Lecture Notes in Computer Science 1935

Edited by G. Goos, J. Hartmanis and J. van Leeuwen

Springer
Berlin
Heidelberg
New York
Barcelona
Hong Kong
London
Milan
Paris
Singapore
Tokyo

Scott L. Delp   Anthony M. DiGioia
Branislav Jaramaz   (Eds.)

Medical Image Computing and Computer-Assisted Intervention – MICCAI 2000

Third International Conference
Pittsburgh, PA, USA, October 11-14, 2000
Proceedings

Springer

Series Editors

Gerhard Goos, Karlsruhe University, Germany
Juris Hartmanis, Cornell University, NY, USA
Jan van Leeuwen, Utrecht University, The Netherlands

Volume Editors

Scott L. Delp
Stanford University
Mechanical Engineering Department
Stanford, CA, USA 94305-3030
E-mail: delp@leland.stanford.edu

Anthony M. DiGioia
Branislav Jaramaz
UPMC Shadyside Hospital and Carnegie Mellon University
Pittsburgh, PA, USA 15232
E-mail: {tony,branko}@cor.ssh.edu

Cataloging-in-Publication Data applied for

Die Deutsche Bibliothek - CIP-Einheitsaufnahme

Medical image computing and computer assisted intervention : third
international conference ; proceedings / MICCAI 2000, Pittsburgh, PA,
USA, October 11 - 14, 2000. Scott L. Delp ... (ed.). - Berlin ;
Heidelberg ; New York ; Barcelona ; Hong Kong ; London ; Milan ; Paris ;
Singapore ; Tokyo : Springer, 2000
  (Lecture notes in computer science ; Vol. 1935)
  ISBN 3-540-41189-5

CR Subject Classification (1998): I.5, I.3.5-8, I.2.9-10, I.4, J.3

ISSN 0302-9743
ISBN 3-540-41189-5 Springer-Verlag Berlin Heidelberg New York

Springer-Verlag Berlin Heidelberg New York
a member of BertelsmannSpringer Science+Business Media GmbH
© Springer-Verlag Berlin Heidelberg 2000
Printed in Germany

Typesetting: Camera-ready by author
Printed on acid-free paper   SPIN: 10781234   06/3142   5 4 3 2 1 0

Preface

MICCAI has become a flagship international conference for medical image computing and computer-assisted interventions. The MICCAI conference was created by merging three closely related and thriving conference series: VBC (Visualization in Biomedical Computing), MRCAS (Medical Robotics and Computer Assisted Surgery) and CVRMed (Computer Vision, Virtual Reality and Robotics in Medicine). MICCAI now provides a single focus for the presentation of high quality research in this important, multi-disciplinary area. The first MICCAI was held in Boston, USA in 1998, and the second meeting was held last year in Cambridge, England. These previous conferences attracted a large number of excellent submissions and were very well attended. These meetings brought together the best theoretical and applied work in this rapidly emerging field, and encouraged constructive dialogue between computer scientists, engineers, and clinicians.

We are delighted to report a high level of interest in MICCAI 2000. A total of 194 papers were submitted, covering a broad range of topics. Of these, 136 were accepted for presentation at the conference – 38 as oral presentations and 98 as posters. All the selected papers appear in these proceedings. Each paper was reviewed by three members of the Scientific Review Committee, selected for technical or clinical expertise of relevance to the subject matter. Final decisions were made by the Program Planning Committee, based on advice from individual reviewers. The result is another volume of high-quality papers that will contribute to the development of this important and exciting area.

We are grateful to the members of the Scientific Review Committee for the time they devoted to the review process. We are also indebted to Branislav Jaramaz, Fredric Picard, and Katharine Cline for superb support in organizing the conference. We are most thankful to the authors of this volume, for submitting a fantastic array of work from basic work on image processing to significant clinical applications of this new technology in orthopaedics, neurosurgery, oncology, surgery, plastic surgery, cardiology, radiology, and other clinical disciplines.

We are pleased to welcome delegates to Pittsburgh and hope that you find MICCAI 2000 an enjoyable and stimulating experience. For readers unable to attend the conference, we hope that you will find this a valuable record of the scientific program, and look forward to meeting you at future MICCAI conferences.

October 2000

Scott L. Delp
Tony DiGioia
Branislav Jaramaz

Conference Organizing Committee

Third International Conference on
Medical Image Computing and Computer-Assisted Intervention
Pittsburgh, Pennsylvania, United States of America
11–14 October 2000

Conference Chairs

Anthony M. DiGioia	UPMC Shadyside Hospital, USA
Scott L. Delp	Stanford University, USA

Co-chairs

Richard D. Bucholz	St. Louis University Health Sciences Center, USA
Alain Carpentier	L'hôpital Broussais, France
Kaigham J. Gabriel	Carnegie Mellon University, USA
W. LeRoy Heinrichs	Stanford University Medical Center, USA
Louis R. Kavoussi	Johns Hopkins Medical Institutions, USA
Michael J. Mack	COR Specialty Associates of North Texas, P. A., USA
Richard M. Satava	Yale University School of Medicine, USA

Program Planning Committee

Alan C.F. Colchester	University of Kent at Canterbury, UK
Takeyoshi Dohi	University of Tokyo, Japan
James S. Duncan	Yale University, USA
Guido Gerig	University of North Carolina, Chapel Hill, USA
Eric Grimson	Massachusetts Institute of Technology, USA
Branislav Jaramaz	UPMC Shadyside Hospital, USA
Frederic Picard	UPMC Shadyside Hospital, USA
Russell H. Taylor	Johns Hopkins University, USA
Jocelyne Troccaz	TIMC - IMAG, France
Max A. Viergever	University Medical Center Utrecht, The Netherlands

Local Organizing Committee

Branislav Jaramaz	UPMC Shadyside Hospital, USA
Frederic Picard	UPMC Shadyside Hospital, USA
Cameron N. Riviere	Carnegie Mellon University, USA

Scientific Review Committee

John R. Adler, Jr.	Stanford University Medical Center, USA
David Altobelli	Private Practice, USA
Licinio Angelini	Universita Degli Studi di Roma La Sapieza, Italy
Takehide Asano	Chiba University, School of Medicine, Japan
Ludwig M. Auer	ISM-International, Germany
Mary Austin-Seymour	University of Washington, USA
Nicholas Ayache	INRIA Sophia-Antipolis, France
William L. Bargar	Private Practice, USA
E. Frederick Barrick	Inova Fairfax Hospital, USA
Harry Bartelink	The Netherlands Cancer Institute, The Netherlands
Andre Bauer	Marbella High Care S.A., Spain
Mitch Berger	University of California at San Francisco, USA
Ulrich Berlemann	Inselspital, Switzerland
Georges Bettega	Hospital A. Michallon, France
Peter Black	Children's Hospital, USA
Dominique Blin	TIMC - IMAG, France
Martin Börner	BGU Unfallklinik, Germany
Jon C. Bowersox	Mt. Zion Medical Center, USA
Richard D. Bucholz	St. Louis University Health Sciences Center, USA
Elizabeth Bullitt	University of North Carolina School of Medicine, USA
Guy-Bernard Cadiere	Centre Hospitalier Universitaire St. Peirre, Belgium
Alain Carpentier	L'hôpital Broussais, France
Steven Charles	The Center for Retina Vitreous Surgery, USA
Kiyoyuki Chinzei	Mechanical Engineering Laboratory, AIST/MITI, Japan
W. Randolph Chitwood	East Carolina University, USA
Philippe Cinquin	TIMC - IMAG, France
Kevin Cleary	Georgetown University Medical Center, USA
Alan C.F. Colchester	University of Kent at Canterbury, UK
Paul Corso	Private practice, USA
Eve Coste- Maniere	INRIA Sophia-Antipolis, France
Hugh Curtin	The Massachusetts Eye and Ear Infirmary, USA
Court Cutting	NYU Medical Center, USA
Ralph Damiano	Washington University St. Louis, USA

Brian Davies — Imperial College of Science, Technology and Medicine, UK

Stephen L. Dawson — Massachusetts General Hospital, USA

Herve Delingette — INRIA Sophia-Antipolis, France

Scott L. Delp — Stanford University, USA

Anthony M. DiGioia — UPMC Shadyside Hospital, USA

Takeyoshi Dohi — University of Tokyo, Japan

Didier Dormont — Hopital de la Pitie-Salpetriere Service de Radiotherapie, France

James S. Duncan — Yale University, USA

Richard Ehman — Mayo Clinic Rochester, USA

Julia Fielding — Harvard Medical School, USA

Elliot K. Fishman — The Russell H. Morgan Department of Radiology and Radiological Science, USA

Kevin Foley — Semmes Murphey Clinic, USA

Thomas Fortin — Hospices Civiles de Lyon, France

Marvin Fried — Albert Einstein College of Medicine, Montefiore Medical Center, USA

Toyomi Fujino — International Medical Information Center, Japan

Toshio Fukuda — Nagoya University, Japan

Kaigham J. Gabriel — Carnegie Mellon University, USA

Barry Gardiner — SGS Surgical Associates, USA

W. Peter Geiss — St. Joseph's Medical Center, USA

Guido Gerig — University of North Carolina, Chapel Hill, USA

Jean-Yves Giraud — TIMC - IMAG, France

Frank Gossé — Hanover Medical University, Germany

Mansel V. Griffiths — University of Bristol, UK

Eric Grimson — Massachusetts Institute of Technology, USA

Gerald Hanks — Fox Chase Cancer Center, USA

Daijo Hashimoto — Tokyo Metropolitan Police Hospital, Japan

Makoto Hashizume — Graduate School of Medical Sciences Kyushu University, Japan

W. LeRoy Heinrichs — Stanford University Medical Center, Japan

Marcel van Herk — The Netherlands Cancer Institute, The Japan

Karl Heinz Höhne — University Hospital Eppendorf, Germany

John Hollerbach — University of Utah, USA

Robert Howe — Harvard University, USA

Koji Ikuta — Nagoya University, Japan

Hiroshi Iseki — Tokyo Women's Medical University, Japan

Cliff Jack — Mayo Clinic Rochester, USA

Kintomo Takakura — Tokyo Women's Medical University, Japan

Mark Talamini — Johns Hopkins University, USA

Shinichi Tamura — Osaka University Medical School, Japan

Charles Taylor — Stanford University, USA

Chris Taylor — University Of Manchester, UK

Russell H. Taylor — Johns Hopkins University, USA

Anthony Timoney — Southmead Hospital, UK

Jun-ichiro Toriwaki — Nagoya University, Japan

Jocelyne Troccaz — TIMC - IMAG, France

Jay Udupa — University of Pennsylvania, USA

Dirk Vandermeulen — University Hospital Gasthuisberg, Belgium

Dirk Vandevelde — University Hospital Antwerpen, Belgium

Michael Vannier — University of Iowa Hospitals and Clinics, USA

Max A. Viergever — University Medical Center Utrecht, The Netherlands

Srinivasan Vijaykumar — University of Illinois at Chicago, USA

Richard Wahl — University of Michigan Medical Center, USA

Eiju Watanabe — Tokyo Metropolitan Police Hospital, Japan

Sandy Wells — MIT, USA

James Wenz — Johns Hopkins University Bayview Orthopaedics, USA

Carl-Fredrik Westin — Harvard Medical School, USA

John Wickham — Guys Hospital, UK

Yasushi Yamauchi — National Institute of Bioscience and Human Technology, Japan

Kazuo Yonenobu — Osaka University Medical School, Japan

Christopher K. Zarins — Stanford University Medical Center, USA

Michael Zelefsky — Memorial Sloan-Kettering Cancer Center, USA

James Zinreich — Johns Hopkins Medical Institutions, USA

MICCAI Board

Table of Contents

Neuroimaging and Neurosurgery

Segmentation

Oncology

Medical Image Analysis and Visualization

Registration

Surgical Planning and Simulation

Endoscopy/Laproscopy

Cardiac Image Analysis

Vascular Image Analysis

Visualization

Surgical Navigation

Medical Robotics

Plastic and Craniofacial Surgery

Orthopaedics

A 3-Dimensional Database of Deep Brain Functional Anatomy, and Its Application to Image-Guided Neurosurgery

K. W. Finnis[1], Y.P. Starreveld[1,2,3], A.G. Parrent[2,3], T.M. Peters[1,2]

Robarts Research Institute[1] and Department of Medical Biophysics[2],
Division of Neurosurgery, London Health Sciences Centre[3]
University of Western Ontario, London, Ontario, Canada

Abstract: We describe a surgical planning environment that permits the determination or refinement of the location of a therapeutic neurosurgical intervention using information derived from an electrophysiological database. Such intraoperative stimulation-response and microelectrode recording data are generated from subcortical exploration performed as part of neurosurgical interventions at multiple centres. We have quantified and nonlinearly registered these intraoperative data, acquired from a large population of patients, to a reference brain imaging volume to create an electrophysiological database. This database can then be nonlinearly registered to future patient imaging volumes, enabling the delineation of surgical targets, cell types, and functional and anatomical borders prior to surgery. The user interface to our system allows the population-acquired physiology information to be accessed in a fully searchable format within a patient imaging volume. This system may be employed in both preoperative planning and intraoperative guidance of stereotactic neurosurgical procedures. We demonstrate preliminary results illustrating the use of this database approach to predict the optimum surgical site for creating thalamic lesions in the surgical treatment of Parkinson's disease.

1. Introduction

Image-guided neurosurgery (IGNS) systems provide neurosurgeons with image-based information that allow neurosurgical sites to be precisely targeted. Such precision is only possible when the target can be seen on the preoperative images of the patient. When desired targets are functionally but not anatomically distinct, as in functional procedures such as pallidotomy or thalamotomy, an approximate target position can only be determined by linear scaling of standardized coordinates [1] or through the use of anatomical atlases fit to anatomical landmarks identified on preoperative scans [2,3]. To refine this initial target into a final target, multiple exploratory trajectories with a recording and/or stimulating electrode are required to characterize the function of the intended target and the tissue surrounding it.

Electrophysiological exploration of the living brain may involve any or all of microelectrode recording, microstimulation, or macrostimulation. Microelectrode recording within brain tissue identifies neurons that produce characteristic firing patterns in response to specific stimuli and allow characterization of function within that particular region of anatomy. All distinct cell-firing patterns encountered along a trajectory, along with the brain coordinates from which they were evoked are entered into the surgical log for later analysis. Micro- and macrostimulation introduce electrical current into a discrete area of brain to produce neuronal membrane depolarization. When stimulation supplies enough current to reach neuronal threshold and excite a small pool of neurons, effects are immediately demonstrated or described by the patient. These stimulation-induced effects may take the form of paresthesias (sensory phenomena), muscle contractions, flashing lights in the visual field, general systemic effects, or alteration of the pathophysiological condition (e.g. cessation of tremor). All verbal responses provided by the patient are entered in the surgical log along with electrode position and the current or voltage settings used to evoke the response.

Many of the surgical targets within the brain, localized through the use of stimulating/recording electrodes, are within anatomical structures that are somatotopically organized. Such structures contain a representation of the human body, which "maps" sub-regions of the deep brain structure to discreet body regions in the same manner that Penfield's homunculus maps the functional organization of the primary somatosensory cortex [4]. Physical responses elicited by the patient, descriptions, and microelectrode recording data obtained during exploration aid the surgeon in mentally reconstructing the somatotopic organization contained within the target structure and in establishing functional borders. Comparison of the patient somatotopic organization with that provided by the literature allows estimation of the location of the probe tip within the target and is used in planning subsequent trajectories.

Each passage of the electrode carries a risk of intracranial hemorrhage. Increasing initial targeting accuracy and decreasing the total number of required trajectories to place a lesion or chronic stimulator is therefore highly advantageous. To accomplish this goal, displaying a map of functional organization with the patient's imaging data set prior to surgery is necessary and has been the subject of several prior endeavors [5,6,7,8].

2. System Description

2.1 Database Construction

During a typical surgical procedure for Parkinson's disease, a preliminary target is indicated on a patient preoperative image, and intraoperative electrophysiological exploration is performed within that anatomical region with the aid of standard stereotactic instrumentation. The function of the region is characterized either by analysis of neural firing patterns obtained during microelectrode recording or by application of electrical stimulation via micro- or macrostimulation electrodes and observing the effect on the patient. Each probe trajectory investigated generates many data elements that must be recorded and interpreted intraoperatively. Historically this has been achieved via dictation or in written format by the surgeon. Our system permits direct entry of these data into the visualization/planning system, although historical data may be entered retrospectively. Prior to inclusion of patient data into a normalized database of function, all electrophysiological data are quantified using a coding structure similar to Tasker's [9], a task which in our system is considerably facilitated through the use of a novel graphical user interface (GUI) as illustrated in Figure 1. The coding structure incorporates the following data:

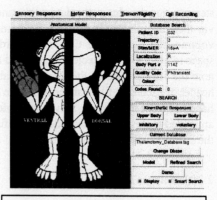

Figure 1. Graphical user interface used to search and enter data into the database. The homunculus figure is employed to define the body part to which the stimulus representation is related.

1. A patient identification number;

2. A trajectory identification number indicating the trajectory from which data were acquired;

3. The method used for data acquisition (i.e. microelectrode recording, microstimulation, or macrostimulation), and if stimulation was applied, the current or voltage, frequency, and duration of the pulse used to evoke the response;

4. The side or sides of the body which evoked the observed modification of cellular firing or experienced effects upon stimulation;

5. A body part number assigned to that discreet region of the body, and;

6. A modifiable response code describing the quality of the response or changes in cell firing patterns. Typical response codes would indicate the involvement of motor, sensory, or autonomic effects; adverse or positive consequences to the physical condition; cognitive changes, or no effect or cellular response.

Within our image-guided neurosurgery system, the coded functional data are plotted directly onto the patient's preoperative image along a virtual trajectory corresponding to the physical probe position and orientation at the location the data were produced. For stereotactic procedures, the data points and trajectory are entered into patient image-space via their coordinates in stereotactic frame space. When tagged to the patient imaging volume, the x, y, and z Cartesian coordinates of the functional data in patient image-space relative to the trajectory origin and the corresponding response codes are saved in a text file for addition to the central database. We also assign a secondary code to the header of each data file that describes the sex, age, pathological condition, procedure and handedness of the patient. Any other information considered surgically relevant may be appended to this secondary code. When fully coded and archived, patient data is added to the central database as described below.

2.2 Non-linear Registration

We employ a nonlinear registration algorithm "ANIMAL" [10], to nonlinearly register (warp) the patient's preoperative image to a standard, high resolution MRI reference brain, which consists of 27 registered T1-weighted MRI scans of the same individual averaged into a single volume. This data set provides excellent clarity of anatomical structure and a high signal to noise ratio [11]. Upon successful nonlinear registration, the deep brain anatomy of the patient image (source volume) appears morphologically identical to the corresponding anatomical region of the reference brain (target volume). Note that since we are only interested in deep brain structures, we do not require that the patient's cortical sulci and gyri be accurately matched.

For each patient image matched to the reference brain, ANIMAL generates a nonlinear deformation grid that accurately describe a one-to-one mapping of the source to the target volume. This three-dimensional deformation grid is used to re-map the coded patient electrophysiologic data to the target volume. This same nonlinear registration technique may be used to map the reference brain to any arbitrary patient's brain. The deformation matrix generated as part of this procedure can then be employed to map points in the population-based electrophysiology database to the anatomy of one brain without the involvement of standard anatomical atlases or linear scaling techniques.

3. Surgical Planning

We have incorporated the electrophysiological atlas described above into our surgical planning environment, so that the atlas information can be used to refine the lesion targets calculated by traditional means (through the use of anatomical atlases or standardized stereotactic coordinates). We believe that it will be possible to use these atlas data as the sole predictor of target loci. Using the inverse of the patient-specific nonlinear transformation matrix generated by nonlinear warping of the patient brain dataset to the reference brain dataset, all data within the functional atlas is transformed into patient image-space as described above. Once transformed, the surgeon may selectively search the database that now conforms to the patient anatomy and obtain electrophysiologic data representative of the desired target. By appropriate selection and display of the functional data from the database, probabilistic functional boundaries within the anatomically homogeneous (MRI) anatomy may be displayed, or alternatively, the locations where the intended therapy was most effective in past patients can be highlighted. Extracted database information is displayed within the IGNS system at the appropriate patient coordinates incorporating user-specified parameters, such as colour, shape representation, and opacity.

The electrical stimulation points extracted from the database are automatically scaled in size to model both the amount of current (microstimulation) or voltage (macrostimulation) used to evoke the response and the resulting spread of current through brain tissue. This scaled representation provides the user with the probable spatial extent of the neurons involved when that particular response was evoked by the stimulation parameters specified in the code. This indicates that while a particular response was evoked at that particular coordinate within the patient brain, neurons responsible for producing the response may have existed anywhere within that volume. For a more comprehensive probabilistic analysis, the surgeon may define a virtual sphere of arbitrary radius at any point in the patient dataset and search the database for all code coordinates that lie within the volume of that sphere. The results of such a search are displayed as a breakdown of all responses that lie within the virtual sphere, as a percentage of the total number of responses. All codes found within the sphere that describe kinesthetic responses are further classified into percentage of responses for specific joints and movements around the joint; muscle contractions are subdivided into discreet muscle groups; and sensory phenomena are arranged into type and their location on the body. This virtual search function is fully customizable by the user to suit individual requirements.

4. Intra-operative guidance

Intraoperatively, comparison of responses observed during electrophysiological exploration, with those predicted by the patient-registered database, provides instantaneous validation of database conformity and displays archived electrophysiologic data in a surgically relevant manner. If lesioning is to be performed, the neurosurgeon can determine the size of the therapeutic lesion or combination of lesions that will optimally destroy or deactivate the anticipated target resulting in minimal trauma to surrounding structures. Accordingly, upon validation of its predictive ability, the database may be used as the sole source of surgical target localization and planning (at the discretion of the surgeon) or used in conjunction with pre-existing localization techniques to refine an approximated target.

5. Initial Clinical Results

The database currently contains information from 75 patients from two clinical sites. At this point all the electrophysiological data have been entered into the database retrospectively, but we are nevertheless in a position to demonstrate the effectiveness of this system to predict the probable location of a lesion to treat Parkinson's disease. Figure 2 demonstrates several clusters of sensory and motor responses positioned in 3-D within patient brain space. a) displays the region of the patient brain (the left thalamus) within which the database points are located; b) shows the location of tremor cells for the population data, and c) demonstrates the trajectory of a lesioning probe that would ablate the region containing the tremor responses. Also shown in c) is the somatotopy exhibited by the thalamic sensory nucleus as predicted by the database. Note that this particular patient's intraoperative data were not contained in the database used in the following demonstration.

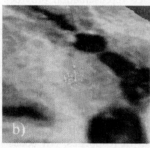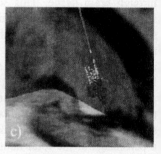

Figure 2: *a) Context for database application; b) Axial: Tremor cells within the Ventralis intermedius (kinesthetic) nucleus; c) Sagittal: Functional border between kinesthetic motor nucleus (small spheres) and Ventralis caudalis sensory nucleus (larger cubes). Somatotopy is evident within the Ventralis caudalis (tongue, fingers and foot sensory regions, from right to left). Trajectory of lesioning probe is also shown in c).*

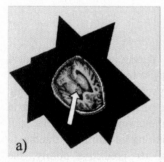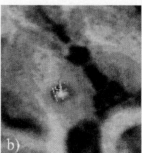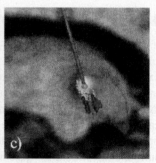

Figure 3: *Post-op MRI. a) Context, showing site of applied lesion. Arrow points to lesion; b) Axial: cluster of tremor cells from database, shown with respect to lesion; c) Sagittal: Tremor cells and sensory nucleus data with respect to lesion. Lesioning electrode modeled to reflect orientation and position of the physical probe that produced the lesion.*

Figure 3 shows the same database information mapped to the post-operative scan of the same patient following surgery. We note that the surgery of this patient was planned and performed using standard techniques, and did not take advantage of the database contents. Nevertheless, it is clearly demonstrated by figure 3b that the cluster of tremor cells lies almost completely within the lesion volume. Fig 3c shows the point clusters as seen in Figure 2, located within the context of the anatomy of the thalamus.

We emphasize that these are preliminary results. Nevertheless, they give us the confidence to proceed to the next stage of this work that will encompass rigorous validation of our registration and targeting strategies, as well as correlating our results with clinical follow-up data.

6. Discussion

This method has several advantages over previous work. Nonlinear registration provides the greatest advantage by accommodating for the neuroanatomical variability found between patients when data are added to the reference brain database and subsequently registered to patient images. None of the earlier work in this field incorporates nonlinear registration (warping) techniques. Further, mapping intraoperative data to the patient image with this technique does not require initial tagging of data to representative slices of digitized anatomical atlases or require anatomical landmarks for the determination of scaling factors to fit the data to images of the patient or reference brain. The improvements offered by our technique can be demonstrated by the fact that functional targets are instantaneously delineated on preoperative images as distinct clusters of functionally related data, without the randomness and scatter seen when linear scaling is employed with anatomical atlas-based approaches [12, 13].

Clustered population electrophysiologic data will display somatotopic organization within proposed targets (Figures 2,3) and may be selectively searched to find specific physiologic regions within the target. Another improvement of this technique is the level of coding detail incorporated. In the past, coding detail contained was neither sufficient to accurately describe an observed response in relation to the parameters that evoked it nor to enable the user to trace the origin of the code back to the original patient or trajectory from which it was produced. Our coding method and database design not only make this possible but permits the user to retrieve data from specific age groups and sex, data evoked from patients with a particular handedness, or codes obtained from patients with specific diseases. By accessing our database through a flexible, user-friendly interface that facilitates detailed searches for any user-defined parameters, our coding technique is ideally suited for intraoperative guidance, preoperative planning and postoperative analysis of electrophysiological data.

A digital probabilistic atlas that utilizes population data will improve in accuracy over time, achieving better statistics with the addition of more data. While only qualitative results were presented in this paper, with 75 patients currently in the database providing over 6100 data points, we are now in a position to commence validation studies that quantitatively assess its design and contents.

7.   References

1.   Favre J., Taha J.M. Nguyen T.T., Gildenberg P.L.,Burchiel K.J. Pallidotomy: A survey of current practice in North America. Neurosurgery 39:883-892, 1996

2.   St-Jean, P., Sadikot, A. F., Collins, D. L., Clonda, D., Kasrai, R., Evans, A. C., and Peters, T. M., Automated atlas integration and interactive 3-dimensional visualization tools for planning and guidance in functional neurosurgery, *IEEE Trans Medical Imaging*, vol. 17, no. 5, pp. 672-680, 1998.

3.   Yeo T.T., Nowinsky W.L. Functional neurosurgery aided by use of an electronic brain atlas. Acta Neurochir [Suppl] 68:93-99, 1997

4.   Penfield W., Rasmussen T. The Cerebral Cortex of Man: A Clinical Study of Localization of Function. Macmillan, New York, 1950.

5.   Thompson, C. J., Hardy, T. L., and Bertrand, G. A system for anatomical and functional mapping of the human thalamus. Comput Biomed Res 10, 9-24. 1977

6.   Tasker R.R., Hawrylyshyn P., Organ W. Computerized graphic display of physiological data collected during human stereotactic surgery. Appl Neurophysiol 41:193-187, 1978.

7.   Yoshida M., Okada K., Nagase A., Kuga S., Watanabe M., Kuramoto S. Neurophysiological atlas of the human thalamus and adjacent structures. Appl Neurophysiol 45:406-409, 1982.

8.   Hardy TL. A method for MRI and CT mapping of diencephalic somatotopography. Stereotact Funct Neurosurg 52:242-249, 1989.

9.   Tasker R.R., Hawrylyshyn P., Organ W. Computerized graphic display of physiological data collected during human stereotactic surgery. Appl Neurophysiol 41:193-187, 1978.

10.   Collins D.L., Peters T.M., and Evans A.C. An automated 3D non-linear image deformation procedure for determination of gross morphometric variability in the human brain. Robb, R. A. 2359, 180-190. 1994. Seattle WA, SPIE. Proceedings of the Third Conference on Visualization in Biomedical Computing.

11.   Holmes C.J., Hoge R., Collins D.L., Woods R., Toga A.W., and Evans A.C., Enhancement of MR images using registration for signal averaging. JCAT 22:324-333, 1998.

12.   Hardy T.L., Bertrand G., Thompson C.J. Organization and topography of sensory responses in the internal capsule and nucleus Ventralis caudalis found during stereotactic surgery. Appl Neurophysiol 42:235-351, 1979.

13.   Tasker R.R., Organ W., Hawrylyshyn P. The Thalamus and Midbrain of Man. Charles C. Thomas Publisher, Springfield, IL. 1982.

Simulation of Corticospinal Tract Displacement in Patients with Brain Tumors

M.R. Kaus[1], A. Nabavi[2,3], C.T. Mamisch[2], W.H. Wells[2,4], F.A. Jolesz[2], R. Kikinis[2], and S.K. Warfield[2]

[1] Philips Research Laboratories, Division Technical Systems, Röntgenstraße 24–26, 22335 Hamburg, Germany,
M.Kaus@pfh.research.philips.com
[2] Surgical Planning Laboratory, Department of Radiology
[3] Department of Neurosurgery, Brigham & Womens Hospital, Harvard Medical School, 75 Francis Street, Boston, MA 02115
[4] Artificial Intelligence Lab, MIT, 545 Technology Square, Cambridge, MA 02139

Abstract. The spatial relationship between the corticospinal tracts and a brain tumor is important for planning the surgical strategy. Although the white matter tracts can be manually outlined from structural MRI this is time consuming and impractical. To enhance the spatial understanding of the distorted pathology we have established a method to retrieve this structural information automatically by registration of a standardized normal brain atlas to the individual pathologic anatomy of brain tumor patients. The skin and the brain were segmented in the patient MRI volume. Subsequently a deformable volumetric atlas of a single normal subject was registered to the patient brain using affine and non-linear registration techniques. The estimated spatial correspondence between atlas and patient brain was used to warp the corticospinal tracts from the atlas onto the patient. The accuracy of the method was evaluated in 5 patients with extrinsic tumors of different histopathology and location by comparing selected anatomical landmark structures from the projected atlas to their manually segmented counterpart in the patient. Our method enables the visualization of complex anatomical information with minimal user interaction for routine surgical planning that would otherwise demand the acquisition of additional imaging modalities.

1 Introduction

Computer assisted surgical planning aids the neurosurgeon in the appreciation of the spatial arrangement of critical brain structures (e.g. motor and sensory cortex) with respect to an adjacent lesion, to define the safest possible surgical approach. If a tumor is located close to the motor cortex, the location of the corticospinal tracts (CST, Figure 2) is particularly important, but unproportionally more difficult to visualize. Nevertheless, connecting functional areas, damage to these white matter tracts results in the same clinical deficits as the destruction of those centers themselves.

Although diffusion tensor MRI provides direct visual information of the white matter tracts, this imaging technique and related image post-processing algorithms are not yet at the stage of providing routinely available studies [1–3]. Manual segmentation of the corticospinal tracts from conventional MRI is difficult if not impossible. Even an experienced, trained radiologist can hardly distinguish different tracts from the bulk of the white matter. In addition, manual segmentation would be time consuming, excluding this approach from clinical routine. Automated segmentation techniques based on image features alone are also inadequate for this task, because the different tracts are almost indistinguishable on conventional MRI.

Non-linear registration of a deformable anatomical template has been used for various image analysis tasks where image information alone is not sufficient. Applications include the identification of cortical sulci or functional areas, and the segmentation of multiple sclerosis lesions or brain tumors [4–8]. Early work of this group reported the volume measurement of basal ganglia schizophrenics by registration of a digital atlas [9]. However, no work has been reported on using an anatomical atlas to retrieve structural information in brain tumor patients where dramatic brain deformation can occur.

In this work we present an automated method which is based on [9] with some variations to retrieve structural information in patients with brain tumors using a standardized normal brain atlas. As an example for a clinically relevant structure, we use our method to extract the corticospinal tracts in 5 patients and validate the results with anatomical landmark structures semi-automatically segmented by trained clinical personnel in MRI and fMRI.

2 Methods

We used a multistage image registration scheme (Figure 1) to align a 3D template of normal anatomy (a deformable volumetric digital brain atlas) to identify structural information in 3D MRI of brain tumor patients. A hierarchical strategy was implemented to avoid local minima in the spatial transform parameter search space and to capture large deformations that can be caused by a brain tumor. The spatial transform that maps the atlas brain onto the patient brain was used to warp the structures of interest (i.e. the corticospinal tracts) from the atlas onto the individual patient dataset. In the following sections we will describe the image data used, the image processing steps, and the validation procedure.

2.1 Image Data

Atlas Data
We used a volumetric brain atlas based of a single normal male where each voxel was labeled (given a number) according to its anatomical membership [10]. For the purpose of hierarchical registration, separate template volumes that contain only one structure (e.g. all segmented structures to form the *head volume*) were

extracted. Here the head volume, the brain volume and the structural volume (containing voxels labeled as thalamus, nucleus caudatus, globus pallidus, pre- and postcentral gyrus, and corticospinal tracts) were formed. The structural volume was projected for validation purposes (see Section 2.3).

Patient Data
5 patients were scanned with a standard protocol (1.5 T Signa GE MRI, post-contrast 3D sagittal spoiled gradient recalled (SPGR) acquisition, $256 \times 256 \times 124$, $0.9375 \times 0.9375 \times 1.5$ mm^3). For one patient (no. 5), a standard fMRI sequence was also available for visualization of the motor cortex. The head, brain and structural volume were semi-automatically segmented in the MRI using a 3D segmentation tool [11].

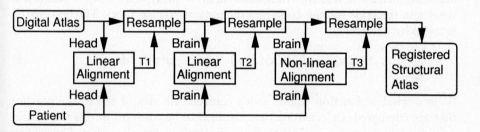

Fig. 1. Schematic description of the registration procedure. *Global shape differences are captured with the global affine transforms T_1 and T_2, while local shape differences are represented with a volumetric deformation vector field T_3.*

2.2 Registration Procedure

Linear Registration The global affine transform $T_1 : I_1(\mathbf{x}) \to I_2(\mathbf{x})$, that roughly aligned the atlas head volume $I_1(\mathbf{x})$ to the patient $I_2(\mathbf{x})$ where $\mathbf{x} = [x, y, z]^T$ was estimated in a 2-steps procedure. First, an initial pose was estimated by aligning the back and the top of the atlas and the patient heads bounding boxes. Second, using a linear registration method for segmented images [12], the head of the atlas and the patient were registered by minimizing

$$S = \sum_{i=1}^{N} f\left(I_2(\mathbf{x}_i), I_1(T^{-1}(\mathbf{x}_i))\right) \tag{1}$$

using Powells method, where N is the total number of voxels and $f(a, b) = 0$ if $a = b$ and $f(a, b) = 1$ otherwise. For binary data f could be efficiently computed using the xor function. To decrease conversion time, a multiresolution image pyramid was formed, subsampling the images by a factor of 2 in each spatial dimension at every resolution level. The minimization was carried from coarse-to-fine resolution on three levels.

The atlas was resampled according to T_1 using nearest neighbor interpolation, the *brain volume* was extracted from the resampled atlas, and the 2-step linear alignment procedure described above was repeated to estimate T_2 to globally align the atlas and the patient brain. It was found experimentally that brains were aligned more precisely when preceded with a global affine head registration.

Non-linear Registration

3D Adaptive Template Matching
We used 3D adaptive template matching to estimate a 3D volumetric deformation vector field based on the work of Dengler *et al.* [13]. Region-based template matching defines a relative translation $\mathbf{u}(\mathbf{x}) = [u(\mathbf{x}), v(\mathbf{x}), w(\mathbf{x})]^T$ between a voxel $\mathbf{x}$ in the atlas volume $I_1(\mathbf{x})$ and the patient volume $I_2(\mathbf{x})$. The best match is found by minimizing

$$S(\mathbf{u}(\mathbf{x})) = \int_{I_2} W(\mathbf{x})\big(I_2(\mathbf{x}) - I_1(\mathbf{x} - \mathbf{u}(\mathbf{x}))\big)^2 dx \quad . \tag{2}$$

W is a window function whose width controls the size of the image patches that are compared. Deformations are assumed to be constant in this region and only estimated locally, such that S is effectively a function of $\mathbf{x}$. Linearizing $I_1(\mathbf{x} - \mathbf{u}(\mathbf{x})) \approx I_1(\mathbf{x}) - \mathbf{u}(\mathbf{x})\nabla I_1(\mathbf{x})$ allows a closed form solution of Equation 2 for each voxel. Setting $\mathbf{S}(\mathbf{u})/d\mathbf{u} = 0$ leads to

$$\mathbf{Qu} = -\mathbf{f}, \quad \mathbf{Q} = \int_{I_2} W\nabla I_1 \nabla I_1^T dx, \quad \mathbf{f} = \int_{I_2} W\nabla I_1(I_2 - I_1)dx \tag{3}$$

which can be directly solved with Cramers rule (dependence on $\mathbf{x}$ omitted for clarity).

Linearization assumes small $\mathbf{u}$. To justify this assumption, the atlas is linearly registered (Section 2.2) to the patient prior to the non-linear registration, and the deformation vector field is estimated on a coarse-to-fine resolution pyramid. Gaussian smoothing and subsampling is efficiently calculated according to Burt *et al.* [14]. Because ∇I_1 is not defined for segmented data, I_1 (and I_2) were smoothed with a 5 point Gaussian prior to the computation of the derivates.

In regions where the gradient ∇I is close or equal to 0, $\mathbf{Q}$ is not invertible and Equation 3 has no solution, i.e. the deformation cannot be estimated in unstructured regions. Therefore, W needs to be large enough to allow computation of ∇I_1 and enable reliable estimation of $\mathbf{u}$ respectively. Because W also determines the variability of $\mathbf{u}$ over the image volume, and we want to allow $\mathbf{u}$ to vary rapidly where necessary, W should be as small as possible and as large as necessary. Thus, we use a spatially varying adaptive window function W

$$W = W(\mathbf{x}) = \sum_{i=1}^{i_{max}} w_i W_i \quad , \quad w_i = \frac{1}{\sum_i \det(\mathbf{Q}_i)}\det(\mathbf{Q}_i) \quad , \tag{4}$$

that takes into account several window functions W_i of size $s_{i+1} = 2s_i$, where s_i is limited by the smallest volume axis length. Qualitatively, the weight w_i is large if the image patch contains large image gradients in a variety of directions. If a small image patch is sufficiently structured then $\mathbf{u}$ can be estimated with a small patch, otherwise larger image patches dominate the estimation. If $\det \mathbf{Q}_i(\mathbf{x}) = 0$, $\forall i$, $\mathbf{u}(\mathbf{x})$ is set to zero.

Our template matching approach results in a smooth deformation field T_3. Since neighboring voxels have similar neighborhoods, the similarity measure (integrated over the window patches) at neighboring voxels is also similar, resulting in a slowly varying or smooth deformation field.

Image Warping

Ideally, to construct the deformed template image $\tilde{I}_1 \leftarrow I_1$, $\mathbf{u}$ is used to translate the value at each location in I_1 according to

$$\tilde{I}_1(\mathbf{x}) = I_1(\mathbf{x} - \mathbf{u}(\mathbf{x})) = I_1(\tilde{\mathbf{x}}). \tag{5}$$

However, our method does not guarantee a one-to-one mapping. In case of e.g. a one-to-many mapping, this resampling method could lead to holes in the deformed template image. Therefore, we estimate the corresponding voxel position in the template image for each voxel in the target image, which essentially means to estimate the inverse of the deformation field $\mathbf{u}(\mathbf{x})$. Assuming that the deformation field is slowly varying, the necessary inverse transform can be approximated with

$$\mathbf{u}(\tilde{\mathbf{x}})^{-1} \approx -\mathbf{u}(\mathbf{x}). \tag{6}$$

Since $\tilde{\mathbf{x}}$ in general will not be a point on the regular grid, $I_1(\tilde{\mathbf{x}})$ is assigned the value $I_1(\bar{\mathbf{x}})$, $\bar{\mathbf{x}}$ being the closest point to $\tilde{\mathbf{x}}$ (nearest neighbor interpolation). This kind of interpolation has the advantage of computational efficiency (as opposed to e.g. tri-linear interpolation) and maintains the original voxel values, which is advantagous when operating on segmented data.

2.3 Validation Experiment

We applied the method to 5 patients with meningiomas and subdural low grade gliomas. Direct validation would be possible by comparing our results to diffusion weighted images. But for the reasons mentioned above their availability is limited. Therefore we chose the approach to a) measure how well our method registers the atlas to the patient brain and b) test the anatomically correct course of the simulated white matter tracts (Figure 2). We validated the CST projections by comparing anatomical landmarks (structures in the brain) which interface with the CST and are unequivocally distinguishable and can thus be segmented from MRI. Such structures are the thalamus, the nucleus caudatus, the globus pallidus, and the pre- and postcentral gyrus. If these control structures from the warped atlas match their manually segmented counterparts, then we infer that the CST are projected correctly.

Table 1. Similarity between manually segmented and projected control structures with respect to the volume of the bounding box. *Reasonable accuracy was achieved for all structures except for the globus pallidus.*

Location	Structure	Case No. (Tumor location)					Mean	SD
		1 (L)	2 (L)	3 (R)	4 (R)	5 (L)		
Left & Right	Brain	92.64	92.29	90.95	92.34	92.81	92.21	0.73
	Thalamus	79.46	83.35	68.24	75.88	74.30	76.25	5.68
Left	Nucleus Caudate	92.71	87.12	76.30	86.94	88.22	86.26	6.04
	Globus Pallidus	45.28	42.32	45.50	42.75	46.66	44.50	1.88
	Precentral Gyri	79.58	79.53	77.81	77.30	93.87	81.62	6.92
	Postcentral Gyri	75.63	83.78	74.06	80.18	91.29	80.98	6.92
Right	Nucleus Caudate	88.28	86.06	76.00	86.01	90.43	85.36	5.54
	Globus Pallidus	45.70	42.48	44.39	41.24	39.05	42.57	2.61
	Precentral Gyri	82.70	80.82	77.90	78.25	88.76	81.87	4.42
	Postcentral Gyri	69.90	82.86	76.12	83.34	89.25	80.29	7.44

The similarity S_s between a manually segmented and a warped structure s was defined as

$$S_s = \frac{\#_{B_s}\text{OverlappingVoxels}}{\#_{B_s}\text{Voxels}}, \qquad (7)$$

where $\#$ is the counting operator applied inside the bounding box B_s that encloses both the manually segmented and the warped control structure. The bounding box was introduced to prevent the similarity measure to be biased towards small structures.

3 Results

Although the brain was significantly deformed by the meningioma, visual validation of the projected corticospinal tracts (CST, Figure 2) correlated well with the anatomical definition. The CST started at the motor cortex (manually segmented from fMRI), passed through the internal capsule (between thalamus and globus pallidus, manually segmented from MRI) and lead into the crus cerebri (part of the brainstem, MRI slice). Thus origin and course towards the brainstem were simulated correctly.

In this image, the size of the motor cortex from fMRI is smaller than the corticospinal tract. This is possibly due to the fact that only a part of the motor cortex has been activated during fMRI.

Table 1 shows the similarity between the manually segmented and the projected structures. The accuracy of the brain match is the accuracy with which

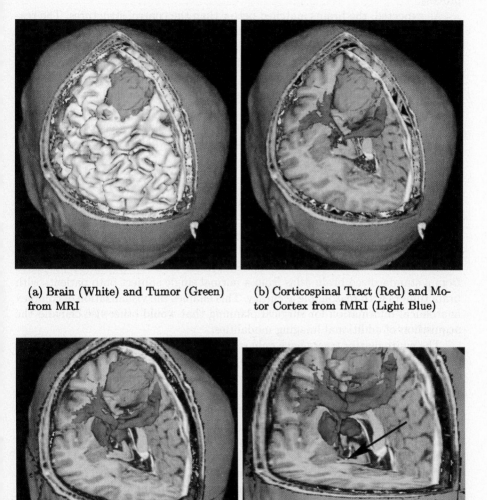

(a) Brain (White) and Tumor (Green) from MRI

(b) Corticospinal Tract (Red) and Motor Cortex from fMRI (Light Blue)

(c) Basal Ganglia (Dark Blue)

(d) Crus Cerebri (Arrow)

Fig. 2. Different views (patient no. 5) for visual verification of the corticospinal tract projection. *Despite a significant brain deformation caused by the meningioma, the corticospinal tract (registered atlas) correctly connects the motor cortex (manually segmented from fMRI, (b)) to the crus cerebri (MRI slice, (c)) and runs between the globus pallidus and the thalamus (manually segmented from MRI, (d)).*

our method maps one structure onto another by means of minimization of a similarity measure. The accuracy of the landmark structure registration describes how well our model describes the brain distortion caused by the pathological process.

As expected, the brains matched better than the control structures. The reason is that the correspondence between the atlas and the patient is established by registration of the atlas to the patient brain. The mapping of the control structures is based on our model of brain deformation. A deformation field based on image structure is only estimated on the the brain surface and the brain-ventricle interface, while the deformation field resulting for the inner brain structures are purely based on interpolation. However, reasonable similarity was established for all control structures. The globus pallidus was aligned with less accuracy, which can be partly explained with an oversegmented globus pallidus in our anatomical atlas.

4 Discussion and Conclusion

We have developed a method for the automated extraction of non-visible inherent information of a structural scan based on non-linear template matching of a brain atlas. It was demonstrated that structural information can be projected from a standardized brain atlas from a normal single subject onto patients with brain tumors with reasonable accuracy. This enables the visualization of complex anatomical information for surgical planning that would otherwise demand the acquisition of additional imaging modalities.

The white matter tracts have only recently been visualized in vivo employing diffusion weighted images [2, 3]. Although their anatomical course has been extensively studied, the surgical anatomy in pathologically deformed brain could not be described presurgically. Since the advent of image guided surgery 3D representation of the anatomy and the co-registration of functional data [16] has increased the interest in precise surgical planning, particularly for tumors adjacent to eloquent areas. The incorporation of functional data has enabled the surgical approach to these lesions. Nevertheless a remaining problem is the displacement of the subcortical fiber connections. The unequivocal identification of functional areas is reported to significantly improve approaches and potential outcome of patients harboring lesions in these areas [16]. Nevertheless it is apparent, that keeping cortical areas intact, while disregarding their connections will result in the same deleterious clinical result, as damaging those cortical areas themselves. Therefore it is of great importance to include the connecting tracts, and particularly the corticospinal tracts.

Particularly the minimal user interaction as well as the utilization of information already available from conventional planning and neuronavigation enhances the potential use of this method. In this paper we used initial semi-automated segmentation, as needed for surgical planning. However, automated methods have become available potentially decreasing user interaction even for this step [15].

Several questions for future work remain. Our method is currently limited to extrinsic tumors because our template matching approach requires equivalent topology. This is not the case when mapping a normal brain atlas to a brain with an intrinsic tumor. However, recent work on atlas deformation in the presence of intrinsic tumors could be investigated in this context [17]. We are currently working on a more solid validation study based on intraoperative verification and preferably on tensor diffusion MRI with a larger number of cases.

Acknowledgments

This investigation was supported (in part) by a Grant from the National Multiple Sclerosis Society (SW). This work was supported (in part) by NIH grants RO1 CA 46627-08, PO1 CA67165-01A1, PO1 AG04953-14, NSF grant BES 9631710, Darpa grant F41624-96-2-0001, Whittaker Foundation RG-96-0440, and DFG (NA 359/1-1).

References

1. C. Pierpaoli, P. Jezzard, P.J. Basser, A. Barnett, and G. Di Chiro. Diffusion tensor MR imaging of the human brain. *Radiology*, 201:637–648, 1996.
2. C.F. Westin, S.E. Maier, B. Khidhir, P. Everett, F.A. Jolesz, and R. Kikinis. Image processing for diffusion tensor MRI. In *MICCAI 99*, pages 441–452, 1999.
3. C. Poupon, J.F. Mangin, V. Frouin, M. Pachot-Clouard, D. Le Bihain and I. Bloch. Regularization of MR diffusion tensor maps for tracking brain white matter bundles. In *MICCAI 98*, pages 489–498, 1998.
4. R. Bajcsy, R. Lieberson, and M. Reivich. A computerized system for the elastic matching of deformable radiographic images to idealized atlas images. *JCAT*, 7(4):618–625, 1983.
5. G.E. Christensen, R.D. Rabbitt, and M.I. Miller. 3D brain mapping using a deformable neuronatomy. *Physics in Medicine and Biology*, 39:609–618, 1994.
6. P. Thompson, D. MacDonald, M.S. Mega, C.J. Holmes, A. Evans, and A.W. Toga. Detection and mapping of abnormal brain structures with a probabilistic atlas of cortical surfaces. *JCAT*, 21(4):467–481, 1996.
7. D.L. Collins, T.M. Peters, W. Dai, and A.C. Evans. Model based segmentation of individual brain structures from MRI data. In *VBC 92*, pages 10–23, 1992.
8. S.K. Warfield, M.R. Kaus, F.A. Jolesz, and R. Kikinis. Adaptive template moderated spatially varying statistical classification. In *MICCAI 98*, Boston, MA, USA, 1998.
9. D.V. Iosifescu, M.E. Shenton, S.K. Warfield, R. Kikinis, J. Dengler, F.A. Jolesz, and R.W. McCarley. An automated measurement of subcortical brain MR structures in schizophrenia. *Neuroimage*, 6:13–25, 1997.
10. R. Kikinis, M.E. Shenton, D.V. Iosifescu, R.W. McCarley, P. Saiviroonporn, H.H. Hokama, A. Robatino, D. Metcalf, C.G. Wible, C.M. Portas, R., Donnino and F.A. Jolesz. A digital brain atlas for surgical planning, model driven segmentation and teaching. *IEEE VCG*, 2(3):232–241, 1996.

11. D.T Gering, A. Nabavi, R. Kikinis, W.E.L. Grimson, N. Hata, P. Everett, F.A. Jolesz and W.M. Wells. An integrated visualization system for surgical planning and guidance using image fusion and interventional imaging. In *MICCAI 99*, pages 809–819, Cambridge, UK, 1999.

12. S.K. Warfield, F.A. Jolesz, and R. Kikinis. A high performance computing approach to the registration of medical image data. *Parallel Computing*, 24:1345–1368, 1998.

13. J. Dengler and M. Schmidt. The dynamic pyramid - a model for motion analysis with controlled continuity. *PRAI*, 2(2):275–286, 1987.

14. P.J. Burt and E.H. Adelson. The laplacian pyramid as a compact image code. *IEEE Transactions on Communications*, 31(4):532–510, 1983.

15. M.R. Kaus, S.K. Warfield, A. Nabavi, E. Chatzidakis, P. Black, F.A. Jolesz and R. Kikinis. Automated brain tumor segmentation in MRI: meningiomas and low grade gliomas. In *MICCAI 99*, pages 1–10, 1999.

16. O. Ganslandt, R. Fahlbusch, C. Nimsky, H. Kober, M. Müller, R. Steinmeier, and J. Romstöck. Functional neuronavigation with MEG: outcome in 50 patients with lesions around the motor cortex. *Neurosurg Focus*, 6(3):Article 3, 1999.

17. B.M. Dawant, S.L. Hartmann, and S. Gadamsetty. Brain atlas deformation in the presence of large space-occupying tumors. In *MICCAI 99*, pages 589–596, 1999.

Registration of 3D Intraoperative MR Images of the Brain Using a Finite Element Biomechanical Model

Matthieu Ferrant[1], Simon K. Warfield[2], Arya Nabavi[2],
Ferenc A. Jolesz[2], and Ron Kikinis[2]

[1] Telecommunications Laboratory, Université catholique de Louvain, Belgium.
[2] Surgical Planning Laboratory, Brigham and Women's Hospital and Harvard Medical School, Boston, USA.
{ferrant,warfield,arya,jolesz,kikinis}@bwh.harvard.edu

Abstract. We present a new algorithm for the non-rigid registration of 3D Magnetic Resonance (MR) intraoperative image sequences showing brain shift. The algorithm tracks key surfaces (cortical surface and the lateral ventricles) in the image sequence using an active surface algorithm. The volumetric deformation field of the objects the surfaces are embedded in is then inferred from the displacements at the boundary surfaces using a biomechanical finite element model of these objects. The biomechanical model allows us to analyse characteristics of the deformed tissues, such as stress measures. Initial experiments on an intraoperative sequence of brain shift show a good correlation of the internal brain structures after deformation using our algorithm, and a good capability of measuring surface as well as subsurface shift. We measured distances between landmarks in the deformed initial image and the corresponding landmarks in the target scan. The surface shift was recovered from up to 1cm down to less than 1mm, and subsurface shift from up to 6mm down to 3mm or less.

1 Introduction

The increased use of image guided surgery systems for neurosurgery has brought to prominence the problem of brain shift, the deformation the brain undergoes after craniotomy, as well as deformations due to tumor resection. These deformations can significantly diminish the accuracy of neuronavigation systems, and it is therefore of great importance to be able to quantify and analyse these phenomena. The subject has recently lead to considerable interest in the medical image analysis community [1–9].

Most of the work that has been done in the field of intraoperative volumetric image alignment is mainly based on image related criteria [10, 8, 4] . Physical deformation models have also been proposed to constrain a deformation field computed from image data using elastic [11, 12] or even viscous fluid deformation models [13, 14]. However, these models do not account for the actual material characteristics of the brain, because the matching is done minimizing an energy measure that consists of a weighted sum of an image similarity term and a relaxation term representing the potential energy of a physical body (e.g. elastic). Therefore, the actual physics of the phenomenon cannot not be properly captured by these models.

There has also been a significant amount of work directed towards simulation using models driven by physics-based forces such as gravity. Skrinjar et al. [15, 7] have proposed a model consisting of mass nodes interconnected by Kelvin models to simulate the behavior of brain tissue under gravity, with boundary conditions to model the interaction of the brain with the skull. Miga et al. [3, 5, 6] proposed a Finite Element (FE) model based on consolidation theory where the brain is modeled as an elastic body with an intersticial fluid. They also use gravity induced forces, as well as experimentally determined boundary conditions.

Even though these models are very promising, it remains difficult to accurately estimate all the forces and boundary conditions that interact with the model.

It is only recently that biomechanical models have been explicitly proposed to constrain the deformation of images [16, 9]. Currently, the drawback of such methods is that they either require user intervention, or another means to compute the forces (or correspondances) applied to the model. Another drawback is that these later methods have only been applied to 2D images thereby limiting the clinical utility and the possibility to efficiently assess the accuracy of the method.

Our ultimate goal is to be able to do prediction of deformation during surgery with the goal of improving intraoperative navigation and tumor resection, and of reducing the amount of intraoperative imaging that is necessary. To be able to do this, one first needs to validate the non-rigid deformation model. Intraoperative MRI provides excellent contrast and spatial resolution, which makes it an ideal testbed for developing and validating nonrigid deformation methods.

We propose a new integrated approach that uses surface-based correspondences to drive a biomechanical model of the brain instead of using estimates of forces that are often difficult to accurately determine. The correspondances are computed using one or multiple active surfaces that are deformed onto the target image. The correspondances between landmark surfaces in the preoperative and intraoperative scans provide an implicit way to compute the forces the model has undergone.

2 Theory

So far, we have limited our model to linear elasticity, as it has been shown that soft tissue deformation can be modeled quite accurately using linear elasticity in the case of small strains [17, 18]. However, our algorithm can easily be used with other constitutive materials such as viscous fluids, etc. The following sections will successively review the theory of finite element modeling and address the FE meshing issue, explain how we have used these principles to solve active surface problems, as well as the way we use it for computing a biomechanical volumetric deformation field.

2.1 Finite element model

Assuming a linear elastic continuum with no initial stresses or strains, the potential energy of an elastic body submitted to externally applied forces can be expressed as [19] [1]:

$$E = \frac{1}{2} \int_\Omega \sigma^T \epsilon \, d\Omega + \int_\Omega \mathbf{F} \mathbf{u} \, d\Omega \tag{1}$$

where $\mathbf{u} = \mathbf{u}(x, y, z)$ is the displacement vector, $\mathbf{F} = \mathbf{F}(x, y, z)$ the vector representing the forces applied to the elastic body (forces per unit volume, surface forces, or

---

[1] Superscript T designs the transpose of a vector or a matrix

forces concentrated at the nodes), and Ω the body on which one is working. ϵ is the strain vector, defined as

$$\epsilon = \left(\frac{\partial \mathbf{u}}{\partial x}, \frac{\partial \mathbf{u}}{\partial y}, \frac{\partial \mathbf{u}}{\partial z}, \frac{\partial \mathbf{u}}{\partial x} + \frac{\partial \mathbf{u}}{\partial y}, \frac{\partial \mathbf{u}}{\partial y} + \frac{\partial \mathbf{u}}{\partial z}, \frac{\partial \mathbf{u}}{\partial x} + \frac{\partial \mathbf{u}}{\partial z}\right)^T = \mathbf{Lu} \qquad (2)$$

and σ the stress vector, linked to the strain vector by the constitutive equations of the material. In the case of linear elasticity, with no initial stresses or strains, this relation is described as

$$\sigma = \left(\sigma_x, \sigma_y, \sigma_z, \tau_{xy}, \tau_{yz}, \tau_{xz}\right)^T = \mathbf{D}\epsilon \qquad (3)$$

where $\mathbf{D}$ is the elasticity matrix characterizing the properties of the material [19].

This equation is valid whether one is working with a surface or a volume. We model our active surfaces, which represent the boundaries of the objects in the image, as elastic membranes, and the surrounding and inner volumes as 3D volumetric elastic bodies.

Within a finite element discretization framework, an elastic body is approximated as an assemblage of discrete finite elements interconnected at nodal points on the element boundaries. This means that the volumes to be modeled need to be meshed, i.e., divided into elements. In [9], Hagemann et al. propose to use the pixels of the image as basic elements of his FE mesh. This approach does not take advantage of the intrinsic formulation of FE modeling, which assumes that the mechanical properties are constant over the element, suggesting that one can use elements covering several image pixels. Also, when performing computations in 3D, which is eventually what is needed for medical applications, the amount of degrees of freedom will be far too large (for a typical 256x256x60 intraoperative MRI, this means about 12 million degrees of freedom !) to perform efficient computations in a reasonable time, even on high performance computing equipment.

Most available meshing software packages do not allow meshing of multiple objects [20, 21], and are usually designed for regular and convex objects, which is often not the case for anatomical structures. Therefore, we have implemented a tetrahedral mesh generator specifically suited for labeled 3D medical images. The mesher can be seen as the volumetric counterpart of a marching tetrahedra surface generation algorithm. A detailed description of the algorithm can be found in [22]. The resulting mesh structure is built such that for images containing multiple objects, a fully connected and consistent tetrahedral mesh is obtained with for every cell, a given label corresponding to the object the cell belongs to. Therefore, different biomechanical properties and parameters can easily be assigned to the different cells or objects composing the mesh. Boundary surfaces of objects represented in the mesh can be extracted from the mesh as triangulated surfaces, which is very convenient for running an active surface algorithm.

The continuous displacement field $\mathbf{u}$ within each element is a function of the displacement at the nodal points of the element $\mathbf{u}_i^{el}$ weighted by its shape functions $N_i^{el} = N_i^{el}(x, y, z)$.

$$\mathbf{u} = \sum_{i=1}^{N_{nodes}} N_i^{el} \mathbf{u}_i^{el} \qquad (4)$$

The elements we use are tetrahedra ($N_{nodes} = 4$) for the volumes and triangles for the membranes ($N_{nodes} = 3$), with linear interpolation of the displacement field.

Hence, the shape function of node i of element el is defined as: $N_i^{el} = K\left(a_i^{el} + b_i^{el}x + c_i^{el}y + d_i^{el}z\right)$, where $K = \frac{1}{6V^{el}}$ for a tetrahedron, and $K = \frac{1}{2S^{el}}$ for a triangle. The computation of V^{el}, S^{el} (volume, surface of el) and other constants is detailed in [19].

For every node i of each element el, we define the matrix $\mathbf{B}_i^{el} = \mathbf{L}_i N_i^{el}$. The function to be minimized at every node i of each element el can thus be expressed as :

$$E(\mathbf{u}_i^{el}) = \int_\Omega \sum_{j=1}^{N_{nodes}} \mathbf{u}_i^{el^T} \mathbf{B}_i^{el^T} \mathbf{DB}_j^{el} \mathbf{u}_j^{el} + \mathbf{F}N_i^{el}\mathbf{u}_i^{el} \, d\Omega \tag{5}$$

We seek the minimum of this function by solving for $\frac{dE(\mathbf{u}_i^{el})}{d\mathbf{u}_i^{el}} = 0$. Equation (5) then becomes :

$$\int_\Omega \sum_{j=1}^{N_{nodes}} \mathbf{B}_i^{el^T} \mathbf{DB}_j^{el} \mathbf{u}_j^{el} \, d\Omega = -\int_\Omega \sum_{j=1}^{N_{nodes}} \mathbf{F}N_i^{el} \, d\Omega \tag{6}$$

This last expression can be written as a matrix system for each finite element:

$$\mathbf{K}^{el}\mathbf{u}^{el} = -\mathbf{F}^{el} \tag{7}$$

Matrices $\mathbf{K}^{el}$ and vector $\mathbf{F}^{el}$ are defined as follows: $\mathbf{K}_{i,j}^{el} = \int_\Omega \mathbf{B}_i^{el^T} \mathbf{DB}_j^{el} \, d\Omega$, $\mathbf{F}_j^{el} = \int_\Omega \mathbf{F}N_i^{el} \, d\Omega$; where every element i, j refers to pairs of nodes of the element el (i and j range from 1 to 4 for a tetrahedron – 1 to 3 for a triangle). $\mathbf{K}_{i,j}^{el}$ and are 3 by 3 matrices, $\mathbf{F}_j^{el}$ is a 3 by 1 vector. The 12 by 12 (9 by 9 for a triangle) matrix $\mathbf{K}^{el}$, and the vector $\mathbf{F}^{el}$ are computed for each element and are then assembled in a global system $\mathbf{Ku} = -\mathbf{F}$, the solution of which will provide us with the deformation field corresponding to the global minimum of the total energy.

We now have constitutive equations that model surfaces as elastic membranes and volumes as elastic bodies.

2.2 Active surface algorithm

The active surface algorithm deforms the boundary surface of an object in one image of the sequence towards the boundary of the same object in the next image of the sequence. The surface is modeled as an elastic membrane, which we deform iteratively onto the target scan by applying image-derived forces. As proposed in [23], the temporal variation of the surface can be discretized using finite differences, provided the time step τ is small enough. Using the previously obtained relation for elastic bodies (see eq. 7), this yields the following semi-implicit iterative equation [2]:

$$\frac{\mathbf{u}^t - \mathbf{u}^{t-1}}{\tau} + \mathbf{Ku}^t = -\mathbf{F}^{\mathbf{u}^{t-1}} \tag{8}$$

Classically, the image force $\mathbf{F}$ is computed as a decreasing function of the gradient so as to be minimized at the edges of the image. To increase the robustness and the convergence rate of the process, we have computed the forces as a steepest gradient descent on a euclidean distance map we efficiently compute from the target object [24]. More details about our active surface algorithm can be found in [25].

---

[2] Superscript t refers to the current temporal iteration.

2.3 Inferring volumetric deformations from surface deformations

The deformation field obtained for the boundary surfaces is then used in conjunction with the volumetric model to infer the deformation field inside and outside the boundary surfaces. The idea is to apply forces to the boundary surfaces that will produce the same displacement field at the boundary surfaces that was obtained with the active surface algorithm. The volumetric biomechanical model will then compute the deformation of the surrounding nodes in the mesh.

Let $\tilde{\mathbf{u}}$ be the vector representing the displacement to be imposed at the boundary nodes. Hence, the forces (see eq. 7) needed to impose these displacements to the volume can be expressed as :

$$\mathbf{F} = \mathbf{K}\tilde{\mathbf{u}} \qquad (9)$$

The solution of the equilibrium equations with these forces will provide us with the displacement at all the nodes in the volumetric mesh with the imposed displacements at the nodes of the boundary surfaces delimiting the objects represented in the mesh. Biomechanical parameters such as the stress tensors can then be derived from the displacements at the nodes using the stress-strain relationship (eq. 3) for every node i:

$$\sigma_i = \sum_{el|i\in el} \mathbf{D}\epsilon_i = \sum_{el|i\in el} \mathbf{D}\mathbf{L}_i N_i^{el} \mathbf{u}_i \qquad (10)$$

3 Experiments

3.1 FE model generation

To build our brain model, we segmented the brain out of the initial intraoperative MRI using our directional watershed algorithm [26]. The volume was further simplified using mathematical morphology to obtain a smooth surface. Figure 1 shows cuts through a sample tetrahedral mesh of the brain overlayed on the corresponding initial image. Note that the mesh has been adaptively refined in the neighborhood of the lateral ventricles, so as to ensure sufficient resolution of the surfaces for the active surface algorithm.

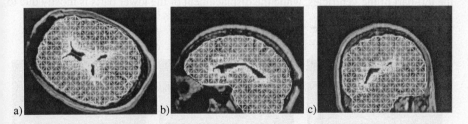

Fig. 1. Axial (a), sagittal (b), and coronal (c) cuts through tetrahedral mesh of the brain overlayed on corresponding cuts through preoperative image.

An isotropic linear elastic material is characterized by two parameters: Young's elasticity modulus E and Poisson's ratio ν [19]. They determine the elastic behavior of the object. The choice of these values is of course critical to the reliability of a physics based deformation model. Their determination has not been addressed very consistently in the literature as the coefficients used often differ significantly from study to study and do not always include the physical units of the values. Recently, Hagemann et. al.[9] published a comparative study of brain elasticity coefficients proposed by different authors, and came to the conclusion that for their application, the only comparable and meaningful values presented by other authors are the ratios of the coefficients for brain and skull. Since we are only interested in modeling the brain, and not the skull, we have chosen to use the parameters Miga et al. [6] obtained with in-vivo experiments instead ($E = 3kPa, \nu = 0.4$).

We have implemented our own FE algorithm, both for the active surface matching and for the volumetric biomechanical deformation. The assembly and solving of the linear matrix systems have been parallelized using the PETSc library [27]. The entire deformation algorithm, using a mesh with approximately one hundred thousand tetrahedra, only takes about 30 minutes on a Sun Ultra 10 440MHz workstation. Using 4 CPUs, the computation time can be reduced to 10 minutes. The average size of the edges of the larger tetrahedra was approximately 15 mm, while the smallest tetrahedra (in the neighborhood of the ventricles) had edges of 1.5 mm. However, it must be noted that the meshing algorithm yields even smaller tetrahedra in the neighborhood of boundary edges. Before we applied our algorithm, the images have been aligned using our rigid registration algorithm based upon maximization of mutual information [28] so as to account for patient movement within the magnet during the operation.

3.2 Active Surface Matching

The active surfaces are extracted from the intraoperative scan at the start of surgery, before opening the dura mater (see Figure 2a), and deformed towards the brain in a later intraoperative image (see Figure 2b). One can very clearly observe that the deformation of the cortical surface is happening in the direction of gravity and is mainly located where the dura was removed. Also, one can observe a shift, as well as a contraction of the lateral ventricles. Figure 4 shows the 3D surface deformation field the brain and the ventricles have undergone. One can very well observe that the shift is mainly affecting the left part of the ventricles, while the displacement of the lower parts is mostly due to volume loss.

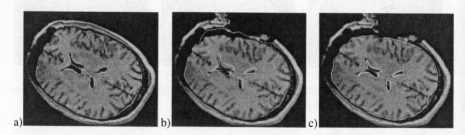

Fig. 2. Axial cut through initial (a and b) and deformed (c) active surfaces overlayed on corresponding slice of (a)initial; b,c)target) intraoperative MR image

3.3 Volumetric FE Deformation

The deformation field obtained with the active surface algorithm is then used as input for our biomechanical FE model. The algorithm yields a deformation vector for every node of the mesh. These displacements can then be interpolated back onto the image grid using the shape functions within every element of the FE mesh (see eq. 4). Figure 3 shows a slice of the deformed image as well as the image of the difference with the target. One can observe that the algorithm captured the surface shift and the ventricular thinning very accurately. The gray-level mean square difference between the target scan and the deformed original scan on the image regions covered by the mesh went down from 15 to 3. However, one can also notice that the left ventricle (lower one on the Figure) was not able to fully capture the thinning. This is due to the approximate model of the lateral ventricles we used in this experiment.

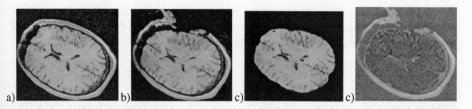

Fig. 3. Slice 29 of a) initial scan b) target scan c)initial scan deformed using our algorithm d)difference between target scan and deformed initial scan.

Figure 6 shows orthogonal cuts through the target intraoperative scan with transparently overlayed color-coding of the intensity of the deformation field. The arrows show the actual displacement of the nodes of the mesh. The extremely dense vector field in the neighborhood of the lateral ventricles is due to the adaptive refinement of the mesh at these locations.

Figure 5a shows the obtained deformation field overlayed on a slice of the initial scan, and Figure 5b shows the same slice of the initial scan deformed with the obtained deformation field. Several landmarks have also been placed on the initial scan (green crosses) and deformed onto the target scan (red crosses), and these last landmarks have also been overlayed on the target scan for comparison with the actual deformed anatomy.

Similar landmarks as those shown on Figure 5 have been placed on 4 different slices where the shift was most visible, and the distance between deformed landmarks and target landmarks (not represented here for better visibility) have been measured. The surface based landmarks on the deformed scan were within 1mm of the landmarks on the target intraoperative scan. The errors between the landmarks placed in between the mid-sagittal plane and the cortical surface were within 2-3mm from the actual landmarks. The largest errors were observed at the level of the mid-sagittal plane and ventricles, which can be explained by the fact that the surface matching of the ventricles was not perfect. Nevertheless, the algorithm reduced the distance between landmarks in the initial and the target scans from up to almost 1cm to less than 1mm for the surface-based landmarks, and from up to 6mm to 3mm or less for the sub-surface landmarks.

4 Discussion and conclusion

We have presented a biomechanical FE deformable model for the registration of image sequences showing brain shift. The biomechanical model is driven by imposing displacements to key boundary surfaces. The displacements at the boundary surfaces are computed using an active surface algorithm.

The algorithm provides us with a physically realistic deformation field and also allows us to inspect the characteristics of the deformed objects. This can be very useful for the inspection of stresses induced by the deformation of certain objects on their surroundings.

Our algorithm was able to track the surface shift the brain undergoes very accurately and partially correct for the subsurface shift. In the experiment we presented, the brain was considered to be an homogeneous elastic body. Further improvements of the algorithm include the modeling of different intracranial structures, and the assignment of the corresponding material properties. Also, we plan to investigate if changing the elasticity coefficients and introducing anisotropy (by modifying the elasticity matrix D appropriately) can improve the non-rigid registration.

Acknowledgments

Matthieu Ferrant is working towards a Ph.D. degree with a grant from the Belgian FRIA. This research was supported (in part) by NIH P41 RR13218-01, NIH P01 CA67165-03, and NIH R01 RR11747-01A. We thank Peter McL. Black for the opportunity and his patience, that enabled us to acquire the data necessary for our study.

References

1. R.D. Bucholz, D.D. Yeh, J. Trobaugh, L.L. McDurmont, C.D. Sturm, C. Baumann, J.M. Henderson, A. Levy, and P. Kessman. The correction of stereotactic inaccuracy caused by brain shift using an intraoperative ultrasound device. In J. Troccaz, E. Grimson, and R. Mösges, editors, *CVRMed-MRCAS '97*, pages 459–466. Springer-Verlag, Berlin, 1997.
2. C.R. Maurer, D.L.G. Hill, A.J. Martin, H. Liu, M. McCue, D. Rueckert, Lloret D., W.A. Hall, R.E. Maxwell, D.J. Hawkes, and C.L. Truwit. Investigation of intraoperative brain deformation using a 1.5 tesla interventional mr system: Preliminary results. *IEEE Transactions on Medical Imaging*, 17:817–825, 1998.
3. D.D. Paulsen, M.I. Miga, F.E. Kennedy, P.J. Hoopes, A. Hartov, and D.W. Roberts. A Computational Model for Tracking Subsurface Tissue Deformation During Stereotactic Neurosurgery. *IEEE Transactions on Biomedical Engineering*, 46(2):213–225, February 1999.
4. D.L.G. Hill, C.R. Maurer, A.J. Martin, S. Sabanathan, W.A. Hall, D.J. Hawkes, D. Rueckert, and C.L. Truwit. Assessment of intraoperative brain deformation using interventional mr imaging. In Berlin Springer-Verlag, editor, *MICCAI '99*, pages 910–919, 1999.
5. M.I. Miga, K.D. Paulsen, J.M. Lemery, S. D. Eisner, A. Hartov, F.E. Kennedy, and D.W. Roberts. Model-updated image guidance: Initial clinical experiences with gravity-induced brain deformation. *IEEE Transactions on Medical Imaging*, 18(10):866–874, October 1999.
6. M.I. Miga, K.D. Paulsen, P.J. Hoopes, F.E. Kennedy, A. Hartov, and D.W. Roberts. In vivo quantification of a homogeneous brain deformation model for updating preoperative images during surgery. *IEEE Transactions on Medical Imaging*, 47(2):266–273, February 2000.
7. O.M. Skrinjar and J.S. Duncan. Real time 3d brain shift compensation. In *IPMI '99*, 1999.

8. N. Hata, A. Nabavi, S. Warfield, W.M. Wells, R. Kikinis, and Jolesz F.A. A volumetric optical flow method for measurement of brain deformation from intraoperative magnetic resonance images. In C.Taylor and A.Colchester, editors, *MICCAI '99*, Lecture Notes in Computer Science, pages 928–935. Springer-Verlag, 1999.

9. A. Hagemann, Rohr K., H.S. Stiel, U. Spetzger, and Gilsbach J.M. Biomechanical Modeling of the Human Head for Physically Based, Non-Rigid Image Registration. *IEEE Transactions on Medical Imaging*, 18(10):875–884, October 1999.

10. N. Hata, R. Dohi, S. Warfield, W.M. Wells, R. Kikinis, and Jolesz F.A. Multimodality deformable registration of pre- and intraoperative images for MRI-guided brain surgery. In *MICCAI '98*, Lecture Notes in Computer Science, pages 1067–1074. Springer-Verlag, 1998.

11. R. Bajcsy and S. Kovacic. Multi-resolution Elastic Matching. *Computer Vision, Graphics, and Image Processing*, 46:1–21, 1989.

12. C. Davatzikos. Spatial Transformation and Registration of Brain Images Using Elastically Deformable Models. *Computer Vision and Image Understanding*, 66(2):207–222, May 1997.

13. G.E. Christensen, S.C. Joshi, and M.I. Miller. Volumetric Transformation of Brain Anatomy. *IEEE Transactions on Medical Imaging*, 16(6):864–877, December 1997.

14. M. Bro-Nielsen and C. Gramkow. Fast Fluid Registration of Medical Images. In *Visualization in Biomedical Computing (VBC '96)*, pages 267–276, 1996.

15. O. Skrinjar, D. Spenser, and J. Duncan. Brain Shift Modeling for use in Neurosurgery. In *MICCAI '98*, pages 641–649. Springer-Verlag, 1998.

16. S.K. Kyriacou, C. Davatzikos, S.J. Zinreich, and R.N. Bryan. Nonlinear elastic registration of brain images with tumor pathology using a biomechanical model. *IEEE Transactions on Medical Imaging*, 18(7):580–592, july 1999.

17. Y.C. Fung. *Biomechanics: Mechanical Properties of Living Tissues*. Springer-Verlag, Berlin, Germany, 1993.

18. E.K. Walsh and A. Schettini. Brain tissue elasticity and CSF elastance. *Neurolog. Research*, 12:123–127, June 1990.

19. O.C. Zienkewickz and R.L. Taylor. *The Finite Element Method*. McGraw Hill Book Co., 1987.

20. Will Schroeder, Ken Martin, and Bill Lorensen. *The Visualization Toolkit: An Object-Oriented Approach to 3D Graphics*. Prentice Hall PTR, New Jersey, second edition, 1998.

21. B. Geiger. Three dimensional modeling of human organs and its application to diagnosis and surgical planning. Technical Report 2105, INRIA, 1993.

22. M. Ferrant, S.K. Warfield, C.R.G. Guttman, F.A. Jolesz, and R. Kikinis. 3D Image Matching Using a Finite Element Based Elastic Deformation Model. In Taylor C. and Colchester A., editors, *MICCAI '99*, volume 1679, pages 202–209. Springer-Verlag, 1999.

23. L.D. Cohen and Cohen I. Finite Element Methods for Active Contour Models and Balloons for 2D and 3D Images. *IEEE Transactions on Pattern Analysis and Machine Intelligence*, 15:1131–1147, 1993.

24. O. Cuisenaire and B. Macq. Fast Euclidean Distance Transformation by Propagation Using Multiple Neighborhoods. *Computer Vision and Image Understanding*, 76(2):163–172, November 1999.

25. M. Ferrant, O. Cuisenaire, and B. Macq. Multi-Object Segmentation of Brain Structures in 3D MRI Using a Computerized Atlas. In *SPIE Medical Imaging '99*, volume 3661-2, pages 986–995, 1999.

26. J.P. Thiran, V. Warscotte, and B. Macq. A Queue-based Region Growing Algorithm for Accurate Segmentation of Multi-Dimensional Images. *Signal Processing*, 60:1–10, 1997.

27. S. Balay, W.D. Gropp, L. Curfman McInnes, and B.F. Smith. PETSc 2.0 for MPI - Portable, Extensible Toolkit for Scientific Computations. http://www.mcs.anl.gov/petsc, 1998.

28. P. Viola and W.M. Wells. Alignment by Maximization of Mutual Information. *International Journal of Computer Vision*, 24(2):137–154, 1997.

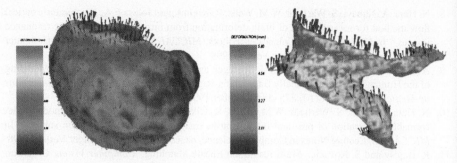

Fig. 4. 3D surface renderings of active surfaces(a) brain surface, b)lateral ventricles) with color-coded intensity of deformation field

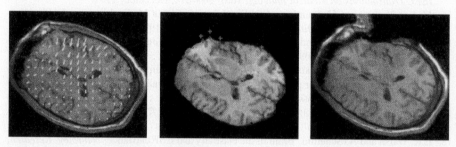

Fig. 5. a)Volumetric deformation field and initial landmarks (green) overlayed on initial intra-operative image slice, b) Same slice of deformed initial image with deformed initial landmarks (red), c) Same slice of target image with deformed landmarks

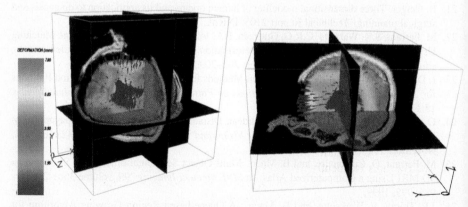

Fig. 6. 3D Volumetric Deformation field (downsampled 12x, scaled 2x) with orthogonal cuts through target intraoperative MR image and transparently overlayed color coded intensity of the deformation field a)Axial view, gravity is downwards. b)Coronal view, gravity goes from left to right

Telesurgery System for Intravascular Neurosurgery

Mitsutaka Tanimoto*1, Fumihito Arai*1, Toshio Fukuda*2
Kouichi Itoigawa*3, Masashi Hashimoto*3
Ikuo Takahashi*4 and Makoto Negoro*5

*1 Dept. of Micro System Engineering, Graduate School of Engineering,
Nagoya University
Furo-cho, Chikusa-ku, Nagoya, 464-8603, JAPAN
*2 Center for Cooperative Research in Advanced Science & Technology,
Nagoya University
*3 TOKAI RIKA Co., LTD.
*4 Anjyo Kousei Hospital
*5 Dept. of Neurosurgery, Nagoya University

Abstract The intravascular neurosurgery is one of the most successful minimally invasive surgery. However, the difficulty to operate a catheter inside the cranial blood vessels because of the complexity and narrowness of blood vessels checks increasing the number of well-skilled doctors. Moreover the X-ray camera, which is used during an operation, causes the X-ray exposure to a doctor. One of the solution to these problems is telesurgery system

A novel telesurgery system for intravascular neurosurgery is described in this paper. This system has three main components; 1) A micro force sensor which can be installed in the catheter tip, and it measures a contact force between the catheter tip and a blood vessel wall. 2) A catheter teleoperation system which enables a surgeon to perform a medical operation to a patient from a distant place. 3) A force display strategy to improve operability and safety during an operation.

An in-vitro teleoperation experiment in the intravascular operating room is also performed and the effectiveness of the telesurgery system is confirmed.

1. Introduction

In recent years, minimum invasive surgery is getting greater attention in the medical field. This technique allows us to reduce physical pain and aftereffects for patients. Here we focus on the intravascular neurosurgery as an example. A catheter is one of the medical tools frequently used in the intravascular surgeries. A catheter is inserted into blood vessels from the groin and navigated far deep into cranial blood vessels from outside of a body through blood vessels.

To operate a catheter inside the cranial blood vessels is very difficult because of the complexity and narrowness of blood vessels. Moreover, this difficulty causes not only an extension of operating time, but also the fatigue of the operators and patients. So there is a problem that there are not so many well-skilled doctors who can operate a catheter appropriately as compared with the number of patients who want to be treated by the intravascular surgery. Moreover, these doctors are always exposed to X-ray radiation whenever they perform an operation. A solution for these problems is a telesurgery system [1]-[4]. So we have studied about telesurgery system for several years, and confirm the possibility of the telesurgery system [1][5].

To perform telesurgery using teleoperation technology, correct and easy information display to understand the situation of slave side (patient side) is very important.

In catheter operation, doctors use only X-ray image as visual information source when they operate a catheter in the patient's blood vessels. The X-ray video is monochrome and it is difficult to understand the condition of catheter and blood vessels especially when the contrast is not clear. So doctors regard their tactile information as important when theyoperate a catheter.

In this paper, we focus on the force display method as an approach of our study. We develop a novel catheter teleoperation system, which is suitable to the catheter operation, and it can realize a fine force display. We also show a new control strategy to improve the performance of safety and operability by two force display methods. One is a multiple force sensor information display method, which displays several forces that are measured by several force sensors on one master device. The other one is a variable impedance characterization of master joystick, which changes the impedance parameter of the master joystick according to the situation of the medical operation. After that a result of the experiment to evaluate these force display method is shown. Finally we show the result of an in-vitro telesurgery experiment in which a doctor is not exposed to the X-ray while the catheter operation.

2. Policy of a System Design

2.1 Problems in the Catheter Teleoperation

Doctors have to maneuver the catheter tip from the inserting hole where is one meter away from the catheter tip. Because the catheter is a very flexible thin tube, control input cannot be transferred to the tip of catheter correctly. Moreover, we can not measure the position of the catheter tip correctly because there is no space to install any position sensor in the catheter tip, and there is also no useful position sensing method from the outside of the body. This means that we cannot use any method, which needs accurate position information.

Doctors regard the tactile information as very important to estimate the situation between a catheter and blood vessels. Especially they want to know the situation between the catheter tip and blood vessel wall because the penetration of a blood vessel wall is the most dangerous situation in accidents.

So we consider that the tactile information is the most important information source for the catheter operation.

2.2 Policy of a System Design

In our system the position information is provided to the doctors by the video. We do not assist the doctors in taking the position information. Of course we do understand the effectiveness of the visual assistant method such as the navigation map, computer graphic overlaying. But in this paper, we focus on the force display approach to take

up the problems, which were mentioned in the previous paragraph. We need the following functions;
1. Fine force reproduction on the master joystick
2. A mechanism which can modify the impedance parameter of the master joystick to change the operability
3. Precise position tracking of slave device

To realize these functions, we employ the force reflection type master slave system as the basic structure.

In the normal force reflection type master slave system, the mechanical impedance of the master arm can not be removed, therefore, this causes bad feeling of the force display. However, to use the impedance control, we can change the mechanical impedance of the master arm and reduce the influence to the force display.

Doctors give a position command to a catheter during a catheter operation. So slave unit must follow the master position as accurate as possible. The position tracking gain should be enough high to make a fine position tracking servo. But if the impedance of the environment is too high, the system become more unstable as the position servo gain become higher. In the case of the catheter operation, the environment is a blood vessel, which has very low impedance, therefore, the system don't become unstable even if we give high position tracking gain to the slave device.

We also take it into account that we should reduce the mechanical impedance of the master joystick when we design the system.

3. Hardware of a Catheter Teleoperation System

We designed a 2 DOF (forward & backward, rotation) master slave system. 2 DOF is necessary and sufficient for the catheter operation. The master joystick has very small mechanical impedance. The slave device has 2-axis force sensor, which measures the contact force between blood vessels and a catheter. There is a micro force sensor in the "tip" of a catheter, which measures the contact force between the catheter tip and the blood vessel wall. The detail is shown in the next section.

3.1 Structure of a Master Joystick

The master joystick is shown in Fig. 1, and motion mechanism is shown in Fig. 2. The first joint is a primastic joint and this joint is connected to the forward and backward motion of a catheter. The second joint is a revolute joint and this joint corresponds to rotating motion of a catheter. Each joint is connected to a motor with a wire mechanism. There is no gear between the motor and the joint to reduce the influence of the friction and backlash. Each joint has a optical rotary encoder, and the resolution of the position measurement is 1.6 micro meter per pulse, 0.023 degree per pulse, respectively.

3.2 Structure of a Slave Device

The slave device is shown in Fig. 3, and the mechanism is shown in Fig. 4. There is an inner unit in an outer shell. The catheter is put between a pair of rollers. A motor is connected to the rollers, and a catheter is driven forward and backward by the rollers. The outer shell rounds to generate the rotating motion of the catheter.

The inner unit is hung from the outer shell with thin beams. There is a strain gauge on the thin beam. This is a force sensor, which measures the pushing component of the contact force between a catheter and a blood vessel. The two rollers are connected by thin beams. There is also a strain gauge on the thin beam. This force sensor measures the rotating component of the contact force. The force measurement mechanism is shown in Fig. 5a and Fig. 5b.

Fig. 1 Master joystick
This joystick has 2 DOF
(forward & backward, rotation).

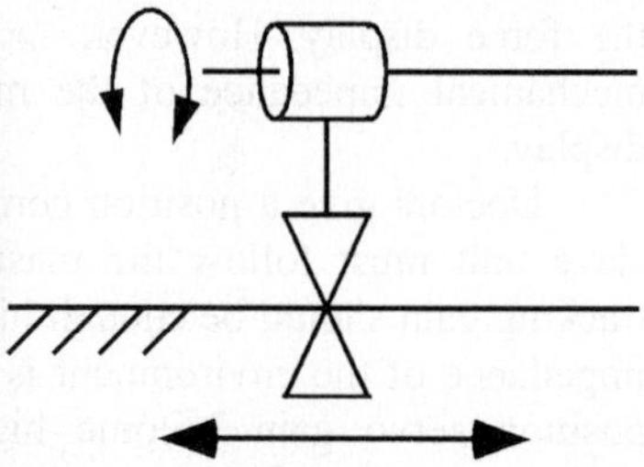

Fig. 2 Mechanism of the master
joystick

Fig. 3 Slave device; The slave device has 2 DOF (forward & backward, rotation) and there is a strain gauge force sensor for each axis. This device was first designed in 1994, and has been improved for several years.

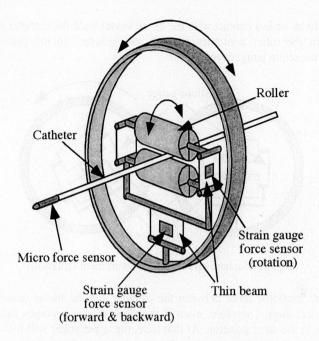

Fig. 4 Mechanism of the slave device

3.3 Micro Force Sensor[6]

A micro force sensor is installed in the tip of a catheter. We have developed this small force sensor for several years and we show the latest version of the micro force sensor here. The diameter and the length is 1.2mm, 5mm, respectively. The sensor is shown in Fig. 6. Doctors can measure the contact force between the catheter tip and the blood vessel well. The resolution is less than 0.5mN, and frequency response is fine up to 2kHz.

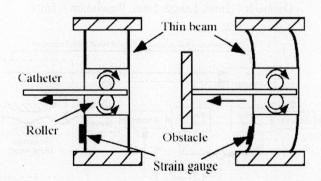

Fig. 5a Mechanism of force measurement (forward & backward)

When the catheter makes contact with the blood vessel wall, the catheter movement is disturbed. But the roller continues to push the catheter. So the pushing force is measured by the strain gauge force sensor.

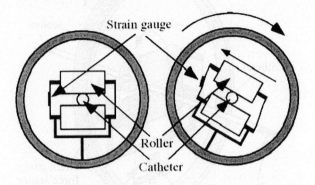

Fig. 5b Mechanism of force measurement (rotation)

There is some frictional force between the catheter and the blood vessel wall while the catheter operation. Therefore, even if the slave device becomes turn round, the catheter stays in the same position. At that time, the upper roller will move to the side direction, and this force is measured by the strain gauge on the thin beam.

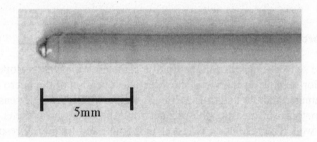

Fig. 6 Micro force sensor
Diameter 1.2mm, Length 5mm, Resolution 0.5mN

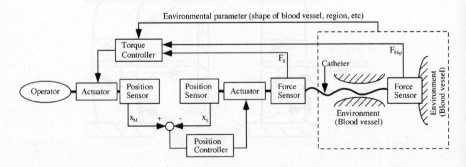

Fig. 7 Catheter teleoperation system

4. Force Display Method for the Catheter Teleoperation System

As we wrote before, we use the force reflection type master slave system as our basic structure. Our system is shown in Fig. 7. We also adopt the impedance control to change the impedance parameter of the master joystick. The target impedance parameter is changed according to the situation of the slave side. We also display 2 forces, which are measured by the strain gauge force sensor at slave device and by the micro force sensor at the tip of a catheter.

4.1 Multiple Force Sensor Information Display Method

Here we focus on the force display of the forward & backward motion. Force information is measured by two force sensors (strain gauge force sensor, micro force sensor). In the usual catheter operation, the force information is transmitted to the operator through the catheter so the operator can't distinguish between those two forces. But we can display these two forces independently because we can measure these two kinds of force respectively.

The force, which is measured by the strain gauge force sensor at the slave device, is always measured whenever the catheter is in the blood vessels. This force is changed according to the movement of the catheter. On the other hand, the force, which is measured by the micro force sensor, does not appear until the catheter tip and the blood vessel wall contact. At that time an operator will feel the contact force as if the catheter contacts to a hard wall. To select an appropriate magnifying coefficient of the contact force at the catheter tip, the operator can distinguish these two forces even if the two forces are displayed on the same master joystick. The formulation of the multiple force sensor information display is written as follows;

$$F_{sensor} = K_{FS} F_S + K_{FStip} F_{Stip} \qquad (1)$$

$$K_{FS}, K_{FStip} \neq 0 \qquad (2)$$

where, F_{sensor} is a total force, F_S, F_{Stip} is a force which is measured by the strain gauge and the micro force sensor, respectively. K_{FS}, K_{FStip} is magnifying coefficient of each force.

4.2 Variable Impedance Characterization (VIC)

In the intravascular surgery, it is difficult to recover from a mistake when a serious situation for a patient occurs. The system must have a function to reduce operator's mistakes. We realize this function by changing the impedance of the master joystick. By selecting a large master joystick impedance, the operator will not be able to move the joystick very fast. In this way the system can reduce the possibility to injure the blood vessel wall by the operator. We also use the impedance changing mechanism to improve the operability of maneuvering the catheter teleoperation system. Generally, there is an appropriate impedance size for each task, and a human naturally changes their arm or finger impedance parameter to match the task to do. But in the teleoperation system, a mechanical system disturbs an operator to get the appropriate impedance for the task. So we make the system to realize an appropriate impedance for the operator. We call this impedance changing mechanism as Variable Impedance Characterization (VIC). The VIC changes the impedance of the master arm dynamically by reflecting the situation of the slave side. The basic concept is written as;

$$F_V = M_V \ddot{x}_M + C_V \dot{x}_M \tag{3}$$
$$M_v = M_v(x_s, D, S, \text{etc.}) \tag{4}$$
$$C_v = C_v(x_s, D, S, \text{etc.}) \tag{5}$$

where, x_M is a position of master arm, F_V is a virtual resistant force against the operator. M_V, C_V are virtual mass and virtual coefficient of viscosity, respectively. x_S is a parameter of the catheter position. In the catheter operation, the two degrees of freedom are needed (motion along axial direction, and rotating motion). Basically the map of the blood vessels is expressed by a parameter of inserted length. To give an appropriate function to x_S, we can express the information of the blood vessel structure such as a position of a branch approximately. D is a diameter of the blood vessel where the tip of a catheter is placed. D will be measured in a prior checkup. If the catheter tip is in a thick blood vessel, an operator can move the catheter quickly, without increasing the risk of injuring blood vessels. But if D is getting smaller, an operator should not move the catheter so fast. So M_V and C_V will be inverse proportion to the diameter with a proper function. S is a parameter which represents the shape of the blood vessels such as a joint angle of a branch or curvature of the vessel.

These concepts are shown in Fig. 8. The virtual resistant force will become strong as the blood vessel is smaller and more complex. To summarize the force display method for the master joystick, it is described as;

$$F_M = -(F_{sensor} + F_v)$$
$$= -(K_{FS} f_S + K_{FSup} f_{Sup} + M_V \ddot{x}_M + C_V \dot{x}_M) \tag{6}$$

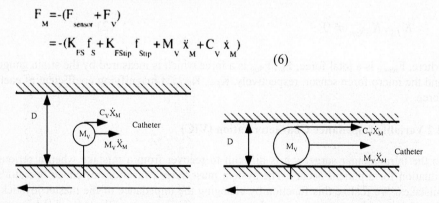

Fig. 8 (a) Virtual resistant force for master arm when the catheter tip is in a thick vessel

Fig. 8 (b) Virtual resistant force when the catheter tip is in a thin vessel or near a branch

5. In-Vitro Telesurgery Experiment without the X-Ray Exposure

An in-vitro telesurgery experiment is performed in the intravascular operating room. An experimental configuration is shown in Fig. 11. The master joystick is placed in the next room of the intravascular operating room. A doctor controls a catheter or a guidewire 5 meter away from a patient (blood vessel model made of soft plastic tube) and the X-ray camera. A doctor can see the X-ray image on the monitor but he can't see the blood vessel model directly. Force information, which is measured by 2 force sensors (micro force sensor in the tip of a catheter and the strain gauge force sensor at the slave device), is displayed on the master joystick. Therefore a doctor can feel force information. A scene of the experiment is shown in Fig. 12-14. Figure 12 shows a doctor's operation. He didn't wear a X-ray protector whole while the operation. Figure 13 is a scene of the patient side. A catheter is inserted into the blood vessel model as same as the doctor's direct operation. Figure 14 is a X ray image which is displayed to the operator. The effectiveness of the catheter teleoperation system has been confirmed through this experiment.

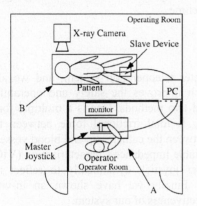

Fig. 11 Experimental Configuration

Fig. 12 Operator side
(from view point A in Fig. 11)

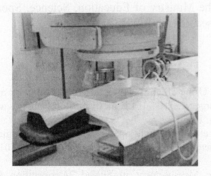

Fig. 13 Patient side
(from view point B in Fig. 11)

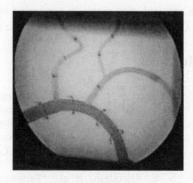

Fig. 14 X-ray image which is
displayed to a operator

6. Toward Practical Use

Easy Sterilization

The part, which contacts with the catheter, should be disposable to realize an easy, cheap but perfect sterilization system. The parts which contact to the catheter are rollers and the guide sheath which will be attached in front and behind of the rollers to support the catheter. By exchanging these parts, we can easily keep the contact point sterilized. We also have to insulate the motor, the rotary encoder and the force sensor from the water or sterilize gas to be able to reuse these parts.

Safety

Automatic emergency stop mechanism should be added to the control law. One of the most dangerous situation is to penetrate the blood vessel wall. The micro force sensor will be able to detect a sign of danger before actual penetration. Some warning sign to the operator should also be needed at the sometime.

7. Conclusion

In this paper, we have shown a new catheter teleoperation system, and we also proposed a new force display method which improves the safety and operability during a catheter teleoperation. We described two methods. One is a multiple force sensor information display (two forces are; 1. total frictional force between the catheter and blood vessels, 2. contact force between the catheter tip and blood vessels) on the master joystick. The other one is a Variable Impedance Characterization (VIC), which changes the impedance of the master joystick dynamically to reflect the situation of a catheter and blood vessels. Finally we have shown an in-vitro telesurgery experiment and confirmed the effectiveness of our system.

Acknowledgement

This research was partially supported by the Ministry of Education, Science, Sports and Culture, Grant-in-Aid for Scientific Research (No.10003371).

References

[1]   F. Arai, et al, "Multimedia Tele-surgery Using High Speed Optical Fiber Network and Its Application to Intravascular Neurosurgery", Proc. IEEE Int'l Conf. on Robotics and Automation ICRA'96, Vol.1, pp.878-883, 1996.4

[2]   M. Mitsuishi, Y. Iizuka, H. Nakanishi, et al., "Remote operation of a micro surgical System", Proc. IEEE International Conference on Robotics and Automation, pp.1013-1019, 1998

[3]   S. E. Salcudean, et al., "Performance Measurement in Scaled Teleoperation for Microsurgery", CVRMed-MRCAS'97, Grenoble, France, pp.789-798, 1997.

[4]   J.W. Hill, "Telepresence Surgery Demonstration System", Proc. 1994 IEEE Int'l Conf. on Robotics and Automation, Vol. 3, pp. 2302-2307, 1994

[5]   M.Tanimoto, et al., "Augmentation of Safety in Teleoperation System for Intravascular Neurosurgery", IEEE Int'l Conf. on Robotics and Automation, pp. 2890 -2895, 1998

[6]   M.Tanimoto, et al., "Micro Force Sensor for Intravascular Neurosurgery", IEEE Int'l Conf. on Robotics and Automation, Vol. 2, pp. 1561-1566, 1997

[7]   K.S. Kwon, et al., "A Robot with Improved Adcolute Positioning Accuracy for CT Guided Stereotactic Surgery", IEEE Trans. on Biomedical Engineering, pp. 153-161, February 1988

[8]   S.E. Salcudean and J. Yen, "Towards a Force-reflecting Motion-Scaling System for Microsurgery", Proc. 1994 IEEE Int'l Conf. on Robotics and Automation, Vol. 3, pp. 2296-2301, 1994

[9]   K.Kaneko, et al., "Impedance Shaping based on Force Feedback Bilateral Control in Macro-Micro Teleoperation System", IEEE Int'l Conf. on Robotics and Automation, Vol. 1, pp. 710-717, 1997

Automatic Recognition of Cortical Sulci Using a Congregation of Neural Networks

Denis Rivière[1], Jean-François Mangin[1], Dimitri Papadopoulos-Orfanos[1], Jean-Marc Martinez[2], Vincent Frouin[1], and Jean Régis[3]

[1] Service Hospitalier Frédéric Joliot, CEA, 91401 Orsay, France
mangin@shfj.cea.fr, http://www-dsv.cea.fr/
[2] Service d'Etude des Réacteurs et de Mathématiques Appliquées, CEA, Saclay
[3] Service de Neurochirurgie Fonctionnelle et Stéréotaxique, La Timone, Marseille

Abstract. This paper describes a complete system allowing automatic recognition of the main sulci of the human cortex. This system relies on a preprocessing of MR images leading to abstract structural representations of the cortical folding. The representation nodes are cortical folds, which are given a sulcus name by a contextual pattern recognition method. This method can be interpreted as a graph matching approach, which is driven by the minimization of a global function made up of local potentials. Each potential is a measure of the likelihood of the labelling of a restricted area. This potential is given by a multi-layer perceptron trained on a learning database. A base of 26 brains manually labelled by a neuroanatomist is used to validate our approach. The whole system developed for the right hemisphere is made up of 265 neural networks.

1 Introduction

The development of image analysis methods dedicated to automatic management of brain anatomy is a widely addressed area of research. A number of works focus on the notion of deformable atlases, which can be elastically transformed to reflect the anatomy of new subjects. An exhaustive bibliography of this approach initially proposed by Bajcsy [1] is largely beyond the scope of this paper (see [23] for a recent review). The complexity and the striking inter-individual variability of the human cortex folding patterns, however, have led several groups to question the behaviour of the deformable atlas framework at the cortex level. The main issues to be addressed are the following:

What are the features of the cortex folding patterns which should be matched across individuals? While some sulci clearly belong to this set of landmark features because they are usually considered as boundaries between different functional areas, nobody knows to which extent secondary folds should play the same role [18]. Some answers to this important issue could stem from foreseeable advances in functional imaging and mapping of cortex connectivity. Deformable atlas methods rely on the optimization of some function which realizes a trade-off between similarity to the new brain and deformation cost. Whatever the approach, the function driving the deformations is non convex. When high-dimensional deformation fields are used, this non-convexity turns out to be particularly problematic since standard optimization approaches are bound to lead to a local optimum. While multi-resolution methods may guarantee that an "interesting optimum" is found, the complexity of the cortical folding patterns implies that a lot of other similar optima exist. An important issue is raised by this observation: is the global optimum the best one according to the pairing of sulcal landmarks?

To overcome some of the difficulties related to the non-convexity of the problem, several teams have proposed to design composite similarity functions relying on manual identifications of the main sulci [23, 4]. These composite functions impose the pairing of homologous sulcal landmarks. While a lot of work remains to be done along this line, this evolution seems required to adapt the deformable atlas paradigm to the human cortex. This new point of view implies a preprocessing of the data in order to extract and identify automatically these sulcal landmarks, which is the subject of our paper.

Our system may be considered as a symbolic version of the deformable atlas approach. The framework is made up of two stages [14, 15]. An abstract structural representation of the cortical topography is extracted first from each new T1-weighted MR image. A symbolic pattern recognition method is then used to label automatically the main sulci. This method can be interpreted as a graph matching approach. Hence the usual iconic anatomical template is replaced by an abstract structural template. The one to many matching between the template nodes and the nodes of one structural representation is simply a labelling operation. This labelling is driven by the minimization of a global function made up of local potentials. Each local potential is a measure of the likelihood of the labelling of a restricted cortex area. This potential is given by a virtual expert in this area made up of a multi-layer perceptron trained on a learning database.

While the complexity of the preprocessing stage required by our method may appear as a weakness compared to the straightforward use of continuous deformations, it results in a fundamental difference. While the evaluation of functions driving continuous deformations is costly in terms of computation, the function used to drive the symbolic recognition relies on only a few hundred labels and can be evaluated at a low cost. Hence stochastic optimization algorithms can be used to deal with the non-convexity problems. In fact, working at a higher level of representation leads to more efficiency for the pattern recognition process, which explains an increasing interest in the community [10, 8, 9].

2 The preprocessing stage

This section describes briefly the robust sequence of treatments that converts automatically a T1-weighted MR image in an abstract structural representation of the cortical topography. The whole sequence requires about one hour on a conventional workstation, including sophisticated triangulation of hundreds of objects. All the steps have been validated with at least 50 different images acquired with 6 different scanners using various MR sequence parameters. The processing sequence is the following:

1. Bias Correction via minimization of the grey level distribution entropy [12];
2. Histogram scale space analysis to assess tissue statistics [13];
3. Brain segmentation using mathematical morphology and Markov fields [13];
4. Hemisphere separation using mathematic morphology [16];
5. Segmentation of grey/CSF using topology preserving deformations [14];
6. Homotopic skeletonization driven by an isophote based erosion [14];
7. Segmentation of the skeleton into topologically simple surfaces [14];
8. Simple surfaces are split according to fold depth local minima, which are considered as clues of a putative burried gyrus [18, 14, 11, 10];
9. The objects provided by the last step are finally gathered in a structural representation which describes their relationships. Three kinds of links are created between these nodes: ρ_T links represent skeleton splits related to the simple surface definition; ρ_P links represent splits related to the presence of a putative burried gyrus; and ρ_C links represent a neighborhood relationship geodesic to the hemisphere hull [14]. The resulting graph is enriched with numerous semantic attributes dedicated to the recognition stage. Nodes are described by their size, maximal depth, gravity center localization, and mean normal. Links of type ρ_T and ρ_P are described by their length, extremity localizations, minimal and maximal depth, and mean direction. Links of type ρ_C are described by their size and the localization of the closest points of the linked nodes. The resulting attributed graph is supposed to include all the information required by the sulcus recognition process.

3 The learning database

Our preprocessing tool can be viewed as a compression system which provides for each individual brain a synthetic description of the cortex folding patterns. A sophisticated 3D browser allows our neuroanatomist to label manually each node with a name chosen in a list of anatomical entities. The lack of a validated explanation of the structural variability of the human cortex is an important problem during this labelling. Indeed, standard sulci are often split into several pieces which leads to ambiguous configurations [17].

It has to be understood that this situation prevents the definition of an unquestionable gold standard to be reached by any sulcus recognition method. Therefore, one of the aims of our research is to favour the emergence of new anatomical descriptions relying on smaller sulcal entities than the usual ones. According to different arguments that would be too long to develop in this paper, these units, the primary cortical folds that appear on the fœtal cortex, are stable across individuals; a functional delimitation meaning is probably attached to them [18]. During ulterior stages of brain growth, some of these sulcal roots merge with each other and form different patterns depending on the subjects. The more usual patterns correspond to the usual sulci. In our opinion, some clues on these sulcal root fusions can be found in the depth of the sulci.

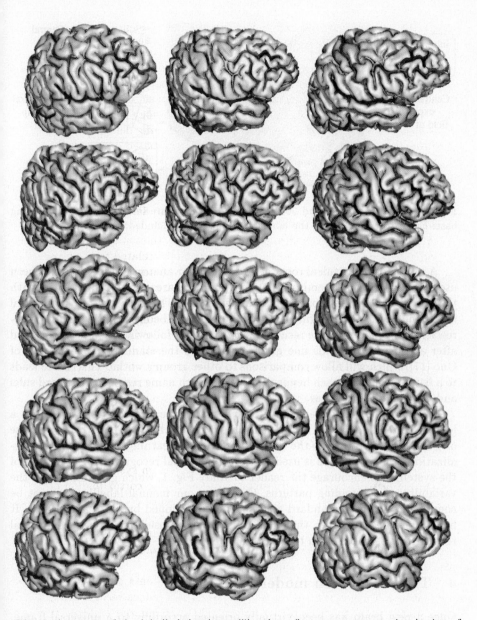

Fig. 1. A survey of the labelled database. The three first rows present nine brains of the learning base, the fourth row presents three brains of the test base, and the last row presents three brains of the generalization base. Each color labels one entity of the anatomical model. Several hues of the same color are used to depict different roots or stable branches of one given sulcus. For instance, color codes of main frontal sulci are: 2 reds = central, 5 yellows = precentral, 3 greens = superior, 2 blues = intermediate, 4 purples = inferior, 8 blues Lateral fissure, red = orbitary, rose = marginal, yellow = transverse...

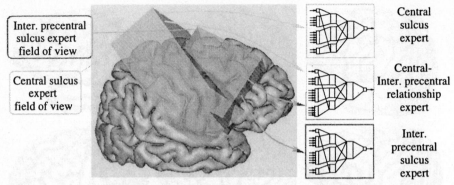

Fig. 2. 60 sulcus experts and 205 relationship experts are inferred from the learning base. Each expert evaluates the labelling of the nodes included in its field of view.

A model of these sulcal roots derived from our anatomical research has been used to label 26 right hemispheres. This model shares striking similarities with the model recently proposed by Lohman in [10]. This new type of anatomical model, however, requires further validations before being properly used by neuroscientists. Therefore, the results described in the following have been obtained after a conversion of this fine grain labelling to the standard nomenclature of Ono [17], which will allow comparisons to other group's works. This choice leads to a list 60 names for each hemisphere, where each name represent standard sulci and usual sulcus branches.

The 26 right hemispheres have been randomly separated into three bases: a learning base made up of 16 brains is used to train the local experts; a test base of 5 brains is used to stop the training before over-learning; and finally, a generalization base of 5 brains is used to assess the actual recognition performance of the system. We encourage the reader to study Fig. 1, which gives an idea of the variability of the folding patterns. Of course, our manual labelling can not be considered as a gold standard and could be questioned by other anatomists. It has to be noted, however, that a lot of information used to perform the manual recognition is concealed in the depth of the sulci.

4   The Markovian model

Once a new brain has been virtually oriented according to a universal frame, for instance the Talairach system, the cortical area where one specific sulcus can be found is relatively small. This localization information can already lead to interesting recognition results [8, 9]. Localization, however, is largely insufficient to perform a complete recognition. Indeed, a lot of discriminating power only stems from contextual information. This situation has led us to introduce a Markovian framework [15]. This framework provides us with a simple way of designing a probability distribution for the labelling: a Gibbs distribution relying on local potentials [6]. These potentials are inferred from the learning base. They embed interactions between the labels of neighboring nodes. These interactions are related to contextual constraints that must be adhered to in order to get anatomically plausible recognitions.

Two families of potential are designed. The first family evaluates the sulcus shapes and the second family evaluates the spatial relationships of pairs of neighboring sulci. Each potential depends only on the labels of a localized set of nodes, which corresponds to the Markov field interaction clique [6]. For a given graph, each clique corresponds to the set of nodes included in the field of view of the underlying expert (see Fig. 2). For sulcus experts, this field of view is defined from the learning base as a parallelepiped of the Talairach coordinate system. The parallelepiped is the bounding box of the sulcus instances in the learning base computed along the inertia axes of this instance set. For sulcus pair relationship experts, the field of view is simply the union of the fields of view of the two related sulcus experts. Pairs of sulci are endowed with an expert if at least 10% of the learning base brains possess an actual link between the two related sulci. For the model of the right hemisphere described in this paper, this rule leads to 205 relationship experts. The whole system, therefore, is made up of a congregation of 265 experts. Each expert potential is defined as the output of a standard multi-layer perceptron [22]. The expert single opinions are gathered by the Gibbs distribution, which gives the likelihood of a global labelling. Hence, the sulcus recognition amounts to minimizing the sum of all of the perceptron outputs.

5 Expert training

Each expert is a multi-layer perceptron which is trained using the usual backpropagation algorithm [22]. Expert inputs are vectors of descriptors of the anatomical feature for which the expert is responsible. These descriptors constitute a compressed code of sulcus shapes and relationships. Sulcus shapes are summarized by 27 descriptors and sulcus relationships by 23 descriptors. These descriptors are computed from a small part of the graph corresponding to one single label (sulcus) or one pair of labels (relationship). A few Boolean logical descriptors are used to inform of the existence of a non-empty instance of some anatomical entity (sulcus, junction with the hemisphere hull, actual link between two sulci...). Continuous semantic descriptors are inferred from the semantic attributes [14] of the subgraph to be analyzed:

Sulcus expert: total size, minimal and maximal depth, gravity center localization, and mean orientation; length, mean direction, and extremity localization of the junction with the hemisphere hull;

Relationship expert: total sizes of both sulci; total sizes of each kind of links; minimal distance between the sulci; semantic attributes of the contact link (junction or buried gyrus [14]), namely junction localization, mean direction, distances between the contact point and the closest sulcus extremities, respective localization of the sulci, and angle between sulcus hull junctions.

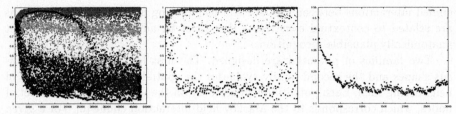

Fig. 3. A survey of the learning of the intermediate precentral sulcus expert. Dark points represent correct examples, light points ambiguous counter-examples, and middle grey points standard counter-examples. The first chart shows the evolution of the perceptron output for the learning base during the training. The second chart is related to the output for the test base. The third chart presents the evolution of the mean error on the test base. A consistent increase of this criterion corresponds to overlearning beginning.

Integer syntactic descriptors of the subgraph topology are also computed:

Sulcus expert: number of connected components using all links or only contact links; number of proximity links between contact related connected components, maximal gap between these components (continuous); number of internal links of "burried gyrus" type;

Relationship expert: numbers of contact related connected components of both sulci, numbers of such components implied in actual links between the sulci, number of contact points, number of links of "burried gyrus" type between the sulci, minimal depth of such links (continuous).

The perceptrons include two hidden layers and one output neuron. The first hidden layer is not fully connected to the input layer. Indeed, various experiments have led to the conclusion that a consistent packaging of the inputs lead to a better generalization power. Hence, the first hidden layer is split in several blocks fed by a specific subset of inputs (see Fig. 2). For instance, syntactic descriptors feed one specific block.

The supervised learning of the experts relies on two kinds of examples. Correct examples extracted from the learning base must lead to the lowest output. Counter-examples are generated from correct examples through random modifications of some labels of the clique nodes. A continuous distance between the correct example and the generated counter-example is used to choose the taught output. Small distances lead to intermediate outputs while larger distances lead to the highest output. This balancing of the counter-example outputs was necessary to overcome learning difficulties fathered by ambiguous counterexamples.

The backpropagation algorithm is iteratively applied to the learning base using a one to ten ratio between correct examples and counter-examples. A stop criterion defined from the test base is used to avoid over-learning. This criterion relies on the sum of two mean errors computed respectively for correct examples and for counter-examples of the test base. The learning is stopped when this criterion presents a consistent increase or after a maximum number of iterations

Base (nb. brains)	Node number			Recognition (%)			U_{base}			$U_{annealing}$		
	min	mean	max	min	mean	max	min	mean	max	min	mean	max
Learning (16):	232	**276**	339	79	**86**	92	-106	**-85**	-75	-108	**-95**	-81
Test (5):	217	**277**	321	73	**80**	85	-68	**-46**	-24	-96	**-85**	-74
Generalization (5):	236	**281**	301	71	**76**	80	-33	**-25**	-20	-74	**-69**	-65

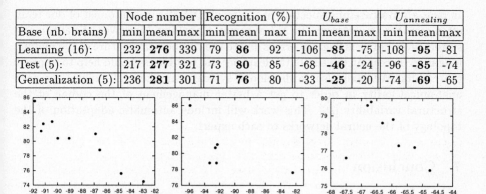

One brain of the learning base   One brain of the test base   One brain of the generalization base

Fig. 4. (top) Node number, recognition rate, energy of the manual labelling (U_{base}), and energy of the automatic labelling for each base. **(bottom)** Recognition rate relative to final energy for ten different minimizations.

(cf. Fig. 3). The minimum value of this criterion is used to get a measure of confidence in the expert opinion. This measure is used to weight the output of this expert during the recognition process.

6 Results

The sulcus recognition process consists of the minimization of the energy made up of the weighted sum of the expert outputs. Each node label is chosen in a subset of the sulcus list corresponding to the expert fields of view which include this node. The minimization is performed using a stochastic algorithm inspired by the simulated annealing principle. This algorithm relies on a dedicated version of the Gibbs sampler [6]. For the following results, the minimization lasts about two hours on a conventional workstation. While an optimized implementation is planned in order to achieve a significant speed-up, it should be noted that the manual labelling work is even slower.

A global measure is proposed to assess the correct recognition rate. This measure corresponds to the proportion of cortical folds correctly identified according to the manual labelling. The contribution of each node to this global measure is weighted by its size (the number of voxels of the underlying skeleton [14]). The mean recognition rate on each of the three bases is proposed in Fig. 4. In order to check the reproducibility of the process, the minimization has been repeated ten times with different initializations for one brain of each base (see Fig. 4). This experiment has shown that the recognition rate is related to the depth of the local minimum obtained by the optimization process.

The recognition rate obtained for the generalization base is 76%, which is very encouraging considering the variability of the folding patterns. As matters stand relative to our understanding of this variability, it should be noted that numerous "errors" of the system correspond in fact to ambiguous configurations.

In fact, after a careful inspection of the results, the neuroanatomist of our team often admits to a preference for the automatic labelling. Moreover, the automatic system often corrects flagrant errors due to the cumbersome nature of the manual labelling. Increasing the size of the learning and test bases should help to improve the results. We plan also to develop a system using several experts for each anatomical entity in order to get a better management of the coding of the structural variability [20]. This work will include automatic adaptation of the topology of the neural networks to each expert.

7 Conclusion

A number of approaches relying on the deformable atlas paradigm consider that anatomical a priori knowledge can be completely embedded in iconic templates. While this point of view is very powerful for anatomical structures presenting low inter-individual variability, it seems insufficiently versatile to deal with the human cortical anatomy. This observation has led several teams to investigate approaches relying on higher levels of representation. All these approaches rely on a preprocessing stage which extracts sulcal related features describing the cortical topography. These features can be sulcal points [3], sulcal lines inferred from skeletons [21, 2], topologically simple surfaces [14], 2D parametric models of sulcal median axis [7, 24, 25], crest lines [5, 11] or cortex depth maxima [10, 19]. In our opinion, this direction of research can lead further than the usual deformable template approach. In fact these two types of work should be merged in the near future. It has to be understood, however, that some of the challenging issues about cortical anatomy mentioned in the introduction require new neuroscience results to be obtained. As such, image analysis teams addressing this kind of research must be responsible for providing neuroscientists with new tools in order to speed-up anatomical and brain mapping research. Our system is used today to question the current understanding of the variability and to help the emergence of better anatomical models. Various direct applications have been developed in the fields of epilepsy surgery planning and brain mapping.

References

1. Ruzena Bajcsy and Chaim Broit. Matching of deformed images. In *IEEE Pric. Sixth Int. Conf. on Pattern Recognition*, pages 351–353, October 1982.
2. A. Caunce and C. J. Taylor. Using local geometry to build 3D sulcal models. In *IPMI'99, LNCS 1613*, pages 196–209. Springer Verlag, 1999.
3. H. Chui, J. Rambo, J. Duncan, R. Schultz, and A. Rangarajan. Registration of cortical anatomical structures via robust 3D point matching. In *IPMI'99, LNCS 1613*, pages 168–181. Springer Verlag, 1999.
4. D. L. Collins, Le Goualher G., and A. C. Evans. Non-linear cerebral registration with sulcal constraints. In *MICCAI'98, LNCS-1496*, pages 974–984, 1998.
5. J. Declerck, G. Subsol, J.-P. Thirion, and N. Ayache. Automatic retrieval of anatomical structures in 3D medical images. In *CVRMed, LNCS-905*, pages 153–162, 1995.

6. S. Geman and D. Geman. Stochastic relaxation, gibbs distributions, and the bayesian restoration of images. *IEEE PAMI*, 6(6):721–741, 1984.

7. G. Le Goualher, C. Barillot, and Y. Bizais. Modeling cortical sulci using active ribbons. *Int. J. Pattern Recognit. Artific. Intell.*, 11(8):1295–1315, 1997.

8. G. Le Goualher, D. L. Collins, C. Barillot, and A. C. Evans. Automatic identification of cortical sulci using a 3D probabilistic atlas. In *MICCAI'98, MIT, LNCS-1496*, pages 509–518. Springer Verlag, 1998.

9. G. Le Goualher, E. Procyk, D. L. Collins, R. Venugopal, C. Barillot, and A. C. Evans. Automated extraction and variability analysis of sulcal neuroanatomy. *IEEE Medical Imaging*, 18(3):206–217, 1999.

10. G. Lohmann and Y von Cramon. Automatic detection and labelling of the human brain cortical folds in MR data sets. In *ECCV*, pages 369–381, 1998.

11. A. Manceaux-Demiau, J.-F. Mangin, J. Regis, O. Pizzato, and V. Frouin. Differential features of cortical folds. In Springer-Verlag, editor, *CVRMED/MRCAS, Grenoble, LNCS-1205*, pages 439–448, 1997.

12. J.-F. Mangin. Entropy minimization for automatic correction of intensity nonuniformity. In *MMBIA, South Carolina*, pages 162–169, 2000.

13. J.-F. Mangin, O. Coulon, and V. Frouin. Robust brain segmentation using histogram scale-space analysis and mathematical morphology. In *MICCAI'98, MIT, LNCS-1496*, pages 1230–1241. Springer Verlag, 1998.

14. J.-F. Mangin, V. Frouin, I. Bloch, J. Regis, and J. López-Krahe. From 3D MR images to structural representations of the cortex topography using topology preserving deformations. *J. Mathematical Imaging and Vision*, 5(4):297–318, 1995.

15. J.-F. Mangin, J. Regis, I. Bloch, V. Frouin, Y. Samson, and J. Lopez-Krahe. A Markovian random field based random graph modelling the human cortical topography. In *CVRMed, Nice, LNCS-905*, pages 177–183. Springer-Verlag, 1995.

16. J.-F. Mangin, J. Régis, and V. Frouin. Shape bottlenecks and conservative flow systems. In *MMBIA, San Francisco*, pages 319–328, 1996.

17. M. Ono, S. Kubik, and C. D. Abernethey. *Atlas of the Cerebral Sulci*. Georg Thieme Verlag, 1990.

18. J. Régis, J.-F. Mangin, V. Frouin, F. Sastre, J. C. Peragut, and Y. Samson. Generic model for the localization of the cerebral cortex and preoperative multimodal integration in epilepsy surgery. *Stereotactic Functional Neurosurgery*, 65:72–80, 1995.

19. M. E. Rettmann, C. Xu, D. L. Pham, and J. L. Prince. Automated segmentation of sulcal regions. In *MICCAI'99, Cambridge, UK, LNCS-1679, Springer-Verlag*, pages 158–167, 1999.

20. D. Rivière, J.-F. Mangin, J.-M. Martinez, F. Chavand, and V. Frouin. Neural network based learning of local compatibilities for segment grouping. In *SSPR'98, LNCS-1451*, pages 349–358. Springer Verlag, 1998.

21. N. Royackkers, M. Desvignes, H. Fawal, and M. Revenu. Detection and statistical analysis of human cortical sulci. *NeuroImage*, 10:625–641, 1999.

22. D. E. Rumelhart, G. E. Hinton, and R. J. Williams. *Learning internal representations by error backpropagation*, pages 318–362. MIT Press, 1986.

23. P. M. Thompson, R. P. Woods, M. S. Mega, and A. W. Toga. Mathematical/computational challenges in creating deformable and probabilistic atlases of the human brain. *Human Brain Mapping*, 9:81–92, 2000.

24. M. Vaillant and C. Davatzikos. Finding parametric representations of the cortical sulci using active contour model. *Medical Image Analysis*, 1(4):295–315, 1997.

25. X. Zeng, L. H. Staib, R. T. Schultz, H. Tagare, L. Win, and J. S. Duncan. A new approach to 3D sulcal ribbon finding from MR images. In *MICCAI'99, Cambridge, UK, LNCS-1679, Springer-Verlag*, pages 148–157, 1999.

BrainSuite: An Automated Cortical Surface Identification Tool

David W. Shattuck and Richard M. Leahy

Signal and Image Processing Institute, Department of Electrical Engineering,
University of Southern California, Los Angeles, California USA 90089-2564
{shattuck,leahy}@sipi.usc.edu

Abstract. We describe a new magnetic resonance (MR) image analysis tool that produces cortical surface representations with spherical topology from MR images of the human brain. The tool provides a sequence of low-level operations in a single package that can produce accurate brain segmentations in clinical time. The tools include skull and scalp removal, image nonuniformity compensation, voxel-based tissue classification, topological correction, rendering, and editing functions. The collection of tools is designed to require minimal user interaction to produce cortical representations. In this paper we describe the theory of each stage of the cortical surface identification process. We then present validation results using real and phantom data.

1 Introduction

Surface representations of the human cerebral cortex are important for visualization and analysis of neuroimaging data [1] and as constraints for localizing functional activation from magneto-encephalography (MEG) and electro-encephalography (EEG) data [2, 3]. MR imaging offers neuroanatomical detail, but extraction of cortical surfaces from MR imagery faces several problems, including measurement noise, partial volume effects, and image nonuniformity due to magnetic field inhomogeneities. In this paper we describe a new tool that provides a comprehensive approach to extracting a representation of the cerebral cortex from T1-weighted MR images.

The problems we address in our work have been addressed by many others. Wells *et al.* presented an Expectation-Maximization (E-M) approach to image nonuniformity and tissue classification [4]. Kapur *et al.* combined the E-M approach with morphological and active contour methods to isolate and segment the brain [5], and later incorporated Gibbs and geometric priors into the E-M method to improve the classifier [6]. Sled *et al.* addressed the problem of nonuniformity using an E-M approach that estimates a gain field to sharpen the histogram of the MR image; this field is kept smooth using a cubic B-spline [7]. Software for this method is publicly available via the Internet (http://www.bic.mni.mcgill.ca/software/N3/) under the name Non-parametric Non-uniform intensity Normalization, or N3.

The general problem of identifying the cortex in MR imagery has been approached from several directions. Atlas-based approaches to segmentation do not typically perform well in the cortex due to intersubject variability. To overcome this problem, Collins *et al.* combined their atlas-based method (ANIMAL) with a low-level method (INSECT) [8]. Active contour methods often perform best after initialization with a low-level segmentation, as described by Xu *et al.* and Pham *et al.* [9, 10]. They find the cerebral cortex using fuzzy C-means classification. The fuzzy membership set is then repeatedly median-filtered until a topologically spherical isocontour is found; this contour represents the boundary between white matter and grey matter. A gradient vector flow method is then used to move this boundary to the medial layer of the cortex. Zeng *et al.* presented an identification sequence that uses a level-set method to find coupled surfaces representing the interior and exterior boundary of the cerebral cortex [11]. Dale *et al.* described a complete method for cortical surface identification that first smoothes image nonuniformities with a cubic spline that normalizes white matter peaks throughout the image, then strips skull and scalp using a deformable template, and finally labels the white matter based on intensity and neighborhood information [12]. Hand-editing is required to correct topological defects in the segmentation prior to tessellation. The tessellated surface is then deformed to refine the cortex.

The tool we describe in this paper is based on a sequence of low-level operations, resulting in a fast yet accurate method to identify the cortex. First, skull and scalp are removed from the image using edge detection and morphological processing [13, 14]. The stripped brain is then processed to remove image nonuniformities using a parametric model that adjusts local intensities of the image to match global properties [14]. Next, the intensity corrected brain is classified at the voxel level into key tissue types: white matter, grey matter, cerebrospinal fluid, and combinations of these [14]. The white matter corresponding to the telencephalon is selected and further processed to ensure that key neuroanatomical structures, such as the ventricles, are interior to the initial white matter volume. This volume is then processed using a graph-based approach to remove topological inconsistencies in the volume, which is the most novel aspect of our approach [15]. The final volume may then be tessellated using the Marching Cubes algorithm [16] to produce a cortical surface that is topologically equivalent to a sphere.

2 Methods

2.1 Removal of Extraneous Tissue

Removal of non-brain tissue from the MR volume facilitates later stages in the algorithm as fewer voxels and fewer tissue types are involved. We refine the brain extraction procedure initially described in [13] to remove skull, scalp, and other tissue from the MR image. This portion of the algorithm has been publicly released as the Brain Surface Extractor (http://neuroimage.usc.edu/bse/), and is currently in use in several neuroanatomical studies. The process begins with an anisotropic diffusion filter [17], which smoothes contiguous tissue regions while respecting the edge boundaries that occur between these regions. We then use a three-dimensional Marr Hildreth edge detector [18], which produces a binary image with edges that are closed boundaries. These edges separate the volume into several distinct objects. We seek the object whose boundary represents the space between the brain and the skull and dura; this object is assumed to be the largest central connected component in the binary edge image.

The use of the anisotropic diffusion filter enhances this boundary, but some connections between the brain and other tissues will still remain due to noise and small anatomical structures such as the optic nerves. Simply finding the largest central connected component in the edge volume will leave several other structures attached to the brain. We use morphological erosion with a 3D rhombus operator of radius one to break these attachments from the brain prior to labeling the connected components. The desired component is then dilated to restore it to approximately its original size. A morphological closing operation with an octagonal element of size four is applied to fill surface pits that may be present due to small errors in the edge detection operation. This operation will also smooth some aspects of the surface detail. However, this will be recovered in later operations when the voxels are classified according to tissue type.

2.2 Image Nonuniformity Compensation

Inhomogeneneity in the magnetic fields during image acquisition and magnetic susceptibility variations in the scanned subjects cause intensity nonuniformities in MR images that prevent characterization of voxel tissue content based solely on image intensity. As a result, segmentation as well as quantitative studies of MR images require compensation for these nonuniformities. We use a parametric tissue measurement model to estimate the local variations in the gain field of the image by comparing the intensity properties of the whole MR volume with the properties of local neighborhoods within the image.

We extend the tissue measurement model of Santago and Gage [19], which describes the probability of measuring a particular intensity given the relative fractions of key tissue types including partial volume types. We incorporate a spatially variant bias term, b_k, which describes the nonuniformity effect at the k-th voxel measurement, x_k:

$$p(x_k|b_k) = \sum_{\gamma \in \Gamma} p(\gamma)p(x_k|\gamma, b_k),$$ (1)

where γ is the tissue class from the set Γ including white matter (WM), grey matter (GM), cerebrospinal fluid (CSF), and partial volume combinations of these (WM/GM, GM/CSF, and CSF/Other). The function $p(x|\gamma, b)$ is determined by the tissue types present in the particular voxel. For voxels composed of a single tissue type (pure voxels),

$$p(x|\gamma, b) = G_{b\mu_\gamma, \sigma}(x),$$ (2)

where $G_{\mu,\sigma}(x)$ is a Gaussian density function of mean μ and variance σ^2. The pure tissue voxels have nonstationary mean values that vary multiplicatively from a global mean value, μ_γ, according to the bias b_k. For mixed tissue types,

$$p(x|\gamma, b) = \int_0^1 G_{\alpha b \mu_a + (1-\alpha) b \mu_b, \sigma}(x) d\alpha, \tag{3}$$

where α is a mixture parameter assumed to be a uniformly distributed random variable as per Santago and Gage, μ_a and μ_b are the global mean values of the pure tissue types associated with the mixture class γ. For each particular image, we compute *a priori* estimates of the global tissue mean values and noise variance from automated analysis of the stripped brain's intensity histogram.

We assume the bias in the image changes very slowly spatially. Within a neighborhood centered about a particular voxel, we approximate the bias as constant. The mixture model (1) then describes the measurements made within the entire region, conditioned on knowing the fractional tissue content and bias of the region.

We select a uniformly spaced lattice of points in the image at which we estimate the bias field. The measurements taken in a neighborhood about each point are described by the mixture model allowing us to fit our model to a normalized histogram of the region. This provides estimates of both the tissue fractions and the bias within the region. We next use a set of outlier rejection steps on the bias estimates to eliminate poor fits between the model and the histogram. Because the bias field is smooth and slowly varying it is sufficient to estimate its values at a coarser sampling than the original image. We then use a tri-cubic B-spline to smooth and interpolate the robust estimate points, providing us with an estimated value for the bias field at each point in the image. We finally remove the bias from the image by dividing the intensity value of each voxel by its corresponding bias estimate.

2.3   Partial Volume Classification

Compensation for the nonuniformity in the MR image greatly simplifies the tissue classification problem. However, noise is still present in the system. For the most part, the brain image can be described by regions of contiguous tissue type. This allows us to use a Gibbs' prior that incorporates a spatial model for the brain tissues into our classification scheme. The model we have selected is quite simple:

$$p(\Lambda) = \frac{1}{Z} \exp\left[-\beta \sum_k \sum_{j \in N_k} \delta(\lambda_k, \lambda_j)\right], \tag{4}$$

where $\Lambda = \{\lambda_1, \lambda_2, \ldots \lambda_{|\Lambda|}\}$ is the set of labels describing the image, Z is a scaling constant to ensure that we have a proper density function and β controls the degree of influence of the prior on the voxel labels. N_k is the D18 neighborhood (neighbors share an edge or face) about the k-th voxel. The δ terms govern the likelihood of different tissue labels being neighbors, hence we set $\delta(\gamma_k, \gamma_j)$ to -2 if labels k and j are identical, -1 if they have a common tissue type, and 1 if they have no common tissues. These scores are scaled according to the inverse of the distance of voxel k to voxel j. In this way, the model penalizes configurations of voxels that are not likely to occur in the brain, e.g., white matter directly adjacent to CSF, while encouraging more likely types, e.g., white matter next to a partial volume mixture of white and grey matter.

We assume that our nonuniformity correction method has performed its task, and remove the bias terms in our image measurement model. We then use Bayes' formula to create a Maximum A Posteriori (MAP) classifier, which maximizes

$$p(\Lambda|X) = \frac{p(X|\Lambda)p(\Lambda)}{p(X)}, \tag{5}$$

where $p(X|\Lambda) = \prod_k p(x_k|\lambda_k)$, with $p(x|\lambda)$ as in equation (1), but with the bias assumed uniform, i.e., $p(x|\gamma) = p(x|\gamma, b = 1)$. Equation (5) is maximized using an iterated conditional modes (ICM) algorithm after initialization with a maximum likelihood classification.

2.4 Constraining the Topology of the Cortical Surface

The cerebral cortex is a single sheet of connected tissue that encloses the telencephalon. This topology begins during the early development of the embryo, and remains during normal development of the cerebrum. We assume the cortical surface to be topologically equivalent to a sphere after closing the cortex at the brainstem. Errors in the tissue classification stage and the limited resolution of the MR image acquisition as compared to the details of the neuroanatomy will result in cortical surface representations that do not have the appropriate topology.

BrainSuite includes the Topological Constraint Algorithm (TCA) presented in [15]. TCA is an iterative correction procedure that decomposes a volumetric object into a graph representation from which topological equivalence to a sphere may be determined. Slices along a particular axis of the volume are examined in turn. First, the foreground connected components in each slice are labeled. Then, connectivity between each of these components and the components in adjacent slices are determined. These connections are used to form a graph, where each connected component is a vertex in the graph and each connection between components is represented as a weighted edge between the corresponding vertices. The weight of each edge represents the strength of connection between two components. The graph captures both topological and geometric information about the object. The process is repeated for the background voxels, creating a second graph. It is our conjecture that if these graphs are trees then the object being analyzed is homeomorphic to a sphere.

TCA performs this graph analysis and then determines a pair of desired trees using a maximal spanning tree algorithm. In this way, the algorithm determines the edges corresponding to the minimal collection of voxels that need to be removed from the graphs to force them to be trees. These edges correspond to regions in the object where small topological handles or holes exist. The algorithm then edits these areas to remove the handles or holes. The process is applied iteratively along each axis, making the smallest changes possible with each iteration. This ensures that the smallest total change to the volume is made. The result of this procedure is an object that, when tessellated with the Marching Cubes algorithm, will be topologically equivalent to a sphere.

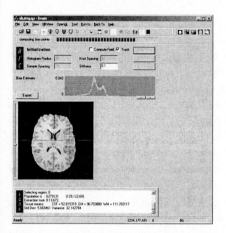 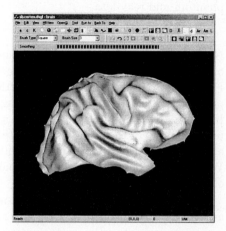

Fig. 1. The BrainSuite graphical user interface during (left) bias correction and (right) surface visualization of a highly smoothed cortical surface.

3   Results and Discussion

3.1   Implementation

We have implemented the sequence of operations into a single, stand-alone interactive tool using Microsoft Visual C++ (see Fig. 1) for use on the Microsoft Windows NT/2000 platforms. The program can be used on common desktop computer hardware to produce cortical volumes within minutes. BrainSuite guides the user through each stage of cortical surface identification and includes hand-editing and connected component tools to allow the user to perform segmentation correction if necessary. Each stage of processing may be performed independently, providing increased flexibility. In addition to the functionality described in this paper, BrainSuite also includes tools for visualization, smoothing, and inflation of cortical surfaces. BrainSuite has been developed to work with BrainStorm, the MEG/EEG Matlab toolbox produced by Baillet et al. [2], and provides the ability to visualize brain activation information on cortical surfaces.

3.2 Validation

The Brain Surface Extractor (BSE) has been used extensively by numerous groups so we do not focus on its accuracy in this paper [20]. BSE was used as the first step for each of the volumes examined in this section. Thus, its performance will directly affect the results of each subsequent stage. In this section we present results for the nonuniformity correction, tissue classification, and topological constraint stages of the algorithm. Validation was performed on phantom and real data.

Nonuniformity Correction Assessing the performance of image nonuniformity correction in real data is problematic due to the lack of a ground truth. For this reason, we test our bias correction algorithm on the BrainWeb Phantom produced by the McConnell Brain Imaging Centre at the Montreal Neurological Institute [21]. The phantom provides a ground truth phantom image and simulated images of the phantom corrupted by noise and RF nonuniformity, available for download via the Internet (http://www.bic.mni.mcgill.ca/brainweb/). We tested our algorithm on the normal brain database with each available set of artifacts.

Ignoring quantization effects, intensity-threshold based classification methods are unaffected by changes of scale and translations of the image intensity. For this reason we compute a Procrustean metric that is the minimum root mean square difference between the two images being compared accounting for all possible translations and scalings of the corrected image intensity,

$$e(\boldsymbol{y}, \tilde{\boldsymbol{x}}) = \min_{a,b} \sqrt{\frac{1}{|\Omega|} \sum_{k \in \Omega} (y_k - (a\hat{x}_k + b))^2}, \tag{6}$$

where $\hat{x}_k$ is the corrected image intensity of the k-th voxel, Ω is the region of interest, and y_k is the intensity of the k-th voxel of the phantom. This metric provides a fair comparison of performance between different bias methods.

| RF Field: | 0% | | | 20% | | | 40% | | |
Bias Correction:	None	N3	BFC	None	N3	BFC	None	N3	BFC
3% Noise	2.98	3.02	3.13	4.13	3.83	2.97	6.23	4.30	3.51
5% Noise	4.88	4.93	4.98	5.50	5.77	4.81	7.05	5.88	4.87
7% Noise	6.65	6.81	6.75	6.96	7.24	6.55	8.05	7.33	6.58
9% Noise	8.30	8.60	8.39	8.41	8.82	8.16	9.15	8.92	8.09

Table 1. Normalized root mean square difference computed between ground truth phantom image and scaled noisy biased phantoms, before and after correction by BFC and N3. Values shown are as a percentage of the ground truth WM intensity. RF field values describe the strength of the simulated nonuniformity field.

To examine the bias correction results specifically, a single mask generated by BSE was used for each image. The stripped brains were then processed by BFC using a single set of parameters for all volumes, and by N3 using the default parameters. We then compared the corrected images to the ground truth using (6) divided by the intensity of WM in the ground truth image. These results are tabulated in Table 1. In the cases where noise was applied with no bias, the normalized RMS difference metric shows that N3 and BFC both left the phantom volumes relatively unchanged. N3 performed slightly better on the 3% and 5% noise phantoms, while BFC performed slightly better on the 7% and 9% cases. In every case with simulated bias fields, the BFC-corrected image was closer to the original than the corresponding N3-corrected image. In most cases the RMS difference was very near to that of the unbiased image with the same level of noise, signifying that we have removed most of the variation attributable to inhomogeneity effects.

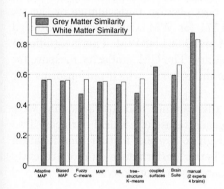 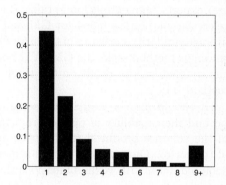

Fig. 2. (left) Similarity comparison computed for several methods using the IBSR dataset. (right) Average fraction of total changes made during each iteration of topological correction. Iteration number represents the largest number of voxels allowed to change to fix a topological problem.

Tissue Classification We tested our tissue segmentation method on data from the Internet Brain Segmentation Repository. The 20 normal MR brain data sets and their manual segmentations were provided by the Center for Morphometric Analysis (CMA) at Massachusetts General Hospital and are available at http://neuro-www.mgh.harvard.edu/cma/ibsr. The data provided by CMA were selected because they have been used in published studies and have various levels of difficulty. A few of these volumes have low contrast and relatively large intensity gradients, and the performance of the tested algorithms is poor on these. The volumes have slice dimensions of 256×256, with resolution of 1 mm$\times$1 mm. Interslice distance is 3 mm, with the number of slices for each volume between 60 and 65.

BrainSuite was used to segment each of the brains in a minimally interactive fashion. We performed skull stripping using one of three parameter settings. Bias correction was performed using the same settings for all twenty brains. The selection of the tissue prior weighting for tissue classification was performed manually, using one of three settings. Settings were only varied when the results were clearly unacceptable. No manual editing of the brain volume or labels was performed. It is possible that more tuning of the parameters to the individual data would produce improved results.

To analyze our performance we use the Jaccard similarity metric, which measures the similarity of two sets as the ratio of the size of their intersection divided by the size of their union. CMA provides reference results for several methods that have been tested using these data; the methods are described in [22]. The results of each method are averaged over the twenty volumes. These averages, our own results, and the grey matter measure provided in [11] are shown in Fig. 2 (white matter metrics are not provided in [11]). Also shown are reference metrics for interoperator variability, 0.876 for GM and 0.832 for WM, proposed by CMA based on an interoperator variability comparison of two experts segmenting four brains. The best average performance of the six reference methods is 0.564 for GM and 0.571 for WM. Our method produced average similarity measures of 0.595 for GM and 0.664 for WM. This is significant since our method seeks the GM/WM boundary. The GM similarity measure for the coupled surface result on the whole brain is 0.657, which outperforms our method [11].

It should also be noted that some of the volumes had artifacts that were beyond the capability of our program to generate useful cortical surfaces. Two of the volumes produced results poor enough to prevent further identification of the cortex. Our methods should achieve better performance on more recently acquired data, with voxel dimensions that are less anisotropic.

Topological Constraint Algorithm We processed 18 of the 20 IBSR brain volumes using the BrainSuite tools to produce inner-cerebral masks. The brainstem and cerebellum were removed manually. Some subcortical structures such as the ventricles and subcortical grey matter are not always well-segmented and were filled using a user-guided flood-filling. This process required approximately three to five minutes per brain. These white matter masks were then processed using the topological constraint algorithm. In each case, TCA successfully produced an object with spherical surface topology, as verified by computing the Euler characteristic on a Marching Cubes tessellation of the object.

The actual changes made to each volume were very minimal, with at most a 0.7% change in membership to the white matter set. This level of change is well within the variability one would expect to see among segmentations by experts. Figure 2 shows the average percentage of changes made with each iteration. The iteration number corresponds to the largest number of voxels changed to fix a specific topological problem. Figure 2 emphasizes that most changes are made in the very early stages of the algorithm and typically correspond to changes

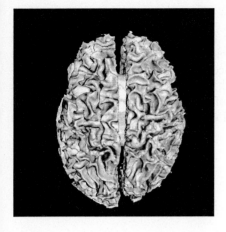 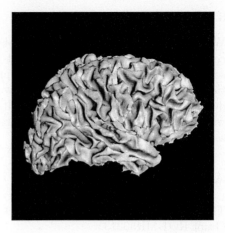

Fig. 3. Axial and sagittal views of a cortical surface with spherical topology, generated using BrainSuite. The surface represents the boundary between white matter and cortical grey matter.

of only a few voxels at a specific location. On newer, higher resolution data our method achieved much better results, and 75% of the topological corrections were achieved in the first iteration [15]. Figure 3 shows renderings of a topologically spherical cortical surface obtained from a high-resolution T1-weighted MR volume using BrainSuite. In this case, less than 0.1% of the initial set membership was changed by the topological constraint algorithm.

Processing Time Table 2 shows the CPU time of each stage of the algorithm applied to the IBSR data. Processing was performed on a 933 MHz Intel Pentium III Xeon with 256K cache and 256MB of RAM. The topological constraint algorithm is the newest component in the toolset, and may be optimized in future work to reduce the total time required for segmentation. High resolution data requires more processing time; the cortical volume in Figure 3 was generated in approximately fifteen minutes. Presently, cortical volumes may be identified from MR images in less than 30 minutes of total operator time using BrainSuite.

	BSE	BFC	PVC	TCA	Total
average	5.1	51.8	3.4	4:01	5:01
best case	3.8	38.7	0.3	2:18	3:06
worst case	8.0	1:11.8	7.8	5:40	6:33

Table 2. CPU times (in minutes:seconds) for each stage of the algorithm applied to the 20 normal MR brain data sets from the IBSR.

3.3 Discussion

We have presented a self-contained toolset that produces topologically spherical cortical surface representations from T1-weighted MR images using an integrated sequence of low-level processing. The processing requires minimal user interaction, and our method was shown to work in very reasonable time on modern desktop computer hardware. The results of our cortical identification technique were validated on both phantom and real data, and were shown to outperform several existing methods.

Acknowledgements

This work was supported in part by the National Institute of Mental Health Grant RO1-MH53213.

References

1. Thompson, P.M., MacDonald, D., Mega, M.S., Holmes, C.J., Evans, A.C., Toga, A.W.: Detection and mapping of abnormal brain structure with a probabilistic atlas of cortical surfaces. Journal of Computer Assisted Tomography **21** (1998) 567–581
2. Baillet, S., Mosher, J., Leahy, R., Shattuck, D.: Brainstorm: A Matlab toolbox for the processing of MEG and EEG signals. NeuroImage **9** (1999) S246
3. Dale, A.M., Sereno, M.I.: Improved localization of cortical activity by combining EEG and MEG with MRI cortical surface reconstruction : a linear approach. Journal of Cognitive Neuroscience **5** (1993) 162–176
4. Wells, W.M., Grimson, W.E.L., Kikinis, R., Jolesz, F.A.: Adaptive segmentation of MRI data. IEEE Transactions on Medical Imaging **15** (1996) 429–442
5. Kapur, T., Grimson, W.E.L., Wells, W.M., Kikinis, R.: Segmentation of brain tissue from magnetic resonance images. Medical Image Analysis **1** (1996) 109–127
6. Kapur, T., Grimson, W.E.L., Kikinis, R., Wells, W.M.: Enhanced spatial priors for segmentation of magnetic resonance imagery. In Wells, W.M., Colchester, A.C.F., Delp, S., eds.: Medical Image Computing and Computer-Assisted Intervention–MICCAI '98. Lecture Notes in Computer Science, Vol. 1496. Springer-Verlag, Berlin Heidelberg (1998) 457–468
7. Sled, J.G., Zijdenbos, A.P., Evans, A.C.: A nonparametric method for automatic correction of intensity nonuniformity in MRI data. IEEE Transactions on Medical Imaging **17** (1998) 87–97
8. Collins, D.L., Zijdenbos, A.P., Baaré, W.F., Evans, A.C.: ANIMAL+INSECT : Improved cortical structure segmentation. In Kuba, A., Sámal, M., Todd-Pokropek, A., eds.: Information Processing in Medical Imaging: Proceedings of the 16th International Conference. Lecture Notes in Computer Science, Vol. 1613. Springer-Verlag, Berlin Heidelberg (1999) 210–223
9. Xu, C., Pham, D.L., Rettmann, M.E., Yu, D.N., Prince, J.: Reconstruction of the central layer of the human cerebral cortex from MR images. IEEE Transactions on Medical Imaging **18** (1999) 467–480
10. Pham, D., Prince, J.: Adaptive fuzzy segmentation of magnetic resonance images. IEEE Transactions on Medical Imaging **18** (1999) 737–752

11. Zeng, X., Staib, L., Schultz, R.T., Duncan, J.: Segmentation and measurement of the cortex from 3-D MR images using coupled-surface propagation. IEEE Transactions on Medical Imaging **18** (1999) 927–937
12. Dale, A.M., Fischl, B., Sereno, M.I.: Cortical surface-based analysis I: Segmentation and surface reconstruction. NeuroImage **9** (1999) 179–194
13. Sandor, S., Leahy, R.: Surface-based labeling of cortical anatomy using a deformable database. IEEE Transactions on Medical Imaging **16** (1997) 41–54
14. Shattuck, D.W., Sandor-Leahy, S.R., Schaper, K.A., Rottenberg, D., Leahy, R.: Magnetic resonance image tissue classification using a partial volume model. Signal and Image Processing Institute Technical Report, University of Southern California **349** (2000)
15. Shattuck, D.W., Leahy, R.M.: Topologically constrained cortical surfaces from MRI. In Hanson, K.M., ed.: Medical Imaging 2000: Image Processing. Proceedings of the SPIE, Vol. 3979. (2000) in press
16. Lorensen, W., Harvey, E.: Marching cubes: A high resolution 3D surface construction algorithm. ACM Computer Graphics **21** (1987) 163–169
17. Gerig, G., Kubler, O., Kikinis, R., Jolesz, F.: Nonlinear anisotropic filtering of MRI data. IEEE Transactions on Medical Imaging **11** (1992) 221–232
18. Marr, D., Hildreth, E.: Theory of edge detection. Proceedings of Royal Society of London **207** (1980) 187–217
19. Santago, P., Gage, H.D.: Quantification of MR brain images by mixture density and partial volume modeling. IEEE Transactions on Medical Imaging **12** (1993) 566–574
20. Rehm, K., Shattuck, D., Leahy, R., Schaper, K., Rottenberg, D.: Semi-automated stripping of T1 MRI volumes: consensus of intensity- and edge-based methods. NeuroImage **9** (1999) S86
21. Collins, D., Zijdenbos, A., Kollokian, V., Sled, J., Kabani, N., Holmes, C., Evans, A.: Design and construction of a realistic digital brain phantom. IEEE Transactions on Medical Imaging **17** (1998) 463–468
22. Rajapakse, J.C., Kruggel, F.: Segmentation of MR images with intensity inhomogeneities. Image Vision Computing **16** (1998) 165–180

Incorporating Spatial Priors into an Information Theoretic Approach for fMRI Data Analysis*

Junmo Kim[1], John W. Fisher III[1,2], Andy Tsai[1], Cindy Wible[3],
Alan S. Willsky[1], and William M. Wells III[2,3]

[1] Massachusetts Institute of Technology,
Laboratory for Information and Decision Systems,
Cambridge, MA, USA
{junmo, atsai, willsky}@mit.edu
[2] Massachusetts Institute of Technology,
Artificial Intelligence Laboratory,
Cambridge, MA, USA
{fisher, sw}@ai.mit.edu
[3] Harvard Medical School,
Brigham and Women's Hospital,
Department of Radiology,
Boston, MA, USA
cindy@bwh.harvard.edu

Abstract. In previous work a novel information-theoretic approach was introduced for calculating the activation map for fMRI analysis [Tsai *et al* , 1999]. In that work the use of mutual information as a measure of activation resulted in a nonparametric calculation of the activation map. Nonparametric approaches are attractive as the implicit assumptions are milder than the strong assumptions of popular approaches based on the general linear model popularized by Friston *et al* [1994]. Here we show that, in addition to the intuitive information-theoretic appeal, such an application of mutual information is equivalent to a hypothesis test when the underlying densities are unknown. Furthermore we incorporate local spatial priors using the well-known Ising model thereby dropping the implicit assumption that neighboring voxel time-series are independent. As a consequence of the hypothesis testing equivalence, calculation of the activation map with local spatial priors can be formulated as mincut/maxflow graph-cutting problem. Such problems can be solved in polynomial time by the Ford and Fulkerson method. Empirical results are presented on three fMRI datasets measuring motor, auditory, and visual cortex activation. Comparisons are made illustrating the differences between the proposed technique and one based on the general linear model.

* J. Kim, J. Fisher, A. Tsai, and A. Willsky supported in part by AFOSR grant F49620-98-1-0349, subcontract GC123919NGD from BU under the AFOSR Multidisciplinary Research Program, and ONR grant N00014-00-1-0089. C. Wible was supported in part under NIMH grants MH40799 and MH52807. W. Wells was supported in part by the Whitaker foundation and by NIH through grant 1P41RR13218.

1 Introduction

In previous work [6], we presented a novel information theoretic approach for calculating ƒMRI activation maps. The information-theoretic approach is appealing in that it is a principled methodology requiring few assumptions about the structure of the ƒMRI signal. In that approach, activation was quantified by measuring the mutual information (MI) between the protocol signal and the ƒMRI time-series at a given voxel. This measure is capable of detecting unknown nonlinear and higher-order statistical dependencies. Furthermore, it is relatively straightforward to implement.

In practice, activation decisions at each voxel are independent of neighboring voxels. Spurious responses are then removed by *ad hoc* techniques (e.g. morphological operators). In this paper, we describe an automatic maximum a posteriori (MAP) detection method where the well-known Ising model is used as a spatial prior. The Ising spatial prior does not assume that the time-series of neighboring voxels are independent of each other. Furthermore, removal of spurious responses is an implicit component of the detection formulation. In order to formulate the calculation of the activation map using this technique we first demonstrate that the information-theoretic approach has a natural interpretation in the hypothesis testing framework and that, specifically, our estimate of MI approximates the log-likelihood ratio of that hypothesis test. Consequently, the MAP detection problem using the Ising model can be formulated and solved *exactly* in polynomial time using the Ford and Fulkerson method [4].

We compare the results of our approach with and without spatial priors to an approach based on the general linear model (GLM) popularized by Friston *et al* [3]. We present results from three ƒMRI data sets. The data sets test motor, auditory, and visual cortex activation, respectively.

1.1 Review of the Information Theoretic Approach

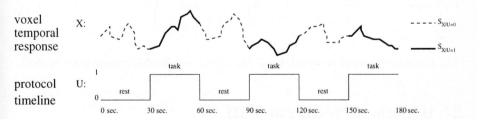

Fig. 1. Illustration of the protocol time-line, $S_{X|U=0}$, and $S_{X|U=1}$.

In [6], each voxel is declared to be active (or not) based solely only the temporal response of that voxel without considering the temporal responses of neighboring voxels. Let $X(\cdot,\cdot,\cdot,\cdot) = \{X(i,j,k,t)|1 \le t \le n\}$ denote the observed fMRI signal, where i, j, k are spatial coordinates and t is the time coordinate. Each voxel (i, j, k) has an associated discrete-time temporal response, $X_1, \ldots, X_n$, where X_t denotes $X(i,j,k,t)$ for convenience.

Figure 1 illustrates the protocol time-line and an associated temporal response. $S_{X|U=0}$ denotes the set of X_i's where the protocol is 0 while $S_{X|U=1}$ denotes the set of X_i's where the protocol is 1. An implicit assumption in our approach is that $S_{X|U=[0,1]}$ are i.i.d. according to $p_{X|U=[0,1]}(x)$. We treat the protocol U as a discrete random variable taking 0 and 1 with equal probability. In this case the MI between X and U is as follows:

$$I(X;U) = H(U) - H(U|X) = h(X) - h(X|U)$$
$$= h(X) - \frac{1}{2}h(X|U=0) - \frac{1}{2}h(X|U=1)$$

where $H(U)$ is the discrete entropy of U and $h(X)$ is the continuous differential entropy of X. It can be shown that $H(U) \le 1$ bit and that $0 \le H(U|X) \le H(U)$ consequently $0 \le I(X;U) \le 1$. Thus, MI as a measure between X and U is normalized.

The differential entropy of X

$$h(X) = -\int_S p_X(x) \log p_X(x)dx$$

is approximated as [6]:

$$\hat{h}(X) \approx -\frac{1}{n}\sum_{i=1}^{n} \log \hat{p}_X(x_i) \tag{1}$$

where $\hat{p}_X(X_i)$ is the Parzen density estimator [5], defined as

$$\hat{p}_X(x_i) = \frac{1}{n\sigma}\sum_j k\left(\frac{x_i - x_j}{\sigma}\right) \tag{2}$$

The kernel $k(x)$ must be a valid pdf (in our case a double exponential kernel).

2 Hypothesis Testing and MI

In order to extend our method to a MAP detection problem using spatial priors it is necessary to formulate a suitable hypothesis test and associated likelihood ratio. Here we show the equivalence of MI to the likelihood ratio of an underlying binary hypothesis testing problem.

2.1 Nonparametric Hypothesis Testing Problem

Consider the following hypothesis test

$$H_0 : x_t, u_t \sim p_X(X)p_U(U) \quad \text{, i.e. } X, U \text{ are independent}$$
$$H_1 : x_t, u_t \sim p_{X,U}(X,U) \quad \text{, i.e. } X, U \text{ are dependent}$$

where the null hypothesis states that the protocol, U, and the *f*MRI time-series, X are statistically independent while the alternative states that they are not. The log-likelihood ratio, which is the optimal test statistic by the Neyman-Pearson lemma, is

$$T_n = \sum_{t=1}^{n} \log \left(\frac{p_{XU}(x_t, u_t)}{p_X(x_t)p_U(u_t)} \right) \tag{3}$$

assuming i.i.d. samples. It can be shown that [1]

$$\lim_{n \to \infty} T_n = nI(X;U) \tag{4}$$
$$= \mathrm{E}\{T_n\} \tag{5}$$

consequently using $I(X;U)$ as the activation test statistic is equivalent to the aforementioned hypothesis test. Since the distribution of U is binomial with equal probability it can be shown that $I(X;U)$ simplifies to

$$I(X;U) = h(X) - \frac{1}{2}h(X|U = 0) - \frac{1}{2}h(X|U = 1) \tag{6}$$
$$= \frac{1}{2}D(p_{X|U=0}\|p_X) + \frac{1}{2}D(p_{X|U=1}\|p_X) \tag{7}$$

where $D(p_1\|p_2)$ is the asymmetric Kullback-Leibler divergence.

2.2 Bias in the Estimate of the Likelihood Ratio

When evaluating the likelihood ratio we substitute 1 into 6. The conditional terms are summed over $S_{X|U=0}$ and $S_{X|U=1}$, respectively. The consequence is that we introduce bias into our estimate of the likelihood ratio or equivalently $I(X;U)$. In order to simplify matters consider the likelihood of the *f*MRI time-series

$$p_{X_1,\ldots,X_n}(x_1, \ldots, x_n) = \prod_{t=1}^{n} p_X(x_t) = \exp(\sum_t \log p_X(x_t))$$
$$= \exp(\sum_t \log \hat{p}_X(x_t) + \sum_t \log \frac{p_X(x_t)}{\hat{p}_X(x_t)})$$
$$= e^{-n[\hat{h}(X) + \frac{1}{n}\sum_t \log \frac{\hat{p}_X(x_t)}{p_X(x_t)}]} \approx e^{-n[\hat{h}(X) + D(\hat{p}_X \| p_X)]}$$

In similar fashion the likelihood ratio is approximately

$$\frac{p_X(x_1,\ldots,x_n|H_1)}{p_X(x_1,\ldots,x_n|H_0)} \approx e^{n(\hat{I}(X;U)-\gamma)} \tag{8}$$

where $\gamma = \frac{1}{2}[D(\hat{p}_{X|U=0}\|p_{X|U=0}) + D(\hat{p}_{X|U=1}\|p_{X|U=1})] - D(\hat{p}_X\|p_X)$,which is nonnegative due to convexity of Kullback-Leibler divergence. More importantly, the divergence terms asymptotically approach zero. Consequently, the approximation approaches the true likelihood ratio.

3 Modeling Voxel Dependency via the Ising Model

We use the Ising model, a simple Markov random field (MRF), as a spatial prior of the binary activation map. Previously, Descombes $et\ al$ [2] proposed the use an MRF (specifically a Potts model) for fMRI signal restoration and analysis. The substantive differences between the proposed method and that of Descombes $et\ al$ include

- Descombes $et\ al$ used simulated annealing to solve the MAP estimation problem with no guarantee of an exact result.
- The MRF prior model was combined with data fidelity terms in a heuristic way.

In contrast, our method combines the likelihood ratio obtained from the data with a spatial prior rigorously within the Bayesian framework leading to an exact solution of a binary MAP hypothesis test.

The Ising model captures the notion that neighboring voxels of an activated voxel are likely to be activated and vice versa. Specifically, let $y(i,j,k)$ be a binary activation map such that $y(i,j,k) = 1$ if voxel (i,j,k) is activated and 0, otherwise. Then this idea can be formulated using Ising model as a prior probability of the activation map $y(\cdot,\cdot,\cdot)$. Let

$$W = \left\{w : w \in \{0,1\}^{N_1 \times N_2 \times N_3}\right\}$$

be the set of all possible [0-1] configurations and let $w(i,j,k)$ be an element of any one sample configuration.

The Ising prior on $y(\cdot,\cdot,\cdot)$ penalizes every occurrence of neighboring voxels with different activation states as follows:

$$P(y(\cdot,\cdot,\cdot) = w) = \frac{1}{Z}e^{-U(w)} \qquad Z = \sum_{w \in W} e^{-U(w)}$$

$$U(w) = \beta \sum_{i,j,k} (w(i,j,k) \oplus w(i+1,j,k) + w(i,j,k) \oplus w(i,j+1,k)$$
$$+ w(i,j,k) \oplus w(i,j,k+1)),$$

where $\beta > 0$. As in [2] we assume that

$$p(X(\cdot,\cdot,\cdot,\cdot)|Y(\cdot,\cdot,\cdot)) = \prod_{i,j,k} p(X(i,j,k,\cdot)|Y(i,j,k))$$

That is, conditioned on the activation map, voxel time-series are independent. The MAP estimate of the activation is then

$$\hat{y}(\cdot,\cdot,\cdot) = \arg\max_{y(\cdot,\cdot,\cdot)} p_Y(y(\cdot,\cdot,\cdot)) p_{X|Y}(x(\cdot,\cdot,\cdot,\cdot)|Y(\cdot,\cdot,\cdot) = y(\cdot,\cdot,\cdot))$$

$$= \arg\max_{y(\cdot,\cdot,\cdot)} \log p_Y(y(\cdot,\cdot,\cdot)) +$$

$$\sum_{i,j,k} y(i,j,k) \log \frac{p_{X|Y}(x(i,j,k,\cdot)|Y(i,j,k) = 1)}{p_{X|Y}(x(i,j,k,\cdot)|Y(i,j,k) = 0)}$$

$$= \arg\max_{y(\cdot,\cdot,\cdot)} \sum_{i,j,k} \lambda_{i,j,k} y(i,j,k) - \beta \sum_{i,j,k} (y(i,j,k) \oplus y(i+1,j,k)$$

$$+ y(i,j,k) \oplus y(i,j+1,k) + y(i,j,k) \oplus y(i,j,k+1)),$$

where $\lambda_{i,j,k} = \ln \frac{p_{X|Y}(x(i,j,k,\cdot)|Y(i,j,k)=1)}{p_{X|Y}(x(i,j,k,\cdot)|Y(i,j,k)=0)} = n(\hat{I}_{i,j,k}(X;U) - \gamma)$ is the log-likelihood ratio at voxel (i,j,k) and $\hat{I}_{i,j,k}(X;U)$ is the MI estimated from time-series $X(i,j,k,\cdot)$. The previous use of MI as the activation statistic fits readily into the MAP formulation.

3.1 Exact Solution of the Binary MAP Estimation Problem

There are 2^{N_v} possible configurations of $y(\cdot,\cdot,\cdot)$ (or equivalently elements of the set W) where $N_v = N_1 N_2 N_3$ is the number of voxels. It has been shown by Greig et al [4] that this seemingly NP-complete problem can be solved exactly in polynomial time (order N_v). Greig et al accomplished this by demonstrating that under certain conditions the binary image MAP estimation problem (using MRFs as a prior) can be reduced to the minimum cut problem of network flow. Consequently, the methodology of Ford and Fulkerson for such problems can be applied directly. We are able to employ the same technique as a direct consequence of demonstrating the equivalence of MI to the log-likelihood ratio of a binary hypothesis testing problem. Details of the minimum cut solution are beyond the scope of this paper and we refer the reader to [4] for further details.

4 Experimental Results

We present experimental results on three fMRI data sets. The protocols are designed to activate the motor cortex (dominant hand movement protocol), auditory cortex (verb generation protocol), and visual cortex (visual stimulation with alternating checkerboard pattern), respectively. Each data set contains 60 whole brain acquisitions taken three seconds apart. We compare the resulting activation map computed by three methods: GLM, nonparametric MI, nonparametric MI with an Ising prior.

We first apply the GLM method to each data set. The coronal slice exhibiting the highest activation for each data set is shown in the first column of figure 2 with the GLM activation map overlaid in white for each data set. The F-statistic threshold for GLM was set such that the visual inspection of the activation map was consistent with our prior expectation of the number of activated voxels. This corresponded to a p-value of 10^{-10}.

In the next column of the figure the same slices are shown using MI to compute the activation map. In this case, the MI threshold γ was set such that all of the voxels detected by the GLM were detected by MI. Consequently figure 2 (b), (e) and (h) contain additional activations when compared to GLM. Some of these additional activations are spurious and some are not.

Finally, the Ising prior was applied to the MI activation map with $\beta = 1$. An intuitive understanding of the relationship of γ and β is as follows. If $\beta = 0$, then there is no prior and the method reduces to MI only. For $\beta \neq 0$ the interpretation is not so simple, but we can consider a special case. Suppose the neighbors of a voxel are declared to be active (in our case there are six neighbors for every voxel), then the effective MI activation threshold γ for that voxel has been reduced by $6\beta/n$. Conversely, if all of the neighbors are inactive then the effective threshold is increased by the same amount. For these experiments, $n = 60$ and $\beta = 1$, this equates to a 0.1 nat (equal to 0.14 bits) change in the MI activation threshold for the special cases described.

Comparison of figures 2 (b), (e) and (h) to figures 2 (c), (f), and (i) shows that many of the isolated activations were removed by the Ising prior, but some of the new MI activations remain. Figure 3 shows the temporal responses of the voxels with the lowest GLM score which were detected by MI with prior but not by GLM. Examination of these temporal responses (with protocol signal overlaid) reveals obvious structure related to the protocol signal.

A reasonable question is whether this result is due to an unusually high threshold set for GLM. In order to address this we next lower the GLM threshold such that the voxels of figure 3 are detected by GLM. We then consider regions of the resulting activation map where new activations have appeared in figure 4. The activations of 4a and 4b (motor cortex, auditory cortex), would be considered spurious in light of the region in which they occur. The result for figure 4c is not so clear as these activations are most likely spurious, but might possibly be related to higher-ordered visual processing.

5 Conclusion

We demonstrated that our previous approach, derived from an information-theoretic perspective, can be formulated in a hypothesis testing framework. Furthermore, the resulting hypothesis test is free of many of the strong assumptions inherent in GLM. As fMRI is a relatively new modality for examining cognitive function we think that it is appropriate to examine nonparametric methods (i.e.

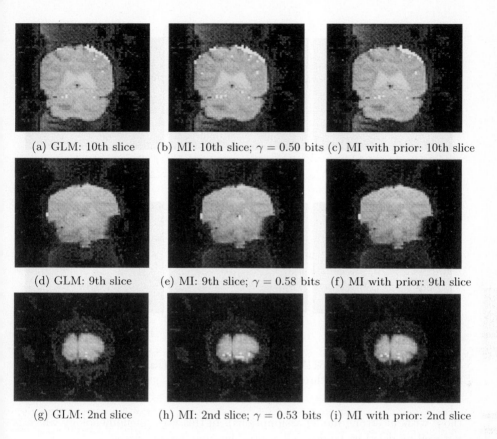

(a) GLM: 10th slice (b) MI: 10th slice; $\gamma = 0.50$ bits (c) MI with prior: 10th slice

(d) GLM: 9th slice (e) MI: 9th slice; $\gamma = 0.58$ bits (f) MI with prior: 9th slice

(g) GLM: 2nd slice (h) MI: 2nd slice; $\gamma = 0.53$ bits (i) MI with prior: 2nd slice

Fig. 2. Comparison of fMRI Analysis results from motor, auditory and visual experiments

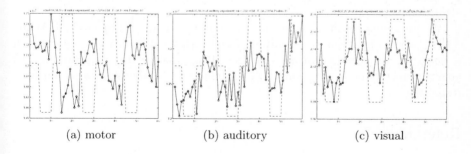

(a) motor (b) auditory (c) visual

Fig. 3. Temporal responses of voxels newly detected by the MI using Ising prior

GLM

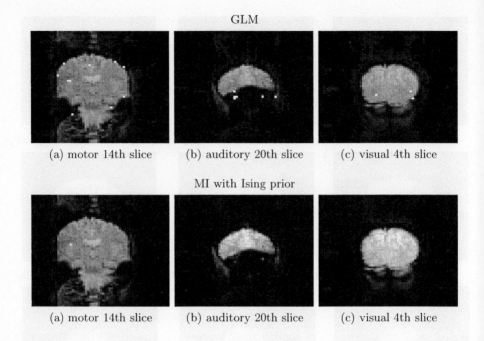

(a) motor 14th slice   (b) auditory 20th slice   (c) visual 4th slice

MI with Ising prior

(a) motor 14th slice   (b) auditory 20th slice   (c) visual 4th slice

Fig. 4. Comparison of fMRI Analysis results from motor, auditory and visual experiments with lowered GLM threshold

those without strong model assumptions). In this way, phenomenology which is not well-modeled by traditional approaches may be uncovered.

We introduced an extension of our method which incorporates spatial priors via the Ising model. A consequence of the hypothesis testing formulation of the original MI-only approach was that the resulting MAP estimation problem (with the addition of spatial priors) could be reduced to a minimum cut network flow problem. Thereby allowing for an exact and relatively fast (polynomial-time) algorithm.

We presented results comparing our approach to the GLM method. While fMRI analysis of patient data is always faced with the difficulty that exact truth is unknown our results indicate that the MI approach with spatial priors was able to detect "true" activations with a significantly smaller number of spurious responses.

References

1. T. M. Cover and J. A. Thomas. *Elements of Information Theory.* John Wiley & Sons, Inc., New York, 1991.
2. X. Descombes, F. Kruggel, and D. Y. von Cramon. Spatio-temporal fmri analysis using markov random fields. *IEEE Transactions on Medical Imaging*, 17(6):1028–1029, Dec 1998.

3. K. J. Friston, P. Jezzard, and R. Turner. The analysis of functional mri time-series. *Human Brain Mapping*, 1:153–171, 1994.
4. D. M. Greig, B. T. Porteous, and A. H. Seheult. Exact maximum a posteriori estimation for binary images. *Journal of the Royal Statistical Society. Series B(Methodological)*, 51(2):271–279, 1989.
5. E. Parzen. On estimation of a probability density function and mode. *Ann. of Math Stats.*, 33:1065–1076, 1962.
6. A. Tsai, J. Fisher III, C. Wible, W. I. Wells, J. Kim, and A. Willsky. Analysis of fmri data using mutual information. In C. Taylor and A. Colchester, editors, *Second International Conference on Medical Image Computing and Computer-Assisted Intervention*, volume 1679 of *Lecture Notes in Computer Science*, pages 473–480. Springer, Sep 1999.

Small Sample Size Learning for Shape Analysis of Anatomical Structures

Polina Golland[1], W. Eric L. Grimson[1], Martha E. Shenton[2], and Ron Kikinis[3]

[1] Artificial Intelligence Laboratory,
Massachusetts Institute of Technology, Cambridge, MA.
{polina,welg}@ai.mit.edu,
[2] Laboratory of Neuroscience, Clinical Neuroscience Division, Department of
Psychiatry, VAMC-Brockton, Harvard Medical School, Brockton, MA.
mshenton@warren.med.harvard.edu
[3] Surgical Planning Laboratory Brigham and Women's Hospital,
Harvard Medical School, Boston, MA.
kikinis@bwh.harvard.edu

Abstract. We present a novel approach to statistical shape analysis of anatomical structures based on small sample size learning techniques. The high complexity of shape models used in medical image analysis, combined with a typically small number of training examples, places the problem outside the realm of classical statistics. This difficulty is traditionally overcome by first reducing dimensionality of the shape representation (e.g., using PCA) and then performing training and classification in the reduced space defined by a few principal components. We propose to learn the shape differences between the classes in the original high dimensional parameter space, while controlling the capacity (generalization error) of the classifier. This approach makes significantly fewer assumptions on the properties and the distribution of the underlying data, which can be advantageous in anatomical shape analysis where little is known about the true nature of the input data. Support Vector Machines with Radial Basis Function kernels are used as a training method and the VC dimension is used for the theoretical analysis of the classifier capacity. We demonstrate the method by applying it to shape classification of the hippocampus-amygdala complex in a data set of 15 schizophrenia patients and 15 normal controls. Using our technique, the separation between the classes and the confidence intervals are improved over a volume based analysis (63% to 73%). Thus exploiting techniques from small sample size learning theory provides us with a principled way of utilizing shape information in statistical analysis of the disorder effects on the brain.

1 Introduction

Statistical shape analysis, or shape based classification, attempts to identify statistical differences between two groups of images of the same organ based on 3D images of the organ. It can be used to study a disease (patients vs. normal controls) or changes caused by aging (different age groups). Size and volume measurements have been widely used for this purpose, but they capture only a small subset of organ shape differences. If utilized properly, shape information can significantly improve our understanding of the anatomical changes due to a particular disorder.

The first step in shape based statistical analysis is extraction of the shape parameters, or a shape descriptor. Most work in constructing models of shape and its variation has been motivated by and used for segmentation tasks, where a generative model of shape variation guides the search over the space of possible deformations of the "representative" shape template, with the goal of matching it to a new input image. Several analytical models for shape variation have been used in conjunction with different shape models [2, 14], followed by introduction of a statistical approach by Cootes and Taylor [5]. They proposed to learn the variation of the shape within a class of examples from the input data by extracting a shape descriptor from every training example and using Principal Component Analysis (PCA) to build a linear model of variation and compress it to a few principal modes of deformation. This method was explored in the domain of image segmentation by several authors using different shape models [9, 11, 15].

When applied to shape analysis, this approach employs PCA to reduce the dimensionality of the shape parameter space and then performs the training in the reduced, low dimensional space [6, 8, 10]. The main reason for dimensionality reduction in shape analysis is the concern that robust parameter estimation is infeasible in the original high dimensional space when only a small number of examples is available. Underlying this concern is the assumption that in order to reliably train a classifier, one has to estimate the density distribution of the data. Using PCA for dimensionality reduction is essentially equivalent to assuming that the examples are generated by a Gaussian distribution, and estimating its parameters – mean and covariance – from the input data. An interesting question then arises: what if the data are not generated by a Gaussian distribution? In the generative case, the Law of Large Numbers justifies using this method for estimating mean and covariance of the class. The estimates are unbiased and as the number of examples grows, they become increasingly accurate. However, the situation is different in the discriminative case. As the number of examples grows, the class mean and covariance estimates converge to their true values, but the Gaussian model based on the estimates does not describe the data more accurately. As a result, a classifier derived from this model does not approach the optimal solution. Since we have no evidence that the shape parameters for any group of subjects (either patients or normal controls) follow a high dimensional Gaussian distribution, it is desirable to avoid making this assumption in the analysis of the data.

We examine an approach to shape classification that makes much weaker assumptions than the ones used by traditional methods. This work is based on small sample size learning theory developed in the Machine Learning community. All approaches to learning from examples share the following very important notion: the trade-off between generalization ability and training error determines the optimal classifier function. The training error, which is equal to the number of misclassified examples in the training set, can be reduced by using increasingly complex classifier functions that can capture the differences between the classes more easily. But the goal of learning is to optimize the classifier performance on new, unseen examples. The expected error rate of the classifier on future examples is called test error, or generalization error. Unfortunately, as the complexity of the classifier increases, its estimation from the training data becomes less robust, and as a result, the test error increases. The optimal classifier function is the one that can capture the class differences in the training data set (low training error) and generalize well the concept of the differences to the examples that had not been seen in the training phase (low expectation of test error). Traditionally, the classifier complexity has been associated with the number of parameters defining it. But the relationship between the complexity, or capacity, of the classifier and the number of parameters is not necessarily a monotonic one (see [3, 17] for examples). Vapnik [16] introduced a better measure of capacity, called VC dimension, and derived theoretical bounds on the generalization error of a classifier based on it. One of the most important observations made within the theoretical analysis is that it is not necessary to solve the density estimation problem as an intermediate step in the learning process. Instead, the optimal classifier can and should be estimated directly from the input data. Based on this theory, a new learning method, called Support Vector Machines (SVMs), was proposed [16, 17]. The main principle of the method is to train the classifier using the input data, while controlling its VC dimension, rather than the number of parameters. In this paper, we demonstrate a framework for statistical shape analysis based on SVMs that for small data sets provides results comparable to other techniques, and is guaranteed to converge to the optimal solution as the number of examples increases. It makes no assumptions on the distribution of the data, and can therefore be applied even when we do not have enough knowledge to estimate the distribution that produced the input data.

SVMs have recently been used in computer vision for classification tasks on grayscale images, such as character recognition [17], face detection [12] and others. While there are similarities between these applications and the problem of anatomical shape analysis, there are also significant differences. Among serious challenges in the general pattern recognition domain are non-linear imaging effects, such as occlusions, or changes in pose and illumination, and creating negative examples for object detections (e.g., a class of "non-faces" in face detection). Moreover, these applications are characterized by a wealth (tens of thousands) of training examples and therefore exploit asymptotic behavior of the learning method, rather than its convergence properties.

The rest of the paper is organized as follows. In the next section, we review Support Vector Machines and the capacity analysis based on VC dimension. It is followed by discussion of shape representation using a distance transform and the description of our algorithm for shape analysis based on SVMs. Section 4 reports the results of the method's application to a data set of 30 examples (15 schizophrenia patient and 15 matched controls) of the hippocampus-amygdala complex and the comparison with a volume based analysis for the same data set. We conclude with the discussion of the results and future research directions.

2 Small Sample Size Learning and Support Vector Machines

In this section, we present a brief overview of Support Vector Machines (SVMs) and the capacity analysis based on the VC dimension. The reader is referred to the tutorial [3] for an extensive introduction and to the book [17] for more details on the theoretical foundation and the proofs.

VC bound. Given a training set of l pairs $\{(\mathbf{x}_i, y_i)\}_1^l$, where $\mathbf{x}_i \in R^n$ are observations and $y_i \in \{-1, 1\}$ are corresponding labels, and a family of classifier functions $\{f_\omega(\mathbf{x})\}$ parametrized by ω, the learning task is to select a member of the family that assigns labels to new, unseen examples while making as few errors as possible. Ideally, we would like to minimize the expectation of the test error, also called *expected risk*,

$$R(\omega) = \int \frac{1}{2}|y - f_\omega(\mathbf{x})| dP(\mathbf{x}, y), \tag{1}$$

where $P(\mathbf{x}, y)$ is the probability distribution that generates the observations. In practice, however, it is difficult to implement, as $P(\mathbf{x}, y)$ is unknown. Instead, the training error, also called *empirical risk*, can be computed:

$$R_{\text{emp}}(\omega) = \frac{1}{2l} \sum_{i=1}^{l} |y_i - f_\omega(\mathbf{x}_i)|. \tag{2}$$

One can show that for most families of classifier functions used in learning, the minimum of empirical risk converges in probability to the minimum of expected risk as the number of training examples increases. Furthermore, for any η ($0 \leq \eta \leq 1$), with probability at least $1 - \eta$, the classifier $f_{\omega^*}(\mathbf{x})$ that minimizes empirical risk $R_{\text{emp}}(\omega)$ on the given training set satisfies

$$R(\omega^*) \leq R_{\text{emp}}(\omega^*) + \sqrt{\frac{h}{l}\left(\log\frac{2l}{h} + 1\right) - \frac{1}{l}\log\frac{\eta}{4}}, \tag{3}$$

where h is a quantity called *VC dimension*. The VC dimension of a family of classifiers is a measure of the complexity of the space of all hypotheses they can generate on any given training set. The right hand side of (3) is often referred to as the *VC bound* and its second term is called *VC confidence*. Note that it is a distribution-free bound, i.e., one does not need to know the distribution of the input data to estimate the convergence rate of the learning algorithm. This also implies that the bound is usually fairly loose for any particular distribution, and tighter bounds could be derived if the data distribution function were known.

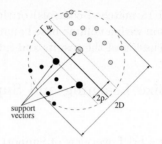

Fig. 1. Linearly separable classes with margin ρ and bounding sphere of radius D.

Support Vector Machines. Let's first consider a situation when the two classes are linearly separable (Fig. 1), i.e., there exists a vector $\mathbf{w}$ and a constant b such that

$$\forall i : \ y_i((\mathbf{w} \cdot \mathbf{x}_i) + b) \geq 1, \tag{4}$$

where $(\cdot)$ denotes a dot product of two vectors. The Support Vector learning machine searches for a $\mathbf{w}$ that maximizes the margin between the two classes

$$\rho(\mathbf{w}) = \frac{1}{2\|\mathbf{w}\|}(\min_{y_i=1}(\mathbf{w} \cdot \mathbf{x}_i) - \max_{y_i=-1}(\mathbf{w} \cdot \mathbf{x}_i)), \tag{5}$$

which can be shown to be equivalent to minimizing $J(\mathbf{w}) = \|\mathbf{w}\|^2$ subject to constraints (4). Using Lagrange multipliers, the solution to this constrained quadratic optimization problem is a linear combination of the training examples $\mathbf{w}^* = \sum_{\mathbf{x}_i} \alpha_i y_i \mathbf{x}_i$ ($\alpha_i \geq 0$) with only a few non-zero coefficients α_i. Moreover, the non-zero α_i's correspond to the training vectors that satisfy the inequality constraint (4) with equality. These are called *support vectors*, as they "support" the separating boundary between the classes (Fig. 1). The resulting classifier $f(\mathbf{x}) = \text{sign}((\mathbf{w}^* \cdot \mathbf{x}) + b) = \text{sign}(\sum \alpha_i y_i (\mathbf{x}_i \cdot \mathbf{x}) + b)$ is a linear function of dot products between the input vector $\mathbf{x}$ and the support vectors and defines a hyperplane in R^n.

To extend this to a non-separable case, we introduce non-negative slack variables ξ_i that measure by how much each training example violates constraint (4). The optimization problem is transformed to minimizing $J(\mathbf{w}) = \|\mathbf{w}\|^2 + C \sum \xi_i$ subject to constraints $\forall i : \ y_i((\mathbf{w} \cdot \mathbf{x}_i) + b) \geq 1 - \xi_i$. The constant C determines a trade-off between maximizing the margin and minimizing the number of errors. The same techniques that were used in the separable case can be applied to this optimization problem as well, and the optimal vector $\mathbf{w}^*$ is still a linear combination of a few support vectors.

This technique can also be extended to non-linear classification by observing that we only use dot products of data vectors to perform training and classification. Consider a *kernel function* $K : R^n \times R^n \mapsto R$, such that for some function $\Psi : R^n \mapsto R^m$ that maps the data into a higher dimensional space, the value of the dot product in R^m can be computed by applying K to the vectors in R^n: $\forall \mathbf{u}, \mathbf{v} \in R^n \ K(\mathbf{u}, \mathbf{v}) = (\Psi(\mathbf{u}) \cdot \Psi(\mathbf{v}))$. We can effectively train a linear classifier in the higher dimensional space R^m without explicitly evaluating Ψ, but rather using K to compute the dot products in R^m. This clas-

sifier produces a non-linear decision boundary back in the original space R^n:
$f(\mathbf{x}) = \mathrm{sign}(\sum \alpha_i y_i (\Psi(\mathbf{x}_i) \cdot \Psi(\mathbf{x})) + b) = \mathrm{sign}(\sum \alpha_i y_i K(\mathbf{x}_i, \mathbf{x}) + b)$.

Several different kernel functions have been proposed in the machine learning community. In this work, we use a family of Gaussian kernel functions $K(\mathbf{u}, \mathbf{v}) = \exp\{-\|\mathbf{u} - \mathbf{v}\|^2/\gamma\}$, where the parameter γ determines the width of the kernel.

Classifier selection. It can be shown that the VC dimension of an optimal hyperplane is bounded by

$$h \leq \min(D^2/\rho^2, n) + 1, \tag{6}$$

where ρ is the margin of the classifier, D is the radius of the smallest sphere that contains all the training examples, and n is the dimensionality of the space (Fig. 1). This bound can also be computed in the non-linear case, as the radius of the bounding sphere in the target space can be estimated using the kernel function. This suggests a method for classifier selection: among optimal classifiers obtained for a hierarchy of function families (e.g., of different kernel width γ), choose the classifier that minimizes the VC bound (3), using the margin and the radius of the bounding sphere to estimate the VC dimension of the classifier.

For problems with a small number of training examples, the VC bound might be too loose to be helpful for classifier selection, and other methods, such as cross-validation, are employed. The relationship between VC dimension and cross-validation is discussed in detail in [17]. The traditional approach to estimating the expected test error from the cross-validation results uses the Law of Large Numbers and De Moivre - Laplace approximation to arrive at the following bound: with probability at least $1 - \eta$

$$|R(\omega^*) - \hat{R}| \leq \Phi^{-1}\left(\frac{1-\eta}{2}\right)\sqrt{\frac{\hat{R}(1-\hat{R})}{l}}, \tag{7}$$

where $\hat{R}$ is the error rate of the cross-validation and $\Phi(x) = \int_0^x \frac{1}{\sqrt{2\pi}} \exp^{-t^2/2} dt$ is the standard error function.

3   Distance Transforms as Shape Descriptors

Numerous models have been proposed for shape description, as discussed in the introduction. In this section, we describe how we extract the shape information from segmented (binary) images using a distance transform and use it for classification.

A distance transform, or distance map, is a function that for any point inside an object is equal to the distance from the point to the closest point on an outline. Since the values of the distance transform at neighboring voxels are highly correlated, using it as a shape descriptor provides the learning algorithm with a lot of information on the structure of the feature space.

Another important property of the distance transform is what we call a *continuity of mapping*. The shape information extraction is an inherently noisy

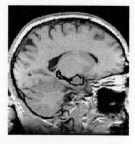  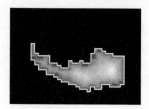

Fig. 2. Data example: an example sagittal slice with the hippocampus-amygdala complex segmented (left), a 3D surface model of the hippocampus-amygdala complex (middle), and a slice of the 3D distance transform (right).

process, consisting of imaging, segmentation and feature extraction, with every step introducing errors. Small changes in the input image (rigid transformation, errors in segmentation, etc.) cause the corresponding feature vector to change as well. The vectors in the training set are therefore computed with uncertainty, and it is important that small errors in any of the steps do not cause significant displacements of the resulting feature vectors. Since the gradient magnitude of the distance map is bounded by 1 everywhere in the image, and changes in the outline have only local effect on the distance map, the distance between the new and the old feature vectors in the feature space is bounded by the "magnitude of transformation" in the image space.

Finally, there is a question of establishing correspondence between the components of different feature vectors, or alignment. We use moments of the distance transform to align the images into a canonical representation, and other techniques, e.g., rigid registration, can further improve the initial alignment of the shapes. This step must use rigid transformations, as the non-rigid deformation of the shape is exactly what should be captured and generalized by the learning algorithm. Moments of the distance map place higher weights on interior points than on points that are closer to the boundary, which reflects our belief that the interior points of the distance transform are estimated more reliably than the outline points. The images are also scaled to be of the same volume, as we consider shape properties independently of the volume.

To summarize, our algorithm takes segmented images, computes a distance transform and its moments, aligns and scales the images so that the object volume and the center of gravity of the distance transform is the same for all example shapes, and the principal axes of the distance transform coincide with the volume sampling directions. Then the distance transform can be either resampled using the same transformation or recomputed using the aligned images. To minimize resampling errors, we perform all of the operations on a sub-voxel level. The resulting distance maps are sampled uniformly in 3D and stored as vectors to be used for learning. Each component of the feature vector corresponds to a particular location in space and is equal to the value of the distance map in that location. Each anatomical structure is processed separately, and the resulting feature vector is a concatenation of the vectors for each structure of interest.

Structure	Right hippocampus		Left hippocampus	
Descriptor	volume	shape	volume	shape
Training accuracy(%)	60.0	83.3	63.3	83.3
Cross-validation(%)	60.0 ± 17.5	66.7 ± 16.9	63.3 ± 17.2	60.0 ± 17.5

Structure	Both hippocampi			
Descriptor	volume	shape	shape & volume	shape & rel. vol.
Training accuracy(%)	66.7	86.7	83.3	83.3
Cross-validation(%)	63.3 ± 17.2	70.0 ± 16.4	$\mathbf{73.3 \pm 15.8}$	70.0 ± 16.4

Table 1. Training and cross-validation accuracy for volume and shape. The results for cross-validation consist of estimated expected error, as well as 95% confidence interval, computed based on Eq. (7). The training accuracy is reported for the parameter setting that yielded the best cross-validation results.

4 Experimental Results

In this section, we report the results of the method applied to a data set that contains MRI scans of 15 schizophrenia patients and 15 matched controls. In each scan, the hippocampus-amygdala complex was manually segmented (Fig. 2a,b). More details on the subject selection and data acquisition can be found in [13]. The same paper reports statistically significant differences in left anterior hippocampus based on relative volume measurements (the volume of the structure normalized by the total volume of intracranial cavity or ICC). For each of the experiments reported in this paper, we systematically explore the space of parameters (the kernel width γ and the soft margin constant C) by sampling it on a logarithmic scale, training a Gaussian kernel classifier and estimate its VC dimension as described in Section 2, and also performing leave-one-out cross-validation. Table 1 contains the summary of training and the cross-validation accuracy for all our experiments.

In order to compare the shape based analysis to the volume based analysis, we first trained a classifier and performed leave-one-out cross-validation based on the relative volume measurements only[1]. We then repeat the experiment with the shape descriptors (distance maps). Fig. 3a,b shows the accuracy sampling for different parameter settings for shape classification, and Table 1 contains the training and the cross-validation accuracy for the best classifier selected based on the cross-validation results. If used on a single side of hippocampus, the shape based results are not significantly different from the volume based result, but there is a significant improvement (63% to 70%) when the two sides are considered together. This suggests that the disease might affect the shape of both hippocampi, and it is the combined shape that is the best indicator of those differences.

---

[1] Note that for volume based classification, the cross-validation accuracy does not differ significantly from the training accuracy, because removing a single training example can only affect the training result if the example is close to the threshold (i.e., it's a support vector), and there could only be very few of those in the low-dimensional space.

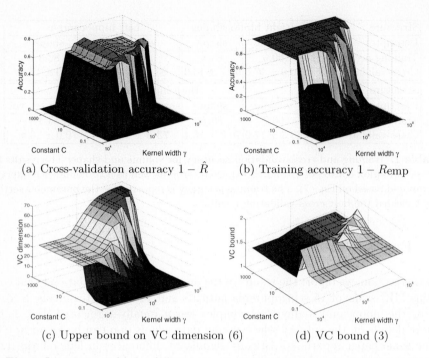

(a) Cross-validation accuracy $1 - \hat{R}$ (b) Training accuracy $1 - R_{\mathrm{emp}}$

(c) Upper bound on VC dimension (6)   (d) VC bound (3)

Fig. 3. Learning results for different parameter settings using both hippocampi.

Given the results reported so far, the natural question to ask is what happens if we combine shape information with volume measurements? If they are decorrelated, better estimates could potentially be obtained by using both. In order to investigate this question, we re-ran the experiment with a slightly modified feature extraction procedure. In the first experiment we kept the volume of the structure unchanged, and then ran the experiment again while scaling the images so that the resulting volume for every image is proportional to the relative volume as reported in the original paper [13]. The results of these experiments are also shown in Table 1. We can see that combining volume and shape information yields additional improvement (cross-validation accuracy 73%).

In order to perform the capacity analysis of the classifier, we estimated VC dimension for each one of the parameter settings and computed the VC bound (3) for the shape based experiment (see Fig. 3). Unfortunately, the bound is too loose to be helpful for classifier selection (it is greater than 1), and we have to resort to cross-validation. Examining the graphs in Fig. 3, we observe that empirical risk R_{emp} and the estimated VC dimension behave in opposite ways: one increases when the other goes down. But when they are combined in (3), the VC confidence dominates the empirical risk term and the sum does not have a distinct minimum. As the number of input examples increases, the estimates of the VC dimension will become more accurate, and the bound can be useful in investigating the functional behavior of the generalization error for different setting of parameters γ and C.

5   Discussion and Conclusion

We have proposed and demonstrated a novel approach to statistical shape analysis of anatomical structures. Our technique is based on the body of knowledge developed in machine learning for situations when the distribution of the data is unknown and the number of examples is limited. It makes no assumptions on the distribution of the data and moreover, is guaranteed to converge to the optimal solution (minimizing the expected error on future examples) as the number of training examples increases. It provides a principled way to attack a problem of modeling statistical shape differences in the medical domain. When applied to the hippocampus study in schizophrenia, shape information allowed us to improve over a purely volume based approach. The results are comparable to the traditional approach (e.g., Csernansky *et al* [6] report 80% accuracy in cross-validation using deformation fields as a shape model and PCA for dimensionality reduction), and if more examples are available, can potentially tighten the bound on the test error.

We used distance maps as shape descriptors in this work, but the learning techniques presented here are applicable to other shape models as well. The main contribution of this paper is in the way we perform learning on the shape features. Comparison between different shape representations and their performance when combined with our analysis is an interesting question that needs to be investigated in the future.

There are several other topics that we would like to explore next, and the most important one is the interpretation of the learning results in terms of shape deformation. Linear models provide a very intuitive interpretation of the parameters as weights assigned to the corresponding vectors (e.g., modes of deformation), but it cannot be extended in a straightforward manner to any family of non-linear models. Finding methods for mapping a non-linear classifier function back into the image domain and providing an interpretation of the results for the medical researches is our current interest.

Another important direction of research in this framework is incorporating invariants into the learning process. To enable efficient learning, information on the structure of the feature space has to be provided to the training algorithm. Some of this is achieved by selecting a good representation. Invariants are important constraints on the feature space that often cannot be explicitly modeled by the shape representation. In some cases, a family of classifiers can be changed to guarantee that the resulting function satisfies the invariance constraint (see [4], for example), or artificial examples can be generated using the invariants and added to the original training set [1]. We plan to explore this direction as a way of further restricting the classifier family and thus improving the accuracy.

To summarize, recent advances in statistical learning theory enabled a new approach to statistical analysis of high dimensional data. As the results presented in this paper demonstrate, the field of statistical shape analysis, and medical research in general, can benefit from these techniques. While there are still many open questions on interpretation of the results and incorporation of prior knowledge into the method, it is a promising direction of research that can help medical researches to get a better insight into various anatomical phenomena.

Acknowledgments. Quadratic optimization was performed using PR_LOQO optimizer written by Alex Smola.
This research was supported in part by NSF IIS 9610249 grant. M. E. Shenton was supported by NIMH K02, MH 01110 and R01 MH 50747 grants, R. Kikinis was supported by NIH PO1 CA67165, R01RR11747, P41RR13218 and NSF ERC 9731748 grants.

References

1. Y.S. Abu-Mostafa. Learning from hints. *J. Complexity*, 10(1):165-178, 1994.
2. F. L. Bookstein. Landmark methods for forms without landmarks: morphometrics of group differences in outline shape. *Medical Image Analysis*, 1(3):225-243, 1996.
3. C. J. C. Burges. A Tutorial on Support Vector Machines for Pattern Recognition. *Data Mining and Knowledge Discovery*, 2(2):121-167, 1998.
4. C. J. C. Burges. Geometry and Invariance in Kernel Based Methods. *In Advances in Kernel Methods: Support Vector Learning*, Eds. B. Schölkopf, C.J.C. Burges, A.J. Smola, MIT Press, 89-116, 1998.
5. T. F. Cootes *et at.* The Use of Active Shape Models for Locating Structures in Medical Images. *Image and Vision Computing*, 12(6):355-366, 1994.
6. J. G. Csernansky *et al.* Hippocampal morphometry in schizophrenia by high dimensional brain mapping. *In Proc. Nat. Acad. of Science*, 95(19):11406-11411, 1998.
7. D. S. Fritsch *et al.* The multiscale medial axis and its applications in image registration. Patter Recognition Letters, 15:445-452 1994.
8. J. C. Gee and R. Bajcsy, Personal communication.
9. A. Kelemen, G. Székely, and G. Gerig. Three-dimensional Model-Based Segmentation. *In Proc. IEEE International Workshop on Model Based 3D Image Analysis*, Bombay, India, 87–96, 1998.
10. J. Martin, A. Pentland, and R. Kikinis. Shape Analysis of Brain Structures Using Physical and Experimental Models. *In Proc. CVPR'94*, 752-755, 1994.
11. M. E. Leventon, W. E. L. Grimson and O. Faugeras. Statistical Shape Influence in Geodesic Active Countours. *In Proc. CVPR'2000*, 316-323, 2000.
12. E. E. Osuna, R. Freund, F. Girosi. Training Support Vector Machines: An Application to Face Detection. *In Proc. CVPR'97*, 130-136, 1997.
13. M. E. Shenton, *et al.* Abnormalities in the left temporal lobe and thought disorder in schizophrenia: A quantitative magnetic resonance imaging study. *New England J. Medicine*, 327:604-612, 1992.
14. L. Staib and J. Duncan. Boundary finding with parametrically deformable models. *IEEE PAMI*, 14(11):1061-1075, 1992.
15. G. Székely *et al.* Segmentation of 2D and 3D objects from MRI volume data using constrained elastic deformations of flexible Fourier contour and surface models. *Medical Image Analysis*, 1(1):19-34, 1996.
16. V. N. Vapnik. The Nature of Statistical Learning Theory. *Springer*, 1995.
17. V. N. Vapnik. Statistical Learning Theory. *John Wiley & Sons*, 1998.

Robust Midsagittal Plane Extraction
from Coarse, Pathological 3D Images

Yanxi Liu, Robert T. Collins and William E. Rothfus*

The Robotics Institute, Carnegie Mellon University, Pittsburgh 15213, USA,
`yanxi,rcollins@cs.cmu.edu`
* University of Pittsburgh Medical Center, Pittsburgh, PA

Abstract. This paper focuses on the evaluation of an ideal midsagittal plane (iMSP) extraction algorithm. The algorithm was developed for capturing the iMSP from 3D normal and pathological neural images. The main challenges are the drastic structural asymmetry that often exists in pathological brains, and the sparse, nonisotropic data sampling that is common in clinical practice. A simple edge-based, cross-correlation approach is presented that decomposes the iMSP extraction problem into discovery of symmetry axes from 2D slices, followed by robust estimation of 3D plane parameters. The algorithm's tolerance to brain asymmetries, input image offsets and image noise is quantitatively measured. It is found that the algorithm can extract the iMSP from input 3D images with (1) large asymmetrical lesions; (2) arbitrary initial yaw and roll angle errors; and (3) low signal-to-noise level. Also, no significant difference is found between the iMSP computed by the algorithm and the midsagittal plane estimated by two trained neuroradiologists.

Keywords: midsagittal plane, asymmetry, robust estimation

1 Introduction

Healthy human brains exhibit an approximate bilateral symmetry with respect to the interhemispheric (longitudinal) fissure plane bisecting the brain, known as the **midsagittal plane** (MSP). However, human brains are almost never perfectly symmetric [4–6]. Pathological brains, in particular, often depart drastically from reflectional symmetry. For effective pathological brain image alignment and comparison (e.g. [4, 10, 11]) it is most desirable to define a *plane of reference* that is invariant for symmetrical as well as asymmetrical brain images and to develop algorithms that capture this reference plane robustly.

We define an **ideal midsagittal plane (iMSP)** as a virtual geometric plane about which the given 3D brain image presents maximum bilateral symmetry. Computationally, this plane is determined by taking the majority votes from both axial and coronal 2D slices based on a sound geometric analysis (Section 2.1). Factors that challenge the robustness of an iMSP algorithm include: (1) the intrinsic factor: the brains being imaged can be either bilaterally symmetric or drastically asymmetric; (2) the extrinsic factor(s): anisotropism, under-sampling, initial transformation errors and artifacts/noise can be introduced during the imaging process.

The well known Talairach framework is an anatomical landmark-based approach to define a 3D brain coordinate system [18]. Patient image to Talairach framework registration is difficult to achieve automatically since it relies on identifying 3D anatomical features that may not be obvious in the image. Furthermore, when the interhemispheric sagittal plane no longer lies on a flat surface due to normal or pathological deformation (Figure 1(a)), the interhemispheric medial plane is ill defined. In contrast, the iMSP is based on global geometry of the head and can be found using low-level image processing techniques. Furthermore, it remains well-defined in pathological cases, forcing a virtual left-right separation consistent with the location where an <u>ideal</u> midsagittal plane would be if not for the presence of local brain deformation (Figure 1(b)).

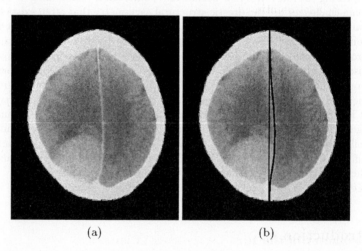

(a)   (b)

Fig. 1. (a) An axial brain slice, the midline is deformed due to a space occupying tumor. (b) The intersection of the extracted iMSP with the same 2D brain slice (straight line), and the deformed midline (curved line) captured by a "snake" active contour.

Designing a robust algorithm that deals with real clinical images originates from our desire to facilitate on-line clinical image database indexing and retrieval for real-time medical consultation [10–13]. We have developed a simple yet robust algorithm that can extract the iMSP of a brain from clinical CT or MR images (Section 2.2). The algorithm has been applied on more than 100 3D clinical images and tested on both synthetic and real images with computed ground truth. Breakdown points of the iMSP extraction algorithm are found by varying brain orientation, lesion size, and noise level, and are compared against a maximization of mutual information method (Section 3). We also consider "ground truth" generated by human experts, and find no significant difference between the orientations estimated by our iMSP algorithm and those hand-picked by two neuroradiologists (Section 3).

The goal of many existing MSP detection algorithms [1–3, 7, 19] is to locate the plane of reflection of a roughly symmetrical brain image. Tolerance of these algorithms to asymmetry is reported to be low [3, 7]. The MSP algorithm developed in [15] is tested on simulated PET images to show its insensitivity to focal asymmetries. More recently, Smith and Jenkinson [17] presented an algorithm for finding symmetry axes in partially damaged images of various modalities. However, no quantitative evaluations have yet been given and the computation is expensive.

2 Ideal Midsagittal Plane Extraction

We define an *ideal head coordinate system* centered in the brain with positive X_0, Y_0 and Z_0 axes pointing in the right, anterior and superior directions respectively (Figure 2, white coordinate axes). With respect to this coordinate system, the iMSP of the brain is defined as the plane $X_0 = 0$. Ideally, a set of *axial (coronal)* slices is cut perpendicular to the $Z_0(Y_0)$ axis, and the intersection of the MSP with each slice appears as a vertical line on the slice[1].

In clinical practice, however, the imaging coordinate system XYZ (Figure 2(a), black coordinate axes) differs from the ideal coordinates due to unintentional positioning errors and/or deliberate realignments introduced so that a desired volume can be better imaged, The orientation of the imaging coordinate system differs from the ideal coordinate system by three rotation angles, pitch, roll and yaw, about the X_0, Y_0 and Z_0 axes, respectively. The imaging coordinate system can also be offset (Figure 2(a)). The goal of an iMSP algorithm is to find the transformation between the two planes: $X_0 = 0$ and $X = 0$.

2.1 Geometry of the Ideal Midsagittal Plane

Under the *imaging coordinate system*, the iMSP can be represented as

$$aX + bY + cZ + d = 0 \tag{1}$$

where (a, b, c) is a vector describing the plane normal. For the rest of this section we assume a nonzero scaling such that $\sqrt{a^2 + b^2} = 1$. Now consider the ith axial slice, represented by the plane equation $Z = Z_i$. The symmetry axis of the ith slice is the intersection of the above two planes: $aX + bY + (cZ_i + d) = 0$. This is the equation of a 2D line (θ_i, ρ_i) in the image XY plane, having line orientation $\theta_i = \arctan(b/a)$ and perpendicular offset from the image origin $\rho_i = c Z_i + d$. By examining this line equation, we can make two immediate observations. First, the orientation angle $\theta_i = \arctan(b/a)$ is the same for the symmetry axes of all slices regardless of their Z_i position. Secondly, the offset

---

[1] The analysis given to the axial slices from now on can be applied to coronal slices (cut along the Y axis) as well with corresponding symbols changed: 'Z' to 'Y'.

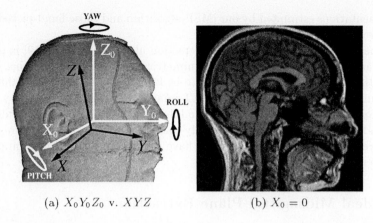

(a) $X_0 Y_0 Z_0$ v. XYZ (b) $X_0 = 0$

Fig. 2. (a) Ideal head coordinate system $X_0 Y_0 Z_0$ vs. the imaging coordinate system XYZ. The iMSP algorithm finds the transformation between planes $X_0 = 0$ and $X = 0$. Rendered head courtesy of the Visible Human Project. (b) A midsagittal plane ($X_0 = 0$) automatically extracted using our iMSP extraction algorithm.

ρ_i of the symmetry axis on slice $Z = Z_i$ is linearly related to Z_i as a function of plane parameters c and d. Therefore, given the translational offset of at least two symmetry axes on different slices, we can estimate c and d by solving a set of linear equations. These observations form the basis of the iMSP extraction algorithm.

Under the ideal coordinate system, the iMSP passes through the origin and has normal vector $(1, 0, 0)$. Due to the scanning geometry, points in the ideal coordinate system are reoriented into the observed imaging coordinate by an unknown rotation $\mathcal{R} = \mathbf{yaw}(\theta)\mathbf{roll}(\phi)\mathbf{pitch}(\omega)$ and displaced by an unknown translation ΔX_0, ΔY_0, and ΔZ_0. Precisely, points in the ideal coordinate system are mapped into the imaging coordinate system by the transformation matrix

$$\begin{bmatrix} X \\ Y \\ Z \end{bmatrix} = \begin{bmatrix} c\phi\, c\theta & c\theta\, s\omega\, s\phi - c\omega\, s\theta & c\omega\, c\theta\, s\phi + s\omega\, s\theta & \Delta X_0 \\ c\phi\, s\theta & c\omega\, c\theta + s\omega\, s\phi\, s\theta & c\omega\, s\phi\, s\theta - c\theta\, s\omega & \Delta Y_0 \\ -s\phi & c\phi\, s\omega & c\omega\, c\phi & \Delta Z_0 \end{bmatrix} \begin{bmatrix} X_0 \\ Y_0 \\ Z_0 \end{bmatrix}$$

where $c\theta \equiv \cos\theta$, $s\theta \equiv \sin\theta$, and so on. After some algebraic manipulation, the iMSP $X_0 = 0$ can be rewritten in terms of the imaging coordinate system as

$$\cos\phi\, \cos\theta\, X + \cos\phi\, \sin\theta\, Y - \sin\phi\, Z - (n \cdot \Delta) = 0 \qquad (2)$$

where $n = (\cos\phi\, \cos\theta, \cos\phi\, \sin\theta, -\sin\phi)^T$ is the unit normal vector of the plane and $\Delta = (\Delta X_0, \Delta Y_0, \Delta Z_0)^T$. Scaling appropriately, and comparing with (1), we find by inspection that

$$a = \cos\theta \qquad b = \sin\theta \qquad c = -\tan\phi \qquad d = -(n \cdot \Delta)/\cos\phi \ . \qquad (3)$$

That is, the shared angle $\theta = \theta_i = \arctan(b/a)$ of each axial slice is actually the yaw angle of the head's imaging coordinate system. Furthermore, the roll angle

ϕ can be determined from the offsets of the 2D symmetry axes on the set of slices by solving a linear system of equations $\rho_i = c Z_i + d$. Note from this equation and equation (3) that when the roll angle ϕ is zero, $c = 0$, and thus all 2D symmetry axes have the same offset $\rho_i = d$; otherwise ρ_i varies linearly from slice to slice. Finally, the quantity $(-d \cos \phi)$ measures the displacement of the imaging coordinate system in the direction normal to the symmetry plane.

In summary, if we can extract the 2D axes of reflection symmetry from a set of axial slices, we can completely determine the geometric equation $aX + bY + cZ + d = 0$ of the ideal MSP. Furthermore, we can infer from this equation some of the 3D pose parameters of the patient's head with respect to the ideal head coordinate system, namely the yaw angle θ, roll angle ϕ and the translational offset along the X_0 axis.

2.2 Midsagittal Plane Extraction Algorithm

The geometric results from the previous section have been used to develop an algorithm for automatically detecting the iMSP of a neural brain scan.

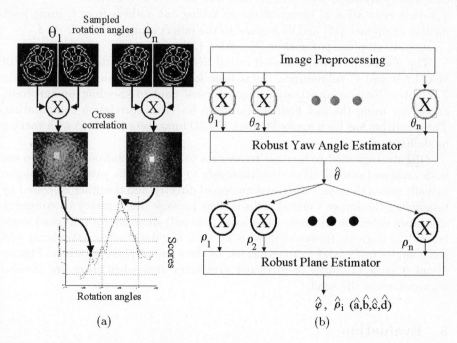

(a)   (b)

Fig. 3. (a) Using pairs of edge images (original image and its rotated reflection w.r.t. $X = 0$ plane) to find the best correlation value for each given rotation angle. The brightest point indicates the highest correlation score. (b) A flow chart of the MSP extraction algorithm, where "X" with a circle around means 2D cross correlation, and with the addition of a square means multiple cross correlations of different rotated images.

Input: a set of brain scans in axial (or coronal) format.

Output: head yaw and roll angles $\hat{\theta}$, $\hat{\phi}$, and translational offset of the iMSP from the $X_0 = 0$ plane.

iMSP Extraction Algorithm[2]

1. isolate the head region and compute binary edge images S_i.

2. find an estimated initial rotation error θ_{init} from one of the lowest 2D brain slices: for each S_i, construct $S'_i = S_i$ reflected w.r.t. $X = 0$ plane.

$\theta_{init} = \arg\max\{C_i(S_i, rot(S'_i, \theta))\}$ where C_i is the cross correlation of S_i and rotated S'_i, θ is sampled in the range of $[-90^o, 90^o]$ or $[-180^o, 180^o]$ if necessary.

3. find symmetry axis orientation θ_i on each S_i (Figure 3(a)):

for each 2D slice S_i,

for j = -10 to 10 step 1 (degrees), $C_i(S_i, rot(S'_i, \theta_{init} + j))$ end;

$\theta_i = \arg\max\{C_i\}$.

end;

4. compute the common yaw (or roll) angle (Figure 3(b)):

$\hat{\theta} = robust(\theta_1, \cdots, \theta_n)$, where function $robust$ eliminates outliers [20] and finds the median from weighted inliers.

5. compute image offsets t_i by finding the maximum cross correlation value of each yaw (roll)-angle-corrected 2D slice and its vertical reflection.

6. robust estimation of image offsets ρ_i: taking out outliers from t_i using least median of squares [16] and fit a plane to the inliers using $\rho_i = c Z_i + d$.

7. compute all plane parameters (Equation (3)).

The algorithm is implemented on an SGI O2 R10000, using a mixture of MATLAB and C subroutines. Total time for all algorithmic steps is roughly 7 minutes. No special attention has yet been paid to speeding up the code, except for using the fast Fourier transform for cross correlation computation. The algorithm has been applied to over 100 3D image sets (Table 1) with varying modalities.

One strategy used to increase robustness to large orientation errors is to use both axial and coronal slices simultaneously to estimate the yaw and roll angles (usually one of these sets of slices is measured directly, and the other is created by resampling the image volume). This is done because the accuracy of symmetry axis offset detection degrades when axial (coronal) slices have a roll (yaw) angle beyond 20 degrees. However, there is no such angular limit to estimating yaw (roll) angles from axial (coronal) slices, as these are in-plane rotations. Figures 4 and 5 show examples of extracted symmetry axes when there are obvious asymmetries in the head.

3 Evaluation

No obvious mishaps have been observed when applying the iMSP extraction algorithm to over 100 image sets with varying modalities and scan geometries. In this section we report a series of experiments that are carried out to test the

---

[2] For further details of the algorithm see [9]

Table 1. A Sample of Input 3D Image Data

Set	Modality	Form	Matrix	Voxel (mm³)	Pathology
5	MR (T1)	coron	256x256x123	0.9375x0.9375x1.5	Normal
17	CT	axial	686x550x9	0.5x0.5x10	Right thalamic acute bleed
58	CT	axial	678x542x17	0.5x0.5x5 (1-9) 0.5x0.5x10 (10-17)	Frontal astrocytoma high grade glial tumor
109	CT	axial	512x512x21	0.4395x0.0.4395x5 (1-10) 0.4395x0.4395x8 (11-21)	Left parietal infarct
110	CT	axial	512x512x24	0.4297x0.4297x5	Normal

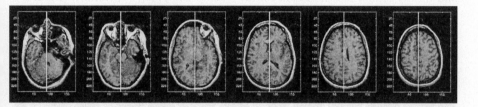

Fig. 4. The symmetry axes extracted from a set of axial slices with a 15 degree roll angle (out of plane rotation).

robustness of the iMSP algorithm. Two ground truth image test sets are created from datasets 5 and 110 respectively (Table 1), one is a sparse, axial CT volume and the other is a dense, coronal MR volume. Each ground truth test set is formed by finding the midsagittal plane by hand, then reflecting one half of the head volume about this midsagittal plane to overlay the other half, producing a perfectly symmetrical volume. Since the constructed test set is perfectly symmetric, the ground truth iMSP is known.

Tolerance to initial offset errors

To evaluate the accuracy of computed roll and yaw angles, the MR test image set was resampled using trilinear interpolation to artificially vary the yaw (roll) angles from -90 to 90 degrees in 5 degree intervals. The algorithm was then run on these sets to determine an estimated yaw and roll angle. In all cases, the error between estimated and actual angles is less than one degree. Given the image sampling interval, these errors are negligable.

Tolerance to asymmetry

The algorithm is tested by superimposing spherical "lesions" of varying density, size and position into the CT test image. The lesions are offset from the midline, resulting in pathological assymetry of the brain. The MSP algorithm's performance starts to decline when the tumor radius reaches 200 pixels (85.94mm),

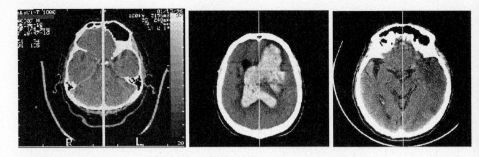

Fig. 5. The symmetry axes extracted from different CT scans where obvious asymmetry is present. Left to right: normal, acute blood, infarct.

and totally fails when the lesion radius reaches 250 pixels (107.4mm) (Figures 6 and 7).

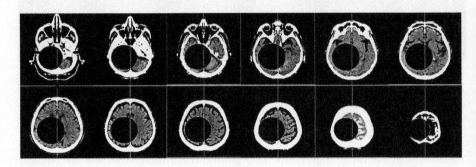

Fig. 6. The iMSP algorithm performs successfully when presented with an artificially grown lesion of 100 pixels (42.97mm) radius.

Tolerance to noise and asymmetry

To study the effects of noise and asymmetry on the iMSP extraction algorithm, we have tested the algorithm on the MR ground truth dataset. The data is artificially degraded by adding different levels of zero-mean Gaussian noise, and by inserting spherical lesions of varying diameters. The algorithm breaking point is determined by incrementally adding noise until the algorithm fails to detect the correct symmetry plane. Each incremental addition of noise corresponds to a loss of 6.02dB of SNR, or roughly 1 bit of information.[3]   Figure 8 shows representative slices, and iMSPs extracted by the algorithm. Naturally, the algorithm is more robust to noise when no lesion is present (Figure 8(d)). But the algorithm can handle very large levels of noise (up to SNR = -10.8dB when no lesion is

---

[3] SNR or Signal to Noise Ratio is defined as 10 * log(var(signal)/var(noise)). An SNR of less than 0 means that the noise has a higher variance than the signal.

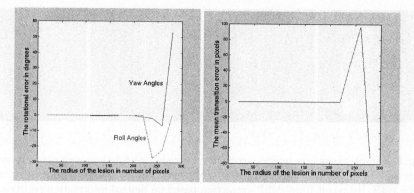

Fig. 7. Yaw and roll angle errors (Left) and translational offset errors (Right) versus the radius of an artificial lesion, in pixels (1 pixel = 0.4297 mm). The spherical lesion is centered at 3D location $X = 310; Y = 210; Z = 350$. The MSP algorithm's performance starts to decline when the tumor radius reaches 225 pixels (96.68 mm), and totally fails when the lesion radius reaches 250 pixels (107.43mm).

present), as well as large lesions (we tested with lesions of radius up to 60 pixels = 56.25mm).

Comparison with a mutual information maximization method

For comparison, an implementation of midsagittal plane extraction based on a general mutual information volume registration algorithm [14] was also tested. Volume registration can be used to identify an iMSP by finding the rigid transformation that best registers a volume with a version of itself reflected about the $X = 0$ plane. The geometry underlying iMSP extraction in this case is analogous to that described in Section 2.2. We choose to compare a volume registration algorithm based on mutual information due to its intensity-based nature, in contrast to our edge-based method. The experimental results show that iMSP extraction based on mutual information registration breaks down at a lower level of noise than our algorithm, as shown in Figure 8. The volume registration algorithm had to be tested on the dense MR dataset, since it could not directly handle sparse, unisotropic clinical CT data.

Comparison with human experts To compare the algorithm with human performance, we had two neuroradiologists (one has 20 years experience, one is an intern) hand-draw the ideal midline on each 2D slice of several randomly chosen brain scans. The radiologists were allowed to view the whole set of the 2D slices from one volumetric image for reference while using a mouse to click on a computer screen directly. Although scan geometry tells us the angles of the symmetry axes on each axial slice should be the same (Section 2.1), there is a variation in the angles determined by the human expert. The standard deviation of the human measurement error on different sets of slices varies from 0.55 to 2.37 degrees (Table 2).

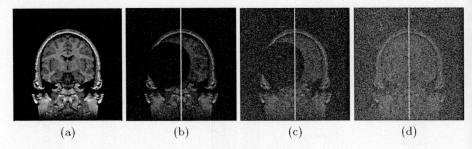

(a)   (b)   (c)   (d)

Fig. 8. Sample results on images artificially degraded with lesions and noise, to determine algorithm breaking points. (a) one slice from original dataset with no noise, SNR = 43.35dB. (b) result from iMSP extraction based on mutual information registration, on a dataset with lesion plus noise. SNR of breaking point is 13.25dB. (c) result from our algorithm, on a dataset with lesion plus noise. SNR of breaking point is -4.82dB. (d) result from our algorithm when run on a dataset with added noise only (no lesion). SNR of breaking point is -10.84dB.

Table 2. Comparison of human vs. computer-estimated yaw angles (in degrees)

Label	Pathology	# Slices	Expert 1 Mean(θ_i)	Expert 2 Mean(θ_i)	Expert 1 STD(θ_i)	Expert 2 STD(θ_i)	Computed Yaw $\hat{\theta}$
CMU121	infarct	17	1.6208	1.5730	1.0290	0.9972	0.8202
CMU126	blood	21	7.1404	6.2472	2.3678	0.5595	6.0347
CMU129	infarct	10	-2.0248	-1.8201	1.1141	1.6304	-2.2318
CMU130	blood	21	1.3257	0.6829	0.9091	0.8967	1.3216
CMU170	normal	20	-4.3187	- 4.5296	1.2781	2.0080	-4.5109
CMU171	normal	22	-1.3028	- 1.7582	1.0626	1.1963	-1.0129

4 Discussion and Future Work

In this paper, we have presented an iMSP extraction algorithm that is capable of finding the ideal MSP from coarsely sampled, asymmetrical neural images, without compromising accuracy on symmetrical ones. We have observed that iMSP computation using our algorithm is not adversely affected by lesions, mass effects, or noise in the images. This may seem strange since cross-correlation is used as a measure for the best matching of two images. It is natural to ask why the algorithm works so well on drastically asymmetrical images. We can provide the following relevant observations: 1. **Majority rules**: For a 3D pathological brain, a lesion only resides on a relatively small number of 2D slices, thus when the iMSP is fit to the whole set of 2D slices, normal slices with prominent bilateral symmetry dominate the iMSP's position. 2. **Edge features**: By using edge features rather than the original intensity images directly, the effect of strong density concentration around lesions is much reduced. 3. **Lower brain slice stability**: Lower brain slices are relatively stable due to the bilateral structure of the skull. In practice, the lower brain slices are given more weight when determining the orientation of the iMSP. 4. **Robust estimators**: Robust estimation techniques are used to remove outliers from computed measurements before combining them to determine plane parameters.

We are currently exploring how to use the iMSP extraction algorithm to facilitate registration of pathological images with other modalities (PET, SPECT), and comparison of brain images (e.g. schizophrenia, acute infarct). Computing similarity among diverse brain images is part of an ongoing effort to study how symmetry-based features can be used for classification and retrieval of medically relevant cases [10–13]. We have also begun to apply the method on 3D pelvis images, to establish a common coordinate system for cross patient comparison, set an initial position for X-ray/CT registration [8], and evaluate left-right abnormality (Figure 9).

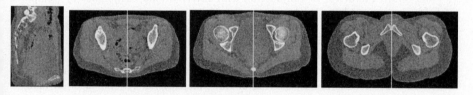

Fig. 9. Results from applying the MSP algorithm to pelvis CT images. Leftmost image is the midsagittal plane automatically extracted.

References

1. P. Allain, J.M. Travère, J.C. Baron, and D. Bloyet. *Quantification of Brain Function. Tracer Kinetics and Image Analysis in Brain PET: Accurate PET positioning with reference to MRI and neuroanatomical data bases,* pages 401,408. Elsevier Science Publishers, B.V., 1993.

2. B Ardekani, J. Kershaw, M. Braun, and I. Kanno. Automatic detection of the mid-sagittal plane in 3-d brain images. *IEEE Transactions on Medical Imaging,* 16(6):947–952, December 1997.

3. M.E. Brummer. Hough transform detection of the longitudinal fissure in tomographic head images. *IEEE Transactions on Medical Imaging,* 10(1):74–81, March 1991.

4. T.J. Crow. Schizophrenia as an anomaly of cerebral asymmetry. In K. Maurer, editor, *Imaging of the brian in psychiatry and related fields.* Springer-Verlag, 1993.

5. R.J. Davidson and K. Hugdahl, editors. *Brain Asymmetry.* MIT Press/Bradford Books, Cambridge, Mass., 1996.

6. N. Geschwind and W. Levitsky. Human brain: left-right asymmetries in temporal speech region. *Science,* 161:186–187, 1968.

7. L. Junck, J.G. Moen, G.D. Hutchins, M.B. Brown, and D.E. Kuhl. Correlation methods for the centering, rotation, and alignment of functional brain images. *Journal of Nuclear Medicine,* 31(7):1220–1226, July 1990.

8. D. LaRose and et al. Post-operative measurement of acetabular cup position using x-ray/ct registration. In *MICCAI2000*, October 2000.

9. Y. Liu, R.T. Collins, and W.E. Rothfus. Robust Midsagittal Plane Extraction from Normal and Pathological 3D Neuroradiology Images. *IEEE Transactions on Medical Imaging*, in revision 2000.

10. Y. Liu and F Dellaert. A Classification based Similarity Metric for 3D Image Retrieval. In *Proceedings of Computer Vision and Pattern Recognition Conference*, pages 800–805, June 1998.

11. Y. Liu and F Dellaert. Classification Driven Medical Image Retrieval. In *Image Understandard Workshop*. DARPA, November 1998.

12. Y. Liu, F. Dellaert, and E. W. Rothfus. Classification driven semantic based medical image indexing and retrieval. Technical Report CMU-RI-TR-98-25, The Robotics Institute, Carnegie Mellon University, Pittsburgh, PA, 1998.

13. Y. Liu, W. Rothfus, and T. Kanade. Content-based 3d neuroradiologic image retrieval: Preliminary results. *IEEE Workshop on Content-Based Access of Image and Video Libraries, in conjunction with International Conference of Computer Vision*, pages 91,100, January 1998.

14. F. Maes, A. Collignon, D. Vandermeulun, G. Marchal, and P. Suetens. Multimodality image registration by maximization of mutual information. *IEEE Transactions on Medical Imaging*, 16(2):187,198, 1997.

15. S. Minoshima, K.L. Berger, K.S. Lee, and M.A. Mintum. An automated method for rotational correction and centering of three-dimensional functional brain images. *Journal of Nuclear Medicine*, 33(8):1579–1585, August 1992.

16. P.J. Rousseeuw. Least median-of-squares regression. *Journal of the American Statistical Association*, 79:871–880, 1984.

17. S. Smith and M. Jenkinson. Accurate robust symmetry estimation. In *Medical Image Computing and Computer-Assisted Intervention - MICCAI'99*, pages 308,317. Springer, 1999.

18. J. Talairach and P. Tournoux. *Co-Planar Steriotaxic Atlas of the Human Brain*. Thieme Medical Publishers, 1988.

19. J.P. Thirion, S. Prima, and G. Subsol. Statistical analysis of normal and abnormal dissymmetry in volumetric medical images. In *Proc. of Workshop on Biomedical Image Analysis*, pages 74–83. IEEE Computer Society, 1998.

20. Z. Zhang. Parameter estimation techniques: a tutorial with application on conic fitting. *Image and Vision Computing*, 15:59–76, 1997.

3-D Reconstruction of Macroscopic Optical Brain Slice Images

Alan Colchester[1], Sébastien Ourselin[2], Yonggen Zhu[1], Eric Bardinet[2], Yang He[1], Alexis Roche[2], Safa Al-Sarraj[3], Bill Nailon[4], James Ironside[4], Nicholas Ayache[2]

[1] KIMHS, University of Kent at Canterbury, UK
{A.Colchester, Y.Zhu, Y.He}@ukc.ac.uk
[2] INRIA, Epidaure Project, Sophia Antipolis, France
{Sebastien.Ourselin, Eric.Bardinet, Alexis.Roche,
Nicholas.Ayache}@sophia.inria.fr
[3] Institute of Psychiatry, London, UK
s.al-sarraj@iop.kcl.ac.uk
[4] National Creutzfeldt-Jakob Disease Surveillance Unit, Western General Hospital,
Edinburgh, UK
{j.w.ironside, william.nailon}@ed.ac.uk

Abstract. We present a method for reconstruction of macroscopic optical images of *post-mortem* brain slices to form a 3-D volume. This forms a key part of a series of procedures to allow *post mortem* findings to be accurately registered with MR images, and more generally provides a method for 3-D mapping of the distribution of pathological changes throughout the brain. In this preliminary work, four brains from a study of Creutzfeldt-Jakob disease were examined. After brain fixation, the pathologist cut coronal slices several mm thick through the brain. The anterior and posterior faces of each slice were photographed. We show that the 2-D co-registration of each such pair of images was most effectively obtained if the slice was placed in a jig before photographing. Fiducials on the jig were detected automatically and point-based rigid registration computed. For co-registration between slices, i.e., across a single cut, an intensity-based method for 2-D non-rigid registration is used which provided satisfactory results. By propagating the 2-D registrations through the volume and using the known slice thickness, the 3-D volume was reconstructed. **Keywords:** registration; magnetic resonance image-to-pathology correlation; Creutzfeldt-Jakob disease.

1 Introduction

Study of the pathological correlates of MR signal abnormalities is important in many neurological conditions. Traditionally, the localisation of both MR and pathological abnormalities is based on naming of anatomical regions by the clinician, radiologist, or pathologist. In some circumstances, this is sufficient to ensure the correspondence between regions, but for many applications more accuracy would be advantageous. We are studying the distribution of MR intensity abnormalities in Creutzfeldt-Jakob disease (CJD), in which various regions of grey matter may show hyperintensity in proton density and T2-weighted images. Different types of CJD appear to show different distributions of change, but more data are needed. Pathology changes revealed on microscopy include varying degrees of neuronal loss, spongiform change, gliosis and prion protein staining. Proper registration between the pathological samples and MR will help to establish the causes of the MR changes. We believe that this will in turn lead to a better understanding of the diagnostic and prognostic implications of *in vivo* imaging and the patterns of spread of disease in the brain, and could help to guide biopsy. In the present paper, we focus on a key process which will aid the co-registration of microscopic data and MR images from the same individual : namely, the reconstruction of whole brain slices into a 3-D volume. This reconstruction also provides a 3-D mapping of the distribution of pathological changes throughout the brain, which is an important goal in its own right.

3-D reconstruction of optical images of large slices can be facilitated if specialised research procedures are used. For example, the optical data in the "Visible Human" were obtained from complete axial sections through the deep frozen cadaver using a macrocryotome [14]. The thin sections themselves were not photographed, the images being acquired from the cut surface of the as-yet-unsectioned cadaver, in a sequence already registered with each other. Such an approach cannot be used for the majority of clinical studies in which conventional pathology procedures must be included. Also, in CJD, any instruments used to cut the brain are likely to become permanently contaminated and can never again be used for any non-CJD studies. Normally, the brain is removed by the pathologist and fixed in preservative. Some weeks later, the brain is cut into coronal slices. From these slices, small rectangular blocks about 1 or 2cm square are cut from selected regions, and from these blocks very thin sections are cut for staining and microscopic examination. We have developed methodology which can be integrated into such pathology routines.

There are several approaches to reconstructing small field-of-view thin sections to form a 3-D volume approximating the block [5,2,7,8,10,13]. Confocal microscopy allows 3-D data to be acquired directly, which is equivalent to 50 or more microscopical sections. However, for human studies, the blocks must be registered with the macroscopic slices, which are at a scale suitable for matching with MR images, and the requirements for reconstructing the macroscopic slices to form a large volume differ from those for reconstructing the small blocks.

2 Methodology

Four fixed *post-mortem* brains were examined for this preliminary study. Using a guide, the pathologist cut coronal slices of constant thickness through the specimen. For brains 3& 4, only one hemisphere was available, and the slices were 12mm thick. For brains 1 & 2, 7mm slices were cut through the whole cerebrum. The next step was carried out in preparation for co-registration of images from the two faces of the same slice (within-slice registration). For the first brain, 6 to 8 needle tracks were made through each slice with a 1mm trochar which had been dipped into Indian ink. A small guide was used to assist in making the track perpendicular to the cut faces of the slice. After photographing one face, the slice was turned over by hand to reveal the other face. For brains 2 - 4, a perspex (Plexiglass) jig was constructed (Fig. 1) to hold each slice during photography. Accurate circular fiducial marks were machined at each corner. Initially, the brain slice was laid on one plate of the jig, and the second plate then screwed down firmly by hand on top of the slice. Photographs were then obtained of both faces. For brains 1 and 2, these were colour transparencies which were subsequently digitised at high resolution (3887 x 2590 pixels, 24-bit colour scale). For brains 3 and 4, a digital camera with a resolution of 1600x1200 pixels was used.

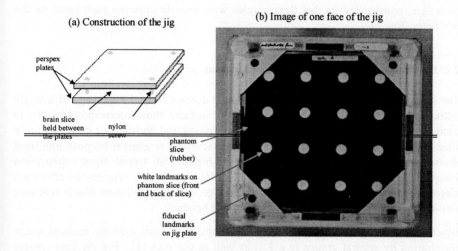

(a) Construction of the jig

(b) Image of one face of the jig

perspex plates

brain slice held between the plates

nylon screw

phantom slice (rubber)

white landmarks on phantom slice (front and back of slice)

fiducial landmarks on jig plate

Fig. 1. Jig to hold each slice during photography of anterior and posterior faces.

To reconstruct the original 3-D relationships, two different types of registration must be computed. The first is to register the anterior (A) and posterior (P) sides of each slice, i.e. within-slice registration. The second is to register the images of surfaces on either side of the cut which separates two adjacent slices, i.e. between-slice registration. Slices were numbered from front to back. Thus "P6" refers to the posterior face of slice 6.

2.1 Within-Slice Co-registration using Artificial Landmarks

The cut surfaces on each side of a slice are separated in the direction perpendicular to the faces by several millimetres (the thickness of the slice). We therefore cannot rely on sufficient similarity between image features to match them correctly. Instead, we explored methods using artificial landmarks, evaluating two options.

Method 1 For the first brain, the needle track landmarks were identified on the images by visual inspection and the centre of the landmarks marked by the pathologist with a cursor. Corresponding marks on the A and P faces were labelled and a standard 2-D rigid co-registration between the sets was computed by minimising the sum of the squared Euclidean distances between the centres.

Method 2 For the second brain, the jig fiducials were visible as circles on a contrasting background at the corners of the image. A program based on a Hough transform was implemented, to detect the circles automatically, to localise their centres to sub-pixel accuracy, and to estimate their radius. 2-D rigid registration between the A and P faces using the circle centres was then calculated. To determine the accuracy and reproducibility of co-registration using the jig, a phantom slice 7mm thick was constructed. This had black rubber faces and 16 circular white discs on each face, constructed so that these circles were exactly opposite each other on the two faces (Fig. 1).

2.2 Between-Slice Intensity-Based Registration

The surfaces on either side of a single cut were derived from the cleavage of a single tissue plane and natural features on the two cut surfaces should correspond exactly to each other. However, distortions occur during cutting and during later movement of a slice, so a non-rigid registration method is essential. A general purpose non-rigid registration framework was used, which computed an initial rigid registration followed by a second affine stage and finally a free-form stage. The results after each stage were examined separately. For each stage, a multi-resolution block matching step was followed by regularisation, as described below.

Block Matching [8] Block matching algorithms have been used for medical image registration by several groups [e.g 3,4] as well as our own [8]. For the first (coarse scale) the image was divided into large square blocks. For each block centre χ_κ in the reference image a corresponding block centre γ_κ in the floating image was found by moving the floating block until a similarity measure was optimised. Large blocks were used in the first iteration (to provide a rough estimate of the necessary large displacements). This was followed by iterations using progressively smaller blocks (down to a minimum of 4x4 pixels) to generate smaller but more accurate displacements. The complexity could thus be kept constant for each iteration. The similarity measure used was the grey level correlation ratio which is an appropriate choice for this application where the true relationship between blocks will tend locally to be at least affine [9]. The output of the block matching was a relatively sparse field of displacement vectors.

Regularisation The hierarchical nature of the block matching algorithm contributed to regularisation of the vector field, but a substantial fraction (about 20%) of vectors were generated from blocks with few features, or with features in the two images that did not represent anatomically corresponding structures. These usually appeared as outliers in the vector field. We therefore adopted robust methods for further regularisation, which was performed following block matching in each of the rigid, affine and free-form stages. For regularisation in the rigid registration stage, a least trimmed squares (LTS) strategy was used to estimate the two-by-two rotation matrix R and a translation vector $t = (t_1, t_2)$ which best superimposed the points $\{\chi_\kappa\}$ and their counterparts $\{\gamma_k\}$. The LTS method is far more robust to outliers than the classical least squares method [12] and can cope with up to 50% of outliers. Following block matching at the beginning of the affine stage, the LTS method was also used for regularisation, but without the rigidity constraints. For the free-form stage, a weighted Gaussian median filter [1] which performs well with outliers was used to convert the sparse deformation field into a dense, regularised field.

2.3 3-D Reconstruction by Propagation of 2-D Registrations through the Volume

The known slice thickness was used to calculate the antero-posterior coordinate ("z" value) of each image. The alternating 2-D within-slice rigid- and between-slice non-rigid- registrations were propagated through the volume to allow each image to be transformed in "x" and "y". After 3-D reconstruction, images of the two faces of the same cut have the same "z" value; we have not at this stage fused, or chosen between, these two coincident images systematically, but we arbitrarily selected one of the pair for display purposes as in Fig. 4.

For propagation using affine non-rigid registrations between slices, the transforms between each image pair were all parametric and propagation was a relatively straightforward analytic process (left half of Fig. 2). It is assumed that all P images have been flipped to bring all images into the same orientation. Let $I(Ai)$ be the A (anterior) image coordinates of the ith slice and the $I(Pi)$ its P (posterior) image. Let $T_R(Pi,Ai)$ be the rigid transformation between these two images and $T_A(P(i+1), Ai)$ be the affine transformation between the P face of slice i and the A face of the next slice i+1. The geometric transformations can be propagated from the A face of slice 1 to the P face of slice k so that $I(Pk)$ is transformed into the coordinate frame of $I(A1)$:

$$I(Pk,A1) = T_R(P1,A1) \times T_A(P2, A1) \times T_R(P2,A2) \dots T_A(P(k-1),A(k)) \times T_R(Pk,Ak) \times I(Pk)$$

However, propagation of free-form registrations required interpolation and image resampling for each new slice (right half of Fig. 2). To reduce the need to extrapolate the free-form vector fields outside the boundary of an image, the propagation direction is normally chosen to run from larger to smaller slices. Free-form propagation involves the following steps (Fig. 2):
(a) Within-slice rigid registration between Ai and Pi
(b) Resampling of Pi to form Pi'
(c) Between-slice non-rigid registration between Pi' and A(i+1), generating a regularised free-form vector field (note that an affine stage is embedded in this).
 (d) Within-slice rigid registration between A(i+1) and P(i+1)
 (e) Transformation of the vector field from (b) using TR(P(i+1),A(i+1)) from step (d)
(f) Warping of P(i+1) to form P(i+1)' which is then in the same 2-D coordinate frame as Ai.
Steps (c) to (f) are repeated, up to the last slice in the propagation sequence.

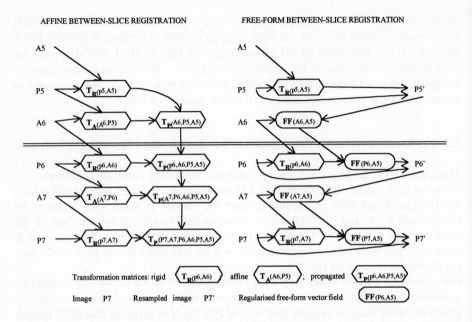

AFFINE BETWEEN-SLICE REGISTRATION   FREE-FORM BETWEEN-SLICE REGISTRATION

Transformation matrices: rigid $\langle T_{R(P6,A6)} \rangle$ affine $\langle T_A(A6,P5) \rangle$; propagated $\langle T_{P(P6,A6,P5,A5)} \rangle$

Image P7   Resampled image P7'   Regularised free-form vector field $\boxed{FF(P6,A5)}$

Fig. 2. Flow chart showing methods for propagating 2-D non-rigid registrations through a sequence of images. As an example, propagation from A5 to P7 is shown. Left: affine propagation (analytical). Right: free-form

3 Results

3.1 Within-Slice Rigid Registration

Method 1 (needle track landmarks) After registration the root mean square and maximum errors (Euclidean distances) between corresponding needle track landmarks were examined as intrinsic measures of the consistency of the results. There was a wide range of errors. For the best slices, the errors were 1.1 mm RMS and 1.6mm maximum. However, for the worst slices, the errors were 13.1 mm RMS and 21.6mm maximum.

Visual inspection of the quality of the registration was carried out, particularly with reference to the silhouette of the slice image, including slice features in the background which protruded beyond the cut surfaces as these should contribute to the silhouette when viewed from the A or the P side. This revealed the large deformations between different parts of the brain slice, which frequently occured when a slice was composed of separated fragments that could move independently of each other.

Method 2 (jig landmarks) Experiments with the phantom slice confirmed good performance of the jig (Fig. 1 and Table 1). Images of both faces of the phantom slice were obtained: (a) with the jig moved and replaced at various orientations relative to the camera (to test the image acquisition, digitisation and image processing steps); (b) after repeatedly dismantling and reassembling the jig (to evaluate mechanical aspects of the jig); and (c) with spacers about 1mm thick at one side of the phantom slice (to simulate the effect of imaging slices having non-parallel faces).

When the jig landmarks were used for registration (the normal situation with brain slices) the error in the phantom slice landmarks (about 0.75mm RMS, Table 1, right hand column) was a measure of the accuracy of the jig. However, the phantom slice itself could not be perfectly accurate, and the intrinsic error of the phantom landmarks (residual error when these same landmarks were used for registration) was about 0.55mm. The intrinsic error of the jig landmarks, calculated in the same way, was about 0.7mm (Table 1).

The intrinsic errors between jig landmarks were also examined when the jig was used for registration of images from brain 2, 3 and 4 (Table 2). These errors were in fact superior to those obtained when the phantom slice was photographed. These results confirmed the reliability of the jig landmark registration when used with real brain slices and also suggested that the image acquisition technique, involving factors such as avoidance of shadows and reflections and use of a digital camera, was improving.

Table 1. Errors in jig landmark positions for phantom slice images averaged across 4 trials of each condition.

		Intrinsic errors[1] (mm)		Accuracy of jig registration[2] (mm)
		Jig landmarks	Phantom landmarks	
Varying jig orientation and position with respect to the cameras	RMS	0.80	0.63	0.73
	max	*1.13*	*1.04*	*1.31*
Effect of dismantling and reassembling	RMS	0.68	0.50	0.61
	max	*1.07*	*1.14*	*1.37*
Effect of non parallel jig faces	RMS	0.67	0.55	0.72
	max	*0.95*	*1.20*	*1.40*

[1] residual errors between landmark positions after registration using the same landmarks
[2] phantom landmark errors after registration using jig landmarks

Table 2. Residual intrinsic errors in jig landmark positions for brain slice images averaged across slices.

		Brain 2	Brain 3	Brain 4
Type of camera		film	digital	digital
Number of slices		3	12	14
Slice thickness, mm		7	12	12
Residual error, mm	(RMS)	0.55	0.19	0.11
	(max)	0.76	0.24	0.16

3.2 Between-Slice Non-Rigid Registration

Between-slice registration was computed in stages (rigid, affine, and free-form) which
were evaluated by visual inspection of subtraction images and of stacked image sets.
Figure 3 shows the successive stages for slices 10 and 11, i.e images P10 and A11, of
brain 1. For this image pair, the subtraction images show progressive improvement
(i.e. increasing uniformity) at each stage. The residual subtraction features (i.e. areas
of non-uniformity) generally were *not* derived from the cut brain surface but from
oblique surfaces or from artefacts. This is a good result because such non-
corresponding features should ideally not influence the registration of those features
which really do correspond.

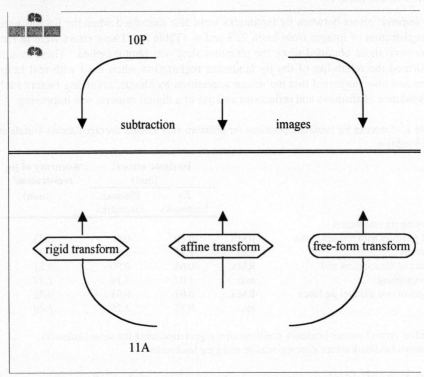

Fig 3: Processing steps for between-slice registration. Example from brain 1, slices 10 and 11.
Middle row: subtraction images to show the residual differences between image P10
(untransformed) and the three stages of transforming flipped slice A11 (rigid, affine and free
form stages).

3.3 3-D Reconstruction by Propagation of 2-D Registrations through the Volume

For brain 3, the alternating within-slice- and between-slice- registrations were
propagated from slice 7 forwards and backwards through the volume. The results

after using the two alternative non-rigid between-slice methods (affine only and free-form following affine) were compared by inspecting the continuity of the outline and of internal image features in the stacks of images. The best results were obtained using the affine method (Fig. 4).

Fig. 4. Completed reconstruction of macroscopic optical brain slice images from brain 3 (only left hemisphere available): surface rendered views. Left image: postero-lateral view; right image: postero-medial view. The methods used were: jig-based (rigid) within-slice registration; affine (non-rigid) between-slice registration; analytical propagation.

4. Discussion

Implications of obtaining surface images from thick slices   The thickness of macroscopic brain slices imposes a wide separation between the images of the anterior and posterior cut surfaces. However, it is wrong to imagine the voxels of the 3-D optical reconstruction as having a large size in the "z" direction. Unlike typical tomographic images, the majority of features visible in our optical images lie on or extremely close to the surface of the slice, i.e. there is very little partial volume effect for such features. The optical data should be visualised as very thin slices with a large interslice gap. Nevertheless, this rule is not true for visible features which do not lie on the cut surface, as inspection of the residual details which did not subtract out after between-slice registration revealed. The difference images after non-rigid registration shown in Fig. 3 showed that the large majority of the residues were derived from such non-corresponding features, confirming the good performance of our method.

Needle track vs. jig methods for within-slice rigid registration   The needle track method avoided the need for gadgetry in the photographing area, an advantage considering the special precautions needed when handling CJD brains. However, making the needle tracks was at least as time consuming as placing slices in the jig. Furthermore, our result showed that it was impossible to avoid substantial deformations when a slice was turned over to photograph its opposite face. In principle, this might have been corrected by using a non-rigid within-slice registration method. However, such a method could not be based on natural image features, because the thick slices do not allow the assumption of correspondence of image features. Also, it would have required a large number of artificial needle track landmarks for accurate estimation and this would have caused an unacceptable number of artefacts in the pathological material. A final disadvantage of the needle track method was the difficulty in localising the needle track landmarks, which tended to have a ragged boundary. In contrast, the jig held the slice stably during photography and provided accurate fiducials, thereby allowing reliable rigid registration.

Affine vs free-form methods for between-slice non-rigid registration The most consistent results in the present study were obtained with affine rather than free-form between-slice registration and this may prove sufficient to initialise a reliable registration procedure with 3-D MR data. The affine transformation was applied globally to the 2-D optical image. This had the advantage of allowing propagation between the complete set of slices to be solved analytically, avoiding the compounding of errors arising in image resampling and regularisation of discrete vector fields. However, a global affine transformation applied to a 2-D image does not allow us to represent independent movement of separated fragments. More detailed evaluation of local matching performance is planned.

Implications of expected physical deformations on choice of between-slice registration algorithms Several factors determine the actual deformations which occur during the *post-mortem* studies. The brain tends to shrink during fixation, and if stood on a surface will flatten towards this. Further deformations occur as a result of compression during cutting. It is likely that such deformations can be represented by an affine transformation, which allows differential stretching in "x" and "y" and skew as well as translation and rotation. However, more complex deformations also occur which are mainly determined by the degree of connectedness between parts. When one part is separated from another by a narrow neck of tissue, for example where the two cerebral hemispheres are joined only by the corpus callosum, bending occurs easily along the join. The narrower the neck of tissue along the join, the more readily bending occurs when the slice is cut or (later) when it is moved. With such connections, stretching will usually be limited. However, when there is complete disconnection, e.g. when a fragment such as the anterior pole of the temporal lobe is left completely separated after cutting, movement relative to other parts is only limited by collision and by the care of the pathologist and photographer in maintaining the original relations. The ideal between-slice matching algorithm thus needs to allow unconstrained deformation between fragments, bending but little stretching along narrow necks of tissue, and more constrained (probably affine) deformations locally within intact blocks of tissue.

Acknowledgments

This work was supported by the EU-funded QAMRIC project BMH4-98-6048 (Quantitative Analysis of MR Scans in Creutzfeldt-Jakob disease).

References

1. D. R. K. Brownrigg. The Weighted Median Filter. *Communications of the ACM*, 27(8):807–818, 1984.
2. F. S. Cohen, Z. Yang, Z. Huang and J. Nissanov. Automatic Matching of Homologous Histological Sections. *IEEE Transaction on Biomedical Engineering*, 45(5):642–649, 1998.
3. D. L. Collins and A. C. Evans. ANIMAL: Validation and Applications of Nonlinear Regsitration-Based Segmentation. *Int. Journal of Pattern Recognition and A.I.*, 8(11):1271–1294, 1997.
4. T. Gaens, F. Maes, D. Vandermeulen and P. Suetens. Non-rigid Multimodal Image Registration Using Mutual Information. In *Proc. MICCAI'98*, vol. 1496 of *Lecture Notes in Computer Science*, pages 1099–1106, Springer.
5. L. S. Hibbard and R. A. Hawkings. Objective image alignment for three-dimensional reconstruction of digital autoradiograms. *Journal of Neuroscience Method*, 26:55–74, 1988.
6. B. Kim, J. L. Boes, K. A. Frey and C. R. Meyer. Mutual Information for Automated Unwarping of Rat Brain Autoradiographs. *Neuroimage*, 5:31–40, 1997.
7. M. S. Mega, S. S. Chen, P. M. Thompson, R. P. Woods, T. J. Karaca, A. Tiwari, H. V. Vinters, G. W. Small and A. W. Toga. Mapping Histology to Metabolism: Coregistration of Stained Whole-Brain Sections to Premortem PET in Alzheimer's Disease. *Neuroimage*, 5:147–153, 1997.
8. S. Ourselin, A. Roche, G. Subsol, X. Pennec and N. Ayache. Reconstructing 3D Structure from Serial Histological Sections. Image and Vision Computing (in press).
9. G. P. Penney, J. W. Weese, J. A. Little, P. Desmedt, D. L. G. Hill and D. J. Hawkes. A Comparison of Similarity Measures for Use in 2D-3D Medical Image Registration. In *Proc. MICCAI'98*, vol. 1496 of *Lecture Notes in Computer Science*, pages 1153–1161, Springer.
10. A. Rangarajan, H. Chui, E. Mjolsness, S. Pappu, L. Davachi, P. Goldman-Rakic and J. Duncan. A robust point-matching algorithm for autoradiograph alignment. *Medical Image Analysis*, 1(4):379–398, 1997.
11. A. Roche, G. Malandain, X. Pennec and N. Ayache. The Correlation Ratio as a New Similarity Measure for Multimodal Image Registration. In *Proc. MICCAI'98*, vol. 1496 of *Lecture Notes in Computer Science*, pages 1099–1106, Springer.
12. Peter J. Rousseeuw and Annick M. Leroy. *Robust Regression and Outlier Detection*. Wiley Series in Probability and Mathematical Statistics, first edition, 1987.
13. T. Schormann, A. Dabringhaus, and K. Zilles. Statistics of Deformations in Histology and Application to Improved Alignment with MRI. *IEEE Transactions on Medical Imaging*, 14(1):25–35, 1995.
14. V. Spitzer, M. J. Ackerman, A. L. Scherzinger and D. Whitlock. The Visible Human Male: a Technical Report. *Journal of American Medical Informatics Association*, 3(2):118-130, 1996.

Ultrasound/MRI Overlay with Image Warping for Neurosurgery

David G. Gobbi[1], Roch M. Comeau[2], and Terry M. Peters[1,2]

[1] Imaging Research Laboratories,
John P. Robarts Research Institute, University of Western Ontario,
London ON N6A 5K8, Canada
dgobbi@irus.rri.on.ca, tpeters@irus.rri.on.ca
http://www.irus.rri.on.ca/Faculty/peters.html
[2] McConnell Brain Imaging Centre, Montreal Neurological Institute,
Montréal QC H3A 2B4, Canada
roch@nil.mni.mcgill.ca

Abstract. Performing a craniotomy will cause brain tissue to shift. As a result of the craniotomy, the accuracy of stereotactic localization techniques is reduced unless the brain shift can be accurately measured. If an ultrasound probe is tracked by a 3D optical tracking system, intra-operative ultrasound images acquired through the craniotomy can be compared to pre-operative MRI images to quantify the shift. We have developed 2D and 3D image overlay tools which allow interactive, real-time visualization of the shift as well as software that uses homologous landmarks between the ultrasound and MRI image volumes to create a thin-plate-spline warp transformation that provides a mapping between pre-operative imaging coordinates and the shifted intra-operative coordinages. Our techniques have been demonstrated on poly vinyl alcohol cryogel phantoms which exhibit mechanical and imaging properties similar to those of the human brain.

1 Introduction

1.1 Stereotactic Surgery and Brain Shift

Stereotactic surgery involves the registration of medical images, typically from volumetric modalities such as MRI and CT, with the head of the patient. With modern computer-aided surgical navigation equipment, the positions of surgical tools are tracked (usually by optical tracking systems) and the registration of the images to the head is accomplished with either bone-mounted or skin-mounted fiducial markers as follows: the markers are first identified in the image and their image coordinates are recorded, then the tracking-system coordinates of the fiducial markers are measured with a tracked pointer.

A basic assumption behind most of the applications of these 'frameless' stereotactic systems that the brain remains stationary during the surgery and that inaccuracies result only from registration and tracking errors. Following a craniotomy, however, brain shift in excess of 10 mm is not uncommon and the shift can be as large as 25 mm [1, 2]. This shift is a nonlinear warping of the brain caused by gravity, pressure inside the brain, and leakage of fluid from the cranial cavity. The extent of brain shift depends primarily on the size of the craniotomy and the duration of the surgery.

Brain shift can be significantly reduced by opening only a small drill-hole in the skull, as is typical of minimally-invasive stereotactic procedures. If stereotaxis is to be used after a craniotomy, however, the brain shift should be measured and corrected for to ensure sufficient positioning accuracy.

This paper discusses the use of intra-operative ultrasound imaging through the craniotomy as a method of quantifying the brain shift. A complementary project within our group is addressing this problem by tracking the cortical surface under the craniotomy [3, 4].

1.2 Stereotactic Ultrasound Imaging

A cluster of infrared LEDs attached to the biopsy needle mount of our ultrasound probe allows the surgical workstation to track the position and orientation of the probe in real-time as the ultrasound video image is captured. Via the tracking system, the slice of patient anatomy that matches the ultrasound video image can be determined. The corresponding oblique slice can be extracted from the pre-operative MRI image volume for comparison. This basic technique has been demonstrated by several groups [5–8]. Our previous contributions include a comprehensive study of this technique involving both phantom work and several clinical procedures [9–11]. At least one company, Sofamor Danek, supports this technique through their SonoNav product (www.sofamordanek.com/sono.htm).

The primary goal of our current study is twofold. First, we are investigating better interactive methods for visualizing the ultrasound video and the corresponding MR slices, in order to provide surgeons or radiologists with tools for very rapid semi-quantitative assessment of brain shift. Second, we are aiming towards rapid and accurate quantification of the brain shift in order to provide a nonlinear warp transformation from pre-operative image space to intra-operative image space. For interactive assessment of brain shift, 2D ultrasound video is used. For quantitative measurement of brain shift over a substantial volume of brain tissue, 3D ultrasound imaging is more suitable.

The creation of the nonlinear warp transformation is accomplished through the identification of several homologous landmarks in the ultrasound image volume and the MR image volume. The warp transformation is interpolated between the landmarks through the use of a radial basis function interpolation technique. We are currently using thin plate splines [12] but are investigating other methods of interpolation (including those which incorporate a mechanical model of brain tissue).

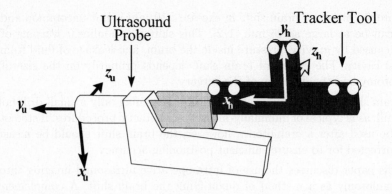

Fig. 1. A POLARIS tracker tool with infrared LEDs is mounted on the ultrasound probe via its biopsy needle mount.

2 Materials and Methods

The primary components of our system are Aloka SSD-2000 and SSD-1700 ultrasound scanners with a 5 MHz burrhole ultrasound probe (Aloka Co., Ltd, Tokyo 181-8522, Japan), a POLARIS optical tracking system (Northern Digital Inc., Waterloo ON N2V 1C5, Canada, www.ndigital.com), and a 450 MHz Pentium II workstation with video input/output capabilities.

A set of infrared LEDs are mounted on the ultrasound probe as shown in Figure 1. The respective coordinate systems of the ultrasound image and of the POLARIS LEDs are labelled in the figure. The coordinate transformation between the ultrasound image and the tracking system was determined through a calibration procedure which we have described in the past [13]. A similar method is used by another group [14], and a set of very different but more widely used methods are described in the literature with reference to freehand 3D ultrasound reconstruction [15].

Our visualization methods and the overall image registration were tested using a special-purpose poly (vinyl alchohol) cryogel (PVA-C) phantom [17, 18]. This phantom is described in detail in [11].

Our software has been developed using the open-source Visualization Toolkit, VTK (http://www.kitware.com/vtk.html) [16]. Several C++ classes which were written for this project have been contributed to the VTK class library. The first, vtkImageReslice, efficently resamples an image based on a linear, perspective, or nonlinear warp transformation. The second, vtkVideoSource, provides a VTK interface for any video digitizer which is compatible with either the Matrox Imaging Library or Microsoft Video for Windows. The third, vtkImageBlend, provides transparency blending of images. We have also contributed significant optimizations for image display in VTK.

Our most significant recent contribution to VTK is a new class heirarchy for geometrical transformations. Two of the several new geometrical transformation classes were developed specifically for this project: the vtkThinPlateSpline-

Transform class provides a landmark-based warp transformation using either a thin-plate spline radial basis function (RBF) or any arbitrary RBF, while the vtkGridTransform defines a warp transformation via a grid of displacement vectors. The tight integration of these classes with VTK provides the following: 1) scalability, in that the image warping in VTK will automatically parallelize over all processors on a multi-processor platform, 2) generality, in that the transformations can be applied to images, vertices, vector fields, or implicit functions and 3) flexibility, in that concatenation and inversion of arbitrary transformations is supported.

In addition to those classes which we have contributed to VTK, we have also developed classes that interface with the POLARIS optical tracking system and which perform 3D ultrasound reconstruction. We have also written a large set of foundation classes for rapid development of visualization applications in the Python programming language. These foundation classes and several example applications will be made available under an open-source license from http://www.atamai.com.

2.1 Image Overlay

Our system provides the means of overlaying both 2D ultrasound video and 3D ultrasound image volumes with 3D MRI volumes. The 2D video overlay is interactive, providing refresh rate of 10 frames per second, and is designed to allow the surgeon or neuroradiologist to rapidly make a semi-quantitative assessment of the brain shift. The 3D overlay is for use in tagging homologous landmarks between a 3D ultrasound volume and an MRI volume. Once a suitable set of landmarks have been identified, a nonlinear warp transformation from preoperative MRI image coordinates to intra-operative surgical coordinates can be generated. The 3D overlay is discussed more fully in the next section.

A video overlay session with a PVA-C phantom is shown in Figure 2. The top window is a rendering of three orthogonal slices through the MRI volume, as well as a virtual model of the ultrasound probe which moves in real-time as the "real" optically-tracked ultrasound probe is moved. The virtual ultrasound probe includes an overlay of the ultrasound video (shown in orange) on the corresponding MRI slice (shown in grey). The 3D scene can be interactively reoriented (rotate, pan, zoom) with the mouse.

A strictly 2D view of the ultrasound/MR overlay is shown in the bottom-left window. This view provides the ultrasound operator with a means of interactively evaluating the brain shift. The ultrasound video is displayed with an orange color scale to provide good contrast with the greyscale MRI. A slider below this window allows the user to control the opacity of the ultrasound overlay.

The raw ultrasound video is also displayed (in the bottom-right window) so that the ultrasound operator will never have a need to divide his or her attention between the computer monitor and the ultrasound machine's CRT display.

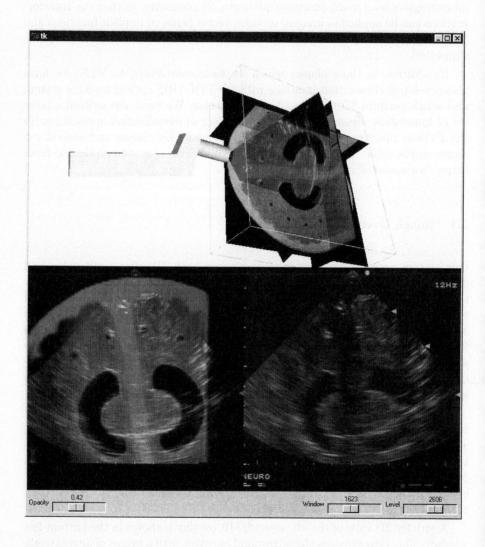

Fig. 2. Overlay of ultrasound video on an MRI volume. These data were obtained from one of our deformable PVA-C phantoms. The rendering frame rate is approximately 10 frames/second.

3 Image Warping

We use a set of homologous landmarks in the ultrasound image volume and the MRI volume to describe the nonlinear warp transformation from pre-operative MRI image space to intra-operative ultrasound image space. Currently, thin plate spline interpolation of the warp transformation is performed between the landmark points. In the future other interpolation methods will be investigated. The pre-operative image is warped, along with any tags or labels on the image, into the intra-operative image space.

The landmarks are identified on a 3D Ultrasound/3D MRI overlay as shown in Figure 3. Figure 3(a) is a volume rendering of the MRI of the phantom, provided to indicate the geometry of the phantom. Figure 3(b) is an MRI volume fused with a 3D ultrasound volume (in orange) from a tracked ultrasound probe. The registration of the volumes was achieved entirely via the tracking of the probe by the POLARIS. The misregistration between the ultrasound and MRI at the bottom of the bean-shaped cavities (which simulate the ventricles in the brain) is due to speed of sound differences between the PVA-C and water. In the future, we will experimentally determine the correct mixture of glycerol and water to obtain the same speed of sound in the cavities as in the phantom material.

To simulate brain shift, we applied a warp to both images as shown by the arrow in Figure 3(c). This simulated warp provides a "gold standard" to which a landmark-driven warp (Figure 3(d)) can be compared.

We used the following method to create the landmark driven warp: Triplanar views of the warped 3D ultrasound volume and non-warped MRI volume were placed side by side. Corresponding landmarks were tagged in the ultrasound and MRI — the landmarks which were used are the small circles which form a ring around the ventricular cavities in the centre of the phantom. The small circles represent large cerebral blood vessels which could be used as landmarks in a real brain. Only the top eight circlular landmarks were used because the others were not within the volume covered by the ultrasound scan.

The landmark coordinates were used to accomplish a thin plate spline warp transformation of the MRI image to match it to the ultrasound image. Comparing Figure 3(c) and (d), the warp transformation matches the simulated ventricles in the MRI images very well. The indentation on the right side of the surface of the phantom is not duplicated, however, because it lies well outside of the cluster of landmarks used for nonlinear registration.

4 Conclusion

Early results of using homologous landmarks to nonlinearly warp an MRI image volume to match a 3D ultrasound volume suggest that good accuracy in the warp is possible as long as the landmarks provide sufficient coverage of the target region of interest within the volumes. In terms of the use of this methodology on patients, this means that landmark must be chosen which lie between the

(a) (b)

(c) (d)

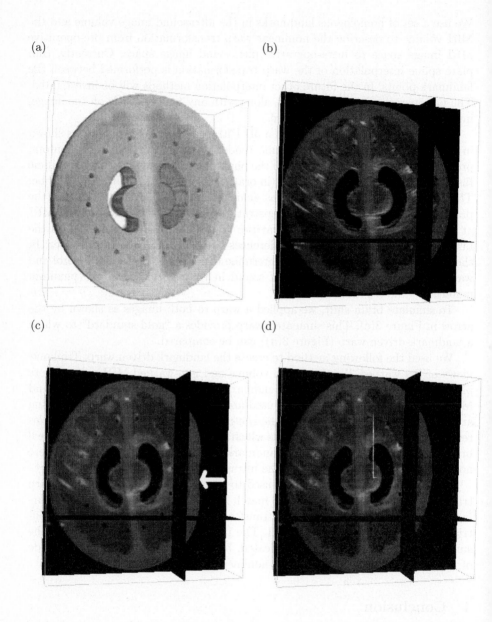

Fig. 3. Orthogonal slice planes through a 3D overlay of ultrasound and MR image volumes.

113

craniotomy and the region of interest, and if possible should surround the region of interest. The ultimate goal is to match the morphology of the pre-operative MRI volume to the intra-operative morphology of brain as demonstrated by 3D ultrasound, using the landmark-based warp scheme described above.

Our support for live video overlay is full-featured and provides an interactive frame rate (>10 frames per second). We aim to further improve the overlay by adding the ability freeze and then annotate the video image. Another important area for future work will be the development of a surgeon-friendly user interface (e.g. by mounting buttons on the ultrasound probe to control the opacity of the overlay or the window/level of the MRI), to make the system as suitable as possible as a routine operating-room tool.

Acknowledgments

PVA-cryogel phantoms used for this project were constructed by Kathleen Surry at Robarts, and our surgical navigation system was developed as a shared effort between the first author and Dr. Yves Starreveld. Funding for this research is provided through grants from the Medical Research Council of Canada and the Canadian Foundation for Innovation. Scholarship funding for David Gobbi is provided by the Ontario Ministry of Education and by the University of Western Ontario.

References

1. D. L. G. Hill, C. R. Maurer Jr., R. J. Maciunas, J. A. Barwise, J. M. Fitzpatrick and M.Y. Wang. Measurement of Intraoperative Brain Surface Deformation under a Craniotomy. *Neurosurgery* **43**:514–528, 1998.
2. D. W. Roberts, A. Hartov, F. E. Kennedy, M. I. Miga and K. D. Paulsen. Intraoperative Brain Shift and Deformation: A Quantitative Analysis of Cortical Displacement in 28 Cases. *Neurosurgery* **43**:749–760, 1998.
3. M.A. Audette and T.M. Peters. Level-Set Segmentation and Registration for Computing Intrasurgical Deformations. *Proc. SPIE Medical Imaging 99, SPIE Vol.* **3661**:110-121, 1999
4. M.A. Audette, K. Siddiqi and T.M. Peters. Level-set Surface Segmentation and Fast Cortical Range Image Tracking for Computing Intrasurgical Deformations. *Medical Image Computing and Computer Assisted Intervention - MICCAI'99*. C. Taylor and A. Cholchester (eds) Lecture Notes in Computer Science **1679**:788–797, Springer-Verlag, Berlin, 1999.
5. R .D. Bucholz, D. Yeh,, J. Trobaugh, L. L. McDurmont, C. D. Sturm, C. Baumann and M. H. Jaimie. The Correction of Stereotactic Inaccuracy Caused by Brain Shift Using an Intraoperative Ultrasound Device. *CVRMed-MRCAS '97 : First Joint Conference Computer Vision, Virtual Reality and Robotics in Medicine and Medical Robotics and Computer-Assisted Surgery*, Troccaz, J., Grimson, E., Mösges, R., (eds.), Grenoble, France, Springer-Verlag, Berlin, 1997.
6. C. Giorgi and D.S. Casolino. Preliminary Clinical Experience with Intraoperative Stereotactic Ultrasound Imaging. *Stereotactic and Functional Neurosurgery* **68**:54-58, 1997.

7. N. Hata, T. Dohi, H. Iseki and K. Takakura. Development of a Frameless and Armless Stereotactic Neuronavigation System with Ultrasonographic Registration. *Neurosurgery* **41**:608–614, 1997.

8. A. Jödicke, W. Deinsberger, H. Erbe, A. Kriete and D.-K. Böker. Intraoperative Three-Dimensional Ultrasonography: An Approach to Register Brain Shift Using Multidimensional Image Processing. *Minim. Invasive Neurosurg.* **41**:13-19, 1998.

9. R.M. Comeau, A. Fenster and T.M. Peters. Integrated MR and Ultrasound Imaging for Improved Guidance in Neurosurgery. *Proc SPIE, Medical Imaging '98, SPIE Vol.* **3338**:474–754, 1998.

10. R.M. Comeau, A. Fenster and T.M. Peters. Intraoperative US in Interactive Image-Guided Neurosurgery. *Radiographics* **18**:1019-1027, 1998.

11. R.M. Comeau, A.F. Sadikot, A. Fenster and T.M. Peters. Intraoperative Ultrasound for Guidance and Tissue Shift Correction in Image-Guided Surgery. *Medical Physics* **27**:787-800, 2000.

12. F.L. Bookstein. Linear Methods for Nonlinear Maps. In *Brain Warping*, A.W. Toga (ed). Academic Press, San Diego, 1999.

13. D.G. Gobbi, R.M. Comeau and T.M Peters. Ultrasound Probe Tracking for Real-Time Ultrasound/MRI Overlay and Visualization of Brain Shift. *Medical Image Computing and Computer Assisted Intervention - MICCAI'99.* C. Taylor and A. Cholchester (eds) Lecture Notes in Computer Science **1679**:920–927, Springer-Verlag, Berlin, 1999.

14. N. Pagoulatos, W.S. Edwards, D.R. Haynor and Y. Kim. Interactive 3-D Registration of Ultrasound and Magnetic Resonance Images Based on a Magnetic Position Sensor. *IEEE Trans. Inf. Tech. in Biomed.* **3**:278–288, 1999.

15. R.W. Prager, R.N. Rohling, A.H. Gee and L. Berman. Rapid Calibration for 3-D Freehand Ultrasound. *Ultrasound in Med. Biol.* **24**:855–868, 1998.

16. W. Schroeder, K. W. Martin and W. Lorensen. *The Visualization Toolkit*, 2nd Edition. Prentice Hall, Toronto, 1998.

17. I. Mano, H. Goshima, M. Nambu and I. Masahiro. New Polyvinyl Alcohol Gel Material for MRI Phantoms. *Magnetic Resonance in Medicine* **3**: 921–926, 1986.

18. K. Chu and B. Rutt. Polyvinyl Alcohol Cryogel: An Ideal Phantom Material for MR Studies of Arterial Flow and Elasticity. *Magnetic Resonance in Medicine* **37**:314–319, 1997.

Model-Updated Image-Guided Neurosurgery: Preliminary Analysis Using Intraoperative MR

Michael I. Miga[1], Andreas Staubert[3], Keith D. Paulsen[1,2], Francis E. Kennedy[1], Volker M. Tronnier[3], David W. Roberts[2], Alex Hartov[1], Leah A. Platenik[1], and Karen E. Lunn[1]

[1] Dartmouth College, Thayer School of Engineering, HB8000, Hanover, NH 03755
{michael.miga, keith.paulsen}@dartmouth.edu
[2] Dartmouth Hitchcock Medical Center, Lebanon, NH 03756
[3] University Hospital, Department of Neurological Surgery, Heidelberg School of Medicine, Im Neuenheimer Feld 400, D-69120 Heidelberg, Germany.

Abstract. In this paper, initial clinical data from an intraoperative MR system are compared to calculations made by a three-dimensional finite element model of brain deformation. The preoperative and intraoperative MR data was collected on a patient undergoing a resection of an astrocytoma, grade 3 with non-enhancing and enhancing regions. The image volumes were co-registered and cortical displacements as well as subsurface structure movements were measured retrospectively. These data were then compared to model predictions undergoing intraoperative conditions of gravity and simulated tumor decompression. Computed results demonstrate that gravity and decompression effects account for approximately 40% and 30%, respectively, totaling a 70% recovery of shifting structures with the model. The results also suggest that a non-uniform decompressive stress distribution may be present during tumor resection. Based on this preliminary experience, model predictions constrained by intraoperative surface data appear to be a promising avenue for correcting brain shift during surgery. However, additional clinical cases where volumetric intraoperative MR data is available are needed to improve the understanding of tissue mechanics during resection.

Keywords: finite element modeling and simulation, image guided therapy, intraoperative image registration techniques

1 Introduction

Over the past 10 years, there has been a signficant effort to understand the nature and extent of brain deformations during neurosurgery. The results of these investigations have suggested that intraoperative tissue movement may compromise the fidelity of preoperative-based image guidance if left unchecked [1], [2]. Recently, there has been a concerted effort by many investigators to augment existing neuronavigation systems to account for intraoperative brain deformation. Several medical centers have adopted the strategy of using intraoperative magnetic resonance (MR) imaging [3], [4]. Recent reports by Knauth et al. have systematically demonstrated in a thirty-eight patient study that intraoperatively updated image guidance significantly increased the amount of complete tumor removal from 37% to nearly 76% [5]. They have also illustrated potential problems with intraoperative MR imaging related to surgically-induced contrast enhancement which can be confused with contrast-enhancing residual tumor [6]. Others continue to question the use of intraoperative MR citing the fate of past experiences with intraoperative computed tomography in the 1980's. Moreover, intraoperative computed tomography did not face the OR-compatibility challenges of MR, yet it has not become a standard component of today's surgical "armamentarium" [7]. In addition, there is significant concern that the tumor-defining boundary shown in MR is not representative of the complete tumor infiltration [7]. Appropriately, investigators are still determining the efficacy of intraoperative MR in order to identify its most important uses [8], [9].

Nonetheless, the need for intraoperative updating seems clear although the paradigm to provide such intraoperative data is somewhat more elusive. As an alternative to full volume imaging in the OR, we are developing a strategy that uses sparse intraoperative data (i.e. cortical shift tracking, coregistered intraoperative ultrasound) to drive a computational model that serves as the source for intraoperative image updates [10]. *In vivo* studies in controlled animal experiments have demonstrated the quantitative feasibility of this approach [11]. Preliminary experiments with humans experiencing intraoperative gravity-induced deformation suggest the model accuracy is comparable to that obtained in animal studies [12]. The advantages of such a strategy are its low cost and minimal impact into an already overcrowded OR environment. Additional impetus for exploiting computational models in the operating room (OR) is provided by the abundance of preoperative data that cannot be updated practically during surgery (i.e. positron emission tomography, electroencephalography, functional MR imaging, and MR spectroscopy). Computational models would be a natural choice to register this data with the updated image database.

A critical step in establishing the utility of computational models for image updating in the OR is to assess model predictions in light of full volume intraoperative image data when available. In this paper, preliminary results are presented from a clinical case having coregistered MR images pre- and immediately-post tumor resection. Results demonstrate the potential predictive capability of models and provide valuable insight regarding the tumor-induced deformation arising from excision.

2 Methods

2.1 Experimental Methods

The patient's history reflected an astrocytoma grade 3 with non-enhancing and enhancing regions in the left pre-central region of the brain. For surgical planning and intraoperative neuronavigation, a preoperative MRI-data set was acquired (T1-FLASH 3D-seq., TR=34.0ms, TE=12.0ms, No.Acq.=1, TA=13:59,

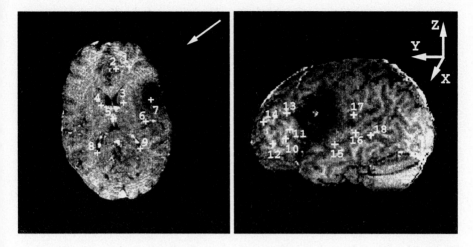

Fig. 1. Subsurface (left) and surface (right) point distribution for deformation tracking. Subsurface points (1,2,3,4) are located more superior at higher elevations relative to the direction of gravity (designated with white arrow in left image), (5,8,9) are located in-plane with tumor and (6,7) are inferior to tumor.

FOV=260, matrix=256*256, No.Sl.=128, Sl.Thk. 1.4 mm) using an open low-field MRI-scanner (Magnetom OPEN 0.2T, Siemens, Erlangen, Germany). The patient was brought to the OR and the patient's head was fixed in a dedicated head-holder coil for intraoperative MRI with head rotated to the right approximately 55 degrees. For neuronavigation the MKM-microscope (Zeiss, Oberkochen, Germany) was used, and the accuracy of patient registration with the preoperative MRI-data set was 1.7 mm. The tumor was resected to the planned extent using the preoperative information. Of note, the margins of the pathologic structure were not clearly recognizable to the surgeon during resection. After total resection according to neuronavigation, the patient was transferred to the intraoperative MRI scanner adjacent to the OR using a dedicated patient transport system [13]. Intraoperative imaging in the Magnetom OPEN was performed using T1- and T2- weighted SE-sequences for diagnostic purposes and a T1-FLASH 3D-sequence (see above) for an intraoperative neuronavigational update [14],[4]. Intraoperatively, no residual enhancing regions of the tumor were found in the post-contrast media T1 images. Qualitative comparison of pre- and intraoperative images also indicated a complete tumor resection, but absolute quantitative estimations were not possible due to a regular brain-shift of approximately 1 cm around the resection cavity. The patient was transferred back to the OR, and the surgical access was closed using standard neurosurgical techniques.

Using this series of scans, we were able to identify 9 subsurface points and 9 cortical surface locations which could be tracked in both preoperative and intraoperative MR scans. Figure 1 illustrates the distribution of subsurface (left) and surface (right) points. We note that the subsurface distribution shown in

#	Descr.	dx	dy	dz	D	#	Descr.	dx	dy	dz	D
1	Ant. Cortex	-4.0	-6.0	-1.3	7.3	10	Ant. Sulcus	-5.0	-7.0	0.0	8.6
2	Ant. Midln Cortex	-1.0	-1.0	-2.6	3.0	11	Ant. Sulcus	-5.0	-8.0	0.0	9.4
3	Ant. L. Lat. Horn	-1.0	-2.0	-1.3	2.6	12	Ant. Sulcus	-3.0	-6.0	1.3	6.8
4	Ant. R. Lat. Horn	-2.0	-1.0	-1.3	2.6	13	Ant. Sulcus	-5.0	-10.0	-2.6	11.5
5	3rd Vent.	-1.0	0.0	0.0	1.0	14	Ant. Sulcus	-4.0	-6.0	0.0	7.2
6	Post. Insular	1.0	2.0	2.6	3.4	15	Post. Sulcus	2.0	3.0	2.6	4.4
7	Post. Insular	-2.0	1.0	2.6	3.4	16	Post. Sulcus	0.0	2.0	2.6	3.3
8	Post. R. Lat. Horn	-1.0	1.0	0.0	1.4	17	Post. Sulcus	1.0	1.0	-1.3	1.9
9	Post. L. Lat. Horn	0.0	-1.0	-1.3	1.6	18	Post. Sulcus	0.0	1.0	0.0	1.0

Table 1. Measured brain feature displacement in millimeters in cartesian coordinates. D is the total displacement magnitude.

Figure 1 is not coplanar but in fact the points are distributed superior, in-plane, and inferior to the tumor location. Table 1 quantifies the deformations of these points as tracked between preoperative and intraoperative MR.

2.2 Computational Methods

We have chosen consolidation physics to represent deformation characteristics of the brain [15]. The governing equation describing mechanical equilibrium is,

$$\nabla \cdot G \nabla \mathbf{u} + \frac{G}{1 - 2\nu} \nabla \varepsilon - \alpha \nabla p + (\rho_t - \rho_f) \mathbf{g} = 0. \tag{1}$$

The description is completed by a mass conservation equation relating volumetric strain to fluid drainage shown here,

$$\nabla \cdot k \nabla p - \alpha \frac{\partial \varepsilon}{\partial t} - \frac{1}{S} \frac{\partial p}{\partial t} = 0, \tag{2}$$

The mathematical framework of coupled equations (1) and (2) has been previously reported in detail [16], [17]. In previous work, material property studies have been reported [18], [19]. In the model's current form, the brain is treated as a homogeneous structure with respect to stiffness properties and heterogeneous with respect to its hydraulic behavior (i.e. $G_w = G_g = 725\ Pa$, $\nu_w = \nu_g = 0.45$, $\alpha_w = \alpha_g = 1.0$, $\frac{1}{S}_w = \frac{1}{S}_g = 0.0$, $k_w = 1.0 \times 10^{-10}\ \frac{m^3 s}{kg}$, $k_g = 5.0 \times 10^{-12}\ \frac{m^3 s}{kg}$, and $\rho_w = \rho_g = 1020\ \frac{kg}{m^3}$ where w and g refer to white matter and gray matter, respectively).

Figure 2a-c illustrate the assumptions regarding boundary conditions and surgical forces. Figure 2a shows the computational geometry which consisted of 21,452 nodes and 113,044 elements. The exterior brain boundary conditions (Figure 2b) are of two types: stress-free (above dashed line) and slippage (below dashed line). Figure 2b also shows the approximate intraoperative brain rotation with gravitational forces represented by a change in surrounding fluid

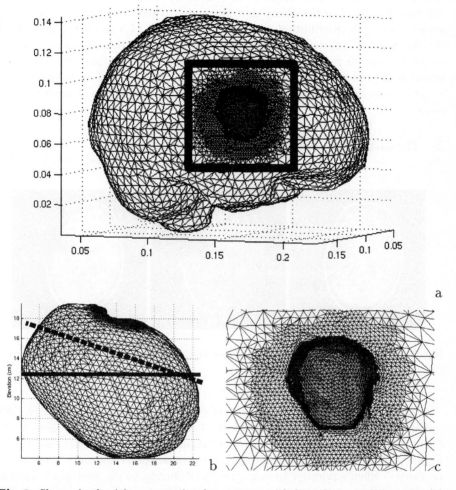

Fig. 2. Shown is the (a) computational geometry with boundary conditions for (b) brain exterior and (c) zoomed view of tumor cavity region designated in (a). In (b), head tilt with respect to gravity shown with solid and dashed line representing air/CSF and free/sliding surfaces respectively. In (c), the subcortical base of the cavity (central area in light grey) has a prescribed normal stress with the walls (dark grey vertical wall perimeter of region) given a larger radially in-plane driving stress distribution.

density for elements above the air/cerebrospinal fluid (CSF) line (solid line). In Figure 2c, we have designated a new boundary condition to simulate the decompression of tissue following the removal of a tumor. Based on the preoperative/intraoperative data, we have approximated the removed tumor mass and deleted this tissue from our model. To simulate decompression, we have applied a non-uniform stress distribution to account for the relaxation of compressive forces previously induced by tumor growth. At the subcortical base of the cavity, the decompressive stress is directed normally and is approximately 5 $mmHg$. Along the perimeter walls of the cavity, stress has been directed radially inward towards the center of the cavity in a plane approximately tangent to the brain suface. The stress distribution values ranged from 5 to 17.5 $mmHg$. The impetus for this non-uniform stress profile is from extensive numerical parameter studies aimed at reproducing the tissue motion observed in the pre/post resection MR scans. We found that a nonuniform distribution characterized by the above description was necessary to explain deformation trajectories seen in measurement data.

3 Results

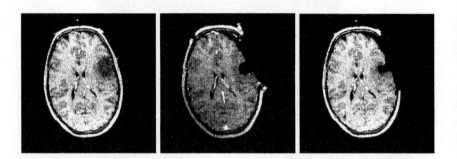

Fig. 3. Co-registered preoperative, intraoperative, and model-based intraoperative update.

Figure 3 is an example of a preoperative image (left), an intraoperative MR update (middle), and a model-based intraoperative update (right). Figure 3 also demonstrates the robustness of the registration algorithm used intraoperatively. Figure 4a.1-a.4 illustrates a point-by-point comparison of anatomic trajectories in Table 1 when subjected to only gravity. Referring to the coordinate system labeled in Figure 1 (right), the largest gravitational component is in the x displacement direction which is also the most satisfying comparison in Figure 4a.1. The most substantial errors occur in surface points along the y displacement direction which corresponds to anterior/posterior movement. In Figure 4b.1-b.4 we have included tumor decompression forces which results in a substantial improvement along the y displacement direction. We see better agreement of surface points both anterior and posterior to the decompressed region.

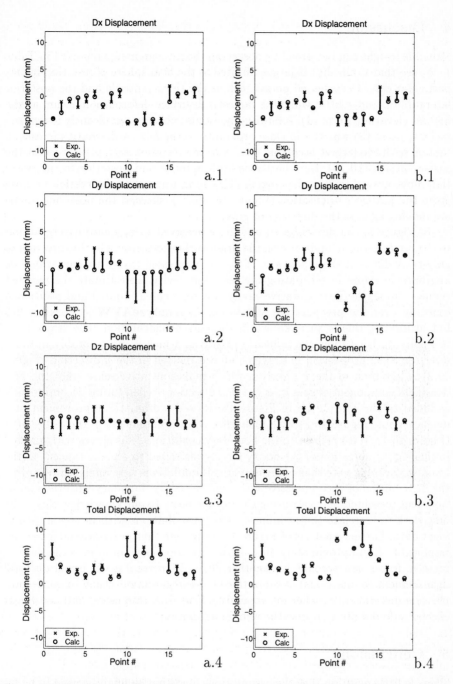

Fig. 4. Tissue trajectory displacements (from Table 1) compared to model calcuations with (a) gravitational forces alone and (b) gravitational forces plus decompression (i.e. indices 1,2,3,4 represent Dx, Dy, Dz, and Total displacements respectively). A good correspondence is achieved when the distance between measured displacement (X) and predicted displacement (O) is small (i.e. line connecting (X) and (O) at each point is small).

4 Discussion

Valuable insight can be gained by observing the measurements reported in Table 1. We see that brain shift is largely limited to the hemisphere of resection in this particular case. Deep seated points such as the third ventricle and the posterior lateral horns move little while the largest subsurface deformation occurs at the highest elevation, point (1). Interestingly we also observe significantly less movement at point (2) which is in close proximity to the falx. A dramatic increase in motion with the largest occurring closest to the resection area is recorded at the surface points (10-18). Here locations both anterior and posterior to the resection move inward toward that region. This is an important observation because given the patient's orientation (Figure 2b) during surgery, the posterior points are moving against the direction of gravity.

In Figure 3, the model-based update is observed to be qualitatively similar to the intraoperative update with the brain surface deforming in the direction of gravity. A different shape and size of the decompressed cavity appears which we attribute to errors in estimating the resection cavity. To estimate the resected tissue volume, the intraoperative cavity was segmented, dilated and positioned centrally over the preoperative MR tumor in **AnalyzeAVW - Version 2.5** but ultimately this is an unknown in the current calculation. The trajectories in Figure 4a.1-a.4 show the influence of gravity which is most notably in the x direction. The proportion of gravity-induced motion in the y direction (Figure 4a.2) is less than in the x direction which makes intuitive sense since the predominant component of gravity is in the x direction. When tumor decompression is added to the calculation in Figure 4b, we see a dramatic improvement in y displacement (Figure 4b.2) with little effect on displacements in the x direction (Figure 4b.1). With respect to the boundary conditions, this agrees with a radial in-plane decompression of the cavity largely restricted to a plane tangent to the brain surface. We note that many numerical simulations were analyzed regarding the distribution of decompressive forces on the cavity boundary with the apportionment presented here capturing the motion best. To our knowledge, this is the first work to correlate tumor-induced stress distributions with *in vivo* measurement data. Further analysis of Figure 4a-b suggests that gravitational forces are responsible for approximately 40% of the intraoperative motion while decompressive forces can account for another 30% (i.e. percentage is based on total displacement comparisons). Assumptions made regarding excision mass, decompression directionality, other intraoperative loads (i.e. retractors) and non-linear effects undoubtedly comprise the remaining motion.

5 Conclusions

There is little question that the assumptions made in this analysis need to be investigated further. However, given these limitations, the model shows significant potential in predicting decompression effects from tumor resection. The clinical data, albeit sparse, highlight intraoperative volume deformations and provide interesting insight to the complexity of tissue movement during surgery. The most

Computational Operation	Time (min)
Problem loading & B.C. deployment	2
Build Preconditioner & 1 Solution	5-6
Successive Solutions w/o Rebuilding Preconditioner	3-4
Output Solutions	0.5

Table 2. Computational burden of FEM for a problem with approximately 21,452 ·nodes and 113,044 elements.

remarkable implication of these results is that with only two temporally-spaced image volumes and with limited knowledge of the actual intraoperative events, a model calculation can predict approximately 70% of the measured intraoperative motion. However, we must temper this enthusiasm with the realization that we are still uncertain as to the quantity and quality of intraoperative data that can be successfully integrated into a model-updating paradigm. In addition, the computational burdens which accompany finite elements must be considered. Table 2 presents the run times associated with different stages of solution evolution for the problem shown in Figures 1-4 on a Pentium III - 550 MHz PC. On the scale of a several hour neurosurgical procedure, we estimate that effective updates need to be performed every 15-20 minutes which appear feasible given the times reported in Table 2. This implies that the remaining steps (i.e. intraoperative data acquisition, data processing, and image updating) would have to be performed in 7-15 minutes which would also seem to be an attainable goal.

Acknowledgments: This work was supported by National Institutes of Health grant R01-NS33900 awarded by the National Institute of Neurological Disorders and Stroke. **AnalyzeAVW - Version 2.5** software was provided in collaboration with the Mayo Foundation, Rochester,MN.

References

1. D. L. G. Hill, C. R. Maurer, R. J. Maciunas, J. A. Barwise, J. M. Fitzpatrick, and M. Y. Wang, "Measurement of intraoperative brain surface deformation under a craniotomy," *Neurosurgery*, vol. 43, no. 3, pp. 514-528, 1998.
2. D. W. Roberts, A. Hartov, F. E. Kennedy, M. I. Miga, and K. D. Paulsen, "Intraoperative brain shift and deformation: a quantitative clinical analysis of cortical displacements in 28 cases," *Neurosurgery*, vol. 43, no. 4, pp. 749-760, 1998.
3. P. M. Black, T. Moriarty, E. Alexander, P. Stieg, E. J. Woodard, P. L Gleason, C. H. Martin, R. Kikinis, R. B. Schwartz, and F. A. Jolesz, "Development and implementation of intraoperative magnetic resonance imaging and its neurosurgical applications," *Neurosurgery*, vol. 41, no. 4, pp. 831-842, 1997.
4. V. M. Tronnier, C. R. Wirtz, M. Knauth, G. Lenz, O. Pastyr, M. M. Bonsanto, F. K. Albert, R. Kuth, A. Staubert, W. Schlegel, K. Sartor, and S. Kunze, "Intraoperative diagnostic and interventional magnetic resonance imaging in neurosurgery," *Neurosurgery*, vol. 40, no. 5, pp. 891-900, 1997.
5. M. Knauth, C. R. Wirtz, V. M. Tronnier, N. Aras, S. Kunze, and K. Sartor, "Intraoperative MR increases the extent of tumor resection in patients with high-grade gliomas," *Am. J. Neuroradiol.*, vol. 20, no. 9, pp. 1642-1646, 1999.

6. M. Knauth, N. Aras, C. R. Wirtz, A. Dorfler, T. Engelhorn, and K. Sartor, "Surgically induced intracranial contrast enhancement: Potential source of diagnostic error in intraoperative MR imaging," *Am. J. Neuroradiol.*, vol. 20, no. 8, pp. 1547–1553, 1999.

7. P. J. Kelly, "Craniotomy for tumor treatment in an intraoperative magnetic resonance imaging unit - COMMENT," *Neurosurgery*, vol. 45, no. 3, pp. 432–433, 1999.

8. C. R. Wirtz, M. M. Bonsanto, M. Knauth, V. M. Tronnier, F. K. Albert, A. Staubert, and S. Kunze, "Intraoperative magnetic resonance imaging to update interactive navigation in neurosurgery: Method and preliminary experience," *Computer Aided Surgery*, vol. 2, pp. 172–179, 1997.

9. V. M. Tronnier, "Craniotomy for tumor treatment in an intraoperative magnetic resonance imaging unit - COMMENT," *Neurosurgery*, vol. 45, no. 3, pp. 431–432, 1999.

10. D. W. Roberts, M. I. Miga, F. E. Kennedy, A. Hartov, and K. D. Paulsen, "Intraoperatively updated neuroimaging using brain modeling and sparse data," *Neurosurgery*, vol. 45, no. 5, pp. 1199–1207, 1999.

11. M. I. Miga, K. D. Paulsen, P. J. Hoopes, F. E. Kennedy, A. Hartov, and D. W. Roberts, "In vivo quantification of a homogeneous brain deformation model for updating preoperative images during surgery," *IEEE Transactions on Biomedical Engineering*, vol. 47, no. 2, pp. 266–273, 2000.

12. M. I. Miga, K. D. Paulsen, J. M. Lemery, S. D. Eisner, A. Hartov, F. E. Kennedy, and D. W. Roberts, "Model-updated image guidance: Initial clinical experiences with gravity-induced brain deformation," *IEEE Trans. Med. Imaging*, vol. 18, no. 10, pp. 866–874, 1999.

13. Wirtz C. R., V. M. Tronnier, F. K. Albert, M. Knauth, M. M. Bonsanto, A. Staubert, O. Pastyr, and S. Kunze, "Modified headholder and operating table for intra-operative MRI in neurosurgery," *Neurol. Res.*, vol. 20, no. 7, pp. 658–661, 1998.

14. Wirtz C. R., V. M. Tronnier, M. M. Bonsanto, A. Staubert, F. K. Albert, , and S. Kunze, "Image-guided neurosurgery with intraoperative MRI: update of frameless stereotaxy and radicality control," *Stereotact. Funct. Neurosurg.*, vol. 68, pp. 39–43, 1997.

15. M. A. Biot, "General theory of three-dimensional consolidation," *J. App. Physics*, vol. 12, pp. 155–164, 1941.

16. K. D. Paulsen, M. I. Miga, F. E. Kennedy, P. J. Hoopes, A. Hartov, and D. W. Roberts, "A computational model for tracking subsurface tissue deformation during stereotactic neurosurgery," *IEEE Trans. Biomed. Eng.*, vol. 46, no. 2, pp. 213–225, 1999.

17. M. I. Miga, K. D. Paulsen, and F. E. Kennedy, "Von neumann stability analysis of Biot's general two-dimensional theory of consolidation," *Int. J. of Num. Methods in Eng.*, vol. 43, pp. 955–974, 1998.

18. M. I. Miga, K. D. Paulsen, P. J. Hoopes, F. E. Kennedy, A. Hartov, and D. W. Roberts, "In-vivo analysis of heterogeneous brain deformation computations for model-updated image guidance," *Computer Methods in Biomechanics and Biomedical Engineering*, vol. 3, pp. 129–146, 2000.

19. M. I. Miga, K. D. Paulsen, P. J. Hoopes, F. E. Kennedy, A. Hartov, and D. W. Roberts, "In vivo modeling of interstitial pressure in the brain under surgical load using finite elements," *ASME J. Biomech. Eng.*, (in press), 2000.

Automatic Landmarking of Cortical Sulci

C. J. Beeston and C. J. Taylor

Division of Imaging Science,
University of Manchester, Stopford Building,
Oxford Rd, Manchester. M13 9PT UK
c.beeston@man.ac.uk, ctaylor@man.ac.uk

Abstract. The positions of the cerebral sulci projected onto a closed hull enclosing the brain tissue provide both a surface description of the brain shape, and meaningful anatomical landmarks. In [1] the sulcal positions were extracted using a thresholding technique, however the use of a single threshold can result in narrower sulci being ignored. We have developed an automatic method of sulcal identification which uses line strength measurement to enhance the contrast of the sulci. The line strength is then projected onto a closed surface surrounding the brain tissue, and line strength filters are run over flat projections of this surface, to give high line strength over the sulcal mouths. Points along the centres of the sulcal mouths are then chosen using non-maximal suppression. We have also experimented with a heat flow model as an alternative to projecting intensities perpendicular to the brain surface. This model allows information to be collected from deeper into the sulci, as it does not rely on the assumption that sulci are perpendicular to the brain surface. This new method is shown to be at least as good for a simple test image as the results from the thresholding technique, and comparisons between the techniques for a real image also indicate greater efficiency.

1 Background

Automatic detection and labelling of the sulci is recognised to be an important problem, since the sulci form anatomical landmarks over the cortex which in some cases indicate functional regions of the brain [2]. However, since the sulcal patterns are extremely variable [3] this is a difficult problem. A large number of examples would be needed to build a sensible statistical model of their variation. Most work on sulcal modelling so far has either used a few hand landmarked examples, where only major sulci are marked for example [4, 5], or used very few brains to build an anatomical atlas [6]. In [7], sulcal patterns are extracted automatically in the form of sulcal basins. However this method relies on a good contrast between grey and white matter, which is not always the case in MR images. Another example of automatic sulcal extraction is given in [8], where lines of positive and negative maximal curvature are found over the brain surface, corresponding to the crests of the gyri, and the troughs of the sulci.

A. Caunce and C. Taylor [1] have already published work on extracting sulcal positions, and building 3-D active shape models (ASMs) from them. In their work, sulcal positions were extracted by segmenting the brain tissue from a 3D image, forming a smooth hull enclosing the brain by a morphological closing operation, and then averaging the image intensities over along the surface normals to this hull, to a depth of 5mm. A threshold was then applied to this projection image, to extract the dark lines formed by the projections of the sulci, and finally thinning to obtain landmark points on the cortical hull over the centres of the sulci. This work complements the work in [1], in that it concentrates on accurate extraction of the sulcal positions from images which have varying degrees of contrast, resolution and sulcal variation, extending the range of data for which sulcal models can be constructed. The method of sulcal extraction we describe is simple, robust and could be applied to a large range of Magnetic Resonance data sets with minimal changes. Thus it allows many examples of the sulcal positions to be collected for use in statistical models of the brain shape and sulcal patterns, without the tedious and time-consuming process of hand landmarking images.

2   Introduction

The initial motivation from this work came from studies of a large set of Magnetic Resonance (IRTSE) images of the brains of people with various types of dementia. The images have very low resolution in z (the voxel dimensions of the images in x, y and z are 0.9, 0.9 and 3.5mm respectively), which causes large partial volume effects, reducing the contrast of the narrow sulci. As well as this, the dementia causes some cerebral atrophy with corresponding sulcal widening, so that the width of the sulci can vary enormously over the brain. Figure 1 shows coronal slices from a normal brain, and from a brain with dementia.

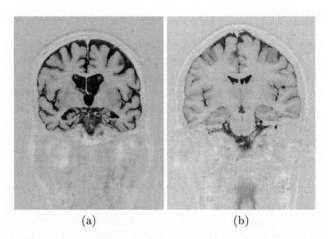

(a)   (b)

Fig. 1. single coronal slices from (a) a normal brain and (b) a brain with cerebral atrophy due to dementia

After the images have been resliced to give cubic voxels, and the brain tissue has been extracted from each image, a 3D morphological closing operation is used to form a smooth hull containing the brain tissue, including the sulci. Figure 2 shows the result of taking each point on the smooth hull, and taking the average of the intensities along a line perpendicular to the hull at this point, to a depth of 5mm. Views of this surface for a normal brain from above and from the side are shown. It can be seen that even for this normal brain, the sulci on the upper surface are wider than those on the lower surface, and that it would be very difficult to choose a single threshold which would preserve all of the sulci.

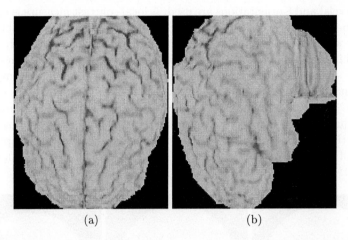

(a)   (b)

Fig. 2. average intensity to 5mm depth perpendicular to the (a) upper and (b) side brain surfaces

It can be seen from figure 2 that when we look at the surface of the brain, the sulci form dark lines against a lighter background. We can also see from figure 1 that since the sulci are planes formed by the infolding of the brain surface, they will appear as lines in at least two orthogonal slices of the image. The next section describes the use of line strength measurement and non-maximal suppression to enhance the sulci, and provide a better estimate of the positions of the centres of the sulci, as projected onto a hull around the brain tissue.

3 Formation of a 3D line strength image

To measure line strength, we have chosen line strength filters at three scales, with line widths equal to 3, 5 and 7 pixels, and a width:length ratio of 1:5. Once we have found the best orientation for a filter of a particular line width, at a particular pixel, we calculate a local contrast measure at the pixel, perpendicular to the best line direction and use this as our measure of best line strength. This reduces the number of pixels which are given an artificially high line strength due to their being in between two dark spots which are bridged by the filter.

To investigate the line strength measurement we use a 100*100*100 pixel test image, containing a sphere of radius 43 pixels centred at (50,50,50), with three surfaces intersecting it to form artificial sulci. The image intensities have been chosen to correspond to the grey and white matter and csf intensities in the real images we are studying. Gaussian noise is then added to the image so that the intensity distributions more closely resemble those in a real image. Figure 3 shows the central plane of the test image. The planes at $x = 27$, $y = x$ and $y = 6 \sin x/5 + 20$ can all be seen.

To make a 3D line strength image, we process each slice of the image, taking the xy, yz and xz projections separately. Line strength filters at 3 scales (widths 3, 5 and 7 pixels) and 8 orientations are used. A local contrast measure is calculated once the best orientation has been found, as described above, this will be referred to as line strength from now on. The line strength images from each scale are combined into one image by taking the highest value at each pixel. Once we have three line strength images, one for each projection, we combine the images into one 3D image by squaring and adding the results at each pixel.

Figure 3 (b) shows the intensity projection of the test image, as described in the introduction, and Figure 3 (c) shows the result of projecting the line strength image in the same way. It is clear that the line strength method improves the contrast of the sulci, at the cost of some apparent widening of the sulci.

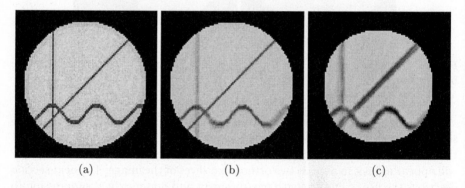

(a) (b) (c)

Fig. 3. (a) central plane of the test image, at z=50. (b) surface projection of the test image, (c) surface projection of the line strength image, inverted for comparison with (b).

4   Locating the sulci using non-maximal suppression

Once we have our 3D line strength image, we project it onto the faces of a cube surrounding the entire image. As the test image is convex, very little information is lost in this way, and the images formed on the 6 planes are sufficient as input to the next stage of processing.

Once again, local contrast images at three scales are calculated, using the best direction of the line strength filter to give the direction in which to calculate the contrast. Non-maximal suppression is then performed on the contrast images from each scale. This involves looking at the local contrast at a fixed distance ($\sqrt{2}$ pixels) from each pixel, perpendicular to the best line strength direction. If the central pixel has a higher value than both of the side pixels, i.e. its contrast is locally maximal, then that pixel position is saved. The resulting networks of points from the three scales are ORed together, and the final network is thinned, to reduce the widths of the lines to one pixel, where this is necessary. A low threshold is also applied, to remove points which are locally maximal but which are in flat regions of the image. The advantage of this method is that it is adaptive, as it relies almost entirely on local contrast measures, with only a very loose global threshold, Thus regions where the sulci are tightly packed are not discriminated against.

Figure 4 shows sulcal positions (light points) obtained using three methods, overlaid on the surface of the test image before noise was added. (a) shows the result of the original method of sulcus detection - a simple threshold is applied, and the resulting lines are thinned to give the final positions. (b) and (c) show the results of the line detection and non-maximal suppression on the images in figure 3 (b) and (c). It can be seen that results from line detection and non-maximal suppression over the brain surface are better than those obtained by thresholding and thinning, particularly around the junctions of sulci. However the points in (c) are further from their ideal positions than those in (b) - for this simple test image, the degradation in the surface image due to widening of the sulcal positions outweighs the improved contrast.

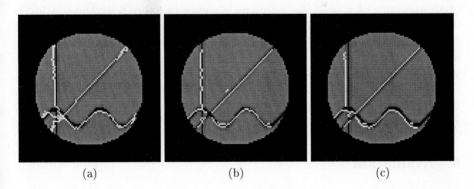

(a) (b) (c)

Fig. 4. Sulcal positions obtained by various methods, overlaid on the surface of the test image before noise was applied. Results from (a) thresholding and thinning, (b) line detection and non-maximal suppression over the intensity projection image, (c) line detection and non-maximal suppression over the line strength projection image

Figure 5 shows the corresponding results for a real image. (a) and (b) show the results from the projected intensity image in figure 2 for a simple threshold, and for line detection. (c) shows the projected 3D line strength image, and (d) shows the results of line detection over this image. No ground truth is available, but the line strength image does appear to give more convincing results than the intensity image.

5 Heat flow model

It seems clear that if we could average the line strength along a sulcus to a depth of greater than 5mm, the definition of the sulci at the brain surface would be further improved. Since the assumption that the sulci are perpendicular to the hull is only good for the outer few millimetres of the brain, a method which allowed the line strength information to be collected along the direction of each sulcus, rather than perpendicular to the surface is desired. To do this, we have implemented a heat flow algorithm, using the heat equation:

$$\Delta T = \kappa \alpha \nabla^2 T \Delta t$$

and assumed that areas of high line strength are both hot and highly conductive: we assume the conductivity κ to be constant and equal to the original line strength, the temperature T is also initialised to the line strength, the time interval for one iteration of the heat flow is one second, and α is chosen so that the average change in T over one iteration is around $1 - 2\%$, for smooth flow. The material outside the line strength image is taken to be cold ($T = 0$) and highly conductive (κ set to higher than any line strength in the image). We apply the algorithm in each orthogonal plane independently at present, and collect the heat flowing out of the surface of the image over several iterations. The result of 10 iterations of flow in the y-z plane is shown in figure 6 for the test image - the resolution of the sulci has improved compared with the surface projection of the linestrength image, because the heat flow model effectively projects the linestrength along the sulci (except, in this image, for the sulci parallel with the y-z plane).

Figures 5 (e) and (f) show corresponding results for flow in the xy plane for the real image. The heat flow projection image (e) appears to have significantly greater contrast than (c), as we had hoped it would.

6 Evaluation

To estimate the efficiency of the different methods, we use the test image before noise has been added, and threshold it to give the ground truth points corresponding to csf, and csf plus grey matter. Table 6 gives the total number of points extracted using each sulcus detection method, the number of points overlapping with the csf ground truth, the number of points within the boundary of csf plus grey matter, and the number of points not within the grey matter.

The methods involving line strength and non-maximal suppression all find fewer points than the thresholding method, mainly because the line strength

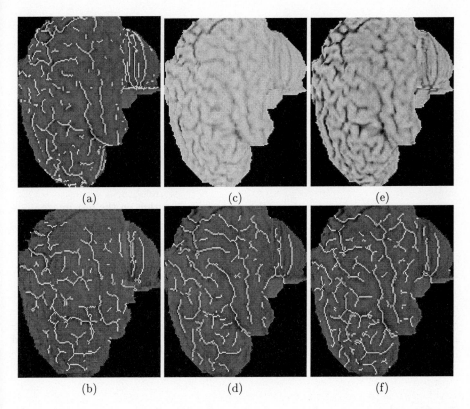

$$(a) \qquad (c) \qquad (e)$$

$$(b) \qquad (d) \qquad (f)$$

Fig. 5. Results from (a) a simple threshold and (b) line detection for the surface projection of the intensity image. (c) surface projection of the 3D linestrength image, and (d) results of line detection over (c). (e) projection of the 3D linestrength image using heat flow (inverted), and (f) results of line detection over (e).

measurement cannot be used too close to the edge of the image, and as we have only dealt with one of the projections of the 3D image, the edges have not been filled in by the results of the other projections. For this test image, the sulci are very distinct, as we have not included any of the partial volume effects we expect from the large voxel dimensions of our real images. Using line detection and non-maximal suppression over the surface of the projected intensity image appears to be a much better method of extracting the sulcal positions than thresholding and thinning.

Looking at the real images, it also appears that using line detection provides a more adaptive method of sulcal extraction, improving both the detection efficiency, and the smoothness of the detected sulci. The heat flow method also looks to have the potential to enhance sulci which are narrow but deep, when results from flowing the line strength ' in the three orthogonal planes are combined.

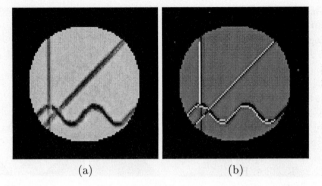

(a)   (b)

Fig. 6. Result of 10 iterations of the heat flow algorithm on the test image.

	(a)	(b)	(c)	(d)
Total points extracted	211	160	154	159
Overlap with csf	125	127	93	106
Overlap with csf or grey matter	208	158	149	159
Points far away from sulci	3	2	5	0

Table 1. Numbers of points extracted using each sulcus detection method: (a) Thresholding and thinning, (b) Line detection over intensity projection image (c) Line detection over line strength projection image (d) Line detection over projection image formed using heat flow model

7 Conclusion

We have shown that projection of a 3D line strength image onto a hull enclosing the brain tissue using a heat flow algorithm, followed by linear feature detection with non-maximal suppression, is an effective method of enhancing the contrast of the cortical sulci in an image, and extracting their positions. Experiments with a test image indicate that the results obtained are at least as good as those obtained using a simple threshold, however a more realistic test image needs to be used to make a more quantitative measure of the efficiency of this method.

Acknowledgements

This work was funded by the UK Engineering and Physical Sciences Research Council. We would also like to thank Professor David Neary of the Manchester Royal Infirmary for the images used in these experiments, and Angela Caunce for the use of her software.

References

1. A Caunce and C J Taylor. Using local geometry to build 3d sulcal models. In *Proc. Information Processing in Medical Imaging*, pages 168–181. Springer-Verlag, 1999.

2. J Rademacher, A M Galaburda, and D N Kennedy et al. Topographical variation of the human primary cortices: Implications for neuroimaging, brain mapping and neurobiology. *Cerebral Cortex*, 3:313–329, 1993.
3. M Ono, S Kubik, and C D Abernathy. *Atlas of the Cerebral Sulci.* Thieme Verlag, New York, 1990.
4. H Chui, J Rambo, and J Duncan et al. Registration of cortical anatomical structures via robust 3d point matching. In *Proc. Information Processing in Medical Imaging*, pages 168–181. Springer, 1999.
5. M Vaillant and C Davatzikos. Hierarchical matching of cortical features for deformable brain image registraion. In *Proc. Information Processing in Medical Imaging*, pages 182–195. Springer, 1999.
6. P Roland and K Zilles. Brain atlases - a new research tool. *Trends in Neurosciences*, 17:458–467, 1994.
7. G Lohmann and Y von Cramon. Automatic detection and labelling of the human cortical folds in magnetic resonance data sets. In *Proc. ECCV*, pages 369–381, 1998.
8. G Subsol, J-P Thirion, and N Ayache. Application of an automatically built 3d morphometric brain atlas: Study of cerebral ventricle shape. In *Proc. Visualization in Biomedical Computing*, pages 373–382. Springer, 1996.

The Skull Stripping Problem in MRI Solved by a Single 3D Watershed Transform

Horst K. Hahn and Heinz-Otto Peitgen

MeVis – Center for Medical Diagnostic Systems and Visualization
Universitätsallee 29, 28359 Bremen, Germany
e-mail: {hahn, peitgen}@mevis.de

Abstract. A robust method for the removal of non-cerebral tissue in T1-weighted magnetic resonance (MR) brain images is presented. This procedure, often referred to as skull stripping, is an important step in neuroimaging. Our novel approach consists of a single morphological operation, namely a modified three-dimensional fast watershed transform that is perfectly suited to locate the brain, including the cerebellum and the spinal cord.

The main advantages of our method lie in its simplicity and robustness. It is simple since neither preprocessing of the MRI data nor contour refinement is required. Furthermore, the skull stripping solely relies on one basic anatomical fact, i.e. the three-dimensional connectivity of white matter. As long as this feature is observed in the image data, a robust segmentation can be guaranteed independently from image orientation and slicing, even in presence of severe intensity non-uniformity and noise. For that purpose, the watershed algorithm has been modified by the concept of pre-flooding, which helps to prevent over-segmentation, depending on a single parameter. The automatic selection of the optimal parameter as well as the applicability are discussed based on the results of phantom and clinical brain studies.

Keywords. Whole brain segmentation, skull stripping,
3D watershed transform, pre-flooding, neurological image processing.

1 Introduction

The skull stripping problem is well-known and has been studied in [1–10]. It is equivalent to the segmentation of the whole brain or the removal of non-cerebral

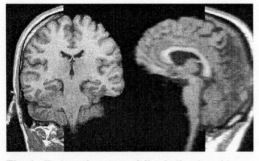

Fig. 1. Fusion of a successfully skull-stripped image with the original data in two projections (healthy volunteer, coronal slicing, thickness 3 mm).

tissue such as skull, scalp, veins or meninges (Fig. 1). Studies of brain anatomy and pathology are most commonly based on magnetic resonance imaging (MRI) due to its good soft tissue separation. Here, T1-weighted protocols, that belong to the fastest MRI protocols available, are often preferred, since they offer a good contrast between gray (GM) and white cerebral matter (WM) as well as between GM and cerebrospinal fluid (CSF).

Whole brain segmentation is often regarded as an essential step in a neurological image processing pipeline, either because the whole brain is the region of interest, such as in studies of Morbus Alzheimer [5], or because the subsequently performed steps benefit from the fact that only a small set of well known tissue types is left over (i.e. WM, GM, CSF and possibly lesions), such as statistical brain tissue segmentation methods [2, 11, 12]. The latter offer a good basis for volumetric or morphometric examinations, e.g. cortex reconstructions [8], and yield promising results in accounting for intensity non-uniformity [11, 12], however, their convergence is distorted by the presence of non-brain tissue [7, 11, 13]. For brain warping techniques, that are used to perform inter-subject studies, it is as well desirable to exclude all non-brain tissue from the matching process [14].

The paper is organized as follows. The next sections will give a brief overview of existing methods and describe the motivation and goals of the novel approach. Then, the method will be derived in detail. An evaluating section assesses the robustness and applicability. The performance as well as the dependency on the parameter introduced below are examined. Finally, extensions and limitations of our approach are discussed.

2 Motivation and Goals

Despite the clear definition of the skull stripping problem, no standardized solution has been published yet. A good survey of the work up to 1996 is given by ATKINS et al. [6]. The image processing techniques found in literature can be divided into three groups:

- *Region-based.* The most common approaches sequentially apply morphological operations and manual editing. First, the white matter gray values are located using thresholding or seeded region growing, followed by a morphological opening that detaches the brain tissue from the surrounding tissue. Morphological dilation and closing are required for the segmentation to cover the whole brain without holes [1, 2, 3, 5, 7, 10].
- *Hybrid methods.* In order to account for the shortcomings of morphological multi-step approaches, they have been combined with edge-based methods [2, 4, 6, 9]. For example, two-dimensional contours such as snakes are applied to the morphological segmentation result in a final step [6].
- *Template-based.* More recently, some investigators succeeded in fitting a balloon-like surface to the intensity-normalized MR data in order to separate the brain from surrounding structures [8]. Another three-dimensional template based on volume data is described in [15].

Our goal in this paper is to develop a skull stripping procedure that is qualified to be the first step in the image processing pipeline. This means, that it should be robust even against considerable radio frequency (RF) non-uniformity as well as noise. This is not the case for most of the existing techniques. Thresholding and region growing based approaches exhibit the problem of *leaking out*, as discussed in [16], and are highly sensible to image non-uniformities. Deformable templates are susceptible to image noise that has to be removed in a preprocessing step.

In order to be applicable both for pathological brain structures and anatomical abnormalities, the algorithm should be as simple as possible: First, in the sense that it consists of one (or a few) well-defined operations, so that segmentation failures can be easily understood and interaction techniques may be properly integrated. Second, in the sense that it should be based on as little assumptions and anatomical a priori knowledge as possible. We do not use deformable templates that require smoothness constraints, which often are not satisfied on the brain boundary, especially in its basal parts.

For clinical purposes it is desirable to provide an automatic segmentation procedure. Finally, the computational costs should be approximately of the order of the number of voxels, to be able to deal with future imaging resolutions.

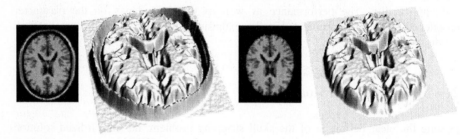

Fig. 2. Axial slice of the MNI phantom (simulated with slice thickness 3 mm, noise level 3 %, RF non-uniformity 20 %) and interpretation of gray value as height information. **left:** original image. **right:** skull stripped with $h_{pf} = 0.16\ I_{max}$ and only one connected hill left, representing the brain.

3   Methods

3.1   Basic Ideas

Similar to other approaches [1, 2, 7], our basic assumption regarding brain anatomy is the *connectivity of white matter*. We define two points of a gray level image to be *connected*, if a path of adjacent voxels exists between them that is at least as bright as the darker one of the two points. This concept can be easily understood by interpreting the image intensity as height information. For each image slice, the topographic interpretation of the gray values consists of several hills corresponding to bright image areas and valleys corresponding to dark image areas (Fig. 2 left). According to the above definition of connectivity, two regions are disconnected, if they are separated by a valley, otherwise they are connected.

Here the significance of the intensity characteristics of T1-weighting for our approach becomes clearly visible: The connected WM is surrounded by darker GM and even darker CSF, and can thus be regarded as the top of a hill. WM regions that are not connected in 2D must be connected in 3D, since the WM is interconnecting all functional parts of the brain. Therefore, we prefer a segmentation procedure that works fully in 3D, rather than 2D techniques that may not account for three-dimensional connectivity.

Assuming that the brain is surrounded by CSF and all non-brain tissue that needs to be removed shows brighter image intensities than CSF, skull stripping becomes equivalent to isolating a single hill in the four-dimensional landscape. (Figures 2 and 3 represent didactical three-dimensional projections of the four-dimensional problem.) This hill then represents the whole brain including the cerebellum and the spinal cord, as long as they are connected within the image. The valley corresponding to CSF and other low intensity tissue such as bone and meninges will then define the accurate border of the segmentation result (Fig. 2 right).

Fig. 3. Same data as Figure 2, but gray level inverted. Hills in the original image are now represented by basins, that can be separated by a watershed transform. **left:** axial, **right:** sagittal view.

3.2 Solution based on a 3D Watershed Transform

We now consider the gray level inverted T1-weighted brain image data. Under this transformation, hills become basins and valleys become crest lines (Fig. 3). The morphological operation that partitions an image into regions, each of them corresponding to a connected basin, is *the watershed transform* [17]. Thus, all that needs to be done for a whole brain segmentation, is to perform *a 3D watershed transform on the inverted original data[1]* and to choose the catchment basin representing the brain.

Before introducing the fast watershed algorithm, we will have a more detailed look at the problem: WM cannot always be regarded as connected in the strict sense defined above, since its image intensity is not constant, even without the presence of RF non-uniformity and noise. Therefore, we have to weaken our criterion for connectivity. We do so by allowing the connecting path to show a lower intensity than the darker of the two connected points up to a maximum difference.

In words of the watershed transform, this is described by the *concept of pre-flooding*: Prior to the transformation, each catchment basin is flooded up to a certain height above its *bottom* (i.e. the darkest voxel) called *pre-flooding height* h_{pf} and will only be regarded as a separate region as long as it holds the water inside. Otherwise, it will be merged with the deepest neighboring basin.

---

[1] It has to be mentioned, that usually the watershed transform is applied to the gradient image.

3.3 Algorithm

The applied fast watershed transform starts by sorting all voxels (of the gray level inverted image, Fig. 3) according to their intensity in an ascending order. Each voxel with regard to its three-dimensional 6-neighborhood is processed exactly once, until the brightest voxels have been processed. If the voxel has some already processed neighbors (i.e. voxels of same or less intensity), *voxel-basin merging* is performed. Otherwise, a new basin is formed since after sorting, an isolated voxel must represent a local intensity minimum. If two or more neighbors have already been processed belonging to different basins, these are tested for *basin-basin merging*. This approach is similar to the one described by MITTELHAEUSSER and KRUGGEL [18], but simpler in that the voxel-basin merging is non-conditional. We now introduce two criteria for the merging procedures (not published in [18]):

- *Voxel-basin merging:* Each voxel will be merged with the deepest neighboring basin, i.e. the basin with the darkest bottom voxel.
- *Basin-basin merging:* All neighboring basins whose depth relative to the current voxel intensity is less or equal to the pre-flooding height h_{pf} will be merged with the same basin as the voxel itself.

After the transform with an appropriate pre-flooding height one basin should exist that represents the whole brain with all parts that are connected via WM, i.e. cerebellum and spinal cord. This basin usually is the largest existing one, unless the field of view (FOV) has been selected too large. In that case, one could choose the basin containing the center of FOV or a manually selected basin.

The number of basins is monotonically decreasing whereas its sizes are increasing with increasing h_{pf}. Figure 4 shows the typical characteristics of that behavior. Here, the region-based nature of our method becomes visible, providing the basis for robustness. The segmentation result does not change continuously with varying h_{pf} but, depending on the image quality, a more or less broad *range of proper h_{pf} values* exists.

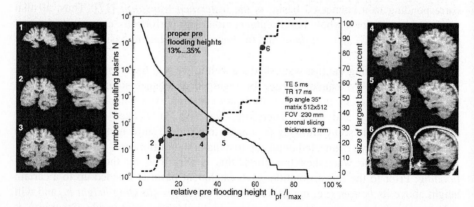

Fig. 4. Algorithmic behavior of the modified watershed transform with increasing pre-flooding height, number of resulting catchment basins (solid line) and size of largest basin (dashed line). For the six given h_{pf} values (filled circles) the result is shown in coronal and axial view. In case no. 5, where under-segmentation did already occur, the largest basin did not contain the brain.

4   Quantitative Evaluation and Results

This section will evaluate the usefulness of the described procedure for a variety of brain images. We use clinical data from different scanners as well as phantom data to answer the following questions:

- Is it possible to successfully segment the whole brain with our method?
- If yes, what is the range of proper pre-flooding heights?

The answer to the second is expected to depend on the image noise level and other aspects of image quality. For each image the level of noise n is measured as the standard deviation of the high-frequency signal present in the image background. In the following, noise and pre-flooding heights h_{pf} will be given in units relative to the maximum image intensity I_{max}.

For clinical MRI segmentation studies no *gold standards* are available. Our "gold standard" was defined by an expert radiologist who classified the segmentation procedure as *unsuccessful* when small parts of cerebrum or cerebellum were excluded or pieces of skull, skin or eyes were included in the segmentation result. On the other hand, parts of meninges or veins, that in the images often seem to be connected to gray matter, were tolerated within a *successful* segmentation.

Moreover, accuracy was assessed by comparing the overlap of a manual expert segmentation with the results of our skull stripping method. On all 7 manually segmented datasets the overlap has been more than 96%.

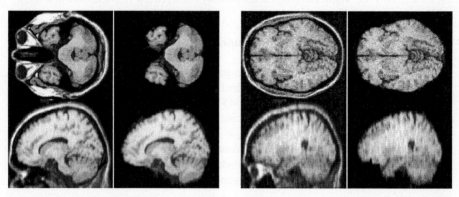

Fig. 5. Two images of the MNI simulated brain database and application of the skull stripping procedure, axial and sagittal view. Image simulation parameters (slice thickness / noise level / RF non-uniformity): **left.** 3 mm / 3% / 20%. **right.** 5 mm / 9% / 40%.

4.1   Phantom Studies

To offer a reproducible data source, we extensively tested our method on the Simulated Brain Database (SBD) [19, 20] from the Montréal Neurological Institute (MNI), McGill University. All available 90 T1-weighted datasets of the normal brain database have been processed. The results were satisfactory: All datasets were successfully segmented by our method, even for extreme slice thickness (9 mm), noise (9%) and RF non-uniformity (40%). Two examples are shown in Figure 5.

140

4.2 Clinical Brain Studies

For further evaluation, 43 clinical brain images of healthy volunteers and patients have been acquired with two MR scanners: A Siemens Magnetom Vision Plus 1.5 T and a Philips S15-ACS 1.5 T, using various 3D T1-weighted acquisition protocols (TR 8.1–17 ms, TE 4–5 ms, flip angle 12°–35°) and different reconstruction parameters (no interslice gap, coronal, axial, and sagittal slicing, slice thickness 1–5 mm). In all 43 cases, the above described method succeeded in segmenting the brain. See Figure 6 for examples. However, in two cases the pre-flooding value h_{pf} for correct segmentation turned out to be unexpectedly high, see Section 5.

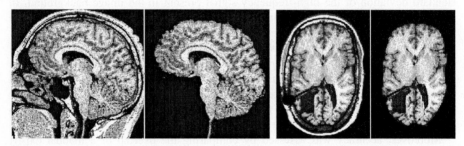

Fig. 6. Clinical brain datasets and application of skull stripping procedure. Two problems are addressed: **left.** extreme image noise level (acquisition parameters: slice thickness 1 mm, TR 8.1 ms, TE 4 ms). **right.** pathological abnormality (enlarged left ventricle, with drainage).

4.3 Robustness and Parameter Dependency

In order to chart the above results and to evaluate the robustness of our method, we plotted the range of proper h_{pf} values against the noise level n, both in units relative to the maximum image intensity I_{max} for 45 of the 90 processed phantom datasets (Fig. 7). The proper h_{pf} values roughly scale linearly with the noise level. The dotted line represents the equation $h_{pf} = 0.11\,I_{max} + 3.5\,n$ chosen such that all 45 datasets are correctly segmented. This behavior makes it possible to automatically select an appropriate pre-flooding height. The broader the h_{pf} range, the more robust is the segmentation to further noise and inhomogeneity (Table 1).

Table 1. Quantitative evaluation of robustness in terms of the h_{pf} range.

Range of proper pre-flooding heights $(h_{pf.max}-h_{pf.min}) / I_{max}$	# datasets	
	phantom	clinical
0 (failure)	0	0
< 5%	6	2
5–10%	25	9
> 10%	59	32

Table 2. Average sorting and transformation times for three different image sizes on two standard SGI machines (Onyx2: 2xR10000, 195 MHz, O2: R5000, 180 Mhz).

# voxels	SGI Onyx2	SGI O2
1 Mio.	2 s / 5 s	5 s / 9 s
4 Mio.	11 s / 21 s	25 s / 43 s
10 Mio.	33 s / 51 s	69 s / 103 s

4.4 Numerical Issues

The complexity of the modified watershed transform is linear in the number of voxels N for the voxel processing, and in the worst case proportional to $N \log N$ for the sorting (about 30 % of the total computing time, see Table 2). In the average case our algorithm shows *linear complexity in N*. All algorithms have been implemented in C++ and are available on the MeVis medical imaging platform ILabMed [21]. For sorting the voxels according to their intensity, we used a QuickSort algorithm [22].

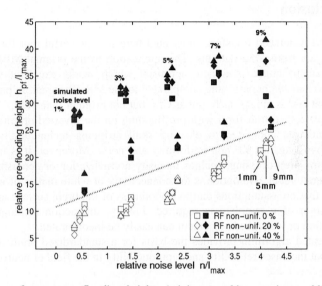

Fig. 7. Range of proper pre-flooding heights (minimum: white, maximum: black markers) plotted against noise level for 45 datasets of the MNI database. For each noise level, slice thickness is increasing from left to right (1 / 5 / 9 mm).

5 Discussion

The described algorithm is able to successfully segment the whole brain in all 133 datasets, without any preprocessing. For any given image the parameter h_{pf} can be varied over a certain range without changing the output, this being a measure for robustness of the method. The sensitivity in comparison to manual expert segmentation is estimated to more than 96 %. Differences (less than 4 %) are mainly located in the dark intensity region at the brain boundary, i. e. the interface between bone, meninges and CSF. Only in two cases, parts of the superior sagittal sinus have been included.

The results were satisfactory even where the range of proper pre-flooding heights h_{pf} has been rather narrow. This has been the case in six of 90 simulated images due to a slice thickness of 9 mm, which is too thick for a good separation of eyes and frontal lobes (compare Figure 5 right), and in two of 43 clinical datasets (see Table 1). When inspecting these two images, we found that some isolated bright voxels existed in different parts of the brain, thus increasing the required h_{pf} value.

The merging rules introduced in Section 3.3 are well-defined and easy to compute. They enabled the watershed algorithm to solve the skull stripping problem in all cases and lead to fast processing times (Table 2). However, further improvements regarding the robustness of the method seem to be possible by modifying the merging rules. The measure for basin depth e.g. could be replaced by a measure that considers all gray values of a basin, not only the darkest one.

6   Conclusion

The modified watershed transform presented here is a powerful tool for segmenting the whole brain from MRI datasets. The application to the original MRI data rather than the gradient image is a new approach which works excellent with images showing an intensity characteristic comparable to the MR T1-signal. In particular, it is important that GM is at least as bright as CSF but not brighter than WM.

The 3D watershed transform with pre-flooding on the inverted original data bears several advantages compared to existing skull stripping techniques, in that it is extremely robust against RF non-uniformity and noise. Moreover, it is fast and does not require any preprocessing, such as intensity normalization or denoising. The only underlying anatomical assumption is the connectivity of white matter. It is model-free in the sense that no assumptions on the smoothness of the brain surface are made and no initialization of a template is required. Finally, interaction techniques such as region selection or watershed placement can easily be incorporated.

The presented algorithm provides the basis for a standardized brain segmentation procedure that increases reliability and reproducibility in the field of neuroimaging.

Acknowledgments

We would like to thank our clinical partners Prof. Dr. B. Terwey, Bremen, and Prof. Dr. P. Stoeter, Mainz, for kindly providing us with brain data and Wolf Spindler, Bremen, for his support on visualization issues.

References

1. R.Kikinis, M.E.Shenton, G.Gerig, J.Martin, M.Anderson, D.Metcalf, C.R.G.Guttmann, R.W.McCarley, W.Lorensen, H.Cline, and F.A.Jolesz. "Routine Quantitative Analysis of Brain and Cerebrospinal Fluid Spaces with MR Imaging" *J. Magnetic Resonance Imaging* 2: 619-629, 1992.
2. T.Kapur, W.E.L.Grimson, W.M.Wells, and R.Kikinis. "Segmentation of Brain Tissue from Magnetic Resonance Images" *Medical Image Analysis* 1(2): 109-127, 1996.
3. S.Sandor and R.Leahy. "Surface-Based Labeling of Cortical Anatomy using a Deformable Atlas" *IEEE Trans. Med. Imaging.* 16(1): 41-54, Feb. 1997.
4. G.B.Aboutanos and B.M.Dawant. "Automatic Brain Segmentation and Validation: Image-Based versus Atlas-Based Deformable Models" *Proc. SPIE Med. Imaging '97*, 299-310, Feb. 1997.

5. P.A.Freeborough, N.C.Fox, and R.I.Kitney. "Interactive Algorithms for the Segmentation and Quantification of 3D MRI Brain Scans" *Comput. Metho. Progr. Biomed.* 53(1): 15-25, May 1997.

6. M.S.Atkins and B.T.Mackiewich. "Fully Automatic Segmentation of the Brain in MRI" *IEEE Trans. Med. Imaging.* 17(1): 98-107, Feb. 1998.

7. A.F.Goldszal, C.Davatzikos, D.L.Pham, M.X.H.Yan, R.N.Bryan, and S.M.Resnick. "An Image Processing System for Qualitative and Quantitative Volumetric Analysis of Brain Images" *J. Comput. Assist. Tomogr.* 22(5): 827-837, 1998.

8. A.M.Dale, B.Fischl, and M.I.Sereno. "Cortical Surface-Based Analysis: I. Segmentation and Surface Reconstruction" *NeuroImage* 9: 179-194, 1999.

9. K.Rehm, D.Shattuck, R.Leahy, K.Schaper, and D.Rottenberg. "Semi-Automated Stripping of T1 MRI Volumes: I. Consensus of Intensity- and Edge-Based Methods" *Proc. 5th Int. Conf. on Functional Mapping of the Human Brain HBM '99*, Düsseldorf, poster no. 86, 1999, abstract pub. in *NeuroImage*.

10. S.A.Hojjatoleslami, F.Kruggel, and D.Y.von Cramon. "Segmentation of White Matter Lesions from Volumetric MR Images" *Proc. Medical Image Computing and Compter-Assisted Intervention MICCAI '99*, Cambridge. Springer LNCS, vol. 1679: 52-61, 1999.

11. W.M.Wells, W.E.L.Grimson, R.Kikinis, and F.A.Jolesz. "Adaptive Segmentation of MRI data" *IEEE Trans. Med. Imaging* 15: 429-443, 1996.

12. D.L.Pham, J.L.Prince. "Adaptive Fuzzy Segmentation of Magnetic Resonance Images" *IEEE Trans. Med. Imaging* 18(9):737-752, Sept. 1999.

13. F.Maes, K.Van Leemput, L.E.DeLisi, D.Vandermeulen, and P.Suetens. "Quantification of Cerebral Gray and White Matter Asymmetry from MRI" *Proc. Medical Image Computing and Compter-Assisted Intervention MICCAI '99*, Cambridge. Springer LNCS, vol. 1679: 348-357, 1999.

14. A.W.Toga. *Brain Warping*, Academic Press, San Diego, 1999.

15. R.S.J.Frackowiak, K.J.Friston, C.D.Frith, R.J.Dolan, and J.C.Mazziotta. *Human Brain Function*, Academic Press, San Diego, 1997. Statistical Parametric Mapping SPM, http://www.fil.ion.ucl.ac.uk/

16. A.J.Worth, N.Makris, V.S.Caviness, jr., and D.N.Kennedy. "Neuroanatomical Segmentation in MRI: Technological Objectives" *Int. J. Pattern Rec. Art. Int.* 11(8): 1161-1187.

17. J.Serra. *Image Analysis and Mathematical Morphology.* London Academic, 1982.

18. G.Mittelhaeusser and F.Kruggel. "Fast Segmentation of Brain Magnetic Resonance Tomograms" *Proc. 1st Conf. on Computer Vision, Virtual Reality and Robotics in Medicine CVR '95.* Springer LNCS, vol. 905, 1995.

19. C.A.Cocosco, V.Kollokian, R.K.-S.Kwan, and A.C.Evans. BrainWeb: "Online Interface to a 3D MRI Simulated Brain Database" *Proc. 3rd Int. Conf. on Functional Mapping of the Human Brain HBM '97*, Copenhagen. *NeuroImage* 5(4): part 2/4, S425, May 1997. http://www.bic.mni.mcgill.ca/brainweb/

20. D.L.Collins, A.P.Zijdenbos, V.Kollokian, J.G.Sled, N.J.Kabani, C.J.Holmes, and A.C. Evans. "Design and Construction of a Realistic Digital Brain Phantom" *IEEE Trans. Med. Imaging* 17(3): 463-468, June 1998.

21. A.Schenk, J.Breitenborn, D.Selle, T.Schindewolf, D.Böhm, W.Spindler, H.Jürgens, and H.-O. Peitgen. "ILabMed-Workstation – Eine Entwicklungsumgebung für radiologische Anwendungen" *Proc. Bildverarbeitung für die Medizin '99*, 238-242, Springer, 1999.

22. W.H. Press, S.A. Teukolsky, W.T. Vetterling, and B.P. Flannery. *Numerical Recipes in C.* Cambridge University Press, Cambridge, 1988, 1995.

A Comparative Statistical Error Analysis of Neuronavigation Systems in a Clinical Setting

Abbasi H. MD PhD, Hariri S. CandMed, Martin D. MD, Kim D. MD,
Adler J. MD, Steinberg G. MD PhD, Shahidi R. PhD

Department of Neurosurgery – Stanford University, 300 Pasteur Dr., R S008,
Stanford CA, 94305
hamid@igl.stanford.edu

Abstract. The use of neuronavigation (NN) in neurosurgery has become ubiquitous. A growing number of neurosurgeons are utilizing NN for a wide variety of purposes, including optimizing the surgical approach (macrosurgery) and locating small areas of interest (microsurgery). The goal of our team is to apply rapid advances in hardware and software technology to the field of NN, challenging and ultimately updating current NN assumptions. To identify possible areas in which new technology may improve the surgical applications of NN, we have assessed the accuracy of neuronavigational measurements in the Radionics™ and BrainLab™ systems. Using a phantom skull, we measured the accuracy of the navigational systems, taking a total of 2616 measurements. We found that, despite current NN tenets, the six marker count does not yield optimal accuracy in either system, and the spreaded marker setting yields best accuracy in both systems. Placing fewer markers around the region of interest (ROI) minimizes registration error, and active tracking does not necessarily increase accuracy. Comparing the two systems, we also found that accuracy of NN machines differs both overall and in different axes. As researchers continue to apply technological advances to the NN field, an increasing number of currently held tenets will be revised, making NN an even more useful neurosurgical tool.

1. Introduction

Neuronavigation (NN) has emerged as an essential tool in neurosurgery. A growing number of neurosurgeons are utilizing NN for a wide variety of purposes. In the early stages of their development, NN machines were very expensive and thus predominantly used only to locate hard-to-find areas of interest (e.g. tumors, aneurysms, abscesses, etc.) in vital regions (e.g. the brainstem, cerebellum, etc). Today, as systems have become less expensive, NN is routinely used to optimize the craniotomy and the neurosurgical approach to the region of interest (ROI).

In macroneurosurgery, neurosurgeons use NN to optimize their approach, minimizing disturbances to surrounding anatomical structures such as the ventricular system. Five mm of accuracy is sufficient for macrosurgery. Microneurosurgery, in contrast, requires much greater accuracy; neurosurgeons must be able to localize extremely small areas of interest in vital regions. All current NN systems have enough accuracy to optimize the approach to the ROI. However, to date, no system has achieved the submillimeter precision truly needed for free-hand stereotaxy and for the location of small areas of interest.

Certain experimental and methodological assumptions have been retained in the use of NN over its development. Current conventional tenets of NN include:

1. Using more markers increases accuracy.
2. Putting more markers around the area of interest increases accuracy.
3. Using 6 markers yields most efficient accuracy.
4. Systems utilizing active tracking [e.g. an active DRF (dynamic reference frame) and probe which emit infrared beams rather than passively reflecting them] have higher accuracy than systems utilizing passive tracking.
5. Minimizing the distance from the marker or marker group to the area of interest increases accuracy.

While rapid advances in hardware and software have emerged in the last few years, there have been only some attempts at challenging the old NN tenets and applying new technology to update these systems. To identify possible areas in which new technology may improve the surgical applications of NN, we conducted accuracy tests of neuronavigational measurements in two currently used systems: Radionics™ and BrainLab™.

The ultimate goal of this project is to give surgeons information about the accuracy of NN machines. Surgeons should be able to estimate the accuracy of images generated by the system to optimize their surgical accuracy. To effectively apply NN to microsurgery, the following questions must be systematically addressed:

1. How many markers efficiently maximize accuracy?
2. What pattern of marker localization maximizes accuracy?
3. Are there significant accuracy differences between various marker arrangements?
4. Are systems significantly different in their level of accuracy?
5. Is accuracy different in the x-, y- and z-axis?

2. Materials and Methods

We built a phantom skull to most realistically simulate the surgical setting. (Had we used a geometric object instead of reproducing the curvature of the skull, our results would not have been applicable to the surgical setting.) We obtained a standard plastic skull, removed the calvaria, and installed 3 Plexiglas square rods of different heights in each of the 3 anatomical fossae (anterior, middle and posterior). We used the edges of these rods as our targets. We installed a Plexiglas ball of known diameter on the phantom's sella turcica (Fig. 1). Replacing the calvaria, we placed a total of 12 markers bilaterally on the exterior of the skull in the following regions: 6 frontal, 2 mastoid, 2 occipital and 2 high parietal (Table 1). We performed a CT of the skull in 1.25 mm slices and sent the data over the network to the two NN machines evaluated in this study. The systems utilize different registration and tracking systems to localize the probe's tip in 3D:

BrainLab VectorVision version 2.3: Passive registration and tracking system: the DRF and probe merely reflect the infrared signal shot out by emitters around the cameras.
Radionics OTS version 2.2: Active registration and tracking system: the DRF and probe contain diodes emitting infrared signals that are collected by two cameras. The DRF and probe are connected to the tracking box that supplies both with electricity.

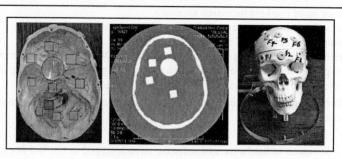

Fig. 1. The Experimental Setup: target points (left); CT of the phantom (middle); the phantom (right).

Table 1: Placement of the markers.

Marker count	Local Setting	Spreaded Setting
4	F1, F3, F4, F6	F1, F3, O1, O6
6	F1, F2, F3, F4, F5, F6,	F1, F3, O1, O3, O4, O6
7 (BrainLab, without O6???) 8 (Radionics)	F1, F2, F3, F4, F5, F6, O1, O6	F1, F3, O1, O2, O3, O4, O5, O6
F1: Frontolateral R below calvaria line	F2: Frontomedian below calvaria line	
F3: Frontolateral L below calvaria line	F4: Frontolateral R above calvaria line	
F5: Frontomedian above calvaria line	F6: Frontolateral L above calvaria line	
O1: Above R mastoid process	O2: R occipital	
O3: High parietal R	O4: High parietal L	
O5: L occipital	O6: Above L mastoid process	

When the probe tip was placed at the edge of a rod, the NN systems visualized the probe's position on their screens in the original axial plane of the CT scans and in the sagital and coronal planes reconstructed from the CT scans (Fig. 2). The exact edge of the rod as displayed on each monitor cross-section was assigned the coordinate values $x=0$ and $y=0$. The x and y coordinates assigned to each cross-sectional view on the monitor were translated as the "real" coordinate axes of the skull (Fig. 2) using Table 2.

In each of the three cross-sections, we measured how far from the actual edge of the rod ($x=0$, $y=0$) the monitor was representing the probe tip. Paging through the images on the monitor to find the largest diameter of the Plexiglas sphere, we used the known diameter of the sphere to establish a system-independent scale for measurements. These measurements were acquired for both the Radionics and BrainLab NN systems. The

Radionics system was analyzed when using local and spreaded marker settings with 4, 6, and 8 markers. The BrainLab system was analyzed when using local and spreaded marker settings with 4, 6, and 7 markers. Table 1 defines the marker locations in these different setups. Thus, we obtained 12 series of measurements, each series consisting of 218 separate measurements.

The absolute distance between the edge of the rod and the NN representation of the probe tip was calculated as SQR $(X^2 + Y^2)$ in each plane. The 3D absolute distance was calculated as SQR $(X^2 + Y^2 + Z^2)$. Because the numbers were squared, the sign of the measurements on each axis was unimportant in these calculations.

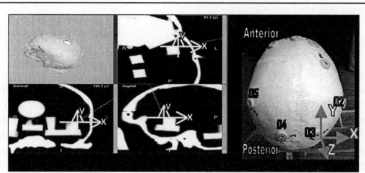

Fig. 2. **Phantom cross-sections as displayed on the NN monitor, dashed lines indicate the probe's approach (left); skull coordinate axes (right).**

Table 2: Relation of the measurements on the monitor to the actual coordinates of the skull as illustrated in *Fig. 2.*

Screen Coordinate	Skull Coordinate	Screen Coordinate	Skull Coordinate
Axial X	+X	Axial Y	+Y
Sagital X	-Y	Sagital Y	+Z
Coronal X	+X	Coronal Y	+Z

3. Results

Radionics

In the local marker setting, the Radionics system was highly sensitive to the number of markers. Specifically, the range of error increased in the z-axis when fewer markers were used. Also, increasing the distance of the probe tip from the marker groups augmented z-axis error. We found that error in reconstructed planes (sagital and coronal) increased rapidly with decreasing marker numbers while error in the original plane (axial) was not very sensitive to the number of markers. In both the spreaded and local marker setting, as expected, error declined as the experimenter went from 4 to 6 to 8 markers. The greatest improvement in accuracy is obtained when increasing from 6 to 8 markers. However, this improvement in accuracy given more markers was not as great as expected.

The findings for the spreaded marker setting were thus similar to the local marker setting, with a few exceptions. In the spreaded marker setting, there were fewer differences in accuracy between the three anatomical fossae. In the local marker setting, placing more markers around the region of interest (ROI) maximizes accuracy. However, in the spreaded marker setting, placing fewer markers around the ROI maximizes accuracy (i.e. localization of the probe tip in the occipital area was most accurate, even though the markers were clustered more frontally).

BrainLab:

In the local marker setting, increasing the distance of the probe tip from the marker groups augmented z-axis error. Increasing the marker count from 4 to 6 decreased accuracy, particularly in the z-axis. The error increased in both the original (axial) and the reconstructed (sagital and coronal) planes, but error in the reconstructed images was greater. Perhaps the algorithm in the procedure registering the various markers can account for this discrepancy. In the local setting, the z-axis contributed most to overall error, however, in the spreaded marker setting, no single axis predominantly contributed to overall error.

In contrast to the Radionics system, where we could improve accuracy with each increase in marker count, in the BrainLab local marker setting, 4 or 7, but not 6 markers, maximize accuracy. Also, in the BrainLab spreaded marker setting, accuracy declines with additional markers. In both systems, the SD and variance of 4 versus 6 markers was not significantly different; this data discounts bias in data collection. In both the BrainLab local and spreaded marker settings, placing more markers around the ROI only somewhat maximizes accuracy.

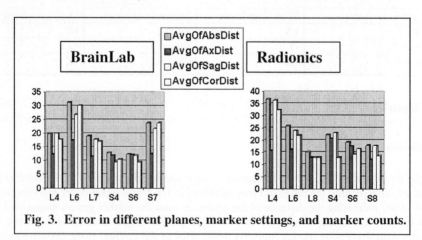

Fig. 3. Error in different planes, marker settings, and marker counts.

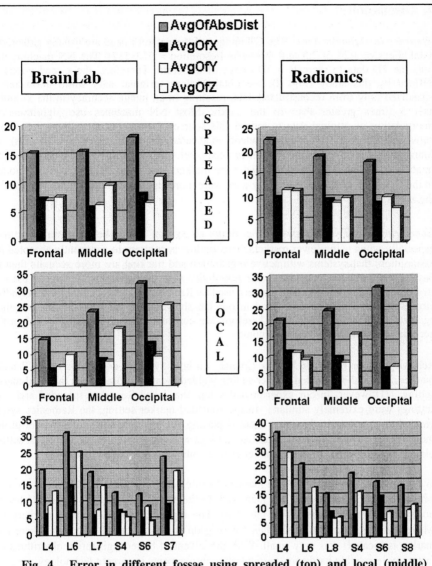

Fig. 4. Error in different fossae using spreaded (top) and local (middle) marker settings. Error in different axes, marker settings, and marker counts (bottom).

4. Discussion

Accuracy in Different Axes. The CT images of the patient's head are usually gathered in axial slices of 1.25 or 2.5 mm thickness (resolution = 512x512). The NN machine then uses the 2D data to recreate a 3D image of the brain. Therefore, the x- and y-axis, as defined by this study (Fig. 2), are obtained from original data while the z-axis is visualized only from reconstructed data, resulting in an image accuracy in the x- and y-axis 5 times greater than in the z-axis. Most NN machines use algorithms for interpolation of original data to reduce this source of error, but there is a compromise between errors in different axes (tradeoff in accuracy between x-/y-axes and z axis). We found poorer accuracy in the Radionics z-axis, but, as expected, we observed an increase in accuracy in this axis when we used more markers. However, in contrast to Radionics, in the BrainLab system, the x- and y-axes were more significant contributors to error than the z-axis.

Comparing the Two Systems. The BrainLab system has a better overall accuracy, especially when using fewer markers. This finding is contrary to a currently accepted NN assumption: that systems with active registration and tracking are more accurate than the passive models. We considered the possibility that the newer NDI camera used by BrainLab (versus the older IGT camera used by Radionics) could account for this finding. However, another study conducted in our lab showed that the IGT camera was more accurate than the NDI camera. Therefore, we concluded that the registration algorithm accounted for BrainLab's better accuracy.

Accuracy Using Different Marker Counts. In both systems, except for in the 7 and 8 marker counts, the spreaded marker setting yielded greater accuracy than the local marker setting. (However, in our experimental setup, the 7 and 8 marker spreaded and local settings were extremely similar.) In the spreaded marker setting, the Radionics system yielded unique data in this experiment -- placing the tip further from the marker groups increased accuracy. In both BrainLab settings and the Radionics local marker setting, placing the tip further from the marker groups somewhat increased error.

We believe that the quality of the markers themselves account for the decreased accuracy when the number of markers are increased. In the early stages of NN, surgeons attached bone screw markers directly to the skull. The registration area in these markers (i.e. where the probe tip is placed when the surgeon is telling the NN machine where each marker is located) was very small. With the advent of better registration algorithms and faster, more reliable hardware, particularly improvements in camera technology, surgeons found that they could obtain sufficient accuracy with less invasive skin markers. The skin marker, however, has a larger registration area (~ 4 mm) and moves in relation to the skull because it is only attached to the skin (Fig. 6). Using the improved registration algorithm, we obtained a visualization of the skull's geometry within ~2-5 mm. While this degree of accuracy is suitable for macroneurosurgery, more delicate microneurosurgery procedures demand much greater accuracy.

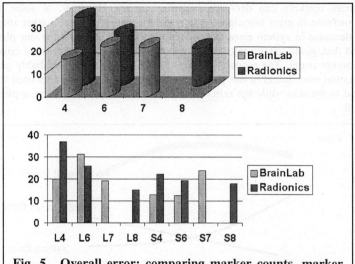

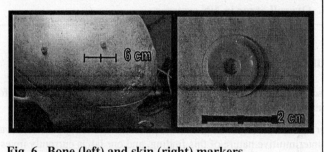

Fig. 5. Overall error: comparing marker counts, marker settings, and NN systems.

Fig. 6. Bone (left) and skin (right) markers

To isolate sources of error, we found that:

$$\text{Overall Error} = \text{Registration Error} + \text{System Error}$$

Registration error is the error generated during the process of telling the NN system where each marker is located. The surgeon attempts to place the probe's tip as close to the center of the marker as possible, registering each marker in the system. However, the diameter of the skin marker's center is relatively large, resulting in part of this registration error as finding the exact center of each marker is virtually impossible. Registration error probably accounts for this experiment's finding that placing fewer markers near the area of interest can maximize accuracy. System error is defined as mechanical, engineering, and software errors, such as machine or camera error.

Having more markers can decrease system accuracy. However, at some point, the marginal increase in error associated with registering an additional marker surpasses the marginal decrease in system error due to more markers (Fig. 7). Using our phantom, we discovered that, given today's registration and system technology, maximal error is found at the 6 marker count. The registration error found in surgeries is probably greater than the registration error found in our phantom model because skin markers used for patients are applied to the skin while the skin markers used for the phantom were applied directly to the skull.

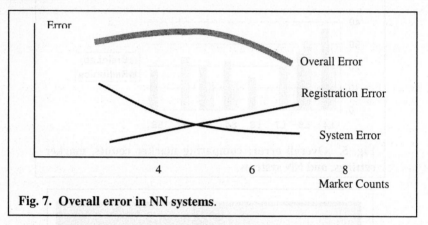

Fig. 7. Overall error in NN systems.

5. Conclusion

This study scrutinized conventional assumptions of neuronavigation and found many of them to be invalid. The new NN tenets supported by our findings are:

1. 4 or 8, but not 6, markers yield most efficient accuracy. We are aware of the counterintuitive nature of this finding, and our lab is currently investigating this result. Additionally, the movement of skin on the skull is not included in this study and may influence the clinical results.
2. Placing fewer markers around the region of interest (ROI) decreases registration error at the ROI.
3. Active tracking does not necessarily increase accuracy.
4. The spreaded marker setting increases accuracy.
5. Accuracy of the NN machines differs both overall and in different axes.

As researchers continue to apply recent developments in hardware and software technology to the NN field, an increasing number of currently held tenets will be challenged and revised, rapidly and dramatically changing the field.

6. Acknowledgements

We thank Bonnie Hale for her invaluable editorship. Sanaz Hariri thanks the Stanford Medical Kolos Scholar Program and the Paul and Daisy Soros Fellowship Program.

7. References

Alberti, O; Dorward, NL; Kitchen, ND; Thomas, DG. "NN--impact on operating time." *Stereotactic and Functional Neurosurgery.* 1997, 68(1-4 Pt 1): 44-8.

Dorward, NL. "NN --the surgeon's sextant [editorial]" *British Journal of Neurosurgery.* Apr 1997, 11(2): 101-3

Germano, IM; Villalobos, H; Silvers, A; Post, KD. "Clinical use of the optical digitizer for intracranial NN." *Neurosurgery.* Aug 1999, 45(2): 261-9, disc. 269-70.

Gumprecht, HK; Widenka, DC; Lumenta, CB. "BrainLab VectorVision NN System: technology and clinical experiences in 131 cases." *Neurosurgery.* Jan 1999, 44(1): 97-104, disc. 104-5.

Kaus, M; Steinmeier, R; Sporer, T; Ganslandt, O; Fahlbusch, R. "Technical accuracy of a NN system measured with a high-precision mechanical micromanipulator." *Neurosurgery.* Dec 1997. 41(6): 1431-6, disc. 1436-7.

Khadem, R; Yeh, CC; Sadeghi-Tehrani, M; Bax, MR; Johnson, JA; Welch, N; Wilkinson, EP; Shahidi, R. "Comparative tracking error analysis of five different optical tracking systems." *Computer Aided Surgery.* 2000, 5(2): 98-107.

Nabavi, A; Manthei, G; Blömer, U; Kumpf, L; Klinge, H; Mehdorn, HM. "Neuronavigation. Computergestütztes Operieren in der Neurochirurgie. [NN. Computer-assisted surgery in neurosurgery]." *Radiologe.* Sept 1995, 35(9): 573-7.

Ostertag, CB; Warnke, PC. "Neuronavigation. Computerassistierte Neurochirurgie. [NN. Computer-assisted neurosurgery]." *Nervenarzt.* June 1999, 70(6): 517-21.

Schönherr, B; Gräwe, A; Meier, U. "Qualitätssichernde Massnahmen bei neurochirurgischen Operationen. Erfahrungen mit intraoperativer Computertomographie und NN. [Quality securing procedures in neurosurgical operations. Experiences with intraoperative CT and NN]" *Zeitschrift für Ärztliche Fortbildung und Qualitatssicherung.* June 1999, 93(4): 273-80.

Schmieder, K; Hardenack, M; Harders, A. "NN in daily clinical routine of a neurosurgical department." *Computer Aided Surgery.* 1998, 3(4): 159-61.

Spetzger, U; Laborde, G; Gilsbach, JM. "Frameless NN in modern neurosurgery." *Minimally Invasive Neurosurgery.* Dec 1995, 38(4): 163-6.

Wirtz, CR; Knauth, M; Hassfeld, S; Tronnier, VM; Albert, FK; Bonsanto, MM; Kunze, S. "NN-- first experiences with three different commercially available systems." *Zentralblatt für Neurochirurgie.* 1998, 59(1): 14-22.

Wirtz, CR; Tronnier, VM; Bonsanto, MM; Hassfeld, S; Knauth, M; Kunze, S. "NN. Methoden und Ausblick. [NN. Methods and prospects]." *Nervenarzt.* Dec 1998, 69(12): 1029-36.

Detection and Quantification of Line and Sheet Structures in 3-D Images

Yoshinobu Sato   Shinichi Tamura

Division of Functional Diagnostic Imaging
Osaka University Graduate School of Medicine
Room D11, 2-2 Yamada-Oka, Suita, Osaka, 565-0871, Japan
yoshi@image.med.osaka-u.ac.jp, http://www.image.med.osaka-u.ac.jp/~yoshi

Abstract. This paper describes a method for the accurate segmentation of line and sheet structures in 3-D images with the aim of quantitatively evaluating anatomical structures in medical images with line-like shapes such as blood vessels and sheet-like shapes such as articular cartilages. Explicit 3-D line and sheet models are utilized. The line model is characterized by medial axes associated with variable cross-sections, and the sheet model by medial surfaces with spatially variable widths. The method unifies segmentation, model recovery, and quantification to obtain 3-D line and sheet models by fully utilizing formal analyses of 3-D local intensity structures. The local shapes of these structures are recovered and quantitated with subvoxel resolution using spatially variable directional derivatives based on moving frames determined by the extracted medial axes and surfaces. The medial axis detection performance and accuracy limits of the quantification are evaluated using synthesized images. The clinical utility of the method is demonstrated through experiments in bronchial airway diameter estimation from 3-D computed tomography (CT) and cartilage thickness determination from magnetic resonance (MR) images.

1   Introduction

With high-resolution 3-D imaging modalities becoming commonly available in medical imaging, a strong need has arisen for a means of accurately quantifying in 3-D anatomical structures of interest from acquired 3-D images. In the 2-D domain, the quantification of vessel diameters from 2-D X-ray projective images has been intensively studied [1],[2] in both the diagnostic imaging and image analysis research communities. As a result, image analysis for 2-D vessel diameter estimation is now recognized as an essential tool for accurate and objective diagnosis. Nevertheless, serious limitations to 2-D quantification as projections of 3-D objects have been pointed out [3]. The goal of this paper is to develop a well-formulated method for such quantification in the 3-D domain.

From the viewpoint of image analysis, the quantification of vessel diameters in 2-D is regarded as the recovery of segmented line descriptions, which typically consist of the medial axes of lines and the width associated with each point on these axes [4]. The 2-D line is naturally extended to two types of 3-D features:

line and sheet [5]. The 3-D line model consists of the *medial axes* of lines [6] and the *cross-sectional shape* associated with each point on these axes, while the 3-D sheet model consists of the *medial surfaces* of sheets and the *width* associated with each point on these surfaces. Examples of line-like structures are ductal structures such as blood vessels and bronchial airways, while structures such as cortices and articular cartilages are regarded as sheet-like.

In this paper, we address the problem of detecting and quantitating line and sheet structures from 3-D images. There has been a considerable amount of work on the detection and description of these structures, which can be categorized into two different approaches, although in general less attention has been paid to sheet-like structures. One approach is to use line and sheet enhancement filtering [5],[7],[8],[9]. Multiscale analysis of 3-D local intensity structures using the eigenvalues of the Hessian matrix has been shown to be effective in discriminating them from other structures as well as in the detection of small structures. However, consideration of 3-D quantification remains insufficient. The other approach is to employ generalized cylinder (GC) recovery [10],[11]. Recovered GCs have rich quantitative parameters, and GC recovery is often applied to successfully extracted contours or surfaces. Nevertheless, the recovery of GCs with arbitrary axes and cross-section is a difficult task, which means that severe restrictions are often imposed on their shapes or user interaction is needed. Unlike these previous approaches, our method unifies segmentation, model recovery, and quantification. While motivated by the GC concept, the method fully utilizes 3-D local intensity structure analysis to obtain explicit 3-D line and sheet models.

The organization of this paper is as follows. In Section 2, we formulate the method for the recovery of line and sheet models, and describe each procedure in detail. In Section 3, we evaluate the basic characteristics and accuracy of the method using synthesized 3-D images, and describe experiments using clinical 3-D data acquired by computed tomography (CT) and magnetic resonance imaging (MRI). Finally, in Section 4 we summarize our conclusions and indicate the directions of future work.

2 Recovery of 3-D Line and Sheet Models

The method is based on a second-order analysis of 3-D local intensity structures. The following is an overview of the method.

Step 1: Existing filtering techniques for line and sheet enhancement are used to extract the initial regions, which should include all potential medial axes and surfaces [7],[9]. These are then used as initial values for the subsequent subvoxel edge localization. The candidate regions, which should include all potential line and sheet regions, are also extracted.

Step 2: The medial axes and surfaces are extracted using local second-order approximation given by the gradient vector and Hessian matrix. The eigenvectors of the Hessian matrix define the moving frames on medial axes/surfaces.

After this, the moving frames are embedded in a 3-D image such that each point within the candidate regions is directly related to its corresponding moving frame.

Step 3: Subvoxel edge localization of the region boundaries is carried out using adaptive 3-D directional derivatives, whose directions are adaptively changed depending on the moving frame, to accomplish accurate segmentation, model recovery, and quantification.

In the following, we begin with a description of Step 2 of the method since Step 1 is described in detail in [7] and [9].

2.1 Detection and Subvoxel Localization of Medial Axes and Surfaces

Let $I(\mathbf{x})$ be a 3-D image in which $\mathbf{x} = (x, y, z)^{\top}$, and $\nabla^2 I(\mathbf{x}; \sigma)$ be the Hessian matrix of the image blurred by the isotropic Gaussian function with a standard deviation σ, which is given by

$$\nabla^2 I(\mathbf{x}; \sigma) = \begin{bmatrix} I_{xx}(\mathbf{x}; \sigma) & I_{xy}(\mathbf{x}; \sigma) & I_{xz}(\mathbf{x}; \sigma) \\ I_{yx}(\mathbf{x}; \sigma) & I_{yy}(\mathbf{x}; \sigma) & I_{yz}(\mathbf{x}; \sigma) \\ I_{zx}(\mathbf{x}; \sigma) & I_{zy}(\mathbf{x}; \sigma) & I_{zz}(\mathbf{x}; \sigma) \end{bmatrix}, \quad (1)$$

where partial second derivatives of the Gaussian blurred image $I(\mathbf{x}; \sigma)$ are represented by expressions like $I_{xx}(\mathbf{x}; \sigma) = \frac{\partial^2}{\partial x^2} I(\mathbf{x}; \sigma)$, $I_{yz}(\mathbf{x}; \sigma) = \frac{\partial^2}{\partial y \partial z} I(\mathbf{x}; \sigma)$, and so on.

The second-order approximation of $I(\mathbf{x}; \sigma)$ around $\mathbf{x}_0$ is given by

$$I_{\mathrm{II}}(\mathbf{x}; \sigma) = I(\mathbf{x}_0; \sigma) + (\mathbf{x} - \mathbf{x}_0)^{\top} \nabla I_0 + \frac{1}{2}(\mathbf{x} - \mathbf{x}_0)^{\top} \nabla^2 I_0 (\mathbf{x} - \mathbf{x}_0), \quad (2)$$

where ∇I_0 and $\nabla^2 I_0$ denote the gradient vector and the Hessian matrix combined with the Gaussian blurrring σ at $\mathbf{x}_0$, respectively. Let the eigenvalues of $\nabla^2 I$ be λ_1, λ_2, λ_3 ($\lambda_1 \geq \lambda_2 \geq \lambda_3$) and their corresponding eigenvectors be $\mathbf{e}_1$, $\mathbf{e}_2$, $\mathbf{e}_3$ ($|\mathbf{e}_1| = |\mathbf{e}_2| = |\mathbf{e}_3| = 1$), respectively. Three types of local structures — line, sheet, and blob — can be classified using these eigenvalues [9]. For the ideal line, $\mathbf{e}_1$ is expected to give its tangential direction and both $|\lambda_2|$ and $|\lambda_3|$, directional second derivatives orthogonal to $\mathbf{e}_1$, should be large on its medial axis, while $\mathbf{e}_3$ is expected to give the orthogonal direction of a sheet and only $|\lambda_3|$ should be large on its medial surface (Fig. 1). Here, structures of interest are assumed to be brighter than surrounding regions.

The initial regions obtained in Step 1 are searched for medial axes and surfaces, which are detected based on the second-order approximation of $I(\mathbf{x}; \sigma_f)$. The medial axis and surface extraction is based on a formal analysis of the second-order 3-D local intensity structure. Here, σ_f is the filter scale used in medial axis/surface detection, and we assume that the width range of structures of interest is around the width at which the filter with σ_f gives the peak response (see [7] and [9] for detailed discussions).

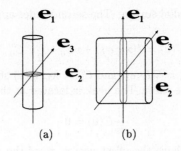

(a)   (b)

Fig. 1. Line and sheet models with the eigenvectors of the Hessian matrix. (a) Line. (b) Sheet.

Line case – medial axis detection – We assume that the tangential direction is given by e_1 at the voxel around the medial axis. The 2-D intensity function, $C(\mathbf{u})$ $(\mathbf{u} = (u,v)^\top)$, on the cross-sectional plane of $I(\mathbf{x};\sigma_f)$ orthogonal to e_1, should have its peak on the medial axis. The second-order approximation of $C(\mathbf{u})$ is given by

$$C(\mathbf{u}) = I(\mathbf{x}_0;\sigma_f) + \mathbf{u}^\top \nabla C_0 + \frac{1}{2}\mathbf{u}^\top \nabla^2 C_0 \mathbf{u}, \qquad (3)$$

where $u e_2 + v e_3 = \mathbf{x} - \mathbf{x}_0$, $\nabla C_0 = (\nabla I \cdot e_2, \nabla I \cdot e_3)^\top$ (∇I is the gradient vector, that is, $\nabla I(\mathbf{x}_0;\sigma_f)$), and

$$\nabla^2 C_0 = \begin{bmatrix} \lambda_2 & 0 \\ 0 & \lambda_3 \end{bmatrix}. \qquad (4)$$

$C(\mathbf{u})$ should have its peak on the medial axis of the line. The peak is located at the position satisfying

$$\frac{\partial}{\partial u}C(\mathbf{u}) = 0 \quad \text{and} \quad \frac{\partial}{\partial v}C(\mathbf{u}) = 0. \qquad (5)$$

By solving Eq. (5), we have the offset vector, $\mathbf{p} = (p_x, p_y, p_z)^\top$, of the peak position from $\mathbf{x}_0$ given by

$$\mathbf{p} = s e_2 + t e_3, \qquad (6)$$

where $s = -\frac{\nabla I \cdot e_2}{\lambda_2}$ and $t = -\frac{\nabla I \cdot e_3}{\lambda_3}$. For the medial axis to exist at the voxel $\mathbf{x}_0$, the peak of $C(\mathbf{u})$ needs to be located in the territory of voxel $\mathbf{x}_0$. Thus, the medial axis is detected only if $|p_x| \leq \frac{1}{2}$ & $|p_y| \leq \frac{1}{2}$ & $|p_z| \leq \frac{1}{2}$. By combining the voxel position $\mathbf{x}_0$ and offset vector $\mathbf{p}$, the medial axis is localized at subvoxel resolution.

Sheet case – medial surface detection – We assume that the direction of the surface normal is given by e_1 at the voxel around the medial surface. The 1-D intensity function, $C(v)$, which is the profile of $I(\mathbf{x};\sigma_f)$ along to e_3, should have its peak on the medial surface. The second-order approximation of $C(v)$ is given by

$$C(v) = I(\mathbf{x}_0; \sigma_f) + vC_0' + \frac{1}{2}v^2 C_0'', \tag{7}$$

where $v\mathbf{e}_3 = \mathbf{x} - \mathbf{x}_0$, $C_0' = \nabla I \cdot \mathbf{e}_3$, and $C_0'' = \lambda_3$. $C(v)$ should have its peak on the medial surface of the sheet. The peak is located at the position satisfying

$$\frac{d}{dv}C(v) = 0. \tag{8}$$

By solving Eq. (8), we have the offset vector, $\mathbf{p}$, of the peak position from $\mathbf{x}_0$ given by

$$\mathbf{p} = t\mathbf{e}_3, \tag{9}$$

where $t = -\frac{\nabla I \cdot \mathbf{e}_3}{\lambda_3}$. The medial surface is detected only if $|p_x| \leq \frac{1}{2}$ & $|p_y| \leq \frac{1}{2}$ & $|p_z| \leq \frac{1}{2}$.

Embedding moving frames The moving frame is defined by the voxel position $\mathbf{x}_0$, the offset vectors $\mathbf{p}$, and the eigenvectors $\mathbf{e}_1$, $\mathbf{e}_2$, $\mathbf{e}_3$ at each detected point of a medial axis or surface. In order to perform the subsequent processes based on moving frames, each voxel within the candidate regions obtained in Step 1 needs to be related to the moving frame. First, we find the correspondences between each voxel and one of the detected points of a medial axis or surface. Once these correspondences are found, each voxel is directly related to its corresponding moving frame. To find the correspondences, we use the Voronoi tessellation of the detected points. The territory of the detected point in the Voronoi tessellation can be regarded as the set of voxels to which each discrete moving frame is applied. This process is identical in both the line and sheet cases.

2.2 Subvoxel Edge Localization and Width Quantification

An adaptive directional second derivative is applied at each voxel based on its corresponding moving frame. The directional derivative is taken along the perpendicular from the voxel to the medial axis or surface. The zero-crossing points of the directional second derivatives are localized at subvoxel resolution to determine the precise region boundaries and quantitate the widths.

Line case At every voxel within the candidate regions, the directional second derivative is calculated depending on its corresponding moving frame. The directional derivative is written as

$$D_{line}(\mathbf{x}; \sigma_e) = \mathbf{r}(\mathbf{x})^\top \nabla^2 I(\mathbf{x}; \sigma_e) \mathbf{r}(\mathbf{x}), \tag{10}$$

where $\mathbf{r}(\mathbf{x})$ is the unit vector whose direction is parallel to the perpendicular from the voxel position $\mathbf{x}$ to the straight line defined by the origin and the medial axis direction of the moving frame. The foot of the perpendicular can be regarded as the corresponding axis position. The origin is given by the voxel position of the medial axis point and the offset vector $\mathbf{p}$. σ_e is the filter scale used in the edge localization; it is desirable that σ_e be small compared with the line width for accurate edge localization.

After the adaptive derivatives have been calculated at all the voxels, subvoxel edge localization is carried out at every voxel in the candidate regions. Let $\mathbf{o}_a$ be the foot of the perpendicular on the axis. Let $\mathbf{r}_a$ be the direction from $\mathbf{o}_a$ to the voxel position $\mathbf{x}_a$. For each voxel, we reconstruct the profiles originating from $\mathbf{o}_a$ in the directions $\mathbf{r}_a$ and $-\mathbf{r}_a$ for $D_{line}(\mathbf{x}; \sigma_e)$ and the initial regions (which we specify as $B_{line}(\mathbf{x})$) obtained in Step 1. The edges are then localized in both directions and the width is calculated as the distance between the two edge locations. The profile is reconstructed at subvoxel resolution by using a trilinear interpolation for $D_{line}(\mathbf{x}; \sigma_e)$ and a nearest-neighbor interpolation for $B_{line}(\mathbf{x})$.

Let $D_+(r)$ and $D_-(r)$ be the profiles of $D_{line}(\mathbf{x}; \sigma_e)$ along the two opposite directions $\mathbf{r}_a$ and $-\mathbf{r}_a$ from $\mathbf{o}_a$. Let $B_+(r)$ and $B_-(r)$ be the profiles of $B_{line}(\mathbf{x})$. Here, r denotes the distance from the foot of the perpendicular on the axis. The localization of edges consists of two steps; finding the initial point for the subsequent search using $B_+(r)$, and then searching for the zero-crossing of $D_+(r)$. The initial point, p_0, is given by r of the first encountered point satisfying $B_+(r) = 0$, starting the search from $r = 0$, that is, the axis point. Given the initial point of the search, if $D_+(p_0) < 0$, search outbound from the axis point along the profile for the zero-crossing position r_+; otherwise, search inbound. After r_- has been similarly estimated, the width (diameter) is given by $|r_+ - r_-|$.

Sheet case At every voxel within the candidate regions, the directional second derivative is taken orthogonal to both $\mathbf{e}_1$ and $\mathbf{e}_2$, that is, along $\mathbf{e}_3$, in its corresponding moving frame. The directional derivative is written as

$$D_{sheet}(\mathbf{x}; \sigma_e) = \mathbf{s}(\mathbf{x})^\top \nabla^2 I(\mathbf{x}; \sigma_e) \mathbf{s}(\mathbf{x}), \tag{11}$$

where $\mathbf{s}(\mathbf{x})$ is the unit vector whose direction is parallel to the medial surface normal of the moving frame. Using a method analogous to that employed in the line case, the profiles of $D_{sheet}(\mathbf{x}; \sigma_e)$ and $B_{sheet}(\mathbf{x})$ are reconstructed for the directions $\mathbf{s}$ and $-\mathbf{s}$. These profiles are then used to determine the edge locations in the two directions and the width (thickness).

3 Experimental Results

3.1 Synthesized Images

We evaluated the medial axis detection performance and width estimation accuracy using synthesized 3-D images of lines and sheets with pill-box cross-sections and bar profiles, respectively. A simulated partial volume effect was incorporated when synthesizing the images. We focused on the effects of the filter scales σ_f

and σ_e, which are used in medial axis/surface detection and width quantification, respectively, on the detection and quantification of various widths of line and sheet structures.

Medial axis detection Synthesized 3-D images of a line with a circular axis were generated. The diameter of the line, D, was varied between 2.0 and $8\sqrt{2}(\simeq 11.3)$ pixels. The radius of the circular axis was proportional to D (we used $4 \times D$). Gaussian noise with 25 % standard deviation in the intensity height of pill-box cross-sections was added to the images. After line and sheet enhancement filtering with integration of scales appropriate for the line diameter [7],[9], the candidate regions were extracted by thresholding and extracting large connective components. The medial axis points were detected within these regions using the procedures described in Section 2.1 with two values for the filer scale σ_f, $\sqrt{2}$ ($\simeq 1.4$) and 4.0 pixels. The same candidate regions were used for both values of σ_f.

Figure 2 shows the volume rendering of the synthesized 3-D images and typical axis detection results. The detection was successful using appropriate combinations of line diameter D and filter scale σ_f (Fig. 2(b)). Many axis points are overlooked when the filter scale is larger than appropriate, while a number of false detections are made when the filter scale is smaller than appropriate (Fig. 2(c)).

Figure 3 shows the performance evaluation results. Detected axis points were evaluated by comparing them with analytically determined axis points. We regarded a detected point as a true detection if the distance between its position and one of the analytically determined points was within two pixels; otherwise, detected points were regarded as false. The false and true positive detection ratios are shown in Figs. 3(a) and (b); the plots verify the observations in Fig. 2. The positions and directions of the detected axis points regarded as true detections were compared with analytically determined ones (Figs. 3(c) and (d)). These graphs clarify the effect of σ_f on accurate and reliable axis detection.

Quantification Synthesized images of a line with a straight axis were generated with and without the same Gaussian noise as that used in the previous experiment. The line diameter D was again varied between 2.0 and $8\sqrt{2}(\simeq 11.3)$ pixels. Medial axis detection was carried out with the filter $\sigma_f = 2D$. The diameter estimation procedures described in Section 2.2 were applied using two values for the filter scale σ_e, $\sqrt{2}$ and 4.0 pixels. The results are shown in Fig 4 (a). Note that the diameter was underestimated when $\sigma_e < \frac{D}{\sqrt{2}}$, which is unlike the case of 2-D line diameter estimation [4], while it was overestimated when $\sigma_e > \frac{D}{2}$ and the error was quite small when $\sigma_e \ll \frac{D}{2}$, both of which are similar to the 2-D line case.

Synthesized images of a sheet with a spherical shape were used with and without Gaussian noise to determine the relationship between the estimated and true thickness. A similar relationship to that in the 2-D line case was observed (Fig 4 (b)).

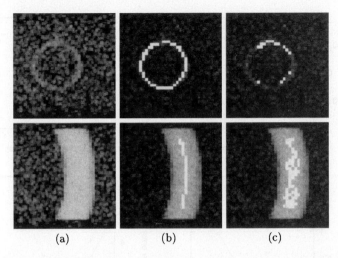

(a)   (b)   (c)

Fig. 2. Medial axis detection from synthesized 3-D images. All units are pixels. (a) Volume-rendered images of original synthesized 3-D images with Gaussian noise. Upper: $D = 2.0$. Lower: $D = 11.3$. (b) Examples of successful axis detection where the filter scale σ_f is appropriate for the line diameter D. The detected axis points are shown as bright points. Upper: $D = 2.0$, $\sigma_f = 1.4$. Lower: $D = 11.3$, $\sigma_f = 4.0$. (c) Examples of undesirable axis detection. Upper: $D = 2.0$, $\sigma_f = 4.0$. When diameter D is smaller than that appropriate for the filter scale σ_f, many true axis points are overlooked. Lower: $D = 11.3$, $\sigma_f = 1.4$. When D is larger than that appropriate for σ_f, many false axis points are detected.

3.2 Clinical Images

Diameter estimation of bronchial airways from CT images In this experiment, we used chest CT images taken by a helical CT scanner. The original voxel dimensions were $0.29 \times 0.29 \times 1.0$ (mm^3). In order to make the voxel isotropic, sinc interpolation was applied along the z-direction. The volume size used in the experiment was $90 \times 70 \times 80$ (voxels) after interpolation.

Fig. 5(a) shows the original CT images. After the initial region extraction by thresholding the line filtered images, the medial axis was detected using $\sigma_f = \sqrt{2}$, 2, and $2\sqrt{2}$ pixels. Fig. 5(b) shows the results of axis detection at the three different scales. Note that the axis points of thin structures were detected only at the smaller two scales while those of large structures (the right segment) were stably extracted at the larger two scales. Fig. 5(c) shows the results of diameter estimation using $\sigma_e = 1.2$ pixels based on the medial axes at these three scales.

Thickness estimation of hip joint cartilages from MR images In this experiment, we used MR images of a hip joint with cartilage enhancement imaging [12]. The original voxel dimensions were $0.62 \times 0.62 \times 1.5$ (mm^3). Sinc interpolation was applied along the z-direction to make the voxel isotropic, and then

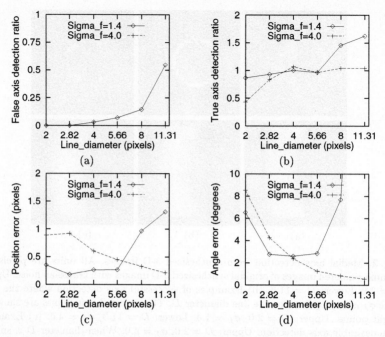

Fig. 3. Performance evaluation of medial axis detection. See text for the definitions of true and false detections. (a) False positive detection ratio, which is the ratio of the number of false detections to all the detections. The ratio was zero for all the line diameters at $\sigma_f = 4.0$ pixels. (b) True positive detection ratio, which is the ratio of the number of true detections to all the analytically determined points. (c) Average position error of axis points regarded as true detections. The distance between detected points and analytically determined points was used as the error. (d) Average angle error of the directions of axis points regarded as true detections.

further applied along all the three directions to make the resolution double. The resultant sampling pitch was 0.31 (mm) in all the three directions. The volume size used in the experiment was $256 \times 256 \times 100$ (voxels) after interpolation.

As shown in Fig. 6(a), cartilages are thin structures; thickness distributions are considered to be particularly important in the diagnosis of joint diseases. The initial cartilage regions were extracted from the enhanced images by the sheet filter. The medial surfaces were extracted using $\sigma_f = 1.4$ pixels. Figs. 6(b) and (c) show the results of thickness distribution estimated using $\sigma_e = 1.2$ pixels. We also obtained the thickness distributions using $\sigma_e = 1.0$ pixel for comparison purposes. The average thickness estimated using $\sigma_e = 1.2$ pixels was $T_f = 4.24$ pixels and $T_a = 3.50$ pixels for the femoral and acetabular cartilages, respectively, compared with $T_f = 4.16$ pixels and $T_a = 3.39$ pixels using $\sigma_e = 1.0$ pixel.

In previous work [13], hip joint cartilages were assumed to be distributed on a sphere approximating the femoral head. The user needs to specify the center of

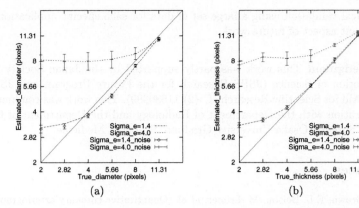

Fig. 4. Accuracy of quantification. The results for the images with 25 % noise are shown by dotted lines. Error bars indicate the standard deviation in the estimated diameters. (a) Diameter estimation of line with a straight axis. (b) Thickness estimation of sheet with a spherical shape.

the sphere, and the cartilage thickness is then estimated along radial directions from the specified center. The method proposed here does not use the sphere assumption, and thus can potentially be applied to badly deformed hip joints as well as to articular cartilages of other joints.

4 Conclusions

The novel method proposed in this paper for detecting and quantitating 3-D line and sheet structures has the following advantages: (1) Well-defined medial axes and surfaces are automatically detected and localized with subvoxel accuracy. (2) The direction along which the width is measured can be automatically estimated. (3) Subvoxel accuracy in width quantification is attainable. These advantages are obtained because the method was formulated based on a thorough analysis of second-order 3-D local intensity structures as a continuous 3-D function. The accuracy limits and the effects of the filter scales on the detection and quantification accuracy were evaluated using synthesized images with various line and sheet structure widths. Finally, the experiments using 3-D CT and MR images demonstrated the applicability of the method to clinical problems.

Future work will address the following issues. Firstly, the behavior of the method at bifurcation and tangency should be examined. Secondly, a method for multiscale integration needs to be developed to combine medial axes and surfaces detected at different scales. Thirdly, the proposed method assumes the isotropy of voxels in 3-D images. However, medical 3-D data generally have lower resolution along the third direction (the direction orthogonal to the slice plane) than within slices, which means that the voxels are essentially anisotropic. Hence, the effects of anisotropic voxels on accuracy requires investigation. Finally, clinical validation using a large set of data for each specific application is an important aspect of future work.

Acknowledgment This work was partly supported by the Japan Society for the Promotion of Science (JSPS Research for the Future Program and JSPS Grant-in-Aid for Scientific Research (C)(2) 11680389). This work was conducted in collaboration with the Department of Radiology and the Department of Orthopaedic Surgery, Osaka University Graduate School of Medicine.

References

1. B.G. Brown, E.L. Bolson, M. Frimer, et al.: Quantitative coronary arteriography, Circulation, **55**, 329-337, (1977).
2. J.H.C. Reiber, C.J. Kooijman, C.J. Slager, et al.: Coronary artery dimensions from cineangiograms — methodology and validation of a computer-assisted analysis procedure, IEEE Transactions on Medical Imaging, **3**(3), 131-141 (1984).
3. Y. Sato, T. Araki, M. Hanayama, et al.: A viewpoint determination system for stenosis diagnosis and quantification in coronary angiographic image acquisition, IEEE Transactions on Medical Imaging, **17**(1), 121–137 (1998).
4. C. Steger: An unbiased detector of curvilinear structures: IEEE Transactions on Pattern Analysis and Machine Intelligence, **20**(2), 113-125 (1998).
5. T.M. Koller, G. Gerig, G. Szekely, et al.: Multiscale detection of curvilinear structures in 2-D and 3-D image data, Proc. Fifth International Conference on Computer Vision (ICCV'95), Boston, MA, 864-869 (1995).
6. B.S. Morse, S.M. Pizer, and A. Liu: Multiscale medical analysis of medical images: Image and Vision Computing, **12**(6): 327–338 (1994).
7. Y. Sato, S. Nakajima, N. Shiraga, et al.: Three dimensional multi-scale line filter for segmentation and visualization of curvilinear structures in medical images, Medical Image Analysis, **2**(2), 143–168 (1998).
8. K. Krissian, G. Malandain, N. Ayache, et al.: Model-based multiscale detection of 3D vessels, Proc. IEEE Conference on Computer Vision and Pattern Recognition (CVPR'98), Santa Barbara, CA, (1998).
9. Y. Sato, C-F Westin, A. Bhalerao, et al.: Tissue classification based on 3D local intensity structures for volume rendering, IEEE Transactions on Visualization and Computer Graphics, **6**(2), 160-180 (2000).
10. T. O'Donnell, A. Gupta, and T. Boult: A new model for the recovery of cylindrical structures from medical image data, Lecture Notes in Computer Science (LNCS), **1205** (CVRMed-MRCAS'97), 223-232 (1997).
11. A.F. Frangi, W.J. Niessen, R.M. Hoogeveen, et al.: Model-based quantification of 3-D magnetic resonance angiographic images, IEEE Transactions on Medical Imaging, **18**(10), 946-956, (1999)
12. N. Nakanishi, H. Tanaka, T. Nishii, et al.: MR evaluation of the articular cartilage of the femoral head during traction, Acta Radiologica, **40**, 60-63 (1999).
13. Y. Sato, T. Kubota, K. Nakanishi, et al.: Three-dimensional reconstruction and quantification of hip joint cartilages from magnetic resonance images, Lecture Notes in Computer Science (LNCS), **1679** (MICCAI'99), 338-347 (1999).

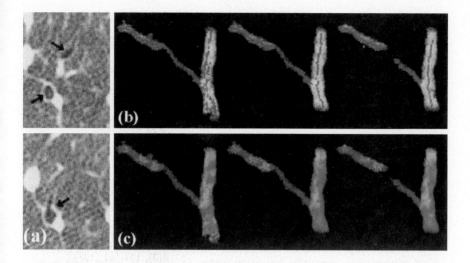

Fig. 5. Diameter estimation of bronchial airways from CT images. (a) Original CT images. The bronchial airway regions, which are darker than the surrounding structures, are shown by arrows. (b) Detection of medial axes at three different scales. Left: $\sigma_f = 1.4$ pixels. Middle: $\sigma_f = 2.0$ pixels. Right: $\sigma_f = 2.8$ pixels. (c) Diameter estimation at the three different scales. The estimated diameter is coded by color: red and blue represent large and small diameters, respectively.

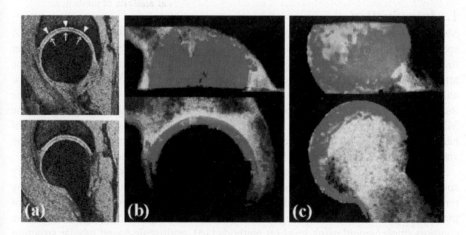

Fig. 6. Thickness estimation of cartilages from MR images. (a) Original MR images. The acetabular (pelvic side) cartilages are shown by arrowheads, and the femoral head cartilages by arrows. (b) Thickness distribution of acetabular cartilages. The estimated thickness is coded by color in the same way as in Fig. 5. The bone regions are volume-rendered in white. (c) Thickness distribution of femoral cartilages.

Fusing Speed and Phase Information for Vascular Segmentation in Phase Contrast MR Angiograms

Albert C. S. Chung[1], J. Alison Noble[1] and Paul Summers[2]

[1]Department of Engineering Science, Oxford University, Oxford, United Kingdom.
[2] Department of Clinical Neuroscience, King's College London, United Kingdom.
Email: {albert, noble}@robots.ox.ac.uk, p.summers@iop.kcl.ac.uk

Abstract. This paper presents a statistical approach to aggregating speed and phase (directional) information for vascular segmentation in phase contrast magnetic resonance angiograms (PC-MRA), and proposes a Maxwell-Gaussian finite mixture distribution to model the background noise distribution. In this paper, we extend our previous work [6] to the segmentation of phase-difference PC-MRA speed images. We demonstrate that, rather than relying on speed information alone, as done by others [12,14,15], including phase information as *a priori* knowledge in a Markov random field (MRF) model can improve the quality of segmentation, especially the region within an aneurysm where there is a heterogeneous intensity pattern and significant vascular signal loss. Mixture model parameters are estimated by the Expectation-Maximization (EM) algorithm [3]. In addition, it is shown that a Maxwell-Gaussian finite mixture distribution models the background noise more accurately than a Maxwell distribution and exhibits a better fit to clinical data.
Keywords: Medical image processing, Statistical segmentation and Medical information fusion.

1. Introduction

Medical diagnosis of vascular diseases is commonly performed on the basis of MRA speed images, which assign high intensity to the moving blood and CSF. The higher the velocity, the higher the image intensity level assigned to it. While most of the arterial anatomy can be shown clearly in the speed images, objects such as intracranial aneurysms containing low and complex flow [17] are poorly represented in the images. The presence of an aneurysm causes significant vascular signal loss in the MRA speed image with some intensity levels approximately equal to those of background tissue, thereby, producing a heterogeneous intensity pattern within the aneurysm. This is problematic in vascular segmentation.

Phase contrast magnetic resonance angiography (PC-MRA) gives not only speed information, but also encodes directional information, which is represented by three separate phase images. Each phase image represents a directional component of the flow. Typical methods for segmentation of MRA images include geodesic active contours implemented using level set methods [14], multiscale-based tubular structure detection [12] and topologically adaptable surfaces [15]. Rather than using just speed information, the directional flow pattern may give additional clues for segmentation. To the best of our knowledge, there is only one related work [19] using both speed and phase images, which uses a multi-resolution, model-based approach to extracting and visualizing vascular flow features.

The method we propose here draws on the fact that the flow pattern in the vasculature is locally coherent. In other words, if blood is flowing in a direction v, neighbouring voxels should have a high probability of exhibiting flow in the same direction v. Local phase coherence measures can be derived to estimate the degree of coherence amongst neighbouring voxels. In this paper, we present a statistical approach, which incorporates the local phase coherence as *a priori* knowledge in a Markov random field (MRF) model to improve the quality of vascular segmentation. Related works can be found in the literature of MRF-based segmentation of conventional MR images, rather than MRA, including mean field approximation [11], ICM [9] and MRF-EM [20].

Moreover, in this paper, we derive the background noise statistical model based on knowledge of the image formation process and use it to derive update equations for Expectation-Maximization (EM) based parameter estimation. In addition, it is shown that the proposed Maxwell-Gaussian finite mixture distribution noise model represents the noise more accurately than a Maxwell distribution used in prior work [1] and shows a better fit to clinical data. It should be noted that a Maxwell distribution describes the probability distribution of the modulus of three Gaussians with zero-mean and equal variance, and the widely known Rayleigh distribution describes the probability distribution of the modulus of two Gaussians with zero-mean and equal variance.

2. Statistical analysis of noise and vascular signals

The phase angle of a complex-valued MR signal S_1 is defined as ϕ_1 and computed by $\arg(S_1) = \tan^{-1}(\text{Im}\{S_1\}/\text{Re}\{S_1\})$, where $\text{Re}\{S_1\}$ and $\text{Im}\{S_1\}$ denote real and imaginary components of the signal respectively, as shown in Figure 1. It is assumed that both real and imaginary components are statistically independent between the two quadrature channels and corrupted by zero-mean Gaussian noise with equal variance σ^2. The probability density function (PDF) of ϕ_1 is given by [13],

$$f_{\phi_1}(\phi_1) = \frac{\exp(-\alpha^2)}{2\pi}\left\{1 + \sqrt{\pi}\alpha \cdot \cos(\phi_1 - \overline{\phi_1}) \cdot \exp\left(\alpha^2 \cdot \cos^2(\phi_1 - \overline{\phi_1})\right)\cdot\left[1 + \text{erf}\left(\alpha \cdot \cos(\phi_1 - \overline{\phi_1})\right)\right]\right\},$$

where $\phi_1 \in [-\pi, \pi)$, $\alpha = A/\sqrt{2}\sigma$, $\text{erf}(x) = (2/\sqrt{\pi}) \cdot \int_0^x e^{-w^2} dw$ is the error function, $\overline{\phi_1}$ is the mean phase and A is the deterministic speed of signal S_1. $f_{\phi_1} = 0$ for $\phi_1 \notin [-\pi, \pi)$. The PDFs at different Signal-to-Noise ratios ($\text{SNR} = A/\sigma = 0, 1, 3 \text{ and } 6$) are plotted in Figure 2. It is noted that the PDF becomes uniformly distributed when the SNR is extremely low, and tends to a Gaussian distribution when the SNR is sufficiently high. To study the goodness-of-fit of a Gaussian approximation at various levels, the PDF was fitted by a Gaussian distribution, where mean and variance were estimated by $\mu = \sum \phi_1 f(\phi_1)$ and $\sigma^2 = \sum (\phi_1 - \mu)^2 f(\phi_1)$ respectively. The absolute difference errors between the PDF and the fitted Gaussian distribution are plotted in Figure 3 at various SNR levels. It is observed that when the SNR is larger than 3, a Gaussian gives a fairly good approximation with absolute difference error less than 8% of the PDF. In this paper, we assume that the SNR is larger than or equal to 3 and the PDF is approximated by a Gaussian. In our experience, this assumption is valid in clinical practice. For a SNR less than 3, a Gaussian approximation is not sufficient, and special attention must be paid to deal with the long and uniform tails at both sides of the PDF, which forms part of our current work [18].

For each velocity component, a phase shift $\Delta\phi$ is produced by the angular difference of the two signal phases [16], i.e. $\Delta\phi = \phi_2 - \phi_1 = \arg(S_2) - \arg(S_1)$, as shown in Figure 1. The two complex-valued signals, S_2 and S_1, are acquired along a specific scanning direction by applying two opposite bipolar gradients. The resulting net phase shift $\Delta\phi$ is directly proportional to the flow rate at this voxels.

As mentioned earlier, the PDFs of ϕ_1 and ϕ_2 are assumed Gaussian. Since the difference between two Gaussians is also a Gaussian, the PDF of $\Delta\phi$, $f_{\Delta\phi}$, follows a zero-mean Gaussian distribution with variance $2\sigma^2$. Figure 4 and Figure 5 show respectively a computer simulated phase image with low SNR and its histogram fitted by a Gaussian distribution.

A MRA speed image is reconstructed on a voxel-by-voxel basis by taking the modulus of the three corresponding phase shift values, $\Delta\phi_x$, $\Delta\phi_y$ and $\Delta\phi_z$, i.e. $i \propto \sqrt{\Delta\phi_x^2 + \Delta\phi_y^2 + \Delta\phi_z^2}$, where i is the image intensity. Since the phase shifts are flow sensitized along the three directional components, x, y and z, and directly proportional to the flow rate, the reconstructed image is called speed image. The distribution described by the modulus of three independent zero-mean Gaussians is a Maxwell distribution [1]. Hence, for the background noise, the noise PDF is given by a Maxwell distribution $f_M(i) = \left(2 \cdot i^2 / \sigma_M^3 \sqrt{2\pi}\right) \cdot \exp\left(-i^2 / 2\sigma_M^2\right)$, where $\sigma_M = \sqrt{2}\sigma$ and $i \geq 0$. Figure 7 and Figure 8 show respectively a computer simulated speed image with low SNR and its histogram fitted by a Maxwell distribution using the relationship $\sigma_M = I_{peak} / \sqrt{2}$, where I_{peak} is the intensity value at which the histogram achieves its maximum, i.e. $df_M(i)/di = 0$ at $i = I_{peak}$. It is observed that the Maxwell distribution fits well in the low intensity region, but not in the intensity region indicated by the arrow in Figure 8. The reason is that both tails of the histogram are not perfectly fitted by a Gaussian distribution (indicated by the arrows in Figure 5). The absolute difference between them is shown in Figure 6 (with a smaller scale on the vertical axis than Figure 5), and reveals one positive residual distribution located at each side. We assume that these residual distributions are non-zero mean Gaussian. Therefore, for each encoding direction, the PDF $f_{\Delta\phi}$ consists of a zero-mean Gaussian (located at the centre) and two non-zero mean Gaussian distributions (located at each side). After the modulus operation, as mentioned earlier, a Maxwell distribution is formed by the modulus of the three zero-mean Gaussian distributions, whereas, the modulus of the residual non-zero mean Gaussian distributions gives a Gaussian distribution [1]. Hence, the noise PDF consists of a linear mixture of a Maxwell and a Gaussian distribution with mean μ_G and variance σ_G^2.

As shown in Figure 9, using the EM algorithm [3], a Maxwell and Gaussian mixture noise model achieves a better fit to the given histogram of a PC-MRA speed image with low SNR. To model the vascular signal with high SNR, we apply the results of our previous work [6] and assume that the signal is uniformly distributed. Thus, we conclude that the overall PDF $f(i)$ of a speed image can be modelled as a Maxwell-Gaussian and uniform finite mixture distribution, which is given by

$f(i) = \underbrace{w_M f_M(i) + w_G f_G(i)}_{\text{Background Noise}} + \underbrace{w_U f_U(i)}_{\text{Vascular Signal}}$, where w_M, w_G and w_U are weights (or

prior probabilities) assigned to the Maxwell, Gaussian and uniform distributions respectively and $w_M + w_G + w_U = 1$.

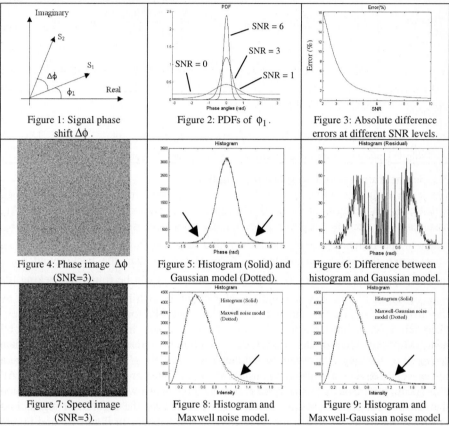

Figure 1: Signal phase shift $\Delta\phi$.	Figure 2: PDFs of ϕ_1.	Figure 3: Absolute difference errors at different SNR levels.
Figure 4: Phase image $\Delta\phi$ (SNR=3).	Figure 5: Histogram (Solid) and Gaussian model (Dotted).	Figure 6: Difference between histogram and Gaussian model.
Figure 7: Speed image (SNR=3).	Figure 8: Histogram and Maxwell noise model.	Figure 9: Histogram and Maxwell-Gaussian noise model

3. Segmentation algorithm

The proposed mixture model has six parameters: w_M, w_G, w_U, σ_M^2, μ_G and σ_G^2, which need to be estimated. The modified EM algorithm can be used to estimate the parameters by maximizing the log-likelihood of the mixture distribution in each iteration [3]. Let N be the total number of voxels and $h(i)$ be the observed histogram distribution. The update equations can be derived and are listed in Table 1.

In our implementation, the initial estimates were found empirically, and convergence was generally reached after about 5 iterations. A typical result is shown in Figure 21. For comparison, a Maxwell and uniform mixture model has been fitted using the EM algorithm [3] in Figure 20, which reveals that the Maxwell model provides a poorer fit than the Maxwell-Gaussian mixture model for background noise. It is also worth noting that, together with the imperfection of the Maxwell noise model, ghosting artifacts and partial volume effect also contribute to the poorer fit of the Maxwell noise model in the clinical data.

Maxwell: (M)	$w_M^{k+1} = \dfrac{1}{N} \cdot \sum_i h(i) \cdot P^k(M\,|\,i)$ and $\left(\sigma_M^2\right)^{k+1} = \dfrac{\sum_i h(i) \cdot P^k(M\,|\,i) \cdot i^2}{3 \cdot \sum_i h(i) \cdot P^k(M\,|\,i)}$, where $P^k(M\,|\,i) = w_M^k f_M^k(i) / f^k(i)$.
Gaussian: (G)	$w_G^{k+1} = \dfrac{1}{N} \cdot \sum_i h(i) \cdot P^k(G\,|\,i)$, $\mu_G^{k+1} = \dfrac{\sum_i h(i) \cdot P^k(G\,|\,i) \cdot i}{\sum_i h(i) \cdot P^k(G\,|\,i)}$ and $\left(\sigma_G^2\right)^{k+1} = \dfrac{\sum_i h(i) \cdot P^k(G\,|\,i) \cdot \left(i - \mu_G^k\right)^2}{\sum_i h(i) \cdot P^k(G\,|\,i)}$, where $P^k(G\,|\,i) = w_G^k f_G^k(i) / f^k(i)$.
Uniform: (U)	$w_U^{k+1} = \dfrac{1}{N} \cdot \sum_i h(i) \cdot P^k(U\,|\,i)$, where $P^k(U\,|\,i) = w_U^k f_U^k(i) / f^k(i)$ and $f_U^k(i) = 1/I_{max}$.

Table 1: Update equations and posterior probabilities at k^{th} iteration for each distribution.

3.1. Segmentation based on speed information

Given an estimated mixture model, a PC-MRA speed image can be segmented statistically on the basis of the MAP (Maximum-A-Posteriori) criterion. Using MAP, assuming the weights (or *prior* probabilities) remain constant, a voxel $x_s \in \{v, b\}$, where v and b denote vessel and background respectively, is classified as a vessel voxel when the vessel probability $w_U f_U(i)$ is greater than the background noise probability $w_M f_M(i) + w_G f_G(i)$. Therefore, a threshold can be found by seeking the intersection of the background and vessel probability distributions. As indicated by the arrows in Figure 20 and Figure 21, the threshold found by using Maxwell-Gaussian noise model tends to more correct and is usually higher than that found by using Maxwell noise model. Figure 22 and Figure 23 show a speed image (data provided by the Department of Clinical Neurosciences, King's College London) and a binary-segmented image produced using Maxwell-Gaussian noise model and MAP classification method. The aneurysm is indicated by an arrow in the figure. It is noted that the resulting segmentation is adversely affected by the significant signal loss inside an aneurysm located at the middle. The same happens in Figure 25 and Figure 26. We will discuss how to overcome this problem in the next subsection.

3.2. Integration of speed and phase information

Note that the flow pattern in the aneurysm appears to show (a) a vortex centred at the singular point, at which the velocity becomes null, and (b) a locally linear (coherent) motion around the neighbourhood of the singular point, which can also be regarded as a deformed circular flow. This is consistent with the results of clinical flow studies [8] and simulations of computational fluid dynamics [5]. The local linear motion exists not only inside the aneurysm, but also in most of the vasculature having laminar flow pattern. We now present a method for the quantification of these local linear motions.

Let $V = \{v_1, \ldots, v_N\}$ be a velocity map, where N is the total number of voxels and the three orthogonal velocity components of a voxel x_s are assigned as the phase shifts in the corresponding phase image, i.e. for $v_s = \left(v_s^x, v_s^y, v_s^z\right)$, $v_s^x = \Delta\phi_x$, $v_s^y = \Delta\phi_y$ and $v_s^z = \Delta\phi_z$. We define a phase feature P as the cosine of the angle between two velocities, v_s and v_t, of the neighbouring voxels, x_s and x_t, where $s,t \in [1,...,N]$. The phase feature P is given by a dot product of the two normalized velocities, $P(v_s,v_t) = v_s \cdot v_t / \|v_s\|\|v_t\|$, where $P \in [-1,1]$. To quantify the local linear motion around a voxels x_s, we measure its local phase coherence (LPC) by applying a 3 x 3 voxel mask to the velocity map V at the voxel within slice. LPC is evaluated by considering the locally coherent motion of its 8 neighbouring voxels, which are numbered in Figure 10, and is defined as a circular addition of the phase features around x_s, i.e.

$$LPC(x_s) = P(v_1,v_2) + P(v_2,v_3) + \cdots + P(v_8,v_1), \tag{1}$$

where $LPC(x_s) \in [-8,8]$. In other words, LPC is a circular addition of dot products of the 8 adjacent normalized velocity pairs. Figure 12 shows a LPC map for the speed image of Figure 22, in which the image intensity value is directly proportional to the strength of LPC. It is observed that the voxels inside the aneurysm and vessels exhibit high LPC and form a piece-wise homogeneous region, whereas the non-vessel voxels have relatively low and random LPC.

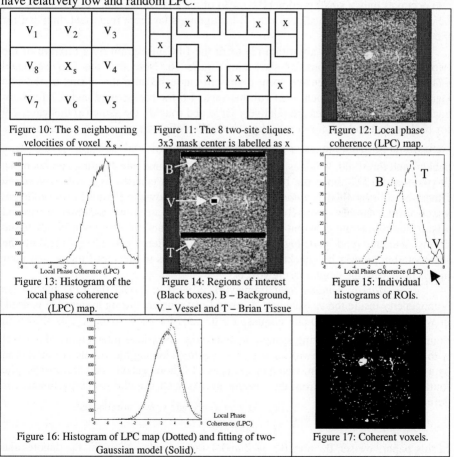

Figure 10: The 8 neighbouring velocities of voxel x_s.

Figure 11: The 8 two-site cliques. 3x3 mask center is labelled as x

Figure 12: Local phase coherence (LPC) map.

Figure 13: Histogram of the local phase coherence (LPC) map.

Figure 14: Regions of interest (Black boxes). B – Background, V – Vessel and T – Brian Tissue

Figure 15: Individual histograms of ROIs.

Figure 16: Histogram of LPC map (Dotted) and fitting of two-Gaussian model (Solid).

Figure 17: Coherent voxels.

We now present an automatic threshold determination method for classification of coherent and non-coherent voxels. A histogram of the LPC map is plotted in Figure 13. It shows that the histogram is right shifted and skewed. In fact, the LPC histogram is constituted by three classes: *background* with low LPC, *brain tissue* with slightly high LPC, and *vessel* with extremely high LPC, which are shown in Figure 14 and Figure 15. Background (B) and brain tissue (T) histograms overlap heavily because of the non-stationary, but slightly coherent, motion of the non-vessel brain tissue. In contrast, the vessel (V) histogram is separated clearly from the background (B) histogram. The point of separation – the desired threshold – is indicated by an arrow in Figure 15. We model the background region and non-background regions with two separate Gaussian distributions. It is worth noting that the theoretical modelling of the LPC histogram is extremely difficult because of the high correlation factor among the velocity random variables, and normalization and dot product operations of the correlated variables in Equation 1. A Gaussian distribution is employed to model the non-background region including both T and V histograms. This is because the V histogram overlaps heavily with the T histogram, and occupies only a small portion (1% - 4%) of the LPC histogram. Moreover, as will be discussed below, estimation for the B histogram is far more important than that for the T and V histograms.

We use the EM algorithm [3] to fit the LPC histogram by a mixture of the two Gaussian distributions, as shown in Figure 16. Note that the two estimated Gaussians merge together and form a smooth curve because the means of the two Gaussians are close to each other (in this case, 1.6 for background and 3.7 for non-background) and variances are relatively large and roughly the same (in this case, 2.9 for background and 2.2 for non-background). We define the mean and variance of the background distribution as μ_B and σ_B^2 respectively, and use $\mu_B + 3 \cdot \sigma_B$ as a background threshold, which is a variant of the background thresholding approach [4]. A voxel with LPC above the background threshold is labelled as a *coherent* voxel. Otherwise, it is labelled as a *non-coherent* voxel. As shown in Figure 17, the coherent voxels form a number of vessel 'clusters', though there are some randomly distributed voxels due to random coherent noise, small coherent motion of the non-vessel tissue during scanning and ghosting artifacts. These 'outliers' can be avoided in the segmentation process by checking their intensity values in the speed image and their interactions with the neighbouring voxels.

A *Markov random field* framework is used to model the LPC piece-wise homogeneity inside the vasculature. A second-order neighbouring system [7] is used, in which we define 8 two-site cliques, $c = \{1, ..., 8\}$, for each voxel x_s, as shown in Figure 11. This neighbouring system describes the immediate interactions of a voxel with its 8 adjacent in slice voxels. These interactions among the voxels are measured by a clique energy function, which encourages LPC homogeneity and discourages the 'outlier' voxels far away from the vasculature. As such, the clique energy function is defined as

$$E_c(x_s, x_i) = \begin{cases} \beta_v, & \text{if } x_s = x_i = v, \text{ and } x_s \& x_i \text{ are coherent.} \\ \beta_b, & \text{otherwise.} \end{cases}$$

In this paper, we set β_v and β_b as 2 and 1 respectively. As such, β_b denotes the number of non-coherent or non-vessel neighbouring voxels. In contrast, β_v denotes twice the number of coherent, vessel voxels in order to favour the coherent, vessel neighbouring voxels. In Section 3.1, the *prior* probability is assumed constant throughout the segmentation process. Now, the total energy function $U(x_s)$ and *prior* probability $P(x_s)$ of voxel x_s are defined as $U(x_s) = \sum_{c=1}^{8} \sum_{i \in c} E_c(x_s, x_i)$, and $P(x_s) = Z^{-1} \cdot \exp\{U(x_s)\}$ respectively, where $Z = \sum_{x_s \in [v,b]} \exp\{U(x_s)\}$ is a normalization constant. It should be noted that the Markov *prior* probability is directly proportional to the number of adjacent, and coherent, vessel voxels. Intuitively, the larger the number of adjacent and coherent, vessel voxels around the voxel x_s, the higher the vessel probability $P(x_s = v)$.

Let $X = \{x_1, ..., x_N\}$, be the true-segmented image, where $x_s \in [v, b]$. Let also $I = \{i_1, ... i_N\}$ be the observed speed image, where $i_s \in [0...I_{max}]$. The Iterated Conditional Modes (ICM) [2] algorithm can be employed to maximize the probability of estimating the true-segmented binary image given the observed image, i.e. $P(X | I)$. Using the parameters estimated by the modified EM algorithm in the previous section, the vessel likelihood $P(i_s | x_s = v)$ and background likelihood $P(i_s | x_s = b)$ are defined as $f_U(i_s)$ and $[w_M f_M(i_s) + w_G f_G(i_s)]/(w_M + w_G)$ respectively. The true segmentation X is initialized by the segmentation results based on speed information, as shown in Figure 23. The ICM algorithm estimates a voxel class by using MAP method, $x_s = \arg \max_{x_s \in [v,b]} P(x_s | i_s)$, to maximize the posterior probability $P(x_s | i_s) \propto P(x_s) \cdot P(i_s | x_s)$ at each voxel iteratively until convergence is reached. The convergent rate is usually around 5 iterations.

4. Results

Intracranial scans (PC-MRA) of two patients were performed using a 1.5T GE MR scanner at the Department of Clinical Neurosciences, King's College London. The volume size was 256x256x28 voxels and voxel size 0.8mm x 0.8mm x 1mm. The Maxwell noise model was compared with the proposed Maxwell-Gaussian noise model by computing the absolute difference error for all slices in the two scans. Results are shown in Figure 18 and Figure 19. This shows that Maxwell-Gaussian model is better than the Maxwell model for describing the background noise distribution in PC-MRA speed images by an average of 9%.

The segmentation algorithm was applied to all slices. Typical segmentation results are shown in Figure 24 and Figure 27. Comparing this with the results of Figure 23 and Figure 26, it is noted that there is a substantial improvement in segmentation, especially in the region of the aneurysm. But still, there are some false negative voxels near the singular point (near the middle of the aneurysm). This is because the level of intensity is very low, and the flow pattern is seriously corrupted by noise. A higher level of understanding of flow topology is required to tackle this problem, which is a subject of our current work.

5. Conclusion

We have derived a Maxwell-Gaussian mixture model for the background noise distribution of PC-MRA images generated by a phase-difference PC-MRA image post-processing algorithm. It has been shown that the Maxwell-Gaussian mixture model fits better than a Maxwell distribution for modelling background noise, which has been used in prior work [1]. Using this mixture model, we have proposed a statistical approach to aggregating speed and phase information available in PC-MRA, and demonstrated that inclusion of phase information as a *priori* knowledge in the MRF model can improve the quality of segmentation, especially in the region within an aneurysm. Future work will include detection of the flow singular point, which may indicate the presence of an aneurysm, perhaps by using knowledge of flow topology [10] and through more detailed studies of application to a large number of aneurysms.

Acknowledgments: We thank Prof. J. Byrne and Dr. A. Martinez at Department of Radiology, Radcliffe Infirmary, Oxford for helpful discussions. A. Chung is funded by a postgraduate scholarship award from the Croucher Foundation, HK.

References

1.   A.H. Andersen & J.E. Kirsch, "Analysis of noise in phase contrast MR imaging", *Med. Phy.*, 23(6), June 1996.
2.   J. Besag, "On the Statistical Analysis of Dirty Pictures", *J. Royal Statistical Society (B)*, 48(3):259-302, 1986.
3.   C.M. Bishop, *Neural Networks for Pattern Recognition*, Clarendon Press, Oxford, 95.
4.   M.E. Brummer, R.M. Mersereau, R.L. Eisner & R.R.J. Lewine, "Automatic Detection of Brain Contours in MRI Data Sets", *TMI* 12(2):153-166, June, 1993.
5.   A.C. Burleson, et. al., "Computer Modeling of Intracranial Saccular and Lateral Aneurysms for the Study of Their Hemodynamics", *Neurosurgery*, 37(4):774-784, 95.
6.   A.C.S. Chung & J.A. Noble, "Statistical 3D vessel segmentation using a Rician distribution", In *MICCAI'99*, pp.82-89 and *MIUA'99*, pp.77-80.
7.   S. Geman and D. Geman, "Stochastic Relaxation, Gibbs Distributions, and the Bayesian Restoration of Images", *PAMI*, 6(6):721-741,1984.
8.   Y.P. Gobin, J.L. Counord, P. Flaud & J. Duffaux, "In Vitro study of haemodynamics in a giant saccular aneurysm model", *Neuroradiology*, 36:530-536, 1994.
9.   K. Held, E.R. Kops, B.J. Krause, W.M. Wells, R. Kikinis & H.W. Muller-Gartner, "Markov Random Field Segmentation of Brain MR Images", *TMI*, 16(6):878-886, 97.
10.   J.L. Helman & L. Hesselink, "Visualization of Vector Field Topology in Fluid Flows", *IEEE Comp. Graphics and Appl.*, 11(3):36-46, 1991.
11.   T. Kapur, W.E.L. Grimson, R. Kikinis & W.M. Wells, "Enhanced Spatial Priors for Segmentation of Magnetic Resonance Imagery", *MICCAI'98*, pp.457-468.
12.   K. Krissian, G. Malandain & N. Ayache, "Model Based Detection of Tubular Structures in 3D Images", *INRIA-Technical Report 3736*, 1999.
13.   B.P. Lathi, *Modern Digital and Analog Communication Systems*, Ch. 11, Hault-Saunders Intl. Ed., 1983.
14.   L.M. Lorigo, O. Faugeras, W.E.L Grimson, R. Keriven, R. Kikinis & C.F. Westin, "Co-dimension 2 Geodesic Active Contours for MRA Segmentation", *IPMI'99*, pp.126-139.
15.   T. McInerney & D. Terzopoulos, "Medical Image Segmentation Using Topologically Adaptable Surface", *CVRMed'97*, pp.23-32.
16.   N.J. Pelc, M.A. Bernstein et. al., "Encoding Strategies for Three-Direction Phase-Contrast MR Imaging of Flow", *JMRI*, 1:405-413, 1991.
17.   H.J. Steiger, et. al., "Computational Simulation of Turbulent Signal Loss in 2D Time-of-Flight Magnetic Resonance Angiograms", *MRM*, 37:609-614, 97.

175

18. P. Summers, A.C.S. Chung & J.A. Noble, "Impact of Image Processing Operations on MR Noise Distributions", In *Proc. of 8th ISMRM*, 2000.

19. P. Summers, A.H. Bhalerao & D.J. Hawkes, "Multiresolution, Model-Based Segmentation of MR Angiograms", *JMRI*, 7:950-957, 1997.

20. Y. Zhang, S. Smith & M. Brady, "Segmentation of brain MR images using Markov random fields", *MIUA '99*, p.65-68.

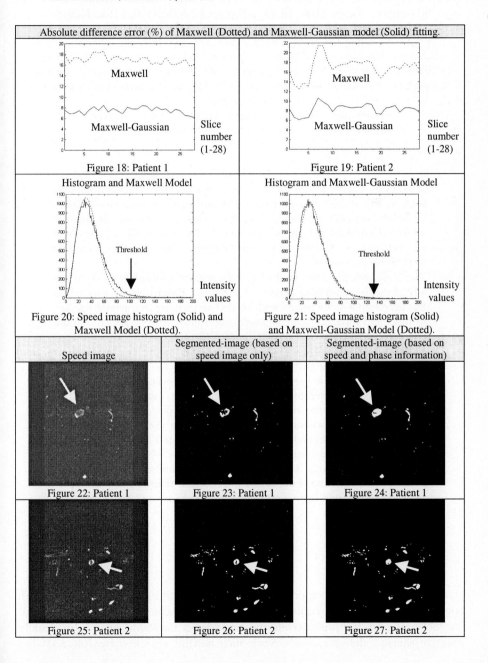

Absolute difference error (%) of Maxwell (Dotted) and Maxwell-Gaussian model (Solid) fitting.

Maxwell

Maxwell-Gaussian

Slice number (1-28)

Figure 18: Patient 1

Maxwell

Maxwell-Gaussian

Slice number (1-28)

Figure 19: Patient 2

Histogram and Maxwell Model

Threshold

Intensity values

Figure 20: Speed image histogram (Solid) and Maxwell Model (Dotted).

Histogram and Maxwell-Gaussian Model

Threshold

Intensity values

Figure 21: Speed image histogram (Solid) and Maxwell-Gaussian Model (Dotted).

Speed image

Segmented-image (based on speed image only)

Segmented-image (based on speed and phase information)

Figure 22: Patient 1

Figure 23: Patient 1

Figure 24: Patient 1

Figure 25: Patient 2

Figure 26: Patient 2

Figure 27: Patient 2

Intraoperative Segmentation and Nonrigid Registration for Image Guided Therapy

Simon K. Warfield[1], Arya Nabavi[1,3], Torsten Butz[4], Kemal Tuncali[2], Stuart G. Silverman[2], Peter McL. Black[3], Ferenc A. Jolesz[1], and Ron Kikinis[1]

[1]Surgical Planning Laboratory http://www.spl.harvard.edu [2]Department of Radiology, [3]Department of Surgery, Harvard Medical School and Brigham and Women's Hospital, 75 Francis St., Boston, MA 02115 USA, [4] Signal Processing Laboratory, Swiss Federal Institute of Technology at Lausanne, 1015 Lausanne, Switzerland.
{warfield,arya,butz,ktuncali,jolesz,kikinis}@bwh.harvard.edu,
pmblack@bics.bwh.harvard.edu, sgsilverman@partners.org

Abstract. Our goal was to improve image guidance during minimally invasive image guided therapy by developing an intraoperative segmentation and nonrigid registration algorithm. The algorithm was designed to allow for improved navigation and quantitative monitoring of treatment progress in order to reduce the time required in the operating room and to improve outcomes.

The algorithm has been applied to intraoperative images from cryotherapy of the liver and from surgery of the brain. Empirically the algorithm has been found to be robust with respect to imaging characteristics such as noise and intensity inhomogeneity and robust with respect to parameter selection. Serial and parallel implementations of the algorithm are sufficiently fast to be practical in the operating room.

The contributions of this work are an algorithm for intraoperative segmentation and intraoperative registration, a method for quantitative monitoring of cryotherapy from real-time imaging, quantitative monitoring of brain tumor resection by comparison to a preoperative treatment plan and an extensive validation study assessing the reproducibility of the intraoperative segmentation. We have evaluated our algorithm with six neurosurgical cases and two liver cryotherapy cases with promising results. Further clinical validation with larger numbers of cases will be necessary to determine if our algorithm succeeds in improving intraoperative navigation and intraoperative therapy delivery and hence improves therapy outcomes.

1 Introduction

Image guided surgical techniques are used in operating rooms equipped with special purpose imaging equipment. The development of image guided surgical methods over the past decade has provided a major advance in minimally invasive therapy delivery. Early work such as that reviewed by Jolesz [1] has established the importance and value of image guidance through better determination of tumor margins, better localization of lesions, and optimization of the surgical approach.

Research in image guided therapy has been driven by the need for improved visualization. Qualitative judgements by experts in clinical domains have been relied upon as quantitative and automated assessment of intraoperative imaging data has not been possible in the past. In order to provide the surgeon or interventional radiologist with as rich a visualization environment as possible from which to derive such judgements, existing work has been concerned primarily with image acquisition, visualization and registration of intraoperative and preoperative data. Intraoperative segmentation has the potential to be a significant aid to the intraoperative interpretation of images and to enable prediction of surgical changes.

Earlier work has been a steady progression of improving image acquisition and intraoperative image processing. This has included increasingly sophisticated multimodality image fusion and registration. Clinical experience with image guided therapy in deep brain structures and with large resections has revealed the limitations of existing rigid registration and visualization approaches [1]. The deformations of anatomy that take place during such surgery are often better described as nonrigid and suitable approaches to capture such deformations are being actively developed by several groups (described below).

A number of imaging modalities have been used for image guidance. These include, amongst others, computed tomography (CT), ultrasound, digital subtraction angiography (DSA), and magnetic resonance imaging (MRI). Intraoperative MR imaging can acquire high contrast images of soft tissue anatomy which has proven to be very useful for image-guided therapy [2]. Multi-modality registration allows preoperative data that cannot be acquired intraoperatively, such as fMRI or nuclear medicine scans, to be visualized together with intraoperative data.

Gering et al. [3] described an integrated system allowing the ready visualization of intraoperative images with preoperative data, including surface rendering of previously prepared triangle models and arbitrary interactive resampling of 3D grayscale data. Multiple image acquisitions were presented in a combined visualization through rigid registration and trilinear interpolation. The system also allows for visualization of virtual surgical instruments in the coordinate system of the patient and patient image acquisitions. The system supports qualitative analysis based on expert inspection of image data and the surgeons expectation of what should be present (normal anatomy, patient's particular pathology, current progress of the surgery etc.)

Several groups have investigated intraoperative nonrigid registration, primarily for neurosurgical applications. The approaches can be categorized by those that use some form of biomechanical model (recent examples include [4–6]) and those that apply a phenomenological approach relying upon image related criteria (recent examples include [7, 8].)

We aimed to demonstrate that intraoperative segmentation is possible and adds significantly to the value of intraoperative imaging. Compared to registration of preoperative images and inspection of intraoperative images alone, intraoperative segmentation enables identification of structures not present in previous images (examples of such structures include the region of cryoablation or radiofrequency treatment area, surgical probes and changes due to resection), quantitative monitoring of the progress of therapy (including the ability to compare quantitatively with a preoperatively determined treatment plan) and intraoperative surface rendering for rapid 3D interactive visualization.

2   Method

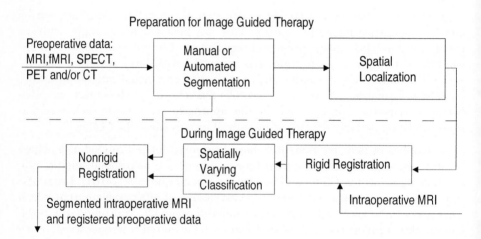

Fig. 1. Schema for Intraoperative Segmentation and Registration

In order to successfully segment intraoperative images, we have developed an image segmentation algorithm that takes advantage of an existing preoperative MR acquisition and segmentation to generate a patient-specific model for the segmentation of intraoperative data. The algorithm uses the segmentation of preoperative data as a template for the segmentation of intraoperative data. Figure 1 illustrates the processing steps that take place before and during the therapy procedure.

We have experimented for several years with a general image segmentation approach that uses a 3D digital anatomical atlas to provide automatic local context for classification [9–11]. The work described here extends our previous work to ensure its suitability for intraoperative segmentation. Rather than a generic digital anatomic atlas we propose here to use a segmented preoperative patient scan to derive a patient-specific anatomical model for intraoperative segmentation of new scans.

Since preoperative data is acquired before surgery, the time available for segmentation is longer. This means we can use segmentation approaches that are robust and accurate but are time consuming and hence impractical to use in the operating room. In our laboratory, preoperative data is segmented with a variety of manual [3], semi-automated [12] or automated [13, 11] approaches. We attempt to select the most robust and accurate approach available for a given clinical application. Each segmented tissue class is then converted into an explicit 3D volumetric spatially varying model of the location of that tissue class, by computing a saturated distance transform [14] of the tissue class. This model is used to provide robust automatic local context for the classification of intraoperative data in the following way.

During surgery, intraoperative data is acquired and the preoperative data (including any MRI/fMRI/PET/SPECT/MRA that is appropriate, the tissue class segmentation and the spatial localization model derived from it) is aligned with the intraoperative data using an MI based rigid registration method [15, 3]. The intraoperative image data then together with the spatial localization model forms a multichannel 3D data set. Each voxel is then a vector having components from the intraoperative MR scan, the spatially varying tissue location model and if relevant to the particular application, any of the other preoperative image data sets. For the first intraoperative scan to be segmented a statistical model for the probability distribution of tissue classes in the intensity and anatomical localization feature space is built. The statistical model is encoded implicitly by selecting groups of prototypical voxels which represent the tissue classes to be segmented intraoperatively (less than five minutes of user interaction). The spatial location of the prototype voxels is recorded and is used to update the statistical model automatically when further intraoperative images are acquired and registered. This multichannel data set is then segmented with a spatially varying classification [10, 13, 16].

Segmentation of intraoperative data helps to establish explicitly the regions of tissues that correspond in the preoperative and intraoperative data. It is then straightforward to apply our previously described [17, 18] and validated [19] multi-resolution elastic matching algorithm. Once the nonrigid transformation mapping from the preoperative to the intraoperative data has been established, the mapping is applied to each of the relevant preoperative data sets to bring them into alignment with the intraoperative scan.

3 Results

In this section illustrative segmentations and two validation experiments are presented. During interventional procedures in the liver and brain, intraoperative MRI (IMRI) data sets were acquired and stored. Our segmentation and nonrigid registration algorithm was applied to these data sets after therapy delivery in order to allow us to assess the robustness, accuracy and time requirements of the approach. In the future we intend to carry out segmentation, nonrigid registration and visualization using the approach described here during the interventional procedures with the goal of improving image guided therapy outcomes.

3.1 Intraoperative Segmentation for Neurosurgery

Figure 2 illustrates the segmentation of six neurosurgery cases using our intraoperative segmentation algorithm. In each case, several volumetric MRI scans were carried out during surgery. The first scan was acquired at the beginning of the procedure before any changes in the shape of the brain took place, and then over the course of surgery other scans were acquired as the surgeon checked the progress of tumor resection. The final scan in each sequence exhibits significant nonrigid deformation and loss of tissue due to tumor resection. In order to test our segmentation approach each subsequent scan was aligned to the first by maximization of mutual information and the first scan was manually segmented to act as an individualized anatomical model. The last scan in each sequence was then segmented with our new segmentation approach. The segmented brain and IMRI is shown in Figure 2. In order to check the quality of the segmentation, each segmentation was visually compared with the MRI from which it was derived. In each case the segmented brain closely matched the expected location.

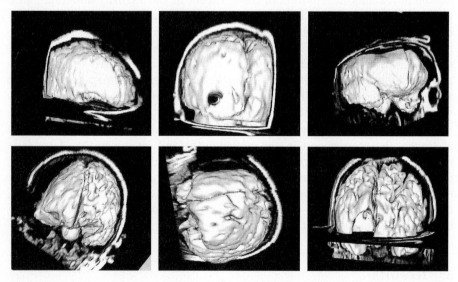

Fig. 2. Visualizations of intraoperative segmentation of brain tissue from six neurosurgery cases. Less than five minutes of user interaction was required for each segmentation. The segmented brain tissue is shown in white with surface rendering and the IMRI is texture mapped in planes along the coordinate axes. This allows ready comparison of the position of the segmented brain and the IMRI (skin appears bright, brain is gray closely matching the segmented brain border).

User Interaction and Computational Requirements Each brain segmentation of Figure 2 involves the segmentation of the entire 3D IMRI scan of 256x256x60 = 3,932,160 voxels (with voxel size 0.9375x0.9375x2.5mm^3.) Such a volume is acquired in approximately 10 minutes intraoperatively (a second scanning mode can acquire a 2D image in approximately 2 seconds). Less than five minutes of user interaction was required for each brain segmentation shown in Figure 2. On a Sun Microsystems Ultra-10 workstation with a 440MHz UltraSPARC-IIi CPU and 512MB RAM each brain tissue segmentation (excluding generation of spatial localization models which requires approximately 200 seconds and can be done preoperatively and excluding rigid registration which requires approximately 30 seconds using maximization of mutual information [15]) required less than 330 seconds to complete. As we have previously described, parallel tissue classification can achieve excellent speedups [20]. On a Sun Microsystems Ultra-80 server with 4 x 450MHz UltraSPARC-II CPUs and 2GB RAM, each brain tissue segmentation required less than 130 seconds to complete. This can be compared to a typical manual segmentation that can take 1800–3600 seconds and has significantly less reproducibility.

3.2 Intraoperative Segmentation for Liver Cryotherapy

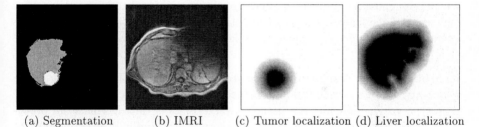

  (a) Segmentation   (b) IMRI   (c) Tumor localization (d) Liver localization

Fig. 3. IMRI of the liver and the segmentation of the liver and tumor. Three of the feature channels used to carry out the segmentation shown in (a) are shown in (b), (c) and (d).

Figure 3 shows IMRI of the liver and the segmentation of the liver and tumor. Intraoperative MRI has been used to guide percutaneous cryotherapy of liver tumors [21]. This figure illustrates the spatial localization of liver and tumor from a 3D volumetric preoperative segmentation (not shown³) and indicates that isointense but different structures in the IMRI can be successfully segmented in the joint feature space formed with the IMRI intensity and spatial localization information.

3.3 Intraoperative Monitoring of Cryotherapy Iceball Formation

Figure 4 shows the intraoperative appearance of an iceball during cryotherapy of another case. Our intraoperative segmentation algorithm allowed rapid, robust

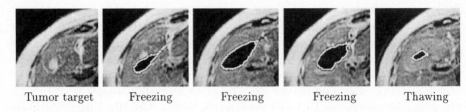

| Tumor target | Freezing | Freezing | Freezing | Thawing |

Fig. 4. Intraoperative imaging of iceball formation. The lesion is bright in the first image and the iceball grows to cover it. The iceball appears as a dark region in the liver, which grows while freezing and shrinks during thawing. The intraoperative iceball segmentation obtained with our method is indicated by the white outline.

and straightforward segmentation of the iceball. By comparing the segmentation of the iceball with a preoperative plan of the desired iceball size and location the therapy progress can be monitored quantitatively.

3.4 Validation Experiments

Key parameters in our segmentation algorithm are the prototype voxels which implicitly model the probability distribution of the intensities of tissues which are to be segmented and the alignment of the spatial localization models which form part of the feature space in which the tissue classification takes place. We studied the effect of variations in these key parameters upon the segmentation.

Reproducibility: Variations in prototype selection Table 1 records the variability of brain segmentation from a single neurosurgery case when the set of prototype voxels modeling the tissue characteristics is varied. The set of prototypes used for the segmentation was subsampled by randomly selecting 90% of the prototypes 100 times. Each of the 100 subsets simulates different user prototype selection. Each subset was used to segment the IMRI using the new method and the volume of the brain segmentation was recorded. The mean, minimum and maximum volume recorded are shown in the table along with the coefficient of variation (C.V.) of the volume of segmentation (which is less than 1%). This indicates the segmentation is extremely robust in the presence of variability in the prototype voxel selection.

Minimum volume	Maximum volume	Mean volume	C.V. (%)
401074 voxels	422440 voxels	414440 voxels	0.97

Table 1. Measures of variability of the volume of the brain (units are voxels) in repeated segmentations, with different selections of prototype voxels, from brain IMRI.

Reproducibility: Variations in preoperative model alignment In order to determine the influence of the alignment of the preoperative segmentation upon the intraoperative segmentation, we selected a neurosurgery case and segmented the brain as described above. We then applied a set of translations and rotations to the preoperative segmentation so that it was no longer correctly registered to the IMRI to be segmented. For each translation and rotation we applied our segmentation method and obtained a segmentation of the brain. We then compared the volume of tissue segmented as brain for each misaligned position with that obtained with the correct alignment and recorded the ratio of the new segmentation volume to the original segmentation volume. The variation in segmentation with translations along each of the coordinate axes is shown in Figure 5. Similar results (not shown) were obtained for rotations around each of the coordinate axes.

Intrapatient registration with maximization of mutual information has a typical accuracy smaller than one voxel (in this case $0.9375 \times 0.9375 \times 2.5 mm^3$). The perturbation of the registration of the preoperative segmentation does not cause a significant change in the intraoperative segmentation until this misalignment reaches around ± 10 mm, which indicates our intraoperative segmentation method is robust to misalignment errors and also to errors in the preoperative segmentation.

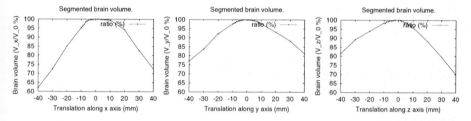

Fig. 5. Reproducibility of brain tissue segmentation as the spatial localization model is translated. The low variability in segmentation around the correct alignment indicates the segmentation is robust to misregistration and spatial localization errors.

4 Discussion and Conclusion

Our early experience with two liver cases and six neurosurgery cases indicates that our intraoperative segmentation algorithm is a robust and reliable method for intraoperative segmentation. It requires little user interaction, is robust to variation in the parameters that require interaction, and is sufficiently fast to be used intraoperatively.

The application of our previously described and validated nonrigid registration algorithm is enabled by intraoperative segmentation establishing the corresponding tissues in the data sets to be aligned. Further work is needed to characterize the accuracy and robustness of our nonrigid registration for intraoperative data, especially in the liver.

Intraoperative segmentation adds significantly to the value of intraoperative imaging. Compared to registration of preoperative images and inspection of intraoperative images alone, intraoperative segmentation enables identification of structures not present in previous images (examples of such structures include the region of cryoablation or RF treatment area, surgical probes and changes due to resection), quantitative monitoring of therapy application including the ability to compare quantitatively with a preoperatively determined treatment plan and intraoperative surface rendering for rapid 3D interactive visualization.

The contributions of this work are an algorithm for intraoperative segmentation and intraoperative registration, a method for quantitative monitoring of cryotherapy from real-time imaging, quantitative monitoring of brain tumor resection by comparison to a preoperative treatment plan and an extensive validation study assessing the reproducibility of the intraoperative segmentation. Empirically the algorithm has been found to be robust with respect to imaging characteristics such as noise and intensity inhomogeneity and robust with respect to parameter selection. Serial and parallel implementations of the algorithm are sufficiently fast to be practical in the operating room.

We have evaluated our algorithm with six neurosurgical cases and two liver cryotherapy cases. Further clinical validation with larger numbers of cases will be necessary to determine if our new approach succeeds in improving intraoperative navigation and intraoperative therapy delivery and hence improves therapy outcomes.

Acknowledgements: This investigation was supported by NIH P41 RR13218, NIH P01 CA67165 and NIH R01 RR11747.

References

1. F. Jolesz, "Image-guided Procedures and the Operating Room of the Future," *Radiology,* vol. 204, pp. 601–612, May 1997.
2. P. M. Black, T. Moriarty, E. Alexander, P. Stieg, E. J. Woodard, P. L. Gleason, C. H. Martin, R. Kikinis, R. B. Schwartz, and F. A. Jolesz, "The Development and Implementation of Intraoperative MRI and its Neurosurgical Applications," *Neurosurgery,* vol. 41, pp. 831–842, April 1997.
3. D. Gering, A. Nabavi, R. Kikinis, W. Grimson, N. Hata, P. Everett, F. Jolesz, and W. Wells, "An Integrated Visualization System for Surgical Planning and Guidance using Image Fusion and Interventional Imaging," in *MICCAI 99: Proceedings of the Second International Conference on Medical Image Computing and Computer Assisted Intervention,* pp. 809–819, Springer Verlag, 1999.
4. A. Hagemann, K. Rohr, H. Stiel, U. Spetzger, and J. Gilsbach, "Biomechanical modeling of the human head for physically based, non-rigid image registration," *IEEE Transactions On Medical Imaging,* vol. 18, no. 10, pp. 875–884, 1999.
5. O. Skrinjar and J. Duncan, "Real time 3D brain shift compensation," in *IPMI'99,* pp. 641–649, 1999.

6. M. Miga, K. Paulsen, J. Lemery, A. Hartov, and D. Roberts, "In vivo quantification of a homogeneous brain deformation model for updating preoperative images during surgery," *IEEE Transactions On Medical Imaging*, vol. 47, pp. 266–273, February 1999.

7. D. Hill, C. Maurer, R. Maciunas, J. Barwise, J. Fitzpatrick, and M. Wang, "Measurement of intraoperative brain surface deformation under a craniotomy," *Neurosurgery*, vol. 43, pp. 514–526, 1998.

8. N. Hata, *Rigid and deformable medical image registration for image-guided surgery.* PhD thesis, University of Tokyo, 1998.

9. S. Warfield, J. Dengler, J. Zaers, C. R. Guttmann, W. M. Wells III, G. J. Ettinger, J. Hiller, and R. Kikinis, "Automatic identification of Grey Matter Structures from MRI to Improve the Segmentation of White Matter Lesions," *Journal of Image Guided Surgery*, vol. 1, no. 6, pp. 326–338, 1995.

10. S. K. Warfield, M. Kaus, F. A. Jolesz, and R. Kikinis, "Adaptive Template Moderated Spatially Varying Statistical Classification," in *MICCAI 98: First International Conference on Medical Image Computing and Computer-Assisted Intervention*, pp. 231–238, Springer-Verlag, Heidelberg, Germany, October 11–13 1998.

11. M. R. Kaus, S. K. Warfield, A. Nabavi, E. Chatzidakis, P. M. Black, F. A. Jolesz, and R. Kikinis, "Segmentation of MRI of meningiomas and low grade gliomas," in *MICCAI 99: Second International Conference on Medical Image Computing and Computer-Assisted Intervention* (C. Taylor and A. Colchester, eds.), pp. 1–10, Springer-Verlag, Heidelberg, Germany, 1999.

12. R. Kikinis, M. E. Shenton, G. Gerig, J. Martin, M. Anderson, D. Metcalf, C. R. G. Guttmann, R. W. McCarley, W. E. Lorenson, H. Cline, and F. Jolesz, "Routine Quantitative Analysis of Brain and Cerebrospinal Fluid Spaces with MR Imaging," *Journal of Magnetic Resonance Imaging*, vol. 2, pp. 619–629, 1992.

13. S. K. Warfield, M. Kaus, F. A. Jolesz, and R. Kikinis, "Adaptive, Template Moderated, Spatially Varying Statistical Classification," *Medical Image Analysis*, vol. 4, no. 1, pp. 43–55, 2000.

14. I. Ragnemalm, "The Euclidean distance transform in arbitrary dimensions," *Pattern Recognition Letters*, vol. 14, pp. 883–888, 1993.

15. W. M. Wells, P. Viola, H. Atsumi, S. Nakajima, and R. Kikinis, "Multi-modal volume registration by maximization of mutual information," *Medical Image Analysis*, vol. 1, pp. 35–51, March 1996.

16. R. O. Duda and P. E. Hart, *Pattern Classification and Scene Analysis.* John Wiley & Sons, Inc., 1973.

17. J. Dengler and M. Schmidt, "The Dynamic Pyramid – A Model for Motion Analysis with Controlled Continuity," *International Journal of Pattern Recognition and Artificial Intelligence*, vol. 2, no. 2, pp. 275–286, 1988.

18. S. K. Warfield, A. Robatino, J. Dengler, F. A. Jolesz, and R. Kikinis, "Nonlinear Registration and Template Driven Segmentation," in *Brain Warping* (A. W. Toga, ed.), ch. 4, pp. 67–84, Academic Press, San Diego, USA, 1999.

19. D. V. Iosifescu, M. E. Shenton, S. K. Warfield, R. Kikinis, J. Dengler, F. A. Jolesz, and R. W. McCarley, "An Automated Registration Algorithm for Measuring MRI Subcortical Brain Structures," *NeuroImage*, vol. 6, pp. 12–25, 1997.

20. S. K. Warfield, F. A. Jolesz, and R. Kikinis, "Real-Time Image Segmentation for Image-Guided Surgery," in *Supercomputing 1998*, pp. 1114:1–14, November 1998.

21. S. G. Silverman, K. Tuncali, D. F. Adams, E. vanSonnenberg, K. H. Zou, D. F. Kacher, P. R. Morrison, and F. A. Jolesz, "MRI-Guided percutaneous cryotherapy of liver tumors: initial experience," *Radiology.* In press.

Efficient Semiautomatic Segmentation of 3D Objects in Medical Images

Andrea Schenk, Guido Prause, and Heinz-Otto Peitgen

MeVis – Center for Medical Diagnostic Systems and Visualization
Universitaetsallee 29, 28359 Bremen, Germany
{schenk,prause,peitgen}@mevis.de

Abstract. We present a fast and accurate tool for semiautomatic segmentation of volumetric medical images based on the live wire algorithm, shape-based interpolation and a new optimization method.
While the user-steered live wire algorithm represents an efficient, precise and reproducible method for interactive segmentation of selected two-dimensional images, the shape-based interpolation allows the automatic approximation of contours on slices between user-defined boundaries. The combination of both methods leads to accurate segmentations with significantly reduced user interaction time. Moreover, the subsequent automated optimization of the interpolated object contours results in a better segmentation quality or can be used to extend the distances between user-segmented images and for a further reduction of interaction time. Experiments were carried out on hepatic computer tomographies from three different clinics. The results of the segmentation of liver parenchyma have shown that the user interaction time can be reduced more than 60% by the combination of shape-based interpolation and our optimization method with volume deviations in the magnitude of inter-user differences.

1 Introduction

Fully automated segmentation is still an unsolved problem for most applications in medical imaging due to the wide variety of image modalities, object properties and biological variability. On the other hand the most general approach, manual contour tracing, is time-consuming, inaccurate and unacceptable for large three-dimensional data sets.

To overcome these problems, a lot of work has been invested in semiautomatic segmentation methods. A popular group attracting considerable attention over the last years are the two-dimensional active contour models or *snakes* introduced by Kass et al. [1]. These algorithms try to minimize an energy functional based on contour deformation and external image forces. For initialization an approximation of the object boundary is required which is usually drawn manually by the user or in some cases is generated from a priori knowledge. However, active contours are sensitive to the settings of their numerous parameters and the quality of the initial contour which limits their applicability to medical images in the clinical routine.

Another promising approach to interactive boundary detection in two-dimensional images is the *live wire* algorithm, also known as *intelligent scissors*, which was first introduced in 1992 [4,5]. With live wire the segmentation process is directly steered by the user who has immediate control over the automatically suggested object contours. The contours are found as minimal paths with respect to a cost function similar to the external energy function of active contours.

Semiautomatic segmentation of three-dimensional objects in volumetric medical images is often based on two-dimensional methods which is prohibitive when each slice needs to be segmented in a larger data set. Although snakes have a direct three-dimensional extension introduced as *balloons* [2] the propagation of an initial two-dimensional snake over the slices and subsequent adjustment is preferred in most applications for reasons of robustness and practicability. For live wire only one approach for segmentation of three-dimensional images has been published so far [3] based on the projection of few interactively defined contours onto orthogonal cross-sections through the volumetric image. With this approach it is difficult to keep track of the topology of complex shaped objects without major user interaction.

The goal of our work was to extend the two-dimensional live wire algorithm to an efficient and easy to use approach for semiautomatic segmentation of three-dimensional objects in medical images. This is achieved by reducing the number of interactively segmented slices and automatic calculation of all missing intermediate contours by an optimized interpolation method which is robust to topological changes. The resulting approach is validated in five hepatic CT scans from three different clinical sites.

2 Methods

2.1 Live Wire

The live wire algorithm is a user-steered segmentation method for two-dimensional images based on the calculation of minimal cost paths by dynamic programming [6] or Dijkstra's graph search algorithm [7]. Several modifications of the basic approach including quality studies have been published [8–11], proving the high accuracy, efficiency, and reproducibility of the algorithm.

The version of live wire we use in our approach is comparable to the algorithm in [11] and we resume the fundamental ideas of this method. The two-dimensional image is transformed into a directed and weighted graph. Vertices of the graph represent image pixels while edges connect neighboring pixels in two directions. The edges are weighted with local cost functions, related to the external image forces of active contours. The cost function used in our implementation is a weighted sum of different gradient functions, the Laplacian function and the gray values at the object boundary.

After the calculation of the cost graph the user starts with the segmentation by setting a first seed point on the boundary and moving the mouse along the outline of the object. Shortest paths – in the sense of paths with lowest accumulated costs – from the seed to at least the current mouse position are computed using Dijkstra's algorithm. Computation and display of the resulting live wire boundary segment is achieved in real-time even on larger images. The live wire path snaps to the boundary while the user moves the mouse over the image and a new seed point has to be set before the path starts to deviate from the desired contour. New shortest paths are computed from the new seed point and the procedure is repeated. A final closing leads to a controlled and piece-wise optimal segmentation result.

Our live wire approach is provided with the additional concepts of boundary snapping, data-driven cooling for automated generation of seed points and learning of cost function parameters [6, 8].

2.2 Shape-Based Interpolation

Interpolation techniques [12] are used for many applications in medical imaging, e.g. image generation, resampling and visualization. They can be broadly divided into two categories: scene-based and object-based interpolation. While scene-based techniques determine the interpolated values directly from the intensities of the given image, object-based interpolation uses additional information about the objects in the image.

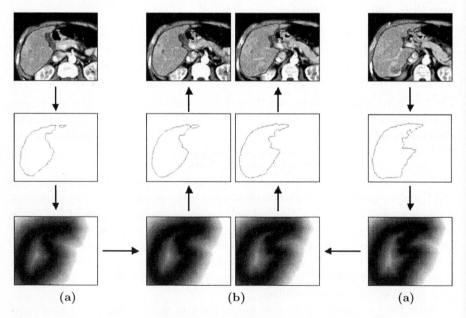

(a) (b) (a)

Fig. 1. Shape-based interpolation: **(a)** user-defined contour in CT images (top), binary scenes (middle), distance images (bottom); **(b)** interpolated distance images (bottom), binary scenes (middle), interpolated contours in CT images (top).

The *shape-based interpolation* method we use in our approach is the first object-based interpolation approach published in the literature [13] and it interpolates between binary images of a three-dimensional data set.

Shape-based interpolation consists of the steps illustrated in Figure 1. In the first step a binary scene is generated from a given object contour. Subsequently, the distance to this boundary is mapped into a new gray-level scene with positive distance values inside and negative distance values outside the object (shown as absolute values in Figure 1). In the third step the distance images are interpolated employing a conventional gray-level interpolation technique such as linear or higher-order interpolation. The interpolated gray-level scenes are converted back to a binary contour image by identifying the zero-crossings.

The distance transformation can be calculated efficiently with a version of the city-block distance [13] or, as in our implementation, with two consecutive chamfering processes [14, 15] realized with 3×3-kernel operations. The chamfering method leads to more accurate results since this transformation is a better approximation of the Euclidean distance.

2.3 Combination of Live Wire and Interpolation

For the segmentation of volumetric objects we have combined the user-steered two-dimensional live wire segmentation and the fully-automated shape-based interpolation. The user starts with the live wire algorithm on individually selected slices. If contours on at least two slices are available, all contours on slices in between can be computed utilizing the shape-based interpolation.

The user has free choice of applying the two methods: either live wire and shape-based interpolation in an alternating fashion, or – in a two-stage approach – interactive contour tracing first and interpolation subsequently. The only restriction is that the first and the last slice of the object of interest must be segmented interactively with the live wire algorithm.

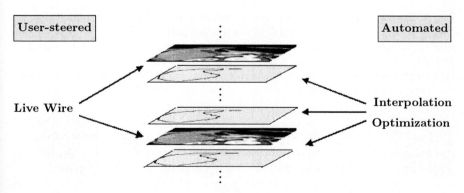

Fig. 2. Combined segmentation scheme: User-steered segmentation of selected slices followed by automatic interpolation and optimization of the intermediate contours.

If the variations of the object boundary between user-selected slices are significant, the interpolation result can be unsatisfying. In this case there is a need for interactive or automatic optimization methods which can be also used for further reduction of the number of interactively segmented image slices.

2.4 Optimization of Interpolated Contours

For the optimization of the interpolated contours we use the basic method of the live wire algorithm and recalculate contour segments as optimal cost paths between seed points with Dijkstra's algorithm.

After the shape-based interpolation the zero-crossings of the interpolated distance images define the boundary points of the new contours. To bring these points into the correct order and to identify separate contours we apply a path search algorithm. Starting with a first contour point (e.g. lower right) we follow neighboring pixels in counterclockwise direction (Figure 3a) until the first point is reached again. The algorithm is repeated until all boundary points are assigned to a contour path and it is provided with a backtracking method to deal with cases of contours connected by a single pixel line (Figure 3b).

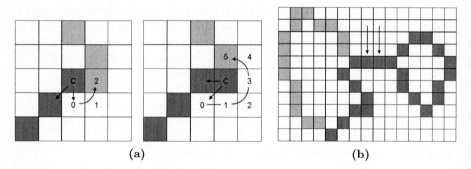

(a)   (b)

Fig. 3. Tracing the contour in the zero-crossings (gray) of the interpolated distance images. a) From the current pixel C of the marked path (dark gray) the next contour point is found as first (light gray) neighbor in counterclockwise direction (numbers describe the search order in the 8-neighborhood). b) If two boundaries are connected via a single pixel line a backtracking algorithm separates the single contours and deletes the connecting pixels (arrows).

To establish a basis for the minimal path search, the seed points from the two adjacent user-defined contours are projected onto the slices with interpolated boundaries. For every user-defined seed point the nearest point on the boundary path is determined. If two points are closer together than a given minimal distance they are merged to one central seed point. The utilization of the user-defined instead of e.g. equally distributed seeds along the contour is suggestive, since these points represent the knowledge of the user in regions of high curvature or weak edges.

The automatic contour optimization starts on all interpolated slices with the calculation of the cost matrix. Based on this data structure and the approximated seed points all contour segments linking two seeds are computed as minimal cost paths.

In the interactive optimization mode the user moves, deletes or inserts new seed points and the piece-wise optimal paths concerning these seeds are recomputed and displayed in real-time.

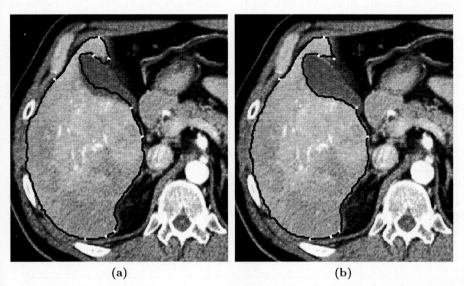

(a)   (b)

Fig. 4. Liver CT with segmented parenchyma: **a)** Image with shape-based interpolated contour and approximated seed points from the two adjacent user-defined slices; larger deviations occur at the gall bladder (dark gray) **b)** contour after automated optimization.

3   Experiments and Results

Our approach has been successfully applied to CT and MR images of the liver and the lung for interactive segmentation of the parenchyma. It is part of an ongoing clinical project for the approximation of patient-specific organ segments and preoperative planning of surgery [17, 18]. Compared with the other image processing steps of this project the segmentation is the most time-consuming part.

The test presented here was designed to quantify the volumetric deviations that have to be expected when the interactive segmentation is reduced and gradually replaced by automated interpolation. For the experiment we used five diagnostic liver CT data sets (I1-I5) from three different hospitals. The underlying examinations are helical CTs during arterial portography (CTAP) (I1-I4) and one biphasic spiral CT scan (I5, data set from the portal venous phase). All images have 512×512 pixels in plane and consist of 34–41 (I1-I4) or 86 (I5) slices with a distance of 4 mm (I1-I4) or 2 mm (I5). Three users segmented the liver parenchyma with the live wire algorithm on all slices and the time needed for the segmentation was recorded. For one image slice an average time of 6-11 seconds and 4-6 seed points were needed depending on the user. The cost function used for the test was identical for all images and was based on a Sobel-like gradient function (feature f_5 in [10] extended to 8-neighborhood) and the gray values at the object border with relative weights of 0.45 and 0.55, respectively. Parameters for both functions (mean and deviation) were determined experimentally.

Table 1. Results for the test images: inter-user variability, volume deviation for the SBI and OPT interpolation methods and saved interaction time (in %).

	Inter-user variability	Automated interpolations SBI	SBI+OPT	Saved interaction time
Image 1	1.68			
8 mm		1.24	0.79	51.0
12 mm		2.03	1.20	65.5
16 mm		3.20	1.95	72.7
Image 2	3.47			
8 mm		2.28	1.83	47.4
12 mm		2.70	1.65	62.3
16 mm		5.70	4.35	72.8
Image 3	1.43			
8 mm		1.57	1.19	51.9
12 mm		2.87	1.98	65.3
16 mm		3.81	2.69	75.0
Image 4	1.24			
8 mm		1.34	1.17	50.1
12 mm		1.89	1.30	67.5
16 mm		2.50	1.92	75.2
Image 5	1.26			
8 mm		1.91	1.85	74.5
12 mm		2.20	1.67	82.9
16 mm		3.17	2.30	86.5

The interactively segmented slices were reduced to data sets with slice distances of 8, 12 and 16 mm and the omitted intermediate slices were automatically calculated with shape-based interpolation (SBI) and, additionally, with the cost-based optimization described above (OPT).

We defined the user-dependent reference volumes as the results of the interactive segmentation on all slices. The differences between two reference volumes (inter-user variability) or between an interpolated volume and its reference volume were calculated with the formula

$$1 - \frac{2|V_i \cap V_j|}{|V_i| + |V_j|}, \qquad i \neq j,$$

which is equal to 1 minus the similarity index [16]. These relative volume differences were averaged over the three users and listed as inter-user variability, respectively volume deviations for the automated interpolations in Table 1.

The results for the sequences I1-I4 show that we can save more than 50% of interaction time with the shape-based interpolation alone if we tolerate deviations in the order of the inter-user variability. Moreover, we can use the optimization method (OPT) for a further reduction of the differences (up to one half) or for further savings of user interaction time.

With an optimized interpolation (SBI + OPT) between all 12 mm slices we have comparable deviations to the SBI method for 8mm slices, but interaction time savings of approximately 66% instead of 50%. The interpolation on the 16 mm distant slices lead to deviations of 2.5-5.7%. With automated optimization these results can be improved, but the inter-user variability can only be reached with manual corrections which require extra interaction time.

With the biphasic image sequence I5 we achieved similar results to those of I1-I4, but due to the smaller slice distance higher savings of interaction time. The interpolation leads to higher deviations compared to those of the other four images, which is also caused by more than twice the number of intermediate slices. Here we have for example a volume deviation of 1.67% and interaction time savings of 82.9% with the shape-based interpolation and automated optimization on 12 mm distant slices (set of every sixth slice) compared to the interactive segmentation on all slices.

If the user verifies the automatically generated contours, the supported manual optimization can be used for further improvements. Corrections of the SBI + OPT results on few slices where the automatically determined lowest cost paths considerably deviate from the object boundary (Figure 5, right), usually lead to a significant reduction of the volume differences.

Fig. 5. Surface rendering of the user-segmented volume in image I1 (left), deviations between this volume and the 12mm SBI volume (middle), deviations for 12mm SBI + OPT (right).

4 Conclusions

In this paper we presented a new approach for the semiautomatic segmentation of three-dimensional objects in medical images. The combination of the live wire algorithm and shape-based interpolation builds an effective and accurate segmentation tool which we extended by an automated optimization method.

· The live wire algorithm provides direct control to the user, is easy to use and offers high flexibility regarding image modalities and segmentation objects. For the segmentation of three-dimensional objects we combined this slice-based method with shape-based interpolation, leading to a significant reduction of interaction time compared to the segmentation on all slices. The applicability of this combination depends on the properties of the object contours, it is less appropriate for strongly undulated boundaries (e.g. white/gray matter in the brain), but a good choice for structures with a more regular surface (e.g. liver, lung). The shape-based interpolation is especially suited for image sequences with a large number of image slices like multi-slice CTs.

We showed in experiments with five hepatic CTs that we can considerably improve the interpolation results or decrease – with constant level of accuracy – the number of user-segmented slices with an automatic optimization method based on the ideas of the live wire algorithm. In the validation we have used equal distances for the interactively segmented images, but we expect even better results with slices individually selected by the experienced user who rates the boundary changes during the segmentation process and specifies the images for contour interpolation.

While the shape-based interpolation and the path-finding algorithm can be calculated efficiently in a few seconds for all interpolated slices, the automated optimization requires more computational time primarily due to the calculation of the cost graph on every interpolated image. Therefore one focus in our current work is the local restriction of cost computation.

Further improvements regarding the accuracy of the interpolated contours can be expected from a better adjustment of the approximated seed points to the underlying image data.

Acknowledgments

The computer tomographies were provided by the Klinikum der Medizinischen Hochschule Hannover, Staedtische Krankenanstalten Krefeld and Klinikum Hof. We would like also to thank D. Boehm, Dr. B. Preim, Dr. R. Rascher-Friesen-hausen and Dr. D. Selle for their support in this project.

References

1. M. Kass, A. Witkin, and D. Terzopoulos. Snakes: active contour models. *International Journal of Computer Vision* 1(4), pp. 321–331, 1988.

2. D. Terzopoulos, A. Witkin, and M. Kass. Constraints on deformable models: Recovering 3D shape and nonrigid motion. *Artificial Intelligence* 36(1), pp. 91–123, 1988.

3. A. X. Falcao and J. K. Udupa. Segmentation of 3D objects using live-wire. In *SPIE on Medical Imaging*, vol. 3034, pp. 191–198, Newport Beach, CA. 1997

4. E. N. Mortensen, B. S. Morse, W. A. Barrett, and J. K. Udupa. Adaptive boundary detection using live-wire two-dimensional dynamic programming. In *IEEE Computers in Cardiology*, pp. 635–638, Durham, North Carolina, IEEE Computer Society Press, 1992.

5. J. K. Udupa, S. Samarasekera, and W. A. Barrett. Boundary detection via dynamic programming. In *Visualization in Biomedical Computing '92*, Chapel Hill, North Carolina, pp. 33–39, 1992.

6. E. N. Mortensen and W. A. Barrett. Intelligent scissors for image composition. In *Computer Graphics (SIGGRAPH '95)*, Los Angeles, CA, pp. 191–198, 1995.

7. D. Stalling and H.-C. Hege. Intelligent scissors for medical image segmentation. In *Digitale Bildverarbeitung fuer die Medizin*, Freiburg, Germany, pp. 32–36, 1996.

8. W. A. Barrett and E. N. Mortensen. Fast, accurate and reproducible live-wire boundary extraction. In *Visualization in Biomedical Computing*, Hamburg, Germany, pp. 183–192, 1996.

9. W. A. Barrett and E. N. Mortensen. Interactive live-wire boundary extraction. *Medical Image Analysis* 1(4), pp. 331–341, 1997.

10. A. X. Falcao, J. K. Udupa, S. Samarasekera, S. Sharma, B. E. Hirsch, and R. A. Lotufo. User-steered image segmentation paradigms: Live-wire and live-lane. *Graphical Models and Image Processing* 60(4), pp. 223–260, 1998.

11. A. X. Falcao, K. Jayaram, J. K. Udupa, and Miyazawa F. K. An ultra-fast user-steered image segmentation paradigm: Live-wire-on-the-fly. In *SPIE on Medical Imaging*, vol. 3661, Newport Beach, CA, pp. 184–191, 1999.

12. T. M. Lehmann, C. Goenner, and K. Spitzer. Survey: Interpolation methods in medical image processing. *IEEE Transactions on Medical Imaging* 18(11), pp. 1049–1075, 1999.

13. S. P. Raya and J. K. Udupa. Shape-based interpolation of multidimensional objects. *IEEE Transactions on Medical Imaging* 9(1), pp. 32–42, 1990.

14. G. T. Herman, J. Zheng, and Bucholtz C. A. Shape-based interpolation. *IEEE Computer Graphics and Applications* 12(3), pp. 69–79, 1992.

15. G. Borgefors. Distance Transformations in Arbitrary Directions. *Computer Vision, Graphics and Image Processing* 27(3), pp. 321–345, 1984.

16. A. P. Zijdenbos, B. M. Dawant, R. A. Margolin, and A. C. Palmer. Morphometric Analysis of White Matter Lesions in MR Images: Method and Validation. *IEEE Transactions on Medical imaging* 13(4), pp. 716–724, 1994.

17. D. Selle, T. Schindewolf, C. J. G. Evertsz, and H.-O. Peitgen. Quantitative analysis of CT liver images. In *Computer-Aided Diagnosis in Medical Imaging*, Chicago, pp. 435–444, Elsevier, 1999.

18. B. Preim, D. Selle, W. Spindler, K. J. Oldhafer, H.-O. Peitgen. Interaction Techniques and Vessel Analysis for Preoperative Planning in Liver Surgery. In *Medical Image Computing and Computer-Assisted Intervention – MICCAI 2000* (this volume).

Space and Time Shape Constrained Deformable Surfaces for 4D Medical Image Segmentation

Johan Montagnat and Hervé Delingette

INRIA Sophia, BP 93, 06902 Sophia-Antipolis Cedex, France
http://www-sop.inria.fr/epidaure/

Abstract. The aim of this work is to automatically extract quantitative parameters from time sequences of 3D images (4D images) suited to heart pathology diagnosis. In this paper, we propose a framework for the reconstruction of the left ventricle motion from 4D images based on 4D deformable surface models. These 4D models are represented as a time sequence of 3D meshes whose deformation are correlated during the cardiac cycle. Both temporal and spatial constraints based on prior knowledge of heart shape and motion are combined to improve the segmentation accuracy. In contrast to many earlier approaches, our framework includes the notion of trajectory constraint. We have demonstrated the ability of this segmentation tool to deal with noisy or low contrast images on 4D MR, SPECT, and US images.

1 Context

Recently, the improvement of medical image acquisition technology has allowed the production of time sequences of 3D medical images (4D images) for several image modalities (CT, MRI, US, SPECT). Tagged MRI is the gold standard of heart motion analysis since it is the only modality permitting the extraction of the motion of physical points located in the myocardium [18]. However, other modalities may be used for meaningful parameters extraction at a lower cost. In particular, the fast development of 3D US imaging is very promising due to its accessibility and low cost [17].

The main target for these new ultra-fast image acquisition devices is to capture and analyze the heart motion through the extraction of quantitative parameters such as volume, walls thickness, ejection fraction and motion amplitude. In order to estimate these parameters, it is necessary to reconstruct the Left Ventricle (LV) motion during a cardiac cycle. Tracking the LV motion in 2D or 3D image sequences has led to several research efforts [10, 9, 2]. Tracking [12, 16] and motion analysis [4, 7] based on deformable models in 4D images take into account time continuity and periodicity to improve their robustness.

In this paper, we propose to track the LV based on 4D deformable models. Our concept of 4D deformable surface models combines spatial and temporal constraints which differs from most previous approaches [12, 16, 4] that decouple them. Furthermore, in contrast to the strategy presented in [7], the motion estimation is not parameterized by a global time-space transformation. It leads to more efficient computation and to greater descriptive ability in motion recovery. Finally, our approach can include the notion of trajectory constraint which is the generalization of the shape constraint.

2   4D Deformable Models

Let I denote a 4D image composed of n volume images corresponding to n different time points $\{t_0, \ldots, t_{n-1}\}$. We define a 4D deformable model $\mathcal{S}$ as a set of n deformable surfaces $\{\mathcal{S}_t\}_{t \in [0, n-1]}$, each surface model $\mathcal{S}_t$ representing a given anatomical structure at time point t. Among the possible geometric representations of deformable surfaces [13], we have chosen the *simplex meshes* [8] discrete surfaces. They are defined by a set of vertices and a constant connectivity function. Their main advantage lies in their simple data structure permitting an efficient implementation both in terms of computational time and memory storage. This is specifically important in the case of 4D deformable models where n surface meshes must be updated at each iteration. Furthermore, simplex meshes are especially well-suited for the computation of curvature-based regularizing forces. All n surface meshes $\mathcal{S}_t$ have the same topology, i.e. there is a one to one correspondence between the d vertices composing each surface. In the rest of the paper, $\mathbf{p}_{i,t}$ denotes the position of vertex number i at time t. While the model undergoes deformations, each surface $\mathcal{S}_t$ evolves in space but it remains at its time step (i.e., t does not change).

A 4D model deforms under the combined action of three forces aiming at recovering the shape and motion of an anatomical structure: (i) the data, or *external*, force attracts each vertex towards the structure boundaries; (ii) the spatial regularizing, or *internal*, force ensures the smoothness of the deforming surface by introducing spatial continuity constraints in the deformation process; (iii) the temporal regularizing force similarly relies on prior knowledge on the time dimension continuity to regularize the deformation process. A second order (Newtonian) evolution scheme discretized using an explicit scheme governs the displacement of each vertex (see [8] for details):

$$\mathbf{p}_{i,t}^{k+\Delta k} = \mathbf{p}_{i,t}^{k} + (1-\gamma)(\mathbf{p}_{i,t}^{k} - \mathbf{p}_{i,t}^{k-\Delta k}) + \alpha_i f_{\text{int}}(\mathbf{p}_{i,t}^{k}) + \delta_i f_{\text{time}}(\mathbf{p}_{i,t}^{k}) + \beta_i f_{\text{ext}}(\mathbf{p}_{i,t}^{k}), \quad (1)$$

where f_{int}, f_{time}, and f_{ext} are the internal, the temporal, and external forces respectively. α_i, β_i, and δ_i are weights including the vertex mass and the iteration step Δk. In all our experiments, the background damping γ is fixed to value 0.35 based on an empirical study showing that this value optimizes the convergence speed in general. The α_i values are always fixed to 1.

Simplex meshes provide a powerful framework for computing internal regularizing forces [8] including smoothing forces without shrinking side effect. External forces are computed as distance functions of the model vertices to the data. This speeds-up the model convergence compared to potential fields approaches and it avoids oscillations [5]. Deformations are computed along each vertex normal direction to avoid creating distorted meshes. Both gradient based and region based criteria are used to determine boundary voxels in images. For the sake of brevity, the external forces computation is not discussed here and the reader may refer to [15] for details.

3 Shape and Temporal Constraints

The main incentive for performing medical image segmentation based on deformable models lies in their ability to incorporate prior knowledge on the data that is being recovered. In most cases, this knowledge is translated mathematically into a set of regularizing constraints that greatly improves the robustness and accuracy of the segmentation process. Indeed, many methods have been proposed to regularize deformations by limiting the model number of parameters [20, 6], or controlling the kind of deformation applied onto the model [3, 11].

We introduce two complementary constraints that are specifically suited for the LV tracking in 4D images. The former consists of a shape constraint that tends to enforce 3D geometric continuity. The latter is a temporal constraint that causes a 4D mesh to rely on prior motion knowledge. It is important to note that, in contrast to many previous works, both constraints are applied simultaneously thus leading to a true 4D approach. Furthermore, each constraint can encapsulate a weak or strong prior knowledge, as summarized in table 1.

Prior knowledge	Spatial constraint	Temporal constraint
Weak	Curvature-based shape smoothing	Temporal position averaging
Strong	Shape constraint	Trajectory constraint

Table 1. Spatial and temporal constraint depending of the amount of prior knowledge.

3.1 Shape Constraints

In the case where no reference shape is known (weak shape constraint), we use the regularizing force defined in [8], that minimizes the variation of mean-curvature over the mesh. Otherwise, we add an additional shape constraint force f_{shape} that is related to a reference shape S' of the anatomical structure. It introduces shape prior knowledge by extending the globally-constrained deformation scheme described in [14] to the 4D case. Let S^k denotes the 4D model after the kth iteration. At initialization, $S'^0 = S^0$. At each iteration, external forces f_{ext} are computed for each vertex so that $\mathbf{p}_{i,t} + f_{\text{ext}}(\mathbf{p}_{i,t})$ corresponds to the myocardium boundary point that best matches $\mathbf{p}_{i,t}$. We estimate a global transformation T^k belonging to a given group of transformations (e.g. affine transformations T_{affine}). T^k approximates the external force field by minimizing the least square criterion:

$$T^k = \arg \min_{T \in \text{T}_{\text{affine}}} \left\{ \sum_{t=0}^{n-1} \sum_{i=0}^{d-1} \| T(\mathbf{p}_{i,t}) - (\mathbf{p}_{i,t} + f_{\text{ext}}(\mathbf{p}_{i,t})) \|^2 \right\}. \quad (2)$$

We then update the reference shape: $S'^{k+\Delta k} = T^k \circ T^{k-1} \ldots \circ T^1(S'^0)$. Thus $S'^{k+\Delta k}$ remains identical to S'^0 up to an affine transformation. A shape force is defined on each vertex of $\mathcal{S}^k$ as a spring-like force towards its updated reference position: $f_{\text{shape}}(\mathbf{p}^k_{i,t}) = \mathbf{p}^k_{i,t}\mathbf{p}'^k_{i,t}$.

Furthermore, a *locality* parameter λ is introduced to weight the influence of the shape force relative to the internal and external forces as described in [14]:

$$\mathbf{p}^{k+\Delta k}_{i,t} = \mathbf{p}^k_{i,t} + (1 - \gamma)(\mathbf{p}^k_{i,t} - \mathbf{p}^{k-\Delta k}_{i,t}) +$$
$$\lambda(\alpha_i f_{\text{int}}(\mathbf{p}^k_{i,t}) + \delta_i f_{\text{time}}(\mathbf{p}^k_{i,t}) + \beta_i f_{\text{ext}}(\mathbf{p}^k_{i,t})) + (1 - \lambda)f_{\text{shape}}(\mathbf{p}^k_{i,t}).$$

When $\lambda = 0$, the 4D model is deformed through the application of a global transformation from its reference shape, thus making the deformation process robust to noise and outliers. Conversely, if $\lambda = 1$, only the weak shape constraint applies and the model shape variability is very high. Any intermediate value of λ produces local deformations combined with a global shape constraint.

3.2 Temporal Constraints

The temporal regularizing force f_{time} is defined as a spring-like force $f_{\text{time}}(\mathbf{p}_{i,t}) = \tilde{\mathbf{p}}_{i,t} - \mathbf{p}_{i,t}$ attracting vertex $\mathbf{p}_{i,t}$ towards a reference point $\tilde{\mathbf{p}}_{i,t}$. When no prior knowledge is used, we define a weak temporal constraint by attracting $\mathbf{p}_{i,t}$ towards the middle position of its two temporal neighbors: $\tilde{\mathbf{p}}_{i,t} = \frac{\tilde{\mathbf{p}}_{i,t-1}+\mathbf{p}_{i,t+1}}{2}$. Applying this force is equivalent to minimizing the speed of each vertex, and therefore to minimizing the kinetic energy of the 4D model.

When using prior information on the trajectory of each vertex, we determine $\tilde{\mathbf{p}}_{i,t}$ such that this point lies on the ideal vertex trajectory. It is important to note that these trajectories usually do not correspond to trajectories of physical points lying on the myocardium but are only used as mathematical constraints. To store prior trajectories, we could store the n vertex positions $\{\mathbf{p}_{i,t}\}_{t\in[0,n-1]}$ of each vertex over time. However, this representation would imply that the trajectory orientation and scale is constant between images, which is not the case. Instead, we choose to store the 3D curve trajectory as a set of geometric parameters $\{\varepsilon_{i,t}, \varphi_{i,t}, \psi_{i,t}\}$ that are invariant to rotation, translation, and scale. The left side of Fig. 1 illustrates the elements composing the trajectory geometry.

Let $\mathbf{p}^{\perp}_{i,t}$ denote the orthogonal projection of $\mathbf{p}_{i,t}$ onto its two temporal neighbors segment $[\mathbf{p}_{i,t-1}, \mathbf{p}_{i,t+1}]$. The position of point $\mathbf{p}_{i,t}$ may be defined through: (i) the *metric parameter* $\varepsilon_{i,t}$ measuring the relative position of $\mathbf{p}^{\perp}_{i,t}$ in segment $[\mathbf{p}_{i,t-1}, \mathbf{p}_{i,t+1}]$; (ii) the angle $\varphi_{i,t}$ measuring the elevation of $\mathbf{p}_{i,t}$ above the segment $[\mathbf{p}_{i,t-1}, \mathbf{p}_{i,t+1}]$ in the plane $(\mathbf{p}_{i,t-1}, \mathbf{p}_{i,t}, \mathbf{p}_{i,t+1})$; and (iii) the angle $\psi_{i,t}$ measuring the discrete torsion. Intuitively, $\varepsilon_{i,t}$, $\varphi_{i,t}$, and $\psi_{i,t}$ correspond to discrete arc length, curvature, and torsion respectively. Let $\mathbf{t}_{i,t}$ denote the discrete tangent, $\mathbf{b}_{i,t}$ the binormal vector, and $\mathbf{n}_{i,t}$ the discrete normal to point $\mathbf{p}_{i,t}$ respectively:

$$\mathbf{t}_{i,t} = \frac{\mathbf{p}_{i,t-1}\mathbf{p}_{i,t+1}}{\|\mathbf{p}_{i,t-1}\mathbf{p}_{i,t+1}\|}, \quad \mathbf{b}_{i,t} = \frac{\mathbf{p}_{i,t}\mathbf{p}_{i,t+1} \wedge \mathbf{p}_{i,t-1}\mathbf{p}_{i,t}}{\|\mathbf{p}_{i,t}\mathbf{p}_{i,t+1} \wedge \mathbf{p}_{i,t-1}\mathbf{p}_{i,t}\|}, \quad \mathbf{n}_{i,t} = \mathbf{b}_{i,t} \wedge \mathbf{t}_{i,t}.$$

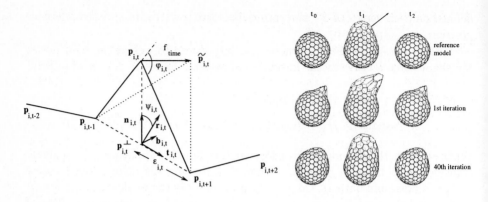

Fig. 1. Left: trajectory geometry and temporal force. Right: temporal force effect.

The metric parameter is defined by $\varepsilon_{i,t} = \frac{\|\mathbf{p}_{i,t}^\perp \mathbf{p}_{i,t+1}\|}{\|\mathbf{p}_{i,t-1}\mathbf{p}_{i,t+1}\|}$, the elevation angle is $\varphi_{i,t} = (\mathbf{p}_{i,t}\mathbf{p}_{i,t+1}, \mathbf{p}_{i,t-1}\mathbf{p}_{i,t})$, and the torsion angle is such that:

$$\mathbf{r}_{i,t} = \frac{\mathbf{t}_{i,t} \wedge (\mathbf{p}_{i,t-2}\mathbf{p}_{i,t-1} \wedge \mathbf{p}_{i,t+1}\mathbf{p}_{i,t+2})}{\|\mathbf{t}_{i,t} \wedge (\mathbf{p}_{i,t-2}\mathbf{p}_{i,t-1} \wedge \mathbf{p}_{i,t+1}\mathbf{p}_{i,t+2})\|} \text{ and } \mathbf{n}_i = \cos(\psi_{i,t})\mathbf{r}_{i,t} + \sin(\psi_{i,t})\mathbf{t}_{i,t} \wedge \mathbf{r}_{i,t}.$$

The position of $\mathbf{p}_{i,t}$ is related to the position of its neighbors and the trajectory parameters by equation:

$$\mathbf{p}_{i,t} = \varepsilon_{i,t}\mathbf{p}_{i,t-1} + (1 - \varepsilon_{i,t})\mathbf{p}_{i,t+1} + \tag{3}$$
$$h(\varepsilon_{i,t}, \varphi_{i,t}, \mathbf{p}_{i,t-1}, \mathbf{p}_{i,t+1})(\cos(\psi_{i,t})\mathbf{r}_{i,t} + \sin(\psi_{i,t})\mathbf{t}_{i,t} \wedge \mathbf{r}_{i,t}),$$

where $h = \|\mathbf{p}_{i,t}\mathbf{p}_{i,t}^\perp\|$. The temporal force is computed with $\tilde{\mathbf{p}}_{i,t}$ defined by equation 3 using the trajectory reference parameters.

The right side of Fig. 1 shows the temporal constraint effect. A spherical 4D model composed of 3 time points (t_0, t_1, and t_2) is shown in the upper row. A single vertex of the model is submitted to an external force at time t_1. The middle row shows the resulting deformation. Surface $\mathcal{S}_1$ is deformed causing surfaces $\mathcal{S}_0$ and $\mathcal{S}_2$ to deform through the temporal constraint, although the deformation is attenuated in time. The bottom row shows the surface converging towards its reference motion after 40 iterations.

3.3   Initialization procedure

In general, to get a first rough position of the 4D LV model, we first proceed by using only highly constrained spatial deformations without any temporal constraint. By using $\lambda = 0$, we basically estimate a set of global affine transformations to align the model with the 4D dataset. Then, we proceed by iteratively increasing the locality parameter λ while adding temporal constraints. This approach allows an evolutional deformation scheme based on a coarse-to-fine strategy.

4   4D Medical Image Segmentation

The 4D model described above has been used to segment 4D MR, SPECT and US images. Figure 2 shows two slices of each image modality at the end of diastole and the end of systole. Due to real time imaging constraints, all images have a sparse resolution. Cardiac MR images have a very high resolution in slice planes. However, the third dimension resolution is much lower ($256 \times 256 \times 9$). SPECT images are sampled on a 64^3 voxel grid. Finally, the 4D US images shown are composed by a set of slices acquired with a rotative probe [15] leading to a low spatial resolution ($256 \times 256 \times 9$, with a 20 degrees angle between two slices).

The 4D model used are made of 500 to 700 vertices per surface. This rough resolution is adapted to the images level of detail. In MR and US images, the internal wall of the LV is reconstructed by a closed surface representing the internal blood volume. In the case of SPECT images, the internal and external walls of the LV appear. A cup shaped surface model is then used. The US image sequences are composed of 8 time points covering only the systole while the MR (13 time points) and the SPECT (8 time points) sequences cover a complete heart cycle.

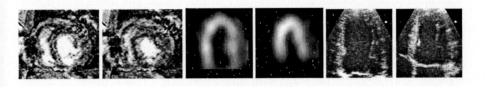

Fig. 2. MR (left), SPECT (centre), and US (right) slices at end diastole and end systole

4.1   MR Images

We show a segmentation experiment on one heart beat sequence. The cardiac MR images contrast varies between slices and the heart boundaries are poorly defined. A 4D model is generated by embedding a set of identical ellipsoids roughly centered on the LV in the first image sequence. Only spatial and temporal smoothing (weak) constraints are used since no relevant prior information was given. The local deformations are constrained by a global affine transformation. The coarse-to-fine deformation algorithm involves two stages composed of 30 iterations each: $\lambda = 10\%$ and $\beta = 0.5$ followed by $\lambda = 40\%$ and $\beta = 0.1$. The weight values $\alpha = 1$ and $\delta = 0.1$ are fixed. The low λ value prevents the surface from being too sensitive to the lack of information in area where the gradient filter gave no response.

Each surface model is composed of 500 vertices and the deformation process for the whole 4D model only takes 1 min 46 s on a 500 MHz Digital PWS with 512 Mb of memory. Four out of the 13 surfaces composing the 4D model are shown in Fig. 3 on the left. The middle graph plots the curve of volume variation through time. It corresponds to an healthy case volume curve.

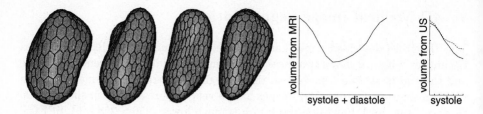

Fig. 3. Left: 4 surfaces from the 4D models deformed in an MRI. Middle: 4D model volume curve. Right: 4D model volume obtained from a US image by 3D (dashed line) and 4D (solid line) segmentation.

4.2 SPECT Images

We have processed a 4D SPECT image database including healthy and pathological patients. The systole is approximatively three time points long whereas the diastole takes the remaining five time points. We compare images of healthy patients with a normal endocardium blood perfusion and pathological patients with an abnormal perfusion due to some ischemic zones. The mean deformation time for all 4D models, made of 700 vertex surfaces, is 2 min 34 s.

The reference model is built from an healthy patient image by 3D segmentation. The high image contrast allows us to use gradient information to compute external forces. The 3D segmentation does not involve any time continuity constraints. A 4D deformation stage with time smoothing forces is therefore needed to obtain a reference model with reliable shape and motion. Shape constraints are especially beneficial for the segmentation of low contrasted images showing pathologies such as ischemia. Due to the similarity between images, the 4D model is roughly initialized in its reference position. Rigid then similarity registration are first used to compensate the differences in location and size between patients ($\alpha = 1$, $\beta = 0.1$, $\delta = 0.2$). Local deformations with an affine constraint are then used. The deformation involves 3 stages (20 iterations each) in the coarse-to-fine algorithm: the locality increases ($\lambda = 20\%, 50\%, 70\%$) while the external force range decreases (range= 4, 2, 2 voxels).

Figure 4 shows a frontal view of the 4D models. Top line displays the reference model obtained by 3D segmentation and revealing poor time continuity. The center line displays the 4D model regularized by time smoothing constraints in the image of a healthy patient. The bottom line shows the model extracted from a pathological case by 4D segmentation. The surface model reveals the pathological heart with weak motion amplitude.

4.3 Ultrasound Images

The speckle noise of US images and the lack of beam reflection on boundaries tangent to the US rays make the segmentation process difficult. A 4D model is built by 3D segmentation of a 8 time points reference image. The model is

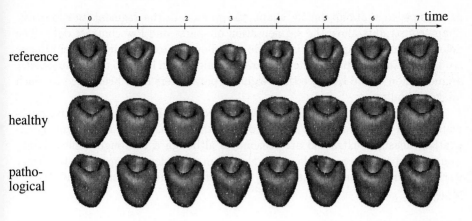

Fig. 4. 4D models of the myocardium: reference model obtained by 3D segmentation (top row), healthy case (middle row), and pathological case (bottom row).

first registered by a similarity transformation to align and adapt its scale to the data. The gradient information is sufficient since the deformations are strongly constrained to speed-up the model convergence. A large force range allows the model to find boundaries far from the initial model position. After registration, the model locally deforms with an affine global constraint ($\alpha = 1$, $\beta = 0.5$, $\delta = 0.1$, range = 20 voxels). Local deformations guided by region based forces are used in the final segmentation. Region-based forces slow down the deformation process and the total reconstruction time was 4 min 42 s.

Right of Fig. 3 shows the evolution of the LV volume through time (solid line). The volume is compared to the result of an iterative 3D segmentation of the same sequence from an earlier study [15] (dashed line). The time regularizing constraints make the 4D curve much more regular. The initial volume value is very close (3% difference) but it grows with time. This is not surprising since the 3D segmentation tends to accumulate errors. Moreover, the 4D curves shows a profile closer from the theoretical line expected. The model volume leads to a 49% ejection fraction. This value compares to the 45% ejection fraction computed from a manual segmentation by a cardiologist on the same sequence.

Figure 5 shows the sequence slices on which are superimposed the model intersections with each plane. The 8 figure columns correspond to the 8 time points. Five rows corresponding to one slice out of two (from top to bottom: 0, 40, 80, 120, and 160 degrees of arc) are shown.

5   Conclusion

We have demonstrated the ability of 4D models to track the LV motion in 4D noisy medical images. The proposed framework relies on complementary spatial and temporal constraints to regularize the deformation while introducing prior knowledge about the LV shape and motion in the segmentation process. Shape constraints allow the segmentation of sparse and low contrast data. The deformable models approach is generic and allows us to deal with different image modalities. In all examples shown above, the algorithm leads to a fully automatic segmentation of the LV once the weighting parameters have been fixed for each image modality.

The resulting surface models are well suited for estimating quantitative parameters such as endocardium volume or wall thickness. Visual results and quantitative measures extracted are reasonable although a thorough clinical study would be necessary to validate the algorithm accuracy. Comparison to earlier work involving 3D segmentation [15] demonstrates the interest of a full 4D approach.

Acknowledgements

We are grateful to GE Medical Systems, Prof. Goris of the University of Stanford Medical School and Dr Lethor of CHU Barbois for MR, SPECT and US image acquisitions respectively. Special thanks go to Michel Audette and Tal Arbel for proof-reading.

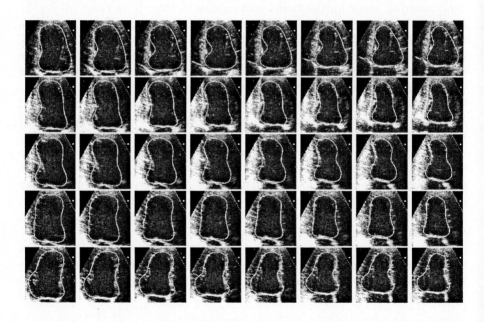

Fig. 5. Set of slices of the US image superimposed with the deformed model intersection. From left to right: cardiac sequence time point. From top to bottom: slices oriented with angle 0, 40, 80, 120, and 160 degrees of arc.

References

1. Bardinet, E., Cohen, L., Ayache, N.: Tracking and motion analysis of the left ventricle with deformable superquadrics. *Med. Image Analysis*, **1(2)** (1996) 129–149
2. Berger, M.-O., Winterfeldt, G., Lethor, J.-P.: Contour Tracking in Echocardiographic Sequences without Learning Stage: Application to the 3D Reconstruction of the 3D Beating Left Ventricle. In *MICCAI*, vol. 1679 of LNCS, (1999) 508–515
3. Besl, P., McKay, N.: A method for registration of 3D shapes. *IEEE Transaction on Pattern Analysis and Machine Intelligence*, **14(2)** (1992) 239–256
4. Clarysse, P., Friboulet, D., and Magnin, I.: Tracking Geometrical Descriptors on 3-D Deformable Surfaces: Application to the Left-Ventricular Surface of the Heart. *IEEE Transaction on Medical Imaging*, **16(4)** (1997) 392–404
5. Cohen, L.: On Active Contour Models and Balloons. *Computer Vision, Graphics, and Image Processing: Image Understanding*, **53(2)** (1991) 211–218
6. Cootes, T., Taylor, C., Cooper, D., Graham, J.: Active shape models, their training and application. *Comp. Vision and Image Understanding*, **61(1)** (1995) 38–59
7. Declerck, J., Feldmar, J., Ayache, N.: Definition of a 4D continuous planispheric transformation for the tracking and the analysis of LV motion. *Medical Image Analysis*, **2(2)** (1998) 197–213
8. Delingette, H.: General Object Reconstruction based on Simplex Meshes. *International Journal of Computer Vision*, **32(2)** (1999) 111–146
9. Giachetti, A.: On-line analysis of echocardiographic image sequences. *Medical Image Analysis*, **2(3)** (1998) 261–284
10. Jacob, G., Noble, A., Mulet-Parada, M., and Blake, A.: Evaluating a robust contour tracker on echocardiographic sequences. *Medical Image Analysis*, **3(1)** (1999) 63–75
11. Lötjönen, J., Reissman, P.-J., Magnin, I., Katila, T.: Model extraction from magnetic resonance volume data using the deformable pyramid. *Medical Image Analysis*, **3(4)** (1999) 387–406
12. McEachen, J., Duncan, J.: Shaped-base tracking of left ventricular wall motion. *IEEE Transaction on Medical Imaging*, **16(3)** (1997) 270–283
13. McInerney, T., Terzopoulos, D.: Deformable models in medical image analysis: a survey. *Medical Image Analysis*, **1(2)** (1996) 91–108
14. Montagnat, J., Delingette, H.: Globally constrained deformable models for 3D object reconstruction. *Signal Processing*, **71(2)** (1998) 173–186
15. Montagnat, J., Delingette, H., Malandain, G.: Cylindrical Echocardiographic Images Segmentation based on 3D Deformable Models. In *MICCAI*, vol. 1679 of LNCS, (1999) 168–175
16. Nastar, C., Ayache, N.: Frequency-Based Nonrigid Motion Analysis: Application to Four Dimensional Medical Images. *IEEE Transaction on Pattern Analysis and Machine Intelligence*, **18(11)** (1996) 1067–1079
17. Papademetris, X., Sinusas, A., Dione, D., Duncan, J.: 3D Cardiac Deformation from Ultrasound Images. In *MICCAI*, vol. 1679 of LNCS, (1999) 420–429
18. Park, J., Metaxas, D., Axel, L.: Analysis of LV motion based on volumetric deformable models and MRI-SPAMM. *Medical Image Analysis*, **1(1)** (1996) 53–71
19. Terzopoulos, D., Witkin, A., Kass, M.: Constraints on Deformable Models: Recovering 3D Shape and Nonrigid Motion. *Artificial Intelligence*, **36(1)** (1988) 91–123
20. Vemuri, B.C., Radisavljevic, A.: From Global to Local, a Continuum of Shape Models with Fractal. In *Computer Vision and Pattern Recognition*, (1993) 307–313

Adaptive-Focus Statistical Shape Model for Segmentation of 3D MR Structures

Dinggang Shen[1] *and* Christos Davatzikos[1,2,3]

[1] Department of Radiology, [2] Department of Computer Science,
[3] Center for Computer-Integrated Surgical Systems and Technology,
Johns Hopkins University
Email: dgshen@cbmv.jhu.edu, hristos@rad.jhu.edu

Abstract. This paper presents a deformable model for automatically segmenting objects from volumetric MR images and obtaining point correspondences, using geometric and statistical information in a hierarchical scheme. Geometric information is embedded into the model via an affine-invariant attribute vector, which characterizes the geometric structure around each model point from a local to a global level. Accordingly, the model deforms seeking boundary points with similar attribute vectors. This is in contrast to most deformable surface models, which adapt to nearby edges without considering the geometric structure. The proposed model is adaptive in that it initially focuses on the most reliable structures of interest, and subsequently switches focus to other structures as those become closer to their respective targets and therefore more reliable. The proposed techniques have been used to segment boundaries of the ventricles, the caudate nucleus, and the lenticular nucleus from volumetric MR images.

1 Introduction

Deformable models have been extensively used as segmentation tools in medical imaging applications. An excellent review of deformable models can be found in [1].

Most boundary-based deformable models adapt to nearby edges under forces emanating from the immediate neighbors and from image gradients. This can cause unrealistic deformations as individual points are pulled towards noisy or fragmented edges. Moreover, it will make the deformable model very sensitive to initialization. To remedy this, prior knowledge in the form of normal statistical variation can be incorporated into the formulation of a deformable model [2], in order to constrain the possible deformations. Chen and Kanada [11] used the statistics of anatomical variations as prior knowledge to guide the process of registering the statistical atlas with a particular subject. Cootes *et al* [3] have developed an active shape model (ASM), based on the statistics of labeled samples. An improvement of ASM has been presented in [12]. Following the seminal work of Cootes *et al*, a flexible Fourier surface model [4] was proposed based on a hierarchical parametric object description rather than a point distribution model. Finally, some researchers consider capturing statistical information from the preprocessed covariance matrix [5].

The aforementioned statistical models can also be implemented hierarchically, since hierarchical implementation usually increases the likelihood of finding the globally optimal match. A review of the hierarchical strategies can be found in [6].

In this paper we present a deformable model for segmentation and definition of point correspondences in brain images, which incorporates geometric as well as statistical information about the shapes of interest, in a hierarchical fashion. Our methodology has three novel aspects, which are briefly described next.

First, an attribute vector is attached to each model point, and it is used to characterize the geometric structure of the model around that point, from a local to a global scale. Our model only deforms the surface segments to the image boundaries with similar geometric structure, based on an energy term that favors similar attribute vectors. The attribute vectors are essential in our formulation, since they distinguish different parts of a boundary according to their shape properties, and therefore they guide the establishment of point-correspondences between the model and an individual anatomy. This is in contrast to most deformable surface models.

The second contribution of our model is in its adaptive formulation. In particular, it can shift focus from one structure to another. The model adaptivity is accomplished both via its hierarchical deformation strategy and via an adaptive focus statistical shape model. This addresses the limitations of previous statistical models. Our adaptive focus statistical model accounts for size differences between different structures when determining the parameters of shape variation. Moreover, it allows the algorithm to selectively focus on certain structures, by biasing the statistics of the model by the statistics of the structures of interest.

Finally, our third contribution is in the training of the statistical shape model. In particular, we build our surface models in a way that point correspondences are defined for the training samples. Although this is readily done in 2D, it is very difficult in 3D. Consequently, other investigators relied on parametric representations that are not necessarily based on point correspondences [4].

2 Adaptive-Focus Deformable Model (AFDM)

In our approach, we first construct a model that represents the typical anatomy of a number of structures by interconnected surfaces (c.f. Fig 1). This is provided in Section 2.1. In Section 2.2, we attach an attribute vector to each vertex of these surfaces. In Sections 2.3 and 2.4, we describe the mechanism that deforms the model to an individual MR volumetric image, and the corresponding energy function being minimized. Finally, in Section 2.5 we present a hierarchical and adaptive-focus strategy for the model deformation.

2.1 Model Description

In the following, we will give definitions for vertices and their neighborhood layers on each surface, and as well as on the interconnections of the surfaces of the model.

Let's assume that there are M separate surfaces V^i ($1 \le i \le M$) in the model, and that the i-th surface V^i has N_i vertices $V_j^i = \begin{bmatrix} x_j^i & y_j^i & z_j^i \end{bmatrix}^T$, where $0 \le j \le (N_i - 1)$.

Vertex V_j^i is assumed to have $R(V_j^i)$ neighborhood layers, and its l-th neighborhood layer has $S_l(V_j^i)$ neighboring vertices, $nbr_l(V_j^i) = \left\{ nbr_{l,m}(V_j^i) \middle| 0 \leq m < S_l(V_j^i) \right\}$. Here, layer l is in the range $1 \leq l \leq R(V_j^i)$. The neighborhood layers are constructed so that no vertex is repeated twice in the neighborhood of another vertex.

In certain applications, some surfaces in the model are in proximity to each other. Therefore, we impose additional constraints that prevent these surfaces from intersecting in the deformation procedure. In particular, if the 3D Euclidean distance between vertices belonging to two different surfaces is below a threshold, the vertices are connected as first-layer neighbors. For example, for each vertex of the boundary of the caudate nucleus that is in contact with the ventricular surface, an additional neighbor is selected as its closest vertex in the ventricular surface (c.f. Fig 1a). If two vertices belonging to the same surface are close enough, based on their 3D Euclidean distance, they are also joined with each other. This helps prevent self-intersections of the surfaces.

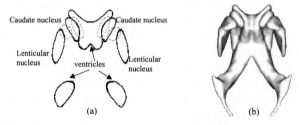

Fig. 1. A model with five surfaces: (a) a cross-section of the 3D model of (b). In (a), any vertex in the dotted ellipses is connected with another vertex of the model.

2.2 Affine-Invariant Attribute Vector

To describe shape characteristics of various scales, an affine-invariant vector of geometric attributes in 2D has been used in [7,8]. Each attribute is the area of a triangle formed by a model point, P_i, and its two neighboring points, P_{i-vs} and P_{i+vs}, in the v_s-th neighborhood layer (c.f. Fig 2a). In the following, we extend the definition of the 2D attribute to the 3D, by using the volume of a tetrahedron (c.f. Fig 2b).

The volume of the tetrahedron, formed by the nearest neighbors of vertex V_j^i, reflects the local structure of the surface around vertex V_j^i. While, the volumes of larger tetrahedrons represent more global properties of the surface around vertex V_j^i. More importantly, even vertices of similar curvatures might have very different attribute vectors, depending on the number of neighborhood layers.

Any four points in the 3D space can establish a tetrahedron. Let's generally denote these four points by $\{W_j | j = 0,1,2,3\}$, where $W_j = [x_j \quad y_j \quad z_j]^T$. The volume of this tetrahedron is:

$$Volume(W_0, W_1, W_2, W_3) = \begin{vmatrix} x_0 & x_1 & x_2 & x_3 \\ y_0 & y_1 & y_2 & y_3 \\ z_0 & z_1 & z_2 & z_3 \\ 1 & 1 & 1 & 1 \end{vmatrix}.$$

If this tetrahedron is linearly transformed by a 4x4 matrix A, then the volume of the new tetrahedron is $|A| \cdot Volume(W_0, W_1, W_2, W_3)$. Thus, it is relatively invariant to linear transformation [7].

The definition of the volume of a tetrahedron can be used to design an attribute vector for each vertex on the model surface. For a particular vertex V_j^i, we can select any three points from the l-th neighborhood layer (see Fig 2b). The volume of the tetrahedron formed by these four vertices is defined by $f_l(V_j^i)$. We compile the volumes calculated for different neighborhood layers into an attribute vector for vertex V_j^i, $F(V_j^i) = [f_1(V_j^i) \quad f_2(V_j^i) \quad \cdots \quad f_{R(V_j^i)}(V_j^i)]$, where $R(V_j^i)$ is the number of neighborhood layers around vertex V_j^i. The attribute vector $F(V_j^i)$ captures different levels of shape information around vertex V_j^i.

The attribute vector can be made affine-invariant, by normalizing it on the whole model, i.e.

$$\hat{F}(V_j^i) = F(V_j^i) \Big/ \sum_{i=1}^{M} \sum_{j=1}^{N_i} \sum_{l=1}^{R(V_j^i)} \left| f_l(V_j^i) \right|,$$

where $\hat{F}(V_j^i) = [\hat{f}_1(V_j^i) \quad \hat{f}_2(V_j^i) \quad \cdots \quad \hat{f}_{R(V_j^i)}(V_j^i)]$. Unlike curvature, the normalized attribute vectors are affine-invariant.

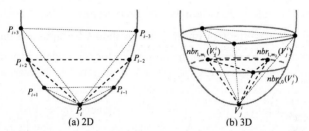

(a) 2D   (b) 3D

Fig. 2. The attribute vector in 2D and 3D.

2.3 Energy Definition

The goal of our deformable model is to define point correspondences, in addition to segmenting structures of interest. Our premise is that the attribute vector, if rich enough, uniquely characterizes different parts of a boundary of a structure. Therefore, in the definition of the energy function to be minimized, we include a term that reflects the difference between the attribute vectors of the model and individual surface. An obvious difficulty in this approach is that the attribute vector of an individual surface cannot be obtained directly from the corresponding MR images, since it is based on a triangularized surface. We overcome this difficulty by deforming our model via a sequence of global and local transformations. Since the attribute vectors are invariant to linear transformation, they remain relatively unchanged in this deformation process. Effectively, this defines a segmentation, but also a set of point correspondences based on a similarity between attribute vectors.

The energy that our deformable model minimizes is defined as follows:

$$E = \sum_{i=1}^{M} \sum_{j=1}^{N_i} \omega_{i,j} E_{i,j} = \sum_{i=1}^{M} \sum_{j=1}^{N_i} \omega_{i,j} \left(E_{i,j}^{\text{model}} + E_{i,j}^{\text{data}} \right). \tag{1}$$

The parameter $\omega_{i,j}$ determines the relative weight for the local energy term $E_{i,j}$. $E_{i,j}$, defined for V_j^i, is composed of two terms: $E_{i,j}^{\text{model}}$ and $E_{i,j}^{\text{data}}$. The term $E_{i,j}^{\text{model}}$ defines the degree of difference between the model and one of its deformed configurations, around vertex V_j^i. The term $E_{i,j}^{\text{data}}$ defines the external energy.

As we elaborated earlier in this section, the term $E_{i,j}^{\text{model}}$ is given by

$$E_{i,j}^{\text{model}} = \sum_{l=1}^{R(V_j^i)} \delta_l \left(\hat{f}_l^{\text{Def}}(V_j^i) - \hat{f}_l^{\text{Mdl}}(V_j^i) \right)^2, \tag{2a}$$

where $\hat{f}_l^{\text{Def}}(V_j^i)$ and $\hat{f}_l^{\text{Mdl}}(V_j^i)$ are, respectively, the components of the normalized attribute vectors of the deformed model configuration and the model at vertex V_j^i. The parameter δ_l denotes the degree of importance of the l-th attribute element in the surface segment under consideration. Notice that $R(V_j^i)$ is the number of the neighborhood layers around vertex V_j^i.

The data energy term, $E_{i,j}^{\text{data}}$, is usually designed to move the deformable model towards an object boundary. Accordingly, for every vertex V_j^i, we require that in the position of V_j^i, the magnitude of image gradient should be high, and the direction of image gradient should be similar to the normal vector of the deformed surface. Since our deformation mechanism, which is defined in section 2.4, deforms a surface segment around each vertex V_j^i at a time, and not just the vertex itself, we want to design an energy term that reflects the fit of the whole segment, rather than a single vertex, with image edges. A surface segment is defined by vertex V_j^i and its neighbors from $R(V_j^i)$ neighborhood layers, where $R(V_j^i)$ can vary throughout the deformation procedure, as detailed in Section 2.5. The l-th neighborhood layer has $S_l(V_j^i)$ vertices, $nbr_{l,m}(V_j^i)$, where $0 \le m < S_l(V_j^i)$. The data energy term $E_{i,j}^{\text{data}}$ is designed as follows.

$$E_{i,j}^{\text{data}} = \sum_{l=1}^{R(V_j^i)} \delta_l \sum_{m=0}^{S_l(V_j^i)} \left(1 - \left| \nabla I \left(nbr_{l,m}(V_j^i) \right) \right| \cdot \left| \vec{h} \left(nbr_{l,m}(V_j^i) \right) \cdot \vec{n} \left(nbr_{l,m}(V_j^i) \right) \right| \right). \tag{2b}$$

2.4 Local Deformation Mechanism

We now describe a greedy deformation algorithm for the minimization of the energy function in equation (1). We suggest considering the deformation of a surface segment as whole, which greatly helps the snake avoid local minima.

Since we deform only one piece of the a model surface at a time, we can introduce discontinuities at the boundary of the segment being deformed. In our 2D model [8], we solved this issue by restricting the local affine transformation so that it leaves the end-points of a deforming segment unchanged. However, for a surface model this is not possible, because the vertices belonging to the $R(V_j^i)$-th neighborhood layer do not necessarily lie on the same plane. Therefore, we cannot necessarily find a local affine transformation that preserves the position of the end-vertices of a deforming segment. To remedy this situation, we used a different form of transformation for each deforming surface segment, which is described next.

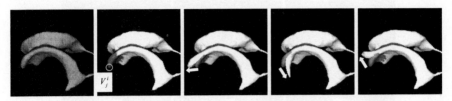

Fig. 3. Demonstration of the proposed deformation mechanism. See text for details.

Let V_j^i be the vertex whose neighborhoods form the surface segment to be deformed at a particular iteration (c.f. Fig 3). The $R(V_j^i)$-th neighborhood layer forms the boundary of the surface segment. Consider a tentative position, $V_j^i + \Delta V$, to which V_j^i is to move during the greedy algorithm. Then, the new position of each vertex, $nbr_{l,m}(V_j^i)$, in the segment is defined as $nbr_{l,m}(V_j^i) + \Delta V \cdot \exp(-l^2/2\sigma^2)$, where σ is a parameter determining the locality of the transformation. We use values of σ that make $\exp(-R(V_j^i)^2/2\sigma^2)$ close to zero, effectively leaving the bounding curve of a deforming segment unchanged, and hence maintaining continuity. The new configuration of the surface segment is then determined by finding ΔV that minimize the sum of two energy terms $E_{i,j}^{data}$ and $E_{i,j}^{model}$.

Fig 3 demonstrates some tentative positions of vertex V_j^i and the corresponding deformations of the surface segment. The gray surface in Fig 3, left, is a ventricular model. The rest of the images show tentative deformations of the ventricular model (white surfaces) overlaid on the undeformed model (gray surface).

2.5 Adaptive-Focus Deformation Strategy

Brain images contain several boundaries. Prior knowledge, in conjunction with the quality of image information (e.g. edge strength), is used in AFDM to guide the deformation of the model in a hierarchical fashion. In particular, surfaces for which we have relatively higher confidence are deformed first. As other surfaces follow this deformation and get closer to their respective targets, they become more reliable features for driving the model's deformation. We demonstrate this scheme using the example of the caudate nucleus (CN), the lenticular nucleus (LN) and the ventricular boundaries. In Fig 4 we show cross-sections of the initial (automatic) placement of a 3D model containing these 5 surfaces, and the deformation of that model after 10 iterations. The one in Fig 4b is the result of AFDM, with the ventricular boundaries deforming first, and the CN and LN boundaries following. In fact, there was a continuous blending in the deformation of the CN and LN as iteration number increased. The result of Fig 4c was obtained via the same model but with a non-adaptive deformation mechanism, i.e. with forces applied to all components of the model simultaneously. In the adaptive focus scheme, the ventricles first pulled the LN close enough to its corresponding boundary in the MR image, before the LN model started deforming. In the non-adaptive scheme, however, the LN deformed towards the wrong boundary.

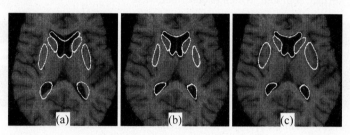

Fig. 4. Demonstration of the adaptive-focus deformation strategy. See text in Section 2.5.

In addition to its cross-component hierarchical formulation, our approach is also hierarchical within components of the model. In particular, the parameter $R(V_j^i)$ that determines the locality of the deformation transformation is typically chosen large in the initial iterations, and is gradually reduced to 1. Therefore, initially, relatively more vertices are involved in the surface segment around vertex V_j^i, and the resulting transformation is of relatively global form. In later stages, the transformation affects the deformable model more locally.

3 Adaptive-Focus Deformable Statistical Shape Model (AFDSM)

In this section, we extend AFDM to incorporate information about the statistical variation of the model. We first describe how we construct models of the training samples using AFDM, while simultaneously establishing point-correspondences. We then extend the statistical shape modeling paradigm of [3-5] to the adaptive focus framework of AFDM. The resulting model is called Adaptive Focus Deformable Statistical shape Model (AFDSM).

3.1 Sample Construction

Statistical shape models have gained popularity in the medical image analysis community after they were first introduced in [3]. One of the difficulties, however, associated with these models is their training, which depends on defining point correspondences in a training sample. This task is fairly straightforward in 2D. However, definition of point correspondences in 3D is a very difficult task. To overcome this difficulty, some investigators have assumed that approximate correspondences can be defined by placing parametric grids on the structure of interest [4,9]. Although this is a convenient way to define correspondences and train a statistical shape model, it is based on only a rough approximation of point correspondences.

In our work we train the deformable model on samples whose point correspondences are defined via AFDM. In particular, images of each training sample are first hand-segmented on a section-by-section basis to the structures of interest (ventricles, LN, CN, in this paper). AFDM is then applied to the segmented images, resulting in a surface representation of each boundary. Notably, since AFDM is based on a similarity between attribute vectors, it determines point-correspondences based on similarity of the geometric structures of the boundaries of interest. We found that AFDM worked very well on these hand-segmented images. In very few cases we had to manually help the algorithm by "pulling" the surface to the boundary.

Fig 5 shows the deformation of the model (initially obtained from a single subject) to a hand-segmented training sample. Cross-sections of the deformed model are shown in black and are overlaid on (gray) sections of the target boundary. Three stages of the process are shown: the initialization, an intermediate result obtained after the model has focused primarily on the ventricles, and the final result.

3.2 Adaptive-Focus Statistical Information

In the previous statistical shape models [3-5], all landmarks in the training samples was given equal weights when calculating shape statistical parameters. In this way, larger features of a shape dominate over relatively smaller, yet important features, merely because their large size influences the measures of shape variability. Furthermore, unreliable features, if they are large, dominate over relatively reliable and important features. To overcome this limitation, in our calculation of the statistical parameters, we weight different vertices of the model differently. In particular, vertices belonging to relatively smaller structures are assigned relatively higher weights and vice versa.

To better explain the importance of variable weighting, we will use the example of a model containing a large structure (ventricles) and a smaller structure (LN). Due to their large size and high variability, the ventricles dominantly affect the statistical shape parameters. In particular, the dominant eigenvectors primarily reflect the variability of the ventricles. Accordingly, a deformation of the ventricles by image-derived forces induces very little deformation on the LN. This is problematic, since the LN should follow the deformation of the ventricles.

Our statistical model is analogous to the one in [3-5]. The variable weighting of the components of the model effectively zooms each component to the same overall size in the space where the statistics are calculated, so that each component is represented in the most important eigenvectors of the corresponding covariance matrix. Each component is then scaled back appropriately to its actual size. More details of the algebraic manipulations involved in these transformations can be found in [8].

4 Experiments and Conclusion

We test the performance of our algorithm in segmenting multiple structures of the human brain with a multi-component model. Fig 1b shows a 5-component model containing the boundaries of the ventricles and the left and right CN and LN. The total number of vertices in this model is 3966. The surfaces of the ventricles, CN and LN have 2399, 760, and 807 vertices, respectively. The total number of triangles in the whole model is 7912. Using the technique in section 2.1, connections among proximal vertices of different components were formed, as in Fig 1a.

As described in Section 3.1, AFDM was applied on hand-labeled images in order to construct the training set. An example of this procedure is shown in Fig 5. The intermediate result was obtained primarily by focusing on the ventricles, while the rest of the components of the model were deformed via a global linear transformation following the ventricular deformation. Fig 6 shows one representative result that was obtained via ADFSM.

We have presented a deformable model for segmenting objects from volumetric MR images. Some extensions of our method are possible. Currently, the vertices are arranged into tetrahedra that represent the 3D structure of objects. An alternative representation could be a medial representation [10]. Combination of our geometric representation with the medial representation should improve the results, since these two methods have complementary merits and weaknesses. Moreover, the creation of our model at multiple resolutions will definitely accelerate the speed of our algorithm.

We finally want to note that all of our experiments have been performed on MR images of elderly individuals, which suffer from reduced white matter/grey matter contrast, and from often extreme atrophy reflected, in part, by very large ventricles. Despite the difficulties imposed by the nature of the data, we have obtained good and robust results in a large set of brain images.

References

1. T. McInerney and D. Terzopoulos: Deformable models in medical image analysis: a survey. Medical Image Analysis, 1(2): 91-108, 1996.
2. L.H. Staib and J.S. Duncan: Boundary finding with parametrically deformable models. IEEE Trans. on PAMI, 14(11):1061-1075, 1992.
3. T.F. Cootes, D. Cooper, C.J. Taylor and J. Graham: Active shape models-their training and application. Computer Vision and Image Understanding, 61(1): 38-59, 1995.
4. A. Kelemen, G. Szekely, and G. Gerig: Elastic model-based segmentation of 3-D neuroradiological data sets. IEEE Trans. on Medical Imaging, 18(10): 828-839, 1999.
5. Y. Wang and L.H. Staib: Elastic model based non-rigid registration incorporating statistical shape information. MICCAI, pages 1162-1173, Cambridge, MA, October 1998.
6. H. Lester, S.R. Arridge: A survey of hierarchical non-linear medical image registration. Pattern Recognition, (32)1, pp.129-149, 1999.
7. Horace H.S. Ip, Dinggang Shen: An affine-invariant active contour model (AI-snake) for model-based segmentation. Image and Vision Computing 16(2):135-146, 1998.
8. Dinggang Shen, C. Davatzikos: A adaptive-focus deformable model using statistical and geometric information. To appear in IEEE Trans. on PAMI, 2000.
9. C. Davatzikos: Spatial transformation and registration of brain images using elastically deformable models. Comp. Vis. and Image Understanding, 66(2):207-222, May 1997.
10. S.M. Pizer, D.S. Fritsch, P.A. Yushkevich, V.E. Johnson, and E.L. Chaney: Segmentation, registration, and measurement of shape variation via image object shape. IEEE Trans. on Medical Imaging, 18(10): 851-865, October 1999.

11. M. Chen, T. Kanade, D. Pomerleau, J. Schneider: 3-D deformable registration of medical images using a statistical atlas. MICCAI, Sept. 1999.
12. N. Duta, M. Sonka: Segmentation and interpretation of MR brain images: An improved active shape model. IEEE Trans. on Medical Imaging, 17(6): 1049-1062, 1998.

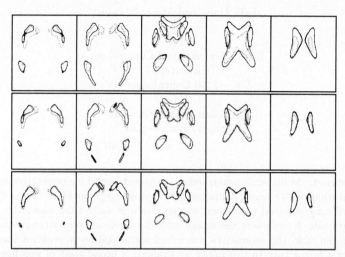

Fig. 5. Deformation of the model to a hand-labeled target image of a training sample, for determining point correspondences. See text in Section 3.1 for details.

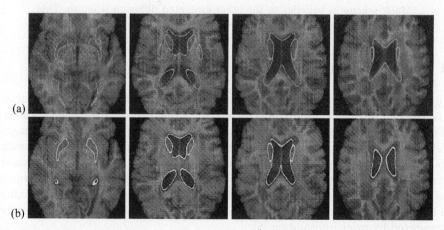

Fig. 6. Example on segmenting multiple structures using AFDSM. (a) Initial position of the model in the four slice images; (b) The final segmentation results corresponding to (a).

Non-linear Diffusion for Interactive Multi-scale Watershed Segmentation

Erik Dam and Mads Nielsen

IT-C
The IT University of Copenhagen
Glentevej 67
DK – 2400 Copenhagen NV
Denmark
{erikdam,malte}@it-c.dk
http://www.it-c.dk

Abstract. Multi-scale watersheds have proven useful for interactive segmentation of 3D medical images. For simpler segmentation tasks, the speed up compared to manual segmentation is more than one order of magnitude. Even where the image evidence does not very strongly support the task, the interactive watershed segmentation provides a speed up of a factor two. In this paper we evaluate a broad class of non-linear diffusion schemes for the purpose of interactively segmenting gray and white matter of the brain in T1-weighted MR images. Through a new scheme GAN, we show that diffusion similar to the nonlinear Perona-Malik scheme is superior to the other evaluated diffusion schemes. This provides a speed up factor of two compared to the linear diffusion scheme.

1 Introduction

In many medical segmentation tasks, the images do not sufficiently clearly outline the relevant anatomical structures for making simple automated segmentations. The counter-example is simple thresholding of bone structures in CT images [21]. However, such techniques do not work for most anatomical structures such as soft tissue in CT (due to varying and indistinguishable attenuation [10]), most structures in MR (due to the image inhomogeneities [10]), and most structures in PET/SPECT (due to noise [10]). Here, the alternatives are either performing a tedious manual outline slice per slice, or creating specialised algorithms heavily supported by prior information [6].

The interactive 3D multi-scale watershed segmentation tool, ∇ *Vision*, may successfully be applied in these situations [8]. The image scale-space is created by Gaussian convolution [35, 17]. The watersheds of the gradient magnitude are computed independently at all scales. A linking procedure gives the simpler large scale watersheds the small scale localisation [26, 20, 15]. The linked watershed regions constitute a multi-scale partitioning of the images.

Ideally, a given anatomical structure may be outlined by a single region. However in most situations, the linked watersheds do not directly compare to the anatomical structures. ∇ *Vision* lets the user arbitrarily change scale and select and deselect regions, and thereby sculpt the anatomical structure. All interaction is geometrical and thereby very intuitive for the clinician. The speed up compared to manual segmentation depends on the interactions required to outline the anatomical structure. Outlining soft tissue in CT may be orders of magnitude faster. Compared to computerised manual segmentation, a complicated task, the masseter (chewing muscle) in MR, shows a speed up factor of two [8, 11].

The Gaussian scale-space is the least committed scale-space. A non-linear scale-space commits itself to certain intensity variations through the non-linear function and to certain local edge shapes through the diffusion structure. This is formalised through the connection between energy minimization methods [23] and non-linear diffusion in the biased non-linear diffusion [25]. In this light, one may argue that the use of the non-linear diffusion schemes is the first step in commitment towards using prior shape and intensity knowledge like in the active contour and core-based segmentation methods [6, 30].

In this paper, we evaluate a large number of non-linear diffusion schemes for the multi-scale watersheds in 2D: non-linear isotropic Perona-Malik [29], non-linear anisotropic schemes [34], and a purely geometrical non-linear scheme (Mean Curvature Motion) [12]. In section 3.4, we argue that these schemes in a natural way span the space of diffusion schemes supporting segmentation.

The flavor of our paper is close to the comparison of diffusion schemes for segmentation performed on the *hyper-stack* [33, 18, 24]. The major differences are that the hyper-stack is based on isophote linking and that it constitutes an automated segmentation algorithm. In this paper we specifically evaluate how the *deep structure* of the various scale-spaces support the segmentation.

The multi-scale watershed segmentation method is introduced in section 2. In section 3 the evaluated diffusion schemes are presented. The evaluation is outlined in section 4 with results in section 5.

2 Multi-scale Watershed Segmentation

During rain the drops gather in pools. The topology of a landscape defines the regions of support for each pool — the *catchment basins*. The boundaries between the catchment basins are termed *watersheds*. On large scale, the watersheds of a landscape are the ridges and the catchment basins are the dales. The geographical concept *watershed* was introduced to mathematicians in [22].

The watersheds allow a simple partitioning of an image. However, for segmentation purposes the regions border should be defined as the watersheds of a dissimilarity measure instead of the original image. A simple, general, and non-committed dissimilarity measure is the gradient magnitude.

The structures that are outlined by this partitioning are defined with respect to the scale at which the gradient is calculated. Different scales are therefore needed to locate objects of different sizes. The theory of scale-space suggests looking at the *deep structure* [17, 35, 32] — how the catchment basins develop over scale.

Each catchment basin corresponds to a local minimum for the gradient magnitude. In [26, 27], the generic events for the gradient magnitude minima are derived. The conclusion is that *fold annihilation, fold creation, cusp annihilation*, and *cusp creation* catastrophes are stable and therefore to be expected for typical images. For the catchment basins, this corresponds to the *annihilation, creation, merge*, and *split* events [26, 7].

Linking of the catchment basins across scale combines the simplification at the detection scale with the fine scale precision at the localisation scale (see figure 1). The segmentation method presented in [26] uses these localised basins as building blocks for the segmentation. The user can shift the detection scale and thereby select building blocks appropriate for sculpting the desired objects.

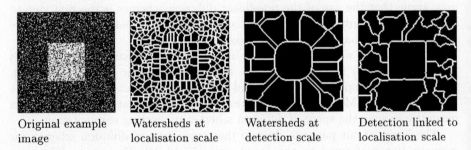

| Original example image | Watersheds at localisation scale | Watersheds at detection scale | Detection linked to localisation scale |

Fig. 1. Linking of catchment basins across scale. The catchment basins at detection scale is linked down to the localisation scale and thereby get fine scale precision.

Similar approaches to the multi-scale watershed segmentation of [26] are presented in [20], [14], and [13]. However, an important contribution of [26] is that an intuitive interface is presented that allows the user to interact directly with the three-dimensional building blocks. This forms the basis for an implementation (the program ∇ *Vision* by *Generic Vision*) that has been tested by clinical researchers with promising results [16, 8]. An important addition in this implementation is an extra "artificial" level of voxel-sized building blocks which allows arbitrary detail in the segmentation.

3 Diffusion Schemes for Multi-scale Watershed Segmentation

The original multi-scale watershed segmentation method relies on the linear Gaussian scale-space to simplify the image. This simplification determines how the catchment basins group into gradually larger building blocks corresponding to image structures at a given scale.

The linear scale-space for an image $I(x)$ is described by the PDE $\frac{\partial L(x;t)}{\partial t} = \Delta L(x;t) = L_{ii}(x;t)$ with the initial condition: $L(x;0) = I(x)$. The Gaussian convolution kernel with standard deviation $\sigma = \sqrt{2t}$ is the Green's function for the PDE.

The Linear Gaussian diffusion scheme (here denoted LG) has extremely nice theoretical properties. In particular, the causality property, the average gray level invariance property, and the fact that the image gets uniform intensity for scale tending to infinity ensures that the linear scale-space is applicable for the multi-scale watershed segmentation method. However, these properties do not ensure that it is an optimal diffusion scheme for the method.

In [26, 27] the generic events for the gradient magnitude minima are derived. When the diffusion scheme is replaced by nonlinear schemes, the analysis of the generic events for the watershed regions is no longer applicable. Some of these nonlinear schemes have been analysed [9]. However, from a practical viewpoint, the linking of the discrete scale levels can handle nearly any diffusion scheme with suitable simplification properties due to robust matching of regions [7].

3.1 Regularised Perona-Malik

The classical Perona-Malik diffusion scheme is designed to preserve edges during the diffusion [29]. The regularisation due to [3] is denoted RPM:

$$\frac{\partial L(\boldsymbol{x}; t)}{\partial t} = div(\ p(|\nabla L_\sigma|^2)\ \nabla L\ ) \quad \text{where} \quad p(|\nabla L_\sigma|^2) = \frac{1}{1 + \frac{|\nabla L_\sigma|^2}{\lambda^2}} \quad (1)$$

The parameter λ is a threshold for the gradient magnitude required to make the scheme preserve an area (an edge). The σ determines the regularisation scale.

3.2 Anisotropic Nonlinear Diffusion

Weickert [34] defines the *anisotropic nonlinear diffusion* equation for a two-dimensional image I by the PDE:

$$\frac{\partial L(\boldsymbol{x}; t)}{\partial t} = div(\ D(J_\rho(\nabla L_\sigma))\ \nabla L\ ) \quad \text{where} \quad L(\boldsymbol{x}; 0) = I(\boldsymbol{x}) \quad (2)$$

The diffusion tensor $D \in C^\infty(R^{2 \times 2}, R^{2 \times 2})$ is assumed to be symmetric and uniform positive definite. The *structure tensor* J_ρ is evaluated at *integration scale* ρ (set to zero in the following), and the gradient ∇L_σ at *sampling scale* σ.

The diffusion equation possesses simplification properties [34] that ensures that the diffusion schemes are applicable for the segmentation method.

For the following diffusion schemes, the diffusion tensor is defined in terms of the eigenvectors $\bar{v}_1 \parallel \nabla L_\sigma$, $\bar{v}_2 \perp \nabla L_\sigma$ and the corresponding eigenvalues λ_1 and λ_2. Furthermore, Weickert presents a diffusivity function w_m designed to preserve edges more aggressively than the Perona-Malik diffusivity function p.

$$w_m(|\nabla L_\sigma|^2) = \begin{cases} 1 & |\nabla L_\sigma| = 0 \\ 1 - exp\left(\frac{-C_m}{\left(\frac{|\nabla L_\sigma|^2}{\lambda}\right)^m}\right) & |\nabla L_\sigma| > 0 \end{cases} \quad (3)$$

Here m determines the aggressiveness of the diffusivity function, and C_m is derived [34, 7] from m such that the flux magnitude function $|\nabla L|\ w_m(|\nabla L_\sigma|^2)$ is increasing for $|\nabla L|^2 < \lambda$ and decreasing for $|\nabla L|^2 > \lambda$.

Isotropic Nonlinear Diffusion

Weickert [34] designs an isotropic nonlinear diffusion scheme (here denoted IND) by the following eigenvalues $\lambda_1 = \lambda_2 = w_m(|\nabla L_\sigma|^2)$. Intuitively, this is an increasingly aggressive version of the Perona-Malik scheme for $m > 0.75$.

Edge Enhancing Diffusion

The anisotropic version is termed *edge enhanced diffusion* and defined by the eigenvalues $\lambda_1 = w_m(|\nabla L_\sigma|^2)$ and $\lambda_2 = 1$. The choice $m = 4$ (which implies $C_m = 3.31488$) is used in [34] with visually appealing results. Here, we also exploit $m = 2$ ($C_m = 2.33666$) and $m = 3$ ($C_m = 2.9183$) for the edge enhancing diffusion scheme.

The isotropic nonlinear scheme enhances edges so aggressively that noise is preserved around edges for a long scale interval. The anisotropic schemes (here denoted EE2, EE3, and EE4 depending on the choice of m) remedy this by smoothing along the edges.

3.3 Mean Curvature Motion

A number of morphological processes possess properties similar to diffusion schemes [2]. *Mean curvature motion* (here denoted MCM) is a special case of the anisotropic nonlinear diffusion scheme [7, 34] where $\lambda_1 = 0$ and $\lambda_2 = 1$. MCM is defined by the following PDE (where $\kappa(L)$ is the mean curvature of the isophote landscape):

$$\frac{\partial L}{\partial t} = -\kappa(L) \, |\nabla L| \tag{4}$$

3.4 Generalised Anisotropic Nonlinear Diffusion

The diffusion schemes previously presented are defined by the diffusivity functions in the gradient direction and the isophote direction. Figure 2 illustrates this "space of diffusion schemes". This inspires the new *Generalised Anisotropic Nonlinear* diffusion scheme (denoted GAN) defined by the following diffusivity functions λ_1 and λ_2:

$$\lambda_1 = w(m, \lambda, |\nabla L_\sigma|^2)$$
$$\lambda_2 = \theta + (1 - \theta) \, \lambda_1 \tag{5}$$

The Weickert diffusivity function (equation 3) is written $w(m, \lambda, s^2)$ instead of $w_m(\lambda, s^2)$ since m is to be perceived as a regular parameter of the diffusivity function. The parameter θ determines the degree of anisotropy.

The GAN scheme is named *Generalised Anisotropic Nonlinear diffusion* since it offers a straigthforward generalisation of the previously presented diffusion schemes. The choice of the parameters, in particular the aggressiveness parameter m and the anisotropy parameter θ, allows the scheme to cover the white area in figure 2. The existing schemes are realised by the following:

Fig. 2. *Space of diffusion schemes.* The gradient direction and isophote direction diffusivity functions determine the positions on the horizontal and vertical axes, respectively. The diffusion schemes with aggressive diffusivity functions are mapped closest to the lower, left corner. The gray area is populated by diffusion schemes that are not suited for segmentation-like purposes — they diffuse more across edges than along them. Thereby the catchment basins merge across possible object borders before merging inside the regions likely to correspond to the desired objects.

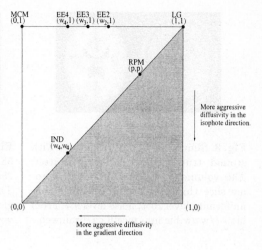

LG	Linear Gaussian diffusion is defined by $\lambda \to \infty$.
IND	Isotropic Nonlinear Diffusion is achived for $\theta = 0$.
RPM	Regularised Perona-Malik scheme is approximated by $\theta = 0$ and $m = 0.75$.
EEx	The Edge Enhancing schemes EE2, EE3, and EE4 are defined by $\theta = 1$ and the corresponding m.
MCM	Mean Curvature Motion is achieved for $\theta = 1$ and $\lambda \to 0$.

4 Evaluation

An obvious evaluation is to let clinicians test the segmentation method on real segmentation tasks with the different diffusion schemes. However, this is not objective and requires extensive work by the clinicians. The alternative is to measure the quality of the segmentations with respect to a "correct segmentation" — the *ground truth*. The quality measure should be general, objective, and quantitative. For specific segmentation tasks, the quality measure could be defined in terms of specific features of the desired segmentations (shape, topology, etc.). However, for a general evaluation method, the measure must be simple and geometric.

In this paper, ground truth segmentations of white and gray matter for both real and simulated MRI brain scans are used (figures 3 and 4). The real data is from the *Internet Brain Segmentation Repository* [1]. The simulated data is from the *BrainWeb* [4, 19, 5]. The quality measure is simply defined by the relative error (misclassified pixels relative to number of pixels in ground truth object, where a pixel is uncorrectly segmented if it is included only in the segmentation or only in the ground truth).

The pixel-sized building blocks allow the user to reach an arbitrary segmentation — therefore any diffusion scheme allow perfect segmentation. The

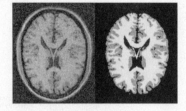
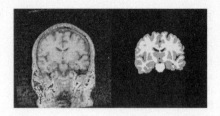

Fig. 3. Simulated T1 MR brain scan with ground truth white and gray matter. The volume is 181x217x181 with coronal slice thickness 1 mm, intensity nonuniformity level 20%, noise level 9%. From http://www.bic.mni.mcgill.ca/brainweb

Fig. 4. Real T1 MR brain scan of a 55 year old male. The set contains 60 256x256 slices with slice thickness $3.0mm$. The slices are coronal with a flip angle of 40 degrees. From http://neuro-www.mgh.harvard.edu/cma/ibsr

evaluation of the semi-automatic method measures the user effort required to reach a specific *quality threshold*. In an evaluation with clinicians, the user effort could be measured as the time needed. For this evaluation, the effort is naturally measured as the minimal number of basic user actions required. The canonical actions are *selection* and *deselection* of the building blocks.

The quantitative evaluation allows automatic optimisation of the parameter sets for the diffusion schemes. For details on the evaluation method, the algorithm for establishing the optimal combinations of building blocks, the optimisation method, and the optimal parameter sets, see the technical report [7].

5 Results

The performance is illustrated by the error as a function of the number of actions for each scheme (with parameters optimised to the specific data set). This is put into perspective by the performance of a *Quad tree* linking scheme [31] (here denoted QT). The number of quad tree blocks required has a close relation to the *box-counting dimension* of the ground truth objects — also denoted the *Hausdorff dimension* [28]. Furthermore, the building blocks of the quad tree are not adapted to the geometry of the image. Thereby the performance of the quad tree linking gives a frame of reference for the performances of the diffusion schemes defined in terms of the complexity of the ground truth objects.

Figures 5 and 6 display the performance on the simulated and the real data, respectively. The best of the existing schemes is the regularised Perona-Malik scheme. The new GAN scheme is slightly better than this. However, the graphs do not give a clear notion of the quantitative differences in performance. Figure 7 delivers the desired relative performance indicator on average for all data sets. The relative performance is determined both for the training set used for optimisation of the parameters for the diffusion schemes and for an independent data set.

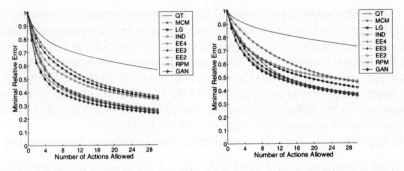

Fig. 5. Evaluation on simulated data from figure 3. White matter is left and gray matter is right. Slices 60, 80, 100, 120, and 140 from the data set are used.

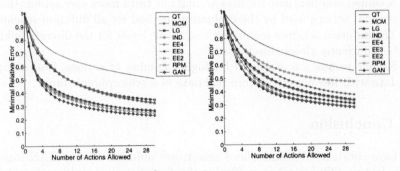

Fig. 6. Evaluation on real data from figure 4. Left is white matter and right is gray matter. Slices 10, 20, 30, 40, and 50 from the data set are used.

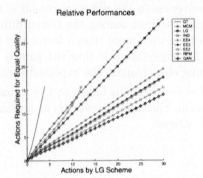

Scheme	Performance	
	Training	Independent
QT	3.83	3.34
MCM	1.18	1.19
LG	1.00	1.00
IND	1.20	1.17
EE4	0.66	0.66
EE3	0.59	0.63
EE2	0.58	0.62
RPM	0.51	0.51
GAN	0.46	0.49

Fig. 7. Average performances with the Gaussian scheme as reference. For a given number of actions the performance of the LG scheme is noted — for each scheme is measured the actions required to get equal quality. The incline of a curve determines the number of actions required to obtain a given segmentation quality relative to the performance of the Gaussian scheme. This performance indicator is displayed in the table for each diffusion scheme. The new GAN scheme requires less than half as many actions compared to the Gaussian scheme for both the training set and an independent data set.

The results show that Perona-Malik is superior among the existing diffusion schemes — somewhat surprising even the anisotropic schemes show inferior performance. Furthermore, the optimal parameter sets for the GAN scheme reveal the optimal degrees of anisotropy and edge preservation aggressiveness. The anisotropy parameter θ is close to zero for all data sets (0.06, 0.12, 0.0, and 0.0). The improved performance compared to RPM is due to slightly increased aggressiveness in the diffusivity function (the parameter m is 1.4, 1.0, 1.1, and 1.0 compared to the approximate value of 0.75 for RPM).

In the technical report [7], a number of other results are documented as well:

- With a tolerance area 1 pixel wide around the borders of the ground truth segments, 97% of the ground truth pixels can be segmented without the use of pixel-sized building blocks.
- A simpel user heuristic requires around one third more user actions than the optimal actions used by the evaluation method for all diffusion schemes.
- The diffusion schemes require up to 30 scale levels for the discrete linking to be sufficiently closely discretised.
- Similar results are measured for higher number of actions, from simulated data with less noise and from real data of a schizophrenic brain.

6 Conclusion

We have constructed a generalised anisotropic diffusion scheme GAN capturing many known diffusion schemes. Tuning this for interactive multi-scale watershed segmentation of white/gray matter in T1-weighted MR slices of the brain shows that diffusion similar to regularised Perona-Malik is superior to the other diffusion schemes. Furthermore, the aggresiveness in the diffusion cut-off is more important than the degree of anisotropy. The best among the tested diffusion schemes yields a decrease in interaction time with more than a factor two compared to linear Gaussian scale-space. Our expectation is that the gain is even higher in a 3D implementation (if nothing else changes, the expected speed up factor in ND is approximately $\sqrt{2}^N$). The conclusion is linked to the segmentation task of white/gray matter in the brain. For cases like vessels or abdominal organs, other diffusion schemes may be optimal.

Ackowledgements

A large part of the work that this paper is based on was conducted at DIKU — the Department of Computer Science, University of Copenhagen, Denmark.

References

1. The internet brain segmentation repository, 1999. MR brain data set 788_6_m and its manual segmentation was provided by the Center for Morphometric Analysis at Massachusetts General Hospital and is available at http://neuro-www.mgh.harvard.edu/cma/ibsr

225

2. L. Alvarez and J.-M. Morel. *Geometry-Driven Diffusion in Computer Vision*, chapter Morphological Approach to Multiscale Analysis: From Principles to Equations. Kluwer, 1994.
3. F. Catté, P.-L. Lions, J.-M. Morel, and T. Coll. Image selective smoothing and edge detection by nonlinear diffusion. *SIAM Journal of Numerical Analysis*, 29:182–193, 1992.
4. C.A. Cocosco, V. Kollokian, R.K.-S. Kwan, and A.C. Evans. Brainweb: Online interface to a 3D MRI simulated brain database. In *3rd Int. Con. on Functional Mapping of the Human Brain*, volume 5 of *NeuroImage*, page 425, 1997. http://www.bic.mni.mcgill.ca/brainweb/.
5. D.L. Collins, A.P. Zijdenbos, V. Kollokian, J.G. Sled, N.J. Kabani, C.J. Holmes, and A.C. Evans. Design and construction of a realistic digital brain phantom. *IEEE Transactions on Medical Imaging*, 17(3):463–468, June 1998.
6. T.F. Cootes, C.J. Taylor, D.H. Cooper, and J. Graham. Active shape models: Their training and application. *CVIU*, (1):38–59, 1995.
7. Erik Dam. Evaluation of diffusion schemes for watershed segmentation. Master's thesis, Uni. of Copenhagen, 2000. Report 2000/1 on http://www.diku.dk/research/techreports/2000.htm.
8. E. Dam, P. Johansen, O.F. Olsen, A. Thomsen, T. Darvann, A.B. Dobrzeniecki, N.V. Hermann, N. Kitai, S. Kreiborg, P. Larsen, M. Lillholm and M. Nielsen. Interactive multi-scale segmentation in clinical use. In *CompuRAD, European Congress of Radiology 2000*, March 2000.
9. James Damon. Generic properties of solutions to partial differential equations. *Arch. Rational Mech. Anal.*, (140):353–403, 1997.
10. Steve Webb (editor). *The Physics of Medical Imaging*. Inst. of Physics Publ., London, 1988.
11. S. Murakami et al. Three and four dimensional diagnosis for TMJ. In *Proc. of CAR 98*, 88–91, 1998.
12. M. Gage. An isoperimetric inequality with applications to curve shortening. *Invent. Math.*, (76):357–364, 1983.
13. John M. Gauch. Image segmentation and analysis via multiscale gradient watershed hierachies. *IEEE Transactions on Image Processing*, 8(1):69 – 79, January 1999.
14. J.M. Gauch and S.M. Pizer. Multiresolution analysis of ridges and valleys in grey-scale images. *IEEE Transactions on Pattern Analysis and Machine Intelligence*, 15(6):635–646, 1993.
15. L.D. Griffin and A.C.F. Colchester. Superficial and deep structure in linear diffusion scale space: isophotes, critical points and separatrices. *Image and Vision Computing*, 13(7):543–557, 1995.
16. Peter Johansen, Mads Nielsen, and Sven Kreiborg. The computation of natural shape, 1999. http://www.diku.dk/research-groups/image/research/NaturalShape/.
17. Jan J. Koenderink. The structure of images. *Biological Cybernetics*, 50:363–370, 1984.
18. André Koster. *Linking Models for Multi-scale Image Sequences*. PhD thesis, Utrecht, 1995.
19. R.K.-S. Kwan, A.C. Evans, and G.B. Pike. An extensible MRI simulator for post-processing evaluation. *Lecture Notes in Computer Science*, 1131:135–140, 1996.
20. Lawrence Lifshitz and Stephen Pizer. A multiresolution hierarchical approach to image segmentation based on intensity extrema. *IEEE PAMI*, 12(6):529 – 540, June 1990.
21. William E. Lorensen and Harvey E. Cline. Marching cubes: A high resolution 3D surface construction algorithm. *Computer Graphics*, 21(4):163–169, 1987.
22. J.C. Maxwell. *On Hills and Dales*. The London, Edinburgh, and Dublin Philosophical Magazine and Journal of Science 4th Series, 40(269):421–425, December 1870.
23. D. Mumford and J. Shah. Boundary detection by minimizing functionals. In *CVPR85*, 22–26, 1985.
24. W.J. Niessen, K.L. Vincken, J.A. Weickert, and M.A. Viergever. Nonlinear multiscale representations for image segmentation. *Comp. Vision & Image Understanding*, 66:233–245, 1997.
25. K.N. Nordstrom. Biased anisotropic diffusion: A unified regularization and diffusion approach to edge detection. *IVC*, 8(4):318–327, 1990.
26. Ole Fogh Olsen. Multi-scale segmentation of grey-scale images. Technical Report 96/30, Department of Computer Science, University of Copenhagen, 1996.
27. Ole Fogh Olsen and Mads Nielsen. Generic events for the gradient squared with application to multi-scale segmentation. In *Scale-Space Theory in Computer Vision*, volume 1252 of *Lecture Notes in Computer Science*, 101–112, Utrecht, The Netherlands, 1997.
28. Edward Ott. *Chaos in Dynamical Systems*. Cambridge University Press, 1993.
29. Pietro Perona and Jitendra Malik. Scale-space and edge detection using anisotropic diffusion. *IEEE PAMI*, 12(7):629 – 639, July 1990.
30. S.M. Pizer, C.A. Burbeck, J.M. Coggins, D.S. Fritsch, and B.S. Morse. Object shape before boundary shape: Scale-space medial axes. *JMIV*, 4:303–313, 1994.
31. H. Samet. The quadtree and related hierarchical data structures. *Surveys*, 16(2):187–260, 1984.
32. Jon Sporring, Mads Nielsen, Luc Florack, & Peter Johansen. *Gaussian Scale-Space Theory*. Kluwer Academic Publishers, Dordrecht, 1997.
33. Koen Vincken. *Probabilistic Multi-scale Image Segmentation by the Hyperstack*. PhD thesis, University of Utrecht, 1995.
34. Joachim Weickert. *Anisotropic Diffusion in Image Processing*. B. G. Teubner, Stuttgart, 1998.
35. Andrew P. Witkin. Scale-space filtering. In *Proceedings of International Joint Conference on Artificial Intelligence*, 1019–1022, Karlsruhe, Germany, 1983.

Medial-Guided Fuzzy Segmentation

George Stetten [1], Stephen Pizer [2]

[1] Department of Bioengineering, University of Pittsburgh,
Robotics Institute, Carnegie Mellon University.

[2] Medical Imaging Display and Analysis Group (MIDAG),
University of North Carolina, Chapel Hill.

Abstract. Segmentation is generally regarded as partitioning space at the boundary of an object so as to represent the object's shape, pose, size, and topology. Some images, however, contain so much noise that distinct boundaries are not forthcoming even after the object has been identified. We have used statistical methods based on medial features in Real Time 3D echocardiography to locate the left ventricular axis, even though the precise boundaries of the ventricle are simply not visible in the data. We then produce a fuzzy labeling of ventricular voxels to represent the shape of the ventricle without any explicit boundary. The fuzzy segmentation permits calculation of total ventricular volume as well as determination of local boundary equivalencies, both of which are validated against manual tracings on 155 left ventricles. The method uses a medial-based compartmentalization of the object that is generalizable to any shape.

1   Introduction

Real Time 3D (RT3D) ultrasound is an imaging modality developed in the early 1990's that electronically interrogates a volume in 3D using a matrix array of transducers instead of a conventional linear array [1]. Unlike 3D ultrasound techniques that rely on physically moving a linear array, the scan rate of RT3D ultrasound is rapid enough (22 frames/second) to smoothly sample heart wall motion of the left ventricle (LV) during a single cardiac cycle. The trade-off for operating at such a high speed in 3D is that the data is very noisy and the resolution is relatively poor, even for ultrasound.

Methods of finding and measuring the LV with conventional ultrasound have concentrated on deformable contours [2-6]. Deformable surfaces have been applied to mechanically scanned 3D echocardiographic data [7]. Attempts at applying deformable methods to Real Time 3D (RT3D) ultrasound, with its lower data quality, have not produced encouraging results [8]. The difficulties encountered when applying deformable models to RT3D ultrasound data have led us to review *bottom-up* approaches based on the measurement of geometric image properties. Successes have been reported in conventional ultrasound using circular arc matched filters to find cross sections of the ventricle [9]. Using templates to find the center of the ventricle and then fuzzy reasoning to find the boundary has also worked quite well [10]. In RT3D data, a Hough transform approach has been developed using circular edge filters, yielding fully automated measurement of balloons [11, 12]. All of these approaches possess fundamentally medial aspects, relating multiple boundary points to common central points deep within the object.

The lineage of the medial approach may be traced to the *medial axis* (otherwise known as the *symmetric axis* or *skeleton*) introduced on binary images by Blum and developed by Nagel, Nackman, and others [13-15]. Pizer has extended the medial axis to gray-scale images producing a graded measure called *medialness*, which links the aperture of the boundary measurement to the radius of the medial axis, to produce what has been labeled a *core*. A core is defined as a height ridge of medialness, a locus in a space of position, radius, and associated orientations [16-17]. Methods involving these continuous loci of medial primitives have proven robust against noise and variation in target shape [18]. We developed a particular method of finding medial loci that analyzes populations of medial primitives called *core atoms*, and uses *medial node models* to identify structures from sets of cores found in this manner. We successfully applied this method to finding the LV in RT3D data by means of a 3-node medial node model (Fig. 1A) demonstrating the automated location of the Apex-to-Mitral-Valve (AMV) axis (Fig. 1B) on 155 ventricles in RT3D ultrasound scans, as described elsewhere [19,20]. We describe here a continuation of that experiment in which a fuzzy segmentation method is applied to the LV, given the initial automated placement of the AMV axis and establishment of its global width. We validate the fuzzy segmentation method using the same 155 ventricles as before.

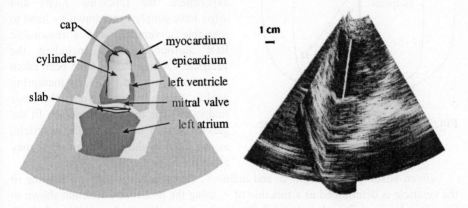

Fig. 1 A. Medial node model for left ventricle. **B.** Apex-to-mitral-valve (AMV) axis

2 Establishing the Probability of a Voxel being in the LV

Given the extremely noisy data in the RT3D images of the heart, we have adopted a statistical approach to segmenting the LV. We assign a probability to individual voxels of being in the LV, as a function of both voxel intensity and location relative to the AMV axis. Dark voxels near the axis are likely to be in the LV. Voxels that are brighter and/or further from the axis are less likely to be in the LV. These two probabilites are combined into a single aggregate probability of being in the ventricle.

The probability $p_L(j)$ that voxel j is within the ventricle due to its location is determined by using a surface model of expected ventricular shape (Fig. 2). The midpoint of the AMV axis is the origin for a spherical coordinate system with the poles of the sphere being the end points of the AMV axis. The angle φ corresponds to *latitude*, with $\varphi = 0°$ at the mitral valve and $\varphi = 180°$ at the ventricular apex. The angle ω corresponds to *longitude*, or rotation around the AMV axis, with $0° < \omega < 360°$. The third coordinate r is the radial distance to the midpoint of the AMV axis.

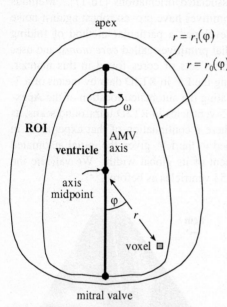

apex

$r = r_1(\varphi)$

$r = r_0(\varphi)$

ω

ROI

AMV axis

ventricle

axis midpoint

φ

r

voxel □

mitral valve

Fig. 2 Model of expected LV shape.

An expected surface for the ventricle is determined by a function $r_0(\varphi)$, such that $r = r_0(\varphi)$ at the LV surface. The function $r_0(\varphi)$ is constant to ω, exhibiting cylindrical symmetry around the AMV axis. A second surface, defining a region of interest (ROI), is specified by the function $r_1(\varphi)$. The ROI surrounds and includes the expected ventricular shape and will be used to generate regional statistics of voxel intensity. For the purposes of this experiment, the functions $r_0(\varphi)$ and $r_1(\varphi)$ have simply been drawn by hand to resemble a ventricle and a reasonable ROI surrounding and including the ventricle. Actual values have been taken from the drawing in Fig. 2, by measuring angles and distances from the axis midpoint. The model is scaled to fit the general size of each particular ventricle, as measured by the core atoms when they establish the AMV axis.

Given voxel j with angle φ and radius r, the probability $p_L(j)$ of it being in the ventricle is determined as a function of r using the probability function shown in Fig. 3A where $r_0(\varphi)$ is the expected distance to the ventricular boundary at latitude φ. Inside the boundary $r = r_0$, the probability is greater than $1/2$, whereas outside it is less than $1/2$.

Ventricular voxels are darker than surrounding voxels, although overall intensity in ultrasound images varies unpredictably from one scan to another. To compensate for this, the probability $p_I(j)$ of voxel j being within the ventricle, given only its intensity, is based on a statistical study of voxels in the ROI for a particular scan. The mean intensity $\bar{I}$ of all voxels in the ROI is computed, weighting each voxel's intensity $I(j)$ by its $p_L(j)$:

$$\bar{I} = \frac{\sum\limits_{j \in ROI} p_L(j) I(j)}{\sum\limits_{j \in ROI} p_L(j)}. \tag{1}$$

Thus ventricular voxels are favored by their tendency to be located near the AMV axis, although other voxels are represented as well. The value of $\bar{I}$ reflects the particular intensities of voxels in the image and is used in the function shown in Fig. 3B to compute the probability $p_I(j)$ for each voxel, given its intensity. The functions in Fig. 3A and Fig 3B have been drawn *ad hoc* for the purposes of this experiment. Determining corresponding functions to optimize the accuracy of the resulting segmentation is beyond the scope of the present research.

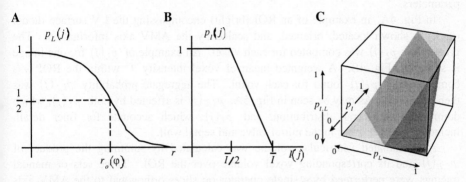

Fig 3 Probabilities $p_L(j)$, $p_I(j)$, and their aggregate probability $p_{I,L}(j)$.

For fuzzy segmentation, we need a single probability $p_{I,L}(j)$ that voxel j is in the ventricle, given both its intensity and location. We define an operator A such that $P_{I,L}(j) = A(P_I(j), P_L(j))$, and design it to exhibit the following behaviors:

(1) It is monotonic with positive slope for both $p_I(j)$ and $p_L(j)$, mapping the domain $[0,1]$ for $p_I(j)$ and $p_L(j)$ into a range $[0,1]$ for $p_{I,L}(j)$.

(2) If either argument $p_I(j)$ and $p_L(j)$ is equal to $1/2$, then $p_{I,L}(j)$ equals the other argument, so that a probability of $1/2$ exerts neither positive nor negative influence.

(3) If either $p_I(j)$ or $p_L(j)$ is 0, then $p_A(j)$ is also 0, so that either can independently exclude any voxel from the ventricle.

A continuous, piece-wise linear function that satisfies the above constraints is shown in Fig. 3C. The aggregate probability $p_{I,L}(j)$ represents a fuzzy segmentation of the LV, as will be validated in the following section.

3   Validating the Fuzzy Segmentation Method

We validated the method for fuzzy segmentation by computing volumes and local boundary equivalencies using $p_{I,L}(j)$ and comparing these results to manual tracings. (A more detailed record of these results and other aspects of this research is available in the author's dissertation [19].) The same 155 ultrasound scans used to establish the AMV axis in the previous experiment [20] were divided blindly into training and test sets. The training set was used to optimize the method's parameters in terms of accuracy of volume measurement and to measure bias using linear regression. Of the many parameters in the overall process, those that were adjusted included the shapes of the surface model in Fig. 2 and the probability curves in Figs. 3A and 3B. The method was then applied to the test set without further adjustment of parameters.

In Fig. 4A, an example of an ROI (bright) encompassing the LV surface model (dark) is shown located, oriented, and scaled by the AMV axis information. The probability $p_L(j)$ was computed for each voxel. An example of $p_L(j)$ for one image is shown in Fig. 4B. A weighted mean of voxel intensity $\bar{I}$ within the ROI was computed and $p_I(j)$ found for each voxel. The aggregate probability $p_{I,L}(j)$ was then computed. As may be seen in Fig. 5A, $p_{I,L}(j)$ is affected by both $p_L(j)$, which dominates the overall distribution, and $p_I(j)$, which accounts for finer detail, including the dark areas in the mitral valve and septal wall.

An automated ventricular volume was calculated by summing the product of $p_{I,L}(j)$ and its corresponding voxel volume over the ROI. Three sets of manual tracings were performed by a single operator on slices orthogonal to the AMV axis (Fig 5B). Voxels within the traces were labeled and used to compute a manual ventricular volume. A comparison between automated and manual volumes is shown in Fig. 6., with an RMS error of 25.9 ml. A reduced RMS error of 9.2 ml was achieved by measuring volume *change* with time for a given heart. This is clinically significant since volume change can yield stroke volume and short-term variation in heart function.

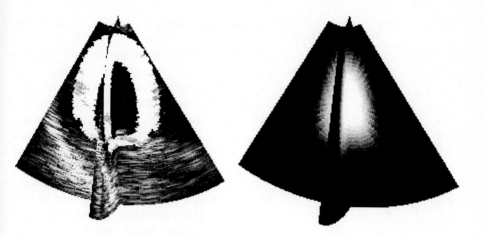

Fig. 4 **A.** ROI (bright) and LV surface model (dark). **B.** $p_L(j)$ for each voxel, bright = 1 and dark = 0

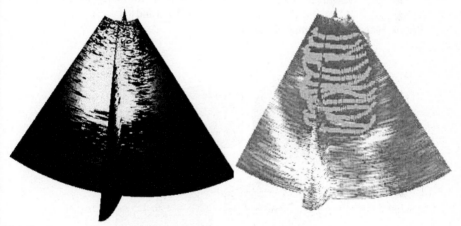

Fig. 5 **A.** Aggregate probability $p_{I,L}(j)$. **B.** Manual tracings of LV.

To localize error, the concept of *truncated wedges* was developed. A full description of truncated wedges may be found elsewhere [19]. The basic concept is to divide an arbitrary shape into compartments that extend from the surface to the medial manifold (Fig. 7). For the particular case of the sphere (Fig. 7a), voxels are sorted by ☐ and☐ into bins and volumes computed for each wedge (solid angle). An equivalent radius is then computed for each wedge given its volume and compared to a radius determined from the manual tracings. The results are shown in Fig. 8.

The local error is expressed in terms of linear distance (cm) as a function of latitude ☐ and longitude ☐. The S.D. for the 3 manual tracings is between 1.5-3.5 mm, while the automated segmentation produces an RMS error generally between 2-6 mm with a spike up to 1.6 cm (which we believe to be caused by the aorta leaving the LV -- a feature of the anatomy not included in the manual tracing).

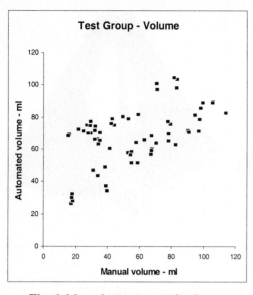

Fig. 6 Manual vs. automated volume.

The accuracy of any segmentation is not well judged solely by comparing total volumes, since an object's volume says nothing about its location and since different shapes may have the same volume. Using the medial framework already established, local error between the manual and automated segmentations was determined in spherical coordinates φ, ω, and r. Longitude ω was not a parameter in the ventricular surface model, but since all the scans were performed with the same transducer orientation, the error could be correlated as a function of both ☐ and ☐.

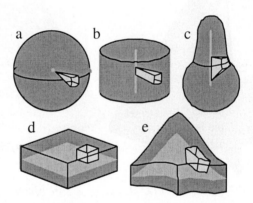

Fig. 7 Truncated wedges based on the medial manifold.

4   Conclusions

Most segmentation routines attempt to locate complete and precise boundaries, which provide explicit parameters for measurements. Perhaps such boundaries seem a more accurate portrayal of physical reality, because, after all, most real surfaces are defined at a microscopic scale and interact with light in a manner easily interpretable by human vision. A probabilistic representation of a surface may be just as useful and more robust for making geometric measurements, but it is not easily visualized or understood. It is clear from Fig. 5A that our fuzzy segmentation method does not yield a reasonable probability for each individual voxel. The fuzzy representation lacks a

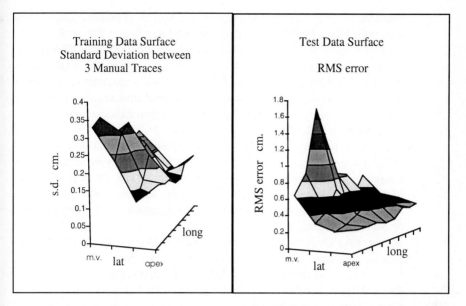

Fig. 8 A. S.D. between manual tracings. **B.** RMS error of automated segmentation.

distinct topology. Only in certain statistical operations, such as the computation of volume, is the fuzzy shape representation appropriate. In those cases, accuracy of measurements based on probabilistic boundaries can be assessed against those from manually determined boundaries.

We believe the inconclusive results from the RT3D ultrasound data may be as much a problem with the manual tracings as with the method itself, and we make no pretense to have optimized the numerous parameters in the method on such a small sample. We merely report an empirical test of a new concept. The use of a medial framework to calculate $p_L(j)$ within an aggregate probability, and to weight an intensity histogram relative to that framework, are promising new avenues of research. In concert with medial node models and truncated wedges, the fuzzy segmentation method can be applied to anatomical targets of any shape.

Support from a Whitaker Biomedical Engineering grant, NSF grant CDR8622201, NIH grants 1K08HL03220, P01CA47982, and HL46242, with data generously supplied by Takahiro Shiota, Department of Cardiology, Cleveland Clinic Foundation, and by Volumetrics Medical Imaging, Inc., Durham, NC.

References

1.   Stetten, G., T. Ota, C. Ohazama, C. Fleishman, J. Castelucci, J. Oxaal, T. Ryan, J. Kisslo, and O.v. Ramm, *Real-Time 3D Ultrasound: A New Look at the Heart.* Journal of Cardiovascular Diagnosis and Procedures, 1998. **15**(2): p. 73-84.

2. Chalana, V., D.T. Linker, D.R. Haynor, and Y. Kim, *A multiple active contour model for cardiac boundary detection on echocardiographic sequence.* IEEE Transactions on Medical Image, 1996. **15**(3): p. 290-298.

3. Hozumi, T., K. Yoshida, H. Yoshioka, T. Yagi, T. Akasaka, T. Tagaki, M. Nishiura, M. Watanabe, and J. Yoshikawa, *Echocardiographic estimation of left ventricular cavity area with a newly developed automated contour tracking method.* Journal of the American Society of Echocardiography, 1997. **10**(8): p. 822-829.

4. Mikic, I., S. Krucinski, and J.D. Thomas, *Segmentation and tracking in echocardiographic sequences: Active contours guided by optical flow estimates.* IEEE Transactions on Medical Imaging, 1998. **17**(2): p. 274-284.

5. Malassiotis, S. and M.G. Strintzis, *Tracking the left ventricle in echocardiographic images by learning heart dynamics.* IEEE Transactions on Medical Imaging, 1999. **18**(3): p. 282-290.

6. Hunter, I.A., J.J. Soraghan, J. Christie, and T.S. Durrani. *Detection of Echocardiographic Left Ventricle Boundaries Using Neural Networks.* in *Computers in Cardiology.* 1993, London, UK, : p. 201-204

7. Coppini, G., R. Poli, and G. Valli, *Recovery of the 3-D shape of the left ventricle from echocardiographic images.* IEEE Transactions on Medical Imaging, 1995. **14**(2): p. 301-317.

8. Stetten, G., T. Irvine, D. Ritscher, O.T.v. Ramm, J. Panza, V. Sachdev, J. Castellucci, M. Jones, and D. Sahn. *Improved accuracy for a semi-automated method for computing right ventricle (RV) cavity volumes from Real Time 3D Echo: Comparison studies to ultrasonic crystals in an open-chest animal model.* in *American College of Cardiology 48th Scientific Sessions.* 1999 (in press), New Orleans

9. Wilson, D.C., E.A. Geiser, and J. Li, *Feature extraction in two-dimensional short-axis echocardiographic images.* Journal of Mathematical Imaging and Vision, 1993. **3**: p. 285-298.

10. Feng, J., W.C. Lin, and C.T. Chen, *Epicardial Boundary Detection Using Fuzzy Reasoning.* IEEE Transactions on Medical Imaging, 1991. **10**(2): p. 187-199, .

11. Stetten, G. and R. Morris, *Shape detection with the Flow Integration Transform.* Information Sciences, 1995. **85**: p. 203-221.

12. Stetten, G., M. Caines, C. Ohazama, and O.T.v. Ramm, *The Volumetricardiogram (VCG): Volume Determination of Cardiac Chambers using 3D Matrix-Array Ultrasound.* Proceedings of the SPIE Symposium on Medical Imaging, 1995. **2432**: p. 185-196.

13. Blum, H. and R.N. Nagel, *Shape description using weighted symmetric axis features.* Pattern Recognition, 1978. **10**: p. 167-180.

14. Nackman, L.R., *Curvature relations in 3D symmetric axes.* CVGIP, 1982. **20**: p. 43-57.

15. Nackman, L.R. and S.M. Pizer, *Three-Dimensional shape description using the symmetric axis transform I:Theory.* IEEE Transactions on Pattern Analysis and Machine Intelligence, 1985. **2**(2): p. 187-202.

16.   Burbeck, C.A. and S.M. Pizer, *Object representation by cores: Identifying and representing primitive spatial regions.* Vision Research, 1995. **35**(13): p. 1917-1930.

17.   Pizer, S.M., D.H. Eberly, B.S. Morse, and D.S. Fritsch, *Zoom invariant vision of figural shape: the mathematics of cores.* Computer Vision and Image Understanding, 1998. **69**(1): p. 55-71.

18.   Morse, B.S., S.M. Pizer, D.T. Puff, and C. Gu, *Zoom-Invariant vision of figural shape: Effect on cores of image disturbances.* Computer Vision and Image Understanding, 1998. **69**: p. 72-86.

19.   Stetten, G., *Automated Identification and Measurement of Cardiac Anatomy Via Analysis of Medial Primitives*, Doctoral Dissertation in Dept. Biomedical Engineering, University of North Carolina, Chapel Hill, 1999, http://www.stetten.com/george/publications/phd

20.   Stetten, G.D. and S.M. Pizer, *Medial Node Models to Identify and Measure Objects in Real-Time 3D Echocardiography.* IEEE Transactions on Medical Imaging, 1999. **18**(10): p. 1025-1034.

Robust 3D Segmentation of Anatomical Structures with Level Sets

C. Baillard and C. Barillot

IRISA-INRIA/CNRS, Campus de Beaulieu,
35042 Rennes cedex, France
{Caroline.Baillard,Christian.Barillot}@irisa.fr

Abstract This paper is concerned with the use of the level set formalism to segment anatomical structures in 3D medical images (ultrasound or magnetic resonance images). A closed 3D surface propagates towards the desired boundaries through the iterative evolution of a 4D implicit function. The major contribution of this work is the design of a robust evolution model based on adaptive parameters depending on the data. First the iteration step and the external propagation force, both usually constant, are automatically computed at each iteration. Additionally, region-based information rather than the gradient is used, via an estimation of intensity probability density functions over the image. As a result, the method can be applied to various kinds of data. Quantitative and qualitative results on brain MR images and 3D echographies of carotid arteries are discussed.

1 Introduction

The 3D segmentation of anatomical structures is crucial for many medical applications, both for visualization and clinical diagnosis purposes. Due to the huge amount of data and the complexity of anatomical structures, manual segmentation is extremely tedious and often inconsistent. Automatic segmentation methods are required to fully exploit 3D data. It is a very challenging task because they can usually not rely on image information only. Anatomical tissues are generally not homogeneous and their boundaries are not clearly defined in the images. It is therefore often necessary to involve prior knowledge about the shape or the radiometric behaviour of the structure of interest.

Deformable models define a powerful tool to accurately recover a structure using very few assumptions about its shape [15]. Such a model iteratively evolves towards the desired location according to a global energy minimization process. The functional energy is based on external forces derived from the data and internal forces related to the geometry of the contour. The limitations of this approach are well-known: the contour must be initialized close to the desired boundaries, and it can not cope with significant protusions nor topological changes. In the last few years, segmentation methods based on level sets have become very popular because they overcome classical limits of deformable models [11, 9, 3]. The evolving surface can change topology and cope with complex geometry, and the result is less dependent on initialization than with any other iterative method. This kind of approach has already been applied within a wide range of applications in computer vision. However, parametrization is still a limitation for practical use. Several evolution models have been proposed, but most of them include many parameters to be tuned: iteration step, weighting parameters, constant propagation term, etc. The tuning of these parameters determines the success of the method.

This paper describes a robust evolution model which enables a volume to be segmented with almost no parameter setting. It involves adaptive parameters depending on the data, and it relies on region-based information rather than gradient, via an estimation of intensity probability density functions over the image. The versatility of the segmentation is demonstrated on both brain structures in MR images and carotid arteries in 3D echography. Our strategy is presented in section 2. The two main stages of the method - intensity distribution analysis and surface evolution - are described in sections 3 and 4. Experimental results are presented and discussed in section 5.

2 Segmentation Strategy Based on Level Sets

Within the level set formulation [11], the evolving surface $S(t)$ is processed as a propagating front embedded as the zero level of a 4D scalar function $\Psi(\mathbf{x}, t)$. This hypersurface is usually defined as the signed distance from $\mathbf{x}$ to the front S (negative *inside* the object). The evolution rule for Ψ is:

$$\frac{\partial \Psi}{\partial t} + F|\nabla \Psi| = 0, \tag{1}$$

where F is a scalar *velocity function* depending on the local geometric properties of the front (local curvature) and external parameters related to the input data (image gradient for instance). The hypersurface Ψ deforms iteratively according to F, and the position of the 3D front $S(t)$ is deduced from Ψ at each iteration step using the relation $\Psi(\mathbf{x}(t), t) = 0$. Practically, the hypersurface Ψ^{n+1} at step $n + 1$ is computed from Ψ^n at step n using the relation:

$$\Psi^{n+1}(\mathbf{x}) = \Psi^n(\mathbf{x}) - \Delta t.F|\nabla \Psi^n(\mathbf{x})|, \quad \forall \mathbf{x} \in \mathbb{R}^3 \tag{2}$$

The design of the velocity function F plays the major role in the evolution process. Several formulations have been proposed [6, 3, 13, 18]. We have chosen the original formulation given by Malladi et.al. [9] for its simplicity:

$$F = h_I(\nu - \rho\kappa) \tag{3}$$

The term ν represents an external propagation force which makes the surface contract or expand. The parameter κ represents the local curvature of the front and acts as a regularization term. The weighting ρ expresses the importance given to regularization. Finally, the term h_I is the data consistency term: it depends on the intensity I of the input data, and acts as a stopping factor at the location of the desired boundaries.

In the following, our definition of the parameters introduced in equations (2) and (3) is presented. They are defined with respect to the intensity probability density functions (PDFs) estimated inside and outside the structure. These PDFs will be respectively denoted $p_i(I)$ and $p_e(I)$, and the prior probability for a voxel to be *inside* the structure will be denoted α_i. Our general strategy is illustrated in figure 1. The PDFs are first estimated by using both the initial segmentation and given statistical distribution models. This allows the parameters of the velocity function to be computed for the evolution of the hypersurface. The two stages are further described in the next two sections.

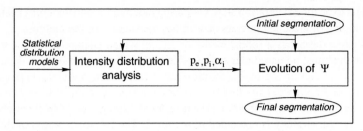

Figure1. Strategy for 3D segmentation. The velocity function controlling the surface evolution is computed according to the estimated intensity density functions inside and outside the structure.

3   Estimation of Intensity Probability Density Functions

We assume that the intensity distribution of the input image can be written as a generalized mixture of n kernels. Furthermore, these kernels do not necessarily have the same form. In this work, we have considered two kinds of laws: the Gaussian distribution and the shifted Rayleigh distribution.

Each component k is associated to a class labelled by λ_k and described by the variable Φ_k, which characterizes the kernel distribution and the prior probability $P(\lambda_k) = \pi_k$ of a voxel belonging to the class λ_k. The distribution mixture is thus described by the variable $\Phi = \{\Phi_k, 1 \leq k \leq n\}$. We begin by estimating Φ using the Stochastic Expectation-Maximisation (SEM) algorithm, which is a stochastic version of the EM algorithm [5]. Whilst the SEM algorithm is less sensitive to initialization than the EM algorithm, it is not guaranteed to find a local maximum of the likelihood surface, and the initialization of Φ_k is still important. For this initialization, we have distinguished two cases according to the value of n.

If $n = 2$, the distribution mixture is bimodal, i.e., the object to be segmented and its background can each be approximated by a monomodal distribution. In this case, the surface initializing the segmentation process is used to produce two independent histograms from the input image: one for the initial segmentation and one for its background. The variables Φ_0 and Φ_1 characterizing the bimodal mixture are independently estimated over these complete data, providing two coarse estimates $\hat{\Phi}_0$ and $\hat{\Phi}_1$. This implicitely assumes that the intensity distribution inside and outside the initial segmentation are somehow representative of the distributions to be estimated. The bimodal approach is depicted in figure 2. Figure 3 shows an example of PDFs estimated with this method for the image shown in figure 6. The bimodal model is not perfect, but it is good enough to separate the two main classes of the image.

If the distribution is not bimodal (case of brain MR images, see figure 5 for instance), the local maxima of the histogram are used as an initial guess for the center of the kernels, and the prior probabilities π_k are initialized equal to each other. This initial guess of Φ is then processed by the SEM algorithm, providing an estimation of π_k and $P(I|\lambda_k)$ for each class λ_k. Among these n classes, only one or two characterize the structure to be segmented; the set of these classes will be noted Λ_i, and the complementary set of the classes exterior to the object will be noted Λ_e. The classes defining Λ_i can be given as input parameters of the algorithm (number of classes and

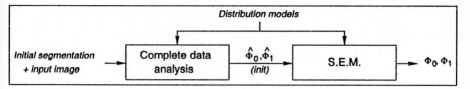

Figure2. Estimation of the mixture model parameters when the distribution is bimodal.

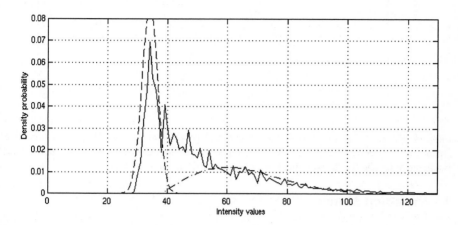

Figure3. Estimation of the intensity mixture model parameters on the ultrasound image of figure 6 (bimodal case). The solid line represents the normalized histogram of the input image. The dash line is the estimated PDF *inside* the carotid arterie and the dashdot line is the estimated PDF *outside* it.

approximative mean value). Alternatively they can be automatically determined from the initialization of the surface (required for the segmentation process): only the classes highly represented *inside the initial surface* are said to belong to Λ_i. Once the sets Λ_i and Λ_e have been defined, the estimated intensity distributions inside and outside the object can be defined as:

$$\begin{cases} p_i(I) = \sum_{k/\lambda_k \in \Lambda_i} \pi_k P(I|\lambda_k) \\ p_e(I) = \sum_{k/\lambda_k \in \Lambda_e} \pi_k P(I|\lambda_k), \end{cases} \tag{4}$$

and the prior probability for a voxel to be inside the structure:

$$\alpha_i = \sum_{k/\lambda_k \in \Lambda_i} \pi_k \tag{5}$$

An example of estimated multimodal distribution mixture is shown in figure 4.

The SEM algorithm does not guarantee the optimal solution, but in practice any initial partition roughly representative of the inner and the outer distribution leads to a correct solution. In particular, the initial surface does not need to be close to the real boundaries.

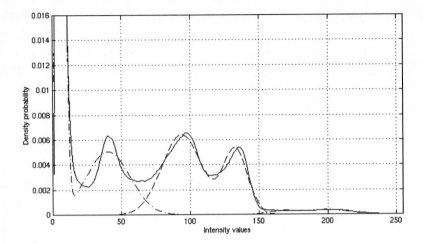

Figure4. Estimation of the intensity mixture model parameters on brain MRI. The solid line represents the normalized histogram of the input image. The dash line is the estimated PDF *inside* the brain and the dashdot line is the estimated PDF *outside* the brain.

4   Evolution Model

4.1   External Propagation Term ν

In equation (3), the sign of ν determines the direction of the external propagation force. Several approaches to 3D segmentation using this evolution model have imposed a one-way propagation force ν, which either contracts or expands the whole surface all along the process [8, 18]. However, when the initial position of the surface can be predicted (by tracking or by registration with an atlas for instance), the predicted and real positions usually overlap. It is therefore necessary to let the surface evolve in both directions. Some propagation models have been designed in order to solve this problem in 2D, by involving a local analysis of intensity [12, 1, 17].

The problem can be expressed as the classification of each point of the current interface $S(t)$. If a point belongs to the object then the surface should locally extend; if it does not, the surface should contract. We perform this classification by maximizing the *a posteriori* segmentation probability $p(\lambda|I)$, where λ denotes the appartenance class of the considered point. According to Bayes rule, the maximization of the posterior distribution $p(\lambda|I)$ is equivalent to the maximization of $p(\lambda)p(I|\lambda)$, where $p(\lambda)$ is the *prior* of the class λ and $p(I|\lambda)$ is the likelihood of intensity. The propagation term ν has then been defined as:

$$\nu = \text{Sign}\{\alpha_i p_i(I) - (1 - \alpha_i)p_e(I)\} \tag{6}$$

The propagation term ν is therefore positive if $p(\lambda_i|I) > p(\lambda_e|I)$. The point is more likely to be *inside* the object than *outside*, and the surface extends. Note, we always have $|\nu| = 1$. Experiments with a continous-varying function have not shown any improvement. Besides, this simple definition needs no tuning.

4.2 Curvature Term $\rho\kappa$

The regularization parameter κ at a point $\mathbf{x}$ is the curvature of the interface, computed at the closest point to $\mathbf{x}$. This curvature is computed using the partial derivatives of Ψ.

The respective roles of the propagation and the regularization terms are entirely determined by the weight ρ, which makes the process very easy to tune. The weighting parameter ρ can be interpreted as the particular curvature radius leading to a stable position ($F = 0$).

4.3 Stopping Factor h_I

The stopping factor is a data consistency term traditionally related to the intensity gradient ∇I of the input image [9]. Since this gradient is only defined for points belonging to the interface (zero level set of the hypersurface), an *extended* gradient function needs to be defined over $I\!R^3$ by the intensity gradient of the closest neighbour on the interface. However, gradient information has no meaning for very noisy and/or low-contrasted images (like ultrasound data). Besides, high gradients do not necessarily indicates a relevant boundary between the structure to be segmented and its background.

Similarly to the approach described in [14] in the 2D case, we have related this term to the *posterior* probability of having a transition between the object and its background. Let $\mathbf{x}$ be a voxel of the current interface, and λ the estimated class of $\mathbf{x}$; the parameter λ is supposed to be known here because it is taken into account via the sign of ν. The posterior probability of $\mathbf{x}$ being a transition, given I and λ, is given by:

$$p_T(\mathbf{x}|I,\lambda) = \begin{cases} p(\mathbf{x} - \mathbf{n} \in \bar{O}|I) & \text{if } \lambda \in \Lambda_i \\ p(\mathbf{x} - \mathbf{n} \in O|I) & \text{if } \lambda \in \Lambda_e, \end{cases} \tag{7}$$

where $\mathbf{n} = \frac{\Delta\Psi}{|\Delta\Psi|}$ is the current normal vector of the hypersurface, O is the object to be segmented and $\bar{O}$ its complementary. If $\nu > 0$, the posterior transition probability of $\mathbf{x}$ is the probability of the neighbouring voxel $\mathbf{x}' = \mathbf{x} - \mathbf{n}$ (located outside the current volume) to be located outside the object to be segmented. Using Bayes rule, this transition probability can be expressed as :

$$p_T(\mathbf{x}|I,\lambda) = \begin{cases} \dfrac{(1-\alpha_i)p_e(I(\mathbf{x}'))}{\alpha_i p_i(I(\mathbf{x}'))+(1-\alpha_i)p_e(I(\mathbf{x}'))} & \text{if } \lambda \in \Lambda_i \\[3mm] \dfrac{\alpha_i p_i(I(\mathbf{x}'))}{\alpha_i p_i(I(\mathbf{x}'))+(1-\alpha_i)p_e(I(\mathbf{x}'))} & \text{if } \lambda \in \Lambda_e \end{cases} \tag{8}$$

The stooping factor h_I at a point $\mathbf{x}$ belonging to the interface is finally defined as a decreasing function of $p_T(\mathbf{x}|I,\lambda)$. Since this probability is only defined on the interface, it is extended to $I\!R^3$ via the closest point on the current interface:

$$h_I(\mathbf{x}) = g(p_T(\tilde{\mathbf{x}}|I,\lambda)) \qquad \forall \mathbf{x} \in I\!R^3, \tag{9}$$

where g is decreasing from $[0;1]$ to $[0;1]$, and $\tilde{\mathbf{x}}$ is the closest point to $\mathbf{x}$ on the interface.

4.4 Iteration Step Δt

The iteration step Δt of equation (2) is usually constant and manually tuned. We propose to compute it automatically at each iteration in order to improve robustness.

The stability of the process requires a numerical scheme for the computation of $\Delta\Psi$, called *upwind sheme*. This scheme induces a limit on the iteration step Δt, called the CFL restriction (Courant-Friedrichs-Levy). More precisely, writing equation (1) as $\Psi_t + H(\Psi_x, \Psi_y, \Psi_z) = 0$, where H is the Hamiltonian defined by:

$$H(u, v, w) = \sqrt{u^2 + v^2 + w^2}.F, \tag{10}$$

the CFL restriction can be expressed in 3D [7] as:

$$1 \geq \Delta t.(\frac{|H_u|}{\Delta x} + \frac{|H_v|}{\Delta y} + \frac{|H_w|}{\Delta z}),$$

where H_u, H_v, H_w denote the partial derivatives of H with respect to u, v, w. Since we work with a regular sampling grid, we can assume $\Delta x = \Delta y = \Delta z = 1$. According to equation (3) and the definition of the parameters involved, the velocity function F is independent from u, v, w. The terms h_I and ν only depends on the image data. The curvature term κ is the curvature of the interface at the closest point, and it does not depend on the local characteristics of Ψ. Therefore the partial derivatives of H can be directly computed from (10) at each iteration of the process, and the best value for Δt which guarantees stability is given by:

$$\tilde{\Delta}t = \min\{\frac{1}{F}.\frac{\sqrt{u^2 + v^2 + w^2}}{|u| + |v| + |w|}\} \tag{11}$$

5 Experimental Results

5.1 Segmentation of Brain MR Images

The segmentation algorithm was run on simulated data provided by the MNI [4]. They consist of a phantom collected with 3 levels of noise and inhomogeneity, associated to a reference classification. The method was applied to segment the whole brain (grey matter + white matter) and numerically assessed with overlapping measures: sensitivity, specificity, and total performance [16]. The segmentation was initialized with a cube $100 \times 70 \times 70$, and the classes of interest (grey matter + white matter) automatically determined as described in section 3. It was compared with a segmentation method based on morphological operators and tuned with a "best practice" parameter set. The numerical results of both methods are summarized in table 1. The total performance achieved with our algorithm is stable around 98.3%, even in critical conditions (9% noise and 40% inhomogeneity). This is far better than the other method which is very dependent on noise. The improvement mainly concerns the sensitivity of the detection.

Experiments have also been run on a database of 18 MR images (volume size $256 \times 256 \times 176$). All the subjects have been processed in the same way, and using exactly the same parameter set. The iteration step is automatically computed, as well as the threshold on intensity values. The weighting parameter on regularization has been set to $\rho = 3$ (see eq. 3), which has appeared to be a good compromise to avoid propagation out of the brain (possible when the cerebro-spinal fluid area is thiner than the image resolution), whilst preserving the highly convoluted characteristics of the cortical surface. Out of the 18 subjects, only 1 segmentation failed, and the failure could be detected because the propagation speed started to increase (it should keep decreasing).

	Levels sets			Morphological filtering		
	Nb iter	Sens.	Spec.	Total	Sens.	Spec.	Total
0% noise, 0% inhom.	710	96.2%	99.0%	98.3%	86.7%	98.7%	95.7%
3% noise, 20% inhom.	680	96.3%	98.9%	98.2%	83.5%	99%	95.2%
9% noise, 40% inhom.	800	95.95%	98.9%	98.2%	69.8%	99.5%	92.3%

Table1. Quantitative assessment on the phantom. The number of iterations, the sensitivity, specificity, and total performance measures are given for three levels of noise and two segmentation methods: the method based on level sets (initialization with a cube $100 \times 70 \times 70$) and a segmentation method based on morphological operators.

All the other 17 subjects were correctly segmented. Figure 5 shows the segmentation results on two different subjects. Protusions of brain and ventricles are properly recovered, despite of the surface being initialized far away from it. The method can naturally cope with change of topology: the ventricles inside the brain have also been recovered. On average, 1000 iterations are necessary to segment the whole brain on real data starting from a $100 \times 70 \times 70$ cube. The number of iterations can be reduced to 300 if the surface is initialized close to the real boundaries by a registration technique (see [2] for details).

5.2 Segmentation of 3D Ultrasound Images

The method has also been applied to 3D ultrasound images of carotid artery. This kind of images is difficult to segment automatically due to the speckle noise. The intensity distribution inside the carotid has been modelized by a Gaussian (shadow area), whereas the exterior is modelized by a Rayleigh distribution (reverberation area) [10]. We have used a weighting value of $\rho = 10$ for regularization. All the other parameters are the same as for experiments on brain MRI. Figures 6 and 7 show the results of segmentation on two different images. They allow us to be very optimistic about the use of this statistical variational approach to segment 3D ultrasound images.

6 Conclusion and Further Work

This paper has presented a robust evolution model for segmenting structures in 3D images using the level set formalism. The level set approach is very appropriate to segment anatomical structures without strong prior information. The statistical analysis of intensity distributions provides relevant information about input data. The design of an adaptive evolution force and an adaptive iteration step provides a good trade-off between convergence speed and stability. Almost no parameter needs to be tuned, only the regularization weight ρ has to be set depending on the data shape and noise. Good quality results have been produced on brain MRI and 3D echographies of carotid. The results demonstrate that a variational approach mixed with region-based analysis (statistical models) significantly improves the sensitivity and the robustness of the segmentation.

Thanks to the genericity of the method, the segmentation can be achieved for various image modalities and various kinds of anatomical structure. Our goal is now to focus on 3D ultrasound. Due to speckle noise, a method mixing variational and statistical models should be particularly appropriate to achieve good automatic segmentation. For

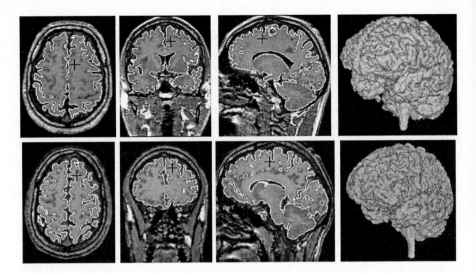

Figure5. Brain segmentation for two different subjects (initialization with a cube of size 70 × 100 × 100 located inside the brain). The first three columns respectively show axial, coronal and saggital planes, the last column shows 3-D views of the segmented brains.

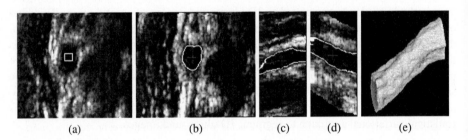

(a)　　　　　　　(b)　　　　　　　(c)　　(d)　　　　(e)

Figure6. Segmentation result on a $256 \times 256 \times 80$ ultrasound image of carotid (isotropic voxels). (a): initialization of the segmentation with a $10 \times 10 \times 25$ cube. (b,c,d): final segmentation (three visualization planes). (e) 3D view of the segmented carotid.

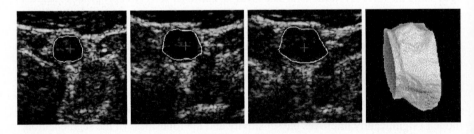

Figure7. Three transaxial slices and a 3D view of a segmented subpart of carotid (non isotropic voxels). The $256 \times 256 \times 36$ input image was initialized with a $50 \times 50 \times 25$ cube inside the carotid.

this purpose, it will be necessary to involve more prior information about the shape of the anatomic structure.

Acknowledgements. The authors would like to thank Drs L. Pourcelot and F. Tranquart from Tours university hospital and Dr A. Fenster from RRI (London, Ontario) for providing the US data, and the GIS "Sciences de la cognition" for granting the MR acquisition project.

References

1. O. Amadieu, E. Debreuve, M. Barlaud, and G. Aubert. Inward and outward curve evolution using level set method. In *ICIP*, Kobe, Japan, oct. 1999.
2. C. Baillard, P. Hellier, and C. Barillot. Segmentation of 3D brain structures using level sets and dense registration. In *IEEE Workshop on Mathematical Methods in Biomedical Image Analysis*, jun. 2000.
3. V. Caselles, R. Kimmel, and G. Sapiro. Geodesic active contours. *IJCV*, 22:61–79, 1997.
4. D.L. Collins, A.P. Zijdenbos, V. Kollokian, J.G. Sled, N.J. Kabani, C.J. Holmes, and A.C. Evans. Design and construction of a realistic digital brain phantom. *IEEE Transactions on Medical Imaging*, 17(3):463–468, jun. 1998.
5. A.P. Dempster, N.M. Laird, and D.B. Rubin. Maximum likelihood from incomplete data via the em algorithm. In *Royal Statistical Society*, pages 1–38, 1976.
6. J. Gomes and O. Faugeras. Reconciling distance functions and Level-Sets. Technical Report 3666, Inria, apr. 1999.
7. R. Kimmel, N. Kiryati, and A. Bruckstein. Analyzing and synthesizing images by evolving curves with the Osher-Sethian method. *IJCV*, 24(1):37–55, 1997.
8. L. M. Lorigo, O. Faugeras, W. E. L. Grimson, R. Keriven, and R. Kikinis. Segmentation of bone in clinical knee MRI using texture-based geodesic active contours. In A. Colchester and S. Delp, editors, *MICCAI*, number 1496 in LNSC, pages 1195–1204, Cambridge, MA, USA, oct. 1998. Springer.
9. R. Malladi, J.A. Sethian, and B.C. Vemuri. Shape modeling with front propagation: A level set approach. *IEEE Tr. on PAMI*, 17(2):158–175, feb. 1995.
10. M. Mignotte and J. Meunier. Deformable template and distribution mixture-based data modeling for the endocardial contour tracking in an echographic sequence. In *CVPR*, pages 225–230, jun. 1999.
11. S. Osher and J.A. Sethian. Fronts propagating with curvature dependent speed: Algorithms based on Hamilton-Jacobi formulation. *J. of Computational Physics*, 79:12–49, 1988.
12. C. Papin, P. Bouthemy, E. Mémin, and G. Rochard. Tracking and characterization of convective clouds from satellite images. In *EUMETSAT Meteo. Satellite Data Users Conf.*, Copenhague, sep. 1999.
13. N. Paragios and R. Deriche. Unifying boundary and region-based information for geodesic active tracking. In *CVPR*, volume 2, pages 300–305, Fort Collins, Colorado, jun. 1999.
14. N. Paragios and R. Deriche. Video and image sequence analysis - geodesic active contours and level sets for the detection and tracking of moving objects. *IEEE Tr. on PAMI*, 22(3):266–280, 2000.
15. D. Terzopoulos. Regularization of inverse visual problems involving discontinuities. *IEEE Tr. on PAMI*, 8(2):413–424, 1986.
16. JH. Van Bemmel and MA. Musen. *Handbook of medical informatics*. Springer, URL : http://www.mieur.nl/mihandbook, 1997.
17. A. Yezzi, A. Tsai, and A. Willsky. Binary and ternary flows for image segmentation. In *ICIP*, Kobe, Japan, oct. 1999.
18. X. Zeng, L.H. Staib, R.T. Schultz, H. Tagare, L. Win, and J.S. Duncan. A new approach to 3D sulcal ribbon finding from MR images. In C. Taylor and A. Colchester, editors, *MICCAI*, number 1679 in LNSC, pages 148–157. Springer, sep. 1999.

A Curve Evolution Approach to Medical Image Magnification via the Mumford-Shah Functional[*]

Andy Tsai[1], Anthony Yezzi Jr.[2], and Alan S. Willsky[1]

[1] Department of Electrical Engineering and Computer Science,
Massachusetts Institute of Technology,
Cambridge, MA, USA
{atsai, willsky}@mit.edu
[2] School of Electrical and Computer Engineering,
Georgia Institute of Technology,
Atlanta, GA, USA
Anthony.Yezzi@ece.gatech.edu

Abstract. In this paper, we introduce a curve evolution approach to image magnification based on a generalization of the Mumford-Shah functional. This work is a natural extension of the curve evolution implementation of the Mumford-Shah functional presented by the authors in previous work. In particular, by considering the image magnification problem as a structured case of the missing data problem, we generalize the data fidelity term of the original Mumford-Shah energy functional by incorporating a spatially varying penalty to accommodate those pixels with missing measurements. This generalization leads us to a PDE-based approach for simultaneous image magnification, segmentation, and smoothing, thereby extending the traditional applications of the Mumford-Shah functional which only considers simultaneous segmentation and smoothing. This novel approach for image magnification is more global and much less susceptible to blurring or blockiness artifacts as compared to other more traditional magnification techniques, and has the additional attractive denoising capability.

1 Introduction

Image magnification is often required to aid in the display, manipulation, and analysis of medical images. It also plays a critical role in computer aided diagnosis and computer assisted surgery. Image magnification involves enlarging a small image to several times its size and often requires some sort of an interpolation scheme. The most straightforward approach for image enlargement is to use a zero-order interpolation technique, commonly known as replication, which may cause the resulting image to appear blocky [2]. Classical enlargement techniques such as bilinear or bicubic interpolation schemes tend to cause blurring across the edges when applied indiscriminantly to the image [2]. More sophisticated

---

[*] This work was supported by ONR grant N00014-00-1-0089, by AFOSR grant F49620-98-1-0349, and by subcontract GC123919NGD from Boston Univ. under the AFOSR Multidisciplinary Research Program on Reduced Signature Target Recognition.

schemes may locate the edges first with local filters prior to interpolation so as to avoid the blurring artifacts [1, 6]. Three important shortcomings are evident in these types of algorithms. One, the interpolation schemes used for magnification are local since they only utilize data values from neighboring pixels. These types of interpolation scheme become problematic when the observed image is noisy. Two, edge detection schemes employed prior to interpolation often only make use of local information (which are very susceptible to noise artifacts) and cannot guarantee continuous closed edge contours. Three, it is unclear in what order the three operations (smoothing, edge detection, and interpolation) should be performed for image magnification since they are not commutative. Our approach for image magnification addresses the first deficiency by using a PDE-based model for interpolating the data which incorporates the use of all data values within each homogeneous region, not just neighboring pixels, to determine the interpolant. As a result, this interpolation scheme is much more robust to noise. The second deficiency is addressed by the use of our active contour model for boundary detection which is more global in nature than local filters (and therefore not as sensitive to noise) and is curve-based (hence providing a continuous closed edge contour). The third deficiency is addressed by our use of the Mumford-Shah model which provides, in a single framework, a tight coupling for *simultaneous* image segmentation, denoising, and magnification. In this manner, the ordering of the different operations is no longer an issue.

Our approach to image magnification is based on the Mumford-Shah active contour model presented earlier by the authors in [8]. In that work, we approached the classic Mumford-Shah problem from a curve evolution perspective. In particular, by viewing an active contour as the set of discontinuities in the Mumford-Shah problem, we used the corresponding functional to determine gradient descent evolution equations to deform the active contour. Each gradient step involved solving an optimal estimation problem for the data within each region. The resulting active contour model offered a tractable implementation of the original Mumford-Shah model to simultaneously segment and smoothly reconstruct the data within a given image in a coupled manner. In this paper, we extend the application of this Mumford-Shah active contour model to one that also handles image magnification. By considering the image magnification problem as a structured case of the missing data problem, we generalize the data fidelity term of the original Mumford-Shah functional by substituting a spatially varying penalty for the traditional constant one (to accommodate those pixels with missing measurements). This generalization leads to a novel approach for simultaneous image magnification, segmentation, and smoothing, thereby providing a new application of the Mumford-Shah functional.

This paper is organized as follows. Section 2 describes the curve evolution implementation of the Mumford-Shah functional proposed in [8] for simultaneous image segmentation and smoothing. In Section 3, we describe the generalization of the Mumford-Shah active contour model for image magnification. Then Section 4 provides some experimental results to illustrate this magnification technique. Finally, we conclude this paper in Section 5 with a summary.

2 The Mumford-Shah formulation as a curve evolution problem

The point of reference for this paper is the Mumford-Shah functional

$$E(\mathbf{f}, \vec{C}) = \beta \iint_{\Omega} (\mathbf{f} - \mathbf{g})^2 \, dA + \alpha \iint_{\Omega \backslash \vec{C}} |\nabla \mathbf{f}|^2 \, dA + \gamma \oint_{\vec{C}} ds \qquad (1)$$

in which $\vec{C}$ denotes the smooth, closed segmenting curve, $\mathbf{g}$ denotes the observed data, $\mathbf{f}$ denotes the piecewise smooth approximation to $\mathbf{g}$ with discontinuities only along $\vec{C}$, and Ω denotes the image domain [4]. The parameters α, β, and γ are positive real scalars which control the competition between the various terms above and determine the "scale" of the segmentation and smoothing. The Mumford-Shah problem is to minimize $E(\mathbf{f}, \vec{C})$ over admissible $\mathbf{f}$ and $\vec{C}$. In our approach, we constrain the set of discontinuities in the Mumford-Shah problem to correspond to evolving sets of curves, enabling us to tackle the problem via a curve-evolution-based approach.

2.1 Optimal image estimation

Any arbitrary closed curve $\vec{C}$ partitions the domain Ω of the image into two regions R and R^c, corresponding to the interior and exterior of curve $\vec{C}$ respectively. If we fix this partitioning, minimizing (1) corresponds to minimizing

$$E_{\vec{C}}(\mathbf{f}_R, \mathbf{f}_{R^c}) = \beta \iint_R (\mathbf{f}_R - \mathbf{g})^2 \, dA + \alpha \iint_R |\nabla \mathbf{f}_R|^2 \, dA$$
$$+ \beta \iint_{R^c} (\mathbf{f}_{R^c} - \mathbf{g})^2 \, dA + \alpha \iint_{R^c} |\nabla \mathbf{f}_{R^c}|^2 \, dA. \qquad (2)$$

The estimates $\hat{\mathbf{f}}_R$ and $\hat{\mathbf{f}}_{R^c}$ that minimize (2) satisfy (decoupled) PDE's which can be obtained using standard variational methods [4]. This method gives the following damped Poisson equation with Neumann boundary condition for $\hat{\mathbf{f}}_R$:

$$\hat{\mathbf{f}}_R - \frac{\alpha}{\beta} \nabla^2 \hat{\mathbf{f}}_R = \mathbf{g} \qquad \text{on } R \qquad (3a)$$

$$\frac{\partial \hat{\mathbf{f}}_R}{\partial \vec{\mathcal{N}}} = 0 \qquad \text{on } \vec{C}. \qquad (3b)$$

In a similar fashion, $\hat{\mathbf{f}}_{R^c}$ is given as the solution to

$$\hat{\mathbf{f}}_{R^c} - \frac{\alpha}{\beta} \nabla^2 \hat{\mathbf{f}}_{R^c} = \mathbf{g} \qquad \text{on } R^c \qquad (4a)$$

$$\frac{\partial \hat{\mathbf{f}}_{R^c}}{\partial \vec{\mathcal{N}}} = 0 \qquad \text{on } \vec{C}. \qquad (4b)$$

Conjugate gradient (CG) method is used to solve the above *estimation PDE's*.

2.2 Gradient flows that minimize the Mumford-Shah functional

With the ability to calculate $\hat{\mathbf{f}}_R$ and $\hat{\mathbf{f}}_{R^c}$ for any given $\vec{C}$, we now wish to derive a curve evolution for $\vec{C}$ that minimizes (1). That is, as a function of $\vec{C}$, we wish to find $\vec{C}_t$ that minimizes

$$E_{\hat{\mathbf{f}}_R, \hat{\mathbf{f}}_{R^c}}(\vec{C}) = \beta \iint\limits_{R} (\hat{\mathbf{f}}_R - \mathbf{g})^2 \, dA + \alpha \iint\limits_{R} |\nabla \hat{\mathbf{f}}_R|^2 \, dA$$

$$+ \beta \iint\limits_{R^c} (\hat{\mathbf{f}}_{R^c} - \mathbf{g})^2 \, dA + \alpha \iint\limits_{R^c} |\nabla \hat{\mathbf{f}}_{R^c}|^2 \, dA + \gamma \oint\limits_{\vec{C}} ds. \tag{5}$$

The curve evolution that minimizes (5) is given by (see [8] for a derivation)

$$\vec{C}_t = \frac{\alpha}{2} \left(|\nabla \hat{\mathbf{f}}_{R^c}|^2 - |\nabla \hat{\mathbf{f}}_R|^2 \right) \vec{\mathcal{N}} + \frac{\beta}{2} \left((\mathbf{g} - \hat{\mathbf{f}}_{R^c})^2 - (\mathbf{g} - \hat{\mathbf{f}}_R)^2 \right) \vec{\mathcal{N}} - \gamma \kappa \vec{\mathcal{N}}. \tag{6}$$

This *Mumford-Shah flow* is implemented via the level set method [5, 7] which offers a natural and numerically reliable implementation of these solutions within a context that handles topological changes in the interface without any additional effort.

2.3 Implementation of the Mumford-Shah active contour model

The algorithm described above requires solving two PDE's (equations (3) and (4)) at every evolution step of the curve making it computationally expensive and impractical. We propose an approximate gradient descent approach to calculate $\hat{\mathbf{f}}_R$, $\hat{\mathbf{f}}_{R^c}$, and $\vec{C}$ to alleviate some of the computational burdens. This approach consists of alternating between these two steps:

- Fix $\hat{\mathbf{f}}_R$ and $\hat{\mathbf{f}}_{R^c}$, and take several gradient descent curve evolution steps to move the curve $\vec{C}$.
- Fix $\vec{C}$, and perform just a few iterations of the conjugate gradients method–without taking it to convergence–to obtain a *rough* estimate of $\mathbf{f}_R$ and $\mathbf{f}_{R^c}$.

Only a rough estimate of $\mathbf{f}_R$ and $\mathbf{f}_{R^c}$ is required to direct the curve to move in the descent direction. The idea is to make the algorithm faster by reducing the number of times $\hat{\mathbf{f}}_R$ and $\hat{\mathbf{f}}_{R^c}$ are calculated and also the amount of time required to calculate each of them. CG method is carried to convergence in the last iteration to obtain an accurate estimate of $\mathbf{f}_R$ and $\mathbf{f}_{R^c}$.

Next, we propose an implementation of our active contour model, building on the preceding modification, to enable our model to handle images with multiple junctions without resorting to more sophisticated level set techniques [3]. Given an image, we apply our Mumford-Shah active contour model for segmentation and smoothing. After segmentation, if any of the resulting subregions require additional segmentation, apply our algorithm again, but this time, restricting the algorithm to operate only in that particular subregion. This approach has the

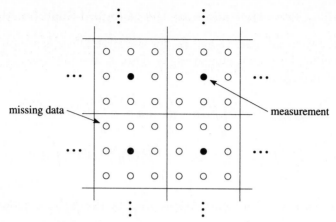

Fig. 1. Diagram showing the locations of missing data in relation to the measurements in our image magnification technique.

natural notion of starting with a crude segmentation and refining the segmentation by telescoping down to the different subregions in order to capture finer and finer details in the image. The attractive feature associated with this hierarchical implementation is that it allows us to handle images with multiple junctions by employing multiple curves to represent such junctions. This implementation also affords us better control as to what details we desire and what objects we would like to capture, in the segmentation and smoothing of our image.

3 Generalization of the Mumford-Shah active contour model for image magnification

Image magnification capability is weaved into the Mumford-Shah active contour model by considering the image magnification problem as a structured case of the missing data problem. Specifically, consider a new grid with three times as many pixels in each direction and assign the value of the original image to the "center" pixel in each 3×3 block on the grid and treat the remaining pixels as missing data points (see Figure 1). From an estimation-theoretic standpoint, we can view these "center" pixels as sparse measurements on a much larger image domain. We then employ the generalized Mumford-Shah curve evolution procedure (described below) to interpolate to this finer grid, using the curve evolution portion of this procedure to partition the domain of the magnified image into different homogeneous subregions so as to provide smooth interpolations where appropriate without blurring across regions of high contrast.

The Mumford-Shah model handles image magnification through the parameter β. In the standard Mumford-Shah formulation (1), β is a constant scalar parameter reflecting our confidence in the measurements. For image magnification in which there are pixels with missing measurements, we replace the constant parameter β by a spatially varying function β whose value at each pixel is inversely proportional to the variance of the measured noise at that pixel. For example, in the situation where the data at pixel (x_o, y_o) is missing, we consider the variance of the data at that pixel as being infinite and accordingly set $\beta(x_o, y_o) = 0$. By introducing this spatially varying β, equation (1) becomes:

$$E(\mathbf{f}, \vec{C}) = \iint_{\Omega} \beta(\mathbf{f} - \mathbf{g})^2 \, dA + \alpha \iint_{\Omega \backslash \vec{C}} |\nabla \mathbf{f}|^2 \, dA + \gamma \oint_{\vec{C}} ds. \tag{7}$$

The gradient flow that minimizes the generalized Mumford-Shah functional, shown in (7), is given by

$$\vec{C}_t = \frac{\alpha}{2} \left(|\nabla \hat{\mathbf{f}}_{R^c}|^2 - |\nabla \hat{\mathbf{f}}_R|^2 \right) \vec{\mathcal{N}} + \frac{\beta}{2} \left((\mathbf{g} - \hat{\mathbf{f}}_{R^c})^2 - (\mathbf{g} - \hat{\mathbf{f}}_R)^2 \right) \vec{\mathcal{N}} - \gamma \kappa \vec{\mathcal{N}} \tag{8}$$

where the optimal estimates $\hat{\mathbf{f}}_R$ and $\hat{\mathbf{f}}_{R^c}$ of (8) satisfy

$$\beta \hat{\mathbf{f}}_R - \alpha \nabla^2 \hat{\mathbf{f}}_R = \beta \mathbf{g} \qquad \text{on } R$$

$$\frac{\partial \hat{\mathbf{f}}_R}{\partial \vec{\mathcal{N}}} = 0 \qquad \text{on } \vec{C}$$

and

$$\beta \hat{\mathbf{f}}_{R^c} - \alpha \nabla^2 \hat{\mathbf{f}}_{R^c} = \beta \mathbf{g} \qquad \text{on } R^c$$

$$\frac{\partial \hat{\mathbf{f}}_{R^c}}{\partial \vec{\mathcal{N}}} = 0 \qquad \text{on } \vec{C}.$$

Over each region of missing data D, the estimation equation reduces to the Laplace equation with the same Neumann boundary condition:

$$\nabla^2 \hat{\mathbf{f}}_D = 0 \qquad \text{on } D$$

$$\frac{\partial \hat{\mathbf{f}}_D}{\partial \vec{\mathcal{N}}} = 0 \qquad \text{on } \vec{C}.$$

As solutions to the Laplace equation, the estimates obtained over any such missing data regions not containing part of $\vec{C}$ take the form of harmonic functions. As such, we can infer much about the smooth nature of these interpolated estimates as they are subject to both a maximum (and minimum) principle as well as the mean value property. However if the curve $\vec{C}$ intersects D, no such smoothing occurs across this boundary, allowing interpolation to be guided by the segmentation defined by $\vec{C}$.

4   Examples

We illustrate our magnification technique by applying it to a noisy microscopic image of red blood cells (Figure 2(a)). The initializing curve is shown in Figure 2(a). Figure 2(b) shows an intermediate step of the Mumford-Shah model.

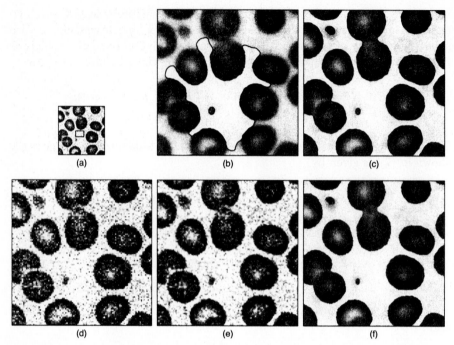

(a)   (b)   (c)

(d)   (e)   (f)

Fig. 2. Three-fold magnification of a noisy 100×100 microscopic image of blood cells. (a) Original image with location of initializing curve. (b) An intermediate step in the magnification and segmentation process. (c) Magnified image with the segmenting curve. (d) Magnified image based on zero-order interpolation. (e) Magnified image based on bilinear interpolation. (f) Magnified image based on the Mumford-Shah model.

Notice the curve automatically proceeds in the outward direction without the aid of any inflationary forces to direct this movement. This is one of the many attractive properties associated with the Mumford-Shah flow (see [8] for more details). The final segmenting curve and the smooth enlarged reconstruction of the image is shown in Figure 2(c). For comparison, in Figure 2(d), we show the blocky and noisy magnification of the original image based on zero-order interpolation. In Figure 2(e), we show a slightly smoother but blurrier magnification of the original image based on bilinear interpolation. In Figure 2(f), we again show the magnified image based on our Mumford-Shah approach but without the segmenting curve. The smooth boundaries of the magnified image are the direct result of the minimum length prior placed on the segmenting curve while the smooth spatially varying structures within each homogeneous regions of the observed image are successfully captured by our PDE-based model for interpolation. It is evident that our magnification approach is better than other conventional image magnification techniques by avoiding many of the processing artifacts such as blockiness and blurring while at the same time, denoising the image. Also, notice in this figure that the starting contour automatically splits into many contours in order to segment this image. This provides a compelling

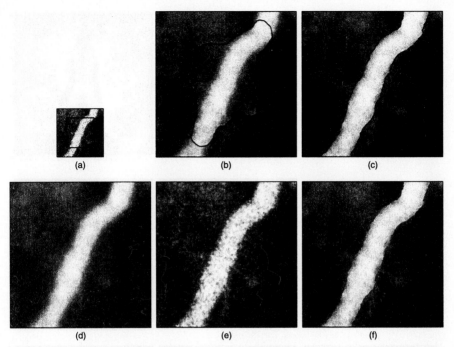

Fig. 3. Three-fold magnification of a noisy 110×110 angiogram. (a) Original image with location of initializing curve. (b) An intermediate step in the magnification and segmentation process. (c) Magnified image with the segmenting curve. (d) Magnified image obtained by isotropic smoothing then replication. (e) Magnified image obtained by replication then isotropic smoothing. (f) Magnified image based on the Mumford-Shah model.

demonstration of the topological transitions allowed by the Mumford-Shah active contour model's level set implementation [5, 7].

We next illustrate our magnification technique by applying it to a noisy angiogram (Figure 3(a)). The initializing curve is shown in Figure 3(a). Figure 2(b) shows an intermediate step of the Mumford-Shah model. Notice the bidirectional flow of the curve as it automatically proceeds toward the boundaries of the image. The final segmenting curve and smooth enlarged reconstruction of the image is shown in Figure 3(c). For comparison, in Figure 3(d), we show a magnified angiogram obtained by first applying isotropic smoothing (i.e. the same smoothing used in the Mumford-Shah framework but applied over the image as one single region) then replication. This image is blurry because the edges of the image were destroyed during the initial smoothing step. In Figure 3(e), we show a magnified angiogram obtained by first replicating the image then applying isotropic smoothing. Notice the magnified image is still noisy because the noise components within the original image have been exaggerated by the zero-order interpolation scheme. In Figure 3(f), we again show the magnified image based on our Mumford-Shah model but displayed without the segmenting curve.

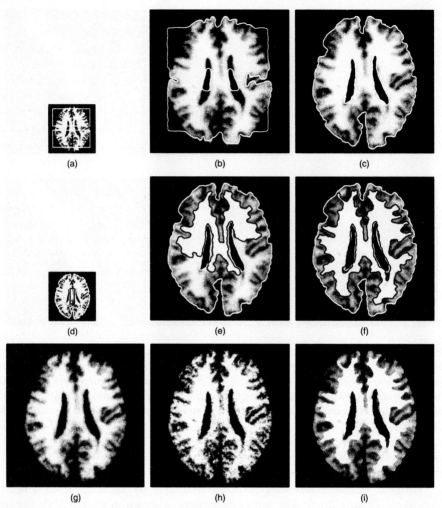

Fig. 4. Three-fold magnification of a noisy 225×225 axial brain pathology section. (a) Original image with the initializing contour for the first curve. (b) An intermediate step in the magnification and segmentation of the image based on the first curve. (c) Magnified image using one segmenting curve which is also displayed. (d) Original image with the segmentation result of the first curve and the initializing contour for the second curve. (e) An intermediate step in the magnification and segmentation of the image based on the second curve. The segmentation result of the first curve is also displayed. (f) Magnified image with the segmentation results from both curves. (g) Magnified image obtained by isotropic smoothing then replication. (h) Magnified image obtained by replication then isotropic smoothing. (i) Magnified image based on the Mumford-Shah model.

We use the noisy brain pathology image shown in Figure 4(a) to demonstrate the hierarchical implementation of our active contour model for image magnification. The segmentation and enlarged reconstruction of the image, shown in Figure 4(c), is obtained using a single curve (which is initialized as shown in Figure 4(a)) and based on the approximate gradient descent approach described earlier. The blurring across the boundary of the white and the gray matter shown in the reconstruction is due to the erroneous implication of this coarse segmentation, namely that the inside of the brain is one region over which smoothing is performed. To provide better details within the brain, we again applied our technique to the interior region of the brain. A second curve, initialized as shown in Figure 4(d), is used in conjunction with the first curve to obtain the segmentation and enlarged reconstruction of the image shown in Figure 4(f). For comparison, in Figure 4(g), we show the magnified image obtained by first isotropically smoothing the original noisy image then magnifying it using zero-order hold. We show in Figure 4(h) the magnified image obtained by first magnifying the original noisy image using zero-order hold then smoothing it isotropically. Finally, in Figure 4(i), we show the magnification results based on our approach without the segmenting curve.

5 Conclusion

The magnification technique we introduced in this paper constitutes a more global approach to interpolating magnified data than traditional bilinear or bicubic interpolation schemes, while still maintaining sharp transitions along region boundaries. The technique is also much less susceptible to blurring artifacts as compared to other more traditional techniques, and has the additional attractive denoising capability. In addition, the curve length penalty in our Mumford-Shah based flow tends to prevent the blocky appearance of object boundaries which is a symptom of replication-based schemes.

References

1. J. Allebach and W. Wong, "Edge directed interpolation," *IEEE International Conference on Image Processing*, vol. 3, pp. 707-711, 1996.
2. J. Lim, *Two-dimensional Signal and Image Processing*, Prentice Hall, 1992.
3. B. Merriman, J. Bence, and S. Osher, "Motion of multiple junctions: A level set approach," *Journal of Computational Physics*, vol. 112, no. 2, pp. 334–363, 1994.
4. D. Mumford and J. Shah, "Optimal approximations by piecewise smooth functions and associated variational problems," *Comm. Pure and Appl. Math.*, vol. 42, 1989.
5. S. Osher and J. Sethian, "Fronts propagation with curvature dependent speed: Algorithms based on Hamilton-Jacobi formulations," *J. of Comp. Physics*, vol. 79, pp. 12–49, 1988.
6. K. Ratakonda and N. Ahuja, "POC based adaptive image magnification," *ICIP*, 1998.
7. J. Sethian, *Level Set Methods: Evolving Interfaces in Geometry, Fluid Mechanics, Computer Vision, and Material Science*, Cambridge University Press, 1996.
8. A. Tsai, A. Yezzi, and A. Willsky, "A curve evolution approach to smoothing and segmentation using the Mumford-Shah functional," *CVPR*, 2000.

Image Segmentation
Based on the Integration of Markov Random Fields
and Deformable Models

Ting Chen[1] and Dimitris Metaxas[2]

[1] Bioengineering Department, University of Pennsylvania, Philadelphia PA 19104, USA.
email: chenting@seas.upenn.edu
[2] VAST Lab, Department of Computer and Information Science, University of Pennsylvania, Philadelphia PA 19104, USA

Abstract. This paper proposes a new methodology for image segmentation based on the integration of deformable and Markov Random Field models. Our method makes use of Markov Random Field theory to build a Gibbs Prior model of medical images with arbitrary initial parameters to estimate the boundary of organs with low signal to noise ratio (SNR). Then we use a deformable model to fit the estimated boundary. The result of the deformable model fit is used to update the Gibbs prior model parameters, such as the gradient threshold of a boundary. Based on the updated parameters we restart the Gibbs prior models. By iteratively integrating these processes we achieve an automated segmentation of the initial images. By careful choice of the method used for the Gibbs prior models, and based on the above method of integration with deformable model our segmentation solution runs in close to real time. Results of the method are presented for several examples, including some MRI images with significant amount of noise.

1 Introduction

Segmentation is the process of assigning pixels in an image to distinct objects or the background. It is one of the fundamental processes for higher-level image analysis. However, it still remains an open research problem. Region-based and boundary-based methods are the two major classes of segmentation algorithms.

In region-based methods, e.g., region growing [1], image pixels are assigned to objects based on homogeneity statistics. If a pixel value is similar to that of its neighbors, then the two pixels will be assigned to the same object. The advantage of this method is that the image information inside the object is considered. The disadvantage is that it may lead to noisy boundaries and holes inside the object because the method does not consider the boundary information explicitly.

In the boundary-based method, e.g., snakes [2], Fourier-Parameterized models [3], a shape model is initialized close to the object boundary and fits to the boundary based on image features. To avoid local minima, most boundary-based methods require that the model be initialized near the solution and it is controlled via an interactive interface. A disadvantage of these methods is that they do not consider pixel information inside the object. For images with low SNR, the boundary-based method may lead to incorrect assignment of pixels unless significant manual intervention is used.

To improve the segmentation results of either of the above two classes of methods a hybrid segmentation framework was proposed in [4] and [5] that combines region-based and boundary-based methods. The algorithm works as follows. In the first step the user selects a single initial pixel inside the object to be segmented and then an affinity operator is applied to the image to estimate the pixels that most likely lie on the boundary of the object. Since the data are usually sparse and/or noisy, in the second step a deformable model is fit to improve the segmentation from the first step that is then used to re-compute the parameters of the affinity operator. The above two steps are then applied recursively to further refine the segmentation results.

A limitation of the affinity operator is that it only considers the local characteristics in the direct neighborhood of every new pixel in order to make a decision as to whether the pixel should be included to the already segmented cluster of pixels or not. The affinity operator does not use any boundary shape or boundary continuity information.

Chan et al. apply the Gibbs prior model in [6] and [7] for image processing based on the work of Besag [8] and Geman and Geman[9]. Compared to the affinity operator, the Gibbs prior models can incorporate much more information. By using specific neighborhood information, the Gibbs prior model can incorporate both boundary and region information during segmentation.

In this paper we develop a new segmentation approach that integrates Gibbs prior models and deformable models for segmenting the boundaries of organs in medical images. The advantages of the method is that:

1. The user only selects a single pixel inside the object,
2. The method can deal with images with significant amount of noise and poorly defined boundaries, and
3. it is computationally efficient and runs in real time (5Hz on SGI R10000 workstation).

1.1 Related Work

As opposed to most previous work on segmentation that falls under either the boundary [14,22] or the region growing methods [1], our method follows a hybrid approach. Recently, there have been several hybrid methods such as the ones developed in [5, 6, 15, 16, 17]. In [18], Zhu et al. develop a unifying framework that generalizes the deformable models, region growing, and prior matching approaches. They define a new energy function that represents them all. By combining the region-based method and the boundary-based method, all these approaches achieve good results. However, they still need good initialization to avoid local minima and they use fixed prior models that may be difficult to compute accurately in advance.

Various successful approaches for object segmentation have been proposed that use the Markov Random Fields property of images in the past few years. [8, 9, 19] used MRF or Gibbsian models that include nearest neighbor correlation information, but do not use object boundary information. [6] and [7] defined higher order neighborhoods in MRF image models that model both region and boundary object information based on theoretical results from [20,21].

Y. Boykov et al. [24], Vittorio Murino and Andrea Trucco [25] use MRF models in their segmentation applications. But neither of them provides a self-adaptive algorithm to refine the prior models. So their methods need an exact prior model to begin with.

J. Dehmeshki et al. developed an automatic adaptive algorithm for MRF image segmentation [26]. In their recursive algorithm, they first segment the image with Gibbs prior models. Then they update the Gibbs prior model based on the segmentation result, and then they segment the image again. They do this repeatedly until a satisfactory result is achieved. However, when they updated the Gibbs prior model, they recalculate parameters by doing statistics on every pixel in the image, which makes the adaptive process computationally expensive. Our method combines the deformable model and the Markov random field model. The deformable model gives a better estimation for the prior used in the MRF since it gives an explicit object boundary estimation. The iteration step between the MRF and deformable model allow us to perform only a few steps for the MRF segmentation, which in turn results in a very efficient segmentation algorithm.

2 Method

2.1 Markov Random Fields and Gibbs Prior for Segmentation

Most medical images are Markov Random Field images, that is, the statistics of a pixel are related to the statistics of pixels in its neighborhood. According to the Equivalent Theorem proved by Hammersley and Clifford in [11], a Markov random field is equivalent to a Gibbs field. Thus for medical images which are MRF images, their joint distribution can be written in the Gibbsian form as follows.

$$\Pi(X) = Z^{-1} \exp\left(-H(X)\right) \quad Z = \sum_{z \in \bar{X}} \left(-H(z)\right) \tag{1}$$

where $H(X)$ models the energy function of image X, $\bar{X}$ is the set of all possible configurations of the image X, Z is a normalizing factor or partition function in statistics terminology, and z is image. The local and global properties of images will be incorporated into the model by designing an appropriate energy function $H(X)$. The lower the value of the energy function, the higher the value of the image joint distribution, the better the image fits to the prior distribution.

We began the establishment of the Gibbs model by designing the energy function as

$$H(X) = H_1(X) + H_2(X) + H_3(X) \tag{2}$$

where $H_1(X)$ models the piecewise pixel homogeneity statistics, $H_2(X)$ models the continuity of the boundary and $H_3(X)$ is the noise model.

To model the piecewise homogeneity, we model $H_1(X)$ as:

$$H_1(X) = \vartheta_1 \sum_{s \in X} \sum_{t \in \partial s} (\Phi(\Delta_{s,t})(1 - \Psi_{st}) + \alpha \Psi_{st}) \quad \text{with} \quad \Psi_{st} = \begin{cases} 0 & \Delta_{s,t} < 0 \\ 1 & \Delta_{s,t} \geq 0 \end{cases} \quad (3)$$

s and t are pixels, pixel t is in the neighborhood of pixel s, ϑ_1 is the weight for $H_1(X)$, α is the threshold for the object boundary, $\Delta_{s,t}$ is the variance between pixels s and t. Φ is a function based on the variance. For simplicity, let $\Phi(\Delta_{s,t}) = \Delta_{s,t}^2$. When we minimize $H_1(X)$, pixels that have similar gray values with their neighbors are considered to be in the same object ($\Delta_{s,t}$ assigned to 0) and will be further smoothed. But for pixels that have different gray values (the variance between two pixels beyond the threshold α) than their neighbors, will be assigned to 1. Thus the variance between these two pixels will remain. Therefore the term will smooth the pixels inside the object and will keep the boundary untouched.

The boundary continuity is modeled as follows:

$$H_2(X) = \vartheta_2 \sum_{s \in X} \sum_{i=1}^{N} W_i(x) \quad (4)$$

where x is a pixel, ϑ_2 is the weight term, N is the number of local configurations that may lie on the boundaries. $W_i(x)$ is the weight function (also called the potential function) for the local configuration of the boundary. For our purpose we consider the potential function on 9 pixels. In our model, the potential functions are defined on a neighborhood system in which the 3 by 3 cliques are used to model the boundary continuity. We depict the clique potential definition in Figure 1. All these local configurations can have different orientations. The vertical and horizontal configurations share the same potential function.

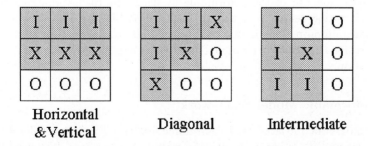

Horizontal &Vertical   **Diagonal**   **Intermediate**

Fig. 1. Figure 1. Pixels labeled "I" are inside the object; pixels labeled "O" are outside the object; pixels labeled "X" all have gray values similar to either "I" or "O". The shading area is the boundary shape when assuming "X" is similar to "I"

Different values are assigned to different local configurations. All three local configurations depicted in Figure 1 most likely exist in an image with smooth and continuous boundary. In $H_2(X)$ we give weight to these local configurations. When we minimize $H_2(X)$, the pixels in the image (especially those near the boundary) will alter their gray values to achieve a smooth and continuous boundary.

To cope with noise in the data, we use as mentioned previously the term $H_3(X)$ in the energy function $H(X)$. For Gaussian Noise, it has the form:

$$H_3(X) = (2\sigma^2)^{-1} \sum_s (y_s - x_s)^2 \tag{5}$$

where y_s is the observed value of pixel s, x_s is the estimated value and σ is the standard deviation.

To minimize $H(x)$ we use a Bayesian framework. According to Bayes's theorem, the posterior distribution of an image is given by

$$P(X|Y) = \frac{\Pi(X)L(Y|X)}{P(Y)} \tag{6}$$

where $\Pi(X)$ is the prior distribution (see (1)) and $L(Y|X)$ is the probability of obtaining an image Y while the actual image is X. Under a Bayesian framework, the segmentation problem can be formulated as the maximization of

$$\Pi(X)L(Y|X) \tag{7}$$

The method requires choosing suitable ϑ_1 and ϑ_2 to achieve a balance between the piecewise homogeneity and boundary continuity. We also need to input the threshold of boundary α and the weight $W_i(x)$ for local characteristics to enable the Gibbs prior model:

We use the ICM method to minimize the energy function $H(X)$, as described in [11].

It is important to note that the result of the Gibbs prior model depends on the proper selection of the respective parameters. In addition, the solution based on the Gibbs prior model can be trapped in a local minimum. Based on our method, the use of the deformable model to fit the estimated boundary at every iteration serves as a way to push the Gibbs prior model solution to the global minimum. This is due to the fact that we re-compute at the end of every iteration the Gibbs prior model parameters based on the deformable model segmentation.

2.2 Deformable Model Framework

We summarize the physics-based deformable model framework developed in [10] that we use and its integration with the estimated boundary data produced by the Gibbs prior model we described above.

Our model is a superellipsoid with local deformations. Given the reference shape s (the superellipsoid) and the displacement d (the local deformations), the position of points p on the model are defined by $p = s + d$. To keep the continuity of the model surface, we impose a continuous loaded membrane deformation strain energy.

$$\varepsilon_m(d) = \int \omega_{10}(\frac{\partial d}{\partial u})^2 + \omega_{01}(\frac{\partial d}{\partial v})^2 + \omega_{00}d^2 du \qquad (8)$$

Where d is a node's local deformation, ω_{00} controls the local magnitude and ω_{10}, ω_{01} control the local variation of the deformation. We can calculate the stiffness matrix K of model based on the strain energy [10].

The model nodes move under the influence of external forces. The model dynamics can be described by the first order Lagrangian method:

$$\dot{d} + Kd = f \qquad (9)$$

where f are external generalized forces.

The external forces we apply in our method are the balloon forces proposed by Cohen[23], based on which the model nodes move outward in the direction of their normal vector. The deformable model is initialized at a pixel inside the object and expands under the influence of the balloon forces. Once the model reaches the estimated boundary based on the MRF segmentation, boundary forces are applied in the opposite direction of the balloon force and the associated nodes stop deforming. Thus the model is aligned with the boundary. The nodal deformations are computed based on the use of finite elements. When most model nodes stop moving, the model stops deforming.

2.3 Integration

Our method is based on the integration of Gibbs Prior models and deformable models that results in efficient and reliable segmentations.

The quality of the Gibbs Prior model segmentation depends on parameters, such as α (the threshold of boundary), the mean value of pixels in the object, etc. In our approach we integrate the deformable and the Gibbs prior models by using the deformable models to compute the parameter values for the Gibbs prior model.

First we have an initial prior model to begin with. This Gibbs prior model will be applied on the image and give the MAP boundary estimation (M step). Then a deformable model is fit to the estimated boundaries (Fit step). Then we recalculate the parameters (such as the mean value of pixels within the object and the boundary threshold) of the Gibbs prior models based on the statistics of pixels inside the boundary of the object that was segmented based on the deformable model. We term this the prior estimation step or E step. The updated prior model parameters based on the deformable model will be more accurate for the object in the given image, since we include both the information inside and on the surface of the object. This way we compute suitable parameters regardless of the chosen initial values.

The recursive approach can be summarized as follows.

1. First use the Gibbs prior model to get an initial segmentation of the object (M step).
2. Once the estimated boundary has been computed, a deformable model is fitted to it (F step).
3. Update the Gibbs parameters based on the property of pixels within and on the object boundary. Use the updated parameters to create a new Gibbs prior model (E step). Then go to step 1.

We apply this three-step procedure recursively until the segmentation results do not change from iteration to iteration.

The advantage of the method is that it is very fast (5Hz on an SGI workstation) and can be used to segment medical images even when there is not enough prior information and the boundaries are not well defined.

3 Experiments

We present several experiments to demonstrate the power of our algorithm. All the experiments run in close to real time (5Hz on an SGI R10000 workstation).

The first experiment shown in Figure 2 shows the various steps of integrating the Gibbs prior model and the deformable model to achieve correct object boundary segmentation. The boundary threshold $\alpha = 6$, the weights for the local configuration $W_i(x) = 5$, there are 100 nodes on the deformable model. A successful segmentation is achieved after one iteration between the deformable and the MRF models.

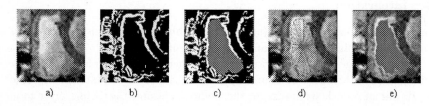

Fig. 2. a) is the MRI image of right ventricular; b) is the result of applying the Gibbs Prior models, the estimated boundary is assigned with white color; c) is the result of fitting a deformable model, the shading area is inside the object; d) shows the deformable model nodes; e) shows the final result after one iteration between the deformable model and the MRF model.

The second experiment shown in Figure 3 shows the effect of the recursive method between deformable model and the MRF model. The initial boundary threshold $\alpha = 6$, the weights for the local configuration $W_i(x) = 5$, there are 100 nodes on the deformable model. A successful segmentation is achieved after four iteration between the deformable and the MRF models.

Experiment three (Figure 4) and four (Figure 5) show the application of our method on visible human image data. In these two experiments, we apply the

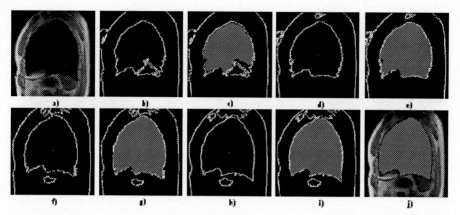

Fig. 3. shows how the recursive method works. a) is the original MRI image of lung; b) is the initial estimation of the boundary using Gibbs filter; c) shows the result of deformable model fitting; d) and f) show the Gibbs estimation after one iteration and two iterations; e) and g) show the reult of deformable model fitting after one iteration and two iterations; h) shows the final Gibbs estimation and i) shows the final deformable model fitting; j) is the final segmentation result in the original image

method to segment the eyeball and a muscle in human head respectively. The boundary threshold $\alpha = 6$. There are 100 model nodes. All the local configuration weights $W_i(x) = 5$. The program stops after one iterations between the deformable and the MRF models.

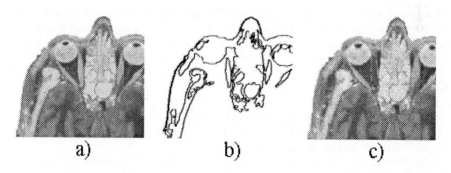

Fig. 4. a) is the original MRI image of human head, and we are going segment the muscle at the center of image; b) is the Gibbs estimation; c) is the final segment result.

These two experiment show the effectness of our method on segmenting subtle humen body structures.

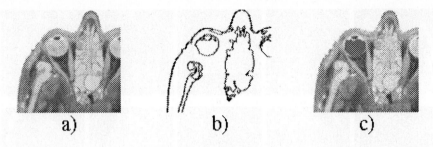

a)   b)   c)

Fig. 5. a) is the original MRI image of human head, and we are going segment the eyeball at the center of image; b) is the Gibbs estimation; c) is the final segment result.

4 Conclusion

In this paper, we described a new approach to the segmentation problem by integrating a probabilistic model and a deformable model. By using the Gibbs prior models, we have a better estimation of the boundary to be used as an initialization for the deformable model. A recursive method was then developed to refine the segmentation results. The algorithm is very fast and is not sensitive to the selection of the initial parameters. The approach is robust to noise and can cope with low SNR. Methods have been developed to make the deformable models work well even in the presence of boundaries with sharp edges. Results of the method have been presented for several examples involving MRI images.

5 Acknowledgement

This investigation is sponsored by NSF-IRI-9624604 NSF Career Award, NIH/NLM 1-RO1-LM-06638-01 and also NIH/NLM Visible Human Image Processing Tools Contract.

References

1. D. H. Ballard and C. M. Brown. Computer vision. Prentice Hall, 1982
2. M. Kass, A. Witkin, and D. Trerzopoulos. Snakes: Active contour models. Intl. J. of Computer Vision, 1(4):321-331, 1988.
3. M. Worring, A. W. M. Smeulders, L. H. Staib, and J. S. Duncan. Parameterized feasible boundaries in gradient vector fields. Computer Vision and Image Understanding, 63(1):135-144, 1996
4. T. N. Jones and D. N. Metaxas. Automated 3D segmentation using deformable models and fuzzy affinity. In J. Duncan and G. Gindi, editors, Lecture Notes in computer Science, Volume 1230: IPMI 97, pages 113-126. Springer, 1997
5. T. N. Jones and D. N. Metaxas., Image Segmentation Based on the Integration of Pixel Affinity and Deformable Models. Proceedings of CVPR 98, June 1998 Santa Barbara, CA.

6. M. T. Chan, G. T. Herman, Emanuel Levitan. A Bayesian Approach to PET Reconstruction Using Image-Modeling Gibbs Prior. IEEE Transaction on Nuclear Science, Vol. 44, No. 3, June 1997

7. M. T. Chan, G. T. Herman, Emanuel Levitan. Bayesian Image Reconstruction Using Image-Modeling Gibbs Prior. 1998 John WileySons, Inc.

8. J. Besag. On the Statistical analysis of dirty pictures. Journal of the Royal Statistical Society Series B, 48:259-302, 1986

9. S. Geman and D. Geman. Stochastic relaxation, Gibbs distributions, and the Bayesian restoration of images. IEEE Trans. Pattern Anal. Mach. Intell. 6, 721-741, 1984

10. D.N. Metaxas. Physics-Based Deformable Models: Application to Computer Vision, Graphics and Medical Imaging. 1996.

11. Gerhand Winkler Image Anlysis, random Fields and Dynamic Monte Carlo Methods, 1995 Springer

12. S. Z. Li Markov Random Field Modeling in Computer Vision, 1995 Springer

13. Hammersley J. M. and Clifford P.: Markov fields on finite graphs and lattices. Preprint University of California, Berkeley.

14. R. Malladi, J.A. Sethian, and B.C. Vemuri. Shape modeling with front propagation: A level set approach. Pattern Matching and machine Intelligence, 17(2), 1997

15. A. Chakraborty and J. S. Duncan. Integration of Boundary finding and region-based segmentation using Game theory. In Y. Bizais et al., editor, Information processing in Medical imaging, pages 189-201. Kluwer, 1995

16. A. Chakraborty, M. Worring, and J.S. Duncan. On multifeature integration for deformable boundary finding. In proc. Intl. Conf. On Computer Vision, pages 846-851, 1995

17. R. Ronfard. Region-based strategies for active contour models. Intl. J. of Computer Vision, 13(2):229-251, 1994

18. S. C. Zhu, T. S. Lee and A. L. Yuille, region competition: Unifying snakes, region growing, and bayes/mdl fro multiband image segmentation. In proc. Intl. Conf. On Computer Vision, pages 416-423, 1995.

19. H.Derin and H. Elliot, Modeling and Segmentation of Noisy and textured images using Gibbs random fields, IEEETrans. Pattern Anal. Mach. Intell. 9, 1987, 39-55

20. V. E. Johnson, W. H. Wong, X. Hu, and C. T. Chen, Image restoration using Gibbs priors: Boundary modeling, treatment of blurring, and selection of Hyperparameters, IEEE Trans. Pattern anal. Mach. Intell. 13, 1991, 413-425

21. D. Geman and G. Reynolds, Constrainted restoration and the recovery of discontinuities, IEEE Trans. Pattern Anal. Mach. Intell. 14, 1992, 367-382

22. M. kass, A. Witkins, D. Terzopoulos: Snakes: Active Contour Models. Int. J. Computer Vision. 1987, Vol. 1, 312-331

23. L. D. Cohen, ON active contour models and balloons. CVGIP: Image understanding, 53(2):211-218:1991.

24. Yuri. Boykov, Olga. Veksler, Ramin. Zabih: Markoc Random fields with efficient Approximation, ICCV 98.

25. Vittorio Murino, Abdrea Trucco, Edge/Region-Based Segmentation and Reconstruction of Underwater Acoustic Image by Markov Random Fields. ICCV 97.

26. J. Dehmeshki, M. F. Daemi, B. P. Atkin and N. J. Miles, Anadaptive Estimation and Segmentation Technique for determination of Major Maceral Groups in Coal, ICSC 95.

Segmentation by Adaptive Geodesic Active Contours

Carl-Fredrik Westin[1], Liana M. Lorigo[2], Olivier Faugeras[2,3], W. Eric L. Grimson[2], Steven Dawson[4], Alexander Norbash[1], and Ron Kikinis[1]

[1] Harvard Medical School, Brigham & Women's Hospital, Boston MA, USA
[2] MIT Artificial Intelligence Laboratory, Cambridge, MA USA
[3] INRIA, Sophia Antipolis, France
[4] Harvard Medical School, Mass. General Hospital, Boston MA, USA
westin@bwh.harvard.edu

Abstract. This paper introduces the use of spatially adaptive components into the geodesic active contour segmentation method for application to volumetric medical images. These components are derived from local structure descriptors and are used both in regularization of the segmentation and in stabilization of the image-based vector field which attracts the contours to anatomical structures in the images. They are further used to incorporate prior knowledge about spatial location of the structures of interest. These components can potentially decrease the sensitivity to parameter settings inside the contour evolution system while increasing robustness to image noise. We show segmentation results on blood vessels in magnetic resonance angiography data and bone in computed tomography data.

1 Introduction

Curve-shortening flow is the evolution of a curve over time to minimize some distance metric. When this distance metric is based on image properties, it can be used for segmentation. The idea of *geodesic active contours* is to define the metric so that indicators of the object boundary, such as large intensity gradients, have a very small "distance" [4, 9]. The minimization will attract the curve to such image areas, thereby segmenting the image, while preserving properties of the curve such as smoothness and connectivity. Geodesic active contours can also be viewed as a more mathematically sophisticated variant of classical snakes [8]. Further, they are implemented with level set methods [15] which are based on recent results in differential geometry [6, 5]. The method can be extended to evolve surfaces in 3D, for the segmentation of 3D imagery such as medical datasets [4, 22]. More recent work developed the level set equations necessary to evolve arbitrary dimensional manifolds in arbitrary dimensional space [1, 13].

This paper introduces the use of spatially adaptive components into the geodesic active contour segmentation framework. These components are derived from local structure descriptors and are used both in regularization of the segmentation and in stabilization of the image-based vector field which attracts the contours to anatomical structures in the images. Local structure can be described in terms of how similar an image region is to a plane, to a line, and to a sphere. Describing the signal in geometrical terms rather than as a neighborhood of voxel values is intuitively appealing. We have applied a tensor-based description method for estimating these geometrical properties to the problem of segmenting bone in computed tomography (CT) data, with a focus on thin bone, which is especially difficult to obtain [18].

2 Background

Local image structure properties can be estimated from derivatives of the intensity image or the responses of directed quadrature filters. We will incorporate them into the geodesic active contour framework, a powerful segmentation technique.

2.1 Local Image Structure

In two dimensions, local structure estimation has been used to detect and describe edges and corners [7]. The local structure is described in terms of dominant local orientation and isotropy, where isotropy means lack of dominant orientation. In three dimensions, local structure has been used to describe landmarks, rotational symmetries, and motion [10, 2, 11, 14, 16]. In addition to isotropy, it describes geometrical properties which have been applied to the enhancement and segmentation of blood vessels in volumetric angiography datasets [12, 17], bone in CT images [21, 20], and to the analysis of white matter in diffusion tensor magnetic resonance imaging [19].

Let the operator $\sum_\Omega$ denote averaging in the local neighborhood Ω about the current spatial location, and assume 3D data. Then the eigenvalues $\lambda_1 \geq \lambda_2 \geq \lambda_3$ of the tensor

$$T_\Omega = \frac{\sum_\Omega \nabla I \nabla I^T}{\sum_\Omega |\nabla I|^2} \tag{1}$$

describe geometrical properties over the neighborhood Ω, such as the following generic cases:

1. $\lambda_1 = 1$, $\lambda_2 = \lambda_3 = 0$, when the signal is locally planar.
2. $\lambda_1 = \lambda_2 = 1$, $\lambda_3 = 0$, when the signal is locally tubular.
3. $\lambda_1 = \lambda_2 = \lambda_3 = 1$, in regions that contain intensity gradients but no dominant orientational structure.

From these cases, we can define scalar measures indicating which local structure is present. An example measure of similarity to thin tubular shapes is

$$c_{linear} = \frac{\lambda_2 - \lambda_3}{\lambda_1}. \tag{2}$$

In higher dimensions, the basic shapes are more complicated than lines and planes, and the possible anisotropies become more complex.

2.2 Geodesic Active Contours, Level Set Methods

The task of finding the curve that best fits an object boundary is posed as a minimization problem over all closed planar curves $C(p) : [0,1] \to \mathbb{R}^2$ [4,3,9]. The objective function is

$$\oint_0^1 g(|\nabla I(C(p))|)|C'(p)|dp$$

where I is the image and g is a strictly decreasing function that approaches zero as image gradients become large. To minimize this weighted curve length by steepest descent, one uses the evolution equation

$$C_t = (g\kappa - \nabla g \cdot \hat{n})\hat{n} \qquad (3)$$

where C_t is the derivative of C with respect to an artificial time parameter t, κ is the Euclidean curvature, and $\hat{n}$ is the unit inward normal.

Level set methods increase the dimensionality of the problem from the dimensionality of the evolving manifold to the dimensionality of the embedding space to achieve independence of parameterization and topological flexibility [6, 5, 15]. For the example of planar curves, instead of evolving the one-dimensional curve, the method evolves a two-dimensional surface. Let u be the signed distance function to curve C; that is, the value of u at each point is the distance to the closest point on C, with interior points designated by negative distances. Consequently, C is the zero level-set of u. Then evolving C according to Equation 3 is equivalent to evolving u according to

$$u_t = g\kappa|\nabla u| + \nabla g \cdot \nabla u \qquad (4)$$

in the sense that the zero level set of u remains the evolving curve C over time. The extension to surfaces in 3D is straightforward and is called *minimal surfaces* [4].

Now presume u is the signed distance function to a surface S in 3D. Expanding ∇g using the chain rule and replacing $\kappa|\nabla u|$ with a more general regularization term λ gives

$$u_t = g\lambda + g'\nabla u \cdot H \frac{\nabla I}{|\nabla I|}. \qquad (5)$$

The *codimension* of a manifold is the difference between the dimension of the embedding space and the dimension of the manifold. We will say that a regularization force for an evolving surface is "codimension-two" if it is based on the smaller principal curvature of the surface; for a tubular surface, this curvature approximates the Euclidean curvature of the centerline of the tube which has codimension two. We will say it is "codimension-one" if it is based on the mean curvature of the surface. If $p_1 \geq p_2$ are the principal curvatures of the surface, then these forces are given by

codimension-one force	codimension-two force
$\lambda = \frac{p_1+p_2}{2}|\nabla u|$	$\lambda = p_2|\nabla u|$

Let $\lambda_1 \geq \lambda_2$ be the eigenvalues of the operator

$$F = P_{\nabla u} \nabla^2 u P_{\nabla u} = (I_d - \frac{\nabla u \nabla u^T}{|\nabla u|^2}) H_u (I_d - \frac{\nabla u \nabla u^T}{|\nabla u|^2}) \tag{6}$$

where I_d is the identity matrix and $P_{\nabla u}$ is the projector onto the plane normal to ∇u. It turns out that $\lambda_1 = p_1|\nabla u|$ and $\lambda_2 = p_2|\nabla u|$, in the limiting case.

In the next section we describe how the local structure tensor (Equation 1) can be incorporated into the geodesic active contour equation (Equation 4). The basic idea that is explored in this paper is how to perform codimension-one regularization in image regions that are similar to a plane, and codimension-two regularization in regions similar to a line.

3 Adaptive Geodesic Active Contours

Observe that the evolution (Equation 4) is based on two terms: (1) a regularization term which ensures smoothness of the evolved surface, and (2) an image term that defines an vector field that is attracting the surface to the image structures of interest. We apply local structure estimation to adaptively modify each of these terms.

3.1 Adaptive Regularization Force

To evolve surfaces in 3D, the minimal surfaces work [4] uses the mean curvature of the surface for regularization, which corresponds to the "codimension-one force". When one wishes to evolve tubular surfaces, the smaller principal curvature is more appropriate to avoid eliminating the high curvatures inherent in the tube-like shape [13], which corresponds to the "codimension-two force."

To illustrate the difference, we created a test volume containing a thin tube connected to a flat sheet. We corrupted this image with a high level of noise. Figure 1 illustrates the result of regularizing this shape using the two different forces. The first image shows the initial shape, obtained by thresholding the volume. The second shows the result of applying the codimension-one force and the third the codimension-two force. Notice that the codimension-two force retained the thin tube while reducing the noise along it, while the codimension-one force deleted the tube but obtained a smoother plane than did the codimension-two force. In both experiments we iterated the evolution equations 20 times. It should be mentioned that with more iterations the result of the codimension-two does get smoother. However our comparison shows that the codimension-one regularization converges faster and therefor is assumed to be more resistant to noise.

This experiment illustrates the potential benefit of choosing the regularization force locally, dependent on the shape of the object to be segmented. One would expect to benefit by choosing a continuous value between the smaller curvature and the mean curvature. We propose to select this value according to the

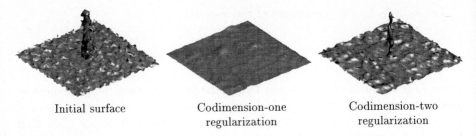

| Initial surface | Codimension-one regularization | Codimension-two regularization |

Fig. 1. Illustration of difference between codimension-one regularization force and codimension-two regularization force: initial surface, result of codimension-one, and result of codimension-two.

local structure estimates. The evolution equation (Equation 5) becomes

$$u_t = g(\gamma \lambda_2 + (1 - \gamma)\frac{\lambda_1 + \lambda_2}{2}) + g' \nabla u \cdot H \frac{\nabla I}{|\nabla I|} \tag{7}$$

where γ is a scalar between 0 and 1 indicating how planar or how tubular the local image structure is. Setting $\gamma = c_{linear}$ (Equation 2) gives the desired result of $\gamma = 1$ for codimension-two regularization of tubular objects and $\gamma = 0$ for codimension-one regularization of planar objects and those without dominant orientational structure.

Referring back to Equation 6 and noting that the *trace* operation yields the sum of the eigenvalues, we obtain another formulation for the mean curvature

$$\frac{\lambda_1 + \lambda_2}{2} = \frac{1}{2|\nabla u|}\text{trace}\left((I_d - \frac{\nabla u \nabla u^T}{|\nabla u|^2})H_u(I_d - \frac{\nabla u \nabla u^T}{|\nabla u|^2})\right). \tag{8}$$

Close to the correct segmentation result, ∇u should be approximately parallel to the image gradient ∇I, so one could project out the ∇I direction instead of the ∇u direction:

$$\frac{\lambda_1 + \lambda_2}{2} \approx \frac{1}{2|\nabla u|}\text{trace}\left((I_d - \frac{\nabla I \nabla I^T}{|\nabla I|^2})H_u(I_d - \frac{\nabla I \nabla I^T}{|\nabla I|^2})\right). \tag{9}$$

The tensor $\frac{\nabla I \nabla I^T}{|\nabla I|^2}$ is a projector onto the direction ∇I; likewise, $I_d - \frac{\nabla I \nabla I^T}{|\nabla I|^2}$ is the projector that removes all components in the direction ∇I. By local averaging of the image-based projectors according to Equation 1, we obtain

$$\lambda_2 \approx \frac{1}{|\nabla u|}\text{trace}\left((I_d - T_\Omega)H_u(I_d - T_\Omega)\right). \tag{10}$$

The local averaging acts to remove all components in all directions around the curve, leaving only those components in the tangent direction. Thus, we have an

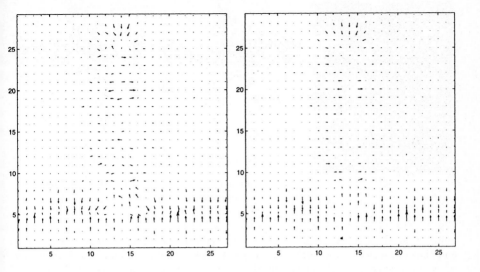

Fig. 2. Visualization of auxiliary vector field. For one slice through the center of our test volume, we show the raw vector field $H\frac{\nabla I}{|\nabla I|}$ (left) and the stabilized vector field $T_\Omega H\frac{\nabla I}{|\nabla I|}$ (right).

approximation to the curvature of the curve, which is the smaller principal curvature of the thin tube around it. Note that we have $\lambda = \lambda_2$ for codimension-two, and for codimension-one $\lambda = \lambda_1 + \lambda_2$. When image noise is high, however, the image gradient estimations will often be less stable than using curvature estimates of the regularized embedding function u. For this reason, our implementation uses Equation 7.

3.2 Adaptive Image Force

Local Image Structure For structures such as vessels in magnetic resonance angiography (MRA) and bronchi CT which appear brighter than the background, a weight on the image term defined by the cosine of the angle between the normal to the surface and the gradient in the image was introduced in [13]. This cosine is given by the dot product of the respective gradients of u and I, so the update equation (Equation 5) becomes

$$u_t = g\lambda + g'(-\frac{\nabla u}{|\nabla u|} \cdot \frac{\nabla I}{|\nabla I|})\nabla u \cdot H\frac{\nabla I}{|\nabla I|}. \tag{11}$$

Similar to the equality $(a^T b)(a^T b) = aa^T \cdot bb^T = A \cdot B$ where A and B are matrices and the dot product of matrices is defined as the sum of the element-wise products, we can write

$$u_t = g\lambda - g'\frac{\nabla u\nabla u^T}{|\nabla u|} \cdot H\frac{\nabla I\nabla I^T}{|\nabla I|^2}. \tag{12}$$

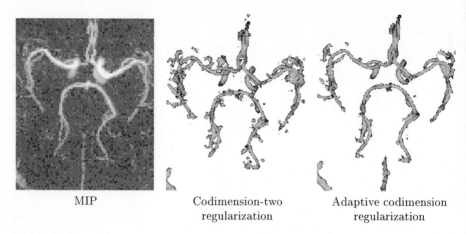

| MIP | Codimension-two regularization | Adaptive codimension regularization |

Fig. 3. The maximum intensity projection (MIP) of an MRA data set of the brain followed by the segmentation obtained with the original auxiliary image vector term and that obtained with adaptive codimension regularization and stabilized image vector term. Notice that the latter captures more vessels while demonstrating increased resistance to noise.

Recall that g must be decreasing function that approaches zero on large image gradients. Choosing $g = \exp(-|\nabla I|)$ as in [13] implies that $\frac{g}{g'} = -1$. Divide both components of the evolution by g to obtain

$$u_t = \lambda + \frac{\nabla u \nabla u^T}{|\nabla u|} \cdot H \underbrace{\frac{\nabla I \nabla I^T}{|\nabla I|^2}}_{T} \tag{13}$$

We observe again the projection operator T as in Equation 9, similar to the differential operator in Equation 1. The same arguments apply: T alone projects onto the direction of the gradient ∇I, but when smoothed, T_Ω has an adaptive behavior. When all gradients in the neighborhood are similar, as in the case of a plane, it projects as T, but when the gradients vary around a thin tubular structure, it projects to the plane orthogonal to the centerline of this structure. One could replace T with T_Ω in Equation 13 to achieve this behavior.

The same intuition can be used, however, to modify the current evolution equation (Equation 11) instead. Since we are interested in only the component of the evolution that is normal to object surface, it is desirable to have stable gradients in this direction; the derived projection operator will have this stabilizing effect. The evolution equation now becomes

$$u_t = \kappa |\nabla u| + (\frac{\nabla u \nabla u^T}{|\nabla u|} \cdot \frac{\nabla I}{|\nabla I|}) \cdot T_\Omega H \frac{\nabla I}{|\nabla I|}. \tag{14}$$

The operator T_Ω can be viewed as an adaptive relaxation of the image-based (auxiliary) vector field based on local image constraints, as illustrated in Figure 2. Notice that the noisy gradients along both the plane and the stick become

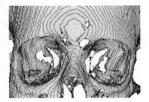  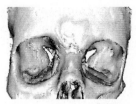

| Initial surface | Codimension-one regularization | Spatially constrained adaptive codimension |

Fig. 4. Incorporation of spatial priors into the auxiliary vector field, in conjunction with an active contour model. The surface used to initialize the evolution is shown first, followed by the results of the unconstrained evolution and of the constrained evolution.

much closer to desired intensity gradients, perpendicular to the image structures. In areas of well-defined gradients the operator T_Ω acts like a projection operator. In unstructured areas the operator will be close to the identity operator.

Figure 3 compares the segmentations obtained by Equation 11 with those obtained by Equation 14 for a cerebral MRA dataset. We corrupted the image with synthetic noise to demonstrate the stabilization of the auxiliary vector field. Notice that the segmentation obtained using the adaptive codimension is smoother without loosing important details which is consistent with our intuition gained from Figure 1 where codimension-one regularization gives a smoother surface, and codimension-two regularization preserves tubular structures.

Incorporating external models External spatial models can also be incorporated through the auxiliary vector field. We represent prior knowledge about the possible locations of the objects to be segmented by a scalar certainty field c. That is, c would be identically zero in regions that could not contain the objects, and could range up to one in areas where the objects are likely to exist. c can be constructed from training datasets, from specification by the user, or from an anatomical atlas. The tensor becomes

$$T_c = cT_\Omega. \tag{15}$$

For the case of bone segmentation in a skull CT image, it is difficult for an active contour model to distinguish the bone boundary from the skin boundary which also has high intensity gradients. However, since bone appears bright in CT, it is unlikely to be present in dark areas. Therefore, we construct a binary mask by thresholding the raw dataset with a threshold that is low enough to capture the entire bone structure. This mask restricts the surface evolution to valid locations. Note that direct thresholding of the raw dataset would create artificial boundaries which would severely corrupt the segmentation; however, acting directly on the auxiliary vector field does not introduce new gradients. Figure 4 shows the ability to segment bone from CT using active contours in conjunction with this spatial prior. The initialization surface is shown first, followed by the unconstrained evolution and the evolution constrained by c. The unconstrained evolution is attracted to all high gradients so finds the skin boundary, while the constrained evolution, is able to obtain the bone surface.

4 Conclusions

We have introduced adaptive capabilities into the evolution equations for level set based active contour models. These capabilities include modification of the regularization force and stabilization of the image-based auxiliary vector field, both of which are based on the codimension of the signal. That is, they depend on how similar the signal is to a line and to a plane. Additionally, we have introduced the option to spatially constrain the surface evolution by altering this vector field. We predict that these modifications can potentially produce more robust segmentations of medical datasets. In this paper we have demonstrated the feasibility of incorporating local structure information into the evolution equations for level set based active contour models. However much more work is needed in order to test the method on large data sets and to improve the current implementation.

Acknowledgments

Carl-Fredrik Westin was funded by CIMIT and NIH grants P41-RR13218 and R01-RR11747. Liana Lorigo was funded by NSF Contract IIS-9610249, NSF Contract DMS-9872228, and NSF ERC (Johns Hopkins University agreement) 8810-274. We gratefully acknowledge Renaud Keriven of ENPC, France for efficient level set prototype code. We thank Peter Everett of Brigham and Women's Hospital for making available the CT dataset.

References

1. Luigi Ambrosio and Halil M. Soner. Level set approach to mean curvature flow in arbitrary codimension. *J. of Diff. Geom.*, 43:693–737, 1996.
2. J. Bigün, G. H. Granlund, and J. Wiklund. Multidimensional orientation: texture analysis and optical flow. *IEEE Transactions on Pattern Analysis and Machine Intelligence*, PAMI–13(8), August 1991.
3. V. Caselles, F. Catte, T. Coll, and F. Dibos. A geometric model for active contours. *Numerische Mathematik*, 66:1–31, 1993.
4. Vicent Caselles, Ron Kimmel, and Guillermo Sapiro. Geodesic active contours. *Int'l Journal Comp. Vision*, 22(1):61–79, 1997.
5. Y.G. Chen, Y. Giga, and S. Goto. Uniqueness and existence of viscosity solutions of generalized mean curvature flow equations. *J. Differential Geometry*, 33:749–786, 1991.
6. L.C. Evans and J. Spruck. Motion of level sets by mean curvature: I. *Journal of Differential Geometry*, 33:635–681, 1991.
7. W. Förstner. A feature based correspondence algorithm for image matching. *Int. Arch. Photogrammetry Remote Sensing*, 26(3):150–166, 1986.

8. M. Kass, A. Witkin, and D. Terzopoulos. Snakes: Active contour models. *Int J. on Computer Vision*, 1(4):321–331, 1988.

9. A. Kichenassamy, A. Kumar, P. Olver, A. Tannenbaum, and A. Yezzi. Gradient flows and geometric active contour models. In *Proc. IEEE Int'l Conf. Comp. Vision*, pages 810–815, 1995.

10. H. Knutsson. Representing local structure using tensors. In *The 6th Scandinavian Conference on Image Analysis*, pages 244–251, Oulu, Finland, June 1989.

11. H. Knutsson, H. Bårman, and L. Haglund. Robust orientation estimation in 2D, 3D and 4D using tensors. In *Proceedings of Second International Conference on Automation, Robotics and Computer Vision, ICARCV'92*, Singapore, September 1992.

12. T.M. Koller, G. Gerig, G. Szekely, and D. Dettwiler. Multiscale detection of curvilinear structures in 2D and 3D image data. In *Proc. ICCV'95*, pages 864–869, 1995.

13. L. Lorigo, O. Faugeras, W.E.L. Grimson, R. Keriven, R. Kikinis, and C.-F. Westin. Codimension-Two Geodesic Active Contours. In *Proc. IEEE Conf. Comp. Vision and Pattern Recognition*, 2000.

14. R. Deriche O. Monga, R. Lengagne. Extraction of zero crossings of the curvature derivatives in volumetric 3D medical images: a multi-scale approach. In *Proc. IEEE Conf. Comp. Vision and Pattern Recognition*, pages 852–855, Seattle, Washington, USA, June 1994.

15. S. Osher and J. Sethian. Fronts propagating with curvature-dependent speed: Algorithms based on Hamilton-Jacobi formulation. *Journal of Computational Physics*, 79(1):12–49, 1988.

16. K. Rohr. Extraction of 3D anatomical point landmarks based on invariance principles. *Pattern Recognition*, 32:3–15, 1999.

17. Y. Sato, S. Nakajima, N. Shiraga, H. Atsumi, S. Yoshida, T. Koller, G. Gerig, and R. Kikinis. Three-dimensional multiscale line filter for segmentation and visualization of curvilinear structures in medical images. *Medical Image Analysis*, 2(2):143–168, 1998.

18. C.-F. Westin, A. Bhalerao, H. Knutsson, and R. Kikinis. Using Local 3D Structure for Segmentation of Bone from Computer Tomography Images. In *Proc. IEEE Conf. Comp. Vision and Pattern Recognition*, pages 794–800, Puerto Rico, June 1997.

19. C.-F. Westin, S.E. Maier, B. Khidhir, P. Everett, F.A. Jolesz, and R. Kikinis. Image Processing for Diffusion Tensor Magnetic Resonance Imaging. In *Medical Image Computing and Computer-Assisted Intervention*, pages 441–452, September 1999.

20. C.-F. Westin, J. Richolt, V. Moharir, and R. Kikinis. Affine adaptive filtering of CT data. *Medical Image Analysis*, 4(2):161–172, 2000.

21. C.-F. Westin, S. Warfield, A. Bhalerao, L. Mui, J. Richolt, and R. Kikinis. Tensor Controlled Local Structure Enhancement of CT Images for Bone Segmentation. In *Medical Image Computing and Computer-Assisted Intervention*, pages 1205–1212. Springer Verlag, 1998.

22. X. Zeng, L. H. Staib, R. T. Schultz, and J. Duncan. Segmentation and measurement of the cortex from 3D mr images. In *Medical Image Computing and Computer Assisted Intervention (MICCAI)*, Boston USA, 1998.

Interactive Organ Segmentation Using Graph Cuts

Yuri Boykov and Marie-Pierre Jolly

Imaging and Visualization Department
Siemens Corporate Research
755 College Road East, Princeton, NJ 08540, USA
[yuri,jolly]@scr.siemens.com

Abstract. An N-dimensional image is divided into "object" and "background" segments using a graph cut approach. A graph is formed by connecting all pairs of neighboring image pixels (voxels) by weighted edges. Certain pixels (voxels) have to be *a priori* identified as object or background *seeds* providing necessary clues about the image content. Our objective is to find the cheapest way to cut the edges in the graph so that the object seeds are completely separated from the background seeds. If the edge cost is a decreasing function of the local intensity gradient then the minimum cost cut should produce an object/background segmentation with compact boundaries along the high intensity gradient values in the image. An efficient, globally optimal solution is possible via standard min-cut/max-flow algorithms for graphs with two terminals. We applied this technique to interactively segment organs in various 2D and 3D medical images.

1 Introduction

Many real world applications can strongly benefit from algorithms that can reliably segment out objects in images by finding their precise boundaries. One important example is medical diagnosis MR or CT where the images are used by the doctor to investigate specific organs of interest. 4D medical images containing information about 3D volumes moving in time are also available nowadays. There is simply too much information in these datasets; the doctors need segmentation tools to be able to concentrate on relevant parts of these images. Precise segmentation of organs would allow accurate measurements, simplify visualization and, consequently, make the diagnosis more reliable.

There are a large number of contour based segmentation tools that were developed for 2D images in the past: snakes (e.g. [17,3]), deformable templates (e.g. [25]), methods computing the shortest path (e.g. [18,6]) and others. Most of these algorithms cannot be easily generalized to images of higher dimensions. To locate a boundary of an object in a 2D image these methods rely on lines ("1D" contours) that can be globally optimized, for example, using dynamic programming [1,23,10]. In 3D images the object boundaries are surfaces and the standard dynamic programming or path search methods cannot be applied directly. Computing an optimal shape for a deformable template of a boundary becomes highly untracktable even in 3D, not to mention 4D or higher dimensional images. Gradient descent optimization or variational calculus methods [3, 4] can still be applied but they produce only a local minimum. Thus, the segmentation results may not reflect the global properties of the deformable contour model. An alternative approach is to segment each of the 2D slices of a 3D volume separately and then glue the pieces together [3]. The major drawback of this approach is that the boundaries in each slice are independent. The segmentation information is not propagated within the 3D volume and the result can be spatially incoherent. In [19] a 3D hybrid model is used to smooth the results and to enforce coherence between the slices. In this case the solution to the model fitting is computed through gradient descent and, thus, may get stuck at a local minimum.

Alternatively, there are many region based techniques for image segmentation: region growing, split-and-merge, and others (see Chapter 10 in [14]). The general feature of these methods is that they build the segmentation based on information inside the segments rather than at the boundaries. For example, one can grow the object segment from given "seeds" by adding neighboring pixels (voxels) that are "similar" to whatever is already inside. These methods can easily deal with images of any dimensions. However, the main limitation of many region based algorithms is their greediness. They often "leak" (i.e. grow segments where they should not) in places where the boundaries between the objects are weak or blurry.

Ideally, one would like to have a segmentation based on both region and contour information. There are many attempts to design such methods. Numerical optimization is the main issue here. General schemes [26] use variational approach leading to a local minimum. In some special cases of combining region and contour information [5, 16] a globally optimal segmentation is possible through graph based algorithms. The main problem in [5] and [16] is that their techniques are restricted to 2D images.

Here we present a new method for image segmentation separating an object of interest from the background based on graph cuts. Formulating the segmentation problem as a two terminal graph cut problem allows for a globally optimal efficient solution in a general N-dimensional setting. Our method has some features of both contour and region based algorithms and it addresses many of their limitations. First of all, our method directly computes the segmentation boundary by minimizing its cost. The only hard constraint is that the boundary should separate the object from the background. At the same time our technique incorporates region information. It is initialized by certain object (and boundary) seeds. There is no prior model of what the boundary should look like or where it should be located. The method can be applied to images of any dimensions. It can also directly incorporate some region information (see Section 4). Our algorithm strongly benefits from both "contour" and "region" sides of its nature. The "region" side allows natural propagation of information throughout the volume of an N-dimensional image while the "contour" side addresses the "leaks".

It should be noted that graph cuts were used for image segmentation before. In [24] the image is optimally divided into K parts to minimize the maximum cut between the segments. In this formulation, however, the segmentation is strongly biased to very small segments. Shi and Malik [21] try to solve this problem by normalizing the cost of a cut. The resulting optimization problem is NP-hard and they use an approximation technique. In [2, 15, 12] graph cuts are applied to minimize certain energy functions used in image restoration, stereo, and other early vision problems. In fact, the optimization scheme that we use in our algorithm is very similar to [12] and [2]. The main contributions of our paper is a new concept of object/background segmentation where a cut must separate the corresponding seed points.

Our method can generate one or a number of isolated segments for the object (as well as for the background). Depending on the image data the method automatically decides which seeds should be grouped into a single object (or background) segment. Our approach also allows effective interaction with a user. Initially, the object and background seeds can be specified manually, automatically, or semi-automatically. In the interactive mode, the user can click in the image to select pixels using a "paint brush" and assign to either the object of interest or the background. After reviewing the corresponding segmentation, the user can specify additional object and background seeds depending on the observed results. To incorporate these new seeds the algorithm can efficiently adjust the current segmentation without recomputing the whole solution.

Interactive segmentation is becoming more and more popular to alleviate the problems inherent to fully automatic segmentation which seems to never be perfect. Our user interface turns out to be identical to the one proposed by Griffin *at al.* [13] in the sense that seeds are marked on the image to impose hard constraints about the object and the background. The difference between their work and ours lies in the underlying segmentation scheme. They use a hierarchical clustering technique instead of graph cuts. Since the segmentation is only region-based, there is no provision to smooth the boundary or minimize its length. Intelligent paint [20] is also region-based. The image is partitioned into small homogeneous regions using a watershed scheme. The user can click the mouse button to select a region where the growing process (paint flow) starts. Other interactive segmentation systems are edge-based. With intelligent scissors [18] and live wire [7], the user draws contours interactively to outline an object of interest in the image. The system computes the best path (sequence of pixels in the image) from the current mouse position to the last mouse button click position according to some energy function based on image gradient. Flickner *et al.* [8] have used the same concept but the contour is parametrized by a spline to produce a smooth contour without outlining all the little nooks of the digitized contour. Vehkomäki *et al.* [22] propose to presegment the image by grouping contour fragments to partition the image into closed cycles. Then, when the user moves the mouse (s)he effectively selects the boundaries between those partitions.

In the next section we explain our segmentation algorithm. Section 3 gives a number of examples where we apply our technique to medical images. In Section 3.1 we show that extracting a single closed contour in a 2D image is a simple special case of our technique. We also demonstrate that with a few simple and intuitive manipulations a user can always segment an object precisely as (s)he wants. More general examples of multiple objects and 3D volumes are considered in Sections 3.2 and 3.3, respectively. Information on possible extensions and future work is given in Section 4.

2 Segmentation Technique

In this section we provide some technical details about our segmentation technique. To segment an image we create a graph with nodes corresponding to pixels (voxels) of the image. There are two additional terminal nodes: an "object" terminal (a source) and a "background" terminal (a sink). The source is connected by edges to all nodes identified as object seeds and the sink is connected to all background seeds. For convenience, all edges from the terminals are referred to as *t-links*. We assign an infinite cost to all t-links between the seeds and the terminals.

Pairs of neighboring pixels (voxels) are connected by weighted edges that we call *n-links* (neighborhood links). Any kind of neighborhood system can be used. Costs of n-links can be based on local intensity gradient, Laplacian zero-crossing, gradient direction, and other criteria (e.g. [18]). The only technical restriction is that the edge costs should be non-negative. Our simplest implementation incorporates undirected n-links $\{p, q\}$ between neighboring pixels p and q (see (a) below) with cost $w_{\{p,q\}} = f(|I_p - I_q|)$ where I_p and I_q are intensities at pixels p and q and $f(x) = K \cdot \exp(-\frac{x^2}{\sigma^2})$ is a non-negative decreasing function.

Such weights encourage segmentation boundaries in places with high intensity gradient. In some examples we also use directed links (p, q) and (q, p) between pixels p and q (see (b)). The weights can be defined as $w_{(p,q)} = max(0, f(|I_p - I_q|) + h \cdot (I_p - I_q))$ where the gradient direction is incorporated. A positive h forces dark pixels to stay inside the segmentation boundary and bright pixels to stay outside. A negative h would achieve the opposite effect.

The general graph structure is now completely specified. Some examples are shown in Figure 2. We draw the segmentation boundary between the object and the background by finding the minimum cost cut on this graph. A cut is a subset of edges that separates the source from the sink. The cost of the cut is the sum of its edge costs[1]. Due to infinite cost of t-links to seeds, a minimum cut

---

[1] In case of directed edges the cost of the cut is the sum of severed edges (p, q) where node p is left in the part of the graph connected to the source and q is in the part connected to the sink.

is guaranteed to separate the object seeds from the background seeds. Note that locations with high intensity gradient correspond to cheap n-links. Thus, they are attractive choices for the optimal segmentation boundary. The minimum cut can be computed exactly in polynomial time using well known algorithms for two terminal graph cuts, e.g. max-flow [9] or push-relabel [11].

It is important that the algorithm efficiently adjusts the segmentation to incorporate any additional seeds that the user might interactively add. We use a max flow algorithm to determine the minimum cut corresponding to the optimal segmentation. When a new seed is added to the image, the corresponding t-link is added to the *residual* graph that was left at the end of the previous cut computation. Then, a new optimal cut can be efficiently obtained without recomputing the whole solution[2]. Deleting a seed from the image is equivalent to adding a t-link to the opposite terminal. Thus, it can also be efficiently implemented.

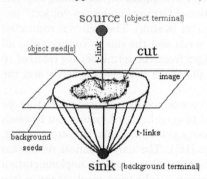

(a) Segmentation of a single object in a 2D image. A cut corresponds to a closed contour separating an object seed from background seeds.

(b) Segmentation of multiple objects in a 3D image. The cut separates the object seeds from the background seeds and creates two isolated object segments.

Fig. 1. Examples of graphs for segmentation of 2D and 3D images.

3 Examples

In this section we consider a number of examples that illustrate how our method can be used to segment medical images. Section 3.1 shows the simplest 2D experiments and explains the main intuitions about our technique. Segmentation of multiple objects is discussed in Sections 3.2. 3D volume segmentation is addressed in Section 3.3.

3.1 Single Object in 2D Images

In this section we show an example of the simplest application for our technique. The goal is to segment an object from the background in a given 2D image. We

---

[2] The exact algorithm is beyond the scope of this paper.

assume that the object appears as one connected blob. The user places object seeds and background seeds to define areas that should be separated by the segmentation.

In Figure 2(a-d) we show the segmentation results for a 2D cardiac MR image. Our interface allows a user to enter seeds with a brush controlled through a mouse. The size of the brush can be changed. Throughout the paper we indicate object and background seeds by bright red and blue colors, respectively. The pixels that the algorithm assigned to the object segment are highlighted in red and background pixels appear bluish. In Figure 2(a) there is only one object segment. In fact, our algorithm is guaranteed to generate a single object segment when the object seeds form one connected group. The segmentation boundary defines a contour with the smallest cost among all contours separating the red (object) seeds from the blue (background) seeds. The computation time on a Pentium II, 333Mhz, for the segmentation of the 128×128 image shown in (a) is 50 milliseconds.

The example in Figure 2(a) shows that in a simple 2D setting our method can be used to extract a single closed contour which is what snakes, deformable templates, or shortest path search algorithms are used for. However, parts (b-d) of the same figure demonstrate the flexibility of our user interface combined with the graph cut segmentation. Our method provides a very simple and intuitive way to adjust the segmentation to the user liking. For example, in (a) the object segment covers only a blood pool. If the user wants to outline the epicardium, as shown in (b), it is enough to draw a few object seeds in the myocardium. Alternatively, if the user wants to get an accurate measurement of the endocardium as in (c), (s)he might have to add a few object seeds in the area of low contrast in the blood pool and a few background seeds in the myocardium. Our technique also allows the user to exclude internal parts of the object from the segmentation. For example, if the user wants to exclude papillary muscles from a volume measurement then (s)he can add background seeds inside these muscles as shown in (d). It should be noted that (b-d) are obtained directly from (a) by adding new seeds and the extra running time is negligible.

Figure 2(e) shows a segmented CT image of the liver. The original data is 512×512 pixels and the initial segmentation takes 1 to 2 seconds. Additional correcting seeds are processed in 100 to 200 milliseconds. This examples highlights the use of directed n-links. When undirected n-links are used as in Figure 2(f), the segmentation boundary oscillates between neighboring edges that are very close together, always choosing the best edge. When directed n-links are used as in Figure 2(g), the boundary is forced to keep brighter pixels inside. Thus, the segmentation boundary coincides with the physical boundary of the organ. In general, the user can achieve any segmentation result that (s)he wants. If the results are not satisfactory in some part of the image, the user can add new object and background seeds providing additional clues where the segmentation was wrong, until all problems are corrected. The exact choice of seed positioning is not relevant. Normally, moving object seeds within a region of "similar" intensity inside one object segment cannot change the optimal segmentation[3]. It can also be shown that adding new object seeds inside the object segment will not change the resulting segmentation. Both properties are equally true for background seeds.

3.2 Multiple Objects

This section gives more general examples illustrating several important properties of our algorithm. The goal is still to segment an object from the background. This time we assume that the image may contain several isolated objects of interest. For example, an MR image may contain ten blood vessels and a doctor may want to investigate two of them. The object seeds provide the necessary clues on what parts of the image are interesting to the user. There are no strict requirements on where the object seeds have to be placed as long as they are inside the object(s) of interest. Such flexibility is justified, in part, by the ability of the algorithm to efficiently incorporate the seeds added later on, when the initial segmentation results are not satisfactory. The object seeds can be placed sparingly and they do not necessarily have to be connected inside each isolated object. Potentially, the algorithm can create as many separate object segments as there are connected components of object seeds. Nonetheless, the isolated seeds (or components of seeds) located not too far from each other inside the same object are likely to be segmented out together. Figure 3(d) illustrates this property of our technique. Figure 3(d) also shows that seeds can be placed to achieve any desired segmentation. We segmented the left kidney as one object by placing a single object "seed" in the middle. We segmented the callices out in the right kidney by placing background seeds in them. The segmentation algorithm automatically decides which object (or background) seeds are grouped into one connected segment and which seeds are placed separately. Note that this property may be also useful in N dimensions when a user does not see how the objects of interest connect when placing the seeds.

The background seeds should provide the clues on what is *not* an object of interest. In many situations the pixels on the image boundary are a good choice for background seeds. If the objects of interest are bright, then background seeds can be spread out in the dark parts of the image. Note that background seeds are very useful when two similar objects touch in some area of the image and one of them is of interest while the other is not. In this case there is a chance that the two objects may be merged into a single segment. This can be avoided by placing a background seed in the undesired object which would force the segmentation to separate the two objects at their merge point. Also, when the segmentation merges two objects of interest that should be separated, the user can add a background seed in between to effectively separate the two objects. This property is illustrated in Figure 3(a-c). At first, in Figure 3(c), the segmentation process grouped the two vessels together. By adding a single

---

[3] This statement can be made precise.

background seed right between the two vessels, as in Figure 3(b), the user forced the process to keep the two objects separated.

3.3 3D Volumes

One example of the general graph structure for a multi-object segmentation in 3D is shown in Figure 2(b). The 3D segmentation has the same properties that we discussed for 2D examples. The main difficulty in 3D is to create a convenient interface for entering seeds. In fact, the ability to scan through a pile of 2D slices was good enough for our purpose. The user can select a few representative slices and enter object and background seeds in these slices. Since all voxels are connected in a 3D graph the information on what parts of the 3D volume are of interest and what parts should be considered background will propagate appropriately. Moreover, upon the initial segmentation the user can scan through the segmented slices and enter correcting seeds in some of the slices where the results are not satisfactory.

Figure 4 shows an example in cardiac MR. We took images of a slice of the heart left ventricle at different time instances and stacked them up into a volume. We then placed seeds in one image in the middle of the volume to indicate that we were interested in the blood pool. The resulting segmentation was almost perfect. We just had to add a couple of background seeds in the first and last slice to fill a small notch in the myocardium. The system was able to also fill the notch in the neighboring slices by propagating the new seeds. This segmentation of 12 256×256 images was done in 5 seconds.

Figure 5 shows a segmentation that we obtained for 3D lung CT data. Each of the lobes and trachea were segmented separately. The segments we combined to obtain multi colored visualization.

4 Conclusions and Future Research

There are several important ideas that we are currently working on. First of all we can incorporate additional regional information by adding finite cost t-links to all non-seed pixels/voxels. Such t-links would connect a pixel to both the object and the background terminals. The costs can reflect how well the pixel's intensity fits into available models of object and background, correspondingly. For example, such models can be represented by intensity histograms of seeded regions. These finite cost t-links are similar to what is used in [2]. Also, we can use approximation multi-way graph cut algorithms [2] to obtain multi-label segmentation. Such segmentation would be able to generate multi colored segmentation similar to what we show in Figure 5 without the need to segment each component independently.

We are also interested to try our segmentation technique on 4D data. To decrease the execution time and increase the memory efficiency of our software, especially for large 3D volumes and 4D datasets, we are considering representing the data using a quad-tree to group very similar pixels and associate groups to single nodes in the graph.

5 Acknowledgements

We would like to thank Bernhard Geiger for providing the 3D visualizations of the segmented volumes. Vladimir Kolmogorov helped us with some of the coding. Finally, discussions with Gareth Funka-Lea and Alok Gupta were very fruitfull.

References

1. A. A. Amini, T. E. Weymouth, and R. C. Jain. Using dynamic programming for solving variational problems in vision. *IEEE Transactions on Pattern Analysis and Machine Intelligence*, 12(9):855–867, September 1990.

2. Y. Boykov, O. Veksler, and R. Zabih. Markov random fields with efficient approximations. In *IEEE Conference on Computer Vision and Pattern Recognition*, pages 648–655, 1998.

3. L. D. Cohen. On active contour models and ballons. *Computer Vision, Graphics, and Image Processing: Image Understanding*, 53(2):211–218, 1991.

4. L. D. Cohen and I. Cohen. Finite element methods for active contour models and ballons for 2-d and 3-d images. *IEEE Transactions on Pattern Analysis and Machine Intelligence*, 15(11):1131–1147, November 1993.

5. I. J. Cox, S. B. Rao, and Y. Zhong. "Ratio regions": A technique for image segmentation. In *International Conference on Pattern Recognition*, volume II, pages 557–564, 1996.

6. M.-P. Dubuisson-Jolly, C.-C. Liang, and A. Gupta. Optimal polyline tracking for artery motion compensation in coronary angiography. In *International Conference on Computer Vision*, pages 414–419, 1998.

7. A. X. Falcão, J. K. Udupa, S. Samarasekera, and S. Sharma. User-steered image segmentation paradigms: Live wire and live lane. *Graphical Models and Image Processing*, 60:233–260, 1998.

8. M. Flickner, H. Sawhney, D. Pryor, and J. Lotspiech. Intelligent interactive image outlining using spline snakes. In *Asilomar Conference on Signals, Systems, and Computers*, volume 1, pages 731–735, 1994.

9. L. Ford and D. Fulkerson. *Flows in Networks*. Princeton University Press, 1962.

10. D. Geiger, A. Gupta, L. A. Costa, and J. Vlontzos. Dynamic programming for detecting, tracking, and matching deformable contours. *IEEE Transactions on Pattern Analysis and Machine Intelligence*, 17(3):294–402, March 1995.

11. A. Goldberg and R. Tarjan. A new approach to the maximum flow problem. *Journal of the Association for Computing Machinery*, 35(4):921–940, October 1988.

12. D. Greig, B. Porteous, and A. Seheult. Exact maximum a posteriori estimation for binary images. *Journal of the Royal Statistical Society, Series B*, 51(2):271–279, 1989.

13. L. D. Griffin, A. C. F. Colchester, S. A. Röll, and C. S. Studholme. Hierarchical segmentation satisfying constraints. In *British Machine Vision Conference*, pages 135–144, 1994.

14. R. M. Haralick and L. G. Shapiro. *Computer and Robot Vision*. Addison-Wesley Publishing Company, 1992.

15. H. Ishikawa and D. Geiger. Segmentation by grouping junctions. In *IEEE Conference on Computer Vision and Pattern Recognition*, pages 125–131, 1998.

16. I. H. Jermyn and H. Ishikawa. Globally optimal regions and boundaries. In *International Conference on Computer Vision*, volume II, pages 904–910, 1999.
17. M. Kass, A. Witkin, and D. Terzolpoulos. Snakes: Active contour models. *International Journal of Computer Vision*, 2:321–331, 1988.
18. E. N. Mortensen and W. A. Barrett. Interactive segmentation with intelligent scissors. *Graphical Models and Image Processing*, 60:349–384, 1998.
19. T. O'Donnell, M.-P. Dubuisson-Jolly, and A. Gupta. A cooperative framework for segmentation using 2D active contours and 3D hybrid models as applied to branching cylindrical structures. In *International Conference on Computer Vision*, pages 454–459, 1998.
20. L. J. Reese. Intelligent paint: Region-based interactive image segmentation. Master's thesis, Brigham Young University, 1999.
21. J. Shi and J. Malik. Normalized cuts and image segmentation. In *IEEE Conference on Computer Vision and Pattern Recognition*, pages 731–737, 1997.
22. T. Vehkomäki, G. Gerig, and G. Székely. A user-guided tool for efficient segmentation of medical image data. In *CVRMed-MRCAS*, pages 685–694, 1997.
23. D. J. Williams and M. Shah. A fast algorithm for active contours and curvature estimation. *Computer Vision, Graphics, and Image Processing: Image Understanding*, 55(1):14–26, 1992.
24. Z. Wu and R. Leahy. An optimal graph theoretic approach to data clustering: Theory and its application to image segmentation. *IEEE Transactions on Pattern Analysis and Machine Intelligence*, 15(11):1101–1113, November 1993.
25. A. Yuille and P. Hallinan. Deformable templates. In Andrew Blake and Alan Yuille, editors, *Active Vision*, pages 20–38. MIT Press, 1992.
26. S. C. Zhu and A. Yuille. Region competition: Unifying snakes, region growing, and Bayes/MDL for multiband image segmentation. *IEEE Transactions on Pattern Analysis and Machine Intelligence*, 18(9):884–900, September 1996.

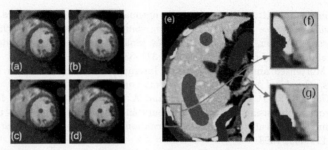

Fig. 2. Single object segmentation. (a-d): Cardiac MRI. (e-g): Liver CT. Directed (f) and undirected (g) n-links.

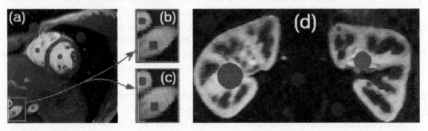

Fig. 3. Segmentation of multiple objects. (a-c): Cardiac MRI. (d): Kidney CE-MR angiography.

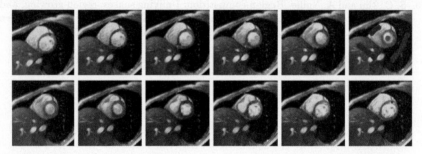

Fig. 4. Volume segmentation of the blood pool in the heart left ventricle in MR.

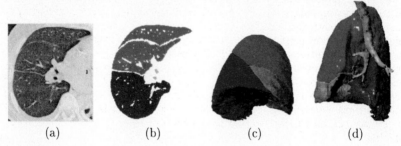

| (a) | (b) | (c) | (d) |

Fig. 5. Segmentation of the right lung in CT. (a): representative 2D slice of original 3D data. (b): segmentation results on the slice in (a). (c-d) 3D visualization of segmentation results.

Distribution of Prostate Cancer for Optimized Biopsy Protocols*

Jianchao Zeng[a], John J. Bauer[b], Ariela Sofer[c], Xiaohu Yao[a],
Brett Opell[d], Wei Zhang[e], Isabell A. Sesterhenn[e], Judd W. Moul[b,f],
John Lynch[d], and Seong K. Mun[a]

[a] Imaging Science and Information Systems Center (ISIS),
Department of Radiology, Georgetown University Medical Center,
Washington,DC 20007
zeng@isis.imac.georgetown.edu (J. Zeng)
http://www.simulation.georgetown.edu
[b] Urology Service, Department of Surgery,
Walter Reed Army Medical Center,
Washington, DC 20307-5001
[c] Department of Operations Research and Engineering,
School of Information Technology and Engineering,
George Mason University,
Fairfax, Virginia 22030
[d] Urology Division, Department of Surgery,
Georgetown University Medical Center,
Washington, DC 20007
[e] Department of Genitourinary Pathology,
Armed Forces Institute of Pathology (AFIP),
Washington, DC 20306-6000
[f] Center for Prostate Disease Research (CPDR), Department of Surgery,
Uniformed Services University of the Health Sciences (USUHS),
Bethesda, Maryland 20814-4799

Abstract. Prostate cancer is the leading cause of death for American men. The gold standard for diagnosis of prostate cancer is transrectal ultrasound-guided needle core biopsy. Unfortunately, no imaging modality, including ultrasound, can effectively differentiate prostate cancer from normal tissues. As a result, most current prostate needle biopsy procedures have to be performed under empiric protocols, leading to unsatisfactory detection rate. The goal of this research is to establish an accurate 3D distribution map of prostate cancer and develop optimized biopsy protocols. First, we used real prostate specimens with localized cancer to reconstruct 3D prostate models. We then divided each model into zones based on clinical conventions, and calculate cancer presence in each zone. As a result, an accurate 3D prostate cancer distribution map was developed using 281 prostate models. Finally, the linear programming approach was used to optimize biopsy protocols using anatomy and symmetry constraints, and the optimized protocols were developed with various criteria.

* This research has been supported, in part, by The Whitaker Foundation, The US Army Medical Research and Matériel Command, and The Center for Prostate Disease Research.

1. Introduction

In the United States, approximately 200,000 new cases of prostate cancer are detected annually, and about 40,000 men die of prostate cancer each year. Transrectal Ultrasonography (TRUS) guided systematic needle biopsy of the prostate has been widely used clinically for the diagnosis and staging of prostate carcinoma. Due to the low resolution of the ultrasound images, however, a urologist can hardly differentiate abnormal tissues from normal ones during the biopsy. Therefore a number of protocols have been developed to help urologists perform prostate needle biopsy. A biopsy protocol designates locations of the needles on the prostate as well as the number of needles to use. The most common protocol currently used clinically is the systematic sextant biopsy protocol[8].

Recent studies, however, have shown that the accuracy of currently used biopsy techniques needs to be improved[1, 4]. These studies show that a significant number of patients who have prostate cancer are not diagnosed at their initial biopsy. As a result, there are a significant number of prostate cancers that are detected on repeat biopsies. If biopsy protocols can be optimized to increase the chances of detecting prostate cancer, the value of these new protocols as a diagnostic tool will be significant.

A number of researchers have investigated techniques for improving the accuracy of biopsy protocols[5, 15]. Eskew et al. introduced a 5-region biopsy protocol in which additional lateral and midline biopsies were added systematically to the traditional sextant biopsy[6]. Our group found that the 5-region protocol showed a statistically significant advantage over the sextant method based on simulation results of 89 patients with cancer[10]. Chang et al. showed that lateral biopsies increase the sensitivity of prostate cancer detection[3]. We have also showed biopsy protocols that use laterally placed biopsies based upon a five region anatomical model are superior to the routinely used sextant prostate biopsy pattern[2]. Goto et al. suggested that new biopsy strategies could be developed based on probability maps of cancer distribution within the prostate[7]. But issues such as how these maps should be built and how new biopsy protocols could be derived from the maps remain to be investigated.

In this research, our goal is to establish an accurate 3D distribution map of prostate cancer, and to develop optimized biopsy protocols. In this paper, we first use real prostate specimens with localized cancer to reconstruct 3D prostate models. We then divide each model into zones based on clinical conventions, and calculate cancer presence in each zone. As a result, an accurate 3D prostate cancer distribution map is developed using 281 prostate models. Further, the linear programming approach is used to optimize biopsy protocols using anatomy and symmetry constraints, and optimized protocols are developed with various clinical criteria.

2. Reconstruction of 3D Prostate Models and 3D Distribution Map of Prostate Cancer

We have developed a deformable modeling technique for the reconstruction of 3D prostate surface models based on step-sectioned whole-mounted radical prostatectomy specimens with localized cancers[14,15,11]. Currently, two hundred and eighty-one 3D prostate models have been so far reconstructed.

In order to establish optimized prostate needle biopsy protocols, we have investigated the spatial distribution of the prostate cancers with the 3D models in such a way that will help directly determine an optimized protocol for improved cancer detection. Based on the low-resolution nature of currently available ultrasound imaging and the current clinical conventions for TRUS guided prostate biopsy, we have divided a prostate gland into different zones (instead of directly dealing with individual voxels) that are accessible by the urologists under the guidance of ultrasound images. By checking cancer presence inside each of these zones using a large number of 3D prostate models, we can develop an overall spatial distribution map of prostate cancer among the various zones. And based on this distribution map, an optimal biopsy protocol can be developed.

This approach is characterized by its precision in determining prostate cancer distribution throughout the whole prostate gland. Since there is no artificial deformation made to the individual prostate models during the process, the resulting information on prostate cancer location and distribution should precisely reflect the original cancer presence. Unlike other deformation-based approaches, our developed approach does not make any approximation and therefore no error is introduced in the results. In addition, the different sizes of individual prostate models will not affect the results of this approach since it uses the same and consistent concept of zones as that used by urologists. That is, although a larger prostate will certainly have a larger size for each of its zones, each corresponding zone in different prostate models carries the same spatial meaning and is treated equally regardless of its size.

(1) Algorithm for dividing zones and detecting cancers in the zones

The algorithm is composed of 2 steps: a 3D-based checking and a 2D-based checking. The 3D-based checking module draws attention to each zone that can precisely identify the presence or non-presence of a cancer inside its scope. The 2D-based checking is used to identify if each zone intersects with cancers or is totally contained inside a large cancer by reviewing 2D section layers of both the cancers and the zone. The 3D-based checking is a quick process dealing with simple situations while the 2D-based checking is a more intensive search process for any complex cases that are left over by the 3D-based checking step.

- *3D-based checking*

The 3D-based checking is performed to quickly make decisions on the presence of a cancer inside a zone using the 3D bounding box information. Only two distinct cases are considered here: a cancer is totally inside a zone, or a cancer is totally outside of a zone. When a cancer completely resides in a zone, every vertex of the cancer bounding box is inside the zone space. In implementation, we have used an 8-bit value to test the *inside* property of each cancer, while every bit corresponds to every vertex (1: inside, 0: outside). Initially the value is set to 0x00000000 and the test is performed for each zone. If the test ends up with a value of 0x11111111, the zone can be marked as positive. Otherwise, further investigation is needed to make a decision. When a cancer is completely outside the zone, we assume that the space is partitioned into two *sides* by each zone wall. All vertices of the cancer bounding box should reside in the same side if the cancer is totally outside of the zone. We need to perform six tests, one for each zone wall. In implementation, we have used a 6-bit value to test the *same-side* property. Each zone is set to 0x000000 initially and every vertex of the cancer bounding box updates this value after the test. The final result of this test is obtained by an "AND" operation of the eight values. If a zone ends up with a value not equal to 0x000000, the cancer is recognized as existing outside of this zone scope, and the zone is thus marked as negative. In the case of multiple cancers, this process is performed with respect to each of them, and all the results will be combined by an "AND" operation. Any result other than 0x000000 will lead to a negative zone.

- *2D-based checking*

The 2D-based checking module deals with those zones that cannot be quickly determined in the 3D-based checking step. For each zone C_i that needs 2D-based checking, first we need to identify which layers of the cancer can possibly intersect with C_i. That is, we need to find those layers of the cancer with a Z coordinate between $C_i.Maximum\_Z\_Value$ and $C_i.Minimum\_Z\_Value$. If we can prove that at least one cancer intersects with the zone in any of these layers, this zone will then be marked as positive for containing cancer. Otherwise, if all applicable layers fail to yield such a proof, this zone will be marked as negative. A *scan line algorithm* is developed to find such a proof fast for each 2D layer.

The *scan line algorithm* starts from the left margin of the zone section, and keeps moving to the right end at a given step. The intersection between the scan line and the polygon edges of the cancer layer can be found quickly since the edges have already been sorted beforehand. If this intersection lies inside the zone section, the 2D-based checking is completed with positive result. Otherwise, the algorithm partitions the layer into two portions: one above and one below the zone. If the scan line has odd number of intersections with the cancer edge in each portion, the cancer polygon edge must contain, at least in part, the zone section, and thus the zone will be marked as positive. If, on the other hand, the intersections are in even number in each portion, the cancer might be in a concave shape here without containing the zone, and the algorithm will continue the scanning process until the end.

(2) A comprehensive 48 division scheme

Obviously, the more divisions the more detailed information we can get for cancer distribution. At the same time, however, in order for the distribution information useful directly to the urologists, a division scheme should allow its zones to be recognizable and accessible by them under the guidance of transrectal ultrasound probe used for prostate biopsy. Under such constraint, we have conducted a comprehensive 48 division of prostates, which is probably the maximum number of division we can get for the purpose of prostate biopsy.

In the 48-zone division, looking at a prostate in axial view, a prostate gland is divided into 4 symmetric compartments sideways, 4 symmetric compartments vertically, and 3 compartments in depth. An example of a prostate model with 48-zone division is shown in Figure 1. The size of the zones will vary with the size of a prostate model. A larger prostate ends up having a larger size for each of its 48 zones. This variation of zone size is the natural reflection of the original prostate, and it does not affect the accuracy of cancer distribution. The labeling of each zone follows the current clinical conventions. The four layers of zones parallel to the YZ plane (sagittal) are labeled along the positive X direction as: left lateral (ll), left mid (lm), right mid (rm) and right lateral (rl), respectively. The three layers of zones parallel to XY plane (axial) are labeled along the positive Z direction as: base, mid and apex, respectively. Similarly, the four layers parallel to the XZ plane (coronal) are labeled along the positive Y direction as: posterior 1 (p1), posterior 2 (p2), anterior 1 (a1) and anterior 2 (a2), respectively. When a specific zone is labeled, the labels of the corresponding three layers are combined in the X-Y-Z order. For example, the zone 'A' in Figure 1 is labeled as llp1-apex.

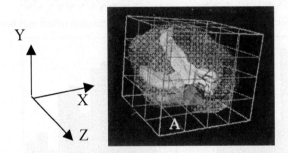

Figure 1. A prostate model with 48-zone division

In order to investigate the cancer distribution inside the prostate gland, we have calculated the appearance of cancers in each zone of each of the 281 individual prostate models using the above algorithm. The results are presented in sagittal layers (Figure 2) and axial layers (Figure 3). From the statistical analysis, we have the following preliminary results: (1) The base zones (36.8%) have a statistically significantly lower cancer distribution than mid zones (56.3%) (p=0.001) and apical zones (53.5%) (p=0.001); (2) There is also a significantly higher cancer distribution in the posterior zones (57.2%) than in the anterior zones (40.5%) (p=0.001); (3) The mid zones (56.3%) have a slightly significantly higher cancer distribution than the apical zones (53.5%) (p=0.032); and (4) There was no significant difference between the left zones (49.2%) and right zones (48.5%) (p=0.494). In order to develop an optimal biopsy protocol using the cancer distribution information, we have used the linear programming approach as described in the next section.

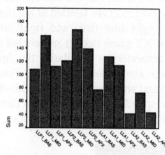

(a) Cancer distribution in left lateral layer

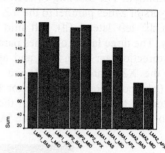

(b) Cancer distribution in left mid layer

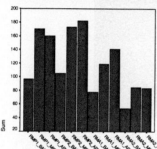

(c) Cancer distribution in right mid layer

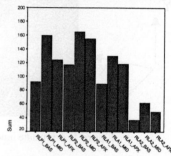

(d) Cancer distribution in right lateral layer

Figure 2. Cancer distribution presented in sagittal layers

lla2 14.95%	lma2 18.51%	rma2 19.22%	rla2 13.17%
lla1 27.76%	lma1 26.6%	rma1 27.76%	rla1 31.67%
llp2 43.06%	lmp2 39.15%	rmp2 37.37%	rlp2 41.64%
llp1 38.43%	lmp1 37.01%	rmp1 34.52%	rlp1 32.74%

lla2 25.98%	lma2 32.03%	rma2 30.25%	rla2 22.06%
lla1 45.20%	lma1 43.77%	rma1 42.35%	rla1 46.26%
llp2 59.79%	lmp2 61.21%	rmp2 61.57%	rlp2 58.72%
llp1 56.58%	lmp1 64.06%	rmp1 60.50%	rlp1 56.94%

(a) Distribution in the base layer

(b) Distribution in the mid layer

lla2 15.66%	lma2 29.18%	rma2 29.89%	rla2 17.44%
lla1 40.93%	lma1 50.89%	rma1 50.18%	rla1 41.99%
llp2 49.47%	lmp2 62.63%	rmp2 64.77%	rlp2 55.16%
llp1 40.21%	lmp1 56.23%	rmp1 56.94%	rlp1 44.13%

(c) Distribution in the apex layer

Figure 3. Cancer distribution presented in axial layers

3. Optimization of Biopsy Protocols

The optimal biopsy protocols can be explored by solving a mathematical optimization problem. Here, we define an m by n matrix $\mathbf{A}$ such that,

$$a_{ij} = \begin{cases} 1, & \text{if prostate i is found to have cancer in zone j} \\ 0, & \text{otherwise} \end{cases}$$

where $m = 281$ is the number of prostate models in the study, and $n = 48$ is the number of zones in the 3D distribution map. We further define variables x_j ($j=1,..,n$) by

$$x_j = \begin{cases} 1, & \text{if a biopsy is taken in zone j} \\ 0, & \text{otherwise} \end{cases}$$

Then the minimum number of zones to be tested that would ensure, in theory, 100% cancer detection for the data set can be found by solving the following problem:

$$\text{Minimize} \quad \sum_{j=1}^{n} x_j$$

$$\text{subject to} \quad \sum_{j=1}^{n} a_{ij} x_j \geq 1, \quad i = 1,\ldots,m \qquad \text{(P1)}$$

$$x_j = 0 \text{ or } 1, \quad j = 1,\ldots,n$$

The problem of minimizing or maximizing a linear function subject to linear constraints is known as a linear program. The special case where the variables are restricted to take on integer values (as we have above) is called an integer program. Linear programs have been studied extensively[12], and sophisticated software packages can often solve problems involving millions of variables and thousand constraints. Although integer programs are usually much harder to solve, recent advances have made it possible to solve many classes of problems effectively[13]. We can also define optimal biopsy protocols as those that provide maximum rate of cancer detection for a given number of biopsies. Again we define variables y_i (i=1,..,m) by

$$y_i = \left\{ \begin{array}{ll} 1, & \text{if cancer is detected for prostate i} \\ 0. & \text{otherwise} \end{array} \right.$$

Then the maximum number of prostates in the data set whose cancer would be detected by a k-biopsy protocol can be found by solving the following integer program:

$$\text{Maximize} \quad \sum_{i=1}^{m} y_i$$

$$\text{subject to} \quad \sum_{j=1}^{n} a_{ij} x_j \geq y_i, \quad i = 1,\ldots,m \qquad \text{(P2)}$$

$$\sum_{j=1}^{n} x_j = k$$

$$x_j, y_i = 0 \text{ or } 1, \quad j =1,\ldots,n, \quad i =1,..,m$$

where the left-hand term of the inequality

$$\sum_{j=1}^{n} a_{ij} x_j \geq y_i$$

represents the number of zones in which cancer is detected for prostate i . If this number is zero, then the inequality forces y_i to be zero. Otherwise, if this number is one or larger, the inequality permits y_i to be either 0 or 1. Since the objective here is to maximize the sum of y_i variables, the optimal solutions will have $y_i = 1$.

Problems (P1) and (P2) can be modified to include various physician preferences. For example, to impose a left-right symmetry in the protocols, we can add the constraints $x_l = x_r$ for every pair of zones l and r that are symmetrical in the distribution map. To restrict biopsies to zones in the posterior area (between urethra and rectum in prostate anatomy), we can impose the condition $x_a = 0$ for all zones in the anterior area. We have investigated a number of strategies for the optimal biopsy protocols, and some examples are shown below. We have used the software package ILOG CPlex 6.5[9] to solve the relevant integer programs. In all cases, the CPlex solved the problems within seconds on a variety of platforms.

(1) Protocols with 100% detection rate
Without constraint:
- *rlp2-base, lma1-base, lmp1-mid, rlp2-mid,*
  lma2-mid, lmp2-apex, rlp2-apex, rma1-apex
- *llp1-base, llp2-base, rma1-base, lmp2-mid,*
  rlp2-mid, lmp1-apex, rlp2-apex, lma1-apex

With constraint (limit to symmetric zones in posterior 1 and 2, and anterior 1):

- *llp2/rlp2-base lmp1/rmp1-mid, lla1/rla1-mid,*
  llp2/rlp2-apex, lla1/rla1-apex
- *lla1/rla1-base lmp1/rmp1-mid, lla1/rla1-mid,*
  llp2/rlp2-apex, lma1/rma1-apex

(b) Protocols with given number of biopsies

With constraint (limit to 8 symmetric zones in posterior 1 and 2, and anterior 1):

- *llp2/rlp2-base lmp1/rmp1-mid, lla1/rla1-mid, llp2/rlp2-apex* (Rate: 280/281)
- *llp2/rlp2-base lma1/rma1-mid, lla1/rla1-mid, llp2/rlp2-apex* (Rate: 280/281)

With constraint (limit to 6 symmetric zones in posterior 1 and 2, and anterior 1):

- *lma1/rma1-base llp2/rlp2-mid, lmp2/rmp2-apex* (Rate: 272/281)

With constraint (limit to 10 symmetric zones in posterior 1 and 2):

- *llp2/rlp2-base, lmp1/rmp1-mid, llp2/rlp2-mid,*
  lmp2/rmp2-mid, llp2/rlp2-apex (Rate: 275/281)

With constraint (limit to 8 symmetric zones in posterior 1 and 2):

- *llp2/rlp2-base, lmp1-rmp1-mid, llp2/rlp2-mid, lmp2/rmp2-apex* (Rate: 274/281)

With constraint (limit to 6 symmetric zones in posterior 1 and 2):

- *llp2/rlp2-base, llp2/rlp2-mid, lmp2/rmp2-apex* (Rate: 271/281)

Some of these results are supporting new protocols recently proposed by urologists. Note that here when there are multiple protocols for the same constraint, it means that they have equal capability of detecting cancer in theory, and it is up to the urologists to decide which one is more appropriate to use with respect to the specific patients. Ideally, an optimized biopsy protocol will achieve maximized detection rate of clinically significant cancer with minimized number of needles at accessible entry locations. At present, we are concentrating only on detecting prostate cancer regardless of its clinical significance. The significance of detected prostate cancer can be later determined in the staging step. In the future, however, both diagnosis and staging of prostate cancer will be addressed by developing unified optimal biopsy protocols. The developed optimal biopsy protocols will be compared and evaluated against those that are currently widely used clinically. The evaluation process will be performed first by computer simulation using 3D prostate models that were not involved during optimization, and then by a limited clinical trial on patients who are recommended for prostate needle core biopsies.

4. Conclusions

Performance of prostate needle biopsy can be significantly improved by providing the urologists with more information on cancer distribution. This research has revealed that such distribution can be precisely calculated using modeling and mapping techniques. We have successfully modeled prostate cancer using real prostate specimens with localized cancer. We have also developed a 3D distribution map of prostate cancer based on which optimized biopsy protocols have been developed. This is the first time ever that prostate needle biopsy protocols have been quantitatively developed and optimized based on a large database of computerized real prostate specimens. Once finally evaluated, the developed biopsy protocols could have great impact on clinical practice for improved outcome of diagnosis of prostate cancer.

Currently, we are also using the developed cancer distribution map and the optimized biopsy protocols to develop an image-guided *in vivo* online system for real patient biopsy. Techniques developed in this research are expected to be applicable to many other clinical procedures such as radiation therapy of prostate cancer.

References

1. Bankhead C. Sextant biopsy helps in prognosis of Pca, but it's not foolproof. Urology Times 1997; 25(8)
2. Bauer JJ, Zeng J et al. 3D Computer simulated prostate models: Lateral prostate biopsies increase the detection rate of prostate cancer. Urology 1999; 53(5):961-967
3. Chang JJ, Shinohara K et al. Prospective evaluation of lateral biopsies of the prostate for cancer detection. J. Urol. 1998; 159(5):179, abstract # 688
4. Daneshgari F, Taylor GD et al. Computer simulation of the probability of detecting low volume carcinoma of the prostate with six random systematic core biopsies. Urology 1995; **45**: 604-609
5. Egevad L, Frimmel H et al. Three-dimensional computer reconstruction of prostate cancer from radical prostatectomy specimens: Evaluation of the model by core biopsy simulation. Urology 1999; 53: 192-198
6. Eskew AL, Bare RL, McCullough DL. Systematic 5-region prostate biopsy is superior to sextant method for diagnosing carcinoma of the prostate. J. Urol. 1997; **157:** 199-202
7. Goto Y, Ohori M et al. Distinguishing clinically important from unimportant prostate cancers before treatment: value of systematic biopsies. J. Urol. 1996; **156:** 1059-1063
8. Hodge KK, McNeal et al. Random systematic versus directed ultrasound guided trans-rectal core biopsies of the prostate. J. Urol. 1989; 142: 71-74
9. ILOG CPLex 6.5 User Manual, ILOG, 1999
10. Kaplan CR, Zeng J et al. Comparison of sextant to 5-region biopsy technique using a 3D computer simulation of actual prostate specimens. J. Urol. 1998; 159(5):179, abstract # 687
11. Lin W, Liang C, Chen C. Dynamic elastic interpolation for 3D medical image reconstruction from serial cross section, IEEE Trans. Medical Imaging 1988; 7(3): 225-232
12. Nash SG, Sofer A. Linear and Nonlinear Programming. McGraw Hill, 1996
13. Nemhauser G, Wolsey L. Integer, Combinatorial Optimization. Wiley, 1988
14. Xuan J, Sesterhenn I et al. Surface reconstruction and visualization of the surgical prostate model. SPIE Medical Imaging 1997, Vol. 3031. San Diego 50-61
15. Zeng J, Kaplan CR et al. Optimizing prostate needle biopsy through 3D simulation. SPIE Medical Imaging 1998, Vol. 3335. San Diego 488-497

Hybrid Classification Approach of Malignant and Benign Pulmonary Nodules Based on Topological and Histogram Features

Y. Kawata, N. Niki, H. Ohmatsu[a], M. Kusumoto[b], R. Kakinuma[a],
K. Mori[c], H. Nishiyama[d], K. Eguchi[e], M. Kaneko[b], N. Moriyama[b]

Dept. of Optical Science, Univ. of Tokushima, Tokushima ,
[a]National Cancer Center Hospital East, [b]National Cancer Center Hospital,
[c]Tochigi Cancer Center, [d]The Social Health Insurance Medical Center,
[e]National Shikoku Cancer Center Hospital

Abstract. This paper focuses on an approach for characterizing the internal structure which is one of important clues for differentiating between malignant and benign nodules in three-dimensional (3-D) thoracic images. In this approach, each voxel was described in terms of shape index derived from curvatures on the voxel. The voxels inside the nodule were aggregated via shape histogram to quantify how much shape category was present in the nodule. Topological features were introduced to characterize the morphology of the cluster constructed from a set of voxels with the same shape category. The properties such as curvedness and CT density were also built into the representation. In the classification step, a hybrid unsupervised/supervised structure was performed to improve the classifier performance. It combined the k-means clustering procedure and the linear discriminate (LD) classifier. The performance of the hybrid classifier was compared to that of LD classifier alone. Receiver operating characteristics (ROC) analysis was used to evaluate the accuracy of the classifiers. We also compared the performance of the hybrid classifier with those of physicians. The classification performance reached the performance of physicians. Our results demonstrate the feasibility of the hybrid classifier based on the topological and histogram features to assist physicians in making diagnostic decisions.

1. Introduction

Lung cancer is the leading cause of cancer death among Japanese men and its incidence continues to increase [1]. In order to improve the recovery rate for lung cancer, detection and treatment at an early stage of growth is necessary. Radiography of the chest is ordinary used for the screening of lung cancer. At present, after screening via chest radiograph to detect suspicious areas, differential diagnosis is ordinarily concluded by histology from biopsy. There are, however, a significant number of malignant cases which should be discriminated as early as possible from benign lesions. Particularity when the peripheral lung for suspicious areas in early development are diagnosed, it is often the case that the differential diagnosis by means of transbronchial or percutaneous biopsies becomes difficult.

There has been a considerable amount of interest in the use of thin-section CT images to observe small pulmonary nodules for differential diagnosis without invasive operation [1]-[3]. In assessing the malignant potential of small pulmonary nodules in thin-section CT images, it is important to examine the condition of nodule interface, the nodule internal intensity, and the relationships between nodules and surrounding structures such as vessels, bronchi, and spiculation [1]-[3]. A number of investigators have developed a feature extraction and a classification methods for characterizing pulmonary nodules. Siegelman et al. investigated CT density in the center of a nodule on two-dimensional (2-D) CT images [4]. Other groups also presented nodule density analysis with a special reference phantom to improve measurement accuracy [5]. Cavouras demonstrated that multiple features including nodule density and texture were useful to classify malignancies from other lesions [6]. Following his work, McNitt-Gray proposed pattern classification approach incorporating multiple features, including measures of density, size measures, and texture of nodules on CT slice images [7]. One promising area of recent researches has been the analysis of three-dimensional (3-D) pulmonary nodule images. We quantified the concave and convex surfaces by using surface curvatures to characterize surface condition of malignant and benign nodules [8],[9]. Hirano presented an index to quantify how a nodule evolved the surrounding vessels [10]. Tozaki proposed a classification approach between pulmonary artery and vein to characterize the relationships between nodules and surrounding structures [11]. Kitaoka developed mathematical models of bronchial displacements caused by nodules to discriminate cancers from inflammatory pulmonary nodules [12]. Although the performances of the computer algorithms are expected to depend strongly on data set, they indicate the potential of using computer aided diagnosis techniques to improve the diagnostic accuracy of differentiating malignant and benign nodules.

This paper focuses on the analysis of the internal structure in the 3-D pulmonary nodules. In previous study [13],[14] we found that curvature indexes such as shape index and curvedness were promising quantities for characterizing the internal structure of nodules. However, there were several distribution patterns of CT density inside the nodule, such as solid or infiltrative types. Therefore, it might be desirable to decompose input samples into classes with different properties to improve classification performance. In this present study, we combine an unsupervised and a supervised model and apply them to classification of malignant and benign nodules. The unsupervised model is based on k-means clustering (KMC) procedure [21] which clustered the nodules into a number of classes by using CT density distribution. A supervised linear discriminate (LD) classifier [21] is designed for each classes by using topological and histogram measures based on curvature indexes. By improving the homogeneity of the nodules, the LD classifier designed may be more robust. The performance of the hybrid classifier will be compared with those of the LD classifier alone and physicians. We will demonstrate that the proposed hybrid structure can improve the accuracy of classification in computer-aided diagnosis applications.

This paper is organized as follows. Section 2 introduces a hybrid classifier and describes the feature extraction schemes of the nodule internal structure. In Section 3, our data set is described and experimental results are presented. Finally, Section 4 concludes this investigation.

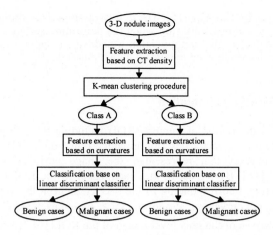

Fig. 1. Block diagram of the hybrid classification of malignant and benign pulmonary nodules.

2. Method

2.1 Overview

We propose to design a hybrid classifier that combines the unsupervised KMC procedure with a supervised LD classifier. The classification procedure of the pulmonary nodules is shown in Fig.1. Since there are several distribution patterns of CT density value inside nodules, the KMC procedure improve the homogeneity of the sample distributions by classifying classes with different properties regarding the distribution patterns of CT density value. In this study the KMC separates the sample data into two classes denoted as class A and B. The class A is the class in which the mean CT density value inside the nodule is high and the class B is the class in which the mean CT density value inside the nodule is low. For each class a LD classifier is designed by using the CT density and the curvature based features inside nodules and then discriminate malignant and benign nodules. The first-stage KMC procedure may improve the performance of the LD classifier if the subclass causes the sample data to deviate from multivariate normal distributions for which the LD classifier is an optimal classifier.

2.2 Nodule Segmentation

The segmentation of the 3D pulmonary nodule image consists of three steps [9] ;1)extraction of lung area, 2) selection of the region of interest (ROI) including the nodule region, 3) nodule segmentation based on a geometric approach. This lung area extraction step plays an essential role when the part of a nodule in the peripheral lung area touches the chest wall. The ROI including the nodule was selected interactively. A pulmonary nodule was segmented from the selected ROI image by the geometric approach proposed by Caselles [15]. The deformation process of the 3-D deformable surface model can automatically stop when the deforming surfaces reach the object's boundary to be detected. In our application we added a stopping condition to exclude vessels and bronchi which were in contact with the nodule [9].

2.3 Curvature based representation

Each voxel in the region of interest (ROI) including the pulmonary nodule was locally represented by a vector description which relied on the CT density value and two curvature indexes that represented the shape attribute and the curvature magnitude [17], [18]. By assuming that each voxel in the ROI lies on the surface which has the normal corresponding to the 3-D gradient at the voxel, we computed directly the principal curvatures κ_1 and κ_2 ($\kappa_1 \geq \kappa_2$) on each voxel from the first and the second derivatives of the gray level image of the ROI [16]. To compute the partial derivatives of the ROI images, the ROI images were blurred by convolving with a 3-D Gaussian function of width σ. In order to take only nonnegative values, we used the shape index defined as

$$S(\mathbf{x};\sigma) = \frac{1}{2} + \frac{1}{\pi} \arctan \frac{\kappa_1(\mathbf{x};\sigma) + \kappa_2(\mathbf{x};\sigma)}{\kappa_1(\mathbf{x};\sigma) - \kappa_2(\mathbf{x};\sigma)} \quad (1)$$

The curvedness of a surface at the voxel $\mathbf{x}$ is defined as

$$R(\mathbf{x};\sigma) = \sqrt{\frac{\kappa_1(\mathbf{x};\sigma)^2 + \kappa_2(\mathbf{x};\sigma)^2}{2}} \quad (2)$$

It is a measure of how highly curved a surface and its dimension is that of the reciprocal of length.

2.4 Curvature based representation

In order to characterize the pulmonary nodule through the local description, we used the shape spectrum which was introduced for object recognition by Dorai and Jain [18]. Using the shape spectrum, we measured the amount of the voxel which had a particular shape index value h. The augment shape spectrum with scale σ is given by

$$H(h;\sigma) = \frac{1}{V} \iiint_O \delta(S(x;\sigma) - h) dO \quad (3)$$

where , V is the total volume of the specified region O, dO is a small region around x and δ is the Dirac delta function. The discrete version of Eq.(1) is derived by dividing the shape index range into B bins and counting the number of voxels falling in each bin k and normalizing it by the total number N of discrete voxels in the specified region. The discrete version is expressed by

$$H(h = \frac{k}{B}) = \frac{1}{N} \sum_{i=1}^{N} \chi_k(S(x_i; \sigma)) \tag{4}$$

with

$$\chi_k(t) = \begin{cases} 1 & \frac{k-1}{B} \le t < \frac{k}{B} \\ 0 & otherwise \end{cases} \tag{5}$$

Here, the segmented 3-D pulmonary nodule image is utilized as the specified region O. The shape index value one is included in the B-th bin. The discrete version of the shape spectrum was called shape histogram. The number of voxel falling in each bin represented the value of the histogram feature. For computational purposes, such as comparing spectra of different nodules, the shape histogram was normalized with respect to the volume of nodule. The normalized number of voxel falling in each bin represents the value of the shape histogram feature. In this study the number of bin B was given the value 100. The similar equations for the curvedness and CT density are obtained in the same manner. The domains of curvedness and CT density were specified to [0,1] and [-1500, 500], respectively. A voxel in which the curvedness value was larger than one was considered as a voxel with curvedness value one. For the CT density the similar process was performed. To classify malignant and benign nodules, we combined a set of histogram features, such as shape, curvedenss, and CT density histogram.

2.5 Topological features

The distribution morphology of the shape category can characterize the internal structure of the nodule. Therefore, we divided the inside of the nodule into four shape categories by the shape index value and then, computed the topological features of each 3-D cluster which constructed from a set of voxels with the same shape category. The four shape categories were peak, saddle ridge, saddle valley, and pit surface types and the interval of the shape index for each shape categories were set [0, 0.25], [0.25, 0.5], [0.5, 0.75], [0.75, 1], respectively. The topological features used here were the Euler number, the number of connected components, cavities, and holes of a 26-connected object [19]. The Euler number of a 3-D digital figure is defined as the following equation,

$$E = b_0 + b_1 + b_2 \tag{6}$$

where E is the Euler number and b_0, b_1, and b_2 respectively represent the number of connected components, holes, and cavities. These valuables, b_0, b_1, and b_2 are called the first, second, an third Betti-number, respectively. Yonekura et al. [19] provided computation schemes for the basic topological properties such as the connected component and the Euler number of 3-D digital object. The Euler number, the number of connected components, cavities, and holes were obtained by their schemes. In addition, we quantified how each shape category distributes inside the nodule by using a technique computing exact Euclidean distance transform [20]. Using the Euclidean distance in the nodule, we measured an amount of voxel which had a particular distance value d in the i-th shape category. The distance spectrum with shape category is given by

$$DH(d;i) = \frac{1}{V} \iiint_O \delta(D(x;i) - d)dO \tag{7}$$

where $D(x; i)$ is the distance value at the voxel x with i-th shape category. The discrete version was derived by the similar manner as the histogram features. In this study the number of bin was given the value 25. The normalized number of voxel falling in each bin represents the value of the shape distribution feature. This feature was included in the topological features.

2.6 Classification

In order to improve the accuracy of a classifier, we combined the unsupervised KMC procedure with the supervised LD classifier. The KMC classified the sample nodules into two classes by using the mean CT density value for three different regions. These different considered regions are as follows: (i)core region shrinking to $T_1\%$ of the maximum distance value of the 3-D nodule image, denoted as R1, (ii) complement of the core region in the 3-D nodule image, denoted as R2, (iii) marginal region extended to $T_2\%$ of the maximum distance value of the 3-D nodule image, denoted as R3. The maximum distance value was obtained by applying the Euclidean distance transformation technique [20] to the segmented 3-D nodule image. In this study T_1 and T_2 values were assigned to 60% and 26%, respectively. For each class a LD classifier was designed by using the topological and histogram features. In order to reduce the number of the features and to obtain the best feature set to design a good classifier, feature selection with forward stepwise selection procedure was applied [22]. In this study the minimization of Wilks' lambda was used as an optimization criterion to select the effective features. A leave-one-out procedure was performed to provide a less biased estimation of the linear discriminate classifier's performance. The discriminant scores were analyzed using receiver operating characteristic (ROC) method [23]. The discriminant scores of the malignant and the benign nodules were used as the decision variable in the ROCKIT program developed by Metz which fits the ROC curve based on maximum likelihood estimation. This program was also used to test the statistical significance of the difference between pairs of ROC curves. The two-tailed p values were reported in the comparison procedure described in the next section.

3. Experimental Results

Thin-section CT images were obtained by the helical CT scanner (Toshiba TCT900S Superhelix and Xvigor). Per patient, thin-section CT slices at 1mm intervals were obtained to observe whole nodule region and its surroundings. The range of pixel size in each square slice of 512 pixels was between 0.3x0.3 mm^2 and 0.4x0.4 mm^2. The 3-D thoracic image was reconstructed by a linear interpolation technique to make each voxel isotropic. The data set in this study included 210 3-D thoracic images provided by National Cancer Center Hospital East and Tochigi

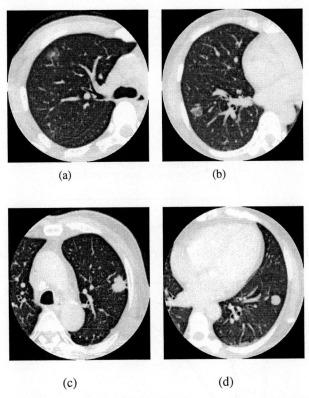

(a) (b)

(c) (d)

Fig. 2. Slice images including pulmonary nodules in class A and class B. (a) Malignant nodule in class A. (b) Benign nodule in class A. (c) Malignant nodule in class B. (d) Benign nodule in Class B.

Cancer Center. Among the 210 cases, 141 contained malignant nodules, and 69 contained benign nodules. Whole malignant nodules were cytologically or histologically diagnosed. In benign cases, lesions showed no change or decreased in size over a 2-year period were considered as the benign nodules. The size of nodules was less than 20 mm in diameter.

The selection of the width σ of the Gaussian function is a critical issue. In previous work [13],[14], we selected the value of the width which provided high accuracy of classification between malignant and benign nodules for discrete width values. From the result, we assigned the value 2.0 to the width. We compared the hybrid classifier with the LD classifier alone. In the hybrid classifier, the KMC procedure classified input samples into two classes denoted as class A and class B. The class A contained 50 cases (13 benign cases and 37 malignant cases) and the class B contained 160 cases (56 benign cases and 104 malignant cases). Fig.2 presents slice images including pulmonary nodules in class A and class B. In comparison with the class B, the nodule of the class A had ill-defined surface and the region with lower CT density value occupied the inside of the nodule. The histogram feature used here was combined with three histogram features such as shape, curvedness, and CT density histograms. The topological features were yield for four clusters. The LD

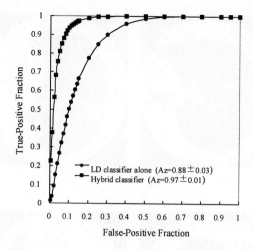

Fig. 3. ROC curves of the hybrid classifier and the LD classifier alone.

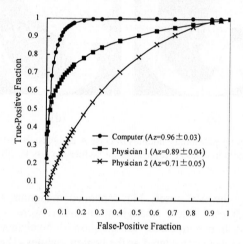

Fig. 4. Comparison of ROC curves obtained by using physicians malignancy rating and the discriminate score of hybrid classifier.

classifiers were designed from the combined topological and histogram features. For the class A and B, the number of the selected features was 9 and 21, respectively. In the LD classifier alone, the number of the selected features was 15. The ROC curves of the hybrid and the LD classifiers were plotted in Fig. 3. The classification accuracy in the hybrid classifier was significantly higher than those in the LD classifier alone ($p<0.01$).

In order to compare the performance of physicians with that of the hybrid classifier, the probability of malignancy of each pulmonary nodule in thin-section CT images which were printed on films, was ranked by eleven physicians on a scale of 1 to 10, where a ranking of 1 corresponded to the nodules with the most benign cases. The number of nodules used in this comparison was 119 cases provided by the National Cancer Center East. Based on the ranking, the ROC curves using three physicians malignancy rating and the computer's discriminate score output in the combined feature space were plotted in Fig. 4. The physicians 1 and 2 respectively have 15 years, and one year of experience in the chest radiology. The difference between the ROC curves of the hybrid classifier and those of two physicians was statistical significance (p < 0.05). These results show that the classification performance of the hybrid classification approach achieved the experienced physician results.

4. Conclusion

In this study, the topological and histogram measures based on curvature information were introduced to characterize the internal structure of 3-D nodule images. A hybrid classifier combining an unsupervised k-means clustering algorithm with a supervised LD classifier has been designed and applied to the classification of malignant and benign nodules. The Az value under the ROC curve for our data set was higher for the hybrid classifier compared to that of the LD classifier alone. A greater improvement was obtained by introducing the k-means clustering procedure. The performance of the hybrid classifier was also compared with those of the experience physicians. The classification performance of the hybrid classifier reached the performance of the experienced physicians. These results indicate that the hybrid classifier is a promising approach for improving the accuracy of classifiers for CAD applications.

Acknowledgments

The authors are grateful to physicians cooperating to the reading test. The authors would like to thank Prof. Charles E. Metz for the ROCKIT program.

References

1. M.Kaneko, K.Eguchi, H.Ohmatsu, R.Kakinuma,T.Naruke, K.Suemasu, N. Moriyama, "Peripheral lung cancer: Screening and detection with low-dose spiral CT versus radiography," Radiology, Vol.201, pp.798-802 (1996)
2. K.Mori, Y.Saitou, K.Tominaga, K.Yokoi, N.Miyazawa, A.Okuyama, M.Sasagawa, "Small nodular legions in the lung periphery: new approach to diagnosis with CT," Radiology, Vol.177, pp.843-849 (1990)
3. C.V. Zwirewich, S.Veda, R.R.Miller, N.L.Muller, "Solitary pulmonary nodule: high-resolution CT and radiological pathologic correlation," Radiology, Vol.179, pp.469-476, (1991)
4. S.S.Siegelman, E.A.Zerhouni, F.P.Leo, N.F. Khouri, F.P. Stitik, "CT of the solitary pulmonary nodule," AJR, Vol.135, pp.1-13 (1980)
5. A.V.Proto and S.R.Thomas, "Pulmonary nodules studied by computed tomography," Radiology, Vol.156, pp.149-153 (1985)

6. D. Cavouras, P. Prassopoulos and N. Pantelidis, "Image analysis methods for solitary pulmonary nodule characterization by computed tomography," European Journal of Radiology, Vol.14, pp.169-172 (1992)

7. M.F. McNitt-Gray, E.M.Hart, J. G. Goldin, C.-W, Yao, and D.R. Aberle, " A pattern classification approach to characterizing solitary pulmonary nodules imaged on high resolution computed tomography," Proc. SPIE, Vol. 2710, pp.1024-1034 (1996).

8. Y.Kawata, N.Niki, H.Ohmatsu, R.Kakinuma, K.Eguchi, M.Kaneko, N.Moriyama, "Shape analysis of pulmonary nodules based on thin-section CT images", Proc. SPIE, Vol. 3034, pp.967-974 (1997)

9. Y.Kawata, N.Niki, H.Ohmatsu, R.Kakinuma, K.Eguchi, M.Kaneko, N.Moriyama, "Quantitative surface characterization of pulmonary nodules based on thin-section CT images", IEEE Trans. Nuclear Science, Vol. 45, pp.2132-2138 (1998)

10. Y.Hirano, Y.Mekada, J.Hasegawa, J. Toriwaki, H.Ohmatsu, and K.Eguchi, "Quantification of vessels convergence in three-dimensional chest X-ray CT images with three-dimensional concentration index", Medical Imaging Technology, Vol.15, pp.228-236 (1997)

11. T.Tozaki, Y.Kawata, N.Niki, H.Ohmatsu, R. Kakinuma, K.Eguchi, N. Moriyama, "Pulmonary organs analysis for differential diagnosis based on thoracic thin-section CT images", IEEE Trans. Nuclear Science, Vol.45, pp.3075-3082 (1998)

12. H. Kitaoka and R. Takaki, "Simulations of bronchial displacement owing to solitary pulmonary nodules," Nippon Acta Radiologica, Vol.59, pp. 318-324 (1999)

13. Y. Kawata, N.Niki, H.Ohmatsu, M. Kusumoto, R. Kakinuma, K. Mori, K. Eguchi, M. Kaneko, N. Moriyama, "Classification of pulmonary nodules in thin-section CT images by using multi-scale curvature indexes" , IEEE Int. Conf. on Image Processing, Vol.2, pp.197-201 (1999)

14. Y. Kawata, N.Niki, H.Ohmatsu, "Curvature based internal structure analysis of pulmonary nodules using thoracic 3-D CT images", IEICE Trans., Vol.J-83-D-II, pp.209-218 (2000).

15. V.Caselles, R.Kimmel, G.Sapiro, and C.Sbert, "Minimal surfaces based object segmentation," IEEE Trans. Pattern Analysis Machine Intelligence, Vol.19, pp.394-398 (1997)

16. J.-P, Thirion and A. Gourdon, "Computing the differential characteristics of isointensity surfaces," Computer Vision and Image Understanding, Vol.61, pp.190-202 (1995)

17. J. J. Koenderink and A.J.V. Doorn, "Surface shape and curvature scales", Image and Vision Computing, Vol.10, pp.557-565 (1992)

18. C.Dorai and A.K. Jain, " COSMOS-A representation scheme for 3D free-form objects", IEEE Trans. Pattern Analysis Machine Intelligence, Vol.19, pp.1115-1130 (1997)

19. T. Yonekura, S. Yokoi, J. Toriwaki, T. Fukumura, "Connectivity and Euler number of figures in the digitized three-dimensional space", Trans. IECE, vol. J65-D, pp.80-87,1982.

20. T. Saito and J. Toriwaki, "Euclidean distance transformation for three-dimensional digital images", Trans. IEICE, Vol.J76-D-II, pp.445-453 (1993)

21. R.O. Duda and P.E. Hart, "Pattern classification and scene analysis", John Wiley, Sons, (1973)

22. M.C. Costanza and A.A. Afifi, "Comparison of stopping rules in forward stepwise discriminate analysis", J. of the American Statistical Association, Vol. 74, pp.777-785, (1979)

23. C.E. Metz, "ROC methodology in radiologic imaging", Investigative Radiology, Vol.21, pp.720-733 (1986)

MRI - Mammography 2D/3D Data Fusion for Breast Pathology Assessment

Christian P. Behrenbruch[1], Kostas Marias[1,2], Paul A. Armitage[1], Margaret Yam[1],
Niall Moore[3], Ruth E. English[4], J. Michael Brady[1]

[1]Medical Vision Laboratory (Robotics), Engineering Science, Oxford University, Parks
Road, Oxford OX1 3PJ, UK
{cpb,jmb}@robots.ox.ac.uk, http://www.robots.ox.ac.uk/~mvl

[2]Department of Surgery, Royal Free and University College Medical School, UCL,
London NW3 2QG, UK

[3]Magnetic Resonance Imaging Centre, John Radcliffe Hospital, Headley Way,
Oxford OX3 9DU, UK

[4]Breast Care Unit, Churchill Hospital, Oxford OX3 7LJ, UK

Abstract. Increasing use is being made of contrast-enhanced Magnetic
Resonance Imaging (Gd-DTPA) for breast cancer assessment since it
provides 3D functional information via pharmacokinetic interaction
between contrast agent and tumour vascularity, and because it is
applicable to women of all ages. Contrast-enhanced MRI (CE-MRI) is
complimentary to conventional X-ray mammography since it is a
relatively low-resolution functional counterpart of a comparatively
high-resolution 2D structural representation. However, despite the
additional information provided by MRI, mammography is still an
extremely important diagnostic imaging modality, particularly for
several common conditions such as ductal carcinoma in-situ (DCIS)
where it has been shown that there is a strong correlation between
microcalcification clusters and malignancy [1]. Pathological indicators
such as calcifications and fine spiculations are not visible in CE-MRI
and therefore there is clinical and diagnostic value to fusing the high-
resolution structural information available from mammography with
the functional data acquired from MRI imaging. This paper presents a
novel data fusion technique whereby medio-lateral (ML) and cranio-
caudal (CC) mammograms (2D data) are registered to 3D contrast-
enhanced MRI volumes. We utilise a combination of pharmacokinetic
modelling, projection geometry, wavelet-based landmark detection and
thin-plate spline non-rigid registration to transform the coordinates of
regions of interest (ROIs) from the 2D mammograms to the spatial
reference frame of the contrast-enhanced MRI volume.

1. Introduction

This paper introduces a system that has been developed to perform data fusion between 2D X-ray mammography and 3D contrast-enhanced MRI of the breast. The objective has been to develop a registration and visualisation framework that makes use of both the high spatial resolution of mammography and the vascular information provided by contrast-enhanced MRI. This is particularly useful in situations where small opacities, microcalcifications and fine spiculations, not visible in an MRI scan, cannot be correlated with voxel enhancement.

The principal concept behind this work is the registration of a highly compressed 2D projective representation of the breast with an uncompressed volume-based acquisition. In the case of the MRI acquisition, the patient is lying facedown in the scanner with the breasts resting pendulously and uncompressed in the breast coils. This is entirely different to the situation in mammography, where the patient normally remains upright with the breasts compressed, (one at a time), between two plates. All of the patients in our study had both a medio-lateral (ML) and a cranio-caudal (CC) mammogram, as is required for breast cancer screening in the UK. It is worth noting that the angular separation between the mammogram view directions depends on the woman's size and shape, but typically varies between 45-60 degrees. More importantly, as regards matching, the compression between the two mammograms differs significantly, often by as much as 1cm.

In order to register the projective X-ray image with the MRI volume, we project the MRI volume in both the CC and the ML direction. However, rather than simply use voxel intensity or approximate T1-correlated X-ray attenuation characteristics [2], we use an enhancement measure based on a two-compartment pharmacokinetic model of contrast uptake [3]. This effectively produces a contrast-based projection that has the visual and structural characteristics of an "enhanced" X-ray mammogram.

The registration technique is outlined in Figure 1. Our registration process actually consists of two separate registrations. The first "partial registration" utilises a curvature measure to correlate boundary points along the film edge of the mammograms with the edge of the volume projection [4]. This curvature measure depends on a good segmentation of the breast edge, which is achieved by an intensity-based search, mathematical morphology for smoothing and spline fitting to produce a smoothed edge profile [5]. The effect of this partial registration is to deform the extremities of the X-ray image to the boundary shape of the MRI contrast projection.

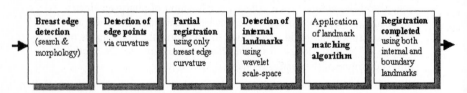

Fig. 1. Overview of X-ray to MRI contrast projection registration process

A second registration is performed based on the selection of internal landmarks using a wavelet-based feature detector. Internal landmarks in both the X-ray image and the MRI projection are matched using the following criteria:

- Scale localisation (i.e. landmarks of similar scale are matched)
- Orientation (via Principal Component Analysis)
- Relative motion between the undeformed and the partially registered (via curvature points) data
- Neighbourhood localisation in the partially registered images

The second registration uses the combination of matched internal landmarks and boundary points (selected on the basis of curvature) to complete the registration. The registration technique adopted to date is thin plate spline warping [6]. This landmark-driven approach is appropriate in applications where we do not necessarily have an intensity correlation between the MRI contrast projection and the X-ray data. It also enables us to perform the partial registration using only boundary points, which is particularly critical in the case where the breast has undergone significant involution[1] and there may be no landmarks or intensities suitable for controlling the internal deformation of the registration process. An example of this 2-stage registration process is shown in Figure 2.

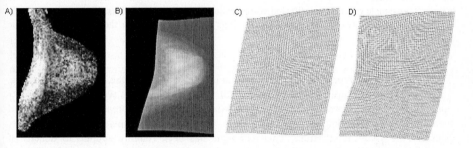

Fig. 2. *A)* shows the functional projection of the MRI volume in the cranio-caudal plane. *B)* shows the deformation of the cranio-caudal X-ray of the breast to the MRI contrast volume projection (*cropped*). *C)* and *D)* show the deformation grid of the partially and fully registered images. The increased level of complexity in the deformation of *D)* can be attributed to the internal landmarks used to complete the registration process

In the following sections, the main components of the registration process are discussed, namely the boundary curvature measure, internal landmark detection and the use of approximating thin-plate splines.

---

[1] The term "involution" refers to the transformation of ductal or parenchymal breast tissue into fat. The incidence of involution increases sharply at the onset of the menopause. The transradiance of fat relative to ductal/parenchymal tissue is the reason screening is typically restricted to women over the age of 50, a heuristic approximation to the age of onset of the menopause.

2. Breast Boundary Feature Detection

The detection of suitable landmarks for boundary deformation between the X-ray mammogram and the MRI volume projection first involves segmenting the breast edge. This is performed using a combination of an intensity-based search, mathematical morphology to smooth the breast edge (closing operator) and finally a B-spline interpolation of the segmented edge for curvature parameterisation [4]. Figure 3 shows the correlation between the spline-parameterised curvature and anatomical landmarks such as the nipple and axilla.

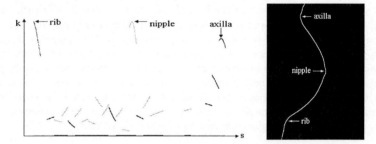

Fig. 3. An example curvature profile and corresponding anatomical landmarks. The curvature profile is separated into positive (*dark*) and negative (*light*) in order to locate the rib/axilla and nipple respectively

These curvature-selected points form the basis of the first step in the registration process, which effectively ensures that the boundary of the X-ray image is warped onto the MRI volume projection. However, this is not a true correction for breast compression and therefore internal landmarks are required to complete the compensation for complex internal soft tissue deformation. This technique has been reliably applied to dozens of mammograms [4,7] and is a sensible approach because while there may be slight changes of compression in successive mammograms (particularly around the time of the menopause), the breast outline shape remains consistent.

3. Partial Registration

Given the edge-localised curvature landmarks, the first registration step is performed (refer to Figure 1) using thin-plate spline approximation. In the thin-plate spline interpolation approach a set of n landmarks (p_i, q_i) is used, where $p_i = (x_{i1}, y_{i1})$ are the co-ordinates of the landmarks of the first image and $q_i = (x_{i2}, y_{i2})$ of the second. The interpolation problem is to find the transformation f that fulfils the condition $f(p_i) = q_i$ for $i=1,\ldots,n$. and minimises a functional $J(f)$. $J(f)$ can be separated in $d=2$ (where d is the dimension of the images) problems for each component f of f. Thus for $d=2$ and $m=2$ (the number of derivatives used we get [6]:

$$J_2^2(f) = \iint_{R^2} \left\{ \left(\frac{\partial^2 f}{\partial x^2}\right)^2 + \left(\frac{\partial^2 f}{\partial y^2}\right)^2 + 2 \cdot \left(\frac{\partial^2 f}{\partial x \cdot \partial y}\right)^2 \right\} dx\,dy \qquad (1)$$

This functional is also known as the bending energy of the deformation. In thin-plate spline theory, the desired function f is the solution of the biharmonic equation $\Delta^2 f = 0$. It can be deformed in certain parts obeying the function $f(x, y)$, as long as the displacements are small and conforms to the minimum bending energy configuration of $f(x, y)$.

4. Wavelet-Based Feature Detection

To complete the registration internal landmarks are required to compensate for the complexity of the soft-tissue deformations. A scale-space approach was chosen, based on tensor wavelet packets [8,9]. Wavelet *packet* decompositions are particularly useful linear superpositions of wavelets that result in large "libraries" of functions that have specific frequency and spatial localisations [10]. The scale space construction follows a conventional dyadic decomposition using quadrature filters (QFs), as expressed in equation(s) 2.

$$\psi_0 \triangleq L_f \psi_0; \qquad \int_R \psi_0(p).dp = 1,$$

$$\psi_{2n} \triangleq L_f \psi_n; \qquad \psi_{2n}(p) = \sqrt{2} \sum_{j \in Z} l_f(j) \psi_n(2p - j), \qquad (2\ a,b,c)$$

$$\psi_{2n+1} \triangleq H_f \psi_n; \qquad \psi_{2n+1}(p) = \sqrt{2} \sum_{j \in Z} h_f(j) \psi_n(2p - j),$$

where ψ_0 is the mother wavelet and ψ_{2n} and ψ_{2n+1} are the resulting wavelet subspaces after convolution and decimation with the low-pass/high-pass equivalent QFs h_f, l_f for pixel p (in one pixel vector). To date, we have used the Coiflet wavelet bases as they feature good spatial localisation (i.e. edge preserving), have compact support [11] and are morphologically relevant to detecting small regions such as microcalcifications.

After wavelet decomposition the scale-space is completed by using an information cost function in the context of a "best basis" algorithm [12] to order wavelet coefficients in such a way that the various subspaces used for the decomposition are ranked by information content:

$$\|f - g\|^2_{L_2(I)} + \lambda \|g\| \qquad (3)$$

evaluated as:

$$\|f - g\|^2_{L_2(I)} \stackrel{\Delta}{=} \left(\int_I |f(p) - g(p)|^2 dx \right)^{1/2} \qquad (4)$$

where $f(p)$ is the original image pixel vector and $g(p)$ is the reconstructed image pixel vector for a given wavelet subspace (packet) in L_2 with respect to the chosen cost function. In this case, we use a second-order approximation to entropy that is relatively invariant to image noise statistics [13]. Each wavelet subspace (filter superposition) is then cumulatively adjointly convolved, in order, with respect to the best-basis assessment of the decomposition. The result is a "stack" of reconstructions from minimum to maximum information content (dependent on the cost function). This analysis is used as the basis of segmentation, similar to the "extremum stack concept" [14], in conjunction with some contour refinement to compensate for smoothing [15].

5. Final Registration

The last step in the registration process is to include both the boundary curvature-based landmarks and detected internal landmarks in the registration process (as outlined in part 3). In addition to the matching criteria discussed previously, the scale localisation information for each internal landmark is used to control the σ_i^2 terms in equation (5). This has the effect of applying a confidence measure to a landmark and adjusting the corresponding level of local deformation caused by each feature point. In this way, features with low-levels of saliency influence the internal deformation to a lesser extent. The implementation involves minimizing the functional [16]:

$$J_\lambda(f) = \sum_{i=1}^{n} \frac{\left|q_i - p_i\right|^2}{\sigma_i^2} + \lambda \cdot J_2^2(f) \tag{5}$$

where λ is the regularisation parameter that controls the amount of smoothness in the deformation and σ_i^2 are the uncertainty terms.

6. Clinical Assessment

In an initial study, five patient datasets demonstrating a wide range of pathology were used to test the procedure[2]. The objective is to gauge how closely a correspondence can be made between the locations of pathology in the X-ray images with the 3D MRI volume representation. Cases were deliberately chosen to encompass a wide range of tissue biomechanical properties. For example, a large, dense ductal carcinoma (as shown in Figure 4, for example) has very different deformation properties than localised fibrosis (a brief numerical assessment of these five cases is presented in Table 1).

For the purpose of presenting the visualisation results shown in this paper, two extreme cases were chosen. The first features a dense circumscribed tumour in a highly involuted breast. This tumour is very large (4.5 centimetres in diameter) but is quite poorly differentiated with a highly fibrous epithelium. This is an interesting case because the effects of compression have quite clearly deformed the shape of the tumour in the mammogram. Figure 4 shows a shape comparison (via our scale-space segmentation) between the mammograms, the MRI contrast projection and the registered images.

The registration shown in Figure 4 is quite good, although the lack of large numbers of internal landmarks has perhaps oversimplified the internal deformation. For further comparison, an averaged centroid coordinate has been taken from both views and used to reconstruct the shape of the "uncompressed" breast using the technique outlined in [17]. This can be compared visually with a segmentation of the tumour in the MRI breast volume using a 3D implementation of the Hayton-Brady pharmacokinetic-based segmentation [3]. This comparison can be seen in Figure 5.

---

[2] The patients were being assessed at the Oxford MRI clinic and volunteered their data in accordance with the local ethical policy.

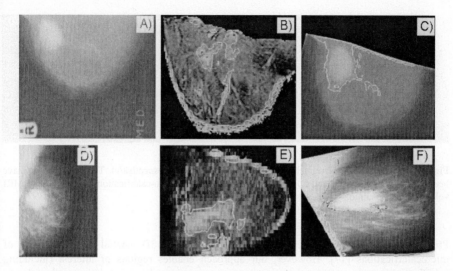

Fig. 4. *A)* and *D)* are the original CC and ML X-rays (*scaled*). *B)* and *E)* are MRI enhancement projections showing segmented regions of interest. *C)* and *F)* are the CC and ML mammograms registered to the MRI projection images with regions of interest segmented

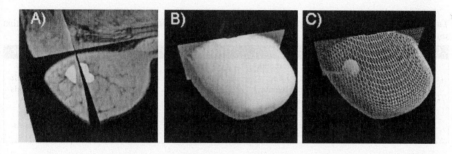

Fig. 5. *A)* shows the pharmacokinetic segmentation of the tumour in 3D from the contrast-enhanced MRI acquisition. Figure *B)* shows the reconstructed shape of the breast from the registered X-ray projections. *C)* shows the location of the tumour in the same registered reference frame

The second case highlights one of the most important applications of this research. It has been observed that certain shapes and distributions of microcalcifications are indicative of ductal carcinoma. However, many cases of DCIS don't enhance particularly well under CE-MRI, nor are microcalcifications visible. However, for surgical planning, it would be useful to be able to correlate the locations of the microcalcification to a 3D position in the breast. Figure 6 shows an example of a patient with severe ductal ectasia ("exploded" ducts) and two coarse calcifications located near the nipple and top-end of the parenchymal tissue. Using exactly the same technique as for the first case, the location of the calcifications in the 3D MRI volume was found.

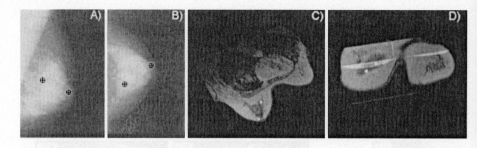

Fig. 6. *A)* and *B)* are the original ML and CC mammograms (*respectively*). The calcifications have been circled for clarity. *C)* and *D)* show the location of two micro-calcifications in the resliced MRI volume after registration

Previous work has enabled a partially compensated 3D spatial reconstruction of microcalcifications [17]. However, our approach enables regions of interest (including calcifications) to be compared against a true uncompressed representation, rather than a simulated and partially uncompressed representation. Table 1 contains a summary of the five cases used for this study. In all cases, clinician-identified regions of interest we registered to within +/- 4 voxels (approx. 4mm using our current MRI volume sequence).

Table 1. The approximate error (with respect to the MRI volume) of each of the cases used in this study

Case	Condition	CC Centroid Error	ML Centroid Error	Avg. Error
1.	Small opacity, infiltrating carcinoma	- 3.5 Voxels (avg)	- 2.5 Voxels (avg)	3 Voxels
2.	Fibroglandular nodule (1.5 cm)	+1.5 Voxels	- 4 Voxels	2.75 Voxels
3.	Large, dense circumscribed tumour (ductal carcinoma)	+3 Voxels	+ 5.5 Voxels	4.25 Voxels
4.	Multiple microcalcifications	-1 Voxel (avg)	- 3 Voxels (avg)	2 Voxels
5.	Spiculated opacity associated with large lymph node	-2 Voxels	+5 Voxels	3.5 Voxels

7. Discussion and Further Work

As with most applications of non-rigid registration, the success of the process outlined in this paper is quite difficult to evaluate. Table 1 shows the estimated error in localising the centroid of the various pathologies in the two views. Clearly, in the case of calcifications, this error estimation is difficult as surrounding tissue, rather than the feature itself must form the basis of the assessment. Additionally, the accuracy of the ML registration must be somewhat less as the cross-plane projection of the MRI volume is limited by the number of slices in the volume data (in this case 32). Further validation is necessary using phantoms, core-biopsy studies and a larger patient data set, however the preliminary results are quite encouraging particularly considering the extent of the deformation required for registration.

Although the principal objective of this research is data fusion for pathology assessment, there are a number of related applications for this work. Firstly, MRI is the most commonly used uncompressed breast imaging modality. By registering X-ray mammography (highly compressed) to MRI, the mammograms are effectively "decompressed". This provides an interesting framework for the study of breast tissue deformation under compression, with MRI effectively providing the uncompressed "gold standard". This would also provide independent validation of Highnam and Brady's h_int model of non-fatty breast anatomy computed from a mammogram [18].

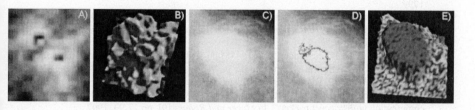

Fig. 7. This sequence of images shows a comparison between the functional projection of the pathology from the contrast-enhanced MRI data set and the X-ray structure in the CC mammogram of the example shown previously. A surface rendering *B)* of the *(scaled)* functional projection *A)* clearly shows two "rings" of contrast enhancement around a low-enhancement centre. This is quite typical of CE-MRI focal enhancement where the necrosis enhancement is relatively low compared with surrounding vascularity. Intensity equalisation and edge detection of the X-ray mass *C,D)* is consistent with the functional information as shown in *D)*. *E)* is a surface rendering of *C)* for comparison with *B)*

Finally, this registration approach may enable a functional representation to be compared with a structural (X-ray) understanding of the pathology. For example, in the case of the large tumour shown in Figures 5 & 6, it is interesting to compare the uptake of contrast in the tumour to the structure shown in the X-ray. This is shown in Figure 7 where the mass actually consists of two enhancing focal regions. With some simple thresholding and edge detection, it can also be shown that the same structure exists in the X-ray, although it would not be immediately apparent to a clinician.

8. Acknowledgements

C. Behrenbruch would like to acknowledge the Commonwealth Scholarship and Fellowship Program for his doctoral funding at Oxford University. The author would also like to acknowledge valuable discussions with Jerome Declerck (MVL) and Dermot Dobson (John Radcliffe Hospital). P. Armitage acknowledges the EPSRC for his post-doctoral grant (GR/M54995). J.M. Brady acknowledges EPSRC support for his senior fellowship.

9. References

1. M. Lanyi, "Diagnosis and differential diagnosis of breast calcifications". *Springer-Verlag*, Berlin, 1986

2. K.G.A. Gilhuijs, "Automated verification of radiation treatment geometry", Universiteit van Amsterdam, *Febo-druk*, Enschede-Utrecht, The Netherlands, 1995

3. P. Hayton, J.M. Brady, L. Tarassenko, N. Moore, "Analysis of dynamic MR breast images using a model of contrast enhancement", *Medical Image Analysis*, 1:3, April, 1997, Oxford University Press

4. K. Marias, J.M. Brady, et. al., "Registration and matching of temporal mammograms for detecting abnormalities", *Medical Imaging Understanding and Analysis*, Oxford 1999

5. K. Marias, C.P. Behrenbruch, J.M. Brady, "Robust Breast Edge Segmentation in Mammography", *Engineering Science Technical Report* 19990805#2, Oxford University, 1999

6. F.L. Bookstein, Principal Warps: "Thin-Plate Splines and the Decomposition of Deformations", IEEE *Transactions on Pattern Analysis and Machine Intelligence*, 11:6, pp 567-585, June 1989.

7. K. Marias, C.P. Behrenbruch, et. al., "Multi-scale Landmark Selection for Improved Registration of Temporal Mammograms", *IWDM (International Workshop in Digital Mammography) 2000*, Toronto, Canada, Elsevier, June 2000.

8. R.R. Coifman, Y. Meyer, S.R. Quake, M.V. Wickerhauser, "Signal processing and compression with wavelet packets". In *Meyer and Roques* [9], pp 77-93, 1992

9. Y. Meyer, S. Roques, (editors), "Progress in Wavelet Analysis and Applications". *Proceedings of the International Conference "Wavelets and Applications"*, Toulouse, France, 8-13 June 1992. Observatoire Midi-Pyrénées de l'Université Paul Sebatier, Editions Frontieres

10. I. Daubechies, "Orthonormal bases of compactly supported wavelets". *Communications on Pure and Applied Mathematics*, XLI: pp 909-996, 1988

11. I. Daubechies, "Ten lectures on wavelets", *CMBS*, SIAM, 61, pp 258-261, 1994

12. R.R. Coifman, & M.V. Wickerhauser, "Entropy based algorithms for best basis selection", *IEEE Transactions on Information Theory*, 32, pp 712-71, 1992

13. G.J.M. Parker, J.A. Schnabel, G.J. Barker, "Nonlinear Smoothing of MR Images Using Approximate Entropy – A Local Measure of Signal Intensity Irregularity". *Technical Report, NMR Research Unit*, Institute of Neurology, University College London, 1998

14. J. Koenderink,, "The structure of images", Biological Cybernetics, vol. 50, pp 363-370, 1984

15. S.C. Zhu, A. Yuille, "Region Competition: Unifying Snakes, Region Growing, and Bayes/MDL for Multiband Image Segmentation", *PAMI* 18:9, September 1996, pp. 884-900.

16. K. Rohr et. al., "Point-based elastic registration of medical image data using approximating thin-plate splines", *Lecture Notes in Computer Science* 1131, Springer, pp. 297-306, 1996

17. M. Yam, J.M. Brady, et. al., "Reconstructing microcalcification clusters in 3-D using a parameterized breast compression model", *Computer Assisted Radiology and Surgery (CARS)*, Paris 1999.

18. R.P. Highnam and J.M. Brady, "Mammographic Image Processing", *Kluwer Academic Press*, 1999

Pre- and Intra-operative Planning and Simulation of Percutaneous Tumor Ablation

Torsten Butz[1,2], Simon K. Warfield[2], Kemal Tuncali[3], Stuart G. Silverman[3],
Eric van Sonnenberg[3], Ferenc A. Jolesz[2], and Ron Kikinis[2]

[1] Swiss Federal Institute of Technology at Lausanne(EPFL), Signal Processing
Laboratoy (LTS), CH-1015 Lausanne, Switzerland
torsten.butz@epfl.ch,
WWW home page: http://ltswww.epfl.ch   FAX: +41-21-693-7600
[2] Surgical Planning Laboratory, Department of Radiology
[3] Department of Radiology, Brigham and Women's Hospital and Harvard Medical
School, 75 Francis Street, Boston, MA 02115, USA

Abstract. We developed a software tool for pre-operative simulation
and planning, and intra-operative guidance, of minimally invasive tumor
ablation, including radiofrequency-, laser- and cryo-therapy. This tool
provides a pre- and intra-operative optimization of the treatment plan,
in order to avoid dangerous probe trajectories, undertreatment of the
tumor, and excessive ablation of healthy tissues.

The simulation is performed within a virtual operating-room consisting in essence of the patient's segmented anatomy from pre- or intraoperatively acquired MR scans. Virtual probes can be placed into this
scene and at the formation of ablated tissue at their tips can be simulated. To verify the simulated treatment plans, we introduced an objective quality measure which also enables a semi-automated optimal probe
placement.

To show the use and to underline the importance of our tool, we investigated a cryo-therapy case which did not succeed. We show that our
software would have predicted the failure of the chosen treatment plan
and how it could have increased the efficacy of the procedure.

1 Introduction

Recent developments in interventional imaging, such as interventional MRI [1],
have opened a vast range of promising medical applications. For example, several
techniques for minimally-invasive percutaneous tumor ablation have been developed; some relying on heating (focused ultrasound, laser or radiofrequency) and
some on freezing (cryo-therapy) the tissue to kill the cells. The immune system
then removes the ablated tissue from the body.

We have placed the emphasis of this paper upon cryo-therapy. Nevertheless,
the generalization to other minimally-invasive ablation techniques is straight forward and some important aspects will be explained for both cryo- and radiofrequency-
ablation.

1.1 Limitations of Conventional Planning

In tumor ablation, the radiologists try to cover the arbitrarily shaped tumor with a given number of discrete objects of approximately known shape, representing the critical ablation temperatures for killing cancerous cells. In cryotherapy, the temperatures within the tumor should nowhere be above, and for thermo-ablation below, this bound. For interstitial tumor ablation, the cooling, respectively heating source is located on the tip of a needle-like probe which is percutaneously inserted through a catheter into the tumor.

The safety and success of a treatment plan depends mainly on the three following points:

1. Choosing secure probe trajectories.
2. Ablating all the cancerous cells.
3. Killing as little of the surrounding healthy tissue as possible.

Unfortunately it is very difficult to find the treatment plan that optimizes these three central factors when the radiologists have to rely only on 2D-slices of pre- and intra-operative MR-scans. In such data, the important 3D shape-information of the different anatomical objects (for example the tumor) and their mutual positions is not efficiently visualized.

1.2 Improving the Planning Process

The presented tool provides several features to improve the planning process for minimally-invasive percutaneous tumor ablation. Segmented MR scans enable a 3D view of the patient's anatomy and virtual needles visualize efficiently the probe's trajectories (optimize point 1 above). We also simulate and visualize the frozen tissue at the tip of the probes. An objective quality measure optimizes and classifies the simulated treatment plans according to point 2 and 3 above.

Results are presented in the context of liver cryo-therapy ([2]). We show how our tool can optimize secure probe trajectories and placements while avoiding undertreatment of the tumor.

2 Methods

The main functions of the tool are a 3D visualization of the patient's anatomy, a simulation of the probes and the frozen tissue, and a semi-automated optimization minimizing an objective quality measure to assess probe placement.

2.1 Visualization

To have a variety of visualization modes at the radiologist's disposal, we added our planning and simulation tool into an existing software package, called the *3D Slicer* ([3]). In this way, our planning tool can be used in conjunction with all the standard functionalities of that program. Let's shortly summarize its main features:

– The visualized 2D slices of 3D MR scans can be arbitrarily oblique, or orthogonal and oriented relative to either the coordinate frame of the scanner or the tracked surgical instrument.
– Different MR scans (e.g. SPGR, T2-weighted MRI, etc.) can be registered and visualized simultaneously or separately.
– Surface models representing the patient's anatomy can be generated from segmentations of MR scans and visualized in a 3D scene.
– Surgical instruments can be tracked and visualized in the 3D scene of the patient's anatomy.

The *3D Slicer* implementation uses primarily the "*Visualization Toolkit*" library (VTK, [7]) and the "*Tool Command Language*/graphical user interface *Toolkit*" scripting language (Tcl/Tk, [8]). We used the same programming environment and simply added the necessary VTK-filters and Tcl-commands.

2.2 Simulation

To simulate the intervention, we combined needle-like VTK-objects representing the probes with VTK-objects representing the frozen or ablated tissue (Figure 1). The radiologist can add these models into the virtual scene of the *3D Slicer* in which the segmented and grayscale data of the patient are also visualized. A powerful visualization is therefore straightforward and, depending on the specific case, it is possible to place any number of probes into the virtual 3D scene (Figure 3).

The virtual probes: So far we have implemented a cryo-probe consisting of one single needle (Figure 1a), a radiofrequency probe with one needle and a radiofrequency probe with three needles, called a cluster (Figure 1b). As *3D Slicer* uses a world coorindate system in millimeter-space, the parameters for the virtual probes are simply given by the parameters of the real probes. This also facilitates the modification of specific simulation-parameters or the addition of other models to the list of virtual probes according to technological and scientific progress.

The frozen, or ablated, tissue: To get a reasonable representation of killed tissue during cryo-ablation, we used the following hypothesis that has historically also guided radiologists carrying out interventions:
Hypothesis: Apart from the 0.5cm outer rim of the formed iceball, all the cells inside the frozen region are killed during cryo-therapy.
The outer cells are simply lying too far away from the cooling source, implying too high temperatures for cell death. Two important statements are equivalent to this hypothesis. First, the shape of the zone representing the killed tissue is the same as the shape of the formed iceball, but smaller. Secondly and equivalently, the radiologist should freeze an additional 0.5cm margin around the tumor to kill all the cancerous cells. We decided to take the second point of view and to simulate the iceballs during cryo-therapy and not directly the killed tissue, but

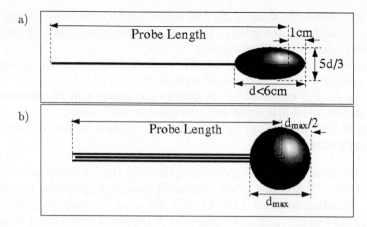

Fig. 1. In a) we show a virtual cryo-probe including the elliptic iceball and a needle. The image in b) shows a radiofrequency cluster consisting of three needles and a spherical ablation model.

to require the coverage of an additional 0.5cm margin around the tumor. This approach is more compatible to MR guidance, as MR images can visualize the iceballs but not the ablated tissue directly.

In order to visualize the additional 0.5cm margin around the tumor, we implemented a VTK filter that dilates the segmented tumor. The result is then added at the appropriate position to the 3D-scene of the *3D slicer*.

We also aimed to simulate the formation of the iceball during cryo-therapy. [4] and [5] have performed studies to describe the iceball-shape during cryo-ablation. Their results as well as our own studies have shown elliptic to teardrop-shaped iceballs for a single probe. Therefore we used ellipsoids to model the frozen zones at the tip of each needle (Figure 1a). The specific parameters were assessed from MR images of several previous cryo-cases showing the iceballs at the end of the freezing cycle. The specific parameters vary for different probes, as e.g. the iceball size depends on the cooling rate and therefore on the probe diameter.

For radiofrequency ablation, we simulate directly the killed tissue and therefore no additional margin has to be simulated. The appropriate model for the ablated zone is spheres, as indicated by previous radiofrequency cases (Figure 1b).

2.3   Optimization

As well as the visual verification of a simulated probe setup, we also provide a more objective quality measure. Such a medically relevant measure has to reflect the following two central factors:

1. How much of the tumor is ablated with a chosen setup. Therefore we have to calculate the percentage of the tumor (including the additional 0.5cm margin) that would be lying within the simulated iceballs.

2. Furthermore it's important to know how much healthy tissue would also be killed with a specific treatment plan. This is done by calculating the volume of the simulated iceball that is not covering cancerous but instead healthy tissue.

A good setup would maximize the first value (the volume of ablated cancerous tissue) and minimize the second (the volume of killed healthy cells). We developed a mathematical expression that reflects these factors and allows the radiologists to use a semi-automated setup-optimization and selection.

Let V_1 denote the volume of killed healthy tissue, V_2 denote the volume of ablated tumor and V_{tumor} denote the volume of the whole tumor (V_1, V_2 and V_{tumor} in ml). x is the vector that represents the positions and orientations of the probes as well as the size of the frozen iceballs at their tips. We can then define the following measure:

$$I(\boldsymbol{x}) = \lambda \cdot \frac{V_1(\boldsymbol{x})}{V_{tumor}} - (1 - \lambda) \cdot \frac{V_2(\boldsymbol{x})}{V_{tumor}} + (1 - \delta_{\boldsymbol{x},\Omega}) \cdot \frac{V_{tumor}}{ml} \tag{1}$$

where we have simply

$$\Omega = \{\boldsymbol{x} \in \mathbf{R}^n \mid \mu_i \leq x_i \leq \sigma_i; \mu_i, \sigma_i \in \mathbf{R}\} \tag{2}$$

and

$$\delta_{\boldsymbol{x},\Omega} = \begin{cases} 1, \text{ if } \boldsymbol{x} \in \Omega \\ 0, \text{ if } \boldsymbol{x} \notin \Omega \end{cases}$$

The potentially optimal setup is found by looking for the vector $\boldsymbol{x}$ that minimizes the measure $I(\boldsymbol{x})$. λ is a very important factor that weighs the two somewhat contradictory terms of minimizing V_1 and maximizing V_2. λ has to be chosen carefully, depending on the specific medical application. For example in brain tumor ablation, it's very important to minimize the ablated healthy tissue, so we would choose a λ close to 1. On the other hand, λ can lie much closer to 0 for liver tumors. Ω defines the bounds of the admissible optimization space and should describe an anatomically possible window for probe trajectories (e.g. we can't pass through any bones, even if the tumor ablation would be more optimal for such a setup).

The optimization algorithm uses primarily a VTK filter for boolean subtraction and reunion of polygonal datasets, a second VTK filter to calculate volumes of polygonal datasets and the "Powell" optimization algorithm from [6].

The combination of visualization, simulation and optimization results in an interactive planning tool for minimally invasive tumor ablation. In practice the radiologists can drag and rotate the probes (so far cryo- or radiofrequency probes) into an initial setup within the patient's segmented anatomy and then start the optimization algorithm to improve the probes' placement. The interventional radiologist can also get an objective quality measure to decide if e.g. an additional probe should be used to increase the likelihood of complete tumor ablation.

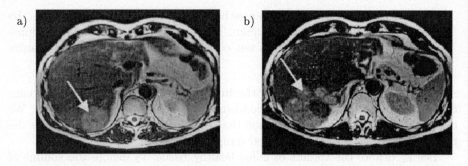

Fig. 2. In figure a) we show the tumor (arrow) in a pre-operative T2-weighted FSE. In b) we can see the ablated area as a dark, ellipsoidal zone. But we also recognize that during the last three months, an outer rim of cancerous cells (arrow) grew around the initial tumor.

3   Results

We applied our planning tool to analyze the failure of a previously performed cryo-therapy case. In Figure 2a we see an axial slice of the pre-operative T2-weighted fast spin-echo (FSE) MR scan of the abdomen. The arrow points at the tumor. In 2b, a slice of the 3-month follow-up scan (same modality) is shown. Unfortunately the tumor had re-appeared (arrow) and had grown very quickly around the effectively ablated zone. The analysis of this case was very important as the radiologists were convinced of having completely frozen the whole tumor including the additional 0.5cm margin. In consequence, some concerns about the frequently used *hypothesis* above arose. So we checked if, because of the failure of this treatment, the *hypothesis* has effectively to be questioned and, if not, to propose a safer treatment plan.

3.1   Simulating the Treatment Plan

In Figure 3a and b, we simulate the setup that was chosen by the radiologists. This is easily possible as the treatment was performed under near-realtime MR guidance and interventional MR scans enable the visualization of the probes and of the growing iceballs.

During pre-operative planning such a 3D view indicates if, for a certain setup, the probes are getting too close to important anatomical structures, such as main vessels or the gallbladder. Furthermore, a chosen probe might be too short to effectively reach its predetermined position in the tumor, and the radiologist can decide to take a longer probe.

In a next step, we wanted to visualize the simulation of ablated tissue as described in section *Simulation*. This is shown in Figure 4a, where the simulated iceball of the chosen setup is in red and the tumor, including the 0.5cm margin, in brown. Volume calculations indicate that just about 63% of the necessary volume are frozen in this setup. To increase this rather poor value, we applied our optimization algorithm for this setup. The frozen volume increases to 74% of the tumor volume (including margin).

A visual interpretation of Figure 4a indicates that the tumor was undertreated with the chosen probe setup. The calculated percentage of killed cancerous tissue (63%) confirms this estimation. Even after optimization, the ablation would still be insufficient (74%). Therefore the use of more cryo-probes is suggested.

3.2 Improving the Treatment Plan

In order to increase significantly the percentage of ablated cancerous tissue, we had to add two more probes to the scene. In Figures 3c, d and Figure 4b, we show the results. For this setup and after optimization, our measure indicated that 92% of the necessary volume would be frozen. The new setup would have increased the likelihood of successful tumor ablation for this patient.

3.3 Checking the Iceball Simulation

To check the quality of our iceball simulation, we superimposed the simulation and the real iceball that was segmented from the intra-operative MR scan at the end of the freezing cycle. The result is shown in Figure 4c (real iceball is shown in white and translucent, and the simulated iceball in red).

The ellipsoids simulate nicely the rough shape of the real iceball. But smaller details of the frozen region are not present in our estimation.

3.4 Post-operative Analysis

At the end of the freezing cycle, the radiologist can check if the performed treatment has most probably been successful. Such an analysis estimates also the quality of our simulation and checks its predictions. We superimposed the segmented iceball at the end of the freezing cycle with the tumor. The result is shown in Figure 4d, and re-confirms the result of our simulation: the tumor has been undertreated.

4 Discussion

The planning tool described here was able to explain the failure of a cryo-therapy case. In particular, even though the *hypothesis* above has still not been proven to be correct, its possible incorrectness can not be inferred from the failure of the presented case. Our tool will help to check and, if necessary, to reformulate this very important *hypothesis*.

We have also presented the different capabilities of the software to minimize the risk of failure:

- Within the visualization package *3D Slicer*, we can provide easily interpretable 3D visualizations of the patients' anatomy.

- Virtual needles and iceballs can be added to the virtual scene. This can ensure the safety of specific insertion directions of the probes and a good choice of the probe (e.g. its length). It improves also the visual quality estimation of a specific setup.
- The addition of an objective quality measure and a semi-automated optimization algorithm minimizes the danger of failure.

With these features, the tool was also able to propose a better probe setup which would have increased significantly the likelihood of success.

There are also some very important improvements necessary to make this tool widely used:

- We are using the models of Figure 1 to simulate the ablated volumes. These rough estimates are based on a limited number of cases and should be refined in the future. For example, the use of several probes does not necessarily result in the same iceball as the union of several single-probe iceballs.
- The optimization algorithm should be improved. For example a parallel implementation would significantly reduce calculation time. So far it takes up to 15 minutes on a 440MHz UltraSPARC-IIi Ultra 10 Workstation with 512MB RAM to optimize the placement of two cryoprobes (presented case). It is also desirable to determine the optimization space Ω (Equation 2) through the anatomical objects in the virtual scene and not explicitly by the radiologists.
- It would be very desirable to combine the tool with a non-rigid registration algorithm ([9]), so that the pre-operative planning can be adjusted to the patient's position in the interventional MR ([11], [10]).

5 Conclusion

The results of this analysis underline the importance of a planning tool having similar functionalities as the presented software. We have shown that this tool has a significant potential to help interventional radiologists before and during image guided interstitial tumor ablation. We have concentrated upon cryotherapy, but other medical applications would be radiofrequency-, laser- or focused ultrasound ablation. The implementation can be done in complete analogy to cryotherapy, and in fact, radiofrequency has already been added to the presented package.

We are convinced that the combination of the presented planning tool with segmentation, non-rigid registration and interventional imaging, has the potential to improve significantly the quality of image guided percutaneous tumor ablation. The general aproach of pre-operative simulation and automated optimization has an even wider range of potential applications. For example for brachytherapy similar approaches have already been developed ([12]) and are used successfully.

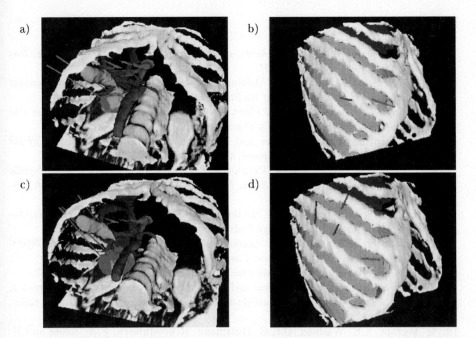

Fig. 3. The virtual cryo-probes, simulating the the needles and the frozen tissue (shown in red), can be set into spatial correspondence to the patient's segmented anatomy and the corresponding intensity scans. The tumor and an additional $0.5cm$ security margin, is shown in brown. The main vessels are in blue, the gallbladder in green and the bones in white. a) and b) show the setup chosen by the radiologists for this specific case, while we present a better setup which would have decreased significiantly the danger of undertreating the cancer in images c) and d).

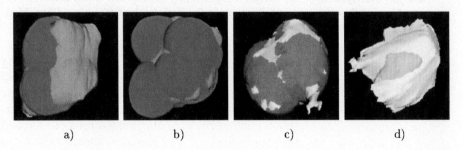

Fig. 4. In a) we show the virtual iceballs (red) according to the real treatment plan. This 3D visualization would indicate undertreatment of the tumor (tumor, including margin, is shown in brown). Volume calculations estimate that only about 63% of the tumor would be killed. In b) we propose a setup that increased the percentage of killed cancerous tissue to about 92% of the whole tumor volume. In image c) we compare our iceball simulation (red) to the real iceball (white and translucent) and see that the simulation is pretty good, though not perfect. In d), we have superimposed the tumor (yellow) and the real iceball at the end of the freezing cycle (white). Even without the 0.5cm margin, the tumor has not been completely frozen.

References

1. F.A. Jolesz; *Image-Guided Procedures and the Operating Room of the Future*; Radiology, 204(3):601-612, May 1997.
2. S.G. Silverman, K. Tuncali, D.F. Adams, et al.; *Percutaneous MR Imaging-guided Cryotherapy of Liver Metastases*; Radiology, 213:122, November 1999.
3. D. Gering; *A System for Surgical Planning and Guidance Using Image Fusion and Interventional MR*; MIT Master's Thesis, 1999
4. J.C. Saliken, J. Cohen, R. Miller and M. Rothert; *Laboratory Evaluation of Ice Formation around a 3-mm Accuprobe*; Cryobiology, 32:285-295, 1995.
5. C.M. Lam, S.M. Shimi and A. Cuschieri; *Thermal Characteristics of a Hepatic Cryolesion Formed in Vitro by a 3-mm Inplantable Cryoprobe*; Cryobiology, 36:156-164, 1998.
6. W.H. Press, S.A. Teukolsky, W.T. Vetterling and B.P. Flannery; *Numerical Recipes in C (Second Edition)*; Cambridge University Press, 1992.
7. W. Schroeder, K. Martin and W. Lorensen; *The Visualization Toolkit: An Object-oriented Approach to 3D Graphics*; Cambridge University Press, 1992.
8. B.B. Welch; *Practical Programming in Tcl and Tk*; Prentice Hall PTR, 1997.
9. S.K. Warfield, M. Kaus, F.A. Jolesz and R. Kikinis; *Adaptive, Template Moderated, Spatially Varying Statistical Classification*; Medical Image Analysis, 4(1):43-55, 2000.
10. M. Ferrant, S.K. Warfield, C.R.G. Guttmann, R.V. Mulkern, F.A. Jolesz and R. Kikinis *3D Image Matching Using a Finite Element Based Elastic Deformation Model*; MICCAI 1999, pp. 202-209.
11. Simon K. Warfield, Arya Nabavi, Torsten Butz, Kemal Tuncali, Stuart G. Silverman, Peter McL. Black, Ferenc A. Jolesz, and Ron Kikinis; *Intraoperative Segmentation and Nonrigid Registration For Image Guided Therapy*; MICCAI 2000, accepted to appear.
12. E.K. Lee, R.J. Gallagher, D. Silvern, Ch.-S. Wuu and M. Zaider; *Treatment Planning for Brachytherapy: An Integer Programming Model, two Computational Approaches and Experiments with Permanent Prostate Implant Planning*; Phys. Med. Biol., 44(1):145-165, 1999.

Volume and Shape Preservation of Enhancing Lesions when Applying Non-rigid Registration to a Time Series of Contrast Enhancing MR Breast Images

C. Tanner[1], J. A. Schnabel[1], D. Chung[1], M. J. Clarkson[1], D. Rueckert[2],
D. L. G. Hill[1], and D. J. Hawkes[1]

[1]Division of Radiological Sciences and Medical Engineering,
The Guy's, King's and St. Thomas' Schools of Medicine and Dentistry,
Guy's Hospital, London SE1 9RT, UK

[2]Department of Computing, Imperial College, London SW7 2BZ, UK

Abstract. In this paper we show first that a non-rigid registration algorithm used to register time-series MR images of the breast, can result in significant volume changes in the region of the enhanced lesion. Since this is physically implausible, given the short duration of the MR time series acquisition, the non-rigid registration algorithm was extended to allow the incorporation of rigid regions. In this way the registration is done in two stages. The enhanced lesions are first detected using the non-rigid registration algorithm in its original form. Secondly, the region of the enhanced lesion is set to be rigid and the new algorithm is applied to integrate this rigid region into the existing registration. By definition, volume and shape will be preserved in this rigid region. Preliminary results of applying this algorithm to 15 datasets are described.

1 Introduction

Breast cancer is the main cause of premature death in women. Screening programs with x-ray mammograms have helped to improve early detection of the disease. However, they are less useful to young women for two reasons. Firstly, the amount of x-ray radiation over the years is in itself a health risk, albeit small. Secondly, the radiopaqueness of dense breast tissue makes the search for a primary tumour difficult, if not impossible [1]. Contrast enhanced magnet resonance (MR) imaging of the breast has shown promising results and is currently under investigation as a screening tool for young women at genetic risk of breast cancer in the UK [2]. In this trial, radiologists evaluate the images on the basis of a standardised protocol, which includes the analysis of Gd-DTPA uptake and washout curves as well as morphological features of suspicious regions. A vital part in this analysis is the production of a subtracted image (post-contrast minus precontrast), which in theory eliminates all unchanged structures while emphasising the enhanced regions. The shape of these enhanced regions are then analysed in an attempt to characterise disease. In practise, however, motion artifacts in the subtracted images can prevent a reliable diagnosis.

Our non-rigid registration algorithm [3] has been shown to significantly reduce the effects of movement artifacts in subtracted contrast-enhanced breast MR images [4] and thus should support the detection of enhanced lesions. Non-rigid transformations can generally change not only the shape but also the volume of image features [5], yet the short MR acquisition time makes volume changes during a scan sequence physically implausible. Studies suggest that breast cancer has a 15-fold increase in elastic modulus in comparison to normal fibroglandular tissue [6]. Therefore, it is reasonable to assume that tumours actually do not change shape during the study, given that little or no force is applied.

Addition of a global volume conserving constraint is physically unrealistic, given that some tissue at the periphery of the field of view may move into or out of the imaged region during the course of the acquisition. While finite element modelling can simulate the elastic tissue properties including volume preservation, it relies on a prior segmentation of the different tissue classes and is extremely computationally expensive. Local volume conservation is a possible solution but will also be computationally very expensive and difficult to implement to achieve good alignment at the skin surface and the chest wall. We therefore propose a two stage algorithm in which we first apply the non-rigid algorithm to provide good quality subtraction images in non-enhancing regions. From these subtracted images we generate a mask which will contain all MR visible enhancing tissue. We then apply a local rigid body constraint to the transformation in this region in order to preserve shape and volume, and combine this with the existing transformation. The final result can then be used for morphometric analysis.

1.1 Related Work

Our non-rigid registration algorithm [3] is based on finding a transformation $\mathbf{T}$ between images A and B, which is composed of a global 3D transformation $\mathbf{T}_g$ and a local transformation $\mathbf{T}_l$, i.e.

$$\mathbf{T}(\mathbf{x}) = \mathbf{T}_g(\mathbf{x}) + \mathbf{T}_l(\mathbf{x}) \tag{1}$$

where $\mathbf{T}(\mathbf{x})$ transforms the position $\mathbf{x} = (x_1 \ x_2 \ x_3)^T$ to $\mathbf{x}'$, which maps the point in the image $A(\mathbf{x})$ into its corresponding point in image $B(\mathbf{x}')$. $\mathbf{T}_g$ is given by the 3D rigid transformation:

$$\mathbf{T}_g(\mathbf{x}) = \begin{pmatrix} \alpha_{11} & \alpha_{12} & \alpha_{13} \\ \alpha_{21} & \alpha_{22} & \alpha_{23} \\ \alpha_{31} & \alpha_{32} & \alpha_{33} \end{pmatrix} \begin{pmatrix} x_1 \\ x_2 \\ x_3 \end{pmatrix} + \begin{pmatrix} \alpha_{14} \\ \alpha_{24} \\ \alpha_{34} \end{pmatrix} \tag{2}$$

where the coefficients α are parameterised by the 6 (or 12) degrees of freedom of a 3D rigid (or affine) transformation. The local transformation $\mathbf{T}_l$ is described by a free-form deformation (FFD) model based on B-splines, which are computationally more efficient than thin-plate splines used in [7]. The object is deformed by manipulating an underlying mesh of control points, yielding a smooth C^2 continuous transformation.

Let $\Omega = \{\mathbf{x} \mid 0 \leq x_i < P_i, \ i \in \{1, 2, 3\}\}$ be the domain of the image volume and let $\Psi = \{\psi_{j_1, j_2, j_3} \mid j_i \in [0, N_i - 1], \ i \in \{1, 2, 3\}\}$ be the mesh of control points with displacements $\phi_{j_1, j_2, j_3}^{(i)}$ and spacing $d_i = \frac{P_i}{N_i - 1}$ for $i \in \{1, 2, 3\}$. $\mathbf{T}_l(\mathbf{x}) = (T_l^{(1)} \ T_l^{(2)} \ T_l^{(3)})^T$ can then be written as:

$$T_l^{(i)}(\mathbf{x}) = \sum_{l=0}^{3} \sum_{m=0}^{3} \sum_{n=0}^{3} B_l(u_1) B_m(u_2) B_n(u_3) \phi_{j_1+l, j_2+m, j_3+n}^{(i)} \qquad i \in \{1, 2, 3\} \tag{3}$$

where $j_i = \lfloor \frac{x_i}{d_i} \rfloor - 1, u_i = \frac{x_i}{d_i} - \lfloor \frac{x_i}{d_i} \rfloor$ for $i \in \{1, 2, 3\}$ and B_l is the l-th basis function of the B-spline defined by:

$$B_0(u) = (1 - u)^3/6 \qquad\qquad B_2(u) = (-3u^3 + 3u^2 + 3u + 1)/6$$
$$B_1(u) = (3u^3 - 6u^2 + 4)/6 \qquad B_3(u) = u^3/6$$

An optimal transformation $\mathbf{T}$ is found by maximising the similarity measure between the overlapping regions of image A and the transformed image of B, i.e $S(A, \mathbf{T}(B))$ by applying a gradient ascent technique. The normalised mutual information (NMI) was chosen as similarity measure, because the contrast enhancement prohibits direct comparison of the intensities, and any dependency on the amount of image overlap is avoided [8], i.e.

$$S(A, B) = \frac{H(A) + H(B)}{H(A, B)} \tag{4}$$

with $H(A)$ being the marginal entropy of image A and $H(A, B)$ being the joint entropy of the images A and B calculated from their joint histogram.

1.2 Objectives

It was shown that non-rigid registration facilitates the detection of enhancing regions [4]. In this paper we investigate to what extent this algorithm changes the volume of an enhanced lesion. This is important since we know that tumours do not shrink or expand during enhancement, and volume changes will produce artifactual changes in the shape of lesions which may hinder attempts to characterise benign and malignant disease. We propose a method to adopt the algorithm to preserve lesion shape and volume.

2 Experiment Design

First we describe our method for assessing volume change. Then we describe an experiment to investigate the effect of control point spacing of the FFD algorithm. In section 2.4 we propose a novel method for conserving volume and shape of enhanced breast lesions.

2.1 Evaluation of Volume Change

First of all, position and extent of the enhancing lesion has to be determined as our ground truth. This has to be based on subtracted contrast enhancing MR images, since malignancies cannot be reliably detected by plain MR imaging [1]. As shown in [4], enhanced lesions can be significantly better detected when non-rigid registration was applied. Thus, our ground truth was determined in accordance with [4] by the following procedure:

1) To reduce motion artifacts, the pre-contrast image ($I1$) was registered to the post-contrast image obtained two minutes after injection ($I2$), resulting in the transformation $\mathbf{T}_{1,2}$.
2) Transformation $\mathbf{T}_{1,2}$ was then used to deform $I1$ to $I1'$ to match $I2$ using tri-linear interpolation.

3) The transformed pre-contrast image $I1'$ was then subtracted from the post-contrast image $I2$ to visualise the region of enhancements.

4) From the subtracted image $I1'$-$I2$, the enhancing lesion was segmented using AN-ALYZE (Biomedical Imaging Resource, Mayo Foundation, Rochester, MN, USA). A mask image M was obtained with intensity values U within the enhancing lesion L and 0 otherwise.

From our ground truth, the volume change induced by each of the tested registrations is evaluated by the following steps:

a) The post-contrast image $I2$ is registered to the pre-contrast image $I1$, resulting in the transformation $\mathbf{T}_{2,1}$.

b) The actual volume change was then determined by transforming the mask image M to M' by $\mathbf{T}_{2,1}$. Tri-linear interpolation was used to transform M to M', so boundary voxels in the transformed image M' will have intensities between 0 and U. Correcting for such partial volume effects, the volume of M' is given by

$$ V_{M'} = \frac{V_{voxel(M')}}{U} \sum_{\mathbf{i} \in M'} M'(\mathbf{i}) \qquad (5) $$

where $V_{voxel(M')}$ is the volume of a voxel in M' and $M'(\mathbf{i})$ is the intensity of voxel $\mathbf{i}$ in M'. The volume change in percent is then

$$ \Delta V = 100 \times \frac{V_{M'} - V_M}{V_M} . \qquad (6) $$

The dependence of the volume change on our ground truth is investigated in section 3.3. Note that $\mathbf{T}$ is in general not invertible since folding leads to violation of the one-to-one mapping property. The ground truth was therefore determined by using $\mathbf{T}_{1,2}$ and not the inverse of $\mathbf{T}_{2,1}$. For the registrations in accordance with [4], the average of the magnitude of the displacement error over all voxel positions in $I1$ is 0.98mm and in the lesion L is 0.81mm. The resultant displacement vector over all voxel positions in $I1$ is (-0.02mm 0.20mm 0.08mm)^T and in L is (-0.17mm -0.19mm -0.03mm)^T.

2.2 Data

The MR scans were acquired with a Siemens 1.5T Impact MR system using a fast gradient echo 3D sequence with TR=12 ms, TE=5 ms, flip angle=35°, FOV=350 mm and axial slice orientation. Five post-contrast scans were obtained after injection of 0.2 mmol Gd-DTPA/kg of body weight at temporal intervals of 1 minute and a voxel size of $1.37 \times 1.37 \times 4.2$ mm^3.

For each registration method, the volume change was evaluated on 15 pairs of pre- and post-contrast images selected from the 54 pairs of pre- and post-contrast breast images used in [4]. Datasets were selected that encompassed a wide range of lesion shapes and sizes.

2.3 Variation of Control Point Spacing

The spacing of the control points determines the flexibility of the non-rigid registration. In [9], the registration results for control point spacing of 20, 15 and 10mm were compared in terms of the squared sum of intensity differences (SSD) and correlation coefficient (CC) of the intensities of the image pair after registration. This showed that decreasing the control point spacing reduced the registration error (excluding the tumour region).

In this paper we investigate the effects on the volume change of the lesion for a control point spacing of 10, 15, 20 and 25mm.

2.4 Coupling of Control Points

The main idea of restricting control points to have the same displacement (i.e. they are coupled) is to produce regions which are only allowed to transform rigidly within the FFD. When control points are coupled, then any region, whose displacement is only controlled by these control points, will have no volume change. To be precise, let an image region $R \subseteq I1$, be under the influence of coupled control points, i.e. for $x \in R$, $\phi_{j_1+l,j_2+m,j_3+n}^{(i)} = K^{(i)}$, a constant, for $l, m, n \in \{0, 1, 2, 3\}$ and $i \in \{1, 2, 3\}$. Then (3) simplifies to

$$T_l^{(i)}(\mathbf{x}) = K^{(i)} \underbrace{\sum_{l=0}^{3} \sum_{m=0}^{3} \sum_{n=0}^{3} B_l(u_1) B_m(u_2) B_n(u_3)}_{1} = K^{(i)}, \quad i \in \{1, 2, 3\}. \quad (7)$$

Thus, the local transformation $\mathbf{T}_l(\mathbf{x})$ for $\mathbf{x} \in R$ is given by the translation vector $(K^{(1)} \ K^{(2)} \ K^{(3)})^T$, which by definition does not introduce shape or volume changes in R.

Let Ψ_{coup} be the set of coupled control points and Ψ_{blend} be the set of control points updated for blending the coupled region into the surrounding. Then the coupling of control points can be incorporated in the registration in following way:

- find $\mathbf{T}(\mathbf{x})$ as in [4] to detect an enhancing lesion L
- determine bounding box $D = [\mathbf{v}, \mathbf{w}]$ of L such that $L \in D$ and D has minimal volume
- include in Ψ_{coup} the adjacent two control points in each direction of D, i.e.
  $\Psi_{coup} = \{\psi_{j_1,j_2,j_3} \mid j_i \in [max(0, \lfloor \frac{v_i}{d_i} \rfloor - 1), min(N_i - 1, \lceil \frac{w_i}{d_i} \rceil + 1)], i \in \{1, 2, 3\}\}$
- include in Ψ_{blend} the adjacent two control points in each direction of Ψ_{coup}, i.e.
  $\Psi_{blend} =$
  $\{\psi_{j_1,j_2,j_3} \mid j_i \in [max(0, \lfloor \frac{v_i}{d_i} \rfloor - 3), min(N_i - 1, \lceil \frac{w_i}{d_i} \rceil + 3)], i \in \{1, 2, 3\}\} - \Psi_{coup}$
- the coupled control point displacements Φ_{coup} are initialised to their mean
- using gradient asscent, the similarity measure (4) is maximised with respect to Ψ_{coup} and Ψ_{blend}, while enforcing the displacement vectors Φ_{coup} to be the same.

This approach only allows a coupled translation. This should be a sufficient approximation for the local transformation of a rigid body, given that the global rotations are covered by $\mathbf{T}_g$. In further work we will investigate whether a local rigid registration (similar to [10]) combined with a FFD improves the registration. Figure 1 shows a 2D example of coupling control points to prevent shape and area change locally.

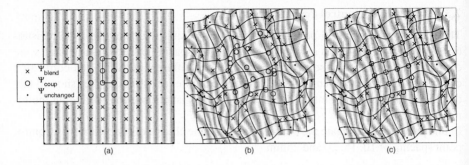

Fig. 1. 2D example of control point coupling in order to prevent shape and area change within the box depicted in (a). The global transformation consists of a 15° rotation. The FFD is based on a control point spacing of 9 pixels. The control point displacements are generated randomly from a normal distribution with a standard deviation of 5 pixels. The deformed grids in (b) and (c) show the transformed position of a mesh drawn through the control points in (a). Note that the deformed grids do not necessarily pass through the displaced control points since the FFD is interpolated using B-spline approximators. (a) sinusoidal source image and the initial distribution of control points, (b) the transformed source image using bi-linear interpolation and the deformed grid and (c) the transformed source image and deformation after coupling the control points Ψ_{coup}. In (c), the image within the corresponding region to the marked box in (a) has not changed shape, while this is not true for (b).

3   Results

In this section we present results of our investigation of the control point spacing in the FFD and of our novel approach to volume and shape preservation. Finally, the dependency of the volume change on the initial lesion segmentation is evaluated.

3.1   Variation of Control Point Spacing

Test were conducted using our original non-rigid registration algorithm with a control point spacing of 10, 15, 20 and 25mm. The summarised results are given in Table 1 and are depicted in Figure 2. Volume change is measured as a percentage relative to the initial volume as defined by (6). The mean volume change over the 15 datasets lie within ±0.3% for all tests with global rigid registration, indicating no global bias towards expansion or contraction. However, volume changes occur within a range of [−17%, 33%] for the non-rigid registration configuration from [4], i.e. a FFD of 10mm after a global affine registration. The standard deviation of the volume change increases as the FFD node spacing decreases from 25mm down to 10mm. This indicates lager changes in volume the smaller the node spacing of the FFD. The trend of the similarity measures shows that at finer node spacing the two images are more similar. The values of the similarity measures are tabulated as negative squared sum of intensity difference (-SSD), correlation coefficient (CC) and normalised mutual information (NMI). This trend is in concordance with the results from [9] and can be explained by the reduced flexibility of the non-rigid registration due to a lower number of control points. The plot in Figure 2 shows that the volume changes induced by the algorithm are not correlated with lesion size.

Registration	Volume Change in %				Mean Similarity		
Method	Min	Max	Mean	Std	-SSD	CC	NMI
affine + FFD 10mm	-16.83	32.05	1.2919	12.0648	-2146.46	0.9847	1.2018
rigid + FFD 10mm	-18.91	17.59	0.1636	10.2618	-2162.61	0.9845	1.2018
rigid + FFD 15mm	-5.77	4.19	0.0072	3.2029	-2430.76	0.9823	1.1985
rigid + FFD 20mm	-2.66	7.29	0.2315	2.3736	-2555.84	0.9812	1.1965
rigid + FFD 25mm	-2.86	1.57	-0.2301	1.4394	-2623.89	0.9804	1.1954

Table 1. Results for the original non-rigid registration algorithm with a control point spacing of 25, 20, 15 and 10mm after global affine or rigid registration. The volume change as well as the similarity between the two images (measured excluding the enhanced lesion) decreases with the increase of node spacing. The evaluated similarity measures are negative squared sum of intensity difference (-SSD), correlation coefficient (CC) and normalised mutual information (NMI).

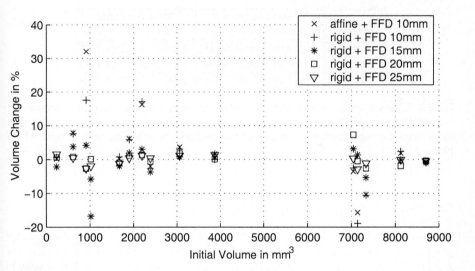

Fig. 2. The volume changes for the 15 lesions and the different methods applied. There is no correlation between initial volume and volume change.

3.2 Coupling of Control Points

The result of the non-rigid registration with control point spacing of 10mm was chosen as the initial transformation, due to its ability to significantly reduce the effects of motion artifacts [4]. This transformation was then optimised for shape and volume preservation of the lesion while blending the lesion region with its surrounding by having two sets of control points, Ψ_{coup} and Ψ_{blend} respectively. As described in Section 2.4, the control points from Ψ_{coup} lie in a dilated box bounding the lesion and are restricted to have the same displacement vectors. Ψ_{blend} covers the zone that is influenced by Ψ_{coup} but which is not in Ψ_{coup}. The results of using coupled control points are shown in Table 2 and Figure 3 in comparison to the global rigid and affine registration. Clearly, the coupling of control points is as good as the global rigid registration in preventing volume changes within the lesion. The observed volume changes indicate the measurement errors, since the rigid transformation is by definition volume preserving. Global affine transformation causes small volume changes, but improves the registration insignificantly in comparison to global rigid registration as shown in [4]. It also leads to higher volume changes when applied before the FFD with 10mm and is therefore not recommended for registrations where volume should be preserved. The mean similarity achieved by coupled control points indicate a registration error similar to a global rigid registration followed by a non-rigid registration with 15mm control point spacing. However, the true registration quality needs to be evaluated in an observer study. The effects of the different registrations on one lesion can be seen in Figure 5.

Registration	Volume Change in %				Mean Similarity		
Method	Min	Max	Mean	Std	-SSD	CC	NMI
coupled	-0.03	0.03	0.0016	0.0155	-2400.42	0.9825	1.1989
rigid	-0.11	0.03	-0.0039	0.0340	-3469.28	0.9725	1.1895
affine	-0.61	1.19	0.0842	0.5230	-3106.54	0.9759	1.1904

Table 2. Results when the control points around the enhanced lesion are coupled in comparison to the global rigid and affine registration. The coupled control points preserve volume within the lesion as good as rigid registration while the mean similarity is within the range of rigid + FFD 15mm.

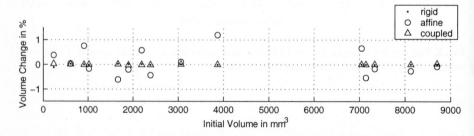

Fig. 3. The volume change of the 15 lesions for global affine and rigid registration and non-rigid registration with coupled control points within the extended region of the lesion.

3.3 Robustness of Results

To analyse the dependency of the previous results on the exact segmentation of the lesions, tests were performed with dilated lesion masks in order to obtain volume change profiles similar to [11]. The mask M was blurred with Gaussian filters with standard deviation of $0.2, 0.4,...1.8$. Then, all intensities greater than zero were set to U. Finally, the dilated masks were transformed using the registrations results from before without changing the sets of coupled and blending control points. Figure 4 shows that for a 5-fold increase of lesion volume, the standard deviation of volume change induced by the algorithm is almost constant for the FFD with node spacings of 15mm and above. There is a small, but significant, reduction in volume change of between 4 and 5% from a FFD of 10mm.

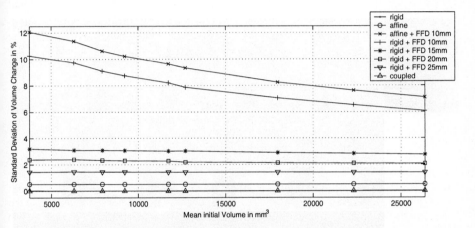

Fig. 4. Dependency of the volume change on the lesion segmentation. The segmented lesion is dilated using Gaussian blurring with a standard deviation of $0, 0.2,...,1.8$ and all intensities greater than zero are set to belong to the lesion. Although the mean initial volume increases 5-folded, the volume change for these new segmentations show clearly the same ranking for the different registration methods. This indicates a weak dependency of the volume change on the lesion's segmentation.

4 Conclusions

In this paper we have shown that the non-rigid registration algorithm assessed in [4] can lead to significant local volume changes. We presented initial results showing that these volume changes are more severe the finer the node spacing in the FFD algorithm that we use.

We proposed a strategy in which a non-rigid registration with 10mm control point spacing is first undertaken. This reduces subtraction artifacts in non-enhancing regions significantly [4] . From the subtracted image the local transformation for the region of the enhancing lesion is then constrained to a translation only. The transformation from the initial non-rigid registration is then optimised to account for this constraint. The final result guarantees volume and shape preservation in the region of the enhanced lesion, within the limits of the interpolation method used in the transformation, while still showing the lesion in the context of its surrounding tissue. In future work, we will extend and improve the proposed method of volume and shape preserving constraints within the framework of a generalised FFD, investigate the rigidity assumption of lesions in the context of contrast enhanced MR breast imaging and undertake an observer study to assess the effect of volume and shape preservation constraints on the quality of the subtracted images. The resulting enhancing region can then be passed on for further morphological analysis in attempt to better distinguish benign from malignant disease.

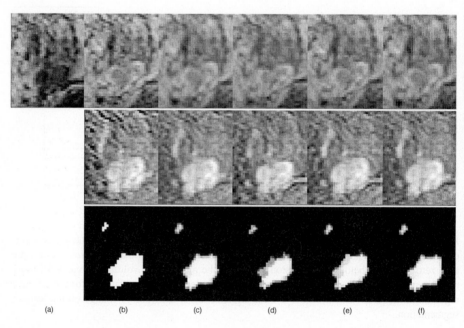

(a)   (b)   (c)   (d)   (e)   (f)

Fig. 5. This figure shows the results of registering the post-contrast image to the pre-contrast image of one patient in 4 different ways. The images display the region of the enhanced lesion on one slice. The top row shows (a) the pre-contrast image, (b) the post-contrast image and the transformed post-contrast images after (c) rigid registration, (d) non-rigid registration with a control point spacing of 10mm as in [4], (e) non-rigid registration with a control point spacing of 25mm after a global rigid registration and (f) non-rigid registration with coupled control points within an extended region of the lesion. The middle row shows the difference images after subtracting the pre-contrast image from the images in the top row. The bottom row pictures the initial and the transformed masks. A clear shrinkage and shape change for the transformed mask of (d) can be observed. The volume changes are (c): 0.00% (d): -16.28% (e): -3.86% (f): 0.00%.

5 Acknowledgments

We would like to thank the Engineering and Physical Sciences Research Council (EP-SRC) for the financial support given by the EPSRC grant GR/M52779 and GR/L08519 and Dr. Sheila Rankin, Dr. Erika Denton and the Radiographers at Guy's and St. Thomas NHS trust for assistance in collection of the data. This work is part of a joint project with the Institude of Cancer Research and we are very grateful to Prof. Martin Leach, Dr. Carmel Hayes and Dr. Andreas Degenhard for useful discussions

References

1. S. Heywang-Köbrunner and R. Beck, *Contrast-Enhanced MRI of the Breast*. Berlin, Germany: Springer-Verlag, 2nd ed., 1995.
2. M. O. Leach, "Assessing Contrast Enhanced MRI as a Method of Screening Women at Genetic Risk of Breast Cancer: Study Design, Methodology and Analysis," in *Proceedings of the Sixth Scientific Meeting of the International Society for MR in Medicine, Sydney, Australia*, p. 226, 1998.
3. D. Rueckert, L. I. Sonoda, C. Hayes, D. L. Hill, M. O. Leach, and D. J. Hawkes, "Non-rigid Registration using Free-Form Deformation: Application to Breast MR Images," *IEEE Transactions on Medical Imaging*, vol. 7, pp. 1–10, August 1999.
4. E. R. E. Denton, L. I. Sonoda, D. Rueckert, S. C. Rankin, C. Hayes, M. O. Leach, and D. J. Hawkes, "Comparison and Evaluation of Rigid, Affine, and Nonrigid Registration of Breast MR Images," *Journal of Computer Assisted Tomography*, vol. 5, pp. 800–805, May 1999.
5. E. A. Stamatakis, J. T. L. Wilson, and D. J. Wyper, "The Effect of Non-Linear Image Registration on Cerebral Lesions," in *Medical Image Understanding and Analysis, Oxford, UK*, pp. 21–24, 1999.
6. A. Sarvazyan, D. Goukassian, E. Maevsky, and G. Oranskaja, "Elastic Imaging as a new Modality of Medical Imaging for Cancer Detection," in *Proceedings of the International Workshop on Interaction of Ultrasound with Biological Media, Valenciennes, France*, pp. 69–81, 1994.
7. C. R. Meyer, J. L. Boes, B. Kim, P. H. Bland, K. R. Zasadny, P. V. Kison, K. Koral, K. A. Frey, and R. L. Wahl, "Demonstration of Accuracy and Clinical Versatility of Mutual Information for Automatic Multimodality Image Fusion using Affine and Thin-plate Spline Warped Geometric Deformations," *Medical Image Analysis*, vol. 1, no. 3, pp. 195–207, 1997.
8. C. Studholme, D. L. G. Hill, and D. J. Hawkes, "An Overlap Invariant Entropy Measure of 3D Medical Image Alignment," *Pattern Recognition*, vol. 32, pp. 71–86, 1999.
9. D. Rueckert, C. Hayes, C. Studholme, P. Summers, M. Leach, and D. J. Hawkes, "Non-rigid Registration of Breast MR Images using Mutual Information," in *Medical Image Computing and Computer-Assisted Intervention, Cambridge, USA*, pp. 1144–1152, 1998.
10. J. Little, D. L. G. Hill, and D. J. Hawkes, "Deformations Incorporating Rigid Structures," *Computer Vision and Image Understanding*, vol. 66, pp. 223–232, 1997.
11. J.-P. Thirion and G. Calmon, "Deformation Analysis to Detect and Quantify Active Lesions in 3D Medical Image Sequences," *INRIA Report*, vol. 3101, February 1997.

Local Integration of Commercially Available Intra-operative MR-Scanner and Neurosurgical Guidance for Metalloporphyrin-Guided Tumor Resection and Photodynamic Therapy

David Dean,[1] Jeffrey Duerk[2], Michael Wendt[3], Andrew Metzger[1], Lothar Lilge[4], Brian Wilson[4], Victor Yang[4], Warren Selman[1], Jonathan Lewin[2], Robert Ratcheson[1]

[1] Department of Neurological Surgery, [2] Department of Radiology, Case Western Reserve University/University Hospitals of Cleveland, 10900 Euclid Ave.
Cleveland, OH 44106 USA
{dxd35, jld3, akm8, wrs, jsl3, rar}@po.cwru.edu
http://neurosurgery.cwru.edu
http://www.uhrad.com

[3] Magnetic Resonance, Siemens, 186 Wood Ave. South
Iselin, New Jersey 08830 USA
michael.wendt@sms.siemens.com

[4]Department of Medical Biophysics, University of Toronto, 610 University Ave., Rm. 7-416
Toronto, Ontario M5G 2M9 Canada
{llilge, wilson, xyang}@oci.utoronto.ca
http://medbio.utoronto.ca

Abstract. We have locally integrated a commercial intra-operative MR-scanner and neurosurgical guidance system in order to conduct a clinical trial testing for improvement in glioma resection following administration of either of two metalloporphyrin drugs, Xcytrin™ or Lutrin™ (Pharmacyclics Inc., Sunnyvale, CA). Like other porphyrins, these two drugs have been shown to bind preferentially to tumor. The metal in Xcytrin is gadolinium. It should enhance intra-operative MR-scans taken during tumor resection. This would overcome problems of non-specific MR-enhancement caused by intra-operative contrast leakage from the vascular compartment. Both drugs provide fluorescence contrast in the presence of 450 nm wavelength light. Intra-operative fluorescence contrast should facilitate Lutrin photodynamic therapy administration to otherwise invisible glioma residual in the walls of the resection cavity. The metal in Lutrin is Lutetium. When illuminated with 732nm wavelength light, Lutrin causes tumor death via release of singlet oxygen (i.e., photodynamic therapy).

1 Introduction

Commercially available intra-operative neurosurgical guidance systems provide for initial tool positioning and craniotomy site placement based on a pre-operative 3D MR or 3D CT image. Currently, neurosurgical image guidance is based primarily on pre-operative images used to optimize craniotomy position and the trajectory used to reach the lesion. Most of the currently available systems use frameless CT and/or MR-visible fiducial markers that are located pre-operatively in the operative suite with a 3D localizer in order to register the pre-operative image to the patient. Once registered, the image may be used to choose a craniotomy (skull opening) site near the lesion and initial surgical tool orientation. Subsequent to surgical entry, the occurrence of "brain shift," surgically-induced swelling, and drop in CSF pressure conspire to reduce the utility of pre-operative images.

Intra-operative MR offers the possibility of image update without interrupting the surgical procedure to take the patient to the Radiology Department. Indeed, intra-operative neurosurgical MR devices are most useful if they do not require interruption of anesthesia. Intra-operative MR-guided neurosurgical procedures began in 1994 with the installation of the GE Medical Systems Signa SP^{TM} "double doughnut," a mid-level field (0.5 Tesla) magnet, at Brigham and Womens Hospital in Boston [7]. Because of the groundbreaking nature of this group's activities, it has developed in-house guidance software, 3D Slicer. The 3D Slicer is available for distribution, but to our knowledge is not FDA-approved.

High field (1-2 Tesla) intra-operative MR-scanners have also been utilized [22], [9]. Intra-operative use of these devices require that all surgical instruments be MRI compatible. The Siemens Open Viva^{TM}, a 0.2 Tesla low field magnet, has been installed and utilized for neurosurgical applications at our site, University of California Medical Center, Los Angeles, and the University Hospitals in Heidelberg and Erlangen, Germany. This device provides excellent images for stereotactic guidance [13] and quantitation [26]. Indeed, Wirtz et al. [26] have found that when this device has been available, brain "tumor removal and [post-operative] survival times were increased significantly." Specifically, they find that intra-operative image updates can compensate for brain shift, and furthermore, unintended residual brain tumor was reduced from 70%, historically, to 30.5% in 242 cases utilizing intra-operative MR-assisted resection.

Wirtz et al.'s [26] findings of improved brain tumor resection efficacy associated with intra-operative MR guidance were tempered when they confirmed an observation made by other groups [12], [1], [6]: in all 66 operations with enhancing lesions, and 20 of 29 with non-enhancing lesions, varying levels of intra-operative enhancement was "induced by the surgical procedure itself". That is, intra-operatively administered intravenous contrast media (gadolinium) leaked into tissue spaces (e.g., parenchymal, CSF, or the resection cavity) where it did not appear on pre-operative scans.

The need for tumor-specific guidance and therapies has motivated us to propose clinical trials of two new investigational drugs, both of which are based on the metalloporphyrin substrate referred to as texaphyrin [28], [15]. We hypothesize either drug would improve intra-operative tumor localization, resection, and/or ablation. These localization methods utilize MR-scans or controlled light emissions. The radio frequency and optically shielded intra-operative suite at University Hospitals of Cleveland provides an optimal setting in which to test these new therapies. In order to improve intra-operative tumor localization in this setting, we have integrated an FDA-approved intra-operative MR system, the Siemens (Munich, Germany) OpenViva^{TM}, and Medtronic's (Minneapolis, MN, USA) neurosurgical guidance system. It is expected that when used with this equipment both drugs would provide tumor localization options that overcome the difficulty of leakage of vascularly administered gadolinium during intra-operative MR-guided tumor resection or ablation.

The first two years of the proposed clinical trial would use the MR-visible porphyrin-based drug, Xcytrin^{TM} (Pharmacyclics Inc., Sunnyvale, CA). It is a porphyrin bound to gadolinium, and has been shown to preferentially bind (specificity) to tumor. Indeed, Phase I clinical trials have already determined details of the toxicity and specificity of this drug [18]. Since Xcytrin is administered pre-operatively, and is expected to provide durable enhancement intra-operatively, it obviates the need for additional intravenous contrast administration. It should prove useful for intra-operative localization of residual tumor and eliminate the occurrence of enhancement induced by surgical disruption of the blood-brain barrier. It is our expectation that resection efficacy will improve if this phenomenon is curtailed. Moreover, it is possible that Xcytrin may provide more contrast enhancement of tumor than is seen following standard vascular gadolinium administration.

The third year of the proposed clinical trial would use a lutetium-bound porphyrin, Lutrin™ (Pharmacyclics Inc., Sunnyvale, CA), for fluorescence-guided localization of tumor. This drug has also been shown to bind specifically to tumor. However, the proposed phase I trial of this drug is necessary [17]. Therefore, we propose to test the specificity of this drug for brain tumor and its sensitivity to photoactivating wavelengths of light (i.e., photodynamic therapy) on resection samples following removal from on study patients. If these tests demonstrate high specificity and sensitivity, a subsequent clinical trial would be proposed to test Lutrin-mediated photodynamic therapy for brain tumors (see Section 6).

2 Metalloporpyrins for Tumor Therapy

The use of metalloporphyrin drugs to locate and ablate cancer cells has been made possible by the work of many photochemists, surgeons, and basic scientists who have pursued the one hundred year old observation that porphyrin molecules bind with high specificity to tumor. There is general speculation on the biochemical mechanism causing tumor's highly selective uptake of porphyrins, but the specific mechanism is unknown. Shibata et al. [20] present evidence that porphyrin uptake by glioma cells is greatest during the mitotic portion of the cell life cycle. Discussion of basic porphyrin structure, including metalloporphyrins such as Xcytrin, and the use of porphyrins for photodynamic therapy is presented in Section 6.

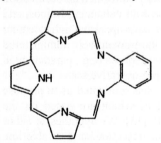

Fig. 1. Texaphyrin.

Porphyrins are intermediate molecules that serve as ligands capable of binding heme, which functions in red blood cells in the mechanism of oxygen distribution. Later, the heme group is recycled in the liver when the red blood cells become too worn. When blood transfusion is used to overcome hemoglobin synthesis deficiency the result is accumulation of non-excretable iron in the liver. The same effect is to be expected with metalloporphyhrins as they are broken down and the associated metal is deposited in fatty organ tissue, rather than excreted. The metal deposited in the case of Xcytrin is gadolinium, and in the case of Lutrin it is lutetium. While the risk of side effects from metal deposition is small with small doses, it would be desirable to determine the minimum Xcytrin or Lutrin dose necessary for MR tumor contrast, photodynamic therapy, and/or fluorescence tumor contrast.

Metalloporphyrins are more hydrophilic than hematoporphyrins (see Section 6). Thus, they move efficiently into plasma and subsequently to tumor. They also wash quickly from the circulation, thus quickly reducing skin and eye photodynamic sensitivity. Prior to 1988, none of the previously created expanded porphyrin systems formed stable complexes with large metal cations, a level of non-lability not equaled in any lanthanide(III) porphyrin complexes [19]. Sessler *et al.* [19] first demonstrated the binding of "Texas-sized" (texaphyrin) porphyrins to lanthanide metal ions, showing that the central binding cavity is expanded 20% over metal-free porphyrins from 4 Nitrogen atoms in hematoporphyrin (i.e., naturally occurring porphyrin) to 5 in texaphyrin (Figure 1). The additional equatorial binding region allows for free interaction of the axial poles of the metal with other molecules. This is important for lanthanide (III) metals such as gadolinium(III) and lutetium(III) which remain free to bind with readily disassociated ligands such as chloride, acetate, and nitrate found at the target cell surface. In distinction to other metalloporphyrins, texaphyrin complexes are more difficult to oxidize and easier to reduce. It is suspected that this enhances their specificity for tumor.

3 Intra-operative MR Xcytrin-based MR Guidance

Why MR and why Xcytrin? The primary advantage of MR over CT is that it usually provides better definition of internal inhomogeneous soft tissue structures like the brain. This is due to the spatial distribution of the MR signal, which is partly determined by the nuclear density of the object imaged. It is also a function of parameters such as the relaxation times T1 and T2 and of fluid flow at each point. MR imaging of the brain primarily measures the distribution and properties of hydrogen, the most abundant element in tissues of living systems. While there can be artifacts from ferrous objects, MR radio frequency electromagnetic radiation

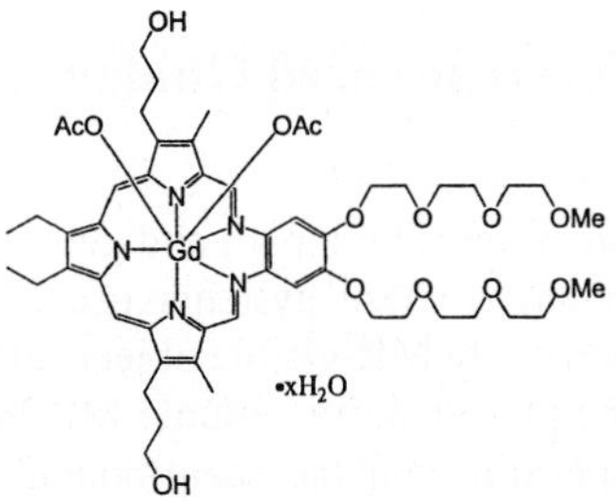

Fig. 2. Xcytrin^{TM} is also referred to as Gadolinium(III)-Texaphyrin.

penetrates boney material like the skull, enabling the brain to be imaged without significant attenuation. Unlike CT, through instrument design and operator control, the influence of all of these parameters on the MR image may be emphasized to give discrimination between tissues with similar electron density. Relevant examples include, first, the strong discrimination seen between gray and white matter on the basis of their significantly different T1 values, and, second, the good contrast commonly obtained between tumor, edema, cerebrospinal fluid, gray matter, and white matter in T2 weighted images.

Damadian [4] found that values of proton relaxation times, T1 and T2, in neoplastic tissue are significantly larger than those in normal brain tissue. Differential proton relaxation times thus potentially provide a valuable neurosurgical pathologic discriminator. However, the shift to MR imaging of brain tumors from CT occurred relatively slowly. It first required demonstration of sufficient MR spatial fidelity [5]. It is generally thought that CT underestimates brain tumor volume, but studies where both CT and MR were available for the same patient at the same time have shown that both modalities have the possibility of detecting a larger volume [23]. Perfusion of the tumor is the basis for gadolinium enhancement, which roughly corresponds to tumor cell proliferation [24]. Therefore, and unfortunately, not all tumors provide good contrast with surrounding tissue following gadolinium administration. It enters the microvasculature of the tumor where the blood brain barrier has been broken, but has no unique specificity for tumor tissue.

Xcytrin^{TM} (Figure 2), referred to as PCI 0120 (Pharmacyclics Inc., Sunnyvale, CA), is a complex of texaphyrin and a trivalent gadolinium. The latter paramagnetic molecule results in Xcytrin's high MR visibility. Selective uptake of Xcytrin by tumor has been associated with long-lasting MR enhancement. This was first observed *in vitro* in isolated tumor cells, murine mammary carcinoma, injected in rat and rabbit thigh and liver, respectively [29]. In clinical trials of radiation therapy sensitization effects, MR enhancement of brain tumor has been reported up to 8 weeks following Xcytrin administration [25]. Deposition of trace amounts of the metallic gadolinium Xcytrin core are also observed in healthy tissues (e.g., kidney, liver, gastrointestinal tract), therefore Xcytrin is not appropriate as a general diagnostic contrast agent. However, if significant therapeutic benefit is expected, such as enhanced glioma resection or radiation therapy sensitization, Xcytrin administration is justified.

4 Intra-operative MR Benefits From Xcytrin-based Guidance

Commercially available neurosurgical navigation systems help plan and guide surgery based on a pre-operative 3D MR image. Primarily, these systems use so called "frameless" fiducial imaging markers, which are usually MR-visible objects affixed to the patient's scalp. Software is used to match the pre-operative volume MR image to the patient on the operating room table by superimposing the scalp-bound fiducial imaging markers seen in the image to their location seen in a coordinate system established in the operating room. The "operating room" location of the markers is obtained by 3D localizer once the patient's head position is fixed for surgery. Surgical entry location and tool trajectory may then be planned and tracked in software on the guidance computer. This procedure assumes that the pre-operative image on the computer is true to the patient's anatomy on the operating room table. Since this condition is known to be violated once the procedure begins, the use of conventional neurosurgical guidance systems is often limited to procedure initiation.

A significant change in spatial relationships occurs following surgical entry. It is referred to as "brain shift." It results from: cerebral edema, release of cerebrospinal fluid (CSF), tissue removal, and gravitationally induced sagging. These factors conspire to reduce the utility of pre-operative images once the procedure has begun [14].

Intra-operative update of the pre-operative MR guidance image can indicate the serial change in lesion and critical brain structure location during surgery. However, as noted in the introduction the value of these images diminished by leakage of intra-operatively administered contrast media can leak from the vascular compartment into other tissue spaces. It is our expectation that resection efficacy will improve if this phenomenon is curtailed. Moreover, it is possible that tumors that do not enhance following vascular gadolinium administration will do so following an Xcytrin regimen. If cases where that occurred we would expect a significant increase in resection efficacy and therapeutic benefit.

5 Intra-operative Metalloporphyrin-based Fluorescence Guidance

Like other porphyrins, both Xcytrin and Lutrin are excited to fluorescence when exposed to light emissions in the 450-480 nm range. High specificity for fluorescent biolocalization of tumor using both drugs has been shown in mice [27], [Woodburn, unpublished data].

The fluorescence of several other photosensitizing porphyrin drugs have been studied in the C6 and 9L mouse and rat glioma models following administration of 3 non-texaphyrin porphyrin drugs: HpD [3], Photofrin [3], [11], ALA [11], and Boronated Protoporphyrin [2]. Stummer et al. [21] found human glioma resection aided by the presence of a fluorescing porphyrin drug, ALA, where "for seven of nine patients, visible porphyrin fluorescence led to further resection of the tumor". The usefulness of Xcytrin and Lutrin for fluorescence contrast localization depends on the specificity of these two drugs for glioma; a determination that is a major goal of the proposed clinical trial.

The hardware necessary to detect residual glioma via fluorescent contrast has been tested in a clinical trial of another porphyrin, Photofrin, at the University of Toronto. The image in Figure 3 was taken after the neurosurgeon had stated that he had completed the brain tumor resection based on information available to him from white light visual inspection of the resection cavity through the operative microscope. Operative white light was then turned off to capture this image through a novel long working distance camera which applies fluorescence excitation light of 450-480 nm wavelength. After capture of the fluorescent light photograph, the operative room white lights are then restored. The fluorescent portion of the image (lighter color in Figure 3) is then overlain on the white light image the surgeon sees in a stereotactic operative microscope. This provides resection guidance for otherwise invisible residual tumor and the surgery continues. It should be noted that because Photofrin has a very low extinction coefficient in the white light used by the surgeon, no light shielding is needed against the 600-650nm range of the Q band that would photoactivate Photofrin and cause tissue necrosis. Thus, it is possible to bring about Photofrin fluorescence without causing inadvertent photodynamic therapy.

We propose to verify the same scenario (i.e., intra-operative fluorescent localization without photoactivation) for Lutrin. This is not necessary for Xcytrin as it cannot photoactivate. If Xcytrin were utilized, we expect resection completion would be additionally sped by detection of larger pieces of residual tumor, including their entire depth, on intra-operative MR images taken after the surgeon had lost all white light visual cues for tumor location in the operative microscope. The use of intra-operative MR-scanning Lutrin or Xcytrin to locate residual tumor would allow the surgeon to complete the resection going back and forth between updated intra-operative MR-scan, white light (for resection), and fluorescent light (to survey for residual tumor). The use of fluorescent contrast for tumor localization could occur as often as desired as long as photo-bleaching of the metalloporphyrin does not occur.

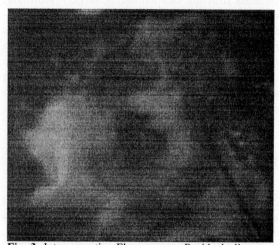

Fig. 3. Intra-operative Fluorescence: Residual glioma at the innermost margin of the resection cavity. This tumor was visualized following 5mg/kg Photofrin intravenous injection and excitation to fluoresce. After the fluorescent picture is taken, the fluorescent portion (lighter color) of this image is superimposed over white light view on screen in the operating room microscope.

As with photodynamic therapy (see Section 6), dosimetry is essential for use of fluorescent activity. At a measurable point the photosensitizer will become "bleached" and fluorescence ceases (i.e., it converts to its inactive form). Fluorescence properties would be a function of the sum photo-activation (fluorescence) and the light intensity used throughout the operative procedure. Stummer et al. [21] found that when using ALA5 as a tumor marker: "Under operating light conditions, fluorescence decayed to 36% in 25 minutes for violet-blue light and in 87 minutes for white light." Photobleaching rates must known a priori; they are clearly a function of the photosensitizer used.

6 Intra-operative MR-guided Photodynamic Therapy

Lutrin (Figure 4), referred to as PCI (Pharmacyclics Inc., Sunnyvale, CA) 0123, has been designed to overcome a number of the limitations of other second-generation photodynamic therapy photosensitization agents. It is a tripyrrolic pentadentate aromatic complex. Its lowest activation wavelength, λ=730nm (range 730-770), allows deep tissue penetration [19]. The administration of PDT activation light delivery and dosimetric accuracy should benefit from intra-operative MR guidance and light shielding of the intra-operative MR suite at University Hospitals of Cleveland. Intra-operative fluorescence of Lutrin should also assist in verifying the location of regions targeted for Photodynamic therapy. It is noteworthy that photons and the optical tools for Lutrin-mediated fluorescent localization and/or photodynamic therapy of brain tumors are inherently MR compatible.

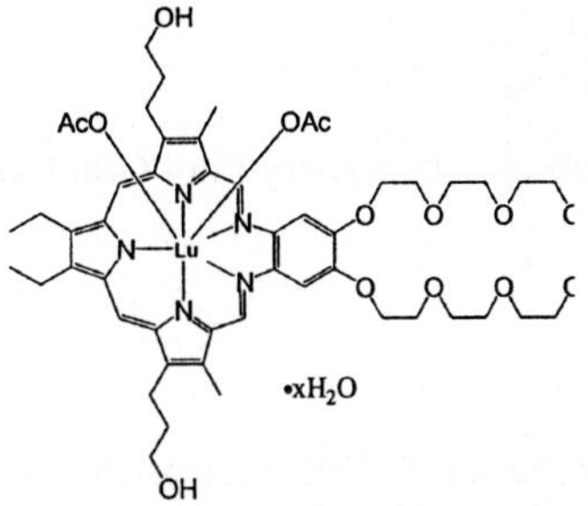

Fig. 4. Lutrin™ is also referred to as Lutetium(III)-Texaphyrin.

Lutrin's high water solubility and aqueous formulation allow it to clear quickly from plasma and other tissues where irreversible toxicity may be a concern and inadvertent photodynamic therapy sensitization is not desired [27]. When irradiated at 732nm Lutrin has a high singlet oxygen ($^1O_2$) yield of >20%, depending on availability of ground state oxygen ($^3O_2$) in local tissues [19], [8].

Selectivity of Lutrin uptake has been demonstrated in two mouse tumor model studies, murine mammary carcinoma [29] and murine melanoma [27]. An expected corollary to the presence of the blood-brain-barrier [16], is a finding of the latter study that lowest levels of Lutrin uptake were in normal brain tissues (i.e., where brain tumor had not broken down the blood-brain-barrier). Lutrin has been found more sensitive to photodynamic therapy illumination than Photofrin in a mouse mammary adenocarcinoma [10]. The proposed study would assay specificity of Xcytrin or Lutrin uptake in normal brain versus brain tumor (glioma).

There are at least two stages of patient care at which there is significant risk for inadvertent and unintended photosensitization to occur. The first is the pre-operative period, between drug administration and surgery. If the drug has high specificity and clears the blood stream reasonably quickly there should be little risk at this stage. The second period where patients are at risk for photoactivation of non-tumor tissue is during surgical treatment and immediately thereafter. It appears for photodynamic therapy of brain tumors there may be two causes of non-specific drug uptake. First, drug may associate with normal tissues because of a natural affinity. Second, drug may spread to the region of normal tissues due to edema associated with surgery [21]. Edema may be controlled by utilizing optimal lag time and appropriate and accurate illumination dosimetry and distribution.

7 Conclusions

Intra-operative MR scanning has been demonstrated to improve glioma and other brain tumor resection efficacy. Metalloporphyrins offer three new opportunities to improve brain tumor resection efficacy when used in conjunction with intra-operative MR. First, an MR-visible contrast agent that binds specifically to tumor, such as Xcytrin, should be able to overcome the non-specific contrast resulting from leakage of vascularly administered gadolinium leakage during intra-operative procedures. Second, Xcytrin or Lutrin fluorescence should provide fluorescent guidance to residual tumor that is not visible under white light. Finally, in future, we expect that

Lutrin-mediated fluorescence will insure maximum tumor resection and photodynamic therapy ablation of unresectable residual.

Acknowledgement

This work was partially supported by an American Cancer Society Pilot Grant to DD.

References

1. Albert, F.K., Forsting, M., Sartor, K., Adams, H.P., Kunze, S.: Early postoperative magnetic resonance imaging after resection of malignant glioma: Objective evaluation of residual tumor and its influence on regrowth and prognosis. Neurosurgery **34** (1994) 45-60

2. Callahan, D.E., Forte, T.M., Afzal, S.M., Deen, D.F., Kahl, S.B., Bjornstad, K.A., Bauer, W.F., Blakely, E.A.: Boronated protoporphyrin (BOPP): localization in lysosomes of the human glioma cell line SF-767 with uptake modulated by lipoprotein levels. Int. J. Radiat. Oncol. Biol. Phys. **45** (1999) 761-71

3. Cubeddu, R., Canti, G., Taroni, P., Valentini, G: Study of porphyrin fluorescence in tissue samples of tumour-bearing mice. J. Photochem. Photobiol. B **29** . (1995) 171-8

4. Damadian, R.: Tumor detection by nuclear magnetic resonance. Science **171** (1971) 1151-3

5. Dean, D., Kamath, J., Duerk, J.L., Ganz, E.: Validation of Object-induced MR Distortion Correction for Frameless Stereotactic Neurosurgery. IEEE Trans. Med. Img. **17** (1998) 810-6

6. Forsting, M., Albert, F.K., Kunze, S., Adams, H.P., Zenner, D., Sartorm, K.: Extirpation of glioblastomas: MR and CT follow-up of residual tumor and regrowth patterns. AJNR Am. J. Neuroradiol. **14** (1993) 77-87

7. Gering, D., Nabavi, A., Kikinis, R., Grimson, W.E.L., Hata, N., Everett, P., Jolesz, F., Wells, W. III.: An Integrated Visualization System for Surgical Planning and Guidance using Image Fusion and Interventional Imaging. In: Taylor, C., Colchester, A. (eds.): Medical Image Computing and Computer-Assisted Intervention--MICCAI'99, Lecture Notes in Computer Science, Vol. 1679. Springer-Verlag Berlin (1999) 809-19

8. Grossweiner, L.I., Bilgin, D., Berdusis, P., Mody, T.D.: Singlet Oxygen Generation by Metallotexaphyrins. Photobiol. Photochem. **70** (1999) 138

9. Hall, W.A., Liu, H., Martin, A.J., Pozza, C.H., Maxwell, R.E., Truwit, C.L.: Safety, efficacy, and functionality of high-field strength interventional magnetic resonance imaging for neurosurgery [In Process Citation]. Neurosurgery **46** (2000) 632-641; discussion 641-2

10. Hammer-Smith, M.J., Ghahranmaniou, M., Berns, MW.: Photodynamic activity of Lutetium-Texaphyrin in a mouse tumor system. Lasers in Surg. & Med. **24** (1999) 276-84

11. Hebeda, K., Saarnak, A., Olivo, M., Sterenborg, H.J., Wolbers, J.G.: 5-Aminolevulinic acid induced endogenous porphyrin fluorescence in 9L and C6 brain tumours and in the normal rat brain. Acta Neurochir (Wien) **140** (1998) 503-12

12. Knauth, M., Wirtz, C.R., Tronnier, V.M., Aras, N., Kunze, S., Sartor, K.: Surgically induced intracranial contrast enhancement: Potential source of diagnostic error in intraoperative MR imaging. AJNR Am J Neuroradiol **20** (1999) 1547-53

13. Lewin, J.S.: Interventional MR Imaging: Concepts, Systems, and Applications in Neuroradiology. AJNR Am J Neuroradiol **20** (1999) 735-48

14. Maurer, C.R.J. Hill, D.L., Martin, A.J., Liu, H. McCue, M., Ruekert, D. Lloret, D., Hall, W.A., Maxwel, R.E., Hawkes, D.J., Truwit, C.L.: Investigative of intraoperative brain

deformation using a 1.5-T interventional MR system: Preliminary results. IEEE Trans. Med. Img. 17 (1998) 817-25.

15. Miller, R. A., Woodburn, K., Fan, Q., Renschler, M.F., Sessler, J.L., Koutcher, J.A. In vivo animal studies with gadolinium (III) texaphyrin as a radiation enhancer. Int. J. Radiat. Oncol. Biol. Phys. 45 (1999) 981-989.

16. Noske, D.P., Wolbers, J.G., Sterenborg H.J.C.M.: Photodynamic therapy of malignant glioma. Clin. Neurol. Neurosurg. 93 (1991) 293-307

17. Renschler, M., Yuen, A.R., Panella, T.J., Wieman, J., Dougherty, S., Esserman, L., Panjehpour, M., Taber, S.W., Fingar, V.H., Lowe, E., Engel, J.S., Lum, B., Woodburn, K.W., Cheong, W.-F., Miller, R.A.: Photodynamic therapy trials with lutetium texaphyrin (Lu-Tex) in patients with locally recurrent breast cancer. In: Dougherty, T.J. (ed.): Optical Methods for Tumor Treatment and Detection. SPIE, Vol. 3247. Bellingham (1998) 35-9

18. Rosenthal, R.I., Nurenberg, P., Becerra, C.R., Frenkel, E.P., Carbone, D.P., Lum, B.L., Miller, R., Engel, J., Young, S., Miles, D., Renschler, M.F.: A phase I single-dose trial of gadolinium enhancing Texaphyrin (Gd-Tex), a tumor selective radiation sensitizer detectable by magnetic resonance imaging. Clin. Cancer Res. 5 (1999) 739-45

19. Sessler, J.L., Dow, W.C., O'Connor, D.O., Harriman, A., Hemmi, G., Mody, T.D., Miller, R.A., Qing, F., Springs, S., Woodburn, K., Young, S.W.: Biomedical applications of lanthanide(III) texaphyrins Lutetium(III) texaphyrins as potential photodynamic therapy photosensitizers. J. Alloys & Compounds 249 (1997) 146-52

20. Shibata Y, A. Matsumura, Yoshida, F, Yamamoto, T, Nakai, K, Nose, T, Sakata, I, Nakajima, S. (1998). Cell cycle dependency of porphyrin uptake in a glioma cell line. Cancer Letter 129 77-85

21. Stummer, W., Stocker, S., Wagner, S., Stepp, H., Fritsch, C., Goetz, C., Goetz, A.E., Kiefmann, R., Reulen, H.J.: Intraoperative detection of malignant gliomas by 5-aminolevulinic acid- induced porphyrin fluorescence. Neurosurgery 42 (1998) 518-25

22. Sutherland, G.R., Taro, K., Louw, D., Hoult, D.I., Tomanek, B., Saunders J.: A mobil high-field magnetic resonance system for neurosurgery. J. Neurosurg. 91 (1999) 804-13

23. Tovi, M., Lilja, A., Bergstrom, A.M., Ericsson, A., Bergstrom, K., Hartman, M.: Delineation of gliomas with magnetic resonance imaging using Gd-DTPA in comparison with computed tomography and positron emission tomography. Acta Radiol. 31 (1990) 417-29

24. Tynninen O., Aronen, H.J., Ruhala, M., Paetau, A., Von Boguslawski, K., Salonen, O., Jaaskelainen, J., Paavonen, T.: MRI enhancement and microvascular density in gliomas. Correlation with tumor cell proliferation. Invest. Radiol. 34 (1999) 427-34

25. Viala, J., Vanel, D., Meingan, P., Lartigau, E., Carde, P., Renschler, M.: Phases IB and II multidose trial of gadolinium Texaphyrin, a radiation sensitizer detectable at MR imaging: Preliminary results in brain metastases. Radiology 212 (1999) 755-9

26. Wirtz, C.R., Knauth, M., Bonsanto, M.N., Sartor, K., Kunze, S., Tronnier, V.M. Clinical evaluation and follow-up results for intraoperative magnetic resonance imaging in neurosurgery. Neurosurgery 46 (2000) 1112-22

27. Woodburn, K.W., Fan, Q., Miles, D.R., Kessel, D., Luo, Y., Young, S.W.: Localization and efficacy analysis of the phototherapeutic lutetium texaphyrin (PCI-0123) in the murine EMT6 sarcoma model. J. Photochem. Photobiol. 65 (1997) 410-5

28. Woodburn, K.W., Qing, F., Kessel, D., Luo, Y., Young, S.W.: Photodynamic therapy of B16F10 murine melanoma with lutetium texaphyrin." J. Invest. Dermatol.110 (1998) 746-51

29. Young, S.W., Qing, F., Harriman, A., Sessler, J.L., Dow, W.C., Mody, T.D., Hemmi, G.W.., Hao, Y., Miller, R.A.: Gadolinium(III) texaphyrin: a tumor selective radiation sensitizer that is detectable by MRI. Proc. Natl. Acad. Sci. USA 93 (1996) 6610-5

Differential Geometry Based Vector Fields for Characterizing Surrounding Structures of Pulmonary Nodules

Y. Kawata, N. Niki, H. Ohmatsu[a], M. Kusumoto[b], R. Kakinuma[a],
K. Mori[c], H. Nishiyama[d], K. Eguchi[e], M. Kaneko[b], N. Moriyama[b]

Dept. of Optical Science, Univ. of Tokushima, Tokushima ,
[a]National Cancer Center Hospital East, [b]National Cancer Center Hospital,
[c]Tochigi Cancer Center, [d]The Social Health Insurance Medical Center,
[e]National Shikoku Cancer Center Hospital

Abstract. This paper presents a scheme to analyze surrounding structures of pulmonary nodules by using differential geometry based vector fields. In this scheme the differential characteristics such as the principal curvatures and directions are computed from the differential values of the isointensity surfaces. Each voxel in the nodule surrounding is described in terms of shape index and curvedness derived from the principal curvatures. Two vector fields are formed from the directions of the maximum principal curvatures of nodule surrounding and gradient vectors of nodule surface, respectively. The gradient vector field is computed by diffusing the gradient vector on the nodule surface. The regions corresponding to the cylindrical or conic figures which are similar to vessel and plural images are segmented by the shape index and curvedness values. Then, the relationship between the segmented regions and the nodule is evaluated by the inner product of the direction of the maximum principal curvature and the gradient vector. We demonstrate the feasibility of the scheme to classify benign and malignant nodules.

1. Introduction

Pulmonary nodules remain a common and difficult diagnostic problem. It is often the case that the differential diagnosis by means of transbronchial or percutaneous biopsies becomes difficult due to the nodule size. There has been a considerable amount of interest in the use of thin-section CT images to observe small pulmonary nodules for differential diagnosis without invasive operation [1], [2]. A number of investigators have developed feature extraction and classification methods for characterization of pulmonary nodules. One promising area of recent researches has been the analysis of three-dimensional (3-D) pulmonary nodule images. In assessing the malignant potential of pulmonary nodules, characterizing the relationships between nodules and their surrounding structures such as vessel, bronchi, and pleura is an important step [1], [2]. In order to analyze the relationships several researches concentrate on evaluating the angle between the nodule and line patterns such as vessel and brachia. There are two major approaches to the development of computing directions of line patterns. One approach uses the center line of the line pattern obtained by a 3-D thinning process [3], [4]. The direction of the line pattern is approximated by the gradient of the center line. The other approach uses the partial derivatives of 3-D gray level images. The direction of the line pattern is approximated by the eigenvectors of Hessian matrix or the direction of principal curvatures [5], [6]. Although the results of these studies were obtained by small data set or simulations, they indicate the potential of using computerized feature extraction techniques to improve the diagnostic accuracy of differentiating malignant and benign nodules.

In our previous work, we found that curvature indexes such as shape index and curvedness was useful for characterizing the internal structure of nodules [7], [8], [9]. In the present study, we introduce the differential geometrical quantities for the structure analysis of the nodule surrounding and explore the feasibility of differential geometry based vector fields in classifying malignancies from other lesions.

2. Methods

2.1 Extraction of nodule and its surrounding region

Thin-section CT images were obtained by the helical CT scanner (Toshiba TCT900S Superhelix). Per patient, about 60 slices at 1mm intervals were obtained to observe whole nodule region and its surrounding. The range of pixel size in each square slice of 512 pixels was between 0.3x0.3 mm^2 and 0.4x0.4 mm^2. The 3D thoracic image was reconstructed by a linear interpolation technique to make each voxel isotropic.

The segmentation of the 3D pulmonary nodule image consists of three steps [7] ;1)extraction of lung area, 2) selection of the region of interest (ROI) including the nodule region, and 3) nodule segmentation based on a geometric approach. The lung area extraction step plays an essential role when the part of a nodule in the peripheral lung area touches the chest wall. The ROI including the nodule was selected interactively. A pulmonary nodule was segmented from the selected ROI image by the geometric approach proposed by Caselles [10]. The region of nodule surrounding was obtained by the subtraction of the segmented 3-D nodule image I_A from its morphological filtering image $I_F = I_A + n_B I_B$ (n_B dilations of the image I_A by structure element I_B). The structure element used was a 3x3x3 non-null uniform square centered in the origin.

2.2 Differential geometry based vector field

Each voxel in the region of interest (ROI) including the pulmonary nodule was locally represented by two curvature indexes that represented the shape index and the curvedness. By assuming that each voxel in the ROI lies on the surface which has the normal corresponding to the 3-D gradient at the voxel, we computed directly the curvatures on each voxel from the first and the second derivatives of the gray level image of the ROI by using an approach proposed by Thirion [11]. Let $I(\mathbf{x}) : \mathbf{R}^3 \to \mathbf{R}$

is a ROI image defined in $\mathbf{R}^3$. To compute the partial derivatives of the ROI images I (x), the ROI images were blurred by convolving with a 3-D Gaussian function with width σ. The computation of the blurred derivatives were efficiently performed by using the recursive implementation of the Gaussian function [12]. Using the formulas proposed by Thirion [11], Gaussian curvature $K(x; \sigma)$ and mean curvature $H(x; \sigma)$ were computed and then the principal curvatures and directions were obtained. The Gaussian and mean curvatures are defined by

$$K(\mathbf{x};\sigma) = \frac{1}{h^2}[I_x^{\,2}(I_{yy}I_{zz} - I_{yz}^{\,2}) + 2I_yI_z(I_{xz}I_{xy} - I_{xx}I_{yz}) \tag{1}$$

$$+ I_y^{\,2}(I_{xx}I_{zz} - I_{xz}^{\,2}) + 2I_xI_z(I_{yz}I_{xy} - I_{yy}I_{xz})$$

$$+ I_z^{\,2}(I_{xx}I_{yy} - I_{xy}^{\,2}) + 2I_xI_y(I_{xz}I_{yz} - I_{zz}I_{xy})]$$

$$H(\mathbf{x};\sigma) = \frac{1}{2h^{3/2}}[I_x^{\,2}(I_{yy} + I_{zz}) - 2I_yI_zI_{yz} + I_y^{\,2}(I_{xx} + I_{zz}) - 2I_xI_zI_{xz} \tag{2}$$

$$+ I_z^{\,2}(I_{xx} + I_{yy}) - 2I_xI_yI_{xy}]$$

where $h = I_x^{\,2} + I_y^{\,2} + I_z^{\,2}$, x, y, and z subscripts denote partial derivatives with respect to the corresponding variable. The principal curvatures $\kappa_1(x; \sigma)$ and $\kappa_2(x; \sigma)$ ($\kappa_1(x; \sigma) \geq \kappa_2(x; \sigma)$)) are obtained by

$$\kappa_i = H(\mathbf{x};\sigma) \pm \sqrt{H^2(\mathbf{x};\sigma) - K(\mathbf{x};\sigma)} \quad (i = 1, 2). \tag{3}$$

The principal directions $\mathbf{v}_1$ and $\mathbf{v}_2$ are defined by

$$\mathbf{v}_i(\mathbf{x};\sigma) = \mathbf{a} \pm \sqrt{H^2(\mathbf{x};\sigma) - K(\mathbf{x};\sigma)}\ \mathbf{b} \quad (i = 1, 2). \tag{4}$$

where $\mathbf{b} = (I_z - I_y, I_x - I_z, I_y - I_x)$ and the x component of the vector $\mathbf{a}$ is obtained by

$$\mathbf{a} \cdot \mathbf{x} = -\frac{1}{2h^{3/2}}[-2I_z^{\,3}I_{xy} + I_y^{\,3}I_{zz} + 2I_y^{\,3}I_{xz} - 2I_y^{\,2}I_zI_{xy} \tag{5}$$

$$+ 2I_z^{\,2}I_xI_{yz} + 2I_z^{\,2}I_yI_{xz} - 2I_y^{\,2}I_xI_{yz} - 2I_zI_xI_yI_{zz}$$

$$+ 2I_xI_yI_zI_{yy} + I_y^{\,2}I_zI_{xx} - 2I_z^{\,2}I_xI_{xz} + I_zI_x^{\,2}I_{zz}$$

$$- I_x^{\,2}I_zI_{yy} + 2I_z^{\,2}I_yI_{yz} - I_zI_y^{\,2}I_{zz} + I_z^{\,3}I_{xx} - I_z^{\,3}I_{yy}$$

$$- I_y^{\,2}I_xI_{xz} + 2I_x^{\,2}I_yI_{yz} - I_y^{\,3}I_{xx} + 2I_xI_z^{\,2}I_{xy} - I_yI_z^{\,2}I_{xx}$$

$$- 2I_zI_y^{\,2}I_{yz} + I_yI_z^{\,2}I_{yy} - 2I_zI_x^{\,2}I_{yz} + 2I_xI_y^{\,2}I_{xy} + I_x^{\,2}I_yI_{zz} - I_x^{\,2}I_yI$$

The y and z components are obtained by circular permutation of x, y, and z. In tube-like structures such as vessel and pleura, axial directions of them are approximately parallel to the direction of maximum principal curvature of the central part of isointensity surface of the structure. Therefore, a vector field of the nodule surrounding was represented by the direction of the maximum principal curvature. We call this vector field the maximum principal curvature vector (MPV) field.

The original shape index defined by Koendrink [13] gives a continuous distribution of surface types between −1 and 1. To introduce the shape spectral function of the objects surface patch, Dorai [14] modified the original definition of the shape index so that the shape index maps the surface types on the interval between 0 and 1. In any definition the distinct surface type corresponds to a unique value of the shape index except the planar surface. Since points on planar surface have an indeterminate shape index, Dorai assigned a symbolic label (a shape index value of 2.0) for the points on planar surface [14]. The definition of the shape index used here was based on the modified definition and [8],[9]. The shape index with the width σ at the voxel x was given by

$$S(x;\sigma) = \frac{1}{2} + \frac{1}{\pi} \arctan \frac{\kappa_1(x;\sigma) + \kappa_2(x;\sigma)}{\kappa_1(x;\sigma) - \kappa_2(x;\sigma)}. \tag{6}$$

The curvedness quantifies how highly curved a surface is, and is inversely proportional to the size of the object. The curvedness $R(x; \sigma)$ was given by

$$R(x;\sigma) = \sqrt{\frac{\kappa_1(x;\sigma)^2 + \kappa_2(x;\sigma)^2}{2}} \tag{7}$$

The selection of the width σ of the Gaussian function is a critical issue. In previous work [8], [9] when the curvature indexes with width σ was utilized to characterize the internal structure of pulmonary nodules, we assigned the value 2.0 to the width, which provided high accuracy of classification between malignant and benign nodules for discrete width values using a data set. In this present study, we also assigned the value 2.0 to the width.

2.3 Gradient vector field

In order to quantify the relationship between the nodule and its surrounding structure such as vessel and pleura, we focus on two indicators of malignancy, which are denoted as vascular convergence and pleural retraction [1], [2]. In the 3-D thoracic CT images, these findings are observed so that the vessel and pleura images are drawn in the nodule. The shape of the vessel and the pleura images are similar to cylindrical or conic structures. Therefore, we measured an amount of the vascular convergence and pleural retraction by computing the absolute value of the inner product of the directions of cylindrical or conic structures and the normal directions of the nodule surface.

The directions of cylindrical or conic structures were obtained by the MPV field. The normal directions of the nodule surface at the vicinity of the nodule was estimated by the following procedure. This procedure was based on the idea which was proposed by Xu [15] to improve the segmentation accuracy of the snake model.

Let $g(\mathbf{x})$: $\mathbf{R}^3 \rightarrow \mathbf{R}$ is an edge map defined in $\mathbf{R}^3$ and $f(\mathbf{x})$: $\mathbf{R}^3 \rightarrow \mathbf{R}$ is a segmented binary image of the nodule defined in $\mathbf{R}^3$. The edge map was derived from the segmented 3-D binary image of the nodule having property that it is larger near the nodule edges. The edge map was represented as

$$g(\mathbf{x}) = G_\sigma * \left| \nabla f(\mathbf{x}) \right| \tag{8}$$

where G_σ is 3-D Gaussian function with the width σ. The edge map has the following three properties; (1) the gradient of an edge map $\nabla g(\mathbf{x})$ has vector toward the edges, which are normal to the edge, (2) these vectors have large magnitudes in the vicinity of the edge, and (3) in the homogeneous region $f(\mathbf{x})$ is nearly constant and $\nabla g(\mathbf{x})$ is nearly zero. Because of the first property, the normal direction of the nodule surface are approximated by the gradient of the edge map in the vicinity of the edges. Since this property is utilized to measure the relationship between nodule surface and surrounding structures, the first property is a desirable property. The capture range will be small due to the second property. The homogenous regions at the surrounding nodule will have zero. The last two properties are undesirable. The approach is to keep the desirable property of the gradients near the nodule edges, but to expand the gradient away from the edges into homogeneous regions of the nodule surrounding using a computational diffusion process. A vector field obtained by the computational diffusion process is denoted as gradient vector (GV) field. The GV field was defined as the vector field $\mathbf{h}(\mathbf{x})$: $\mathbf{R}^3 \rightarrow \mathbf{R}^3$ that minimized the energy functional

$$E = \int_{\mathbf{R}^3} \mu \left| \nabla \mathbf{h} \right|^2 + \left| \nabla g \right|^2 \left| \mathbf{h} - \nabla g \right|^2 d\mathbf{x} \tag{9}$$

where the gradient operator ∇ is applied to each component of $\mathbf{h}$ separately. Minimizing Eq. (9) makes the vector field smooth when there is no data. When $\left| \nabla g \right|$ is small, the first term dominates the energy. On the other hand the second term dominates the energy as $\left| \nabla g \right|$ is large. The parameter μ is a regularization parameter controlling the tradeoff between the first term and the second term. Using the calculus of variations, it is found that the GV field must satisfy the Euler equation. We implemented the iteration technique presented by Xu [15] to solve the Eq. (9).

2.4 Characterization of the nodule surrounding based on vector fields

Each voxel in the nodule surrounding was represented by the following six attributions; (1) the shape index value (denoted as SH), (2) the curvedness value (denoted as CV), (3) the CT density value (denoted as D), (4) the direction of maximum principal curvature (denoted as MPV), (5) the gradient vector (GV), and (6) the absolute value of inner product of MPV and GV (denoted as F). To select the central part of cylindrical or conic structures, we specified two threshold values T_{SH} and T_{CV} for SH and CV, respectively. We then obtained the region consisting of voxels which satisfied with a condition A of SH $\geq T_{SH}$ and CV $\geq T_{CV}$. This process

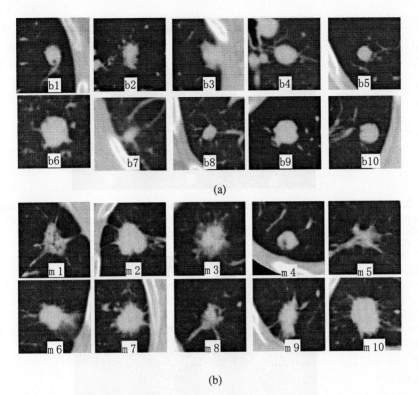

(a)

(b)

Fig. 1. Slice images of the ROIs of benign and malignant nodules. (a) Benign cases. (b) Malignant cases.

means that the vessel and pleura regions with cylindrical or conic structures are segmented and the central parts of these regions are selected because the curvedness value of the central part of cylindrical or conic structures become larger than that of the marginal part of them. We finally computed the ratio of the number of voxel concentrating on the nodule to the total number of voxel in the selected region. This ratio was represented as feature $F1$. The feature $F1$ was used as a measure of the ratio of vessels or pleura concentrating on the nodule

3. Experimental results

The data set in this study included twenty 3-D thoracic images acquired from National Cancer Center Hospital East. Among the 20 cases, 10 contained malignant nodules, and 10 contained benign nodules. Whole malignant nodules were cytologically or histologically diagnosed. Fig. 1 presents slice images of the ROIs of benign and malignant nodules used here. In malignant cases, m7 and m10 were diagnosed as squamous cell carcinoma and others were diagnosed as adenocarcinoma. In benign cases, b1, b2, b3, b5, and b7 were diagnosed as inflammatory and others were diagnosed as tumor-like lesion.

Fig. 2 shows the distribution of the attributions on a slice of the ROI of a benign

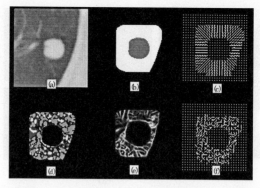

Fig. 2. Computation results of a benign case. (a) Slice image of the ROI. (b) Nodule surrounding region. (c) Gradient vector field. (d) Distribution of shape index. (e) Distribution of curvedness. (f) Vector field of the direction of the maximum principal curvature.

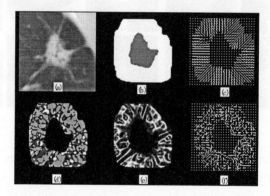

Fig. 3. Computation results of a malignant case. (a) Slice image of the ROI. (b) Nodule surrounding region. (c) Gradient vector field. (d) Distribution of shape index. (e) Distribution of curvedness. (f) Vector field of the direction of the maximum principal curvature.

case. Fig.2 (a) shows the distribution of CT density value on the slice of the ROI and (b) presents the slice image of nodule surrounding colored white and the segmented nodule colored gray. Fig.2 (c) demonstrates the distribution GV field in the nodule surrounding. Fig. 2 (d), (e), and (f) demonstrate the distribution of the shape index value, the curvedeness, and the vector field of the maximum principal curvature in the slice of the nodule surrounding, respectively. In the shape index distribution, the defined color code corresponding to each surface type is as follows; peak surface is white, saddle ridge surface is red, saddle valley surface is blue, and pit surface is green. In the curvedness distribution the defined gray scale corresponding to the curvedness value is as follows; the value zero is black, the value lager than one is white, and the interval from 0 to 1 is represented by gray. In Fig.2 (c) and (f) the length of each vector was normalized to one. Fig. 3 shows the distribution of the attributions on a slice of the ROI of a malignant case. The figure arrangement is the same as that of Fig.2. In Fig.3 (a) it is observed that vessel and plural images are presented in the nodule surrounding. It is also observed that the cylindrical or conic structures such as vessel and plural image are represented as peak surface or saddle

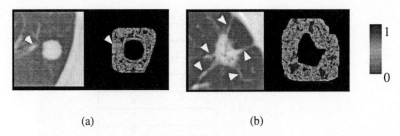

(a)   (b)

Fig. 4. Distribution of F value of the nodule surrounding. (a) Benign case. (b) Malignant case. Left: Slice image of the ROI. Right: Distribution of F value represented by a color code. The range of F value is between 0 and 1.

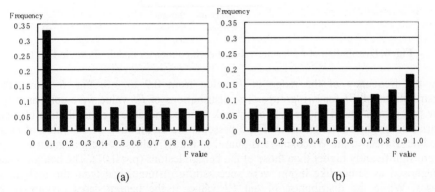

(a)   (b)

Fig. 5. Histogram of F value. (a) Benign case. (b) Malignant case.

ridge surface types in Fig.2(d) and Fig.3 (d), and the center part of the structures have high curvature value in Fig.2 (e) and Fig.3 (e). In Fig.2 (f) and Fig. 3 (f) it is observed that directions of the maximum principal curvatures in the part corresponding to the cylindrical or conic structures are parallel to the axial directions of the structures.

Fig. 4 shows the distribution of F value of the nodule surrounding. The F value is represented by a color code as shown in Fig.4. The color code is varied from red to blue as F value becomes zero to one. The vessel region converging on the nodule and pleura region radiating from the nodule denoted by the arrows are colored blue. From these results it can be expect that the frequency of high F value is depended on the existence of vessel or pleura converging on the nodule.

Fig. 5 shows a histogram of the F value of the region constructed by a set of voxels which are satisfied with the condition A. In this experiment, the values of T_{SH} and T_{CV} were set to 0.5 and 0.9, respectively. The frequency was normalized by the total number of voxels of the constructed region. Compared the distribution of the histogram of the benign case with that of the malignant case, it is observed that the voxels with higher value of F occupy in the nodule surrounding. From these results it can be expect that the frequency of high F value is depended on the existence of vessel or pleura converging on the nodule.

To analyze the statistical significance of the data, we divided the patients into two groups; patients with malignant and benign nodules. The feature $F1$ value was given

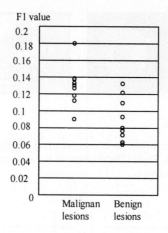

Fig. 6. Distribution of *F1* feature value of each pulmonary nodule shown in Fig. 1.

by the frequency of the range of F value from 0.9 to 1.0. Fig. 6 shows the distributions of *F1* for two groups. The distributions of *F1* value were compared for the two groups based on the Wilcoxon rank sum test because the distribution of their variables might be nongaussian. Fig.6 presents the distribution of *F1* value of each nodule. Based on the Wilcoxon rank sum test, the *F1* values of the malignant lesions were significantly higher than those of the benign lesions (p < 0.01). The benign cases diagnosed as tumor-like lesion were successfully distinguished from the malignant cases. While the distribution of the *F1* value of the benign cases diagnosed as inflammatory overlapped with that of the malignant cases. There is a benign lesion in which the vascular convergence and pleural retraction are observed in the nodule surrounding [1], [2]. Therefore, the potential of the feature *F1* for nodule classification may be limited.

4. Conclusion

In this study, we have introduced the differential geometrical quantities for the structure analysis of nodule surrounding and evaluated the feasibility of differential geometry based vector fields in classifying malignancies from other lesions. While the number of studied nodules in our data set was too small to achieve a definitive conclusion, the preliminary result from this work was very encouraging, demonstrating the potential for vector field analysis used in the computerized characterization to discriminate benign from malignant pulmonary nodules.

We have demonstrated classification strategy of benign and malignant nodules based on the differential geometrical quantities in the nodule surrounding. The classification strategy was designed by an interactive analysis of the local intensity structures. Further work should therefore include the development of an automated method of the parameter adjustment such as the width of the Gaussian smoothing kernel. In addition, we will also investigate classification schemes considering the multidimensional feature space.

References

1. K.Mori, Y.Saitou, K.Tominaga, K.Yokoi, N.Miyazawa, A.Okuyama, M.Sasagawa, "Small nodular legions in the lung periphery: new approach to diagnosis with CT," Radiology, Vol.177, pp.843-849 (1990)
2. K. Kuriyama, R. Tateishi, O. Doi, K. Kodama, M. Tatsuta, M Matsuda, T. Mitani, Y. Narumi, M. Fujita, "CT-pathologic correlation in small peripheral lung cancers", AJR, Vol.149, pp.1139-1143 (1987)
3. T.Tozaki, Y.Kawata, N.Niki, H.Ohmatsu, R. Kakinuma, K.Eguchi, N. Moriyama, "Pulmonary organs analysis for differential diagnosis based on thoracic thin-section CT images", IEEE Trans. Nuclear Science, Vol.45, pp.3075-3082 (1998).
4. Y.Hirano, Y.Mekada, J.Hasegawa, J. Toriwaki, H.Ohmatsu, and K.Eguchi, "Quantification of vessels convergence in three-dimensional chest X-ray CT images with three-dimensional concentration index", Medical Imaging Technology, Vol.15, pp.228-236 (1997)
5. J. Williams, L. Wolff, "Analysis of the pulmonary vascular tree using differential geometry based vectored fields", Computer Vision and Image Understanding, Vol. 65, pp.226-236 (1997)
6. Y. Sato, S. Nakajima, N. Shiraga, H. Atsumi, S. Yoshida, T. Koller, G. Gerig, and R. Kikins,"Three-dimensionla multi-scale line filter for segmentation and visualization of curvlinear structures in medical images", Medical Image Analysis, Vol.2, pp.143-168 (1998)
7. Y.Kawata, N.Niki, H.Ohmatsu, R.Kakinuma, K.Eguchi, M.Kaneko, N.Moriyama, "Quantitative surface characterization of pulmonary nodules based on thin-section CT images", IEEE Trans. Nuclear Science, Vol. 45, pp.2132-2138 (1998)
8. Y. Kawata, N.Niki, H.Ohmatsu, M. Kusumoto, R. Kakinuma, K. Mori, K. Eguchi, M. Kaneko, N. Moriyama, "Classification of pulmonary nodules in thin-section CT images by using multi-scale curvature indexes", IEEE Int. Conf. on Image Processing, Vol.2, pp.197-201, (1999)
9. Y. Kawata, N.Niki, H.Ohmatsu, "Curvature based internal structure analysis of pulmonary nodules using thoracic 3-D CT images", IEICE Trans., Vol.J-83-D-II, pp.209-218 (2000)
10. V.Caselles, R.Kimmel, G.Sapiro, and C.Sbert, "Minimal surfaces based object segmentation", IEEE Trans. Pattern Analysis Machine Intelligence, Vol.19, pp.394-398 (1997)
11. J.-P, Thirion and A. Gourdon, "Computing the differential characteristics of isointensity surfaces," Comput. Vision and Image Understanding, Vol.61, pp.190-202 (1995)
12. R. Deriche, "Recursively implementing the gaussian and its derivatives", INRIA Research Report, 1893 (1993)
13. J. J. Koenderink and A.J.V. Doorn, "Surface shape and curvature scales", Image and Vision Computing, Vol. 10, pp.557-565, (1992)
14. C.Dorai and A.K. Jain, " COSMOS-A representation scheme for 3D free-form objects", IEEE Trans. Pattern Analysis Machine Intelligence, Vol.19, pp.1115-1130 (1997)
15. C. Xu and J.L. Prince, "Snake, shape, and gradient vector flow", IEEE Trans. Image Processing, Vol. 7, pp.359-369 (1998)

Non-distorting Flattening for Virtual Colonoscopy

Steven Haker[1], Sigurd Angenent[2], Allen Tannenbaum[3], and Ron Kikinis[4]

[1] Departments of Computer Science and Diagnostic Radiology,
Yale University, New Haven CT 06520.
haker@cs.yale.edu
[2] Department of Mathematics,
University of Wisconsin, Madison WI 53705.
[3] Departments of Electrical and Computer Engineering and Biomedical Engineering,
Georgia Institute of Technology, Atlanta GA 30332-0250.
tannenba@ece.gatech.edu
[4] Harvard Medical School, Brigham and Women's Hospital,
Harvard University, Boston MA 02115.

Abstract. In this paper, we consider a novel 3D visualization technique based on conformal surface flattening for virtual colonoscopy. Such visualization methods could be important in virtual colonoscopy since they have the potential for non-invasively determining the presence of polyps and other pathologies. Further, we demonstrate a method which presents a surface scan of the entire colon as a cine, and affords a viewer the opportunity to examine each point on the surface without distortion. From a triangulated surface representation of the colon, we indicate how the flattening procedure may be implemented using a finite element technique. We give a simple example of how the flattening map can be composed with other maps to enhance certain mapping properties. Finally, we show how the use of curvature based colorization and shading maps can be used to aid in the inspection process.

1 Introduction

Three dimensional visualization is becoming an increasingly important technique in surgical planning, non-invasive diagnosis and treatment, and image-guided surgery. Surface warping and flattening, which allow the easy visualization of highly undulated surfaces, are methods that are becoming increasingly widespread. For example, flattened representations of the brain cortical surface are essential in functional magnetic resonance imaging since one wants to show neural activity deep within the folds or sulci of the brain. 3D visualization is also of great importance in virtual colonoscopy in which one can non-invasively determine the presence of pathologies.

Virtual colonoscopy is currently an active area of research by radiologists as a minimally invasive screening method for the detection of small polyps (see [8] and the references therein). In the colon, this has become possible because of imaging devices which allow single breath hold acquisitions of the entire abdomen at acceptable resolutions. Most reports have focused on methods which use computer graphics to simulate conventional colonoscopic procedures [8, 13, 14]. Virtual colonoscopy has some fundamental problems, which it shares with conventional colonoscopy. The most important one is that the navigation using inner views is very challenging and it happens frequently that sizable areas are not inspected at all, leading to incomplete examinations. An alternative approach for the inspection of the entire surface of the colon is to simulate the approach favored by pathologists, which involves cutting open the tube represented by the colon, and laying it out flat for a comprehensive inspection. In some recent work [11], a visualization technique is proposed using cylindrical and planar map projections. It is well-known that such projections can cause distortions in shape as is discussed in [11] and the references therein.

In this paper, we take another approach. We present a method for mapping the colon onto a flat surface in a conformal manner. A *conformal mapping* is a one-to-one mapping between surfaces which preserves angles, and thus preserves the local geometry as well. Our approach to flattening such a surface is based on a certain mathematical technique from Riemann surface theory, which allows us to map any highly undulating tubular surface without handles or self-intersections onto a planar rectangle in a conformal manner. There is some related work in the interesting paper [15] on the topological flattening of a tube onto the plane and its application to virtual colonoscopy. In [15], an electro-magnetic field is simulated by placing charges along a fly-through path. The resulting field lines which emanate from a point on the path define a surface whose intersection with the colon surface forms a loop which is flattened into the plane.

Our approach differs in that no flight path needs to be calculated, and the conformal nature of our flattening map allows us to enhance mapping properties and correct for distortion.

From a triangulated surface representation of the colon, we indicate how the flattening function may be found by solving a second order elliptic partial differential equation (PDE) using finite element techniques. Once the colon surface has been flattened onto a rectangular region of the plane, we utilize a method by which the entire colon is presented as a cine, and which allows the viewer to examine each surface point *without distortion* at some time in the image sequence. Thus in this sense 100% view with 0% distortion can be achieved. Moreover, we explicitly show how various structures of the colon may be studied using this approach. We demonstrate the use of shading maps and colorization as a function of surface curvatures to enhance visualization and inspection.

2 Analytical Approach to Colon Flattening

We first consider a mathematical model for the colon surface. See [3, 6] for the full details. Let $\Sigma \subset \mathbf{R}^3$ represent a smooth embedded surface (no self-intersections) which is topologically an open-ended cylinder. The boundary of Σ consists of two topological circles, which we will call σ_0 and σ_1. We want to introduce a cut C on Σ from end to end, and construct an angle preserving one-to-one map $f : \Sigma \backslash C \rightarrow \mathbf{R}^2$, which sends $\Sigma \backslash C$ to a rectangle R such that σ_0 and σ_1 are mapped to the left and right hand edges of R respectively, while the cut C is mapped to the upper and lower edges of R. The construction of f begins by finding, before the cut is made, a solution u to the Dirichlet problem $\Delta u = 0$ on $\Sigma \backslash (\sigma_0 \cup \sigma_1)$, with boundary conditions $u = 0$ on σ_0, and $u = 1$ on σ_1. Here Δ is the Laplace-Beltrami operator [12] on the surface Σ.

The cut C on Σ is then determined as the trace of the smooth curve obtained by following the gradient of u from a point on σ_0 to σ_1. We next compute the harmonic function v which is conjugate to u by specifying boundary conditions on the cut surface and again solving a Dirichlet problem; see [6]. The mapping $f = (u, v)$ sends the surface Σ to a rectangle R, as desired. The mapping can easily be extended across the cut and thus the cut need not hinder visualization.

3   Approximating the Flattening Function

In the previous section, we outlined the analytical procedure for finding the flattening map f. Here we will discuss the finite element method for finding an approximation to this mapping. See [9] for details about this method. In [1], we described a related method for brain flattening. However, because of the differences in topology between the brain and colon surface, the boundary conditions for the flattening maps are quite different. We now assume that Σ is a triangulated surface, and we look for a flattening map f which is continuous on Σ and linear on each triangle. Here, we will concentrate on finding u, the method for its conjugate v being similar.

It is a classical result [12] that the harmonic function u is the minimizer of the Dirichlet functional

$$\mathcal{D}(u) := \frac{1}{2} \int \int_{\Sigma} |\nabla u|^2 dS, \quad u|_{\sigma_0} = 0, \quad u|_{\sigma_1} = 1, \tag{1}$$

where ∇u is the gradient with respect to the induced metric on Σ. Let $PL(\Sigma)$ denote the space of piecewise linear functions on Σ. For each vertex $V \in \Sigma$, let ϕ_V be the continuous function, linear on each triangle, such that

$$\begin{aligned} \phi_V(V) &= 1, \\ \phi_V(W) &= 0, \ W \neq V. \end{aligned} \tag{2}$$

This set $\{\phi_V\}$ forms a basis for $PL(\Sigma)$, and so any $u \in PL(\Sigma)$ can be written as

$$u = \sum_{V \in \Sigma} u_V \ \phi_V. \tag{3}$$

To approximate the solution to the PDE, we minimize $\mathcal{D}(u)$ over all $u \in PL(\Sigma)$ which satisfy the boundary conditions. To minimize $\mathcal{D}(u)$, we introduce the matrix

$$D_{VW} = \int \int \nabla \phi_V \cdot \nabla \phi_W \, dS, \tag{4}$$

for each pair of vertices V, W. Since $D_{VW} \neq 0$ if and only if V, W are connected by some edge in the triangulation, D is sparse. Simple formulas exist [1, 6, 9] for the calculation of the elements of D; these formulas involve only the angles between adjacent surface edges. One can show that $u = \sum_V u_V \phi_V$ minimizes the Dirichlet functional over $PL(\Sigma)$ with the boundary conditions, if for each vertex $V \in \Sigma \backslash (\sigma_0 \cup \sigma_1)$,

$$\sum_{W \in \Sigma \backslash (\sigma_0 \cup \sigma_1)} D_{VW} \, u_W = - \sum_{W \in \sigma_1} D_{VW}. \tag{5}$$

This is simply a matrix equation. One can quickly solve for the unknown $\{u_W\}$ using standard tools from linear algebra such as the conjugate gradient method.

4   Inspection and Distortion Removal

In practice, once the colon surface has been flattened into a rectangular shape, it will need to be visually inspected for various structures. In this section, we present a simple technique by which the entire colon surface can be presented to the viewer as a sequence of images or cine. In addition, this method allows the viewer to examine each surface point without distortion at some time in the cine. Here, we will say a mapping is without distortion at a point if it preserves the intrinsic distance there. It is well known that a surface can not in general be flattened onto the plane without some distortion somewhere [4]. However, it may be possible to achieve a surface flattening which is free of distortion along some curve. The Mercator projection of the earth does this along the equator. See [11] for a nice discussion of the classical geographic projections and their application to virtual colonoscopy. In our case, the distortion free curve will be a level set of the harmonic function u described above (essentially a loop around the tubular colon surface), and will correspond to the vertical line through the center of a frame in the cine. Specifically, suppose we have conformally flattened the colon surface onto a rectangle $R = [0, u_{max}] \times [-\pi, \pi]$. Let F be the inverse of this mapping, and let $\phi^2 = \phi^2(u, v)$ be the amount by which F scales a small area near (u, v), i.e. let $\phi > 0$ be the "conformal factor" for F. Fix $w > 0$, and for each $u_0 \in [0, u_{max}]$ define a subset $R_0 = ([u_0 - w, u_0 + w] \times [-\pi, \pi]) \cap R$ which will correspond to the contents of a cine frame. We define a mapping

$$(\hat{u}, \hat{v}) = G(u, v) = \left(\int_{u_0}^{u} \phi(\mu, v) d\mu, \int_{0}^{v} \phi(u_0, \nu) d\nu \right) \tag{6}$$

from R_0 to $\mathbf{C}$ which has differential

$$dG(u, v) = \begin{pmatrix} \hat{u}_u & \hat{u}_v \\ \hat{v}_u & \hat{v}_v \end{pmatrix} = \begin{pmatrix} \phi(u, v) & \int_{u_0}^{u} \phi_v(\mu, v) d\mu \\ 0 & \phi(u_0, v) \end{pmatrix} \tag{7}$$

and in particular $dG(u_0, v) = \phi(u_0, v) \times \begin{pmatrix} 1 & 0 \\ 0 & 1 \end{pmatrix}$. The conformality of the flattening map, together with this value for $dG(u_0, v)$, implies that the composition of the flattening map with G sends the level set loop $\{u = u_0\}$ on the colon surface to the vertical line $\{\hat{u} = 0\}$ in the $\hat{u}$–$\hat{v}$ plane without distortion. In addition, it follows from the formula for dG that lengths measured in the $\hat{u}$ direction accurately reflect the lengths of corresponding curves on the colon surface.

5 Application

We tested our algorithms on $256 \times 256 \times 124$ CT colon data sets provided to us by the Surgical Planning Laboratory of Brigham and Women's Hospital. Two slices from one such data set are shown in Fig. 1. First, using the fast segmentation methods of [10] we found the colon surface. Unfortunately, the segmentation algorithm itself does not guarantee that the surface found will be a topological cylinder. In fact, it may contain numerous minute handles which arise because the boundary of the colon, as represented in the data set, may not be sharp. We used a morphological based method [7] by which these handles can be effectively removed and a surface which has the the topology of a closed-ended cylinder can be extracted. This is done in such a way that the large-scale geometry of the surface is not adversely affected.

Our segmentation method is a so called "level-set" method, in which the colon surface to extract is defined implicitly as the zero level set $\{(x, y, z) \mid \Psi(x, y, z) \equiv 0\}$ of a function Ψ defined on the 3D volume. It is well known that the Gaussian and mean curvatures of such an iso-surface can be calculated from the derivatives of Ψ; doing so allows us to avoid having to make these calculations on the triangulated surface after extraction. In fact, we may use the function Ψ to determine the entire second fundamental form for the iso-surface, using the formula

$$II = \frac{-1}{||\nabla\Psi||} \, T^{\,t} \, H_\Psi \, T \tag{8}$$

where H_Ψ is the 3×3 Hessian matrix of second derivatives of Ψ, and T is a 3×2 matrix whose columns are arbitrary orthonormal vectors perpendicular to the surface normal $N = \frac{\nabla\Psi}{||\nabla\Psi||}$. The eigenvectors e_1 and e_2 of II yield the principal directions (the directions in which the degree of surface bending is extremal) as Te_1 and Te_2, while the eigenvalues k_1, k_2 are the corresponding principal curvatures. See [5] for applications of principal direction vectors to surface visualization. The Gaussian curvature can then be found by $K = k_1 k_2 = Det(II)$, the mean curvature by $H = \frac{1}{2}(k_1 + k_2) = \frac{1}{2}Trace(II)$.

One would expect polyps to have relatively high Gaussian curvature as compared to the flatter surrounding colon surface. Further, these areas should be convex with respect to the colon interior, and thus have positive mean curvature with respect to the outward surface normal. This suggests that for visualization, we color the flattened surface according to the Gaussian curvature of the colon surface, but only where both the Gaussian and mean curvatures of the colon surface are positive. The other areas may be colored with some neutral color. This alone, however, is not satisfactory for visualization of the flattened colon because the folds of the colon will not be represented. One solution to this problem is to use *shading maps*, an idea from computer graphics. The idea is to translate surface normals from the original surface to the corresponding point on the flattened surface. When rendered under identical lighting conditions, the original surface and the flattened surface with these "pseudo-normals" will have similar appearances, due to similar shading. This allows us to color the surface any way we wish, and still have the surface folds distinguishable in the flattened representation.

In Fig. 2, two views of the extracted colon surface and the corresponding flattened surface are shown using the coloring and shading scheme described above. We broke the flattened surface into two pieces to fit the page. The flattened surface is not exactly rectangular because we cut the colon surface along triangle edges rather than following the gradient of u exactly as described in Section 2. Fig. 3 shows more detailed views. On the left is an exterior view of a piece of the colon surface. In the center is a "fly-through" view of the same region, and on the right is the corresponding flattened region, corrected for distortion along the vertical center line as described in Section 4. In practice this image would be a single frame of a cine. Notice that the entire section of the colon is visible in the flattened version, while the coloring and shading scheme indicate the convex areas and surface folds. We are currently investigating the usefulness of this visualization scheme for the detection of polyps in a clinical setting.

The distortion correction method described in Section 4 is an example of how the conformal flattening map may be composed with another map to obtain desired mapping properties. Another simple example of this sort of enhancement is shown in Fig. 4. Here, we have composed the flattening map with another one-to-one conformal mapping from the plane to itself. This second conformal mapping was chosen to minimize the overall distortion of area in a least squares sense. However, it is not possible to correct for all distortion this way. Other such enhancing compositions are a current area of research [2].

References

1. S. Angenent, S. Haker, A. Tannenbaum, R. Kikinis, "Laplace-Beltrami operator and brain surface flattening," *IEEE Trans. Medical Imaging*, **18** (1999), pp. 700-711.
2. S. Angenent, S. Haker, A. Tannenbaum, R. Kikinis, "On area preserving mappings of minimal distortion," in *System Theory: Modeling, Analysis, and Control*, edited by T. Djaferis and I. Schick, Kluwer, Holland, 1999, pp. 275-287.
3. H. Farkas and I. Kra, *Riemann Surfaces*, Springer-Verlag, New York 1991.
4. M. P. Do Carmo, *Riemannian Geometry*, Prentice-Hall, Inc. New Jersey, 1992.

5. A. Girshick, V. Interrante, S. Haker, T. Lemoine. "Line direction matters: an argument for the use of principal directions in 3D line drawings," *First International Symposium on Non Photorealistic Animation and Rendering (NPAR 2000)*, to appear.

6. S. Haker, S. Angenent, A. Tannenbaum, R. Kikinis, "Nondistorting flattening maps and the 3D visualization of colon CT images," Technical Report, Department of Electrical Engineering, Georgia Tech, November 1999. Submitted for publication to *IEEE Trans. of Medical Imaging*.

7. S. Haker, "Extracting simply connected iso-surfaces from volumetric data," in preparation.

8. A. Hara, C. Johnson, J. Reed, R. Ehman, D. Ilstrup, "Colorectal polyp detection with CT colography: two- versus three-dimensional techniques," *Radiology* **200** (1996), pp. 49-54.

9. T. Hughes, *The Finite Element Method*, Prentice-Hall, New Jersey, 1987.

10. S. Kichenasamy, P. Olver, A. Tannenbaum, A. Yezzi, "Conformal curvature flows: from phase transitions to active contours," *Archive Rational Mechanics and Analysis* **134** (1996), pp. 275-301.

11. D. Paik, C. Beaulieu, R. Jeffrey, C. Karadi, and S. Napel, "Visualization modes for CT colonography using cylindrical and planar map projections," Technical Report, Department of Radiology, Stanford University School of Medicine, Stanford, CA, 1999.

12. J. Rauch, *Partial Differential Equations*, Springer-Verlag, New York 1991.

13. G. Rubin, S. Napel, and A. Leung, "Volumetric analysis of volumetric data: achieving a paradigm shift," *Radiology* **200** (1996), pp. 312-317.

14. D. Vining, "Virtual endoscopy: is it reality?," *Radiology* **200** (1996), pp. 30-31.

15. G. Wang, E. McFarland, B. Brown, and M. Vannier, "GI tract unraveling with curved cross sections," *IEEE Trans. Med. Imaging* **17** (1998), pp. 318-322.

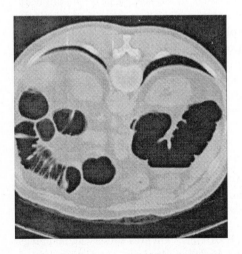 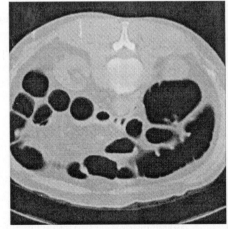

Fig. 1. Two Slices from CT Colon Data Set

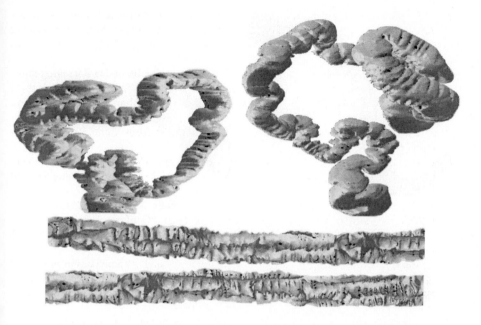

Fig. 2. Two Views of Colon Surface and Flattened Representation

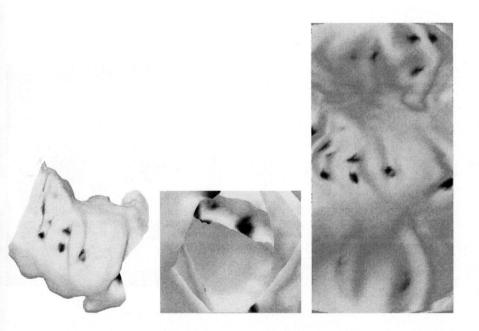

Fig. 3. Exterior, Fly-Through and Flattened Views

Fig. 4. Improved Conformal Mapping with Detail

HeartPerfect: Data Mining in a Large Database of Myocardial Perfusion Scintigraphy

Bernard Hotz[1] and Jean-Philippe Thirion[1]

[1] HealthCenter Internet Services, 449 route des crêtes,
06560 Sophia-Antipolis, France
bernard.hotz@healthcenter.com

Abstract. We are presenting a method to obtain diagnosis and prognosis information by searching similar images into a large database of Myocardial Perfusion Scintigraphy (MPS) cases for which diagnosis is known. We are applying similarity measures to cardiac images pre-registered with a template. Our database is composed of 1430 patient cases with associated clinical information. For each new case, we sort all the patients of the database from most to less similar ones and compute a severity criterion, based on a statistical analysis of normal and diseased most similar patients. By varying a threshold on the severity criterion and testing the classification of controlled cases, we have measured the operational characteristic of this test (ROC curves), and shown increased performance in sensitivity and specificity for disease detection with respect to clinicians and to experts in consensus. Through the extension of database to patients' outcome information, we expect to extend this method to prognosis.

1   Introduction

The interpretation of medical images is often based on a comparison with a virtual normal reference image. This reference image can be the average of images of presumed normal subjects. This approach is restrictive because the comparison is made only with respect to a normal model. As there are many possible locations, extensions and degrees for cardiac abnormalities, it is not possible to generate a set of synthetic images representing abnormal cases, which is at the same time compact and extensive. All that we can do is to perform comparison relatively to a very large panel of presumed abnormal subjects, coming from a lot of different sources and supposed to recover the most frequent coronary artery disease patterns. We have designed a prototype, HeartPerfect, which is implementing such concept and relies on a database of 1430 rest and stress Myocardial Perfusion Scintigraphy (MPS) exams.

We first describe the method used to pre-process the data and then give the definition of the similarity measures that we are using. We are then showing one result of most similar patients search using our criteria. We define a severity measure of Coronary Artery Disease (CAD) from most similar patient profiles. At last, we compute the operational characteristic of our method for disease detection by comparing results of severity measure on controlled patients (i.e. patients with known clinical information or angio-verified cases.

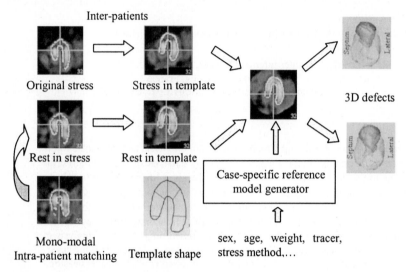

Figure 1: statistical stress and rest perfusion defects computation using CardioMatch® method and a case-specific reference model.

2   Pre-processing of the database

We are using 1430 stress/rest myocardial perfusion studies (MPS) from five institutions[1] collaborating in a prospective clinical trial to evaluate the performance of the CardioMatch® software (see [2]). All 1430 rest and stress images are registered into a common template geometry using an elastic spline 3D transformation to eliminate size and shape differences (see [1]). After this transformation stress or rest 3D images can be compared between each other at a voxel to voxel basis. The intensities in the image are normalized with respect to a common reference image presenting a normal perfusion, by using a linear regression and outliers rejection. A case specific reference model (CSRM) depending on a database of 100 normal cases as well as on the particular clinical parameters of that case is computed using multi-linear regression. It allows generating an infinity of adapted models without storing a too large amount of data. Parameters such as patient sex, weight and age, stress tracer and stress method are used to compute case-specific normal average and standard deviation images, which are then used to determine statistically significant 3D stress defects (see [3], as well as Figure 1). All cases of the database are pre-processed, once for all, using this method.

---

[1] Oakland, Santa Clara, South San Francisco, Stanford University and Walnut Creek, Kaiser trial.

3   Similarity criteria

We have defined three possible ways to obtain lists of similar profiles, with respect to:
1. similar regression parameters.
2. similar stress perfusion images.
3. similar stress perfusion defects.

For a given case, we generate the list of subjects presenting similar clinical parameters. Such parameter-based similarity measure can be used to restrict further exploration to only a subset of the database better suited to the patient's case. This is especially useful when large numbers of subjects are used, which is not our case here, as we have "only" about 1500 cases.

We started with a very simple image-based similarity criterion, which is linear correlation:

$$Linear\ correlation = \frac{\sum_{myocardium}(i_1 - \overline{i_1})(i_2 - \overline{i_2})}{\sqrt{\sum_{myocardium}(i_1 - \overline{i_1})^2 \sum_{myocardium}(i_2 - \overline{i_2})^2}}$$

More evolved similarity measures can be found in [6] or [4] for brain applications or [7], [8] for non-medical applications. However, the advantage of this criterion is that it is independent of intensity normalization, which reduces one possible cause of variability. It is also relatively fast to be computed: 2" on a 400MHz PC for a comparison with the 1430 cases of the database.

We have also developed a defects-based similarity criterion, being the overlap percentage of the defects between the reference patient case and the case to compare. The overlap is the number of voxels significantly abnormal in both cases divided by the number of voxels significantly abnormal in at least one case:

$$Defects\ overlap = \frac{_{card}\{v \mid v \in (defects_1 \cap defects_2)\}}{_{card}\{v \mid v \in (defects_1 \cup defects_2)\}}$$

4   Most similar patients search for a typical pathological case

To illustrate our method, we are presenting the 4 most image-based and defects-based similar patients obtained for a typical pathological case (see Figure 2 and Figure 3 respectively).

4.1 Image-based similar cases

In Figure 2 (right column) we present also the average image obtained from all cases whose correlation level is higher than 0.8 (16 cases in this particular example), each profile being weighted by a function of its correlation coefficient. Although this image is very smooth, we can clearly see hypo-perfusion in the infero-lateral location similar to the studied patient's case. Such image can be used for visual quality control to verify that the cases, which are found, are indeed similar.

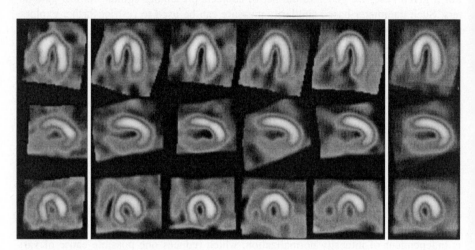

Figure 2: VLA, SA and HLA slices of stress images of the studied case (left). Middle are the 4 most similar images (from most to less similar). Right is the average image for correlation > 0.8.

	Correl.	Sex	Age	Weight	Tracer	Stress	Risk	History	Clinicians
Case	1.000	M	63	90.0	Thal	exercise	0.943	CABG	Fixed
Sim.1	0.875	M	64	95.0	Tetro	exercise	0.082	CABG	Mixed
Sim.2	0.874	M	61	81.8	Mibi	persantine	0.123	MI	Mixed
Sim.3	0.869	M	76	72.7	Tetro	exercise	0.791	CABG	Fixed
Sim.4	0.848	M	78	68.2	Tetro	persantine	0.943	MI	Ischemic

Table 1: clinical parameters associated with the similar images displayed in Figure 2. Stratification corresponds to Diamond and Forrester (D&F) pre-test probability (see [5]). CABG stands for Coronary Artery Bypass Graft, MI stands for Myocardial Infarction.

We are also giving the clinical information associated to these cases (see Table 1). We will see later on that such information, measured from a larger set of similar patients, can allow us to infer diagnosis and prognosis information.

4.2 Defects-based similar cases

Such measurement makes sense only for patients presenting large perfusion defects (size > 10% of the total myocardium volume, i.e. diseased cases). Results are presented in Figure 3. One interesting finding is that defects computed on the average image obtained with the image-based similarity measure (Figure 2, right column) have generally a larger defects overlap than all other images (Figure 3, right column).

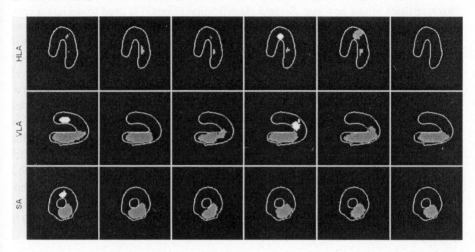

Figure 3: HLA, VLA and SA slices of stress perfusion defects images of the considered case (left), the 4 most similar defects images, from most to less similar (middle) and the defects image which corresponds to average image of Figure 2.

	Overlap	Sex	Age	Weight	Tracer	S method	Risk	History	Clinicians
Case	100%	M	63	90.0	Thal	exercise	0.943	CABG	Fixed
Sim.1	68.7%	M	76	72.7	Tetro	exercise	0.791	CABG	Fixed
Sim.2	64.0%	M	72	97.3	Tetro	exercise	0.609	Blank	Ischemic
Sim.3	63.7%	M	54	77.3	Tetro	exercise	0.998	CABG	Ischemic
Sim.4	62.8%	M	60	86.8	Tetro	adenosine	0.943	ANGIO +	Mixed

Table 2: parameters of similar defect images displayed in Figure 3

5   CAD severity measure based on similar patients' statistics

The main interest for building lists of similar profiles is to compile clinical information available on similar patients to extrapolate a risk of having a CAD for the considered case. In our database, each MPS was provided with the following information concerning CAD: *CAD risk stratification*, according to both Diamond/Forrester and Framingham methods. *"Second classification"* according to known history of CAD (10 possible values: angiography (negative, positive or minor), no history of CAD, transplant, evaluation of pre-operative risk, myocardial infarction, coronary artery bypass graft or percutaneous angioplasty, history of mediastinal radiation, percutaneous myocardial reperfusion). *Clinical interpretation* of MPS studies from Nuclear Physicians. *Quantitative results of CardioMatch®* defects analysis (location, relative size and degree of severity of abnormalities as a fraction of the myocardial mass).

In our experiments, we have used only information 2 and 3 to measure the operational characteristics of our new test. Second classification (Class2) is used to declare a case as "positive" (i.e. diseased) if it has a positive angiography, a Myocardial Infarction or a D&F pre-test probability larger than 0.9. It is declared as "negative" (i.e. normal) if it has a negative angiography or a D&F probability smaller than 0.1. The selection based on D&F stratification applies only to profiles without history of CAD. Out of our 1430 MPS cases, it remains 363 positive and 180 negative cases. Clinical interpretation is an alternative to "Class2". A case is classified as "positive" if it is considered as "ischemic", "fixed" or "mixed" by the nuclear physician and as "negative" if it is considered as "normal". In this way, we got 747 positive and 662 negative cases. Our severity criterion is the sum of all positive cases weighted by the correlation coefficient, normalized between a minimal correlation value and 1, minus the weighted negative cases. The minimal correlation level shall not be too high otherwise too few cases are defined as similar to perform statistical analysis. We have determined empirically that stable severity criteria can be obtained for a minimal correlation of 0.6 (see Figure 4).

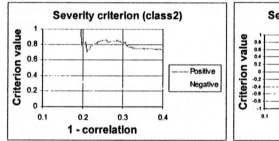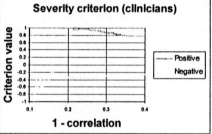

Figure 4: determining a minimal correlation level to define which cases to use in severity criteria.

6 Measuring ROC curves of severity criteria based on similarity

To measure the operational characteristics of our severity criteria, we are measuring results only for cases with an established ground truth, that is, 131 angio-verified cases. Of course, a given case is not used in the computation of its own severity criteria to avoid bias. The same data set of 131 angio-verified cases was also blindly classified by 3 experts in a consensus session (Dr. G. DePuey, St. Lukes Hospital, New York, Dr. J. Vansant, Emory University, Atlanta, Dr. H. Yasuda, Mass General, Boston). We also compare the results with the performance of the clinicians that were studying these cases on a regular basis as well as with CardioMatch® software results.

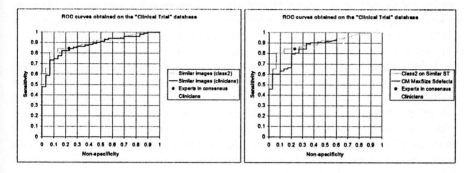

Figure 5: Left: comparing ROC for severity based on Class2 and severity based on Clinical diagnosis. Right: ROC curves to compare results obtained by clinicians, by experts in consensus, by CardioMatch® and by severity based on most similar patients (Class2).

We have first compared our two severity criteria to conclude that there is a small advantage in using Class2 classification instead of clinical diagnosis classification (see Figure 5). We have then compared results obtained by clinicians, by experts in consensus, by CardioMatch® and by severity based on most similar patients (Class2). For the same sensitivity than the clinicians (89%), both CardioMatch® and HeartPerfect are showing a much larger specificity (53% for CardioMatch®, 57% for HeartPerfect against 34% for clinicians). When compared to Experts in consensus (sensitivity = 84%, specificity = 77%), CardioMatch® is almost equivalent to experts (sensitivity = 84%, specificity = 73%, i.e. one case difference with respect to experts). For 84% sensitivity, HeartPerfect is reaching 87% specificity, which means a gain of 3 cases with respect to experts. We have also computed the distribution of negative and positive cases with respect to severity criteria (see **Figure 6**). It shows the discriminating power of the tests and also a slight advantage of the "class2" criterion over the criterion based on clinical information.

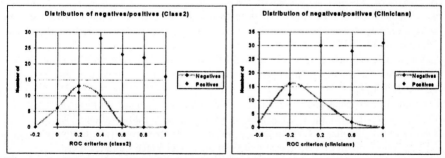

Figure 6: distribution of negative and positive cases with respect to severity criteria. Class2 (left) and clinical information (right).

7 Conclusion and perspectives

We have described in this paper HeartPerfect, a prototype for data mining based on image similarity measures for the exploration of large databases of MPS studies. It relies on the geometric 3D normalization of all images performed in a pre-processing step. From a given patient case, it can output cases in the database presenting either the same clinical parameters, similar images based on image correlation measurements, or patients with similar perfusion defects. By computing statistics on most similar patients, we demonstrate that such a system is able to achieve a classification into CAD diseased or normal with a better performance than physicians in clinical routine and slightly superior performance than experts in consensus. The interest of HeartPerfect is that it can evolve simply through the enrichment of the database; more cases would allow restricting the search of similar images to subclasses of patients presenting similar clinical parameters. Up to now, we have used our system for Computer Aided Diagnosis only. In the future, we plan to enrich the database with information about the different outcomes of the patients, which would allow deriving statistics on possible outcome and ultimately, could allow inferring prognosis information.

8 References

1. J. Declerck, J. Feldmar, M.L. Goris and F. Betting, "Automatic Registration and Alignment on a Template of Cardiac Stress & Rest SPECT Images", in Mathematical Methods in Biomedical Image Analysis, June 1996.
2. M.L. Goris, W. Pace, M. Petersen and A. Kwan, "The Effect of Lesion Modulation on the Operating Characteristics of Myocardial Perfusion Studies", EANM'98, Berlin, Germany, August 1998.
3. M.L. Goris, W.M. Pace, J-P. Thirion, B. Hotz and P. Similon, "Case specific reference images for patient and acquisition variables in Myocardial Perfusion Analysis", EANM'99, Barcelona, Spain, Vol. 26, Num. 9, September 1999.
4. A. Guimond, G. Subsol and J-P. Thirion, "Automatic MRI Database Exploration and Applications", in Int. J. on Pattern Recognition and Artificial Intelligence (IJPRAI), Vol. 11, Num. 8, pages 1345—1365, December 1997.
5. G.A. Diamond and J.S. Forrester, "Analysis of Probability as an Aid in the Clinical Diagnosis of Coronary-Artery Disease", in the New England Journal of Medicine, Vol. 300, Num. 24, pages 1350—1358, June, 1979.
6. A. Venot, J.F. Lebruchec, J.L. Golmard and J.C. Roucayrol, "An Automated Method for the Normalization of Scintigraphic Images", in the Journal of Nuclear Medicine, Vol. 24, pages 529—531, 1983.
7. C. Nastar, M. Mitschke, and C. Meilhac. "Efficient query refinement for image retrieval". In Computer Vision and Pattern Recognition (CVPR '98), Santa Barbara, June 1998.
8. S.Aksoy and R.Haralick, "Textural features for image database retrieval", in Proceedings of CVPR, Workshop on Content-Based Access of Image and Video Libraries, 1998.

Retrospective Correction of MR Intensity Inhomogeneity by Information Minimization

Boštjan Likar[1], Max A. Viergever[2], and Franjo Pernuš[1]

[1] University of Ljubljana, Department of Electrical Engineering
Tržaška 25, 1000 Ljubljana, Slovenia
bostjan.likar@fe.uni-lj.si
franjo.pernus@fe.uni-lj.si

[2] Image Sciences Institute, University Hospital Utrecht
P.O. Box 85500, 3508 GA Utrecht, The Netherlands
max@isi.uu.nl

Abstract. In this paper the problem of retrospective correction of intensity inhomogeneity in MR images is addressed. A novel model-based correction method is proposed, based on the assumption that an image corrupted by intensity inhomogeneity contains more information than the corresponding uncorrupted image. The image degradation process is described by a linear model, consisting of a multiplicative and an additive component which are modeled by a combination of smoothly varying basis functions. The degraded image is corrected by the inverse of the image degradation model. The parameters of this model are optimized such that the information of the corrected image is minimized while the global intensity statistic is preserved. The method was quantitatively evaluated and compared to other methods on a number of simulated and real MR images and proved to be effective, reliable, and computationally attractive.

1 Introduction

In magnetic resonance imaging (MRI), image intensity inhomogeneity is an adverse phenomenon which manifests itself by slow intensity variations of the same tissue class over the image domain. Intensity inhomogeneity may be caused by a number of factors including poor radio frequency (RF) coil uniformity, static field inhomogeneity, RF penetration, gradient-driven eddy currents, and overall patient anatomy and position [1], [2]. Spurious intensity variations, which may reach up to 30% of the image intensity amplitude [3], [4] usually do not affect the visual impression of the image significantly, but may have serious implications for MR image analysis, e.g., in segmentation, registration, or quantification.

Methods for correction of intensity inhomogeneities may be prospective or retrospective. The former require an acquisition protocol tuned to inhomogeneity correction, while the latter can be applied to any MR image, since they only use the information naturally occurring in an image. A number of prospective methods have been introduced in MRI, which either use phantom acquisitions [1], [2], [5]-[8] or measure the excitation field [9] or the static uniformity of the reception coil [10] *in vivo*. However, the patient-independence requirements and/or extended scan time makes them impractical for clinical use. As an alternative, a number of retrospective methods have been proposed. The most intuitive retrospective methods for correcting multiplicative and smooth intensity variations are homomorphic filtering (see e.g. [11]), image blurring [7], smoothing [13], averaging [14], Fourier domain filtering [15], and homomorphic unsharp masking [16]. The histogram matching method was also applied [17]. A number of methods were proposed that fit polynomials or thin-plate splines to manually or automatically selected points [12], to regions defined by inhomogeneity-tolerant preliminary segmentation [3], or directly to image data by using multiple-valley criterion functions instead of least squares [18]. Sled *et al.* presented an iterative optimization method for MRI, which seeks the smooth multiplicative field that maximizes the frequency content of the distribution of tissue intensity [19]. A segmentation-based method, using the expectation-minimization algorithm to iteratively classify and correct the image, was introduced in [20], improved in [4], and recently further extended in [21]. Segmentation-based methods that use fuzzy C-means clustering can be found in [22]-[24], while the usage of Markov random fields was considered in [25]. Filtering out and weighted re-integration of well-characterized local derivatives to estimate the bias field was proposed by Vokurka *et al.* [26]. In similar vein, in [27] a new variational shape-from-orientation approach was suggested. A review and evaluation of MRI non-uniformity corrections for brain tumor response measurements was provided by Velthuizen *et al.* [28], reporting that different methods give significantly different correction images and thus concluding that non-uniformity correction is not yet well understood.

In a recent paper [29], we described a retrospective shading correction method based on entropy minimization, designed for correcting the intensity inhomogeneity in 2D images. In this paper, we generalize the method to 3D MR images. The derivation of our algorithm is based on the assumption that an image corrupted by intensity inhomogeneity contains more information than the corresponding uncorrupted image. The method will be quantitatively evaluated and compared with two recently proposed methods, [19] and [26], on a large number of simulated and real MR images.

2 Theory

2.1 Problem formulation

Let $v(x)$ denote the acquired image and let $u(x)$ denote the "true" image of the imaged object. The two images are related by:

$$v(x) = f(u(x)) \quad \text{or} \quad u(x) = f^{-1}(v(x)) \tag{1}$$

with f denoting the image degradation model (bias field) that introduces a spatially dependent intensity degradation to the true image $u(x)$ and f^{-1} representing the inverse of the degradation model. The problem of retrospective correction of intensity inhomogeneity is to find the true image $u(x)$ from the acquired image $v(x)$.

2.2 Correction strategy

The proposed correction strategy is based on the assumption that because of the image degradation process the information content of the acquired image will be higher that the information content of the true image:

$$I[v(x)] = I[f(u(x))] > I[u(x)] \tag{2}$$

The information I of any image $v(x)$ can be quantitatively expressed by the Shannon-Wiener entropy $H[\,v(x)\,]$ as:

$$I[v(x)] = H[v(x)] = -\sum_{n} p(n) \log p(n) \tag{3}$$

where $p(n)$ is the probability that a point in image $v(x)$ has value n.

Suppose a correction model $\tilde{f}^{-1}$, which is constrained in such a way that it can not change the mean intensity of the input image $v(x)$ and can not transform the input image to a uniform image, is given. Then, an estimate $\tilde{u}(x)$ may be obtained as:

$$\tilde{u}(x) = \tilde{f}^{-1}(v(x)) \tag{4}$$

According to the assumption made in (2), transforming the image $v(x)$ by the correction model $\tilde{f}^{-1}$ (4) in such a way that the information I of the transformed image $\tilde{u}(x)$ is minimized should lead to the optimal correction model $\tilde{f}_o^{-1}$:

$$\tilde{f}_o^{-1} = \arg\min_{\tilde{f}^{-1}}\left\{ I\left[\tilde{f}^{-1}(v(x))\right] \right\} \tag{5}$$

which defines the transformation of the image $v(x)$ to the corrected image $\tilde{u}_o(x)$:

$$\tilde{u}_o(x) = \tilde{f}_o^{-1}(v(x)) \tag{6}$$

3 Correction by a linear model

3.1 Modeling

The linear model of image degradation consists of a multiplicative $m(x)$ and an additive $a(x)$ intensity degradation component:

$$v(x) = u(x)\, m(x) + a(x) \tag{7}$$

By the inverse of the degradation model, the estimation $\tilde{u}(x)$ is obtained as:

$$\tilde{u}(x) = v(x)\, \tilde{m}^{-1}(x) + \tilde{a}^{-1}(x) \tag{8}$$

where

$$\tilde{m}^{-1}(x) = \frac{1}{\tilde{m}(x)} \quad \text{and} \quad \tilde{a}^{-1}(x) = -\frac{\tilde{a}(x)}{\tilde{m}(x)} \tag{9}$$

are the multiplicative and additive correction components, respectively. These two components are described by smoothly varying basis functions $s_i(x)$:

$$\tilde{m}^{-1}(x) = \sum_{i=1}^{K} {}^m b_i \, {}^m s_i(x) \quad \text{and} \quad \tilde{a}^{-1}(x) = \sum_{i=1}^{K} {}^a b_i \, {}^a s_i(x) \tag{10}$$

that are uniquely defined by parameters ${}^m b$ and ${}^a b$, respectively.

To neutralize the global transformation effect (8) the correction components have on the acquired image $v(x)$, we introduce the mean-preserving condition:

$$\frac{1}{\Theta} \int_\Omega v(x) d\Omega = \frac{1}{\Theta} \int_\Omega \left(v(x)\tilde{m}^{-1}(x) + \tilde{a}^{-1}(x) \right) d\Omega \tag{11}$$

which ensures that the mean intensity values of $v(x)$ and $\tilde{u}(x)$, defined over the relevant image domain Ω of size Θ; $\Theta = \int_\Omega d\Omega$, will be the same.

To equalize the mean absolute contributions of individual basis functions $s_i(x)$ to the global intensity transformation, the parameters ${}^m b$ and ${}^a b$ are normalized:

$$\frac{1}{\Theta} \int_\Omega |v(x)^m s_i(x)| d\Omega = 1 \quad \text{and} \quad \frac{1}{\Theta} \int_\Omega |{}^a s_i(x)| d\Omega = 1 \quad \text{for all } i \tag{12}$$

As the smoothly varying basis functions $s_i(x)$ we use polynomial terms $q_i(x)$ in the following form:

$$^m s_i(x) = \frac{q_i(x) - {}^m c_i}{{}^m d_i} \quad \text{and} \quad {}^a s_i(x) = \frac{q_i(x) - {}^a c_i}{{}^a d_i} \tag{13}$$

with neutralization constants c_i, needed to fulfil the mean-preserving condition (11), and normalization constants d_i, needed to fulfil (12). The two correction components $\tilde{m}^{-1}(x)$ and $\tilde{a}^{-1}(x)$, which define the correction model (8), are modeled by the linear combination of neutralized and normalized polynomials that fulfil (11) and (12):

$$\tilde{m}^{-1}(x) = 1 + \sum_{i=2}^{K} {}^m b_i \frac{q_i(x) - {}^m c_i}{{}^m d_i} \quad \text{and} \quad \tilde{a}^{-1}(x) = \sum_{i=2}^{K} {}^a b_i \frac{q_i(x) - {}^a c_i}{{}^a d_i} \tag{14}$$

3.2 Correction

The optimal parameters ${}^m b_o$ and ${}^a b_o$ are found by Powell's multi-dimensional directional set method and Brent's one-dimensional optimization algorithm [30]:

$$\left\{ {}^m b_o, {}^a b_o \right\} = \arg \min_{\{^m b, ^a b\}} \left\{ I\left[v(x) \, \tilde{m}^{-1}(x) + \tilde{a}^{-1}(x) \right] \right\} \tag{15}$$

The optimal parameters ${}^m b_o$ and ${}^a b_o$ define the optimal components $\tilde{m}_o^{-1}(x)$ and $\tilde{a}_o^{-1}(x)$, respectively, which transform the acquired image $v(x)$ into the optimally corrected image $\tilde{u}_o(x)$:

$$\tilde{u}_o(x) = v(x) \, \tilde{m}_o^{-1}(x) + \tilde{a}_o^{-1}(x) \tag{16}$$

3.3 Implementation details

Three variations of the proposed information minimization method are implemented. In the first method, named MA2, the multiplicative and additive component are modeled by a second order polynomial. The other two methods, named M2 and M4, consist solely of a multiplicative component, which is modeled by a second and fourth order polynomial, respectively.

The correction domain Ω is extracted by thresholding the image and subsequently eroding the obtained non-background area, using a simple six-voxel structuring element. Prior to the optimization process, the neutralization constants c_i and normalization constants d_i of the correction components (14) are calculated.

Effective calculation of the entropy H is crucial for the performance of the proposed method. The entropy (3) is defined from a set of probabilities $p(n)$ that can be obtained from the intensity histogram of the current estimate $\tilde{u}(x)$. Since $\tilde{u}(x)$ is obtained by an intensity transformation applied to a given image, an integer gray value g is transformed to a new real value g', which in general lies between two integer values, say k and $k+1$. We use partial intensity interpolation by which the histogram entries $h(k)$ and $h(k+1)$ are fractionally updated by $k+1-g'$ and $g'-k$, respectively. Prior to the calculation of the set of probabilities $p(n)$, the histogram $h(n)$ is slightly blurred to reduce the effects of imperfect intensity interpolation. To provide a high statistical power, which is proportional to the number of samples used to form a 1D histogram from a 3D image, we use 8 bit intensity quantization. Due to the high statistical power, histograms may be formed by substantially sub-sampling the image data without affecting the method's performance.

4 Experiments and results

4.1 Validating the correction strategy

In this section we tested the proposed correction strategy using the MA2 method. For this purpose, two simulated MR T1-weighted images [31]-[33], one with 0% (uniform image) and the other with 40% intensity non-uniformity, were used. In Fig. 1 the recovery of the intensity distribution can be observed by comparing the histograms of the uniform, non-uniform, and corrected image.

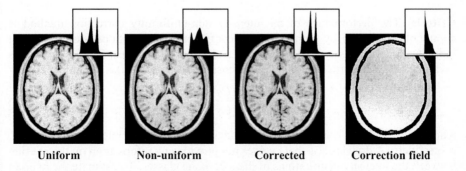

| Uniform | Non-uniform | Corrected | Correction field |

Fig. 1. Simulated uniform and non-uniform (40% intensity inhomogeneity) images, and the corresponding retrospective correction of the non-uniform image.

4.2 Evaluation

Methods. Five methods were included in the quantitative evaluation, i.e. the so-called MA2, M2, and M4 variations of the information minimization method described in this paper, the nonparametric nonuniform intensity normalization (N3) method proposed by Sled *et al.* [19], and the fast model independent (FMI) method introduced by Vokurka *et al.* [26]. The N3 and FMI methods were implemented at our site by using the code available via internet at [34] and [35], respectively. The implementations were validated. The parameters of the N3 and FMI methods were set to their default values.

Experimental data. One set of simulated and four sets of real MR data were used:

Set 1: Consisted of 12 (4 x T1, T2, and PD-weighted) MRI volumes from the BrainWeb MR simulator (181x217x181voxels, 8 bit, 1mm slice thickness, 3% noise) [31]-[33].

Set 2: Six (2 x T1, T2, and PD-weighted) MRI volumes, 3 of a normal volunteer (256x256x25 voxels, 8bit) and 3 of a brain tumor patient with multiple malignant meningioma lesions (256x256x20 voxels, 8 bit), were used. A complete gray and white matter manual segmentation was available for each volume [36].

Set 3: The set contained 33 (11 x T1, T2, and PD-weighted) real MR images (181x217x181 voxels, 8 bit) with manual labeling of the regions of pure gray and white matter, acquired at 11 different sites from 11 multiple sclerosis patients having a moderate number of white matter lesions [19].

Set 4: The set contained 18 normal coronal three-dimensional T1-weighted spoiled gradient echo MRI scans acquired on two different imaging systems. A complete segmentation was available [37]-[39]. The images were all 8 bit with the resolution of either 128x128 or 160x160 and they had 50 to 58 slices.

Set 5: A breast T1-sagittal image (1T Siemens Magnetom Harmony, body coil, 4 mm slice thickness, 256x256x20 voxels, 8 bit) and a prostate T1-axial image (1T Siemens Magnetom Harmony, phased array, 3 mm slice thickness, 256x256x20 voxels, 8 bit) were used.

Criteria. The performance of an intensity non-uniformity correction method is commonly evaluated by comparing the coefficients of variations (*cv*) within the individual tissue classes in the original and corrected images. The *cv*, which is invariant to the uniform multiplicative intensity transformation, is computed as:

$$cv(C) = \frac{\sigma(C)}{\mu(C)} \tag{17}$$

where $\sigma(C)$ and $\mu(C)$ are the standard deviation and mean intensity of class C, respectively. A drawback is that *cv* does not provide any information on the overlap between the intensity distributions of distinct tissue classes. To estimate the overlap we introduce the so-called coefficient of joint variations $cjv(C_1, C_2)$:

$$cjv(C_1,C_2) = \frac{\sigma(C_1)+\sigma(C_2)}{|\mu(C_1)-\mu(C_2)|} \qquad (18)$$

which is the sum of the standard deviations of two distinct classes; C_1 and C_2, normalized by the difference of means. The cjv is invariant to the uniform linear, i.e. multiplicative and additive, intensity transformation and is thus a more general quantitative measure than the cv.

Results. Quantitative evaluation of the MA2, M2, M4, N3, and FMI methods was performed by computing the cjv(GM,WM) of the gray and white matters for all images from the first four experimental sets.

Table 1 contains the results of non-uniformity correction of the images from the first two sets. Ideally the cjv of the simulated uniform and corrected uniform images of set 1 should be identical. This was approximately true for all methods except the FMI method, which corrupted the images with 0% non-uniformity. The corrected images with simulated 40% intensity non-uniformity indicate that all methods, except the FMI, can efficiently reduce the overlap of their intensity distributions. The cjv-s of all corrected images with 40% non-uniformity are similar to the cjv-s of the images with 0% non-uniformity for all methods, except for the FMI method and the N3 method when applied to a normal T2 image. The methods applied to the real normal and tumor volumes (set 2) yielded similar reductions of the non-uniformity criteria, with the exception of M4 method, which slightly increased the cjv of T2 tumor volume, and the FMI method, which increased the cjv of the three normal volumes.

Table 1. Correction of simulated and real data of image sets 1 and 2.

Set	Mod	Bias	cjv(GM,WM) in [%]					
			Orig.	MA2	M2	M4	N3	FMI
1 - Normal	T1	0%	51.6	52.0	52.1	52.1	51.9	72.5
(simulated)		40%	69.3	51.6	51.5	51.7	51.8	96.4
	T2	0%	83.2	82.1	82.6	82.5	83.1	125.5
		40%	106.4	82.3	82.4	82.4	103.9	116.8
	PD	0%	64.9	64.7	64.7	65.0	66.4	93.5
		40%	163.0	64.3	64.5	64.1	66.7	107.9
1 - MS lesions	T1	0%	50.9	51.4	51.5	51.5	51.3	88.0
(simulated)		40%	68.0	50.8	50.8	51.1	51.1	65.3
	T2	0%	74.9	74.1	74.2	74.3	74.9	150.3
		40%	123.8	73.7	74.1	74.2	77.8	98.3
	PD	0%	66.9	66.7	66.7	67.0	68.7	117.3
		40%	195.6	66.5	66.6	66.1	68.7	125.9
2 - Normal	T1		138.0	131.4	133.0	134.0	127.0	144.3
(real)	T2		90.7	88.4	90.4	89.8	90.7	105.0
	PD		77.8	70.7	72.1	72.1	73.2	87.1
2 - Tumor	T1		175.7	167.5	168.6	170.4	168.1	173.6
(real)	T2		169.3	164.7	166.6	175.5	161.3	163.8
	PD		133.8	118.0	115.2	118.2	114.6	121.3

Table 2 summarizes the mean changes of the cjv(GM,WM) for the image sets 3 and 4. The FMI method performed poorly on nearly all images from the two sets. On T1 and T2 images from set 3 the cjv were best reduced by the M4 method. The MA2 method was the most effective in correcting the PD images from set 3, while the T1 images from set 4 were again best corrected by the M4 method.

Table 2. Mean changes of cjv(GM,WM) in [%] for the images from sets 3 and 4.

Set	Mod	MA2	M2	M4	N3	FMI
3	T1	-4.39	-2.78	-5.92	-0.28	44.62
3	T2	-18.10	-23.65	-29.58	-21.69	17.47
3	PD	-35.76	-27.97	-30.66	-28.90	-2.19
4	T1	-0.14	-0.04	-3.54	-0.12	5.56

In general, the M4 method performed the best from the methods tested and was also faster than the N3 and FMI methods. A successful application of the M4 method to the non-brain images from set 5 is illustrated in Fig. 2.

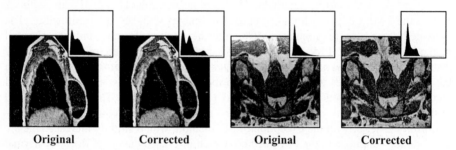

Original   Corrected   Original   Corrected

Fig. 2. Correction of the breast and prostate image by the M4 method.

5. Conclusion

A novel information-theoretic approach to retrospective correction of intensity inhomogeneities in MRI, based on modeling the bias field and minimizing the information of the acquired images by the parametric polynomial model, was proposed. The modeling approach offers the possibility of incorporating knowledge about the image formation process in the correction algorithm, while the global intensity uniformity criterion, i.e. the entropy, enables the optimization of arbitrarily complex image formation models. The method can be widely applied to different types of MR images because it solely uses the information that is naturally present in an image, without making assumptions on its spatial and intensity distribution. The proposed method requires no preprocessing, parameter setting, nor user interaction and my thus be a valuable tool in MR image analysis.

Acknowledgements

This work was supported by the Ministry of Science and Technology of the Republic of Slovenia under grant J2-0659-1538 and by the IST-1999-12338 project, funded by the European Commission. The authors express their sincere thanks to Robert P. Velthuizen, John G. Sled, Elizabeth A. Vokurka, Chris Stork, and Andrew J. Worth for their generous help and cooperation.

References

1. B. R. Condon, J. Patterson, D. Wyper, A. Jenkins, and D. M. Hadley, "Image non-uniformity in magnetic resonance imaging: its magnitude and methods for its correction." *The British Journal of Radiology,* 60:83-87, 1987.
2. A. Simmons, P. S. Tofts, G. J. Barker, and S. Arridge, "Sources of intensity nonuniformity in spin echo Images at 1.5 T." *Magnetic Resonance in Medicine,* 32:121-128, 1994.
3. C. R. Meyer, P. H. Bland, and J. Pipe, "Retrospective correction of intensity inhomogeneities in MRI." *IEEE Transactions on Medical Imaging,* 14:36-41, 1995.
4. R. Guillemaud and M. Brady, "Estimating bias field of MR images." *IEEE Transactions on Medical Imaging,* 16:238-251, 1997.
5. D. A. G. Wicks, G. J. Barker, and P. S. Tofts, "Correction of intensity nonuniformity in MR images of any orientation." *Magnetic Resonance Imaging,* 11:183-196, 1993.
6. M. Tincher, C. R. Meyer, R. Gupta, and D. M. Williams, "Polynomial modeling and reduction of RF body coil spatial inhomogeneity in MRI." *IEEE Transactions on Medical Imaging,* 12:361-365,1993.
7. L. Axel, J. Costantini, and J. Listerud, "Intensity correction in surface-coil MR imaging." *American Journal of Radiology,* 148:418-420, 1987.
8. E. R. McVeigh, M. J. Bronskill, and R.M. Henkelman, "Phase and sensitivity of receiver coils in magnetic resonance imaging." *Medical Physics,* 13:806-814, 1986.
9. R. Stollberger and P. Wach, "Imaging of the active B_1 field *in vivo*." *Magnetic Resonance in Medicine,* 35:246-251, 1996.
10. P. A. Narayana, W. W. Brey, M. V. Kulkarni, and C. L. Sivenpiper, "Compensation for surface coil sensitivity variation in magnetic resonance imaging." *Magnetic Resonance Imaging,* 6:271-274, 1988.
11. B. Johnson, M. S. Atkins, B. Mackiewich, and M. Anderson, "Segmentation of multiple sclerosis lesions in intensity corrected multispectral MRI." *IEEE Transactions on Medical Imaging,* 15:154-169, 1996.
12. B. M. Dawant, A. P. Zijdenbos, and R. A. Margolin, "Correction of intensity variations in MR images for computer-aided tissue classification." *IEEE Transactions on Medical Imaging,* 12:770-781, 1993.
13. W. W. Brey and P. A. Narayana, "Correction for intensity falloff in surface coil magnetic resonance imaging." *Medical Physics,* 15:241-245, 1988.
14. A. Koivula, J. Alakuijala, and O. Tervonen, "Image feature based automatic correction of low-frequency spatial intensity variations in MR images." *Magnetic Resonance Imaging,* 15:1167-1175, 1997.
15. J. Haselgrove and M. Prammer, "An algorithm for compensation of surface-coil images for sensitivity of the surface coil." *Magnetic Resonance Imaging,* 4:469-472, 1986.
16. B. H. Brinkmann, A. Manduca, and R. A. Robb, "Optimized homomorphic unsharp masking for MR greyscale inhomogeneity correction." *IEEE Transactions on Medical Imaging,* 17:161-171, 1998.
17. L. Wang, H. M. Lai, G. J. Barker, D. H. Miller, and P. S. Tofts, "Correction for variations in MRI scanner sensitivity in brain studies with histogram matching." *Magnetic Resonance in Medicine,* 39:322-327, 1998.
18. C. Brechbühler, G. Gerig, and G. Székely, "Compensation of spatial inhomogeneity in MRI based on a parametric bias estimate.", in *Visualisation in Biomedical Computing, Proceedings of VBC'96, Lecture Notes in Computer Science,* 1131:141-146, 1996, Springer-Verlag.
19. J. G. Sled, A. P. Zijdenbos, and A. C. Evans, "A nonparametric method for automatic correction of intensity nonuniformity in MRI data." *IEEE Transactions on Medical Imaging,* 17:87-97, 1998.

20. W. M. Wells III, W. E. L. Grimson, and F. A. Jolesz, "Adaptive segmentation of MRI data." *IEEE Transactions on Medical Imaging*, 15:429-442, 1996.

21. K. van Leemput, F. Maes, D. Vandermeulen, and P. Suetens, "Automated model-based bias field correction of MR images of the brain." *IEEE Transactions on Medical Imaging*, 18:885-896, 1999.

22. S. K. Lee and M. W. Vannier, "Post-acquisition correction of MR inhomogeneity." *Magnetic Resonance in Medicine*, 36:275-286, 1996.

23. D. L. Pham and J. L. Prince, "An adaptive fuzzy C-means algorithm for image segmentation in the presence of intensity inhomogeneities." *Pattern Recognition Letters*, 20:57-68, 1999.

24. D. L. Pham and J. L. Prince, "Adaptive fuzzy segmentation of magnetic resonance images." *IEEE Transactions on Medical Imaging*, 18:737-752, 1999.

25. J. C. Rajapakse and F. Kruggel, "Segmentation of MR images with intensity inhomogeneities." *Image and Vision Computing*, 16:165-180, 1998.

26. E. A. Vokurka, N. A. Thacker, and A. Jackson, "A fast model independent method for automatic correction of intensity nonuniformity in MRI data." *Journal of Magnetic Resonance Imaging*, 10:550-562, 1999.

27. S. H. Lai and M. Fang, "A new variational shape-from-orientation approach to correcting intensity inhomogeneities in magnetic resonance images." *Medical Image Analysis*, 3:409-424, 1999.

28. R. P. Velthuizen, J. J. Heine, A. B. Cantor, H. Lin, L. M. Fletcher, and L. P. Clarke, "Review and evaluation of MRI nonuniformity corrections for brain tumor response measurements." *Medical Physics*, 25:1655-1666, 1998.

29. B. Likar, J. B. A. Maintz, M. A. Viergever, and F. Pernuš, "Retrospective shading correction based on entropy minimization." *Journal of Microscopy*, 197:285-295, 2000.

30. W. H. Press, B. P. Flannery, S. A. Teukolosky, and W.T. Vetterling, "Numerical Recepies in C." Second Edition, *Cambridge University Press*, 1992.

31. C. A. Cocosco, V. Kollokian, R. K.-S. Kwan, and A. C. Evans, "BrainWeb: Online Interface to a 3D MRI Simulated Brain Database." *NeuroImage* 5:S425, 1997. Available: *http://www.bic.mni.mcgill.ca/brainweb/*

32. R. K.-S. Kwan, A. C. Evans, and G. B. Pike, "An Extensible MRI Simulator for Post-Processing Evaluation." *in Visualisation in Biomedical Computing, Proceedings of VBC'96, Lecture Notes in Computer Science*, 1131:135-140, 1996, Springer-Verlag.

33. D. L. Collins, A. P. Zijdenbos, V. Kollokian, J. G. Sled, N. J. Kabani, C. J. Holmes, and A. C. Evans, "Design and Construction of a Realistic Digital Brain Phantom." *IEEE Transactions on Medical Imaging*, 17:463-468, 1998.

34. *ftp://ftp.bic.mni.mcgill.ca/pub/mni_n3/*

35. *http://www.niac.man.ac.uk/*

36. R. P. Velthuizen, L. P. Clarke, S. Phuphanich, L. O. Hall, A. M. Bensaid, J. A. Arrington, H. M. Greenberg, and M. L. Silbiger. "Unsupervised measurement of brain tumor volume on MR images." *Journal of Magnetic Resonance Imaging* 5:594-605, 1995.

37. J. Talairach and P. Tournoux, "Co-planar stereotaxic atlas of the human brain." *New York: Thieme*, 1988.

38. P. A. Filipek, D. N. Kennedy, and V. S. Caviness, "Volumetric analysis of central nervous system neoplasm based on MRI." *Pediatric Neurology*, 7:347-351, 1991.

39. D. N. Kennedy, P. A. Filipek, and V. S. Caviness, "Anatomic segmentation and volumetric calculations in nuclear magnetic resonance imaging." *IEEE Transactions on Medical Imaging*, 8:1-7, 1989.

Comparision of Three High-End Endoscopic Visualization Systems on Telesurgical Performance

David Mintz[1], Volkmar Falk, MD[2], and J. Kenneth Salisbury, Jr[3]

[1] DM is a Senior Systems Analyst with Intuitive Surgical Inc,
1340 W. Middlefield Rd., Mountain View, CA 94043
david_mintz@intusurg.com
http://www.intusurg.com

[2] VF is with the Department of Cardiothoracic Surgery Stanford School of Medicine

[3] KS is a Scientific Advisor at Intuitive Surgical Inc. and a member of the faculty in Computer Science at Stanford University

Abstract. Previous studies and surgeon interviews have shown that most surgeons prefer quality standard definition (SD)TV 2D scopes to first generation 3D endoscopes. The use of a telesurgical system has eased many of the design constraints on traditional endoscopes, enabling the design of a high quality SDTV 3D endoscope and an HDTV endoscopic system with outstanding resolution. The purpose of this study was to examine surgeon performance and preference given the choice between these. The study involved two perceptual tasks and four visual-motor tasks using a telesurgical system using the 2D HDTV endoscope and the SDTV endoscope in both 2D and 3D mode. The use of a telesurgical system enabled recording of all the subjects motions for later analysis. Contrary to experience with early 3D scopes and SDTV 2D scopes, this study showed that despite the superior resolution of the HDTV system surgeons performed better with and preferred the SDTV 3D scope.

1 Background

Various studies have examined the effect of 3D versus 2D imaging systems on performance for a variety of laparoscopic tasks or procedures. These studies led to inconclusive results. While some reports demonstrate a clear benefit for 3D vision systems, with a reduction in overall procedure time [1] [2] [3], others could not confirm these results [4] [5] [6].

These studies were performed using first generation 3D systems which have lower resolution than standard 2D systems. The first generation 3D endoscopes used were primarily single channel scopes with minimal stereo separation, resulting in limited stereopsis. These stereo scopes also all used single chip CCD cameras with limited video resolution. Most of the 3D camera systems used frame interlaced displays with shutter glasses to present the 3D images to the surgeons.

Despite these limitations, a number of conclusions can be drawn from the studies referenced above. Depth perception is regarded to be superior with 3D by most users. With first generation systems, however, this did not always result in improved in performance, particularly for more complex tasks and for users with high levels of endoscopic experience. This is supported by our interviews of experienced endoscopic surgeons, where we found that they prefer quality 2D scopes with 3 CCD cameras to low resolution 3D endoscopes. The current lack of commercial promotion of these first generation 3D systems by endoscope manufacturers is anecdotal evidence which further supports this view.

The design of the first generation endoscopes used in these studies was severely constrained by the size and weight that a surgeon is willing to support throughout a surgical procedure. In contrast, the *da Vinci*™ telesurgical system used for this study employs a robot arm, with a comparatively large payload capability, to hold the endoscope during the procedure. The easing of size and weight constraints afforded by the robot arm allowed the development of an endoscope with superior stereo effect and three-chip CCD cameras resulting in a system with resolution on par with current 2D scopes. These less stringent design specifications also enabled the use of a very high resolution 2D scope attached to a three-chip HDTV camera that would be prohibitively large for use in a standard endoscopic case.

The primary motivation for our study was to look at surgeon performance and preference given the choice between an endoscope with superior 3D effect and three-chip CCD's, and a 2D scope with resolution superior to anything currently available. We seek to determine whether surgeons would continue to value additional resolution above depth perception in a camera system, or if passing a certain threshold in quality for the 3D scope has changed this rule.

2 Methods

In this study we compared the perceptions and performance of 15 experienced endoscopic surgeons ($>$ 100 endoscopic cases a year) when using three different endoscopic viewing systems with the *da Vinci*™ telesurgical system. The three camera systems were: the 3D system that is standard with the *da Vinci*™ system; this same system in 2D mode; and a high quality 2D scope and prototype HDTV camera system. Prior to participating in the study, each of the surgeons went through a standard visual acuity test (eye chart). Only subjects with a corrected acuity of 20/20 or better were included. The study consisted of several perceptual tasks, followed by a group of visual-motor tasks, and finally a questionnaire.

2.1 The Telesurgical system

The *da Vinci*™ system is a surgical teleoperator developed by Intuitive Surgical Inc. of Mountain View, CA. The *da Vinci*™ system creates an immersive operating environment for the surgeon by providing both high quality stereo visualization and a man-machine interface that directly connects the surgeon's hands to the motion of his surgical tool tips inside the patient's body [7]. The registration, or alignment, of the surgeon's hand motion to the motion of the surgical tool tips is both visual and spatial. The system projects the 3D image of the surgical site atop the surgeon's hands (via two CRT's and mirrored overlay optics), restoring hand-eye coordination and providing a natural correspondence in motions. Furthermore, the controller transforms the spatial motion of the tools into the camera frame of reference, so that the surgeon feels as if his hands are inside the patient's body. Lastly, the *da Vinci*™ system restores the degrees of freedom lost in conventional laparoscopy by placing a 3 degree-of-freedom wrist inside the patient and controlling this wrist naturally, bringing a total of seven freedoms of motion to the control of the tool tip (3 orientation, 3 translation and grip). The system also uses its digital controller to filter out surgeon tremor, making the tool tips steadier than the unassisted hand. Additionally, the system allows for variable motion scaling from the masters to the slaves. For example, a 3:1 scale factor maps 3 cm of translation on the masters into 1 cm of translation at the slaves.

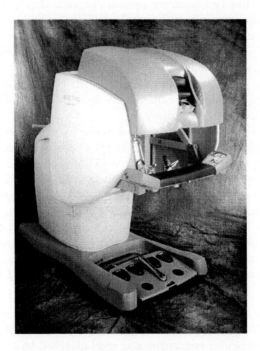

Fig. 1. *da Vinci*™ System Surgeon console

The surgeon's console on the *da Vinci*™ system, Fig. 1, transmits the image of the surgical site to the surgeon through a high-resolution stereo display, which uses two medical grade CRT monitors to display one image to each of the surgeon's eyes [7]. The virtual image plane of the stereo viewer is placed conincident with the range of motion of the masters using two sets of mirriors. This stereo viewer supports standard definition television (NTSC or PAL) signals. For this study NTCS was used. In order to facilitate the use of the HDTV scope, a custom surgeon's console was built that uses an HDTV monitor and two mirrors to place the high resolution 2D image in the workspace of the masters, thereby replicating the immersive sensation of the production console.

2.2 Endoscopic Camera systems

Two vision systems were used for this study. The *InSite*™ system was used to provide both the 2D and 3D enviornments in NTSC format. A second vision system, including a high quality monocular endoscope, an HDTV camera head and a custom built stereo viewer, was used to provide the 2D HDTV view.

InSite™ **Vision system.** The *InSite*™ vision system, provided standard with *da Vinci*™, includes a 3D-NTSC endoscope, developed by Precision Optics Corp. This scope has two separate optical channels that provide wide stereo separation while maintaining symmetric optical distortion in each eye. The payload of the *da Vinci*™ system camera arm enabled the scope and camera head designers to opt for higher quality whenever there was a design trade-off between quality and size or weight. The result is a scope with a stereo separation of 5.5 mm, giving good stereo effect, and optical resolution of 3.3 mrad/lp for each channel. Two 3-chip CCD cameras are used to capture the full resolution provided by the optics of the endoscope.

The stereo viewer in the surgeon console is specially designed to match the endoscope. The nominal working distance for the endoscope, when used in cardiac surgery is 38 mm. Given the 5.5 mm stereo separation of the optical channels, the angle of convergence with the tissue plane is 8.3 deg. Human interoccular spacing is approximately 6.5 cm. The stereo viewer was designed such that the image plane is placed 43 cm away from the eye piece. The angle subtended by the surgeon's eyes is the 8.6 deg, resulting in a realistic stereo effect. This scope was used in 2D mode for a portion of the study by providing the image captured by one of its optical chains to both monitors in the surgeon console. This allows a direct comparison of 2D vs. 3D performance independent of resolution.

HDTV System. The prototype HDTV endoscopic system used in this study has exceptional resolution (1.4 mrad/lp) throughout its field of view. A high quality 10 mm monocular endoscope was used attached to a broadcast quality 3-chip CCD HDTV camera to obtain this resolution. Because the stereo viewer included in the production *da Vinci*™ surgeon console cannot display HDTV signals, a custom HDTV-capable console was built. The viewer in this console consisted of an HDTV monitor and a pair of mirrors used to project the image into the master manipulator workspace. The design of this console mimicked that of the SDTV stereo viewer described above.

2.3 Experiments

All subjects underwent a structured introduction to the system and were allowed to become acquainted with its features. The two scopes were set at the same fixed distance from the tasks.

Six tasks were performed using each visualization system. Subjects completed each of the tasks with a given visualization system before moving on to the next. The order in which the visualization systems were presented to the surgeons was randomized in order to minimize any bias due to learning. The tasks were always presented in the same order.

The first two tasks were purely perceptual in nature while the remaining four involved manipulations with the telesurgical system. This permitted a separation of the subjects' perceptual and sensorimotor skills under each viewing condition. The six tasks are outlined below.

1. **Resolution Estimation.** Subjects were presented standard resolution patterns and asked to determine the level of resolution they could discriminate in both the horizontal and vertical directions.
2. **Relative Distance Estimation** Subjects were presented with a set of six pegs of varying height positioned at varying distances from the scope. This was designed to cause confusion between distance and size. They were told that the pegs were each in one of three rows. The subjects were asked to classify each peg as near, medium, or far.
3. **Switch Touching.** Each subject was given control of a telesurgical manipulator with his dominant hand and told to touch each of four switches placed at varying heights about the viewing field. The state of the switches was recorded, along with the position and orientation of the manipulator tip.
4. **Peg Placement.** Using the same manipulator subjects was directed to pick up cylindrical beads from a starting location and place them on the pegs viewed in task 2. The number of beads dropped as well as the total time to place all six beads was measured. Dropped beads were returned to the starting position for the subjects to retrieve and continue with the task.
5. **Suturing.** Subjects were asked to perform a running suture (4-0 silk) connecting printed dots (distance 4 mm horizontally, 5 mm vertically) on a fixed rubber glove using two telesurgical instruments (needleholders). Time to complete the task and errors (stitch outside a 1mm circle around the marked dot) were recorded.
6. **Knot Tying.** Using the same two instruments, each subject was instructed to tie three instrument knots using a 4-0 Silk suture that was previously stitched into a rubber glove. Time to perform the task was recorded.

2.4 Questionnaire

After completion of the tasks, all subjects were interviewed with regard to their performance under the various viewing conditions using a standardized questionnaire. System-related questions included subjective assessment of resolution, depth perception and color perception with the different set-ups.

3 Results

The results of the study are summarized below. A full statistical analysis has not been provided, but can be found in Falk et al. [8]

3.1 Experiments

1. **Perceived Resolution.** As expected the HDTV system showed significantly better perceived resolution both horizontally and vertically when compared to both of the 2D and 3D NTSC setups. Interestingly, there was a perceived resolution difference between the 2D and 3D NTSC. Although the difference was small, both the horizontal and vertical resolution was perceived as better in the 3D mode. This may perhaps be explained by the fact that various cerebral summation effects (by cortical vision centers) enhance among other factors acuity and pattern recognition beyond what can be expected from two equal monocular inputs [9]. According to Cagnello and Rabin, acuity is enhanced by roughly 10% when comparing binocular to monocular vision [10]. Given the same resolution, a 3D-system may therefore not only enhance depth perception, but also give better visualization.

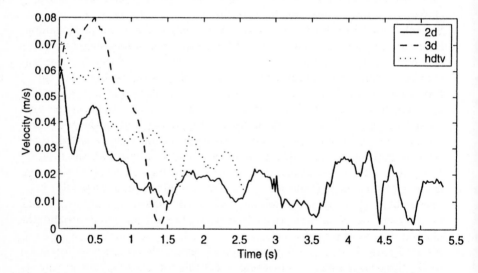

Fig. 2. Example velocity profile. (Subject #3 moving from switch 1 to 2)

2. **Switch Touching Task.** The large amount of raw data taken during this task has allowed a significant amount of post processing and analysis. In the interest of space, not all the metrics studied will be reported here. A more complete analysis of these data is available in Falk et al. [8]. The 3D NTSC system performed as well or better than both of the 2D systems for all metrics

considered. For several of the metrics calculated (total time between switch contacts, mean velocity, and mean acceleration, in particular) performance for the 3D system was significantly better ($p < 0.03$) than for either of the two 2D setups. Additionally the length of the deceleration phase, defined as time from peak velocity to contact with the target switch, was significantly shorter for the 3D system than either 2D system. A typical velocity profile is shown in Fig. 2.

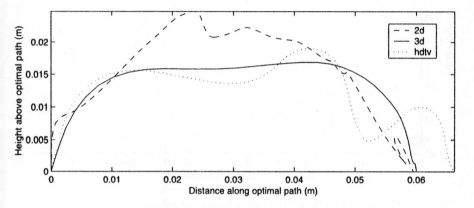

Fig. 3. Movement from switch 1 to 2 for subject #3

Inspection of the raw data supplied some other interesting insights, which are difficult to quantify. For instance, Fig. 3 shows the motion of the instrument tip mapped onto a plane defined by the straight line distance from switch to switch, and the maximum deviation from that line. What is interesting about this view of the data is that it supports the claim that motion under 3D visualizationwas made more confidently and efficiently than with the 2D setups.This is can be seen in the smoothness and directness of the path taken. This was typical of many of the subjects. It is also interesting to note that in the case pictured, the motion under HDTV viewing did not finish at the same point as the motions under the other viewing systems. This is because the instrument tip was past the target switch when the switch was contacted, presumably by the wrist or shaft of the tool. This is a further indication of the difficulty encountered in determining depth while using a 2D system.

3. **Relative Distance Estimation and Peg Placement Tasks.** All subjects were able to assign correctly the pins to their corresponding row under all viewing conditions. Interestingly, using the same pegboard patterns, but now requiring placement of objects on the pins, the results from task 4 were markedly different. With the involvement of motor-skill (acquisition and placement of objects), performance was clearly improved under 3D vision. Although the difference in time to complete the task was only statistically

significant between the 2D-NTSC and the 3D-NTSC systems, the task was performed faster under 3D than under HDTV by 12 of the 14 subjects, as seen in Fig. 4.

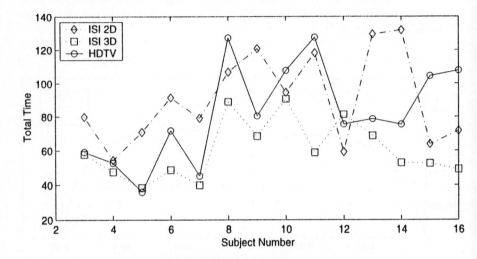

Fig. 4. Total time for peg placement

4. **Suturing and Knot Tying Tasks.** Results for both the suturing drill and knot tying tasks were again in favor of the 3D-setting [8]. Most subjects experienced difficulties with needle orientation under 2D vision.

3.2 Questionnaire

Asked for the best visualization, 53% of the surgeons preferred the HDTV-set-up followed by the 3D-NTSC set-up (33%), while 13% did not see a difference. Eighty percent of the subjects evaluated their own performance under 3D viewing conditions as superior to either 2D setting. As expected, 93% of subjects felt depth perception was best with the 3D-set-up. In terms of resolution the majority favored the HDTV system (HDTV 73%, 3D 13%, undetermined 13%). When asked if problems with depth arose during performance, 87% of the subjects found problems with the 2D-NTSC setting and 47% also for the 2D-HDTV setting (3D: 0%). Asked for the overall effectiveness of working under 2D and 3D conditions, 87% voted for 3D. Eighty percent of the subjects rated the relative importance of depth perception more important than resolution. When asked about specific problems with 2D vision, subjects complained about incorrect needle positioning and inability to grasp suture. On a verbal rating scale (1-10) of subjective importance for endoscopic work, depth perception averaged at a level of 8.4, demonstrating its perceived importance for endoscopic work. This was despite the fact that 93% of all subjects worked with a 2D-scope in their daily endoscopic practice and stated in the questionnaire that they felt that 3D vision was not an improvement prior to the trial.

4 Discussion

Despite a predisposition to 2D systems, the surgeons participating preferred, and performed best when using, the 3D system even when presented with a 2D system with truly superior resolution. This is in contrast with many of the earlier studies and our previous experience with first generation 3D endoscopic systems compared to standard resolution 2D systems. We conjecture that there is a threshold of quality, in either resolution or 3D effect, when the addition of depth cues provided by a 3D system becomes more valuable than better resolution. This assertion is reasonably supported by the results of the tasks and the questionnaire, and warrants further investigation.

The telesurgical platform provides several design advantages that have allowed this 3D endoscope quality threshold to be reached. First the scope and camera are held by a robot rather than by a surgeon, so they can be designed larger and heavier than those used in standard endoscopic surgery. Secondly the surgeon is seated and immersed in the endoscopic view with his head in a fixed position. This relieves the constraint for mobility and peripheral vision that is imposed when the surgeon is working at the operating table. This allows the use of two properly aligned medical grade monitors to present the three-dimensional view, rather than shutter glasses or heads up displays which produce an inferior 3D image.

It is reasonable to expect that surgeons will welcome the better visual resolution that will come with next-generation HDTV-based stereo endoscopes. However, to make cost-effective design choices, it will be important to better understand the tradeoffs between depth perception and spatial resolution touched upon in this study and their effect upon surgical task performance.

References

1. Dion, Y.M., Gaillard, F.: Visual integration of data and basic motor skills under laparoscopy. Influence of 2-D and 3-D video-camera systems. Surg Endosc 1997;11:995-1000
2. Buess, G.F., van Bergen, P., Kunert, W., Schurr, M.O.: Comparative study of varios 2D and 3D vision systems in minimally invasive surgery. Chirurg 1996;67:1041-6
3. Von Pichler, C., Radermacher, K., Boeckmann, W., Rau, G., Jakse, G.: Stereoscopic Visualization in Endoscopic Surgery: Problems, Benefits and potentials. Perception 1997;6:198-217
4. Hanna, G.B., Shimi, S.M., Cuschieri, A.: Randomised study of influence of two dimensional versus three dimensional imaging on performance of laparoscopic procedures. Lancet 1998;351:248-51
5. Jones, D.B., Brewer, J.D., Soper, N.J.: The influence of three-dimensional video systems on laparoscopic task performance. Surg Laparosc Endosc 1996;6:191-7

6. Chan, A.C., Chung, S.C., Yim, A.P., Lau, J.Y., Ng, E.K., Li, A.K.: Comparison of two-dimensional vs three-dimensional camera systems in laparoscopic surgery. Surg Endosc 1997;11:438-40

7. Guthart, G.S., and Salisbury, J.K.:The Intuitive Telesurgery System: Overview and Application, Proc. of the IEEE International Conference on Robotics and Automation (ICRA2000), San Francisco CA, April 2000.

8. Falk, V., Mintz, D., Reitz, B.A.: Influence of 3D-Vision on Telemanipulator Performance. Submitted for publication 2000

9. Frisen, L., Lindblom, B.: Binocular summation in humans: Evidence for a hirachic model. J Physiol 1988;402:773-782

10. Rabin, J.: Two eyes are better than one: binocular enhancement in the contrast domain. Opthalmic Physiol Opt 1995;15:45-48

Three-Dimensional Slice Image Overlay System with Accurate Depth Perception for Surgery

Ken Masamune,Yoshitaka Masutani*,Susumu Nakajima+,Ichiro Sakuma**,
Takeyoshi Dohi**, Hiroshi Iseki*** and Kintomo Takakura***

Dept. of Biotechnology, Tokyo Denki University
*Graduate school of Engineering, the University of Tokyo,
**Graduate school of Frontier Science, the University of Tokyo
+Dept. of Orthopedic surgery, the University of Tokyo
***Dept. of Neurosurgery, Tokyo Women's Medical University
Ishizaka, Hatoyama, Hiki, Saitama, 350-0394 Japan
masa@b.dendai.ac.jp

Abstract.
In this paper, we describe a three-dimensional (3-D) display, containing a flat two-dimensional (2-D) display, an actuator and a half-silvered mirror. This system creates a superimposed slice view on the patient and gives accurate depth perception. The clinical significance of this system is that it displays raw image data at an accurate location on the patient's body. Moreover, it shows previously acquired image information, giving the capacity for accurate direction to the surgeon who is thus able to perform less-invasive therapy. Compared with conventional 3-D displays, such as stereoscopy, this system only requires raw 3-D data that are acquired in advance. Simpler data processing is required, and the system has the potential for rapid development. We describe a novel algorithm, registering positional data between the image and the patient. The accuracy of the system is evaluated and confirmed by an experiment in which an image is superimposed on a test object. The results indicate that the system could be readily applied in clinical situations, considering the resolution of the pre-acquired images.

1. Introduction

Recently, many researches described the use of three-dimensional (3-D) medical images for surgical simulation and planning, before and after surgery from the end of 1980's[1]. Ordinarily, these 3-D medical images are shown on the flat two-dimensional (2-D) monitor of a computer. In general, 3-D images play an important role in allowing the surgeon to recognize the region of interest and to coordinate the spatial relationships between the patient and the images. However, surgeons are constrained by having to watch a display in a corner of the operating room. Thus, their actions are not truly objective and it is not possible to register the images with the patient in an accurate manner. This problem is particularly acute for 2-D slice images.

To increase comprehension of pre-acquired images, 3-D displays are becoming more popular in the medical field, with various modes of visualization on offer. The most popular method of observing such images is to use the principle of stereoscopic display, for example, using a polarized shuttering system or reticular lens system. Such 3-D display systems are very convenient in the operating theater, especially in the field of Virtual Reality (VR).

However, there are problems in the use of binocular stereoscopic methodology, especially from the point of registration of 3-D medical images. In a stereoscopic 3-D display, we normally use two images: right and left. Thus, the surgeon looks at the right image with the right eye and at left image with the left eye. In general, we continuously tune our eyes' convergence and focus to recognize 3-D objects. In the case of stereoscopic displays, we can only observe pre-fixed 2-D images, which create quasi 3-D images with distortion. Subjectively we perceive it as 3-D, but there may be significant inaccuracies in registration. Thus, stereoscopic displays are not good enough for surgery, where accurate registration of image and subject is essential.

To overcome the above problems, we have developed a novel 3-D display called 'Slice display', which contains one flat 2-D display, an actuator and a half-silvered mirror. Compared with conventional 3-D displays, this system is so simple that it needs less heavy computation in image processing and has the potential for rapid development with an easy method of image-object registration. In Section 2, we describe the basic requirements for 3-D displays in surgery. The system we have developed is described in Section 3. Section 4 describes and defines the accuracy of the method of registration, which is the most important practical element. Phantom testing was performed to evaluate not only the mechanical accuracy, but also the depth perception of this system.

2. Basic requirements of 3-D slice displays for surgery

Three-dimensional displays will provide the most valuable information for surgeons if they meet the following conditions:
1. The surgeon should be able to recognize the 3-D position of any surgical devices on the pre-operative images with accuracy.
2. Image quality should be high.
3. There should be minimal eyestrain and fatigue during long surgical procedures.
4. The surgeon should be able to observe the neighborhood of the diseased target in 3-D.
5. Set-up, registration and sterilization of equipment should be straightforward.
Of course, the actual content and quality of the 3-D image should also be considered. The above conditions are basic points to apply in the operating theater. In particular, we put great emphasis on conditions 1, 2 and 3. And more, displaying a slice is considered clinically significant as a 3-D visualization addressed in [2].

3. System Description [3]

Our technique of slice-display has one degree of freedom of movement. Its principle is shown in Fig. 1. In this case, the surgeon can see the patient's head directly. In addition, the reflected image of the flat 2-D display set is visible above the patient in the half-silvered mirror. Finally, the surgeon can observe 2-D slice image data apparently floating "inside" the patient's head. By moving the 2-D display in the direction of the line of vision (Fig. 1), the virtual floating image is also moved in the same direction. We put a linear sensor on the 2-D display to measure the display's location. Using the measured value, we can calculate the appropriate slice image, once registration is performed. The registration method is described below. This floating 2-D slice is located relative to the 2-D display, so that the surgeon always has the correct view through the half-silvered mirror and the image will follow actual operating movements. Thus, it provides a natural view, minimizing fatigue during the operation.

Fig. 2 is the inside view of the Volume display. We use an AC motor (15 W, 100 V) and a ball screw shaft to actuate the 2-D display. This is a 13.3-inch Super TFT Color LCD flat display screen (Selectop, Hitachi Co. Ltd., Japan). The linear encoder measures increments of 0.01 mm. A computer (Apple Macintosh LCIII: 32MB RAM) is used to create and display the slices. The maximum size of the system is 1,000 × 340 × 390 mm and its weight is 12 kg.

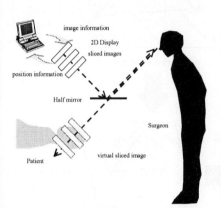 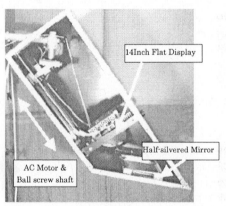

Fig. 1. Principle of Slice display;　　　　　　**Fig.2** Inside the Slice display.

4. Registration

Registration is the most important problem for display. Although many registration methods have been studied [4][5], most methods are inconvenient for practical use because they usually require complicated instruments, procedures and calculation. Considering clinical use, we applied a simple rigid matching using fiducial markers on the patients and in the images. A flow chart of the registration system is shown in Fig. 2. The fiducial marker was a triangular board and three cones shown on Fig. 3. As markers, we used capsules containing liquid Vitamin E, to enhance resolution in magnetic resonance imaging (MRI) images. The procedure was as follows:

1. Three fiducial markers were placed on the patient's head; 3-D images were obtained and converted for left-right image direction (mirroring).
2. Three markers were identified in the 3-D images.

3. A new slice was generated, containing three markers simultaneously. This image was called the "fiducial slice image".
4. Re-slicing was carried out perpendicular to the fiducial slice image.
5. The floating image was adjusted to the real image using three markers and the surgeon's own perception.

After registration, we could observe a series of sliced images and relate them to the correct positions. All the above procedures were executed by a macro in "NIH-Image" software on the computer [6][7]. Display position was measured by linear digital scale and the data were transferred from the multiplexer to the computer by RS232C at a speed of 1200 bps.

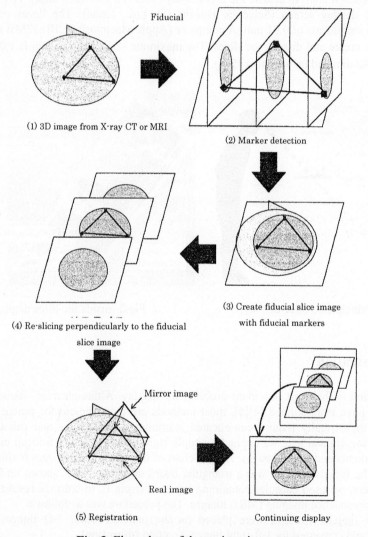

Fiducial

(1) 3D image from X-ray CT or MRI

(2) Marker detection

(3) Create fiducial slice image
with fiducial markers

(4) Re-slicing perpendicularly to the fiducial
slice image

Mirror image

Real image

(5) Registration

Continuing display

Fig. 3. Flow chart of the registration system.

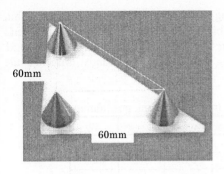
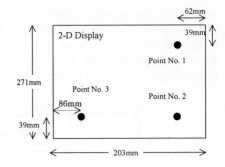

Fig.4 Triangular marker for registration. **Fig. 5** Three points measured on the display.

5. Evaluation of accuracy

We evaluated the mechanical and depth perception accuracy of the system. The latter factor is important because, even if we produce an accurate system, we will still have errors in the surgeon's perception of depth.

Mechanical accuracy

Firstly, the mechanical accuracy was evaluated. The 2-D display was moved by 50-mm increments, and three points on the display (Fig. 5) were measured in Table 1., using a 3-D digitizer (Micro Scribe 3-D,Immersion company, U.S.A.), with an accuracy of less than 0.51 mm.

Table 1 Measured coordinate values of three points.

Position of 2D Display		0.00(mm)	50.00(mm)	100.00(mm)	150.00(mm)	200.00(mm)	250.00(mm)	300.00(mm)
Point No.1	x(mm)	33.07	31.02	28.99	26.94	24.94	22.89	20.88
	y(mm)	-21.06	-23.33	-285.58	-27.82	-30.09	-32.33	-34.59
	z(mm)	34.15	38.1	42.15	46.13	50.12	54.12	58.05
Point No.2	x(mm)	25.9	23.86	21.79	19.74	17.72	15.65	13.65
	y(mm)	-28.72	-30.95	-33.17	-35.45	-37.72	-39.89	-42.18
	z(mm)	26.36	30.35	34.37	38.35	42.31	46.3	50.22
Point No.3	x(mm)	35.34	33.32	31.3	29.27	27.22	25.19	23.18
	y(mm)	-37.6	-39.8	-42.13	-44.39	-46.6	-48.91	-51.15
	z(mm)	26.29	30.3	34.31	38.28	42.26	46.21	50.18

From these results, the maximum distortion of the display is 0.39 mm, as seen at 300 mm. This value is less than the error of the 3-D digitizer, and it will thus not be a significant source of error in this system.

Accuracy of 3-D perception

In the next stage, the accuracy of 3-D perception using this system was evaluated. We prepared three points on the 2-D display as described above. Nine volunteers attempted to point the floating display image in the vacant space under the display system. We first measured human trembling while pointing the 3-D digitizer within one second, and the results are shown in Table 2. The trembling error was not depend on the position of the display. Considering the accuracy of the 3-D digitizer (0.51 mm), this indicated that human trembling would not significantly influence the positioning procedure during the evaluation.

Table 2. Measurement of hand trembling during positioning.

Display position		Average	Maximum	Minimum	SD
	Point No.1	0.13	0.35	0.02	0.052
0	Point No.2	0.14	0.37	0.02	0.052
	Point No.3	0.13	0.34	0.04	0.047
	Point No.1	0.12	0.37	0.02	0.052
100	Point No.2	0.13	0.35	0.01	0.050
	Point No.3	0.12	0.34	0.03	0.049

(mm)

For evaluation, three positions of the display were measured 10 times, and the difference between the actual and theoretical point was measured and calculated. Fig. 6 shows the positioning task by volunteer. Table 3 shows the average error for each volunteer sorted by result.

Table 3. Average positioning error.

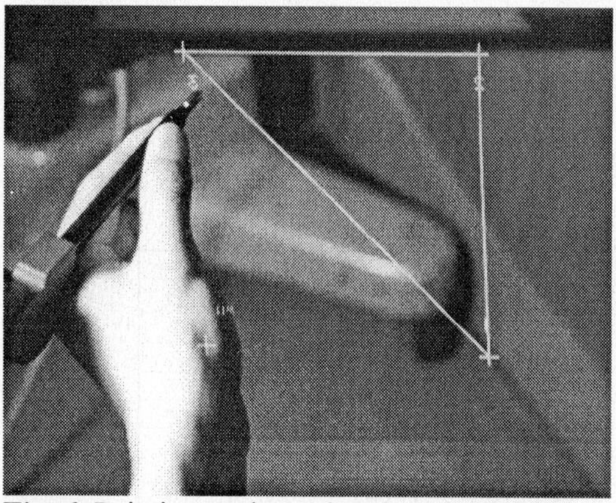

Fig. 6. Pointing testing using the 3-D digitizer.

	Average	SD
Average	2.13	1.671
Volunteer 1	0.98	0.221
Volunteer 2	1.03	0.142
Volunteer 3	1.06	0.139
Volunteer 4	1.08	0.141
Volunteer 5	1.09	0.130
Volunteer 6	1.43	0.567
Volunteer 7	3.25	2.145
Volunteer 8	3.83	2.580
Volunteer 9	5.43	3.793

(mm)

We evaluated whether the display position depends on pointing accuracy. Table 4 and the Fig. 7. show the results. "Display position" means the distance between the first position and measuring position of the display. It increases with the distance between the surgeon's eye and the floating 2-D slice display. Most volunteers were able to point within an accuracy of 1 mm. Considering the resolution of MRI, we thus appear to have sufficient accuracy for clinical use. However, some volunteers had much greater errors; over 10 mm when the display was most displaced. This error arose mainly from using 2-D slice images and the human eye. If we present the surgeon with a 3-D volume data set, this human error will be reduced because there will be one more degree of freedom of movement to perceive three-dimensional space.

Table 4. Pointing accuracy testing.

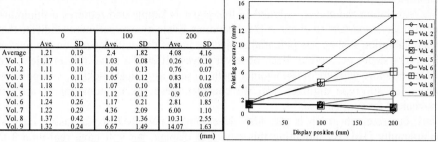

	0		100		200	
	Ave.	SD	Ave.	SD	Ave.	SD
Average	1.21	0.19	2.4	1.82	4.08	4.16
Vol. 1	1.17	0.11	1.03	0.08	0.26	0.10
Vol. 2	1.11	0.10	1.04	0.13	0.76	0.07
Vol. 3	1.15	0.11	1.05	0.12	0.83	0.12
Vol. 4	1.18	0.12	1.07	0.10	0.81	0.08
Vol. 5	1.12	0.11	1.12	0.12	0.9	0.07
Vol. 6	1.24	0.26	1.17	0.21	2.81	1.85
Vol. 7	1.22	0.29	4.36	2.09	6.00	1.10
Vol. 8	1.37	0.42	4.12	1.36	10.31	2.55
Vol. 9	1.32	0.24	6.67	1.49	14.07	1.63
					(mm)	

Fig.7. Graph of the data in Table 4 (sorted by distance from the display).

The above experiments verify the pointing accuracy of this system. As a phantom experiment using this system, we used the MRI volume of a volunteer's head as overlay. Fig.8 shows the result. A floating 2-D re-slice image of MRI is observed superimposed on the patient's head. Fig. 7 (R.) shows the image of the eyeballs, and confirms the position of the eyes three-dimensionally.

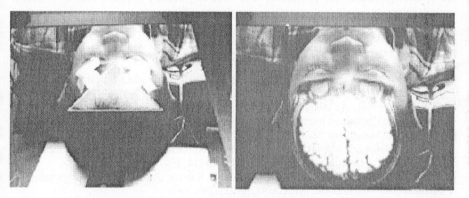

Fig.8. Image overlay experiment Left: adjusting the triangle marker to obtain registration. Right: the 2-D slice image is floating, with accurate depth perception.

6. Discussion

In this system, we used a simple raw intersection image, which is familiar to surgeons. In the future, we could also apply more useful pre-acquired data, such as 3-D organ models and structural information on blood vessels. The concept of using a 3-D image visualized in a half-silvered mirror is not an original idea, and similar research is presented in references 1, 2, 3, 4, 8 and 9. The difference in this system lies in the use of a simple system to create the 3-D image, which would be useful in extending the expression of the 3-D image. This system would be very useful to use in real-time image updating, such as X-ray computerized tomography (CT) or

interventional magnetic resonance (MR) systems. It will solve the problems of deformation of organs inside the patient.
The registration method we applied in this paper is simple and requires a minimum set of fiducial markers to obtain 3-D registration. Evaluation experiments indicate that this registration system has potential subjective errors that will depend on the human observer's depth perception. Each observer's eyes differ. To solve this problem, we could add more markers or use another method, such as optical tracking, 3-D-position sensor, or arm-type digitizer, and we consider these special components are not convenient for surgical use.
Thus, our research group is also now developing a new 3-D display using integral photography [10]. By mounting this new 3-D display, we will have real-time 3-D images in the operating theater in the near future.

7. Conclusion

A 2-D sliced medical image overlay display with accurate depth perception is described. Real depth perception, simple registration, and the capability to be used in the operating theater are presented. The accuracy results indicate that it is useful enough to apply in clinical situations considering the resolution of the pre-acquired images.

References
[1] R. Taylor, et al., Computer Integrated Surgery - Technology and Clinical Application - , MIT Press, 1995
[2] D.T. Gering, et al., An Integrated Visualization System for Surgical Planning and Guidance Using Image Fusion and Interventional Imaging, Proc. MICCAI99, pp.809-819, 1999
[3] T. Ono, K. Masamune, M. Suzuki, T. Dohi, H. Iseki and K. Takakura, Study on three-dimensional surgical support display using enhanced reality, Proc. of 7th annual meeting of Japan Society of Computer Aided Surgery, pp.113-114, 1997 (in Japanese)
[4] M. Blackwell, C. Nikou, A.M. DiGioia and T. Kanade, An Image Overlay System for Medical Data Visualization, MICCAI98, pp.232-240, 1998
[5] W.E.L. Grimson, et al., An Automatic Registration Method for Frameless Stereotaxy, Image Guided Surgery, and Enhanced Reality Visualization, IEEE Trans. Med. Imag., Vol.15, No.2, 1996
[6] Y. Masutani, H. Iseki, et al., NIH Image for surgical operations - Guide in 3-D medical image processing, Nakayama Syoten Press, 1996
[7] NIH Imagae Home page   http://rsb.info.nih.gov/nih-image/
[8] Y. Masutani, T. Dohi, Y. Nishi, M. Iwahara, H. Iseki and K. Takakura, VOLUMEGRAPH An integral photography based enhanced reality visualization system, Proc. of CAR'96, p.1051, 1996
[9] Y. Masutani, M. Iwahara, O. Samuta, Y. Nishi, N. Suzuki, M. Suzuki, T. Dohi, H. Iseki and K. Takakura, Development of integral photography based enhanced reality visualization system for surgical support, Proc. of ISCAS'95, pp.16-17, 1995
[10] S. Nakajima, K. Masamune, I. Sakuma, T. Dohi, Development of a 3-D Display System for Surgical Navigation,Trans. of the Institute of Electronics, Information and Communication Eng. D-II, Vol. J83-D-II, No.1, pp.387-395, 2000

Surgical Navigation System
with Intuitive Three-Dimensional Display

Susumu Nakajima[1], Sumihisa Orita[2], Ken Masamune[3],

Ichiro Sakuma[2], Takeyoshi Dohi[2], and Kozo Nakamura[1]

[1]Department of Orthopaedic Surgery, The University of Tokyo,
[2]Faculty of Engineering, The University of Tokyo,
7-3-1 Hongo, Bunkyo-ku, Tokyo, 113-8655, Japan
susumu@miki.pe.u-tokyo.ac.jp
[3]Department of Biotechnology, Tokyo Denki University
Hatoyama Hiki-gun, Saitama, 350-0394, Japan

Abstract. This paper describes a newly developed surgical navigation system that superimposes the real, intuitive 3-D image of the patient's internal structure on the patient's body, and helps surgeons to perform surgery. The system consists of a personal computer, a lens array, a supporting stage, a liquid crystal display and a half-silvered mirror. The 3-D images are generated by real-time computer-generated integral photography, and superimposed on the patient's body via a half-silvered mirror, as if they could be seen through the body. The differences between the theoretical positions of the projected 3-D images and the recognized ones were found to be less than 3.8 mm, within an area of 100 × 100 × 40 mm in the vicinity of the operation. The location of the point in the superimposed 3-D image could be recognized within an error of 3 to 4 mm. A 3-D bone image of a wrist joint superimposed on the corresponding part could be observed with correct motion parallax by more than one person and appeared to the observers as if it could be seen through the body. Because of the simplicity and the intuitiveness of the navigation image, this system will become applicable to clinical use in the near future.

1. INTRODUCTION

Surgical navigation systems have come to be widely used for many surgical fields. They usually show three-dimensional (3-D) information in pre-operation images to surgeons, as a set of two-dimensional (2-D) sectional images displayed away from the surgical area. Therefore, surgeons have to follow some extra procedures to perceive 3-D information of the patient during surgery. Firstly, they need to turn their gaze to the computer display showing the navigation images. Next, they must reconstruct the 3-D information in their minds, with the help of experience and anatomical knowledge. Lastly, they must turn their eyes back to the surgical area and register the 3-D information reconstructed in their minds with the actual patient's body. These procedures interrupt the flow of surgery and the reconstructed 3-D information sometimes differs between individual surgeons. To solve these problems, 3-D images should be displayed three-dimensionally in the space that coincides with the patient's body.

The following requirements should be met for surgical navigation using 3-D displays:

1. Geometrical accuracy over the projected objects (especially depth perspective);
2. Visibility of motion parallax over a wide area;
3. Simplicity of application.
4. Avoidance of the need for extra devices, such as the wearing of special glasses;
5. Simultaneous observation by many people;
6. Absence of visual fatigue;

All of these requirements are important, especially 1, 2, and 3.

To meet all the requirements, we used the principle of integral photography (IP) proposed by Lippmann in 1908 [1]. Integral photography records and reproduces 3-D objects using a "fly's eye" lens array and photographic film (Fig. 1). Igarashi proposed a computer-generated IP method that constructs an image by transforming the 3-D information about the object into 2-D coordinates on the computer display, instead of taking a photograph [2]. However, the computing time for this method of creating IP was too costly and the system was not practical for rendering 3-D medical objects. To solve this problem, we have devised a new rendering algorithm for computer-generated IP and have developed a practical 3-D display system using it [3]. The 3-D location of the projected 3-D objects could be recognized within an error of 2.0 mm.

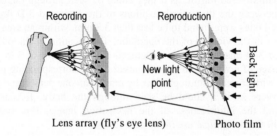

Fig. 1. How to record and reproduce a 3-D object by integral photography. Light rays reflected at a point on the 3-D object pass through the centers of all lenses on the array and are recorded on film. When the film is lit from behind, light rays intersect at the original point to become a new light point.

To superimpose the 3-D images on the patient, we used a half-silvered mirror that both reflects light rays and allows them to pass through it. The patient's body can be seen through the half-silvered mirror, and the projected 3-D image is reflected and can be seen at the corresponding position. This principle has been used for a similar purpose by Masutani et al. [4,5], who used stationary IP images of medical 3-D objects recorded on film and then superimposed them on the patient during surgery via a half-silvered mirror. Their method required several hours to record an IP image, and the image could not be altered, since conventional photographic film was used to record the images. Blackwell [6] used a binocular stereoscopic vision display system to show 3-D images. This display method reproduces the depth of projected objects by the fixed binocular disparity of the images. However, this does not always give the observer a sense of depth, as this is created by the spacing of the observer's eyes, which is not always the same as that of the fixed lenses. Motion parallax cannot be reproduced without wearing a tracking device. As well as limiting the number of observers, this also causes visual fatigue, because although there are 2-D images on the displays, the observer has a remote 3-D sensation. Consequently, systems based on binocular stereoscopic vision are not suitable for surgical navigation display.

This paper presents a novel surgical navigation system with an intuitive 3-D display that can reduce the burden of procedures required to recognize 3-D information during surgery. The 3-D display designed for this purpose and the registration method for computer-generated IP are presented in Section 2. The accuracy of the recognized location of the projected 3-D image to the system and that of the registered image to the object are presented in Section 3. Associated problems and the feasibility of the system are discussed in Section 4.

2. METHODS

2.1 System Configuration

The surgical navigation system developed in this study consists of a personal computer (Dual Pentium II 400 MHz), a hexagonal lens array (lens diameter: 2.32 mm), a supporting stage, a liquid crystal display (LCD; 6.0 inch, XGA, 213 ppi) and a half-silvered mirror (Fig. 2). The source image on the LCD is generated from the 3-D data by the algorithm described later in this paper. Any kind of 3-D data source, such as magnetic resonance (MRI) or computerized tomography (CT) images, can be processed.

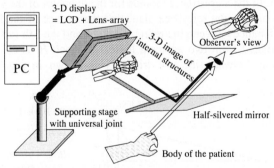

Fig. 2. System configuration of the surgical navigation system with the 3-D display.

2.2 Three-dimensional display with computer-generated integral photography

In the proposed computer-generated IP, each point in a 3-D space is reconstructed by the convergence of rays from pixels on the computer display, through lenses in the array. The observer can see any point in the display from various directions as if it were fixed in 3-D space. Each point appears as a new light source. A 3-D object can thus be constructed as an assembly of such reconstructed light sources.

Because the resolution of LCDs is much lower than that of photographic film, there is a decreased amount of information that can be displayed. Thus, not all points on a 3-D object can be displayed on the LCD, and only the 3-D information of the points most suitable to each pixel of the LCD can be processed. Coordinates of those points in the 3-D object that correspond to each pixel of the LCD must be computed for each pixel on a display. Fig. 3 shows the fast 3-D rendering algorithm for IP that computes coordinates of one point for each pixel of the computer display.

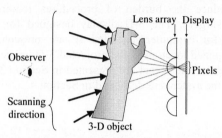

Fig. 3. Schema of a new, fast, 3-D rendering algorithm for IP. The line that occurs at the center of a pixel is extended through the center of the lens corresponding to the pixel. Then, the first point seen from the observer's side where the line intersects the 3-D object is displayed on the pixel.

Unlike natural objects, information about any point in the original object can be directly acquired in the case of medical 3-D images, making this method free from the pseudoscopic image problem that is peculiar to the original principle of IP.

The LCD is used to display the image without distortion. The plane that contains all the centers of the lenses on the lens array is used as the reference plane in the direction of depth. In the algorithm, D stands for the diameter of the lenses on the lens array; h the gap between the reference plane and the surface of the LCD; r the pixel pitch and Hi the distance between the reference plane and the i-th farthest point where two light rays radiated from two adjacent lenses intersect. Hi is calculated from the parallel relationship between the two triangles as shown in Fig.4, formulated as follows.

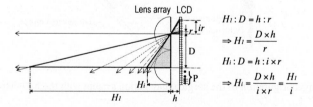

$$H_1 : D = h : r$$
$$\Rightarrow H_1 = \frac{D \times h}{r}$$
$$H_i : D = h : i \times r$$
$$\Rightarrow H_i = \frac{D \times h}{i \times r} = \frac{H_1}{i}$$

Fig. 4. Relation between Hi and other dimensions. P: pixels on LCD

Hi becomes the distance of the i-th farthest light point from the reference plane that is created by two adjacent lenses as an intersection of two rays, while i is a natural number less than or equal to the number of pixels included in the radius of a lens, $D/2r$.

$$H_i = \frac{D \times h}{i \times r} = \frac{H_1}{i} \qquad (1)$$

In the same way, given that Hi/j is the distance of the i-th farthest light point created by two lenses, the distance between their centers is j-times the diameter of the lenses.

$$H_{i/j} = \frac{D \times j \times h}{i \times r} = H_1 \times \frac{j}{i} \qquad (2)$$

Thus, the spatial resolution in depth is equal to H_{iMax} ($iMax$ is the largest number of i). As described above, the extent of the projection space can theoretically be much farther than H_1, although the further the distance between a point and the reference plane becomes, the worse the quality of the point.

Fig. 5 shows a 3-D stationary image generated from a CT of the head. The image could be observed from various viewpoints with correct motion parallax.

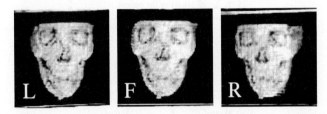

Fig. 5. Photos of projected 3-D stationary skull images by IP taken from different directions. They demonstrate the ability of the system to display motion parallax. The letter denotes the position of the observer relative to the 3-D image: F: front, L: left, R: right.

2.3 Registration method between the 3-D image and the patient's body

Because the relationship between the location of the projected 3-D image and that of the navigation system can be found from calculation, registration between the 3-D image and the patient's body could be performed by registration between the navigation system and the patient's body. The registration was, therefore, performed as follows (Fig. 6).

(1) Coordinates of the point are measured by an optical tracking system (POLARIS, Northern Digital Inc., Ontario, Canada) and the measured value is directly used for calculation. Four points (P_{sys}) are pre-defined on the surface of the navigation system for calibration.

(2) MRI of the object is obtained, using four MRI markers attached to the surface of the object.

(3) The 3-D image of the object is rendered from the MRI by the above-mentioned method and projected by the navigation system.

(4) The position of the navigation system is adjusted manually so that the projected 3-D image roughly coincides with that of the object. Subsequently, the coordinates of P_{sys} are measured by the tracking device, and those of the images of the markers can be found by calculation.

(5) The coordinates of the markers on the object are measured by the tracking device. The position of the projected 3-D image is then transformed, as the four coordinates of the marker images coincide with the measured coordinates of the corresponding ones on the object. This procedure allows theoretical registration of the projected 3-D image with the object.

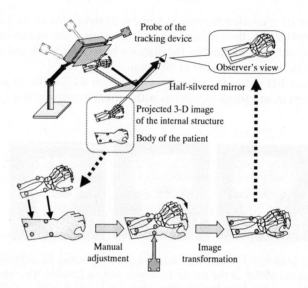

Fig. 6. Scheme of the registration method. The coordinates of the markers in the projected 3-D image can be found from the measured position of the navigation system by the optical tracking system. The position of the projected 3-D image can be adjusted roughly by manual movement of the system, and precisely by image transformation, so that the observer can see the 3-D image at the corresponding position.

3. EXPERIMENTS

3.1. Accuracy of the recognized location of the projected point to the system

Because geometrical accuracy is the most important consideration for medical imaging, the accuracy of the recognized location of the projected point to the navigation system within the objective space was measured. A square with sides of 100 mm was displayed at a distance of 10 to 40 mm from the surface of the 3-D display and projected into the objective space away from the displayed location by the half-silvered mirror. The coordinates of the recognized location of the vertices and the center of the square were measured by the optical tracking system. Five observers repeated the measurements three times for each vertex. The mean differences between the measured and the theoretical coordinates were less than 3.8 mm, within the space of $100 \times 100 \times 40$ mm (Table 1).

Table 1. Mean error of the measured coordinates.

Distance from display	Mean value (S.D.)
10.0	2.5 (1.4)
20.0	1.9 (0.6)
30.0	2.5 (1.3)
40.0	3.8 (2.5)

Unit:mm

3.2. Accuracy of the recognized location of the registered image to the object

After registration, the position of the 3-D image theoretically coincides with that of the object. However, some errors in the recognized positions of the points in the image are inevitable. We therefore measured the accuracy of the recognized location of the registered image to the real object. A 50-mm cubic phantom object with four MRI markers (a vitamin E capsule filled with vegetable oil with a diameter of about 6 mm) on its surface and the same two markers inside the phantom was made of plate acrylic resin (Fig. 7). First, MRI of the phantom with six markers was taken and two inside markers were removed. Next, the 3-D image generated from the MRI was superimposed on the phantom with the above-mentioned registration method. Lastly, the position corresponding to the two markers in the phantom were displayed in the 3-D image and their coordinates were measured three times by the optical tracking system. The mean differences between the measured and the true coordinates of the two markers are shown in Table 2. The errors were nearly equal to the radius of the marker.

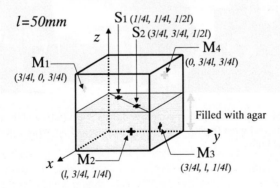

Fig. 7. Scheme of the cubic phantom made of plate acrylic resin. The lower half was filled with agar. Mn (n=1~4) and Sn (n=1,2) denote the position of the marker attached to the surface and the inside, respectively.

Table 2. The mean differences between the measured and the true coordinates of the inside marker.

marker	mean error of the coordinates	mean error in length
S1	(1.58, 2.52, 2.83)	4.1
S2	(1.42, 2.34, 2.72)	3.2

Unit:mm

3.3. Trial of the procedure from data acquisition to superimposition of the 3-D image

To examine the practicality of the system, we applied our procedure, from data acquisition to superimposition of a 3-D image, in a continuous manner. The wrist of a healthy volunteer was used as the object. An MRI image of the wrist was taken first. Images of the bones of the wrist and hand were segmented and superimposed on the corresponding part of the volunteer. It took about 2 h for the whole procedure, including data acquisition, segmentation, 3-D rendering and superimposition. The structures of the superimposed bone could be observed by many persons at the same time with correct motion parallax and appeared subjectively as if they could be seen through the body (Fig. 8).

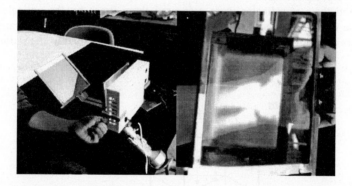

Fig. 8. Position of the surgical navigation system and the subject (left) and the superimposed 3-D wrist joint on the real wrist (right). The internal structure of the wrist could be observed as if seen through the body.

4. DISCUSSION

From the results of the accuracy test, it appears that observers can be guided by the navigation system within an error of 3 to 4 mm. This is insufficient, but allows for surgical navigation, because the surgeon can observe not only the navigation image, but also the real area during surgery. However, to minimize surgical invasion, navigation must be made more accurate. The spatial resolution of the 3-D image projected, using the principle of computer-generated IP, is proportional to the ratio of the lens diameter in the lens array to the pixel pitch of the computer display. Thus, both the lens diameter and, in particular, the pixel pitch need to be made much smaller. Judging from recent progress in the technology of manufacturing flat-panel LCD displays, this will soon be realized. Because the navigation system with this 3-D display is very simple and easy to use, we intend to apply this to such clinical uses as surgical planning and surgical navigation.

To conclude, we have developed a novel surgical navigation system with an intuitive 3-D display and have examined the errors in the recognized location of the projected object by using this system. The projected objects were geometrically accurate. Because of its simplicity and the accuracy of the image projections, this system will be able to be applied to clinical use.

Acknowledgments

We would like to thank Ryo Mochizuki (NHK Engineering Service) for lending us the super high-resolution LCD. This study was partly supported by the Research for the Future Program (JSPS-RFTF 96P00801).

REFERRENCES

1. Lippmann, M. G.: Epreuves reversibles donnant la sensation du relief. J. de Phys., 7, 4th series (1908) 821-825
2. Igarashi, Y., Murata, H., Ueda, M.: 3-D display system using a computer generated integral photograph. Japan J. Appl. Phys. 17 (1978) 1683-1684
3. Nakajima, S., Masamune, K., Sakuma, I., Dohi, T.: Three-dimensional display system for medical imaging with computer-generated integral photography. Proc. of SPIE Vol.3957 (2000) 60-67
4. Masutani, Y., Iwahara, M., Samuta, O., Nishi, Y., Suzuki, N., Suzuki, M., Dohi, T., Iseki, H., Takakura, K.: Development of integral photography-based enhanced reality visualization system for surgical support. Proc. of ISCAS'95 (1995) 16-17
5. Masutani, Y., Dohi, T., Nishi, Y., Iwahara, M., Iseki, H., Takakura, K.: VOLUMEGRAPH. An integral photography-based enhanced reality visualization system. Proc. of CAR'96 (1996) 1051
6. Blackwell, M., Nikou, C., Digioia, A. M., Kanade, T.: An image overlay system for medical data visualization. Proc. of MICCAI'98 (1998) 232-240

Functional Analysis of the Vertebral Column Based on MR and Direct Volume Rendering

P. Hastreiter[1], C. Rezk-Salama[2], K. Eberhardt[3], B. Tomandl[3], T. Ertl[4]

[1] Neurocenter, University of Erlangen-Nuremberg, Germany
hastreiter@nch.imed.uni-erlangen.de
[2] Computer Graphics Group, University of Erlangen-Nuremberg, Germany
[3] Division of Neuroradiology, University of Erlangen-Nuremberg, Germany
[4] Visualization and Interactive Systems Group, University of Stuttgart, Germany

Abstract. Degenerative diseases of the vertebral column are mainly combined with misalignments of the intervertebral discs and deformations of the spinal cord. For the investigation of severe spinal stenosis and herniated intervertebral discs x-ray myelography is required providing important functional information about the motion of vertebral segments. To avoid this time-consuming procedure and the injection of contrast agent, we introduce a non-invasive approach using direct volume rendering of a novel T2 weighted MR sequence. It provides meaningful representations of extreme spinal pathologies. Above all, the essential functional information is obtained using an adjustable couch directly applied within the MR scanner. Due to the distribution of the grey values, we suggest a coarse segmentation of the vertebral column using a fast sequence of filtering operations and volume growing. Further on, details are delineated implicitly using interactive adjustment of transfer functions. Thereby, the CSF, the roots of the spinal nerves and the vertebrae are clearly visible. Additionally, we propose an approach for volumetric measurement providing the functional dependence of the CSF volume. The presented non-invasive strategy gives a comprehensive understanding of spatial and functional information including spinal instability and may thus lead to further reduce the need of x-ray myelography.

Keywords: Visualization, Functional Analysis, Spine, Therapy Planning

1 Introduction

Degenerative discogenic diseases of the vertebral column are mainly caused by misalignments of the intervertebral discs and deformations of the spinal cord. For their investigation a clear representation of the spinal cord in relation to the roots of the nerves and the surrounding bone structures is required. For these examination there has been a great change of diagnostic methods. This is mainly related to the enormous development of magnetic resonance *(MRI)* and x-ray computed tomography *(CT)*. Above all, MRI gains of increased importance since the parameters of the imaging sequence can be optimized according to the application and the applied post-processing procedure [1]. As a further advantage of tomographic imaging techniques three-dimensional *(3D)* image data is produced which provides spatial understanding. However, severe cases such as extreme spinal stenoses and herniated intervertebral discs with unconfirmed root compressions still require to perform x-ray myelography. So far, due to its detailed imaging capabilities it is the only imaging approach giving an exact correlation of the roots of the spinal nerves, the vertebral segments and surrounding bone structures. Further on, it provides valuable functional information which is essential for a comprehensive examination. This is of great importance since the location of vertebral segments and the representation of the spinal cord changes considerably during motion. However, as major drawback the injection of a non-ionic contrast agent is required which is applied intrathecally. In general four standard projections are taken with anterior-posterior, lateral and left-anterior-oblique (LAO) or right-anterior-oblique (RAO) orientation. This is a time-consuming procedure which is also combined with a certain risk.

The idea of the presented approach is to introduce a new strategy for a comprehensive analysis of discogenic diseases within the vertebral column based on MR image data exclusively. After a short overview in section 2 explaining the main issues of the applied image data, a newly developed MR compatible couch is discussed in section 3 which allows to position the spine. Then, section 4 demonstrates which anatomical structures have to be separated explicitly and how to segment them efficiently. In order to visualize the available 3D image data comprehensively an appropriate visualization approach is indispensable. Therefore, we suggest to use interactive direct volume rendering which allows to show all available information including vertebral instability. The applied hardware accelerated approach is briefly illustrated in section 5. It is also explained how to efficiently achieve a meaningful representation of the vertebral segments, the spinal cord and the roots of the spinal nerves. Additionally, in case of an arteriovenous malformation a fusion strategy is introduced which allows to integrate a representation of dural vessel information as presented in [2]. Since volumetric information is of major interest section 6 presents an approach which allows to analyze the functional dependence of the CSF volume. Finally, section 7 presents several clinical examples demonstrating the value of the suggested approach.

2 Image Data

The advances of MR scanning techniques produced a variety of special MR sequences which give more detailed insight to the different structures of the vertebral column. As a basis of our approach MEDIC *(Multi Echo Data Image Combination)* data is applied which is a strongly T2-weighted MR sequence. It clearly distinguishes the CSF, the vertebral discs and the roots of the spinal nerves. As demonstrated in figure 1 *(left)* they are mapped to high intensities whereas bone structures have low intensities. Since parts of the surrounding tissue is in the same range of data values simple visualization strategies like maximum intensity projection *(MIP)* provide insufficient 3D representations. Contrary to that explicit segmentation of the image data is difficult due to partial volume effects. Therefore, interactive direct volume rendering is applied based on its capability to perform implicit segmentation using transfer functions.

As an advantage over the FISP *(Fast Imaging with Steady State)* sequence which was previously [1] evaluated using two-dimensional *(2D)* visualization, the *MEDIC* sequence allows to perform measurements with an improved in-slice resolution. Thereby, smaller intra-dural structures become visible much better and any post-processing like segmentation or volume visualization is considerably simplified. As presented in [2], MR-CISS *(Constructive Interference in the Steady State)* data, shown in figure 1 *(right)*, is optimally suited to detect structures within the dura. Thereby, tiny vessel malformations are clearly distinguished from CSF. Contrary to that, the applied *MEDIC* sequence hardly delineates these vessel structures. However, it optimally shows the intervertebral discs and differentiates the CSF from epidural fat.

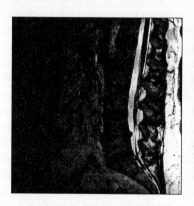 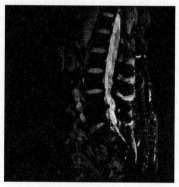

Fig. 1. MR-*CISS (left)* and MR-*MEDIC (right)* difference of bone structures, discs and CSF.

3   Functional Analysis

Functional information about the motion dependence of the vertebral column represents a fundamental prerequisite for the investigation of degenerative discogenic diseases. Using an adjustable couch which is applied directly in a standard MR scanner, different positions are achieved simulating the movement of the vertebral column. Depending on the state of the disease a "normal" position is defined for every individual patient according to the normal position used for standard MRI scans. This is in correspondence with the system described in [3]. Then, using the device the patient is successively lifted at the head and at the back in order to perform inclination and reclination relative to the normal position. Thereby, changes of the vertebrae and the spinal cord are clearly detected allowing for geometric and volumetric assessment. As major advantage of this approach, it is applicable with an arbitrary MR scanner. Further on, the couch gives the necessary assistance which is essential for older patients or in case of severe diseases.

Other approaches suggesting an investigation with open MR scanners have severe disadvantages. The main drawback is the restricted availability of these systems since they are installed at very few sites. Due to the low magnetic field of such scanners only image data of lower resolution is provided. However, high resolution is essential for the analysis of intra-dural structures such as the roots of the spinal nerves and for volumetric measurements related to the CSF. Above all, the lack of an assisting device supporting the movement of the patient is the most important drawback. Thereby, longer scanning times are prohibited and older patients or patients with extreme spinal pathologies are excluded from an examination.

4 Explicit Segmentation

According to section 2 the MR-*MEDIC* shows high contrast between CSF and surrounding tissue. However, both the CSF and the intervertebral discs are in the same range of intensity values. This makes an implicit delineation difficult using a lookup table which affects the data globally. Therefore, explicit segmentation is performed with fast filtering and volume growing (see figure 2). Thereby, the CSF including the roots of the spinal nerves and the vertebrae are clearly separated.

Initially, noise reduction is performed to optimize the data for further pre-processing. Successively, the CSF including the nerve roots is extracted by semi-automatic volume growing. As a basis of this strategy flexible bounding boxes are applied which are interactively adjusted to the underlying anatomy. Thereby, minimal post-processing is required and the target structure is optimally separated. In the same way the area is isolated which includes the bone structures and the intervertebral discs.

Based on the segmentation results, the original image data is attributed using unique tag numbers for the CSF including the nerve roots, the bone area and the surrounding tissue. Thereby, separate transfer functions are assigned which locally manipulate the look-up tables for color and opacity values. Due to the immediate visual control and the low computational expense this strategy is much faster and more robust than an explicit segmentation of details which might easily miss important features.

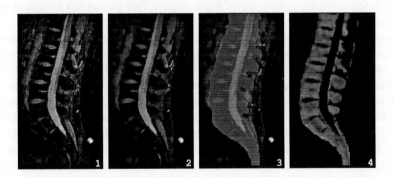

Fig. 2. Segmentation of MR-*MEDIC*: (1) original image, (2) noise reduction , (3) volume growing for the separation of CSF and bone structures, (4) inversion of segmented bone structures.

5 Visualization

For the analysis of medical image data it is indispensable to choose an appropriate visualization approach. Compared to indirect strategies [4] which rely on polygonal representations, direct volume rendering has proved to be superior [5]. Since the information of all data samples is used, image characteristics like partial volume effects are better taken into account. However, interactive manipulation is an essential feature for an application in a clinical environment [6].

All visualizations of the presented approach are based on direct volume rendering using 3D-texture mapping [7]. The basic idea of the algorithm is to define the image

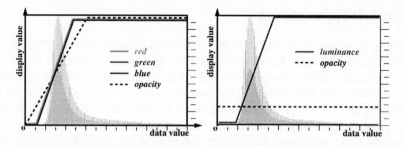

Fig. 3. Intensity histogram and transfer functions for Figure 6: *(left)* Setting for the CSF including the roots of the spinal nerves — *(right)* Setting applied to the background.

data as a 3D-texture. Then, equidistant slices parallel to the image plane are cut out of the volume. Thereby, the texture mapping hardware is exploited by performing trilinear interpolation in order to reconstruct every slice. Finally, the 3D representation is obtained by blending the textured slices back-to-front onto the viewing plane.

As an advantage of this approach, interactive manipulation of any viewing parameter is achieved even if applied to volumes of high resolution. Due to the applied trilinear interpolation scheme and the number of sampling points, which are appropriately adjusted, the resulting images are of high quality. This guarantees to reproduce fine structures and ensures their clear delineation.

Using the segmentation of the *MEDIC* data as described in section 4, detail information is delineated by adjusting the respective transfer functions. For every tagged subvolume four separate curves are used describing the correlation between the original data values and the displayed color components (RGB) and opacity. In order to speed up the process of adjusting these transfer functions, pre-defined templates are used.

In Figure 3 the setting of the transfer functions for color and opacity values is shown which leads to the visualization presented in Figure 6. Although arbitrary transfer functions are applicable, a piecewise linear mapping is sufficient. The visualization of the CSF (tag 1) requires to use low opacity values for low data values and high opacity values for high data values. Further on, a linear ramp of high gradient is used in between in order to produce a smooth transition. The transfer function for color values is positioned within the transition from low to high opacity values. This leads to a good impression of depth and enhances the representation of the nerve roots. To reveal the vertebrae and the intervertebral discs (tag 2) the opacity is set to its maximum. Just for low data values a steep linear ramp is applied in order to suppresses artifacts remaining from the coarse segmentation. This results in a fully opaque representation which supports the anatomical orientation. For the background region (tag 3) the opacity is set to full transparency since there is no further information of importance.

6 Volumetric Measurement

The MR compatible device discussed in section 3 allows to investigate different positions of the vertebral column: normal position, inclination, reclination. Depending on the selected position the representation of the CSF volume changes. Therefore, knowledge of this functional dependence is of major interest for the examination. An essential prerequisite for this volumetric measurement is the explicit segmentation of the CSF as presented in section 4. Including the roots of the spinal nerves this is easily achieved with MR-*MEDIC* data due to the clear delineation from surrounding structures.

With respect to denerative discogenic diseases of the vertebral column it is important to assess the functional dependence of the CSF volume in the vicinity of the respective intervertebral discs. Therefore, as shown in figure 4, a strategy is suggested which allows standardized measurements using a cylindrical bounding box. Having set its extent to a user defined value, the bounding box is applied to the volume data of all functional states which are investigated. Additionally, within each data set the bounding box is rotated appropriately. Using its centerline as a reference it is then centered at the selected intervertebral disc. As a result a subvolume of the segmented CSF including the nerve roots is specified which is compared to the volume of the other positions.

According to section 2 the *CISS* sequence differentiates between CSF and the roots of the spinal nerves. In contrast to that both structures are represented as a unique volume with MR-*MEDIC*. However, as presented in figure 5, the combined volume of the *MEDIC* data is identical to the sum of the subvolumes obtained with *CISS* data. In consequence there is a clear dependence of the CSF volume and the volume of the roots exiting a specific segment as well as the respective parts of the descending roots. Therefore, the *MEDIC* sequence is sufficient to evaluate the degree of compression. This was confirmed by neurophysiological examinations with 20 patients.

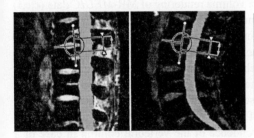

Position	CSF+roots *(MEDIC)*	roots *(CISS)*	CSF *(CISS)*
inclinat.	40374	10750	29624
normal	44773	10488	34295
reclinat.	35325	11562	23763

Fig. 4. Volumetric measurement.   **Fig. 5.** Volumetric measurements in mm^3.

7   Results and Discussion

So far, the presented approach was applied to 50 healthy volunteers and 80 patients with severe spinal stenosis and spondylo listeses. In all cases the diagnosis was performed in comparison with x-ray myelography. All MR-MEDIC volumes were acquired with a Siemens MR Magnetom Symphony 1.5 Tesla scanner which provides the necessary resolution to resolve the relevant structures appropriately. In all cases volumes were used consisting of images with a 256^2 matrix and 30–50 slices. The size of the voxels was set to $1.17 \times 1.17 \times 1.17$ mm^3. In order to achieve high frame rates, all direct volume rendering was performed with a SGI Onyx2 (R10000, 195MHz) with BaseReality graphics hardware providing 64MB of texture memory.

In all cases the investigation was performed with a standardized protocol. Using the MR-MEDIC sequence three scans were performed showing the normal position, inclination and reclination of the vertebral column. On the basis of this image data, direct volume rendering was performed of every data set. In order to give a good overview of the lesion, the selected directions of view comprise posterior-anterior, oblique posterior and lateral. The following examples demonstrate the value of the presented approach:

- In figure 6 the image data of a healthy volunteer is shown. The example demonstrates the strategy of the applied visualization technique and illustrates the capabilities of the presented approach with respect to intra-dural substructures such as the roots of the spinal nerves.
- The case of a multi segment stenosis in levels L2–L5 is presented in figure 7. Applying direct volume rendering the degree of the lesion is precisely determined in all levels. Due to osteochondrosis disc L5/S1 is not visible. The comparison of inclination and reclination illustrates the extremely restricted range of movement which is the case for the majority of older patients.
- The example of figure 8 shows a multi segment spinal stenosis. Using x-ray myelography the contrast agent stops completely at the stenosis just after intrathecal application (a). About 30 minutes later the situation changes completely after the contrast agent has past the stenosis (b,d). However, the representation of the lesion prohibits to determine its exact level which is essential information to minimize the risk of surgery. In contrast, direct volume rendering of MR-*MEDIC* data (c,e,f) clearly shows the location of all segments with different directs of view. This allows to evaluate the compression of the nerve roots at every level.
- In figure 9 the compression of both nerve roots in level L5 is presented. Based on an additional segmentation of the vertebrae a semi-transparent visualization tremendously assists the investigation of the lesion from all directions. The aliasing artifacts within the lateral view are caused by a sub-optimal field of view (FoV). This was required to minimize the time of examination to prevent breathing artifacts.
- Figure 10 shows the functional dependence of the vertebral column in case of a healthy volunteer. Direct volume rendering was performed after explicit segmentation of the bone structures.

As demonstrated in these examples, our strategy based on non-invasively acquired MR-*MEDIC* data allows to analyze degenerative diseases of the vertebral column in a comprehensive way. Compared to the normal position, the CSF and the intervertebral discs are better delineated during inclination of the vertebral column. Additionally, any instability and movement of vertebral segments are clearly represented with the functional image data. In all cases with extreme spinal stenosis and a reduced CSF volume of more than 70 % the level of the lesion was clearly determined. Contrary to that x-ray myelography prohibited a secure determination in these cases.

Due to the distribution of the grey values and the complexity of the target structures, the explicit segmentation of the CSF including the roots of the spinal nerves and the surrounding vertebrae requires a time-consuming process. Therefore, the approach based on filtering and semi-automatic volume growing provides a more convenient and robust way for segmentation. Since the structures are clearly delineated within the image data, this process takes approximately 5–10 minutes. For the successive direct volume rendering of the attributed volume interactive adjustment of transfer functions is important. Thereby, direct visual feedback is guaranteed. Based on pre-defined lookup tables for color and opacity values, the adjustment of the respective transfer functions takes another 5 minutes. Applying 3D texture mapping, the trilinear interpolation scheme provides excellent image quality and the hardware acceleration ensures high frame rates.

8 Conclusion

The presented approach introduces a non-invasive strategy in order to investigate severe degenerative diseases of the vertebral column. Using MR-*MEDIC* images the important structures are clearly delineated. Thereby, only the CSF including the roots of the spinal nerves and in selected cases the bone structures have to be segmented explicitly. This allows convenient volumetric measurements for a more detailed analysis of the lesion. A detailed representation of the volume data is then obtained using implicit delineation on the basis of direct volume rendering. The essential functional information about the motion dependence of the vertebral column is obtained with a MR compatible device. The presented clinical examples demonstrate the value of our approach. In consequence, the pre-operative investigation of severe degenerative diseases within the vertebral column is effectively assisted and the need of x-ray myelography is considerably reduced.

References

1. K. Eberhardt, H. Hollenbach, B. Tomandl, and W. Huk. Three–Dimensional MR myelography of the lumbar spine: comparative case study to X–ray myelography. In *European Radiology*, volume 7, pages 737–743, 1997.
2. C. Rezk-Salama, P. Hastreiter, K. Eberhardt, B. Tomandl, and T. Ertl. Interactive Direct Volume Rendering of Dural Arteriovenous Fistulae. In *Proc. MICCAI'99 (2nd Int. Conf on Med. Img. Comput. and Comp.–Assis. Interv.)*, volume 1679 of *Lect. Notes in Comp. Sc. 1679*, pages 42–51. Springer, 1999.
3. J. Dvorak and M. Panjabi et al. Functional radiographic diagnosis of the lumbar spine. *SPINE*, 16:562–571, 1991.
4. S. Nakajima, H. Atsumi, A. Bhalerao, F. Jolesz, R. Kikinis, T. Yoshimine, T. Moriarty, and P. Stieg. Computer-assisted Surgical Planning for Cerebrovascular Neurosurgery. *Neurosurgery*, 41:403–409, 1997.
5. B. Kuszyk, D. Heath, D. Ney, D. Bluemke, B. Urban, T. Chambers, and E. Fishman. CT Angiography with Volume Rendering : Imaging Findings. *American Jour. of Radiol. (AJR)*, pages 445–448, 1995.
6. L. Serra, R. Kockro, C. Guan, N. Hern, E. Lee, Y. Lee, C. Chan, and W. Nowinsky. Multimodal Volume–Based Tumor Nerurosurgery Planning in the Virtual Workbench. In *Proc. Med. Img. Comput. and Comp.–Assis. Interv. (MICCAI)*, volume 1496 of *Lec. Notes in Comp. Sc.*, pages 1007–1015. Springer, 1998.
7. B. Cabral, N. Cam, and J. Foran. Accelerated Volume Rendering and Tomographic Reconstruction Using Texture Mapping Hardware. *ACM Symp. on Vol. Vis.*, pages 91–98, 1994.

Note to reviewers: the color pages can also be found at
http://www9.informatik.uni-erlangen.de/eng/research/vis/Spine/miccai.html

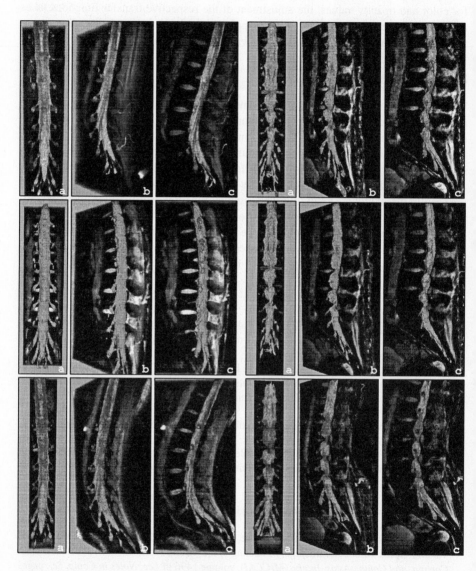

Fig. 6. Visualization showing normal position *(top)*, inclination *(middle)* and reclination *(bottom)* of healthy volunteer with standardized views: *(a)* posterior-anterior, *(b)* oblique posterior and *(c)* lateral.

Fig. 7. Visualization showing normal position *(top row)*, inclination *(middle row)* and reclination *(bottom row)* of patient with spinal stenosis with standardized views: *(a)* posterior-anterior, *(b)* oblique posterior and *(c)* lateral.

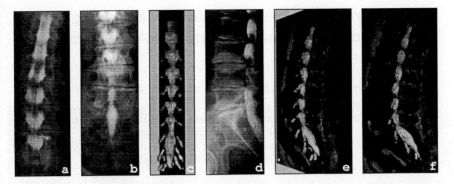

Fig. 8. Patient with multi-segment spinal stenosis: comparison of x-ray myelography (a,b,c) and examination based on MR-*MEDIC* data and direct volume rendering with different directions of view: posterior-anterior *(c)*, oblique posterior *(e)* and lateral *(f)*.

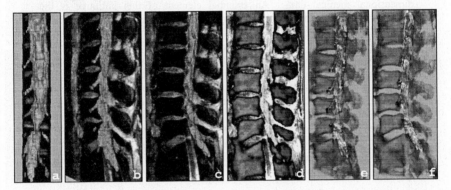

Fig. 9. Direct volume rendering of MR-*MEDIC* data showing a mono-segment stenosis in combination with a spondylo listesis in L4 and L5: 3D representation without *(a,b,c)* and with *(d,e,f)* explicit segmentation of bone structures.

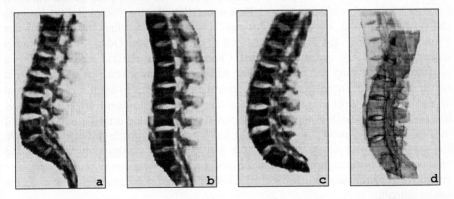

Fig. 10. Direct volume rendering of functional MR-*MEDIC* data: Healthy volunteer showing the bone structures of the vertebral column: normal position (a), inclination (b), reclination (c) and superposition of inclination and reclination (d).

Toward a Common Validation Methodology for Segmentation and Registration Algorithms

Terry S. Yoo [1], Michael J. Ackerman [1], Michael Vannier [2]

[1] National Library of Medicine, National Institutes of Health, Bethesda, MD, 20894, USA
{yoo, ackerman}@nlm.nih.gov
[2] Department of Radiology, College of Medicine, University of Iowa, Iowa City, IA, USA
michael.vannier@uiowa.edu

Abstract. The National Library of Medicine and its partners are sponsoring Insight, a public software toolkit for segmentation and registration of high dimensional medical data. An essential element of this initiative is the development of a validation methodology, a common means of comparing the precision, accuracy, and efficiency of segmentation and registration methods. The goal is to make accessible the data, protocol standards, and support software necessary for a common platform for the whole medical image processing community. This paper outlines the issues and design principles for the test and training data and the supporting software that comprise the proposed Insight Validation Suite. We present the methods for establishing the functional design requirements. We also present a framework for the validation of segmentation and registration software and make some suggestions for validation trials. We conclude with some specific recommendations to improve the infrastructure for validating medical image processing research.

Introduction

The National Library of Medicine (NLM), with its partner institutes: the National Institute of Dental and Craniofacial Research (NIDCR), the National Eye Institute (NEI), the National Institute of Mental Health (NIMH), the National Science Foundation (NSF), the National Institute of Deafness and Other Communication Disorders (NIDCD), and the National Cancer Institute (NCI), are sponsoring a program to develop an application programmer interface (API) and first implementation of a segmentation and registration toolkit called Insight [6]. The goal of this initiative is to create a self-sustaining code development effort to support image analysis research in segmentation, classification, and deformable registration of medical data. The intent is to amplify the investment being made through the Visible Human Project and future programs for medical image analysis by reducing the reinvention of basic algorithms. We are also hoping to empower young researchers and small research laboratories with the kernel of an image analysis system in the public domain.

The Insight Software Research Consortium is a team from academia and industry. The prime contractors are: General Electric Corporate R&D, Kitware, Inc., MathSoft, Inc., the Univ. of North Carolina at Chapel Hill, the Univ. of Pennsylvania (the VAST

Lab and Radiology), and the Univ. of Tennessee. Subcontracts have been extended to: Harvard Brigham and Women's Hospital, U. Penn's GRASP Lab, the Univ. of Pittsburgh, and Columbia and the Univ. of Utah.

As part of the software initiative, the Insight Team is creating a software toolkit, and an algorithm validation methodology. These developments have begun to raise and consolidate difficult research questions. The medical imaging community as a whole is beginning to recognize the difficulties in obtaining definitive ground truth with regard to questions of segmentation and registration. How do we validate segmentation and registration methods in the absence of firm control data? How do we generate a model for ground truth? In the absence of ground truth, software researchers often compare the output of one algorithm with that of a known algorithm in the literature. What metrics do we choose as comparison points between output images? What relevance do these metrics have on clinical outcomes?

Background

The Visible Human Project was initially formed to collect data from human subjects to serve as a guidebook and baseline dataset in modern anatomy research and education. Data from two subjects, one male and one female, were collected through a variety of methods including the standard radiological techniques of X-ray CT studies, magnetic resonance imaging, and plain film radiographs. In addition to these conventional clinical studies, the subjects were frozen and sectioned at 1 mm (male subject) and 1/3 mm (female subject) intervals. The exposed surfaces were photographed with 35 and 70 mm film and digitized with an electronic camera. The resulting data has entered into broad use in education and in medical research [5].

In February 1998, a workshop sponsored jointly by NLM and NIDCD explored the growing needs of the research and education community for more powerful digital tools and higher resolution models of human anatomy. Among their many recommendations, the workshop participants recommended the pursuit of advanced image analysis software tools to accommodate future higher resolution data [4]. The demand for more powerful segmentation and registration tools was confirmed through panel discussions and other exchanges during the Second Visible Human Conference, held at NIH in Bethesda, MD in October 1998 [1].

The Insight Team convened its first organizational meeting in November of 1999. At that time, the software consortium opened the discussions on the difficult issues of validating segmentation and registration algorithms. A meeting of the Insight Subcommittee on Validation was scheduled for March 2000 to explore the breadth of these topics and draft an approach for solving them.

Design Principles

The first task before the Subcommittee on Validation was to establish some basic design requirements for the Insight Validation Suite. Since the consortium and its related professionals represent a diverse set of users of medical image data, the task of achieving agreement on the basic issues was not simple. The committee used Quality function deployment (QFD), a management tool to capture the needs and priorities of the Insight user. The QFD process, originally applied in industrial circles, helps streamline production and decrease time-to-market by targeting specific customer desires [2].

Method: QFD Analysis

The QFD process is comprised of several steps, each driving towards the goal of having a list of specific, prioritized features for a segmentation and registration validation suite. In the first step, Insight developers acting as primary users, answered a series of questions. For example, workshop participants answered questions such as, "What do you like about validation software as it exists today?" and "If cost were not an issue, what capability would you ask for and why would you want it?" The questions are designed to provoke brain-storming and to encourage descriptive, free-form responses. Multiple responses to each question are allowed and, in fact, encouraged. Each separate idea is recorded anonymously on a separate index card.

The next step establishes categories into which the users' needs can be classified. All index cards from the previous step are placed face-up on a table, and the participants are asked to help sort the cards into categories with common attributes. For example, cards stating "no technical support necessary" and "complicated to use" could both be grouped together because they both address the ease-of-use of a system. Any participant can group, ungroup, or regroup cards as he or she sees fit, and all grouping is done with minimal discussion. Grouping continues until a small, pre-determined number of categories (or card groupings) is established.

After the groupings are established, they must be named. In this step, it is important to let the users define the name for each category. This allows the users' linguistics to enter the design process from the very beginning. For each stack of index cards grouped together in the previous step, all cards in that stack are read aloud. The participants discuss the group's common characteristics and come to a consensus on what the category name should be. During this process, groupings may be combined or divided as necessary to capture all main ideas presented in the index cards. The category name of each grouping is recorded. In the committee meeting, a reduced set of thirteen groupings defined by the sorting process resulted in eighteen unique categories after the naming process was completed.

Finally, a voting process is used to prioritize the categories. Each participant is given a certain number of ratings to distribute among the categories. The ratings vary in value from nine (the best rating) down to one, and each rating can be given to a fixed number of categories. The votes are tallied, and the categories are ranked from highest to lowest priority according to points received.

Results: Affinity Diagram

The results of the QFD Analysis from the meeting of the Insight Subcommittee on Validation can be seen in Figure 1. The category names were determined by workshop participants and are listed on the left. The right-most column reflects the total score as determined by the voting process.

Category	Score
1 . Software Issues	144
2 . Consensus Acceptibility	123
3 . Statistical Foundation	120
4 . Ground Truth	110
5 . Quantitative evaluation	107
6 . Robustness	101
7 . Extensible Databases / data quality	68
8 . Registration	65
9 . Automation	61
10 . Efficiency	57
11 . Application	43
12 . Multimodality	36
13 . Resolution	26

Figure 1. Results from QFD Voting process.

The principle user of the Insight Toolkit and its validation suite is expected to be the software tool designer. Insight is an application programmer's interface (API), and it will support access and hooks for user interface programming; however, it will not contain tools for user interfaces nor visualization and graphics software. Insight is expected to provide mathematical and statistical rigor to advance medical applications through a compact, portable toolkit. In this context, issues of portability, software extensions, robustness, and other Software Issues are expected to top the list of issues. As a public resource Consensus Acceptability is also expected as a top priority. However, our committee does not explicitly have a professional statistician, yet questions of Statistical Foundations, Ground Truth, and Quantitative evaluation are among the five most important issues reported through the QFD analysis.

While not overly surprising, these results have indicated possible gaps in the expertise of the Insight Team. Moreover, the current initiative does not support a strong statistical component. Recommendations for bridging these gaps are made in a later section.

Experimental Design of Validation Trials

The design of blind evaluation for computer algorithm validation is an area in need of more study. The committee considered the work of Yu as one possible model [7].

Figure 2. shows the basic structure of a blind trial designed by Yu and his team. Separate evaluators were selected, and no evaluator knew whether the output decisions were made by a medical student, by any of eight physicians (including the patient's actual diagnosing physician), or a computer system. Figure 3 shows a modified view of a proposed automated structure for validating segmentation and registration algorithms. The costs of truth generation of the blind evaluation are often

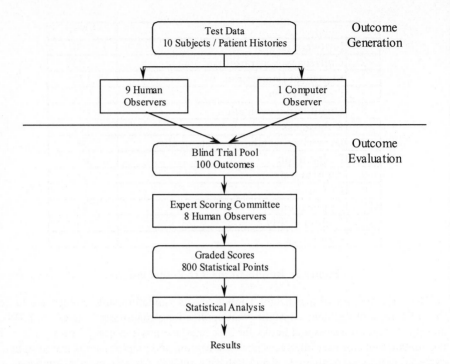

Figure 2. Simplified clinical trial model for evaluating computer performance.

prohibitive to the average algorithm designer. Insight is attempting to provide these elements as part of its work.

Design elements

Some essential design elements that are common to both structures include *truth generation*, the concept of *blind evaluation or scoring*, the use of *multiple observers* to capture the variation in the human decision process, and the analysis of the output through statistical analysis.

Some of the differences include the absence of multiple input datasets in the Insight model arising from a lack of sufficient data. This poverty of input information may introduce significant bias into the test structure.

The output of the outcome generation is handled differently. In the human trials case, a blind evaluation committee is used. In the Insight model, the blind committee is replaced with an automated scoring system. The input is also modified slightly. While truth is generated from collected human data in both cases, the Insight model attempts to pre-collect truth data for re-use in multiple trials. This must be done carefully, separating test data from training data to keep from affecting the experiment. The output of the automated scoring system will be differences in "figures of merit," a term coined by Udupa. These elements suggest that the automated scoring system itself must be validated and shown to be free of

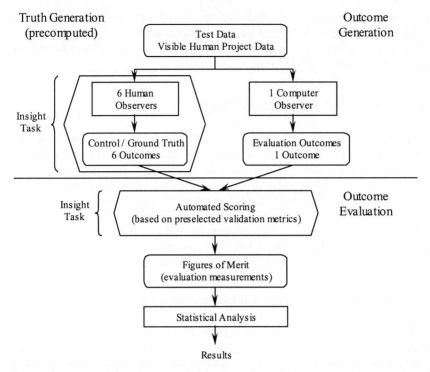

Figure 3. Modified Insight clinical trial model for evaluating segmentation and registration algorithms.

experimental bias. The figures of merit should also be demonstrated to be clinically useful indicators of outcome.

Validation Requirements

The committee iterated over several issues that are considered requirements for a validation suite. They include the removal of experimental bias (including human bias and software training bias). The blind evaluation of the outcomes is also considered essential. The separation of training data from testing data (including training truth and testing truth) is required. Many methods require a large body of training data to begin to make associations, and these data must not be shared with the test data. In addition, one model for software evolution is the tweaking of software parameters. To keep from customizing a method to a particular dataset or group of datasets, a large body of data should be available both for training and testing. Also, the periodic renewal of the testing suite to keep from incremental bias encroaching from repeated testing.

The statistical analysis of segmentation algorithms should study the precision of the method (how well it corresponds with ground truth), the accuracy of the method (how consistent it is), and the efficiency of the method (including the amount of expert user interaction it requires). The committee listed the difference in human vs. computer tasks as an essential element of study. Any validation study should recognize computers are much better at quantitative measurements of images than humans. These effects are of great interest in any validation study undertaken.

Researchers working in the field of segmentation and registration should consider these issues carefully when performing experiments.

Committee Policy Recommendations

The emerging results of this study are beginning to provide a principled infrastructure upon which we can build a validation suite for the evaluation of segmentation and registration software tools. The goal is to archive and make accessible the data, standard protocols, and support software necessary to serve as a foundation for a common platform for the whole medical image processing community.

However, all of the building blocks for such a foundation do not presently exist. There are still significant scientific concerns that must be addressed. To contend with these issues, we are making the following recommendations to the scientists of the medical image processing research community and the programs that support them.

Recommendation: A broad mission for data collection

Common data, accessible across the community, fosters collaborative research and comparative analysis across research groups. The Visible Human Project data is a clear example of how data can create a rendezvous for scientific research. However, the current Visible Human Project data itself is not sufficient to support a common validation effort in segmentation and registration research. The Insight Subcommittee on Validation recommends that the National Institutes of Health and other related agencies embark on a mission to collect comparable data from a wide cross-section of the population as an ongoing effort to support medical imaging into the next century.

Multiple datasets: As a single data point, the Visible Human Project data cannot begin to address the issues of human variation. More importantly, there are dangers in using a limited collection of data to serve for both the training as well as the testing of algorithms and software systems. Significant biases will be introduced into the research and development process; new work will grow to be based on flawed science. The existing data can be used in a limited fashion. For instance, segmentation software can be trained on various portions of the anatomy and tested on others. However, many applications are anatomy specific, and much of the anatomy of humans has only one or two examples per subject.

Multiple modalities: Any data collection initiative should emphasize multiple datasets and multiple modalities. The growth of imaging modalities and the growth of computing in medicine has created an increasing need for the fusion of multiple sets of volume data. The intent is to bring as much information together as possible through advanced computing to create greater insight and better illuminate the human condition.

Clinical Data: The subcommittee recommends that any data collection initiative consider clinical data sources as primary means of obtaining medical information.

The existing Visible Human Project data is limited to post mortem images. These data do not accurately reflect the types of images that pervade the field of medical imaging. Moreover, the Visible Human Project as a study in macro anatomy of embalmed cadavers is in some risk of distancing itself from mainstream medicine. New imaging techniques in functional imaging are providing powerful vehicles for studying human physiology at a cellular level.

Registration datasets: Data supporting the validation of registration algorithms should be collected; it may differ from the type of data needed for segmentation experiments. Fiducial information and landmarks play an important role in registration techniques and should be included in any data collection effort.

Continuing effort: The subcommittee recommends that the mission in medical image data collection be framed as a continuing effort. Imaging technology evolves constantly, improving its ability to resolve fine structures as well as visualize previously inaccessible information. Magnetic resonance functional imaging and ultrasound doppler imaging are two examples of new types of information rather than just improvements in existing modalities. Legacy images stored in databases will lose their relevance rapidly due to the pace of change in imaging systems.

The issues of Consensus Acceptability and Extensible Databases were prominent among the elements of the QFD affinity diagram. All data archived in the recommended effort should be publicly available, indexed via a database system, capable of supporting the anticipated growth suggested by the need for a continuing effort in data collection.

Recommendation: A program in statistics

The Insight Subcommittee on Validation recommends that the National Institutes of Health and related funding agencies undertake an initiative in the support of statistical research in validation science. Several issues surrounding experimental design, the control of human and software development bias, and the quantification of ground truth repeatedly appeared in the planning process. The need for a strong statistical foundation was one of the highest priorities discovered in the QFD analysis.

Questionable truth: A comprehensive and consistent treatment of comparing multiple random variables in the absence of imperfect truth would be an important research resource for the imaging community. The intractable issues of Ground Truth were a high priority in the committee's QFD analysis. Unlike many other scientific research areas, most questions of segmentation and registration can only be compared with the performance of skilled human observers. Ground truth assembled from the responses of multiple people, will be subject to the random variations of human decisions. Published methods for receiver operating characteristic (ROC) analysis for segmentation and registration and other common statistical means of managing such data should be sought.

Limiting or understanding the free imaging variables: The committee recommends the study of imaging protocols to quantify and normalize the variation among imaging systems. No two scans taken of the same patient correspond absolutely. Variation among manufacturers, among technique used at different institutions, and even arising from the relative age of different medical scanners yield differences that are difficult to characterize. Without better understanding of these differences, comparing two output segmentations may be subject to independent variables that are not relevant to the clinical questions being asked. The committee suggests further study in how to normalize the differences among scanners, placing some emphasis on common radiology and pathology technique for data acquisition.

Understanding and reducing experimental bias: The committee recommends an exploration of the sources of bias in validation studies. Much of the significance of a scientific study can be lost if overwhelmed by flawed experimental design. In particular, many segmentation techniques such as Bayesian analysis, genetic programming, and neural network processing require that representative training data be provided to create the statistical framework necessary for their computations. It is essential to separate the test data from the training data, otherwise the tests will be biased and the technique being studied will falsely report superior performance. Additionally, the test data itself must periodically be renewed, otherwise the effects bias collected over time through repeated testing with the same data will indirectly adversely influence the development of segmentation and registration methods. Human bias must also be considered since the human visual system will be used to generate much of the ground truth used in performance experiments. There is much uncharted scientific study in how to create good experiments in imaging software validation.

Figures of merit – evaluation and grading vs. clinical relevance: The committee has also recommended that a study of validation metrics be undertaken with regard to their clinical relevance. Specifically, we recognize that different metrics (e.g., minimum cross sectional area, total surface area, or total volume to name a few) may have different levels of importance depending on the presented pathology, the procedure under consideration, the age of the subject, or any of a number of variables. Characterization of the validation metrics in how well the reflect clinical conditions is necessary to enable useful predictions and directions for future research. What is desired is a measurement that is easy to quantify in a digital imaging context that reflects clinical outcome. Such metrics can then become valid measurements of performance for our segmentation and registration systems.

Conclusions and Future Work

The Insight Software Consortium Subcommittee on Validation is assembling the foundations of a public resource for validating algorithms for segmentation and validation of medical image data. A requirements analysis process was used to generate some common ground among the variety of interests represented among the consortium of anatomists, computer scientists, surgeons, radiologists, psychologists and other related professionals. A basic framework has been proposed as a model for discussing and refining the design of validation experiments and the infrastructure necessary to support them. Based on its findings, the committee is recommending to the Visible Human Project software sponsors and other NIH programs the creation of a long term data collection initiative. In addition, the committee recommends a related program to establish a strong statistical foundation for clinical trials and validation studies in segmentation and registration research.

The QFD Voting process forced the participants to work together to define and name categories of importance from a medical point-of-view, ensuring that the focus of the analysis was correctly targeted to the end-users. Data for a separate analysis based on the Kano method [3] was collected during the same workshop; that data is under study and the results will be available at MICCAI 2000.

Acknowledgements

This work would also not be possible without the support and dedication of the sponsoring agencies and the program officers whose vision launched and continue to sustain this initiative. This distinguished group includes Dr.'s Lindeberg,. Slavkin,. Oberdorfer, Jacquet, Griffin, Clarke, Croft, Gulya, and Huerta. We would like to thank the Insight Team for their continuing participation in this development effort. Many of the ideas published here reflect the collective thinking of the group rather than the particular expertise of the authors. In addition, we would especially like to thank Dr.'s Molholt and Imielinska and Columbia University for organizing and hosting the Insight Software Consortium Subcommittee Meeting on Validation. We'd also like to thank Hillary Schmidt for both her participation and her particular contribution to the validation model presented in this paper.

References

1. R. A. Banvard and P. Cerveri, eds., 1998, Proceedings of the Second Visible Human Project Conference. October 1-2, 1998, Bethesda, MD: US Dept. of Health and Human Services, Public Health Service, NIH.
2. J. Hauser and D. Clausing, 1988, The House of Quality, Harvard Business Review, May-June 1988, pp. 63-73.
3. N. Kano, 1993, A Perspective on Quality Activities in American Firms, California Management Review, 35/3 (Spring 1993): 12-31.
4. NIDCR, 1998, Summary Report: Virtual Head and Neck Anatomy Workshop. February 25, 1998, Bethesda, MD: US Dept. of Health and Human Services, Public Health Service, NIH. (http://www.nidcr.nih.gov/news/strat%2Dplan/headneck/contents.htm).
5. V. Spitzer, M. J. Ackerman, A. L. Scherzinger, and D. Whitlock, 1996, The Visible Human Male: A Technical Report, J. of the Am. Medical Informatics Assoc., 3(2) 118-130.
6. T. S. Yoo and M. J. Ackerman, 2000, A New Program in Medical Image Data Processing, *Medicine Meets Virtual Reality 2000* (Proceedings of the 8th Annual Medicine Meets Virtual Reality Conference), J. Westwood, et al. eds., IOS Press, Amsterdam: pp. 385-391.
7. V. L. Yu, et al.. 1979, Antimicrobial Selection by a Computer, J. of the Am. Medical Assoc., 242(12) 1279-1282..

An Improved Metric Space for Pixel Signatures

A. S. Holmes and C. J. Taylor

Imaging Science, Stopford Building, Oxford Road,
University of Manchester, Manchester M13 9PT, UK
ash@sv1.smb.man.ac.uk,
http://www.isbe.man.ac.uk

Abstract. Our previous work in computer-aided mammography has used scale orientation pixel signatures to provide a rich description of local structure. However, when treated as vectors for statistical classification, the Euclidean space they define has unsatisfactory metric properties. We have also described a novel measure of signature similarity known as best-partial-match (BPM) distance that recognises similar structures whilst remaining robust to background variability and the presence of other structures. In this paper we describe a scheme that makes use of the BPM similarity measure to define a non-linear transformation of pixel signatures into a space with improved metric properties. Using BPM distance we select a set of prototype signatures. We apply multidimensional scaling to these prototypes to construct a new space in which a Euclidean metric behaves in the same way as the BPM distance measure. Support vector regression is then used to learn the non-linear transformation between the new and original spaces permitting a run-time method of transforming any signature into an improved metric space. We use mammographic data to test the performance of our transformed signatures and compare the results with those produced by the raw signatures. Our initial results indicate that our scheme provides an efficient run-time method of transforming signatures into a space suitable for statistical analysis.

1 Introduction

It has been shown that prompting in mammography can improve a radiologist's performance, even if the prompting system makes errors [1]. Prompting involves using computer-based image analysis to locate potential abnormalities so that the radiologist's attention can be drawn to them. Recently, non-linear scale-orientation signatures have been used for the detection of malignant tumours in mammographic images [2]. The approach can in principle be extended to detect other normal and abnormal structures. However, when the signatures are treated as vectors for statistical classification, the Euclidean vector space they define has unsatisfactory metric properties - a small change in underlying structure may produce a large change in the vector. This provides an unsatisfactory basis for statistical analysis. We have previously introduced a novel measure of signature similarity known as the BPM distance that is robust to small changes in scale and orientation of signature structure as well as intrinsic variability and the presence of other potentially confounding structures [3]. The distance measure is based upon the transportation ('earth-mover') algorithm [4]. In this paper, we present a scheme that uses the BPM distance to define a non-linear transformation of pixel signatures into a space with improved metric properties. Firstly, k-means clustering is used to select a set of signature prototypes. A distance matrix is then constructed by measuring the BPM distance from every signature prototype to every other signature prototype. Multidimensional scaling is then used to construct a new space in which measurement of the inter-point *Euclidean* distances produces an approximation to the *BPM* distance matrix. Finally, support vector regression is used to learn the mapping between the original and new spaces. The regression parameters may be applied to any signature to transform it into the new space.

2 Scale-orientation pixel signatures

Morphological M and N filters belong to a class of filters, known as sieves, that remove peaks or troughs smaller than a specified size [5]. By applying sieves of increasing size at a number of orientations, a scale-orientation signature can be constructed for each pixel in an image. The signature is a 2-D array in which columns are values for the same orientation, rows are values for the same scale and the values themselves are the grey-level change at the pixel, resulting from the application of the filter at a particular scale and orientation. Sieves have been shown to have desirable properties when compared to other methods of constructing scale-orientation signatures [2, 6]. In particular, the results at different positions on the same structure are similar (local stationarity) and the interaction between adjacent structures is minimised. Figure 1 shows examples of scale-orientation signatures for the centre pixel of some simple structures.

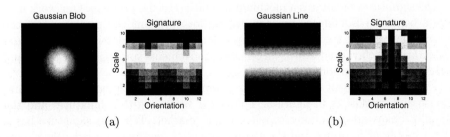

Fig. 1. Scale-orientation signatures for (a) a Gaussian blob and (b) a Gaussian line.

3 Measuring signature similarity using BPM distance

In real applications, signatures may be obtained from one or more structures embedded in a variable mammographic background, resulting in partial similarity between signatures. This suggests that a fraction of a given signature should match to another obtained from the same structure. Therefore, a useful measure of similarity should be able to recognise when similar structures are represented in two signatures despite intrinsic variability and regardless of the presence of other background structure. We have addressed this problem using a *transportation problem* framework [3]. Transportation problems are a class of linear programming problems in which one attempts to minimise the cost of delivering integral quantities of goods produced at a number of warehouses to a number of markets whilst balancing supply and demand [4], as shown in figure 2. There is a cost associated with every possible route from warehouses to markets.

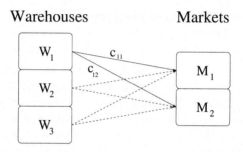

Fig. 2. Representation of a transportation problem.

Signature elements can be thought of as warehouses, each containing goods proportional to the pixel intensity of the element. Alternatively, the elements can be thought of as markets, each requiring goods proportional to the pixel intensity. Choosing a suitable set of costs to capture the two-dimensional nature of the signatures facilitates a meaningful solution to the problem. Our work uses costs based on Euclidean distance, taking into account the periodicity of the orientation axis of the signatures. Thus, localised movement in both scale and orientation is favoured above larger scale movements and signature scale and orientation information is preserved.

We can implement partial signature similarity by supplying a dummy warehouse and market as in figure 3. The larger boxes are the signatures and the smaller boxes are a dummy warehouse and market. The first signature has a total supply of S_1 goods and the second signature requires S_2 goods. We choose a specified fraction of the first signature, fS_1, to describe the second signature. This dictates what the values in the dummy warehouse and market should be, as shown in figure 3. We can solve the problem of finding an appropriate value for the partial match fraction f by considering the transportation solution unit

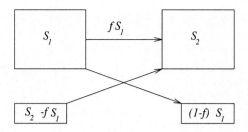

Fig. 3. Partial matching using the transportation problem.

cost as a function of f. We observe a minimum in the transportation cost at the value of f that results in the best description of one signature in terms of the other. This is termed the best-partial-match (BPM) fraction. For example, when a structure is common to both signatures, the BPM fraction corresponds to a solution that finds the common structure but imports and/or exports the remaining material in the two signatures. We use a polynomial fit iterative search to find the BPM fraction at which the minimum unit cost occurs. This minimum unit transportation cost is termed the BPM distance.

4 Signature transformation scheme

Our transformation scheme consists of three separate components; selecting a set of signature prototypes, constructing a new space and learning the transformation into the new space.

4.1 Selecting signature prototypes

Standard k-means clustering is a method of selecting a set of n prototypes from an underlying probability distribution of N samples. In our case, we choose to find a set of 256 prototype pixel signatures that are representative of pixel signatures formed from a complete range of mammographic structure. 256 initial prototypes are randomly selected and then a large number of signatures (2^{16}) are examined in turn. Each examined signature is assigned to its nearest prototype and the mean of that prototype is adjusted accordingly. This continues for all N signatures, resulting in a set of prototypes that are representative of the examined signatures. Figure 4 shows a mammogram containing a wide variety of structures. For each of the prototype signatures obtained by k-means clustering, the image pixel with the most similar signature is highlighted, given a visual representation of the prototype distribution. As expected, relatively few prototypes are required to represent background tissue. In contrast, the prototype density is higher in regions containing more structure. Now that we have a generated a representative set of signature prototypes, we may use multidimensional scaling to construct an improved metric space.

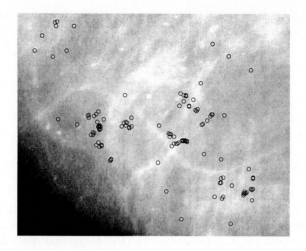

Fig. 4. A mammogram containing a variety of structure. The markers indicate the closest image signatures to the set of signature prototypes.

4.2 Constructing a new space

There are many types of multidimensional scaling but they all attempt to solve the following problem; given a set of distances, d_{rs}, find a configuration of points that will reproduce an approximation to these distances. The distances may or may not be *Euclidean* and they need *not* satisfy the triangle inequality

$$d_{rt} \leq d_{rs} + d_{st} \tag{1}$$

satisfied if they were produced by a metric. Our current work uses an analytical technique although we plan to investigate the use of an iterative scaling algorithm as this may offer an improvement to our scheme. Our analytical method is briefly presented below. The reader is referred to Ripley [7] for a more detailed explanation.

A configuration X is an $n \times p$ matrix of data, considered as n points in p dimensions. We note that the distance matrix may be constructed from X as follows

$$\text{if } M = XX^T, \tag{2}$$

$$d_{rs}^2 = M_{rr} + M_{ss} - 2M_{rs}. \tag{3}$$

This observation leads to the following implementation, known as classical or metric multidimensional scaling. If L is the given distance matrix, we define

$$T = L^2. \tag{4}$$

Next, we remove row and column means and add back the overall mean. This follows from equation 3.

$$T' = -\frac{1}{2}\left[T - \frac{(T\mathbf{1})\mathbf{1}^T}{n} - \frac{\mathbf{1}(T\mathbf{1})^T}{n} + \frac{\mathbf{1}^T T\mathbf{1}}{n^2}\right]. \tag{5}$$

The rank of T' gives the number of dimensions that the constructed configuration will have. Next, we find the eigendecomposition of T' using schur decomposition

$$T' = CD^2C^T \tag{6}$$

and find D by taking the matrix square root. To finish, we must find the order of the eigenvalues i.e. the diagonal values of D with the largest first and take the corresponding columns of CD to be our configuration X. Given a Euclidean distance matrix, this technique produces a perfect configuration i.e. the given distance matrix is perfectly reproduced. If the distance matrix is non-Euclidean, the technique produces an approximation to the distance matrix.

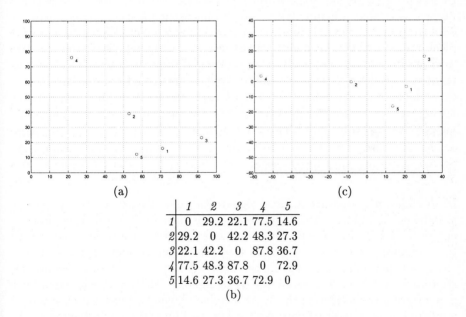

(a)

(c)

	1	2	3	4	5
1	0	29.2	22.1	77.5	14.6
2	29.2	0	42.2	48.3	27.3
3	22.1	42.2	0	87.8	36.7
4	77.5	48.3	87.8	0	72.9
5	14.6	27.3	36.7	72.9	0

(b)

Fig. 5. (a) a random configuration of points, (b) the distance matrix generated by the configuration and (c) a configuration reconstructed from the distance matrix.

Figure 5(a) is a randomly generated set of 2D points. The distance matrix in figure 5(b) is generated from the data points by measuring the inter-point

Euclidean distances. The distance matrix is symmetric and has a zero diagonal. Multidimensional scaling then generates a new configuration from the distance matrix, as shown in figure 5(c). If the distance matrix is Euclidean, that is it contains Euclidean distances, then the new configuration will exactly reproduce the generating distance matrix when the inter-point distances are measured. However, if the distance matrix is not Euclidean, the new configuration can only reproduce an approximation to the generating distance matrix. It is interesting to note that the distance matrix is invariant to translation, rotation and reflection of the configuration of data points. In our case, a distance matrix is constructed by measuring the inter-prototype BPM distances. Multidimensional scaling then transforms each of these prototypes into a new space, from the distance matrix.

4.3 Learning the space transformation

The use of support vector machines (SVM) in empirical data modelling is becoming widespread as they generalize well and address the curse of dimensionality. They provide promising empirical performance and have many attractive features. They can be used in both classification (SVC) and regression (SVR) problem domains. The reader is referred to Vapnik [8] for a more exhaustive treatment of SVMs and some recent applications. When the input space is not linearly separable, a mapping into a high dimensional feature space, where the data is linear, is required. However, a kernel function can be used to avoid explicit evaluation of this computationally intensive mapping. The kernel function used in our work is the popular Gaussian radial basis function. The regression input space is defined as the original prototype signatures, formed using k-means clustering and the output space is defined as the reconstructed prototypes in the new space. After performing SVR, the regression parameters can be applied to any signature to transform it into the new space.

5 Experimental evaluation

In the following experiment, pixel signatures were constructed by applying a morphological M filter, centred on every image pixel, at different scales and orientations. Ten scales were used that increased logarithmically and were chosen to encompass the range of structure sizes that occur in mammograms. As orientation is periodic, for each scale the filter was applied at twelve orientations, separated by fifteen degrees. A region of interest (ROI) was selected from a mammogram (mdb245ls) in the MIAS database [9] and a pixel was chosen from a salient linear structure. Figure 6(a) shows the ROI and the selected pixel. The ROI exhibits a variable background and contains a variety of structures that are commonly found in mammograms. Euclidean and BPM distances were then measured between the chosen pixel's signature and every pixel signature in the ROI using raw or transformed signatures. Plotting the similarity distance at each pixel forms a similarity image for each method. These images were inverted so that bright regions correspond to similar pixels and dark regions corresponded to pixels that were less similar. Three similarity images were generated. Firstly, Euclidean distance was measured between raw signatures. This allows us to see whether our scheme offers any improvement in metric properties compared to the original Euclidean vector space. Secondly, BPM distance was measured between raw signatures and finally, Euclidean distance was measured between transformed signatures.

6 Results and Discussion

Figures 6(b), (c) and (d) show the similarity images generated by the experiment described in section 5.

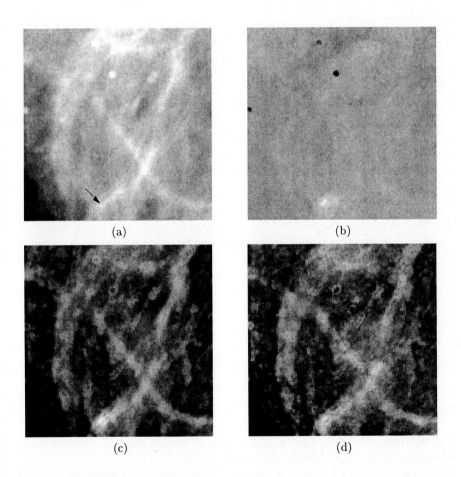

(a)   (b)

(c)   (d)

Fig. 6. Pixel signature similarity images are generated from (a) a pixel from a linear structure in the original image using (b) Euclidean distance between raw signatures, (c) BPM distance between raw signatures and (d) Euclidean distance between transformed signatures.

Figure 6(b) shows the similarity image generated by measuring Euclidean distance between raw signatures. This image clearly demonstrates the poor metric properties of the vector space defined by the raw signatures. The method has not been able to segment other linear structures in the image, although it has emphatically rejected blob-like structures. This image confirms the need for the transformation scheme described in this paper. Figure 6(c) shows the similarity image generated by explicitly measuring the BPM distance between raw signatures. This image took approximately a day to generate. We have previously discussed the suitability of the BPM measure, as indicated by this result [3] but for completeness we note that the method has successfully detected linear structures and suppressed other structures and background, thus measuring similarity in a practically useful way. This result encourages us to examine the performance of our transformation scheme as shown in figure 6(d). This figure shows the similarity image generated by transforming all image signatures using the scheme described in section 4 and measuring the Euclidean distance between the line pixel signature and every other image pixel. Having already constructed the new space and learnt the mapping from the original to the new space 'off-line', this image took a quarter of a second to generate, showing one of the advantages of our scheme. It is readily apparent that as well as offering significant improvement in computational efficiency, the scheme also offers a high degree of accuracy in approximating the BPM distance measurement.

7 Conclusions

The results described in section 6 indicate that our scheme provides a comprehensive and efficient method of transforming pixel signatures into a space with improved metric properties. Future work will investigate ways to improve the scheme and provide quantitative results. We will then examine its use in classification performance when detecting focal abnormalities and normal structure in mammograms.

8 Acknowledgements

Anthony Holmes is funded by the EPSRC on a quota studentship.

References

1. I. W. Hutt. *The Computer-aided Detection of Abnormalities in Digital Mammograms.* PhD thesis, University of Manchester, 1996.
2. R. Zwiggelaar, T. C. Parr, J. E. Schuum, I. W. Hutt, S. M. Astley, C. J. Taylor, and C. R. M. Boggis. Model-based detection of spiculated lesions in mammograms. *Medical Image Analysis*, 3(1):39–62, 1999.
3. A. S. Holmes and C. J. Taylor. Developing a measure of similarity between pixel signatures. In *British Machine Vision Conference*, pages 614–622, 1999.

4. F. L. Hitchcock. The distribution of a product from several sources to numerous localities. *J. Math. Phys.*, 20:224–230, 1941.

5. J. A. Bangham, P. D. Ling, and R. Young. Multiscale recursive medians, scale-space and transforms with applications to image processing. *IEEE Transactions on Image Processing*, 5(6):1043–1048, 1996.

6. R. Harvey, A. Bosson, and J. A. Bangham. The robustness of some scale-spaces. In *British Machine Vision Conference*, pages 11–20, 1997.

7. B. D. Ripley. *Pattern recognition and neural networks*. Cambridge University Press, 1996.

8. V. Vapnik. *The nature of statistical learning theory*. Springer-Verlag, New York, 1995.

9. J. Suckling, J. Parker, D. Dance, S. Astley, I. Hutt, C. Boggis, I. Ricketts, E. Stamatakis, N. Cerneaz, S. Kok, P. Taylor, D. Betal, and J. Savage. The mammographic images analysis society digital mammogram database. In *Exerpta Medica. International Congress Series*, volume 1069, pages 375–378, 1994. mias@sv1.smb.man.ac.uk.

Evaluation of an Automatic System for Simulator/Portal Image Matching

F.M. Vos[1], J. Stoeckel[1,3], P.H. Vos[2], A.M. Vossepoel[1]

[1] Pattern Recognition Group, Delft University of Technology
Lorentzweg 1, 2628 CJ Delft, The Netherlands
{frans,jonathan,albert}@ph.tn.tudelft.nl

[2] Bernard Verbeeten Institution
Brugstraat 10, 5000 LA Tilburg, The Netherlands
vos.ph@bvi.nl

[3] EPIDAURE Project, INRIA
2004 route des Lucioles, BP 9306902, Sophia Antipolis Cedex, France
Jonathan.Stoeckel@sophia.inria.fr

Abstract. User guided systems for X-ray/portal image matching are commonly employed in radiation therapy for position verification and correction. To enable extended clinical support a fully acceptable solution must be generic, accurate, fast and automatic. In this paper a new registration system using the structure tensor is compared with the performance of experienced radiographers. It is shown that local extremum lines (emanating from bone ridges) are invariant with respect to the photon energy and that they can be detected without a priori knowledge. Registration based on such creases is validated with a large database of images from anterior irradiation fields for prostate treatments. It appears that the precision of the structure tensor approach is significantly better than human observers. The execution time is less than 10 seconds while human supervision can be restricted to detect outliers only.

1 Introduction

In the European Union, cancer is the second leading cause of death. Radiotherapy is one of the most effective means for its treatment [1]. The objective of radiation therapy is to deliver a prescribed radiation dose to the tumor site to eradicate the cancer cells. Unfortunately, incorrect dose delivery may easily result in complications or recurrence of the tumor. For this reason it is of paramount importance to verify the patient's position during treatment.

The boundary of a treatment area is often prescribed with the aid of a classic X-ray image (henceforth we will refer to it as the *simulator image*, see Figure 1). Alternatively, such a 'reference image' may be reconstructed from CT-data (through ray-tracing) resulting in a digitally reconstructed radiograph, DRR [2]. During treatments a so-called *portal image* is acquired with the treatment beam using an electronic portal imaging device (EPID). Discrepancies in the patient setup can be derived from the comparison of

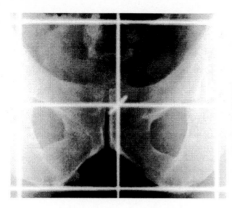 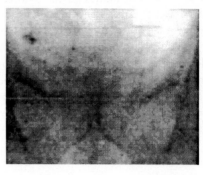

Fig. 1. Examples of a simulator (left) and portal image (right) of the pelvis. Both images were linearly stretched for clarity. The outer white 'rectangle' in the simulator image represents the treatment area.

the prescribed and given treatment field through image matching. Consequently, the patient's position can be corrected in subsequent treatments.

In the 'Bernard Verbeeten Institution', the reference image consists of an simulator image because the DRR modality is not available. To do the image registration, a human operator manually delineates appropriate bone features in the simulator image. Subsequently, the template is positioned by hand at the corresponding location in the portal image through rotation and translation. Thus, the 'prescription image' (the simulator image) is related to the treatment area (depicted by the portal image) via the projected anatomy of the patient.

Although the sketched approach is generally accepted in clinical practice, it is well known to be sensitive to observer variability and time consuming due to the necessary user interaction. An ideal solution should meet the following requirements:

- fast
- automatic,
- accurate as well as precise,
- generic, i.e. not use explicit knowledge about the visualized anatomy.

In the literature, not many articles describe *X-ray - portal* image matching. In [3] [4] fiducial marks are extracted manually from the simulator image and a top-hat filter is used to recover them from the portal image. The registration is achieved through chamfer matching. A knowledge based approach was proposed in [5] applying a snake fitting algorithm to segment bone crests from images of the pelvis. At last [6] introduces a matched filter approach to emphasize bone features. Subsequently, the filtered images are registered through correlation. Although good results are given (a success rate of 95%), the experimental outcome is incomparable for the lack of a well described criterion of 'success'.

Clearly, none of the proposed solutions completely satisfies the requirements mentioned. In this paper a new technique will be described in which features are automatically extracted from both the simulator and the portal image without using *a priori* knowledge.

The paper is structured as follows. Section 2 gives a global description of our technique. Then, in Section 3 the effectiveness specifically regarding the accuracy and precision will be evaluated. For testing we will use images of the pelvic area made for prostate cancer treatment. Finally, in Section 4 we will summarize our findings and draw conclusions.

2 Approach

2.1 Outline

The image processing involved in the verification process comprises a number of stages as illustrated in Figure 2.

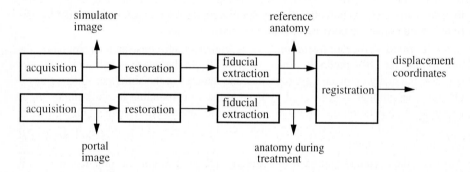

Fig. 2. Overview of the position verification system. The output consists of the displacement between two images.

The simulator and portal images are acquired with beams of very different photon energies (approximately 100 kV and 10 MV respectively). It can be seen in Figure 1 that portal images lack contrast and sharpness in the projected patient anatomy compared to low energy X-ray images. Moreover, they contain disturbing artefacts requiring restoration (for an elaborate description we refer to [7]). Also, corrections have to be included for the crosswires in the simulator image (Figure 1). It is unavoidable that by their presence image areas are occluded. The previously noticed difference in portal/simulator image structure has proven to be too large for correlation based methods [6][8]. For this reason we have opted to use so called fiducial 'structures' in the image that are assumed to be immobile with respect to the tumor (that often cannot be detected).

In the following sections we will investigate two types of fiducial features, describe techniques to extract them from the images and elaborate on the registration.

2.2 Definition of fiducial features

To allow proper verification of the patient's position fiducial features should satisfy the following prerequisites:

- detectable in both simulator and portal image,
- fixed with respect to the treatment target,
- in a reproducible position with respect to the radiation energy.

The most prominent aspects of portal images are the projections of bony structures, air cavities and body outlines. It has been proven that local extremum lines (ridges in the simulator image, troughs in the portal image) correspond to lines of maximum bone thickness [9]. In the remainder we will refer to such features as 'creases'. In the context of prostate treatments, it is generally assumed in clinical practice that the pelvic bone satisfies the second condition [10][11]

Let us consider the influence of photon energy on the position of two types of image features, namely *creases* and *edges* (i.e. transitions in image intensity). To this end consider a mono-energetic source and an image line detector of which the measured signal is linear with the number of incoming photons. Additionally, let us assume that the treatment site consists only of soft tissue and bone. Then, the measured signal on position x at the detector can be modelled by:

$$I(x) = I_0 \cdot e^{-\mu_b(\varepsilon)d_b(x) - \mu_s(\varepsilon)d_s(x) - \mu_a(\varepsilon)d_a(x)}, \tag{1}$$

where I_0 reflects the source intensity and the sensitivity of the detector, $\mu_y(\varepsilon)$ represents the attenuation coefficient of a material y as a function of energy, and $d_y(x)$ the pathlength through the material. The subscripts are b for bone, s for soft tissue and a for air.

The position of *creases* in the image (i.e. lines of maximal bone thickness) are found where the *first derivative* of I is zero. Furthermore, *edges* (due to bone/soft tissue transitions) are at locations where the *second derivative* has a zero crossing.

By definition:

$$C_{sd} = d_b(x) + d_s(x) + d_a(x) \tag{2}$$

is constant, representing the distance between source and detector.

To find an expression for both types of features, we will assume that the thickness of the patient is a linear function of the position x. Consequently:

$$d_a(x) = C + kx, \tag{3}$$

with k a constant.

Setting the first derivative of $I(x)$ (Equation 1) to 0 and solving for $d\,d_b(x)/dx$ we find (using Equation 2 and Equation 3):

$$\frac{d}{dx}d_b(x) = k\frac{(\mu_s(\varepsilon) - \mu_a(\varepsilon))}{(\mu_b(\varepsilon) - \mu_s(\varepsilon))}. \tag{4}$$

Similarly, setting the second derivative of $I(x)$ to zero results in:

$$\frac{d}{dx}d_b(x) = k\frac{(\mu_s(\varepsilon) - \mu_a(\varepsilon))}{(\mu_b(\varepsilon) - \mu_s(\varepsilon))} \pm \frac{1}{\sqrt{\mu_b(\varepsilon) - \mu_s(\varepsilon)}}\sqrt{\frac{\partial^2}{\partial x^2}d_b(x)}. \tag{5}$$

Note that Equation 4 recurs in Equation 5.

The right hand side in *Equation 4* is approximately constant (1.04 at 100 kV and 1.10 at 10 MV with k = 1). Consequently, there is practically no dependence on the energy level. The position of ridges is determined merely by the first derivative of d_b. The second term in *Equation 5*, however, may change dramatically as a function of energy as the factor before the square root varies from 2.5 to 7.1. Thus, the remaining influence of the attenuation coefficients may easily result in displacement of edge features. This comparison illustrates the superiority of creases over transitions as fiducial marks.

2.3   Fiducial feature extraction

In previous work we have tested several automatic crease extraction techniques [12]. Following this study we have opted to use a crease measure based on the structure tensor[13]. The structure tensor (F) is defined by the dyadic product of the gradient vector with itself:

$$\overline{F} = \overline{(\nabla f)(\nabla f)^t} = \begin{bmatrix} \overline{f_x f_x} & \overline{f_x f_y} \\ \overline{f_x f_y} & \overline{f_y f_y} \end{bmatrix}. \tag{6}$$

Here, f_x and f_y denote the partial derivatives of an image function $f(x,y)$. They are calculated through Gaussian derivatives. Also, a Gaussian is used to 'smoothen' the structure tensor (depicted by the overbar).

Let $\lambda_1 \geq \lambda_2$ be the eigenvalues of the structure tensor at a certain point on f, and v_1, v_2 the corresponding eigenvectors. A point is qualified as a ridge point if the first derivative in the direction of v_1 is zero.

In the description above, it can be seen that the width of the Gaussian derivative introduces a (first) scale parameter. This parameter is used only to suppress the effect of spatially uncorrelated noise on the derivatives. Therefore, it is chosen small, to correspond as much as possible to the inner scale of the image (i.e. the smallest features of interest). A second parameter is introduced by the Gaussian tensor smoothing. This parameter is tuned to prevent spurious creases.

It should be noticed that in our approach, changing the amount of tensor smoothing does not change the position of creases. It only influences the scale on which they are found. If a scale space of the input image were used instead, creases are found in different positions for different scales of the image [14].

In addition to the structure tensor approach, we have implemented the technique proposed by Gilhuijs [4] for comparison. Here, a top-hat transformation [15] is applied to the input image:

$$f_{out} = f_{in} - (f_{in} \bullet K) \quad, \tag{7}$$

where the operator • represents greyscale opening (i.e. morphological erosion followed by dilation). The kernel is rectangular of small size, typically 7x7 pixels. To identify the creases the transformed image (f_{out}) is merely thresholded. Both the exact kernel size and threshold level are to be determined experimentally.

In Section 3 we will evaluate both approaches.

2.4 Registration

To match the crease images, standard chamfer matching [16] was modified as follows. Instead of taking the distance transform to calculate the offset between two images, the 'reference-ridge image' is inverted after applying a Gaussian filter. Firstly, this allows the match to take into account extra information about the strength of individual points. Secondly, the influence of found crease points is limited locally, which is controlled by the width of the Gaussian. This is important in case of both over- and under-detection.

While performing the registration we will only consider translations and in-plane rotations. Thus, out-of-plane rotations are neglected. (Earlier, only poor results have been reported regarding the estimation of out-of-plane rotations based on two-dimensional reference images [9]).

From the preceding description it can be seen that our technique does not require explicit knowledge about the patient anatomy (conform requirement 3 in the introduction). Moreover, the user interaction is limited to a posteriori check of the result.

The remaining requirement (accuracy) will be explored in the next section.

3 Results

To evaluate our method we will focus here on the accuracy and the precision. It may be expected that they differ with the principal direction, i.e. transversal or longitudinal, due to differences in orientation information. For instance, in the images of the pelvis vertical structures dominate. Also, outliers will effect the outcome as a consequence of the occlusion of key bone structures by the crosswires. At last estimating in-plane rotation in addition to translation may contribute to a larger standard deviation due to the surplus in degrees of freedom.

To monitor the effect of such influences we conducted the following experiments. Over one year we randomly selected 57 patients visiting our institution for prostate cancer treatment. For each patient an anterior simulator image was obtained as well as several corresponding portal images (portal images were acquired in approximately 11 treatments per patient). Two experienced clinical physicists aligned each portal-simulator image pair manually. The averaged results of these two matches defined our gold standard. This ground truth was compared with the outcome by radiographers in clinical practice, the top-hat based technique and our method. The parameters of the latter two approaches were experimentally determined by minimizing their standard deviation.

In Table 1: the mean distances and corresponding standard deviations are tabulated between the ground truth and each of the three alternatives. To this end 549 images were included without in-plane rotation according to the gold standard. Consequently, they were aligned using translation only. The field dimensions typically were $(100 \text{ mm})^2$

digitized at a resolution of 75 dpi. The execution time for one match was on the order of a few seconds.

Table 1: mean distances and corresponding standard deviations between matches of experts (defining the gold standard) and three alternative 'systems'.

matching by translation only		radio-graphers	tophat	structure tensor
# of images		549	549	549
transversal:	mean (mm)	-0.08	0.11	0.27
	st.dev (mm)	1.28	1.87	1.61
longitudinal:	mean (mm)	0.52	0.49	0.55
	st.dev (mm)	1.33	2.04	1.87

All three matching 'systems' appear about equally *accurate* and not significantly different from the gold standard. The systematic difference in transversal and longitudinal *precision* (standard deviation) is caused by the shape of the anatomy: vertical structures dominate. Also, the precision of both the tophat as well as the orientation tensor based technique is notably worse than the human observers. Further analysis showed that only a few outliers are at the basis of this outcome. As was noticed previously, the cross hairs in the imaging device may block the central part of the portal image. In that case the absence of key bone structures such as the symphysis contributes to a larger standard deviation.

Most of the outliers will be detected immediately by a human observer. For this reason it seems obvious to exclude the matches that exceed a minimum distance from the gold standard. Table 2: shows the result when matches are ignored that deviate more than 7.5 mm. In this case, apparently no large differences in accuracy are found with the gold

Table 2: As in Table 1:, but leaving out those matches that are more than 7.5 mm from the gold standard for each system separately.

matching by translation only		radio-graphers	tophat	structure tensor
discarded outliers (n=549)		0.2%	2%	2.2%
transversal:	mean (mm)	-0.08	0.05	0.19
	st.dev (mm)	1.28	1.42	0.90
longitudinal:	mean (mm)	0.5	0.45	0.57
	st.dev (mm)	1.27	1.43	0.98

standard. The precision of the structure tensor technique may not be dramatically smaller than both the radiographers or the tophat approach, but it is significantly better by the Fisher test.

To study the influence of rotation, 78 images from 8 patients were analyzed in which small in-plane revolutions were indicated by the gold standard (all smaller than 3°). Table 3: shows the outcome. It may come as a surprise that the radiographers matched the

Table 3: as in Table 2:, now also matching by rotation by radiographers and the structure tensor approach

matching by translation and rotation		radio-graphers	structure tensor
# of images		78	76
transversal:	mean (mm)	-0.28	0.11
	st.dev (mm)	1.49	1.11
longitudinal:	mean (mm)	0.42	0.42
	st.dev (mm)	1.45	1.05
rotation:	mean ($^{\circ}$)	-	0.36
	st.dev ($^{\circ}$)	-	0.89

images using translation only in *all* cases. As a logical consequence, the registration turned out to be less precise (with standard of deviations approximately 1.4 mm). In our view this only emphasizes the difficulty of retrieving in-plane rotations by humans. Contrasting to this, the rotation-angle appeared well estimable by the automatic techniques (with a precision on the order of 0.9°). The effect of the additional parameter on the transversal/longitudinal accuracy and precision was negligible.

4   Discussion/conclusion

Comparing our results using the tophat technique with those reported in the literature, we noticed remarkably worse performance in our outcome. In [4] the precision of both the radiographers as well as the tophat technique is approximately twice as good (with standard deviations of about 0.5 mm). This difference might be attributed to the quality of the portal images. It is well known that our fluoroscopic portal imaging systems yield worse signal to noise ratio than liquid filled ionisation chambers employed by [4] (see also [3]). Another remarkable difference is that in [4] the reference template is drawn manually.

From the preceding sections we conclude that a registration system for portal/simulator image matching should be generic, accurate, automatic and fast. Local extremum lines can be detected without using a priori knowledge; their position is invariant with the photon energy. The accuracy of an automatic system based on the structure tensor is not significantly different from experienced physicists as well as radiographers (the latter doing the work in clinical practice). The precision is significantly better than that of human observers when outliers are discarded. Moreover, in-plane rotations are taken into account *without* sacrificing accuracy or precision. These results are obtained when user control is admitted to detect outliers only. The execution time of our system is less than 10 seconds which is sufficient for clinical application.

References

[1]   P. Rubin, "Clinical Oncology - A multidisciplinary approach for physicians and students", 7th edition, Saunders & Co, Philadelphia, 1993.

[2]   R. Bansal, L. Staib, Z. Chen, A. Rangarajan, J. Knisely, R. Nath, J. Duncan, "A minimax entropy registration framework for patient setup verification in radiotherapy", Computer Aided Surgery, vol. 4, pp. 287-304, 1999.

[3]   K.G.A. Gilhuijs, J. Bijhold, M. van Herk, H. Meertens, "Automatic on-line inspection of patient set-up in radiation therapy using digital portal images", Medical Physics, vol. 20, pp. 667-677, 1993.

[4]   K.G.A. Gilhuijs, A. Touw, M. van Herk, R.E. Vijlbrief, "Optimization of automatic portal image analysis", Medical Physics, vol. 22, pp. 1089-1099, 1995.

[5]   F. Yin, K. Nie, L. Chen, C.W.Chen, "Automated extraction of pelvic features in portal and simulator images", SPIE Conference on Medical Imaging and Image Processing, Newport Beach, California, Vol. 2710, pp. 1020-1023, 1996.

[6]   F. Kreuder, B.Schreiber, C. Kausch, O. Doessel, "A structure based method for online matching of portal images for an optimal patient set-up in radiotherapy", Philips Journal of Research, vol. 51, no. 2, pp. 317-337, 1998.

[7]   J. Stoeckel, "Simulator portal image matching - a patient position verification system for radiotherapy", graduation report, Pattern Recognition Group, Technical University Delft, 1999.

[8]   J. Moseley, P. Munro, "A semiautomatic method for registration of portal images", Medical Physics, vol. 21, no. 4, pp. 551-557, 1994.

[9]   K.G.A. Gilhuijs, P.J.H. van de Ven, M. van Herk, "Automatic three-dimensional inspection of patient setup in radiation therapy using portal images, simulator images, and computed tomography data", Medical Physics, vol. 10, no. 6, pp. 389-399, 1996.

[10]   J.A. Antolak, I.I. Rosen, C.H. Childress, G.K. Zagars, A. Pollack, "Prostate target volume variations during a course of radiotherapy", International Journal of Radiation Oncology, Biology and Physics", vol. 42, no. 3, pp. 661-672, 1998.

[11]   M.J. Zelefsky, D. Crean, G.S. Mageras, O. Lyass, L. Happerset, "Quantification and predictors of prostate position variability in 50 patients evaluated with multiple CT scans during conformal radiotherapy", Radiotherapy and Oncology, vol. 50, no. 2, pp. 225-234, 1999.

[12]   J. Stoeckel, F.M. Vos, P.H. Vos, A.M. Vossepoel, "An evaluation of ridge extraction methods for portal imaging", submitted for publication to ICPR conference 2000.

[13]   M. Kass, W.Witkin, "Analyzing oriented patterns", Computer Vision, Graphics and Image Processing, vol. 37, pp. 362-385, 1987.

[14]   D.S. Fritsch, E.L. Chaney, A. Boxwala, M.J. McAuliffe, S. Raghavan, A. Thall, J.R.D. Earnheart, "Core-based portal image registration for automatic radiotherapy treatment verification", International Journal for Radiation Oncology, Biology and Physics, vol. 33, no. 5, pp. 1287-1300, 1995

[15]   J. Serra, "Image analysis and mathematical morphology", 2nd edition, Academic Publishers, San Diego, Ca, 1988.

[16]   G. Borgefors, "Hierarchical chamfer matching: a parametric edge matching algorithm", IEEE Transactions on Pattern Analysis and Machine Intelligence, vol. 10, no. 6, pp. 849-865, 1988.

Image Registration by Maximization of Combined Mutual Information and Gradient Information

Josien P. W. Pluim, J. B. Antoine Maintz, and Max A. Viergever

Image Sciences Institute, University Medical Center Utrecht
Utrecht, The Netherlands
{josien,twan,max}@isi.uu.nl

Abstract. Despite generally good performance, mutual information has also been shown by several researchers to lack robustness for certain registration problems. A possible cause may be the absence of spatial information in the measure. The present paper proposes to include spatial information by combining mutual information with a term based on the image gradient of the images to be registered. The gradient term not only seeks to align locations of high gradient magnitude, but also aims for a similar orientation of the gradients at these locations.

Results of combining both standard mutual information as well a normalized measure are presented for rigid registration of three-dimensional clinical images (MR, CT and PET). The results indicate that the combined measures yield a better registration function than mutual information or normalized mutual information per se. The accuracy of the combined measures is compared against a screw marker based gold standard, revealing a similar accuracy for the combined measures to that of the standard measures. Experiments into the robustness of the measures with respect to starting position imply a clear improvement in robustness for the measures including spatial information.

1 Introduction

Mutual information is currently a popular registration measure, which has been shown to form the base of accurate and robust registration methods by several independent studies [1, 2, 3, 4, 5], and in particular by the Retrospective Registration Evaluation Project (RREP), an international study comparing the accuracy of sixteen registration methods against a screw marker gold standard [6]. However, failure of the measure in certain situations has also been reported [7, 8, 9, 10]. Such situations can arise when the images are of low resolution, when the overlapping part of the images is small or as a result of interpolation methods. A possible solution to failure of mutual information may be to include spatial information, something that is not contained in the measure. Rueckert et al. [11] recently proposed 'higher-order mutual information', which incorporates spatial information by forming four-dimensional intensity histograms. We propose to include spatial information by combining mutual information with gradient information. Image gradients by themselves have been shown to be useful registration criteria [10, 12].

2 Method

2.1 Mutual Information

The mutual information of two images is the amount of information that one image contains about the other or vice versa. When transforming one image with respect to the other such that their mutual information is maximized, the images are assumed to be registered.

The mutual information I of two images A and B can be defined as

$$I(A, B) = H(A) + H(B) - H(A, B). \tag{1}$$

Here, $H(A)$ and $H(B)$ denote the separate entropy values of A and B respectively. $H(A, B)$ is the joint entropy, i.e. the entropy of the joint probability distribution of the image intensities. In this paper, we use the Shannon measure of entropy, $-\sum_{p \in P} p \, \log p$ for a probability distribution P. The joint probability distribution of two images is estimated by calculating a normalized joint histogram of the grey values. The marginal distributions are obtained by summing over the rows, resp. the columns, of the joint histogram.

Recently, it was shown that the mutual information measure is sensitive to the amount of overlap between the images and *normalized* mutual information measures were introduced to alleviate this problem. Examples of such measures are the normalized mutual information $Y(A, B)$ introduced by Studholme et al. [7]

$$Y(A, B) = \frac{H(A) + H(B)}{H(A, B)}$$

and the entropy correlation coefficient $ECC(A, B)$ used by Maes et al. [2]

$$ECC(A, B) = \frac{2\, I(A, B)}{H(A) + H(B)}.$$

These two measures have a one-to-one correspondence and we will therefore only use $Y(A, B)$ in this paper.

2.2 Incorporating Gradient Information

Image locations with a strong gradient are assumed to denote a transition of tissues, which are locations of high information value. The gradient is computed on a certain spatial scale. We have extended mutual information measures (both standard and normalized) to include spatial information by multiplying the mutual information with a gradient term. The gradient term is based not only on the magnitude of the gradients, but also on the orientation of the gradients.

The gradient vector is computed for each sample point $\mathbf{x} = \{x_1, x_2, x_3\}$ in one image and its corresponding point in the other image, $\mathbf{x}'$, which is found by geometric transformation of $\mathbf{x}$. The three partial derivatives that together form the gradient vector are calculated by convolving the image with the appropriate first derivatives of a Gaussian kernel of scale σ. The angle $\alpha_{\mathbf{x},\mathbf{x}'}(\sigma)$ between the gradient vectors is defined by

$$\alpha_{\mathbf{x},\mathbf{x}'}(\sigma) = \arccos \frac{\nabla \mathbf{x}(\sigma) \cdot \nabla \mathbf{x}'(\sigma)}{|\nabla \mathbf{x}(\sigma)||\nabla \mathbf{x}'(\sigma)|}, \tag{2}$$

with $\nabla \mathbf{x}(\sigma)$ denoting the gradient vector at point $\mathbf{x}$ of scale σ and $|\cdot|$ denoting magnitude.

For multimodal images, the different imaging techniques can lead to a tissue having different intensities in either image. As a result, the gradients of the images can point in different directions. However, since the images fundamentally depict the same anatomical structures, gradients in two multimodal images – at least in principle – will have the same orientation and either identical or opposing directions. Consequently, we use the following weighting function w, which favours both very small angles and angles that are approximately equal to π (see figure 1(a)):

$$w(\alpha) = \frac{\cos(2\,\alpha) + 1}{2}. \tag{3}$$

Furthermore, the different imaging processes of different modalities imply that multimodal images do not necessarily depict the same tissue transitions. Hence, strong gradients that emerge with a certain imaging technique may be absent or less prominent with another technique. Since we are only interested in including strong gradients that appear in *both* images, the angle function is multiplied by the minimum of the gradient magnitudes. Summation of the resulting product for all samples gives us the gradient term with which we multiply the mutual information measure. Two examples of the gradient measure (before summation) for different combinations of multimodal images can be found in figure 1(b). Tissue transitions that are depicted in both modalities are emphasized.

The proposed registration measure becomes

$$I_{new}(A, B) = G(A, B)\, I(A, B) \tag{4}$$

with

$$G(A, B) = \sum_{(\mathbf{x},\mathbf{x}') \in (A \cap B)} w(\alpha_{\mathbf{x},\mathbf{x}'}(\sigma))\, \min(|\nabla \mathbf{x}(\sigma)|, |\nabla \mathbf{x}'(\sigma)|). \tag{5}$$

Similarly, the combination of normalized mutual information and gradient information is defined: $Y_{new}(A, B) = G(A, B)\, Y(A, B)$.

3 Results

We illustrate the behaviour of the known and the newly proposed measures by showing registration functions for two multimodal matching problems: MR to

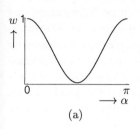

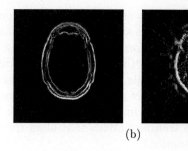

(a)   (b)

Fig. 1. (a) Weighting function for gradient angles, (b) Examples of the gradient function G per pixel for MR-T1/CT (left) and MR-T1/PET (right).

CT and PET to MR. We make use of the datasets from the aforementioned RREP study, which consist of pairs of either CT or PET and MR (PD, T1 and T2) images, seven pairs each. For most MR images, a *rectified* version is also available, corrected for scaling and intensity inhomogeneity. Following the examples of registration functions, we evaluate the accuracy of the new measures by comparing the registration results against a screw marker based gold standard. Finally, we give an indication of the behaviour of the registration functions in six dimensions, by traversing the search space from different starting positions and comparing the number of local maxima and the errors encountered for the different measures.

In the computation of the gradients, our choice of scale was motivated by past research on edge based measures for image registration [12], which demonstrated the best performance of edge based measures at smaller scales. Searching a trade-off between small scale and image resolution, we have opted for a σ of 1.5 mm for all images.

3.1   Registration Functions

MR-T1 and CT

Although MR images depict different anatomical details than CT images, there generally are corresponding structures – and hence corresponding gradients – in both images. Figure 2 shows some examples of MR-CT registration functions, with the zero position corresponding to the gold standard solution. We first show an example of a well-defined mutual information function (rotation around an in-plane axis, top row) and find that the function is not significantly altered by the inclusion of gradient information. As mutual information is sensitive to the number of samples used, the registration function is generally less smooth for images of lower resolution, for example, images that have been downsampled for use in a multiresolution registration method. Indeed, when subsampling the images by a factor of three (middle row), both standard and normalized mutual information functions deteriorate, while the functions for the combined measures are virtually unchanged.

In the bottom row of figure 2 the behaviour of the measures for translation in the slice direction is shown. The local minima in the mutual information and normalized mutual information functions are a result of interpolation. The interpolation method used (linear interpolation) influences the entropy measures by blurring noise and other small structures. Fluctuations in the registration function occur as a result of grid-aligning transformations, for which interpolation is not applied. Here the images have equal slice thicknesses and the local minima correspond to transformations that align the image slices (see [14] for a more detailed explanation).

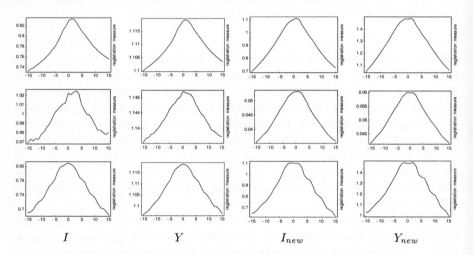

$$I \qquad Y \qquad I_{new} \qquad Y_{new}$$

Fig. 2. Registration functions for MR-T1 and CT matching. From top to bottom: (i) rotation around an in-plane axis, (ii) *idem*, images subsampled by a factor of three in each dimension and (iii) translation in slice direction.

MR-T1 and PET

Registration of MR and PET images is more likely to end in misregistration, both because of the fewer similarities between the image contents and because of the lower intrinsic resolution of PET images. The RREP PET images have a relatively low sampling resolution in the slice direction. As a result, the registration functions are particularly ill-defined for the out-of-plane rotations, as can be seen in the top two rows of figure 3. Since it is (at least partly) an overlap problem, the normalized mutual information measure performs rather better. However, for both standard and normalized mutual information, it is obvious that optimization of such functions will not be robust. By including gradient information the registration functions for out-of-plane rotation are vastly improved (rightmost two columns).

In the third row (in-plane rotation), the position of the global optimum for the combined measures is closer to the gold standard solution compared to the global optimum of standard mutual information. The registration functions for in-plane translations (rows four and five) are well-defined for all measures. Interpolation-induced local minima are found in the registration functions for translation in the slice direction (bottom row). The inclusion of gradient information reduces the artefacts, as can be seen in the rightmost two functions.

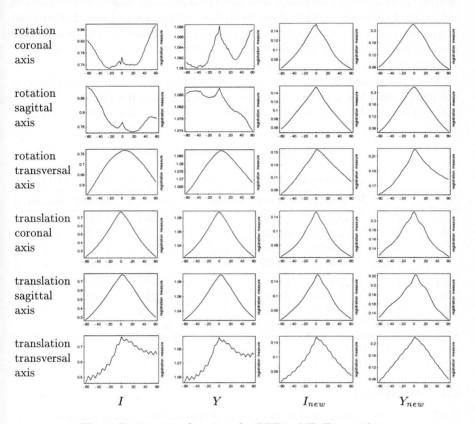

rotation coronal axis

rotation sagittal axis

rotation transversal axis

translation coronal axis

translation sagittal axis

translation transversal axis

I Y I_{new} Y_{new}

Fig. 3. Registration functions for PET to MR-T1 matching.

3.2 Accuracy

The accuracy of the proposed measures has been tested by comparison of the registration results against a screw marker based gold standard. Each pair of either CT or PET and MR (PD, T1 and T2, both rectified and nonrectified) from the RREP study was registered using the four measures discussed in this paper. The number of datasets for registration of CT to MR nonrectified and rectified was 21 and 20, respectively. For PET and MR, the numbers were 21 and 14. Screw marker based solutions were available for these registration cases. We define the accuracy measure as the averaged Cartesian distance between the positions of the eight corner points of the image volumes with our solution and with the gold standard solution. The measure is in millimetres. To minimize the dependence of the results on the optimization method (Powell's direction set method [13]), the starting position of all registration experiments was the gold standard transformation. Table 1 summarizes the registration results for all four measures. The average accuracy, standard deviation and number of mismatches are given. To avoid distortion of the results by outliers, image pairs that resulted in misregistration with any of the measures were excluded for *all* measures for computation of the results. This was the case for two MR and CT image pairs and two MR and PET image pairs.

From the results in table 1 can be concluded that the combined measures do not compromise the accuracy of mutual information based methods. Moreover, optimization of the combined measures did not result in any misregistrations.

Table 1. Registration results (in mm)

		nonrectified MR (n=21/21)			rectified MR (n=20/14)		
		average	std dev	failures	average	std dev	failures
MR/	I	2.4451	1.1737	1	1.3009	0.5254	1
CT	Y	2.1876	0.9485	0	1.1316	0.5410	0
	I_{new}	2.0410	0.8599	0	1.4318	0.4313	0
	Y_{new}	2.0737	0.7109	0	1.8048	0.7616	0
PET/	I	3.9304	1.8102	2	3.8436	1.8426	0
MR	Y	3.6407	2.0821	1	3.2228	1.4035	0
	I_{new}	3.6917	1.4735	0	2.6733	1.6286	0
	Y_{new}	3.5443	1.4355	0	3.2845	2.0767	0

3.3 Robustness

To illustrate the behaviour of the various registration functions in six dimensions, we have investigated the search space of a PET and an MR image. A gradient ascent optimization was started from 64 different positions: the corner points of a 6D hypercube with sides of 10 millimetres in length. The number of different maxima found was counted, where two end positions were considered identical maxima when the difference between each of the six transformation parameters was not more than the step size of the gradient ascent method (here: 1 mm or degree). The registration error of a maximum was identical to the measure defined in the previous section. Figure 4(a) shows the number of maxima found for each of the four registration measures, while (b) shows the mean (horizontal line) and standard deviation (vertical line) of the registration errors of all maxima. The number of maxima is considerably lower for normalized mutual information compared to the standard measure, while it decreases even further when applying the combined measures. Similar behaviour is found for the mean and standard deviations of the registration errors.

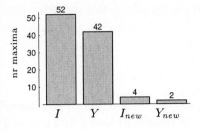 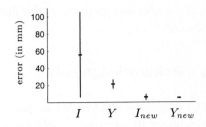

Fig. 4. Left: number of maxima; right: mean (horizontal line) and standard deviation (vertical line).

4 Discussion

We have proposed the adaptation of mutual information measures, by incorporating spatial information. The measure combines either standard or normalized mutual information with gradient information. The essence of the gradient information is that, at registration, locations with a large gradient magnitude should be aligned, but also that the orientation of the gradients at those locations should be similar.

The results presented in this study indicate that the combined measures yield registration functions outperforming both the standard mutual information function with respect to smoothness and attraction basin as well as a normalized mutual information measure. The functions of the combined measures are better defined, containing fewer erroneous maxima and leading to the global maximum from larger initial misregistrations. The measures perform better for low resolution images and can decrease interpolation induced local minima. Well-defined mutual information registration functions are not significantly altered by the inclusion of gradient information.

The accuracy of the combined measures was shown to be similar to that of standard and normalized mutual information. Image pairs that yielded mismatches with (normalized) mutual information, were accurately registered with either of the combined measures. The robustness of standard and normalized mutual information was studied by Studholme et al. [7], showing poor robustness for mutual information and good performance for normalized mutual information. However, we *have* encountered a mismatch and ill-defined registration functions for normalized mutual information (only for the nonrectified MR images, which were not included in Studholme's study). Experiments into the robustness of the measures with respect to starting position implied that including spatial information considerably improves robustness.

Several issues of the method can be improved upon or should be investigated further, including the dependence of the method on the scaling parameter in the gradient computation and the matter of differences in intrinsic resolution. PET images have a significantly lower intrinsic resolution than MR images and it is possible the method can be improved upon by taking this difference into account.

5 Acknowledgments

The images were provided as part of the project, "Evaluation of Retrospective Image Registration", National Institutes of Health, Project Number 1 R01 NS33926-01, Principal Investigator Prof. J. Michael Fitzpatrick, Vanderbilt University, Nashville, TN. We thank the Laboratory for Medical Imaging Research in Leuven (especially Dr Frederik Maes) for kindly supplying us with their software for mutual information based registration.
This research was funded by the Netherlands Organization for Scientific Research (NWO).

References

[1] W. M. Wells III, P. Viola, H. Atsumi, S. Nakajima, and R. Kikinis. Multi-modal volume registration by maximization of mutual information. *Medical Image Analysis*, 1(1):35–51, 1996.

[2] F. Maes, A. Collignon, D. Vandermeulen, G. Marchal, and P. Suetens. Multi-modality image registration by maximization of mutual information. *IEEE Transactions on Medical Imaging*, 16(2):187–198, 1997.

[3] C. R. Meyer, J. L. Boes, B. Kim, P. H. Bland, K. R. Zasadny, P. V. Kison, K. Koral, K. A. Frey, and R. L. Wahl. Demonstration of accuracy and clinical versatility of mutual information for automatic multimodality image fusion using affine and thin-plate spline warped geometric deformations. *Medical Image Analysis*, 1(3):195–206, 1997.

[4] C. Studholme, D. L. G. Hill, and D. J. Hawkes. Automated three-dimensional registration of magnetic resonance and positron emission tomography brain images by multiresolution optimization of voxel similarity measures. *Medical Physics*, 24(1):25–35, 1997.

[5] P. Viola and W. M. Wells III. Alignment by maximization of mutual information. *International Journal of Computer Vision*, 24(2):137–154, 1997.

[6] J. West *et al.* Comparison and evaluation of retrospective intermodality brain image registration techniques. *Journal of Computer Assisted Tomography*, 21(4):554–566, 1997.

[7] C. Studholme, D. L. G. Hill, and D. J. Hawkes. An overlap invariant entropy measure of 3D medical image alignment. *Pattern Recognition*, 32(1):71–86, 1999.

[8] P. Thévenaz and M. Unser. Spline pyramids for inter-modal image registration using mutual information. In A. Aldroubi, A. F. Laine, and M. A. Unser, editors, *Wavelet Applications in Signal and Image Processing*, volume 3169 of *Proc. SPIE*, pages 236–247. SPIE Press, Bellingham, WA, 1997.

[9] C. E. Rodríguez-Carranza and M. H. Loew. A weighted and deterministic entropy measure for image registration using mutual information. In K. M. Hanson, editor, *Medical Imaging: Image Processing*, volume 3338 of *Proc. SPIE*, pages 155–166. SPIE Press, Bellingham, WA, 1998.

[10] G. P. Penney, J. Weese, J. A. Little, P. Desmedt, D. L. G. Hill, and D. J. Hawkes. A comparison of similarity measures for use in 2D-3D medical image registration. *IEEE Transactions on Medical Imaging*, 17(4):586–595, 1999.

[11] D. Rueckert, M. J. Clarkson, D. L. G. Hill, and D. J. Hawkes. Non-rigid registration using higher-order mutual information. In *Medical Imaging: Image Processing*, Proc. SPIE. SPIE Press, Bellingham, WA, 2000. In press.

[12] J. B. A. Maintz, P. A. van den Elsen, and M. A. Viergever. Comparison of edge-based and ridge-based registration of CT and MR brain images. *Medical Image Analysis*, 1(2):151–161, 1996.

[13] W. H. Press, B. P. Flannery, S. A. Teukolsky, and W. T. Vetterling. *Numerical recipes in C*. Cambridge University Press, Cambridge, UK, 1992.

[14] J. P. W. Pluim, J. B. A. Maintz, and M. A. Viergever. Interpolation artefacts in mutual information based image registration. *Computer Vision and Image Understanding*, 77(2):211–232, 2000.

An Image Registration Approach to Automated Calibration for Freehand 3D Ultrasound

J.M. Blackall[1], D. Rueckert[2], C.R. Maurer, Jr.[3], G.P. Penney[1], D.L.G. Hill[1], D.J. Hawkes[1]

[1]Division of Radiological Sciences and Medical Engineering
The Guy's King's and St. Thomas' Schools of Medicine and Dentistry
Guy's Hospital, London SE1 9RT, UK

[2]Department of Computing, Imperial College, London SW7 2BZ, UK

[3]Department of Neurological Surgery and Biomedical Engineering
University of Rochester, Rochester, NY 14642, USA

Abstract. This paper describes an image registration approach to calibration for freehand three-dimensional (3D) ultrasound. If a conventional ultrasound probe is tracked using a position sensor, and the relation between this sensor and the two-dimensional (2D) image plane is known, the resulting set of B-scans may be correctly compounded into an image volume. Calibration is the process of determining the transformation (rotation, translation, and optionally image scaling) that maps B-mode image slice coordinates to points in the coordinate system of the tracking sensor mounted on the ultrasound probe. A set of 2D ultrasound images of a calibration phantom is obtained using a tracked ultrasound probe. Calibration is performed by searching for the calibration parameters that maximise the similarity between a model of the calibration phantom, which can be an image volume or a geometrical model, and the ultrasound images transformed into the coordinate space of the phantom. Validation of this calibration method is performed using a gelatin phantom. Measures of the calibration reproducibility, reconstruction precision and reconstruction accuracy are presented for this technique, and compared to those obtained using a conventional cross-wire phantom. Registration-based calibration is shown to be a rapid and accurate method of automatic calibration for freehand 3D ultrasound.

1 Introduction

Three-dimensional ultrasound addresses some of the disadvantages of conventional two-dimensional ultrasound imaging, reducing subjectivity inherent in the mental transformation of the 2D images into an anatomical volume; reducing the problems of speckle, shadowing and enhancement; and improving diagnostic (e.g., obstetrics) and therapeutic (e.g., biopsy, brachytherapy, cryosurgery, RF ablation) decisions [4]. One approach to 3D ultrasound uses a 2D phased array of elements, which are used to transmit a broad beam of ultrasound diverging away from the array and sweeping out a pyramidal volume. Another approach is to control the position of the transducer using stepping motors in the transducer's scan head or a translational or rotational mechanical device. The technique under consideration in this paper is that of sensed freehand acquisition.

Freehand 3D ultrasound is based on tracking the arbitrary motion of a conventional B-mode probe. The motion of the probe in all six degrees of freedom is tracked using a sensor that provides position and orientation, which is commonly magnetic [1, 3, 6, 7], but may also be optical, mechanical or acoustic. The probe is moved slowly over the region of interest. The B-scans are digitised by a frame grabber, tagged with the measurements from the position sensor, and stored as 2D arrays of pixels. Providing that the probe has been calibrated, the set of B-scans may then be assembled into a 3D data set, using the associated position data. The scan planes may intersect each other so that the same region in space is sampled from more than one direction, but gaps may remain where no samples are taken. The data may then be visualised by reslicing, or by creating 3D surface or volume renderings [8].

To calibrate the ultrasound probe, the position and orientation of the B-scan plane with respect to the sensor mounted on the probe must be determined. The calibration transformation consists of six rigid-body transformation parameters (three for rotation and three for translation) and optionally two image scale factors. These are combined with the position sensor measurements to calculate the correct positions of the B-scans during reconstruction. Accurate calibration is essential for a consistent reconstruction that preserves true anatomical shape and must be repeated every time a sensor is mounted on a probe [7, 8].

An estimate of the six rigid-body calibration parameters can be obtained by external measurements of the probe and position sensor [5]. However, the origin of the sensor coordinate system is not well defined with respect to the external cases of the sensor and probe. Calibration is usually performed by imaging an artificial object with known physical properties or dimensions, known as a phantom. Measurements obtained by scanning the phantom, combined with its known physical properties, can be used to determine the six rigid body parameters and two scaling factors [8]. Several common approaches use an invariant point, line, or plane. An invariant point is typically obtained with the cross-wire phantom [1, 3, 7], but a small metal sphere can also be used [6]. Another calibration phantom is the three-wire phantom [2], which consists of three orthogonal wires mounted in a water bath. An invariant plane can be the floor of a water bath [7, 8], or the 'virtual plane' provided by the "Cambridge phantom" [7]. A comparison of existing calibration techniques has been carried out by Prager et al [7].

In this paper we propose a new calibration algorithm for freehand ultrasound images. The calibration algorithm is based on the assumption that correct calibration parameters should allow the registration of a set of tracked 2D ultrasound images of a phantom with a digital model of its structure in 3D.

2 Method

The calibration algorithm is based on the assumption that the correct calibration parameters will provide the optimal registration of a set of tracked 2D ultrasound images of a phantom with a 3D model of the phantom, which can be an image volume generated using CT or MR, or a geometrical model. Figure 1 illustrates the four different coordinate systems used by the calibration algorithm. The ultrasound coordinate system describes points in the ultrasound image. The probe coordinate system describes points relative to a tracking sensor attached to the ultrasound probe. The world coordinate system forms the reference frame and describes points in physical space, i.e., relative to the origin of the tracking device. The transformation between probe coordinates and world coordinates is the output of tracking the position sensor relative to the origin of the tracking device. The model coordinate system describes points in model space. The transformation between model and world coordinates can be determined by attaching fiducial markers to the phantom.

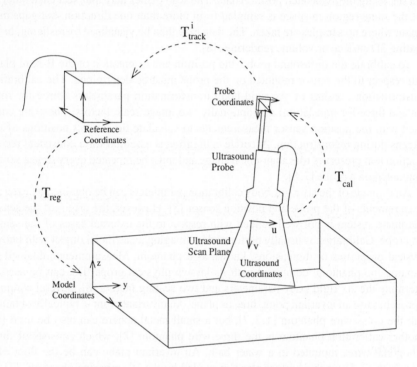

Fig. 1. Coordinate systems used in calibration algorithm.

The transformation between coordinates in the ultrasound image and coordinates in the model can be expressed as follows:

$$
\begin{pmatrix} x_i \\ y_i \\ z_i \\ 1 \end{pmatrix} = \mathbf{T}_{reg} \, \mathbf{T}^i_{track} \, \mathbf{T}_{cal} \begin{pmatrix} u_i \\ v_i \\ 0 \\ 1 \end{pmatrix} \tag{1}
$$

where x_i, y_i, z_i are model coordinates; u_i, v_i are coordinates in the ith ultrasound image; $\mathbf{T}_{reg}$ is the rigid-body registration transformation, from world to model coordinates; $\mathbf{T}^i_{track}$ is the tracking transformation, from probe to world coordinates, for the ith image; and $\mathbf{T}_{cal}$ is the rigid body (and optionally image scaling) calibration transformation, from ultrasound to probe coordinates. In this equation only the calibration transformation is unknown, all the other transformations are known.

2.1 Similarity Measure

A 3D digital model of the calibration phantom may be generated by imaging the phantom, e.g., in CT or MR. Since the properties of the imaging modality used to derive the model data may be quite different from those of ultrasound, the similarity measure must be able to deal with these. In this paper we use normalised mutual information (NMI) as a similarity measure which has been shown to align images from different modalities accurately and robustly and is invariant to the amount of overlap [11]:

$$I(A, B) = \frac{H(A) + H(B)}{H(A, B)} \tag{2}$$

Here $H(A), H(B)$ denote the marginal entropies of both images and $H(A, B)$ denotes their joint entropy. The marginal and joint entropies are both calculated from the joint histogram of A and B. If the correct calibration parameters are found the ultrasound images and model images should be correctly aligned and the normalised mutual information should be maximal.

The similarity measure is calculated by taking each 2D ultrasound image in turn. For each pixel in the ultrasound image the corresponding location in the model data is calculated using equation (1) and the intensity of the model at this location is calculated from the eight neighbouring voxels using trilinear interpolation. The pixel intensity of the original ultrasound image and the interpolated model intensity are used to add a single value to the joint histogram, $H(A, B)$. This approach avoids compounding a 3D ultrasound image before evaluating the similarity measure, as would be necessary if a traditional image-to-image registration were performed.

2.2 Optimisation

Given an initial estimate of the calibration transformation $\mathbf{T}_{cal}$, an improved estimate may be determined by finding the calibration transformation $\mathbf{T}$ which maximises the similarity measure I as a function of the set of ultrasound images B and the digital model A:

$$\hat{\mathbf{T}}_{cal} = \arg \max_{\mathbf{T}} \{I(A, \mathbf{T}(B))\} \tag{3}$$

In this paper we use a simple iterative optimisation algorithm. At each iteration the algorithm evaluates the similarity measure for the current estimate of the calibration parameters as well as for the current estimate with increments and decrements of step size $\pm \delta$. At the end of each iteration the current estimate of the calibration parameters is updated with those estimates which have maximised the similarity measure.

For computational efficiency we have implemented a multi-resolution search strategy which searches for the calibration parameters at different levels of image resolution and with different step sizes [11]. The algorithm starts with an initial step size of $\delta = 4$mm for the translation parameters, $\delta = 4°$ for rotation parameters, and $\delta = 4\%$

for scale changes. If no further improvement of the similarity measure can be achieved, the step size is reduced by a factor of two and the process is repeated. The algorithm stops if the step size reaches a lower limit, having been halved eight times.

3 Calibration Experiments

The ultrasound machine used in our experiments is a Siemens Sonoline Versa Pro. Calibration experiments were performed on a 10 MHz linear array probe (Siemens 10.0 L 25). A depth setting of 40 mm was used, and controls such as time-gain compensation, number of focal zones and overall gain were set appropriately and kept constant throughout all experiments. Twenty infrared LEDs were attached to the ultrasound probe. The probe was tracked during image acquisition using an optical tracking system (Optotrak 3020, Northern Digital) which tracks infrared LEDs to a 3D spatial accuracy of approximately 0.2mm.

3.1 Calibration Phantoms

Gelatin Phantom A gelatin phantom, pictured in Figure 2, was constructed with which to carry out registration-based calibration. This multimodality phantom was formed using gelatin powder (G-2500, Sigma Chemicals) dissolved in water at a concentration of 15% by weight. A 2% by weight concentration of silica (S-5631, Sigma Chemicals) was added to generate speckle texture and also provide MR contrast within included structures, as shown in Figure 3. These materials yield acoustic and mechanical properties which are similar to those of human tissue [9] and so allow correct image scaling factors to be determined. The phantom was constructed in several stages. A block of gelatin containing silica was allowed to set, then a selection of shapes were cut from this block, and placed onto a base layer of set gelatin in a rigid acrylic container. This was then covered with more molten gelatin, leaving the structures of interest embedded at various depths and orientations. An MR scan of the gelatin phantom was acquired to act as its digital model. Eight fiducial markers filled with a dilute aqueous solution of Gadolinium were attached to the acrylic container. The spatial resolution of the MR scan was 0.869mm × 0.869mm × 1.165mm.

Ten sets of calibration data, each consisting of 100 ultrasound images, were acquired with a layer of water covering the top of the phantom to provide acoustic coupling whilst allowing free movement of the probe. During scanning, an effort was made to move the probe in all six degrees of freedom while imaging the regions of interest. Prior to ultrasound image acquisition the fiducial markers were accurately localised using an Optotrak ball-tipped probe, allowing the registration transformation, T_{reg}, between physical and model coordinates to be determined. For each data set, the algorithm was tested with 64 starting estimates, based on a preliminary point-based calibration, but with all combinations of ±5mm and ±5° added to each of the six rigid-body parameters.

Cross-wire Phantom Point-based calibration is a very widely used technique in freehand 3D ultrasound and has been shown to yield the most precise and accurate recon-

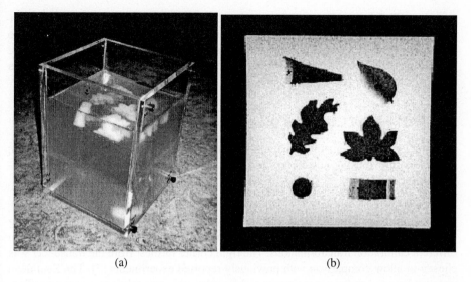

(a)   (b)

Fig. 2. (a) The gelatin phantom used for registration-based calibration experiments, (b) MR slice through gelatin phantom showing shapes of internal structures. Calibration is achieved by automatic registration of a set of tracked 2D ultrasound images to this MR image volume.

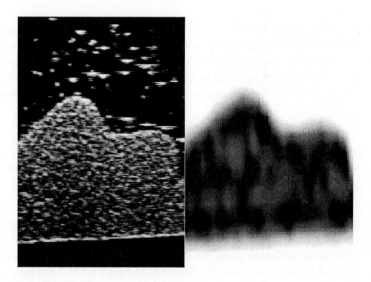

Fig. 3. Corresponding US and MR image slices through gelatin phantom.

structions [7]. It is therefore useful to compare the point-based method to our registration-based approach to calibration. A cross-wire phantom was constructed by crossing two cotton threads in a water bath, as described in [1, 3, 7]. Ten cross-wire data sets were then acquired, each consisting of 50 ultrasound images, and the point at which the wires cross was manually segmented from each.

3.2 Assessment of Calibration Quality

The quality of our registration-based calibration was assessed using several criteria, based on a previously published protocol [7], and compared to a conventional cross-wire calibration.

Calibration Reproducibility Reproducibility of calibrations was assessed by looking at the position of a single point in the ultrasound image in probe coordinates under two different calibration transformations. The right corner furthest from the transducer was chosen to allow comparison with previously reported experiments [7]. The Euclidean distance between these two points in probe coordinates was taken as a measure of reproducibility, and this was calculated for each pair of calibrations.

$$\Delta \mathbf{x}^r = \|\mathbf{T}_{cal\,1}\,\mathbf{x}^c - \mathbf{T}_{cal\,2}\,\mathbf{x}^c\| \qquad (4)$$

where $\Delta \mathbf{x}^r$ is the reproducibility measure; $\mathbf{x}^c$ is the corner point of the ultrasound image; and $\mathbf{T}_{cal\,1}$, $\mathbf{T}_{cal\,2}$ are the two calibration transformations.

Reconstruction Precision Reconstruction precision was assessed by scanning a cross-wire from 50 different viewpoints. The cross was manually located in each image and then transformed to reconstruction space using the associated tracking data, and the calibration on test. The change in 3D position was calculated as a measure of reconstruction precision. This was calculated for each pair of views, using all calibrations.

$$\Delta \mathbf{x}^p = \left\| \mathbf{T}_{track}{}^{view\,1}\,\mathbf{T}_{cal}\,\mathbf{x}^{view\,1} - \mathbf{T}_{track}{}^{view\,2}\,\mathbf{T}_{cal}\,\mathbf{x}^{view\,2} \right\| \qquad (5)$$

where $\Delta \mathbf{x}^p$ is the precision measure; $\mathbf{T}_{track}{}^{view\,i}$ is the tracking transformation corresponding to the ith view of the cross-wire; $\mathbf{x}^{view\,i}$ is the ith cross-wire point in ultrasound image coordinates; and $\mathbf{T}_{cal}$ is the calibration transformation.

Reconstruction Accuracy Reconstruction accuracy was assessed by scanning 8 ball bearings, 10 times each. The centre of the ball was located in each image, then transformed to reconstruction space using the associated tracking data and the calibration on test. These ball bearings were 3mm in diameter, and were set in 3mm hemispherical divots, which were accurately machined in a sheet of acrylic. The divots were arranged in two rows of four, successive divots were separated by approximately 100mm. The position of the centre of each divot was determined using an Optotrak pointer with a 3mm diameter ball-point tip and thus corresponds to centre of the ball bearing. The Euclidean distance between this gold-standard position and the position determined by manual identification in the ultrasound image was taken as a measure of point reconstruction accuracy, and was calculated for each view of each ball bearing, using all calibrations.

$$\Delta \mathbf{x}^a = \left\| \mathbf{T}_{track}{}^{view\ 1}\, \mathbf{T}_{cal}\, \mathbf{x}^{view\ 1} - \mathbf{x}^{gs} \right\| \tag{6}$$

where $\Delta \mathbf{x}^a$ is the point reconstruction accuracy measure; $\mathbf{T}_{track}{}^{view\ i}$ is the tracking transformation corresponding to a view of the ith ball bearing; $\mathbf{x}^{view\ i}$ is the ith ball bearing centre point in ultrasound image coordinates; $\mathbf{x}^{gs}$ is the gold-standard position of the centre of the divot; and $\mathbf{T}_{cal}$ is the calibration transformation on test.

Similarly, the difference between the gold-standard separation and the separation determined from the ultrasound images was taken as a measure of distance reconstruction accuracy, and calculated for each pair of views of adjacent ball bearings, using all calibrations.

$$\Delta \mathbf{x}^d = \left\| \left(\mathbf{T}_{track}{}^{view\ 1}\, \mathbf{T}_{cal}\, \mathbf{x}^{view\ 1} - \mathbf{T}_{track}{}^{view\ 2}\, \mathbf{T}_{cal}\, \mathbf{x}^{view\ 2} \right) \right\| - \left\| \left(\mathbf{x}^{gs\ 1} - \mathbf{x}^{gs\ 2} \right) \right\| \tag{7}$$

where $\Delta \mathbf{x}^d$ is the distance reconstruction accuracy measure; $\mathbf{T}_{track}{}^{view\ i}$ is the tracking transformation corresponding to The ith view of a ball bearing; $\mathbf{x}^{view\ i}$ is the ith ball bearing centre point in ultrasound image coordinates; $\mathbf{x}^{gs\ i}$ is the gold-standard position of the centre of the ith divot; and $\mathbf{T}_{cal}$ is the calibration transformation on test.

4 Results

The results of the calibration quality experiments are summarised in table 1. The calibration reproducibility measure $\Delta \mathbf{x}^r$ indicates the extent to which repeated experiments using the same technique yield the same calibration parameters. The calibration precision measure $\Delta \mathbf{x}^p$ indicates the precision with which a single physical point can be reconstructed from a number of ultrasound images. The calibration point reconstruction accuracy measure $\Delta \mathbf{x}^a$ indicates the accuracy with which points can be reconstructed in the correct positions and similarly the distance reconstruction accuracy measure $\Delta \mathbf{x}^d$ indicates the extent to which distances can be correctly reconstructed.

Quality Measure	$\Delta \mathbf{x}^r$		$\Delta \mathbf{x}^p$		$\Delta \mathbf{x}^a$		$\Delta \mathbf{x}^d$	
Calibration Method	Point	Registration	Point	Registration	Point	Registration	Point	Registration
Mean (mm)	1.05	1.84	0.80	1.15	1.15	1.16	-0.00019	-0.025
Std. Dev. (mm)	0.43	1.26	0.46	0.62	0.40	0.45	0.60	0.69
Max (mm)	2.00	4.91	3.52	5.15	3.10	3.18	1.74	2.44
Min (mm)	0.19	0.00	0.02	0.02	0.32	0.24	-2.57	-2.97

Table 1. Summary of results of calibration validation experiments. Four measures of calibration quality, $\Delta \mathbf{x}^r$, $\Delta \mathbf{x}^p$, $\Delta \mathbf{x}^a$, and $\Delta \mathbf{x}^d$ are shown for both the conventional Point-based method and our Registration-based technique.

These results indicate that the registration-based calibration approach yields similarly impressive point and distance accuracy results to the conventional point-based method. On average, both techniques allow point reconstruction to within just over 1mm of the true position in world coordinates, and distance reconstruction to within 0.025%. Point-based calibration performs slightly better than registration-based in terms of reproducibility and precision, tending to produce more stable calibration parameters, and more precise reconstructions, but in all quality measures the differences are fairly small.

5 Discussion

In this paper we have proposed a novel registration-based method for 3D calibration of a 2D ultrasound probe. The method entails establishing the 3D rigid-body transformation and 2 scaling factors between the 2D ultrasound image coordinates and the probe coordinates. This is achieved by scanning an object of known dimensions and finding the calibration transformation that maximises the normalised mutual information between a set of 2D ultrasound images and a 3D model of the scanned object.

We have shown that the accuracy, precision and reproducibility of registration-based calibration is comparable with that obtained using the established point-based method. Our new method has the significant advantage that it takes about 2 minutes to acquire the data, and subsequent processing of the data is fully automatic, requiring no segmentation. The optimisation algorithm currently takes about 15 minutes to run on a Sun Ultra 10 workstation, which is practical for most applications, but we believe that improvements to the algorithm could allow running time to be reduced even further. In comparison, point-based calibration requires about 30 minutes of careful scanning, followed by 30 minutes of user interaction with the data, to manually identify the cross-wire point in each image.

The current method of phantom construction is sub optimal in that the gelatine used may change shape over time and may gain or lose water during the scanning process. Also, the MR image used as our 3D digital model of the phantom may also introduce errors due to geometric distortion and uncertainty in image scaling. If the geometry of the phantom was determined using a more accurate scan, for example CT, or an accurately machined ultrasound compatible phantom were constructed, we would expect a corresponding increase in accuracy, precision, and reproducibility.

Values of calibration quality measures resulting from our registration-based calibration experiments compare favourably with equivalent measures reported previously in other studies of calibration techniques for 3D freehand ultrasound [7]. However, these values are not directly comparable, due to the differences in accuracy of the tracking devices used in the two studies.

Acknowledgements
We would like to thank EPSRC for financial support, and the members of the Computational Imaging Science Group, for their assistance and encouragement.

References

1. C. D. Barry, C. P. Allott, N. W. John, P. M. Mellor, P. A. Arundel, D. S. Thomson, and J. C. Waterton. Three-dimensional Freehand Ultrasound: Image Reconstruction and Volume Analysis. *Ultrasound in Medicine and Biology*, 23(8):1209–1224, 1997.
2. J. Carr. Surface Reconstruction in 3D Medical Imaging. PhD Thesis, University of Canterbury, Christchurch, New Zealand, 1996.
3. P. R. Detmer, G. Bashein, T. Hodges, K. W. Beach, E. P. Filer, D. H. Burns, and D. E. Strandness, Jr. 3D Ultrasonic Image Feature Localization based on Magnetic Scanhead Tracking: In Vitro Calibration and Validation. *Ultrasound in Medicine and Biology*, 20(4):923–936, 1994.
4. A. Fenster and D. B. Downey. 3D Ultrasonic Imaging: A Review. *IEEE Engineering in Medicine and Biology*, 15(6):41–51, 1996.
5. S. W. Hughes, T. J. D'Arcy, D. J. Maxwell, W. Chiu, A. Milner, J. E. Saunders, and R. J. Sheppard. Volume Estimation from Multiplanar 2D Ultrasound Images Using a Remote Electromagnetic Position and Orientation Sensor. *Ultrasound in Medicine and Biology*, 22(5):561–572, 1996.
6. D. F. Leotta, P. R. Detmer, and R. W. Martin. Performance of a Miniature Magnetic Position Sensor for Three-dimensional Ultrasound Imaging. *Ultrasound in Medicine and Biology*, 23(4):597–609, 1997.
7. R. W. Prager, R. N. Rohling, A. H. Gee, and L. Berman. Rapid Calibration for 3-D Freehand Ultrasound. *Ultrasound in Medicine and Biology*, 24(6):855–869, 1998.
8. R. N. Rohling. 3D Freehand Ultrasound: Reconstruction and Spatial Compounding. PhD Thesis, University of Cambridge, England, 1998.
9. L. K. Ryan and F. S. Foster. Tissue Equivalent Vessel Phantoms for Intravascular Ultrasound. *Ultrasound in Medicine and Biology*, 23(2):261–273, 1997.
10. C. Studholme, D. L. G. Hill, and D. J. Hawkes. Automated 3-D registration of MR and CT images of the head. *Medical Image Analysis*, 1(2):163–175, 1996.
11. C. Studholme, D. L. G. Hill, and D. J. Hawkes. An overlap invariant entropy measure of 3D medical image alignment. *Pattern Recognition*, 32(1):71–86, 1999.

Symmetrization of the Non-rigid Registration Problem Using Inversion-Invariant Energies: Application to Multiple Sclerosis

Pascal Cachier and David Rey

INRIA Sophia - Epidaure Project
2004 Route des Lucioles BP 93
06902 Sophia Antipolis Cedex, France
{Pascal.Cachier, David.Rey}@sophia.inria.fr

Abstract. Without any prior knowledge, the non-rigid registration of two images is a symmetric problem, i.e. we expect to find inverse results if we exchange these images. This symmetry is nonetheless broken in most of intensity-based algorithms. In this paper, we explain the reasons why most non-rigid registration algorithms are asymmetric. We show that the asymmetry of quadratic regularization energies causes an oversmoothing of expending regions relatively to shrinking regions, hampering in particular registration-based detection of evolving processes. We therefore propose to use an inversion-invariant energy to symmetrize the registration problem. To minimize this energy, two methods are used, depending on whether we compute the inverse transformation or not. Finally, we illustrate the interest of the theory using both synthetic and real data, in particular to improve the detection and segmentation of evolving lesions in MR images of patients suffering from multiple sclerosis.

1 Introduction

Non-rigid registration of two images I and J consists in finding a transformation between the two sets of points of both images. If we make the assumption that every point of an image has an homologous point in the other, and that the topology has been conserved, the result of the algorithm should be a smooth homeomorphic transformation $T(I, J)$, going from I to J.

Among non-rigid registration algorithms, intensity-based algorithms are driven by a similarity energy E_{sim} (e.g. the sum of square differences (SSD)) that attracts both images toward each other using their intensity values [1]. At the same time, the transformation $T(I, J)$ is kept smooth either by constraining $T(I, J)$ to belong to some space $\mathcal{T}$ of regular transformations (e.g. splines with few control points [2]), or by choosing a regularization energy E_{reg} (e.g. the linear elastic energy), or both [3]. This energy E_{reg} can be minimized with E_{sim} in a weighted sum $E_{sim} + \lambda E_{reg}$ [4,5], or minimized alternatively with E_{sim} [6]. See [7] for a review of registration techniques.

1.1 Asymmetry and related problems

Without any other prior knowledge, the registration problem is symmetric and we expect to obtain the inverse transformation if we exchange I and J: $T(J, I) = [T(I, J)]^{-1}$. However, this symmetry is usually broken in practice, giving birth to the terminology "reference" or "template" image and "study" or "floating" image (here $T(I, J)$ goes from the reference image I to the study image J). Most, if not all, of the energies used in medical image non-rigid registration leads to asymmetrical registration. This includes the usual similarity energies and quadratic regularization energies such as linear elastic or thin plate energies. Even the general, non-linear elasticity is conceptually asymmetric since one of the images is supposed to be in a no-stress state and not the other; see [8] for an example showing the importance of this consideration.

The first obvious problem with asymmetry is that we have to choose arbitrarily a reference image, and thus the results of measures relying on this registration depend on how the reference image is chosen.

However, there can be a more subtle problem than the previous one. Indeed, using standard similarity measures and quadratic regularization energies will lead to systematically biased solutions: they will smooth the transformation on expanding regions more than on shrinking regions. Both $T(I, J)$ and $T(J, I)$ will have more difficulty to describe a local growth than a local shrink.

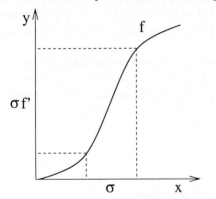

Fig. 1. *Smoothing f by a kernel of size σ is roughly the same as smoothing f^{-1} by a kernel of size $\sigma f'$*

In order to understand this, let us present a 1D example where we smooth a function f using a convolution with a kernel of some finite extension σ: we use values of f in an interval of length σ to smooth f. As shown on fig. 1, the same smoothing is approximately obtained on f^{-1} if it is convolved with a kernel of size $\sigma f'$. Thus, the transformation around an expanding region ($f' > 1$) would be less smoothed if the images were exchanged and the inverse transformation smoothed with the same kernel of size σ.

This explains why the detection of evolving brain lesions using non-rigid registration has appeared easier on shrinking lesions than on expanding lesions in our previous work [9]. We thus detected expanding and shrinking regions by working on shrinking regions of both $T(I, J)$ and $T(J, I)$.

1.2 Towards symmetry of registration

We have found that the asymmetry of intensity-based non-rigid registration is due to one or several of the following reasons:

1. Order non-preservation of the energies: An energy $E(I, J, T)$ being used for registration, we may have $E(I, J, T_1) < E(I, J, T_2)$ but not $E(J, I, T_1^{-1}) < E(J, I, T_2^{-1})$ for two estimates T_1 and T_2 of the transformation.

2. Non-stable transformation space: The space of allowed transformation $\mathcal{T}$ may not be a group and thus not stable by inversion, i.e. we may have $T \in \mathcal{T}$ but not $T^{-1} \in \mathcal{T}$.

3. Local Minima: The optimization algorithm may get stuck in different local minima of the energy when the images are exchanged.

A previous work on the symmetrization of the registration has been done by Christensen [5]. In his work, two asymmetric registrations are made simultaneously, but are constrained to be close to the inverse of each other. This constraint is efficient at smoothing the registration. However, the algorithm is only asymptotically symmetric when the weight of this constraint tends towards infinity. Also, it relies on the computation of an inverse transformation whose existence is not ensured by linear elasticity. This work is nice but we felt that the symmetrization could be handled theoretically in a more efficient way.

In this paper, we propose to tackle the first point of the previous list and to ensure that the minima of the energy are invariant when exchanging the images, by forcing the similarity energy E_{sim} and the smoothness energy E_{reg} to be symmetric, i.e. invariant if we exchange I and J and invert the transformation: $E_{sim}(I, J, T) = E_{sim}(J, I, T^{-1})$ and $E_{reg}(T) = E_{reg}(T^{-1})$. This is close to a very recent work by Trouvé and Younes [10] where the authors consider distances of two parametric closed curves verifying some properties, including this kind of symmetry.

In section 2 we present a way to construct symmetric energies. In section 3 we present two optimization procedures, depending on whether we compute the inverse transformation or not. We illustrate the theory with a synthetic example in section 4, and in section 5 we show the interest of this method to detect and segment evolving lesions in multiple sclerosis.

2 Symmetric Energies

2.1 Symmetrization of regularization energies

Let $\mathbf{x}$ be a point of I, and let $d^n T(\mathbf{x})$ be the n-th differential tensor of T at $\mathbf{x}$. Let $E_{reg}(T)$ be a regularization energy that can be written as an integral of some function e_2 of the differentials of T:

$$E_{reg}(T) = \int e_2(dT(\mathbf{x}), d^2 T(\mathbf{x}), ...) \, d\mathbf{x} \qquad (1)$$

E_{reg} is symmetric (or inversion invariant) if $E_{reg}(T) = E_{reg}(T^{-1})$. If we change variables in $E_{reg}(T^{-1})$

$$E_{reg}(T^{-1}) = \int e_2(d(T^{-1})(\mathbf{y}), d^2(T^{-1})(\mathbf{y}), ...) \, d\mathbf{y}$$

$$= \int e_2(d(T^{-1}) \circ T(\mathbf{x}), d^2(T^{-1}) \circ T(\mathbf{x}), ...) |dT(\mathbf{x})| \, d\mathbf{x}$$

it appears that a sufficient condition for E_{reg} to be symmetric is that

$$e_2(dT(\mathbf{x}), d^2T(\mathbf{x}), ...) = e_2(d(T^{-1}) \circ T(\mathbf{x}), d^2(T^{-1}) \circ T(\mathbf{x}), ...)|dT(\mathbf{x})|$$

where $|dT(\mathbf{x})|$ is the Jacobian of T at point $\mathbf{x}$.

There are many such symmetric energies. Among them, we are more particularly interested in energies that are somehow linked to the asymmetric energies we are used to. There are several possibilities to symmetrize an asymmetric energy. Among them, the most natural is perhaps to take the mean of $e_2(dT, d^2T, ...)$ and $|dT|e_2(d(T^{-1}) \circ T, d^2(T^{-1}) \circ T, ...)$, i.e. the symmetrization E^*_{reg} of a regularization energy E_{reg} can be defined by

$$E^*_{reg}(T) = \frac{1}{2}(E_{reg}(T) + E_{reg}(T^{-1})) \tag{2}$$

The trick is that $e_2(d(T^{-1}) \circ T, d^2(T^{-1}) \circ T, ...)$ can be calculate (more or less easily) from the derivatives of T. Thus, $E^*_{reg}(T)$ is indeed of the form (1) and can be computed without inverting T. For example, let us consider the following quadratic energy in 2D: $E_{reg}(T) = \int_{\mathbf{R}^2} ||dT||^2 = \int_{\mathbf{R}^2} ||\partial_x T||^2 + ||\partial_y T||^2$. Since $d(T^{-1}) \circ T = (dT)^{-1}$, we find that the symmetrization of this energy using eq. (2) is:

$$E^*_{reg}(T) = \frac{1}{2} \int (1 + \frac{1}{|dT|}) \left(||\partial_x T||^2 + ||\partial_y T||^2 \right) \tag{3}$$

Note that the integrand tends to $+\infty$ when dT tends to 0: in practice, this will force the transformation to be one-to-one.

2.2 Symmetrization of Similarity Energies

Similarity measures can be symmetrized using the same technique. If $E_{sim}(I, J, T)$ is a similarity energy, a symmetric similarity energy can be deduced by setting:

$$E^*_{sim}(I, J, T) = \frac{1}{2}(E_{sim}(I, J, T) + E_{sim}(J, I, T^{-1})) \tag{4}$$

Here again, changing the variables inside $E_{sim}(J, I, T^{-1})$ usually eliminates T^{-1} from the formulas, and therefore the inversion of T can be avoided. For example, if $E_{sim}(I, J, T) = \int (I - J \circ T)^2$ is the standard SSD criterion between two images, then its symmetrization using (4) is

$$E^*_{sim}(I, J, T) = \frac{1}{2} \int (1 + |dT|)(I - J \circ T)^2 \tag{5}$$

3 Optimization of the Symmetric Energy

3.1 Finite element implementation without inversion

Our first implementation uses the fact that the symmetric energies can be computed and optimized without inverting the displacement field. To produce the results presented in the next section, we have minimized in 2D the symmetric energy (7) which is the symmetrization of (6):

$$E(I, J, T) = \int (I - J \circ T)^2 + \lambda \int ||dT||^2 \tag{6}$$

$$E^*(I, J, T) = \frac{1}{2} \left[\int (1 + |dT|)(I - J \circ T)^2 + \lambda \int (1 + \frac{1}{|dT|}) ||dT||^2 \right] \tag{7}$$

In our experiences, the accuracy of the symmetry has been very sensitive to the discretization of the continuous optimization problem. We found out that results were significantly better using a finite element discretization of the energy rather than a finite difference scheme. A triangularization of the rectangular grid defined by the pixels is used. The images as well as the transformation are defined as piecewise linear functions on this triangularization. To minimize the energy, we use a simple gradient descent coupled with a multiresolution scheme.

We encountered two types of problems with the previous algorithm:

1. The energy is difficult to minimize, mainly because it includes the term $1/|dT|$ that raises numerical problems, especially when the smoothness constraint is set low ($\lambda \ll 1$).
2. The optimization by gradient descent is asymmetric (point 3 of the list given in introduction): for example, the derivative of the energy (7) with respect to T uses the derivative of J and not the derivative of I. This hampers a perfect symmetry of the result.

3.2 Alternate minimization with inversion

These last problems can be avoided if we compute the inverse transformation. Our second method uses a very simple yet efficient alternated-minimization method: instead of minimizing $E^*(I, J, T) = E(I, J, T) + E(J, I, T^{-1})$, we minimize alternatively the asymmetric energies $E(I, J, T)$ and $E(J, I, T^{-1})$ (given by eq. (6)) using respectively I and J as the reference image. This requires an inversion of the transformation every time we change of reference image.

For a better comparison with the previous algorithm in 2D we have used the same finite element discretization. However, it is not necessary, which is one advantage of this method: faster optimizations are possible with the same quality of symmetry.

The inverse transformation is computed by finding the zero of $f_{\mathbf{y}}(\mathbf{x}) = \mathbf{y} + T(\mathbf{x})$ using a Newton scheme. Interpolation of the transformation is necessary; we have used a bi- or trilinear interpolation. The inverse is found with an error usually less than 0.1 voxel. Note that we cannot have an arbitrarily small error because the inverse of a piecewise bi/trilinear transformation is not piecewise bi/trilinear.

We have to be careful with the computation of the inverse. Indeed, $E(I, J, T)$ does not guarantee that the transformations is invertible. Forcing the corrections brought at each iteration to be small, and/or smoothing them, often helps the transformation to be invertible at every iteration, especially during the first iterations at full resolution when only a small number of points are moving. Furthermore, imposing small corrections seems to give more symmetrical results.

4 Results on a Synthetic Example

In this synthetic example we have two 520×280 images containing two discs, one multiplying its radius by 2 and the other shrinking symmetrically (fig. 3). We have registered these images using the asymmetric energy (6) (Asym algorithm), and using the algorithms described in sec. 3.1 (Sym1) and sec. 3.2 (Sym2).

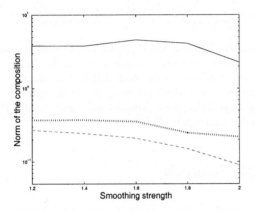

Fig. 2. *Mean error in pixels of* $T(I, J) \circ T(J, I)$ *for Asym (plain), Sym1 (dots) and Sym2 (dashes).*

When we register these images using Asym, we are confronted to the problem mentioned in sec. 1.1 that shrinkage is more easily recovered than growth. We can analyse it using the absolute value of the logarithm of the Jacobian (AVLJ) of the transformation, which expresses the local increase or decrease of area and we used for the detection of brain evolution in [9]. If the growth and the shrinkage were recovered equally, we would find equal AVLJ inside both circles, and if they were furthermore exactly recovered, these numbers would be equal to $\log\left[(\pi(2r)^2)/(\pi r^2)\right] = \log(4) \approx 1.39$. Instead, we find that both AVLJ are increasingly underestimated with the strength of the smoothing, which is expected, but also that the AVLJ of the expanding region is systematically underestimated relatively to the AVLJ of the shrinking region (fig. 3). This shortcoming is reduced with Sym1, and virtually eliminated with Sym2. We have also registered the images in both directions and computed $T(I, J) \circ T(J, I)$, which should be equal to identity in the case of a perfect symmetry. Mean errors of this composition relatively to the identity have been computed (fig. 2) and confirm the previous results.

We would like to emphasize that for the same amount of smoothing, Sym1 and Sym2 do *not* find a solution closer to the real transformation than the asymmetric algorithm for both circles: the small circle is registered better but the big circle is less shrinked. The distance to the real transformation is not uniformly decreased; errors are just shared more equitably.

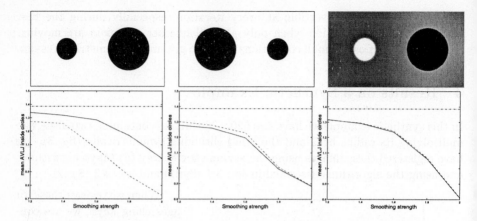

Fig. 3. *Top: The two images to register (left and middle), and an example of a log-Jacobian image on the right (positive numbers are white, negative are black). Bottom: Graphs showing the mean absolute value of the log-Jacobian (AVLJ) inside the big circle (in plain) and inside the little circle (in dashed) relatively to the strength of the smoothing (x-axis), using respectively Asym (left), Sym1 (middle) and Sym2 (right). The real AVLJ is $log(4) \approx 1.39$ (dashed-dotted line) for both circles.*

5 Application to Multiple Sclerosis

In this section we would like to show that the method described in this paper can significantly improve results in the detection and segmentation of evolving lesions in multiple sclerosis [11, 9]. In a previous article [9] we explain that small expanding lesions can be difficult to detect with temporal analysis of apparent deformations: if there is an important expansion locally between images I and J, we would need a one to many mapping due to limited resolution of the image. To avoid this, we considered only shrinking regions from I to J, and then shrinking regions from J to I. By thresholding shrinking areas we obtained the sought segmentations.

However it is possible to avoid the computation of 2 vector fields (direct and reverse) between images I and J, by computing only one vector field thanks to a symmetrical approach. Here, we have used the alternated minimization method (sec. 3.2), without using a finite element scheme but the fast scheme of [6].

We have made some experiments on an evolving lesion extracted from two T2-weighted MRI of a patient with multiple sclerosis. Time between the two acquisitions is 8 weeks. The resolution of the images is really poor as we can see on the first two columns of figure 4 (voxel size is $0.9 \times 0.9 \times 5.5mm^3$). We have computed four vector fields for our experiment: asymmetric and symmetric fields from I to J, and asymmetric and symmetric fields from 2 to 1. We compute the Jacobian on each field, and we manually search for the Jacobian isovalue (in red on fig. 4) that segments the lesion. As explained in [9], it does not exist any efficient automatic method to find the best threshold at the moment.

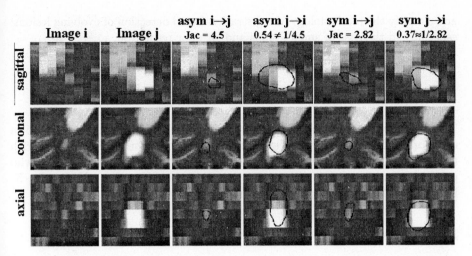

Fig. 4. *Manual segmentations were done independently. Values are not inverse in the case of asymmetrical computation, and inverse in the case of symmetrical approach. Furthermore, the segmentation of evolving areas better corresponds to reality.*

A symmetrical computation of the deformations significantly improves the detection and segmentation of evolving lesions. First it makes it possible to use the same value v to threshold the Jacobian (Jac$> v > 1$ for expansion, and Jac$< 1/v < 1$ for shrinking). With an asymmetrical approach we had to compute the direct and reverse field and only search for shrinking areas. In figure 4 manual segmentations were done independently. Values are not inverse in the case of asymmetrical computation, and inverse in the case of symmetrical approach. Furthermore, the segmentation of evolving areas seems to better correspond to reality as we can see in figure 4.

6 Conclusion

In this paper we have listed the reasons of the asymmetry of intensity-based registration algorithms. We have shown that asymmetry may hamper the equal retrieval of expanding and shrinking areas, perturbing registration-based detection of evolving processes. To reduce the asymmetry of registration, we have introduced inversion-invariant similarity and smoothness energies, and a way to symmetrize the widely used asymmetric energies.

We have worked on a particular asymmetric energy E^* that is easily expressed as a function of the transform T. We have minimized E^* with and without computation of the inverse transform. On a synthetic example, we have shown that the use of a symmetric energy was reducing considerably the shortcomings of asymmetric energies, and that the computation of the inverse transform, if it exists, is helpful. Finally, a symmetric approach improves segmentation results and simplify the methodology of registration-based detection of evolving lesions, especially in the case of small lesions and/or poor resolution.

Acknowledgment

This work was supported by the EC-funded BIOMORPH project 95-0845, a collaboration between the Universities of Kent and Oxford (UK), ETH Zürich (Switzerland), INRIA Sophia Antipolis (France) and KU Leven (Belgium) who provided us with multiple sclerosis images time series. Thanks to N. Ayache, X. Pennec and J. Stoeckel for proofreading and useful comments.

References

1. A. Roche, G. Malandain, N. Ayache, and S. Prima. Towards a Better Comprehension of Similarity Measures Used in Medical Image Registration. In *Proc. of MICCAI'99*, pages 555 – 566, 1999.
2. O. Musse, F. Heitz, and J.-P. Armspach. 3D Deformable Image Matching using Multiscale Minimization of Global Energy Functions. In *Proc. of CVPR'99*, 1999.
3. D. Rueckert, C. Hayes, C. Studholme, P. Summers, M. Leach, and D. J. Hawkes. Non-Rigid Registration of Breast MR Images Using Mutual Information. In *Proc. of MICCAI'98*, pages 1144 – 1152, 1998.
4. M. Ferrant, S. K. Warfield, C. R. G. Guttmann, R. V. Mulkern, F. A. Jolesz, and R. Kikinis. 3D Image Matching using a Finite Element Based Elastic Deformation Model. In *Proc. of MICCAI'99*, pages 202 – 209, 1999.
5. G. E. Christensen. Consistent Linear-Elastic Transformations for Image Matching. In *Proc. of IPMI'99*, pages 224 – 237, 1999.
6. X. Pennec, P. Cachier, and N. Ayache. Understanding the "Demons" Algorithm: 3D Non-Rigid Registration by Gradient Descent. In *Proc. of MICCAI'99*, pages 597 – 605, 1999. http://www.inria.fr/RRRT/RR-3706.html.
7. J. B. A. Maintz and M. A. Viergever. A Survey of Medical Image Registration. *Medical Image Analysis*, 2(1):1–36, 1998.
8. S. K. Kyriacou and C. Davatzikos. A Biomechanical Model of Soft Tissue Deformation, with Applications to Non-Rigid Registration of Brain Images with Tumor Pathology. In *Proc. of MICCAI'98*, pages 531 – 538, 1998.
9. D. Rey, G. Subsol, H. Delingette, and N. Ayache. Automatic Detection and Segmentation of Evolving Processes in 3D Medical Images: Application to Multiple Sclerosis. In *Proc. of IPMI'99*, pages 154–167, 1999. http://www.inria.fr/RRRT/RR-3559.html.
10. A. Trouvé and L. Younes. Diffeomorphicc Matching Problems in One Dimension: Designing and Minimizing Matching Functionals. In *Proc. of ECCV'00 Part 1*, pages 573–587, 2000.
11. G. Gerig, D. Welti, C. Guttmann, A. Colchester, and G. Székely. Exploring the Discrimination Power of the Time Domain for Segmentation and Characterization of Lesions in Serial MR Data. In *Proc. of MICCAI'98*, pages 469–480, Boston, USA, 1998.

A APPENDIX

For a better comprehension, we develop the mathematics for obtaining the symmetric energies (3) and (5). If $T : \mathbf{R}^2 \to \mathbf{R}^2$ is a diffeomorphic 2D transformation, we can compute its Jacobian matrix at a point $\mathbf{p} = (x, y)$:

$$dT(\mathbf{p}) = \left(\frac{\partial T}{\partial x}(\mathbf{p}) \frac{\partial T}{\partial y}(\mathbf{p}) \right) = \begin{pmatrix} \frac{\partial T_x}{\partial x}(\mathbf{p}) & \frac{\partial T_x}{\partial y}(\mathbf{p}) \\ \frac{\partial T_y}{\partial x}(\mathbf{p}) & \frac{\partial T_y}{\partial y}(\mathbf{p}) \end{pmatrix}$$

Since $d(T^{-1}) \circ T = (dT)^{-1}$, we have

$$d(T^{-1})(\mathbf{q}) = \frac{1}{|dT(\mathbf{p})|} \begin{pmatrix} \frac{\partial T_y}{\partial y}(\mathbf{p}) & -\frac{\partial T_y}{\partial x}(\mathbf{p}) \\ -\frac{\partial T_x}{\partial y}(\mathbf{p}) & \frac{\partial T_x}{\partial x}(\mathbf{p}) \end{pmatrix} \tag{8}$$

with $\mathbf{q} = T(\mathbf{p})$. Now, if

$$E_{reg}(T) = \int_{\mathbf{R}^2} \left(\frac{\partial T_x}{\partial x}(\mathbf{p}) \right)^2 + \left(\frac{\partial T_x}{\partial y}(\mathbf{p}) \right)^2 + \left(\frac{\partial T_y}{\partial x}(\mathbf{p}) \right)^2 + \left(\frac{\partial T_y}{\partial y}(\mathbf{p}) \right)^2 d\mathbf{p}$$

then, using (8),

$$E_{reg}(T^{-1}) = \int_{\mathbf{R}^2} \frac{1}{|dT(\mathbf{p})|^2} \left[\left(\frac{\partial T_x}{\partial x}(\mathbf{p}) \right)^2 + \left(\frac{\partial T_x}{\partial y}(\mathbf{p}) \right)^2 + \left(\frac{\partial T_y}{\partial x}(\mathbf{p}) \right)^2 + \left(\frac{\partial T_y}{\partial y}(\mathbf{p}) \right)^2 \right] d\mathbf{q}$$

$$= \int_{\mathbf{R}^2} \frac{1}{|dT(\mathbf{p})|} \left[\left(\frac{\partial T_x}{\partial x}(\mathbf{p}) \right)^2 + \left(\frac{\partial T_x}{\partial y}(\mathbf{p}) \right)^2 + \left(\frac{\partial T_y}{\partial x}(\mathbf{p}) \right)^2 + \left(\frac{\partial T_y}{\partial y}(\mathbf{p}) \right)^2 \right] d\mathbf{p}$$

Finally we sum this last equation to $E_{reg}(T)$ to find the symmetrization of $E_{reg}(T)$ (eq. (3)).

Similarly, if $E_{sim}(I, J, T) = \int_{\mathbf{R}^2} (I(\mathbf{p}) - J \circ T(\mathbf{p}))^2 d\mathbf{p}$,

$$E_{sim}(J, I, T^{-1}) = \int_{\mathbf{R}^2} (J(\mathbf{q}) - I \circ T^{-1}(\mathbf{q}))^2 d\mathbf{q}$$

$$\overset{\mathbf{q}=T(\mathbf{p})}{=} \int_{\mathbf{R}^2} (J \circ T(\mathbf{p}) - I(\mathbf{p})) |dT(\mathbf{p})| d\mathbf{p}$$

We find eq. (5) by summing this last equation to $E_{sim}(T)$.

Modeling and Analysis of Ultrasonic Echoes Reflected from a Surface under Two Layers

Joyoni Dey and Takeo Kanade

Carnegie Mellon University

Abstract. Fat and muscle layers in the human soft-tissue causes degradation of resolution and distortions of shape in ultrasound images of organs. Novel ultrasonic applications such as registration between CT and ultrasound images, requiring comparable levels of accuracies between modalities could suffer from the distortions in ultrasound images. In this work, we describe a possible method to compensate or correct for these distortions in ultrasound images. This can be used as a pre-processing step before registration with other modalities. We derived a model of ultrasound field after reflection from an interface embedded in two layers. Experiments were performed using single-element transducers and custom made tissue mimicking phantoms of fat and muscle and a steel-block. We fit the model to the experimental data using Levenberg-Marquardt inversion.

Keywords: Medical ultrasound, biomedical modeling and simulation

1 Introduction

In addition to diagnostic applications, ultrasound is becoming increasingly popular for novel applications such as using intra-operatively obtained ultrasound images for registration with other modalities for computer-aided surgery. In Fig-

Intra-operative Ultrasound Image Pre-operative CT Data

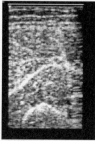 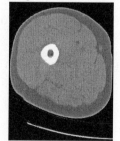

Fig. 1. Ultrasound image and CT slice of human femur

ure 1 we show an ultrasound image of a human femur bone (the semi-circular reflection at the bottom) and a CT image of the femur. (Note that Figure 1 is shown for the purpose of illustration only – the images belong to different patients). It is evident that the bone outline can be used for the registration of the images. The problem with registration between these two modalities is that while CT-data is very accurate, ultrasound has imaging artifacts. In much of the human body there are layers of fat and muscle. These intervening layers can cause degradation of resolution, shape distortion and other artifacts.

We describe a method to compensate for these artifacts for the case of imaging the bone under fat and muscle layers. This method can be used as a preprocessing step to modify the ultrasound images before proceeding with the registration. We developed a forward ray-tracing model of the received ultrasound waves after propagation into two layers and reflecting back from a highly reflecting interface. We chose to model this particular set-up because it closely resembles the bone under layers of fat and muscle. Our derived model can handle arbitrary interface geometries and aperture configurations. We also describe an inversion process to estimate the parameters of the forward model. The main parameters of interest in our application are the depths of the interfaces for the case of two layers on a reflective interface.

We performed real-life experiment on uniform thickness tissue-mimicking phantoms with known thicknesses, placed on a steel-block. This set up mimics a fat and muscle layer on the highly reflective interface, the bone. We used a circular aperture transducer operating at 7.5 MHz. We then fit our forward model (for a special case where the layers are planar and parallel to the aperture) to this experimental data by iteratively updating the model parameters (the depths of the interfaces in this case) so as to minimize the least square-errors between the model and the data.

2 Related Work

Several authors have investigated the phase aberrations and amplitude distortions in ultrasound wave caused by layers of fat and muscle in the human body wall [1]- [11]. Some of them account for diffraction effects [1]- [4] while others consider ray-tracing approximations, which account for only reflection and refraction at the boundaries. When the organs of interest are large compared to the wavelength, ray-tracing can be used. Effects of composite layers on beam-formation such as defocusing and beam-shifts and imaging artifacts such as split-image, shadowing, and enhancements have been shown in the literature [5, 6, 7]. In these works, ray-tracing has been used to explain these effects. While it does not consider diffraction, the advantage of ray-tracing based modeling is that it is highly parallelizable and fast. The cost of the implementation is particularly important if we are considering inversions of the model. It facilitates the inversion further if we have a closed form forward model. In a related work [12], we attempted to derive a fully analytical model for the propagation of a monochromatic spherical wave through general media geometries and arbitrary apertures under ray-tracing approximations. The modeling is similar to the work of Odegaard et al [10] – a main point of difference is that in the simulations described by Odegaard et al, the amplitude spreading factors were obtained numerically by perturbing rays in two directions, while we derived these factors in closed form by mathematically obtaining the ratios of differential flux-tube areas [12]. In this paper, when modeling the special case of imaging the bone under layers of fat and muscle, we used parts of our above mentioned results for propagation of the monochromatic spherical ultrasound wave through layered media.

3   Theory : Forward Model

In Figure 2(a), we show an ultrasonic aperture and a reflective interface (Plane 2) under two media. We have broken up the interface surfaces into small planar segments. The dimensions have to be of the order of the wavelength of the ultrasonic wave, about 0.2 to 0.3 mm. We want to model the reflected waveform received at the aperture. First we consider a spherical wave emitted by a point in the aperture and observe the reflected waves for that source. The response for the entire aperture is the sum of the reflected waveforms received corresponding to each point source in the aperture. For a focussed aperture we can add the appropriate transmit and receive focusing delays before adding the responses from the different parts of the aperture. To derive our model, we used ray-tracing

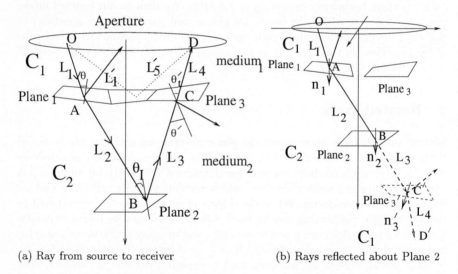

(a) Ray from source to receiver   (b) Rays reflected about Plane 2

Fig. 2. Ray Tracing from source to receiver

assumption. Geometrically this implies that a spherical wave remains spherical after reflection. But refraction introduces a change in the shape of the waveform. Calculations show that for the parameters under consideration, the ray tracing approximation is not very restricting [12]. Other main assumptions are longitudinal propagation (valid for soft-tissues). We also considered absorption loss. Ignoring multiple reflections, the received wave at D due to a point source at O, $f(P_D, t)$ would have two terms, $f_1(P_D, t)$ and $f_2(P_D, t)$. The first term, $f_1(P_D, t)$ is due to pure reflection from the interface between the first and second layer. The second term, $f_2(P_D, t)$ is the reflected wave from the second layer (it goes through refraction at the points A and C as well).

We can straightforwardly write the first term, for which the relevant ray-paths are those shown in dashed lines in Figure 2(a). $f_1(P_D, t) = \sum_{rays'} e^{-(\mu_1(L_1' + L_5'))}$ $\frac{1}{(L_1' + L_5')} V(\theta) \mathbf{p}[t - (\frac{(L_1' + L_5')}{C_1})]$ where inverse of the total path-length $L_1' + L_5'$ appears as the intrinsic attenuation factor (since this is pure reflection, the wave is spherical and for a spherical wave, the attenuation is $1/r$, where r is the equivalent radial distance covered from the source) ; $V(\theta)$ is the reflection coefficient, where θ is the angle of incidence of the ray. $rays'$ denotes the set of rays that reached from O to D on reflection at first interface. The exponent term is the term due to absorption loss. μ_1 is the attenuation coefficient in the first medium.

The ray O-A-B-C-D, is the ray to be considered for the second term. This set is sufficient because of the linearity [12]. The segments OA and CD (with path lengths L_1 and L_4) are in the first medium with propagation velocity C_1. AB and BC (of lengths L_2 and L_3) are in the second medium, with propagation velocity C_2. $f_2(P_D, t) = \sum e^{-(\mu_1 L_1 + \mu_2(L_2 + L_3) + \mu_1 L_4)} |f_D| \mathbf{p}[t - (\frac{L_1}{C_1} + \frac{L_2 + L_3}{C_2} + \frac{L_4}{C_1})]$ where L_i are the lengths of the path in each media for the ray (shown in solid lines) from each source to the point of interest in Figure 2(a). The first exponent is the absorption loss attenuation factor [10, 13, 14]. μ_i being the attenuation coefficients inside each medium. $|f_D|$ is the intrinsic attenuation factor, which is not exactly $1/r$ anymore because of the refraction and so on.

It is an elaborate process to calculate this factor $|f_D|$, and is out of scope of this paper except for a brief outline. We can first simplify the rays – reflect ray BC and CD in Figure 2 (a) about the Plane 2, and imagine it as continuing along AB as shown in Figure 2(b). Now we see that as far as the derivation of $|f_D|$ is concerned, it is the same as if the wave was propagating into three-layers, except here the first and the third layers happen to be the same. In [12] we have formulated in detail the intrinsic attenuation factors after propagation through three (or more) media. We show the final result here, $|f_D| = R(\theta_I)|1 + V(\theta)||1 + V'(\theta')|\frac{1}{L_1}\sqrt{\frac{\Delta a a_2 A_1}{\Delta D_1 s_1 S_1}}\sqrt{\frac{\cos\theta_I}{\cos\theta'}}$. In this equation, $V(\theta)$ and $V(\theta')$ are the reflection coefficients at the first and the third interface. The angles are shown in Figure 2(a). The first factor $R(\theta_I)$ is a function of the incident angle θ_I corresponding to the reflection at the second interface (plane B in Figure 2(b)). For a perfect reflector, this factor is unity. The area ratios $\sqrt{\frac{\Delta a a_2 A_1}{\Delta D_1 s_1 S_1}}$ are the ratios of flux-tube areas that constructed perpendicular to the ray (part of the wavefront). Calculation of these areas in close-form are shown in [12]. Briefly, the approach taken is to change the ray differentially in two independent directions and calculate the sides of the triangular areas by taking derivative of the points of intersection of the rays and the interfaces. The areas of the triangles are then the magnitudes of cross-products of the sides.

The final received signal at point D in the aperture due to the source at O is the sum of $f(P_D, t) = f_1(P_D, t) + f_2(P_D, t)$. Finally, we can apply the aperture forward transmittance (for focusing, steering etc), if any, at each point and add the signals arriving at D for all the point sources. Then to obtain the total response of the entire aperture, we would add the arriving wave at all the points in the aperture, appropriately modified by the received focusing delays, if any. For the *unfocused* transducer case, we would simply add the response arriving at all the points of the transducer, giving the final received time signal as

$$f(t) = \sum_{P_D} \sum_{P_O} \sum_{rays'} e^{-(\mu_1(L_1' + L_5'))} \frac{1}{(L_1' + L_5')} V(\theta) \mathbf{p}[t - (\frac{(L_1' + L_5')}{C_1})]$$

$$+ \sum_{P_D} \sum_{P_O} \sum_{rays} e^{-(\mu_1 L_1 + \mu_2(L_2 + L_3) + \mu_1 L_4)} |f_D| \mathbf{p}[t - (\frac{L_1}{C_1} + \frac{L_2 + L_3}{C_2} + \frac{L_4}{C_1})]$$ (1)

where the integral over P_O indicates we sum the waves arrived at D for all the point sources, the integral over P_D indicates that we then add the received waveform at each such point in the aperture.

4 Inversion of the Model

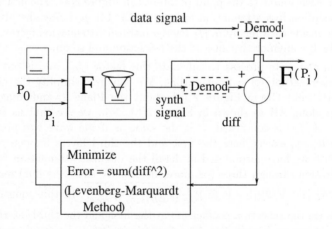

Fig. 3. Iterative Inversion

We wrote a simulator to obtain a synthetic output of the received signal given in Equation 1, given the input parameters of the receiver, and the media parameters – the interface geometries and the physical variables. This is the forward model, denoted by block diagram F in Figure 3. We use this block diagram to show the inversion of the model (ignoring the dotted blocks for the present). We start with an initial estimate of the parameter values, P_0 (in this case the depth of the two interfaces from the aperture). The initial estimate is chosen by a simple thresholding operation on the experimental data signal. The parameters are used by the forward model F to get a synthesized signal output (or an ultrasound image in general). The error difference between the synthesized output and the actual experimental data signal (or images) drives the parameters to their next values. We adopt the Levenberg-Marquardt minimization for this purpose. The simulated output signal from the receiver in our case is a time signal, $y = y(n; \mathbf{p})$ where n is the time sample index and $\mathbf{p}$ is the parameter set used. In this case, the parameter set consists of the two depths of the planar interfaces, given by D_1 and D_2. The scalar error that is minimized is given by $E(\mathbf{p}) = \sum_n (y_n - y(n; \mathbf{p}))^2$ where y_n and $y(n; \mathbf{p})$ are the data signal and the synthesized signal respectively.

5 Experiments and Model Fitting

5.1 Experimental Equipment and Procedure

The ultrasonic transducer used is a 7.5 MHz, single element circular aperture transducer. We use a RiTec control box with variable amplitude and frequency to trigger the transducer. We excite the transducer with a short square pulse of time-duration half of the time-period at resonance, (about 67 nano-secs). We place the transducer over the fat-phantom and obtain the return-echoes. The echoes are captured, and then filtered by an anti-aliasing low pass (cut off 10 MHz) and a high pass filter (to remove the DC bias, cut off at 20 KHz). The filtered signal is sampled at 25 MHz by an Sonix analog to digital converter. The Sonix board is triggered by a synchronizing trigger source from the control box. We use "fat" and "muscle" mimicking phantoms. These physical objects designed from water-agar-gel mixtures mimic some ultrasound characteristics (such as propagation velocity) in human fat and muscle. We also use a steel-block about 5 cm thick. To create a highly reflective interface under the two layers (fat and muscle) the phantoms are stacked on the steel-block so that the muscle-phantom lies on the steel-block and the fat-phantom lies on top of the muscle. To obtain good coupling between the phantom layers, they are wetted with water.

In order to invert the forward model F in Equation 1, we need to determine the pulse $\mathbf{p}(t)$ for this transducer. We do so this experimentally by a standard method of obtaining an echo from a steel-block under-water. The pulse obtained is observed to be nearly a Gaussian. We fit a Gaussian curve (in amplitude, mean, standard deviation, and the carrier frequency) using the simplex algorithm. We obtain the pulse to within a scale. The phantom parameters such as mu_i, C_i, ρ_i are provided by the manufacturer.

5.2 Results of Inversion

A straightforward inversion of the forward model has the problem that, due to the modulation of the signals at the carrier frequency, f_0 there would be local

Table 1. Convergence Data

Using Demodulated Signals			Using Modulated Signals		
	Start	End		Start	End
Error	45306	6203	Error	32651	12702
Depth D_1	15.0000mm	14.9568mm	Depth D_1	14.9568mm	14.9727mm
Depth D_2	35.3000mm	35.4342mm	Depth D_2	35.4342mm	35.4220mm

minima at regular intervals of roughly λ_0 from the global minimum [12]. To obtain a smoother error bowl, we first need to add a demodulator shown in dotted lines in Figure 3, so that the envelopes of the signals should be involved in the minimization, rather than the modulated signals themselves. The signals are real and typically the energy is concentrated (in "humps") around the carrier f_0. There are many techniques to demodulate signals such as using the Hilbert Transform. We adopted simple intuitive steps described briefly as follows. The receiver signals are real, hence the real part of their FFT is symmetric and the imaginary part, antisymmetric. All the information of the signal is then entirely available on one side of the spectrum. We extract the real and imaginary "humps" of the spectrum from the positive side of frequency (humps centered around the positive f_0). We shift each of them by f_0 (to baseband) to get $G_r(f)$ and $G_i(f)$. The signal $g(n) = IFFT(G_r(f) + jG_i(f))$ therefore has all the information of the original signal and is a baseband signal. But $g(n)$ is complex [12] and difficult to use in the minimization as such. Hence we use the real function $s_n = \sqrt{2g(n)g^*(n)}$. The $\sqrt{2}$ factor is needed such that the original modulated signal and the baseband "envelope" signal s_n has the same energy. Some examples are given in [12].

After convergence with the envelope signals, we are roughly within $\lambda/2$ of the global minimum, we revert back to using the raw (modulated RF) signals again (increased sensitivity) and minimize the difference between those to do fine-tuning of the parameter fitting.

From a rough estimate of the location of the echoes from the experimental "data" signal, we start the iteration at $D_1 = 15.0mm$ and $D_2 = 35.3mm$. The starting and final errors and converged parameters are given in the second and third columns of Table 1. After this stage the echoes seem to be slightly out of phase (particularly the first one). This is because so far we used the envelope of the signals for the error minimization rather than using the signals themselves. Hence, the next obvious fine-tuning step is to eliminate the demodulation step and run the inversion on the raw signals. The error and parameter values are shown in the last three columns in Table 1. The starting parameters for these sets of iterations (using the raw signals) are the end parameters arrived at by using the demodulated signals (third column, Table 1. We observe that for the same parameters, the starting error is much larger for the modulated case than for the demodulated case in Table 1. And the final error is higher than what we obtained at the end of the runs for the demodulated case. This is not an unexpected result. Matching of envelopes is a different criterion from matching of the signals themselves.

As a last step, we do the scale correction between simulation and experiment. We assume that the initial estimate of the scale (by energy ratio of the signals) was near enough to the correct scale so as not to hamper the convergence of the other parameters. Once we obtain a good estimate of the parameters $\mathbf{p} = [D_1, D_2]$, we can correct for the scale factor analytically. At the correct parameters, $\mathbf{p_c}$, the energy is given by $E(\mathbf{p}) = \sum_n (y_n - Ay(n; \mathbf{p_c}))^2$. The right scale, at which this error is minimum, can be calculated by setting the derivative of $E(\mathbf{p})$ with respect to A to zero. This is given by $A_c = \dfrac{\sum_n s_n s(n; \mathbf{Pc})}{\sum_n s^2(n; \mathbf{Pc})}$. The original scale was $A = 4.4904$. After the correction, the scale is $A_c = 3.9139$. The correctly-scaled simulation and the experiments are shown in Figure 4.

Figure 4 show a qualitatively good match between experiments and simulations. The first echo seems to have matched well except toward end of the pulse. The second echo seems to have matched in shape well in the midsection but not in the beginning and the end. We note that we have not modeled the noise in our simulations. When the signal is small (towards the beginning/end of the echoes) the effect of the noise is higher. We calculate some quantitative ratios between the compared experimental and simulation echoes shown in Figure 4. Considering the first echo, the RMS amplitude ratio of the experimental one to that of the the fitted model is close to unity, 1.0353 (or 0.1507 db). For the second echo, the RMS amplitude ratio of the experimental to that of the fitted model is 0.9564 (or -0.1936 db). The final parameters are obtained as $D_1 = 14.9727$ and $D_2 = 35.4220$. Hence the layer thicknesses are within $0.03mm$ and $.45mm$ of the manufacturer specified nominal values of $15mm$ and $20mm$.

6 Conclusions

In this work, we have modeled ultrasound echo-field reflected from a interface embedded in two layers. We performed experiments on tissue mimicking phantoms and fit our model to the data. For model-fitting, we have minimized the mean-squared-error using the Levenberg-Marquardt minimization technique. We obtained good quantitative matches.

It is noted that ultrasound speckle has not be considered here in simulations or experiments. The speckle is expected to add noise to the error function to be minimized, thereby making the inversion more prone to false minima. The effect can be reduced to some extent perhaps by using multiresolution approach – first using smoothing functions on the envelopes and obtaining a minima roughly close to the global minimum ; and then using the end parameters of that step as the initial guess into the next resolution step; and so on until finally using the original signals to fine tune the minima location.

Finally, it is noted even though we presented results for a single-element ultrasound transducer, the method is not fundamentally different for a linear array

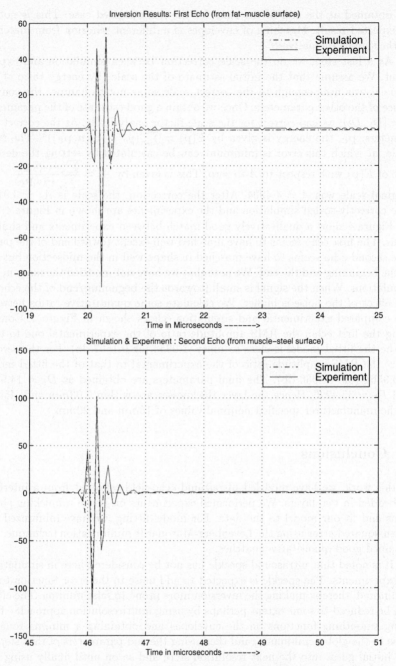

Fig. 4. Comparison of Experiment and Simulation: first (top) and second (bottom) echo

aperture. Each rectangular element of the array can be first discretized to point sources emitting spherical waves. Distortion of each emitted wave due to layered inhomogeneity can be calculated as shown here. Then transmit/receive focusing delays can be applied to the rectangular elements of the array and the result summed to obtain each of the scan lines.

Acknowledgments: We would like to thank Rajesh Gopakumar and Farhana Kagalwala for many helpful discussions and suggestions.

References

1. T. D. Mast, L. M. Hinkelman, L. A. Metley, M. J. Orr, R. C. Waag, "Simulation of ultrasonic pulse propagation, distortion, and attenuation in the human chest wall", J. Acoust. Soc. Am. **106**, 3665-3677, December 1999
2. L. M. Hinkelman, T. D. Mast, L. A. Metley, and R. C. Waag, "The effect of abdominal wall morphology on ultrasonic pulse distortion. Part I: Measurements", J. Acoust. Soc. Am. **104** 3635-3649 (1998)
3. T. D. Mast, L. M. Hinkelman, M. J. Orr, and R. C. Waag, "The effect of abdominal wall morphology on ultrasonic pulse distortion. Part II: Simulations", J. Acoust. Soc. Am. **104** 3651-3664 (1998)
4. C. W. Mantry and S. L. Broschat, "FDTD Simulations for Ultrasound Propagation in a 2-D Breast Model", Ultrason. Imaging **18** 25-34 (1996)
5. G. Kossoff, D.A. Carpenter, D.E. Robinson, D. Ostry and P.L. Ho, "A sonographic technique to reduce beam distortion by curved transducers," Ultrasound in Med. and Biol. **15**, 375-382(1989).
6. D. E. Robinson, L. S. Wilson, and G. Kossoff, "Shadowing and enhancement in ultrasonic echograms by reflection and refraction", J. Clin. Ultrasound, **9**, 181-188 (1981)
7. E. E. Sauerbrei, " The split image artifact in pelvic ultrasonography: the anatomy and physics," J. Ultrasound Med. **4**, 29-34 (1985)
8. D. A. Carpenter, D. E. Robinson, P. L. Ho, D.C.C. Martin and P. Isaacs, "Body wall aberration correction in medical ultrasonic images using synthetic-aperture data," IEEE 1993 Ultrason. Symp. Proc. **3**, 1131-1134(1993).
9. D.A. Carpenter, G. Kossoff and K. A. Griffiths, "Correction of distortion in US images caused by subcutaneous tissues: Results in tissue phantoms and human subjects," Radiology **195** 563-567(1995).
10. L. Odegaard, S. Holm, F. Teigen and T. Kleveland, "Acoustic field simulation for arbitrarily shaped transducers in a stratified medium," IEEE 1994 Ultrason. Symp. Proc. 1535-1538 (1994).
11. L. Odegaard, S. Holm and H. Torp, "Phase aberration correction applied to annular array transducers when focusing through a stratified medium," IEEE 1993 Ultrason. Symp. Proc. 1159-1162(1993).
12. J. Dey, Modeling and analysis of ultrasound propagation in layered medium. Doctoral Dissertation, Department of ECE, Carnegie Mellon University, 1999, CMU-RI-TR-99-26.
13. L. M. Brekhovskikh, *Waves in layered media* (Academic Press, 1980).
14. A. D. Pierce, *Acoustics: An introduction to its physical principles and applications* (McGraw Hill, Inc, 1981).

Localization of 3D Anatomical Point Landmarks in 3D Tomographic Images Using Deformable Models*

Sönke Frantz, Karl Rohr, and H. Siegfried Stiehl

Universität Hamburg, Fachbereich Informatik, Arbeitsbereich Kognitive Systeme
Vogt-Kölln-Str. 30, 22527 Hamburg, Germany
{frantz,rohr,stiehl}@informatik.uni-hamburg.de
http://kogs-www.informatik.uni-hamburg.de/PROJECTS/imagine/Imagine.html

Abstract. Existing differential approaches to the localization of
3D anatomical point landmarks in 3D tomographic images are relatively
sensitive to noise as well as to small intensity variations, both of which
result in false detections as well as affect the localization accuracy. In
this paper, we introduce a new approach to 3D landmark localization
based on deformable models, which takes into account more global image information in comparison to differential approaches. To model the
surface at a landmark, we use quadric surfaces combined with global
deformations. The models are fitted to the image data by optimizing
an edge-based fitting measure that incorporates the strength as well as
the direction of the intensity variations. Initial values for the model parameters are determined by a semi-automatic differential approach. We
obtain accurate estimates of the 3D landmark positions directly from
the fitted model parameters. Experimental results of applying our new
approach to 3D tomographic images of the human head are presented.
In comparison to a pure differential approach to landmark localization,
the localization accuracy is significantly improved and also the number
of false detections is reduced.

1 Introduction

In this contribution, we address the problem of extracting 3D anatomical point
landmarks from 3D tomographic images, focusing on anatomical structures of
the human head. The driving task is landmark-based 3D image registration,
which is fundamental to computer-assisted neurosurgery. Existing work on the
automated extraction of 3D point landmarks is based on differential approaches
(e.g., [17],[11]). However, while being computationally efficient, differential approaches are relatively sensitive to noise as well as to small intensity variations,
both of which result in false detections as well as affect the localization accuracy.
In this paper, we introduce a new approach to 3D landmark localization based on

---

* This work was supported by Philips Research Hamburg, project IMAGINE (IMage-
and Atlas-Guided Interventions in NEurosurgery).

deformable models, which takes into account more global image information in comparison to differential approaches and thus opens the possibility of increasing the robustness as well as the accuracy in landmark extraction. Previously, deformable models have been applied to segmentation, tracking, and image registration (see [8] for a survey), whereas the localization of 3D point landmarks based on deformable models has not been considered so far.

Exemplarily, we here focus on two different types of landmarks, namely, salient surface loci (curvature extrema) of *tips* and *saddle structures*. Examples of corresponding 3D point landmarks of the human head are the tips of the ventricular horns or the saddle points at the zygomatic bones (see Fig. 1). To represent such structures, we utilize 3D surface models. In the literature, a variety of 3D surface models has been used (e.g., [15],[13],[3],[14],[7],[16],[19],[1]). Note that in comparison to earlier work on deformable models, we are here interested in the accurate localization of salient surface loci. Central to an efficient solution of this specific problem is that the model surface exhibits a unique point whose position can be directly computed from the model parameters. As a compromise between generality and efficiency, we here use quadric surfaces as shape prototypes, which are combined with additional global deformations to enlarge the range of shapes (Sect. 2).

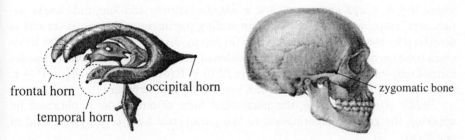

Fig. 1. Ventricular horns of the human brain (from [12]) and the human skull (from [2]). Examples of 3D point landmarks are indicated by black dots.

Model fitting is formulated as an optimization problem where a suitable fitting measure is optimized w.r.t. the model parameters. Here, we use an edge-based fitting measure that incorporates the strength as well as the direction of the intensity variations. This measure, which is described in Sect. 3, is a 3D generalization of a 2D fitting measure suggested in [20]. Usually, only the strength of the intensity variations is incorporated (e.g., [19],[14]). To determine initial values for the model parameters, we have developed a semi-automatic differential approach (Sect. 4). Experimental results of applying our new approach to 3D tomographic images of the human head are presented in Sect. 5. In particular, we analyze the localization accuracy of our new approach and compare it with that of an existing differential approach.

2 Geometric Models of Tips and Saddle Structures

As 3D shape prototypes, we here use quadric surfaces, namely, ellipsoids for 3D tip-like structures and hyperboloids of one sheet for 3D saddle structures. However, real structures in general show deviations from these prototypes (e.g., the ventricular horns generally have a bended shape and partly show a tapering). To take into account bending in the case of tip-like structures, we here additionally apply a quadratic bending deformation along the centerline of the ellipsoid (i.e., the z-axis) [3]: $\mathcal{B}(\mathbf{x}) = (x + \delta \cos v \, z^2, y + \delta \sin v \, z^2, z)^T$, where $\mathbf{x} = (x, y, z)^T$ denotes an arbitrary surface point and $\delta \geq 0$ determines the strength and v the direction of bending. To transform the object-centered model coordinate system to the image coordinate system, we here use a rigid transformation, $\mathcal{R}(\mathbf{x}) = \mathbf{R}\mathbf{x} + \mathbf{t}$, where $\mathbf{t} = (X, Y, Z)^T$ denotes the translation vector and $\mathbf{R}$ the rotation matrix depending on the rotation angles α, β, γ.

Tips. The parametric form of our model is obtained by applying the bending deformation and the rigid transformation to the parametric form of an ellipsoid:

$$\mathbf{x}_{tip}(\theta, \phi) = \mathbf{R} \begin{pmatrix} a_1 \cos\theta \cos\phi + \delta \cos v \, (a_3 \sin\theta)^2 \\ a_2 \cos\theta \sin\phi + \delta \sin v \, (a_3 \sin\theta)^2 \\ a_3 \sin\theta \end{pmatrix} + \mathbf{t}, \tag{1}$$

where $0 \leq \theta \leq \pi/2$ and $-\pi \leq \phi < \pi$ are the latitude and longitude angle parameters, resp., and $a_1, a_2, a_3 > 0$ are scaling parameters. Hence, the model is described by the parameter vector $\mathbf{p} = (a_1, a_2, a_3, \delta, v, X, Y, Z, \alpha, \beta, \gamma)$. The landmark position of our model, i.e., the position of the curvature extremum of the deformed ellipsoid, is given by $\mathbf{x}_l = \mathbf{x}_{tip}(\pi/2, 0) = \mathbf{R}(\delta \cos v \, a_3^2, \delta \sin v \, a_3^2, a_3)^T + \mathbf{t}$. Figure 2 (left) shows an example of a bended tip-like structure.

Saddle structures. Here, the parametric form of our model is obtained by applying the rigid transformation to the parametric form of a hyperboloid of one sheet:

$$\mathbf{x}_{saddle}(\theta, \phi) = \mathbf{R} \begin{pmatrix} a_1 \sec\theta \cos\phi \\ a_2 \sec\theta \sin\phi \\ a_3 \tan\theta \end{pmatrix} + \mathbf{t}, \tag{2}$$

where $|\theta| < \pi/2$ and $0 \leq \phi \leq \pi$. Thus, the model is described by the parameter vector $\mathbf{p} = (a_1, a_2, a_3, X, Y, Z, \alpha, \beta, \gamma)$. The landmark position is here given by $\mathbf{x}_l = \mathbf{x}_{saddle}(0, \pi/2) = \mathbf{R}(0, a_2, 0)^T + \mathbf{t}$. Figure 2 (right) shows an example of a saddle structure.

3 Model Fitting Using an Edge-Based Fitting Measure

The geometric models introduced in Sect. 2 are fitted to the image data by optimizing an edge-based fitting measure w.r.t. the model parameters. Our fitting measure, which is a generalization of the 2D fitting measure in [20], exploits (a) the similarity between the directions of the intensity gradient and the normals of the model surface as well as (b) the strength of the intensity variations.

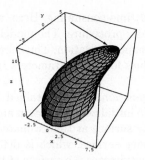 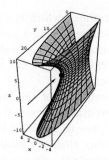

Fig. 2. Geometric models based on quadric surfaces as 3D shape prototypes. The landmark positions are indicated by a black dot.

As a result, the influence of neighboring structures during fitting is diminished significantly, which increases the robustness of model fitting.

During fitting, we consider the contributions of the intensity gradient in the direction of the normal of the model surface, utilizing the projection of the intensity gradient onto the unit normal of the model surface:

$$M_{fit}(\mathbf{p}) = \pm \iint \ < \nabla g(\boldsymbol{x}(\theta, \phi; \mathbf{p})), \frac{\partial \boldsymbol{x}(\theta, \phi; \mathbf{p})}{\partial \theta} \times \frac{\partial \boldsymbol{x}(\theta, \phi; \mathbf{p})}{\partial \phi} > \ d\theta \, d\phi \ \to \ min., \quad (3)$$

where g denotes the intensity function, $\boldsymbol{x}$ is the parametric form of the respective geometric model, which depends on θ, ϕ, and the model parameter vector $\mathbf{p}$, and $< \cdot, \cdot >$ denotes the inner product. The choice of the sign $\pm$ of the fitting measure depends on the appearance of the landmark at hand in the image: In the case of a dark structure, where the intensity gradient points outward, the sign is positive, whereas the sign is negative in the case of a bright structure. Thus, the better the similarity between the directions of the intensity gradients and the normals of the model surface is and the stronger the intensity variations along the model surface are, the smaller is the measure in Eq. (3). It is worth noting that in our implementation, only those surface points in Eq. (3) are considered where $\pm < \nabla g(\boldsymbol{x}(\theta, \phi; \mathbf{p})), \frac{\partial \boldsymbol{x}(\theta, \phi; \mathbf{p})}{\partial \theta} \times \frac{\partial \boldsymbol{x}(\theta, \phi; \mathbf{p})}{\partial \phi} >$ is less than zero. That is, surface points where the direction of the intensity gradient differs significantly from the surface normal are excluded. As a result, the influence of neighboring structures is further reduced. We optimize the fitting measure in Eq. (3) w.r.t. the model parameters by applying the conjugate gradient method ([10]).

A similar 3D fitting measure was suggested in [5]. However, in [5] a discrete formulation of Eq. (3) was used. Also, different geometric models based on Fourier surfaces were used in [5], and the approach was applied to segmented data only, while here we do not require segmented data.

4 Initialization of the Model Parameters

A central issue of fitting deformable models to the image data is the determination of suitable initial values for the model parameters. Often, initial values are manually determined, which is tedious and time-consuming. Here, we initialize the model parameters using a semi-automatic procedure. The model for tip-like structures is initialized using an ellipsoid as approximation. Thus, for both models from Sect. 2, we have to find initial values for nine parameters (translation, rotation, and scaling).

An initial estimate of the landmark position, $\hat{\mathbf{x}}_l$, is obtained by a semi-automatic differential approach ([11],[6]). To initialize the rotation angles α, β, γ, we exploit the pose of the local isointensity surface at the estimated landmark position, assuming that the local isointensity surface, which is defined by the implicit equation $g(\mathbf{x}) - g(\hat{\mathbf{x}}_l) = 0$, well approximates the surface of the anatomical structure at hand. The rotation angles are then determined by the direction of the intensity gradient $\nabla g(\hat{\mathbf{x}}_l)$ (estimate of the normal) as well as the principal curvature directions of the local isointensity surface at $\hat{\mathbf{x}}_l$ (see, e.g., [4],[17] for computing the curvature of isointensity surfaces). The scaling parameters a_1, a_2, a_3 are initialized based on the principal curvatures κ_1, κ_2 of the local isointensity surface at the estimated landmark position. For example, in the case of a tip, we have the relations $\kappa_1 = a_3/a_1^2$ and $\kappa_2 = a_3/a_2^2$. Note, however, that we have only two principal curvatures, while we have three scaling parameters. To cope with this problem, we here initialize one scaling parameter manually.

5   Experimental Results for 3D Tomographic Images

We have applied our approach to 3D synthetic data as well as to 3D tomographic images of the human head. In this section, we present experimental results of applying our approach to a 3D MR image and a 3D CT image of one patient. In Sect. 5.1, we describe the parameter setting, and in Sect. 5.2 we present the results obtained for different anatomical landmarks of the human head.

5.1   Parameter setting

In the case of the 3D MR image (T1-weighting, voxel size $\approx 0.86 \times 0.86 \times 1.2 \text{mm}^3$), we considered the tips of the frontal and occipital ventricular horns as well as the saddle points at the zygomatic bones (see Fig. 1, where the landmarks are indicated). The field-of-view of the CT image captures only a part of the ventricular horns, and therefore we here considered only the zygomatic bones. Instead of the original, anisotropic CT data, we used interpolated data with an isotropic voxel size of 1.0^3mm^3 (for interpolation we applied the approach in [9]). Initial estimates of the landmark positions were determined by applying the semi-automatic differential approach in [11],[6], where partial derivatives of the intensity function were estimated using cubic B-spline image interpolation ([18]) and Gaussian smoothing. The scale of the Gaussian filters was coarsely adapted to the scale of the respective landmark: For the ventricular horns we used $\sigma = 1.5 \text{mm}$, while for the zygomatic bones we used $\sigma = 1.0 \text{mm}$ (in our implementation of derivative computation, the filter scales are specified in units of mm and are afterwards converted to voxel units based on the respective image resolution). In case of several detections, we selected the candidate with the maximal operator response. For computing the curvature of the local isointensity surface, the scale of the Gaussian filters was the same as that used for landmark detection.

The fitting measure $M_{fit}(\mathbf{p})$ as well as its derivative w.r.t. the model parameter vector $\mathbf{p}$ ($\nabla M_{fit}(\mathbf{p})$ is required for optimization) involve (a) the parametric forms of the models as well as partial derivatives of the parametric forms w.r.t. θ, ϕ, and the model parameters and (b) image derivatives. Expressions involving the parametric forms were determined analytically. Image derivatives were computed using Gaussian filters with $\sigma = 1.0$mm. For numerical evaluation of the integral in the fitting measure in Eq. (3) and in its derivative, we adopted a scheme based on equidistant sampling of the two-dimensional parameter space (θ, ϕ) and cubic interpolation ([10]). The image derivatives were trilinearly interpolated. To diminish the influence of neighboring structures, model fitting was restricted to a spherical region-of-interest (ROI) centered at the estimated landmark position, where the ROI radius was set to 15 voxels.

5.2 Results for the ventricular horns and the zygomatic bones

Ventricular horns We considered four different landmarks, namely, the tips of the frontal and occipital ventricular horns in both hemispheres. For each landmark, the semi-automatic differential approach in [11],[6] yielded a reasonable initial estimate of the landmark position. The rotation angles α, β, γ as well as the scaling parameters a_1 and a_2 were automatically determined based on the differential characteristics of the local isointensity surfaces at the position estimates (see Sect. 4). Only the scaling parameter a_3 was manually initialized. The bending parameters were initially set to zero. Figure 3a visualizes the initialization result obtained for the left frontal ventricular horn; for the other landmarks, we obtained similar initialization results.

Given the relatively large number of parameters, model fitting was then performed in two steps for reasons of robustness: To achieve a coarse adaption, we first fitted only the six parameters of the rigid transformation, while the other parameters were kept constant. In the second step, all parameters (translation, rotation, scaling, and bending) were considered during optimization. Model fitting took in total between 34 and 184 seconds (SUN SPARC Ultra 2) and succeeded in all cases. Figure 3b exemplarily shows the fitting result obtained for the left frontal ventricular horn. One can see that the fitted model surface well agrees with the ventricle surface.

The localized landmark positions derived from the fitted model turned out to be good. Figures 3c and 4 show the localization results obtained for the tips of the left frontal and occipital horn, resp. Please note that for visualization purposes, the model surfaces as well as the landmark positions are represented by voxel positions only, while the fitting results yield subvoxel positions.

498

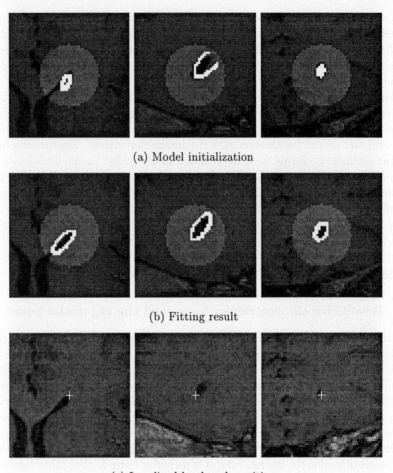

(a) Model initialization

(b) Fitting result

(c) Localized landmark position

Fig. 3. Localization of the tip of the left frontal ventricular horn in a 3D MR image. Orthogonal sections at the ROI center depicting (a) the surface initialization and (b) the fitting result. The considered spherical ROI is highlighted. (c) Orthogonal sections at the localized landmark position based on the fitted model (white cross).

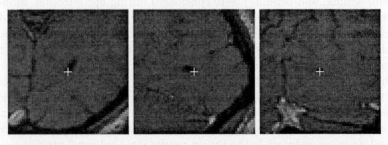

Fig. 4. Localization of the tip of the left occ. ventr. horn in a 3D MR image. Orthogonal sections at the localized landmark position based on the fitted model (white cross).

Zygomatic bones In both modalities, we obtained reasonable initial values for the model parameters. Only the scaling parameters were manually coarsely initialized. In contrast to the experiments using the ventricular horns, we here performed model fitting in a single step in which all parameters were adapted. Model fitting took between 35 and 117 seconds and gave in all cases good results. The localized landmark positions derived from the fitting results are in all cases satisfying as visual inspection revealed. Figure 5 shows the localization results for the saddle point at the left zygomatic bone in MR and CT.

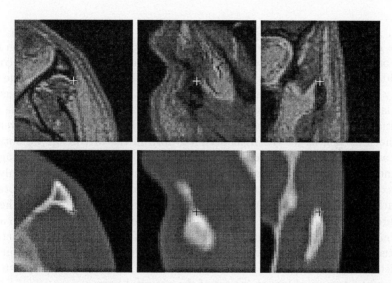

Fig. 5. Localization of the saddle point at the left zygomatic bone in a 3D MR image (top) and a 3D CT image (bottom). Orthogonal sections at the localized landmark position based on the fitted model (white (black) cross in the MR (CT) image).

Localization accuracy We now analyze the localization accuracy of our new approach in the case of the MR image, using as ground truth positions that were manually determined in agreement with up to four persons. Note that manually only voxel positions were determined, while our new approach yields subvoxel positions. For comparison, we use the results obtained with a differential approach ([11],[6]), which was here also used to determine initial estimates of the landmark positions. Table 1 summarizes the computed Euclidean distances to the ground truth positions for six landmarks. One can see that for each landmark, the locus obtained with our new approach based on deformable models is better (i.e., closer to the reference position) than the locus obtained with the differential approach. The mean Euclidean distance from the positions localized with our new approach to the reference positions is $\bar{e}_{new} = 1.22$mm, while for the differential approach we have $\bar{e}_{differential} = 2.11$mm. Thus, the localization accuracy was improved by 0.89mm.

500

	$e_{\text{differential}}$	e_{new}		$e_{\text{differential}}$	e_{new}
Left frontal horn	1.92mm	0.90mm	Right frontal horn	1.72mm	1.28mm
Left occipital horn	3.32mm	1.23mm	Right occipital horn	2.58mm	1.61mm
Left zygomatic bone	0.86mm	0.78mm	Right zygomatic bone	2.26mm	1.52mm

Table 1. Comparison of the localization accuracy of a differential approach to landmark localization ($e_{\text{differential}}$) and our new approach based on deformable models (e_{new}) for six landmarks in a 3D MR image. $e_{\text{differential}}$ and e_{new} denote the Euclidean distances to the reference positions, which were determined manually.

False detections One problem with differential approaches is that often more than one landmark candidate is detected, i.e., we have to ensure that a correct candidate is selected for model initialization. To this end, we studied the suitability of using the fitting results to automatically identify false detections. For each landmark from above, we used *all* detected candidates to determine a set of initial values for the model parameters: For the left and right frontal ventricular horn as well as for the right occipital horn in the MR image we obtained two candidates, while for the left (right) zygomatic bone in the MR image we obtained three (five) candidates. In the case of the other landmarks, only one correct candidate was detected. We then compared the fitting results obtained for each candidate based on the value of the fitting measure divided by the surface area (the normalization was done to avoid a bias due to the surface area). We found that in all cases but one, the selection of a correct candidate actually resulted in the best fitting result. For the right occipital horn, it turned out that the detected two candidates are both correct in the sense that they refer to two prominent anatomical loci at the tip of the occipital horn.

6 Conclusion

In this paper, we introduced a new approach to 3D landmark localization based on deformable geometric models. By fitting these models to the surface at the landmark at hand, we obtain accurate estimates of the 3D landmark positions. Experimental results using 3D MR and CT images demonstrated the applicability of our approach. The utilization of differential approaches for determining initial values for the model parameters resulted in reasonable initializations. Using manually determined positions as ground truth, we analyzed in detail the accuracy of the estimated 3D landmark positions obtained from the fitting results. In particular, we compared the localization results with those of a differential approach. It turned out that our new approach based on deformable models significantly improves the localization accuracy and also reduces the number of false detections. Future work will include further experiments using clinical data. Also, extensions of the geometric models will be studied to capture a broader range of shapes.

References

1. E. Bardinet, L.D. Cohen, and N. Ayache. Superquadrics and Free-Form Deformations: A Global Model to Fit and Track 3D Medical Data. In N. Ayache, ed., *Proc. CVRMed'95*, LNCS 905, pp. 319–326. Springer-Verlag, Berlin, 1995.
2. R. Bertolini and G. Leutert. *Atlas der Anatomie des Menschen. Band 3: Kopf, Hals, Gehirn, Rückenmark und Sinnesorgane.* Springer-Verlag, Berlin, 1982.
3. K. Delibasis and P.E. Undrill. Anatomical object recognition using deformable geometric models. *Image and Vision Computing*, 12(7):423–433, 1994.
4. L.M.J. Florack, B.M. ter Romeny, J.J. Koenderink, and M.A. Viergever. General Intensity Transformations and Differential Invariants. *Journal of Mathematical Imaging and Vision*, 4(2):171–187, 1994.
5. L. Floreby, L. Sönmo, and K. Sjögreen. Boundary Finding Using Fourier Surfaces of Increasing Order. In A.K. Jain, S. Venkatesh, and B.C. Lovell, eds., *Proc. ICPR'98*, pp. 465–467. IEEE Computer Society Press, Los Alamitos, CA, 1998.
6. S. Frantz, K. Rohr, and H.S. Stiehl. Improving the Detection Performance in Semi-automatic Landmark Extraction. In C.J. Taylor and A.C.F. Colchester, eds., *Proc. MICCAI'99*, LNCS 1679, pp. 253–262. Springer-Verlag, Berlin, 1999.
7. A. Kelemen, G. Székely, and G. Gerig. Three-dimensional Model-based Segmentation of Brain MRI. In B. Vemuri, ed., *Proc. IEEE Workshop on Biomedical Image Analysis*, pp. 4–13. IEEE Computer Society Press, Los Alamitos, CA, 1998.
8. T. McInerney and D. Terzopoulos. Deformable Models in Medical Image Analysis: A Survey. *Medical Image Analysis*, 1(2):91–108, 1996.
9. E.H.W. Meijering, K.J. Zuiderveld, and M.A. Viergever. Image Reconstruction by Convolution with Symmetrical Piecewise nth-Order Polynomial Kernels. *IEEE Trans. on Image Processing*, 8(2):192–201, 1999.
10. W.H. Press, B.P. Flannery, S.A. Teukolsky, and W.T. Vetterling. *Numerical Recipes in C.* Cambridge University Press, 1988.
11. K. Rohr. On 3D differential operators for detecting point landmarks. *Image and Vision Computing*, 15(3):219–233, 1997.
12. J. Sobotta. *Atlas der Anatomie des Menschen. Band 1: Kopf, Hals, obere Extremität, Haut.* Urban & Schwarzenberg, München, 19th edition, 1988.
13. F. Solina and R. Bajcsy. Recovery of Parametric Models from Range Images: the Case for Superquadrics with Global Deformations. *IEEE Trans. on Pattern Analysis and Machine Intelligence*, 12(2):131–147, 1990.
14. L.H. Staib and J.S. Duncan. Model-based Deformable Surface Finding for Medical Images. *IEEE Trans. on Medical Imaging*, 15(5):720–730, 1996.
15. D. Terzopoulos, A. Witkin, and M. Kass. Constraints on deformable models: recovering 3D shape and nonrigid motion. *Artificial Intelligence*, 36(1):91–123, 1988.
16. D. Terzopoulos and D. Metaxas. Dynamic 3D Models with Local and Global Deformations: Deformable Superquadrics. *IEEE Trans. on Pattern Analysis and Machine Intelligence*, 13(7):703–714, 1991.
17. J.-P. Thirion. New Feature Points based on Geometric Invariants for 3D Image Registration. *Internat. Journal of Computer Vision*, 18(2):121–137, 1996.
18. M. Unser, A. Aldroubi, and M. Eden. B-Spline Signal Processing: Part I—Theory. *IEEE Trans. on Signal Processing*, 41(2):821–833, 1993.
19. B.C. Vemuri and A. Radisavljevic. Multiresolution Stochastic Hybrid Shape Models with Fractal Priors. *ACM Trans. on Graphics*, 13(2):177–207, 1994.
20. M. Worring, A.W.M. Smeulders, L.H. Staib, and J.S. Duncan. Parameterized feasible boundaries in gradient vector fields. In H.H. Barrett and A.F. Gmitro, eds., *Proc. IPMI'93*, LNCS 687, pp. 48–61. Springer-Verlag, Berlin, 1993.

Fiducial Registration from a Single X-Ray Image: A New Technique for Fluoroscopic Guidance and Radiotherapy

T. S. Y. Tang   R. E. Ellis   G. Fichtinger

Computing and Information Science, Queen's University at Kingston, Canada
CISST/ERC Johns Hopkins University, Baltimore, Maryland, USA
contact: ellis@cs.queensu.ca

Abstract. Fiducial registration is useful both in applications where other registration techniques have poor performance and for validation of new registration techniques. Registration of 3D CT or MR images to 2D X-ray images is particularly difficult, in part because automated contour extraction from 2D images is not yet well solved and in part because of the considerable computational expense in matching the contours to the 3D images.

This work addresses the problem of fiducial registration from a single X-ray image. We have developed an algorithm for fast, efficient registration of 3D fiducial locations to the lines cast from the X-ray source to the 2D projective image that is 60 times faster than the popular iterated closest-point algorithm. The algorithm has been tested on fluoroscopic images from portable C-arms and on portal images from a radiotherapy device. On these images, six or seven fiducials can be registered within seconds to an absolute accuracy of about one millimeter and two degrees.

1 Introduction

Image-guided minimally invasive therapy and conformal radiotherapy can benefit from registration of a patient or tool to a preoperative image. In the case of image-guided therapy, registration of a rigid (or near-rigid) tissue to a preoperative image permits intraoperative guidance; for example, in many fluoro-guided percutaneous procedures the soft target volume is determined in preoperative CT or MR scans but the target is not visible intraoperatively. In the case of conformal radiotherapy and fractionated radiosurgery, it is essential that the patient is registered on the coach of the linear accelerator in the position and orientation that was planned in pre-operative CT and/or MRI images.

These functions can be performed by registering multiple projected X-ray images, e.g., from several positions or with several imaging systems. However, multi-image registration not only increases the ionizing radiation dose to the patient but also requires additional time (with the risk of patient motion occurring between the imaging instants) or additional resources in the form of additional imaging systems (which increases the cost). The ideal registration technique would use an automatic registration of contours or intensities from the intraoperative image. However, automatic procedures are not yet sufficiently fast and robust for reliable therapeutic use so an alternative technique is worthy of examination.

This work addresses the problem of how to estimate the pose of an known object with a single 2D image using fiducial markers. The location of the fiducial markers in the anatomy is known from a preoperative 3D image, and the specific problem is to estimate the pose of the markers in the coordinates of the 2D imaging system. Our pose estimation process has two steps: the calibration of the image and imaging device, and the reconstruction of the pose from the image and the known spatial distribution of the fiducial markers. We will describe an accurate algorithm for estimating the pose by minimizing the distance between the set of 3D fiducial points and the lines that project from the X-ray source to the 2D image sensor. By taking advantage of the geometric constraints inherent in the problem we have developed an algorithm that is more than 60 times times faster than the iterative-closest-point algorithm and that can achieve submillimeter registration accuracy from a single fluoroscopic image.

2 Background

Single-image registration requires a calibrated image and a registration algorithm. Here, we briefly review the main issues and previous work on these issues.

2.1 Image Calibration

We are interested in using a single-X-ray image for registration, concentrating on C-arm fluoroscopy and portal imaging. For portal images only minimal calibration is needed, as the images are practically free of distortion and the machine's internal parameters (such as the source-to-screen distance and pixel size) are readily determined from routine procedures. But it is critical that a C-arm fluoroscope be calibrated properly, as there is a substantial amount of distortion in fluoroscopic images and substantial mechanical deformation of the C-arm beam. Various methods for determining global [10, 12] and local [3, 9] spatial transformations for "unwarping" image distortion have been devised, typically using a grid of lines or points with known spatial coordinates. The deformation for mechanical beams, which is dependent on the C-arm orientation, can be determined beforehand [13]. For finding the internal parameters of the imaging device, X-ray calibration methods [6, 11] and camera calibration methods [5, 14] can be applied.

2.2 Point-Based Image Registration

Point-based 3D/3D registration of a pair of corresponding point sets, also known as the absolute orientation problem, has been well addressed. Arun et al. [1] provided a least-squares solution based on singular value decomposition, and Horn [7, 8] gave closed-form solutions using quaternions and orthonormal matrices respectively.

If the correspondence between points is not known, or if one of the sets is not a point set, the iterative closest point algorithm (ICP) of Besl and McKay [2] can be used. ICP is a general algorithm for registering a point set to a geometric model, which may be a set of points, lines, or surfaces. Typically the model is constructed preoperatively and the data are gathered intraoperatively. ICP can also be accelerated [2] to speed the iteration in regions where the error gradient is smooth.

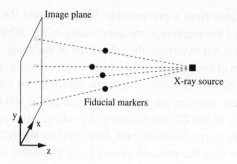

Fig. 1. Imaging geometry.

By reversing the temporal sequence, i.e., by gathering the points preoperatively and constructing a line set model from the intraoperative image, one can use ICP to solve the registration problem. From the X-ray image of the object, which is a 2D point set, a set of *back-projected lines* can be constructed from the 2D image points to the 3D X-ray source location (which is known from the device calibration). ICP can then be applied between this set of 3D lines and the 3D point set of the object.

However, as we will show below, accelerated ICP needs thousands of iterations to converge and often yields incorrect answers as it converges to a local non-global minimum. A main reason for this is that ICP may start with a wrong correspondence as a point could correspond to two or more lines. Even if ICP is modified to force a unique correspondence, results are not satisfactory because of long computation time and high failure rate when the number of points are small. (ICP takes 172 CPU seconds on a SUN Ultra 60 to register just five points, with a failure rate of more than 16%.)

3 Computational Methods

The main weakness of ICP in this application is the incorrect correspondence that is found from the various initial pose estimates. The initial estimate is very important in this application, because a good initial estimate should lead to a correct correspondence and thus the correct pose estimate. The algorithm described here is in two parts. First the algorithm quickly finds a set of initial estimates of the pose, each of which should be very close to the correct estimate; then, for each initial estimate, the algorithm refines the estimate by using iterative gradient descent to minimize the total least-squares distances from the points to the corresponding model lines.

Formally, the registration problem is defined here as
Given:

- An object O that is a set of 3D points $\mathcal{P}^B = \{p_1^B, \ldots, p_n^B\}$ in frame **B**;
- A set of n coplanar image points $\mathcal{J}^C = \{j_1^C, \ldots, j_n^C\}$ in frame **C**;
- An X-ray source location s^C;
- A distance metric $\rho(\cdot, \cdot)$ that measures the distance between a point z and the closest point on line $L(t) = a + tb$; and

– An objective function $F(\cdot)$ of a set of values;

Construct:

the set of back-projected lines from the image points to the X-ray source $\mathcal{L}(t)^C$, where $L_i(t) = j_i^C + t(s^C - j_i^C)$, then

Find:

The rotation R_B^C and translation t_B^C that minimizes $F(\rho)$, where

$$\rho_i = \rho \left(R_B^C p_i^B + t_B^C ,\; j_i^C + t(s^C - j_i^C) \right)$$

is minimized.

The distance metric used here is the squared Euclidean distance between a point z and a line $L(t) = a + tb$, and $F(\rho)$, so finding the registration is a least-squares minimization problem.

3.1 Finding Initial Estimates

We have previously established that the algorithm for finding initial estimates that Besl and McKay originally proposed [2], which was based on symmetries in SO(3) orientations, is inadequate [13].

To find good initial estimates, consider the problem of registering three points to three lines. Suppose there is no symmetry in $\mathcal{P}$, i.e., that by accident or design there is no rotation and/or reflection of $\mathcal{P}$ similar to the original. Pick three points in $\mathcal{J}^C$ that are not collinear and use these three back-projected lines to try to register with all non-collinear 3-permutations of the n points in $\mathcal{P}^B$. The idea is that, for each permutation, we generate the (two) initial estimates by finding the orientation and position of the triangle formed by the three points, say p_i^B for $i = 1 \ldots 3$. For brevity, denote $L_{j_i^C}(t)$ be the line formed by j_i^C and s^C, and let $a_i = L_{j_i^C}(t_i)$ and $a_i' = L_{j_i^C}(t_i')$ for some t_i and t_i'.

Let the triangle with vertices a_i be denoted as $\mathcal{A}$. There are two possible registrations, so there are two such triangles $\mathcal{A}$. The problem is to find some $a_i, i = 1 \ldots 3$, such that the triangle $\mathcal{A}$ is congruent to the triangle $\mathcal{P}$ formed by p_i. To find the orientation of the triangle, observe that if a point a_1' is arbitrarily chosen then there is a triangle $\mathcal{A}'$ that contains a_1' and that is a scaled version of the registration $\mathcal{A}$. The triangle $\mathcal{A}'$ is therefore similar to the triangle $\mathcal{P}$, so

$$\frac{\|p_1 - p_2\|}{\|a_1' - a_2'\|} = \frac{\|p_1 - p_3\|}{\|a_1' - a_3'\|} = \frac{\|p_2 - p_3\|}{\|a_2' - a_3'\|} \tag{1}$$

Two constants, r and s can be defined as

$$r = \frac{\|p_1 - p_2\|}{\|p_1 - p_3\|} = \frac{\|a_1' - a_2'\|}{\|a_1' - a_3'\|} \qquad s = \frac{\|p_1 - p_2\|}{\|p_2 - p_3\|} = \frac{\|a_1' - a_2'\|}{\|a_2' - a_3'\|} \tag{2}$$

Rewriting Equations (1) with the terms from Equation (2) gives two equations of constraint for the points a_2' and a_3':

$$f(t_2', t_3') = r^2 \|a_1' - a_3'\|^2 - \|a_1' - a_2'\|^2 = 0 \tag{3}$$

$$g(t_2', t_3') = s^2 \|a_2' - a_3'\|^2 - \|a_1' - a_2'\|^2 = 0$$

Partial derivatives of these two non-linear functions can be found, so standard optimization methods for solving two equations with two unknowns can be applied. In this work the Levenberg-Marquardt algorithm was used. Because $L_{j_i^C}(t)$ is defined such that $t = 0$ is at the source, and $t = 1$ is on the image plane, all t_i are between 0 and 1. If t_1' is fixed at 0.5, by using all combinations of 0 and 1 as initial values for t_2' and t_3' two solutions were always found.

The registration $\mathcal{A}$ can be determined by scaling the solution $\mathcal{A}'$ so that $\mathcal{A}$ is congruent to $\mathcal{P}$.

Two points, say a_u' and a_v', and the source s form a triangle on the plane containing the two lines $L_{j_u^C}$ and $L_{j_v^C}$. We want to find the triangle $\triangle sa_u a_v$ that is similar to $\triangle sa_u' a_v'$ and $\|a_u - a_v\| = \|p_u - p_v\|$. The two similar triangles yield the constraint

$$\frac{\|s - a_u\|}{\|s - a_u'\|} = \frac{\|a_u - a_v\|}{\|a_u' - a_v'\|} \tag{4}$$

From the line $L_{j_u^C}(t)$, it can be determined that

$$\|s - a_u\| = \|s - L_{j_u^C}(t_u)\| = \frac{\|a_u - a_v\| \cdot \|s - a_u'\|}{\|a_u' - a_v'\|} \tag{5}$$

Expanding this equation and solving for t_u we find

$$t_u = \frac{\|a_u - a_v\| \cdot \|s - a_u'\|}{\|a_u' - a_v'\| \cdot \|s - j_u\|} \tag{6}$$

Using this algorithm, all t_i can be found and thus the simplified problem is solved. We call this algorithm the *3-point-line* algorithm (3PL).

3.2 The 3PLFLS Registration Algorithm

For a 3-permutation $\mathcal{M}^B$ of the point set $\mathcal{P}^B$ there are two sets of approximate solutions, $\hat{\mathcal{R}}_1^C$ and $\hat{\mathcal{R}}_2^C$, that are found from 3PL. Each solution consists of three points that lie on the back-projected lines. The transformations $\hat{T}_{iB}^C$ from $\mathcal{M}^B$ to $\hat{\mathcal{R}}_i^C$ can be determined by any solution of the absolute-orientation problem [1, 7, 8]. Applying transformation $\hat{T}_{iB}^C$ to all points in $\mathcal{P}^B$ transforms the points to $\hat{\mathcal{Q}}^C$. For each transformation, the correspondence is chosen by matching a point to the closest line.

Unlike in ICP, this correspondence is now *fixed* for all subsequent iterations.[1] We call this the *Fixed Least Squares* algorithm, abbreviated as FLS. Using $\hat{T}_{iB}^C$ as the initial estimate, the least-squares error between the points and their corresponding lines is minimized by a Levenberg-Marquardt optimization. The final solution is the correspondence that gives the least RMS error from FLS.

We call the registration algorithm that uses three point-line correspondences as initial estimates, and then fixes the correspondences for subsequent least-squares error minimization, the 3PLFLS algorithm. Figure 2 summarizes this algorithm in pseudocode.

This algorithm is efficient because it is polynomial in the number of markers used. For n points, there are $2n(n-1)(n-2)$ initial 3PL estimates to be computed. Because it would be uncommon to use a large number of fiducial markers, the number of initial estimates is always small (even for 10 markers there are only 1440 initial estimates). Because the optimization algorithm uses fixed correspondences, the computationally expensive nearest-neighbor calculation that tends to dominate ICP calculations is avoided. If the structure of the fiducial points is known beforehand it may be possible to limit the number of Levenberg-Marquardt iterations, so the computation time can indeed be seen to grow cubically with the number of fiducial markers.

4 Experimental Methods

The 3PLFLS algorithm was tested on fluoroscopic and portal images. The fluoroscopic tests were conducted at Kingston General Hospital (Canada) and the portal tests were conducted at Johns Hopkins University (USA). The fluoroscopic tests were performed in conjunction with highly accurate independent 3D detection methods that provided both a relative and an an absolute measurement of registration accuracy. The portal images were evaluated with a relative accuracy measurement only.

4.1 Fluoroscopic Methods

The fluoroscopic tests were conducted on two single-plane C-arms (models BV25 and BV26, Philips). Two objects, plastic models of a femur and tibia, were instrumented with six 0.8 mm tantalum fiducial markers each. The location of the fiducial markers were determined by Roentgen stereophotogrammetry and cross-validated with CT, as we have previously reported [4]; the maximum diameter of the point sets was 72 mm. Arrays of infrared light-emitting diodes (IRED's) were attached to the C-arm and the test object; the markers were contacted with an optically tracked probe (Optotrak 3020, Northern Digital) and Horn's algorithm was used to register the location of the markers to the C-arm coordinate system via the local IRED array.

The C-arms were calibrated using our previously reported methods [13]; the image-to-source distance was found to be approximately 900 mm for each C-arm. Images of each test object were taken and the video signals from the C-arm were digitized directly

---

[1] The original ICP algorithm reselects the correspondence in each iteration.

Algorithm 3PLFLS

$\mathcal{L}^C \leftarrow \{L_{j_i^C} | 1 \leq i \leq n\}$

$\mathcal{L}_3^C \leftarrow \{L_{j_i^C} | i = i_1, i_2, i_3 \text{ where } j_i^C \text{ are not collinear}\}$

for $\mathcal{M}^B$ = permutation of 3 points in $\mathcal{P}^B$ **do**

  if $\mathcal{M}^B$ is collinear **then**

    continue

  end if

  $(\hat{\mathcal{R}}_1^C, \hat{\mathcal{R}}_2^C) \leftarrow 3PL(\mathcal{M}^B, \mathcal{L}_3^C)$

  foreach $\hat{\mathcal{R}}_i^C$ **do**

    $\hat{T}_B^C \leftarrow AO(\mathcal{M}^B, \hat{\mathcal{R}}_i^C)$

    $\hat{Q}^C \leftarrow Transform(\hat{T}_B^C, \mathcal{P}^B)$

    $Cor(\cdot) \leftarrow$ correspondence between $\hat{Q}^C$ and $\mathcal{L}^C$

    $(\tilde{T}_B^C, \epsilon_{rms}) \leftarrow FLS(Cor(\mathcal{P}^B), \mathcal{L}^C, \hat{T}_B^C)$

    if $\tilde{T}_B^C$ has the least RMS error **then**

      $T_B^C \leftarrow \tilde{T}_B^C$

      $BestCor(\cdot) \leftarrow Cor(\cdot)$

    end if

  end for

end for

$(T_B^C, \epsilon_{rms}) \leftarrow FLS(BestCor(\mathcal{P}^B), \mathcal{L}^C, T_B^C)$

$Q^B \leftarrow Transform(T_B^C, \mathcal{P}^B)$

end algorithm

Fig. 2. The 3PLFLS algorithm for finding correspondence and registration. $T_{F_1}^{F_2} = AO(\mathcal{P}^{F_1}, \mathcal{P}^{F_2})$ calculates the transformation using an absolute orientation solution such that $T_{F_1}^{F_2}$ transforms a point set $\mathcal{P}^{F_1}$ in frame F_1 to a point set $\mathcal{P}^{F_2}$ in frame F_2. The function 3PL() calculates the two solutions with the three-point-line back-projection algorithm.

into a PC. The pose of each object was estimated with the accelerated ICP algorithm and with the 3PLFLS algorithm; the ICP algorithm was initialized, as recommended by Besl and McKay [2], from a set of 312 distinct poses. Registration accuracy was tested by using the ground-truth data provided by the optical tracking system.

4.2 Portal-Imaging Methods

Portal-imaging presents significant challenges due to its relatively poor spatial resolution and high X-ray energy. Pixels are large (0.9 mm) and the "hard" beam easily penetrates small steel or tantalum objects, leaving no visible trace in the image.

The portal-imaging tests were conducted on a conventional radiotherapy linear accelerator (Varian Systems). A plastic vertebral phantom was instrumented with sharp metal screws of 9-10 mm length, gradually widening, with 3 mm diameter at the end. We selected screws, rather than spherical objects, to better simulate actual clinical circumstances. The phantom was cased in a rigid transparent plastic box that had a thin cross-hair on the top. This arrangement allowed us to set the phantom box first in the center of the CT gantry, then in the isocenter of the linear accelerator, using the laser set-up mechanisms of the therapy device.

The phantom was placed on the patient table and imaged from a nominal AP view. The gantry was then rotated 45° and imaged again. The phantom was slightly rotated and translated, then the imaging protocol was repeated. The four images were processed by manually selecting the center locations of the markers on the portal-imaging console.

5 Experimental Results

The fluoroscopic results were statistically indistinguishable for the two C-arms tested, so the results were pooled. The ICP algorithm was deemed to have failed if it converged to a point-line correspondence that was known *a priori* to be incorrect. The performance of results calculated by ICP and 3PLFLS are shown in Table 1. Relative and absolute errors of registration using 3PLFLS are shown in Table 2.

Table 1. Performance of ICP and 3PLFLS on fluoroscopic images. n is the number of fiducial markers involved; time is in CPU seconds on a SUN Ultra 60. A registration is considered as *failed* if the correspondence of the fiducial markers is incorrect.

Algorithm	$n = 4$		$n = 6$	
	CPU Time	Failure Rate	CPU Time	Failure Rate
ICP	88.52	36.7%	154.58	0
3PLFLS	0.53	0	2.19	0

Table 2. Accuracy of registration for fluoroscopic images. $\delta\theta$ is the rotation error in degrees, δd is the translation error in millimeters. SD is the standard deviation of the errors. MAX is the maximum error.

Object	RMS Error	Absolute Errors	
		$\delta\theta$ (SD; MAX)	δd (SD; MAX)
Femur	0.10 (0.02/0.12)	0.86 (0.40; 1.75)	1.57 (0.77; 3.29)
Tibia	0.06 (0.02/0.09)	0.76 (0.42; 1.77)	1.40 (0.69; 3.40)

Table 3. Root-mean-square (RMS) errors, in millimeters, and angular errors, in degrees, of registration for portal images of an instrumented lumbar vertebral phantom. For each pose of the phantom a portal image was taken at $0°$ and at $45°$ orientation of the accelerator beam. The relative angular error is the estimated pose difference from $45°$.

Phantom	RMS error, 0° image	RMS error, 45° image	Angular error
Pose 1	0.47	0.43	$+1.44°$
Pose 2	0.46	0.34	$-0.66°$

6 Discussion and Conclusions

We have developed an efficient and effective algorithm for registering a set of 3D points to a single 2D X-ray image. The algorithm is faster than the well known ICP algorithm by more than a factor of 60. The algorithm makes use of geometric constraints that are natural in the problem domain and exploits the combinatorics of point-line matching. Efficacy of the algorithm has been established on two fluoroscopic C-arms and one portal imaging system.

510

The errors in estimating the pose are extraordinarily small, given the difficulty of the problem. The point sets were of only 70 mm in diameter, yet could be registered from a single fluoroscopic image to within 2 mm absolute accuracy for an image-to-source distance of over 900 mm. This result includes all sources of error, including unwarping of the fluoroscopic imaging physics, estimation of the 3D X-ray source location with respect to the image, and estimation of the projected fiducial centroid in the image. In particular, we found that we could estimate the pose within ±0.5 mm in the plane parallel to the image, and that the translational component normal to the image was the primary contributor to the registration error.

This problem is much harder than the problems encountered in virtual fluoroscopic navigation, where an instrument is tracked optoelectronically and then virtually superimposed on the image by performing forward calculations of perspective projection. In virtual navigation the error in locating the X-ray source location are effectively cancelled by perspective, whereas here such cancellation is not inherent in the problem. Cancellation does, however, occur if *two* point sets are tracked. In such a case the relative pose of the objects can be very accurately estimated, particularly in planes parallel to the image.

The algorithm scales as a moderate polynomial in the number of markers. If there are n markers, then the number of triplets of markers to be exhaustively searched is $O(n^3)$ complexity. If a large number of markers is required by the surgical application, alternative registration algorithms (such as geometric hashing [15]) could be considered. However, the purpose of the proposed algorithm is to accurately solve the registration problem for surgical guidance in which there are not many markers. In particular, the algorithm works very well when the number of markers is minimal (four) or nearly minimal.

There are many potential clinical applications of single-image registration. The technique is directly applicable to conformal radiotherapy, where the technique can localize target anatomy that has been previously instrumented. For percutaneous RF ablation of liver, radio-opaque markers can be placed around a tumor and subsequently be used in fluoroscopic guidance of an ablative instrument (the liver parenchyma is relatively stiff and locally moves as a rigid body). For cancer in the pelvic region, the pelvic bone can be instrumented and nearby soft tissues can potentially be treated percutaneously. In these and other applications, the speed and convenience of single-image registration may be able to improve the therapeutic outcome and improve the quality of life of the patient.

Acknowledgments

This research was supported in part by Communications and Information Technology Ontario, the Institute for Robotics and Intelligent Systems, and the Natural Sciences and Engineering Research Council. Fluoroscopic and CT images were acquired with the assistance of the Department of Surgery and the Department of Diagnostic Radiology at Kingston General Hospital, Canada. Portal and CT images were acquired with the assistance of the Department of Radiology and the Department of Radiation Oncology at Johns Hopkins university, USA.

References

1. K. S. Arun, T. S. Huang, and S. D. Blostein. Least-squares fitting of two 3-D point sets. *IEEE Transactions on Pattern Analysis and Machine Intelligence*, PAMI-9(5):698–700, 1987.

2. Paul J. Besl and Neil D. McKay. A method for registration of 3-D shapes. *IEEE Transactions on Pattern Analysis and Machine Intelligence*, 14(2):239–256, February 1992.

3. John M. Boone, J. Anthony Seibert, William A. Barrett, and Eric A. Blood. Analysis and correction of imperfections in the image intensifier-TV-digitizer imaging chain. *Medical Physics*, 18(2):236–242, 1991.

4. R. E. Ellis, S. Toksvig-Larsen, M. Marcacci, D. Caramella, and M. Fadda. Use of a biocompatible fiducial marker in evaluating the accuracy of CT image registration. *Investigative Radiology*, 31(10):658–667, 1996.

5. Keith D. Gremban, Charles E. Thorpe, and Takeo Kanade. Gemoetric camera calibration using systems of linear equation. In *Proceedings IEEE International Conference on Robotics and Automation*, pages 562–567, Philadephia, PA, 1988.

6. Kenneth R. Hoffmann, Jacqueline Esthappan, Shidong Li, and Charles A. Pelizzari. A simple technique for calibrating imaging geometries. In *Medical Imaging: Physics of Medical Imaging*, pages 371–376. SPIE, 1996.

7. Berthold K. P. Horn. Closed-form solution of absolute orientation using unit quaternions. *Journal of the Optical Society of America A*, 4(4):629–642, April 1987.

8. Berthold K. P. Horn, Hugh M. Hilden, and Shahriar Negahdaripour. Closed-form solution of absolute orientation using orthonormal matrices. *Journal of the Optical Society of America A*, 5(7):1127–1135, July 1988.

9. Laurent Launay, Catherine Picard, Eric Maurincomme, René Anxionnat, Pierre Bouchet, and Luc Picard. Quantitative evaluation of an algorithm for correcting geometric distortions in DSA images: applications to stereotaxy. In *Medical Imaging: Image Processing*, volume 2434, pages 520–529. SPIE, 1995.

10. Ewa Pietka and H. K. Huang. Correction of aberration in image-intensifier systems. *Computerized Medical Imaging and Graphics*, 16(4):253–258, 1992.

11. Anne Rougée, Catherine Picard, Cyril Ponchut, and Yves Trousset. Geometrical calibration of X-ray imaging chains for three-dimensional reconstruction. *Computerized Medical Imaging and Graphics*, 17(4/5):295–300, 1993.

12. Stephen Rudin, Daniel R. Bednarek, and Roland Wong. Accurate characterization of image intensifier distortion. *Medical Physics*, 18(6):1145–1151, 1991.

13. Thomas S Y Tang. Calibration and point-based registration of fluoroscopic images. Master's thesis, Queen's University, Kingston, Ontario, Canada, 1999.

14. Roger Tsai. A versatile camera calibration technique for high-accuracy 3D machine vision metrology using off-the-shelf TV cameras and lenses. *IEEE Journal of Robotics and Automation*, RA-3(4):323–344, August 1987.

15. H. J. Wolfson and I. Rigoutsos. Gemoetric hashing: An overview. *IEEE Computational Science and Engineering*, 4(4):10–21, 1997.

Multimodal Non-rigid Warping
for Correction of Distortions in Functional MRI

Pierre Hellier, Christian Barillot

IRISA, INRIA-CNRS unit, Campus de Beaulieu, F-35042 Rennes cedex, France.
e-mail : {phellier,barillot}@irisa.fr

Abstract. This paper deals with the correction of distortions in EPI acquisitions. Echo-planar imaging (EPI) data is used in functional resonance imaging (fMRI) and in diffusion tensor MRI (dMRI) because of its impressive ability to collect data rapidly. However, these data contain geometrical distortions that degrade the quality of the scans and disturb their interpretation. In this paper, we present a fully automatic 3D registration algorithm to correct these distortions. The method is based on the minimization of a cost function (including mutual information and regularization) with a hierarchical multigrid optimization scheme. We present a numerical evaluation on simulated data and results on real data.

keywords: registration, multimodal fusion, EPI distortions, fMRI, unwarping.

1 Introduction

1.1 Context

To explore the brain, on its anatomical and functional sides, different modalities are now commonly used in clinical diagnosis and therapy planning. We distinguish two types of images: anatomical images (MRI, CT, angiography) and functional images (fMRI, PET, SPECT, MEG, EEG). These acquisitions measure different anatomical or physiological properties within the patient that are not redundant but complementary. Therefore, these images must be aligned (registered) so that no information is excluded from the diagnosis and therapeutic processes. To register these images, a rigid (translation, rotation and scaling) transformation is generally sufficient because the volumes are acquired from the same patient (see [13, 14] for tutorials). Nevertheless, we might have to perform non-rigid registration if there are distortions in one acquisition.

Among the functional images of the brain, fMRI is an appealing technique because it offers a good tradeoff between spatial and temporal resolution. To increase its temporal resolution, echo-planar imaging (EPI) is used because it makes possible to collect at least five slices per second at a reduced spatial resolution. The drawback of this impressive acquisition rate is that it may introduce artefacts and distortions in the data. More details about these distortions can be found in [11].

If the distortions do not vary during the time series, they will not affect much the detection of subtle signal changes, but they will perturb the localization of the functional activity once being overlapped to the anatomical volume. It becomes necessary to correct these geometrical distortions in order to accurately identify activated areas.

1.2 Related work

Many research has been made to develop algorithm that perform automatic rigid alignment of multimodal images. Earliest methods rely on the extraction and matching of singular structures (fiducials, curves or surfaces). The major problem is the extraction of the attributes to be matched, and the precision/robustness with respect to this extraction. "Voxel-based" methods are overwhelming "feature-based" methods as they are automatic and more accurate [22]. "Voxel-based" methods rely on the maximization of a similarity measure between two volumes.

Several similarity measures have been proposed: Woods et al. [23] proposed a similarity measure based on the local comparison of $2nd$ order moments. Collignon et al. [3] and Viola et al. [21] proposed simultaneously to use mutual information, a statistical measure of the dependence between two distribution based on information theory. Studholme et al. [18] presented an overlap invariant version of the mutual information. More recently, Roche et al. [17] presented a method based on the maximization of the correlation ratio.

Although many efforts have been made to perform rigid registration, as far as we know, there has been few research concerning non-rigid multimodal registration. Two radically different approaches have been proposed:

The first approach consists in correcting the inhomogeneity field with the phase information [8, 1, 10]. The phase of the raw MR signal (k-space signal) is generally not used since we only need the amplitude to construct the images. The dominant distortion caused by eddy currents may be considered to be a scaling and, in the phase encoding direction, a shear and a translation. This information can then used to correct the recorded signal. These methods require to have the phase information, which is a constrain for almost all MR equipments. Furthermore, these methods are designed to correct only eddy currents-induced artefacts, which is not the only source of distortions.

Another way of considering the problem is the computer vision point of view, where the goal is to design a non-rigid multimodal registration method that can compensate for the EPI distortions. In this category, we distinguish two approaches: Maintz et al. [12] and Gaens et al. [6] proposed an algorithm that seek a non-rigid transformation by maximization of mutual information. They use a "block-matching" minimization scheme with a gaussian filtering of the estimated deformation field to avoid blocky effects. On local windows, the estimation does not take into account the spatial context of the deformation field and only a translation is estimated. Furthermore, these methods are only performed in $2D$.

An interesting approach is described in [7]. This method considers the multimodal registration problem as a monomodal registration problem, and therefore estimates alternatively an intensity correction and a monomodal registration. This method dramatically depends on the intensity correction scheme and for these reasons, the iterative algorithm is not proved to be stable.

1.3 General description of the method

In this paper we propose a method that does not require pre-processing, nor phase acquisition, and we do not estimate any intensity correction. After rigid registration, we estimate a deformation field by minimizing a cost function that is composed of two terms: a similarity measure and a regularization term in order to ensure spatial coherence of the deformation field. We also use a minimization procedure described in [9]. We design a multigrid minimization scheme that is flexible, efficient and simple.

This paper is organized as follows: we describe successively the rigid registration step, the formulation of the problem and the multigrid minimization scheme. Then we present an evaluation of the method on simulated data and results on real data.

2 Non-rigid multimodal registration method

2.1 Rigid registration step

To initialize the algorithm, we perform a rigid registration step. We estimate a rigid transformation that maximizes the mutual information. Given two discrete random variables A and B and their marginal probability distribution $p_A(a)$ and $p_B(b)$, let us note $p_{A,B}(a,b)$ the joint distribution. Mutual information $I(A,B)$ is then defined as [3, 21]:

$$I(A,B) = \sum_{a,b} p_{A,B}(a,b) \frac{p_{A,B}(a,b)}{p_A(a)p_B(b)} = H(A) + H(B) - H(A,B),$$

with

$$H(A) = -\sum_a p_A(a) \log_2(p_A(a)) \text{ and } H(A,B) = -\sum_{a,b} p_{A,B}(a,b) \log_2(p_{A,B}(a,b)).$$

We choose an arbitrary world coordinate system, whose anatomical orientation is known, and in which the center of the axis correspond to the center of the volume, with a voxel size of $1mm$. The transformation T that maps the floating volume B (EPI acquisition) onto the reference volume A (anatomical volume) is estimated in the world coordinate system.

The registration is performed through a multiresolution optimization scheme (construction of a pyramid of volumes by successive isotropic filtering and subsampling in each direction). At each resolution level, the similarity $I(A, T(B))$ is maximized w.r.t. the parameters of the transformation using a Powell's algorithm [16]. We calculate the joint histogram on the overlapping part of A with $T(B)$ by partial volume interpolation, the latter being known as providing a smooth cost function. Let us note T_0 the final rigid transformation.

2.2 An energy-based formulation

To compensate for local geometrical distortions, a $3D$ deformation field w must then be estimated. Let us note T_w the transformation associated with the deformation field w. The total transformation $T_w \circ T_0$ maps the floating volume onto the reference volume A. The field w is defined on S_B, where S_B denotes the voxel lattice of volume B.

In a Bayesian context, we formulate the registration problem with gibbs prior as the minimization of a cost function:

$$U(w; A, B, \mathsf{T}_0) = -I(A, (\mathsf{T}_w \circ \mathsf{T}_0)(B)) + \alpha \sum_{<s,r> \in \mathcal{C}_B} ||w_s - w_r||^2, \quad (1)$$

where $\mathcal{C}_B$ is the set of neighboring pairs of volume B (if we note $\mathcal{V}$ a neighborhood system on S_B, we have: $< s, r > \in \mathcal{C}_B - s \in \mathcal{V}(r)$), and α controls the balance between the two energy terms. The second term is a spatial regularization term that ensures the coherence of the deformation field.

2.3 Multigrid minimization

Motivations The direct minimization of equation (1) is impossible for different reasons: if we estimate iteratively the deformation field on very small regions (the region could eventually be reduced to a voxel), mutual information will be inadequate, because the entropy measure is only meaningful for large groups of voxels. Furthermore, the algorithm will be extremely time-consuming, because the propagation of the regularization will be limited to small regions, and thus very slow. Finally, we need to specify a model to be estimated for the deformation field.

To overcome these difficulties (that are classical in computer vision when minimizing a cost function involving a large number of variables), multigrid approaches have been designed and used in the field of computer vision [5, 15, 19]. Multigrid minimization consists in performing the estimation through a set of nested subspaces. As the algorithm goes further, the dimension of these subspaces increases, and the estimation becomes more and more accurate. In practice, the multigrid minimization usually consists in choosing a set of basis functions and estimating the projection of the "real" solution on the span of the set of basis functions.

Description We use a multigrid minimization based on successive partitions of the initial volume, which is an extension of our previous work [9]. At each grid level ℓ, corresponding to a partition of cubes, we estimate an incremental deformation field dw^ℓ that refines the previous estimation w^ℓ. Let us note T_ℓ the transformation associated with the incremental deformation field dw^ℓ.

At grid level ℓ, $\Xi_\ell = \{\Xi_n, n = 1 \cdots N_\ell\}$ is the partition of the volume B into N_ℓ cubes Ξ_n. A 12-dimension parametric deformation field is estimated on each cube Ξ_n, therefore the total increment deformation field dw^ℓ is piecewise affine. Contrary to block-matching algorithms, we have an interaction between the cubes of the partition, so that we do not have "block-effects" in the estimation.

At coarsest level ℓ_c, the partition is a regular one, with cubes of size $2^{3\ell_c}$. When we change of grid level, each cube is regularly divided and we stop at grid level ℓ_f. Please note that there is no need to have a regular subdivision, it may be adaptive (see [9]) and constrained by functional ROI for instance. The final transformation $\mathsf{T}_{\ell_f} \circ \ldots \circ \mathsf{T}_{\ell_c} \circ \mathsf{T}_0$ expresses the hierarchical decomposition of the deformation field.

Estimation At grid level ℓ and on each cube Ξ_n, we estimate an affine displacement increment defined by the parametric vector Θ_n^ℓ: $\forall s = (x, y, z) \in \Xi_n, dw_s = P_s \Theta_n^\ell$, with $P_s = \mathbb{I}_3 \otimes [1 \ x_s \ y_s \ z_s]$ (operator $\otimes$ denotes the Kronecker product). Let us note

$T_{\Theta_n^\ell}$ the transformation associated with the parametric field Θ_n^ℓ. We have $T_\ell = T_{dw^\ell}$ and $T_{\Theta_n^\ell} = T_{dw^\ell|\Xi_n}$, where $T_{dw^\ell|\Xi_n}$ denotes the restriction of $T_{\Theta_n^\ell}$ to the cube Ξ_n.

A neighborhood system V^ℓ on the partition Ξ_ℓ derives naturally from $\mathcal{V}$ (section 2.2):

$\forall n, m \in \{1 \cdots N_\ell\}, m \in V^\ell(n) - \exists s \in \Xi_n, \exists r \in \Xi_m/r \in \mathcal{V}(s)$. $\mathcal{C}$ being the set of neighboring pairs on S^k, we must now distinguish between two types of such pairs: the pairs inside one cube and the pairs between two cubes:

$\forall n \in \{1 \ldots N_\ell\}, < s, r > \in \mathcal{C}_n^\ell \Leftrightarrow s \in \Xi_n, r \in \Xi_n$ and $r \in \mathcal{V}(s)$.

$\forall n \in \{1 \ldots N_\ell\}, \forall m \in V^\ell(n), < s, r > \in \mathcal{C}_{nm}^\ell \Leftrightarrow m \in V^l(n), s \in \Xi_n, r \in \Xi_m$ and $r \in \mathcal{V}(s)$.

With these notations, at grid level ℓ, the cost function (1) can be modified as:

$$\overset{\star}{U}(\Theta^\ell; A, B, T_0, w^\ell) = -\sum_{n=1}^{N_\ell} I(A, (T_{\Theta_n^\ell} \circ T_{w^\ell} \circ T_0)(B_{|\Xi_n}))$$

$$+\frac{\alpha}{2}\sum_{n=1}^{N_\ell}\left[\sum_{m\in V^\ell(n)}\sum_{<s,r>\in\mathcal{C}_{nm}^\ell}\|(w_s^\ell + P_s\Theta_n^\ell) - (w_r^\ell + P_r\Theta_m^\ell)\|^2\right]$$

$$+\alpha\sum_{n=1}^{N_\ell}\left[\sum_{<s,r>\in\mathcal{C}_n^\ell}\|(w_s^\ell + P_s\Theta_n^\ell) - (w_r^\ell + P_r\Theta_n^\ell)\|^2\right], \tag{2}$$

where $B_{|\Xi_n}$ denotes the restriction of volume B to the cube Ξ_n. The minimization is performed with an ICM algorithm (each cube is iteratively updated while its neighbors are "frozen"). On each cube, Powell's algorithm is used to estimate the parametric affine increment.

3 Results

3.1 Simulated data

To evaluate the multimodal registration method, we use the simulated database of the MNI (Brainweb:http://www.bic.mni.mcgill.ca/brainweb)[4]. The T1-weighted MR volume is the reference volume (3% noise and 9% inhomogeneity), whereas the T2-weighted MR volume is the floating volume.

From the T2-weighted MR volume, we extract a sub-volume and we apply a rigid transformation (3 rotations and 3 translations). To simulate local geometrical distortions, we apply a Thin Plate Spline [2] deformation to the volume. The thin plate deformation is computed by choosing one point in the volume and a displacement for this point. We choose a displacement of magnitude 5 voxels, with no privileged direction. Furthermore, the thin-plate deformation field is constrained to be naught at the border of the volume.

After rigid registration (see figure 1), distortions are clearly visible. On the axial view, ventricles are not well registered ; on the sagittal and coronal view, the sagittal mid-plane is not well aligned. We then perform the multigrid non-rigid registration from grid level 7 until grid level 5 to avoid useless computational efforts. After non-rigid registration, the internal structures are accurately registered (see ventricles on the axial view for instance).

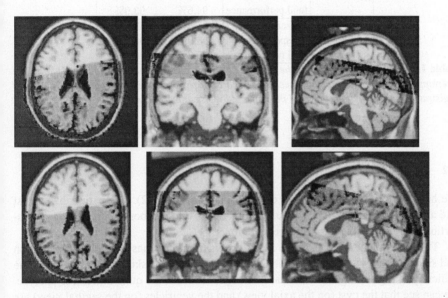

Fig. 1. *Results of the registration on simulated data. Top: results after rigid registration. Distortions are visible on axial view (ventricles) and on coronal view (sagittal mid-plane). Bottom: results after performing a 3D multimodal non-rigid registration. Anatomical structures are well registered, see ventricles on axial view for instance.*

As we have the segmentation of the phantom (grey matter and white matter classes), we can evaluate objectively the registration process. We deform the segmentation volumes as it is described at the beginning of section (3.1). We can assess the quality of the registration by computing overlapping measures (specificity, sensitivity and total performance, see [20] for tutorial) between the initial classes and the deformed classes, once registered with the estimated deformation field. These numerical results are shown on table (1). At the end of grid level 5, we manage to recover up to 95% of the segmentation, which is a satisfactory result because we use only binary classes (and not fuzzy classes) and a simple linear interpolation scheme. Accordingly to the overlapping measures, we verify that the similarity is increasing as the registration process goes further.

Registration	Overlap measure	Grey matter	White matter
Rigid	sensibility	74.7%	76.6%
	specificity	93.0%	92.8%
	total performance	87.0%	87.6%
Non-rigid grid level 7	sensibility	84.7%	86.0%
	specificity	97.2%	96.2%
	total performance	93.2%	92.9%
Non-rigid grid level 6	sensibility	86.6%	86.8%
	specificity	98.5%	97.3%
	total performance	94.6%	93.9%
Non-rigid grid level 5	sensibility	87.5%	87.0%
	specificity	98.9%	98.0%
	total performance	95.8%	95.3%

Table 1. *Numerical evaluation of the multimodal registration method on simulated data. The overlapping measures (specificity, sensitivity and total performance) are computed after rigid registration and at each grid level of the non-rigid registration process.*

3.2 Real data

We have performed the algorithm on real data (see figure (2). The patient has a cyst and a bone tumor, therefore the multiple interfaces (air/cyst/bone) introduce large distortions that are visible after rigid registration.

There are many artafactsin the fMRI acquisition: there has been signal saturation and signal drops (visible in the cyst and in the border of the skull). That illustrates the difficulty of registering clinical data. Although the results are quite difficult to analyze, we can see that the cyst (on the axial view) and the ventricles (on the sagittal view) are better aligned after non-rigid registration.

4   Conclusion

In this paper we presented a new method for $3D$ multimodal non-rigid registration. After rigid registration, we estimate a deformation field with a hierarchical multigrid algorithm. The estimation is performed by minimizing a cost function that is composed of a similarity measure (mutual information) and a regularization term. We have presented results on real data an a numerical evaluation of the algorithm on simulated data.

In the future, we intend to investigate the influence of the similarity measure on the non-rigid registration process. Normalized mutual information [18] and correlation ratio [17] are appealing measure that may give slightly different results. Another perspective is to perform the objective evaluation of this algorithm on a large set of clinical data and study the influence of non-rigid registration on the localization of activated areas.

Acknowledgments. This work has been partly supported by the Brittany Country Council under a contribution to the student grant. The authors would like to thank the SIM laboratory and the Pontchaillou hospital for providing the data.

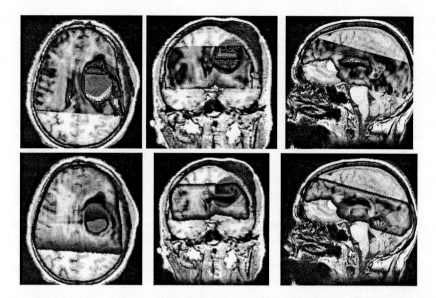

Fig. 2. *Results on real data. Top: results of the rigid registration. The multiple artefacts are visible: distortions, signal saturation, signal drops. Bottom: Results of the non-rigid registration. The registration is more accurate, in particular for the ventricles and for the cyst.*

References

1. M. Bastin. Correction of eddy current-induced artefacts in diffusion tensor imaging using iterative cross-correlation. *Magnetic Resonance Imaging*, 17(7):1011–1024, 1999.
2. F. Bookstein. Principal warps: Thin plate splines and the decomposition of deformations. *IEEE PAMI*, 11(6):567–585, 1989.
3. A. Collignon, D. Vanderneulen, P. Suetens, and G. Marchal. 3D multi-modality medical image registration using feature space clustering. In *Proc. of CVRMed*, pages 195–204, 1995.
4. D.L. Collins, A.P. Zijdenbos, V. Kollokian, J.G. Sled, N.J. Kabani, C.J. Holmes, and A.C. Evans. Design and construction of a realistic digital brain phantom. *IEEE TMI*, 17(3):463–468, 1998.
5. W. Enkelmann. Investigations of multigrid algorithms for the estimation of optical flow fields in image sequences. *CVGIP*, 43(2):150–177, 1988.
6. T. Gaens, F. Maes, D. Vandermeulen, and S. Suetens. Non-rigid multimodal image registration using mutual information. In *Proc. of MICCAI*, pages 1099–1106, October 1998.
7. A. Guimond, A. Roche, N. Ayache, and J. Meunier. Multimodal brain warping using the demons algorithm and adaptive intensity correction. Technical Report 3796, INRIA, http://www.inria.fr/RRRT/RR-3796.html, November 1999.
8. J. Haselgrove and J. Moore. Correction for distortion of echo-planar images used to calculate the apparent diffusion coefficient. *Magnetic Resonance in Medecine*, pages 960–964, 1996.
9. P. Hellier, C. Barillot, E. Mémin, and P. Pérez. Medical image registration with robust multigrid techniques. *Proc. of MICCAI*, pages 680–687, September 1999.
10. M. Horsfield. Mapping eddy current induced fields for the correction of diffusion weighted echo planar images. *Magnetic Resonance Imaging*, 17(9):1335–1345, 1999.

11. P. Jezzard and S. Clare. Sources of distortions in functional MRI data. *Human Brain Mapping*, 8:80–85, 1999.

12. J. Maintz, E. Meijering, and M. Viergever. General multimodal elastic registration based on mutual information. *Proc. of Medical Imaging : Image Processing*, number 3338 in SPIE Proceedings, pages 144–154, April 1998.

13. J. Maintz and MA. Viergever. A survey of medical image registration. *Medical Image Analysis*, 2(1):1–36, 1998.

14. CR. Maurer and JM. Fitzpatrick. A review of medical image registration. In *Interactive image guided neurosurgery*, pages 17–44. American Association of Neurological Surgeons, 1993.

15. E. Mémin and P. Pérez. Dense estimation and object-based segmentation of the optical flow with robust techniques. *IEEE Trans. Image Processing*, 7(5):703–719, 1998.

16. M. Powell. An efficient method for finding the minimum of a function of several variables without calculating derivatives. *The Computer Journal*, pages 155–162, 1964.

17. A. Roche, G. Malandain, X. Pennec, and N. Ayache. The correlation ratio as a new similarity measure for multimodal image registration. In *Proc. of MICCAI*, pages 115–1124, October 1998.

18. C Studholme, D. Hill, and D. Hawkes. An overlap invariant measure of 3D medical image alignment. *Pattern Recognition*, 32:71–86, 1999.

19. D. Terzopoulos. Image analysis using multigrid relaxation methods. *IEEE PAMI*, 8(2):129–139, 1986.

20. JH. Van Bemmel and MA. Musen. *Handbook of medical informatics*. Springer, URL : http://www.mieur.nl/mihandbook, 1997.

21. P. Viola and W. Wells. Alignment by maximisation of mutual information. In *Proc. ICCV*, pages 15–23, 1995.

22. J. West, J. Fitzpatrick, *et al.*. Comparaison and evaluation of retrospective intermodality brain image registration techniques. *Journal of Computer Assisted Tomography*, 21(4):554–566, 1997.

23. R. Woods, JC. Mazziotta, and SR. Cherry. MRI-PET registration with automated algorithm. *Journal of Computer Assisted Tomography*, 17(4):536–546, 1993.

Robust 3D Deformation Field Estimation by Template Propagation

P. Rösch[1], T. Netsch[1], M. Quist[2], G. P. Penney[3], D. L. G. Hill[3], and J. Weese[1]

[1] Philips Research Laboratories, Division Technical Systems,
Röntgenstraße 24–26, D-22335 Hamburg, Germany
[2] MIMIT Advanced Development, Philips Medical Systems Nederland B.V.,
Veenpluis 4–6, NL-5680 DA Best, The Netherlands
[3] Radiological Sciences and Biomechanical Engineering.
The Guy's, King's and St Thomas' School of Medicine,
King's College London, London SE1 9RT, U.K.

Abstract. A new robust method to automatically determine a 3D motion vector field for medical images in the presence of large deformations is proposed. The central idea of this approach is template propagation. Starting from an image position where valid starting estimates are known, small sub-volumes (templates) are registered rigidly. Parameters of successfully registered templates serve as starting estimates for its neighbors. The registration proceeds layer by layer until the relevant image volume is covered. Based on this principle, a template-based registration algorithm has been implemented. Using the resulting set of corresponding points, the parameters of a non-rigid transformation scheme are determined. The complete procedure has been validated using four MR image pairs containing considerable deformations. In order to obtain an estimate for the accuracy, homologous points determined by template propagation are compared to corresponding landmarks defined by an expert. For landmarks with sufficient structure, the average deviation is well below the voxel size of the images. Because of the larger number of homologous points available, transformations incorporating the output of template propagation yielded a larger similarity between the reference image and the transformed image than an elastic transformation based on landmark pairs.

1 Introduction

For both medical and non-medical applications the generation of motion and deformation fields is a prerequisite for the automated analysis of image pairs. In medical applications it is desirable to distinguish between deformations caused by bending joints and moving organs on the one hand and pathologies on the other hand. Examples are following the size and shape of tumors, compensating respiratory motion in cardiac perfusion studies and comparison between pre- and post treatment.

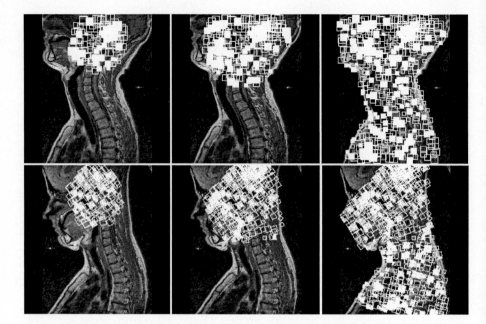

Fig.1 Process of 3D template propagation starting at one location where valid starting estimates are known (left) propagate (center) until the whole volume of overlap is covered with templates (right). The top row shows a slice of the reference image, the bottom row represents a slice of the target image.

One possibility for deformation field estimation is to select a set of corresponding landmarks manually and to determine the motion vectors at the other locations by interpolation [1]. The disadvantage of interpolation functions based on sparse data is that the properties of tissue are not taken into account and that the shapes of rigid bodies (e.g. bones) are affected by the transformation. In order to avoid this, the image can be segmented into parts that are deformed while other parts are kept rigid [2].

The methods described in the following attempt to reduce user interaction by automatically establishing correspondence between image locations. One class of automatic algorithms extract and match crest lines [3] or region boundaries [4]. The procedures yield transformation parameters for the selected feature lines only and the deformation field has to be extended to the whole volume afterwards.

Another method is iterative gray-value based elastic registration where the full image content is used. Motion field estimation is an optimization procedure aiming at the deformation field yielding maximum similarity between the images. Global optimization schemes for elastic registration change the whole deformation field during each refinement step. Thus, a regularization step e.g. by including biomechanical models [5] is essential to avoid local optima resulting from the large number of parameters. Another possibility is to vary a mesh of points controlling a free-form deformation and to introduce a penalty function to constrain the deformation to be smooth [6].

An alternative to global optimization is block-matching. The algorithm presented in [7] starts by registering the whole image rigidly. Afterwards the image is successively split into smaller and smaller portions which are again registered rigidly using the results of the "parent" block as starting estimates. Although this method works for images with relatively small deformations, the presence of large deformations in 3D image pairs (like fig.1) does not allow even for an approximate rigid matching of image parts that are larger than a few cm so that the initial step of this procedure can not be applied in the case of large deformations. The algorithm closest to the approach presented here is the one in [8] where displacement vectors at the nodes of a 3D grid are varied and the correlation of image features in the environment of the nodes is used as similarity measure. Registration starts at a coarse resolution and the resulting translation vectors are refined in subsequent steps at finer resolutions.

Rather than applying a multi-resolution strategy to determine local translation vectors, the procedure described in this contribution establishes a chain of successfully registered small sub-volumes (templates) from an image position with known translation and rotation parameters to the other locations. The volume of the templates is so small that the structures contained are not significantly affected by deformations. Thus, a rigid registration yields valid local translation and rotation parameters at the template center. In contrast to current block matching procedures, the template to be registered next is chosen based on the success of previous registration steps in its neighborhood. As a result, image regions that can not be "trespassed" by template propagation because they do not contain sufficient structure are circumvented. Apart from the propagation strategy, the choice of a similarity measure that is applicable to small volumes (e.g. 5x5x3 voxels) is crucial for the success of this method. Two measures that fulfill these requirements, standard cross-correlation and local correlation (LC) [9] gave similar results for the example images. However, the applicability of cross-correlation is limited to single modality registration, in the case of MR data it is even limited to images acquired with the same protocol. As LC has been successfully applied for single- and multi-modality registration [10] and future evaluations will include MR image pairs with different protocols and images originating from different modalities, only results based on LC are reported here.

The algorithm is described in section 2. Experimental results are given in section 3 and discussed in section 4. Finally, conclusions are drawn in section 5.

2 Algorithm

This algorithm finds an elastic registration in three consecutive steps. First, templates are selected from the reference volume. Secondly, these templates are rigidly registered to the deformed volume by template propagation. Finally, an elastic transformation according to [2] based on the corresponding points originating from the previous step is performed.

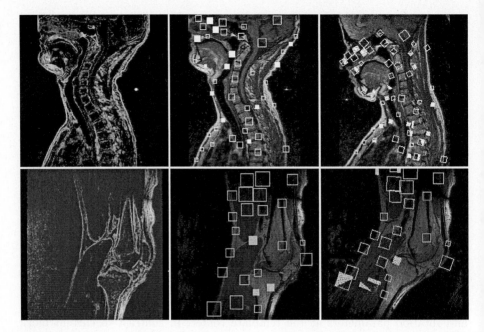

Fig. 2 Local variance distribution (left), subset of selected templates (center) and corresponding positions found by template matching (right) of the neck image (256x256x40 voxels of size 1.25x1.25x2.5 mm) and knee image pair (256x256x125 voxels, 0.98x0.98x1 mm). The figure shows saggital slices of the 3D data sets close to the center.

Template selection addresses the following considerations:

- selection of image features that are relevant for the registration method
- determination of template size such that rigidity may be assumed within the template but that it still contains sufficient structure for the rigid registration process
- homogeneous distribution of templates within the volume of interest

An image feature that is closely related to LC is the local variance. At image location x local variance is defined as $\sigma_L^2(x,r) = \sum_{u \in S(x,r)} (I(u) - \bar{I}_S)^2$ where $S(x,r)$ is a sphere of radius r around x, $I(u)$ is the grey value at position u and $\bar{I}_S$ denotes the average gray value within the sphere. The amount of local variance a template encloses serves as criterion to determine its size. If a template becomes too large for the rigidity assumption, it is rejected. The distance between template centers should be related to the scale of the deformation present in the images. Large deformations might require overlapping templates.

The idea of template propagation is to establish a chain of successfully registered templates from a starting point to the most remote template in the image. The selection of the template to be registered next as well as its starting estimates are based on the distance to previously registered templates and on the confidence in the registration result. As similarity measure for registration as well as for rating the registration result LC [10] is used. A LC value of almost one indicates a successful registration while values approaching zero correspond to poor registrations.

After all selected templates have been registered, a sub-set of template pairs is chosen as input for the elastic transformation scheme. The selection of corresponding points is performed in analogy to the template selection. Instead of a minimum amount of local variance, a minimum LC value is defined. This step removes misleading registration results which go along with small LC values.

3 Experiments and results

The algorithm has been applied to four 3D MR image pairs, three of the human neck and one of the knee. As the results of the neck images are similar, only one neck and the knee example are reported. Image sizes and resolutions are given in the caption of fig. 2. Template selection and registration has been performed with the parameters given in tab. 1. In order to visualize the results of template selection and template matching, the outlines of template volumes are plotted into the image volume (fig. 2). Although it is possible to reveal gross registration errors by visual inspection of template positions in the reference and target image (fig. 2), a quantitative evaluation of registration accuracy requires results obtained by an independent process. For the data sets used here, pairs of corresponding points picked by an expert were available. As the landmark based deformation (fig. 4a) requires corresponding points in all image parts, the expert indicated corresponding points like "anterior thigh" in regions where no salient 3D anatomical points are available. In the following, the validation scheme is described and results are presented.

During template selection, templates are centered at landmark positions in the source image, even if these positions do not meet the selection criterion based on local variance. The remaining templates are selected according to section 2 and template propagation is performed. In order to investigate the accuracy of the algorithm, positions found by the expert and by template propagation have been compared by calculating the Euclidean distance $\|\Delta r\| = \|r_{\text{template,target}} - r_{\text{landmark,target}}\|$. This procedure has been performed for the neck and knee images using 22 and 17 landmarks respectively. Results are compiled in tab. 2.

Finally, corresponding points found by template matching have been used to apply an elastic transformation incorporating rigid structures to the "neck down" image. To allow a comparison with results based on manually selected landmarks [2], information about the rigidly registered vertebra C1–C7 has been added and templates in related areas have been discarded. Transformed images are shown in fig. 4.

Tab. 1 Experimental parameters. Calculation times refer to a 400 MHz SUN UltraSPARC with 512 MB memory. $d_{t,min}$ denotes the minimum distance between template centers, l_{max} is the maximum template size.

name	image	starting estimates given for	$d_{t,min}$ mm	l_{max} mm	number of templates	time for selection min	time for registration min
exp.1	neck	voxel of C1	5	15	3879	6	31
exp.2	neck	voxel of C1	4	25	7844	9	68
exp.3	knee	patella	10	29	564	4	28
exp.4	knee	patella	4	25	5884	32	271

Tab.2 Euclidean distance between positions of corresponding points specified by an expert user and positions resulting from template matching. LC denotes the maximum local correlation value found for a template centered at the landmark position. Parameter settings for the experiments are given in tab. 1

neck landmark name	exp.1 $\|\Delta r\|$ mm	LC	exp.2 $\|\Delta r\|$ mm	LC	knee landmark name	exp.3 $\|\Delta r\|$ mm	LC	exp.4 $\|\Delta r\|$ mm	LC
Fourth ventricle			1.59	0.34	Attach.				
L arytenoid	10.39	0.48	1.21	0.73	patella ten.	1.91	0.53	2.02	0.53
L carotid aorta Jn	4.48	0.47	3.38	0.5	Ant. Tibial a.	13.89	0.39	13.1	0.39
L mandible	1.57	0.53	3.43	0.49	Anterior thigh	1.69	0.52	1.61	0.55
L orbit	0.45	0.77	0.45	0.78	Common				
L straight sinus	4.07	0.3			peroneal n.	1.28	0.32	1.16	0.36
L T5 rib	1.63	0.59	1.81	0.59	Gastrocnemius	7.05	0.23	9.66	0.29
Lower sternum	6.28	0.67	6.56	0.71	Genicular a.	8.8	0.29	2.51	0.54
Mandible	8.36	0.47	4.7	0.57	Patella lig.	1.07	0.48	2.01	0.51
Pituitary	1.74	0.53	1.99	0.5	Patella tendon	7.53	0.3	2.35	0.41
R arytenoid	11.94	0.26	17.19	0.3	Popliteal v.	1.88	0.33	1.85	0.34
R carotid Innom Jn	0.99	0.57	2.25	0.58	Post. calf			4.77	0.26
R mandible	2.04	0.69	9.69	0.44	Quad. femoris	0.75	0.25	4.64	0.24
R orbit	1.13	0.87	1.21	0.87	Quadriceps				
R straight sinus	9.14	0.31	17.6	0.26	fem.	7.53	0.35	2.12	0.41
R T5 rib	0.79	0.59	0.69	0.59	Supf. thigh v.	1.72	0.43	4.11	0.41
Sphenoidal sinus	3.53	0.45	3.5	0.38	Trib. G.				
Sternum	2.31	0.6	2.6	0.61	saphenous v.	1.57	0.54	1.57	0.54
T2 rib left	1.58	0.62	2.18	0.63	Trib.2 G.				
T2 rib right	2.15	0.6	0.95	0.62	saphenous v.	1.51	0.49	1.42	0.52
T2 spinous process	1.15	0.55	0.81	0.55	Trib.3 g.				
Trachea bification	5.52	0.38	5.98	0.38	saphenous v.	1.14	0.58	1.1	0.65

4 Discussion

In this section, results of template selection, template registration and the non-rigid transformation are interpreted. Fig. 2 shows that templates are positioned preferably at borders between different tissue types or at the air/tissue boundary which correspond to large values of local variance. The sizes of templates located in areas with more or less uniform gray values are considerably larger than those templates positioned on tissue borders. This is particularly visible for the knee image where the local variance of muscular tissue hardly exceeds that of the background (fig. 2).

The large amount of deformation present in the neck image is reflected in the variation of local transformation parameters. For example, the rotation angle of templates around the axis perpendicular to the plane of the slices shown in the upper part of fig. 2 varies from $+34°$ at C1 to $-6°$ at the center of the image. In the knee image pair, this rotation angle varies between $+8°$ and $-28°$.

It follows that the propagation of starting estimates is essential for a successful registration of small templates in these cases. Using e.g. starting estimates that are valid for the C1 region for templates at the center of the neck image does not result in successful registrations due to the limited capture range of the optimization procedure. Fig. 2 indicates that the propagation approach led to correct template alignment even in the lung area of the neck image where little structure is present. This visual impression is confirmed by comparing registered templates with manually picked landmarks (tab. 2). Tab. 2 shows that generally $\|\Delta r\|$ decreases with increasing LC value as assumed before. When interpreting the contents of tab. 2 it must be taken into account that the landmarks have not been selected with respect to their suitability for template matching. In accordance with the template propagation procedure, the similarity measure can, however, be used to distinguish successfully registered templates from others. For successfully registered templates, i.e. templates with high LC values $\|\Delta r\|$ is comparable or smaller than one voxel diameter (about 3 mm for the neck and about 1.7 mm for the knee data sets). These deviations are of the same order as typical errors in manual landmark picking. In conclusion, template propagation was successful and produced accurate results.

The landmark "lower sternum" shows a deviation of about two voxel diameters despite its LC value of almost 0.8. Fig. 3 shows enlarged views of this landmark position and its environment. In the coronal and axial slices, motion artifacts in the skin region manifest themselves as a "saw-tooth" pattern. The reason for these artifacts is most likely the regular movement of the chest due to breathing that interferes with the time interval required for the acquisition of an image slice. This explains the registration result obtained for the "lower sternum" landmark: rather than corresponding positions on the sternum, corresponding points with respect to the "saw-tooth" pattern have been detected by template matching.

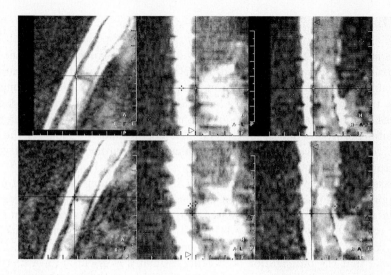

Fig. 3 Motion artifacts at the position of the "lower sternum" landmark in the "head up" image (top) and the "head down" image (bottom). From left to right, sagittal, coronal and axial slices are shown. The cross-hairs indicate corresponding positions found by template matching. Details are given in the text.

A comparison between exp.1 and exp.2 (and between exp.3 and exp.4) shows that the increase of template density and overlap increased the registration accuracy of some templates at the cost of higher computation time. The computation times between 40 min and 300 min are closely related to the number of templates and hence to the large amount of deformations considered by the algorithm.

Finally, we compare the results of two elastic transformations of the "neck down" image according to [2]. The first is based on the manually picked landmarks and the other is based on the corresponding points found by template propagation. Ideally, an elastic transformation of the "head down" image would yield the "head up" image. Thus, the performance of elastic transformation results can be evaluated by an investigation of its differences to the "head up" image. Bright areas in the difference images fig. 4b and fig. 4d indicate locations where misregistration has occured.

The average absolute gray value of the difference image fig. 4d is by about 30% smaller than that in fig. 4b. This indicates an improvement of non-rigid registration results by using template matching rather than manually picked landmarks. The gray values in both difference images at the positions of segmented vertebra are almost zero. This follows directly from the procedure described in [2]. The improvements are particularly evident for the vertebra Th1-Th6 that have not been segmented and for the chest region.

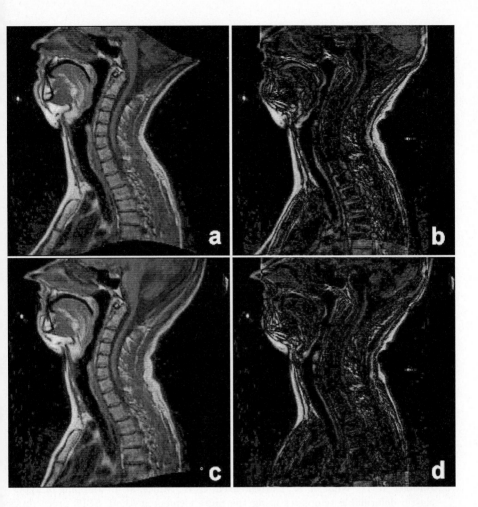

Fig. 4 Result of elastic transformations according to [2] (a,c) and difference images (b,d) between transformed image and the "head up" image. The elastic transformation is based on 22 landmarks picked manually (a,b) and on 178 corresponding points determined by template matching (c,d).

It should be noted that the back of the head is only visible in the "head up" image. It follows that for several templates selected in this image, a corresponding position in the "head down" image does not exist. However, results of template registration have been used as input for deformation estimation. Although the outer shape of the occiput seems to match better in fig. 4c, closer inspection reveals the presence of artifacts due to the different field of view in both images. These artifacts could be avoided easily by restricting the volume of interest to structures present in both images.

5 Conclusions and future work

The performance of a new automatic algorithm for robust 3D deformation field estimation has been demonstrated and tested. Given approximate rigid registration parameters at one image location only, the template propagation paradigm results in a large, evenly distributed set of homologous points between two volumes despite the presence of large deformations. These points can be the input for various elastic interpolation methods to obtain motion field estimations for an image pair. To test template propagation, deviations between corresponding points found by template registration and landmarks picked by an expert have been investigated. The average deviation for successfully registered templates are significantly smaller than one voxel diameter. Results of a registration algorithm which allows pre-segmented rigid bodies to be incorporated into a non-linear deformation have been significantly improved by using homologous points determined by template registration rather than a smaller number of manually picked landmarks. The properties of the LC measure allow for an extension to multi-modality elastic registration which will be addressed in our future work.

References

1. Bookstein, F. L.: Principal warps: Thin-plate splines and the decomposition of deformations, IEEE Trans. Patt. Anal. Mach. Intell **11** (1989) 567–585
2. Little, J. A., Hill, D. L. G., Hawkes, D. J.: Deformations Incorporating Rigid Structures. Computer Vision and Image Understanding **66** (1997) 223–232
3. Declerck, J., Subsol, G, Thirion, J.-P., Ayache, N.: Automatic retrieval of anatomical structures in 3d medical images, Lecture Notes in Computer Science **950** (1995) 153–162
4. D.Davatzikos, J.L.Prince, R.N.Bryan: Image Registration Based on Boundary Mapping, IEEE Trans. Medical Imaging **15** (1996) 112–115
5. Rabbitt, R. D., Weiss, J. A., Chreestensen, G. E., Miller, M. I.: Mapping of hyperelastic deformable templates using the finite element method, SPIE proceedings **2573** (1995) 252–265
6. Rueckert, D. Sonoda, L.I., Hayes, C., Hill, D. L. G., Leach, M. O., Hawkes, D. J.: Nonrigid Registration Using Free-Form Deformations: Application to Breast MR images, IEEE Trans Medical Imaging **18** (1999) 712–721
7. Kostelec, P. J., Weaver, J. B., Healy, D. M. Jr.: Multiresolution elastic image registration, Med. Phys. **25** (1998) 1593-1604
8. Collins, D. L., Holmes, C. J., Peters, T. M., Evans, A. C.: Automatic 3-D Model-Based Neuroanatomical Segmentation, Human Brain Mapping **3** (1995) 190–208
9. Rösch, P., Blaffert, T., Weese, J.: Multi-modality registration using local correlation, in: H. U. Lemke, M. W. Vannier, K. Inamura, A. G. Farman (Eds) CARS'99 Proceedings, Elsevier (1999) 228–232
10. Weese, J., Rösch, P., Netsch, T., Blaffert, T., Quist, M: Gray-Value Based Registration of CT and MR Images by Maximization of Local Correlation, MICCAI'99, Lecture Notes in Computer Science, **1679** (1999) 656–663
11. Studholme, C., Hill, D.L.G., Hawkes, D.J.: Automated 3-D registration of MR and CT images of the head. Med. Image Anal. **1** (1996) 163–175

Tetrahedral Mesh Modeling of Density Data for Anatomical Atlases and Intensity-Based Registration

Jianhua Yao, Russell Taylor

Computer Science Department, The Johns Hopkins University, Baltimore, MD

Abstract:
In this paper, we present the first phase of our effort to build a bone density atlas. We adopted a tetrahedral mesh structure to represent anatomical structures. We propose an efficient and automatic algorithm to construct the tetrahedral mesh from contours in CT images corresponding to the outer bone surfaces and boundaries between compact bone, spongy bone, and medullary cavity. We approximate bone density variations by means of continuous density functions in each tetrahedron of the mesh. Currently, our density functions are second degree polynomial functions expressed in terms of barycentric coordinates associated with each tetrahedron. We apply our density model to efficiently generate Digitally Reconstructed Radiographs. These results are immediately applicable as means of speeding up 2D-3D and 3D-3D intensity based registration and will be incorporated into our future work on construction of atlases and deformable intensity-based registration.

1. Introduction and Background

One of the most critical research problems in the analysis of 3D medical images is the development of methods for storing, approximating and analyzing image data sets efficiently. Many groups are developing electronic atlases for use as a reference database for consulting and teaching, for deformable registration-assisted segmentation of medical images and for use in surgical planning. Researchers at INRIA have built atlases based on surface models and crest lines [1-3]. Cutting *et al* [4] have built similar atlases of the skull. These atlases are surface models with some landmarks and crest lines and don't contain any volumetric density information. Chen [5] and Guimond [6] have built an average brain atlas based on statistical data of the voxel intensity values from a large group of MRI images. Their atlases only have intensity information and don't describe the structural information of the anatomy. Pizer *et al* have proposed a medial model representation called M-rep to represent the shapes of 3D medical objects [7, 8]. A CSG scheme allows M-reps to describe complicate shapes. One advantage of this approach is that it is easy to deform medial atoms to accommodate shape variations. One drawback is that the current representational scheme is primarily useful for exterior shapes, and more work will be needed to extend this work to volumetric properties.[1]

One goal of our current research is to construct a deformable density atlas for bone anatomy and to apply the atlas to different applications. A related goal is to provide an efficient representation for 2D-3D and 3D-3D intensity based registration. Our intent is to use intensity-based deformable registration methods both in the

---

[1] The authors have had very useful discussions with Dr. Pizer about the issues involved, and are looking forward to exploring these issues with the UNC group.

construction of the atlas and in exploiting it for patient-specific procedure planning and intraoperative guidance. It may be possible in many cases to dispense with patient CT in favor of radiographs. The atlas will provide a basis for representing "generic" information about surgical plans and procedures. The atlas will provide infrastructure for study of anatomical variation. The techniques should be extendable to a wide variety of bony anatomies. It may also provide an initial coordinate system for correlating surgical actions with results, as well as an aid in postoperative assessment.

This atlas should include: 1) model representations of "normal" 3D CT densities and segmented surface meshes; 2) 3D parameterization of surface shape & volumetric properties; and 3) statistical characterization of variability of parameters.

We adopt a tetrahedral mesh model to represent volumetric properties for several reasons. Tetrahedra provide great flexibility and other representations can be converted into tetrahedral meshes relatively easily. Tetrahedra are easy to deform and are easy to compute with. It is convenient to assign properties and functions to the vertices and tetrahedra. Computational steps such as interpolation, integration, and differentiation are fast and often can be done in closed form. Finite element analysis is often conveniently performed on tetrahedral meshes.

This paper reports current progress in building suitable tetrahedral density models from patient CT scans and illustrates the usefulness of this representation in performing an important calculation in 2D-3D registration. In section 2, we elaborate the method to construct the tetrahedral mesh from bone contours. We outline the key steps of the method, including contour extraction, tetrahedron tiling, branching and continuity preserving. In section 3, we present the method to compute and assign the density function to the tetrahedron and evaluation the accuracy of the density function. Then in section 4, we show the results using our density model to generate Digitally Reconstructed Radiographs (DRRs). Finally section 5 discusses current status and future work.

2. Construction of Tetrahedral Mesh Models from Contours

There are several techniques to construct the tetrahedral meshes. The easiest way first divides the 3D space into cubicle voxels that can be easily divided into four tetrahedra [9]. The drawback of this method is that the mesh generated is far too dense and doesn't capture any shape property of the model. A variation of this method first merges similar voxels into larger cubes or uses oct-tree techniques to subdivide the space, then performs the tetrahedronization. 3D Delaunay triangulation is another method. But this method is time consuming and requires some post-processing since the basic algorithm produces a mesh of the convex hull of the anatomical object. Boissonnat *et al* [10] proposed a method to construct tetrahedral mesh from contours on cross section. They computed 2D Delaunay triangulation on each section, then tiled tetrahedral mesh between sections and removed external. They also solved complex branching problem by contour splitting. But they didn't consider the intercrossing and continuity between tetrahedra, so their tetrahedral mesh may not be valid for volumetric analysis.

Our tetrahedral mesh construction algorithm is derived from the surface mesh reconstruction algorithm of Meyers, *et al* [11]. The tetrahedral mesh is constructed slice by slice and layer by layer based on bone contours extracted in a separate algorithm. The algorithm is straightforward and fast. The running time is $O(n)$, where *n* is the total number of vertices in the model. We only sketch the algorithm here. More details can be found in our research report [12].

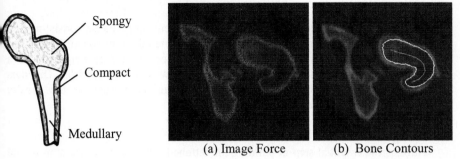

(a) Image Force   (b) Bone Contours

Figure 1. Bone Structure

Figure 2   Contour Extraction

2.1 Contour Extraction

Bones contain two basic types of osseous tissue: compact and spongy bone [13]. Compact bone is dense and looks smooth and homogeneous. Spongy bone is composed of little beams and has a good deal of open space. The shaft of long bones is constructed of a relatively thick collar of compact bone that surrounds a medullary cavity. Short, irregular and flat bones share a simple design: they consist of thin plates of compact bone on outside and spongy bone within. Figure 1 illustrates this structure.

We extract both the outer contour and the inner contour of the bone. If the shell between contours is too thin, only outer contour is extracted[2]. In our algorithm, we apply a deformable "snake" contour model to extract the bone contours [14]. The forces that drive the deformable contour model can be expressed as: $F = F_{internal} + F_{image} + F_{external}$, where $F_{internal}$ is the spline force of the contour. F_{image} is the image force generated from the iso-value. And $F_{external}$ is an external force to allow the contour shrink or expand relative to the medial axis of the contour. Provided the initial value of the contour, the algorithm can automatically converge to the contour of the bone. The algorithm can also be applied repeatedly using the outer contour as the initial value to find the inner contour. It should be mentioned that the users only need to provide the initial contour for the first slice. For any following slice, the contour generated on previous slice can be used as the initial value. So our method is highly automatic. Figure 2a shows the image force used in our algorithm to extract the inner and outer contour of the bone.

The tetrahedral model is a solid model. So after obtaining the contour of the bone, we also compute the medial axis of the inner contour (or of the outer contour if no inner contour has been extracted) and treat it as the innermost layer. Currently, we use Lee's method to extract the medial axis [15]. Hence we can tile the medial axis and the bone contour to build a solid model.

The contours are fitted into B-spline curves and re-sampled at desired resolution for the follow up tiling. During sampling, the segments with larger curvature are

---

[2] In the current implementation, this decision is made manually on a slice by slice basis, but it will be easy to automate.

sampled at higher resolution, while those with smaller curvature at lower resolution. Figure 2b shows the contours and the medial axis extracted by our algorithm.

2.2 Data Structure of Tetrahedral mesh

We chose a simple data structure to represent the tetrahedral mesh, consisting of a list of vertex coordinates and a list of tetrahedra. Each tetrahedron contains links that reference its four vertices and the four face neighbors. The face neighbors of a tetrahedron are those tetrahedra share a common face with this tetrahedron. This data structure is easy to maintain and update. Other information such as the density function is also stored in the tetrahedron.

2.3 Tiling

Tiling is the essential step in building the tetrahedral mesh. The idea is to divide the space between adjacent slices into tetrahedra by the aid of the bone contours, then connect the tetrahedra into a mesh. The tiling operations are based on local information, so the algorithms are fairly efficient even on large data set. The tiling can be expressed as the traversal of four ordered contours on two adjacent slices (two on each slice). At each step, one or two contours advance. Once the contours advance, new tetrahedra are generated. The tiling continues until all contours are traversed.

2.3.1 Tiling Patterns: There are 32 distinct tiling patterns for each single step [12]. There are two categories of the tiling patterns: advance-one-contour and advance-two-contours. Figure 3 shows two tiling pattern examples. In order to understand the tiling pattern, we give some definitions. In Figure 3, the tiling happens between adjacent slice N and slice N+1. The current vertices of the contours are called the front vertices. The next vertices of the front vertices are called the candidate vertices for next advance. In Figure 3a, $a1, b1, c1, d1$ are the front vertices and $a2\ b2\ c2\ d2$ are the candidate vertices of contour $a\ b\ c\ d$ respectively. The front faces are those triangles composed by the front vertices. The front faces are the connections between the tetrahedra generated in last tiling and those to be generated in current tiling. In Figure 3a, triangle $a1b1c1$ and triangle $b1c1d1$ are the front faces. The pivot vertices are those vertices chose for current tiling. We can have one or two pivot vertices in each tiling. The selection of pivot vertices determines the tiling pattern. In Figure 3a, $b2$ and $d2$ are the pivot vertices. And in Figure 3b, a2 is the pivot vertex. In general, the pivot vertices and the front faces together decide the tiling pattern. In Figure 3a, the tiling pattern produces three new tetrahedra. They are $b2a1b1c1$, $b2b1c1d1$ and $b2c1d1d2$ (we denote the tetrahedron by its four vertices).

There are totally 32 tiling patterns according to the selection of pivot vertices and the fact of the front faces. Among these, $\binom{4}{2} \times 2 \times 2 = 24$ cases are advance-two-contours patterns and $\binom{4}{1} \times 2 = 8$ cases are advance-one-contour patterns.

2.3.2 Metric Functions: As shown in last section, we have 32 tiling patterns. In order to choose the best pattern for tiling, a metric function must be evaluated on each candidate pattern. And we find out that most metric functions used in surface tiling are also good for tetrahedron tiling [11]. The metric function used in our algorithm is a combination of minimizing span length, matching direction and matching normalized arc length. Details of the metric functions are in our research report [12].

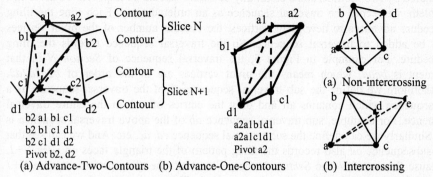

b2 a1 b1 c1
b2 b1 c1 d1
b2 c1 d1 d2
Pivot b2, d2
(a) Advance-Two-Contours

a2a1b1d1
a2a1c1d1
Pivot a2
(b) Advance-One-Contours

Figure 3. Tiling between slices

(a) Non-intercrossing

(b) Intercrossing

Figure 4. Intercrossing between tetrahedra

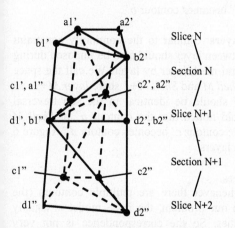

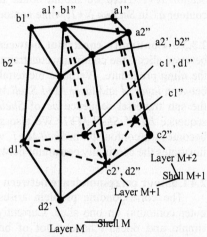

Figure 5. Continuity constraint between slices
Traversal sequence of section N is *b2c2a2d2*, and that of section N+1 is *d2a2b2c2*. (For clarity, all internal crossing edge are omitted)

Figure 6. Continuity constraint between layers
Traval sequence of Shell M is *b2c2a2d2*, and that of Shell M+1 is *d2c2a2b2*. (For clarity, all internal crossing edge are omitted)

2.3.3 Constraints: The tiling problem seems very unconstrained due to the 32 tiling patterns we mentioned in section 2.3.1. But in order to form a reasonable tetrahedral mesh, some constraints must be imposed.

2.3.3.1 **No intercrossing between tetrahedra:** This constraint is obvious for a valid subdivision of the volume. The case that causes intercrossing is shown on figure 4. Figure 4.a shows a non-intercrossing case. But in Figure 4.b, line *ae* pierces triangle *bcd* and causes the intercrossing of two tetrahedra *abcd* and *abce*. So when we select a tiling pattern, those newly generated tetrahedra shouldn't intercross with each other and shouldn't intercross with the old tetrahedra.

2.3.3.2 **Continuity constraint between slices:** On a valid tetrahedral mesh, all triangular faces except those on the boundary should have two neighbor tetrahedra. Because we build the mesh slice by slice, the triangle faces between sections should be shared by two sections and a continuity constraint should be imposed. To solve this problem, we record the traversal sequence as an ordered array. Each time the tiling procedure advances to new pivot vertices, the contour number of the pivot vertices will be added to traversal sequence. So the traversal sequence records the tiling procedure. For example in Figure 5, the traversal sequence of *Section N* in that segment is *bcad*, which means the pivot vertices are in the order of *b2c2a2d2*. Furthermore we define the sub traversal sequence *ab* of the traversal sequence as a subsequence which contains all and only the entries of *a b* in the entire traversal sequence. For example, sub traversal sequence *ab* of the above traversal sequence is *ba*. Similarly we can define the sub traversal sequence *cd, ac*, etc. And we can see that the subsequence *cd* also records the tiling pattern of the triangle faces on *slice N+1*. Because *Section N+1* and *Section N* share *slice N+1*, the tiling pattern on *slice N+1* should be preserved while tiling *Section N+1*. This means that the sub traversal sequence *cd* of *Section N* should be identical to the sub traversal sequence *ab* of *Section N+1*. Here we should notice that the contour *c'* in *Section N* becomes the contour *a"* in *Section N+1*, while contour *d'* becomes contour *b"*.

2.3.3.3 **Continuity constraint between layers**: Similar to the continuity constraint between slices, the continuity constraint between layer should also be imposed during the tiling procedure. We build the tetrahedral mesh layer by layer. We call the space between *layer M* and *layer M+1 Shell M*. *Shell M* and *Shell M+1* share *layer M+1*. So the sub traversal sequence *ac* of *Shell M* should be identical to the sub traversal sequence *bd* of *Shell M+1*. We also should notice that the contour *a'* in *Shell M* becomes the contour *b"* in *Shell M+1*, while contour *c'* becomes contour *d"*. Figure 6 illustrates the continuity constraint between layers.

2.4 Contour correspondence between slices

The corresponding problem arises whenever there are multiple contours (the outer contours) on one slice. Currently in our problem, the bone model is usually simple and doesn't have a lot of branches. So the correspondence is not very complicate. We solved the correspondence problem by simply examining the overlap and distance between contours on adjacent slices.

2.5 Branching problems: Problems associated with branching structures have been studied intensively in surface mesh construction [11]. Some of the techniques are suitable for tetrahedral mesh construction. There are two kinds of branching problems: branching between layers and branching between slices.

2.5.1 Branching between layers: The layer branching occurs when the numbers of layers of corresponding contours between adjacent slices are different. The branching between layers can be easily converted to the tiling of three contours (one on one slice, two on the other slice), which is a special case of the tiling of four contours.

Branching between layers usually happens at the boundary between spongy bone (2 layers) and compact bone (3 layers).

2.5.2 Branching between slices: The branching between slices occurs when the numbers of contours between adjacent slices are different. We adopt the methods used in the surface mesh construction [11]. We construct a composite contour that connects the adjacent contours at the closest points. (Figure7a). This composite contour is then tiled with the single contour from the adjacent slice. In Figure 7, we show the case of *1:2* branching. This method can handle more general cases, i.e. the *m:n* branching cases. In the general case, the method can produce composite contours for both slices and then conduct the standard tiling procedure.

2.6 Results

Figure 8 shows wire frame renderings of two tetrahedral mesh models produced by our method. Figure 8.a is a femur model. Figure 8.b is a half pelvis model. Table 1 lists some facts about these two models.

Model	Num of Vertices	Num of Tetrahedra	Num of Slices	Num of Voxels inside	Volume (mm^3)	Area (mm^2)
Femur	1400	7815	46	859503	144720	17593
Pelvis	3887	18066	89	873494	497664	83453

Table 1. Facts about two tetrahedral model

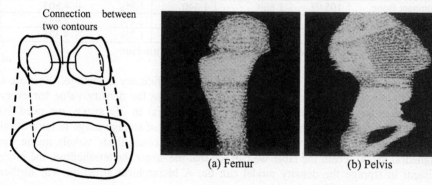

(a) Femur   (b) Pelvis

Figure 7 Branching between slices

Figure 8. Tetrahedral Mesh Model

3. Density Function

We assign an analytical density function to every tetrahedron instead of storing the density value of every pixel in the model. The advantage of such a representation is that it is in explicit form and is a continuous function in 3D space. So it is easy to compute the integral, to differentiate, to interpolate, and to deform.

We build our density function in the barycentric coordinate base. The definition of barycentric coordinate in a tetrahedron is as following. For any point K inside a tetrahedron $ABCD$, there exists four masses w_A, w_B, w_C, and w_D such that, if placed at the corresponding vertices of the tetrahedron, their center of gravity (barycenter) will coincide with the point K. (w_A, w_B, w_C, w_D) is called the barycentric coordinate of K.

Furthermore barycentric coordinates are in a form of homogeneous coordinates where $w_A+w_B+w_C+w_D=1$, and the barycentric coordinate is defined uniquely for every point.

Currently we define the density function as a quadratic polynomial. For each tetrahedron, we first get a sample of the pixels inside the tetrahedron, and obtain the pixel density via the CT data set, then fit a polynomial function of the density in the barycentric coordinates of those sampled pixels. After getting the density function, we can compute the density of any point inside the tetrahedron by their barycentric coordinate. In our initial experience quadratic polynomials have worked well, although we are considering the use of higher order polynomials in barycentric Bernstein form with adaptively simplified tetrahedral meshes.

In order to test the accuracy of the density function, we randomly chose sample points $\{p_i\}$ inside the tetrahedron. The number of sample points of each tetrahedron for testing is the same as the number used to generate the density function. Then we compute the densities $\{c_i\}$ of $\{p_i\}$ by interpolation using CT data set and compute the densities $\{d_i\}$ of $\{p_i\}$ using the density function of the tetrahedron. Table 2 lists the comparison of $\{d_i\}$ and $\{c_i\}$. We present the results in three bone categories: compact bone, spongy bone and medullary cavity. We got the results from the femur density model we built in last section. From the results we can learn that the density distributions in compact bone and medullary cavity area are more homogeneous.

	Avg Density $Avg(c_i)$	Avg Density Diff $Avg(|c_i\text{-}d_i|)$	Std Dev $(|c_i\text{-}d_i|)$	Avg Density Diff $Avg(|c_i\text{-}d_i|/c_i)$	Max Density Diff $Max(|c_i\text{-}d_i|)$
Compact Bone	105.98	1.869	1.440	1.9%	4.803
Spongy Bone	77.9	2.309	2.046	2.4%	7.437
Medullary Cavity	70.956	1.783	1.354	1.3%	4.325

Table 2. Accuracy of Density Function

The other advantage of our density model is its efficiency in storage. We store a density function for each tetrahedron instead of storing the density value for every voxel inside the tetrahedron. For quadratic polynomial in 3D space, we need 10 parameters to describe it. Table 3 gives the comparison of storage usage in tetrahedral density model and regular CT image (here we only count those voxels inside the tetrahedral mesh). From the table we can see that the larger the tetrahedron, the more efficient in storage the density model can be. A hierarchical structure will further improve the storage efficiency.

	Num of Tetra	Avg Volume (mm³)	Avg Num of voxels inside	Storage per tetra (bytes)	Avg Storage in CT image (bytes)
Compact Bone	5110	13.958	75.6	Min 72	151
Spongy Bone	2504	26.745	181.1	Max 84	362
Medullary Cavity	164	34.228	200.4	Avg 75	401

Table 3. Storage of Tetrahedral Density Model

4. Computing DRRs from Tetrahedral Mesh Density Model

We can employ our tetrahedral mesh density model to support fast computation of DRRs. When a ray passes through a 3D data set, it may go though hundreds of voxels but only a few tetrahedra. Because the density function is in an explicit form, the integral along a line can be computed in close form. Furthermore the neighborhood information is stored in the mesh, so it is fast to get the next tetrahedron hit by the ray from current hitting tetrahedron. And it is also fast to find the entry point of the casting ray from the neighborhood information. In contrast generating DRR from CT data require computing partial voxel crossing, which is a time consuming operation. So we have an efficient way to generate the DRR from our tetrahedral density model, which is an important technique in 2D-3D intensity-based registration. The computation can be even faster if we have a hierarchical model to represent different level of details.

	Running time	Avg. elems Passed through	Avg Density	Avg Density Diff $Avg(|c_i-d_i|/c_i)$	Std Dev of Density Diff $(|c_i-d_i|/c_i)$
CT Data set	45 s	65.6 voxels	142.5	4.9%	3.7%
Density Model	14 s	18.3 tetras	142.3		

Table 4. Comparison of DRR using CT data set and Tetrahedral Density Model

Figure 9a is a DRR generated from a CT data using the ray casting algorithm. Figure 9b is a DRR generated from our tetrahedral density model. The image size is 512*512. Table 4 shows the comparison of the two DRRs in Figure 9. We compare the pixel values $\{c_i\}$ of the DRR generated from CT data and the pixel values $\{d_i\}$ of the DRR generated from density model (We only compare those non-zero values). We find out that the running time for an un-optimized implementation using density model is only about 1/3 of the running time using CT data set. Figure 10 illustrates the tetrahedra passed by a ray.

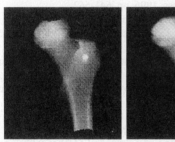 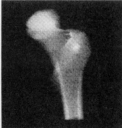 

(a) DRR from CT   (b) DRR from density model

Figure 9. DRR of femur

Figure 10. Tetrahedra passed by a ray

5. Discussions and Future work

In this paper we have presented the first phase of our density atlas research. We proposed a very efficient and automatic method to construct the tetrahedral mesh from bone contours. We assigned a density function to the tetrahedron and showed that it is reasonably accurate. We also presented the fast generation of DRRs from our density mesh, which is an essential technique for 2D-3D intensity based registration and volume visualization.

We will continue our work on building the density atlas. We are investigating the technique to simplify the tetrahedral mesh based on face collapsing and build a hierarchical data structure to represent multiple levels of details of the density model

[9]. We will also investigate the deformation rule of the tetrahedral mesh and make the density atlas deformable. We will build an average density model from a large group of data by incorporating the statistical information in the model. And we will apply finite element technique to study the biomedical and bio-mechanical properties of the anatomy. Ultimately we will apply the density atlas to various application such us 2D-3D registration, surgical planning etc.

Acknowledgements

This work was partially funded by NSF Engineering Research Center grant EEC9731478. We thank Integrated Surgical System (Davis, CA) for providing the femur data and the Shadyside Hospital for the pelvis data. We specially thank Stephen Pizer, Sandy Wells, Branislav Jaramaz, Robert VanVorhis, Jerry Prince, and Chengyang Xu for their useful suggestions. Some of their suggestions have been included in this paper; more of them are still under investigation.

References

[1] G. Subsol, J.-P. Thirion, and N. Ayache, "First Steps Towards Automatic Building of Anatomical Atlas," INRIA, France 2216, March, 1994.

[2] G. Subsol, J.-P. Thirion, and N. Ayache, "A General Scheme for Automatically Building 3D Morphometric Anatomical Atlas: Application to a Skull Atlas," INRIA 2586, 1995.

[3] J. Declerck, G. Subsol, J.-P. Thirion, and N. Ayache, "Automatic Retrieval of Anatomical Structures in 3D medical Images," INRIA, France 2485, Feb. 1995.

[4] C. B. Cutting, F. L. Bookstein, and R. H. Taylor, "Applications of Simulation, Morphometrics and Robotics in Craniofacial Surgery," in *Computer-Integrated Surgery*, R. H. Taylor, *et al*, Eds. Cambridge, Mass.: MIT Press, 1996, pp. 641-662.

[5] M. Chen, T. Kanade, *et al*, "3-D Deformable Registration of Medical Images Using a Statistical Atlas," Carnegie Mellon University, Pittsburgh, PA CMU-RI-TR-98-35, December, 1998.

[6] A. Guimond, J. Meunier, and J.-P. Thirion, "Average Brain Models: A Convergence Study," INRIA, France 3731, July, 1999.

[7] S. M. Pizer, A. L. Thall, and D. T. Chen, "M-Reps: A New Object Representation for Graphics," Univ. of North Carolina, Chapel Hill 1999.

[8] S. M. Pizer, D. S. Fritsch, *et al*, "Segmentation, Registration, and Measurement of Shape Variation via Image Object Shape," *IEEE Trans. Medical Imaging*, 1999.

[9] I. J. Trotts, B. Hamann, and K. I. Joy, "Simplification of Tetrahedral Meshes with Error Bounds," *IEEE Transactions On Visualization and Computer Graphics*, vol. 5, pp 224-237, 1999.

[10] J.-D. Boissonnat and B. Geiger, "Three dimensional reconstruction of complex shapes based on the Delaunay triangulation," INRIA, France 1697, May, 1992.

[11] D. Meyers, S. Skinner, and K. Sloan, "Surfaces from Contours," *ACM Transactions on Graphics*, vol. 11, pp. 228-258, 1992.

[12] J. Yao, "Building Tetrahedral Mesh Model from Bone Contours," ERC, Johns Hopkins Univ. ERC-2000-01, 2000.

[13] E. N. Marieb, *Human Anatomy and Physiology*, second edition, 1992.

[14] M. Kass, A. Witkin, and D. Terzopoulos, "Snakes: Active Contour Models," *International Journal of Computer Vision*, pp. 321-331, 1988.

[15] D. T. Lee, "Medial Axis Transformation of a Planar Shape," *IEEE Transactions On Pattern Analysis and Machine Intelligence*, vol. 4, pp. 363-369, 1982.

Nonrigid Registration of 3D Scalar, Vector and Tensor Medical Data[*]

J. Ruiz-Alzola[1,2], C.-F. Westin[2], S.K. Warfield[2], A. Nabavi[2,3], and R. Kikinis[2]

[1] Dep. Señales y Comunicaciones. University of Las Palmas de Gran Canaria, Spain
[2] Dep. Radiology. Harvard Medical School and Brigham & Women's Hospital, USA
[3] Dep. Neurosurgery. University of Kiel, Germany
{jruiz,westin,warfield,arya,kikinis}@bwh.harvard.edu

Abstract. New medical imaging modalities offering multi-valued data, such as phase contrast MRA and diffusion tensor MRI, require general representations for the development of automatized algorithms. In this paper we propose a unified framework for the registration of medical volumetric multi-valued data. The paper extends the usual concept of similarity in intensity (scalar) data to vector and tensor cases. A discussion on appropriate template selection and on the limitations of the template matching approach to incorporate the vector and tensor reorientation is also offered. Our approach to registration is based on a multiresolution scheme based on local matching of areas with a high degree of local structure and subsequent interpolation. Consequently we provide an algorithm to assess the amount of structure in generic multi-valued data by means of gradient and correlation computations. The interpolation step is carried out by means of the Kriging estimator that outperforms conventional polynomial methods for the interpolation of sparse vector fields. The feasibility of the approach is illustrated by results on synthetic and clinical data.

1 Introduction

While there is a large amount of research done on the registration of scalar datasets provided by different types of intensity-based MRI scans and other medical imaging modalities, with a proliferation of algorithms and a solid theoretical background [1], this does not seem to be the case for non-scalar datasets despite their increasing clinical relevance. For example, techniques offering higher dimensional output fields are *Phase Contrast Angiography MRI (PCA-MRI)*, which provides a vector field of blood velocities, or *Diffusion Tensor MRI (DT-MRI)*, which provides a second order symmetric-tensor field description of the

---

[*] This work has been partially funded by the Spanish Gov. (MEC), visiting research fellowship FPU PRI1999-0175 for the first author, jointly by the European Commission and the Spanish Gov. (CICYT), research grant 1FD97-0881-C02-01, by grant RG 3094A1/T from the National Multiple Sclerosis Society (SKW) and by US grants NIH NCRR P41 RR13218, NIH P01 CA67165 and NIH R01 RR11747. A. Nabavi has been supported by the DFG (NA 365/ 1-1)

local water diffusion in each tissue. The major aim of this paper is to develop a common framework for the three-dimensional registration of scalar, vector and tensor fields that can be readily embedded in medical imaging applications. In particular our goal is to map a reference anatomy, depicted by the signal $S_r(\mathbf{x})$, onto a deformed one, represented by the signal $S_d(\mathbf{x})$. Both signals are the output of scanning devices and could correspond to different studies of the same patient or to different patients. Equation (1) describes a model to characterize the relationship between both datasets – see also Fig. (1) –, where D models the deformation applied to the reference signal, and both H and the noise v model the inter-scan differences.

$$S_d(\mathbf{x}) = H\left[D\left[S_r(\mathbf{x})\right]\right] + v(\mathbf{x}) \qquad (1)$$

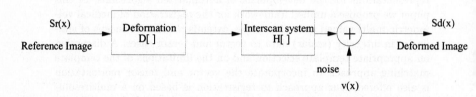

Fig. 1. Model relating two different scans

The deformation D represents a space-variant shift system and, hence, its response to a signal $S(\mathbf{x})$ is $D[S(\mathbf{x})] = S(\mathbf{x}+\mathbf{d}(\mathbf{x}))$. With regard to the inter-scan differences, we are considering H to be a non-memory, possibly space-variant, system depending on a set $\mathbf{h}(\mathbf{x}) = (h_1(\mathbf{x})\dots h_p(\mathbf{x}))^t$ of unknown parameters and the noise to be spatially white and with zero mean. With these simplifications and defining $S_r^H(\mathbf{x}) = H\left[S_r(\mathbf{x})\right]$, the model (1) reduces to:

$$S_d(\mathbf{x}) = S_r^H(\mathbf{x} + \mathbf{d}(\mathbf{x})) + v(\mathbf{x}) \qquad (2)$$

The goal of registration is to find the displacement field $\mathbf{d}(\mathbf{x})$ that makes the best match between $S_r(\mathbf{x})$ and $S_d(\mathbf{x})$ according to (2). The approach proposed in this paper is based on template matching by locally optimizing a similarity function (Sect. 2). A local structure detector for generic tensor fields (Sect. 3) allows to constrain the matching to areas highly structured. In order to obtain the deformation field in the remaining areas we propose (Sect. 4) an interpolation scheme whose key feature is the probabilistic weighting of the samples using a *Kriging Estimator* [4] as an alternative to global polynomial models. The whole approach is embedded in a multiresolution scheme using a Gaussian pyramid in order to deal with moderate deformations and decrease the influence of false optima. We also present (Sect. 5) some illustrative results carried out on synthetic and clinical data.

2 Template Matching

Several schemes can be used to estimate the displacement field in (2). When there is no *a priori* probabilistic information about the signal and noise characterization, a *Least-Squares* [3] approach is a natural choice. For this all that is required is a suitable definition of an inner product and, thereafter, an induced norm. Note that scalar, vector and tensor fields are applications of a real domain onto Euclidean vector spaces and which allows us to define the inner product between fields by means of the integral over the whole domain of the inner products between their values. Let us consider the functional set $\mathcal{F} = \{f : D \longrightarrow V\}$ where D is a real domain and V is an Euclidean space. Then an inner product can be defined on $\mathcal{F}$ as $< f_1, f_2 > = \int_D w(\mathbf{x}) < f_1(\mathbf{x}), f_2(\mathbf{x}) > d\mathbf{x}$, where $w(\mathbf{x})$ is a weighting function for the inner product. Note that the inner product in the left-hand side is defined between fields and in the right-hand side, inside the integral, is defined between values of the field.

The least squares estimator is obtained minimizing a cost function (3) that consists of the squared norm of the estimation error.

$$C^*(\mathbf{d}(\mathbf{x}); \mathbf{h}(\mathbf{x})) = \|S_d(\mathbf{x}) - S_r^H(\mathbf{x} + \mathbf{d}(\mathbf{x}))\|^2 \tag{3}$$

The dependency on the unknown parameters can be removed by estimating them using some constraints. For example, if the parameters are assumed to be constant, a least-squares estimation can be obtained, $\hat{\mathbf{h}}(\mathbf{d}) = \hat{\mathbf{h}}(S_d(\mathbf{x}), S_r(\mathbf{x} + \mathbf{d}(\mathbf{x})))$, and substituted in C^* to obtain a new cost function (4) that only depends on $\mathbf{d}(\mathbf{x})$.

$$C(\mathbf{d}(\mathbf{x})) = C^*(\mathbf{d}(\mathbf{x}); \hat{\mathbf{h}}(\mathbf{x})) \tag{4}$$

The optimization of $C(\mathbf{d}(\mathbf{x}))$ in order to obtain the displacement field $\mathbf{d}(\mathbf{x})$ is a daunting task that requires additional constraints to make it feasible. *Template Matching* trades off accuracy and computational burden to approximate a solution for this optimization problem. In particular it assumes that (2) holds with a uniform displacement field (i.e. translational) and constant parameters in a neighborhood of the point $\mathbf{x}_0$ to be corresponded. This prevents the template matching method to be successfully used in the presence of significant rotational deformation fields. There is a fundamental trade-off to be considered in the design of the neighborhood: it must be non-local, and hence large in size, in terms of the $S_d(\mathbf{x})$ space-frequencies to avoid the *ill-posedness* arising from the lack of discriminant structure (aperture problem [2]), and it must be local, and hence small in size, in terms of the unknown displacement field spatial-frequencies to guarantee invariance of the transformation parameters. If the template is too small, the problem remains *ill-posed* and the contribution of noise leads to false solutions. Conversely, if the template is too large the estimated parameters for the transformation are a smoothed version of the actual ones. Adaptive templates with different sizes and weights can help to deal with this problem.

Let $T(\mathbf{x} - \mathbf{x}_0)$ be a window function centered in a generic point $\mathbf{x}_0$ in the deformed dataset and designed following the previous remarks. The template matching assumptions transform (2) into (5), that holds for every point $\mathbf{x}_0$ in the deformed dataset.

$$T(\mathbf{x} - \mathbf{x}_0)S_d(\mathbf{x}) = T(\mathbf{x} - \mathbf{x}_0)S_r^H(\mathbf{x} + \mathbf{d}(\mathbf{x}_0)) + v(\mathbf{x}) \tag{5}$$

Equation (5) has an intuitive interpretation: for every point $\mathbf{x}_0$ in the deformed dataset a template $T(\mathbf{x} - \mathbf{x}_0)S_d(\mathbf{x})$ is constructed and compared against every possible template around it in the transformed (by H) reference dataset. This means that a complex global optimization problem has been split into many simple local ones. In order to constrain the variability of the estimated displacement field it can be smoothed after the matching using a linear filter.

2.1 Similarity Functions

A *Similarity Function* is any convenient monotonic function of the cost (3), $S(\mathbf{d}(\mathbf{x})) = F[C(\mathbf{d}(\mathbf{x}))]$, that leaves the locations of the optima unchanged. Under the template matching assumptions the similarity function $S(\mathbf{d}; \mathbf{x}_0)$ is local and, hence, it depends on $\mathbf{x}_0$. Here the least-squares method referred to above is used to obtain suitable local similarity functions for the template matching of generic tensor fields.

Let us first consider that H is the identity mapping and, hence, (3) and (4) are equal. Direct use of (4) provides the similarity function (6) that corresponds to the well-known *Sum of Squared Differences*.

$$S_{SSD}(\mathbf{d}; \mathbf{x}_0) = \|T(\mathbf{x} - \mathbf{x}_0)\left(S_d(\mathbf{x}) - S_r(\mathbf{x} + \mathbf{d})\right)\|^2 \tag{6}$$

The extension of (6) using inner products and assuming that $\|S_r(\mathbf{x} + \mathbf{d})\|^2$ is almost constant for all possible $\mathbf{d}$ leads to an alternative similarity that corresponds to the *Correlation* measure.

$$S_C(\mathbf{d}; \mathbf{x}_0) = \;<S_d(\mathbf{x}), S_r(\mathbf{x} + \mathbf{d})> \tag{7}$$

Let us now consider that H is an affine transformation of the intensity. In this case $S_r^H(\mathbf{x} + \mathbf{d}) = aS_r^H(\mathbf{x} + \mathbf{d}) + b$ and the cost (3) turns out to be (8)

$$C^*(\mathbf{d}; a, b; \mathbf{x}_0) = \|T(\mathbf{x} - \mathbf{x}_0)\left(S_d(\mathbf{x}) - aS_r(\mathbf{x} + \mathbf{d}) - b\right)\|^2 \tag{8}$$

Minimizing (8), expanding the inner product, it is not difficult to find an estimator for the parameters a and b. Substituting them into (8) and making some manipulations its possible to derive the well-known *correlation coefficient* similarity measure (9), whose absolute value is to be maximized and where $\mathbf{s} = T(\mathbf{x} - \mathbf{x}_0)S_d(\mathbf{x})$, $\mathbf{p}_1 = T(\mathbf{x} - \mathbf{x}_0)S_r(\mathbf{x} + \mathbf{d})$, $\mathbf{p}_2 = T(\mathbf{x} - \mathbf{x}_0)$ and $\hat{\mathbf{p}}_2 = \frac{\mathbf{p}_2}{\|\mathbf{p}_2\|}$. The correlation coefficient provides a geometric interpretation as the cosine of the angle between the normalized versions of $T(\mathbf{x} - \mathbf{x}_0)S_d(\mathbf{x})$ and $T(\mathbf{x} - \mathbf{x}_0)S_r(\mathbf{x} + \mathbf{d})$, obtained removing the mean value and making the signals to have unit norm.

$$S_\rho(\mathbf{d}; \mathbf{x}_0) = \left\langle \frac{\mathbf{s} - <\mathbf{s}, \hat{\mathbf{p}}_2>}{\|\mathbf{s} - <\mathbf{s}, \hat{\mathbf{p}}_2>\|}, \frac{\mathbf{p}_1 - <\mathbf{p}_1, \hat{\mathbf{p}}_2>}{\|\mathbf{p}_1 - <\mathbf{p}_1, \hat{\mathbf{p}}_2>\|} \right\rangle \tag{9}$$

The application of the equations above requires a proper definition of the inner product (10) and its induced norm (11). We assume that the tensors are cartesian (defined with respect to an orthonormal basis) and we are using the Einstein notation for sums (any repetition of an index entails a summing over this index). Note that with these definitions the elements with the form $\mathbf{f}- <\mathbf{f},\hat{\mathbf{p}}_2>$ in (9) are the mean value of $\mathbf{f}$ weighted by $T(\mathbf{x}-\mathbf{x}_0)$.

$$< S_1(\cdot), S_2(\cdot); \mathbf{x}_0 > = \int_D S_{1_{i_1\ldots i_n}}(\mathbf{x})S_{2_{i_1\ldots i_n}}(\mathbf{x})d\mathbf{x} \tag{10}$$

$$\|S(\cdot); \mathbf{x}_0\|^2 = \int_D S_{i_1\ldots i_n}(\mathbf{x})S_{i_1\ldots i_n}(\mathbf{x})d\mathbf{x} \tag{11}$$

2.2 Warped Vectors and Tensors

Vector and tensor data are linked to the body under inspection and, thereafter, any warping of the supporting tissue will lead to a consequent warping or reorientation of these data. In fact, as far as the function and the domain share the same reference system (or any rigidly related one), any transformation applied to the domain will affect to the function too. In particular, we are considering the domain transformation to be analytic and, hence, to preserve local topology, i.e., differential line elements map onto differential line elements, the same being true for differential surface and volume elements; moreover, differential ellipsoids are mapped onto differential ellipsoids. The warping of the domain can be expressed by the transformation $\mathbf{x} = \mathbf{T}(\mathbf{x}')$, where $\mathbf{x}$ stands for points in the reference dataset and $\mathbf{x}'$ for points in the deformed one. The differential of the transformation is

$$d\mathbf{x} = [\nabla \otimes \mathbf{T}(\mathbf{x}')]\,d\mathbf{x}' \tag{12}$$

where the tensor product $\otimes$ has been used to define the *deformation gradient* $\mathbf{A}(\mathbf{x}') = [\nabla \otimes \mathbf{T}(\mathbf{x}')]$, which can be recognized as the Jacobian matrix of the mapping. Consequently, two vectors $\mathbf{v}$ and $\mathbf{v}'$ and two second order tensors $\mathbf{P}$ and $\mathbf{P}'$ are locally related as follows:

$$\mathbf{v} = [\nabla \otimes \mathbf{T}(\mathbf{x}')]\,\mathbf{v}' \tag{13}$$

$$\mathbf{P} = [\nabla \otimes \mathbf{T}(\mathbf{x}')]^{-1}\mathbf{P}'[\nabla \otimes \mathbf{T}(\mathbf{x}')] \tag{14}$$

The *Polar Decomposition Theorem* [5] states that for any non-singular square matrix, such as the *Deformation Gradient* $\mathbf{A}(\mathbf{x}')$, there are unique symmetric positive definite matrixes $\mathbf{U}(\mathbf{x}')$ and $\mathbf{V}(\mathbf{x}')$ and also an unique orthonormal matrix $\mathbf{R}(\mathbf{x}')$ such that $\mathbf{A}(\mathbf{x}') = \mathbf{R}(\mathbf{x}')\mathbf{U}(\mathbf{x}') = \mathbf{V}(\mathbf{x}')\mathbf{R}(\mathbf{x}')$. This leads to important geometric interpretations of the geometric mapping. For example, notice that a sphere is first stretched by the mapping in the directions of the eigenvectors of $\mathbf{U}(\mathbf{x}')$ and then rotated by $\mathbf{R}$. Thereafter, a transformation such that locally $\mathbf{R}(\mathbf{x}') = \mathbf{I}$ is said to be a *Pure Strain* at $\mathbf{x}'$ while if $\mathbf{U}(\mathbf{x}') = \mathbf{V}(\mathbf{x}') = \mathbf{I}$ it is said to be a *Rigid Rotation* at that point.

It was said above (Sect. 2) that the matching approach to data registration relies on the fact that the displacement field is constant inside the matching template. This imposes a hard local constraint on the deformation gradient, since it must locally be close to the identity matrix. Therefore, unless the template is reduced to a single point (for example, by smoothing and downsampling) one should not include the local directional change of the vector and tensor fields due to the domain warping in the matching problem. Nevertheless, once the displacement field has been estimated, it is possible to use (13) and (14) to compute the reorientation of the vector or tensor field.

3 Structure Measures

Matching must be constrained to areas with local high discriminant structure in order to be successfully applied, making sure that the local similarity function are narrow around the optima. We propose to threshold a convenient measure of cornerness to identify the appropriate areas. In the scalar case some of these measures are based on identifying points corresponding to curved edges by means of the locally averaged outer product (i.e. the correlation matrix) of the gradient field [6]. The gradient of a tensor field is expressed in invariant form with a tensor product $\nabla \otimes S(\mathbf{x})$. Using an orthonormal basis, the gradient of a scalar field in any point is a vector whose components are the partial derivatives of the function. For a vector field, the gradient turns out to be the Jacobian matrix with the partial derivatives for each component and for a second order tensor field it is a $3D$ array with the partial derivatives of every component of the tensor. The *comma* [5] convention becomes very handy to represent gradients in component form and it has been used in (15), where $, k$ indicates an indexing of the partial derivatives and n is the order of S. Note that for each component of a generic field we obtain a gradient vector that is arranged into the overall gradient tensor.

$$\nabla \otimes S(\mathbf{x}) = S_{i_1 \ldots i_n, k}(\mathbf{x}) \tag{15}$$

In the scalar case, the *correlation matrix* of the gradient vector, i.e., the symmetric, positive semidefinite second order tensor formed by the mathematical expectation of the outer product of the gradient vector (16) provides the basis to assess cornerness analyzing its associated quadratic form (ellipsoid): The rounder the ellipsoid the bigger the cornerness while a very elongated ellipsoid would indicate a straight edge.

$$\mathbf{R}_{\nabla S}(\mathbf{x}) = E\left\{\nabla S(\mathbf{x}) \otimes \nabla S(\mathbf{x})\right\} \tag{16}$$

In order to extend this idea to the vector and tensor cases, note first that (16) can be directly extended into the *correlation tensor* (17).

$$\mathbf{R}_{\nabla S}(\mathbf{x}) = E\left\{\nabla \otimes S(\mathbf{x}) \otimes \nabla \otimes S(\mathbf{x})\right\} \tag{17}$$

For example, if we had a second order tensor field we would obtain a sixth order correlation tensor. The *generalized correlation matrix* (18) is defined as the contraction of all the non-differential components in (17) with respect to an orthonormal basis.

$$\mathbf{R}_{\nabla \mathbf{S}}(\mathbf{x}) = E \left\{ S_{i_1 \ldots i_n, k}(\mathbf{x}) S_{i_1 \ldots i_n, l}(\mathbf{x}) \right\} \qquad (18)$$

Due to the linear nature of the expectation operator, the generalized correlation matrix of the gradient of a generic tensor field is the sum of the correlation matrices of the gradient vectors of each coordinate. This is consistent with the fact that each component adds its own edge and, therefore, the ellipsoid associated to the generalized correlation matrix get rounder as new components are added unless they do not provide additional directional information.

Matching is therefore constrained to points where the ellipsoids associated to the generalized correlation matrices are large and round enough, meaning strong and curved edge.

4 Interpolation: the Kriging Estimator

The structure analysis carried out in the previous section provides a sparse set of clusters of $3D$ points on which a local estimation of the displacement can be performed by template matching. Two comments must be made in order to interpolate the displacement field in the remaining areas. First of all the field can be discontinuous due to the relative motion between different bodies; this prevents the direct use of any global polynomial interpolation approach since no single model can fit everywhere in the dataset. Second, the displacement field components on the boundaries of touching organs must have continuous components in the directions normal to the surface interfaces, though the field can be discontinuous in the orthogonal tangent plane. This simply indicates that the motion at the boundaries can be decomposed in a discontinuous sliding component and in a continuous pushing one.

The *Kriging Estimator (KE)*, which originated in *geostatistics* [7, 8], is a method to deal with spatially dependent data. Essentially it provides a linear estimator of an unknown sample of a random field using a set of known samples. The estimator is designed to minimize the mean squared error under the constraint of the the estimator to be unbiased, i.e., the weights must sum up one. From a practical point of view, the KE weighs the contribution of each known sample according to its distance to the one to be estimated. This means that distant samples lose importance and eventually can be ignored in the estimator while closer samples, which are more likely to be part of the same body and to have similar realization values, increase their relative importance. Its application to the interpolation of $3D$ scalar medical images has been referred elsewhere [9]. Here the method is used in the interpolation of displacement fields considering each spatial component independently. A hypothesis is made about a so-called *variogram* function, which is the mean squared error between two samples at a distance r. See [4] for a discussion on different variograms. Essentially a parametric model is proposed as theoretical variogram and parameter estimation from the known data using, for example, least-squares, leads to the experimental variogram, which is the one actually used.

The *KE* interpolator provides an estimator for the components $d_i(\mathbf{x}_0)$ of the displacement field in $\mathbf{x}_0$ as a linear combination of the corresponding components matched at the high structure locations $d_i(\mathbf{x}_j)$, as it is indicated in (19).

$$d_i(\mathbf{x}_0) = \sum_{j=1}^{N} h_j d_i(\mathbf{x}_j) \qquad (19)$$

The weights of the estimator are obtained solving the linear system of equation (20), where $\gamma(\|\mathbf{x}_i - \mathbf{x}_j\|)$ is the evaluation of the variogram between point $\mathbf{x}_i$ and $\mathbf{x}_j$.

$$\begin{pmatrix} 0 & \gamma(\|\mathbf{x}_2 - \mathbf{x}_1\|) & \dots & \gamma(\|\mathbf{x}_N - \mathbf{x}_1\|) & 1 \\ \gamma(\|\mathbf{x}_2 - \mathbf{x}_1\|) & 0 & \dots & \gamma(\|\mathbf{x}_N - \mathbf{x}_2\|) & 1 \\ \vdots & \vdots & \vdots & \vdots & \vdots \\ \gamma(\|\mathbf{x}_N - \mathbf{x}_1\|) & \gamma(\|\mathbf{x}_N - \mathbf{x}_2\|) & \dots & 1 & 0 \\ 1 & 1 & \dots & 1 & 0 \end{pmatrix} \begin{pmatrix} h_1 \\ h_2 \\ \vdots \\ h_N \\ \mu \end{pmatrix} = \begin{pmatrix} \gamma(\|\mathbf{x}_0 - \mathbf{x}_1\|) \\ \gamma(\|\mathbf{x}_0 - \mathbf{x}_2\|) \\ \vdots \\ \gamma(\|\mathbf{x}_0 - \mathbf{x}_N\|) \\ 1 \end{pmatrix}$$

$$(20)$$

5 Results

The framework presented in this paper is under evaluation in a number of clinical cases at Brigham & Women's Hospital. The implementation relies on sampled data and therefore the integrals in (10) and (11) become sums. The window function $T(\mathbf{x})$ is defined to be the unity inside a cube centered in the origin and zero outside. The local structure detection is based on (18) where the expectation is computed from the sample mean. Since $\det(\mathbf{R}_{\nabla S}(\mathbf{x})) = \prod \lambda_i$ and $\mathrm{tr}(\mathbf{R}_{\nabla S}(\mathbf{x})) = \sum \lambda_i$ (λ_i are the eigenvalues of $\mathbf{R}_{\nabla S}$), only points above a threshold in $t_1(\mathbf{x}) = \frac{\det(\mathbf{R}_{\nabla S}(\mathbf{x}))}{\mathrm{tr}(\mathbf{R}_{\nabla S}(\mathbf{x}))}$ are considered. False detections due to high gradients are avoided thresholding $t_2(\mathbf{x}) = \frac{\det(\mathbf{R}_{\nabla S}(\mathbf{x}))}{\mathrm{tr}(\frac{1}{N}\mathbf{R}_{\nabla S}(\mathbf{x}))^N}$ that varies between zero and one depending on the shape of the associated ellipsoid, with one meaning a perfect sphere. The structure detector is only applied to the deformed dataset, providing clusters of points in which the displacement vector is to be estimated using template matching around a neighborhood. The warp is interpolated in a generic point of low structure using Kriging derived from the N closest highly structured points with known displacement. The estimated warp is finally smoothed with a linear filter. The full approach is embedded in a gaussian pyramid so that warps estimated at coarser resolution levels are linearly interpolated in order to be used as initial displacement in the next higher resolution level.

Figure (2.a) shows a 32×32 MRI slice of the corpus callosum that is deformed by a synthetic gaussian field as depicted in Fig. (2.b). In Fig. (2.c) it is shown the correctly estimated displacement field in areas of high structure (blue arrows) and the interpolated field (red) using the Kriging estimator with a linear variogram $\gamma(r) = ar$. All experiments carried out have confirmed the superior performance of the Kriging estimator over polynomial interpolators both with synthetic and real deformations.Figure (3.a) shows an axial slice of a DT-MRI dataset [1] of the corpus callosum where the principal eigenvectors directions have been represented using a color coding ranging from blue (in-plane projection) to red (orthogonal to plane) [10]. The whole approach has been applied to warp this dataset into another corresponding to a different individual, shown in Fig. (3.b), using three levels of a gaussian pyramid, local templates of dimension 3×3 and a linear variogram for the Kriging interpolator that is limited to take into account the 20 closest samples. Figure (3.c) shows a T2W zoomed version of the righthand side of the former, corresponding to the posterior corpus callosum and the estimated deformation field.

6 Conclusions

We have presented a unified framework for non-rigid registration of scalar, vector and tensor medical data. The approach is local, since it is based on template-matching, and resorts to a multirresolution implementation using a Gaussian pyramid in order to provide a coarse-to-fine approximation to the solution which allows to deal with moderate deformations and avoids false local solutions. The method does not assume any global *a priori* regularization and, therefore, avoids the computational burden associated to those approaches. We also have extended the concept of discriminant structure to the tensor case, providing a new operator to detect it. The *Kriging Estimator* outperforms polynomial approaches for the interpolation of sparse displacement fields and to the best of our knowledge this is the first time to be used with this purpose in medical image analysis. The whole approach is under evaluation in a number of clinical cases at Brigham & Women's Hospital with preliminary promising results.

References

1. A. Roche et al. Towards a Better Comprehension of Similarity Measures Used in Medical Image Registration. *Proc. MICCAI'99*, pp. 555-566, Cambridge, UK, 1999
2. T. Poggio, V. Torre, and C. Koch. Computational vision and regularization theory. *Nature*, vol. 317, pp. 314-319, sept. 1985
3. T. K. Moon and W. W. Stirling. *Matehematical Methods and Algorithms for Signal Processing*, Prentice-Hall, NJ, 2000
4. J.L. Starck, F. Mrtagh and A. Bijaoui. *Image Processing and Data Analysis. The Multiscale Approach*, Cambridge University Press, UK, 1998

---

[1] GE Signa 1.5 Tesla Horizon Echospeed 5.6 scanner, Line Scan Diffusion technique,1 min/slice, no averaging, effective TR=2.4 s, TE=65 ms, $b_{high} = 750s/mm^2$, fov=22 cm, voxel size $4.8 \times 1.6 \times 1.5mm^3$, 6 kHz readout bandwidth, acquisition matrix 128×128

5. L.A. Segel. *Mathematics Applied to Continuum Mechanics*, Dover, NY, 1987
6. K. Rohr. On 3D Differential Operators for Detecting Point Landmarks. *Image and Vision Computing*, 15:219-233, 1997
7. D. Krige. *A Statistical Approach to Some Mine Valuation and Allied Problems on the Witwatersrand*, Master Thesis, University of Witwatersrand, 1951
8. N. Cressie. Kriging Nonstationary Data. *Journal of the American Statistical Association*, 81:625-634, 1986
9. R. W. Parrott et al. Towards Statiscally Optimal Interpolation for 3D Medical Imaging. *IEEE Engineering in Medicine and Biology*, pp. 49-59, sept. 1993
10. S. Peled et al. Magnetic Resonance Imaging Shows Orientation and Asymmetry of White Matter Tracts. *Brain Research*, vol. 780, pp. 27-33, jan. 1998

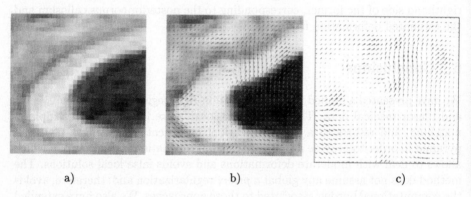

a)   b)   c)

Fig. 2. Displacements fields (b) real (c) blue: estimated, red: Kriging interpolated

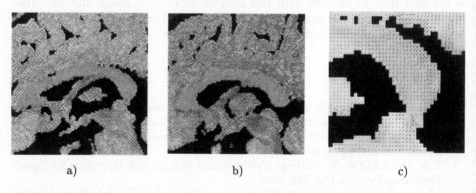

a)   b)   c)

Fig. 3. DT-MRI interpatient warping. a, b) Axial DTMRI of different individuals. c) zoomed T2W of the posterior corpus of a) and estimated deformation

A Fluoroscopic X-Ray Registration Process for Three-Dimensional Surgical Navigation

Robert Grzeszczuk[1], Shao Chin[1], Rebecca Fahrig[2], Hamid Abassi[1], Hilary Holz[1], Daniel Kim[1], John Adler[1], Ramin Shahidi[1]

[1]Image Guidance Laboratories, Stanford University
[2]Department of Radiology, Stanford University

Abstract

We describe a system involving a computer-instrumented fluoroscope for the purpose of 3D navigation and guidance using pre-operative diagnostic scans as a reference. The goal of the project is to devise a computer-assisted tool that will improve the accuracy, reduce risk, minimize the invasiveness, and shorten the time it takes to perform a variety of neurosurgical and orthopedic procedures of the spine. For this purpose we propose an apparatus that will track surgical tools and localize them with respect to the patient's anatomy and pre-operative 3D diagnostic scans using intraoperative fluoroscopy for *in situ* registration. The resulting system leverages equipment already commonly available in the Operating Room (OR), allowing us to provide important new functionality that is free of many current limitations, while keeping costs contained.

1. Introduction

Computer-assisted spine surgery is an active area and numerous systems, both commercial and research, have been described in the literature. Most of these systems provide similar functionality and methodology. The majority employ optical trackers for the purpose of registration and localization. Typically, the vertebral body of interest is fully exposed intra-operatively and a small number of distinct anatomical landmarks are digitized for the purpose of coarse registration. Subsequently, a larger number of points are digitized on the surface of the vertebrae to refine the registration with a surface matching technique. The procedure is often cumbersome, time consuming, and of limited accuracy. This is mainly due to difficulties in identifying characteristic anatomical landmarks in a reproducible fashion (particularly in older patients and pathologically changed bone structures) and inherent inaccuracies of surface matching techniques. While dynamic reference frames (DRFs) are commonly used to monitor target movement, any safeguarding against DRF misregistration requires the entire process, including the laborious manual digitization part, to be repeated. This gets even more cumbersome in procedures involving multiple vertebrae (e.g., cage placements) requiring multiple DRFs to be maintained. Finally, aforementioned techniques cannot be applied in the context of percutaneous procedures, because they rely on the target structure being directly visible to the optical tracking device.

Recently, several academic researchers [1] and commercial vendors (e.g., FluoroNav, Medtronic,Minneapolis, MN,USA) began experimenting with fluoroscopy as an intra-operative imaging modality. The relative low cost and pervasiveness of C-Arm devices in modern ORs can explain this increased interest. Most of these attempts focus on improving conventional 2D navigation techniques via tracking of the C-Arm and re-projecting pre-operative CT data onto multiple planes. Such techniques are helpful in lowering the amount of ionizing radiation delivered to the patient and the OR staff during free-hand navigation and also in providing more information to the surgeon about the relative position of the surgical tools with respect to the patient's anatomy. However, they essentially automate and streamline the current workflow and rely on the surgeon's ability to create a complex spatial model mentally. In contrast, our approach employs stereoscopic registration in order to relate patient's anatomy to pre-operative diagnostic scans in 3D.

2. Materials and Methods

2.1. Overview

A mobile fluoroscopic device (C-Arm, GE, Milwaukee, WI, USA) is used for instrumented intra-operative free hand imaging of skeletal anatomy. The position and orientation of its camera is tracked by an optoelectronic device (Optotrack, Northern Digital, Waterloo, Ontario, Canada). NTSC video output of the X-ray camera as well as its position and orientation are processed by an SGI 320 (SGI, Mountain View, CA, USA) NT workstation. The image data is used by the system to register a pre-operative CT data set (512x512 130 slices, 1.0 mm thick) to the patient's reference frame. Tracking data is used to monitor the relative position changes of the camera with respect to the patient during free-hand navigation.

During the course of the procedure, the system operator performs an initial registration of the patient's anatomy to the pre-operative CT study by taking two protocoled fluoroscopic views (e.g., AP and lateral) of the target vertebrae. These images are used to compute the C-Arm-to-CT registration using a fully automatic technique described in detail below. As a result, we obtain the position and orientation of the C-Arm's camera in the reference frame of the CT study. We can then use this information, together with the position and orientation of the camera in the tracker reference frame to follow the future relative motions of the camera, assuming static system. Misregistration, due to either patient movement or system error, can be detected at any time by comparing the predicted DDR to the actual fluoroscopic image, at which point the objects can be re-registered within seconds. A surgical tool visible in at least two fluoroscopic views can be back-projected into the CT reference frame using standard stereoscopic techniques well known in the computer vision community. Alternatively, we can track a surgical tool externally (e.g., using the optical tracking device already employed, or a robotic interface) to facilitate a variety of procedure types.

2.2. Intrinsic Camera Calibration

Before the system can be used effectively, it needs to be calibrated to eliminate imaging distortions and extract parameters required to characterize the imaging system. During the process of intrinsic camera calibration we determine the set of parameters that characterize the internal geometry of our imaging system. All of these parameters are well documented in the computer vision literature [2]: effective focal length (i.e., the exact distance from the X-ray source to the image plane), focal spot (i.e., the exact place where the optical axis pierces the image plane), magnification factors, and pincushion distortions. It is also well understood that many of C-Arm's intrinsic parameters vary with orientation. For example, the focal length and focal spot will be different for vertical and horizontal positioning, due to mechanical sag of the arm. Similarly, the pincushion distortion depends on the orientation of the device with respect to the Earth's magnetic field, and external source of EMFs. For these reasons we fix a calibration jig to the face of the image intensifier, in order to adaptively calibrate the system every time an image is acquired. The intrinsic parameters thus produced are used during the registration phase.

2.3. Extrinsic Camera Calibration and Registration

Once the intrinsic camera parameters are measured, we can proceed to register our imaging system to the patient. The problem can be posed as follows: given a CT study of the relevant anatomy and a fluoroscopic image of the same anatomy, find the position and orientation of the camera in the reference frame of the CT study that produced the image. For this purpose, we adopted a technique used by a commercial frameless image-guided radiosurgery system (Cyberknife, Accuray Inc., Sunnyvale, CA, USA) [3]. In this approach, the system hypothesizes about the camera's actual six degrees of freedom (DOF) position by introducing a virtual camera and computing a simulated radiograph (Digitally Reconstructed Radiograph, or DRR). The radiograph represents the actual patient position. If the DRR matches the radiograph exactly, the virtual camera position identifies the actual camera position and thus the registration is found, otherwise another hypothesis is formed. In practice, the accuracy of such basic radiograph-to-DRR registration method can be substantially improved if two or more views of the same anatomy can be used. In this scenario, we will assume that the relative positions of the cameras are known (e.g., from an optical tracking device), so instead of finding two or more independent camera positions [4], we can reformulate our problem as that of finding the pose of CT study with respect to the rigid configuration of multiple cameras. In essence, we can view this registration process as an extrinsic calibration of an abstract imaging system consisting of multiple, rigidly mounted cameras.

The radiograph-to-DRR registration problem has three parts. First, we need to develop the means of exploring the range of possible patient positions to be represented in the DRRs (i.e., hypothesis formation strategy), second, identify those elements in the images (image features) that most effectively capture information about the patient pose, then finally, third, use a comparison statistic or cost function for the two sets of image feature data to indicate when the best match has been achieved.

There are three fundamental ways of generating hypotheses for possible patient placement. One approach, called matched filter area correlation, generates a single reference image from a DRR representing the desired patient position from the camera point of view. This reference image, or part thereof, is shifted and rotated relative to the acquired image until the best fit is achieved, in a manner of a sliding window matched filter. A second approach, referred to here as interpolative area correlation, consists of calculating a set of reference DRRs that samples a full range of possible patient positions and orientations, and then making an interpolative comparison of the acquired fluoroscopic image with each of the DRRs. Using two cameras, either of the aforementioned methods can accurately measure all threes translational and one rotational degree of freedom, providing there is no out-of-plane degrees of freedom [6]. A third method consists of interactively re-projecting DRRs while perturbing the pose of the patient image in the CT study, until a DRR is made that matches the fluoroscopic image. The iterative re-projection technique can accurately measure all six DOFs and is the approach we have taken here.

Comparison of the DRRs and acquired radiographs is a problem in pattern recognition. The sets of image data used to compare two images are called feature vectors. The most primitive feature vector is simply the complete set of pixel gray-scale values, where each pixel's brightness is the magnitude of a vector component. More sophisticated feature vectors are usually sought to emphasize the important large-scale structure in the image and minimize the extraneous information and noise. Formal systems of feature extraction involve recasting the original grayscale feature vector on an orthogonal system of eigenvectors that relate to large-scale patterns that are not necessarily physical structures. Heuristic feature vectors identify the positions and shapes of physical edges, boundaries, and other discernible structures.

The DRR and radiograph feature vectors **A** and **B** are compared using a similarity statistic or cost function. This can be simply the cross-correlation coefficient $r = \mathbf{A} \cdot \mathbf{B} = cos\theta$., where **A** and **B** have been normalized to unit length. (The vectors can be centered on their means before normalization, which gives Pearson's correlation coefficient.) A more general and flexible comparison can be made with the chi-squared statistic: $\chi^2 = \Sigma(A_i - B_i)^2 / w_i^2$, where each vector component is weighted according to both its reliability and usefulness, and the vectors are not necessarily normalized. When the vectors are normalized, and all vector components carry equal weight, χ^2 is proportional to $1 - r$.

3. Results

For fiducial-based registration, the fluoroscope image contrast is expanded and then thresholded to highlight the fiducial shadows. The image-plane coordinates of the fiducials are automatically extracted using one of three methods: (1) A Sobel edge filter is convolved across the image to further enhance the fiducials, the image is thresholded again, and then x and y moments are computed in the neighborhood of the bright fiducial edge features; (2) a matched filter representing the shadow of the fiducial is convolved across the image and the convolution maxima are isolated by thresholding; (3) if spherical fiducials have been used, a circular Hough transform is applied to the image, resulting in a bright maximum at the center of each fiducial shadow. Registration of skeletal landmarks is accomplished by edge-filtering the fluoroscope

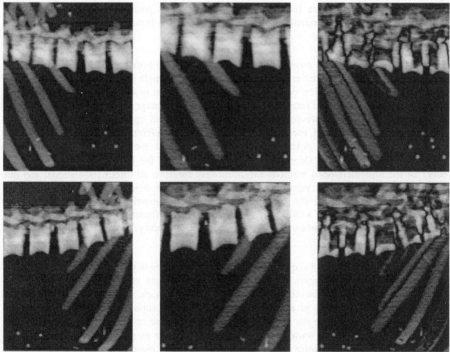

Figure 1. Registration in progress for a rigid torso phantom. Top row shows data for camera position A, the bottom for position B. Left column shows test DRRs, center the actual fluoroscopic images, while the right one shows the results of the correlation (in a mis-registered configuration).

and DRR reference images and then locating the points where the anatomical edges intersect line segments at fixed positions in the images. These points make up the feature vectors for the radiographs and DRRs. With eight to ten feature points in each fluoroscope view, the translational registration can again achieve ± 0.2 mm and ± 0.5 degrees precision.

If the fluoroscope images have a pixel pitch of 1.0 mm, the position of a 2 - 3 mm diameter spherical fiducial can be found with a precision of ± 0.2 mm. This yields a translational registration precision of 0.2 mm or better. The rotational precision depends on the fiducial spacing and the angle of projection of the fiducial configuration in the fluoroscope image. For fiducials spaced 25 mm apart a typical projection angle will resolve out-of-plane rotations with a precision of ± 0.5 degrees.

4. Discussion

The main factors differentiating our solution from that of others include: selecting fluoroscopy for the *in situ* imaging technique, using stereo photogrammetry to extract 3D information from projective images, as well as using a robust, precise and practical registration method. Unlike all currently available 3D spinal navigation packages, which require full exposure of the vertebral body for the sake of registration and real-time optical tracking, our method employs fluoroscopic imaging and registration using percutaneously implanted markers or skeletal anatomy as a minimally invasive approach. This allows us to help guide surgical tools using pre-operative 3D diagnostic scan. While our method does not permit a real-time DRF for the sake of target movement monitoring, periodic re-registration is much more practical than in conventional approaches: misregistration can be detected and eliminated by simply re-acquiring two new fluoroscopic images and running a fairly automatic procedure that requires little or no intervention by the operator. The registration technique has been adopted from Accuray's Cyberknife radiation treatment methodology, which has been shown to register either fiducial markers or skeletal anatomy with sub-millimeter and sub-degree precision in all six degrees of freedom and computation efficiency leading to a time scale of approximately one second to complete the entire registration process. Additional proprietary improvements in digital re-projection techniques using off-the-shelf computer graphics hardware will further enhance the robustness, accuracy, and performance of the registration method.

Similarly, our registration method coupled with more sophisticated visualization and biomechanical modeling techniques can potentially be generalized to handle non-rigid deformations resulting from inter-vertebral displacement. This would allow us to apply our methodology in clinical scenarios that involve more than a single vertebral body, without cumbersome patient fixation or making assumptions about unchanged pose between the time of the scan and intra-operative positioning. The ability to intra-operatively register articulated, deformable spinal anatomy in near real-time and with high accuracy would be a critical improvement over existing systems.

An important differentiating factor is our ability to track surgical tools with respect to the patient's anatomy as defined by a pre-operative diagnostic scan. Unlike traditional approaches where optical tracking is used to follow the surgical tools, we employ fluoroscopy for both registration and tool tracking. This permits us to apply our system in the context of minimally invasive percuteaneous procedures where the tool may not be exposed and visible to the tracking CCD camera. This is particularly advantageous for flexible and/or articulated effectors, which cannot be tracked optically. Finally, an additional benefit of our approach is more effective use of the fluoroscope with less exposure to ionizing radiation on the part of the patient as well as the surgeon, because instrumented fluoroscopy can be used in a more controlled manner than during conventional free hand imaging.

References

1. R. Hofstetter, M. Slomczynski, M. Sati and L.-P. Nolte, CAS, 4:65-76, 1999.
2. R.Y. Tsai, "An Efficient and Accurate Camera Calibration Technique for 3D Machine Vision", Proceedings of IEEE Conference on Computer Vision and Pattern Recognition, Miami Beach, FL, 1986, pages 364-374.
3. M. J. Murphy, "An automatic six-degree-of-freedom image registration algorithm form image-guided frameless stereotaxic radiosurgery," in Medical Physics 24(6), June 1997.
4. J. Weese, G.P. Penny, T.M. Buzug, C. Fassnacht and C. Lorenz "2D/3D registration of pre-operative CT images and intra-operative X-ray projections for image guided surgery," in CARS97, H.U. Lemke, M.W.Vannier and K Inamura ed., pages 833-838, 1997.
5. M. Roth, C. Brack, R. Burgkart, A. Zcopf, H. Gotte and A. Schwiekard "Multi-view contourless registration of bone structures using single calibrated X-ray fluoroscope," CARS99, pages 756-761, 1999.

Block Matching:
A General Framework to Improve Robustness
of Rigid Registration of Medical Images

S. Ourselin, A. Roche, S. Prima, and N. Ayache

INRIA Sophia - Epidaure Project
2004 Route des Lucioles BP 93
06902 Sophia Antipolis Cedex, France
{Sebastien.Ourselin, Alexis.Roche, Sylvain.Prima,
Nicholas.Ayache}@sophia.inria.fr

Abstract. In order to improve the robustness of rigid registration algorithms in various medical imaging problems, we propose in this article a general framework built on block matching strategies. This framework combines two stages in a multi-scale hierarchy. The first stage consists in finding for each block (or subregion) of the first image, the most similar subregion in the other image, using a similarity criterion which depends on the nature of the images. The second stage consists in finding the global rigid transformation which best explains most of these local correspondances. This is done with a robust procedure which allows up to 50% of false matches. We show that this approach, besides its simplicity, provides a robust and efficient way to rigidly register images in various situations. This includes for instance the alignment of 2D histological sections for the 3D reconstructions of trimmed organs and tissues, the automatic computation of the mid-sagittal plane in multimodal 3D images of the brain, and the multimodal registration of 3D CT and MR images of the brain. A quantitative evaluation of the results is provided for this last example, as well as a comparison with the classical approaches involving the minimization of a global measure of similarity based on Mutual Information or the Correlation Ratio. This shows a significant improvement of the robustness, for a comparable final accuracy. Although slightly more expensive in terms of computational requirements, the proposed approach can easily be implemented on a parallel architecture, which opens potentialities for real time applications using a large number of processors.

1   Introduction

Numerous intensity-based methods have been successfully devised for rigid registration of medical images [7]. For the monomodal as well as the multimodal case, the general approach consists in assuming a global relationship between the intensities of the images to register (e.g., affine, functional, statistical, etc.) [12], and then deriving and maximizing a suitable similarity measure (resp. correlation coefficient [1], correlation ratio [13], mutual information [16, 6], etc.). Several issues are common to these methods.

First, all the similarity measures are known to be highly non-convex with respect to the transformation parameters. Thus, their global maximization is seldom straightforward. In most implementations, for sake of low computation time, this latter is performed using a standard deterministic optimization technique (simplex or Powell's method, gradient descent, etc.) which is easily trapped in local maxima. Moreover, the non-convexity of the criterion is enhanced by the interpolation procedure which is used when considering sub-voxel transformations [10].

Second, the assumption of a global relationship between the image intensities may be violated by the presence of various image artefacts. For instance, intensity inhomogeneities are found in several image modalities (bias field in MR, shadowing in ultrasound). Also, biological changes occur in temporal studies (evolutive lesions, tumors, ischemia, etc.) and, more generally, the anatomical structures to be matched may not be strictly equivalent (alignment of serial sections, mid-sagittal plane computation, etc.). These artefacts or anatomical disparities can bias a global criterion and hamper its maximization.

In this paper, we address these problems using a block matching strategy interleaved with a robust transformation estimator. Block matching techniques have already been used in non-rigid medical image registration [2, 4, 8], but very rarely in rigid registration. Instead of maximizing a global similarity measure with respect to the transformation parameters, they treat the problem in a sequential manner. The first step is to build a displacement field by locally translating image regions (or blocks) so as to maximize a given similarity measure; thus, the global maximization is replaced by several *local* maximizations. The second step is to determine the spatial transformation by regularizing the previous displacement field. This two-step procedure is generally applied iteratively in a multi-scale hierarchy.

When using this approach to estimate a rigid transformation, the major difficulty is that the displacements to be found are generally much larger than in a non-rigid context. Therefore, the blocks have to be moved in wider neighborhoods, which may cause the resulting displacement field to contain some severe outliers. These outliers may strongly affect the estimation of the rigid transformation. For that reason, the robustness of the transformation estimator is a key issue.

Our implementation of block matching in rigid mode is described in Section 2, while Section 3 presents results obtained in three different rigid registration problems: alignment of histological sections, computation of the mid-sagittal plane in brain images, and multimodal registration of CT with MR images.

2 Description of the method

The algorithm takes two images as input: a reference image I and a floating image J. The output will be the transformation T and the image $J' = J \circ T^{-1}$, which is aligned with I. The whole process follows from an iterative scheme where, at each step, two successive tasks are performed. The first is computing a displacement field between I and the current floating image J'; this is done through a block matching strategy. The second is gathering these displacements to determine a rigid transformation S using a robust estimator. Updating the current transformation according to $T \leftarrow S \circ T$, we get the new floating image J' by resampling only once the image J in terms of the new T. This two-step procedure is integrated within an iterative multi-scale scheme.

In this section, we have chosen to describe the 3D implementation of the method. However, it should be clear to the reader that the 2D adaptation is straightforward.

2.1 Computation of correspondences by a block matching strategy

The basic principle is to move a block $\mathcal{A}$ of the floating image in a neighborhood Ω, and to compare it to the blocks $\mathcal{B}$ that have coincident positions in the reference image (see Figure 1). The block size and the size of the search neighborhood are constant at a given scale level; they will be respectively denoted $N_x \times N_y \times N_z$ and $\Omega_x \times \Omega_y \times \Omega_z$. The best corresponding block $\mathcal{B}$, with respect to a given similarity measure, allows one to define a match (a_i, b_i) between the centers of the blocks $\mathcal{A}$ and $\mathcal{B}$. In order to improve the computation time, we subsample the search set Ω by considering tentative matches every Σ_x (resp. Σ_y, Σ_z) voxels along the x (resp. y, z) direction. These parameters determine the *resolution* of the displacement field. Also, the centers of the test blocks $\mathcal{A}$ are taken from a sub-lattice of the floating image. We thus introduce another set of subsampling factors $(\Delta_x, \Delta_y, \Delta_z)$, which determines the *density* of the displacement field.

Since the $(N, \Omega, \Sigma, \Delta)$ parameters are voxel distances they are always integers. To take into account possible image anisotropy, their relative values are chosen so as to correspond roughly to the same *real* distances.

Contrary to methods based on a global criterion, the block matching furnishes explicit point correspondences. However, we should suspect some of them to be unreliable. Bad matches often occur when the displaced block is uniform or when it includes structures that are absent from the reference image. To handle outliers, we first reject matches corresponding to similarity measures that are below a given threshold. In this manner, a primary set of outliers is detected. The remaining outliers will be rejected during the robust estimation of the transformation parameters (see Section 2.2).

Multi-scale scheme When choosing large search areas Ω, we expect the algorithm to recover gross displacements while accurate matches may be achieved by assigning low values to Δ and Σ. Moreover, the complexity of our block matching

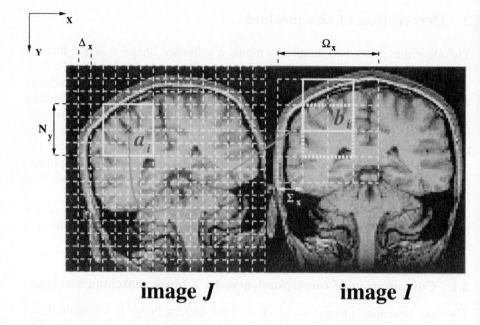

image *J*   image *I*

Fig. 1. *Illustration of the block matching strategy. In a given direction, the parameters have the following meaning: N is the block size; Ω is the size of the search set; Σ is the resolution of the displacement field; Δ is the density of the displacement field.*

procedure is proportional to $(N \times \Omega)/(\Delta \times \Sigma)$. To optimize the trade-off between performance and computational cost, we make use of a multi-scale scheme. We start out with a coarse scale, large values for N, Ω, Δ, and Σ. When refining the scale, all of these parameters are halved; in this manner, the complexity is independent of the scale level. The details of this multi-scale implementation will be found in [9].

Choice of the similarity measure We have implemented several similarity measures. In practice, the correlation coefficient (CC) is well suited for monomodal registration, whereas the correlation ratio (CR) and the mutual information (MI) are better adapted to multimodal registration.

An important fact to bear in mind is that the cost of the block matching procedure is strongly dependent on the time required to evaluate the similarity measure between two blocks. In particular, the MI measure yields computation times that are often prohibitive. Contrary to the case of CR, computing MI between two blocks requires first evaluating their joint histogram [13]. Then, one still has to perform a considerable number of operations, mainly because $n_a \times n_b$ logarithm computations are needed, n_a and n_b being the respective number of intensity classes in the blocks $\mathcal{A}$ and $\mathcal{B}$ to be compared. As a consequence, the computational cost of the block matching procedure using MI is one order of magnitude greater than when using CR.

2.2 The least trimmed squares minimization

A least trimmed squares (LTS) regression is used to find the rigid transformation T that best fits the displacement field computed through block matching. This minimization scheme has been proven to be far more robust to outliers than the classical least squares (LS) method [15]. It consists in solving the following minimization problem,

$$\min_{T} \sum_{i=1}^{h} \|r_{i:n}\|^2,$$ (1)

where $\|r_{i:n}\|^2$ are the squared ordered Euclidean norms of the residuals, $r_i = a_i - T(b_i)$, n is the total number of displacement vectors, and h is set to $h = \lfloor n/2 \rfloor$ to achieve a 50% breakdown point.

As discussed in [14], the LTS criterion may be minimized by means of a simple iterative LS estimation. This technique, generally requiring no more than 10 iterations, has the advantage of being computationally very efficient owing to the fact that the solution of one LS minimization is analytical [3]. We notice, however, that this is not guaranteed to converge to the global minimum of the LTS criterion, but only to a local minimum. Although we have not yet observed cases where the iterative LS was trapped in aberrant solutions, more robust LTS minimization strategies [14] will be looked into in future implementations.

3 Results and discussion

We have applied this block matching algotrithm to three different medical image registration problems. First, we present the histological slices alignment [9]. Second, we present the computation of the mid-sagittal plane [11]. Finally, we have evaluated the robustness of our method with the CT/MR registration problem.

3.1 Histological slices alignment

A similar block matching scheme was applied to estimate a 2D rigid transformation between successive anatomical brain slices [9] (see Figure 2 left). In particular, we realigned several datasets of rat's brains containing about 25 sections each with a resolution of 768×576 pixels ($0.03mm \times 0.03mm$) and an intersection gap of $0.4mm$. The maximization of the global CC [5] were also applied. Our method yields the best visual results, in particular for the inner strucures (see Figure 2 right).

3.2 Mid-sagittal plane computation

We present in Figure 3 the computation of the mid-sagittal plane for MR and SPECT images. The MR image has a size of 256^3, with a voxel size of $0.78mm^3$. The SPECT image has a size of 64^3, with a voxel size of $3.12mm^3$. In [11], an analysis of the accuracy and the robustness of our method has been presented

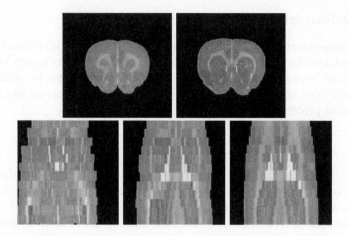

Fig. 2. *Two successive slices (top line) and the median perpendicular view of the section set (bottom line): left, initial data; middle, after registration using global CC; right, after registration using the block matching.*

using artificial and real images. Several image databases have been created in order to evaluate the influence of anatomical differences. A number of experiments have demonstrated the robustness and the subvoxel accuracy of our algorithm.

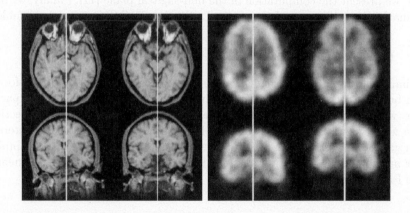

Fig. 3. *axial (top) and coronal (bottom) views, for the initial MR and SPECT and the reoriented and recentered image.*

3.3 CT/MR rigid registration

We have applied our block matching algorithm to the 3D registration of CT with MR (T1 weighted) images. We used eight brain image pairs from the Vanderbilt database (see Figure 5), each ground-truth transformation being known thanks to a prospective, marker-based method [17]. As we are faced with a multimodal registration problem, we have chosen CR as the similarity measure to drive the block matching. Another choice, such as MI, is theoretically possible but, as discussed in Section 2.1, this is prohibitive in terms of computation time.

In these experiments, the block matching algorithm is compared with two methods based on the maximization of a global similarity measure, respectively CR and MI. In these cases, the CR or MI measure was computed using partial volume interpolation (PV) [6], and the maximization was undertaken using Powell's method in a two-step multiresolution scheme (the CT was subsampled by factors $4 \times 4 \times 1$ in the first step, and not subsampled at the second).

In all the $(3 \times 8 = 24)$ registration experiments, the initial transformation between the CT and the MR was set to the identity. The registration errors we obtained with the different methods are represented in Figure 4 and Table 1. On the one hand, we consider the *center error*, i.e., the error made on the translation of the CT image center. However, since this error does not reflect the errors in the estimation of the rotation, we also define the *corner error* as the RMS error computed over the eight vertices of a bounding box centered in the CT image, with a relative size of $33\% \times 33\% \times 100\%$. This box is roughly tangential to the patient's head. Of course, to be meaningful, these errors have to be compared with the images resolution: $1.25 \times 1.25 \times 4$ mm^3 for the MR, and $0.65 \times 0.65 \times 4$ mm^3 for the CT.

Table 1. Mean and median registration errors found using three different algorithms in CT/MR registration.

Method	Center errors		Corner errors	
	Mean (mm)	Median (mm)	Mean (mm)	Median (mm)
Global CR	6.57	6.29	7.45	5.80
Global MI	5.08	3.11	5.07	3.08
Block Matching with CR	2.39	2.03	3.70	3.32

These comparative performances lead us to the following observations. First, it is clear that the global methods have both failed in one case (actually, for the same patient dataset). Notice, it was possible to obtain better results by manually resetting the initial transformation parameters, but these are not presented here since our goal is to compare fully automatic methods. We warn the reader that some other results obtained with the global methods might correspond to a local, but non-global, maximum of the similarity measure.

As may be observed in Table 1, the block matching algorithm has provided the best average and median center errors. However, the trend is different when

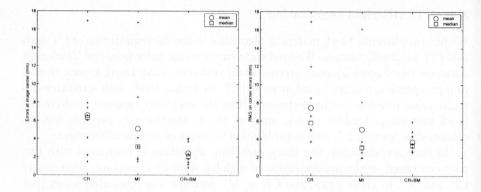

Fig. 4. Registration errors of three different algorithms in CT/MR registration: CR (global maximization), MI (global maximization), and CR+BM (the present block matching algorithm using CR as local similarity measure).

looking at the corner errors, as the performance of global MI and block matching are actually comparable (better for the average for the block matching, while better for the median for global MI). This means that the rotation was best estimated when using global MI: mean rotation errors are 2.85° (global CR), 0.75° (global MI), and 1.68° (block matching).

Another observation is that the performances of the block matching algorithm are remarkably stable across experiments. This is well reflected in Figure 4, where one sees that the error measures corresponding to the block matching algorithm are more concentrated than those corresponding to global CR and global MI. We may thus conclude that the block matching algorithm has provided more robust, if not accurate, estimates of the actual transformations than the global approaches based on MI and CR.

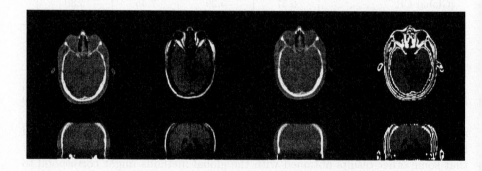

Fig. 5. *CT/MR registration. Left: original CT; Center left: original MR; Center right: registered CT to MR; Right: MR with registered CT contours superimposed*

4 Conclusion

In this paper, we have presented a general strategy for the rigid registration of medical images which is based on the combination of a block matching technique with a robust transformation estimator. We have presented experiments of histological slices alignment, computation of the mid-sagittal plane in brain images, and multimodal CT/MR registration. Our results suggest that, in some cases, the block matching approach can significantly outperform global approaches in terms of robustness.

Concerning the first two experiments, previous analysis [9, 11] has demonstrated the accuracy and robustness of our method. Further analysis is needed in order to better demonstrate its interest in other multimodal registration problems. Moreover, a comparative study between mutual information and correlation ratio within the block matching approach will be investigated.

5 Acknowledgments

The authors would like to thank Elf and Sanofi-Research, Monptellier (France), for the rat's brain datasets presented in Section 3.1.

The MR image presented in Section 3.2 has been provided by Dr. Neil Roberts, Magnetic Resonance and Image Analysis Research Centre (University of Liverpool, UK). The SPECT image presented in the same Section has been provided by Pr. Michael L. Goris, Department of Nuclear Medicine (Stanford University Hospital, USA).

The images and the standard transformations we used in Section 3.3 were provided as part of the project, "Evaluation of Retrospective Image Registration", National Institutes of Health, Project Number 1 R01 NS 33926-01, Principal Investigator, J. Michael Fitzpatrick, Vanderbilt University, Nashville, TN.

References

1. G.L. Brown. A Survey of Image Registration Techniques. *ACM Computing Surveys*, 24(4):325–376, 1992.
2. D. L. Collins and A. C. Evans. ANIMAL: Validation and Applications of Nonlinear Registration-Based Segmentation. *International Journal of Pattern Recognition and Artificial Intelligence*, 11(8):1271–1294, 1997.
3. D.W. Eggert, A. Lorusso, and R.B. Fisher. Estimating 3D Rigid Body transformations: A Comparison of Four Major Algorithms. *Machine Vision Applications, Special Issue on Performance Characteristics of Vision Algorithms*, 9(5/6):272–290, 1997.
4. T. Gaens, F. Maes, D. Vandermeulen, and P. Suetens. Non-rigid Multimodal Image Registration Using Mutual Information. In *Proc. MICCAI'98*, volume 1496 of *Lecture Notes in Computer Science*, pages 1099–1106, Cambridge, Massachusetts (USA), October 1998. Springer.
5. L.S. Hibbard and R.A. Hawkings. Objective image alignment for three-dimensional reconstruction of digital autoradiograms. *Journal of Neuroscience Method*, 26:55–74, 1988.

6. F. Maes, A. Collignon, D. Vandermeulen, G. Marchal, and P. Suetens. Multimodality Image Registration by Maximization of Mutual Information. *IEEE Transactions on Medical Imaging*, 16(2):187–198, 1997.

7. J. B. A. Maintz and M. A. Viergever. A survey of medical image registration. *Medical Image Analysis*, 2(1):1–36, 1998.

8. J.B. A. Maintz, E.H.W. Meijering, and M.A. Viergever. General multimodal elastic registration based on mutual information. In K. M. Hanson, editor, *Medical Imaging: Image Processing*, volume 3338 of *Proc SPIE*, Bellingham, WA, 1998. SPIE Press.

9. S. Ourselin, A. Roche, G. Subsol, X. Pennec, and N. Ayache. Reconstructing a 3D Structure from Serial Histological Sections. *Image and Vision Computing*, 2000. In press.

10. J. P. W. Pluim, J. B. A. Maintz, and M. A. Viergever. Mutual information matching and interpolation artefacts. In K. M. Hanson, editor, *Medical Imaging: Image Processing*, volume 3661 of *Proc SPIE*, Bellingham, WA, 1999. SPIE Press.

11. S. Prima, S. Ourselin, and N. Ayache. Computation of the Mid-Sagittal Plane in 3D Images of the Brain. In *Sixth European Conference on Computer Vision, ECCV'2000*, Lecture Notes in Computer Science, Dublin, Ireland, June 2000. Springer.

12. A. Roche, G. Malandain, N. Ayache, and S. Prima. Towards a Better Comprehension of Similarity Measures used in Medical Image Registration. In *Proc. MICCAI'99*, volume 1679 of *LNCS*, pages 555–566, Cambridge (UK), October 1999.

13. A. Roche, G. Malandain, X. Pennec, and N. Ayache. The Correlation Ratio as a New Similarity Measure for Multimodal Image Registration. In *First International Conference on Medical Image Computing and Computer-Assisted Intervention*, volume 1496 of *Lecture Notes in Computer Science*, pages 1115–1124, Cambridge (USA), October 1998. Springer.

14. P.J. Rousseeuw and K. Van Driessen. Computing LTS Regression for Large Data Sets. Technical report, Statistics Group, University of Antwerp, 1999.

15. Peter J. Rousseeuw and Annick M. Leroy. *Robust Regression and Outlier Detection*. Wiley Series in Probability and Mathematical Statistics, first edition, 1987.

16. P. Viola. Alignment by Maximisation of Mutual Information. *International Journal of Computer Vision*, 24(2):137–154, 1997.

17. J. West and al. Comparison and evaluation of retrospective intermodality brain image registration techniques. *Journal of Comp. Assist. Tomography*, 21:554–566, 1997.

Generalized Correlation Ratio for Rigid Registration of 3D Ultrasound with MR Images

A. Roche[1], X. Pennec[1], M. Rudolph[2], D. P. Auer[3], G. Malandain[1],
S. Ourselin[1], L. M. Auer[2], and N. Ayache[1]

[1] INRIA, Epidaure Project, Sophia Antipolis, France
[2] Institute of Applied Sciences in Medicine, ISM, Salzburg, Austria
[3] Max Planck Institute for Psychiatry, Munich, AG-NMR, Germany

Abstract. Automatic processing of 3D ultrasound (US) is of great interest for the development of innovative and low-cost computer-assisted surgery tools. In this paper, we present a new image-based technique to rigidly register intra-operative 3D US with pre-operative Magnetic Resonance (MR) data. Automatic registration is achieved by maximization of a similarity measure that generalizes the correlation ratio (CR). This novel similarity measure has been designed to better take into account the nature of US images. A preliminary cross-validation study has been carried out using a number of phantom and clinical data. This indicates that the worst registration errors are of the order of the MR resolution.

1 Introduction

Over the last years, the development of real-time 3D ultrasound (US) imaging has revealed a number of potential applications in image-guided surgery. The major advantages of 3D US over existing intra-operative imaging techniques are its comparatively low cost and simplicity of use. However, the automatic processing of US images has not gained the same degree of development as other medical imaging modalities, probably due to the low signal-to-noise ratio of US images.

The registration of US with pre-operative Magnetic Resonance (MR) images will allow the surgeon to accurately localize the course of instruments in the operative field, resulting in minimally invasive procedures. At present, few papers have been published on this particular registration problem [5]. Most of the approaches that have been proposed are based on stereotactic systems. For instance, in [8] registration is achieved by tracking the US probe with a DC magnetic position sensor. Existing image-based methods match homologous features extracted from both the US and MR data. Features are user-identified in [1], while semi-automatically extracted in [2]. More recently, Ionescu *et al* [3] registered US with Computed Tomography (CT) data after automatically extracting contours from the US using watershed segmentation.

The present registration technique expands on the correlation ratio (CR) method [12]. It is an intensity-based approach as it does not rely on explicit feature extraction. In a previous work [11], we reported preliminary results of US/MR registration by maximization of CR and mutual information (MI). While results obtained using CR were more satisfactory than when using MI, the method was still lacking precision and robustness with respect to the initialisation of the transformation parameters.

In this paper, we have improved the CR method following three distinct axes: (1) using the gradient information from the MR image, (2) reducing the number of intensity parameters to be estimated, and (3) using a robust intensity distance. These extensions are presented in the following section, while section 3 proposes an original evaluation of the method accuracy using phantom and clinical data.

2 Method

2.1 Correlation ratio

Given two images I and J, the basic principle of the CR method is to search for a spatial transformation T and an intensity mapping f such that, by displacing J and remapping its intensities, the resulting image $f(J \circ T)$ be as similar as possible to I. In a first approach, this could be achieved by minimizing the following cost function:

$$\min_{T,f} \quad \sum_k \left[I(x_k) - f(J(T(x_k)))\right]^2, \tag{1}$$

which integrates over the voxel positions x_k in image I. In the following, we will use the simplified notations $i_k \equiv I(x_k)$, and $j_k^{\downarrow} \equiv J(T(x_k))$, where the arrow expresses the dependence in T. This formulation is asymmetric in the sense that the cost function changes when permuting the roles of I and J. Since the positions and intensities of J actually serve to predict those of I, we will call J the "template image". In the context of US/MR registration, we always choose the MR as the template.

In practice, the criterion defined in eq (1) cannot be computed exactly due to the finite nature of the template image. One obvious problem is that the transformed position of a voxel will generally not match a grid point of J, such that the corresponding intensity $j_k^{\downarrow}$ is unknown. A classical approach is then to linearly interpolate $j_k^{\downarrow}$ using the eight neighbours of $T(x_k)$ in the grid of J. However, instead of interpolating the image intensity, we may directly interpolate the incremental contribution of x_k, i.e., $[i_k - f(j_k^{\downarrow})]^2$. The difference between these two approaches is illustrated in figure 1. In fact, the last method is equivalent to the so-called partial volume (PV) interpolation, originally proposed by Maes *et al* [4] in the context of joint histogram computation. We have found PV to generally outperform classical linear interpolation in terms of smoothness of the resulting registration criterion.

Another difficulty to compute eq (1) is that some points x_k may transform outside the template domain and lack eight grid neighbours. We decide not to take into account such points in the computation of the registration criterion.

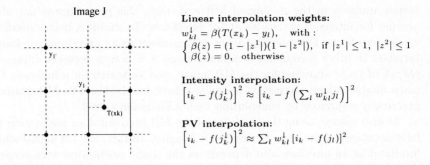

Fig. 1. Illustration of linear interpolation in the 2D case, and two related strategies of interpolating the registration criterion.

Doing so without particular attention, the criterion would become zero when every point x_k transforms outside J. Hence, in order to avoid an absolute minimum when the image overlap is small, we impose the additional constraint that the variance of I be large in the overlapping region. Justifications of this particular normalization strategy will be found in [12], while related normalization issues are discussed in [15, 16].

These practical considerations lead us to the following modification of eq (1):

$$C(T, f) = \frac{\sum_{k,l} w_{kl}^{\downarrow} [i_k - f(j_l)]^2}{n^{\downarrow} \mathrm{Var}(I^{\downarrow})}, \tag{2}$$

where j_l is the intensity of a voxel with coordinates y_l in the coordinate system of J. The terms depending on T are marked with an arrow: $w_{kl}^{\downarrow}$ are the linear interpolation weights, $n^{\downarrow}$ is the number of points x_k such that $T(x_k)$ has eight neighbours in the template grid, and $\mathrm{Var}(I^{\downarrow})$ is the intensity variance computed over these points.

If no constraint is imposed to the intensity mapping f, an important result is that the optimal f at fixed T enjoys an explicit form that is very fast to compute [12]. The minimization of eq (2) may then be performed by travelling through the minima of $C(T, f)$ at fixed T. This yields the correlation ratio, $\eta_{I|J}^2(T) = 1 - \min_f C(T, f)$, a measure that reaches its maximum when $C(T, f)$ is minimal. In practice, the maximization of η^2 is performed using Powell's method.

2.2 Bivariate correlation ratio

Ultrasound images are commonly said to be "gradient images" as they enhance the interfaces between anatomical structures. The physical reason is that the amplitudes of the US echos are proportional to the *difference* of acoustical impedance caused by successive tissue layers. Ideally, the US signal should be high at the interfaces, and zero within homogeneous tissues.

As stated above, the CR method tries to predict the intensities of the US by remapping those of the MR. Hence, uniform regions of the original MR will remain uniform in the remapped MR and, thus, this procedure is not able to account for intensity variations at the interfaces. To enable a better prediction, we propose to use the modulus of the MR gradient as an additional explanatory variable. In other terms, our template image J is now a vectorial image, $J = (M, \|\nabla M\|)$, M standing for the MR image, and we search for a function f that maps double-valued intensities to single-valued intensities. The MR gradient is practically computed by convolution with a Gaussian kernel.

At first glance, using the modulus of the MR gradient does not appear to be fully acceptable from the physics of US imaging. In fact, the US signal which is produced at an interface also depends on the tissue orientation with respect to the scan line. Thus, perhaps a more appropriate choice than $\|\nabla M\|$ would be the dot product, $\nabla M.\mathbf{u}$, where $\mathbf{u}$ is the scan direction. The main difficulty in using this last expression is that $\mathbf{u}$ is unknown before registration since it depends on the position of the US probe in the MR coordinate system.

We have not yet studied the effect of using the projected MR gradient versus the modulus. Still, we believe that there are good reasons not to take into account information from the gradient orientation, at least as a first-order approximation. Through diffraction of the ultrasound beam on interfaces, the received echo is actually less dependent on the direction propagation than would be the case with perfectly specular reflection. This, added to log-compression, tends to equalize the response values corresponding to different tissue orientations.

2.3 Parametric intensity fit

If we put no special constraint on the mapping f to be estimated, then f is described by as many parameters as there are distinct intensity values in the template image [11]. That approach makes sense as long as the number of intensity classes in J is small with respect to the number of voxels used to draw an estimate. In our case, J is a double-valued image (with, in general, floating precision encoding of the MR gradient component), and the number of parameters to be estimated becomes virtually infinite.

We will therefore restrict our search to a polynomial function f. Let m_l and g_l denote the intensity of the voxel with coordinates y_l, respectively in the original MR, M, and in $\|\nabla M\|$. We are searching for a mapping of the form:

$$f(m_l, g_l) = \sum_{p+q \leq d} \theta_{pq} \, m_l^p g_l^q, \tag{3}$$

where d is the specified polynomial degree. The number of parameters describing f then reduces to $(d+1)(d+2)/2$. In all the experiments presented below, the degree was set to $d = 3$, implying that 10 coefficients were estimated. It is shown in [13] that minimizing eq (2) with respect to the polynomial coefficients brings us to a weighted least square (WLS) linear regression problem. As is standard, this is solved by the method of singular value decomposition (SVD).

This polynomial fitting procedure, however, has significant extra computational cost with respect to the unconstrained fitting. Recall that, in the basic version of the CR method, f is updated for each transformation trial. Such a strategy is no longer affordable when estimating a polynomial function. Instead, the minimization of eq (2) may be performed alternatively along T and f, resulting in the following algorithm: (1) given a current transformation estimate T, find the best polynomial f and remap J accordingly; (2) given a remapped image $f(J)$, minimize $C(T, f)$ with respect to T using Powell's method; (3), return to (1) if T or f has evolved.

2.4 Robust intensity distance

Our method is based on the assumption that the intensities of the US may be well predicted from the information available in the MR. Due to several ultrasound artefacts, we do not expect this assumption to be perfectly true. Shadowing, duplication or interference artefacts may cause large variations of the US intensity from its predicted value, and this even when the images are perfectly registered. Such bad intensity matches are false negative.

The sensitivity of the registration criterion to false negative may be reduced by replacing the expression $(1/n^\downarrow) \sum_{k,l} w_{kl}^\downarrow [i_k - f(j_l)]^2$ in eq (2) with a robust scale estimate. A similar idea was developed in [6]. We propose here to build such an estimate from a one-step S-estimator [14]:

$$\hat{S}^2(T, f) = \frac{S_0^2}{Kn^\downarrow} \sum_{k,l} w_{kl}^\downarrow \rho \left(\frac{i_k - f(j_l)}{S_0} \right),$$

where ρ is the objective function corresponding to a given M-estimator, K is a normalization constant to ensure consistency with the normal distribution, and S_0 is some initial guess of the scale. The new registration criterion is then: $C(T, f) = \hat{S}^2(T, f)/\mathrm{Var}(I^\downarrow)$.

This criterion implies few modifications of our alternate minimization strategy. As a function of T, it may still be minimized by means of Powell's method. As a function of f, the solution is found by a simple iterative WLS procedure as shown in [13], generally requiring no more than 5-6 iterations.

In our implementation, we have opted for the Geman-McClure ρ-function, $\rho(x) = \frac{1}{2}x^2/(1 + \frac{x^2}{c^2})$, for its computational efficiency and good robustness properties, to which we always set a cut-off distance $c = 3.648$ corresponding to 95% Gaussian efficiency. The normalization constant is then $K = 0.416$.

Initially, the intensity mapping f is estimated in a non-robust fashion. The starting value S_0 is then computed as the weighted median absolute deviation of the corresponding residuals, $\{|i_k - f(j_l)|\}$ (see [13] for details). Due to the initial misalignment, S_0 tends to be overestimated. Thus, it may not allow to reject efficiently bad intensity matches. For that reason, we reset S_0 at each new iteration, i.e., after completing one minimization along T and one minimization along f.

3 Experiments

The experiments related in this section were performed within the framework of the European project ROBOSCOPE[1]. The goal is to assist neuro-surgical operations using real-time 3D ultrasound images and a robotic manipulator arm. The operation is planned on a pre-operative MRI and 3D US images are acquired during surgery to track in real time the deformation of anatomical structures. In this context, the rigid registration of the pre-operative MR with the first US image (dura mater still closed) is a fundamental task to relate the position of the surgical instruments with the actual anatomical structure. This task being determinant for the global accuracy of the system, different datasets were acquired to simulate the final image quality and to perform accuracy evaluations.

It should be emphasized that all the US images provided in this project were stored in Cartesian format, which means that the actual (log-compressed) ultrasound signal is resampled on a regular cubic lattice. As a consequence, the images undergo sever interpolation artifacts (blurring) in areas which are distant from the probe. In the following, we will refer to US images as cubic images, but one has to keep in mind that this is somewhat artificial. Notably, the voxel size in Cartesian US images should not be confused with the real spatial resolution, which is in fact spatially dependent.

We computed all the MR/US registrations using the previously described algorithm. The location of the US probe being linked to the pathology and its orientation being arbitrary (the rotation may be superior to 90 degrees), it was necessary to provide a rough initial estimate of the transformation. Here, this was done using an interactive interface that allows to draw lines in the images and match them. This procedure was carried out by a non-expert, generally taking less than 2 minutes. However this user interaction could be alleviated using a calibration system such as the one described in [8]. After initialization, we observed that the algorithm found residual displacements in the range of 10 mm and 10 degrees. In all the experiments, the gradient norm of the MR image was computed by linear filtering using a Gaussian kernel with $\sigma = 1$ voxel.

3.1 Principle of the accuracy evaluation

To estimate the accuracy of the algorithm, one should ideally compare the result of a registration with a gold-standard. Up to our knowledge, there is no such gold-standard for MR/US registration. To get around this problem, our main idea is to use several MR and/or US images to compute registration loops and test for the residual error on test points. What we call a registration loop is a succession of transformation compositions that sould ideally lead to the identity transformation. A typical loop is a sequence of the form $US_i \rightarrow MR_i \rightarrow MR_j \rightarrow US_j \rightarrow US_i$ in the case of the Phantom data described below.

If we were given perfectly registered images within each modality, this loop would only be disturbed from the identity by errors on the two MR/US registrations. As the variances are additive, the variance of the observed error should

---

[1] http://www.ibmt.fhg.de/Roboscope/home.htm

roughly be: $\sigma^2_{loop} = 2\sigma^2_{MR/US}$. Unfortunately, we are not provided with a ground truth registration within each modality: we need to estimate it. This time, as we are combining one MR/MR, one US/US and two MR/US registrations, the variance of the loop error will be roughly: $\sigma^2_{loop} \simeq 2\sigma^2_{MR/US} + \sigma^2_{MR/MR} + \sigma^2_{US/US}$. The *expected* MR/US accuracy is then $\sigma_{MR/US} \simeq \sqrt{(\sigma^2_{loop} - \sigma^2_{MR/MR} - \sigma^2_{US/US})/2}$.

However, what we really measure is the maximum or *conservative* MR/US accuracy, $\sigma_{MR/US} \simeq \sigma_{loop}/\sqrt{2}$. In order to minimize the influence of intra-modality registrations errors in this figure, we need to provide very accurate MR/MR and US/US registrations. For that purpose, we designed the following algorithm.

Multiple intra-modality registration To relate n images together, we need to estimate $n - 1$ rigid registrations $\bar{T}_{i,i+1}$. To obtain a very good accuracy, we chose to register all image pairs, thus obtaining $n(n - 1)$ transformations $\bar{T}_{i,j}$, and estimate the transformations $\bar{T}_{i,i+1}$ that best explain our measurements in the least-square sense, i.e. that minimizes the following criteria:

$$C(\bar{T}_{1,2}, \ldots \bar{T}_{n-1,n}) = \sum_{i \neq j} \text{dist}^2(\bar{T}_{i,j}, T_{i,j}),$$

where $\bar{T}_{i,j}$ is recursively defined by $\bar{T}_{i,j} = \bar{T}_{j-1,j} \circ \ldots \circ \bar{T}_{i,i+1}$ if $j > i$, and $\bar{T}_{i,j} = \bar{T}^{(-1)}_{j,i}$ if $j < i$.

We used a robust variant of the left invariant distance on rigid transformations introduced in [10]: let σ_r and σ_t be typical scales on the rotation angle and on the translation magnitude and χ^2 a threshold. If (r, t) are the rotation vector and the translation of transformation T, the distance between two transformations is

$$\text{dist}^2(T_1, T_2) = \min \left(\|r_2^{(-1)} \circ r_1\|^2/\sigma_r^2 + \|t_1 - t_2\|^2/\sigma_t^2 , \ \chi^2 \right).$$

The standard deviations σ_r and σ_t are manually adjusted to correspond roughly to the residual rotation and translation error after convergence. To obtain the minimum, we used a Newton gradient descent similar to the one described in [10], but on all transformations $\bar{T}_{i,i+1}$ together (see [13] for details).

3.2 Data

MR and US compatible Phantom We developed for the European Project ROBOSCOPE an MR and US compatible phantom made of two balloons that can be inflated with known volumes in order to simulate deformations. In this experiment, we used 8 acquisitions with different balloons volumes, each acquisition consisting of one 3D MR and one 3D US image. However, we cannot directly compare the MR/US registrations as the phantom is moved between the acquisitions. Thus, the first step is to rigidly register all the MR images together and similarly for the US images.

The main problem for the multiple intra-modality registration of the phantom images is that the balloons deform inbetween acquisitions. The only rigid part is the outer part of the container. Thus, intra-modal registrations of MR images were carried out using a feature-based registration algorithm known to handle a large amount of outliers [9]. For the US images, since it is very difficult to extract meaningful features on these images, we used the robust block matching technique proposed in [7].

As we are testing the *rigid* registration, we cannot register MR and US images across acquisitions. Thus, the simplest loops we can use for accuracy estimations are the $n(n-1)$ following loops: $US_i \rightarrow MR_i \rightarrow MR_j \rightarrow US_j \rightarrow US_i$. Of course, only $n-1$ loops are independent but since the ideal value is known (the identity) there is no need to correct the estimation for the number of free parameters.

Baby dataset with degraded US images This dataset was acquired to simulate the degradation of the US images quality with respect to the number of converters used in the probe. Here, we have one MR T1 image of a baby's head and 5 transfontanel US images with different percentages of converters used.

As we have no or very few deformations within the images, we can rigidly register all the US images onto our single MR and test the 30 following loops $US_i \rightarrow MR \rightarrow US_j \rightarrow US_i$ (only 5 of them being independent). For that, we still need to register the US images together. Since there are no deformations between the acquisitions (we only have a motion of the probe and a degradation of the image quality) the algorithm is much more efficient and accurate than for the Phantom.

Patient dataset during tumor resection This dataset is an actual surgical case: two MR T1 images with and without a contrast agent were acquired before surgery. After craniotomy (dura mater still closed), a set of 3D US images was acquired to precisely locate the tumor to resect. In this experiment, we use the three US images that are large enough to contain the ventricles.

The two MR images were registered using the feature based method with a very high accuracy (probably overestimated as we only have two images) and we tested the loops $US_i \rightarrow MR_0 \rightarrow MR_1 \rightarrow US_i$. As the acquisition cone of the US probe is completely within the Cartesian image (see Fig. 4), the region of interest is much smaller than the images size: we took our typical points at the corners of a $80 \times 80 \times 80$ mm^3 cube centered in the image.

3.3 Results and discussion

We put in table (1) the standard deviations of the residual rotation, of the residual translation and of the displacement of the test points for the different registration involved. Since we took the origin of the images at the center, the σ_{trans} value corresponds to the mean error at the center of the image while σ_{test} corresponds to the maximum registration error within the US image (except for the patient experiment, the test points are taken at the corners of the image).

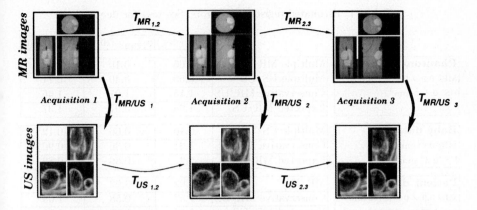

Fig. 2. MR and US images of the Phantom and the rigid registrations that are involved.

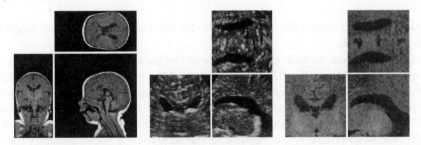

Fig. 3. Example registration of the MR and US images of the baby. From left to right: original MR T1, original US and registered MR.

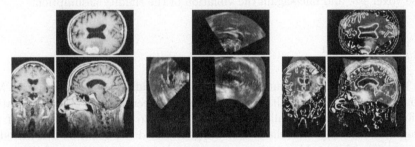

Fig. 4. Example registration of the MR and US images of the patient. From left to right: MR T1 image with a contrast agent, manual initialisation of the US image registration, and result of the automatic registration with the MR contours superimposed.

The results on the Phantom show that the MR/US registration accuracy is of the order of the MR resolution. One could probably expect a better conservative accuracy by acquiring larger US image including some rigid landmarks for multiple US/US registration. One finds the same type of results for the other

Table 1. Estimated registration errors. See text for details.

		σ_{rot} (deg)	σ_{trans} (mm)	σ_{test} (mm)
Phantom dataset	Multiple MR/MR	0.06	0.10	0.13
MR: 0.9 × 0.9 × 1 mm^3	Multiple US/US	0.60	0.40	0.71
US: 0.41^3 mm^3	Conservative MR/US	1.15	1.01	1.46
	Expected MR/US	1.06	0.97	1.37
Baby dataset	Multiple US/US	0.10	0.06	0.12
MR: 0.9^3 mm^3	Conservative MR/US	1.21	0.36	0.90
US: 0.3^3 mm^3	Expected MR/US	1.21	0.36	0.89
Patient dataset	MR/MR	0.02	0.02	0.03
MR: 0.9 × 0.9 × 1.1 mm^3	Conservative MR/US	1.57	0.58	1.65
US: 0.63^3 and 0.95^3 mm^3	Expected MR/US	1.57	0.58	1.65

datasets: slightly under the MR voxel size for the baby data and a bit larger for the patient data.

However, when we look more carefully at the patient results, we find that the loop involving the smallest US image (real size $150 \times 85 \times 100$ mm, voxel size 0.63^3 mm^3) is responsible for a corner error of 2.61 mm ($\sigma_{trans} = 0.84$ mm) while the loops involving the two larger US images (real size $170 \times 130 \times 180$, voxels size 0.95^3 mm^3) do have a much smaller corner error of about 0.84 mm ($\sigma_{trans} = 0.39$ mm). We suspect that a non-rigidity in the smallest US could accounts for the registration difference between the two MR images. Another explanation could be a misestimation of the sound speed for this small US acquisition leading to a false voxel size and once again the violation of the rigidity assumption.

4   Conclusion

We have presented a new automated method to rigidly register 3D US with MR images. It is based on a multivariate and robust generalization of the correlation ratio (CR) measure that allows to better take into account the nature of US images. Incidentally, we believe that the generalized CR could be considered in other registration problems where conventional similarity measures fail.

Testings were performed on several phantom and clinical data, and accuracy was evaluated using an original method that does not require the knowledge of a ground truth. We estimated the worst registration errors (errors at the Cartesian US corners) to be of the order of 1 millimeter.

In our experiments, registration was tested with US images stored in Cartesian format. This obviously does not help registration owing to the fact that: 1) intensities of voxels distant from the probe are unreliable, 2) resampling on a cubic lattice artificially increases the concentration of such voxels, consequently increasing their influence on the registration criterion. The present algorithm may be straightforwardly extended to use polar US images as inputs (without interpolation of the US signal, see section 2). We believe that this could significantly improve both the accuracy and the robustness of registration.

Further developments also include non-rigid registration, in order to correct for false distance artefacts in US, as well as for tissue deformations due to brain shift and operative manipulations.

Acknowledgements This work was partially supported by la Région PACA (France), and by the EC-funded ROBOSCOPE project HC 4018, a collaboration between The Fraunhofer Institute (Germany), Fokker Control System (Netherlands), Imperial College (UK), INRIA (France), ISM-Salzburg and Kretz Technik (Austria).

References

1. H. Erbe, A. Kriete, A. Jödicke, W. Deinsberger, and D.-K. Böker. 3D-Ultrasonography and Image Matching For Detection of Brain Shift During Intracranial Surgery. In *Appl. of Comp. Vision in Med. Image Processing*, volume 1124 of *Excerpta Medica – Int. Congress Series*, pages 225–230. Elsevier, 1996.
2. N. Hata, M. Suzuki, T. Dohi, H. Iseki, K. Takakura, and D. Hashimoto. Registration of Ultrasound echography for Intraoperative Use: A Newly Developed Multiproperty Method. In *Visualization in Biomedical Computing*, volume 2359 of *Proc. SPIE*, pages 252–259, Rochester, MN, USA, October 1994. SPIE Press.
3. G. Ionescu, S. Lavallée, and J. Demongeot. Automated Registration of Ultrasound with CT Images: Application to Computer Assisted Prostate Radiotherapy and Orthopedics. In *Proc. MICCAI'99*, volume 1679 of *LNCS*, pages 768–777, Cambridge, UK, 1999. Springer Verlag.
4. F. Maes, A. Collignon, D. Vandermeulen, G. Marchal, and P. Suetens. Multimodality Image Registration by Maximization of Mutual Information. *IEEE Transactions on Medical Imaging*, 16(2):187–198, 1997.
5. J. B. A. Maintz and M. A. Viergever. A survey of medical image registration. *Medical Image Analysis*, 2(1):1–36, 1998.
6. C. Nikou, F. Heitz, J.-P. Armspach, and I.-J. Namer. Single and multimodal subvoxel registration of dissimilar medical images using robust similarity measures. In *Conf. on Med. Imaging*, volume 3338 of *SPIE*, pages 167–178. SPIE Press, 1998.
7. S. Ourselin, A. Roche, S. Prima, and N. Ayache. Block Matching: a General Framework to Improve Robustness of Rigid Registration of Medical Images. In *MICCAI'2000*, Pittsburgh, USA, October 2000.
8. N. Pagoulatos, W.S. Edwards, D.R. Haynor, and Y. Kim. Interactive 3-D Registration of Ultrasound and Magnetic Resonance Images Based on a Magnetic Position Sensor. *IEEE Trans. on Information Technology in Biomed.*, 3(4):278–288, 1999.
9. X. Pennec, N. Ayache, and J.P. Thirion. Chap. 31: Landmark-based registration using features identified through differential geometry. In *Handbook of Medical Imaging*. Academic Press, 2000. In press.
10. X. Pennec, C.R.G. Guttmann, and J.P. Thirion. Feature-based registration of medical images: Estimation and validation of the pose accuracy. In *Proc. MICCAI'98*, number 1496 in LNCS, pages 1107–1114, Cambridge, USA, 1998. Springer Verlag.
11. A. Roche, G. Malandain, and N. Ayache. Unifying Maximum Likelihood Approaches in Medical Image Registration. *Int. J. of Imaging Systems and Technology*, 11:71–80, 2000.
12. A. Roche, G. Malandain, X. Pennec, and N. Ayache. The Correlation Ratio as a New Similarity Measure for Multimodal Image Registration. In *Proc. MICCAI'98*, volume 1496 of *LNCS*, pages 1115–1124, Cambridge, USA, 1998. Springer Verlag.
13. A. Roche, X. Pennec, M. Rudolph, D.P. Auer, G. Malandain, S. Ourselin, L.M. Auer, and N. Ayache. Generalized Correlation Ratio for Rigid Registration of 3D Ultrasound with MR Images. Technical report, INRIA, 2000. In press.
14. Peter J. Rousseeuw and Annick M. Leroy. *Robust Regression and Outlier Detection*. Wiley Series In Probability And Mathematical Statistics, first edition, 1987.
15. C. Studholme, D. L. G. Hill, and D. J. Hawkes. An overlap invariant entropy measure of 3D medical image alignment. *Pattern Recognition*, 1(32):71–86, 1998.
16. P. Viola and W. M. Wells. Alignment by Maximization of Mutual Information. *Intern. J. of Comp. Vision*, 24(2):137–154, 1997.

Simulating Minimally Invasive Neurosurgical Interventions Using an Active Manipulator

Arne Radetzky[1], Michael Rudolph[1], Werner Stefanon[1], Stephen Starkie[2], Brian Davies[2], and Ludwig M. Auer[1]

[1] Institute of Applied Sciences in Medicine, ISM-Austria, Jakob-Haringer Str. 3, A-5020 Salzburg, Austria
{A.Radetzky, Michael.Rudolph, Werner.Stefanon, Auer}@ism-austria.at
http://www.ism-austria.at

[2] Department of Mechanical Engineering, Imperial College of Science, Technology and Medicine, London, SW7 2BX, UK
{s.starkie, b.davies}@ic.ac.uk

Abstract. This application report describes the software system ROBO-SIM, which is a planning and simulation tool for minimally invasive neurosurgery. Using actual patients' datasets, ROBO-SIM consists of a planning unit and a simulator for microsurgical manipulations. The planning steps are 1. definition of the trepanation for entry into the intracranial space and virtual craniotomy, 2. the target point within the depth of the brain, 3. control of the surgical track, 4. definition of go-areas for use with an intra-operative active manipulator. The simulator allows neurosurgeons to perform virtual surgical interventions using actual patient data and the same instruments as for the real operation.

1 Introduction

With the advent of computer-assisted methods in surgery, image-guided planning has become an increasingly accepted procedure in neurosurgery under the term of 'neuronavigation' [1-4]. As soon as robotic manipulators are considered to augment precision of microsurgical procedures [5-8], complex preoperative planning and simulation becomes mandatory and represent an important part of the total duration of a robot-assisted operative procedure. An approach to combine a newly developed planning and simulation platform with the capability for robot-assisted neurosurgery is presented in the following. The integral setup called ROBO-SIM is designed for manipulator-assisted virtual procedures through a trepanation in the skull of 1-2cm diameter and a miniaturized approach of a few millimeters diameter to target areas in the depth of the brain and its ventricular system. ROBO-SIM is part of the operating system ROBOSCOPE (project of the EU-Telematics program, No. 4.018), including a robot arm, NEUROBOT, on which the actual endoscope is mounted. The surgeon who plans and performs a virtual surgical procedure thus works with the real instruments also used during actual surgery, while looking at a virtual scenario of the operating field created by aid of a 3D-MRI-dataset of an actual patient.

2 NEUROBOT

NEUROBOT (see Fig. 1), developed by Fokker control systems b.v., is used by the surgeon as an active manipulator with inbuilt robotic capabilities such as active constraints, within permitted regions, precise pattern control and the ability to automatically track features as they move and deform. NEUROBOT provides four degrees of freedom around a pivot point, which is the point of trepanation into the skull. Mounted to the robot is a force sensor with the ability to hold a standard neuro-endoscope. With the force sensor, every movement of the endoscope causes an active movement of the robot allowing a low-force input. In addition, rigid borders, e.g. pre-defined go-areas and no-go-areas, can be simulated using the active constraint mode. A foot switch controls the mode during operation. If it is activated, the robot can be moved; otherwise, the robot remains at its current position.

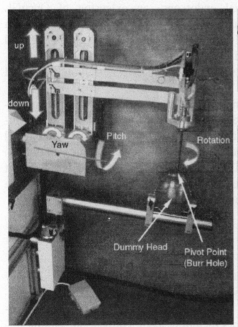
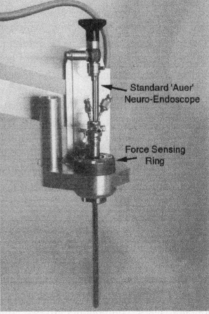

Fig. 1. Left: The active manipulator NEUROBOT provides four degrees of motion around a pivot point (burr hole). A dummy head can be clamped to the robot for simulated interventions. Right: A standard neuro-endoscope attached to a force sensing ring can be used to guide the robot. (Courtesy of Fokker Control Systems, B.V.)

NEUROBOT will be used either for real or simulated interventions. Thus, neurosurgical interventions can be trained using the same equipment as in real operations.

For simulation, a dummy head can be clamped to the robot, which can be adjusted for different trepanation points.

3   ROBO-SIM

To reduce the required amount of graphical power and nevertheless enable planning and simulation of neurosurgical procedures on actual patients' datasets, the use of a combination of volume- and surface-rendered data for visualization and simulation seems most practical. 3D-MRI datasets of the brains from actual patients are used to simulate views of the outer surface of the head as well as of inner surfaces, such as the ventricular system or cystic brain lesions, by aid of virtual endoscopy. For the simulations, 3D-MRI-datasets from a 1,5T Sigma echo-speed whole body scanner (GE Medical Systems, Milwaukee, USA) are used. The computer platform for the development and use of the system is an SGI Onyx2 Infinite Reality as well as SGI O2-workstations and conventional PCs in several different program versions. The rendering engine is based on OpenGL and OpenGL Volumizer [9]. To speed-up the rendering process especially for virtual endoscopy, an extension of OpenGL Volumizer, called Flight-Volumizer, was developed [10]. Flight-Volumizer uses a field of view rendering facility, which allows only the data visible through the field of view of an endoscope to be rendered. In addition, simulations of deformations, fragmentations and coagulations are possible using surface rendering on segmented models of the ventricle.

ROBO-SIM consists of two main components: the planning tool and the simulation tool. Using the planning tool, all the steps necessary to preplan a minimally invasive neurosurgical intervention are included. These steps are 1. 'Define Entry- and Targetpoint' for planning of an approach and target point for the trepanation, 2. 'Check Surgical Track' for checking and changing the trajectory from entry to target point, 3. 'Virtual Craniotomy' for simulating the trepanation and 4. 'Define Active Constraints' for defining go- and no-go-areas for robotic surgery.

3.1   Define Entry and Target Point

Usually the pre-operative planning begins with the definition of the point of entry into the skull for the trepanation (entry point) and the target point. Several pre-defined planning screens are included in ROBO-SIM consisting of a number of components. These components are two editors including axial, coronal and sagittal images of the patient for planning the entry and the target point, a virtual endoscopic view, a planning view, a main (global) view and several views with slices along the axis of the virtual endoscope. All screens are interactive and new screens can be defined by the neurosurgeon using the main components. Fig. 2 shows an example of such a planning screen. Moving the crosses for the entry or target point results in changing other slices and the virtual endoscopic view. The main or global view shows the position of the endoscope with respect to the patient's skull.

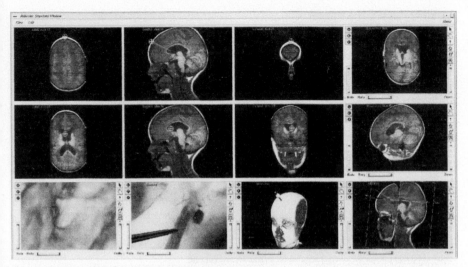

Fig. 2. An example screen of ROBO-SIM: The upper row of images allows the definition of the point of entry into the skull (trepanation). The middle row allows the definition of a requested target point. All images are primarily interactive; interaction may, however, be blocked. The image on the lower left shows a Flight-Volumizer virtual endoscopic view from a frontal approach into the left lateral ventricle. Lower middle images: view with a remote virtual endoscopic camera along the ventricular lumen shows the position of the virtual transendoscopic instrument and an outside view with a virtual endoscope in position for a left frontal approach. Right column: MR images along the endoscope and a virtual 3D view with the trajectory from entry to target point.

3.2 Check Surgical Track and Virtual Craniotomy

The next step is to check if the surgical track – i.e. the line from entry to target point - avoids areas of the brain that should not be transgressed by the endoscope. The standard screen to check the surgical track consists of a main view and a virtual endoscopic view. The axial, sagittal and coronal images are also displayed, with marked red dots where the surgical track intersects the slices (see Fig. 3). These images are used to outline the track of the endoscope. The surgical track can be moved by dragging with the computer mouse.

After checking the surgical track, a virtual trepanation can be performed. The size, depth and position of the trepanation can be planned. The trepanation is directly and interactively visualized in the volume rendered outside view of the patient's skull (see Fig. 3).

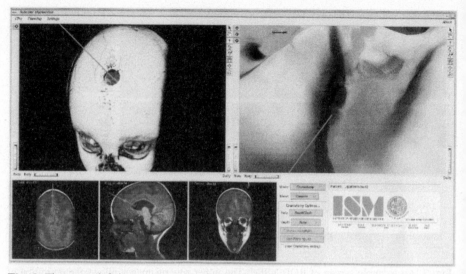

Fig. 3. The upper left image shows a volume-rendered outside view of the patient's dataset and the virtual trepanation at the entry point. The upper-right image shows the end of the surgical track. The surgical track can be moved with the mouse. The lower row shows axial, sagittal and coronal images of the patient's dataset overlaid with the surgical track and the point of intersection of the track and the image.

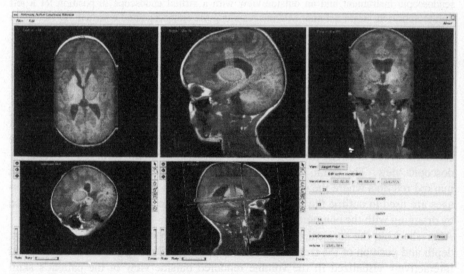

Fig. 4. The 'Active Constraint Screen' is used to define an arbitrary ellipsoid indicating the go-area for the active manipulator developed within the ROBOSCOPE project. The active manipulator restricts the surgeon performing a real operation from leaving this predefined area. The ellipsoid can be changed in size, orientation and position by the sliders on the lower right. A cylinder shows the direction from the entry to the target point. The radius of the cylinder indicates the radius of the trepanation.

3.3 Define Active Constraints

The next step is to define the target area (see Fig. 4). This could be, for example, a brain tumor. The target area is used as active constraint for the active manipulator during an operation. To prevent the neurosurgeon from accidentally damaging healthy brain areas, physical resistance is provided to the surgeon by the motors embodied in the NEUROBOT. At present, arbitrary ellipsoids are allowed to define the target area.

3.4 Simulation

After finishing the planning steps, the simulation can be performed using the same active manipulator, which is used during a real operation. A volume rendered view of the ventricle can be used for real-time virtual endoscopy.

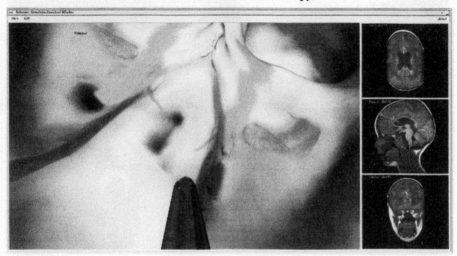

Fig. 5. Simulation screen of ROBO-SIM: Several tissue fragments were removed in center of the virtual endoscopic view by using a biopsy forceps. The right column shows the position of the endoscope in relation to the MR dataset and the surgical track.

In addition to simulation of virtual endoscopy, a full simulation tool based on deformable surface meshes was integrated in ROBO-SIM (see Fig. 5). The surface meshes are created by general object reconstruction based on simplex meshes [11]. For the physical-based simulation of these surface meshes, neuro-fuzzy systems [12] are used. These special neuro-fuzzy systems can emulate the real deformation even without a properly defined mesh structure [13]. Thus, volumetric meshes are not needed for realistic impressions of virtual deformations. In addition to the simulation of deformations, tools for coagulation and fragmentation of tissue are integrated in ROBO-SIM (see Fig. 6). Dynamic collision detection between the ventricle and the endoscope is performed using a cylinder-triangle collision test [14].

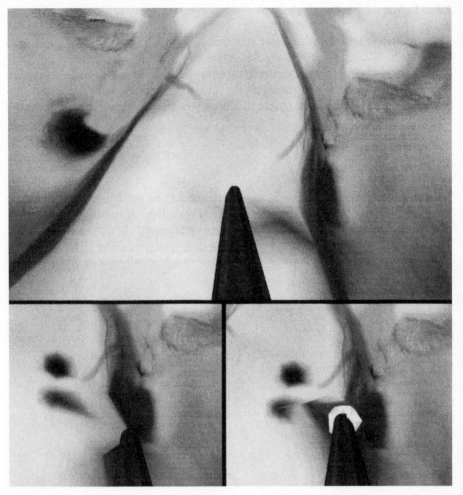

Fig. 6. Upper image: simulation of a deformation caused by a collision between the endo-scopic instrument and the surface of the ventricle. Lower left image: parts of the tissue were already removed by the biopsy forceps (dark spots) and another part is gripped. Lower right image: The tissue is fragmented and the part that was ripped out, still remains in the biopsy forceps.

Using the real instruments mounted onto the robot arm, as well as the visual impression from the simultaneously moving virtual instruments and the virtual endoscopic anatomical landscape and its changes during manipulations, surgeons are able to simulate real procedures such as the removal of a tumor.

A first evaluation of ROBO-SIM on an actual patient dataset was already performed by an expert surgeon with very good results because of the high realism of the visualization and simulation. The only criticized points were the lack of blood flows and membranes. A more thoroughly test comparing the simulation results and the subjective opinions of a group of experienced neurosurgeons from all over Europe will follow. The neurosurgeons will fill in a questionary on subjective parameters concerning practicability, user friendliness and the quality of the simulation.

4   Conclusions

The development of surgical simulators, comparable to flight simulators, has been initiated in a number of institutions (see, for example, [15-20]). Most of these simulators are working with simplified models of the human anatomy instead of using the anatomy of a real patient. ROBO-SIM is able to use directly the digital imaging datasets of the actual patient's neuro-anatomy. A simulation of minimally invasive neurosurgical procedures is considerably more complex than in most other areas, because the creation of the illusion of reality is required at a microsurgical level. Simulations, such as the real-time visualization of movements during manipulations, or the transfer of tactile sensations to the surgeon, or the visualization of the effect of robotic activities, provide a formidable challenge for high-end graphical computing and other disciplines.

To reduce the required amount of graphical power and nevertheless enable planning and simulation of neurosurgical procedures, the use of a combination of volume- and surface-rendered data for visualization and simulation seemed most practical in the hands of this group. 3D-MRI datasets of the brains from actual patients are used to simulate views of the outer surface of the head as well as of inner surfaces, such as the ventricular system or cystic brain lesions, by aid of virtual endoscopy. Although a direct manipulation of the MRI datsets is possible on high-end graphic systems [10], it has been proven that this approach is not useful simulating minimally invasive neurosurgical procedures. The MRI datasets show to less details and resolution for realistic impressions of deformations and fragmentation. Instead, the simulation works with surface models that represent a very exact model of the actual patient's ventricle.

At present ROBO-SIM is implemented on an Onyx2 IR (Silicon Graphics) with two MIPS1000 CPUs using the graphic library Open Inventor [21]. However, all of the algorithms, except the volume rendering, are already ported to a PC. In a first comparison, a two processor PC (Pentium III 800 MHz) with a fast 3D graphics board based on a NVIDIA Geforce2 chip was capable of a 50 percent higher simulation speed than the Onyx. The volume rendering is still a problem on the PC. However, the recently introduced VolumePro500 board allows orthographic real-time volume rendering on a PC [22]. Thus, porting ROBO-SIM to a PC seems a most promising task.

Acknowledgements

This project was partly supported by the Austrian Ministery of Research (grant to ISM) and by the European Commission, programme telematics (no. 4.018). The segmentations of the ventricles are done by algorithms provided by INRIA, Sophia Antipolis.

References

1. Auer LM, Auer D, Knoplioch JF. Virtual Endoscopy for Planning and Simulation of Minimally Invasive Neurosurgery. Lecture Notes in Computer Science 1205, Springer-Verlag 1997; 315 – 18

2. Goradia TM, Kumar R, Taylor R, Auer LM. True Volumetric Stereotaxy for Intracerebral Hematomas. MMVR: Global Healthcare Grid: Transformation of Medicine through Communication. Amsterdam:IOS Press, 1999.

3. Auer DP, Auer LM. Virtual Endoscopy. A new Tool for Teaching and Training in Neuroimaging. Int.J. of Neuroradiol. 4, 1998, 3-14.

4. Auer LM, Auer, DP. Virtual Endoscopy for Planning and Simulation of Minimally Invasive Neurosurgery. Neurosurgery 43, 1998, 529-548.

5. Auer LM. Robots for Neurosurgery. In: Minimally Invasive Techniques for Neurosurgery. Current Status and Future Perspectives, eds.: Hellwig DH, Bauer BL, Springer Verlag 1998; 243-249

6. Benabid AL; Cinquin P; Lavalle S; Le-Bas JF; Demongeot J; De-Rougemont J. Computer-driven robot for stereotactic surgery connected to CT scan and magnetic resonance imaging. Technological design and preliminary results. APPL-NEUROPHYSIOL. 50/1-6 (153-154) 1987

7. Davies BL, Ng WS, Hibberd RD, "Prostatic Resection; an example of safe robotic surgery." Robotica, Cambridge University Press, Vol 11, pp561-566, 1993.

8. Goradia, T.M., R Taylor, Auer LM. Robot-Assisted Minimally Invasive Neurosurgical Procedures: First Experimental Experience. Lecture Notes in Computer Science 1205, Springer 1997, 319-22

9. Grzeszczuk R, Henn C, Yagel R. Advanced Geometric Techniques for Ray Casting Volumes. In: Course Notes, 4, SIGGRAPH'98, 1998

10. Radetzky A, Schröcker F, Auer LM. Improvement of Surgical Simulation using Dynamic Volume Rendering. In: Medicine Meets Virtual Reality, Studies in Health Technology and Informatics, Amsterdam:IOS Press, 2000, 272-278

11. Delingette, H. General Object Reconstruction Based on Simplex Meshes. In: International Journal of Computer Vision, Boston, MA, 1999, 1-32.

12. Nauck D, Klawonn F, Kruse R. Foundations of Neuro-Fuzzy Systems, New York: John Wiley & Sons Inc., 1997.

13. Radetzky A, Nürnberger A, Teistler M, Pretschner DP. Elastodynamic shape modeling in virtual medicine, In: International Conference on Shape Modeling and Applications, IEEE Computer Society Press, 1999, 172-178.

14. Karabassi E-A, Papaioannou G, Theoharis T und Boehm A. Intersection test for collision detection in particle systems. Journal of Graphics Tools 4(1), 1999, 25-37.

15. Jambon, A-C, Dubois P, Karpf S. A Low-Cost Training Simulator for Initial Formation in Gynecologic Laparoscopy. Troccaz J. Grimson E, eds. In: Proc. of CVRMed-MRCAS'97, Lecture Notes in Computer Science 1205, Berlin:Springer, 1997; 347-355.

16. Suzuki N, Hattori A, Kai S. Surgical Planning System for Soft Tissues Using Virtual Reality. In: Morgan KS, eds. Medicine Meets Virtual Reality: Global Healthcare Grid, Amsterdam: IOS Press, 1997; 39:159-163.

17. Bockholt U, Ecke U, Müller W, Voss G. Realtime Simulation of Tissue Deformation for the Nasal Endoscopy Simulator (NES). Westwood JD, Hoffman HM, Robb RA, Stredney D, eds. In: Medicine Meets Virtual Reality, Studies in Health Technology and Informatics 62, Amsterdam:IOS Press, 1999; 74-75.

18. Székely G, Bajka M, Bechbühler C, Dual J, Enzler R et al. Virtual Reality Based Surgery Simulation for Endoscopic Gynaecology. Westwood JD, Hoffman HM, Robb RA, Stredney

D eds. In: Medicine Meets Virtual Reality, Studies in Health Technology and Informatics 62, Amsterdam:IOS Press, 1999; 351-357.

19. Bro-Nielson M, Tasto JL, Cunningham R, Merril GL. PreOp™ Endoscopic Simulator: A PC-Based Immersive Training System for Bronchoscopy. Westwood JD, Hoffman HM, Robb RA, Stredney D, eds. In: Medicine Meets Virtual Reality, Studies in Health Technology and Informatics 62, Amsterdam:IOS Press, 1999; 76-82.

20. Ursino M, Tasto JL, Nguyen BH, Cunningham R, Merril GL. CathSim™: An Intravascular Catheterization Simulator on a PC. Westwood JD, Hoffman HM, Robb RA, Stredney D eds. In: Medicine Meets Virtual Reality, Studies in Health Technology and Informatics 62, Amsterdam:IOS Press, 1999; 360-366.

21. Wernecke J. The Inventor Mentor. New York: Addison-Wesley, 1994.

22. Pfister H, Hardenberg J, Knittel J, Lauer H, Seiler L. The VolumePro Real-Time Ray-Casting System. In: Computer Graphics Proceedings, Annual Conference Series, ACM SIGGRAPH, 1999.

Bayesian Estimation of Intra-operative Deformation for Image-Guided Surgery Using 3-D Ultrasound

A. P. King, J. M. Blackall, G. P. Penney, P. J. Edwards, D. L. G. Hill, D. J. Hawkes

Division of Radiological Sciences and Medical Engineering
The Guy's, King's and St. Thomas' Schools of Medicine and Dentistry
Guy's Hospital, London SE1 9RT, UK

email: andrew.king@kcl.ac.uk

Abstract. This paper describes the application of Bayesian theory to the problem of compensating for soft tissue deformation to improve the accuracy of image-guided surgery. A triangular surface mesh segmented from a pre-operative image is used as the input to the algorithm, and intra-operatively acquired ultrasound data compounded into a 3-D volume is used to guide the deformation process. Prior probabilities are defined for the boundary points of the segmented structure based on knowledge of the direction of gravity, the position of the surface of the surgical scene, and knowledge of the tissue properties. The posterior probabilities of the locations of each of the boundary points are then maximised according to Bayes' theorem. A regularisation term is included to constrain deformation to the global structure of the object.

The technique is demonstrated using a deformable phantom designed to have similar properties to human tissue. Results presented demonstrate that the algorithm was able to recover much of the deformation for a number of objects at varying depths from the source of deformation.

This technique offers a convenient means of introducing prior knowledge of the operative situation into the problem of soft tissue deformation and has the potential for greatly improving the utility of image-guided surgery.

1 Introduction

Image-guided surgery systems register pre-operative images such as magnetic resonance (MR) or computed tomography (CT) to the physical space of the patient in the operating theatre. Typically surgeons place a tracked pointer into the surgical scene and view the position of the tip of this pointer as three orthogonal slices through the pre-operative image on a computer monitor. Recently we described a system which uses an augmented reality (AR) approach to image-guidance [1]. By integrating the surgical microscope into the image-guidance process it is possible to produce accurately aligned overlays in the eye-pieces of the microscope showing critical structures such as arteries or nerves. The surgeon is then able to visualise information from the pre-operative data without having to look away from the surgical scene.

A common problem in any image-guidance system is that of tissue deformation [2][3]. If tissue has moved, deformed or been resected intra-operatively, the pre-operative image will no longer be an accurate representation of the real surgical scene. Therefore, overlaying information from the image will contain errors. The aim of this paper is to present a novel technique for updating the shape and location of a structure segmented from a pre-operative image based on ultrasound data acquired intra-operatively. This will allow the accurate overlaying of information from pre-operative images in the presence of deformation. Previous work related to visualising or compensating for deformation using ultrasound data includes [4] and [5].

Bayesian approaches have been applied in computer vision as a means of introducing prior knowledge to object segmentation and recognition problems which are ill-posed due to high levels of noise (e.g. [6]). A parameterisation of a known object is defined, together with prior probability distributions for the parameter values which indicate the amount of deformation allowed away from the prototypical shape. Bayes' theorem is used to maximise the posterior probability of the object position and shape given noisy image data.

In medical imaging Bayesian techniques have been previously applied in, for example, segmentation of MR images [7]. However, the high levels of noise in ultrasound images would seem to make them ideal data for the application of such techniques. Previous work with this concept includes Kao et al. [8], who used a Bayesian approach for edge detection in single B-mode ultrasound images, including a derivation of a discrete ultrasound imaging model. Lin et al. [9] applied a Bayesian technique to the problem of boundary estimation in compounded 3-D ultrasound volumes, with the aim of producing realistic renderings for visualisation.

This paper describes the application of a Bayesian approach to the problem of compensating for intra-operative tissue deformation using information from a 3-D ultrasound volume. The problem is regularised by the use of a boundary rigidity term. The technique is demonstrated using a deformable phantom designed to have similar elastic properties to human tissue.

2 Method

The problem can be stated as follows: given an initial segmentation of an object, in the form of a triangular surface mesh, and intra-operative 3-D ultrasound data of the object, what is the location and shape of the object which maximises the sum of the posterior probabilities of the boundary points subject to the regularisation constraint? In the following text, $v_i \in \Re^3, i = 1 \ldots N$ denotes the initial locations of the boundary points, and the updated positions are denoted by $v_i' \in \Re^3, i = 1 \ldots N$.

2.1 3-D Ultrasound

True 3-D ultrasound systems directly acquire a 3-D volume using a rectangular array [10]. However, it is thought that the cost and size of the probes currently make their use impractical for intra-operative use. Another approach is to compound a 3-D volume from a set of 2-D ultrasound images. This can be achieved using a number of different methods [10]. All of these approaches require calibration of the ultrasound probe and volume compounding. Several approaches exist for performing the calibration and compounding, and good reviews are provided in [11] and [12] respectively.

For the work described in this paper a freehand 3-D ultrasound system was used. A Siemens Sonoline Versa Pro ultrasound machine with a 10MHz probe was calibrated using a cross-wire phantom. The residual error from the calibration process was 0.67mm. The probe was tracked during acquisition using an Optotrak 3020 optical tracking system from Northern Digital Inc, which tracks infra-red LEDs to a 3-D spatial accuracy of ~ 0.2mm. The 2-D slices were compounded into a 3-D volume using the pixel nearest neighbour technique. This algorithm takes every pixel intensity in each 2-D ultrasound image, and places it into the nearest voxel in the final volume image. If more than one intensity is placed into the same voxel, the conflict is resolved by averaging the values concerned.

2.2 Bayesian Estimation

The problem of updating the position of each boundary point in a segmented structure to compensate for intra-operative deformation can be expressed in Bayesian terms. Given prior probabilities $P(v_i' = x)$ for each boundary point v_i', and the ultrasound data $I(x)$, the problem becomes one of maximising the posterior probability of a boundary being present given the data. Using Bayes' theorem, this posterior probability for a boundary element i is given by

$$P((v_i' = x)|I(x)) = \frac{P(I(x)|v_i' = x)P(v_i' = x)}{P(I(x))} \tag{1}$$

The prior probability of the image data $P(I(x))$ is assumed to be uniform. The probability of the data given that there is a boundary present, $P(I(x)|v_i' = x)$, is the imaging model, which can be expressed as a function of the 3-D ultrasound image $f_{US}(x)$. Depending on the type of tissue involved, either the image intensity, the gradient of the image intensity, or a weighted combination of the two may be more appropriate for this model. For example, gradient information should be used at boundaries between tissue types which have different degrees of scatter but similar acoustic impedance, whereas at the boundaries between tissue types with similar scatter but differing acoustic impedance, the ultrasound image intensity will be more appropriate. Prior knowledge is introduced here in the form of a weighting factor k_m which determines the proportions of intensity and gradient information used.

$$P(I(x)|v_i' = x) = k_m f_{US}(x) + (1 - k_m)\|\nabla f_{US}(x)\| \tag{2}$$

In many applications, the maximum value for the prior probability $P(v_i' = x)$ might be taken to be at the initial boundary location, with the probability falling away by some function of distance from this initial position. However, this is a term in which significant prior knowledge of the operative situation can be utilised. As well as the initial location of the boundary point, it is not unreasonable to suppose that knowledge of the direction of gravity and the position and orientation of the surface of the operative scene will also be available. These values can be used to bias the prior probability field. For example, deep structures are likely to shift less than superficial ones after a craniotomy [13], and deformation is most likely to occur in the direction of gravity.

Therefore, the prior probability field for a given boundary point is given by

$$P(v_i' = x) = F_{prior}(g, v_i', v_i, s_p, s_n, d_e) \tag{3}$$

where g is a vector indicating the direction of gravity. The surface of the surgical scene is modelled as a plane defined by the point s_p and the normal s_n. The amount of deformation observed at the surface in the direction of gravity is denoted by d_e.

Hence the maximum prior probability will occur beneath the initial position of the point in the direction of gravity, at a displacement dependent on its distance from the surface and the amount of deformation observed at the surface.

2.3 Regularisation

If the above approach were used in isolation there would be no constraint to preserve the global structure of the object and each boundary point would be free to move to its most probable location individually. To constrain the solution so that excessive deformation does not occur, a regularisation term is included and the problem expressed as one of energy minimisation. The total energy of the system is given by

$$E = (1 - \lambda)E_i + \lambda E_r \tag{4}$$

where $E_i = -P(v_i' = x)|I(x))$ as given in (1), and λ is the regularisation weighting.

The rigidity term E_r of the energy equation represents the degree to which the global shape of the structure has deformed. The degree of deformation which is permitted will vary depending on the tissue type of the structure. For example, soft tissue such as white matter in the brain should be allowed to deform more than some tumours which will have a greater degree of rigidity. Therefore, prior knowledge of the tissue type can be utilised here in the form of the weighting factor λ in (4). The rigidity energy is given by

$$E_r = \sqrt{\sum_{i=1}^{N} \| (v_i' - v_i) - \frac{1}{N} \sum_{i=1}^{N} (v_i' - v_i) \|^2} \tag{5}$$

This expression is the standard deviation of each boundary point's movement away from the mean movement over the whole structure. Hence a shift of the entire structure is not penalised, whereas structural deformation is. This expression allows a certain amount of deformation away from the initial shape, but penalises excessive deformation.

2.4 Search Strategy

The global energy minimum represents the state of the object boundary which maximises the sum of the posterior probabilities of its boundary points, subject to the regularising rigidity constraint. This minimum is found using a gradient descent search strategy: the change in posterior probability resulting from each point v_i' moving in each of 6 directions is calculated. The movement that causes the maximum increase in the posterior probability for each point is chosen. Each point is allowed to move if it decreases the energy E after the regularisation constraint is considered.

Since segmented surface meshes typically contain a large number of boundary points, each of which must be allowed to seek its maximum posterior probability, the search for the optimal position and shape takes place in a high-dimensional space which may have many local minima. To reduce the search time, a two-phase technique is employed: each boundary point is first allowed to maximise its prior probability without reference to the data, then each point maximises its posterior probability according to (1). Both phases use the rigidity constraint described in Section 2.3. In addition to this, to enable a larger capture range without a significant increase in computational complexity, a multiresolution approach is used. The ultrasound image volume is smoothed with a $5 \times 5 \times 5$ Gaussian filter and subsampled to produce a series of successively coarser representations of the image volume. The search begins with a relatively large step size at a coarse scale. The step size and coarseness of the ultrasound data is gradually decreased to enable the boundary estimate to be refined through scale space. The search stops when the boundary points have reached a stable energy state.

3   A Deformable Phantom

To demonstrate the technique a deformable phantom was constructed (see Figure 1). The phantom material used was gelatine (15% gelatine - 85% water by weight) as this has been demonstrated to have similar elastic properties to human tissue [14]. Regions of internal structure showing different MR and ultrasound contrast to the structural gelatine were made by adding silica to the gelatine mixture. These regions were immersed into the structural gelatine which was set and placed in a perspex box, with a moveable perspex plate at one end. By turning a number of perspex screws the plate moves and deforms the gelatine block. MR images of the phantom were acquired before deformation (for segmenting the initial structures) and afterwards (to measure the deformation). The phantom was scanned freehand with a tracked calibrated ultrasound probe after deformation.

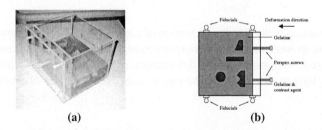

(a)   (b)

Fig. 1. (a) The deformable gelatine phantom, (b) diagram showing plan view of phantom

A number of fiducial markers were attached to the phantom prior to MR imaging. These contained a gadolinium solution which gives high image contrast in MR. These imaging markers were replaced with physical localisation caps which were localised with a tracked pointer before ultrasound scanning to obtain an MR-physical registration. This registration, together with the tracking and calibration data for the ultrasound probe enabled the ultrasound volume to be aligned with the MR images.

In (3) a general term for the prior probabilities for boundary points was given. The precise form of these priors is dependent on the tissue type. Observations of the nature of deformations present in 15% concentration gelatine suggested that the amount of deformation has an approximately linear relationship with distance from the source of deformation. Hence the prior probability for boundary point i being in location x is defined by

$$P(v_i' = x) = e^{-(\|(v_i'-v_i)-g(d_e-d_s/k_d)\|)^2} \tag{6}$$

where $d_s = (v_i' - s_p) \cdot s_n$ is the distance of v_i' beneath the surface of the surgical scene, and k_d is a constant which was chosen so that the expected deformation dropped to zero at the wall of the phantom furthest from the perspex plate. The distribution of the prior probabilities is Gaussian, with its peak displaced by the linear term. Work is in progress to model brain deformation more accurately [2][15] and it will be possible to incorporate more realistic models as they become available. In the meantime the simple linear model is compatible with linear compression of an elastic material.

For the purposes of this demonstration, the gravity direction and surface position and normal were all defined manually from the MR images. The direction of gravity was defined as the direction of movement of the perspex plate.

4   Results

Figure 2(a) shows a sample slice through the pre-deformed MR image of the phantom overlaid with contours extracted from the deformed MR image. The MR images have an in-plane resolution of 1mm × 1mm, and a slice thickness of 1.5mm. The deformation due to movement of the perspex plate (which was at the bottom of the slice) is apparent. Three structures were segmented from the undeformed MR image using the Analyze software package (Biomedical Imaging Resource, Mayo Foundation, Rochester, MN, USA) and triangle meshes constructed using the marching cubes algorithm. Renderings of these are shown in Figure 2(b). These three structures were chosen because one is close to the deforming plate, and so the deformation would be expected to be large, one was far from the plate, and one object was elongated and covered a range of depth from the plate. Figures 2(c) and (d) show sample slices through the ultrasound volume, and its gradient magnitude image respectively. These images have a voxel size of 1mm × 1mm × 1mm.

Figures 3, 4 and 5 show overlays (in dark outlines) onto the post-deformed MR (i.e. the gold standard) of the pre- and post-deformed structures. Sample slices are shown but the deformation is computed in 3-D. It can be seen that in all cases the algorithm has recovered much of the deformation that has occurred. Note that whereas in Figures 3 and 4 the movement of the shape could be approximated by a rigid body motion, in Figure 5 the structure of the shape has deformed significantly owing to the fact that it

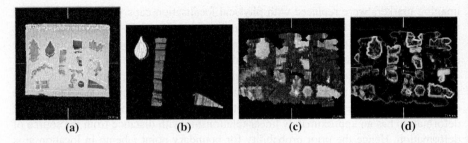

(a)   (b)   (c)   (d)

Fig. 2. The deformable phantom, (a) sample slice through pre-deformation MR image overlaid with contours extracted from post-deformation MR image, (b) renderings of 3 segmented structures from pre-deformation MR image, (c) sample slice through the compounded ultrasound volume, (d) sample slice through the gradient magnitude of the ultrasound volume

extends over a range of depth from the plate. Even in this case the errors in boundary alignment appear to be only $\sim 1 - 2$mm. However, residual errors do remain. These could arise from errors incurred while calibrating the ultrasound probe, tracking the probe, in the MR-physical registration, or from the assumptions used in calculating the priors.

5 Discussion

This paper has outlined a framework for the combination of information from an intra-operative modality with clinical knowledge of the expected surgical deformation to automatically compensate for soft tissue deformation. The technique has its theoretical grounding in Bayesian theory. The results presented here have demonstrated that the technique works well on phantom data. The Bayesian framework is a convenient means of introducing prior knowledge into the problem of deformation compensation, so with more clinically based priors it is not unreasonable to expect good results from clinical data too. This is an area for future investigation. These priors could be taken from measurements of intra-operative deformation, e.g. by using an intra-operative MR scanner [13]. Alternatively, this technique could be combined with a predictive model to provide more realistic priors. For example, finite element models (FEMs) offer a means of constructing a physical model of tissue, which can be used to predict deformation fields given information about the surgical situation [16], such as gravity direction and resection of tissue.

It should be emphasised that the technique described here only compensates for deformation of structures segmented from a pre-operative image modality. No interpolation between these structures is performed. For applications such as pre-operative visualisation or intra-operative augmented reality overlays, this is acceptable. Indeed, it can be seen as an advantage in that relatively sparsely sampled intra-operative image data can be used effectively. However, a more general deformation algorithm would need to provide a deformation field that covers the whole of the surgical scene. Although the current technique does not provide this, the deformation field it provides

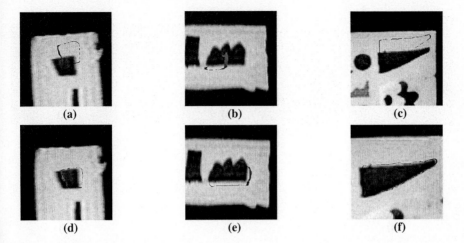

Fig. 3. Results for shape 1 (a structure near the source of deformation), (a), (b), (c) outlines of undeformed shape on orthogonal slices through the post-deformation MR image, (d), (e), (f) outlines of deformed shape on the post-deformation MR image

could be used as input to an algorithm which does provide such information, such as an FEM [16].

Intra-operative deformation is often caused by resection of tissue, e.g. a tumour. Using the technique described in this paper, information regarding the extent of this resection could be incorporated into the prior probabilities. For example, the probabilities for nearby points could be set relatively high in the area of resection, allowing them to move into it without penalty.

Another important issue in deformation is volume change. Some types of tissue are incompressible, i.e. their volume remains constant under deformation. Other tissue types may undergo volume change under certain conditions. For example, cerebrospinal fluid may drain away, or aedematous regions may increase in volume. Knowledge of the likely compressibility of a structure could be incorporated into the model by including it in the regularisation term.

Another area for future research is the imaging model. The current model defined in (2) does not include knowledge of the ultrasound probe orientation at the time of acquisition. By using this knowledge a more realistic ultrasound imaging model could be incorporated into the framework described here.

The current implementation runs in approximately 5 minutes on a SUN Ultra 10 for an object containing $\sim$ 1600 boundary points. Compounding the 3-D ultrasound volume and calculating its gradients is performed separately and takes $\sim$ 60 minutes, but there are many ways in which this process could be speeded up. This technique therefore offers the potential for fast intra-operative updates, which give it a significant advantage over many predictive models such as FEMs.

In conclusion, the technique described here is a novel application for Bayesian theory, and has the potential to be able to deal with many of the important issues in defor-

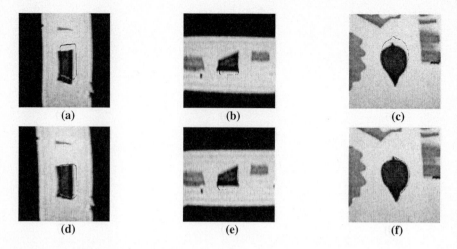

Fig. 4. Results for shape 2 (a structure far from the source of deformation), (a), (b), (c) outlines of undeformed shape on orthogonal slices through the post-deformation MR image, (d), (e), (f) outlines of deformed shape on the post-deformation MR image

mation. It offers the possibility of improving the utility of image-guidance systems. As such, it represents a highly promising area for future investigation.

Acknowledgements

We would like to thank the EPSRC for funding this project. We are also grateful to the radiology and radiography staff, and the staff of the medical physics workshop at Guy's hospital for their cooperation.

References

1. P. J. Edwards, A. P. King, C. R. Maurer, Jr, D. A. de Cunha, D. J. Hawkes, D. L. G. Hill, R. P. Gaston, M. R. Fenlon, S. Chandra, A. J. Strong, C. L. Chandler, A. Richards, and M. J. Gleeson. Design and evaluation of a system for microscope-assisted guided interventions (MAGI). In *Proceedings MICCAI '99*, pages 842–851, 1999.
2. D. L. G. Hill, C. R. Maurer, R. J. Maciunas, J. M. Barwise, J. A. Fitzpatrick, and M. Y. Wang. Measurement of intraoperative brain surface deformation under a craniotomy. *Neurosurgery*, 43(3):514–528, 1998.
3. D. W. Roberts, A. Hartov, F. E. Kennedy, M. I. Miga, and K. D. Paulsen. Intraoperative brain shift and deformation: A quantitative analysis of cortical displacement in 28 cases. *Neurosurgery*, 43(4):749–760, 1998.
4. R. D. Bucholz, D. D. Yeh, J. Trobaugh, L. L. McDurmont, C. D. Sturm, C. Baumann, J. M. Henderson, A. Levy, and P. Kessman. The correction of stereotactic inaccuracy caused by brain shift using an intraoperative ultrasound device. In *Computer Vision, Virtual Reality, and Robotics in Medicine 1997*, pages 459–466, 1997.

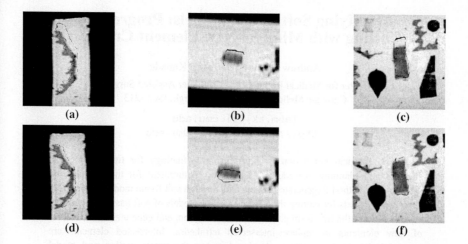

Fig. 5. Results for shape 3 (a structure covering a range of depth), (a), (b), (c) outlines of undeformed shape on orthogonal slices through the post-deformation MR image, (d), (e), (f) outlines of deformed shape on the post-deformation MR image

5. D. G. Gobbi, R. M. Comeau, and T. M. Peters. Ultrasound probe tracking for real-time ultrasound/MRI overlay and visualization of brain shift. In *Proceeding MICCAI '99*, pages 920–927, 1999.
6. A. L. Yuille, P. W. Hallinan, and D. S. Cohen. Feature extraction from faces using deformable templates. *International Journal of Computer Vision*, 8:99–111, 1992.
7. T. Kapur, E. L. Grimson, R. Kikinis, and W. M. Wells. Enhanced spatial priors for segmentation of magnetic resonance imagery. In *Proceedings MICCAI '98*, pages 457–468, 1998.
8. C-M. Kao, X. Pan, E. Hiller, and C-T. Chen. A Bayesian approach for edge detection in medical ultrasound images. *IEEE Transactions on Nuclear Science*, 45:3089–3096, 1998.
9. W. J. Lin, S. M. Pizer, and V. E. Johnson. Boundary estimation in ultrasound images. In *Information Processing in Medical Imaging*, pages 285–299, 1991.
10. A. Fenster and D. B. Downey. 3-D ultrasound imaging: A review. *IEEE Engineering in Medicine and Biology Magazine*, 15(6):41–51, 1996.
11. R. W. Prager, R. N. Rohling, A. H. Gee, and L. Berman. Rapid calibration for 3-D freehand ultrasound. *Ultrasound in Medicine and Biology*, 24(6):855–869, 1998.
12. R. N. Rohling, A. H. Gee, and L. Berman. Radial basis function interpolation for 3-D ultrasound. Technical report, Cambridge University Engineering Department, 1998.
13. C. R. Maurer, Jr., D. L. G. Hill, A. J. Martin, H. Liu, M. McCue, D. Rueckert, D. Lloret, W. A. Hall, R. E. Maxwell, D. J. Hawkes, and C. L. Truwit. Investigation of intraoperative brain deformation using a 1.5 Tesla interventional MR system: Preliminary results. *IEEE Transactions on Medical Imaging*, 17:817–825, 1998.
14. L. K. Ryan and F. S. Foster. Tissue equivalent vessel phantoms for intravascular ultrasound. *Ultrasound in Medicine and Biology*, 23(2):261–273, 1997.
15. T Hartkens, D. L. G. Hill, C. R. Maurer, Jr, A. J. Martin, W. A. Hall, D. J. Hawkes, D. Rueckert, and C. L. Truwit. Quantifying the intraoperative brain deformation using interventional MR imaging. In *Proceedings ISMRM*, 2000.
16. K. D. Paulsen, M. I. Miga, F. E. Kennedy, P J. Hoopes, A. Hartov, and D. W. Roberts. A computational model for tracking subsurface tissue deformation during stereotactic neurosurgery. *IEEE Transactions on Biomedical Engineering*, 46:213–225, 1999.

Modifying Soft Tissue Models: Progressive Cutting with Minimal New Element Creation

Andrew B. Mor and Takeo Kanade

Center for Medical Robotics and Computer Assisted Surgery
Carnegie Mellon University, Pittsburgh, PA 15213

{abm,tk}@ri.cmu.edu
http://www.mrcas.ri.cmu.edu

Abstract. Surgical simulation is a promising technology for training medical students and planning procedures. One major requirement for these simulation systems is a method to generate realistic cuts through soft tissue models. This paper describes methods for cutting through tetrahedral models of soft tissue. The cutting surface follows the free form path of the user's motion, and generates a minimal set of new elements to replace intersected tetrahedra. Intersected elements are progressively cut to minimize the lag between the user's motion and model modification. A linear finite element model is used to model deformation of the soft tissue. These cutting techniques coupled with a physically based deformation model increases the accuracy and applicability of a surgical simulation system.

1. Introduction

Surgical simulation is getting increased attention due to its promise of increasingly realistic modeling. It will be used by medical students when learning new procedures and will give surgeons the ability to simulate patient specific data when planning complex procedures. Two important parts of any surgical simulation system are soft tissue modeling and the ability to modify the simulated anatomical structures. Modifying the simulated tissue, as a cut is taking place, in a physically realistic and consistent fashion is an important part of any surgical simulation. Physically accurate real-time models are required to give the user a correct feel of how the soft tissue responds.

Previous work has been done on physically based tissue modeling and on techniques required to realistically modify an object's topology. But no work had addressed both topics simultaneously. Because the modification of the tissue model directly impacts the accuracy and speed of the soft tissue simulation, it is important to address both issues simultaneously. There has been no work addressing the problem of how to modify the topology of individual elements while a cut is taking place, an important issue if the nominal element size is easily discernible.

1.1 Prior Work

Soft tissue simulation can be implemented utilizing either surface or volumetric models. Volumetric models can simulate objects with interior structure that surface models can not encode. [5] gives a good overview of different methods used to simulate deformable objects. Finite element methods are based on a continuum model and generate physically based results. [9] introduced models based on elasticity theory to the computer graphics community. [3] modeled human muscle using an FEM mesh. [2] developed a system utilizing classical, three dimensional solid finite element models using condensation, precalculation of the inversion and exploitation of the sparse structure of the force vector to accelerate the calculation of static deformations. [4] demonstrated a representation they named tensor mass based on linearly elastic continuum mechanics that are solved in a dynamic fashion over time. [11] recently introduced a non-linear finite element method that handles large-scale deformations better than linear methods.

There has been less attention applied to the problem of modifying deformable objects. Modification precludes any precomputation for the soft-tissue simulation. Cutting of surface-based finite element models was demonstrated by [8]. [6] created a static cutting surface by specifying the beginning and end points of a cutting edge, and then interpolating between them. Intersected elements were split according to the number of intersected edges. [1] made free-form cuts through an object by tracking the tip and the direction of a cutting edge. A small cutting plane was generated at every time stop between the previous edge position and the current edge position, and any element that is intersected is split into 17 smaller elements once the cutting tool leaves the element.

This paper is organized in the following manner: Section 2 states the proposed methods and contains an overview of the system, Section 3 describes minimal set cutting, Section 4 details the method for progressive cutting, Section 5 describes the soft tissue model, and Section 6 describes our results. The results and future work are then discussed.

2. Soft Tissue Simulation and Object Modification

To provide a convincing visual and haptic experience for surgical simulation, the simulation of the soft tissue must be performed in as realistic and accurate a manner as possible. Finite elements models generate the most accurate results, and are becoming more acceptable for simulation use as computing power increases. We use a linear finite element utilizing lumped parameters. This model is quickly and easily updated when changes to the topology of the model are made.

Cutting tissue is one of the main activities performed during surgery, and the results of such modification in a simulator can greatly affect both the sense of realism of the cutting process and the computational load on the host computer. Three items can greatly affect the cutting process. First, the simulation should allow cuts along free-form surfaces that the user traces out with a cutting tool. If the user is constrained to cut along regular grid lines, or only along boundaries between elements, the sense of realism will be greatly compromised. Cuts will not appear where the user expects them to be, and the resulting surface will either be artificially smooth or stair-cased in shape. Second, cuts should occur as the user moves the cutting tool through the object. If the cut lags behind, a disconnect will form between the user and the simulation. Third, if the cutting process generates a large number of smaller elements, the computational cost of the new elements may be so great so as to adversely affect the update rate of the simulation. To address these issues, we propose the use of a method that generates a minimal set of new elements that follows the motion that the user traces out. We also propose generating a progressive, temporary cut, as the user moves the cutting tool within an element. Once the cut leaves an element, it is cut permanently.

2.1 System Overview

The soft tissue simulation system includes an input device, finite element modeler, and object modification routines. The general flow of the system, after start-up when the model is loaded and the individual element parameters are first determined, begins with gathering the position of the tool that the user is holding. The current implementation uses keyboard input for simplicity, but the system is easily extendable to using a 6DOF haptic device. Next, we determine the intersection state between the tool and the user. If any elements are partially intersected, where the cutting edge is still within the interior of the element, then temporary elements are created using the progressive cutting method. If any elements are fully intersected, the element is fully split into its minimal subset. Lastly, the internal elastic forces are determined, and then nodal positions are updated. The scene is then rendered to the user.

3. Minimal Set Cutting

The described cutting method creates a minimal set of new tetrahedra that contain the cut surfaces that follow the user's motions. There are two main activities required for cutting tetrahedral models: the detection and storage of intersections, both of edges and faces; and the subsequent subdivision of the intersected element when a cut is completed. First, intersections between the cutting tool and the model are detected and stored. Once an element is completely cut through, it is then subdivided based on the number and type of intersections among its edge and faces.

3.1 Intersection Detection

Cutting with a scalpel can be viewed as the motion of a finite length cutting edge passing through an object. If the body of the blade is ignored, and the edge is taken to be infinitely sharp, then the problem is reduced to tracking the passage of a line segment moving in time and space through a tetrahedral mesh [1]. Figure 1 shows the path of a cutting edge from time t_i to t_{i+1}, as it creates two face intersections and one edge intersection.

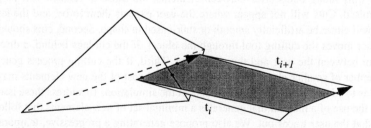

Figure 1. Cutting edge intersection with a tetrahedron.

The swept surface created by the path of the cutting edge must be tested at every time step for intersections with the model. Two tests are required: the intersection of the path of the tip of the cutting tool and the faces of the tetrahedron, and the intersection of the swept surface and the edges of the tetrahedron. These two tests generate, respectively, face intersections, which mark the base of the cut, and edge intersections, which occur where the model is split in two by the cutting edge.

The procedure for updating the intersection state of the model starts with a global search to determine if the path of the blade has intersected the model. A global search is inefficient, but guarantees that all intersections will be detected. Other initial collision detection techniques can also be used. If any tetrahedron has been intersected, then all 6 of the tetrahedron's edges and all 4 of its faces are tested against the swept surface and cutting tip path. The current, deformed, positions of the vertices are used when testing for intersections. When an intersection occurs, the barycentric coordinates of the intersection are determined. The corresponding rest position of the intersection is then used when updating the soft tissue parameters of the new elements. After all the tetrahedra are tested, if any were intersected, then the model is marked as having an intersection present.

Next, if the model has been intersected, all of the intersected elements are checked to see if the cutting instrument has either passed through any non-intersected faces or edges, or has left the tetrahedron. If a new intersection occurs, then the intersection information for that tetrahedron is updated. Neighboring elements that also contain the newly intersected edge or face are updated, thereby using spatial coherency to propagate the cutting motion through the model. If the cutting edge no longer passes through an intersected element, then the user has completed the cut, and the element will be subdivided.

3.2 Element Subdivision

Tetrahedral elements cut by planar, or near-planar, surfaces will fall into one of five different topological cases, based on the number of cut edges and intersected faces. There are two different cases where the tetrahedron is completely cut through into two pieces, and three cases where the element is cut, but not completely through. Figure 2 shows the different cases.

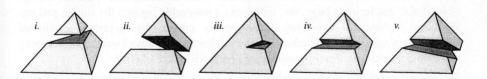

Figure 2. The five cases of tetrahedron subdivision after a completed cut.

Within the framework of the tetrahedral mesh numbering, there are multiple orientations of the cut element based on the ordering of the vertices and cut edges. When the element is completely cut into two, there are four different permutations when three edges are cut, as in case *i*, and three different permutations in case *ii*, when four edges are cut. The four different permutations of case *i* correspond to each of the four vertices being cut away from the remaining three. When only one edge is cut, case *iii*, there are six permutations, and there are twelve permutations for both cases *iv* and *v*.

Individual procedures were implemented for each type of intersection, so that no excess elements would be created. Only five to nine new elements are created for each cut element, depending on the type of intersection. This minimal subdivision uses only the original vertices of the element and vertices created due to the cutting action: one vertex at the point of each face intersection, and two vertices at the point of each edge intersection. Figure 3 demonstrates how each half of the cut element from case *ii* in Figure 2 is minimally subdivided. In this case, six elements are created to replace the original one.

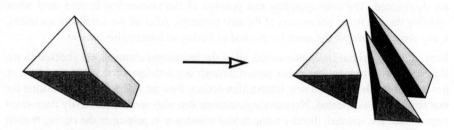

Figure 3. Minimal element subdivision.

Each procedure uses a lookup table to determine how to mirror or rotate the vertices to fit the default orientation. After the ordering is determined, any new edges that are needed are created. Then the new tetrahedra are created and the original tetrahedron is removed.

4. Progressive Cutting

Previous methods of modifying objects, and the basic technique described above, do not split an element until the cut has been completed. With current models, where the average element size can be quite large, this introduces a noticeable lag into the cutting process. We have implemented a method of progressive cutting that generates a minimal subdivision of a partially cut tetrahedron. The subdivision is always based on the geometry of the original element, thereby minimizing a potential source of error.

The general procedure for progressive cutting utilizes a temporary subdivision of each partially cut element. An example cut is shown in Figure 4. First, any temporary face intersections caused by the cutting edge are updated for each partially cut tetrahedron. A temporary face intersection occurs when the cutting edge intersects a face. This type of intersection does not occur for the permanent intersections described previously.Then, the modified topology of the partially cut element is checked for any changes. A change occurs when a new intersection is created: for example, when the element is first cut into, or when the cutting edge or tip passes through another edge or face. If the topology has changed, a new minimal set of temporary tetrahedra are created and all the old temporary tetrahedra are removed. If the modified topology has not been changed, then the temporary elements are updated using the new positions of any temporary face intersections. Once the cutting edge leaves an element and the cut is completed, the temporary elements are removed, and a final subdivision is created.

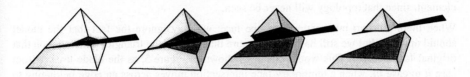

Figure 4. Progressive cutting example.

There are eleven different combinations of intersected edges, faces, and temporary face intersections, which are enumerated in Table 1. Momentarily marking temporary intersections as permanent, many of the cases can directly use the procedures described in the previous section on minimal cutting. The cases which are unique to progressive cutting were implemented in a similar fashion as described above, with a minimum number of new elements created for each cut.

Table 1. Enumeration of different cases for temporary intersections.

Case:	1	2	3	4	5	6	7	8	9[a]	10	11
Edge Ints.	0	0	1	1	1	2	2	2	2	3	3
Face Ints.	1	2	0	1	2	0	1	2	2	0	1
Temp. Face Ints.	1	2	2	1	2	2	1	2	2	2	1
Interior Ints.	1	0	0	1	0	0	1	0	0	0	1

a. Same number of intersections as case 8, but case 9 has a different subset of edges intersected.

There are two important issues unique to progressive cutting that must be addressed: how to deal with the tip of the cutting edge being within the interior of an element; and how to deal with two intersections on one face, which happens when a face is intersected permanently by the passage of the tip of the blade and temporarily by the cutting edge.

When the tip is within the interior of an element, ideally we would want the model to be able to open up along the cut of the blade, so that the user could see all the way up to the base of the cut, where the tip is located. But given the nature of the subdivision for a generic cut, this would not be possible. An example of this is shown in Figure 5, case i, which corresponds to Case 1 from Table 1. In this case, the tip of the blade is inserted fully through one face, with the tip of the cutting edge within the interior of the tetrahedron. I_i is the tip of the cutting edge, I_f is a permanent face intersection, and I_{tf} is a temporary face intersection. There are now two intersections on one face. Ideally, as described above, the object would be able to open up along the edges between the two face intersections. But if no intermediate nodes are inserted between the two face intersections, a straight line will always connect them, and the model will not be able to open up. The fact that the model can not open up along these edges allows us to ignore the location of the tip within the model, and, in fact, to generate an arbitrary topology within the interior of the original element, since that topology will never be seen.

When there are two intersections on one face, we may ignore the fact that the model should open up, but we still have to make sure that none of the triangles generated on that original face overlap. This would occur, as shown in Figure 5, as the blade travels from case *ii* to case *iii*, when a temporary face intersection moves across an edge belonging to the other face intersection. The shaded area shows the overlapping area of the two triangles. If this occurs, then the modified topology of the partially cut tetrahedron has changed, and a new set of tetrahedra will be created, as shown in case *iv*.

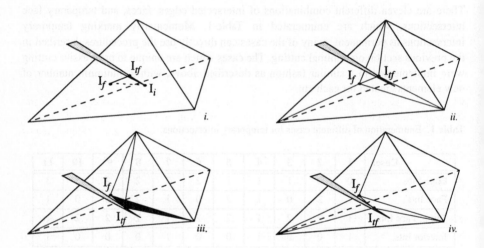

Figure 5. Temporary subdivision, with two intersections on one face.

5. Soft Tissue Modeling and Linear Finite Elements

The problem of modeling soft tissue can be approached in many different ways, depending on the trade-off of speed vs. accuracy. We implemented a linear finite element model due to its increased realism when compared to mass spring or surface based models. The mass and damping terms are diagonalized and lumped at the nodal positions. The system is explicitly solved to integrate the Newtonian dynamics of the nodal masses.

The linear finite element model was generated in a manner similar to [4]. The global stiffness matrix is set up, then individual nodal and edge terms are broken out of the global matrix, and stored with the individual nodes and edges. The stiffness matrix is calculated in the standard manner [12].

Given an isoptropic solid, the equation for the element stiffness matrix, K^e, simplifies to:

$$K^e = V^e B^T DB .$$ (1)

where the matrix D is determined by material properties and the matrix B encodes the shape functions of the element. For a tetrahedron, this 12x12 stiffness matrix can be viewed as being composed of four 3x3 nodal displacement stiffness matrices, K^e_{ii}, and six edge displacement stiffness matrices, K^e_{ij}. The first stiffness matrix relates to the force felt by a node due to its own displacement from its home position, while the second matrix determines the force felt by the node i due to the displacement of node j from node j's home position; this can be viewed as an edge effect since it occurs between two nodes.

To calculate the internal elastic force acting on a node i, the contributions from all the tetrahedra that node i belongs to are summed:

$$f_i = K_{ii}\overrightarrow{P_i^o P_i} + \sum_{j \in N(P_i)} K_{ij}\overrightarrow{P_j^o P_j} .$$ (2)

where f_i is the nodal force, K_{ii} is the sum of the K^e_{ii} tensors associated with all the tetrahedra that node i belongs to, P_i^o is the home position of node i, the tensor K_{ij} is the sum of the K^e_{ij} tensors associated with the edge from node i to node j, and $N(P_i)$ is a list of all of the nodal neighbors of the node i.

6. Results

Figure 6 shows the results of cutting a section off the left lobe of a liver model. Figure 7 shows the results of cutting through a rectangular object that is under tension. There were 576 tetrahedra in the rectangular object before cutting, on a 4x6x4 cubical lattice, and 954 afterwards. 60 elements were cut, removed, and replaced by a new set of 312 tetrahedra, an average of 5.2 elements added for every element removed. This compares very favorably to a similar cut demonstrated in [1], where 110 elements were cut and 1870 new elements were created.

There is one major limitation with the current implementation. When the user first starts cutting into an element, or when a final cut creates an element that is very small, the model can become unstable. This occurs because the stiffness matrices for the small elements are quite large, and even small displacements can generate very large forces. This is not a problem with the cutting techniques, only with the soft tissue modeling. Because of this instability, the examples demonstrated were performed on a graphical simulator, without haptics. The examples were also run with a very small time step, to insure the stability of the soft tissue model, and were not real-time.

Figure 6. Cutting off a section of the liver. (Model courtesy of Project Epidaure, INRIA.)

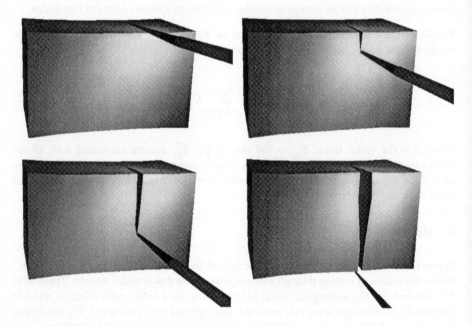

Figure 7. Cutting of a rectangular object under tension.

7. Discussion

We have presented a technique for realistically modifying the topology of simulated objects in an efficient manner. The technique of minimal cutting generates a minimum set of new elements, and therefore is useful in a wide range of applications where the computational load and update rates are factors in the realism of the simulation. Progressive cutting is useful in situations where the size of individual elements is noticeable to the user and any apparent lag between the motion of the cutting tool and the modification of the model would be distracting. The described techniques work on any tetrahedral mesh, irrespective of the overall shape of the model. The soft tissue simulation will run at update rates around 100Hz on the liver model, given the size of the model and results demonstrated in [4]. This will be sufficient to interpolate for output through a force feedback device.

Future work will be directed towards fixing the limitations imposed by the use of the linear finite element model. Linear models are only valid up to approximately 10% deformation, so a non-linear model will be investigated. Also, the problems caused by small tetrahedra will be addressed. One technique currently under investigation is to check the length of all new edges and the height of the new tetrahedron with respect to a minimum edge length. If the new element does not pass the check, then snap the position of the new vertex to the nearest original edge.

Acknowledgments

We would like to thank the Project Epidaure Team at INRIA for the model of the liver.

References

[1]   D. Bielser, V. A. Maiwald, M. H. Gross. "Interactive Cuts through 3-Dimensional Soft Tissue." Proceedings of the Eurographics '99 (Milano, Italy, September 7-11, 1999), Computer Graphics Forum, Vol. 18, No. 3, C31-C38, 1999.

[2]   M. Bro-Nielsen and S. Cotin, "Real-time Volumetric Deformable Models for Surgery Simulation Using Finite Elements and Condensation", Proceedings of Eurographics '96.

[3]   D. T. Chen and D. Zeltzer. "Pump It Up: Computer Animation of a Biomechanically Based Model of the Muscle Using the Finite Element Method." Computer Graphics (SIGGRAPH), num 26, July 1992.

[4]   Cotin, S., Delingette, H., Ayache, N., "Efficient Linear Elastic Models of Soft Tissues for Real-Time Surgery Simulation", INRIA T.R. No. 3510, October 1998.

[5]   S. Gibson and B. Mirtich. "A Survey of Deformable Models in Computer Graphics." TR-97-19, Mitsubishi Electric Research Laboratories, Cambridge, MA, 1997.

[6]   Mazura, A., Seifert, S., "Virtual Cutting in Medical Data", Medicine Meets Virtual Reality, San Diego, CA, 1997.

[7]   Reznik and Laugier, "Dynamic Simulation and Virtual Control of a Deformable Fingertip", Proceedings of IEEE International Conference on Robotics and Automation, Minneapolis, MN, 1996.

[8]   Song, G., Reddy, N., "Towards Virtual Reality of Cutting: A Feasibility Study", Proceedings of 16th Annual International Conference of the IEEE Engineering in Medicine and Biology Society, Baltimore, MD, 1994.

[9]   Terzopolous, D., Platt, J., et al. "Elastically Deformable Models", Computer Graphics Proceedings, Annual Conference Series, Proceedings of SIGGRAPH 87, pp 205-214, 1987.

[10]   Terzopolous, D. and Waters, K. "Physically-Based Facial Modeling, Analysis, and animation", Journal of Visualization and Computer Animation, 1:73-80, 1990.

[11]   Y. Zhuang and J. Canny. "Real-time Simulation of Physically Realistic Global Deformation." IEEE Vis'99. San Francisco, California. October 24-29, 1999.

[12]   Zienkiewicz, O., Taylor, R., The Finite Element Method, McGraw-Hill Book Co., London, Fourth Edition, 1988.

Interaction Techniques and Vessel Analysis for Preoperative Planning in Liver Surgery

Bernhard Preim[1], Dirk Selle[1], Wolf Spindler[1], Karl J. Oldhafer[2],
Heinz-Otto Peitgen[1]

[1]MeVis – Center for Medical Diagnostic Systems and Visualization, Universitätsallee 29, 28359 Bremen, Germany, Email: {preim, selle, spindler, peitgen}@mevis.de

[2]Universitätsklinikum Essen, Klinik für Allgemein- und Transplantationschirurgie Hufelandstr. 55, 45122 Essen, Germany, Email: karl.oldhafer@uni-essen.de

Abstract. We present visualization and interaction techniques for preoperative planning in oncologic liver surgery. After several image processing steps a 3d visualization of all relevant anatomic and pathologic structures is created. In this 3d visualization a surgeon can flexibly specify resection regions with resection tools which can be applied selectively to different structures. The combination of several views which can be synchronized makes it easy to compare different views on the resection plan.

In addition, we present the application of vessel analysis techniques in order to make suggestions for optimal resections according to guidelines for liver surgery. The basic idea for these suggestions is to define the region which has to be removed in order to resect a lesion with a given tumor free margin. For this purpose, the vessels involved and the region supplied by them is estimated. It turned out that the resections suggested provide a reasonable and useful basis for preoperative planning. This contribution presents novel methods which have not been evaluated thoroughly yet.

Keywords: computer-assisted surgery, vessel analysis, visualization

1. Introduction

Liver carcinoma belong to the most wide-spread malignant diseases world-wide. Among the well-established therapies surgical intervention is the only one which can have a curative effect (5-year survival rate $\approx$ 30 %). Long-term survival depends on whether all lesions have been removed entirely with a sufficient tumor free margin. Therefore it is essential to decide on a reliable base whether a tumor can be treated surgically and which region should be resected. Currently, this decision is made on the base of planar slices of CT- and MR-images. This raises several problems: the spatial relationships between vessels and lesions are difficult to judge. The volume of lesions, the vessels involved in the resection of a lesion and the region which is supplied by the involved vessels can only roughly be estimated. Furthermore, it is often difficult to decide whether multiple lesions should be resected separately.

In this paper, we present the SURGERYPLANNER which is dedicated to liver surgery and provides decision support for the above-mentioned questions. The SURGERY-PLANNER is based on our long-term project on image processing and analysis of CT liver images [4], [9], [10]. The SURGERYPLANNER contains three main components:

- a flexible 3d visualization for the exploration of the previously identified structures,
- a resection planning module which provides resection tools in order to try a resection strategy (e.g. to simulate the removal of parts of the liver) and
- a module which suggests resection regions on the background of guidelines for tumor-free margins and an analysis of the involved intrahepatic vessels.

Besides supporting the surgeon to preoperatively decide on the optimal resection strategy the SURGERYPLANNER is useful to discuss the intervention and to explain it to a patient.

2. Medical Background

The main issue of liver surgery is to take into account the individual intrahepatic vessel anatomy. The anatomy of the liver is characterized by four hierarchical vessel systems: the portal venous system, the liver veins, the arteries and the bile ducts. For pre-operative planning the portal venous system plays a key role as it defines the functional units of the liver. Hepatic veins are also essential as they drain the liver. Bile ducts and arteries are very close to the portal veins and are of minor importance for surgery planning.

Following the Couinaud model [2] which is wide-spread in Europe the liver is divided into segments which are defined according to the branching structure of the portal vein. A segment is supplied by a third-order portal vein branch. The hepatic veins proceed between the different segments. Liver segments are highly variable from patient to patient in shape and size [4]. Even the number of segments may be different. As the boundaries between the segments can not be localized by external landmarks it is impossible to identify segments exactly during an operation. In this paper we describe methods to support liver surgery by preoperatively identifying liver segments and subsegments as suitable resection regions. The prerequisites for this purpose are the segmentation of the liver and of all lesions as well as the segmentation of the portal venous system and of hepatic veins.

In order to get a better understanding of the requirements surgeons actually have in practice, we have polled liver surgeons by sending out some 30 questionaires from which 11 have been returned (all 11 are male surgeons, average age 40 years, experience in tumor surgery on the average 10 years). All surgeons indicated that the spatial relations between vessels and lesions are difficult to judge and that a 3d visualization and quantitative analysis would be helpful for this purpose. Almost all, 10 of the 11 surgeons indicated they would appreciate trying out resection strategies preoperatively which reveals that it is often not obvious how to resect a lesion. Based on the survey we developed the following scenario which guides our development: a surgeon selects a lesion and specifies a safety margin (in liver surgery 10 or 15 mm are considered appropriate). The system visualizes the margin and highlights all vessels involved. Beside the directly involved vessels the peripheral part of this vessel system (starting from the directly involved vessels) is highlighted to indicate that these would be destroyed. Finally, it is estimated which region of the liver parenchyma is supplied by the involved vessels because this region should be removed. In addition, the volume of the resection region (percentage of the total liver volume) is displayed. If too much tissue would be removed the safety margin can be decreased to find out whether the resection is possible with a 5 mm margin (which reduces the long-term survival expectance but is still considerably better than no operation at all).

3. Prior and Related Work

Computer support for planning liver resections is challenging. The identification and delineation of the liver and of the lesions inside it, as well as the vessel segmentation requires excellent radiological data and a variety of dedicated and robust methods which still form an area of active research. Several complex image processing steps are needed to come up with a 3d model for surgery planning. In [5] the problems are described and a particular problem – transfer function design to highlight the liver and the vessels in a volume visualization – has been dealt with. Another particular problem, the liver vessel analysis, has been tackled in [3] where a model-based approach has been used. A sophisticated system for liver surgery simulation has been developed at INRIA [6] which includes advanced 3d interaction techniques. The Visible Human dataset was used for the creation of their model. Furthermore, experiences with a deformable volumetric model of the liver parenchyma have been described in [1]. With this model, cutting procedures can be simulated.

The use of the SURGERYPLANNER requires image processing with our HEPAVISION system which integrates the algorithms for liver and tumor segmentation, vessel segmentation, vessel analysis and liver segment approximation [4], [8-10]. For the liver and tumor segmentation a live wire segmentation and a modified watershed transform are available. For the analysis of the vessel systems the result of the vessel segmentation is skeletonized and transformed into a graph which represents the branching structure of the vessel system. The portal venous system and the hepatic veins are separated automatically. Furthermore, the individual branches of the vessel system are detected. Based on this analysis liver segments and subsegments are approximated. We compared different methods for this purpose. The easiest way to approximate liver segments is the nearest neighbor approach. With this approach each voxel of the liver parenchyma is supposed to be supplied by the portal venous branch which is nearest to it (in the Euclidean metric). The HEPAVISION system has been used in cooperation with 4 clinical partners.

4. Interaction and Visualization Techniques

In the image processing stage, carried out with HEPAVISION, the liver, the lesions and liver segments are identified and delineared. A *tagged volume* represents for each voxel to which object(s) it belongs and is used as input for the SURGERYPLANNER.

4.1 Visualization of Anatomic and Pathologic Structures

The SURGERYPLANNER uses the information which results from the image processing to allow the user to design a 3d visualization. The user can select a subset of the identified objects and can assign different viewing parameters (e.g. colors) to each object (see Fig. 1 and Fig. 2). These objects are rendered in an OPENINVENTOR viewer with the usual facilities for camera control to explore the visualization. The SURGERY-PLANNER is based on a volume renderer developed at MEVIS. The renderer provides various direct volume rendering (DVR) as well as surface rendering techniques. It turned out to be useful to present the segmented structures as surface rendering (including shading effects which reveal the spatial relations more clearly) and to embed the segmented structures in a DVR of the surrounding bones to provide the visual context to assess the viewing direction. The viewer used in the SURGERYPLANNER is a special viewer which extends the OPENINVENTOR facilities in two ways: it provides a shadow projection on a camera-fix plane for better perception of spatial relations and supports 3d interaction with a SpaceBall. With these facilities users can recognize and explore spatial relations to become familiar with them.

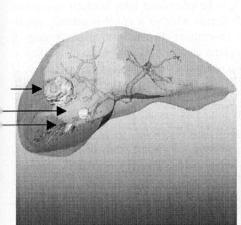 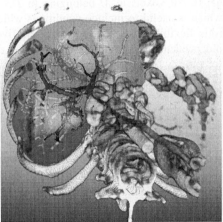

Fig. 1: Semi-transparent liver segments with three lesions (see arrows) and the portal venous system inside.

Fig. 2: The liver with arteries, veins and the portal vein embedded in the volume-rendered surrounding.

4.2 Cutting Arbitrary Regions

As surgeons want to be able to try resection strategies we developed techniques to cut arbitrary regions from a volume. Efficient removal of arbitrary regions from a volume has been described in [11]. We follow a different approach using implicit functions to define resection regions analytically. A large variety of 3d objects, like wedges and cylinders, can be conveniently defined with implicit functions.

The strategy of our algorithm is as follows: a convex resection region R is defined by an implicit function. With a transformation T the resection region is mapped into a discrete mask volume M, the size of which corresponds to the data volume V. For each voxel $v = \{v_x, v_y, v_z\}$ in V the corresponding voxel $m = \{m_x, m_y, m_z\}$ in M is TRUE if m belongs to $T(R)$. In order to identify quickly the voxels which belong to $T(R)$, an additional data structure, a brick volume B, is introduced. In B one item represents a brick of M with the original size of 12×12×12 voxel. B is used to record which region in M has been processed. Since R is convex it is in most cases sufficient to check the vertices of a brick whether they belong to $T(R)$. If the test yields TRUE for all vertices the brick is filled iteratively. If the test yields FALSE for all vertices the whole brick is outside $T(R)$. Only if the test yields TRUE for some vertices and FALSE for others the brick is subdivided. The recursive subdivision may be finished by examining 3×3×3-sized cubes (which are filled completely depending on the test of the central voxel) or by examing individual voxels. The latter high-quality mode is only applied when the user stops moving the resection tool. Like a recursive fill, the algorithm starts at a voxel inside of $T(R)$, considers its brick and recursively visits neighboring bricks (with subdivision if required) until no neighboring bricks are found which belong to $T(R)$.

With this approach the resection region is explicitly represented in M and can be modified with morphologic image processing techniques (dilatation and erosion, close gaps). All voxels in M belonging to $T(R)$ can be combined with boolean operations which is useful to include conditions which decide whether a voxel is actually drawn (this flexibility is used later for selective resections). In an 8-bit mask volume, 8 independent (and overlapping) resection regions can be managed which is sufficient for surgery planning. If R is moved to R' two options are available: (1) the old mask M is cleared and only $T(R')$ will be cut, or (2), $T(R)$ remains marked in M as deleted. In the latter case the movement of R defines a trace of arbitrary shape. Note, that this trace is not restricted to be convex, thus every possible resection can be defined.

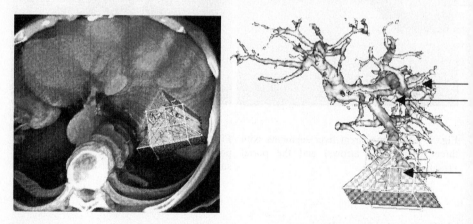

Fig. 3: A wedge-shaped resection tool is used for virtual resection in a CT data set. Both views are synchronized with respect to their viewing direction. The user can virtually resect in either view. In the left view the user gets an impression of the resection while the right view is used for orientation as it contains the vessels to be saved if possible and all lesions to be removed.

4.3 Resection Tools for the Specification of Resection Regions

For the trial of resections different resection tools are available: wegdes, clip planes, cylinders and spheres. Resection tools can be parameterized within appropriate dialoges and can be transformed in a 3d visualization by means of manipulators provided by the Graphics Library OPENINVENTOR. The properties of a resection tool define its visualization (color, rendering style) and its initial orientation and position. Resection tools can be applied selectively to different structures so that certain objects are visible even if they belong to the resection region. Such visualizations have been succesfully used for anatomy education and are described in [12]. In the context of surgery planning it is helpful to see a lesion and major blood vessels in a region where other structures are removed (see Fig. 4). The user can thus assess the distance between the tumor and the boundary of the resection region and the vessels involved in this resection.

4.4 Visualization of and Interaction with Resection Tools

Concerning the usability of resection tools the visualization and the direct-manipulative movement is crucial. Resection tools and their manipulators should be recognizable but should not occlude the resection region too heavily. A trade-off between these goals can be achieved with semi-transparent resection tools or tools which are visualized via their outlines or a combination thereof. The color must differ strikingly from that of the data values. Manipulators should be provided to support all transformation tasks: translation, rotation and scaling. After the size has been adjusted a resection tool is primarily translated and less often rotated. Based on this observation resection tools should be equipped with a manipulator dedicated for translation. The OPENINVENTOR Jack Manipulator is appropriate for this task. With this manipulator translations in the orthogonal planes are explictly supported. The Jack Manipulator can also be used to rotate an object. For this purpose a second mode exists with handles for translations temporarily hidden. The interactive resection is typically carried out in the following way: the user starts with a medium-sized resection tool and moves it through the data in order to remove a lesion. In this process, it is often necessary to rotate the visualization to evaluate what has been removed. After a rough boundary has been specified it is refined with a smaller-sized resection tool. Gaps in the resection region may be ignored as these may be filled automatically. A two-handed interaction with the system is possible with one hand to control the resection (with a 3d input device) and the other to simultaneously control the camera (with a 2d mouse).

4.5 Synchronization of Different Views

In the process of surgery planning a variety of different visualizations are generated: the viewing direction changes, different objects are visible, and resection tools are applied. For many visualization goals no single visualization is appropriate and even the interactive handling is insufficient, e.g. to compare different views under certain aspects. Therefore, it is crucial to have multiple views which can be flexibly parametrized. In the SURGERYPLANNER the user can freely add viewers which are named automatically. In an overview each viewer is represented by an icon and its name in order to switch quickly between different viewers (see Fig. 4).

In each viewer not only the viewing direction may be altered but also the subset of objects displayed, the appearance of objects and the resection tools applied. To support the comparison of different views, the user can define synchronizations between selected viewers. Viewers may be synchronized for example concerning the application of resections, or concerning rotations of the scene, or filter operations. Moreover, in every dialog which affects the content of a viewer it can be specified whether these settings shoul d be applied to all viewers (regardless whether synchronizations apply).

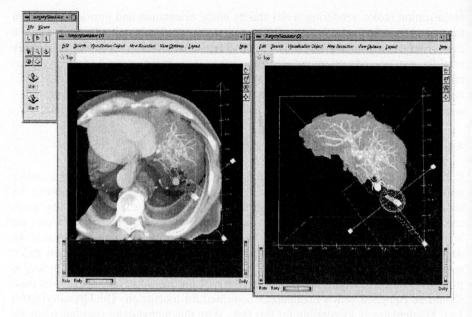

Fig. 4: Two synchronized views for the specification of a resection. The left view provides the visual context (bones, heart), while the right view (with the unoccluded liver) is used to move the resection tool.

5. Suggestions for Optimal Resections

With direct-manipulative resection tools the user has full control to flexibly develop and explore resection strategies. However, some experience with 3d interaction and a few minutes to specify a resection are required. An alternative approach to preoperative planning is to let the system emphasize what should be removed given a lesion and the location and branching structure of the portal venous system and taking into account how the resection influences vessels and the supplied liver parenchyma. We developed such a high-level support based on the scenario described in Sect. 2: starting from a selected lesion and a given tumor free margin, the involved parts of the vessel system are identified and the supplied liver parenchyma is calculated.

We started to investigate this approach by working with CT images of 8 corrosion casts of human cadavers which have been kindly supplied by PD Dr. Fasel (University Genf, Department for Morphology). With these models the portal vein could be reconstructed up to the 6th branching order – the overall length of the extracted vessels is an order of magnitude larger than in clinical data. In these models we included a sphere with 1 cm diameter as model of a focal lesion and calculated the liver parenchyma to resect for different safety margins (see Fig. 5 and Fig. 6). Safety margins are defined by a sequence of dilatations of the lesion. The calculation of resction regions also considers large hepatic veins which should not be damaged. For each corrosion cast this calculation has been carried out for 1500 tumor locations summarized in a video which we carefully discussed with our clinical partner Prof. Oldhafer. It turned out that the resection regions proposed by the system correspond well to surgical practice in shape and size. In the large majority of the cases, the highlighted regions could have been resected. In a second experiment we analyzed a clinical data set with 3 lesions. This CT data set allowed us to extract 4th order brancing of the portal vene. With this prerequisite suggestions for resections could be defined (see Fig. 7 and Fig. 8).

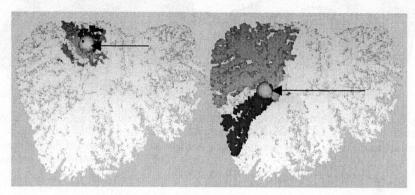

Fig. 5: A sphere has as a model for a focal lesion been moved through a model of a corrosion cast with the portal vein. The vessels not involved in the resection are displayed white. Different colors indicate the vessels involved in the resection with different margins (0.5 cm, 1.0 cm, 1.5 cm, 2.0 cm, 2.5 cm).

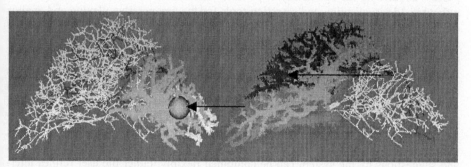

Fig. 6: The same experiment as in Fig. 5 with a model with hepatic veins and the portal venous system. The vessels not involved in the suggested resection are displayed skeletonized to better reveal the branching pattern. Different colors again indicate the vessels involved in the resection with different margins. The hepatic vein is considered in the suggested resection.

The way by which the tumor appears in a CT image and thus the size and shape of the segmented tumor is not perfectly reliable. Moreover, the surgeon cannot exactly follow a planned resection. Suggestions for resections have to consider the limited accuracy of the data and the procedure. Therefore the calculation is carried out for different margins: the red one being the inner region which must be definitely resected, the other colors representing more distant zones. The visualization thus indicates the risk involved in the procedure for the damage of major blood vessels (which are in the orange or yellow region). The resection regions suggested can be fine-tuned interactively. For this purpose the user selects a region which is subsequently transformed in the mask volume for manipulation with a resection tool.

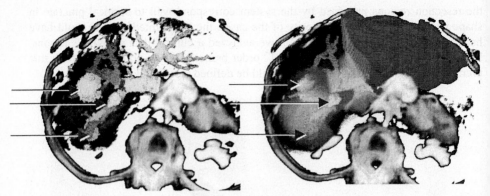

Fig. 7: In a case with three lesions (see the arrows) the dark vessels are involved in the removal whereas gray vessels are not affected.

Fig. 8: The regions supplied by the vessels involved in the resection are rendered semitransparent. The resection of the left hemi-liver (49,4 % of the liver parenchyma) is indicated.

6.   Conclusion and Future Work

The SURGERYPLANNER provides a reliable base to aid in the complex decision making process concerning the operability of patients with focal liver lesions. For this purpose a flexible 3d visualization and resection tools for the preoperative planing and analyis of a resection are provided. An efficient algorithm for cutting arbitrary regions is used for interactive resection. With these methods, several patients have been virtually resected using the data processed with our HEPAVISION system. The handling of the resetion tools turned out to be difficult, without stereo rendering. Once a resection is defined, however, it helps considerably to judge an operation. The selective application of resection tools and synchronization mechanisms are essential for the usefulness of the system. Based on established guidelines and a user-defined margin suggestions for resections are generated as an alternative approach to surgery planning. With this unique feature the risk for long-term survival can be estimated. Interactive resections and suggestions by the system can be combined flexibly: either the user refines interactively a resection area suggested by the system or the system analyzes a user-defined resection area. The combination of flexible interaction techniques with high-level support is not bound to liver surgery but can be applied, e.g. to lung surgery.

The support provided by the SURGERYPLANNER is considerable but also has limitations. The system does not guide the surgeon during the operation. Intraoperative support, as in neurosurgery, is difficult to realize in liver surgery due to the surmounting problems with intraoperative registration. Moreover, the decision whether a patient can tolerate a resection is not fully supported. In particular, primary liver carcinoma are often inoperable due to the extent of liver cirrhosis which accompanies the cancer disease. In some cases patients are inoperable due to lesions outside the liver or due to the genral state of the patient. The SURGERYPLANNER can only be used to decide whether patients are locally operable (that is whether liver lesions might be resected).

Future work will concentrate on the following problems: an in-depth evaluation in the preoperative use of the SURGERYPLANNER, the use of MR data which requires to fuse several MR sequences which selectively highlight the relevant structures and the refinement of the suggestions for resections. In rare cases where a lesion is centrally located at the back side ($\approx$ 5 percent) the suggestions generated require to resect the liver ex-situ (which is an extremely demanding procedure carried out only at a few hospitals). In these cases our suggestions should consider the access to the resection region to find out whether an alternative access is possible. Currently, the user can treat these cases with interactive resection.

Acknowledgments

We would like to thank our colleagues A. Schenk, D. Böhm and Dr. T. Schindewolf for their contribution to the HEPAVISION system. Furthermore, we thank Prof. Galanski, Dr. Högemann, and Dr. Stamm (Department of Diagnostic Radiology, Medical School Hannover) as well as Prof. Klose and his team (Department of Diagnostic Radiology, Philipps-University Marburg) for fruitful discussions. Fig. 1, 3, 5, 6, 8 and 9 are based on data kindly provided by Prof. Klose. Fig. 2 and Fig. 4 are renditions of datasets kindly provided by Prof. Galanski. We thank Felix Ritter, University of Magdeburg, who provided us with a special viewer which supports shadow projection and the use of 3d input devices. We acknowledge the German Reseach Council (DFG) for support of our project under the number PE 199/9-1.

References

1. Boux de Casson F, and Laugier C (1999) "Modeling the Dynamics of a Human Liver for a Minimally Invasive Surgery Simulator", *Proc. of MICCAI*, Springer, LNCS, Vol 1679, pp. 1156-1165
2. Couinaud, C. *Le foie – Etudes anatomiques et churgicales*, Paris, Masson, 1957
3. Dokladal, P, Lohou C, Perroton L, and Bertrand G (1999) "Liver Blood Vessel Extraction by a 3d toplogical approach", *Proc. of MICCAI* (Cambridge, UK, 19.-22. Sept.), Springer, LNCS, Vol 1679, pp. 98-104
4. Fasel JHD, Selle D, Gailloud P et al. (1998) "Segmental Anatomy of the Liver: Poor Correlation with CT", *Radiology*, Vol 206 (1), pp. 151-156
5. Fishman, EK, Kuszyk BS, Heath DG, Gao L, and Cabral B (1996) "Surgical Planning for Liver resection", *IEEE Computer*, January 1996, pp. 64-72
6. Marescaux J, Clement JM, Tassetti V et al. (1998) "Virtual reality applied to Hepatic Surgery Simulation: The Next Revolution", *Annals of Surgery*, Vol 228 (5), pp. 627-634
7. Mazziotti A and Cavallari A (1997) *Techniques in Liver Surgery*, Greenwich Medical Media
8. Oldhafer KJ, Högemann D, Stamm G, Raab R, Peitgen HO, and Galanski M (1999) "Dreidimensionale Visualisierung der Leber zur Planung erweiterter Leberresektionen", *Der Chirurg*, Vol 70, Springer, pp. 233-238
9. Schenk, A, Prause G and Peitgen HO (2000) "Efficient Semiautomatic Segmentation of 3D Objects in Medical Images", *Proc. of MICCAI* (Pittsburgh, 11.-14. Okt.), Springer, LNCS (to appear in this volume)
10. Selle D, Schindewolf T, Evertsz CJG, and Peitgen HO (1999) "Quantitative analysis of CT liver images", *Excerpta Medical International Congress*, Vol 1182, Elsevier, pp. 435-444
11. Udupa JK and Odhner D (1991) "Fast Visualization, Manipulation and Analysis of Binary Volumetric Objects", *IEEE Comput. Graphics Appl.*, Vol 11 (6), pp. 53-62
12. Tiede, U, Bomans, M, Höhne KH et al. (1993) "A computerized three-dimensional atlas of the human skull and brain", *Am. J. Neuroradiology*, Vol 14 (3), pp. 551-559

First Steps in Eliminating the Need for Animals and Cadavers in Advanced Trauma Life Support®

Christoph Kaufmann, MD, MPH, FACS
Scott Zakaluzny, BA
Alan Liu, PhD

Surgical Simulation Laboratory
National Capital Area Medical Simulation Center
Uniformed Services University of the Health Sciences

Abstract. The Advanced Trauma Life Support® course is designed to provide for optimal initial resuscitation of the seriously injured patient. The surgical skills component of this course requires the use of cadavers or anesthetized animals. Significant anatomical differences and ethical issues limit the utility of animals. The cost and difficulty in procuring cadavers makes widespread use of this option impractical. A combination of mannequin and computer-based surgical simulators is being developed to replace these animals and cadavers. The first of these simulators, a pericardiocentesis trainer, is complete.

1. Introduction

The Advanced Trauma Life Support course (ATLS®) provides the doctor with effective strategies for initial management of the trauma patient. The ATLS® course is strongly recommended for all physicians and surgeons involved in the care of seriously injured patients. Globally, an average of 19,000 doctors are trained each year [1]. The entire ATLS® course focuses on rapid assessment and treatment to afford the best possible patient outcome. The ATLS® course includes a surgical skills practicum in addition to skills stations teaching accurate patient assessment, triage, establishing and maintaining the airway, using imaging to identify various injuries, and trauma care decision making [1]. The procedures in the surgical skills portion of the course include: tube thoracostomy, surgical airway, pericardiocentesis, and diagnostic peritoneal lavage. Today, ATLS® students learn these procedures on anesthetized animals or cadavers. Further consideration is needed since anesthetized animals do not present the correct anatomy for realistic training, are not reusable, are expensive, and raise ethical issues. Cadavers, which have the correct anatomy, still pose problems because they are expensive and can be difficult to procure. Tissue degradation and frozen cadaver specimens present additional problems.

Recent advances in computer simulation technology permit many of these procedures to be practiced in a virtual environment. In this paper, we provide introduction to our work in developing an ATLS® surgical skills lab that eliminates the use of animals and cadavers.

2. Methods

The National Capital Area Medical Simulation Center is a state-of-the art medical education facility at the Uniformed Services University of the Health Sciences (USUHS). It consists of three functional areas: clinical examination rooms, computer lab, and surgical simulation lab. USUHS has received permission from the American College of Surgeons to examine the utility of using computer-based surgical simulators in teaching the ATLS® surgical skills practicum. The surgery simulation lab includes the MedSim patient simulator [2], a desktop 3D haptic reach in virtual environment, a BDI vascular anastomosis simulator, CathSim IV insertion simulators, Ultrasim ultrasound simulator, HT Medical PreOp broncoscopy simulator, and other simulation technologies.

The MedSim patient simulator is a computer-controlled mannequin designed to simulate a human patient. Among the various features on this device are reactive pupils, arm motion response to painful stimulus, and both radial and carotid pulses. Two ATLS® surgical skills can be taught on the simulator. The simulator can accept and respond to chest tubes used in tube thoracostomy. The simulator also permits the creation of a surgical airway and responds with the appropriate physiological reaction to improved ventilation and oxygenation. The mannequin's airway is connected to a computer which analyzes the percent oxygen in inhaled air, displays blood oxygen saturation results on a pulse-oxygen monitor, and exhales carbon dioxide. The mannequin allows for the delivery of and detection of eighty-five drugs. This is a very abbreviated list of features of this highly useful teaching tool. The simulator can reproduce physiologic responses consistent with various forms of injury by using the computer interface to control forty different patient attributes from age to insulin production and invoke up to thirty medical complications.

One of the conditions which the mannequin can simulate is cardiac tamponade, a diagnosis which mandates pericardiocentesis, although this procedure cannot be performed on the MedSim mannequin. Pericardiocentesis is a simple needle-based procedure and lends itself to virtual reality simulation. Pericardiocentesis involves the needle aspiration of excess fluid (e.g., blood) from the pericardial sac to permit the heart to fill and function properly. Since this is an emergency procedure, external guidance (e.g., fluoroscopy) is not available. The physician must be familiar with acceptable entry locations and angles. The physician must also know how far to advance the needle.

We have developed a virtual reality based simulator for pericardiocentesis. The simulation is based on a desktop 3D virtual environment similar to that described in [3]. A flat panel liquid crystal display (LCD) is used to generate stereoscopic images of the environment. The LCD display incorporates a grid to focus light rays from alternate pixels to each eye, permitting a different view to be generated for each eye generating the 3D perspective [4]. Special eyewear is not required. This display facilitates a more comfortable viewing of the virtual environment over extended periods of time.

The pericardiocentesis simulator is running on a PC with a 550 MHz Intel Pentium III Xeon® processor, a half a gigabyte of memory, and an Evans and Sutherland Tornado 3000® video card. The image on the flat screen LCD is reversed so that it can be viewed in a mirror situated horizontally below and in front of the monitor. (Fig. 1) The student looks down at the horizontally situated mirror and sees a 3-D rendered field with an external view of a virtual thorax and a virtual cannula. A Phantom haptic interface [5] controls the position and orientation (pose) of this virtual cannula. In order to increase the realistic feel of the simulation, the Phantom is set-up in such a way that the arm extends underneath the viewing mirror so that the stylus on the Phantom is moving in three-space with the same orientation and position as the cannula it represents in the virtual image being viewed. (Fig. 2) This preserves the hand eye axis of the operator. Using the Phantom, the student selects an entry location and angle, then inserts the cannula into the chest. (Fig. 3) Haptic feedback is provided to simulate the effect of both an increase and a decrease in resistance associated with that of the cannula meeting and penetrating the skin and chest wall. The program provides both positive and negative feedback by having the cannula fill with blood from the pericardium and by displaying a warning message, respectively. (Fig. 4 &5) If the student chooses either the wrong entry point and/or advances the needle in the wrong direction an obvious warning message appears (allowing for correction).

3. Discussion

Medical simulators provide significant advantages over animals and cadavers. Simulators generate repeatable scenarios that are impossible to predictably simulate with cadavers or animals. Not only can difficult situations be simulated, they can be repeated as many times as needed since no specimen is destroyed. By using the same simulator and same loaded scenario, the computer simulation provides a means of ensuring uniformity in training and evaluating performance. In addition, rare anatomic variations or pathology can be easily emulated. These provide exposure to unusual cases hands-on as contrasted with a description in a book or lecture. The program's warnings also act as an additional way of instructing students. The warnings provide the immediate feedback that allows the student to correct their actions and learn to do the procedure properly. The obvious advantage of simulators is that they are reusable and, following initial start-up cost, they require only minimal maintenance compared to purchasing new cadavers or animals. This makes simulators more cost-effective over the long term. Most importantly, no ethical issues are involved when using computer-based simulation.

Computer-based and mechanical medical simulators do have disadvantages. Limited visual and tactile realism can still be a problem (e.g., bleeding is difficult to simulate). Simulators require high-end equipment and new programming which can create a high initial cost. In addition, the field of medical simulators is still in its infancy and is pushing the envelope of the present technology. Many useful procedures currently do not have adequate simulations because the technology does not exist to support

Figure 1- Demonstrates the flat screen LCD and mirror used to view the 3D rendered field.

Figure 2- Demonstrates the flat screen LCD, mirror, and Phantom haptic feedback device.

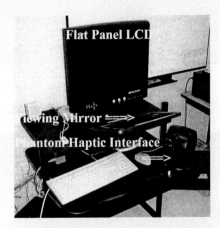

Figure 3- Screen capture demonstrates the field of view and the cannula. Viewed here in 2D and black and white as opposed to 3D and color.

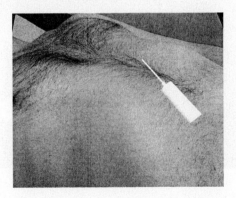

Figure 4- Screen capture demonstrates the cannula filling with blood in a successful procedure.

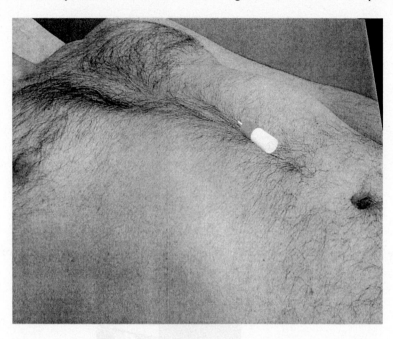

Figure 5- Screen capture demonstrates example of a warning message.

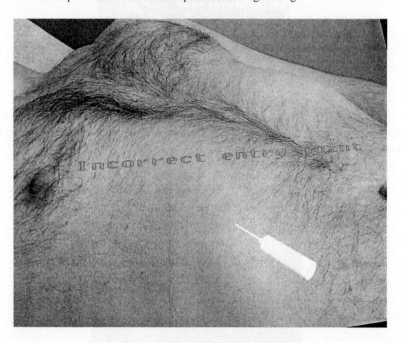

them. The simulators that are being developed now are used to assure that specific tasks can be completed properly, but many other surgical procedures cannot be simulated as of yet. This is primarily because of the complexity of human anatomy, the need to represent physiologic processes such as bleeding, and the lack of accurate haptic representation of various human tissues. [6]

4. Conclusion

Computer-based medical simulators are a viable alternative to animals and cadavers for teaching ATLS® surgical skills. A skill-specific surgical simulator has been developed for pericardiocentesis. The basic aspects of correct location, angle, and depth for cannula insertion can be demonstrated and evaluated. Work is on-going to evaluate the efficacy of this device and to develop a second surgical simulator, diagnostic peritoneal lavage.

Most recently, the pericardiocentesis simulator has been moved to a lower end platform that does not require the flat screen LCD or Phantom haptic device. This move was made to make the program more portable and fiscally feasible for wide-spread use. The development and testing of this system is on-going.

5. References

1. American College of Surgeons Committee on Trauma, "Advanced Trauma Life Support® for Doctors", 6th edition, Chicago, IL (1997)
2. MedSim Advanced Medical Simulations, Ltd. Product literature. URL: www.medsim.com
3. Poston T., Serra L., "Dextrous virtual work", *Communications of the ACM v. 39 (May) p. 37-45, 1996*
4. Kaufmann, C., Liu A., Burris D., "DTI Autostereographic Display: Initial Evaluation," *Medicine Meets Virtual Reality 2000*, J.D. Westwood et al (Eds.) IOS Press, Amsterdam (2000) 156-158
5. Massie T.H., Salisbury J.K., "The PHANToM Haptic Interface: A Device for Probing Virtual Objects." *Proceedings of the ASME Winter Annual Meeting, Symposium on Haptic Interfaces for Virtual Environment and Teleoperator Systems, Chicago, IL, Nov. 1994*
6. Kaufmann, C., "Role of Surgical Simulators in Surgical Education," *Asian J. Surgery* 1999; 22 (4): 398-401

Planning and Simulation of Robotically Assisted Minimal Invasive Surgery[*]

Louaï Adhami, Ève Coste-Manière, Jean-Daniel Boissonnat

INRIA Sophia-Antipolis B.P. 93 FR-06902 Sophia-Antipolis, France.
{**Louai.Adhami, Eve.Coste-Maniere,
Jean-Daniel.Boisonnat**}**@sophia.inria.fr**
http://www.inria.fr/chir

Abstract. This paper proposes a framework for pre-operative planning and simulation of robotically assisted Minimal Invasive Surgery (MIS). The design of an integrated system is presented for cardiovascular interventions. The approach consists of a planning, validation and simulation phase. The goals of each phase being, respectively, to propose suitable incision sites for the robot, to validate those site and to enable realistic simulation of the intervention. With the patient's pre-operative data, we formulate the needs of the surgeon and the characteristics of the robot as mathematical criteria in order to optimize the settings of the intervention. Then we automatically reproduce expected surgeons' movements and guaranty their feasibility. Finally we simulate the intervention in real-time, paying particular attention to potential collisions between the robotic arms.

Keywords: Surgical simulators, robotics and robotic manipulators, therapy planning.

1 Introduction

The extension of MIS techniques to use robotics came from the limitations imposed on the surgeon by manually controlled MIS instruments. Namely, the surgeon finds his movements, vision and tactile sensing reduced and severely altered ([9],[3]). The introduction of robotic manipulators would remedy the loss of dexterity in the movements by incorporating additional degrees of freedom at the end of the tools; i.e., in the part that would be inside the patient. In addition, more precision and better tactile feedback can be achieved by proper design of the system, which can also improve the hand eye alignment of the surgeon and offer better 3d visualization.

However, this innovation has its own limitations and problems. Beginning with the limitations, and despite the increased dexterity, the region reached from a set of incision sites will remain restrained, thus these sites have to be carefully chosen for each patient depending on his anatomy. Moreover, the forces that can

---
[*] This work is partially supported by the Télémédecine project of the French Ministry of Research.

be delivered by a robotic manipulator may vary significantly with the position of the latter, which stresses even more on the choice of the incision sites or *ports*. Now moving to the problems introduced by the use of a robot, and setting aside classical control and liability concerns, the main handicap of such systems is the issue of potential collisions between the manipulator arms. Again this stresses on the proper positioning of both the incision ports and the robot. Finally, and no matter how intuitive the controlling device is made, the surgeon will need time and proper training before using his new "hands" in the most efficient way. Therefore simulation would be used to rehearse the intervention, validating the planned ports and helping the surgeon gets accustomed to both his tools and his patient.

In short, a strong need of planning has been pointed out for robotically assisted MIS, as well as suitable simulation. These, along with necessary intermediate steps will be described in this paper as a generic methodology for robotically assisted MIS. However, and in order to have an illustrative example, the MIDCAB (minimally invasive direct coronary artery bypass) intervention was chosen as a test bed for our work. The MIDCAB intervention is a closed chest cardiac intervention where damaged portions of cornary vessels are bypassed. This choice is based on the importance of the MIBCAD intervention, and on the sub-millimeter precision needed to perform it, which makes the use of a robotic manipulator very attractive. Likewise, a prototype version of the daVinci^{TM} robot[1] was selected for the implementation of our approach.

The proposed work is part of a more complete effort of computer assistance for MIS interventions which comprises, in addition to the pre-operative phase presented here, an augmented reality intra-operative guidance system. Thus during the intervention, the surgeon will be guided in accordance with his pre-operative plans, and online validation will insure a safe, collision free manipulation of the robot.

Related work The use of computer assisted systems in the preparation and simulation of minimally invasive techniques is being recognized as an efficient tool in surgery. Various successful systems have been proposed and are already operational. Some examples are the haptic system based on the work of [4], the more general purpose system in [7], and [2] in neurosurgery. These systems do not handle teleoperated robotics, with the notable exception of the KISMET simulation tool that is being merged with the ARTEMIS robotic environment (see [1]), but which lacks systematic planning.

In the case of robotically assisted MIDCAB intervention, existing planning attempts are rather empirical ([9] and [10]). A more systematic effort is found in [6], where pre-operative data is used to measure the distance between the concerned vessels. However, no robotic system is used, and a mini-thoracotomy is performed. To the best of our knowledge, there presently is no system that integrates the planning, simulation and online control of a robotic surgery system for laparoscopic or cardiac MIS interventions.

---

[1] The da Vinci surgical system is a trademark of Intuitive Surgical Inc.

2 Overview

Our approach, schematically summarized in figure 1, can be subdivided in four main parts outlined below.

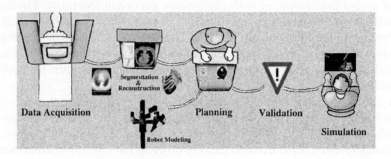

Fig. 1. General Approach.

1. **Preliminary processing**: Pre-operative radiological data is often difficult to handle; therefore, suitable pre-processing such as segmentation and 3d reconstruction should first be carried out. Since the data used is patient dependent, it is also desirable to have a completely automatic process to handle the output of a radiological imaging device directly. Moreover the robot system that will be used has to be modeled.
2. **Planning**: The goal of the planning phase is to determine the best incision sites for the introduction of the robot arms and endoscope. For that, three sources of information are combined in an optimization algorithm where some predefined mathematical criteria have been integrated. The criteria translate features such as collision avoidance between the manipulator arms or reachability of targeted organs.
3. **Validation**: Once a set of ports is obtained, the feasibility of the intervention is checked by automatically reproducing the expected movements of the surgeon and looking for collisions or other problems, such as an out of reach condition.
4. **Simulation**: Finally, the surgeon is presented with the incision ports and he is now ready to test them, and may require another set if he judges the proposed one not satisfactory. In addition, he would be able to "rehearse" his intervention, thus comparing different strategies and getting accustomed to the use of the robot.

3 Preliminary steps

Radiological data must first be processed and combined with the robot model in order to carry on with the planning. A set of anatomical entities has to be isolated for every type of surgical intervention. Moreover, some special handling may be required for the port placement to allow or forbid access to certain physiological parts.

3.1 Segmentation

The patients' anatomical entities that are of interest are isolated using a novel knowledge based segmentation technique that will be the subject of a future publication. The process is invariant from one patient to another, in particular it does not have any parameters to be tuned by the radiologist. However, it should be noted that manual interventions are still required at this stage.

For all robotically assisted MIS intervention, we define the *admissible points* as those sites that allow the introduction of the robot arms and endoscope. They are coupled with *admissible directions* pointing outward and perpendicular to the skin, which will be used during the planning phase. Figure 2 shows a segmentation for the MIDCAB intervention, where the admissible sites are the intercostal spaces. Additional algorithms were added to compute the density of the traversed ray from the admissible point to the skin in the normal direction, in order to determine whether a proposed port is acceptable or not; e.g, a ray passing through a bony structure is eliminated.

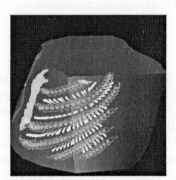
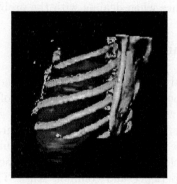

Fig. 2. (a) The admissible ports and their directions, the rib centers are shown as spheres. (b) Using **nuages** to reconstruct in 3d the anatomical entities needed for the MIDCAB intervention.

3.2 Reconstruction

3d reconstruction is essential for efficient visualization and manipulation of the segmented data. The software used for the reconstruction is called **nuages** and is described in [5]. The underlying algorithm is based on projected Voronoï diagrams, where the input is a set of closed non-intersecting contours, and the output is a mesh of triangles representing the reconstructed surface in 3d. This algorithm has the advantages of outputting a low and controllable number of triangles, and of not being prone to distortive effects such as the staircase effect observed in marching cubes algorithms. On the other hand, its input is somewhat restrictive, as only non-intersecting slice-wise closed curves are allowed. Figure 2 (b) depicts a reconstruction for the MIDCAB intervention.

3.3 Robot Modeling

Modeling the robot is an essential step that will gain in importance as we advance through the stages of our approach. In the planning phase, only the Denavit-Hartenberg (DH) models are needed. For validation, an implementation of the kinematics have to be carried out, and finally the simulation would incorporate as much as possible the dynamics of the system.

To carry out the modeling, we developed a generic C++ library, where OpenGL^R output and collision detection are implemented. Two primitives were retained for the modeling of the robot body, namely rectangular parallelepipeds and cylinders. Moreover, inverse kinematics are carried out either analytically or numerically. An analytic solution is used when there is the same number of degrees of freedom (dofs) and constraints, whereas a numerical solution is used when there are more dofs than constraints. In the latter case, artificial constraints are added to reflect the proximity between the arms, which would be of great significance when dealing with the problem of collision avoidance.

4   Planning

As already mentioned, the choice of the incision points for the robot arms and endoscope are critical to the success of the intervention. Therefore, some sort of optimization is desired in order to guarantee the best possible locations of the ports, if such positions exist. Naturally, the best way to insure the feasibility of the intervention is through simulation, which is the object of section 6. Nevertheless, one cannot try out all possible configurations, even if driven by some experience and common sense. The proposed solution is to choose from the set of all possible candidates, the *admissible* points, a triplet that optimizes a set of predefined criteria. These criteria are based on the descriptions of the surgeons' difficulties and priorities when performing MIS interventions, whether robotically assisted or not.

4.1   Criteria

The first step is for the surgeon to specify his points of interest, or *target* points, along with desired *attack direction and area* (see figure 4 (b) for an example). The target points are representatives of the areas on which the surgeon wants to operate, whereas the attack direction specifies the orientation of the surgical tool with respect to the surface at the attack point. This defines the criteria considered in the optimization process, which are:

1. The *target area*, which is the area chosen by the surgeon to translate the importance of the site on which he wishes to work.
2. The *attack* angle, which is the angle between the attack direction at the target point on the one hand, and the straight line connecting the latter to the admissible point on the other. It reflects the ease with which the surgeon can operate on a given location with respect to the ideal inclination he had specified.
3. The *dexterity* parameter, which is proportional to the angle between the surface normal at the admissible point and the straight line connecting the latter to the target point. This measure of dexterity has to be interpreted in accordance with the robot capabilities.

4.2 Optimization

In order to isolate the most interesting triplet according to the above enumerated criteria, the following steps are carried out.

1. Eliminate admissible candidates that are not fully accessible from the ports by rendering a cone with its apex at the admissible point and its base covering the target area (see figure 3 (a)). The technique of rendering parts of scene to detect interference was inspired from [8].
2. For the endoscope, sort the admissible sites in a way to minimize the angle between the target normal and the line connecting the admissible point to the target point. This step privileges a direct view over the areas of interest. It should also be noted that the targets are weighted in accordance to their area. Moreover points with angles greater than the camera angle used during the intervention, are eliminated.
3. For the arms, sort in the same way as for the endoscope, but with the angle limitation relaxed. Moreover, eliminate admissible candidates that make to obtuse an angle this time with the normal to the skin. For example 60° is the limit in the MIDCAB case to avoid excessive stress on the ribs. The different angles are shown in figure 3 (b). Moreover, a check is performed on the different dimensions of the links to insure that no movement will cause an out of reach condition.
4. Finally choose the triplet that compromises the score of each site according to the above criteria with maximizing the distance between the ports. The latter requirement is a natural remedy against the problem of collision between the robot arms.

4.3 Results

The results obtained on a test case for the MIDCAB intervention are shown in figure 4 (b), next to those recommended by [9] in (a). The optimization is exhaustive, although hierarchical elimination is carried out to speed up the process. The evaluation of the results is ultimately through clinical trials; however, a first

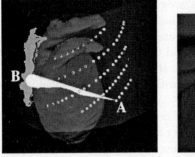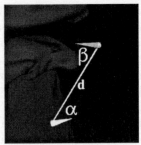

Fig. 3. (a) An example of a partially occluded area. Imagine having a flash light at point A, the targeted area B is not fully illuminated. (b) The measures used for optimization: α is the angle introduced in the 2nd step, β in the 3rd and d the distance in 4th step.

estimation will be through virtual validation and simulation discussed in section 5 and 6. These evaluations were very positive: typically one of the first three triplets passed the validation test that we now present.

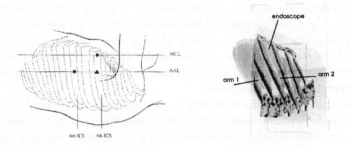

Fig. 4. Port placement results for the MIDCAB intervention: (a) Surgeons' recommendations (reproduced from [9]), and (b) optimized ports. Circle/yellow for camera port, square/triangle/cyan for tools ports.

5 Validation

Once a suitable triplet has been identified, the robot is placed in the proposed position (see section 3.3 for the way inverse kinematics are carried out). In order to guarantee the feasibility of the required surgical act, the current configuration is tested for a collision free path connecting the different target area. The details of interference detection are presented in section 6.1.

The trajectory generated between two target areas is a straight line, as this is the way the surgeon is expected to navigate. The possibility of collisions between discrete time steps are handled by the interference detection algorithm that tests sweeps the volume covered by the manipulator arms. In addition to interference detection, out of reach conditions and possible singularities are monitored and signaled. Finally, the endoscope is positioned relative to the tips of the tool arms at a predefined distance, in a way to guarantee a good visibility at all times. Figure 5 depicts a test for the MIDCAB intervention.

If no problem is detected, then the triplet is accepted and the surgeon can proceed with manual verification, else the triplet is refused and another one is computed. If no suitable ports can be found, then the robot is placed in a predefined nominal position and the surgeon would try to find one manually.

Fig. 5. A validation sequence. Note the colliding state and the out of reach condition in the last image.

6 Simulation

The simulator has the double aim of offering the surgeon a realistic environment to develop good control over the robot, and of validating the suggested incision ports.

6.1 Collision detection

Exact real-time collision detection is an essential requirement for a realistic simulation and a safe validation of the strategy that will be used in the intervention. Collisions are stratified as external (with the anatomical entities) and internal (between the manipulators), the latter being further subdivided into static and dynamic (continuous movement). For the latter types, an algorithm to detect interferences between rectangular parallelepipeds and cylinders was written. It is capable of handling both static and dynamic collisions, thus considerably reducing the sensitivity to the discretization step. On the other hand, external collisions are detected using the method suggested in [8], which makes use of the graphics hardware. Both methods run at frame rate for typical data.

6.2   The Interface

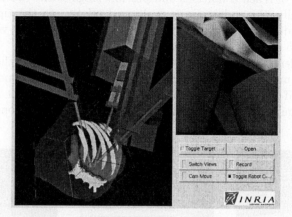

Fig. 6. A single interface is used for both planning and simulation.

Figure 6 shows a snapshot of the interface developed for surgeon interaction. Besides the usual 3d navigation and all the functionalities included for the planning phase, a realistic control of the robot has been implemented using two 3d pointing devices[2]. The models used are those of the prototype of the daVinci™ system. As is the case with the actual robot, the surgeon manipulates the tools simultaneously in the camera coordinate frame, and may also control the camera using a special switch. Automatic validation may be replayed, and a new optimization may be requested at any time.

7   Conclusion and Future Trends

A framework for pre-operative planning and simulation of robotically assisted MIS intervention has been proposed. The three main steps are planning, validation and simulation. The results obtained in the planning phase are very encouraging and seem coherent with those proposed manually. Moreover, the interface was well accepted by the surgeons.

The validation step forms a solid guaranty of the results proposed in the planning. Efficient collision detection methods make it possible to try out different setups in a few seconds. Finally the simulation interface offers an intuitive tool to rehearse the intervention and comfort the surgeon. Current efforts are directed towards the following extensions of the system:

1. Merging the planning step with the validation step so as to directly generate the desired set of ports.

---

[2] Magellan Space Mouse, http://www.logicad3d.com.

2. The integration of the present work with the intra-operative part to enable clinical validation of the results on the MIDCAB and other interventions.
3. More realistic simulation by the use of deformable models to represent soft tissues.
4. Improvement of the segmentation technique until no manual intervention is required.
5. More realistic representation of the organs; e.g, integrating a coronary map on the heart.

Acknowledgements

This work would not have been possible without the help of Prof. Alain Carpentier and Dr. Didier Loulmet from the Hôpital Broussais in Paris. We would also like to acknowledge the Intuitive Surgical Inc. team, especially Dr. Kenneth Salisbury and Dr. Gary Guthart.

References

1. ARTEMIS homepage. http://wwwserv2.iai.fzk.de/artemis/.
2. L. Auer, A. Radetzky, C. Wimmer, G. Kleinszig, H. Delingette, and B. Davies. Visualisation for planning and simulation of minimally invasive procedures. *Lecture Notes in Computer Science*, 1679:1199–1209, Sept. 1999.
3. G. B. Cadière and J. Leroy. *Principes généraux de la chirurgie laparoscopique. Encycl Méd Chir (Techniques chirurgicales - Appareil digestif)*, volume 40, page 9. Elsevier-Paris, 1999.
4. S. Cotin, H. Delingette, and N. Ayache. Real-time elastic deformations of soft tissues for surgery simulation. *IEEE Transactions On Visualization and Computer Graphics*, 5(1):62, 1999.
5. B. Geiger. Three dimentional modeling of human organs and its application to diagnosis and surgical planning. Technical Report 2105, INRIA-Sophia, 1993.
6. H. Gulbins, H. Reichenspurner, C. Bechker, D. Boehm, A. Knez, C. Schmitz, R. Bruening, R. Harbrel, and B. Reichart. Preoperarive 3d-reconstruction of the ultrafast-CT images for the planning of the minimally invasive direct coronary artery bypass operation (MIDCAB). In *Second World Congress of Minimally Invasive Cardiac Surgery, Minneapolis, Minnesota*, June 1998.
7. U. Kühnapfel, H. Çakmak, and H. Maaß. 3Dmodeling for endoscopic surgery. In *Proc. IEEE Symposium on Simulation*, pages 22–32, Delft University, Delft, NL, Oct. 1999.
8. J.-C. Lombardo, M.-P. Cani, and F. Neyret. Real-time collision detection for virtual surgery. In *Computer Animation, Geneva*, May 1999.
9. D. Loulmet, A. Carpentier, N. d'Attellis, A. Berrebi, C. Cardon, O. Ponzio, B. Aupècle, and J. Y. M. Relland. Endoscopic coronary artery bypass grafting with the aid of robotic assisted instruments. *The journal of thoraic and cardiovascular surgery*, 118(1), July 1999.
10. H. Tabaie, J. Reinbolt, P. Graper, T. Kelly, and M. Connor. Endoscopic coronary artery bypass graft (ECABG) procedure with robotic assistance. *The Heart Surgery Forum (http://www.hsforum.com)*, 2(0552), Sept. 1999.

A Framework for Predictive Modeling of Intra-operative Deformations: A Simulation-Based Study

Stelios K. Kyriacou[1,2], Dinggang Shen[1], and Christos Davatzikos[1,2]

[1] Department of Radiology,
[2] Center for Computer-Integrated Surgical Systems and Technology (ERC),
Department of Computer Science,
Johns Hopkins University
kyriacou@cbmv.jhu.edu, dgshen@cbmv.jhu.edu, hristos@rad.jhu.edu

Abstract. Deformations that occur between pre-operative scans and the intra-operative setup can render pre-operative plans inaccurate or even unusable. It is therefore important to predict such deformations and account for them in pre-operative planning. This paper examines two different, yet related methodologies for this task, both of which collect statistical information from a training set in order to construct a predictive model. The first one examines the modes of co-variation between shape and deformation, and is therefore purely shape-based. The second approach additionally incorporates knowledge about the biomechanical properties of anatomical structures in constructing a predictive model. The two methods are tested on simulated training sets. Preliminary results show average errors of 9% (both methods) for a simulated dataset that had a moderate statistical variation and 36% (first method) and 23% (second method) for a dataset with a large statistical variation. Use of the above methodologies will hopefully lead to better clinical outcome by improving pre-operative plans.

Keywords: finite element modeling and simulation, registration techniques, deformable mapping

1 Introduction

A fundamental problem encountered in several kinds of surgical procedures is that pre-operative plans cannot be accurately executed, due to deformations that occur between the pre-operative setup and the intra-operative environment. For example, in prostate therapy, patients are often imaged in the subpine position and operated on in the lithotomy position. Moreover, anatomical deformations can also be caused by the surgical instruments themselves. This problem is very significant in robotically assisted percutaneous therapy involving needle and catheter insertion, since a pre-operative plan based strictly on pre-operative image coordinates cannot be accurately executed by a robotic system, unless soft tissue deformation is predicted and/or tracked during the procedure.

A number of investigators have used biomechanical models to predict intraoperative deformations in particular with respect to the brain shift during image-guided neurosurgery [1–3]. Nevertheless, there have been few attempts to use statistical training sets in addition to the biomechanical models (c.f. [1]). In this paper we present steps towards the development of a framework for predicting soft tissue deformation, assuming that the deformation of interest is observed in a training set, from which a predictive model is constructed. We investigate two different, yet related frameworks. The first one is referred to as *shape-based estimation (SBE)* and it extends ideas that have been used in statistical shape models [4]. In particular, from the training set we find the principal modes of co-variation between shape and deformation, by applying a principal component analysis on vectors that hold jointly landmark coordinates and their respective deformed coordinates vectors. When presented with a new shape, which corresponds to the anatomy of the individual patient, we express it in terms of the principal eigenvectors via an optimization procedure. We thus simultaneously obtain the most likely deformation of the individual anatomy. Our goal here is to find the component of the deformation that can be predicted from the patient's shape, based on the premise that anatomy (e.g. bone, muscle, ligaments) to some extent determines or constrains possible deformations.

The second approach that we examine is referred to as *force-based estimation (FBE)*. It is based on the premise that often there is additional knowledge about the biomechanical properties of the deforming anatomy. Hence, this knowledge should be utilized. Accordingly, we find the modes of co-variation between shape and forces, rather than shape and deformation. (Forces are calculated from the observed deformation and elastic properties; the latter can be optimized for best prediction of deformation in the training set.) From the resulting forces, we then find the deformation via the biomechanical model, using finite elements.

At this stage of our work, we have not worked on a particular application, but rather we created simulated shapes and deformations, in order to test and compare our methodologies. We plan to use this method for prostate therapy.

2 Methods

In this section we first describe how we created a training set by sampling probability distributions for the shape, elastic properties, and force parameters, and feeding them to a finite element model. We then describe SBE and FBE. All of our experiments are on 2D shapes. However, the principles are applicable to 3D.

Creating a statistical sample. In order to develop and test our methodology, we created training sets using a biomechanical model. The training samples were created by loading a plain-strain 2D shape with a uniform pressure at the top and fixed boundary conditions at the bottom (see figure 1, left panel). The shape is comprised of five ellipse-like areas. Each ellipse represents a different anatomical region with possibly different elastic properties. The material behavior is assumed to be linear elastic with a Poisson ratio of 0.48, which gives an almost incompressible behavior typical of soft tissues. The Young's moduli E_i

(where i may refer to ellipse i) are within the range of the Young's moduli of very soft tissue (order of 10,000 N/m^2). Linear elasticity is not necessary but is just a convenient first approximation. ABAQUS/CAE [5] is used for the automatic creation of the geometry, the application of elastic properties and boundary conditions, meshing, and the solution of the resulting finite element model. Although linear materials were used, large deformation mechanics rendered the problem nonlinear. Gaussian distributions were sampled for all major parameters of the simulated training samples: pressure at the top, major and minor diameters for all ellipses, the Young's moduli, and positions of the 3 small ellipses. We created two sets of training samples, having relatively lower and relatively higher variation in parameters. The first training set included 20 samples and the second one included 40 samples.

Extracting Statistical Parameters. Both of the methods that we examine in this paper, namely SBE and FBE, use principal component analysis, although they apply it in a somewhat different way. Hence, we will describe the basic principle of both methods first.

Assume that a collection of points, perhaps landmarks, defining a shape are arranged in a vector $\mathbf{s}$, and another collection of vectors are arranged in the vector $\mathbf{q}$. The vector $\mathbf{q}$ represents displacements in SBE and forces in FBE. In the case of SBE the displacements correspond to all the points in $\mathbf{s}$, while in the case of FBE the forces correspond to points in $\mathbf{s}$ that are on the boundary of the shape. Consider the vector $\mathbf{x}$, created by concatenating $\mathbf{s}$ and $\mathbf{q}$. We apply principal component analysis in order to determine the modes of variation of $\mathbf{x}$ from the training set, in a way similar to [4]. This procedure is summarized next:

– Align the training samples.

– Create the vector $\mathbf{x}_i$ for each sample i out of n samples.

– Calculate the mean shape vector $\mathbf{x}_{mean}$ and covariance matrix $\mathbf{C}$ and extract its eigenvalues matrix $\mathbf{D}_{full}$ (with the eigenvalues, λ_i, sorted by decreasing size) and eigenvectors matrix $\mathbf{V}_{full}$ (with eigenvectors $\mathbf{v}_i$ being the columns of the matrix).

– A new vector $\mathbf{x}$ can be created within this Statistical Shape Model:

$$\mathbf{x} = \mathbf{x}_{mean} + \mathbf{V}\mathbf{a} \qquad (1)$$

where $\mathbf{V}$ is the part of $\mathbf{V}_{full}$ that corresponds to m largest eigenvalues and $\mathbf{a}$ is an m-dimensional **coefficients vector** (with elements a_i). In this work we use all the eigenvectors that have a non-zero eigenvalue.

Predicting q for a new shape. The procedure described above effectively determines the modes of covariation between the shape vector, $\mathbf{s}$, and the vector to be predicted, $\mathbf{q}$. When presented with a patient's pre-operative images (target), we know the vector $\mathbf{s}_t$ (t stands for target), which represents the patient's undeformed anatomy. Our goal is then to predict the vector $\mathbf{q}$, which will determine the patient's deformed anatomy. However, the eigenvectors in $\mathbf{V}$ have dimensionality higher than that of $\mathbf{s}_t$, and therefore since we don't know $\mathbf{q}$, we cannot find $\mathbf{a}$ via projection on $\mathbf{v}_i$. Instead, we solve an optimization problem in which the vector $\mathbf{a}$ is found so that it yields a shape that best fits $\mathbf{s}_t$, while being most likely. The probability distribution of **a** is determined from the training set.

Specifically, for a given target shape, s_t, we find the vector **a** that minimizes the following objective function:

$$\mathcal{E}(\mathbf{a}) = \|\mathbf{s} - \mathbf{s}_t\|^2 + \mu \left(e^{0.5 \sum_i^m \frac{a_i^2}{\lambda_i}} \right)^2, \tag{2}$$

where μ is a relative weighting factor and

$$\mathbf{s} = \mathcal{T}(\mathbf{x}_{mean} + \mathbf{Va}),$$

with $\mathcal{T}(\cdot)$ representing the operation of truncating the second half of a vector (the components corresponding to **q**). The first term in (2) seeks vectors that get as close as possible to the patient's undeformed anatomy. The second term is proportional to $1/Prob(\mathbf{a})$ and penalizes vectors **a** that are unlikely. The solution is found using the Levenberg-Marquardt [6] (nonlinear) optimization scheme. We start with a guess vector $\mathbf{x}_{guess}$ based on eq. 1 and $a_i = k\sqrt{\lambda_i}$, where k is a constant e.g. 0.1. This procedure results in an estimate for the "missing part" of **x**, namely **q**.

SBE and FBE. The predicted vector **q** is different in SBE and FBE methods. In particular, in SBE, we set **q**=displacements, thus predicting deformation directly from the patient's undeformed anatomy. This is accomplished by defining point-correspondences in the undeformed and deformed configurations of the training samples.

In FBE, we set **q**=boundary forces. Since boundary force measurements are not directly available, this method is less straightforward and is described in more detail next. We start with some nominal elastic parameters, which in a real case scenario would be initially taken from the literature or perhaps from experimental data. From the displacement vectors that are available for each training sample and from the nominal elastic parameters, we calculate a forces vector **f** for each training sample, via a finite element solution. We then use these forces in place of the vector **q** and calculate the statistics of **x**, as described previously. In principle, we can iterate this procedure, by varying the elastic parameters and each time evaluating the accuracy of the predictive model via jack-knifing, an approach which we plan to use in future work.

Having established the predictive model, when presented with a new target (i.e. the pre-operative anatomy of the patient), we use the predictive model to obtain force estimates. We do this by minimizing $\mathcal{E}(\mathbf{a})$, as described previously. We then feed these estimated forces into a finite element mesh created for that patient and, by using the nominal elastic parameters (or those determined from the training set, in future work) we calculate the displacement field, and therefore the deformed (intra-operative) anatomy. Note that if no additional constraints are imposed, the problem is ill-posed, in that there are infinite number of solutions differing by a rigid-body motion. Since rigid-body motion is not of concern in our application, and in order to make the solution unique, we applied 3 additional displacement constraints on arbitrarily selected points.

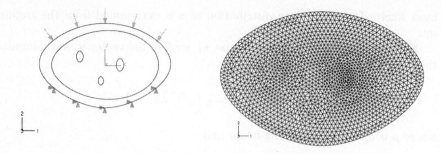

Fig. 1. Left: The model used in the phantom study. Uniform pressure is applied at the top and fixed boundary conditions are applied at the bottom of the large ellipse. Each ellipse may have its own elastic properties. Right: A sample finite element mesh with approximately 1500 nodes and 3000 linear triangular elements.

3 Results

Our first training set was comprised of 20 samples. The set was created by varying the Young moduli etc. according to Gaussian distributions, as explained in Methods. The standard deviation of each parameter was set equal to 1/20 of the absolute of its mean. Four typical sample shapes from this sample set are shown in figure 2, top left, showing the extent of variation in the samples.

In order to test our predictive model, we used jack-knifing. In particular, we sequentially used 6 of the 20 training samples as test samples (targets), and for each target the remaining 19 samples were used to build the predictive model. The following estimates were used for the elastic properties for FBE: $E_1 = E_{1mean} * 2, E_2 = E_{2mean} * 0.5, E_3 = E_{3mean} * 2, E_4 = E_{4mean} * 0.5$. By E_{1mean} etc. we denote the mean values used in the Gaussian distribution to create the training dataset, and 1 denotes the 2 largest ellipse regions, while 2, 3, and 4 denote the small lower right, lower left, and upper left ellipse regions respectively. By using material properties different than the true (mean) ones, we wanted to test the robustness of our predictive model to non-perfect estimates for the elastic parameters. Accordingly, we built 6 different predictive models, from which we calculated an average prediction error for each target. A typical example is shown in Fig. 2, which includes the results of both SBE and FBE. Note that each shape consists of 5 ellipses. The target undeformed shape is represented by a "." and the target deformed shape by an "o". The result of SBE is represented by an "x" while the result of BFE by a "+". In this example, both methods achieved very good predictions, since the "+", "x", and "o" practically coincide.

In order to examine the performance of these methods under higher variability in the sample set, we also created a second sample set with twice as high standard deviation. Fig. 3 shows a typical result from that sample set, with the same notation as in Fig. 2. Here, FBE outperformed SBE.

Tables 1 and 2 summarize the results for the first and second training datasets respectively. The error measure was obtained by calculating the root mean square error throughout all nodes on the boundaries of the five ellipses. The errors were normalized against the root mean square of the displacement vector for the same nodes.

Error in SBE	Error in FBE
0.115256	0.054368
0.047286	0.066952
0.100564	0.153132
0.065547	0.056310
0.081303	0.070515
0.116419	0.153702

Table 1. Normalized errors obtained for various targets for the low variation training dataset. The mean for SBE was 0.0877 and the mean for FBE was 0.0925.

Error in SBE	Error in FBE
0.304303	0.053248
0.126730	0.073419
0.346008	0.566100
0.532299	0.387028
0.512355	0.048511

Table 2. Normalized errors for various targets for the high variation training dataset. The mean for SBE was 0.3643 and the mean for FBE was 0.2257.

In order to examine the error introduced in FBE by the inherently approximating nature of finite element modeling, we tested the behavior of the two methods using a statistical sample set composed of 20 *identical* samples. SBE gave an error of 0.0000, as expected, while FBE gave an error of 0.0032. We conclude that the approximating nature of FEM is not a serious source of error in the methodology.

Computational time requirements. The most CPU intensive operation was the solution of the finite element problems. Each solution needed an average of 2 minutes on a 225 MHz Octane SGI with 128 MB RAM. In contrast, a statistical solution was on the order of 10 seconds and an optimization solution was on the order of 60 seconds. Note that FBE required the solution of a finite element problem for each sample and the target so this method was slower than SBE by an order of magnitude.

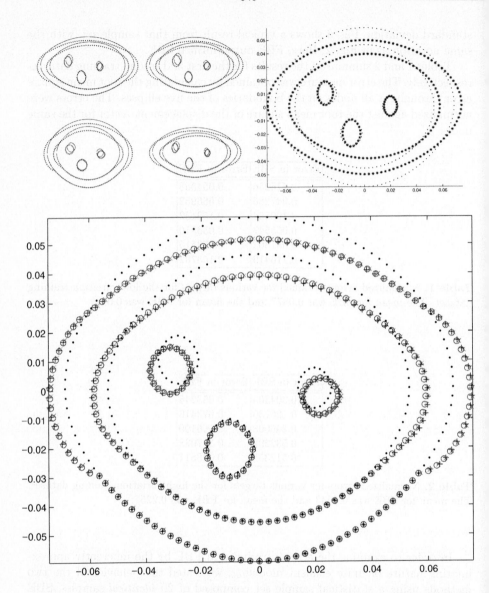

Fig. 2. Optimization results from a typical experiment in the low variation training dataset. Top Panel Left: Four sample shapes, showing the extent of variation in the samples. Solid and dotted lines depict the undeformed and deformed configuration respectively. Top Right: The initial undeformed shape used to initialize the Levenberg-Marquardt method is represented by "*", and this target's undeformed shape is represented by ".". Bottom Panel: The target deformed shape predicted by the two methods. SBE: "+". FBE: "x". True target deformed shape: "o". True target undeformed shape: ".".

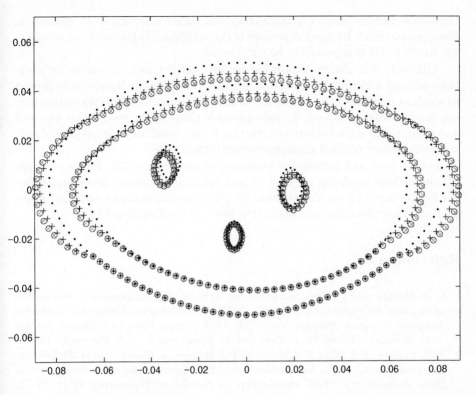

Fig. 3. Optimization results from a typical experiment in the high variation training dataset. SBE: "+". FBE: "x". Target deformed shape: "o". Target undeformed shape: ".".

4 Discussion

We presented a methodological framework for predicting deformations, in part using information obtained from a training set. We examined two methods. The first method (SBE) is purely shape-based and it constructs a predictive model based on the primary modes of covariation between anatomy and deformation. That is, it finds the component of deformation that can be predicted purely from knowledge of a patient's anatomy. The second method (FBE) utilizes knowledge about the biomechanical properties of the deforming anatomy, leaving the rest to a model analogous to the one used in SBE. Our preliminary results indicate that FBE outperforms SBE when large variations in anatomy, forces, and elastic parameters are present. However, both methods performed equally well under relatively lower variability. This is expected, since large deformation mechanics deviate from linearity, and therefore a linear statistical model should be expected to be relatively less accurate in that scenario.

FBE has the advantage that it utilizes knowledge about the biomechanical properties of anatomical structures. Moreover, it does not only estimate deformations on a discrete number of landmarks, but rather it determines a continuous displacement field. Its main drawback is the additional computational complexity, which in 3D is expected to be substantial.

Although it is clearly impossible to fully predict intra-operative deformations without additional intra-operative data, it is certainly beneficial to predict as much as possible of the deformation, and to account for this deformation during pre-operative planning. As pre-operative plans might need to be adjusted and recomputed intra-operatively, starting from a good initial guess should substantially reduce optimal planning computational time.

Future work will include the extension of our work to 3D. Potential applications include modeling of prostate and spine deformation, predicting deformations induced by needle insertion by using intra-operatively obtained force-feedback measurements, and predicting prostate swelling in gene therapy.

References

1. O.M. Skrinjar and J.S. Duncan. Real time 3D brain shift compensation. *Proceedings of the 16th International Conference, IPMI'99 (Information Processing in Medical Imaging), Visegrad, Hungary, June/July 1999, Lecture Notes in Computer Science 1613, Springer. (Edited by A. Kuba and M. Samal and A. Todd-Pokropek)*, 1999.

2. K.D. Paulsen, M.I. Miga, F.E. Kennedy, P.J. Hoopes, A. Hartov, and D.W. Roberts. A computational model for tracking subsurface tissue deformation during stereotactic neurosurgery. *IEEE Transactions on Biomedical Engineering*, 46(2):213–25, 1999.

3. C.R. Maurer Jr., D.L.G. Hill, A.J. Martin, H. Liu, M. McCue, D. Rueckert, D. Lloret, W.A. Hall, R.E. Maxwell, D.J. Hawkes, and C.L. Truwit. Investigation of intraoperative brain deformation using a 1.5-T interventional MR system: preliminary results. *IEEE Transactions on Medical Imaging*, 17(5):817–25, 1998.

4. T.F. Cootes, A. Hill, C.J. Taylor, and J. Haslam. Use of active shape models for locating structures in medical images. *Image and Vision Computing*, 12(6):355–65, 1994.

5. Abaqus version 5.8. *Hibbit, Karlsson, and Sorensen, Inc., USA*, 1998.

6. D.W. Marquardt. An algorithm for least-squares estimation of nonlinear parameters. *J. Soc. Indust. Appl. Math.*, 11:431, 1963.

Real-Time Large Displacement Elasticity for Surgery Simulation: Non-linear Tensor-Mass Model

G. Picinbono, H. Delingette, and N. Ayache

INRIA Sophia - Epidaure Project
2004 Route des Lucioles BP 93
06902 Sophia Antipolis Cedex, France
{Guillaume.Picinbono, Herve.Delingette, Nicholas.Ayache}@sophia.inria.fr
http://www-sop.inria.fr/epidaure/Epidaure-eng.html

Abstract. In this paper, we describe the latest developments of the minimally invasive hepatic surgery simulator prototype developed at INRIA. A key problem with such a simulator is the physical modeling of soft tissues. We propose a new deformable model based on non-linear elasticity and the finite element method. This model is valid for large displacements, which means in particular that it is invariant with respect to rotations. This property improves the realism of the deformations and solves the problems related to the shortcomings of linear elasticity, which is only valid for small displacements. We also address the problem of volume variations by adding to our model incompressibility constraints. Finally, we demonstrate the relevance of this approach for the real-time simulation of laparoscopic surgical gestures on the liver.

1 Introduction

A major and recent evolution in abdominal surgery has been the development of laparoscopic surgery. In this type of surgery, abdominal operations such as hepatic resection are performed through small incisions. A video camera and special surgical tools are introduced into the abdomen, allowing the surgeon to perform a procedure less invasive. A drawback of this technique lies essentially in the need for more complex gestures and in the loss of direct visual and tactile information. Therefore the surgeon needs to learn and adapt himself to this new type of surgery and in particular to a new type of hand-eye coordination. In this context, surgical simulation systems could be of great help in the training process of surgeons.

Among the several key problems in the development of a surgical simulator [1, 13], the geometrical and physical representation of human organs remain the most important. The deformable model must be at the same time very realistic (both visually and physically) and very efficient to allow real-time deformations. Several methods have been proposed: spring-mass models [8, 12], free form deformations [2], linear elasticity with finite volume method [9] or various finite element methods [6, 10, 15, 4].

In this article we propose a new real-time deformable model based on non-linear elasticity and a finite element method. We first introduce the linear elasticity theory and its implementation through the finite element method, and we then highlight its shortcomings when the "small displacement" hypothesis does not hold. Then we focus on our implementation of St Venant-Kirchhoff elasticity and incompressibility constraints.

2 Shortcomings of the linear elasticity

Linear elasticity is often used for the modeling of deformable materials, mainly because the equations remain quite simple and the computation time can be optimized.

The physical behavior of soft tissue may be considered as linear elastic if its displacement and deformation remain small [11, 14] (typically less than 10% of the mesh size). We represent the deformation of a volumetric model from its rest shape $\mathcal{M}_{\text{initial}}$ with a *displacement vector* $\mathbf{U}(x, y, z)$ for $(x, y, z) \in \mathcal{M}_{\text{initial}}$ and we write $\mathcal{M}_{\text{deformed}} = \mathcal{M}_{\text{initial}} + \mathbf{U}(x, y, z)$.

From this displacement vector, we define the linearized *Green-St Venant strain tensor* (3×3 symmetric matrix) E_l and its principal invariants l_1 and l_2:

$$E_l = \frac{1}{2} \left(\nabla \mathbf{U} + \nabla \mathbf{U}^t \right) \qquad l_1 = tr\, E_l \qquad l_2 = tr\, E_l^{\,2}. \qquad (1)$$

The linear elastic energy $W_{Elastic}$, for homogeneous isotropic materials, is defined by the following formula (see [5]):

$$W_{Elastic} = \frac{\lambda}{2} \left(tr E_l \right)^2 + \mu\, tr E_l^{\,2} = \frac{\lambda}{2} (div\, \mathbf{U})^2 + \mu \|\nabla \mathbf{U}\|^2 - \frac{\mu}{2} \|rot\, \mathbf{U}\|^2, \quad (2)$$

where λ and μ are the *Lamé coefficients* characterizing the material stiffness.

Equation 2, known as *Hooke's law*, shows that the elastic energy of a deformable object is a quadratic function of the displacement vector.

2.1 Finite element method

Finite element method is a classical way to solve linear elasticity problems. Its most interesting property is to provide a continuous description of physical equations. We chose to use P_1 finite elements where the elementary volume is a tetrahedron with a node defined at each vertex. At each point $\mathbf{M}(x, y, z)$ inside tetrahedron $\mathbf{T}_i$, the displacement vector is expressed as a function of the displacements $\mathbf{U}_k$ of vertices $\mathbf{P}_k$. For P_1 finite elements, interpolation functions Λ_k are linear ($\{\Lambda_k; k = 0, ..., 3\}$ are the barycentric coordinates of $\mathbf{M}$ in the tetrahedron):

$$\mathbf{U}(x,y,z) = \sum_{j=0}^{3} \mathbf{U}_j \Lambda_j(x,y,z) \qquad \Lambda_j(x,y,z) = \alpha_j.X + \beta_j$$

$$\alpha_j = \frac{(-1)^j}{6V(\mathbf{T}_i)}(\mathbf{P}_{j+1} \times \mathbf{P}_{j+2} + \mathbf{P}_{j+2} \times \mathbf{P}_{j+3} + \mathbf{P}_{j+3} \times \mathbf{P}_{j+1}),$$

where $\times$ stands for the cross product between two vectors, and $V(\mathbf{T}_i)$ is the volume of the tetrahedron.

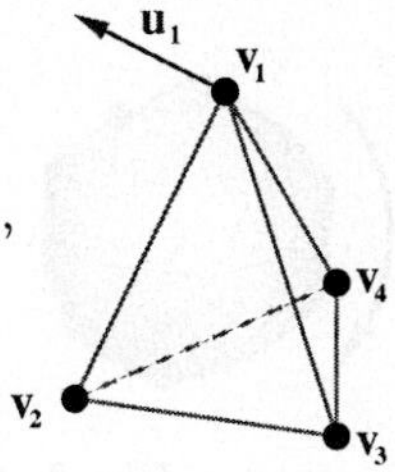

Fig. 1. P_1 finite element

Using this equation for the displacement vector $\mathbf{U}$ leads to the finite element formulation of linear elastic energy in the tetrahedron $\mathbf{T}_i$ [10]:

$$W_{Elastic}(\mathbf{T}_i) = \sum_{j,k=0}^{3} \mathbf{U}_j^t [\mathcal{B}_{jk}^{\mathbf{T}_i}] \mathbf{U}_k$$

$$\mathcal{B}_{jk}^{\mathbf{T}_i} = \lambda(\alpha_j \otimes \alpha_k) + \mu [(\alpha_k \otimes \alpha_j) + (\alpha_j.\alpha_k) Id_3], \qquad (3)$$

where $[\mathcal{B}_{jk}^{\mathbf{T}_i}]$ is the tetrahedron contribution to the stiffness tensor of the edge $(\mathbf{P}_j, \mathbf{P}_k)$ (or of the vertex $\mathbf{P}_j$ if $j = k$), $\{\alpha_j, k = 0, .., 3\}$ are the shape vectors of the tetrahedron and $\otimes$ stands for the tensor product of two vectors.

Finally, to obtain the force $\mathbf{F}_p^{\mathbf{T}_i}$ applied by the tetrahedron $\mathbf{T}_i$ on the vertex $\mathbf{P}_p$, we derive the elastic energy with respect to the vertex displacement $\mathbf{U}_p$:

$$\mathbf{F}_p^{\mathbf{T}_i} = 2\sum_{j=0}^{3}[\mathcal{B}_{pj}^{\mathbf{T}_i}]\mathbf{U}_j. \qquad (4)$$

We have been using this linear elasticity formulation for several years through two deformable models, the **pre-computed model** [6] and the **tensor-mass model** [10, 7]. Furthermore, it can be extended to anisotropic linear elasticity [15], which allows to model fiber-reinforced materials, very common within biological tissues (blood vessels, tendons, muscles, ...).

2.2 The problem of rotational invariance

The main limitation of the linear model is that it is not invariant with respect to rotations. When the object undergoes a rotation, the elastic energy increases, leading to a variation of the volume (see figure 2). In the case of a global rotation of the object, we could solve the problem with a specific change of the reference frame.

But this solution proves itself to be ineffective when only one part of the object undergoes a rotation (which is the case in general). This case is presented by the cylinder of figure 3: the bottom face is fixed and a force is applied to the central top vertex. Arrows show the trajectory of some vertices, which are

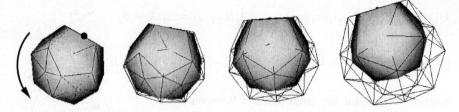

Fig. 2. *Global rotation of the linear elastic model (wire-frame)*

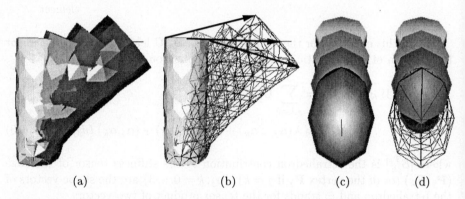

(a)   (b)   (c)   (d)

Fig. 3. *Successive deformations of a linear elastic cylinder. (a) and (b): side view. (c) and (d): top view*

constrained by the linear model to move along straight lines. This results in the distortion of the mesh. Furthermore, this abnormal deformation is anisotropic since the object only deforms itself in the rotation plane (figure 3(c) and 3(d)).

This unrealistic behaviour of the linear elastic model for large displacements leads us to consider different models of elasticity.

3   St Venant-Kirchhoff elasticity

A model of elasticity is considered as a large displacement model if it derives from a strain tensor which is a quadratic function of the deformation gradient. Most common tensors are the **left** and **right Cauchy-Green strain tensors** (respectively $B = \nabla\phi\nabla\phi^t$ and $C = \nabla\phi^t\nabla\phi$, ϕ being the deformation).

The St Venant-Kirchhoff model is a generalization of the linear model for large displacements, and is a particular case of hyperelastic materials. The basic energy equation is the same (equation 2), but now E stands for the complete **Green-St Venant strain tensor**:

$$E = \frac{1}{2}(C - I) = \frac{1}{2}\left(\nabla\mathbf{U} + \nabla\mathbf{U}^t + \nabla\mathbf{U}^t\nabla\mathbf{U}\right). \tag{5}$$

Elastic energy, which was a quadratic function of $\nabla\mathbf{U}$ in the linear case, is now a polynomial of order four with respect to $\nabla\mathbf{U}$:

$$W = \frac{\lambda}{2}\left[(div\ \mathbf{U}) + \frac{1}{2}\|\nabla\mathbf{U}\|^2\right]^2 + \mu\|\nabla\mathbf{U}\|^2 - \frac{\mu}{2}\|rot\ \mathbf{U}\|^2$$

$$+ \mu(\nabla\mathbf{U} : \nabla\mathbf{U}^t\nabla\mathbf{U}) + \frac{\mu}{4}\|\nabla\mathbf{U}^t\nabla\mathbf{U}\|^2, \tag{6}$$

where $A : B = tr(A^tB) = \sum_{i,j} a_{ij}\ b_{ij}$ is the dot product of two matrices.

3.1 Finite element modeling

With the notations introduced in section 2.1, we express the St Venant-Kirchhoff elastic model with finite element theory as:

$$W(\mathbf{T}_i) = \sum_{j,k} \mathbf{U}_j^t \left[\mathcal{B}_{jk}^{\mathbf{T}_i}\right] \mathbf{U}_k + \sum_{j,k,l} \left(\mathbf{U}_j.\mathcal{C}_{jkl}^{\mathbf{T}_i}\right)(\mathbf{U}_k.\mathbf{U}_l) \tag{7}$$

$$+ \sum_{j,k,l,m} \mathcal{D}_{jklm}^{\mathbf{T}_i}\,(\mathbf{U}_j.\mathbf{U}_k)\,(\mathbf{U}_l.\mathbf{U}_m),$$

where:

- $\mathcal{B}_{jk}^{\mathbf{T}_i}$ is the (3x3) symmetric matrix of the linear elastic model (equation 3),
- $\mathcal{C}_{jkl}^{\mathbf{T}_i}$ is a vector: $\mathcal{C}_{jkl}^{\mathbf{T}_i} = \frac{\lambda}{2}\,\alpha_j\,(\alpha_k.\alpha_l) + \frac{\mu}{2}\,[\alpha_l\,(\alpha_j.\alpha_k) + \alpha_k\,(\alpha_j.\alpha_l)]$,
- and $\mathcal{D}_{jklm}^{\mathbf{T}_i}$ is a scalar: $\mathcal{D}_{jklm}^{\mathbf{T}_i} = \frac{\lambda}{8}\,(\alpha_j.\alpha_k)\,(\alpha_l.\alpha_m) + \frac{\mu}{4}(\alpha_j.\alpha_m)(\alpha_k.\alpha_l)$.

The force applied at each vertex inside a tetrahedron is derived from the elastic energy $W(\mathbf{T}_i)$:

$$\mathbf{F}^p(\mathbf{T}_i) = \underbrace{2\sum_{j}\left[\mathcal{B}_{pj}^{\mathbf{T}_i}\right]\mathbf{U}_j}_{\mathbf{F}_1^p(\mathbf{T}_i)} + \underbrace{\sum_{j,k}2\,(\mathbf{U}_k \otimes \mathbf{U}_j)\,\mathcal{C}_{jkp}^{\mathbf{T}_i} + (\mathbf{U}_j.\mathbf{U}_k)\,\mathcal{C}_{pjk}^{\mathbf{T}_i}}_{\mathbf{F}_2^p(\mathbf{T}_i)} \tag{8}$$

$$+ \underbrace{4\sum_{j,k,l}\mathcal{D}_{jklp}^{\mathbf{T}_i}\,\mathbf{U}_l\,\mathbf{U}_k^t\,\mathbf{U}_j}_{\mathbf{F}_3^p(\mathbf{T}_i)}.$$

The first term of the elastic force ($\mathbf{F}_1^p(\mathbf{T}_i)$) corresponds to the linear elastic case presented in section 2.1. The next part of the article deals with the generalization of the tensor-mass model to large displacements.

3.2 Non-linear Tensor-Mass Model

The main idea of the tensor-mass model is to split, for each tetrahedron, the force applied at a vertex in two parts: a force created by the vertex displacement and forces produced by the displacements of its neighbours:

$$\mathbf{F}_1^p(\mathbf{T}_i) = [\mathcal{B}_{pp}^{\mathbf{T}_i}]\mathbf{U}_p + \sum_{j\neq p}[\mathcal{B}_{pj}^{\mathbf{T}_i}]\mathbf{U}_j. \tag{9}$$

This way we can define for each tetrahedron a set of **local stiffness tensors** for vertices ($\{\mathcal{B}_{pp}^{\mathbf{T}_i}; p = 0, ..., 3\}$) and for edges ($\{\mathcal{B}_{pj}^{\mathbf{T}_i}; p, j = 0, ..., 3; p \neq j\}$). By doing this for every tetrahedron, we can accumulate on vertices and edges of the mesh the corresponding contributions to the **global stiffness tensors**:

$$\mathcal{B}_{pp} = \sum_{\mathbf{T}_i \in N(\mathbf{V}_p)} \mathcal{B}_{pp}^{\mathbf{T}_i} \qquad \mathcal{B}_{pj} = \sum_{\mathbf{T}_i \in N(\mathbf{E}_{pj})} \mathcal{B}_{kl}^{\mathbf{T}_i}.$$

These **stiffness tensors** are computed when creating the mesh and are stored for each vertex and edge of the mesh.

The same principle can be applied to the quadratic term ($\mathbf{F}_2^p(\mathbf{T}_i)$ of equation 8) and the cubic term ($\mathbf{F}_3^p(\mathbf{T}_i)$). The former brings **stiffness vectors** for vertices, edges, and triangles, and the latter brings **stiffness scalars** for vertices, edges, triangles, and tetrahedra. The following array (table 1) summarizes the stiffness data stored on each geometrical primitive of the mesh:

Stiffness data distribution	Tensors	Vectors	Scalars
Vertex $\mathbf{V}_p$	$\mathcal{B}^{pp}$	$\mathcal{C}^{ppp}$	$\mathcal{D}^{pppp}$
Edge $\mathbf{E}_{pj}$	$\mathcal{B}^{pj}$	$\mathcal{C}^{ppj}\ \mathcal{C}^{jpp}$ $\mathcal{C}^{jjp}\ \mathcal{C}^{pjj}$	$\mathcal{D}^{jppp}\ \mathcal{D}^{jjjp}\ \mathcal{D}^{jpjp}$ $\mathcal{D}^{pjjp}\ \mathcal{D}^{jjpp}$
Triangle $\mathbf{F}_{pjk}$		$\mathcal{C}^{jkp}$ $\mathcal{C}^{kjp}$ $\mathcal{C}^{pjk}$	$\mathcal{D}^{jkpp}\ \mathcal{D}^{jpkp}\ \mathcal{D}^{pjkp}$ $\mathcal{D}^{jjkp}\ \mathcal{D}^{jkjp}\ \mathcal{D}^{kjjp}$ $\mathcal{D}^{kkjp}\ \mathcal{D}^{kjkp}\ \mathcal{D}^{jkkp}$
Tetrahedron $\mathbf{T}_{pjkl}$			$\mathcal{D}^{jklp}\ \mathcal{D}^{jlkp}\ \mathcal{D}^{kjlp}$ $\mathcal{D}^{kljp}\ \mathcal{D}^{ljkp}\ \mathcal{D}^{lkjp}$

Table 1. *Storage of the stiffness data on the mesh*

Given a tetrahedral mesh of a solid —in our case an anatomical structure— we build a data structure incorporating the notion of vertices, edges, triangles, and tetrahedra, with all the necessary neighbours. For each vertex, we store its current position $\mathbf{P}_p$, its rest position $\mathbf{P}_p^0$, and its stiffness data. For each edge, we store stiffness data. Finally for each tetrahedron, we store the Lamé coefficients λ and μ, the four shape vectors α_k, and the stiffness data.

During the simulation, we compute forces for each vertex, edge, triangle, and tetrahedron, and we update the vertex positions from the differential equations of continuum mechanics [3]:

$$\mathbf{M\ddot{U}} + \mathbf{C\dot{U}} + \mathbf{F(U)} = \mathbf{R}. \tag{10}$$

Following finite element theory, the mass $\mathbf{M}$ and damping $\mathbf{C}$ matrices are sparse matrices that are related to the stored physical properties of each tetrahedron. In our case, we consider that $\mathbf{M}$ and $\mathbf{C}$ are diagonal matrices, i.e., that mass and damping effects are concentrated at vertices. This simplification called *mass-lumping* decouples the motion of all nodes and therefore allows us to write equation 10 as the set of independent differential equations for each vertex.

Furthermore, we choose an *explicit integration scheme* where the elastic force is estimated at time t in order to compute the vertex position at time $t + 1$:

$$\left(\frac{m_i}{\Delta t^2} - \frac{\gamma_i}{2\Delta t}\right) \mathbf{P}_i^{t+1} = \mathbf{F}_i + \frac{2m_i}{\Delta t^2}\mathbf{P}_i^t - \left(\frac{m_i}{\Delta t^2} + \frac{\gamma_i}{2\Delta t}\right) \mathbf{P}_i^{t-1}.$$

One of the basic tasks in surgery simulation consists in cutting soft tissue. With our deformable model, this task can be achieved efficiently. We simulate the action of an electric scalpel on soft tissue by successively removing tetrahedra at places where the instrument is in contact with the anatomical model.

When removing a tetrahedron, about a hundred update operations are performed to suppress the tetrahedron contributions to the stiffness data of the surrounding vertices, edges, and triangles. By locally updating stiffness data, the tissue has exactly the same properties as if we had removed the corresponding tetrahedron at its rest position. Because of the volumetric continuity of finite element modeling, the tissue deformation remains realistic during the cutting.

4   Incompressibility constraint

Living tissue, which is essentially made of water, is nearly incompressible. This property is difficult to model and leads in most cases to instability problems. This is the case with the St Venant-Kirchhoff model: the material remains incompressible when the Lamé constant λ tends towards infinity. Taking a large value for λ would force us to decrease the time step and therefore to increase the computation time. Another reason to add an external incompressibility constraint to our model is related to the model itself: the main advantage of the St Venant-Kirchhoff model is to use the strain tensor E which is invariant with respect to rotations. But it is also invariant with respect to symmetries, which could lead to the reversal of some tetrahedra under strong constraints.

We choose to penalize volume variation by applying to each vertex of the tetrahedron a force directed along the normal of the opposite face $\mathbf{N}_p$ (see figure on the right), the norm of the force being the square of the relative volume variation:

$$\mathbf{F}^p_{incomp} = \left(\frac{V - V_0}{V_0}\right)^2 \mathbf{N}_p.$$

These forces act as an pressure increase inside the tetrahedron. This method is closely related to Lagrange multipliers, which are often used to solve problem of energy minimization under constraints.

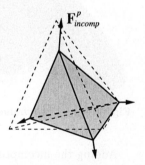

Fig. 4. *Penalization of the volume variation*

5   Results

In the first experiment, we wish to highlight the contributions of our new deformable model in the case of partial rotations. Figure 5 shows the same experience as the one presented for linear elasticity (section 2.2, figure 3). On the left we can see that the cylinder vertices can now follow trajectories different from straight lines (figure 5(a)), leading to much more realistic deformations than in the linear (wire-frame) case (figures 5(b) and 5(c)).

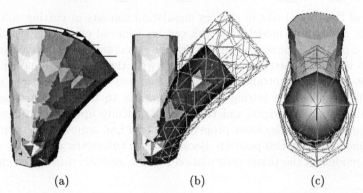

(a)   (b)   (c)

Fig. 5. (a) Successive deformations of the non linear model. Side (b) and top (c) view of the comparison between linear (wire-frame) and non linear model (solid rendering)

In the second example (figure 6), we apply a force to the right lobe of the liver (the liver is fixed on its central back part, and Lamé constants are: $\lambda = 4.10^4$ and $\mu = 10^4$). Using the linear model, the right part of the liver undergoes a large (and unrealistic) volume increase, whereas with non-linear elasticity, the right lobe is able to rotate, giving a much more accurate deformation.

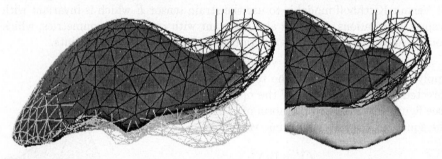

Fig. 6. Linear (wireframe), non-linear (solid) liver models, and rest shape (bottom)

Adding the incompressibility constraint on the same examples decreases the volume variation even more (see table 2[1]), and also stabilizes the behaviour of the deformable models in strongly constrained areas.

The last example is the simulation of a typical laparoscopic surgical gesture on the liver. One tool is pulling the edge of the liver sideways while a bipolar cautery device cuts it. During the cutting, the surgeon pulls away the part of

---

[1] For the cylinder: left, middle and right stand for the different deformations of figures 3 and 5(a)

Volume variations (%)			Linear			Non-linear			Non-linear incomp.		
Cylinder	left	middle	right	7	28	63	0.3	1	2	0.2	0.5	1
Liver			9			1.5			0.7		

Table 2. *Volume variation results*

the liver he wants to remove. This piece of liver undergoes large displacements and the deformation appears fairly realistic with this new non-linear deformable model (figure 7).

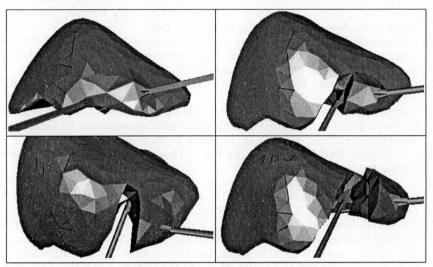

Fig. 7. *Simulation of laparoscopic liver surgery*

Obviously, the computation time of this model is larger than for the linear model because the force equation is much more complex (equation 8). With our current implementation, simulation frequency is four times slower than with the linear model. Nevertheless, with this non-linear model, we can reach a frequency update of 30Hz on meshes made of about 2000 tetrahedra (on a PC Pentium PIII 500MHz). This is sufficient to reach visual real-time with quite complex objects, and even to provide a realistic haptic feedback using force extrapolation as described in [15].

6 Conclusion

We proposed in this article a new deformable model based on large displacement elasticity, a finite element method, and a dynamic explicit integration scheme. It solves the problem of rotational invariance and takes into account the incompressibility properties of biological tissues. Including this model into our laparoscopic surgery simulator prototype improves its bio-mechanical realism and thus increases its potential use for learning and training processes.

Our future work will focus on generalizing large displacement elasticity to anisotropic materials. Also, the large displacement model requiring more computation time, it will be usefull to develop a "hybrid" model: linear elasticity for the strongly constrained parts of the organ and large displacement elasticity for the free parts that can undergo large deformations. This will be done following the experience we had with previous "hybrid" models of this type [7].

References

1. N. Ayache, S. Cotin, H. Delingette, J.-M. Clement, J. Marescaux, and M. Nord. Simulation of endoscopic surgery. *Journal of Minimally Invasive Therapy and Allied Technologies (MITAT)*, 7(2):71–77, July 1998.
2. C. Basdogan, C. Ho, M. A. Srinivasan, S. D. Small, and S. L. Dawson. Force Interaction in Laparoscopic Simulation: Haptic Rendering Soft Tissues. In *Medecine Meets Virtual Reality (MMVR'6)*, pages 28–31, San Diego CA, january 1998.
3. K.-L. Bathe. *Finite Element Procedures in Engineering Analysis*. Prentice-Hall, 1982.
4. Morten Bro-Nielsen and Stephane Cotin. Real-time Volumetric Deformable Models for Surgery Simulation using Finite Elements and Condensation. In *Eurographics '96. ISSN 1067-7055*, pages 57–66. Blackwell Publishers, 1996.
5. P. G. Ciarlet. *Mathematical elasticity Vol. 1: Three-dimensional elasticity*. Elsevier Science Publishers B.V., 1988.
6. S. Cotin, H. Delingette, and N. Ayache. Real-time elastic deformations of soft tissues for surgery simulation. *IEEE Transactions On Visualization and Computer Graphics*, 5(1):62–73, January-March 1999.
7. S. Cotin, H. Delingette, and N. Ayache. A Hybrid Elastic Model allowing Real-Time Cutting, Deformations and Force-Feedback for Surgery Training and Simulation. *The Visual Computer*, 2000. to appear (see INRIA research report RR-3510).
8. F. Boux de Casson and C. Laugier. Modeling the Dynamics of a Human Liver for a Minimally Invasive Surgery Simulator. In *MICCAI'99*, Cambridge UK, 1999.
9. G. Debunne, M. Desbrun, A. Barr, and M.-P. Cani. Interactive multiresolution animation of deformable models. In *10th Eurographics Workshop on Computer Animation and Simulation (CAS'99)*, September 1999.
10. H. Delingette, S.Cotin, and N.Ayache. A Hybrid Elastic Model allowing Real-Time Cutting, Deformations and Force-Feedback for Surgery Training and Simulation. In *Computer Animation*, Geneva Switzerland, May 26-28 1999.
11. Y. C. Fung. *Biomechanics - Mechanical Properties of Living Tissues*. Springer-Verlag, second edition edition, 1993.
12. D. Lamy and C. Chaillou. Design, Implementation and Evaluation of an Haptic Interface for Surgical Gestures Training. In *International Scientific Workshop on Virtual Reality and Prototyping*, pages 107–116, Laval, France, June 1999.
13. J. Marescaux, J-M. Clément, V. Tassetti, C. Koehl, S. Cotin, Y. Russier, D. Mutter, H. Delingette, and N. Ayache. Virtual Reality Applied to Hepatic Surgery Simulation : The Next Revolution. *Annals of Surgery*, 228(5):627–634, 1998.
14. W. Maurel, Y. Wu, N. M. Thalmann, and D. Thalmann. *Biomechanical Models for Soft Tissue Simulation*. Springer-Verlag, 1998.
15. G. Picinbono, J.C. Lombardo, H. Delingette, and N. Ayache. Anisotropic Elasticity and Force Extrapolation to Improve Realism of Surgery Simulation. In *IEEE International Conference on Robotics and Automotion: ICRA 2000*, pages 596–602, San Francisco, CA, April 2000.

Multi-DOF Forceps Manipulator System for Laparoscopic Surgery

Ryoichi Nakamura[1], Etsuko Kobayashi[2], Ken Masamune[3], Ichiro Sakuma[2],
Takeyoshi Dohi[2], Naoki Yahagi[2], Takayuki Tsuji[2], Daijo Hashimoto[4], Mitsuo
Shimada[5], Makoto Hashizume[6]

[1] Dept. of Precision Engineering, Graduate School of Engineering,
[2] Institute of Environmental Studies, Graduate school of Frontier Science,
the Univ. of Tokyo, 7-3-1, Hongo, Bunkyo-ku, Tokyo 113-8656 Japan
{ryoichi, etsuko, sakuma, dohi, yahagi, tsuji}@miki.pe.u-
tokyo.ac.jp
[3] Dept. of Bio-Technology, College of Science and Engineering, Tokyo Denki University,
Ishizaka, Hatoyama-cho, Hiki, Saitama, 350-0395, Japan
masa@b.dendai.ac.jp
[4] Department of Surgery, Saitama Medical Center, Saitama Medical School,
1981, Tsujido, Kamoda, Kawagoe-shi, Saitama 350-8550 Japan
[5] Department of Surgery and Science, Graduate School of Medical Sciences,
[6] Dept. of Disaster and Emergency Medicine, Grad. School of Medical Science,
Kyushu University, 3-1-1, Maidashi, Higashi-ku, Hukuoka 812-8582 Japan
mshimada@surg2.med.kyushu-u.ac.jp
mhashi@dem.med.kyushu-u.ac.jp

Abstract. Many problems in laparoscopic surgery are due to the poor degrees
of freedom of movement (DOF) in controlling forceps and laparoscopes. The
Multi-DOF forceps manipulator we have newly developed has two additional
DOF of bending on the tip of forceps, and provides new surgical fields and
techniques for surgeons. The most remarkable characteristics of the prototype
described in this paper are: **1)** the *small* diameter (ϕ 6 mm) and the *small* radius
of curvature of bending; **2)** the *large* force generated on each bending and
grasping axis; **3)** the confirmation of *perfect cleanness and sterilization* of this
manipulator. The effectiveness of these characteristics was confirmed in actual
testing by surgeons. This manipulator can solve problems met with in
laparoscopic surgery, and will establish new standards for laparoscopic surgery
with higher effectiveness and safety.

1 Introduction

Laparoscopic surgery is now widely established as minimal invasive surgery (MIS).
However, it has some problems and difficulties due to the "minimal invasive"
method. The forceps and laparoscope are inserted into the visceral cavity through the
trochars that are fixed on the abdominal wall. These tools have only four degrees of
freedom of movement (DOF) and small working areas (Fig. 1).

Using conventional tools (forceps), surgeons can approach an operation point only
through a single approach path. However, in some cases, various alternative approach

paths for an operation area are needed. The tools thus require additional DOF to solve these problems (Fig. 2).

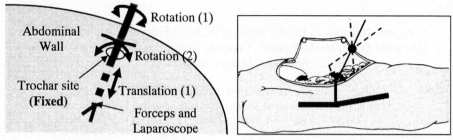

Fig. 1. Four DOF of surgical tools in laparoscopic surgery

Fig. 2. Approach path of surgical tools
◄——— is approach path of conventional tools.
◄- - - are approach paths of tools with additional DOF.

Surgical robots, which can help surgeons' manipulation such as steadier holding, more precise operating, and remote control, are being studied and developed. However, there are many problems in the use of robots in medical fields. The most important consideration must be safety, both in terms of engineering and medical effectiveness.

Recently, forceps with additional DOF for MIS have been studied in many institutes and applied to many actual cases [1-5], but in laparoscopic surgery, there are very few such cases. This is mainly due to the small force of additional DOF and reduced safety. We must develop new mechanisms with additional DOF, taking these considerations into account.

In this paper, we present a novel forceps manipulator with Multi-DOF. This manipulator has important features for laparoscopic surgery. The methods used in order to construct the forceps manipulator with the desired features are presented in Section 2. The experimental results are presented in Section 3, including the measurement of mechanical characteristics and the performance of this manipulator in surgical environments. Using this manipulator, problems in laparoscopic surgery are clearly solved, and new standards with higher effectiveness and safety are achieved.

2 Method

2.1 Requirements

When considering the need for increased force delivery and safety, and the basic requirement for new surgical tools for laparoscopic surgery, we defined the characteristics of the forceps we have developed as follows:

(1) DOF: two bending and one grasping (or scissoring or another DOF for end effector) on the tip of the forceps;
(2) Radius of curvature of bending: as small as possible (for safety);

(3) Diameter of the forceps: about 5 mm;
(4) Generated force: about 0.5 kgf for bending / about 1 kgf for grasping.

From the viewpoint of expanding the number of approach paths and the fields of operation of forceps, those with additional DOF and a large radius of curvature of bending would seem to be very useful. However, an instrument with a wide operating area also risks large and fatal errors if it is mishandled. Achieving large bends in the limited space of the visceral cavity is both difficult and dangerous. We decided to develop the first prototype with only two additional DOF and a small radius of curvature of bending.

It is difficult to determine the requirement of force generated accurately, because the measurement of the handling force of surgical tools in surgery is difficult. We decided the requirement from an experimental study using PVC [6].

2.2 Bending Mechanism

We developed the Multi-DOF forceps manipulator, which provides two additional DOF of bending. The bending mechanism is composed of four stainless steel rings with a coupling giving one DOF, stainless wire, and Teflon tubes (shown in Fig. 3 and 4). This mechanism is driven by four stainless steel wires (Fig. 5). The ranges of bending motion are 0 to 90 degrees for each two DOF.

Fig. 3. Stainless steel ring joint

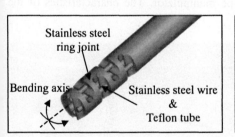

Fig. 4. Bending mechanism

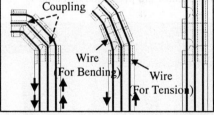

Fig. 5. Wire drive of bending mechanism

This mechanism is very simple, with high rigidity. It can easily be miniaturized. A wide range of bending provides a wide range of operations, but it increases the risk of error. We therefore developed this mechanism to be small enough to avoid fatal errors in operation. The diameter of the first prototype is 6 mm, and the minimum radius of curvature of bending is about 8.5 mm.

2.3 Mechatoronical Control

This bending forceps (Forceps with Bending Mechanism) needs very large forces driving the wires to generate similar forces on the tip of the forceps (bending and grasping forces). Moreover single-handed manual control of three DOF (two bending and one grasping) is very difficult and inadequate for this forceps. This is why we use mechatronic components for this bending forceps to drive the bending and grasping DOF.

The motors, which cannot be sterilized, are placed sufficiently far from the surgical field for maintenance of sterility, and stainless steel wires transmit the forces produced by the motors. In this system, since there are no electronic components (motors, sensors, etc.) on the tip of the forceps manipulator (which is inserted into the patient), this manipulator is perfectly sterilizable. This arrangement, with remote electronic components, and the Multi-DOF forceps manipulator system, which is lightweight, simple, and perfectly clean, can easily be assembled (Fig. 6).

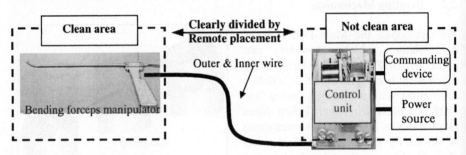

Fig. 6. Total system components of the Multi-DOF forceps manipulator.

Fig. 7. A photograph of the first prototype manipulator. The characteristics of the manipulator are presented in Table 1.

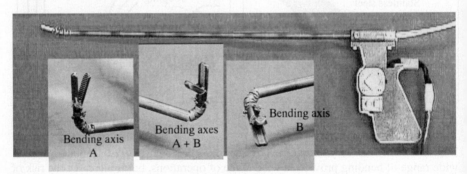

Fig. 7. First Prototype of Multi-DOF forceps manipulator

Table 1. Specification of the bending forceps manipulator system

Range of actions (DOF)	2 bending and 1 grasping
Moving range	0-90 °(bending) / 0-60 °(grasping)
Diameter	6 mm
Size	455 mm (l) ×130 mm(h) ×15 mm(d)
Weight	270g (Multi-DOF forceps)
Material	Stainless (SUS304)

3 Experimental Results

3.1 Mechanical Characteristics

Firstly, we checked the characteristics of bending movements. The bending angles are shown in Fig. 8 and 9. The movement characteristics have hysteresis, with large backlash due to the stretching and friction of the wires. The difference between bending drive and extension drive (on drive speed and backlash) is mainly caused by the difference in the driving forces of wire (for bending and for extension). The difference between bending axes A and B may be caused by the differences in the loss of drive power, due to the different placement of the hinges of each axis. The accuracy of the movement is within 4.3 degrees for each axis.

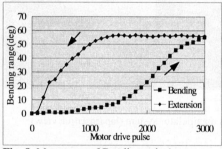

Fig. 8. Movement of Bending axis A   **Fig. 9.** Movement of Bending axis B

The generated forces, bending and grasping, are measured using weights. The results are shown in Table 2. The generated forces of bending (driving and holding) and grasping are sufficient, but the forces of extension are notably inadequate. This is because of the low power of the spring that controls the tension of the wire for extension.

Table 2. Generated force of Multi-DOF forceps manipulator

	Generated force	
	Driving	Holding
Bending	0.45 kgf	0.50 kgf
Extension	0.05 kgf	0.08 kgf
Grasping	0.85 kgf	

3.2 *In vivo* Testing

We tested the clinical performance of this Multi-DOF forceps manipulator by using it in a semi-clinical situation. The surgeon carried out trial operations (grasping, bending, extension, etc.) around the liver after ordinary laparoscopic cholecystectomy on a pig (Fig. 10 and 11).

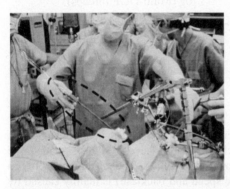
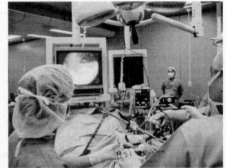

Fig. 10. *In vivo* testing of the Multi-DOF forceps manipulator system (overview)

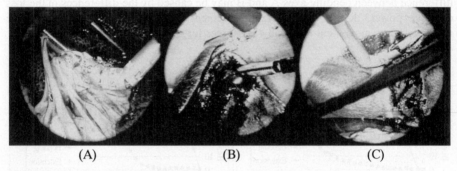

(A)   (B)   (C)

Fig. 11. *In vivo* testing of the Multi-DOF forceps manipulator system (laparoscopic view)

The results show that the bending force (driving and holding), the grasping force, and the radius of curvature of bending are nearly adequate. "User-friendliness", which mainly depends on the size, weight, and man-machine interface of this forceps manipulator, is satisfactory.

Using the Multi-DOF forceps manipulator, an operation that is impossible using current straight, rigid forceps, was carried out. In Fig. 11(B), although inserted into the visceral cavity through the left upper trochar, the manipulator works as if it had been inserted through the right upper trochar. In Fig. 11(C), the bent part of the forceps was effectively used to control the position of the liver (pushing). However, the extension force (driving and holding) was insufficient, and the stick-slip movement appeared because of this. The poor responses of movement (due to the backlash of the wire drive system) are not a problem in bending and extension. However, in grasping, a high response of movement is needed. We plan to solve these problems in extension forces and the response of grasping movements in the next prototype.

In this experiment, although there were indeed problems caused by the extension drive, the basic functions of this Multi-DOF forceps manipulator were confirmed as sufficiently useful for laparoscopic surgery.

4 Discussion

4.1 Multi-DOF Forceps Manipulator

Using simple and rigid components to achieve a bending mechanism, this manipulator established a high level of performance, with large output based on small components. The remote placement of electronic components ensured the clean, simple and lightweight nature of the Multi-DOF forceps manipulator. In the near future, this equipment promises to overcome the problems with current forceps and will raise the standards of laparoscopic surgery with higher effectiveness and safety.

The most important problems of this manipulator are that of extension force and response of movement. However, this problem may be solved if we can develop intelligent control mechanisms for force sensing and feedback systems.

4.2 Future Work ~ New System Concept ~

The conventional laparoscope, with a rigid scope, provides a limited and small workspace to the surgeon. Flexible laparoscopes solve this problem in part, but controlling the line of sight in a flexible laparoscope is more complex. The new forceps manipulator we are developing now integrates the bending forceps with a fiberscope (Fig. 12). Using this manipulator, the surgeon can see the workspace from two different lines of sight (one is that of the laparoscope, and the other is that of the fiberscope) without controlling the view-line of an additional endoscope (fiberscope). We think surgeons can efficiently operate using both a laparoscope view ("macro" view) and a fiberscope view ("micro" view). Especially when operating on the far side of organs, the fiberscope view will be very helpful for surgeons.

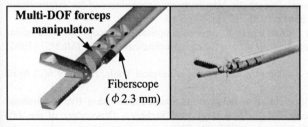

Fig. 12. New Multi-DOF forceps manipulator integrated with a fiberscope

Using the Multi-DOF forceps manipulator, we will construct a new surgical manipulator system concept. We have already developed a laparoscope manipulator using a five-bar linkage system [7]. Integrating the forceps manipulator and five-linkage manipulator, we will develop new surgical manipulators for laparoscopic surgery. Both of these system components (Multi-DOF forceps manipulator and five-bar linkage manipulator) were developed to achieve safety and reliability.

5 Conclusion

We have developed a Multi-DOF forceps manipulator with bending mechanism for laparoscopic surgery. This mechanism achieved two additional DOF of bending, with small radius and a large driving force on the tip of the slim forceps. Through the *in vivo* testing, although some problems were found, the effectiveness of this manipulator was clearly demonstrated. This manipulator can solve many problems in laparoscopic surgery due to the poor DOF of controlling surgical tools. We are currently developing a new surgical manipulator for laparoscopic surgery using this Multi-DOF forceps manipulator.

6 Acknowledgements

This study was partly supported by the Research for the Future Program JSPS-RFTF 96P00801 and JSPS-RFTF 99I00904.

References

1. Rininsland H.: ARTEMIS. A telemanipulator for cardiac surgery, *European Journal of Cardio-Thoracic Surgery* 16(2) (1999) 106-111
2. Shennib H., Bastawisy A., McLoughlin J., Moll F.: Robotic computer-assisted telemanipulation enhances coronary artery bypass, *Journal of Thoracic & Cardiovascular Surgery* 117(2) Feb (1999) 310-313
3. Cohn, M., L. S. Crawford, J. M. Wendlandt, S. S. Sastry: Surgical Applications of Milli-Robots, *Journal of Robotic Systems* 12(6) (1995) 401-416
4. Steve Charles, Hari Das, Tim Ohm, Curtis Boswell, Guillermo Rodriguez, Robert Steele, Dan Istrate: Dexterity-enhanced Telerobotic Microsurgery *Proceedings of 8th International Conference on Advanced Robotics (ICAR `97)* (1997) 5-10
5. Ikuta K., Kato T., Nagata S.: Micro active forceps with optical fiber scope for intra-ocular microsurgery, *Proceedings of the IEEE Micro Electro Mechanical Systems (MEMS)* (1996) 456-461
6. Gupta V., Reddy NP., Batur P.: Forces in Laparoscopic Surgical Tools, *Presence* 6(2) April (1997) 218-228
7. Kobayashi E. et al.: Development of a laparoscope manipulator using five-bar linkage mechanism, *CAR97 Computer Assisted Radiology and Surgery - Proceeding of the 11th International Congress and Exhibition* – (1997) 825-830
8. R. Nakamura et al.: Development of forceps manipulator system for laparoscopic surgery, CARS2000 Computer Assisted Radiology and Surgery - Proceeding of the 14th International Congress and Exhibition - (2000) 105-110
9. Hashimoto D.: *Gasless Laparoscopic Surgery* World Scientific Publishing (1995)

A Wide-Angle View Endoscope System Using Wedge Prisms

Etsuko Kobayashi[1], Ken Masamune[2], Ichiro Sakuma[1] and Takeyoshi Dohi[1]

[1]Institute of Environment Studies, Grad. School of Frontier Sciences,
The University of Tokyo,7-3-1 Hongo Bunkyo-ku, Tokyo, 113-8656, Japan
{etsuko, sakuma, dohi}@miki.pe.u-tokyo.ac.jp
[2]Depatment of Bio-technology, Tokyo Denki University,
Ishizaka, Hatoyama-cho, Saitama, 350-0394, Japan
masa@b.dandai.ac.jp

Abstract. We describe a novel robotic endoscope system. It can be used to observe a wide area without moving or bending the endoscope. The system consists of a laparoscope with zoom facility and two wedge prisms at the tip. This new concept produces excellent characteristics as follows. Firstly, it can change the field of view even in a small space. Secondly, it is safe because it avoids the possibility of hitting internal organs. Finally, because it does not require a large mechanism for manipulation of the endoscope, it does not obstruct the surgeon's operation. During evaluation, we confirmed that the range of view and levels of image deformation were acceptable for clinical use.

1. Introduction

Endoscopic surgery is becoming increasingly popular as a form of minimally invasive surgery. In such surgery, smooth manipulation of the endoscope is essential. However, it is difficult for surgeons and camera assistants to manipulate the endoscope steadily and smoothly, because of the restricted space and degrees of freedom of movement, and because of limitations in hand-eye coordination.

Robotic endoscopic systems have the potential to solve these problems [1][2][3]. They can hold the endoscope steadily and the surgeon can exert remote control using a computer display. For these reasons, many robotic endoscopic systems have been developed. Most of them manipulate existing endoscopes, of which there are two types: rigid and flexible. Rigid endoscopes need to rotate the insertion point to observe a wide range of views, whereas flexible types rotate the tip of the endoscope to achieve the same end. Thus, to obtain a wide field of view a large operating area is required.

However, in endoscopic fields such as neurosurgery, where there is normally only a limited operating area, it is difficult to obtain a sufficiently wide field of view. It is therefore desirable to develop a new type of endoscope that can achieve a wide range of views without moving the endoscope itself.

There is a commercially available laparoscopic system that alters the field of view by trimming the field viewed by the laparoscope and moving the cutting tool area appropriately [4]. With this system, the surgeon can move the field of view without moving the endoscope. However, because the image uses only part of the whole, the quality is poor.

With this in mind, we propose a new type of robotic wide-view endoscope that does not require rotation or bending of the endoscope to move the field of view. This is clinically significant as it allows the surgeon to make extensive observations in a small space and is safe because it avoids the possibility of accidentally hitting organs.

In Section 2, we propose a design for a novel robotic endoscope, with wedge prisms for moving the field of view. Theoretical formulae for controlling the image are presented in Section 3. In Section 4, we validate the basic specification of this endoscope system.

2. System description

2.1 System requirements

To develop a first prototype of a wide-view endoscope system, we decided to focus on the rigid type laparoscope, since it is ubiquitous in minimally invasive surgery and provides high quality images.

The requirements of this system are as follows.

Movement of the laparoscopic view

In laparoscopic surgery, the laparoscope should be able to rotate around the insertion point and to move back and forth. Thus, the wide-view laparoscope should have at least two degrees of angular freedom and be able to zoom in and out.

Sterilization

Because the tip of the endoscope is inserted in the abdomen, it needs to be sterilized. To achieve this, the mechanism must be simple.

High quality image

As for all medical endoscopes, the image quality must be good.

2.2 System description

To achieve the above requirements, we used two wedge prisms and a laparoscope with zoom capacity. Figure 1 depicts a wedge prism. It has a vertex angle of (θ_w) to bend the light axis θ_d. The relationship between θ_d and θ_w is as follows:

$$\theta_w = \arctan\left[\frac{\sin\theta_d}{n - \cos\theta_d}\right]$$

where 'n' is the refractive index of the prism (Fig. 1a).

If the wedge prism is put in the tip of a laparoscope and rotated around its axis, the light axis turns around the axis with an angle of θ_d (Fig. 1b).

Another wedge prism that has the same wedge vertex angle (θ_w) as the first is mounted close behind the first (Fig. 1c). When the two prisms are rotated around the light axis of the laparoscope, the light axis can bend to any direction in a cone. If the wedge vertex angle (θ_w) is small, the maximum angle of such bending is $2\theta_d$ (Fig. 1a).

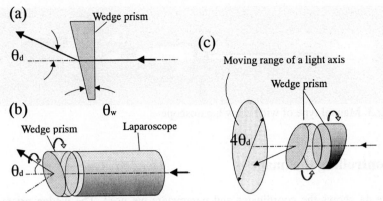

Fig.1. Wedge prism

a) Cross-section of a wedge prism. The wedge prism has a wedge vertex angle (θ_w) to refract light at an angle of θ_d ; b) Laparoscope with a wedge prism; c) Combination of two wedge prisms.

Figure 2 shows the construction of the wide-view endoscope system. We used an automatic micro-zoom laparoscope (Shinko Optical Co. Ltd, Tokyo, Japan). It consists of a short (155-mm) laparoscope with one lens at its tip, an automatic zoom component and a charge coupled device (CCD) camera[5]. Two wedge prisms are mounted in the tip, and each prism is set on a sleeve. These two sleeves rotate independently around within the laparoscope, and are connected to motors by gears. The system is controlled by the surgeon, using a man-machine interface.

The wedge vertex angle of the wedge prism that we used was 18°8' and the light bending angle (θ_d) was 10°. Therefore, the system can bend the light axis in a cone with a vertex angle of 40°. In this mechanism, only two sleeves rotate within the laparoscope, so the mechanism can be simple and small, and yet can observe a wide range of view.

Figure 3 shows a manual version of the wide-view endoscope. The diameter of the prism was 25 mm. We are constructing a smaller motorized prototype with a smaller (12-mm) prism.

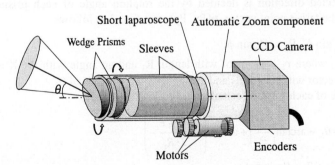

Fig.2. Construction of the wide-range endoscope system

Fig.3. Manual type of wide-view laparoscope

3. Controlling the image

Figure 4a shows the coordinates and parameters we used. The wedge prism that is nearest the tip of the laparoscope is named Prism 1. The other prism is named Prism 2. Figure 4b shows a cross-section in the YZ plane. The origin is set at a node of the light axis and Prism 1. As we decided that positioning accuracy was not required for the final endoscopic image, we used approximate values for some parameters and we considered the optical model to be that of a simple pinhole camera model. The distance between the virtual screen and the focal point is Z_f. The distance between the virtual screen and the origin is Z_1 and the distance between the virtual screen and the node is Z_2. In this model, because of limited space in the abdomen, we assumed that all the targets were in the virtual screen plane.

Figure 4b shows the refracting angle of each prism. For Prism 2, to simplify the formula, we assumed that the angle of incidence (equal to the refracting angle of Prism 1) was zero degrees [6].

$$\phi_1 = \arcsin\{n\sin(D(\theta_w))\}$$
$$\phi_2 = D(\theta_w)$$
$$D(\theta_w) = -\theta_w + \arcsin(n\sin\theta_w)$$

The relation between the target point and the rotation angle of the prisms is shown in Fig. 4c. In the virtual screen, the reflected light axis of Prism 1 appears as a circle with a radius of R_1. The reflected light beam of Prism 2 is a circle with a radius of R_2.

The reflected direction is decided by the rotation angle of each prism (θ_1, θ_2). Therefore, the target point vector, $\mathbf{p}(Px, Py)$, was found as follows:

$$R_1 = Z_1 \sin\phi_1, R_2 = Z_2 \sin\phi_2$$

$\mathbf{p} = \mathbf{r}_1 + \mathbf{r}_2$, where $\mathbf{r}_1$ is a vector with length R_1 and an angle with the X axis of θ_1, and $\mathbf{r}_2$ is a vector with length R_2 and an angle with the X axis of θ_2.

The angle of each prism is calculated as follows (Fig. 4d).

$$\theta_1 = \theta_a + \theta_b = \arctan\left(\frac{P_y}{P_x}\right) + \arccos\left(\frac{\sqrt{P_x^2 + P_x^2}}{2 \cdot R_1}\right)$$

$$\theta_2 = \arctan\left(\frac{P_y - R_1\sin\theta_1}{P_x - R_1\cos\theta_1}\right)$$

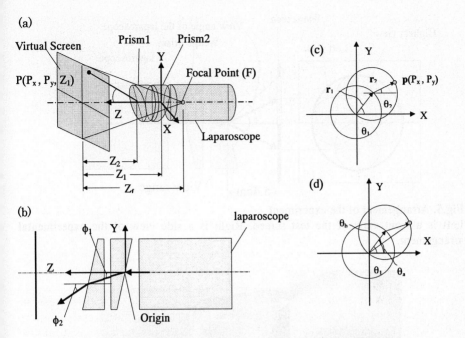

Fig. 4. Controlling the image, (a) Optical model of the wide-view laparoscope; (b) Cross-section of the system; (c) Virtual screen; (d) Virtual screen.

4. Evaluation

4.1 Measurement of the range of view

Firstly, we measured the range of view of the endoscope. Figure 5 shows the arrangement of the experiment. The test screen was set 5.4 cm in front of the tip of the laparoscope. We measured the center points (the node of the light axis and the screen) of the highest, the lowest, and the extreme left and right views. As a result, the range of view was 41° in height and 39° in width. Figure 6 shows the highest and lowest endoscopic images of a life-size liver model. The liver model was set 7.5 cm in front of the tip. You can see that the surgeon can observe the almost whole liver by the wide-view endoscope.

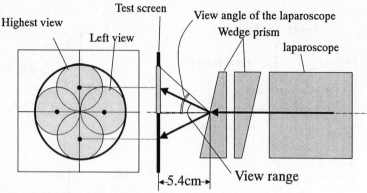

Fig.5. Arrangement of the experiment
Left is a front view of the test screen. Right is a side view of the experimental arrangement.

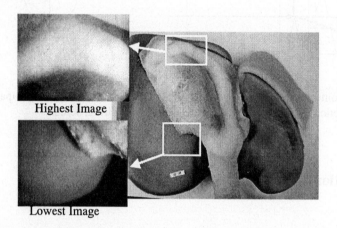

Fig.6. Wide-view endoscopic image of a life-size liver model

4.2 Evaluation of image deformation

We evaluated the image deformation caused by the prisms. Table 1 shows the ratio of the height to the width of a 5-mm square. We measured this for the highest, lowest, and extreme left and right images. For comparison, we measured the deformation of the periphery of the normal laparoscopic image.

Table 1. Comparison of deformation of the image between the experimental system and the unaltered laparoscope.

	Images	Ratio of height to width
Wide-view laparoscope	Highest image	0.85
	Lowest image	0.98
	Left end	0.97
	Right end	1.05
Ordinary laparoscope		1.02

5. Discussion

In this system, the surgeon can move the field of view without rotating or bending the endoscope. Therefore, this system offers excellent advantages.

-It can move the field of view, even in a small space.

-It is completely safe because it also avoids the possibility of hitting organs.

-Because it does not require a large mechanism for moving the endoscope, it does not obstruct the surgeon's operation.

In the evaluation, we confirmed the wide range of movement of our system. Because the view angle of the laparoscope is 60°, we can observe a circle with a diameter of 17 cm when only 5 cm in front of the tip of the endoscope. By contrast, the ordinary laparoscope can only observe a circle of 5.6 cm without being moved. From this result, for most of the organs the whole structure can be observed without moving the endoscope. Therefore, we can say it has a sufficiently wide range for clinical use.

There was some deformation of the image. However, it was considered acceptable, because it was similar to that seen in the periphery of the ordinary laparoscopic image. Color aberration is also inevitable because of the prisms. These aberrations and deformations could be simply corrected by an image processing board.

6. Conclusions

We propose a new type of endoscope system that can allow observation over a wide angle and provide high quality images without moving the endoscope itself. This has great significance for future clinical use.

Acknowledgement

This study was partly supported by the Research for the Future Program (JSPS-RFTF 96P00801) and (JSPS-RFTF 99I00904)

References

1. Taylor RH, Funda J, Eldridge B, Gomory S, Gruben K, LaRose D, Talamini M, Kavossi L, Anderson J A Telerobotic Assistant for Laparoscopic Surgery. Computer Integrated Surgery: The MIT Press, (1995) pp 581-592
2. Sackier JM, Wang Y Robotically Assisted Laparoscopic Surgery: From Concept to Development. Computer Integrated Surgery: The MIT Press, (1995) pp 577-580
3. Finlay PA, Ornstein MH Controlling the Movement of a Surgical Laparoscope. IEEE Engineering in Medicine and Biology Vol.14 No.3 (1995) pp.289-299
4. Olympus co.ltd. Home page, http://www.olympus.co.jp/
5. Etsuko Kobayashi, Ken Masamune, Ichiro Sakuma, Takeyoshi Dohi, Daijo Hashimoto A New Safe Laparoscopic Manipulator System with a Five-Bar Linkage Mechanism and an Optical Zoom Journal of Computer Aided Surgery, 4(4), (1999) pp.182-192
6. Faugeas:Three-Dimentional Computer Vision, MIT Press, 1993

Computer-Assisted TURP Training and Monitoring

M.P.S.F. Gomes and B.L. Davies

Mechatronics in Medicine Laboratory, Department of Mechanical Engineering,
Imperial College of Science, Technology and Medicine, Exhibition Road,
London SW7 2BX, UK
{p.gomes, b.davies}@ic.ac.uk
http://www.me.ic.ac.uk/case/mim/

Abstract. A computer-assisted surgical trainer for Transurethral Resection of the Prostate (TURP) is described. The surgeon resects a rubber prostate phantom using real surgical tools, under endoscopic visual guidance. Additionally, a computer display provides navigational information and shows the progression of the resection. The core of the trainer consists of a low-cost PC and a commercial optical tracker. The extension of the trainer to the operating room, to be used as an *in vivo* monitoring system, is proposed. This novel two-stage approach aims to bridge the gap between training on a simulator and performing a real procedure in the operating room.

1 Introduction

There is an increasing need for validated training aids for endoscopic surgery that do not require the presence of a qualified surgeon, the involvement of a patient, or the use of live animals [1-12]. One of the key requirements of minimally invasive endoscopic procedures is the surgeon's ability to locate anatomical features in 3D space usually hidden behind an access port, and to position tools in specific locations whilst avoiding others. This requires mastering the anatomy, as well as an ability to visualise a 3D mental picture of it, from a combination of theoretical knowledge, experience, preoperative examinations, ultrasound scans, and a localised view given by an endoscopic camera.

In the specific case of Transurethral Resection of the Prostate, the goal is to use a minimally invasive resectoscope to remove adequate prostatic tissue, without resecting distal to the verumontanum, damaging the ureteric orifices, or perforating the capsule (Fig.1).

The traditional training follows the typical pattern of i) observing an experienced surgeon doing a number of procedures, and ii) performing a number of procedures, closely monitored by an experienced surgeon. Courses are conducted at the Royal College of Surgeons of England, London, where trainee urological surgeons can resect a prostate phantom but this is usually a one-off session and not all trainees attend.

The urologist conventionally carries out a TURP using a resectoscope inserted through the urethra. The resectoscope (Fig.3) performs three functions:

1. It allows the urologist to visualise the area to be resected (the prostatic urethra). This is accomplished with the help of a high intensity light source, a camera and a

series of lenses. The camera is usually linked to a monitor for more convenient visualisation (Fig.2).

2. It resects prostatic tissue, by utilising a wire loop through which a high frequency AC current is passed. A different waveform is used to coagulate bleeding arteries and veins.

3. It has a separate channel for irrigation with a sterile hypo-osmolar solution. Irrigation is used to improve the viewing of the operating field.

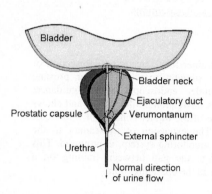

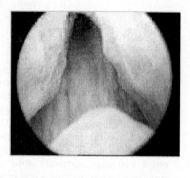

Fig. 1. Anatomy of the prostate gland

Fig. 2. Endoscopic view of the prostate gland, showing the verumontanum and the bladder neck

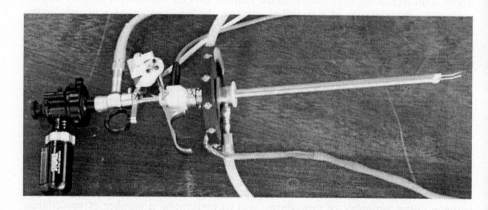

Fig. 3. Resectoscope elements (the ring and the potentiometer arrangement attached to the working element have been specifically developed for the trainer)

The main objectives of the computer-assisted TURP trainer presented are:
• To provide a realistic hands-on trainer.

A physical phantom of the prostate gland, of similar "feel" to real tissue, is resected, using very similar surgical equipment and procedure as those used in the

operating room. The nature of the procedure, and the way it is perceived, must not be significantly affected by the trainer.

- To provide objective data to enable measurement of surgical skills and performance.

The amount of tissue resected, perforation or damage to high-risk areas, and the time taken to complete the procedure can all be logged.

- To provide extra guidance for navigation in 3D space.

A computer graphical interface is provided that depicts a scaled 3D model of the prostate phantom, with the current position of the resectoscope superimposed in real time, so that the trainee always knows the position of the cutting loop within the prostate. Visual warnings are also given when the loop is close to high-risk areas. The important anatomical landmarks and the tissue resected are also shown, allowing the progress of the resection to be monitored.

2 Materials and Methods

The system comprises a mock-up abdomen (a perspex box with a rubber membrane) and a rubber bladder. For each resection, a rubber prostate is inserted into the rubber bladder and clamped. The box has an outlet for the irrigant and sits in a tray, in a supporting fixture. The position and orientation of the box, and hence of the phantom, is monitored using a camera-based tracker. The artificial prostate is resected by the surgeon, using a resectoscope instrumented for tracking. Two screens are provided, one depicting the endoscopic view, the second depicting 3D navigational information in relation to the prostate and the resectoscope. A snapshot of the trainer in operation is given in Fig.4.

Fig. 4. Computer-assisted TURP trainer in use. The optical tracker is is mounted on a tripod and cannot be seen in this image.

2.1 Instrumentation

Resectoscope motion The resectoscope's position and orientation is tracked using a purpose-built IRED ring clamped to its sheath. A camera-based system (Flashpoint 5000, Image Guided Technologies Inc., Colorado, USA) is used to track the 100mm diameter ring, with 12 equally spaced IREDs (Fig.5). This novel IRED tool is necessary because the resectoscope can be rotated about the axis of its sheath by more than ±180°.

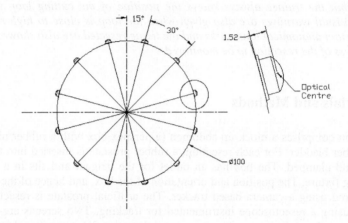

Fig. 5. IRED arrangement for optical tracking

A tool definition file is supplied to the optical tracker, containing the relative positions of all the IREDs and the tool tip. During the operation, the optical tracker calculates the tool tip's location from the positions of the visible IREDs.

There are many constraints to the shape of the ring, e.g. it has to allow the use of a conventional resectoscope and hence allow the operation of the irrigation inlet and outlet taps; it cannot obstruct the surgeon's movements or interfere with the procedure; it should not significantly change the "feel" of the tool; it has to be sterilisable; and it should also be easily adapted to standard instruments. The current design complies with these constraints.

A Dynamic Reference Frame (a special IRED probe, used as the origin of a coordinate system for the readings from the optical tracker) is attached to the perspex box, in a fixed location. This makes the readings, provided by the camera-based tracking system, independent from the position of the camera sensor.

Cutting loop travel The prostate phantom is resected by an electrical cutting loop. The loop moves along the inside of the resectoscope shaft, in a linear path of 25mm extension. The resected chip separates from the rest of the phantom when the cutting loop fully enters the resectoscope's body. The surgeon moves the loop in and out of the resectoscope by operating a standard "scissor grip" on the resectoscope handle.

A potentiometer has been coupled to the resectoscope's trigger by a gear mechanism to track the cutting loop travel. The potentiometer is connected to the low cost PC by means of an A/D board, using a cable with detachable plugs, similar to those for the IRED ring.

Diathermy activation The cutting action of the diathermy loop is activated by pressing a foot pedal connected to the diathermy unit. This action is sensed by a switch attached to the foot pedal and this is used to appropriately modify the computer display. The switch is connected to the PC, via the same data acquisition A/D board.

2.2 Software

Platform The software for the trainer has been developed using Microsoft Visual C++, Microsoft Foundation Classes, and OpenGL, for the Windows 9x platform. The software runs on a Pentium III 500 MHz machine.

Graphical user interface Whilst operating conventionally, the surgeon views the endoscopic camera display, and uses both hands to operate the resectoscope and ancillary equipment, as well as one foot to activate the diathermy unit. For the trainer, extra intervention or information has to be kept simple and to a minimum. Essentially, what the computer display of the trainer provides is i) information about the resectoscope's position in three-dimensional space, and ii) information about the tissue already resected.

In order to convey information in a quick and simple way, different views can be shown (Fig.6):
1. A 3D rendering of the model and of the resectoscope. The user can decide between displaying the whole model or displaying a partial view by cutting away the top part, which lies above the resectoscope.
2. A 2D view of the section through the cutting loop position, perpendicular to the base of the model, showing the model's outline, the resectoscope's position, and the area resected.
3. A collection of 2D views (similar to that described in (b) above), shown as thumbnail images, at pre-specified positions, equally spaced between the bladder neck and the verumontanum.

The user can chose which views are shown at any one time, as well as wireframe or solid models. The display of the resectoscope changes colour to provide simple and immediate visual feedback about the state of the instrument (green in normal mode; red if cutting; grey if obscured). The background of the 2D enlarged view also changes from light grey to yellow or red if the resectoscope's tip is too close to the bladder neck of the verumontanum, respectively.

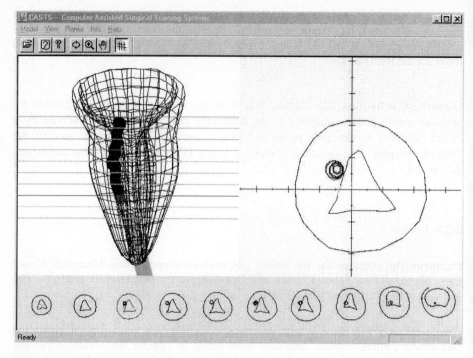

Fig. 6. The graphical user interface

Model representation The graphics model of the prostate phantom is a set of NURBS (Non-Uniform Rational B-Splines), stored in the IGES (Initial Graphics Exchange Specification) standard format, and loaded at the start-up of the trainer's software.

The data for the model was obtained by measuring hard casts of the rubber prostate, using the optical tracker, and fitting NURBS to these points, using Rhino (Robert McNeel and Associates, Washington, USA).

Cuts representation A separate graphical model is created, containing the cuts made on the phantom, without changing the original prostate model. This allows clear visualisation of the resected material and the progress of the resection.

Soft tissue deformation During a resection, the phantom moves and deforms, just like the real prostate, by the compressing and cutting action of the resectoscope, in a way related to the nature of the phantom's material and the phantom's fixture. In order to provide a realistic graphical display of this deformation, the NURBS control points are displaced and rendered. The displacements are precalculated and stored in look-up tables for speed.

Deformations caused by prolonged cutting and swelling due to the absorption of the irrigant are also observed. Clamps are used to partially constrain these deformations and make them predictable without detracting from the realistic 'feel' of the prostate which moves, *in vivo*, within the pelvic girdle attachments.

3 Discussion and Conclusions

Preliminary tests by surgeons have shown the *in vitro* system to be realistic and a useful addition to training. The computer display provides additional visual confirmation of the resectoscope cutting relative to the critical prostatic features. The use of a resectable physical phantom of the prostate provides a realistic sense of "feel" on the handle of the resectoscope when pressing and cutting the prostate. This approach has been found to provide a simpler and lower-cost system than using a totally virtual reality (VR) system in which the haptic sense at the tool handle is not provided by resecting a physical phantom but is provided by a complex series of force profiles applied to the tool using motorised torque control.

However, several problems have been found which need tackling before widespread use of the trainer can be beneficial. Inaccuracies of several milimeters have been observed when the traditional resectoscope's cutting loop bends and also when the phantom swells by absorbing irrigant. Also, with the progress of the resection, the system slows down due to increased graphical rendering (of the cuts model) and some cuts made are not recorded or are displayed late. This however can be compensated for, to some degree, by increased computing costs.

The solution to improve accuracy lies in closing the loop, i.e., in verifying what tissue is actually resected. This can be accomplished by using ultrasound to image the resection. However, this would change the procedure and, therefore, would not comply with our first objective listed in section 1. Additionally, the current phantom cannot be imaged with ultrasound and an alternative, which would still provide similar "feel" to human tissue, would have to be found. Ideally, the phantom would be non-absorbent and made to tighter manufacturing tolerances.

For the proposed *in vivo* monitoring system (Fig.7), pre-operative transrectal ultrasound scans of the patient's prostate (taken just before the resection procedure) are to be used to build up a model of the specific 3D anatomy. Per-operative rectal ultrasound scanning, using an ultrasound probe fitted with infra-red emitter diodes (IREDs), allows the progress of the resection to be monitored during the procedure.

This attempt to improve the quality of prostatectomy training has given a number of insights into the problems of CAS (Computer-Assisted Surgery) *in vitro* trainers for general use in soft tissue surgery. It is felt that the ability to link the *in vitro* approach to an *in vivo* training and monitoring system is potentially rewarding. The need to bridge the wide gap between simple training systems and patient trials by trainees is of great importance.

Acknowledgements

The collaboration of Mr. A.G. Timoney, from the Department of Urology, Southmead Hospital, Bristol, United Kingdom, and of Mr. P.V.S. Kumar, from the Taunton and Somerset Hospital, United Kingdom, is gratefully acknowledged. This project is sponsored by the Engineering and Physical Sciences Research Council (EPSRC) of the UK. Karl Storz Endoscopy (UK) has kindly loaned medical equipment. Limbs & Things (UK) provided moulds for the prostate model.

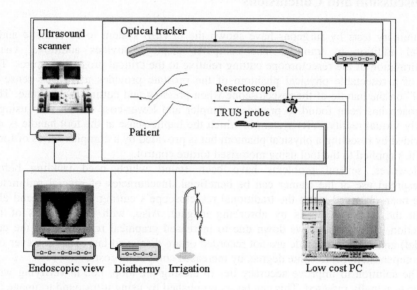

Fig. 7. The proposed *in vivo* TURP monitor

References

1. http://www.cine-med.com/
2. http://www.ht.com/products_endosim.htm
3. http://www.vrweb.com/WEB/DEV/MEDICAL.HTM#MIST
4. Ayache, N., Cotin, S., Delingette, H., Clement, J. M., Y.Russier and Marescaux, J. (1998) Simulation of endoscopic surgery, Minimally Invasive Therapy & Allied Technologies 7 (2), 71-77
5. Bauer, A., Soldner, E. H., Ziegler, R. and Muller, W. (1995) Virtual reality in the surgical arthroscopic training, Second Annual International Symposium on Medical Robotics and Computer Assisted Surgery, Baltimore, Maryland, USA, 350-355
6. Gomes, M. P. S. F., Barrett, A. R. W., Timoney, A. G. and Davies, B. L. (1999) A computer-assisted training/ monitoring system for TURP - Structure and design, IEEE Transactions on Information Technology in Biomedicine 3 (4), 242-251
7. Kleinszig, G., Radetzky, A., Auer, D. P., Pretschner, D. P. and Auer, L. M. (1998) ROBO-SIM: Simulation of Endoscopic Surgery, Society for Minimally Invasive Therapy 10th Anniversary International Conference, London
8. Logan, I. (1995) Virtual environment knee arthroscopy training system, http://www.enc.hull.ac.uk/CS/VEGA/medic/surgery.html

677

9. McCarthy, A. D. and Hollands, R. J. (1998) A commercially viable virtual reality knee arthroscopy training system, Medicine Meets Virtual Reality: 6, San Diego, USA, 302-308
10. Schill, M.A., Wagner, C., Hennen, C., Bender, H-J. and Manner, R. (1999) EyeSi – A simulator for intra-ocular surgery, Medical Image Computing and Computer-Assisted Intervention - MICCAI'99, Cambridge, UK, 1166-1174
11. Szekely, G., Brechbuhler, C., Hutter, R., Rhomberg, A. and Schmid, P. (1998) Modelling of soft tissue deformation for laparoscopic surgery simulation, Medical Image Computing and Computer-Assisted Intervention - MICCAI'98, Boston, USA, 550-561
12. Xuan, J., Wang, Y., Sesterhenn, I. A., Moul, J. W. and Mun, S. K. (1998) 3-D model supported prostate biopsy simulation and evaluation, Medical Image Computing and Computer-Assisted Intervention - MICCAI'98, Boston, USA, 358-367

Estimating 3D Strain from 4D Cine-MRI and Echocardiography: In-Vivo Validation

Xenophon Papademetris[1], Albert J. Sinusas[23], Donald P. Dione[2],
R. Todd Constable[2] and James S. Duncan[13]

[1] Departments of Electrical Engineering, [2] Diagnostic Radiology, and [3] Medicine,
Yale University New Haven, CT 06520-8042
papad@noodle.med.yale.edu

Abstract. The quantitative estimation of regional cardiac deformation from 3D image sequences has important clinical implications for the assessment of myocardial viability. The validation of such image-derived estimates, however, is a non-trivial problem as it is very difficult to obtain ground truth. In this work we present an approach to validating strain estimates derived from 3D cine-Magnetic Resonance (MR) and 3D Echocardiography (3DE) images using our previously-developed shape-based tracking algorithm. The images are segmented interactively and then initial correspondence is established using a shape-tracking approach. A dense motion field is then estimated using a transversely linear elastic model, which accounts for the fiber directions in the left ventricle. The dense motion field is in turn used to calculate the deformation of the heart wall in terms of strains. The strains obtained using our algorithm are compared to strains estimated using implanted markers and sonomicrometers, which are used as the gold standards. These preliminary studies show encouraging results.

1   Introduction

It is the fundamental goal of many forms of cardiac imaging and image analysis to measure the regional function of the left ventricle (LV) in an effort to isolate the location and extent of ischemic or infarcted myocardium. In addition, the current management of acute ischemic heart disease is directed at establishing coronary reperfusion and, in turn, myocardial salvage. Regional function may also be related to the degree of salvage achieved.

However, while there have been many methods proposed for the estimation of 3D left ventricular deformation (e.g. [1, 6, 7, 16]), the validation of these results is an extremely important and often neglected aspect of work in this area. Often *phantoms* are used with known shapes and displacements usable as ground truth (e.g. Kraitchman[8]). In Young [19] it was shown that away from the free surfaces of the gel-phantom, a Rivlin-Mooney [9] analytic model accurately reproduced the 2-D displacement of magnetic tags. This showed agreement between the theory (model) and the image-derived displacements. However, *in vivo* measurements of the beating heart usually present additional complexities. An alternative validation method is to use *simulated images* with known ground truth (e.g.

Amini[1], Prince [14] and Haber [6]). One example[1] uses a kinematic model of the left ventricular motion by Arts [2] within an MR tag image simulator [18] to generate synthetic images with known displacements. In the shape-tracking work of Shi[17], *implanted markers* are used as the gold standard. These markers are physically implanted on the myocardium before the imaging. Here, algorithm generated displacements are compared to the marker-displacements (these are easily identifiable from the images). This technique has the disadvantage of comparing trajectories for a small number of points, however, it is done on *real* data as opposed to simulations. However, in all previous efforts including those noted above, to the best of our knowledge, there has been no validation of estimates of *in-vivo* cardiac deformation (Shi et al[17] only validate displacements.)

In this paper we describe the experiments used to validate strains estimates derived from our algorithm, which uses a transversely isotropic model for the left ventricle and shape-based displacements [12]. We first briefly review the algorithm used to estimate the deformation and then describe the methodology used for the validation. We then present results for the validation of these deformation estimates derived from Magnetic Resonance (MR) and 3D-Echocardiography (3DE) image sequences.

2 Our algorithm

2.1 Image Acquisition

Canine MR-images: ECG-gated magnetic resonance imaging was performed on a GE Signa 1.5 Tesla scanner. Axial images through the LV were obtained with the gradient echo cine technique. The imaging parameters were: section thickness=5 mm, no intersection gap, 40 cm field of view, TE 13 msec, TR 28 msec, flip angle 30 degrees, flow compensation in the slice and read gradient directions, 256 x 128 matrix and 2 excitations. The resulting 3D image set consists of sixteen 2D image slices per temporal frame, and sixteen temporal 3D frames per cardiac cycle, with an in-plane resolution of $1.6mm$ and a slice thickness of $5mm$.

3D Echocardiography (3DE): The 3DE images were acquired using an HP Sonos 5500 Ultrasound System with a 3D transducer (Transthoracic OmniPlane 21349A (R5012)). The 3D-probe was placed at the apex of the left ventricle of an open-chest dog using a small ultrasound gelpad (Aquaflex) as a standoff [12]. Each acquisition consisted of 13–17 frames per cardiac cycle depending on the heart rate. The angular slice spacing was 5 degrees resulting in 36 image slices for each 3D frame.

2.2 Segmentation and Shape-Based Tracking

The endocardial and epicardial surfaces were extracted interactively[11] and then sampled to 0.5 voxel resolution. Next, curvatures are calculated and the shape based tracking algorithm applied to generate a set of initial matches and confidence measures for all the points on the surface. Given a set of displacement vector measurements u^m and confidence measures c^m we model theses estimates probabilistically by assuming that the noise in the individual measurements as normally distributed with zero mean and a variance σ^2 equal to $\frac{1}{c^m}$. These assumptions result in a measurement probability of the form:

$$p(u^m|u) = \frac{1}{\sqrt{2\pi\sigma^2}}\, e^{-\frac{(u-u^m)^2}{2\sigma^2}} \tag{1}$$

2.3 Modeling the myocardium

The left ventricular myocardium is modeled using a biomechanical model. We use a transversely linear elastic model which allows us to incorporate information about the preferential stiffness of the tissue along fiber directions from [5]. These fiber directions are shown in figure 1. The model is described in terms of an internal or strain energy function of the form:

$$W = \epsilon' C \epsilon \tag{2}$$

where ϵ is the strain and C is the 6×6 matrix containing the elastic constants which define the material properties (see [12, 13] for the details.)

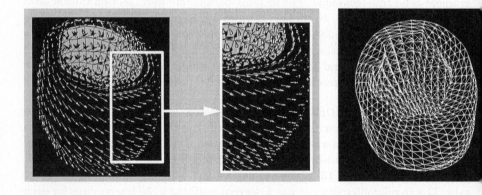

Fig. 1. Left: Fiber direction in the left ventricle as defined in Guccione[5]. **Right:** Volumetric Model of the LV consisting of hexahedral elements.

The left ventricle of the heart is specifically modeled as a transversely elastic material to account for the preferential stiffness in the fiber direction, using the following material matrix C:

$$
C^{-1} =
\begin{bmatrix}
\frac{1}{E_p} & \frac{-\nu_p}{E_p} & \frac{-\nu_{fp}}{E_f} & 0 & 0 & 0 \\
\frac{-\nu_p}{E_p} & \frac{1}{E_p} & \frac{-\nu_{fp}}{E_f} & 0 & 0 & 0 \\
\frac{-\nu_{fp}E_f}{E_p} & \frac{-\nu_{fp}E_f}{E_p} & \frac{1}{E_f} & 0 & 0 & 0 \\
0 & 0 & 0 & \frac{2(1+\nu_p)}{E_p} & 0 & 0 \\
0 & 0 & 0 & 0 & \frac{1}{G_f} & 0 \\
0 & 0 & 0 & 0 & 0 & \frac{1}{G_f}
\end{bmatrix}
\tag{3}
$$

where E_f is the fiber stiffness, E_p is cross-fiber stiffness and ν_{fp}, ν_p are the corresponding Poisson's ratios and G_f is the shear modulus across fibers. ($G_f \approx E_f/(2(1+\nu_{fp}))$). The fiber stiffness was set to be 3.5 times greater than the cross-fiber stiffness [5]. The Poisson's ratios were both set to 0.4 to model approximate incompressibility. Using a Markov Random Field analogy, we can describe the model probabilistically using an equivalent prior probability density function $p(u)$ of the Gibbs form:

$$
p(u) = k_1 \exp(-W(C, u))
\tag{4}
$$

2.4 Integrating Model and Data

Having defined both the data term (equation 1) and the model term (equation 4) as probability density functions we naturally proceed to write the overall problem in a Bayesian estimation framework. Given a set of noisy input displacement vectors u^m, the associated noise model $p(u^m|u)$ (data term) and a prior probability density function $p(u)$ (model term), find the best output displacements $\hat{u}$ which maximize the posterior probability $p(u|u^m)$. Using Bayes' rule we can write:

$$
\hat{u} = \frac{\arg\max}{u} \, p(u|u^m) = \frac{\arg\max}{u} \left(\frac{p(u^m|u)p(u)}{p(u^m)} \right)
\tag{5}
$$

Taking logarithms in equation (5) and differentiating with respect to the displacement field u results in a system of partial differential equations, which we solve using the finite element method [3]. To achieve this, a volumetric model of the LV is constructed using hexahedral elements as shown in figure 1. For each frame between end-systole (ES) and end-diastole (ED), a two step problem is posed: (i) solving equation (5) normally and (ii) adjusting the position of all points on the endocardial and epicardial surfaces so they lie on the endocardial and epicardial surfaces at the next frame using a modified nearest-neighbor technique and solving equation (5) once more using this added constraint. This ensures that there is a reduction in the bias in the strain estimates.

3 Experimental Results

In this section we present validation of the image derived strains using implanted markers and sonomicrometers as *gold standards*.

3.1 Implanted Image-Opaque Markers:

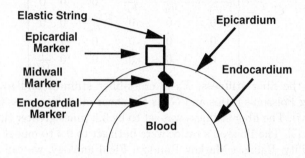

Fig. 2. Implantation of Image-Opaque Markers. This figure shows the arrangement of markers on the myocardium. First a small bullet-shaped copper bead attached to an elastic string was inserted into the blood pool through a needle track. Then the epicardial marker was sutured (stitched) to the myocardium and tied to the elastic string. Finally, the midwall marker was inserted obliquely through a second needle track to a position approximately half-way between the other two markers.

Methodology: To validate the image-derived strains, markers were implanted on canine hearts as follows: First the canine heart was exposed through a thoraco-tomy. Arrays of endocardial, midwall and epicardial pairs of markers were then implanted as shown in figure 2. They were loosely tethered, combinations of small copper beads (which show up *dark* in the MR images) at the endocardial wall and the midwall region and small plastic capsules filled with a 200:1 mixture of water to Gd-DTPA at the epicardial wall (which show up *bright* in the MR images). Marker arrays were placed in two locations on the canine heart wall. The location of each implanted marker is determined in each temporal frame by first manually identifying all pixels which belong to the marker area (as the marker 'image' extends to more than one voxel) and then computing the 3D centroid of that cluster of points, weighted by the grey level[1]. figure 3 shows a short-axis MR slice of the heart with the identified marker pixels shown in blue (left). The marker centroids are shown on the right.

Results: The image-derived strains, estimated using the algorithm described in section 2, were compared to strains derived from implanted markers[2]. In the case of the markers the strains were computed as follows using only the epicardial and endocardial markers. In each region of the LV that contained markers, groups of either 6 or 8 markers (depending on the geometry) were connected to form wedge

---

[1] In the case of dark markers the image is first inverted.

[2] Note that in the estimation of the image-derived strains, we do not use any infor-mation regarding the position of the implanted markers.

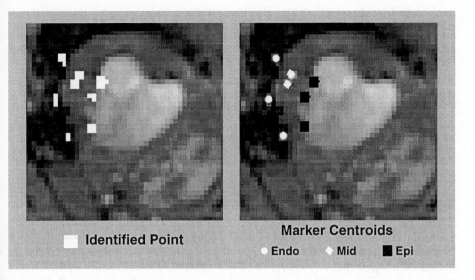

Identified Point

Marker Centroids

⬤ Endo ◆ Mid ■ Epi

Fig. 3. Localization of implanted markers. Arrays consisting of 12 markers each were placed at two positions on the left ventricle. In this figure, we show the portion of one marker array as it intersected a short-axis MR image slice. A human observer identified the pixels corresponding to each marker (left) and the marker positions (right) were found by calculating centroids of these points.

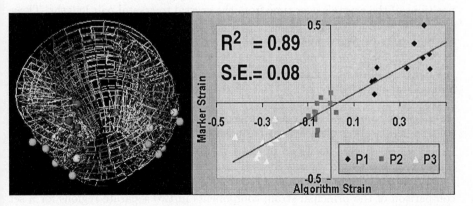

Fig. 4. Algorithm-derived Strains vs. Implanted Marker-derived Strains. **Left:** Reconstructed LV volume from cine-MRI at ED with marker positions noted as spheres. **Right:** Scatter plot of principal strains derived from baseline and post-infarction cine-MRI studies using algorithm vs. same strains derived from implanted marker clusters at two positions in the LV wall for $N = 4$ dogs (There was a total of 12 usable extracted marker arrays).

or hexahedral elements. Given the known displacements, we then calculated the strains between these markers. These strains were compared to the average strains in the elements contained within each marker array. We used principal strains, (these are the eigenvalues of the strain tensor[9]), as the marker arrays where large and included elements where the cardiac-specific directions varied widely.

Comparison results are shown in figure 4 for $N = 4$ dogs (2 acquisitions per dog, one pre-occlusion and one post-occlusion). We observe a strong correlation of the principal strain values ($r^2 = 0.89$).

3.2 Sonomicrometers

Methodology: In the case of the 3DE images we validate the strain estimates using implanted sonomicrometers (Sonometrics Corporation, London Ontario, Canada.) The canine heart is again first exposed through a thoracotomy. With the aid of an implantation device constructed in our laboratory, two crystal-arrays each consisting of 12 crystals (3 sub-epicardial, ∼2.0 mm, 6 mid-wall and 3 sub-endocardial, ∼0.75 mm diameter) were placed in the heart wall. Finally, to define a fixed coordinate space, 3 crystals attached to a plexi-glass frame were secured in the pericardial space under the right ventricle.

Digital sonomicrometry employs the time of flight principal of ultrasound to measure the distance between a transmitter and a receiver. A total of 27 crystals are used in each study. The distances between all possible pairs of crystals are recorded along with LV and aortic pressure at a sampling frequency of greater than 125 Hz. There are a number of preprocessing steps involved in obtaining the positions of the crystals over time from the crystal to crystal pair lengths. These are described by Dione[4] (see also Ratcliffe[15].) The efficacy of this technique was illustrated by additional work [10] that showed that the distances obtained with sonomicrometers compared favorably ($r = 0.992$) with those obtained using the more established technique of tracking implanted bead displacements using biplane radiography.

Results: We compared our image-derived strains, estimated using the algorithm described in section 2, to concurrently-estimated strains derived from sonomi-crometers at several positions in the LV myocardium in the same dogs. The sonomicrometers were visually located from the images and the two nearest sec-tors [12] of algorithm-derived strains were selected for comparison purposes. The comparison of the principal strain components in two separate regions for a set of 3 studies showed a strong correlation ($r^2 = 0.80$). Here we again compare the principal strains as in the last section. A scatter plot of algorithm-derived princi-pal strains versus sonomicrometer-derived principal strains is shown in figure 5.

4 Conclusions

In this work, we describe methodology to validate 3D image-derived cardiac de-formation estimates, and we use it to compare the output of our algorithm[12]

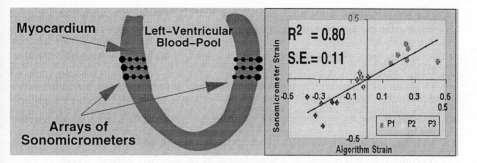

Fig. 5. 3DE Algorithm-Derived Strains vs. Sonomicrometer-derived Strains. **Left:** Placement of arrays of sonomicrometers in the Left Ventricular Wall (the crystals are shown enlarged, they are barely visible in the images.) **Right:** Scatter plot of principal strains derived from these 3DE studies using the algorithm vs. same strains derived from sonomicrometer arrays (12 crystals in each cluster) at two positions in the left ventricular wall. Note the high correlation between the two sets of strain values ($r^2 = 0.80$).

to implanted markers and sonomicrometers. We demonstrate good correlation of the image-derived estimates of strain, from both cine-MRI and 3D Echocardiography, to strains estimated from implanted markers and sonomicrometers, used as *gold standards*. At this point we only compare transmurally averaged principal strains between end-diastole and end-systole. In the future we hope to extend the validation to account for more time points and also to non-transmurally averaged cardiac specific strains.

References

1. A. A. Amini, Y. Chen, R. W. Curwen, V. Manu, and J. Sun. Coupled B-snake grides and constrained thin-plate splines for analysis of 2D tissue defomations from tagged MRI. *IEEE Transactions on Medical Imaging*, 17:3:344–356, June 1998.
2. T. Arts, W. Hunter, A. Douglas, A. Muijtens, and R. Reneman. Description of the deformation of the left ventricle by a kinematic model. *J. Biomechanics*, 25(10):1119–1127, 1992.
3. K. Bathe. *Finite Element Procedures in Engineering Analysis*. Prentice-Hall, New Jersey, 1982.
4. D. P. Dione, P. Shi, W. Smith, P. De Man, J. Soares, J. S. Duncan, and A. J. Sinusas. Three-dimensional regional left ventricular deformation from digital sonomicrometry. In *19th Annual International Conference of the IEEE Engineering in Medicine and Biology Society*, pages 848–851, Chicago, IL, March 1997.
5. J. M. Guccione and A. D. McCulloch. Finite element modeling of ventricular mechanics. In P. J. Hunter, A. McCulloch, and P. Nielsen, editors, *Theory of Heart*, pages 122–144. Springer-Verlag, Berlin, 1991.
6. E. Haber, D. N. Metaxas, and L. Axel. Motion analysis of the right ventricle from MRI images. In *Medical Image Computing and Computer Aided Intervention (MICCAI)*, pages 177–188, Cambridge, MA, October 1998.
7. W. S. Kerwin and J. L. Prince. Cardiac material markers from tagged MR images. *Medical Image Analysis*, 2(4):339–353, 1998.

8. D. L. Kraitchman, A. A. Young, C. Chang, and L. Axel. Semi-automatic tracking of myocardial motion in MR tagged images. *IEEE Transactions on Medical Imaging*, 14(3):422–433, September 1995.

9. L. E. Malvern. *Introduction to the Mechanics of a Continuous Medium*. Prentice-Hall, Englewood Cliffs, New Jersey, 1969.

10. D. Meoli, R. Mazhari, D. P. Dione, J. Omens, A. McCulloch, and A. J. Sinusas. Three dimensional digital sonomicrometry: Comparison with biplane radiography. In *Proceedings of IEEE 24th Annual Northeast Bioengineering Conference*, pages 64–67, 1998.

11. X. Papademetris, J. V. Rambo, D. P. Dione, A. J. Sinusas, and J. S. Duncan. Visually interactive cine-3D segmentation of cardiac MR images. *Suppl. to the J. Am. Coll. of Cardiology*, 31(2. Suppl. A), February 1998.

12. X. Papademetris, A. J. Sinusas, D. P. Dione, and J. S. Duncan. 3D cardiac deformation from ultrasound images. In *Medical Image Computing and Computer Aided Intervention (MICCAI)*, pages 420–429, Cambridge, England, September 1999.

13. X. Papademetris, A. J. Sinusas, D. P. Dione, and J. S. Duncan. Estimation 3D left ventricular deformation from echocardiography. *Medical Image Analysis*, in-press.

14. J. L. Prince and E. R. McVeigh. Motion estimation from tagged MR image sequences. *IEEE Transactions on Medical Imaging*, 11:238–249, June 1992.

15. M.B. Ratcliffe, K.B. Gupta, J.T. Streicher, E.B. Savage, D.K. Bogen, and L.H. Edmunds. Use of sonomicrometry and multidimensional scaling to determine the three dimensional coordinates of multiple cardiac locations: feasibility and initial implementation. *IEEE Trans Biomed Eng*, 42:587–598, 1995.

16. P. Shi, A. J. Sinusas, R. T. Constable, and J. S. Duncan. Volumetric deformation analysis using mechanics–based data fusion: Applications in cardiac motion recovery. *International Journal of Computer Vision*, 35(1):65–85, November 1999.

17. P. Shi, A. J. Sinusas, R. T. Constable, E. Ritman, and J. S. Duncan. Point-tracked quantitative analysis of left ventricular motion from 3D image sequences. *IEEE Transactions on Medical Imaging,*, 19(1):36–50, January 2000.

18. E. Waks, J. L. Prince, and A. Douglas. Cardiac motion simulator for tagged MRI. In *Mathematical Methods in Biomedical Imaging Analysis*, pages 182–191, 1996.

19. A. A. Young, L. Axel, and et al. Validation of tagging with MR imaging to estimate material deformation. *Radiology*, 188:101–108, 1993.

Automating 3D Echocardiographic Image Analysis

Gerardo I. Sanchez-Ortiz*, Jérôme Declerck,
Miguel Mulet-Parada and J. Alison Noble

Medical Vision Laboratory, Department of Engineering Science,
University of Oxford, Parks Road, Oxford OX1 3PJ, UK.
http://www.robots.ox.ac.uk/~mvl

Abstract. 3D echocardiography is a recent cardiac imaging method actively developed for quantitative analysis of heart function. A major barrier for its use as a quantitative tool in routine clinical practice is the absence of accurate and robust segmentation and tracking methods necessary to make the analysis fully automatic. In this article we present a fully-automated 3D echocardiographic image processing protocol for assessment of left ventricular (LV) function. We combine global image information provided by a novel multi-scale fuzzy-clustering segmentation algorithm, with local boundaries obtained with phase-based acoustic feature detection. We fit and track the LV surface using a 4D continuous transformation. To our knowledge this is the first report of a completely automated method. The protocol is viable for clinical practice. We exhibit and compare qualitative and quantitative results on three 3D image sequences that have been processed manually, in semi-automatic manner, and in fully automated fashion. Volume curves are derived and the ejection fractions errors with respect to manual segmentation are shown to be below 5%.

1 Introduction

The last few years have seen the emergence of 3D echocardiography acquisition systems in the market. Methods of acquisition are improving (in terms of spatial and temporal resolution), moving now towards real-time volumetric acquisition. However, interpretation and analysis of the data is more complex and time consuming than for conventional 2D echocardiography. As recent research studies have shown [1–3], the use of three-dimensional data provides more precise information on the pathophysiology of the heart than conventional analysis of 2D views ([4–6] and references therein), especially for volume and ejection fraction calculation.

Previous work has shown the feasibility of reconstructing a three-dimensional surface of the heart from sparse views [2]. However, in that work, the amount of interaction required to obtain a reconstruction of the endocardium was very large. In [1], a 3D finite element mesh of the left ventricular (LV) myocardium is computed and used to perform strain analysis. The approach is interesting,

---

* E-mail: giso@robots.ox.ac.uk

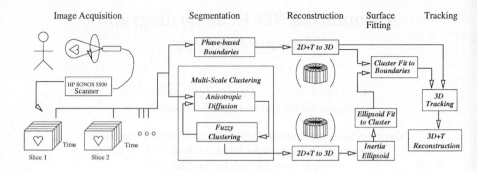

Fig. 1. Diagram shows the process for automatic analysis of 3D echocardiography.

however the analysis depended on the high quality images acquired from open-chest and is obviously not viable for routine protocol.

In this paper we present a framework for fully-automated 3D transthoracic echocardiographic image analysis, viable for routine clinical practice. The image data is acquired using a transthoracic 3D probe, the final result of the processing is a segmented left ventricular cavity over the whole cardiac cycle. No user interaction is necessary for this process. Previous work [3] from our laboratory has shown that it is possible to perform automatic tracking of the LV cavity, given a good segmentation of the image sequence and a correct initialisation for the tracker. This paper addresses especially these two issues. First, we initialise a 4D surface tracker [7] using a surface estimate of the LV cavity. The cavity surface estimate is obtained automatically with a novel method for multi-scale fuzzy-clustering based on combined ideas from [8, 9]. We then use the tracker to follow LV features detected with an acoustic-based edge detector for 2D cardiac sequences [6]. In the following sections we present the details of the method, and the experiments which have been conducted on three human healthy subjects. Left ventricular cavity volume curves are computed and compared to those of a manual segmentation performed by an expert.

2 Materials and Methods

The method consists of the following steps: first, feature points are extracted from the image to get a precise localisation of the endocardial boundary. Secondly, an approximate LV cavity is estimated using an iterative process of anisotropic diffusion and fuzzy clustering. Finally, a series of matching steps are required to a) generate a model of the initial endocardial surface and b) map this model through the cardiac sequence. Successful matching over the sequence requires a good initialisation shape, which is provided by the first matching. The procedure is schematised in Figure 1 and detailed in the following.

2.1 Image Acquisition

Digital 3D echocardiographic data was acquired on a HP SONOS 5500 ultrasound machine using a 3-5MHz rotating transducer, and stored as a sequence of

2D echograms (2D+T image), one for each probe angle (see Figure 2 for examples of 2D+T images). For normal studies we used 12 coaxial planes, one every 15 degrees (see Figure 3(a)); for one "dense" dataset we acquired 60 slices, one every 3 degrees. Data was acquired at a frame rate of 25 frames per second, the pixel size being 0.5mm x 0.33mm. Scanning was performed using ECG and respiration gating, on an apical view (*i.e.* the probe was located at the apex and roughly aligned with the LV long axis).

2.2 Phase-based Image Features Extraction

A phase-based spatio-temporal feature detection method is used to find candidate endocardial border points. An early implementation of the method used is described in [6] and the performance of the technique is tested in detail on clinical data in [10] within the context of 2D echocardiographic image tracking. Briefly, the idea is to detect endocardial border points according to their phase signature (edge shape) rather than intensity gradient information. The reason for this choice is that the acoustic reflection from the endocardial border varies according to the relative angle between the boundary and the transducer. The net effect is to produce a border with variable intensity contrast around its length. This makes it difficult to detect the endocardial border using an intensity-gradient based approach. This problem is even more of an issue in 3D echocardiography, since some 2D acquisition planes might be non-optimal because they are determined automatically once the first plane position has been chosen. In [6], feature asymmetry is proposed as an alternative measure for endocardial border detection and a spatio-temporal (2D+T) version of this idea is developed.

2.3 Scale-space Clustering for LV Region Extraction

In order to have an automated surface fitter that reliably finds and tracks the LV wall without human intervention we need boundary points to be present at reasonable intervals all over the LV surface. Because ultrasound images have a very low signal-to-noise ratio in regions of the LV wall parallel to the insonification beam, obtaining boundary points all over the endocardium is difficult even with the phase-based method. However, this problem can be overcome if the fitter is initialized with a surface that lies reasonably close from the target LV surface.

In order to extract an estimate of the wall in every region of the LV cavity, we use a multi-scale fuzzy clustering algorithm as a complementary segmentation method. This algorithm does not rely exclusively on the local differential structure of the data but takes into account the global characteristics of the image. In this way, a continuous approximation of the LV cavity boundary is provided even in regions of the images with low contrast and low signal-to-noise ratio. This also diminishes the effect of outliers detected by the phase-based method that correspond to noise and other anatomical structures. The results of the clustering method are then used to initialize the surface fitter in the manner described in Section 2.4.

Fuzzy Clustering We depart from Bezdek's fuzzy c-means clustering algorithm [11] which uses information about image attributes (like intensity) to divide the image domain into a pre-determined number of regions (clusters) and assigns every pixel in the image a degree of membership to the clusters, *i.e.* a probability of belonging to each of the regions. The algorithm groups the attributes by iteratively approaching a minimum of an energy function that measures the dissimilarity between the pixel attributes and those attributes of the cluster centres of each region (the energy function is typically a distance in the attributes' vector space).

The image attributes we use are intensity and position in an elliptic-cylindrical coordinate system, which is a natural choice for the 2D+T LV long-axis images (in [8] we investigated a similar approach for 3D MR images, with an elliptic coordinate system). The attributes are rescaled to homogenize their value ranges in the attribute space.

The origin of the elliptic-cylindrical coordinate system is first placed in the centre of the image (placing the left ventricle near the centre of the imaging window is therefore an image acquisition requirement, although anyway this is normally the case given that the left ventricle is the imaging subject), and then the position is refined by computing the centre of mass of the LV cluster. The cluster corresponding to the LV cavity is automatically identified as the one with the lowest intensity and position. In case some pixels belonging to this cluster are scattered around in the image (which rarely happens after the image has been smoothed as described in the next section), the largest connected component is computed to select only the points of the cluster belonging to the LV cavity.

None of the only two parameters to which clustering could be sensitive proved to be crucial to get a correct estimation of the LV cavity. These two parameters are, the number of clusters, and the geometry-intensity weight, which determines how to weight the geometric position with respect to the image intensity in the attributes space. For all the datasets (around 100 2D+T images corresponding to 3 studies), using 6 clusters and a weight of 1 gave satisfactory results. Small variations to these values only modify the quality of the segmentation results (*i.e.* how close the cluster approximates the LV cavity, but they do not, for instance, fuse the cavity and the background clusters). If precise identification of the LV is required, these values can be useful to weight geometry v. intensity in images with poor contrast or boundaries definition. However, a cluster that approximates the LV cavity is sufficient for our protocol.

Anisotropic Diffusion for Scale-Space Generation In order to overcome the problematic effect of intensity fluctuations of the noisy ultrasound images, the clustering process is performed at different levels of resolution in a scale-space of the image [12]. The scale-space is generated using the knowledge-based anisotropic diffusion (KBAD) algorithm [9]. Anisotropic diffusion algorithms smooth the image intensity (I) while preserving sharp edges by using the heat diffusion equation $\frac{\partial I(x,y)}{\partial \tau} = \nabla \cdot (c(x,y) \nabla I(x,y))$, where τ is the diffusion time (related to the scale of resolution) and the conductance c is normally a monotonically decreasing function of the magnitude of the intensity gradient

$(c = c(\|\nabla I\|))$. In the KBAD scheme the conductance term is a tensor and an explicit function of the the position $\mathbf{p}$ (in 2D+T in our image domain), the image intensity and its gradient, *i.e.* $\mathbf{C} = \mathbf{C}(\mathbf{p}, I, \nabla I)$. It can therefore incorporate *a priori* and *a posteriori* information of the geometric and dynamic characteristics of the image. It can also be used to introduce a probabilistic measure of the image intensity distribution [8,13], as explained in the next section.

Multi-Scale Fuzzy-Clustering The KBAD scheme used for generating the scale-space gets feedback from the clustering in progress. The fuzzy classification of the image domain provides a measure of the *a posteriori* probability that neighbouring pixels belong to the same tissue type, and is therefore incorporated into the diffusion process by means of the conductance function, penalizing or encouraging diffusion between pixels depending on the probability of them to belong to the same cluster. *A priori* knowledge about the system is also introduced by choosing the elliptic-cylindrical coordinate system for the clustering attribute space.

The clustering is updated at regular intervals during the diffusion process (*i.e.* at different levels of the scale-space), and the initially coarse segmentation of the image is gradually improved until it converges to a meaningful region partition, as the smoothing action of the diffusion process clears the image from noise. The first clustering is done after some iterations of the diffusion, then repeated at regular intervals. The computational expense of repeating the clustering at different scales of resolution is not high because energy minimisation is faster in the lower dynamic range of the smoothed image. Since the segmentation does not have to be precise, the process is performed on subsampled images (reduced by a factor 8) making processing time shorter (a 2D+T image can, for instance, be processed in under half a minute in a O2 SGI work station).

The combined diffusion-clustering algorithm penalizes inter-cluster diffusion and encourages intra-cluster diffusion, resulting in homogeneous intensity clusters with high contrast between them. The two driving mechanisms of the diffusion process are, on the one hand, the gradient based function governed by the differential structure of the image intensity function, and on the other hand, the intensity and spatial based clustering through which knowledge has been introduced by the elliptic-cylindrical symmetry of the image. Since the clustering scheme uses non-local information in order to perform the classification, the diffusion process is enriched with information about the global characteristics of the image.

2.4 Surface Fitting and Motion Tracking

The phase-based feature points obtained for each of the 2D+T images are reconstructed in 3D (*i.e.* slices placed coaxially). The LV cluster is reconstructed in 3D in the same manner, and then used to create an ellipsoid of inertia by computing the eigenvectors of its mass (points) distribution. The surface of the LV cavity at end diastole is obtained in three steps: i) reconstruction of the ellipsoid of inertia; ii) matching of this ellipsoid surface to the cluster points in order to get a first approximation of the shape of the cavity; and iii), matching of the latter to the feature points detected with the feature asymmetry algorithm.

This LV cavity shape is then sequentially fitted to the feature points at all time frames. A 3D matching is initialised with the result obtained at the previous time frame. Each of these matching steps is processed using a variant of the method described in [7]: a surface S_1 in an Image1 (either the ellipsoid, the manually segmented shape or the cluster shape) is deformed using B-spline tensor products to some feature points extracted from an Image2 (these being the boundary of the cluster or the points obtained as described in the segmentation section). The geometric transformation $\mathbf{f}$ that maps the points in Image1 to Image2 is such that the cost function defined by $C(\mathbf{f}) = \sum_{\mathbf{M_i} \in S_1} d[\mathbf{f}(\mathbf{M_i}) - CP_2(\mathbf{f}(\mathbf{M_i}))]^2$ is minimized, where $\mathbf{M_i}$ is a feature point from S_1, and the function CP_2 finds the closest feature point in Image2. d is a distance measure between two points. Computations of the local B-spline transformation is iterated until convergence is achieved. Details of the subset selection and minimisation process are outlined in [7]. As a final refinement, the control points of the 3D B-spline tensor products that define the deformations at each time frame are interpolated over time using a periodic temporal B-spline. The final deformation of the surface over time is therefore a continuous 4D (3D+T) tensor product of B-splines from which valuable dynamic information can be computed.

3 Results and Discussion

3.1 Phase-based Feature Points Extraction

An example of feature detection on a slice at one time frame is shown in Figure 2(b), and a representation of the detected points over a spatio-temporal sequence is shown in Figure 2(c). Points are generally well defined over most of the endocardium, but a significant amount of spurious edges belonging to other anatomical structures or produced by noise and image acquisition artifacts are also detected and might perturb the motion tracking process. For instance, in Figure 2(g) we can see some boundary outliers in an image with a strong intensity inhomogeneity at the centre of the LV cavity (the inhomogeneity is due to a bad gain setting during image acquisition). For these reasons a careful initialisation of the surface fitting is essential to avoid tracking boundary outliers.

3.2 LV Cavity Cluster

The 2D+T sequence is partitioned into clusters, as shown in Figure 2(d). The boundary of the central cluster (which corresponds to the LV cavity) is shown in Figure 2(e) and (f). Only the end-diastolic cluster is used for initialisation, but the clustering has been performed in 2D+T for more robustness. In Figure 2(h) we can see the clustering results on the image with the inhomogeneity artifact shown in Figure 2(g), and in Figure 2(i) we can see volume rendering results using maximum intensity projection of the same LV cluster in 2D+T space. Notice

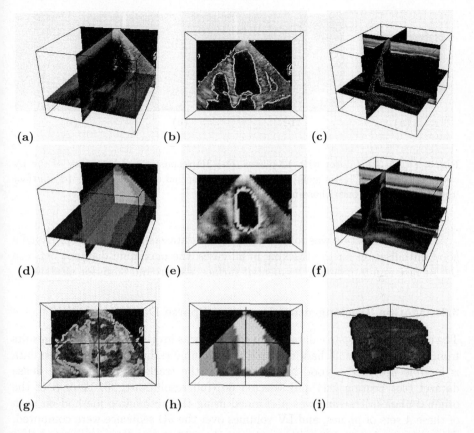

(a) (b) (c)

(d) (e) (f)

(g) (h) (i)

Fig. 2. (a) Example of original 2D+T image (depth is time). Feature detection overlaid on the original 2D+T image, at a particular time frame (b) and over the sequence (c). (d) Clustering results (depth is time), showing the LV cavity cluster in yellow (light grey). The boundary of the LV cluster is overlaid on the subsampled image at a particular time frame (e) and over the sequence (f). (g) Example of low quality ultrasound image with phase-based edges overlaid. (h) Clustering results of the same image (showing that the LV cluster is a good approximation to the cavity), and (i) volume rendering (maximum intensity projection) of the LV cluster in 2D+T space.

that the clustering algorithm works fine in spite of the intensity inhomogeneity (because it uses geometric as well as intensity information).

Once the 2D+T clusters have been computed for all planes of the image (all probe angles as shown in Figure 3(a)), the clusters are reconstructed in 3D (Figure 3(b) shows the boundaries of the reconstructed 3D cluster). The ellipsoid of inertia is computed from these points. This computed ellipsoid is then deformed onto the cluster points as shown in Figure 3(c). Finally, this deformed shape is deformed again onto the feature asymmetry points (Figure 3(d)). This last surface defines the shape of the LV cavity which is used for motion tracking.

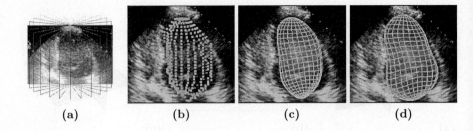

  (a)   (b)   (c)   (d)

Fig. 3. (a) A 3D dataset with 12 planes. (b) 3D reconstructed boundary of the LV cavity cluster. (c) Cluster's ellipsoid of inertia deformed onto the cluster. (d) Surface from (c) deformed onto phase-based edges.

The reconstruction was robust on the three datasets processed. It provided a good initialisation for the tracking in all cases: the maximum distance between the surface and a manually segmented surface was around 8mm for all data.

3.3 Motion Tracking using Dense vs. Sparse Data Acquisition

The effect of using sparse data on our protocol was investigated (previous results from our laboratory [3] had indicated that volume estimation from sparse data is feasible provided a good initialization for the tracker). We used the dense dataset (see Section 2.1) and selected 60 (full resolution), 30, 12 or 6 of the original planes. Tracking was performed using the automated method on each of these 4 sets of planes, and LV volumes over the 3D sequence were computed. The maximum volume variation between the sets at any given time was below 6% of the volume. In order to be consistent with the other acquisitions, all other results using this dataset (patient (1)) were obtained using only 12 planes.

3.4 Comparing User Guided and Fully Automatic Processing

Four tracking processes are compared on each of the three datasets: a) Manual segmentation performed by an expert for the complete 3D sequence; b) automatic tracking using manual segmentation of the end-diastolic surface as initialisation; c) automatic tracking using as an initial surface an ellipsoid manually placed by an expert with the help of a 3D graphic interface; and d) the fully automatic method, *i.e.* initialised using clustering as described in Sections 2.3 and 3.2. Figure 4 shows the volume curves obtained with these four methods on the three datasets. If the initialisation is not sufficiently close to the target LV cavity (method (c)), the tracking fails and the volumes are meaningless. On the other hand, the trackings (b) and (d) give similar results, showing that clustering based automatic initialisation gives as good results as initialisation from manual segmentation.

While initialising the tracker with a surface close to the actual LV reduces errors due to the sparsity of the data (see [3]), our results show a systematic

695

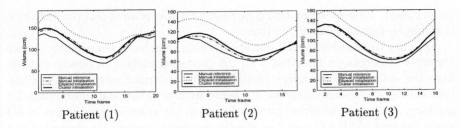

Patient (1)   Patient (2)   Patient (3)

Fig. 4. The plots show volume curves for three different patients. In each of them, the full line shows results of manual segmentation. Using the automatic tracking, three initialisation shapes were used: a manual segmentation (dashed lines), a manually selected ellipsoid (dotted lines), and the automatic cluster's inertia-ellipsoid deformed to the cluster (thick lines).

Fig. 5. The table shows the ejection fractions computed for the three patients using all four reconstruction methods described in relation to Figure 4.

Method	Ejection Fraction (%)		
	Patient (1)	Patient (2)	Patient (3)
Manual segmentation	51	45	54
Manual initialisation	46	43	52
Ellipsoid initialisation	38	37	46
Cluster initialisation	46	41	54

overestimation of the volume. The main reason is that endocardial boundaries are not clearly defined in some regions of the lateral wall, often shadowed by the rib cage. In these regions the tracker followed spurious edges located outside the cavity. However, the computed ejection fractions are, in the case of the fully automatic method, less than 5% over the value obtained from the manual segmentation (see table in Figure 5).

4 Conclusions and Future Work

We demonstrated the feasibility of an automatic method to track endocardial boundary from 3D transthoracic ultrasound data. We first use *multi-scale clustering* to provide an approximate segmentation of the LV cavity. This estimated shape is used to initialise the *3D fitter and tracker*, which then follows the precise boundary candidates obtained with the *phase-based* method. Using local and global information from the 3D image sequences, the combination of these three techniques overcomes the problem of tracking sensitivity to the initial shape: we obtained very similar tracking results using a manually defined surface and the automatically defined one. Ejection fractions computed with the fully automatic method were less than 5% over the value obtained using manual segmentation by an expert. No user interaction is necessary for this process and the technique is viable for routine clinical practice.

In order to improve the LV surface tracking (and volume estimate) we are currently developing a filter based on the clustering results, to remove outliers from the endocardial boundary candidates. Also, normal vectors to the edges provided by the feature detector are being used as additional information for tracking. We have also investigated segmenting the reconstructed image in 3D space (instead of 2D+T) using an elliptic coordinate system for clustering and a 3D version of the feature detector. Results show that the method could also be used on 3D images obtained with a different acquisition protocol (*e.g.* using a real time 3D transducer-array).

Results from a clinical study comparing volumes obtained with our protocol and those from SPECT Multi-gated Acquisition (MUGA) will be available by the time of the meeting. This research is funded by grants from the EPSRC (grant GR/L52444) and the MRC (grant G9802587).

References

[1] X. Papadimetris, A. Sinusas, D. Dione, and J. Duncan. 3D cardiac deformation from ultrasound images. In *MICCAI*, pages 420–429, Cambridge, UK, Oct 1999.

[2] F. Sheehan, E. Bolson, R. Martin, G. Bashein, and J. McDonald. Quantitative three-dimensional echocardiography: methodology, validation and clinical applications. In *MICCAI*, pages 102–109, Boston, MA, USA, Oct. 1998.

[3] G.I. Sanchez-Ortiz, J.A. Noble, G.J.T. Wright, J. Feldmar, and M. Mulet-Parada. Automated LV motion analysis from 3D echocardiography. In *MIUA*, Oxford, UK, July 19-20 1999.

[4] A. Giachetti. On-line analysis of echocardiographic image sequences. *Medical Image Analysis*, 2(3):261–284, 1998.

[5] G. Jacob, A. Noble, M. Mulet-Parada, and A. Blake. Evaluating a robust contour tracker on echocardiographic sequences. *Medical Image Analysis*, 3(1):63–76, 1999.

[6] M. Mulet-Parada and J.A. Noble. 2D+T Boundary Detection in Echocardiography. In *MICCAI*, pages 186–196, Cambridge, Mass. U.S.A, Oct. 1998.

[7] J. Declerck, J. Feldmar, M.L. Goris, and F. Betting. Automatic registration and alignment on a template of cardiac stress and rest reoriented SPECT images. *IEEE Transactions on Medical Imaging*, 16(1):1–11, 1997.

[8] G.I. Sanchez-Ortiz. Fuzzy Clustering Driven Anisotropic Diffusion: Enhancement and Segmentation of Cardiac MR Images. In *IEEE Nuclear Science Symposium and Medical Imaging Conference*, volume 3, pages 1873–1875, Toronto, Canada, Nov. 1998.

[9] G.I. Sanchez-Ortiz, D. Rueckert, and P. Burger. Knowledge-based tensor anisotropic diffusion of cardiac MR images. *Medical Image Analysis*, 3(1):77–101, 1999.

[10] M. Mulet-Parada. *Intensity independent feature extraction and tracking in echocardiographic sequences*. PhD thesis, University of Oxford, 2000.

[11] J.C. Bezdek and P.F.Castelaz. "prototype classification and feature selection with fuzzy sets". *IEEE Trans. on systems, man and cybernetics*, SMC-7:87–92, 1977.

[12] B.M. ter Haar Romeny (Ed.). *Geometry-Driven Diffusion in Computer Vision*. Computational Imaging and Vision. Kluwer Academic Publishers, 1994.

[13] S.R. Arridge and A. Simmons. Application of multi-spectral probabilistic diffusion to dual echo mri. In *MIUA*, Leeds, UK, 1998.

MyoTrack: A 3D Deformation Field Method to Measure Cardiac Motion from Gated SPECT

Jean-Philippe Thirion[1] and Serge Benayoun[1]

[1] HealthCenter Internet Services, 449 route des crêtes,
06560 Sophia-Antipolis, France
jean-philippe.thirion@healthcenter.com

Abstract. We are presenting a new method for Gated SPECT analysis. We segment the myocardium walls in the first time frame (end diastole) and then compute volumetric motion fields between time frames for tracking. The application of dense motion field to segmented walls provides us with individual 3D trajectories for endo- and epicardium wall points. Besides traditional measurements like ejection fraction, radial motion and thickness variations, it allows us to derive original measurements such as focus of contraction, axial contraction, elevation and twist, and compute an average image where cardiac motion blur is suppressed. The method is packaged in a comprehensive research prototype, MyoTrackVB, which is currently evaluated by medical partners with more than 200 cases already processed.

1 Introduction

Many different imaging modalities have been used to study human heart motion: Nuclear Medicine (first path, 2D and 3D blood pool, gated SPECT), MRI (Gated MRI, Tagged MRI, Phase contrast MRI), Gated CT, Ultrasound, angiography.... SPECT imagery suffers from low resolution and high level of noise in the images and medical researchers interested in heart motion analysis generally disregard Gated SPECT to the benefit of other more contrasted but most costly or invasive modalities. However, from a clinical standpoint, Gated SPECT is one of the most widespread exam for wall motion analysis. It can be performed simultaneously to a Myocardial Perfusion Scintigraphy (MPS) study without much additional cost. One advantage is that wall motion information is obtained in the same reference frame than perfusion. Our method, *MyoTrack*, is the adaptation of a generic volumetric motion field technique (see [16]) to the case of Gated SPECT images and is based on tracking. In contrast, the majority of existing Gated SPECT techniques are based on independent segmentations of time frames or measurements dedicated to very peculiar indices such as the "ejection fraction" (EF). In that sense, we are much closer to methods developed for MRI or CT. We describe first the physical particularities of Gated SPECT as well as the clinically relevant measurements used to analyze Cardiac Motion. We summarize the principles of existing analysis methods. We then describe MyoTrack and explain how physical constraints of Gated SPECT are taken into account. Finally, we present quantitative results obtained from diseased and normal patients.

2 Specificity of Gated SPECT image sequences

Gated SPECT is a 3D Computed Tomography technique where acquisition is gated with the ECG. As images are obtained by averaging a lot of cardiac cycles, the image sequence is cyclic, first frame being the successor of the last one. It is measuring photon emitted by the destruction of radionuclides whose concentration can be considered as constant during the acquisition and proportional to the perfusion of the heart muscle (mainly the myocardium of the left ventricle). The heart being a muscle, it is incompressible which means that the divergence of the deformation field is null within the myocardium. As intensity in the images is proportional to perfusion, it means that we should measure constant intensity within the myocardium.

One major drawback of Gated SPECT is that the acquisition device has a large point spread function, which means limited resolution but also changes in image maxima when the myocardium is contracting: a larger wall appears brighter in the image. A normally perfused myocardium has a uniform perfusion distribution, but unfortunately diseased patients are generally presenting large perfusion defects where very little information is available. As noise level is high and as extra activities in bowels or liver can blend with the myocardium, automatic segmentation is really challenging (Figure 1).

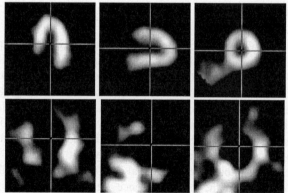

Figure 1: Horizontal long, vertical long and short axes of a normal subject (top) and a diseased subject (bottom)

3 Existing methods for Cardiac Dynamic analysis

One useful clinical index is the Ejection Fraction or EF, which is the ratio of the difference between End Diastolic Volume (EDV) and End Systolic Volume (ESV), divided by the EDV. The Stroke Volume (SV) is EDV-ESV. A lot of clinical papers are dealing with semi-automatic or automatic determination of EF (see for example [1],[2],[3]). Other useful clinical measures are radial motion and wall thickening (see [4],[5],[6]). Existing Gated SPECT analysis techniques can be classified into physical measurement of wall boundary positions, and into evaluation of mid-wall and wall thickness. Most methods rely on a frame by frame analysis based on independent segmentation of the myocardium. In [5] temporal analysis of intensity variations is related to width variations by empirical studies on point spread function effects. MyoTrack is based on tracking and is using comprehensively the whole sequence. It is much closer to methods developed for other modalities such as Gated MR ([7]), Tagged MR ([8], [9],[10]), Phase Contrast MR ([11]), Gated CT ([12],[13]), echocardiography ([14]) or PET ([15]). Tracking methods can be classified into 3D surface-based tracking and 3D dense motion field computation. MyoTrack is closer to motion field methods that we are now detailing.

The work of Song and Leahy [12] is the application of 2D+T optical flow techniques to 3D+T and is heavily relying on fluid mechanics. Two constraints are of particular interest to our problem, which are the conservation equation and the incompressibility constraint. The method is iterative (conjugate gradient descent) and solve for the two constraints (conservation and incompressibility) as well as a smoothness constraint. It is applied to Gated CT images. In the same spirit, Benayoun and Ayache [13] are measuring 3D dense deformation field. Aside from smoothness, the other constraint is the tracking of differential singularities. Although it is less physical than density conservation, it allows providing a better matching with respect to motion in the direction tangent to object boundaries. The cardiac application is also 3D Gated CT (dog heart). In the work of Klein [15], conservation and smoothness are solved for and the method is applied to 3D Gated PET images.

4   MyoTrack, a new method for Gated SPECT analysis

We have designed an iterative and multi-resolution generic method of motion field computation described in [16]. We have shown that 3D deformation computation can be also viewed as the "diffusion" of a deformable model through the contours of a reference object and that image driven constraints as simple as inward/outward unitary displacements along normals to contours, determined from tissue characterization, is sufficient to perform matching. We have also shown that image smoothing applied to the 3D deformation fields is an effective way to perform regularization. In [16], we have briefly described brain matching and gated SPECT analysis as potential applications, with some feasibility results. Since then, we have refined our technique in order to achieve a real clinical application for gated SPECT analysis that we call MyoTrack and which is presented here.

4.1 General principle

The general principle of MyoTrack is to separate 3D motion computation from myocardium segmentation (

Figure 2). The myocardium is segmented in the first image of the Gated SPECT sequence (generally the end diastole time frame) and then tracked throughout the whole sequence. The segmentation is based on template matching. The interest is twofold: first, it allows performing inter-patients comparisons as well as to design reference population models. It also helps providing a-priori knowledge for specific case interpretation as any pre-computation performed in the template's shape can be transferred to the specific case.

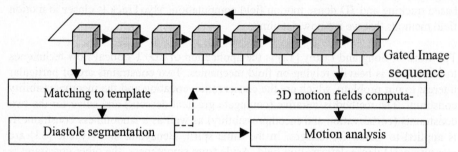

Figure 2: General principle. Segmentation and 3D image tracking are separated

4.2 Myocardium segmentation for end diastole time frame

Segmentation of the myocardium is a 4-steps method:
1. During the initialization phase an affine transformation is found between the end diastole image and a reference template shape. We are using an Iterative Closest Point method (ICP, see [17]) and are searching for an affine transformation. Two 3D deformable parameterized ellipsoids are defined, once for all in the template's reference frame.
2. The coupled ellipsoids are iteratively deformed to match to the edge points as well as to ensure smoothness constrains and anatomical a-priori constraints (for example, myocardium approximate width conservation). Sample points for regular angular subdivision in polar coordinates are used to represent deformed ellipsoids.
3. The base hole is segmented using a deformable 2D contour in polar representation.
4. A 3D closed surface representation in Cartesian coordinates is computed from the radius maps of endo- and epicardium positions and base definition.

4.3 Dense motion field computation

We only describe here the adaptation necessary from [16], which are intensity conservation, cyclicity, incompressibility and adaptive rigidity constraints. The output is, for each time frame, the node positions of a deformed 3D grid. The reference is a regular 3D grid defined in end diastole time frame. The method is iterative and multi-resolution, ending with a resolution as thin as voxel sizes.

1. Intensity conservation is ensured by a formula derived from optical flow and used for motion computation. s and m are the respective intensities in the reference and model images and $(m-s)^2$ is used to avoid singularities when $\vec{\nabla}s$ is small:

$$\vec{v} = \frac{(m-s)\vec{\nabla}s}{(\vec{\nabla}s)^2 + (m-s)^2}$$

2. To ensure cyclicity as well as regularity through time, we are restricting at each iteration the $x(t)$, $y(t)$, and $z(t)$ displacements of each node of the deformable grid to first harmonics.
3. Incompressibility corresponds to null divergence of the deformation field. We are measuring, at each iteration, the volume element of each cell of the grid (within the Myocardium). As the grid in the end diastole image is regular, this volume should remain one voxel cube. Homothetic variations are applied as additional constraints to make cell volumes tend toward unit values.
4. Our rigidity constraint is adaptive. The endocardium cavity is much more deformable than epicardium boundary thanks to the use of adaptive filtering to regularize deformation fields.

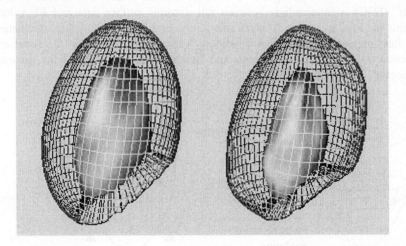

Figure 3: End diastole and end-systole surface representation (endo- and epicardium), with perfusion information.

4.4 Dynamic 3D surface representations

As they are defined from a regular grid in the end diastole frame, deformation fields can be applied to 3D surface representation of the myocardium walls. Each vertex of the surface representation follows a 3D closed trajectory deduced from the 3D deformed grids via tri-linear interpolation within the grid cells (see Figure 3). On the contrary to all traditional methods of gated SPECT analysis, and much closer to Gated MR or CT techniques, we are able to analyze other aspects than simply surface radial motion.

5   Motion analysis methods

MyoTrackVB has various outputs which are: dynamic 3D surface representation of the myocardium walls in Cartesian coordinates, dynamic Bull's eye representations (or polar maps) of wall positions and perfusion, measures obtained from dense 3D deformation field for all time frames....

5.1   Dynamic 3D surfaces, EF,...

Dynamic 3D surface representations can be manipulated and visualized using MyoTrackVB (see Figure 3) and convey important clinical information about myocardium function. To get more tractable information, we can also compute the endocardium volume for each time frame, which is giving EDV, ESV, SV and EF values. The "Focus of Contraction" (FOC) is intermediate between full dynamic surface data and EF value. We define the FOC as the locus which minimizes, in a least square sense, the distance to all lines supporting instantaneous velocity vectors. We found that the focus of contraction trajectory is very compact and situated near the center of the myocardium "equator" (see results section).

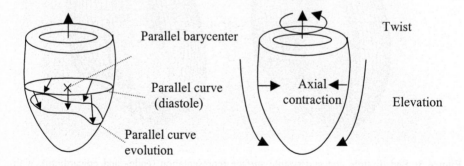

Figure 4: evolution of parallel lines to measure axial contraction, elevation and twist

MyoTrackVB can present also the amplitude and phase of radial motion and thickness variations in Bull's eye representation (Figure 5). Contrary to other Gated SPECT methods, each vertex of the surface representation is free to move in 3D, except for regularization constraints, which is changing definitions such as width variation which becomes the evolution of the distance between identified pairs of points instead of surface motion. 3D trajectories allow us to study other motion, such as axial contraction, elevation and twist. To define more precisely such quantities (see Figure 4), we are considering the evolution of "parallels", that is, sets of points having the same angle θ in polar coordinates in end diastole image and forming a closed curve. Parallel curve barycenter motion along the axis of the myocardium is the elevation, the square root of the surface variation of the minimal surface defined by the curve gives the axial contraction ratio and the "rotation" of the curve around the myocardium axis gives the twist component. These values are defined for all "latitudes" θ ranging from $-\pi/2$ to $+\pi/2$.

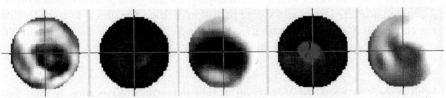

Figure 5: Left to right, thickness variation amplitude and phase, radial motion amplitude and phase, and perfusion.

5.2 Using the dense motion field computation

To reduce noise level, all time frames are summed to give a single static representation for MPS studies. However, cardiac motion blurs the image, which reduces the signal. As suggested in [16] or [15], we are using the computed motion field to compensate for cardiac motion to get a static image of the heart with a much higher contrast to noise ratio (see
Figure 6). The deformation fields are applied to all time frame images to make them superimposable to the end diastole image, and all deformed images are then summed. This operation is especially useful in the case of low count data.

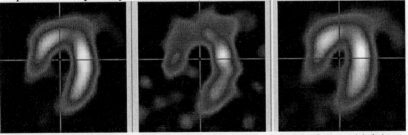

Figure 6: Diastole frame (middle), direct average of 8 time frames (right), average with motion compensation (left).

6   Results

We have performed several validation experiments of the Ejection Fraction computation with different medical partners. The Hôpital Henri Mondor has performed a blind validation on 41 severely diseaded patients, imaged two times with Thallium, at 20 minutes and at 4 hours (see [18]). Comparison is performed with ERNA (2D blood pool) and QGS ([2]). St Luke's Hospital, New York, as performed a blind validation with 57 single head MIBI data, with comparison with ERNA and with SPECT EF. We have also performed a validation (non-blind) with Emory University Hospital on 22 cases with comparison with gated MRI and with QGS. EF correlation between MyoTrack and other methods are ranging between 80% and 95%. In total, MyoTrack has been run fully automatically on more than 200 cases without major failure of segmentation or tracking (visual assessment with contours overlay). We are presenting results for 5 normal cases with a normal angiogram and good perfusion and 5 diseased cases with a severe infarct and visible perfusion deficits. The aim is to compare different outputs of MyoTrack in various conditions in order to evaluate, beside EF computations, which measurements has to be investigated with respect to further clinical validation.

With respect to Focus of contraction, a first interesting observation is that, for all cases (normal as well as abnormal) the trajectory of the FOC is a double loop, the larger one being covered during diastole and the smaller one during systole. We have computed the barycenter of each FOC trajectory as well as the average distance to the barycenter (compactness) and the Residual Mean Square (RMS) distance between the instantaneous velocity direction lines and the FOC, normalized using EDVs (focalization).

For Axial contraction, elevation and twist 1D profiles, we are presenting these profiles in Figure 7. The pattern is more coherent for normal than for diseased subjects. The clinical usefulness of these indices has still to be evaluated. It might be that, thanks to perfusion defects, the twist motion could be retrieved with more accuracy for diseased than for normal cases, which would be troublesome for clinical application, but the two other indices are very promising.

7 Conclusion

We have developed a new technique for the interpretation of Gated SPECT image sequences based on flow field techniques and inspired from methods developed for other kind of modalities such as Gated MRI or Gated CT. We have packaged the whole method into a fully automated research prototype called MyoTrackVB. MyoTrackVB is taking as input raw tomograms and has a large variety of possible outputs: average image with Cardiac motion blur compensation, 3D dynamic surfaces interactive display, amplitude and phase Bull's eyes for radial motion and thickness, axial contraction, elevation and twist 1D profiles, focus of contraction and Ejection Fraction and endocardium volumes (EDV, ESV) computations. These measures are now studied in collaboration with medical partners to determine their clinical relevance. The automation of the processing has already permitted to process more than 200 cases.

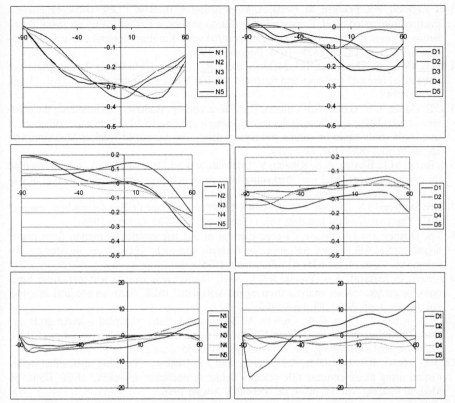

Figure 7: Top to bottom: axial contraction, elevation and twist for (left) normal subjects and (right) disease subjects.

Acknowledgments
Thanks to Drs. M. Meignan, J. Rosso and E. Itti from Hôpital Henri Mondor, Creteil, France, Drs. G. Depuey and M. Kamran from St'Lukes Hospital, New York, USA and Dr. J. Vansant and Mr. M. Blais from Emory University Hospital, Atlanta, USA, for collaborating with us on EF validation, and to Dr. Martin Kernberg, from the Department of Medicine, UCSF, California, and Dr. Michael Goris, Stanford University, for stimulating discussions about cardiac anatomy and SPECT.

8   References

1. E.G. Depuey, K. Nichols, and C. Dobrinsky, ``Left ventricular ejection fraction assessed from gated technetium-99m-sestamibi SPECT'', J. of Nuclear Medicine, vol. 34, no. 11, pp 1871--1876, November 1993.
2. Germano, H. Kiat, P.B. Kavanaugh, M. Moriel, M. Mazzanti, and H.T. et al. Su, ``Automatic quantification of ekection fraction from gated myocardial perfusion spect'' The Journal of Nuclear Medicine, vol. 36, pp. 2138--2147, 1995.

3. H. Everaert, P.R. Franken, M. Goris, A. Momen, and A. Bossuyt, ``Left ventricular ejection fraction from gated SPECT myocardial perfusion studies: a method based on radial distribution of count rate density across the myocardial wall'' European Journal of Nuclear Medicine, vol. 23, no. 12, pp. 1628--1633, December 1996.

4. M.L. Goris, C. Thompson, L.J. Malone, and P.R. Franken, ``Modelling the integration of myocardial regional perfusion and function'' Nuclear Medicine Communication, vol. 15, pp. 9--20, 1994.

5. C.D. Cooke, E.V. Garcia, S.J. Cullom, T.L. Faber, and I.P. Roderic, ``Determining the accuracy of calculating systolic wall thickening using a fast Fourier transform approximation: A simulation study based on canine and patient data'', The Journal of Nuclear Medicine, vol. 35, no. 7, pp. 1185--1192, July 1994.

6. I. Buvat, M.L. Bartlett, A.N. Kitsiou, V. Dilsizian, and S.L. Bacharach, ``A "hybrid" method for measuring myocardial wall thickening from gated pet/spect images,'' The Journal of Nuclear Medicine, vol. 38, no. 2, pp. 324--329, February 1997.

7. L.H. Staib and J.S. Duncan, ``Model-based deformable surface finding for medical images,'' IEEE Trans. on Medical Imaging, vol. 15, no. 5, pp. 720--731, October 1996.

8. T.S. Denney and J.L. Prince, ``Reconstruction of 3-d left ventricular motion from planar tagged cardiac MR images: An estimation theoretic approach,'' IEEE Trans. on Medical Imaging, vol. 14, no. 4, pp. 625--635, December 1995.

9. J. Park, D. Metaxas, A.A. Young, and L. Axel, ``Deformable models with parameter functions for cardiac motion analysis from tagged MRI data,'' IEEE Trans. on Medical Imaging, vol. 15, no. 3, pp. 278--289, June 1996.

10. A.A. Amini, Y. Chen, R.W. Curwen, V. Mani, and J. Sun, ``Coupled b-snake grids and constrained thin-plate splines for analysis of 2-d tissue deformation from tagged mri,'' IEEE Trans. on Medical Imaging, vol. 17, no. 3, pp. 344--355, June 1998.

11. F.G. Meyer, R.T. Constable, A.J. Sinusas, and J.S. Duncan, ``Tracking myocardial deformation using phase contrast MR velocity fields: A stochastic approach,'' IEEE Trans. on Medical Imaging, vol. 15, no. 4, pp. 453--465, August 1996.

12. S.M. Song and R.M. Leahy, ``Computation of 3-d velocity fields from 3-d cine CT images of a human heart,'' IEEE Trans. on Medical Imaging, vol. 10, no. 3, pp. 295--306, September 1991.

13. S. Benayoun and N. Ayache, ``Dense non-rigid motion estimation in sequences of medical images using differential constraints,'' International Journal of Computer Vision, vol. 26, no. 1, pp. 25--40, January 1998.

14. F.H. Sheehan, M.S. Bolson, R.W. Martin, G. Basheim, and J. McDonald, ``Quantitative three-dimensional echocardiography: Methodology, validation, and clinical application,'' in MICCAI'98, October 1998, vol. 1496 of Lecture Notes in Computer Science, pp. 102--109, Springer.

15. Gregory J. Klein, ``Forward deformation of pet volumes using material constraints,'' in IEEE Workshop on Biomedical Image Analysis (WBIA'98), B. Vemuri, Ed., Santa Barbara, California, June 1998, pp. 64--71, IEEE.

16. Jean-Philippe Thirion, ``Image matching as a diffusion process: an analogy with Maxwell's demons,'' the journal of Medical Image Analysis (MEDIA), vol. 2, no. 3, pp. 243--260, 1998.

17. J. Declerck, J. Feldmar, M.L. Goris, and F. B etting, ``Automatic Registration and Alignment on a Template of Cardiac Stress \& Rest SPECT Images,'' in Mathematical Methods in Biomedical Image Analysis, June 1996.

18. E. Itti, J. Rosso, M. Meignan, S. Benayoun and J-P. Thirion, "MYOTRACK: a new method to measure ejection fraction from Gated SPECT: preliminary evaluation of precision and accuracy with 201Tl", American Society of Nuclear Medicine SNM'2000, saint Louis, USA.

An Adaptive Minimal Path Generation Technique for Vessel Tracking in CTA/CE-MRA Volume Images

Brian B. Avants, James P. Williams

Imaging & Visualization Department - Siemens Corporate Research
755 College Road East, Princeton, NJ 08540
avants, jwilliams@scr.siemens.com

Abstract. We present an efficient method for the segmentation and axis extraction of vessels and other curvilinear structures in volumetric medical images. The image is treated as a graph from which the user selects seed points to be connected via 1-dimensional paths. A variant of Dijkstra's algorithm both grows the segmenting surface from initial seeds and connects them with a minimal path computation. The technique is local and does not require examination or pre-processing of the entire volume. The surface propagation is controlled by iterative computation of border probabilities. As expanding regions meet, the statistics collected during propagation are passed to an active minimal-path generation module which links the associating points through the vessel tree. We provide a probabilistic basis for the volume search and path-finding speed functions and then apply the algorithm to phantom and real data sets. This work focuses on the contrast-enhanced magnetic resonance angiography (CE-MRA) and computed tomography angiography (CTA) domains, although the framework is adaptable for other purposes.

1 Introduction

The domain for this investigation is the volumetric images produced by contrast-enhanced magnetic resonance angiography (CE-MRA) and computed tomography angiography (CTA.) In the CE-MRA imaging protocol, a contrast agent, usually based on the rare-earth element Gadolinium (Gd) (a highly paramagnetic substance), is injected into the bloodstream. In such images, blood vessels and organs perfused with the contrast agent appear substantially brighter than surrounding tissues. In CTA, a contrast agent is injected which increases the radio-opacity of the blood making the vessels appear dense.

The goal of the majority of CTA/CE-MRA examinations is diagnosis and qualitative or quantitative assessment of pathology in the circulatory system. The most common pathologies are aneurysm and stenosis caused by arterial plaques.

The modern clinical workflow for the reading of these images increasingly involves interactive 3D visualization methods, such as volume rendering, for quickly pinpointing the location of the pathology. Once the location of the pathology is determined, quantitative measurements can be made on the original 2D slice data or, more commonly, on 2D multi planar reformat (MPR) images produced at user-selected positions and orientations in the volume.

In the quantification of stenosis, it is desirable to produce a cross-sectional area/radius profile of a vessel so that one can compare pathological regions to patent (healthy) regions of the same vessel. In order to accurately measure cross-sectional area or radius a definition of the cross-sectional plane is required.

Segmentation research in this area has focused in recent years with the application of curve propagation methods (such as snake and active contours), region competition and mathematical morphology to analysis of the vessel tree [4,8]. Sethian [5,6] provides a fast marching method for vessel tree segmentation that requires pre-processing of the image. Kimia [7] and others have used balloons. Morphological methods, combined with the eikonal approximation, have also proven fruitful [3]. However, our approach uses segmentation simply as a learning procedure that provides both a metric and bounds for an explicit vessel centerline search. Vessel axes encode the vessel tree more sparsely. Our method makes the extraction of such abstractions sufficiently fast to make axis creation and manipulation part of a real-time, interactive clinical workflow.

The foundation given here provides a clinical framework that is related to the above work but novel in its design and in its efficiency. No pre-processing of the images is required. The interface is semi-automated in which the image reader places seed points within the vessel tree. A max-probability surface propagation algorithm then makes a dynamic segmentation of the vessels containing the seeds. Parameters are also set for the subsequent minimal path calculation used to approximate the vessel centerlines. This piece-wise linear approximation can then be used as a visual handle to the vessel region, as a navigation tool for automatic placement of MPR's orthogonal to the vessel and also as a starting point for further segmentation or automated quantification algorithms.

2 Methods: Segmentation Module

2.1 Graph Based Approach

In our application, we view the image volume as a k-regular graph (where 'k' denotes the number of edges at each node/voxel) through which we would like to compute a small set of the minimal connectivity. The user contributes the location of nodes to be connected through the image volume. The algorithm then consists of two parts: a volume search and a path search. The first part searches for the subset of the full image graph that is most likely to contain the paths between the user's seed points. The associated volumes are also given a statistical characterization. The edge cost function for the second phase is determined from the statistical distribution of the voxels explored during the first phase. The second phase computes the minimal connectivity of the selected vessel tree points according to parameters set by the partial segmentation results.

The constraint of the minimal path computation to only those regions that contain the objects to be connected is an important part of the efficiency and correctness of the algorithm if it is to operate on volumetric data. If unnecessary (non-object) regions of the image are searched, not only does the execution time rapidly increase (as the algorithm is O(NlogN)), but the chance that the path will mistakenly cross a object boundary increases. It is imperative, then, to restrain the search to only those regions in which it is probable to have a true path.

2.2 Surface Expansion

Our approach to the segmentation aspect of this problem is motivated by Sethian's fast marching method. He observes in [6] that the fast marching method, which is a level set surface propagation tool, is fundamentally the same as Dijkstra's algorithm in a regular graph. The former, however, computes velocities (inversely proportional to edge costs) incrementally according to the local solution to the eikonal partial differential equation (pde). The latter is defined, usually, over a simple network with fixed edge costs.

Sethian describes his method as producing a gradient field for the arrival time of the front such that:

$$|\nabla T| = \frac{1}{F} \tag{1}$$

In his application of the fast marching method on medical images [2], the velocity function is given by:

$$F = \frac{1 - \varepsilon\kappa}{(1 + \nabla G * I)} \tag{2}$$

The first term is an expansion term; the second term slows expansion where curvature, κ, is large. ∇G is the gradient of the image pre-processed with a gaussian filter. Time is zero at all initialized source points and arrival time is calculated such that T>0 at all successive iterations. The proof of the algorithm's correctness is given in [5].

We incorporate a similar approach to Sethian's, but do not focus entirely on segmentation. The surface propagation is simultaneously used to learn an appropriate data based cost function for calculating the minimal path. Also, in our case, the velocity model is probabilistic and non-viscous thereby allowing velocity to be determined by data at a particular voxel rather than the voxel's neighborhood. This probabilistic model effectively learns the intensity distribution of the segmented region and is subsequently used to generate a data-based closed form for the edge cost function.

We redefine the speed function such that we are examining the probability of the local data being included in a known set. Then, the speed function for each node is determined by maximizing the probability of advancement in each possible direction of new exploration. We sample the data locally according to Dijkstra's method (analogous to the eikonal model of front advancement, if connectivity is k-regular) and then define speed to be:

$$F_{ijk} = V * P(\text{data included in the known set}) \tag{3}$$

If V is 1 (expansive), then F_{ijk} can be written:

$$F_{ijk} = 1 - P(\text{data not included in the known set}) \tag{4}$$

and we see the parallel between this and Sethian's purely geometric $F = 1 - \varepsilon \kappa$. The change in time is then given by:

$$\frac{dT}{ds} = \frac{1}{\max(F_{ijk}(P))} \tag{5}$$

Initially the front will expand most rapidly in the most probable directions. After long times, however, lower probability regions may be explored. This fits with the intuitive notion that probable events occur often and improbable ones only at large intervals.

If the probability calculation varies only according to data at each voxel, and the exploration will execute in O(NlogN). If probability changes according to the delivery from the min-heap, the entire boundary should be recomputed (as it must be with pure level-set methods.) This guarantees the maximum probability exploration, given the elapse of time, but reduces efficiency with a worst-case bound of N^2.

Instead, we use a bounded-error approach where recomputation is performed only when potential maximum error exceeds a given threshold. This generates an error-bounded approximation of the maximum probability exploration with relatively little loss in efficiency.

2.3 Edge Cost / Velocity Model

The probability model given for the data determines our speed function. In vessel path finding for CTA/CE-MRA we are interested in quickly connecting vessel-like regions without crossing neighboring non-vessel regions. Stated another way, we are interested in discrimination between energy levels, vessel membership being analogous to a particular energy level. Therefore, we want rapid propagation through regions of similar energy and slow movement elsewhere.

2.3.1 Partitioning Distributions

We formulate the vessel search in terms of a thermodynamic partitioning problem. Sections of the image are allowed two possibilities, vessel and non-vessel, where the existence of one state excludes the other. Each image region, whether it is a single voxel or collection of voxels, is then treated as a canonical ensemble. The comparison of two regions is performed with the partition function [1] given by:

$$Q = \sum_{all-states} \exp (E_i / kT) \qquad (6)$$

Where each state is described by its energy E and all states share the constants kT.

Equilibrium requires that, given a system, we have a time independent minimum of information to describe it. That information for equilibrium systems (and our image) is given by N, the number of particles (proportional to the number of voxels), volume V (the sum of voxel volumes) and temperature T, which we can set. We assume that, from the macroscopic perspective, we can also measure the system's average energy.

2.3.2 The Fermi-Dirac Distribution

These assumptions lead directly to the Fermi-Dirac probability function for two states [1]:

$$P_{FD} = \frac{1}{1 + \exp\left(\dfrac{E_1 - E_0}{kT}\right)} \qquad (7)$$

P_{FD} is a step function with slope determined by the size of kT. The inflection point occurs where E_1 (the ensemble temperature) equals E_0, beyond which probability approaches one.

Therefore, the kT term determines the parsimony in the allocation of probability. Increasing kT makes the slope gentle, supplying slowly changing probabilities to a broad range of energies about E_1. As the temperature approaches absolute zero, the function approaches zero slope. Higher temperatures result in steep vertical slopes.

2.3.3 Implementation

In our framework, given fixed temperature, we can examine the change of an equilibrium system surrounded by a narrow border of ensembles that may or not contain the vessel. We choose a low temperature (generally, such that kT is within an order of magnitude of unity) to make propagation through high probability energy regions fast, and low probability regions slow.

Initialization occurs when the user selects seeds. Each seed is considered a vessel containing system for which we have N, V, T and the energy,

$$E_1 = \frac{1}{N} \sum_{\substack{ensemble\ members}}^{N} E_i(I) \tag{8}$$

which is the standard averaging procedure, where N=1 and E(I) is the intensity. These points are treated as ensembles as stated above. For each source, surrounding regions are explored and compared point-by-point to the expanding ensemble < E > with which they are in contact. The border point with the highest probability of being vessel tissue will have the smallest arrival time. That voxel is added to its source ensemble such that the latter becomes a system of size N+1. < E > is then re-measured and the process repeats.

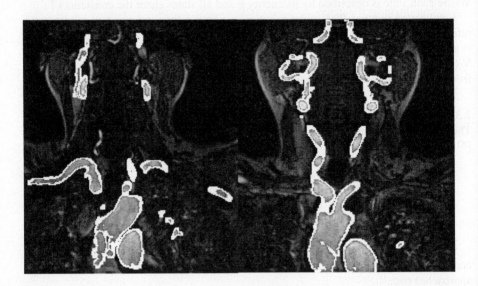

Figs. 1a (left),b (right). Slices from a segmented 3-D carotid CE-MRA data set illustrate the region explored and segmented by phase 1 of the algorithm. The explored region is highlighted by a white boundary.

2.4 Approximation Methods

We could simplify our calculation by weighting the source points with large N values and giving all border systems N = 1 such that adding a new voxel won't appreciably change E_1, the mean of the ensemble values. However, for calculating the average and standard deviation of the vessel intensity (needed for the path finding parameters), we would like more exact discrimination. We therefore may be forced to recompute border probabilities when we re-measure E_1. This increases the order by a factor of N. However, we can use the monotonicity of the p.d.f. as well as the relative stability of < E > over time to limit the frequency of re-computation. The derivative of P_{FD} is:

$$\frac{d}{dE} P_{FD} = \frac{-m}{(m + \exp(E_0 - E_1)(m + \exp(E_0 - E_1))} \tag{9}$$

where m = 1/kT. We first note that the derivative of our p.d.f. approaches the delta function if temperature decreases. As temperature increases, the derivative spreads and its magnitude decreases. Generally speaking, the error is only appreciable along the region of the p.d.f where there is a significant slope. For the low temperature fermi function, this region is narrow.

We can illustrate the nature of the measurement error as follows

$$P_{FD}(E_{+error}) = \frac{1}{1 + \exp(E_1(1+\varepsilon) - E_0)} \tag{10}$$

We see that our error can be thought of as occurring on the measurement of E_1. We can, then, limit our error concern to only those values of E_1 where the error might be appreciable. We expect to encounter three cases:

1) $E_0 > (E_1 + |\Delta E|)$, error rapidly diminishing as E_0 increases
2) $(E_1 - |\Delta E|) < E_0 < (E_1 + |\Delta E|)$, error significant
3) $E_0 < (E_1 - |\Delta E|)$, error rapidly diminishing as E_0 decreases

We store, for each E_0 probability computed, the value of E_1 against which it was compared. Then, at the time of delivery, we compute

$$Error \approx P_{FD}(E_0 \mid E_{1old}) - P_{FD}(E_0 \mid E_{1new}) \tag{11}$$

If this value is greater than χ, the error tolerance, we re-compute the probability and re-insert the voxel. This process is repeated for all delivered nodes with error greater than the tolerance. Generally, we apply a 5% rule to case 1 and 2 errors and 10 % to case 3 errors that could benefit from updating. Fortunately, not only is the range that we need to recompute small, but the value < E > stabilizes over time. In practice this means that the vast majority of border recomputations occur early in the expansion process when the boundaries contain relatively few nodes.

3 Methods: Path Finding Module

3.1 Path Finding Speed Function

The speed function for the path finding derives naturally from our former statements. Here, we do not have to choose between two states. The only state we are concerned with is the vessel state. Due to our surface exploration, then, we know the degeneracy (histogram) of energy levels in the vessel. So, we use that knowledge along with the canonical distribution to compute our speed function:

$$F_{ijk} = L * W_j * P_{canonical}(E_j) = L * W_j * \exp(-E_j / kT) \tag{12}$$

Where Wj is the degeneracy of energy Ej and L is an euclidean term that depends on the distance in patient space. The advantage that we have here is that we know all energies locally and can normalize the computation by the 'known' natural temperature of the distribution. It is proportional to the standard deviation of the intensities in the region, which we measured in our surface propagation algorithm. This guarantees that we have a provably minimal path that will move through the vessel region with a very high probability.

Even if there is error in the segmentation, the use of the canonical distribution for the speed function makes it highly unlikely the actual minimal path will deviate into

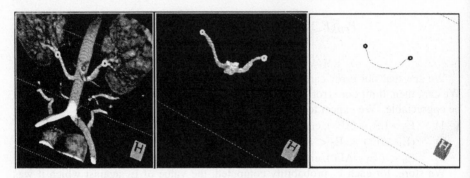

Fig. 2 (left), **Fig. 3 (center)**, and **Fig. 4 (left)** Volume-rendered MR angiography data. Renal arteries have been tracked by the selection of two seeds, one near each kidney. The segmented search region and extracted path are shown.

the erroneously segmented parts of the volume. This is because the canonical distribution peaks around < E > due to the large degeneracy of that energy level. [1]

3.2 Active Backtracking and System Equilibration

Although we instantiate a min-heap for each source, we compute the minimum over all heaps for the expansion of exploration. In this way, we have a virtual single min-heap. The movement of all surfaces, then, is governed by the interaction of the set of heaps. This, naturally, leads to surfaces in more homogeneous regions of the image propagating more rapidly and thus initiating meetings with other surfaces first.

Surface meetings occur when different surfaces share osculating border points. This is the condition for producing a path. We set the parameters for the path finding by measuring the degeneracy of the energy levels within each surface and computing the (approximate) standard deviation of the energies. An instance of Dijkstra's algorithm with the above speed function is then initiated from the osculating points backward into each surface thus producing a minimal path as it goes. The stopping condition for this search is the discovery of either an original source or a minimal path computed at a previous meeting.

After the paths are computed, the two osculating regions then share statistical information and merge into a single heap. Then, the algorithm continues as before, but with one less surface. The final termination condition is that all surfaces have met and all paths connecting them have been computed.

The data structures containing the vessel boundaries, paths and sources are maintained. This allows subsequent sources to be added to the image volume.

The algorithm will then begin a new set of surface searches and meetings will be handled in the same way as before, except that if the old surface is met, active backtracking is not symmetric. Instead, the new surface is allowed to find paths both backward into its volume and forward towards the paths computed in prior executions of the algorithm.

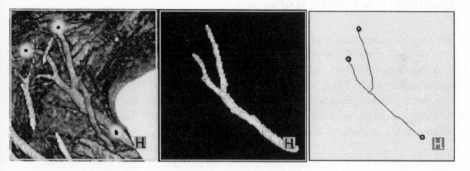

Fig. 5 (left) Volume-rendered CTA of the coronary arteries and 3 seed points. **Fig 6 (center)**, Segmented exploration region **Fig 7 (right)** Axes obtained from the search.

4 Performance Evaluation

4.1 Narrow Vessels – Coronary CTA, Renal MRA

The segmentation of the narrow vessels is very rapid and the path is smooth and remains within the boundaries of the volume-rendered vasculature. Computation times for the renal paths are about 1-2 seconds on a 733MHz Pentium III machine. These times vary with the spatial resolution of the images.

4.2 Large Vessels – Carotid/Upper Aorta CE-MRA

Larger vessels with a larger enhanced region require the algorithm to do more exploration. There is, then, greater possibility for deviation of the minimal path relative to the narrow vessels. For aorta tracking in 256x256x64 volumes, typical times for tracking and segmentation of the aorta from above the kidneys to the illiac branch takes 2-3 seconds. Re-centering the path with a distance transform takes and additional 2-3 seconds on a 73MHz Pentium III machine.

4.3 Segmentation Results

[Fig. 1a,b] Shows original slice data from a CE-MRA carotid/upper aorta study. This data was seeded with four points, two at the visible extrema of the subclavian arteries and two in vessel extrema near the top of the head, one on each side. The images depict the region explored by the phase 1 expansion in light gray with a white border.

5 Summary & Future Work

We have demonstrated a physics-based framework for the extraction of vessel paths from CE-MRA and CTA volumes. The two-phase computation is adaptive and rapid. It is not dependent on arbitrary thresholds or similar parameters. With minimal optimization, we feel that this algorithm is capable of real-time operation. In current tests the algorithm designed for vessels is being successfully used for segmenting and tracking the large intestine CT colonography studies and for delineating bone in peripheral CT studies.

References

[1] Andrews, Frank C., *Equilibrium Statistical Mechanics* (2nd edition), John Wiley & Sons, Inc. New York, NY, 1975.
[2] Malladi R., Sethian J.A., "A Real-Time Algorithm for Medical Shape Recovery," *Proceedings of International Conference on Computer Vision,* pp. 304-310, Mumbai, India, January 1998.
[3] Masutani Y., Schiemann Th., Hohne K. "Vascular Shape Segmentation and Structure Extraction Using a Shape-Based Region-Growing Model", *MICCAI '98,* pp. 1242-1249, Cambridge, MA, USA, October 1998.
[4] McInerney T. et al. "Medical Image Segmentation Using Topologically Adaptable Surfaces," LNCS Vol. 1205, *Proceedings of Computer Vision Virtual Reality and Robotics in Medicine '97,* pp 23-32, 1997.
[5] Sethian J. "A Fast Marching Level Set Method for Monotonically Advancing Fronts," *Proceedings of the National Academy of Sciences,* Vol. 93, pp. 1591-1595, 1996.
[6] Sethian J. *Level Set Methods and Fast Marching Methods: Evolving Interfaces in Computational Geometry, Fluid Mechanics, Computer Vision, and Materials Science,* Cambridge University Press, 2nd Edition, 1999.
[7] Tek H., Kimia B. "Volumetric Segmentation of Medical Images by Three-Dimensional Bubbles," CVIU, 64(2):246-258, 1997.
[8] Zhu S., Yuille A., "Region Competition: Unifying Snakes, Region Growing, and Bayes/MDL for Multiband Image Segmentation" *IEEE Transactions on Pattern Analysis and Machine Intelligence,* Vol. 18, No. 9, September 1996.

Exploiting Weak Shape Constraints to Segment Capillary Images in Microangiopathy

M Rogers[1], J Graham[1] and R A Malik[2]

[1]Imaging Science and Biomedical Engineering, University of Manchester, Manchester M13 9PT, U.K.
[2]Department of Medicine, Manchester Royal Infirmary, Manchester, U.K.
mdr@server1.smb.man.ac.uk
www.isbe.man.ac.uk

Abstract. Microangiopathy is one form of pathology associated with peripheral neuropathy in diabetes. Capillaries imaged by electron microscopy show a complex textured appearance, which makes segmentation difficult. Considerable variation occurs among boundaries manually positioned by human experts. Detection of region boundaries using Active Contour Models has proved impractical due to the existence of confusing image evidence in the vicinity of these boundaries. Despite the fact that the shapes have no identifying landmarks, the weak constraints imposed by statistical shape modelling combined with genetic search can provide accurate segmentations.

1. Diabetic Nerve Capillaries

Peripheral neuropathy is an important and debilitating symptom of diabetes. Among the pathological manifestations of neuropathy is microangiopathy (disease of small blood vessels) which affects the capillaries in the endoneurium - the interstitial connective tissue in peripheral nerves separating individual nerve fibres. The two main effects are:

1. a thickening of the basement membrane or an accumulation of basement membrane material in the capillaries causing an apparent contraction of the luminal area, and
2. a proliferation of endothelial cell material together with a thickening of the basement membrane, which manifests itself in arteries, arterioles and occasionally venules.

The aetiology of the condition is unknown. Quantitative studies of the variation in shape and size of the Basement Membrane (BM), Endothelial Cell (EC) and lumen region may shed light on the progression of nerve capillary damage [1]. The structure of two endoeneurial capillaries as they appear in electron micrographs is shown in Fig. 1.

Currently segmentations of these areas for the purpose of measurement are taken by hand [1], which is a time consuming process and restricts the quantity of samples that can be analysed. There is a requirement for an automated approach, both to reduce the labour required and to make the measurements more objective.

The nerve capillary structures have a complex appearance, containing no consistent structural features and wide variation in shape and structure. The image evidence defining the required boundary is extremely variable from image to image and is often highly ambiguous.

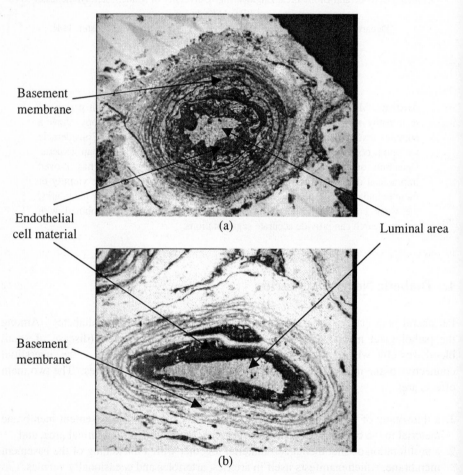

Fig. 1. Two examples of nerve capillary images showing the large variation in appearance and structure. Image (a) exhibits atypical texture around the EC/BM boundary. Image (b) shows an area of locally confusing image evidence at the top of the capillary structure.

2. Segmentation of Capillaries

Byrne and Graham [2] applied an Active Contour Model to this difficult segmentation problem and achieved encouraging results however, the active contours were initiated by manual positioning, and often became 'trapped' on confusing image evidence (see Fig. 1(b)), which provided good but incorrect local boundaries. In this study we seek to obtain more accurate segmentations by:

1. using genetic search to overcome the problem of local maxima in hill climbing methods,
2. constraining the solutions using a model of capillary shape.

There is considerable variability in the shape of capillaries so shape constraints will not be as powerful as those that can be exploited, for example, in the detection of organs in anatomical images. However, capillary shapes are not totally unconstrained either, and conducting a search within the statistical limits imposed by a training set should contribute to better solutions than might be obtained with a totally data-driven approach.

Active Shape Models (ASMs) have been applied successfully to analysis of radiological images [3,4]. The power of this method derives from the statistics of consistent landmark positions on training objects (see section 3 below). The shape representation is then manipulated by a hill climbing search mechanism. Capillaries do not have recognisable landmarks. However, it is still possible to use boundary points to represent the shape, and the resulting descriptions are easily manipulated by Genetic Algorithms (GAs).

3. Materials and Methods

3.1. Data

Our data set consists of 44 electron microscope images (8-bit grey scale, 768x575 pixels) of nerve capillaries. Significant variation in the quality of the images has been introduced by inconsistencies in the image acquisition process. A set of 10 images to be used as the training set for grey level modelling has been chosen manually. The selection was made based on the perceived quality of the images.

Each capillary image has been annotated three times by experts who marked the boundaries between the various structures within the capillaries. Even for expert human annotation, the boundary positions are difficult to judge and there is considerable variation in the position of manually placed boundaries even for the cleanest images. Table 1 includes the mean point-to-line distances between different expert annotations of the same image for the 10 images in the 'good' image set. The average closest annotated boundary from any sampled point is 7.8 pixels. Many of the differences between boundaries are small and represent variation in positioning the same perceived boundary. However, some of the larger distances represent different interpretations of the positions of the boundaries between the relevant structures, revealing a genuine ambiguity of interpretation. Fig. 2 shows examples of capillary images with two possible interpretations of the EC/BM boundary position from the three annotations.

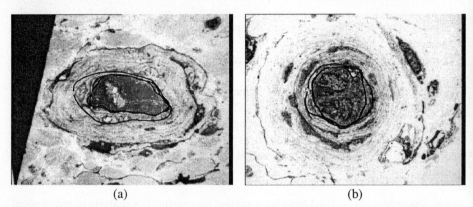

(a)   (b)

Fig. 2. Two examples of nerve capillaries with multiple expert annotations (marked as solid black lines) of different positions for the Endothelial Cell/Basement Membrane boundary

3.2. Texture Discrimination

As we wish to describe texture boundaries, we transform the grey level images into a texture image. We use the method described by Byrne and Graham [2] in which Laws texture filters [5] are applied to a set of training images to give a set of texture features. These are combined using linear discriminant analysis to provide a texture discrimination function. This function is then used to generate the texture images used in GA search. Fig. 3 shows an example of a texture image generated in this way.

3.3. Active Shape Models

Active Shape Models [6] are generated using a statistical analysis of shape and local grey level appearance over a training set of images. For a detailed description of ASM search see [7]. The training images are labelled with a set of landmark points marking consistent features throughout the set of images. Further evenly spaced model points between chosen features are often required to provide an adequate representation of the shape of a structure. The local grey level appearance is modelled over a patch at each model point.

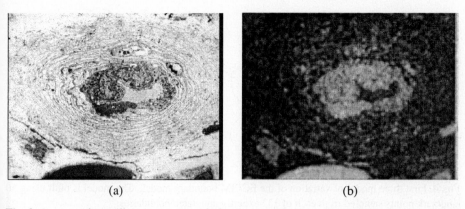

Fig. 3. An example capillary image (a) with the corresponding texture image (b) generated using the texture discrimination function.

The variation in shape across the set is described by applying Principal Component Analysis (PCA) to the landmark points, resulting in a Point Distribution Model (PDM) 7. In this way any valid example of the shape being modelled can be approximated using:

$$\mathbf{x} = \overline{\mathbf{x}} + \mathbf{Pb} \qquad (1)$$

where $\overline{\mathbf{x}}$ is the mean shape vector, $\mathbf{P}$ is a set of orthogonal modes of variation and $\mathbf{b}$ is a vector of shape parameters. Conversely, for any shape $\mathbf{x}$ the parameter vector $\mathbf{b}$ can be calculated:

$$\mathbf{b} = \mathbf{P}^T (\mathbf{x} - \overline{\mathbf{x}}) \qquad (2)$$

The model can be used as a representation for image search. By constraining the shape to lie within a specified range of 'allowed' shapes determined by the training set, solutions can be found which are guaranteed to have 'legal' shapes.

In the capillary images there are no consistently identifiable 'landmark' points. To take account of this it was necessary to modify the ASM method. The training points were provided by sampling evenly spaced points from each boundary in the set of expert annotations. The first point on each boundary was defined to lie at one end of the major axis of the best fitting ellipse to the original boundary. Each training example contains 50 boundary points. The shape model training set contains all three annotations from each of the 44 capillary images giving 132 sets of boundary points. The first three modes of variation from the shape model built in this way are shown in Fig 4.

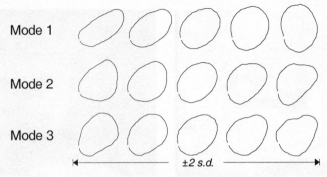

Mode 1

Mode 2

Mode 3

|◄─────────── ±2 s.d. ───────────►|

Fig. 4. First three modes of variation of the EC/BM boundary model. The model is built using 50 landmark points sampled from each of 132 expertly annotated boundaries.

The grey level appearance in the capillary images is very variable across the entire set, and this can lead to a model that fits well to flat image data. In order to produce good models capable of discriminating between regions, a subset of 10 images with good discrimination between endothelial area and basement membrane were manually chosen to build the grey level models. An area of 20x3 pixels around each landmark was modelled. Since the model points do not correspond to any particular anatomical feature, there is no reason to believe that the grey level appearance at one model point should be different from that at any other point. Any difference between the grey level models is a feature of the small sample rather than any real property of the data. Therefore a single local model, calculated from all model points was used for all points in the PDM.

A further adjustment to the standard ASM grey level modelling scheme was introduced to utilize our prior knowledge of how the texture images were formed. An ASM was built and an iteration of search was carried out using the original landmarks as a starting point. This has the effect of finding the local best fit of the profile model to the image data, and relocating landmarks closer to areas of high texture gradient. The new point positions after search are then used to build a new grey level model that will find areas of high texture gradient more robustly. This is only appropriate as we want image search to identify areas of high texture gradient as good model matches. The process can be thought of as removing small errors from the annotated landmark positions. Comparative results between a standard ASM with a combined grey level model and an ASM built in this manner are given in section 4.

3.4. Genetic Algorithms

In the standard implementation ASMs perform a hill-climbing search. Small areas of locally difficult texture evidence in the capillary images cause many local maxima to be present in an objective function of the quality of fit of an ASM (see Fig. 5). It is necessary to apply a search method capable of overcoming local maxima to correctly determine the global solution. Genetic Algorithms [8] are a commonly used stochastic search mechanism which can find good solutions in the presence of numerous local maxima in the objective function. In general, the fittest individuals of a population of candidate boundaries tend to reproduce and survive to the next generation.

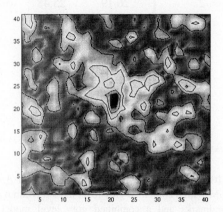

Fig. 5. A plane through the search space defined by the objective function (eqn. 3), produced by varying translation parameters shows many local maxima. Axis values are in pixels.

We use as the objective function:

$$f = \sum_i \exp\left(\frac{-m_i}{2}\right) \qquad (3)$$

where m_i is the Mahalanobis distance for the fit of model point i from the mean position. Hill et al. [9] show that convergence of GAs is accelerated by performing an iteration of hill-climbing ASM search at the end of each generation. We can think of the GA locating hills in the objective function and the ASM reaching the top of them. This approach has been shown to speed up GA convergence [9].

4. Results

GA searches were run on the set of 10 texture processed nerve capillary images that had been used to build the ASM grey level model in a leave-one-out cross validation, using a population size of 100 individuals and a maximum of 25 generations. The individual with the highest fitness from the final population was taken as the boundary found by the search. Result boundaries were evaluated against the expert annotations available for the corresponding image by calculating the point-to-line distance for each landmark point. Several annotated boundaries exist for each image giving a set of point-to-line distances for each landmark. Table 1 gives details of the point-to-line distances obtained for each of the 10 image searches.

Image	A E_{bm}	B E_{bm}	C E_{bm}
1	5.98	6.76	10.34
2	26.19	25.48	6.71
3	15.28	14.52	8.22
4	6.73	3.84	5.86
5	12.60	12.94	10.18
6	5.97	5.43	6.23
7	6.32	5.65	6.60
8	16.70	14.79	6.91
9	6.42	4.94	6.48
10	129.3	84.62	8.72
Average	23.15	17.90	7.62

Table 1. Search results from leave-one-out GA searches. Column A shows results from GA search with a standard ASM with a combined grey level model; column B shows results using an ASM with optimised landmarks and a combined grey level model and column C shows differences between expert annotations.

	%	E_{bm}	$E_{b,\sigma}$
Successful	72.	17.1	12.5
(E_{bm}<35)	73	4	7
Failure	27.	62.9	28.5
(E_{bm}>35)	27	9	3
Total	100	29.6	32.3
		4	0

Table 2. Robustness results from the set of 44 capillary images, both combined and seperated into successful and unsuccessful search categories. E_{bm} is the mean point-to-line distance of each landmark to the closest single annotation, $E_{b,\sigma}$ is the standard deviation of the values

The searches of all the images have found good approximations to the annotated boundaries except for image number 10 which has a far larger error than the rest of the set for all three methods. Of the two GA experiments, the performance of the ASM with an optimized grey level model is slightly better throughout the set with a mean point-to-line error E_{bm} of 17.90 against the standard ASM error of 23.15. Ignoring the results from image 10, average E_{bm} errors become 11.36 and 10.48 for the standard and optimised landmark models respectively, which are comparable to the error in expert annotation. Fig. 6 shows examples of GA search results together with the closest expert annotation.

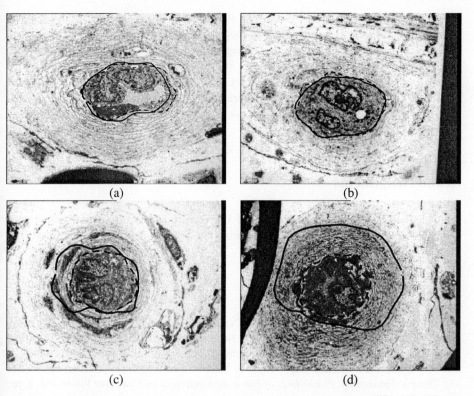

Fig. 6. GA search results: solid lines are automatic segmentations and dashed lines show the closest expert annotation. (a) shows the result for image 1 from Table 1; (b) shows the result for a successful search from the robustness set of 44 images; (c) show a search result that has been influenced by local image evidence; (d) shows the results from image 10 in Table 1 which is an example of a failed search.

Table 2 presents an extension of the evaluation in which leave-one-out tests were carried out on the entire set of 44 usable capillary images. In each case 43 images were used to train the shape model, but the training of the texture model was limited to the 'good' set of 10 images. The search results have been classified into successful searches, with a mean point-to-line distance less than 35 pixels, and unsuccessful searches. The GA search technique is successful in 72.7% of cases. The overall point-to-line error throughout the entire set was 29.6 pixels.

5. Conclusions and Discussion

Actives Shape Models manipulated by GA search have been shown to produce results that have comparable accuracy to human experts. However, the method is inadequately robust at its current state of development; just over 25% of the searches converge on totally misleading evidence. Inspection of the images on which the search failed suggests that this occurred in cases where the textured appearance of the boundary is significantly different from those in the model, including image 10 in the 'good' set. Fig. 1(a) gives an example of such an image with unusual texture appearance. This indicates that the most likely approach to improving robustness is in development of more adaptive texture models. It is not yet clear whether the very wide variation in texture in some images is a genuine feature of the capillary images or if it has been caused by some error in the imaging process. Analysis of a larger image set will allow investigation of this.

Acknowledgements

We would like to thank Dave Walker of the Department of Medicine, Manchester Royal Infirmary for his assistance in annotating the capillary images used in this work.

References

1. R A Malik, S Tesfaye, S D Thompson, A Veves, A K Sharma, A J M Boulton, J D Ward. Microangiopathy in Human Diabetic Neuropathy: Relationship between Capillary Abnormalities and the Severity of Neuropathy. *Diabetologia*, vol. 30, pp. 92-102, 1989.
2. M J Byrne and J Graham. Application of Model Based Image Interpretation Methods to Diabetic Neuropathy. *Proceedings of European Conference on Computer Vision*, vol. 2, pp. 272-282, 1996.
3. T F Cootes, A Hill, C J Taylor, J Haslam. The Use of Active Shape Models for Locating Structures in Medical Images. *Image and Vision Computing* vol.12, no.6, pp.355-366, 1994.
4. T F Cootes, A Hill, C J Taylor, Medical Image Interpretation Using Active Shape Models: Recent Advances. *14th International Conference on Information Processing in Medical Imaging*, pp. 371-372, 1995.
5. K I Laws. Textured Image Segmentation. *University of Southern California*, 1980.
6. T F Cootes, C J Taylor, D H Cooper and J Graham. Active shape models – their training and application. In *Computer Vision and Image Understanding*, vol. 61, no. 1, pp. 38-55, 1995.
7. T F Cootes, C J Taylor, D H Cooper and J Graham. Image search using trained flexible shape models. *Advances in Applied Statistics*, pp. 111-139, 1994.
8. L Davis. Genetic Algorithms and Simulated Annealing. *Pitman, London*, 1987.
9. A Hill, T F Cootes and C J Taylor. A Generic System for Image Interpretation Using Flexible Templates. *BMVC*, pp. 276-285, 1992.

Guide Wire Tracking
During Endovascular Interventions

S.A.M. Baert, W.J. Niessen, E.H.W. Meijering, A.F. Frangi, M.A. Viergever

Image Sciences Institute, University Medical Center Utrecht
Rm E 01.334, P.O.Box 85500, 3508 GA Utrecht, The Netherlands
shirley@isi.uu.nl

Abstract. A method is presented to extract and track the position of a guide wire during endovascular interventions under X-ray fluoroscopy. The method can be used to improve guide wire visualization in the low quality fluoroscopy images. A two-step procedure is utilized to track the guide wire in subsequent frames. First a rough estimate of the displacement is obtained using a template matching procedure. Subsequently, the position of the guide wire is determined by fitting the guide wire to a feature image in which line-like structures are enhanced. In this optimization step the influence of the scale at which the feature is calculated and the additional value of using directional information is investigated. The method is applied both on the original and subtraction images. Using the proper parameter settings, the guide wire could successfully be tracked based on the original images, in 141 out of 146 frames from 5 image sequences.

Keywords: guide wire tracking, medical image processing, interventional radiology

1 Introduction

Endovascular interventions are rapidly advancing as an alternative for conventional invasive open surgical procedures. During interventions a guide wire is advanced under fluoroscopic control. Accurate positioning of the guide wire and the catheter with regard to the vasculature is a prerequisite for a successful procedure. Owing to the low dose used in fluoroscopy in order to minimize the radiation exposure of the patient and radiologist, image quality is often limited. Additionally, motion artifacts owing to patient motion and guide wire motion, further limit image quality. Therefore, a method to extract and track guide wires is presented which can deal with the low signal-to-noise ratio inherent to fluoroscopic images, and disappearance of the guide wire in a few frames owing to motion blur. The method can be used to improve guide wire visualization, potentially enabling a reduction in radiation exposure. It can also be used to detect the position of the guide wire in world coordinates for registration with preoperatively acquired images as a navigation tool for radiologists.

2 Methods

In order to represent the guide wire, a spline parameterization is used. For all experiments in this paper, we used a third order B-spline curve defined by:

$$\mathbf{C}(u) = \sum_{i=0}^{p} N_{i,3}(u)\mathbf{P}_i \qquad 0 \leq u \leq 1 \tag{1}$$

where $\mathbf{P_i}$ denote the control points (forming a control polygon) with p the number of control points and $N_{i,3}(u)$ are the third degree B-spline basis functions defined on the non-periodic knot vector

$$U = \{0, \dots, 0, u_4, \dots, u_{m-4}, 1, \dots, 1\} \tag{2}$$

where m is the number of knots.

In order to find the spline in frame $n+1$ if the position in frame n is known, a two-step procedure is introduced. First, a rigid translation is determined to capture the rough displacement of the spline. Next, a spline optimization procedure is performed in which the spline is allowed to deform for accurate localization of the guide wire. These steps can be understood as a coarse-to-fine strategy, where the first step ensures a sufficiently good initialization for the spline optimization.

2.1 Rigid transformation

In order to obtain a first estimate of the displacement, a binary template is constructed based on the position in the present frame. The best location of this template in the new frame is obtained by determining the highest cross correlation with a certain search region in this image (or features derived from it; see section 2.3). This template matching technique is based on the assumption that a local displacement, $\mathbf{d} = (d_x, d_y)$, of a structure in one image I_0 can be estimated by defining a certain window W (say $K \times L$ pixels in size) containing this structure, and by finding the corresponding window in a second image I in the sequence by means of correlation. The correlation coefficient (CC) is computed as:

$$CC(\mathbf{d}) = \frac{\sum_{\mathbf{x} \in W}[I_0(\mathbf{x}) - \langle I_0 \rangle_W][I(\mathbf{x} + \mathbf{d}) - \langle I \rangle_{W,\mathbf{d}}]}{\sqrt{\sum_{\mathbf{x} \in W}[I_0(\mathbf{x}) - \langle I_0 \rangle_W]^2 \sum_{\mathbf{x} \in W}[I(\mathbf{x} + \mathbf{d}) - \langle I \rangle_{W,\mathbf{d}}]^2}} \tag{3}$$

where

$$\langle I_0 \rangle_W = \frac{1}{KL} \sum_{\mathbf{x} \in W} I_0(\mathbf{x}) \quad \text{and} \quad \langle I \rangle_{W,\mathbf{d}} = \frac{1}{KL} \sum_{\mathbf{x} \in W} I(\mathbf{x} + \mathbf{d}) \tag{4}$$

denote the mean values of the image intensities in the respective windows.

2.2 Spline optimization

After the first step is carried out, the spline is optimized under internal and external forces. The internal constraints are related to the geometry of the curve and influence the length (first derivative of the B-spline) and the bendedness (second derivative of the spline). This curvature (K) is computed as:

$$K = \int_0^1 \frac{|\mathbf{C}'(s) \times \mathbf{C}''(s)|}{L^3} ds \qquad (5)$$

where L is the length of the curve given by

$$L = \int_0^1 \mathbf{C}'(s) ds \qquad (6)$$

The parameters for the curvature are set sufficiently large to avoid strange shapes of the spline and sufficiently small to ensure that the internal forces only have a small influence on the total spline energy. For the external forces, the image intensity or a feature image derived form it (see section 2.3) is used. The spline contains four or five control points and one hundred sample points. The spline is optimized using Powell's direction set method [1].

2.3 Feature image

It has previously been shown [2] that using original images for the matching and optimization steps, the guide wire could not effectively be tracked due to presence of other objects in the image. Therefore a feature image in which line-like structures are enhanced is used to determine the optimal spline position.

In order to enhance elongated structures in the image, the eigenvalues of the Hessian matrix are calculated at scale σ. This Hessian matrix is defined as:

$$H = \begin{pmatrix} L_{xx} & L_{xy} \\ L_{xy} & L_{yy} \end{pmatrix} \qquad (7)$$

where L_{xy} represents the convolution with the scaled Gaussian derivative

$$L_{xy} = L * \frac{\partial^2}{\partial x \partial y} G(\mathbf{x}, \sigma) \qquad (8)$$

and $G(\mathbf{x}, \sigma)$ is given as

$$G(\mathbf{x}, \sigma) = \frac{1}{2\pi\sigma^2} e^{-\left(\frac{x^2+y^2}{2\sigma^2}\right)} \qquad (9)$$

The corresponding eigenvalues are given by:

$$\lambda_{1,2}(\mathbf{x}, \sigma) = \frac{1}{2}\left(L_{xx} + L_{yy} \pm \sqrt{(L_{xx} - L_{yy})^2 + 4L_{xy}^2} \right) \qquad (10)$$

Let λ_1 denote the largest absolute eigenvalue. On line-like structures λ_1 has a large output. Since we are interested in dark elongated structures on a brighter background, only positive values of λ_1 are considered; pixels with negative values of λ_1 are set to zero. The feature image is subsequently constructed by inverting this image since the optimization is based on a minimum cost approach, see Figure 1. Next to the eigenvalues, the eigenvectors give insight in the orientation

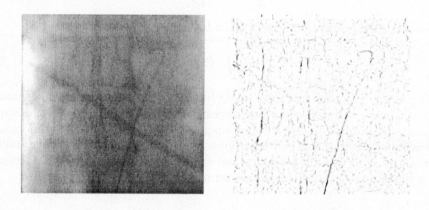

Fig. 1. *The original image (left) and the feature image computed with the eigenvalues of the Hessian matrix with $\sigma = 1.5$ (right).*

of the line structure. Therefore, in the optimization scheme, we also investigated the use of this directional information to effectively attract the guide wire only to line structures with similar orientation. Hereto, the inner product between the spline and the orientation of the feature is used given by

$$O(\hat{x}_i) = \lambda_1(\hat{e}_2 \cdot \hat{x}_i) \tag{11}$$

where $\hat{e}_2$ is the normalized eigenvector corresponding to λ_2 and $\hat{x}_i$ is the normalized first derivative of the spline in sample point i.

To enable a coarse-to-fine strategy and to reduce sensitivity to noise, the feature image can be calculated at multiple scales (σ). We compare the use of the eigenvalue image calculated on the original image with computing the feature image on subtraction images. These images are obtained by subtracting the first frame from frame n. The first frame is used to ensure sufficient guide wire movement so as to make it clearly visible in the subtraction image. In our experiments we used image sequences with a maximum of 49 frames. For longer sequences it will be better to subtract for example frame $n - 20$ from the present frame, to limit the effects of background motion. Alternatively, methods for motion correction in subtraction techniques can be used [3].

2.4 Images

The method was applied on five image sequences, three sequences of the thorax and two abdominal image sequences, with a sequence length between 14 and 49 frames. An overview of these sequences is given in Figure 2. Only J-tipped guide wires were used during the interventions. The image series were acquired on a H5000 (4 sequences) and a H3000 (1 sequence) X-ray fluoroscopy system (Philips Medical Systems, Best, the Netherlands).

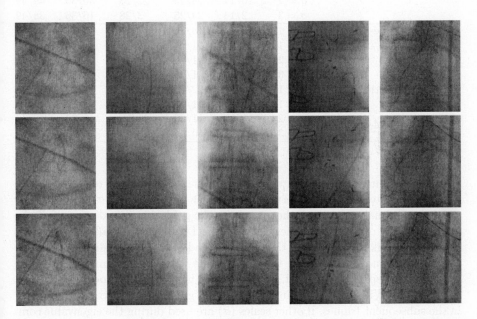

Fig. 2. *Three non-subsequent frames of the sequences I to V (left to right) used during the evaluation of the method.*

3 Results

We compared the performance of the method on the Hessian feature image calculated on the original image with the one calculated on the subtraction image using different scales (σ). In the spline optimization step, we compared the use of the scalar feature image and the vector feature image (see section 2.3). Evaluation of the tracking algorithm is difficult since no ground truth is available. In order to determine the effectiveness of the guide wire tracking procedure, the number of frames that could not be tracked was determined based on visual inspection of the results. When the spline model lies on top of the tip of the guide wire, the result was rendered as correct.

Table 1 summarizes the tracking results. Results without first applying a rough registration are not listed. These results were worse for every image sequence, which shows the need for a rough initialization before spline optimization.

Table 1. Number of frames correctly tracked

			Seq. I	Seq. II	Seq. III	Seq. IV	Seq. V
	Hessian	$\sigma = 1$	12/14	24/25	14/28	28/30	49/49
		$\sigma = 1.5$	12/14	25/25	25/28	30/30	49/49
Original		$\sigma = 2$	9/14	24/25	23/28	29/30	49/49
image		$\sigma = 3$	1/14	14/25	9/28	13/30	2/49
	Orientation	$\sigma = 1$	12/14	24/25	17/28	27/30	49/49
		$\sigma = 1.5$	12/14	25/25	25/28	30/30	49/49
		$\sigma = 2$	10/14	25/25	22/28	30/30	49/49
		$\sigma = 3$	2/14	24/25	3/28	16/30	38/49
	Hessian	$\sigma = 1$	9/14	18/25	17/28	23/30	49/49
		$\sigma = 1.5$	9/14	11/25	20/28	28/30	49/49
Subtraction		$\sigma = 2$	8/14	11/25	22/28	28/30	48/49
image		$\sigma = 3$	2/14	1/25	22/28	28/30	48/49
	Orientation	$\sigma = 1$	9/14	22/25	14/28	13/30	49/49
		$\sigma = 1.5$	10/14	11/25	16/28	29/30	49/49
		$\sigma = 2$	11/14	11/25	20/28	29/30	48/49
		$\sigma = 3$	1/14	10/25	20/28	27/30	48/49

Using the feature image computed on the original image for the registration and optimization step, we can observe from Table 1 that the scale $\sigma = 1.5$ appears to be the optimal scale. With these settings, the guide wire could successfully be tracked in 141 out of 146 frames. In sequence I the tracking failed in 2 frames owing to line-like structures in the neighborhood of the guide wire. Motion blur due to a fast movement of the tip of the guide wire caused a mismatch in the registration step in 3 frames of sequence III. These failures appeared only in a single frame. Without manual intervention, the guide wire was tracked correctly in the subsequent frames. If other scales (σ) are used during the eigenvalue computation, the performance of the tracking method becomes worse. For smaller σ, the tracking results, especially the matching step, is influenced by image noise. Conversely, at larger σ, broader line-like structures are more enhanced, so that the guide wire is not always detected correctly.

If we use orientation information instead of scalar information in the optimization step, the same results are obtained for $\sigma = 1.5$. The failures on this scale were obtained in the matching step and they could not be corrected using different external energy forces in the spline optimization step. On a larger scale ($\sigma = 3$), the tracking results improved for sequence II and V. Using orientation information, the spline is less attracted to enhanced structures with an orientation that differs form the orientation of the spline. The optimization improves the localization of the spline which subsequently influences the registration step in the next frame.

If the eigenvalues of the Hessian matrix are calculated for the subtraction image, good results are obtained in the last two sequences, but the tracking failed most of the times in sequence I, II and III. These sequences are very noisy and by performing digital subtraction, the noise level is amplified. Using the orientation instead of scalar information in the optimization step for the subtraction images, the performance of the method hardly changes. An example of the good tracking results for sequence III and IV obtained with $\sigma = 1.5$ is shown in Figure 3 and 4.

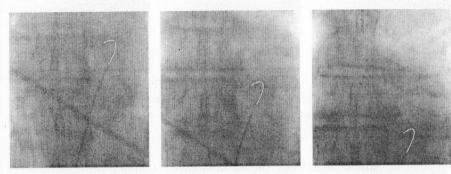

Fig. 3. *Three frames (frame 2, 10 and 25) of sequence III, which gives an impression of the tracking results. The method was applied on the feature image in which the eigenvalues of the Hessian matrix were calculated with $\sigma = 1.5$ on the original images of the sequence.*

4 Discussion

There is relatively little literature on tracking guide wires from 2D fluoroscopy images. In [4], guide wire tracking is used to evaluate the possibility of extracting myocardial function from guide wire motion. However, guide wire tracking is only performed in a single frame and not in time. Some other work is aimed at the tracking of guide wires during endovascular interventions to control the position of a catheter inside the human body with external devices [5, 6] (active tracking), or to reconstruct 3D catheter paths [7]. The presented method is a new approach to improve visualization and localization during endovascular interventions.

Based on the tests in the previous section, a number of conclusions can be drawn. First, the use of the cross-correlation step is helpful, as it supplies a good initialization for the subsequent fitting procedure. It is a known drawback from snake algorithms [8] to be sensitive to the initialization, and using the first rough alignment this is circumvented. Second, the scale parameter at which the feature image was computed is essential. It has to be selected sufficiently large to reduce sensitivity to noise, and sufficiently small to ensure that other line-like structures at a larger scale are not enhanced.

734

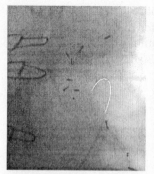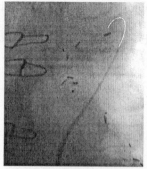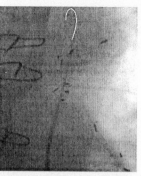

Fig. 4. *Three frames (frame 4, 11 and 22) of sequence IV, which gives an impression of the tracking results. The method was applied on the feature image in which the eigenvalues of the Hessian matrix were calculated with $\sigma = 1.5$ on the subtraction images of the sequence.*

For the best parameter settings, the feature image containing the eigenvalues of the Hessian matrix gave the best overall results. The guide wire was successfully tracked in 141 out of 146 frames from 5 sequences. Subtracting a previous image from the current frame can help if there are elongated structures in the neighborhood of the spline, but the movement of the guide wire tip has to be large enough to get a good result. Using orientation information instead of scalar information in the optimization step gave some improvement on larger scales.

References

1. W.H. Press, S.A. Teukolsky, W.T. Vetterling, B.P. Flannery, Numerical Recipes in C: The art of scientific computing, 1992.
2. S.A.M. Baert, W.J.Niessen, E.H.W. Meijering, A.F. Frangi, M.A. Viergever, Guide wire tracking in interventional radiology, Proceedings of the 14th International Congress and Exhibition CARS 2000, 537-542.
3. E.H.W. Meijering, W.J. Niessen, M.A. Viergever, Retrospective Motion Correction in Digital Subtraction Angiography: A Review, IEEE Transactions on Medical Imaging 1999, Vol.18, No.1: 2-21.
4. D. Palti-Wasserman, A.M. Brukstein, R.P. Beyar, Identifying and Tracking a Guide Wire in the Coronary Arteries During Angioplasty from X-Ray Images, IEEE Transactions on Biomedical Engineering 1997, Vol.44, No.2: 152-164.
5. H. Starkhammar, M. Bengtsson, D.A. Kay, Cath-Finder Catheter Tracking System: a new device for positioning of central venous catheters. Early experiments from implantation of brachial portal systems. Acta Anaesthesiol. Scand. 1990, No. 34: 296-300.
6. J. Ragasa, N. Shan, R.W. Watson, Where antecubital catheters go: a study under fluoroscopic control, Anesthesiology 1989, No. 71: 378-380.
7. H.-J. Bender, R. Männer, C. Poliwoda, S. Roth, M. Walz, Reconstruction of 3D Catheter Paths from 2D X-ray Projections, Proceedings of the Second International Conference MICCAI'99: 981-989, Lecture Notes in Computer Science Vol.1679.
8. M. Kass, A. Witkin, D. Terzopoulos, Snakes: Active contour models. Int. Journal of Computer Vision 1987, Vol. 1, No. 4, 321-331.

A Deformable Vessel Model with Single Point Initialization for Segmentation, Quantification and Visualization of Blood Vessels in 3D MRA

Marcela Hernández-Hoyos, Alfred Anwander, Maciej Orkisz,
Jean-Pierre Roux, Philippe Douek, Isabelle E. Magnin

CREATIS, CNRS Research Unit (UMR 5515) affiliated to INSERM, Lyon, France
CREATIS, INSA 502, 20 av. Albert Einstein, 69621 Villeurbanne cedex, France
marcela.hernandez@creatis.insa-lyon.fr
http://www.creatis.insa-lyon.fr

Abstract. We deal with image segmentation applied to three-dimensional (3D) analysis of of vascular morphology in magnetic resonance angiography (MRA) images. The main goal of our work is to develop a fast and reliable method for stenosis quantification. The first step towards this purpose is the extraction of the vessel axis by an expansible skeleton method. Vessel boundaries are then detected in the planes locally orthogonal to the centerline using an improved active contour. Finally, area measurements based on the resulting contours allow the calculation of stenosis parameters. The expansible nature of the skeleton associated with a single point initialization of the active contour allows overcoming some limitations of traditional deformable models. As a result, the algorithm performs well even for severe stenosis and significant vessel curvatures. Experimental results are presented in 3D phantom images as well as in real images of patients.

1. Introduction

Atherosclerosis is the principal acquired affection of the vascular wall. It remains the number one cause of death among people older than 60 years. Its major complication is the arterial stenosis. Precise stenosis quantification is necessary for diagnosis purposes and treatment planning.

Accurate quantitative measurements of vessel morphology require previous vessel segmentation. Two main approaches for vessel segmentation can be distinguished. The first one relies on purely photometric criteria, mainly on thresholding and region-growing techniques [1]-[4]. Its major advantage is its generally simple implementation. However, a further modelling step is necessary to extract meaningful measurements from thus segmented images. The second approach exploits the geometrical specificity of the vessels, in particular the notions of orientation and tubular shape. Most of these approaches consists in vessel-tracking [5]-[9] and use (often implicitly) a generalized-cylinder model, *i.e.* an association of an axis (centerline) and a surface (vessel wall). Consequently, the segmentation process involves two tasks: centerline extraction and vessel contour detection (most often using a 2D deformable model) in the planes usually perpendicular to the axis. This procedure results in a stack of 2D contours along the vessel, allowing quantitative cross-section measurements and visualization by means of triangulation-based rendering. Other more recent approaches use 3D deformable models of surface [10][11]. Although deformable models have been shown to be useful for segmenting images from various imaging modalities most of them, for good convergence, require an initialization close to the final solution.

In this context, we propose a novel implementation for vessel segmentation using deformable models for vessel centerline and contours. The principal contribution of our approach compared to the others resides in the simplicity of the model initialization and in the flexibility of its evolution. The algorithm starts from a single point inside the vascular segment to extract, and a simulated propagation of this point along the vessel defines its skeleton. In a similar way, the active model for the contour detection starts from the single point determined by the axis position in the orthogonal plane.

2. Method

On the basis of the discrete generalized-cylinder model [12], we propose a segmentation technique divided into two steps. First, the extraction of the vessel central axis is performed using an expansible vessel skeleton. Contours are then detected in the planes locally orthogonal to the centerline with an improved 2D active contour.

2.1. Expansible Vessel Skeleton

The vessel axis extraction is achieved by an expansible skeleton method. It is based on a tracking strategy, which starts from a user-selected point within the vessel, and then iteratively estimates the subsequent axis points. Geometrically, the axis is defined by a list of points embedded in the image volume $(x,y,z) \in \Re^3$. We consider the expansion of the axis as an iterative process where at each iteration, a new point is added to the model. At iteration k, the axis is represented by:

$$A_k = \{ \mathbf{p}_i, i = 1..k \} \qquad (1)$$

where $\mathbf{p}_i$ is the position of point i and k is the total number of points at this iteration. Point generation is a two-step procedure. First, a prediction of the new point position is obtained, based on the vessel local orientation at the last axis point. This position is then corrected under the influence of image forces and shape constraints.

Prediction. Let $\mathbf{p}_i$ be the point added to the axis during the last iteration, the position of the next point is predicted by applying a "constant velocity" displacement of $\mathbf{p}_i$ along the vessel local orientation. The vessel local orientation is estimated by inertia moment minimization for a small volume (box or cell) centered on the current point

p_i. Let us recall that, for an axis χ passing through the gravity center, the inertia moment of the box can be expressed as:

$$M_\chi = e_\chi^T \mathbf{M} e_\chi \qquad (2)$$

e_χ is a unit vector corresponding to χ, $\mathbf{M}$ is the inertia matrix defined with respect to any coordinate system whose origin is the gravity center. The vessel orientation corresponds to the axis χ for which M_χ is minimum. It is given by the eigenvector $e_{\chi min}$ associated with the smallest eigenvalue of the inertia matrix $\mathbf{M}$. The expression of the new predicted point $\hat{p}_{i+1}$ is:

$$\hat{p}_{i+1} = p_i + \mu\, s_i\, e_{\chi min} . \qquad (3)$$

Since $e_{\chi min}$ is only an orientation unit vector, s_i and μ are included in order to control respectively the direction and the magnitude of the movement. The latter is a fixed parameter, while s_i is computed using the displacement vector d_i from the previous iteration:

$$s_i = \text{sign} \left(< e_{\chi min}, d_i > / \|e_{\chi min}\| . \| d_i \| \right), \text{ where } d_i = p_i - p_{i-1} . \qquad (4)$$

Correction. The inertia-based approach described above provides a convenient way to automate the axis extraction of most regular vascular structures. Moreover it offers a good noise-robustness and gives satisfactory results even for small vessels [13]. It has been shown however, that the algorithm constructs an absurd axis for some structures such as the one presented in Fig. 1 [7].

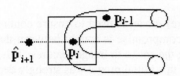

Fig. 1. Erroneous computation of the new point based on the main axis of inertia alone (the rectangle represents the cell used for the inertia moments computation)

We thus propose to estimate the position of the new point by adding a correction to the predicted position $\hat{p}_{i+1}$. This correction results from submitting the point $\hat{p}_{i+1}$ to two kinds of elementary forces: external and internal ones. The external forces are used to attract the point toward a position which has a high likelihood of lying along the vessel centerline. Assuming that the maximum of the MR signal is reached at the vessel axis, we propose two external forces, which respectively attract the point towards a local maximum of intensity and towards the gravity center of a box centered in $\hat{p}_{i+1}$. The total displacement due to these forces can be written as:

$$\Delta_{i+1}^{ext} = \Delta_{i+1}^{gc} + \Delta_{i+1}^{max} , \text{ where:} \qquad (5)$$

- $\Delta_{i+1}^{gc} = -\eta \, (\mathbf{p}_{i+1}^{gc} - \hat{\mathbf{p}}_{i+1})$ is the displacement due to the attraction force exerted by the local gravity center $\mathbf{p}_{i+1}^{gc}$, weighted by the coefficient η.

- $\Delta_{i+1}^{max} = -\kappa \, (\mathbf{p}_{i+1}^{max} - \hat{\mathbf{p}}_{i+1})$ is the displacement due to the attraction force exerted by the local maximum intensity point $\mathbf{p}_{i+1}^{max}$, controlled by κ.

The internal forces are used to impose smoothness and continuity constraints, in order to limit oscillations and reduce the noise-sensitivity. These forces provide the internal constraints over the first and second order discrete derivatives and act like the regularization terms used in the classical snakes internal energy [14]. The expression of the total displacement due to the internal forces is:

$$\Delta_{i+1}^{int} = -\alpha_a \, (\hat{\mathbf{p}}_{i+1} - \mathbf{p}_i) - \beta_a \, (\mathbf{p}_{i-1} - 2\mathbf{p}_i + \hat{\mathbf{p}}_{i+1}) . \tag{6}$$

Two coefficients dictate the physical characteristics of the axis: α_a controls its elasticity while β_a controls its flexibility. Taking into account the displacements (5) and (6), the estimated new point position is:

$$\mathbf{p}_{i+1} = \hat{\mathbf{p}}_{i+1} + \Delta_{i+1}^{ext} + \Delta_{i+1}^{int} . \tag{7}$$

When the axis has been tracked, we proceed to the vessel lumen boundary extraction using 2D active contours applied orthogonally to the centerline. For this purpose, the centerline is interpolated using a B-spline curve and the image volume is re-sliced.

2.2. Vessel Contour Extraction

The active contour (snake) model is used to ensure continuity and smoothness of the detected boundaries. A compromise is sought between these internal constraints and external forces attracting the snake to image features. This concept was introduced by [14], but convergence of this initial model was strongly dependent on its initialization. Much work was then devoted to overcome this problem. We first recall basic concepts concerning the active contours, then we propose a new implementation, which only needs a single point as initialization. The contour is automatically initialized by the cross-section between the vessel axis and the orthogonal image plane.

Active Contour Model. A deformable contour model is defined by a parametric curve $v(s,t) = (x(s,t), y(s,t))^T$, which evolves in time t and space, and its energy $E(v) = E_{int}(v) + E_{ext}(v)$, where $s \in [0,1]$ is the arc length. The energy $E_{int}(v)$, corresponding to internal forces, imposes constraints on the first and second derivatives of the curve:

$$E_{int}(v) = E_{elast}(v) + E_{flex}(v) = \alpha_c \int_0^1 \left| \frac{\partial v(s,t)}{\partial s} \right|^2 ds + \beta_c \int_0^1 \left| \frac{\partial^2 v(s,t)}{\partial s^2} \right|^2 ds \qquad (8)$$

and the coefficients α_c and β_c play the same role as α_a and β_a in (6). There are two kinds of external forces represented by a potential energy $E_{ext}(v) = \int_{s=0}^1 P(v(s,t)) ds$: image forces and a balloon force. The link between potentials $P_i(v)$ and forces is given by : $F_i(v) = -\nabla P_i(v)$. The image forces are usually designed to attract the snake toward strong intensity gradients. One can also use prior knowledge about the vessel outline as function of the intensity, so that the potential energy function has a minimum at the vessel boundary. With an initialization far from any edge, the active contour would collapse. One solution consists in adding an external balloon force F_b [15], which acts like an inflating pressure, and makes the active contour model move without any image gradient. It is expressed as $F_b(v) = b\mathbf{n}(s,t)$, where $\mathbf{n}(s,t)$ is the normal unitary vector oriented outward at the point $v(s,t)$, and b is a coefficient which controls the inflation. The total external force $F_{ext}(v)$ is a weighted sum of these forces.

The parametric curve $v(s)$ is fitted to the image by minimization of the energy $E(v)$. The minimization is a temporal and spatial discrete process, using a finite number N of snake points $v(i,k)$ that approximate the snake as a polygon. The partial derivatives in (8) are approximated by finite differences $\partial v(s,t)/\partial s \approx (v((i+1),k) - v((i-1),k))/2h$, with h the discretization step of the snake points on the contour. The snake is iteratively deformed, explicitly using the external forces $F_{ext}(v)$ at each point of the snake, while implicitly minimizing the internal energy. The new position $v(i,k)$ at time step t is computed by solving the associated Euler equations. These equations can be solved by matrix inversion:

$$v(i,k) = [\gamma \mathbf{I} + \mathbf{A}]^{-1} \cdot [\gamma \cdot v(i,(k-1)) + F_{ext}(v(i,(k-1)))] . \qquad (9)$$

This equation is referred to as the evolution equation of the snake [14]. The matrix $\mathbf{A}$ represents a discretized formulation of the internal energy. The damping parameter γ controls the snake deformation magnitude at each iteration.

Proposed Numerical Implementation and Parameterization. The discretization step h depend on the number of snake points N and, initially, is equal to the distance between the snake points. The evolution equation (9) is only valid if the snake points are equally distanced and the step h is unchanged. After each iteration however, the length of a snake grows due to external force, and h does not correspond to the real distance between the snake points. The internal energy, especially the term $E_{elast}(\mathbf{v})$ associated with the contour's tension, grows with the snake length and stops the snake before the boundary is reached. This is particularly annoying if the initialization is far from the final position. In this case, the model needs to be resampled with a new (higher) number of snake points N, or/and a new step h. The N by N matrix $[\gamma\mathbf{I}+\mathbf{A}]$, as well as its inverse have to be recomputed. This is a time-consuming task which limits the classical snake model application in many cases.

To make the snake scale independent and allow an initialization by a single pixel in the image, we propose a new numerical scheme for the snake energy minimization. It was designed to preserve the validity of the equation (9) in spite of the evolution of the snake's size. It only needs a single computation of the matrices $[\gamma\mathbf{I}+\mathbf{A}]$ and $[\gamma\mathbf{I}+\mathbf{A}]^{-1}$, and uses a fixed number of points for all the iterations. We propose to scale the snake size at each iteration, in order to normalize the internal energy of the model with a fixed discretization step $h'=1$. The scaled snake $\mathbf{v}'(i,(k-1))$ has a normalized length equal to the number of its points. It is obtained from the actual snake using the average distance $\overline{h}_{k-1}$ between its points:

$$\mathbf{v}'(i,(k-1))=\mathbf{v}(i,(k-1))/\overline{h}_{k-1} . \tag{10}$$

The scaled snake is deformed according to the evolution equation (9), using the external forces F_{ext} from the non-scaled snake. Hence, the new (scaled) positions for the snake points $\mathbf{v}'(i,k)$ are:

$$\mathbf{v}'(i,k)=[\gamma\mathbf{I}+A]^{-1}\cdot[\gamma\cdot\mathbf{v}'(i,(k-1))+F_{ext}(\mathbf{v}(i,(k-1)))] . \tag{11}$$

Let $\Delta\mathbf{v}'(i,k)$ be the snake deformation vectors for each snake point:

$$\Delta\mathbf{v}'(i,k)=\mathbf{v}'(i,k)-\mathbf{v}'(i,(k-1)) . \tag{12}$$

The deformation vectors are computed with a normalized internal energy, and the deformation is directly applied on the non-scaled snake:

$$\mathbf{v}(i,k)=\mathbf{v}(i,(k-1))+\Delta\mathbf{v}'(i,k) . \tag{13}$$

In this new position, the snake points are redistributed along the snake polygonal outline, to ensure an equal distance h_k between all snake points for the next iteration.

3. Results and Applications

3.1. Stenosis Quantification

In order to assess the precision of the methods described above, we tested them on phantom images. Our Computer Assisted Design (CAD) phantoms are cylinders whose internal surface represents the endoluminal shape of the vessel (Fig. 3). Their internal reference diameter is 6 mm and each of them comprises 2 stenosis. We tackled the problem of area measurements for stenosis quantification. Cross-sectional areas from the normal ("healthy") parts of the vessel were used to calculate the average normal area $\overline{S}_n$ and the corresponding coefficients of variation defined as:

$$CV_\sigma = \frac{\sigma_s}{\overline{S}_n} \times 100 \qquad CV_{max} = \frac{\max\left(S_{max} - \overline{S}_n, \overline{S}_n - S_{min}\right)}{\overline{S}_n} \times 100 \qquad (14)$$

where σ_s is the standard deviation of the measured area, S_{max} and S_{min} are respectively the maximum and minimum normal area. The degree of stenosis was computed as the pathological cross-sectional area S_m and the $\overline{S}_n$ ratio, as indicated in Fig. 2.

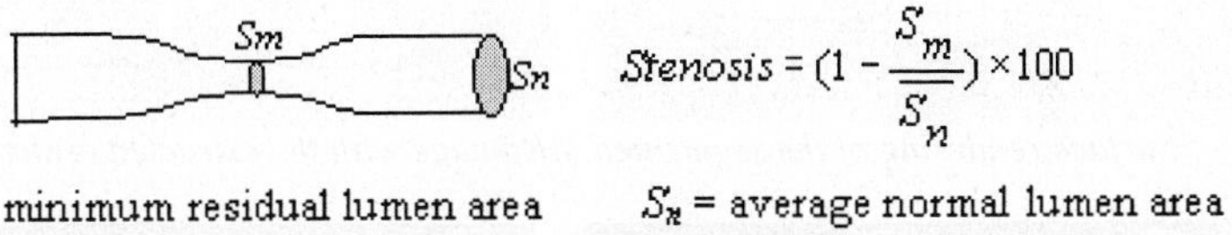

$$Stenosis = \left(1 - \frac{S_m}{\overline{S}_n}\right) \times 100$$

S_m = minimum residual lumen area $\qquad S_n$ = average normal lumen area

Fig. 2. Stenosis quantification

The stenosis severity estimation was quantitatively evaluated by comparing it to the true theoretical stenosis values of the phantoms. A summary of quantitative results for these data is given in Fig. 3. The CV coefficients ($CV_\sigma < 2$ %, $CV_{max} < 5$ %) indicate a good stability of the vessel contour extraction results. The fluctuations can be attributed to the partial volume effect. Indeed, let us consider the case of the phantom (a) which presents the greatest CV. The cross-sectional area from a normal part (diameter = 6 mm) is 28.27 mm. Its voxel size is 0.78 mm x 0.78 mm x 0.76 mm. It means that each pixel missing (or exceeding) in the extracted surface, represents an error of 2,15 %. This shows that our results contain not more than one or two misclassified pixels, which is a correct value if one takes into account the partial volume effect. Moreover, the stenosis quantification absolute errors are not more significant than the above mentioned variations and could be explained by the same reasons. The slightly larger errors for the phantom (b) may be due to a more important partial volume effect, since the image, in this case, was acquired with a coarser resolution.

3.2. Qualitative Results

The axis position as well as the cross-sectional contour shape have been qualitatively evaluated on data from 7 patients with stenosis located in different arteries: carotid, renal and lower limbs. According to a visual appreciation, the results were satisfactory for large vessels as well as for small and low intensity ones. Fig. 4 shows examples of successful centerline extractions for vessels of different sizes in patient images.

(a)   *(b)*

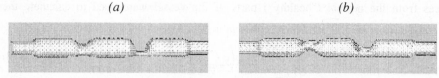

CAD images of the phantom

Maximum Intensity Projection (MIP)

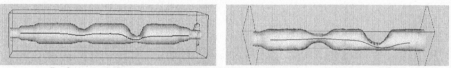

Surface rendering of the segmented MR image with the extracted centerline

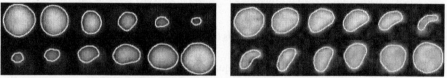

A few planes orthogonal to the extracted centerline.
The extracted contours are superimposed onto the images

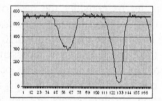 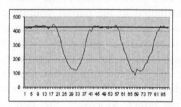

Cross-sectional area. The average normal cross-sectional area is indicated by the bold line

Stenosis	Estimated
50%	48.33%
95%	94.33%

Stenosis	Estimated
75% concentric	70.53%
75% eccentric	73.80%

$$CV_\sigma = 1.92\%$$
$$CV_{max} = 4.91\%$$

$$CV_\sigma = 1.98\%$$
$$CV_{max} = 4.29\%$$

Fig. 3. Results from phantom data. Column (a): Phantom with a 50% and a 95% stenosis, both elliptic and eccentric. Column (b): Phantom with a 75% semi-lunar, eccentric stenosis and a 75% semi-lunar concentric stenosis.

3.3. Visualization Application

Virtual endoscopic or fly-through methods, combining the features of endoscopic viewing with cross-sectional and volumetric imaging, are of great interest for the physicians. At present, some of clinical workstations provide this functionality, but their major drawback is that the user needs to define the path manually. This manual planning of flight paths is a tedious, imprecise and time-consuming task. A direct application of our vessel axis extraction method in the field of visualization is the automation of the path construction for guiding virtual endoscopic exploration of vessels. Fig. 5 shows an example of aorta exploration: the vessel axis is visualized at the same time as the vascular walls. In this case, the translucent wall allows the operator to see the collateral small vessels like the renal arteries, which may be used as reference points for trajectory location.

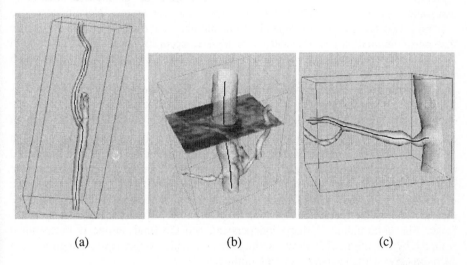

(a)　　　　　　　　(b)　　　　　　　　(c)

Fig. 4. Vessel axis extraction results. (a) Carotid artery and its axis. (b) Aorta central axis and orthogonal plane (c) Severely stenosed renal artery and its extracted axis

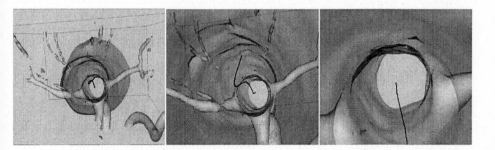

Fig. 5. Aorta virtual endoscopy guided by the detected centerline

4. Discussion and Conclusion

We devised a vessel segmentation method applied to stenosis quantification in 3D MRA images. The method was qualitatively and quantitatively evaluated. The visual appraisal of the results shows a precise estimation of the vessel axis position and orientation as well as of contour location. Quantitative results show that the stenosis severity was estimated with good accuracy even for severe stenosis.

Concerning the vessel axis extraction, our method differs from other vascular network extraction techniques founded on second derivatives [7][16] or on curvatures [17], proved to be noise-sensitive. Hence our first objective was to reduce the noise-sensitivity in order to cope even with very small and low contrast vessels. Due to its expansible nature, the centerline extraction algorithm overcomes the initialization limitations of deformable models as well as some of geometrical flexibility constraints such as the adaptation to significant curvatures. Nevertheless, we remarked that the results are dependent on the size of the cell used for the inertia moments computation and it needed to be adjusted according to the size to the vessel to be extracted. A multiscale approach could be considered to cope with vessel width variability.

Concerning the vessel contour detection, its advantages can be summarized as follows. The contour motion is independent of the scale of the snake. With the damping parameter γ (eq. 11), the snake can move with a controlled speed of about one pixel/ iteration throughout its evolution, and thus avoids missing energy minima on its way. Furthermore, the method does not need much computation time compared to the classical snake model: the scaling of the control points is a simple multiplication for each point and the points redistribution is a fast operation. The normalized snake with a fixed number of control points does not need to change the matrix of the snake model, and is initialized only once. This makes the snake much faster. The deformation is shape independent, and the final outline is independent from the initialization. This allows an initialization with a single pixel, and growing in each direction while keeping a smooth outline.

Nevertheless, the proposed approach shows some limitations. Based on a generalized-cylinder model, it cannot deal with bifurcations. At present, the algorithm succeeds in extracting the vessel axis even at the joint points, but it tracks only one branch at a time. Additional work is necessary to process the whole vessel tree. Furthermore, reliable boundary detection should not be limited to 2D contours in the planes locally orthogonal to the axis, but should be carried out in a 3D neighborhood.

Acknowledgements

This work is in the scope of the scientific topics of the GDR-PRC ISIS research group of the French National Center for Scientific Research (CNRS).

References

1. X. Hu, N. Alperin, D. Levin, K. Tan, M. Mengeot - Visualization of MR angiographic data with segmentation and volume-rendering techniques, *J Magnetic Resonance Imaging* 1, 539-546, 1991
2. D.L. Wilson, J.A. Noble - Segmentation of cerebral vessels and aneurysms from MR angiography data, *Information Processing in Med Imaging. 15th Int. Conf,* Springer, Berlin, 423-428, 1997
3. Y. Masutani, T. Schiemann, K-H. Höne, Vascular shape segmentation and structure extraction using a shape-based region-growing model, *MICCAI'98*, Boston USA, 1242-1249, 1998
4. A.C.S. Chung, J.A. Noble, Statistical 3D vessel segmentation using a Rician distribution, *MICCAI'99*, Cambridge UK, 82-89, 1999
5. B. Verdonck, I. Bloch, H. Maître - Accurate segmentation of blood vessels from 3D medical images, *ICIP'96*, Lausanne , 311-314, 1996
6. B. Nazarian, C. Chédot, J. Sequeira, Automatic reconstruction of irregular tubular structures using generalized cylinders, *Revue de CFAO et d'informatique graphique*, 11(1-2), 11-20, 1996
7. C. Lorenz, I-C. Carlsen, T.M. Buzug, C. Fassnacht, J. Weese - Multi-scale line segmentation with automatic estimation of width, contrast and tangential direction in 2D and 3D medical images, *CVRMed/MRCAS'97*, Grenoble, France, 233-242, 1997
8. O. Wink, W.J. Niessen M.A. Viergever, Fast quantification of abdominal aorta aneurysmes from CTA volumes, *MICCAI'98*, Boston USA, 138-145, 1998
9. KC. Wang, R.W. Dutton, C.A. Taylor, Improving geometric model construction for blood flow modeling, *IEEE Engineering in Medicine and Biology*, 18(6), 33-39, 1999
10. A.F. Frangi, W.J. Niessen et al, Model-Based quantitation of 3-D magnetic resonance angiographic images, *IEEE Transactions on Medical Imaging*, 18(10), 946-956, 1999
11. A. Bulpitt, E. Berry, An automatic 3D deformable model for segmentation of branching structures compared with interactive region growing, *Med Image Understanding Anal*, Leeds UK, 25-28, 1998
12. U. Shani, D. Ballard, Splines as embeddings fot generalized cylinders, *CVGIP'84*, 27(2), 1984.
13. M. Hernández-Hoyos, M. Orkisz et al, Inertia-based vessel axis extraction and stenosis quantification in 3D MRA images, *CARS'99*, Paris, 189-193,1999
14. M. Kass, A. Witkin, D. Terzopoulos, Active contour models, *Int. J Computer Vision*, 1, 321-331, 1988
15. L. D. Cohen, On Active Contour Models and Balloons, *Computer Vision, Graphics and Image Processing: Image Understanding*, 53(2), 211-218, 1991
16. Y. Sato, S. Nakajima, *et al.* - 3D Multi-scale line filter for segmentation and visualization of curvilinear structures in medical images, *CVRMed/MRCAS'97*, Grenoble, France, 213-222, 1997
17. V. Prinet, O. Monga - Vessels representation in 2D and 3D angiograms, *CARS'97*, 240-245, 1997

Development of a Computer Program for Volume Estimation of Arteriovenous Malformations from Biplane DSA Images

Roger Lundqvist[1], Michael Söderman[2], Kaj Ericson[2], Ewert Bengtsson[1], Lennart Thurfjell[1]

[1] Centre for Image Analysis, Uppsala University,
Uppsala, Sweden
[2] Dept. of Neuroradiology, Karolinska Hospital
Stockholm, Sweden

Abstract. The goal of this work was to develop a computer program for accurate estimation of size and location of intracranial arteriovenous malformations (AVM) from biplane digital subtraction angiography (DSA) images. The program will be used for diagnostic purposes to predict the outcome of Gamma knife radiosurgery and thus give basic data for the optimization of the management and comparison of different treatment modalities. Our solution is based on the so called intersecting cone model (ICM). Volume measurements with this model have been shown to correlate well to volume data from the dose-planning equipment (Leksell GammaPlan®, Elekta, Sweden). The method described in this paper produces a fast and accurate implementation of the ICM suitable for clinical practice. The implementation has been validated from phantom images and simulated images and the errors have been shown to be small compared to the approximations made by the ICM.

1 Introduction

Accurate volume estimation of intracranial arteriovenous malformations (AVMs) is important in radiosurgical treatment. The risk for complications is directly related to the irradiated volume and the radiation dose given, as are the chances of cure [1], [2], [3]. In Gamma Knife radiosurgery (GKRS) of AVM, the dose planning (Leksell GammaPlan®, Elekta, Sweden) is based on biplane digital subtraction angiography (DSA) images, preferably integrated with magnetic resonance imaging (MRI) [4] and [5]. Such dose planning is usually only available in institutions where GKRS is performed. In the clinical situation knowledge about the predicted outcome of GKRS is valuable in patient management, even if the facilities for GKRS are lacking [6]. For example centres performing AVM embolisation may find it valuable to accurately estimate AVM volume from diagnostic angiography before and after procedures, for an objective evaluation of the results. They may also compare their results with the radiosurgical outcome model and optimize the combined treatment [7]. A potentially hazardous continued endovascular treatment may be halted when the volume reduction has reached a level where radiosurgery can safely be applied. This assessment may be done by placing a stereotactic box over the head of the patient at the end of the diagnostic angiography and perform volume measurement, [6] and [7].

Some years ago a computer program for measurement of AVM volumes was developed at the Karolinska Hospital, in collaboration with General Electric Medical Systems. It was based on two or more most often perpendicular DSA images, with the AVM contour delineated in each of them. The volume was approximated with the interior of the intersection of the two X-ray cones intersecting the AVM contours in stereotactic space. The approach was named the intersecting cone model (ICM). The values obtained with ICM is an overestimation of the AVM volume, but it has been shown that the difference is small when compared to data from the dose planning [7]. The program developed at that stage was, however, closely connected to an old version of the GE Advantage Windows system and not easily applicable for further improvements. The measurement algorithms used internal functions in the GE Advantage Windows system, which made them hard to port to a new program. Finally, the algorithm for the volume measurements was very slow when the sizes of the AVM volumes were large.

These limitations of the old program gave rise to the decision to develop a new program, independent of the old one and with new algorithms to solve the volume measurement problem, but still based on the ICM. The goal was to create a fast and flexible system suitable for clinical practice and further development. The result of this work, including validation and description of the methods used, is presented in this paper.

2 Methods

All DSA images used in this work were acquired on a General Electric Medical Systems Advantx®, angiography equipment. The image file format for all images followed the DICOM 3.0 standard. The DSA images were corrected for geometric distortion by an algorithm developed in collaboration with General Electric Medical Systems [8], before the volume measurement process was started.

2.1 Stereotactic box calculations

The first step is to register the different DSA projections into the same coordinate system. This is made from a stereotactic plexiglass box (Elekta, Sweden) with engraved fiducials, which is used during the angiography. It is essential that the box is retained in the same position relative to the head during acquisition of the lateral and PA projections, but this is easily achieved with biplane angiography without the need to fix the box to the head of the patient. The detection of the stereotactic box in the images is performed by manual marking of the fiducials by the mouse, see Fig. 1.

We consider the stereotactic box coordinate system as fixed and the positions of the X-ray origin X_0 and projection image plane P_p as moved relative to

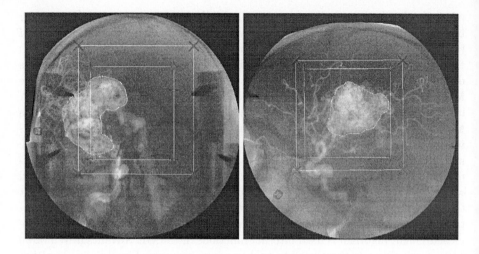

Fig. 1. In the images two different DSA projections with a large visible AVM are shown. The stereotactic box has been defined and the AVM has been delineated by the user

this coordinate system, see Fig. 2. The projection image plane can be located anywhere along the X-ray direction. In our solution, however, we regard it as located at the specific position where the pixel size in the projected image corresponds to some specific predefined value. This allows us to couple distances and positions in the projected 2D-image to positions in stereotactic 3D-space.

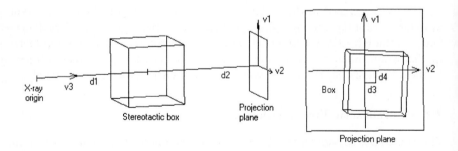

Fig. 2. The left part of the figure shows a sideview of the angiography examination. To the right we see the same scene viewed in the direction from the X-ray source in direction of the point in the middle of the projection image matrix

We need to know a number of variables to be able to do the registration of the images into stereotactic space. For each projection, the distance d_1, from the box centre to X_0, and the distance d_2, from box centre to P_p, have to be found.

Also the relative rotation around the three box axes of the vector $\overline{v_3}$, pointing in the X-ray direction, has to be found. These rotations we denote as r_x, r_y and r_z. Finally, we also have to find the translation of the point in the middle of the P_p image matrix, which we denote as p_0, relative to the box centre. This is expressed as a two component translation, d_3 and d_4, along the two base-vectors $\overline{v_1}$ and $\overline{v_2}$ spanning the projection image plane P_p.

From the three rotations we can build a rotation matrix $\mathbf{R}$. If we denote the base vectors $\overline{b_1}$, $\overline{b_2}$ and $\overline{b_3}$ in the stereotactic space, which are the unit vectors in respective direction, we can express the base-vectors in the X_0-P_p coordinate system according to:

$$\overline{v_1} = R\overline{b_1} \qquad\qquad \overline{v_2} = R\overline{b_2} \qquad\qquad \overline{v_3} = R\overline{b_3} \qquad (1)$$

This leads to the following equations for the points X_0 and P_p :

$$\overline{X_0} = -d_1\overline{v_3} + d_4\overline{v_1} + d_3\overline{v_2} \qquad\qquad \overline{p_0} = d_2\overline{v_3} + d_4\overline{v_1} + d_3\overline{v_2} \qquad (2)$$

If we use the cosinus-theorem it is possible to calculate a projected point p' on P_p for every 3D-point p in the stereotactic space, since we know the base-vectors and origins for the two coordinate systems. In the same way it is possible to express the equation for an arbitrary ray through 3D-space from point X_0 to a point p' on P_p.

There are a number of different approaches to find the 7 parameters (r_x, r_y, r_z, d_1, d_2, d_3, d_4) needed to perform these calculations. We have chosen to solve this problem by using a numerical method, since an analytical approach would lead to an overdetermined equation system and still result in an approximation of the real solution. We use an iterative optimization method based on Powell's algorithm implemented as it is described in [9], where a cost function expressing how well a specific parameter set solves the projection problem is minimized. In our solution this cost function calculates the sum of the distances between the box fiducials projected down on P_p and the corresponding points marked by the user directly in the image. For each projection there are totally eight fiducials used. The optimization process results in one optimal parameter configuration for each projection image which defines the images relation to the stereotactic space.

2.2 Volume measurement calculations

The next step is to draw the AVM contours in at least two DSA projections by the mouse, see Fig. 1. After that a result volume is created in the computer memory to be used when calculating the AVM volume. The center of it corresponds to the middle of the stereotactic coordinate system. In the next step, rays are cast through this volume from X_0 to all points lying on the drawn AVM contour on P_p. All voxels passed by the rays are set to a non-zero value in the result volume. This results in a hollow cone for each projection, which is filled through a floodfill algorithm. A different bit is set for each projection, which makes it possible to find the voxels that are set in both projections. In the next step the number of voxels set in both projections are counted and multiplied by the volume of an individual voxel to find an estimation of the total AVM volume.

There is a problem if we want to use the full image resolution in this process, since typically DSA images have a matrix of 1024x1024, leading to a result volume of a size in the order of 1024x1024x1024 voxels. Even if only 2 data bits per voxel is enough to handle the storage of the resulting volume, it would still be a too large volume for most computers. The solution to this is to perform the volume measurement in a two step multi-resolution process. First we use a low resolution result volume, 128x128x128 voxels, and perform the whole process. The boundaries of the AVM volume are found and a new result volume, containing only the part with the AVM, with 8 times higher resolution is created. This results in a volume with manageable size, which is used in the final step where the whole process of casting rays through the AVM contours is repeated once again. It can also be noticed that the edge voxels of the cones are not completely belonging to the AVM. Since the number of such voxels becomes large at the finest resolution, we make the assumption that they contribute with approximately half their size to the measured volume. For this reason the edge voxels are counted separately and only half of their voxel volume are added to the final result.

3 Experiments

3.1 Phantom data

Some geometrical object phantoms were constructed and mounted inside the stereotactic box. These were two cylinders of different sizes and one rectangular box shaped object. The objects were measured manually and their "true" volumes were calculated based on their geometrical forms. Two different image series were acquired to test for different positions of the objects. In both series the objects were positioned with their sides in approximately perpendicular angles towards the DSA projections to test the ability of the program to make an accurate volume estimation, Fig. 3. The only difference between the series was a translation of the objects relative to the X-ray sources. The objects were redrawn and processed three times for each series and the results were compared to the true volumes.

3.2 Simulated data

A validation of the combined errors from the manual marking of the stereotactic box together with the numerical solution of the box equation system was also made. A test program was written in C++ using OpenGL to create simulated images of the box fiducials, with known rotation angles, translations and distances relative to the imaging equipment. First the known positions of the

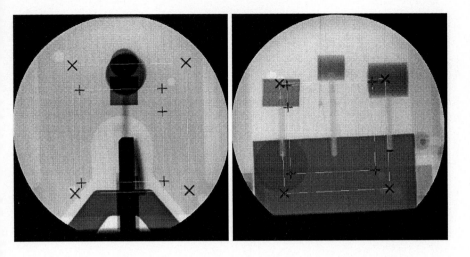

Fig. 3. In this figure two approximately perpendicular projections of the phantom images used in the experiments are shown

fiducials were defined in a 3D coordinate system and then a random configuration for the parameter transformations were chosen within reasonable limits. These transformations were applied to the 3D coordinate system through OpenGL:s internal functions for rotations, translations and perspective projection and the fiducials were drawn in 3D space and projected onto the image plane, Fig. 4.

After this procedure the image was saved and loaded into the volume measurement program and processed in the same way as the real DSA images, hence resulting in an estimated solution of the true parameter configuration used when creating the simulated image. A total of 50 simulated images were constructed in this way and processed by the program and maximal, absolute mean and standard deviation for the errors in the transformation parameters were calculated.

Finally, the contribution of these errors on the outcome of the actual volume measurement were studied. This was accomplished by randomly adding errors to the parameters, based on the knowledge of the error size, and compare the volume measurements with parameter errors included to those without. The parameter errors, however, were not independent of each other, since different transformations compensate for errors made by others. For this reason we made an inspection of the distributions of the errors and found that an assumption of a multivariate normal distribution seemed justifiable. According to that assumption the random errors were produced by generating multivariate random numbers based on the covariance matrix of the errors. This was repeated for some AVM volumes of different sizes and for each volume a total of 100 volume measurements with included errors were made. The distribution of these measurements compared to the values without added errors was then examined.

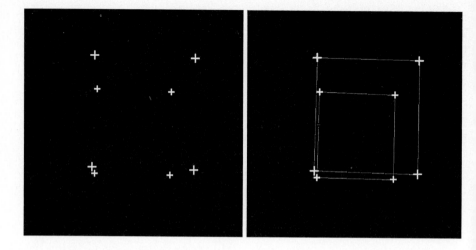

Fig. 4. In this figure a simulated image of the stereotactic fiducials before and after definition of the stereotactic box is shown

3.3 Comparisons to old program

The old program has been validated earlier against the radiosurgical dose planning system (LGP) and showed good correlation with that [7]. For that reason the volume estimates of the new program also has been compared with results from the old program. A perfect comparison is impossible to accomplish since the file formats for saving the manual marks of the fiducials and drawn AVM contours are not compatible between the systems. This means that the manual interaction has to be performed twice, independently of each other. Nevertheless, a number of measured volumes from different examinations have been compared to verify that the two programs produce similar results.

4 Results

4.1 Phantom data

Due to the specific geometrical forms of the phantoms, the measured volumes with the ICM should be very close to the true volumes, provided that they are positioned with their sides at right angles to the two DSA projections. In Table 1 the results from the estimates of the phantom volumes based on the ICM compared to the volumes calculated from the geometrical forms of the objects are presented. In the comparisons we can see that the spread for the same objects within series is small compared to the between series measurements, although the angles of the objects were very similar. Wee can also study the accuracy of the method. For all objects the series which produces the best result gave a very good approximation of the volume. However, we should also notice that the measurements are actually smaller than the reference volume for some series. The best explanation to this is that the reference value also is an estimation of the volume since it is based on manual measurements. The edges of the objects were also not perfectly sharp in the images which could lead to some errors when drawing the contours.

Table 1. Volume measurements from phantom images. In each series three measurements were made based on differently drawn AVM contours

Object	Ref. value	Series 1 Mean	Series 1 Range	Series 2 Mean	Series 2 Range
Small cylinder	5.840	5.745	0.114	5.565	0.022
Large cylinder	15.303	16.037	0.030	15.287	0.086
Rectangular box	11.088	10.540	0.059	11.076	0.120

4.2 Simulated data

In Table 2 the results from the error estimates of the transformation parameters based on the simulated image experiments are presented. It is clear that the errors are not particularly large except for parameters d_1 and d_2. These errors, however, compensate for each other to a large extent and the correlation factor was 0.93. For this reason the overall error caused by these parameters are small compared to other sources.

Table 2. Distribution of volume measurements when random errors are added. A total of 100 measurements for each volume were made

Measures	No error	Mean	Min	Max	Stdev	Range
Volume 1	1.230	1.228	1.183	1.257	0.015	0.074
Volume 2	2.190	2.205	2.153	2.303	0.031	0.150
Volume 3	3.162	3.154	3.129	3.187	0.011	0.058
Volume 4	4.564	4.552	4.421	4.616	0.029	0.196
Volume 5	6.238	6.232	6.135	6.328	0.037	0.193
Volume 6	8.383	8.385	8.265	8.471	0.034	0.206
Volume 7	11.815	11.835	11.707	12.062	0.077	0.355

The results from the study of how the parameter errors effect the volume measurements are presented in Table 3. We can see from this table that the spread of the measurements for the same volume gets larger with larger AVM volume size. However, these deviations should be compared to the size of the overestimation error with the ICM. For instance if the AVM has an ellipsoid shape the overestimation is about 20%. This can be compared with the maximal deviation from the mean value in these measurements, which never exceeds 5%. For this reason a conclusion is that the errors made by the program, caused by approximations in the calculations and manual interaction, are not decisive when studying the accuracy of the method.

Table 3. Parameter errors from simulated image experiments. A total of 50 images were used

Errors	r_x	r_y	r_z	d_1	d_2	d_3	d_4
Abs. mean	0.108	0.096	0.130	4.837	3.983	0.124	0.115
Stdev	0.153	0.122	0.158	5.868	4.879	0.114	0.092
Max	0.381	0.308	0.371	13.873	11.606	0.366	0.299
Range	0.750	0.568	0.674	26.271	19.940	0.459	0.340

4.3 Comparisons to old program

The volume estimates calculated with the new program are in agreement with the old program. Some minor differences can be noticed, which can be explained by the fact that the manual interaction is performed independently two times. The differences in results are within the error range found from the simulated image experiments and they are also very small compared to the assumptions made by the ICM.

The processing time with the new program is much shorter than the old program. For example a calculation of a large AVM takes less than 5 seconds with the new program using a PC with an Intel Pentium III 500MHz CPU. A similar volume could take several minutes to calculate with the old program. Even though the computers are not directly comparable a significant improvement in performance can be noticed with the new algorithms, which makes the program much more suitable for clinical use.

5 Discussion and conclusions

We have developed a computer program for estimation of AVM volumes based on biplane DSA images. The program has been thoroughly validated against phantom data and simulated images to estimate the errors in the volume measurement. It has been clearly shown that the errors made by using the program are not large compared to the assumptions of the ICM. The implementation is also fast which makes it very useful for clinical practice. The system will be extended with improved functionality for image database support and there are several clinical applications for it. One is the possibility to calculate already at the diagnostic angiography and decide if a newly detected AVM is suitable for radiosurgical treatment. An early decision may then be taken to treat the patient with endovascular means or open surgery in the first place. A repeated volume measurement after this treatment will show if the AVM now is treatable with radiosurgery. An obvious application is, of course, as quality control of endovascular procedures. The results of embolisations are usually estimated in very vague terms. A simple and relatively accurate volume measurement will objectively show how much that has been accomplished with embolisation.

6 Acknowledgments

This project was funded by the Swedish Foundation for Strategic Research through the VISIT-program.

References

1. Karlsson B., Lax I., Söderman M.: "Factors influencing the risk for complications following Gamma Knife radiosurgery of cerebral arteriovenous malformations", Radiotherapy and Oncology 1997;43(3):275-80
2. Karlsson B., Lax I., Söderman M.: "Can the probability for obliteration of arteriovenous malformations following radiosurgery be accurately predicted?", Int Journ Rad Onc Biol Phys 1999;43(2):313-319
3. Karlsson B., Lax I., Söderman M.: "Risk for haemorrage in the latency period following Gamma Knife radiosurgery for cerebral arteriovenous malformations", Int. J. Rad. Onc. Biol. Phys. 2000;Submitted
4. Guo W., Lindqvist M., Lindqvist C., et al.: "Stereotaxic Angiography in Gamma Knife Radiosurgery of Intracranial Arteriovenous Malformations", AJNR 1992;13:1107-1114
5. Guo W-Y., Nordell B., Karlsson B., et al.: "Target delineation in radiosurgery for cerebral arteriovenous malformations. Assessment of the value of stereotaxic MR imaging and MR angiography", Acta Radiol 1993;34(5):457-63
6. Ericson K., Söderman M., Karlsson B. and Lindquist C.: "Volume determination of intracranial arteriovenous malformations prior to stereotactic radiosurgical treatment", Interventional Neuroradiology 2 1996;271-275
7. Söderman M., Karlsson B., Launnay L., Thuresson B. and Ericson K.: "Volume measurement of cerebral arteriovenous malformations from angiography", Neuroradiology 2000, In press.
8. Söderman M., Picard C. and Ericson K.: "An algorithm for correction of distortion in stereotactic digital subtraction angiography", Neuroradiology 40: 1998;277-282
9. Press W.H., Flannery B.P., Teukolsky S.A., Vetterling W.T.: "Numerical recipes in C", Cambridge: Cambridge University Press, 1988.

Geometrical and Morphological Analysis of Vascular Branches from Fundus Retinal Images

M. Elena Martínez-Pérez[1], Alun D. Hughes[2], Alice V. Stanton[2], Simon A. Thom[2], Neil Chapman[2], Anil A. Bharath[1], and Kim H. Parker[1]

[1] Department of Biological and Medical Systems.
Imperial College of Science, Technology and Medicine.
Prince Consort Road, London, SW7 2BY, UK
[2] Department of Clinical Pharmacology, NHLI,
Imperial College of Science, Technology and Medicine, St Mary's Hospital
South Wharf Road, London W2 1NY, UK
elena.martinez@ic.ac.uk

Abstract. A semi-automatic method to measure and quantify geometrical and topological properties of complete vascular trees in fundus images is described. The method is validated by comparing automatic *vs.* manual measurements in 17 individual bifurcations. We also compared automatic analyses of complete vascular trees from 4 pairs of red-free and fluorescein images. Preliminary results comparing 10 hypertensive and 10 normotensive subjects show changes in geometrical properties similar to those reported in previous studies. Several topological indices show differences in the arterial, but not venous, trees in hypertensive subjects. This suggests that a combination of geometrical and topological measurements of the whole vascular network may provide more sensitive indicators of morphological changes due to diseases.

1   Introduction

The basic geometrical properties in branching trees are lengths and diameters of the branches and the angles at which branches meet at bifurcations. Generally both the arteries and veins of the retina are binary trees, whose properties can be considered either locally or globally. The local properties are, for example, the relationships between the branching angles and the ratios of diameters at individual bifurcations. The global properties focus on ordering and relationships between branches. As an example, the orientation of minor branches relative to the anatomical features of the retina could be an interesting topological property.

Retinal blood vessels from fundus images have usually been studied in terms of individual bifurcations, measuring a few of the most clearly visible bifurcations in an image. Differences in geometrical properties in retinal blood vessel have been found as a result of different diseases. Hypertension, for example, causes: reduction in the vessel diameter [8], reduction in the number of vessels present per unit volume of tissue [10], reduction of the branching angle [9] and increase in the length to diameter ratio [3].

In order to study the global properties of branching structures, it has been usual to group bifurcations into schemes based on order or generation. Some of these models are ordered from the root of the tree toward the leaves, others from the leaves toward the root. Some of these schemes assume all branches are symmetrical, incorporating diameters and length but not branching angles. Others assume all branches with the same order have the same diameter and length and bifurcate with the same asymmetry. Several studies have been carried out formulating models of the branching geometry, particularly of the human conducting airways, following these different schemes (discussed by [7]). There are no studies, as far as we are aware, related to retinal blood vessels from this global point of view.

In this work a semi-automatic method to measure and quantify geometrical and topological properties of continuous vascular trees in fundus images is described. The combination of geometrical with topological properties in the study of retinal vascular structures could yield more reliable indices to differentiate pathologies in the vascular system.

2 Vessel segmentation

Blood vessels are segmented using a previously described algorithm based on scale-space analysis [5]. Two geometrical features based upon the first and the second derivative of the intensity image along the scale-space, maximum gradient and principal curvature, are obtained by means of Gaussian derivative operators. A multiple pass region growing procedure is used which progressively segments the blood vessels using the feature information together with spatial information about the 8-neighboring pixels. The algorithm works with red-free as well as fluorescein retinal images. Figure 1(a) shows the scanned negative of a red-free retinal photograph and (b) the segmented binary image where the optic disc region is marked in grey, vessels are tracked from this area outwards.

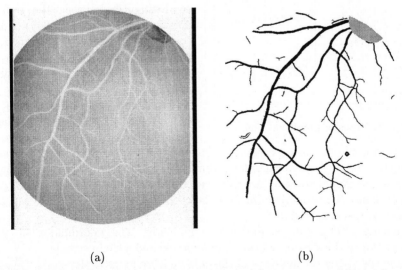

(a)   (b)

Fig. 1. (a) Red-free fundus photograph negative, (b) the segmented binary image with the optic disc marked in grey, vessels are tracked from this area outwards.

3   Tree labelling

Labelling each tree of vessels involves three steps: thinning the segmented binary image to produce its skeleton, detecting significant points and tracking the skeleton of the tree.

Thinning. The skeleton of the vascular tree is obtained from the segmented binary image by a thinning process where pixels are eliminated from the boundaries towards the centre without destroying connectivity in an 8-connected scheme. This is an approximation to the medial axis of the tree. A pruning process is applied to eliminate short, false spurs, due to small undulations in the vessel boundary.

Detecting significant points. Three types of significant points in the skeleton must be detected: terminal points, bifurcation points and vessel crossing points.

In a first pass, skeleton pixels with only one neighbor in a 3×3 neighborhood are labelled as terminal points and pixels with 3 neighbors are labelled as candidate bifurcation points. Because vessel crossing points appear in the skeleton as two close bifurcation points, a second pass is made using a fixed size window centered at the candidate bifurcations. The number of intersections of the skeleton with the window frame determine whether the point is a bifurcation or a crossing. Figure 2 summarises the three possible cases.

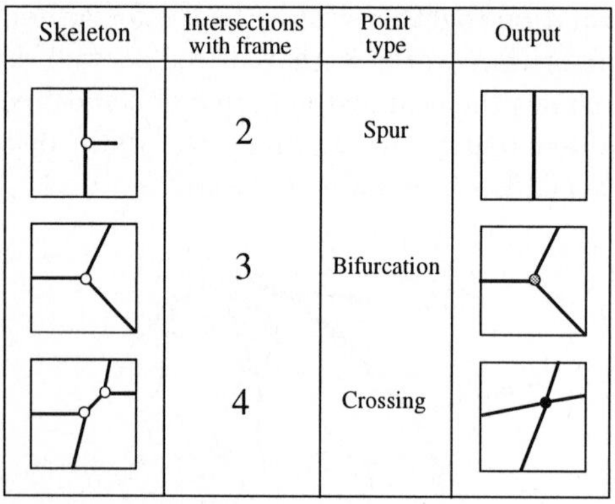

Skeleton	Intersections with frame	Point type	Output
	2	Spur	
	3	Bifurcation	
	4	Crossing	

Fig. 2. Types of significant points.

Once all of the significant points are identified, bifurcation points are labelled as $-r$ where r is the radius of the maximum circle centered on that point that fits inside the boundary of the bifurcation. The sign is used to distinguish between the radius and chain code numbers that are described in the following section. For the crossing points the skeleton is modified to define a single crossing point.

Until this stage the process is fully automatic, but it fails when two true bifurcation points are very close and are merged into a crossing point or when two vessels cross at a very acute angle so that the two candidate bifurcation points fall outside the window frame and are thus defined as two bifurcation points. These cases must be corrected by hand. The complete image is labelled and normally contains several independent vascular trees.

Tracking. Each tree is tracked individually. The coordinates of the starting point of the skeleton of the root are saved and a chain code is generated specifying the direction of the next skeleton point as shown. When the first bifurcation point is reached the chain of the current branch is ended and the coordinates of the starting points of the two daughters

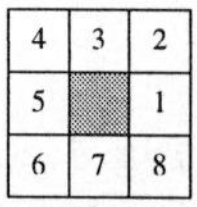

4	3	2
5		1
6	7	8

are saved. The process is iteratively repeated for every branch of the tree until a terminal point is reached and all daughters have been numbered. Using a binary scheme, the root is labelled 1 and thereafter the daughters of parent k_0 are labelled $k_1 = 2k_0$ and $k_2 = 2k_0 + 1$. With this numbering scheme we are able to track the tree in either direction.

4 Measurement of tree properties

Using the data generated in the labelling stage and the binary image, three types of geometrical features are automatically measured: lengths, areas and angles.

Lengths. Two lengths are measured directly from the skeleton. The first is the true length of the branch from the starting to the end points of the skeleton computed as $L_t = N_o + \sqrt{2}N_e$ where N_o and N_e are the number of pixels with odd and even direction codes along the skeleton. The second measure of length is the end-to-end distance L_s.

Areas. Using the value of the radius defined at each bifurcation point, a circle and a line tangent to this circle and perpendicular to the skeleton are drawn on the border image at each end of the branch. This is done to close the region to be measured as illustrated in Figure 3(a). This ensures that ambiguities in the area near bifurcation points are excluded. The number of vessel pixels

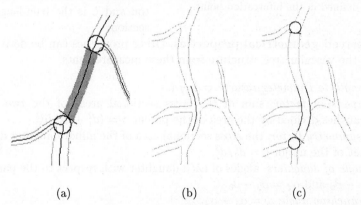

(a)	(b)	(c)

Fig. 3. (a) The area A is the number of vessel pixels bounded by the two perpendicular lines to the skeleton and L_a is its true length. (b) and (c) the borders at crossing points are closed using a displaced section of the skeleton until maximum correlation of pixels with the borders is found.

bounded by these tangents is defined as the area A. The average diameter of the selected branch is calculated as $d = A/L_a$, where L_a is the true length of the segment measured along the skeleton.

When two blood vessels cross, the borders in the crossing region for both vessels are opened. These are closed by taking a section of the skeleton centered at the crossing point and displacing it parallel to the skeleton in both directions until maximum correlations of pixels with the border are found (Figure 3(b,c)).

This maintains the natural shape of the curvature of the vessel in the crossing region.

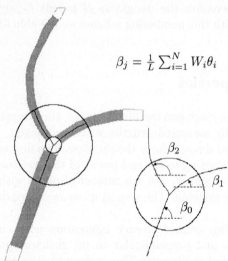

$$\beta_j = \tfrac{1}{L} \sum_{i=1}^{N} W_i \theta_i$$

Fig. 4. Three average angles β_j are defined and measured within a distance of 5 times the radius defined in the bifurcation point.

Angles. A bifurcation angle is associated with a parent vessel. Because a vessel branch can be very tortuous, and because daughters can be parents themselves, three different angles are defined and measured: the *total* average angle of the branch, a *head* angle and a *tail* angle. The latter two are defined within a distance of 5 times the radius of the particular bifurcation point (Figure 4). An average angle is defined as $\beta_j = \tfrac{1}{L} \sum_{i=1}^{N} W_i \theta_i$, where θ_i is the associated angle of the pixel i (45° steps in the chain code), W_i is the associated weight in the chain code $W = 1$ for odd and $W = \sqrt{2}$ for even directions, N is the number of pixels in the section of the skeleton and L is the true length of the section.

Derived geometrical properties. Other properties can be derived to describe the vascular tree structure from these measurements.

- *Length to Diameter ratio*: $\lambda = L_a/d$.
- *Expansion factor*: sum of the cross sectional areas of the two daughter branches divided by the area of the parent $\gamma = (d_1^2 + d_2^2)/d_0^2$.
- *Asymmetry factor*: the cross sectional area of the minor daughter divided by that of the major $\zeta = d_2^2/d_1^2$.
- *Angle of daughters*: angles of each daughter with respect to the parent $\alpha_1 = \beta_1 - \beta_0$ and $\alpha_2 = \beta_2 - \beta_0$.
- *Branching angle*: $\omega = \alpha_1 + \alpha_2$.
- *Angular asymmetry*: the angle of the major daughter respect to the parent divided by that of the minor $\eta = \alpha_1/\alpha_2$.
- *Tortuosity*: $T = L_t/L_s$.
- *Junction exponent*: the exponent x relating the parent and daughter vessel diameters $d_0^x = d_1^x + d_2^x$. Murray [6] suggested that $x = 3$ for maximum efficiency in arteries.

5 Preliminary Results and Conclusions

Validation of measurements. The true values of geometrical properties in retinal blood vessels remain unknown. Instead indirect methods are used and evaluated by comparison with an established technique rather than with the true quantity. With this purpose two validation studies were undertaken. Automatic measurements of individual bifurcations were compared with manual measurements for 17 randomly chosen bifurcations from red-free retinal images. The mean and the standard deviation of the normalised differences of diameter were -11 ± 14 % ($n = 51, p < 0.001$), *i.e.* automatic were smaller than manually measured diameters. The normalised differences of branching angles were 13 ± 12.3 % ($n = 17, p < 0.006$), *i.e.* automatic were larger than manual angles. The p values were calculating using the two sided Wilcoxon signed-rank test [2]. A Runs test for the differences as function of parent vessel diameters showed no systematic tendencies with vessel width. Since the manual measurements involved the average of 5 diameters measured close to the bifurcation and the angles between straight lines fitted by eye, these differences can not be taken as error but as indications of the variability of different measurement techniques.

We also compared automatic measurements of the clinically more common red-free images with automatic measurements of fluorescein images for the same eye in 4 subjects. The normalised differences for the three geometrical features diameters d, branching angle ω and true length of the parent vessel L_t (excluding the roots) are shown in Table 1, some of the derived geometrical properties such as length to diameter ratio λ, tortuosity T, expansion factor γ and asymmetry factor ζ are also shown.

Feature	Differences (%)	n	p
d	2.5 ± 24.5	277	0.012
ω	-0.1 ± 12.6	128	0.317
L_t	-0.3 ± 9.1	107	0.026
λ	-7.8 ± 25.8	107	< 0.001
T	0.4 ± 2.3	107	0.073
γ	-11.7 ± 31.8	128	< 0.001
ζ	-19.3 ± 30.5	128	< 0.001

Table 1. Mean and standard deviations of the normalised differences of several geometrical features between red-free and fluorescein measurements. Values of p were calculated using the two sided Wilcoxon signed-rank test. Diameter d, branching angle ω, true length L_t, length to diameter ratio λ, tortuosity T, expansion factor γ and asymmetry factor ζ. Measures with $n = 107$ were done in parent branches excluding roots.

The p values of the Wilcoxon test show that there is no a significant difference between ω and T. The rest of the features were significantly different but the normalised differences are all small and significance is due to the large number of vessels measured. For example, the normalised difference in d corresponds to only 2.5%. The Runs test again showed no significant tendencies, except for the diameters in which there is a tendency to under estimate diameter in vessels with diameters less than 12 pixels. Most of these vessels are terminal branches which were more difficult to detect in the lower contrast red-free images. It should also be noted that the number of bifurcations analysed using the automatic analysis is significantly higher than in any study involving manual measurements.

Structural analysis. A table of measurements is automatically obtained with the basic geometrical features described above for each tree in an image. At the moment, arterial and venous trees are identified by an expert. Figure 5 shows (a) one arterial tree with its keys marked and (b) one venous tree, both from the image in Figure 1(b).

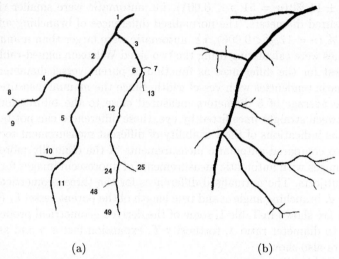

(a)   (b)

Fig. 5. Selected trees from Figure 1(b). (a) arterial tree with its vessels numbered and (b) the associated venous tree.

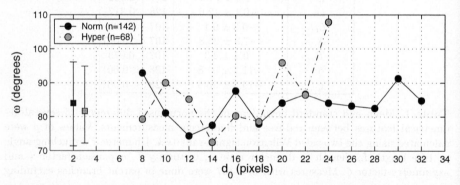

Fig. 6. Median of the branching angle ω as a function of the parent diameter d_0 in arterial trees. Normotensive and Hypertensive groups, where n = number of bifurcations. Grand medians and quartiles are to the left (square symbols).

Preliminary results for 10 hypertensive and 10 normotensive subjects using red-free images (2000×2300 pixels) have been obtained. A diameter-based classification system, originally used for lung airways [7], is used for our comparisons. Figure 6 shows the branching angle ω of arterial trees as a function of the parent diameter d_0. The overall median shows that ω for hypertensives is more acute than for normotensives as reported in [9], although the difference is not significant. In the normotensive group ω is relatively constant as the parent diameter increases, whereas the hypertensive group shows an increase in ω for parent vessels ≥ 14 pixels. A pixel corresponds to approximately 4 μm based on the diameter of the optic disc for this scale of images, although due to inter-individual variations in refractive indices of the eye it is not possible to establish an absolute scale.

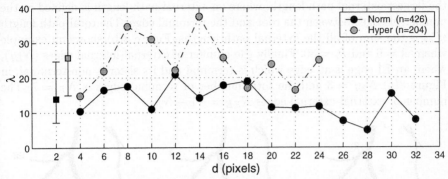

Fig. 7. Median of the length to diameter ratio λ as a function of diameter d in arterial trees. Normotensive and Hypertensive groups, where $n = $ number of vessels. Grand medians and quartiles are to the left (square symbols).

Figure 7 shows the median of the length to diameter ratio λ of the arterial tree vessels as a function of the diameter d. The overall median is significantly larger in hypertensives than normotensives in agreement with [3]. For the normotensive group λ is nearly constant over all diameters whereas in the hypertensive group λ seems to be larger in the medium sized vessels $7 < d < 16$ pixels. Note also that the number of vessels in the normotensive group is much higher than in hypertensive, as reported in [10], and that there are no vessels with $d > 25$ pixels in the hypertensive group, in agreement with the reduction in vessel diameters [8].

Topological analysis. Since we also keep the information about connectivity between branches in the table of measurements that is automatically generated, we are able to make topological analyses of the vascular trees. Vascular trees are branching structures which are treated as binary rooted trees, a special type of graph. A graph is a set of points (or vertices) which are connected by edges. A tree is a graph with no cycles. A rooted tree is a tree in which one vertex is distinguished as the root and a binary tree is one in which at most three edges are contiguous to any vertex [4].

Some important topological features of biological trees are those which measure the symmetry and elongation of a given branching structure. The topological indices we examined in this work are the most commonly used indices within biological applications [1]. (a) Strahler branching ratio (R), (b) maximum tree exterior path length or altitude (A), (c) total exterior path length (P_e), and (d) number of external-internal edges (N_{EI}), and the total number of external (or terminal) edges (N_T).

The Strahler branching ratio (R) is calculated by ordering all the edges within a given tree using Strahler ordering. This ordering scheme assigns all external edges an order of one. Where two edges of order m come together, the third edge is assigned to order $m + 1$. Where an edge of order m meets an edge of order n, the third edge is assigned to order $max(m, n)$. The Strahler branching ratio (R) is defined as $R = N_m/N_{m+1}$, where N is the number of edges and m is the order, Figure 8(a) shows this ordering scheme. The altitude (A) is computed as the largest external path length, where an external path length is defined as the number of edges between the root and the terminal edge. The total path length (P_e) is the sum of all the external path lengths. Figure 8(b) shows an example where $A = 6$ and $P_e = 30$. Finally the number of external-internal edges (N_{EI}), where an EI edge is an edge which is terminal and its sister is non-terminal, in Figure 8(c) $N_{EI} = 3$ and the total number of terminal edges is $N_T = 7$. The number of external-external edges $N_{EE} = N_T - N_{EI}$.

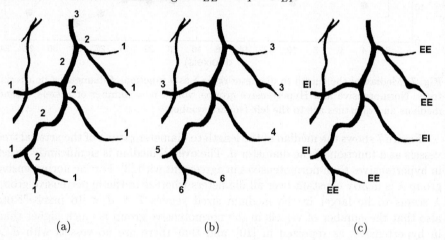

(a)   (b)   (c)

Fig. 8. Topological indices: (a) Strahler order scheme where the branching ratio is: $R = N_m/N_{m+1}$, (b) Altitude $(A = 6)$ and total path length $(P_e = 30)$, and (c) number of external-internal edges $(N_{EI} = 3)$, where the total number of terminal edges is $N_T = 7$.

Using the same set of normotensive and hypertensive subjects as before, we calculated each of these topological indices for all trees with $N_T > 4$, for both artery and vein trees. Table 2 shows the p values for the t-test between groups, where we can see that all indices for arteries were significantly different except for the altitude (A), whereas for veins all indices were not significantly different. These results suggest that arteries from the hypertensive group are topologically different from normal. Comparing R for different Strahler orders showed that neither arterial nor venous trees were self-similar.

The purpose of this analysis is not to reach any conclusions about physiology, but to show the kind of analysis that could be usefully applied to provide insights about the changes in the geometry of the whole retinal blood vasculature in diseases such as hypertension. We have shown that the (nearly) automatic method of geometrical quantification proposed in this paper is sensitive enough to show differences in geometrical changes already reported in the literature. It also sug-

Index	Arteries $(n = 21)$	Veins $(n = 25)$
NT	$p = 0.013$	$p = 0.501$
A	$p = 0.247$	$p = 0.385$
P_e	$p = 0.049$	$p = 0.989$
EI	$p = 0.043$	$p = 0.189$

Table 2. Comparison of topological indices between normotensive and hypertensive groups for arteries and veins separately. p values for the t-test between groups. Number of terminal edges NT, altitude A, total path length P_e and number of external-internal edges EI.

gests that analysis of the geometry and topology of the whole vascular network may provide more sensitive indicators of changes due to disease. Future work will include fractal analysis and other combinations of geometrical features and comparisons with other ordering schemes in order to quantify retinal vascular trees more fully.

References

1. G.M. Berntson. The characterization of topology: a comparison of four topological indices for rooted binary trees. *J. Theor. Biol.*, 177:271–281, 1995.
2. J. M. Bland and D. G. Altman. Statistical methods for assessing agreement between two methods of clinical measurement. *The Lancet*, 1(8):307–310, February 1986.
3. L.A. King, A.V. Stanton, P.S. Sever, S. Thom, and A.D. Hughes. Arteriolar length-diameter (l:d) ratio: A geometric parameter of the retinal vasculature diagnostic of hypertension. *J. Hum. Hypertens.*, 10:417–418, 1996.
4. N. MacDonald. *Trees and Networks in Biological Models*. John Wiley & Sons, New York, 1983.
5. M.E. Martínez-Pérez, A.D. Hughes, A.V. Stanton, S.A. Thom, A.A. Bharath, and K.H. Parker. Retinal blood vessel segmentation by means of scale-space analysis and region growing. In C. Taylor and A. Colchester, editors, *MICCAI-99*, volume 1679 of *Lectures Notes in Computer Science*, pages 90–97. Springer-Verlag, 1999.
6. C.D. Murray. The physiological principle of minimum work. i. the vascular system and the cost of blood volume. *Proc. Nat. Acad. Sci.*, 12:207–214, 1926.
7. C.G. Phillips, S.R. Kaye, and R.C. Schroter. A diameter-based reconstruction of the branching pattern of the human bronchial tree. part i. description and application. *Respir. Physiol.*, 98:193–217, 1994.
8. A.V. Stanton, P. Mullaney, F. Mee, E.T. O'Brien, and K. O'Malley. A method of quantifying retinal microvascular alterations associated with blood pressure and age. *J. Hypertens.*, 13:41–48, 1995.
9. A.V. Stanton, B. Wasan, A. Cerutti, S. Ford, R. Marsh, P.P. Sever, S.A. Thom, and A.D. Hughes. Vascular network changes in the retina with age and hypertension. *J. Hypertens.*, 13:1724–1728, 1995.
10. H.A.J. Struijker, J.L.M. le Noble, M.W.J. Messing, M.S.P. Huijberts, F.A.C. le Noble, and H. van Essen. The microcirculation and hypertension. *J. Hypertens.*, 10(7):S147–S156, 1992.

Computer-Based Assessment of Body Image Distortion in Anorexia Nervosa Patients

Daniel Harari[1], Miriam Furst[1], Nahum Kiryati[1],
Asaf Caspi[2], and Michael Davidson[2]

[1] Dept. of Electrical Engineering – Systems, Tel Aviv University
Ramat Aviv 69978, Israel
{danielh, mira, nk}@eng.tau.ac.il
[2] Dept. of Psychiatry and Eating Disorders Center, Sheba Medical Center
Tel Hashomer 52621, Israel

Abstract. A computer-based method for the assessment of body image distortions in anorexia nervosa and other eating-disordered patients is presented. At the core of the method is a realistic pictorial simulation of lifelike weight-changes, applied to a real source image of the patient. The patients, using a graphical user interface, adjust their body shapes until they meet their self-perceived appearance. Measuring the extent of virtual fattening or slimming of a body with respect to its real shape and size, allows direct, quantitative evaluation of the cognitive distortion in body image. In a preliminary experiment involving 20 anorexia-nervosa patients, 70% of the subjects chose an image with simulated visual weight gain of about 20% as their "real" body image. None of them recognized the original body image, thus demonstrating the quality of the transformed images. The method presented can be applied in the research, diagnosis, evaluation and treatment of eating disorders.

1 Introduction

Anorexia nervosa (AN) is a psychiatric disorder, characterized by low body weight (less than 85% of the optimal weight), refusal to gain weight, an intense fear of gaining weight or becoming obese, amenorrhea (absence of the monthly period) and perceptual distortions or disturbances of body image or weight [1]. Disordered persons may also exhibit the practice of ingesting large quantities of food and then voiding it through self-induced vomiting, also known as bulimia. Bulimia depletes the body of fluids and of potassium, adversely affecting heart function. The disorder, therefore, involves high incidence of both mental and physical morbidity and high rates of mortality (5-18% of known AN patients) [18].

Body image distortion is a major symptom of AN and other eating disorders. It is relatively resistant to treatment and can be perceived as a cognitive and perceptual core of the disorder [20]. The assessment of body image is, therefore, a central and reliable diagnostic measure, and is also important for the assessment of the changes during treatment [21]. Since the perceptual distortion allows a reliable representation of the pathology, a way of documenting and quantifying the perceptual distortion is of major necessity [30]. Evaluating the existence and severity of the distortion as a direct measure for the perceptual pathology, is also important for research purposes. It may contribute to the understanding of the etiology and the cognitive basis of the disorder, to the evaluation of different treatment methods and to research on the epidemiology of the disorder. For instance, quantifying the distortion in body image at different time points along the course of treatment will give a valid indication for the efficiency of the treatment used. A sensitive measure of body image distortions may be used for early detection, where a clinical intervention may avoid potentially chronic effects of adapting to eating disorders habits (as in anorexia and bulimia nervosa) and other cases of sub-clinical eating disorders.

The increasing incidence of eating disorders and the rising acknowledgment of the central role of the perceptual distortion of the body, lead to the development of various methods for the assessment of body size self-estimation in AN. The first to be employed was the *movable caliper* or *visual size estimation* method, which involved the manual manipulation of the distance between two markers until it was perceived to match the particular body part size to be estimated [24]. Various methods, based on the same principle of estimating body part widths, are referred to as *body part* or *body-site* methods [2, 22, 25]. One drawback of these methods is that body parts are considered in isolation and the context of the whole body is missing. Furthermore, in these methods an external mediator is involved in conducting the measurements. This can adversely affect the reliability of quantitative measurement of the subject's self-perception and self-assessment.

Other methods focus on estimating the body as a whole [3, 15, 13, 19]. In these procedures, the participant views an image and adjusts its size until it matches his or her view of his or her body size [14]. Among the most commonly used procedures is the *distorting mirror* in which an adjustable mirror distorts the size of the whole body both vertically and horizontally [29]. These *whole body* distortion methods lead, in most cases, to unnatural, unrealistic images. The distortion level is technically limited and the distortion of specific body parts cannot be controlled.

Computer-based techniques that involve body part and whole body distortions were presented in [12, 23]. Both methods provide body outlines or silhouettes, not a realistic recognizable image of the participant.

In this paper we present a new diagnostic tool that facilitates evaluation of the self perception of the subject's body image and provides a quantitative measure of body image distortion. Starting with a real image of the subject, highly realistic fattened and slimmed images are synthesized. Body parts are independently modified while maintaining the natural appearance and global characteristics of the body shape. User interactivity is simple and friendly, allowing direct manipulation by an untrained subject without an external mediator.

2 The Algorithm for Weight-Change Simulation

Fattening and slimming processes have different effects on different body parts. The body weight-change simulation algorithm first subdivides the body image into principal body-parts, that have different patterns of shape-change, when a weight gain or loss process takes place. Each body part is separately and independently processed. Then, the altered body parts are merged back into a whole body, with special attention to joint areas. The whole process is based on a method developed for the representation of biological shape modifications (such as growth and evolution) in living organisms.

2.1 Image Acquisition and Preprocessing

The subjects (eating disorders patients) are dressed in pink or purple colored full-body leotards. Front-view body images are captured with a Sony digital still camera (model MVC-FD73), with the subjects' arms raised by about $25°$, having their fists clenched and their legs parted by about 50cm. A uniform matte blue screen is used as background. The pixel resolution is 640x480.

The identification of body parts is semi-automatic. The user points at seven different landmarks on the source body image. Slightly misplaced landmarks are corrected automatically. These seven landmarks are used by the algorithm to isolate ten different body parts (left hand, right hand, head, upper chest, lower chest, belly, hips, pelvis, left leg, and right leg). Each body part is processed independently, according to its shape-change characteristics.

The extraction of the body figure from the background of the image is automatic. First the image is transformed from the RGB color space to the HSV (Hue, Saturation, brightness Value) color space. Since the background is uniform matte blue, hue-thresholding (supported by an adaptive brightness threshold for reliable segmentation of hair) allows sharp and exact extraction of the body figure from the source image (see the middle column in Fig. 1). The segmented body shape, together with the seven user marked landmarks, is used to create a binary mask for each of the ten different body parts. These masks are essential for the shape transformation algorithm, which is applied to each body part independently.

2.2 The Method of Transformation Grids

There exist several techniques and methods for the representation of shape and shape change. However, most of these techniques refer to shapes as still objects and to shape changes as partial displacements governed by external forces. Since in our case the shapes to be modified are human body parts, that have their own pattern of growth and deformation, a special kind of shape change method is required. Such a method was presented by Bookstein [6] who has found, in his research, a way to express biological shape differences using a mathematical approach based on D'Arcy Thompson's distorted grid graphs [28].

Thompson invented the method of *transformation grids* to represent the relationship between a pair of *homologous* shapes throughout their interiors. A *homology* between two organisms is defined to be a "maximally inclusive scheme of pairs of parts, organs or structures that manifest the same positional relations among themselves in both organisms" [6]. Thompson has drawn grid intersections at roughly homologous points on two organisms, and then showed how the square grid for one was distorted to produce the grid of the other. He suggested to interpret the pair of diagrams as a transformation of the whole picture plane which maps the points of one diagram into corresponding points in the other, while varying smoothly in between. His suggestion accorded with his belief that constituent parts of an organism could never evolve quite independently [6].

While exploring a mathematical model to produce Thompson's distorted grids, Bookstein [6] argued that Thompson was wrong in the construction of pairs of diagrams which were unsymmetrically specified: a rectangular grid on one diagram and an unrestricted grid on the other. Instead, Bookstein suggested the use of biorthogonal grids for the transformation as a representation of general lines of growth. He produced a canonical coordinate system which reduces all change of shape to gradients of differential directional growth [6], enabling the measure of shape change without measuring shape at all. In computing the biorthogonal curve systems, Bookstein assumed correspondence between boundaries and extended it to the interiors. His strategy was to describe the homology between two images by discrete pairs of homologous landmarks. He chose a complex biharmonic function, with only isolated "sources" and "sinks" of distortion as a convenient model for the homologous correspondence between the two images, from which the mapping function could be defined. Considering a mapping of a small patch, Bookstein defined the *roughness* of the transformation as the squared distance between the mapping of the centroid of the patch and the centroid of the mapped patch. Further, he has shown that if the *roughness* of the transformation is minimized, the non-linear mapping function will be uniquely defined, and the biological characteristics of the shape will be preserved [6–8].

Formally, suppose we are given a shape $\mathcal{P}$ that should be transformed into a shape $\mathcal{Q}$. Let $\{P_i = (X_i, Y_i)\}$ be a subset of N points on the outline of $\mathcal{P}$, and suppose that their corresponding points on the outline of $\mathcal{Q}$ are respectively $\{Q_i = (U_i, V_i)\}$. Based on these homologous (corresponding) pairs, Bookstein [6] defines the following transformation from *any* point $P = (X, Y) \in \mathcal{P}$ to a point $Q = (U, V) \in \mathcal{Q}$:

$$U = a_u \cdot X + b_u \cdot Y + c_u + \sum_{i=1}^{N} \omega_i^u \cdot F\{\delta(Q_i, P)\}$$

$$V = a_v \cdot X + b_v \cdot Y + c_v + \sum_{i=1}^{N} \omega_i^v \cdot F\{\delta(Q_i, P)\} \,, \tag{1}$$

where $\delta(Q_i, P)$ is the Euclidean distance between Q_i and P and

$$F(\delta) \equiv \delta^2 \cdot \log(\delta^2) \,.$$

This transformation is parameterized by $2N + 6$ parameters, namely a_u, b_u, c_u, $\{\omega_i^u\}_{i=1}^N$, a_v, b_v, c_v, $\{\omega_i^v\}_{i=1}^N$. The transformation can be made unique by requiring that its *roughness* be minimized, where the roughness R is defined as

$$R = \int\int [(\nabla^2 U)^2 + (\nabla^2 V)^2]\, dU\, dV \tag{2}$$

Differentiating R and equating to zero, leads to the following six constraints on the parameters:

$$\sum_{i=1}^N \omega_i^u = \sum_{i=1}^N \omega_i^v = 0$$

$$\sum_{i=1}^N \omega_i^u \cdot U_i = \sum_{i=1}^N \omega_i^v \cdot U_i = 0$$

$$\sum_{i=1}^N \omega_i^u \cdot V_i = \sum_{i=1}^N \omega_i^v \cdot V_i = 0 \tag{3}$$

Now, substituting the N corresponding pairs $P_i \to Q_i$ in Eqs. 1 and Eqs. 3 yields $2N + 6$ equations from which the unknown parameters can be obtained. Using these parameters, the transformation given by Eqs. 1 can now be applied to all points $P \in \mathcal{P}$.

2.3 Body Shape Transformation

The outcome of the preprocessing steps (subsection 2.1) are a segmented body image and the shapes of the ten body parts represented by binary masks. In order to simulate weight-change in each body part according to Bookstein's model, there is a need for pairs of corresponding landmarks in the original and modified shapes.

For each body part, a set of three arbitrary points $P_i = (X_i, Y_i)$, located far from each other on the outline, is selected. These three points are mapped to points $Q_i = (U_i, V_i)$ on the outline of the modified shape by scaling the magnitudes of the vectors connecting them to the centroid. The scaling factor is is specific to the body part and proportional to the global change required.

Substituting for the three pairs of points $P_i \to Q_i$ and for $N = 3$ in Eqs. 1 and Eqs. 3, yields 12 independent linear equations, from which the coefficients a_u, b_u, c_u, $\{\omega_i^u\}_{i=1}^3$, a_v, b_v, c_v, and $\{\omega_i^v\}_{i=1}^3$ can be obtained. Using these coefficients, Eqs. 1 are now used to transform the complete outline of the original body part into a weight modified outline. Repeating this procedure for all body parts, yields a set of independently transformed outlines.

The next step is to merge the modified outlines of all the body-parts into a single outline of a whole body. Adjacent pairs of body parts are merged, subject to constraints that preserve the natural appearance of the modified body shape and global characteristics such as body height and skeleton integrity.

Having obtained the whole outline of the weight-modified body, N pairs of corresponding points from the original and modified outlines are selected (N is about 10% of the total number of points on the outline). Substituting for these pairs of points in Eqs. 1 and Eqs. 3, we obtain $2N + 6$ equations that uniquely define the coefficients in Eqs. 1. These coefficients specify the transformation of the whole body. Eqs. 1 can now map *all* points (X, Y) in the original body image onto points (U, V) in the weight modified body image.

The ratio between the 2-D body areas in the modified and original images is a useful quantitative index for body image distortion evaluation. Using a cylindric approximation for the 3-D structure of body parts, weight ratio is roughly the square of the area ratio. A better approximation could be obtained by monitoring the weight and 2-D body-shape area of humans during weight change processes.

2.4   Postprocessing

The non-linear transformation maps each point in the original body image onto the target image plane. Due to discretization effects, some points in the modified body image have no corresponding source points in the original image, and therefore lack color properties. These discontinuities are eliminated using 3×3 median-based color interpolation around colorless pixels. This procedure may create some artifacts on the external side of the body outline. These are masked using the (known) modified outline. The result is a clear, sharp and realistic color body image, in which the body figure is fatter or slimmer by the specified weight-change factor. Typical results are shown in Fig. 1.

3   Implementation and Performance Evaluation

The suggested method for the assessment of body image distortions in eating-disordered patients is implemented in the MATLAB environment, and is executable on computers with a wide variety of operating systems. For each subject, a set of weight-modified body images is generated offline and stored as a virtual album. A graphical user interface is provided for controlling and monitoring the automated body-image album generation.

Convenient access to the stored images is facilitated by a graphical browser, allowing immediate retrieval and display of the weight-modified body images.

To evaluate the quality of the weight-modified images, an experiment was conducted among 30 healthy people (not suffering from eating disorders). The purpose of the experiment was to determine whether the simulated body images were realistic enough, to confuse an original body image with an artificially generated one.

The experiment consisted of 45 trials. In each two body images were presented: an original image and an artificial one. A total of 15 simulated image albums of teenage and young female adults were used in the experiment. The weight change factor of the simulated image in each trial was chosen in the range {-16%,+28%}. The position (right/left) of the two images on the screen

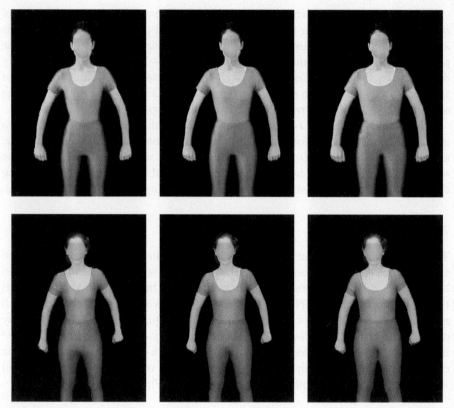

Fig. 1. *Middle column:* Body images of two anorexia-nervosa patients, extracted (segmented) from the original images (not shown). *Left column:* Computer generated body images with about 5% weight loss. *Right column:* Computer generated body images with about 15% weight gain.

was random. In each trial the subject was asked to choose which of the two displayed images was the original body image. A graphical user interface was used to control the experiment and store its results automatically.

The results of the experiment show that in trials with a simulated weight change of up to $^+_-8\%$, the subjects confusingly chose the artificial body image as the "original" one in 44% of the cases. This proves that the artificially generated body images are highly realistic. As could be expected, larger simulated weight changes appear less natural and are easier to detect. Over the full weight-change range of {-16%,+28%} the error rate of subjects was about 30%.

4 Discussion

We presented a method for the assessment of body image distortions in eating disorders patients, that improves upon previous techniques. Realistic simulation of the human body weight-change process is achieved via independent manipulation of separate body parts, followed by the presentation of the subject's natural-looking weight-modified body image.

The weight change simulation algorithm is based on a unique biological shape modification technique presented by D'Arcy Thompson and Bookstein [28, 6]. Other methods for shape deformation and modification do exist. *Snakes* or *active contour models* are energy-minimizing splines guided by external constraint forces and influenced by image forces that pull them toward features such as lines and edges [16]. Some active-contour based models suggest shape deformation by making the shape's contour behave like a balloon which is inflated by an additional force [10]. An active simulate the changes in shape and cross-sectional area occurring during the contraction of isolated muscle fibre [17].

Caricaturing algorithms [9, 26] exaggerate features on the shape's contour. The resulting shapes resemble the original ones but are not natural looking. Image *morphing* techniques [11, 27] are defined between at least two shapes or objects. However, in the case of weight change simulation only the original body image is given, hence we remain with the problem of creating a synthesized target body image to which the metamorphosis process can be targeted.

Based on facial image caricaturing [5], an interesting method for the study of body size perception and body types was very recently presented [4]. Using clinical categories of body types (BMI - Body Mass Index) as prototypes, two approaches to changing the body shape were considered. The first approach alters the body shape by exaggerating or minimizing differences of feature points on the original body and its corresponding prototype. The second approach alters the body shape by referring to all prototypes simultaneously, following evolutionary paths of feature points among the prototypes. A large image-database (hundreds of images) is necessary for the extraction of stable prototypes with minimal variability within each prototype and with similarity between the BMI category prototypes [4]. The user is required to manually delineate hundreds of feature points, used by the system to locate and extract the body and the different body parts, and to refer to the category prototypes.

The method presented here was developed as the first phase of a two-phase project. In the next phase of the project, the body-image distortion as assessed by the method presented, will be compared with the distortion as assessed by several common and clinically validated assessment tools, such as: (a) structured interview of anorexia and bulimia syndromes, (b) eating disorder inventory, and (c) body shape questionnaire. In addition, the algorithm's assessment results will be compared to common clinical measures of diagnosis and follow-up: (a) body weight and BMI, (b) menstrual cycle, and (c) stage of therapy.

The proposed body image distortion assessment tool is currently being deployed and tested at the Eating Disorder Center of Sheba Medical Center. A group of 20 admitted patients suffering from AN participated in an experiment, in which they where asked to choose an image from a 24-picture album of their body (at various simulated weight-change levels), that corresponds to their body size as they perceive it. A high percentage of the subjects (70%), both youth and young adults, chose an image in which a visual weight gain of about 20% was simulated, as their "real" body image. None of them recognized their true source body image. This demonstrates the quality of the transformed body images. The suggested method is expected to be a valuable tool for diagnosis, treatment and follow-up in patients with eating disorders.

References

1. American Psychiatric Association (1994). "Diagnostic and Statistical Manual of Mental Disorders", 4th ed.
2. Askevold F. (1975). "Measuring Body image". *Psychotherapy and Psychosomatics*, Vol 26, pp. 71-77.
3. Bell C., Kirkpatrick S. W., & Rinn R. C. (1986). "Body image of anorexic, obese, and normal females". *Journal of Clinical Psychology*, Vol. 42, pp. 431-439.
4. Benson P. J., Emery J. L., Cohen-Tovee E. M., & Tovee M. J. (1999). "A comuter-graphic technique for the study of body size perception and body types". *Behavior Research Methods, Instruments & Computers*, Vol. 31, pp. 446-454.
5. Benson P. J., & Perrett D. J. (1991). "Synthesizing continuous-tone caricatures". *Image & Vision Computing*, Vol. 9, pp. 123-129.
6. Bookstein F. (1978). "The measurement of biological shape and shape change". *Lecture Notes in Biomathematics*, Springer-Verlag New York.
7. Bookstein F. L., & Green W. D. K. (1993). "A feature space for derivatives of deformations". *Information Processing in Medical Imaging*, 13th International Conference, IPMI '93 Proceedings, Springer-Verlag, Berlin, Germany, pp. 1-16.
8. Bookstein F. L. (1989). "Principal warps: Thin plate splines and the decomposition of deformations". *IEEE Transactions on Pattern Analysis & Machine Intelligence*, Vol. 11, pp. 567-585.
9. Brennan S. E. (1985). "Caricature generator: The dynamic exaggeration of faces by computer." *Leonardo*, Vol. 18, pp. 170-178.
10. Cohen L. D. (1991). "On active contour models and balloons". *CVGIP - Image Understanding*, Vol. 53, pp. 211-218.
11. Daw-Tung L., & Han H. (1999). "Facial expression morphing and animation with local warping methods". Proceedings 10th International Conference on *Image Analysis and Processing, IEEE Computer Society*, Los Alamitos, CA, USA, pp. 594-599.
12. Dickson-Parnell B., Jones M., Braddy D., & Parnell C. P. (1987). "Assessment of body image perceptions using a computer program". *Behavior Research Methods, Instruments, & Computers*, Vol. 19, pp. 353-354.
13. Freeman R. J., Thomas C. D., Solyom L., & Hunter M. A. (1984). "Video-camera for measuring body image distortion". *Psychological Medicine*, Vol. 14, pp. 411-416.
14. Gardner R. M. (1996). "Methodological issues in assessment of the perceptual component of body image disturbance". *British Journal of Psychology*, Vol. 87, pp. 327-337.
15. Garner D. M., Garfinkel P. E., & Bonato D. P. (1987). "Body image measurement in eating disorders". *Advances in Psychosomatic Medicine*, Vol. 17, pp. 119-133.
16. Kass M., Witkin A., & Terzopoulos D. (1987). "Snakes: active contour models". *International Journal of Computer Vision*, Vol. 1, pp. 321-331.

17. Klemencic A., Pernus F., & Kovacic S. (1999). "Modeling morphological changes during contraction of muscle fibres by active contours". *Computer Analysis of Images and Patterns*, 8th International Conference, CAIP'99. Proceedings, Springer-Verlag, Berlin, Germany. (Lecture Notes in Computer Science, Vol. 1689, pp. 134-141).

18. Neumaker K. J. (1997). "Mortality and sudden death in anorexia nervosa". *International Journal of Eating Disorders*, Vol. 21, pp. 205-212.

19. Probst M., Vandereycken W., & Van Coppenolle H. (1997). "Body size estimation in eating disorders using video distortion on a life-size screen". *Psychotherapy & Psychosomatics*, Vol. 66, pp. 87-91.

20. Rosen J. C. (1990). "Body image disturbance in eating disorders". in T. F. Cash: *Body images: development, deviance and changes*, New York: Guilford, pp. 190-214.

21. Rosen J. C. (1996). "Body image assessment and treatment in controlled studies of eating disorders". *International Journal of Eating Disorders*, Vol. 20, pp. 331-343.

22. Ruff G. A., & Barrios B. A. (1986). "Realistic assessment of body image". *Behavioral Assessment*, Vol. 8, pp. 235-251.

23. Schlundt D. G., & Bell C. (1993). "Body image testing system: a microcomputer program for assessing body image". *Journal of Psycho-pathology & Behavioural Assessment*, Vol. 15, pp. 264-285.

24. Slade P. D., & Russell G. F. M. (1973). "Awareness of body dimensions in anorexia nervosa: Cross-sectional and longitudinal studies". *Psychological Medicine*, Vol. 3, pp. 188-199.

25. Thompson J. K., & and Spana R. E. (1988). "The adjustable light beam method for the assessment of size estimation accuracy: Description, psychometrics and normative data". *International Journal of Eating Disorders*, Vol. 7, pp. 521-526.

26. Steiner A. (1995). "Planar shape enhancement and exaggeration". M.Sc. thesis, Department of Electrical Engineering, Technion - Israel Institute of Technology, Haifa, Israel.

27. Tal A., & Elber G. (1999). "Image morphing with feature preserving texture". *Blackwell Publishers for Eurographics Association. Computer Graphics Forum*, Vol. 18, pp. 339-348.

28. Thompson D. W. (1917). "On growth and form" (abridged edition, 1966). *Cambridge University Press*, London, Great Britain.

29. Traub A. C., & Orbach J. (1964). "Psychophysical studies of body image. I. The adjustable body-distorting mirror". *Archives of General Psychiatry*, Vol. 11, pp. 53-66.

30. Williamson D. A., Cubis B. A., & Gleaves D. H. (1993). "Equivalence of body image disturbances in anorexia and bulimia nervosa". *Journal of Abnormal Psychology*, Vol. 102, pp. 177-180.

A Realistic Model of the Inner Organs from the Visible Human Data

K. H. Höhne[1], B. Pflesser[1], A. Pommert[1], M. Riemer[1], R. Schubert[1], T. Schiemann[1], U. Tiede[1], U. Schumacher[2]

[1]Institute of Mathematics and Computer Science in Medicine,
[2]Institute of Anatomy
University Hospital Eppendorf,
D–20246 Hamburg, GermanyHamburg, Germany
{hoehne, pflesser, pommert, riemer, schubert, tiede,
schumacher}@uke.uni-hamburg.de

Abstract. The computer-based 3D models of the human body reported to date suffer from poor spatial resolution. The Visible Human project has delivered high resolution cross-sectional images that are suited for generation of high-quality models. Yet none of the 3D models described to date reflect the quality of the original images. We present a method of segmentation and visualization which provides a new quality of realism and detail. Using the example of a 3D model of the inner organs, we demonstrate that such models, especially when combined with a knowledge base, open new possibilities for scientific, educational, and clinical work

1. Introduction

While in classical medicine, knowledge about the human body is represented in books and atlases, present-day computer science allows for new, more powerful and versatile computer-based representations of knowledge. Their most simple manifestations are multimedia CD-ROMs containing collections of classical pictures and text, which may be browsed arbitrarily or according to various criteria. Although computerized, such media still follow the old paradigm of text printed on pages accompanied by pictures. This genre includes impressive atlases of cross-sectional anatomy, notably from the photographic cross-sections of the Visible Human Project (1-3). In the past years, however, it has been shown that pictorial knowledge, especially about the structure of the human body, may be much more efficiently represented by computerized *3D models* (4) , which can be constructed from cross-sectional images generated by computer tomography (CT), magnetic resonance imaging (MRI), or histologic cryosectioning, as in the case of the Visible Human Project. Such models may be used interactively on a computer screen or even in "virtual reality" environments. If such models are connected to a data base of descriptive information, they can even be interrogated or disassembled by addressing names of organs (4-6). They can thus be regarded as a "self-explaining body".

Until now, the Visible Human Project has not reported 3D models that reflect the rich anatomical detail of the original cross-sectional images. This is due to the fact that, for the majority of anatomical objects contained in the data, the cross-sectional images could not be converted into a set of coherent realistic surfaces. If, however, we succeed in converting all the detail into a 3D model, we gain an unsurpassed representation of human structure that opens new possibilities for learning anatomy and simulating interventions or radiological examinations. This paper presents a new technique for segmenting and modeling the 3D anatomy of the Visible Human data that leads to a nearly photorealistic computer based model.

2. Earlier Work

Work on 3D anatomical reconstruction from spatial sequences of cross-sectional medical images goes back to the late seventies (7). Since then, techniques have been improved continuously (8-10). However, the impact of these techniques on practical medicine remains limited, because the procedure for creation of 3D models is rather complex. The process of 3D visualization requires - in contrast to classical 2D imaging - a processing step that identifies the regions in the data volume that correspond to anatomical objects. Otherwise we could not unveil their surfaces or remove objects obscuring the ones behind. This step is called *segmentation*. A common approach to 3D visualization which tries to avoid this step is *volume rendering*. Instead of an explicit identification of organs, transparency values are assigned to the image volume according to the color values or color changes at the object borders. In the case of the Visible Human data, this method (11) yields semitransparent views with acceptable quality e. g. for the outer surface, the muscles and the skeleton. It fails, however, to display internal structures properly. In addition, organ borders are not explicitly indicated, thus making the removal or exclusive display of an organ impossible.

Segmentation, i. e. the exact determination of the surface location of an organ is crucial to its realistic display. Complete automatic segmentation using methods of computer vision fails, since the capabilities of computers are far inferior to those of a human observer. The brute force approach to segmentation is manual outlining of objects on the cross-sections (11). Besides the fact that this procedure is tedious and very time consuming, it is largely observer-dependent and thus does not yield exact and coherent surfaces. Furthermore, despite the high resolution of the data set, important details such as nerves and small blood vessels cannot clearly be identified, because their size and contrast is too small even at the high resolution of the Visible Human data set.

3. Material and Methods

We therefore aimed at a method that yields surfaces for the segmentable organs that are as exact as possible and textured with their original color. In order to arrive at a complete model, we decided to model non segmentable objects like nerves and small blood vessels artificially on the basis of landmarks present in the image volume.

Data. Photographic cross-sectional images from the male Visible Human were used to create models of the head, neck, and torso as well as the corresponding inner organs. The original data set consists of 1871 photographic cross-sections with a slice distance of 1 mm. The cross-sections themselves have a spatial resolution of 1/3 mm. For the model of the inner organs, the resolution was reduced to 1mm for reasons of data storage and computing capacity. From 1049 such slices, an image volume of 573x330x1049 volume elements ("voxels") of 1 mm^3 was composed, where each voxel is represented by a set of red, green and blue intensities ("RGB-tuple"). The Visible Human data set also includes two sets of computer tomographic images of 1mm slice distance, one taken from the fresh, the other (like the photographic one) from the frozen cadaver. Both were transformed into an image volume congruent with the photographic one.

Segmentation. The image volume thus created was segmented with an interactive tool, which is an extension of an earlier development for RGB data (13). It can be regarded as an "electronic sculpturing tool". On a cross-section, an expert "paints" a typical region of the organ under consideration. All voxels in the volume with similar RGB-tuples are then collected by the program to form a "mask". This mask usually needs to be refined by repeating this procedure in order include the target organ completely. The resulting cluster in RGB space has an ellipsoid shape, due to the slight correlation of the colors; this cluster is then converted into the mathematical form of an ellipsoid, which facilitates subsequent computations. There are, of course, other regions with the same characterization present in the volume. If they are not connected to the target organ, it can be isolated easily; if not, 3D interactive cutting tools are used.

Graphic modeling. There are many anatomical constituents, such as nerves and small vessels, which are too small to be segmented adequately, but need to be included in a comprehensive anatomical model. For these, we have developed a modeling tool that allows us to include tube-like structures into the model. Ball-shaped markers of variable diameter are imposed by an expert onto the landmarks still visible on the cross-sections or on the 3D image; these markers are automatically connected to form tubes of varying diameter. Unlike the segmented objects, which are represented as sets of voxels, the modeled structures are represented as composed of small triangles.

Knowledge modeling. Each voxel of the spatial representation of anatomical data thus gained is connected to a knowledge base containing object descriptions within the structure of a semantic network (4,14,15). Different networks were created for different views (e. g. systematic or topographic anatomy). Within the views, the anatomical constituents are linked by relations like *part of, has part, branching from* or *branching to.*

Visualization. The 3D visualization algorithm we have developed is characterized by the fact that it renders *surfaces* from *volume* data. This is different from the volume rendering typically used for visualizing the Visible Human data set. As in volume rendering, the program casts rays from an image plane onto the image volume in viewing direction. However, the rays stop at the first encountered object. Surface texture and surface inclination (important for proper computation of light reflection) are calculated from the RGB-tuples at the segmented border line. The decisive quality improvement is achieved by determining the object surfaces with a spatial resolution higher than the resolution of the original voxels. While such subvoxel resolution may be achieved easily by sampling with high rates in volume rendering, the determination of distinct adjacent *surfaces* in subvoxel resolution is a major problem that has been solved only recently (16). This makes the high resolution visualization of distinct surfaces possible, which is the characteristic feature of our approach.

The objects modeled with the surface modeler are visualized with standard computer graphics methods within the context of the volume objects. The visualization program, an extended version of the VOXEL-MAN system (4), runs on UNIX workstations. Because of the high resolution and the sophisticated algorithms, the computation of a single image may take several minutes, even on a high end workstation. The model is therefore being made available for interaction and exploration via precomputed "Intelligent QuickTime VR Movies" (17), which may be viewed on any personal computer.

4. Results

Using the methods described above, we have built a model of the organ systems within the torso of the male Visible Human. It contains more then 850 *three-dimensional* anatomical constituents (*not* just labeled cross-sections) and more than 2000 relations between them.

The following features of this model represent innovations:

- Because of the exact segmentation and the visualization method developed, the visual impression is one of unsurpassed realism;
- There is, to date, no computer model of human structure that contains so many *three-dimensional* organs;
- The model is space-filling i. e. any voxel is labeled as an element of a 3D object;
- The integrated formal organization of pictorial and descriptive information allows a virtually unlimited number of ways of using the model.

The model is a general-knowledge representation of gross anatomy, from which all classical representations (pictures, movies, solid models) may be derived via mouse click. In addition, the model may be used interactively for teaching and simulation of diagnostic and therapeutic procedures. While the underlying principle was reported earlier (4), the model we describe is the first to offer sufficient detail and comprehensiveness to serve these purposes seriously. FIG. 1 gives an impression of image quality and the degree of detail.

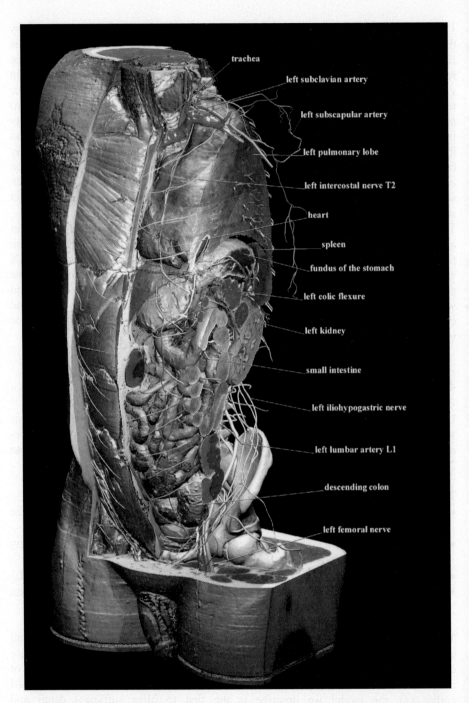

Fig. 1. The model can be viewed from any direction, cuts may be placed in any number and direction and objects may be removed or added. Annotations may be called by mouse click.

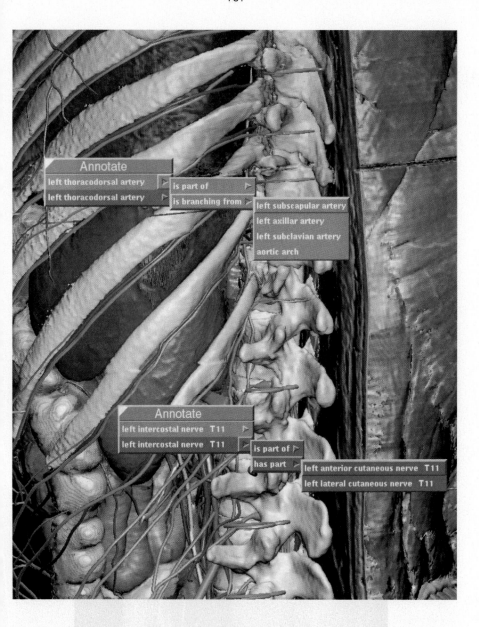

Fig. 2. Inquiring the semantic network behind the pictorial model according to different "views". The user has clicked onto a blood vessel and a nerve and received information about systematic(red) and topographic anatomy(blue).

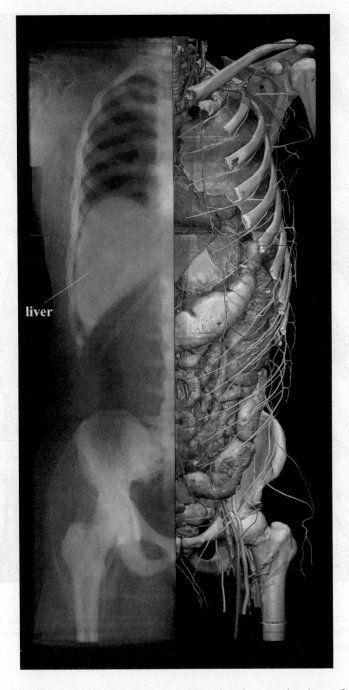

Fig. 3. Different viewing modes such as X-ray imaging may be chosen from any direction and for any part of the image. The X-ray image may be interrogated as to which object contributes to the image.

Since the model is volume-based, cut planes (which can be placed in any number and direction) show the texture of the original cross-sectional images and thus look realistic (FIG. 1). This virtual dissection capability not only allows an interactive dissection for learning purposes, but also simulates the rehearsal of a surgical procedure. In addition, the image of a "self-explaining body" allows us to inquire about complex anatomical facts. FIGures 1 and 2 show different possibilities for presenting these explanations. The more traditional way of annotating structures of interest is demonstrated within the user-specified 3D scene in FIG. 1. These annotations can be obtained simply by pointing and clicking with the mouse on the structure of interest. However, pressing another button of the mouse will result in popup menus shown in FIG. 2; these offer structured knowledge about anatomy and function. Such information is available because every voxel, and therefore any visible point of any user-created 3D-scene, is linked to the knowledge base.

A special feature of the model involves the possibility of simulating radiological examinations (FIG. 3). Since the absorption values for every voxel are available in the original tomographic data, artificial X-ray images from any direction can be computed. Based on the information of the model, both the contributing anatomical structures and the extent of their contribution to the final absorption can be calculated. Similarly, the information present in cross-sectional radiological images (Computer Tomography, Magnetic Resonance Imaging, Ultrasound) can be clarified by presenting them in the corresponding 3D context.

5. Discussion

Computer-based models of the human anatomy created to date are more like sophisticated video games than they are professional tools for medical education and practice. We have presented a general approach for creating models that have the spatial resolution, the degree of detail, and the organization suitable for useful application in educational, scientific, and clinical work. This has been demonstrated with the example of a 3D model of the inner organs derived from the Visible Human data set. The versatility of the approach makes it suitable for anatomy and radiology teaching as well as for simulation of interventional procedures. Last, but not least, it can be used to provide patients with necessary information. A first version of a 3D anatomy atlas based on the described concept has been published recently (18).

Yet there are still improvements to be made. Like all postmortal images, those of the Visible Human suffer from postmortal artifacts, such as collapsed blood vessels. These artifacts cannot be corrected completely. One more severe limitation is the fact that the data is derived from one single individual. The interindividual variability of organ shape and topology in space and time is thus not yet part of the model. Inclusion of variability into 3D models is a difficult problem not yet generally solved. So far most progress has been achieved for 3D atlases of the brain (19).

However, the current model should be an excellent basis for further developments. One such development is the inclusion of physiology, e. g. the modeling of blood flow or propagation of electrical fields throughout the body. Applications such as the computation of body surface potential maps (20) should profit from an increased level of detail. Furthermore, because of the more detailed characterization of tissues, surgical simulation involving soft tissue deformation could be made much more realistic (20). This approach is thus an important, albeit early step towards computer models that not only look real, but also act like a real body.

References

1. Spitzer, V., Ackerman, M. J., Scherzinger, A.L., Whitlock, D.: The visible human male: a technical report. J-Am-Med-Inform-Assoc 1996; 3(2):118-30.
2. Ackerman MJ. The Visible Human Project: a resource for anatomical visualization. Medinfo. 1998; 9 Pt 2:1030-2.
3. Spitzer VM, Whitlock DG. The Visible Human Data Set: the anatomical platform for human simulation. Anat-Rec. 1998 Apr; 253(2):49-57.
4. Höhne, K. H., Pflesser, B., Pommert, A., Riemer, M., Schiemann, T., Schubert, R., Tiede, U. A new representation of knowledge concerning human anatomy and function. Nature Med. 1995; 1(6):506-511.
5. Rosse C, Mejino J, Modayur B, Jakobovits R, Hinshaw K, Brinkley JF. Motivation and organizational principles for anatomical knowledge representation: the digital anatomist Symbolic Knowledge Base. J-Am-Med-Inform-Assoc 1998; 5(1): 17-40.
6. Golland P, Kikinis R, Halle M, Umans BA, Grimson WEL, Shenton ME et al. AnatomyBrowser, A novel approach to visualisation and integration of medical information. Computer Assisted Surgery 1999; 4: 129-143.
7. Herman GT, Liu HK. Display of three-dimensional information in computed tomography. Journal of Computer Assisted Tomography 1977;1:155-160.
8. Höhne, K. H., Bernstein R. Shading 3D-images from CT using gray level gradients", IEEE Trans. Med. Imaging 1986; 5(1):45-47.
9. Levoy M. Display of surfaces from volume data. IEEE Comput Graphics and Appl.1988; 8(3):29-37
10. Tiede, U., Bomans, M., Höhne, K. H., Pommert, A., Riemer, M., Wiebecke,G.: Investigation of medical 3D rendering algorithms. IEEE Comput. Graphics. Appl. 1990;10(2):41-53.
11. Stewart, J. E., Broaddus, W.C., Johnson, J.H.: Rebuilding the Visible Man. In: Höhne, K. H. and Kikinis R, editors. VBC 96. Proceedings of the Conference on Visualization in Biomedical Computing; 1996 Sep 22—25; Hamburg, Germany. Berlin Heidelberg: Springer-Verlag 1996. Lecture Notes in Computer Science 1131, p. 81-85
12. Mullick, R., Nguyen, H.T.: Visualization and Labeling of the Visible Data Set: Challenges and Resolves. In: K. H. Höhne and R. Kikinis, editors. VBC 96. Proceedings of the Conference on Visualization in Biomedical Computing; 1996 Sep 22—25; Hamburg, Germany. Berlin Heidelberg: Springer-Verlag 1996. Lecture Notes in Computer Science 1131, p. 75-80.
13. Schiemann, T., Tiede, U., Höhne, K. H.: Segmentation of the Visible Human for high quality volume based visualization. Med. Image Anal. 1997; 1(4):263-271.
14. Schubert, R., Höhne, K. H., Pommert, A., Riemer, M., Schiemann, T., Tiede, U.: Spatial knowledge representation for visualization of human anatomy and function. In: Barrett, H. H., Gmitro, A. F., editors. IPMI 93. Proceedings of the 13th Conference on Information

Processing in Medical Imaging; 1993 Jun 14–18; Flagstaff, Arizona. Berlin Heidelberg: Springer-Verlag 1993. Lecture Notes in Computer Science 687, p. 168-181.

15. Pommert, A., Schubert, R., Riemer, M., Schiemann, T., Tiede, U., Höhne, K. H.: Symbolic modeling of human anatomy for visualization and simulation. In: Robb R, editor. VBC 94. Proceedings of the Conference on Visualization in Biomedical Computing; 1994 Oct 4—7, Rochester, MN. Bellingham Washington 1994: SPIE Proceedings 2359. p. 412-423.

16. Tiede, U., Schiemann, T., Höhne, K. H.: High quality rendering of attributed volume data. In: Ebert D, et al., editors. Vis 98. Proceedings of the Conference IEEE Visualization; 1998 Oct 18–23; Research Triangle Park, NC. Los Alamitos, CA: IEEE Computer Society Press; 1998. p. 255-262.

17. Schubert, R., Pflesser, B., Pommert, A., Priesmeyer, K., Riemer, M., Schiemann, T., et al. Interactive volume visualization using "intelligent movies". In: Westwood JD et al., editors. MMVR 99. Proceedings of the Confenrence Medicine Meets Virtual Reality; 1999 Jan 20–23; San Francisco, CA. Amsterdam: IOS Press 1999. Studies in Health Technology and Informatics 62. p. 321-327.

18. Höhne, K. H., Pflesser, B., Pommert, A., Priesmeyer, K., Riemer, M., Schiemann, T., Schubert, R., Tiede, R., Frederking, H.-C., Gehrmann, S., Noster, S., Schumacher, U.: VOXEL-MAN 3D-Navigator: Inner Organs. Regional, Systemic and Radiological Anatomy. Springer-Verlag Electronic Media, Heidelberg, 2000. (3 CD-ROMs, ISBN 3-540-14759-4

19. Mazziotta, J. C., Toga, A. W., Evans, A., Fox, P., Lancaster, J.: A probabilistic atlas of the human brain: theory and rationale for its development. The International Consortium for Brain Mapping (ICBM). Neuroimage.1995; 2(2): 89-101

20. Sachse, F. B., Werner, C. D., Meyer-Waarden, K., Dössel, O.: Applications of the Visible Human Male Data Set in Electrocardiology: Calculation and Visualization of Body Surface Potential Maps of a Complete Heart Cycle. In Proceedings of the 2nd Users Conference of the National Library of Medicine's Visible Human Project, Bethesda, Maryland 1998 p. 47-48.

21. Cotin, S., Delinguette, H., Ayache, N.: Real-time elastic transformations of soft tissues for surgery simulation. IEEE Trans. Visualization Comp. Graph. 1999. 5: 62-73.

Acknowledgements

We thank Victor Spitzer and David Whitlock (University of Colorado) and the National Library of Medicine for providing the Visible Human data set. We are also grateful to Jochen Dormeier, Jan Freudenberg, Sebastian Gehrmann, and Stefan Noster for substantially contributing to the segmentation work. The knowledge modeling work was supported by the German Research Council (DFG) grant # Ho 899/4-1.

Automatic Centerline Extraction for 3D Virtual Bronchoscopy

Tsui Ying Law and Pheng Ann Heng

Department of Computer Science and Engineering
The Chinese University of Hong Kong,
Shatin, N.T., Hong Kong
{tylaw, pheng}@cse.cuhk.edu.hk

Abstract

Centerline extraction is the basis to understand three dimensional structure of the lung. In this paper, an automatic centerline determination algorithm for three dimensional virtual bronchoscopy CT image is presented. This algorithm has two main components. They are *end points retrieval algorithm* and *graph based centerline algorithm*. The end points retrieval algorithm first constructs a binary tree which links up all necessary center points of each region in each slice from segmented lung airway tree volume data. Next, it extracts the end points of the lung airway tips from the binary tree. The graph based centerline algorithm reads the end points and applies distance transform to yield a distance map which shows all shortest paths from the start point to those end points. Then, modified Dijkstra shortest path algorithm is applied in the centerline algorithm to yield the centerline of the bronchus. Our algorithm is tested with various CT image data and its performance is encouraging.

1 Introduction

The idea of using virtual reality is currently of great interest to the medical imaging community. Virtual reality system can provide an initial assessment of the condition of patients. Bronchoscopy is a medical diagnosis for evaluating the endobronchial anatomy. Virtual reality and volumetric imaging develop so rapidly that their technologies are matured enough to create a simulated (virtual) environment for medical surgery, which takes less cost and risks. Virtual bronchoscopy [7] is a new concept to combine volumetric imaging and virtual reality technology for simulating bronchoscopy. It is gaining public attention because of its potential for decreasing discomfort and inconvenience, considerably lower cost and risks, in comparison with routine bronchoscopic screening procedures.

Simulation technology makes it possible for an user to experience adverse scenarios without risk to human life or damage to expensive equipment [2]. In addition, it is useful for training medical students or physicians to achieve better surgery skills. Comparing the virtual and real surgery, although it has the physical biopsy, color/texture limitations, virtual one takes the advantages of measurements providing, reproducibility, flexibility and non-invasiveness.

We are currently developing a virtual bronchoscopy system for training and diagnosis purposes. The system allows healthcare providers to practise procedures in an environment where mistakes do not have serious consequences. In addition, it lowers risk associated with training on human patients, avoids the use of animals for training, and establishes standards and optimization of specific procedures.

One required function of the virtual bronchoscopy system is to extract and recognize the 3D structure of bronchus in lung area. The automated extraction algorithm of the bronchus area [5] is an enhancement of algorithm described in [8] and [9]. It is for the recognition of bronchus in three-dimensional CT images, which is based on region growing method. Region growing method [11] is the most popular and widely used method for the detection of tree structured objects such as bronchus and blood vessels [4]. The algorithm is applied to real 3D CT images and experimental results have been given to demonstrate its ability to extract bronchus area and the result is satisfactory.

We can extract the bronchus. However, they are just a simple set of voxels. To understand their 3D structure, additional analysis is necessary. Centerline extraction [4] is the basis to understand 3D structure. To extract centerlines, thinning processing is usually applied. However, it produces skeletons which may not reflect the true shape of the original pattern. Therefore, a more accurate algorithm for extracting the centerline of bronchus is desired.

The organization of this paper is given as follows. After giving an introduction in Section 1, the detailed description and experimental results of our centerline algorithm are presented in Sections 2 and 3 respectively. Conclusion is given in Section 4.

2 Automatic Centerline Determination Algorithm

Our automatic centerline determination algorithm is composed of two main components. The first component is *end points retrieval algorithm* which converts segmented lung airway tree volume data into a set of end points. The second component is *graph based centerline algorithm*. The algorithm read the end points and it yields a distance map which shows all shortest paths from the start point to those end points. Those end points can be used to construct a set of centerlines of the bronchus.

2.1 End Points Retrieval

Assume that the slices of the volume data are in axial format, we first determine all the center points of every region in each slice. In each slice, every pixel is scanned once to perform region growing searching. Segmented volume data can be divided into two categories: feature or non-feature pixel. The feature pixels represent the lung tree airway extracted, while the non-feature pixels represent those remaining areas. When a feature pixel is being visited, region growing is started from the pixel and search its eight neighbors recursively. Pixels are marked when they have been visited. Those unmarked feature pixels will start region growing to form other regions. After the whole slice is scanned, a few regions should be extracted. The center of mass of each region is determined and it is treated as the center point of that region.

After the center points of all slices are determined, they will be linked up by our center points linking algorithm. Based on the bifurcation characteristics of the lung tree volume, we link up those center points starting from the main bronchus to tiny airway tips. By the algorithm in [5] which adopts genetics algorithm to determine the start point, a seed point which belongs to the main bronchus is generated. The center point of the corresponding region that contains the seed point is then used as the initial point in the center points linking algorithm.

The initial point is inserted into a linked list which will contain subsequent center points for processing. Let L be the linked list, p be the first element of the list and $\{c_1, c_2, \ldots c_n\}$ be the set of unlinked center points which are in the previous or next slices of p. Distance between p and all elements in $\{c_1, c_2, \ldots c_n\}$ will be calculated so that any point will be linked up to p if it is close to p.

To measure the closeness between two points, there are two threshold values $distance\_thershold$ and $possible\_distance\_thershold$ used in our algorithm. Let x_l and x_{l+1} be two center points in l^{th} and $l+1^{th}$ slices respectively. If the distance between x_l and x_{l+1} is not larger than $distance\_thershold$ and x_l has less than 2 children, x_l is linked to x_{l+1} which represents that x_l is the parent of x_{l+1}. If the distance between x_l and x_{l+1} is larger than $possible\_distance\_thershold$, x_l and x_{l+1} do not have child-parent relationship. If the distance falls between $distance\_thershold$ and $possible\_distance\_thershold$, they may have parent-children relationship and further test is needed. In this case, x_l is inserted into the list L so that all its preceding points in the list have the same slice number but with smaller minimum distance. When x_l is extracted again and it has one child or less, x_l is linked to x_{l+1} if x_{l+1} has not yet been selected from other points. Then x_l is the parent of x_{l+1}. If x_{l+1} is a child point of x_l, it is then inserted into the end of L so that its children will be found later. Every point should have only one parent. Once it is selected from a point, the child-parent relationship is established and no other point can be its parent therefore.

The center points linking algorithm is shown in Fig. 1. After the algorithm is completed, a binary tree which links up all necessary center points is yielded. In the binary tree, end points are the points which do not have any child. It implies that they are corresponded to the lung airway tree tiny tips. Since the binary tree is created, those end points can be easily extracted by the traversal of the tree.

2.2 Graph Based Centerline Algorithm

Based on chamfer distance transform and Dijkstra's single source shortest path algorithm [10] [3], Blezek [1] applies a centerline algorithm to virtual endoscopy. However, he mainly concerns one-start-point-one-end-point endoscopy such as colonoscopy and user is required to supply the start and end points. His algorithm cannot be applied directly to virtual bronchoscopy, which has many branches. In order to develop an automatic algorithm for centerline extraction for virtual bronchoscopy, we introduce the following modifications.

Our centerline algorithm first calculates a three dimensional chamfer 3-4-5 distance map of the volume data which is a two-pass procedure. All voxels in the interior of the object are labelled with the distance to the nearest background voxel. In other word, a voxel in the center interior of the object should have a larger distance value as it is farther apart from the background comparing with those boundary voxels. From the distance map, the maximum distance value V_{Max} is determined. All voxel distance values are squared after subtracted from V_{Max} to achieve the result that more interior voxels have smaller distance values and closer to boundary voxels have larger distance values relatively. Thus, voxel has zero distance value is as far from the boundaries of the object as possible. The squaring is performed to encourage the path to maintain a medial path.

A weighted directed graph $G = (V, E)$, where V is the set of vertices and E is the set of edges in the graph is constructed. All voxels are

```
LinkCenterPoint(list L){
1.    While list L not empty{
2.          Extract the first member p of the list L
3.          Locate all unlinked center points {c₁, c₂, ...cₙ} in the previous and
            next slices of p
4.          Calculate the distance from {c₁, c₂, ...cₙ} to p as [d₁, d₂, ...dₙ]
5.          Sort the center points [c₁, c₂, ...cₙ] by the corresponding [d₁, d₂, ...dₙ]
6.          if d₁ ≤ distance_threshold
7.                link up p to c₁
8.                insert c₁ into list L
9.          if d₂ ≤ distance_threshold
10.               link up p to c₂
11.               insert c₂ into list L
12.         if d₁ or d₂ > distance_threshold and ≤ possible_distance_threshold
13.               if p not yet appear in the linked list L
14.                     insert p into the list L in a suitable position
15.               else link up p to the corresponding point (c₁ or c₂)
            }
      }
```

Figure 1: Center Points Linking Algorithm

considered as nodes (vertice). Edges are created between a vertex and its 26-connected neighbour. The weight of a edge is assigned as the distance value of the voxel in in-direction. For example if the distance value of voxel V_1 is four, then the weight of edges (V_1, V_0) and (V_1, V_2) are four.

A general method solving the single-source shortest-path problem is known as Dijkstra's algorithm. It is a prime example of greedy algorithms. Greedy algorithms generally solve problems in stages by doing what appears to be the best thing at each stage. Similarly Dijkstra's algorithm proceeds in stages. In each stage, Dijkstra's algorithm selects a vertex v which has the smallest distance among all the unknown vertices, and declares that the shortest path from source to the vertex v is known. The remainder of a stage consists of updating the values of the neighbours of the vertex v. A priority queue Q that contains all the unknown vertices keyed by their distance values is maintained. It is practical to implement the priority queue Q with a binary heap [10] [3]. The resulting algorithm is sometimes called the modified Dijkstra algorithm. Binary heaps have two properties, namely, a structure property and a heap-order property. A heap is a binary tree that is completely filled, with the possible exception of the bottom level, which is filled from left to right. Such a tree is known as a complete tree. It can be represented in an array and no

```
dijkstra(G, StartPt){
1.    initial distance[i] = ∞ and previous vertices are NULL
2.    set distance[StartPt] = 0
3.    build a binary minimum heap which is keyed to distance[]
4.    while heap is not empty {
5.          retrieve the minimum item v from the heap
6.          for each adjacent w to v
7.                if w not out of bound and w is a feature point
8.                     update w distance and previous vertex if required
9.                     update heap if w information is modified
      }
}
```

Figure 2: Modified Dijkstra Shortest Path Algorithm

links are necessary. For any element in array position i, the left child is at position $2i$, the right child is in the cell after the left child $(2i + 1)$, and the parent is at position $i/2$. Thus not only are links not required, but the operations required to traverse the tree are extremely simple and can be very fast on most computers. The heap-order property allows operations to be performed quickly. In a heap, for every node X, the key in the parent of X is smaller than or equal to the key in X, with the exception of the root which has no parent. By the heap-order property, the minimum element can always be found at the root. Thus, we can retrieve the minimum element quickly.

When the Dijkstra algorithm as shown in Fig. 2 is completed, the previous vertex in the shortest path from the source vertex $StartPt$ to each vertex is determined. The previous vertex should be a 26-neighbour of the current vertex. Therefore, three dimensional 26-connected chains are generated from the source vertex $StartPt$ to each feature vertex. To improve the efficiency of the procedure, only those feature vertices are considered. If the vertices are non-feature (background) one, their previous vertex should be remained $NULL$ which is not changed since initialization.

The centerline required is constructed by extracting the shortest path from each end point to the source point. It is accomplished by following the chain code from end points back to source point. The voxels visited are marked. Combining all the visited voxels will form the centerline of the object.

3   Experiment and Discussion

The automatic centerline extraction algorithm is implemented and tested
on real 3D CT data which is obtained from a dog lung. Seven sets of
lung airway tree segmented data are generated by the method mentioned
in [5]. Among the seven image sets a reduced resolution image set that
is composed of 64 images, where each of them is a 128x128 8-bit gray-
scale image, are prepared for comparison. Other six image sets contain
around 185 images where each of them is a 256x256 8-bit gray-scale image. None of pre-processing operations such as smoothing, interpolation
is applied. All experiments are performed on Sun UltraSparc 5/270 workstation with Solaris 2.6 OS, 512MB RAM and 100Mbps network speed.
Table 1 presents the running time of the algorithm. The calculation time
is approximately linear with number of object voxels.

Table 1: CPU Running Time of End Points Retrieval Algorithm, Distance
Transform and Dijkstra Shortest Path Algorithm

Data	Volume Size(Voxels)	Object Voxels	End Point Time(s)	DT Time(s)	Dijkstra Time(s)	End Point Voxels	Center -line Voxels
set1	256x256x185	54398	4.3	5.79	3.85	91	1490
set2	128x128x64	9657	0.39	0.53	0.43	47	465
set3	256x256x185	60025	4.32	5.7	3.69	93	1356
set4	256x256x180	72378	4.3	5.7	4.42	129	1784
set5	256x256x185	98885	4.43	5.99	4.95	162	1664
set6	256x256x180	91697	4.42	5.98	5.36	206	2565
set7	256x256x200	99094	4.88	6.61	5.78	98	1466

distance_threshold and possible_distance_threshold are set to be 3 and
26 respectively in the end points retrieval algorithm. User can alter the
threshold whenever necessary. For a 12MB lung airway tree volume data,
it spends about 4 seconds to generate over 90 end points. The graph based
centerline algorithm takes about 10 seconds to complete, and one third
of the execution time is contributed to the completion of the modified
Dijkstra shortest path algorithm. One of the results in Blezek [1] which
consists of 99054 object voxels with a 256x256x256 volume size requires
20 seconds to complete. The algorithm in Blezek [1] requires user to input
a start point and an end point to his shortest path searching algorithm.
On the contrary, our algorithm includes end points searching function
that does not require user to input the start point. In the case of similar
object voxel number, our algorithm requires 12 seconds including end

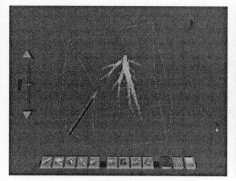
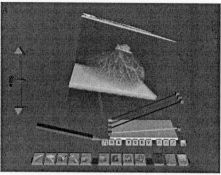

(a) Transparent Lung Tree Airway Data and Centerline Generated

(b) Transparent Lung Volume Data and Centerline Generated

(c) Visualization of Lung Volume in Cutbox View

(d) Visualization of Segmented Lung Tree in Cutbox View

Figure 3: Various Visualization Results

points searching, distance transform and Dijkstra shortest path searching. Our algorithm is much more efficient than Blezek's algorithm.

The extracted centerline is visualized with 3D texture mapping techniques [6]. This method is a direct data visualization technique that is similar to ray casting. Three dimensional textures are a logical extension of 2D textures. In 3D textures, texels become unit cubes in texel space. The 3D texture is used as a voxel cache, processing 2D layer each time by all rays simultaneously. Fig. 3 show the images of the reconstructed centerlines. The centerlines are shown to be extracted satisfactorily.

4 Conclusion

An automatic centerline determination algorithm from CT images for three dimensional virtual bronchoscopy is presented. The end points retrieval algorithm extracts end points of the lung airway tips. Distance transform and modified Dijkstra shortest path algorithm are then applied in the centerline algorithm which yields the centerline of the bronchoscopy. Experimental results show that our algorithm is more efficient than algorithm in Blezek [1] which is the basis of our centerline algorithm. Our test cases include various CT image data sets. For a typical 256x256x180 segmented lung tree airway volume data, it requires about 12 seconds for the whole centerline determination procedure. The experimental results show that the performance of our algorithm is encouraging.

Acknowledgments

This work is supported by Hong Kong RGC Earmarked Research Grants CUHK 4167/97E.

References

[1] Daniel J. Blezek, Richard A. Robb, *Centerline Algorithm for Virtual Endoscopy based on Chamfer Distance Transform and Dijkstra's Single Source Shortest Path Algorithm*, Part of the SPIE Conference on Physiology and Function from Multidimensional Images, San Diago, California, SPIE Vol. 3660, Feb 1999.

[2] Richard D. Bucholz, MD, FACS, *Advances in Computer Aided Surgery*, CAR'98.

[3] Thomos H. Cormen, Charles E. Leiserson, Ronald L. Rivest, Introduction to Algorithms, The MIT Press, McGraw-Hill Book Company, 1990.

[4] Fumikazu Iseki, Tsagaan Baigalmaa et al., *Extraction of 3D Tree Structure of Blood Vessels in Lung Area from Chest CT Images*, CAR'98, 1998.

[5] Tsui-Ying Law, Pheng-Ann Heng, *Automated Extraction of Bronchus from 3D CT Images of Lung*, pp. 906-916, Medical Imaging 2000: Image Processing, Proceedings of SPIE Vol. 3979(2000), 2000.

[6] T. McReynolds et al., *Advanced Graphics Programming Techniques Using OpenGL*, SIGGRAPH 98 Course Notes 17, July 1998.

[7] J. R. Mellor, G. Z. Yang et al., *Virtual Bronchoscopy*, Image Processing and its Applications, 4-6 July 1995, Conference Publication No. 410, IEE, 1995.

[8] Kensaki Mori, Jun-ichi Hasagawa et al., *Automated Extraction and Visualization of Bronchus from 3D CT Images of Lung*, Lecture Notes in Computer Science 905, Computer Vision, Virtual Reality and Robotics in Medicine, First International Conference, CVRMed'95, Nice, France, Springer, pp. 542-548, April 1995.

[9] Kensaku Mori, Jun-ichi Hasegawa et al., *Recognition of Bronchus in Three-Dimensional X-ray CT Images with Applications to Virtualized Bronchoscopy System*, Proceeding of ICPR'96, IEEE, 1996.

[10] Mark Allen Weiss, Data Structures & Algorithm Aanlysis in C++, Addison Wesley Books, Second Edition, 1999.

[11] Terry Yoo, *Segmentation and Classificaiton*, Siggraph 93 Course Notes 21, pp. 12-15 , Aug. 1993.

Mixed Reality Merging of Endoscopic Images and 3-D Surfaces

Damini Dey[1], Piotr J. Slomka[1,2], David G. Gobbi[1], Terry M. Peters[1]

[1]Imaging Research Laboratory, John P. Robarts Research Institute, London, Canada
[2]Diagnostic Radiology and Nuclear Medicine, London Health Sciences Center, London, Canada

Abstract. In image-guided neurosurgery, "mixed reality" merging has been used to merge video images with an underlying computer model. We have developed methods to map intra-operative endoscopic video to 3D surfaces derived from pre-operative scans for enhanced visualization during surgery. We acquired CT images of a brain phantom, and digitized endoscopic video images from a tracked neuro-endoscope. Registration of the phantom and CT images was accomplished using markers that could be identified in both spaces. The endoscopic images were corrected for radial lens distortion, and mapped onto surfaces extracted from the CT images via a ray-traced texture-mapping algorithm. The localization accuracy of the endoscope tip was within 1.0 mm. The mapping operation allows the endoscopic images to be permanently painted onto the surfaces. Our method allows panoramic and stereoscopic visualization from arbitrary perspectives (though the original endoscopic video was monoscopic) and navigation of the painted surface after the procedure.

1   Introduction

The term "Mixed reality" was introduced by Milgram in 1994 in an attempt to classify merging of real images with virtual images [1]. In industry, particularly in telerobotics, mixed reality imaging has been employed to combine real and virtual worlds. This has become important when there is a time-delay in the telerobotic system, that prevents the operator from real-time manipulation of the final task. This situation can be somewhat improved through the use of a local computer model that is matched to the manipulated environment. The operator can perform the operations based on the behaviour of the model, and through merging the model image with the (delayed) video representation of the actual manipulated site, can optimize the teleoperation of the remote robot.

In medicine, mixed reality imaging has been employed to merge real-time video images with an underlying computer model (based on MRI or CT scans) for the purpose of surgical guidance. Grimson et al. projected surfaces extracted from pre-operative MRI data onto patients during neurosurgery to provide the surgeon with "x-ray" vision [2]. Konen et al. mapped anatomical landmarks derived from pre-operative MRI images to live video sequences acquired with a optically tracked endoscope [3]. In contrast to these approaches, Jannin et al. merged microscope images with surfaces that were derived from anatomical images [4], while Clarkson et al. described a method to texture-map a single 2D stereoscopic video image onto CT surface images of the skull [5].

In some general image visualization applications, multiple photographic or video images have been "stitched" together to form visual panoramas [6-8]. However, in most of these methods, the camera acquiring the images is constrained in some manner, e.g to rotate about a fixed central axis or to follow a pre-defined path [7,8]. The acquired camera images are mapped back to planar, spherical or cylindrical surfaces to form visual panoramas [6-8]. To our knowledge, for applications in medical imaging, multiple 2D images have not previously been accurately texture-mapped to 3D surfaces from arbitrary camera viewpoints.

One of the most widely used sources of video images in medicine is the endoscope. Endoscopes are employed in many surgical procedures, including orthopedic, cardiac, neuro and abdominal surgery [9-11]. In recent years virtual endoscopy has emerged as a means of examining complex 3-D data sets [12-13]. The virtual endoscope is "placed" within the 3-D digital volume, and images that represent what an actual endoscope would see are generated. It has been common to combine such virtual endoscopic navigation with a standard display of a set of orthogonal planes [12, 13].

In this paper, we present a new means of combining the video endoscopic images with the models of the surfaces from which the images originated. This approach permanently maps the acquired endoscopic images onto the surface, regardless of the viewing pose of the endoscope, and provides the three-dimensional (3D) relative context of the endoscopic image to the underlying structure in an intuitive, visual manner. We have achieved this by developing methods to register the 2D images acquired by a tracked endoscope to the 3D object surfaces, and by implementing an automatic painting algorithm via ray-traced texture mapping. This work also entails accurate modeling of the endoscope optics.

1.1 Geometrical Framework

In general, video image is a 2D projection of the 3D scene. The imaging geometry can be described by the pinhole camera model [14]. Within this model, a 3x4 perspective transformation matrix relates an arbitrary point in the 3D scene to a point in the 2D image. A point x in 3D, defined by the homogeneous co-ordinates $x = (x,y,z,1)$, is related through the 3x4 transformation matrix T to a 2-D point $u = (u,v,1)$:

$$wu^T = Tx^T$$

(1)

The matrix T represents a rigid body transformation from 3D scene (or world) co-ordinates to 3D camera co-ordinates, followed by a projective transformation onto the 2D imaging plane. w is a scaling factor for homogeneous co-ordinates [14]. If the camera characteristics are known, a virtual camera view can be generated from any arbitrary position in the 3D scene. The transformation matrices, obtained via careful calibration of the endoscope, are employed during image rendering.

Pasting the 2D image acquired by the camera back to the 3D surface, involves further calculations to determine the 2D-3D mapping transformation for each pixel in the 2D image. Since the mapping described in Equation (1) is generally not invertible, every point on the camera image maps in 3D to an infinitely long line [14-15].

We have adopted a computer graphics-based approach to merge multiple 2D images back to the 3D surface. Our system involves calibrating the endoscope and deriving an optical model, as well as tracking the endoscope in real time with an optical tracking tool. Through knowledge of this optical model, we can identify the surface patch viewed by the endoscope. The final mapping transformation of each pixel in the endoscopic image is determined by ray-traced texture mapping.

2 Methods

We perform the following steps to acquire and display the endoscopic images:

2.1 Image Acquisition

Images may be acquired using either CT or MRI. In the preliminary work presented here, we employed a 3-D CT scan volume of an anatomical head phantom. The CT data was segmented by 3D region-growing and a polygonal surface of the phantom was extracted via the Marching Cubes algorithm [16].

2.2 Optical Modeling

We derived an optical model of the endoscope by imaging a simple calibration pattern at several known distances from the endoscope tip and determining from the resulting images a model of the endoscopic viewing cone [17]. From the same set of calibration pattern images we subsequently extracted a model of the radial (or barrel) distortion of the endoscope lens [18], which is characterized by a 4th order polynomial.

2.3 Endoscope Tracking

For this work we used a 2.7 mm straight AESCULAP neuro-endoscope, tracked by the POLARIS Optical Tracking System (Northern Digital Inc., Waterloo, Canada). This system consists of a T-shaped assembly of infra-red light-emitting diodes (LEDs), and a position-sensor consisting of two cylindrical lenses coupled to video cameras mounted on a bar. The LED assembly was attached to the shaft of the endoscope via a plexiglass mount, and the endoscopic video images were digitized using a Matrox Corona framegrabber board (Matrox Inc, St Laurent, Québec) in a Pentium II 450-MHz PC. The anatomical phantom used in this work was registered to its 3D image by using a least-squares point-matching algorithm.

The LEDs, mounted at the end of the endoscope, were continuously tracked by the POLARIS during the experiments, with the 3-D position, and orientation of the tip being computed in real time.

2.4 Surface patch identification

Texture mapping is a computer graphics technique commonly used to map 2D images (textures) to the 3D surfaces represented in the computer by a set of polygons (triangles in our case) [20]. In order to map a texture to a surface, the correct texture co-ordinates for the vertex of each polygon must be computed. This is quite simple for mapping a texture to a regular surface such as a plane, cylinder or a sphere. However, to project a texture to an arbitrary 3D surface from an arbitrary 3D position without visual distortions, is not straightforward. For example, if texture co-ordinates for an arbitrary 3D surface are computed based on the assumption that the surface is cylindrical or spherical, the texture can show severe distortions. Therefore, in our application, the texture co-ordinates and scaling have to be computed accurately.

2.5 Surface Painting

For each endoscopic view, we know the position, rotation and orientation of the endoscope tip in 3D. We place a modeled endoscopic "viewing cone" at the position of the endoscope tip, and extract the surface patch of the viewed object that falls with this cone (Figure 1 (a)). This surface is then intersected with 5 rays whose intersections correspond to the 3D positions to which the center and 4 equally-spaced points around edges of the texture patch are mapped (Figure 1(b)). Finally we calculate the texture co-ordinates of each polygon by tracing virtual rays from the vertex of each triangle, through the texture, to the localized endoscope tip (Figure 1 (c)) [19]. These co-ordinates are used to stretch and "pin" the 2D texture to the surface.

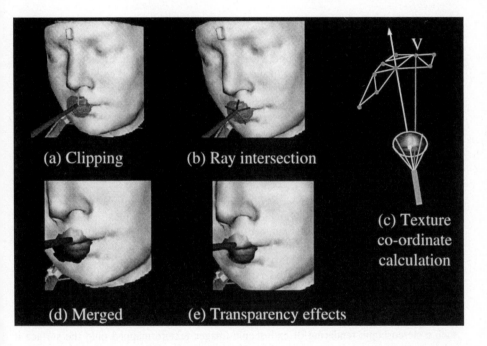

Figure 1. Our texture mapping method. (a) The endoscope tip is localized in the pre-operative image and the surface patch is extracted by clipping. (b) The surface is intersected by 5 rays corresponding to the center and the 4 edges of the texture map. (c) Texture co-ordinates for each polygon are calculated by tracing virtual rays from each vertex V of each polygon to the localized endoscope tip. (d) The texture co-ordinates are used to "pin" the texture to the surface.(e) The transparency of darker pixels can be adjusted so that they do not occlude the surface.

As the endoscope sweeps across the object surface, cutout patches corresponding to the intersection of the endoscopic viewing cone with the entire surface are extracted. The texture co-ordinates for the new endoscopic position are computed, and the endoscopic images are texture-mapped to this surface. For each pose of the endoscope, a texture-mapped cutout patch is added to the rendered scene. The surface patch corresponding to the most recent endoscopic pose is added last.

This software is written primarily in C++, and the Python Programming Language, and is interfaced with the graphics toolkit *The Visualization Toolkit (VTK)* [16].

As the distance between the surface of an object and the endoscope tip increases, the light intensity decreases. In our application we have the option to transparently render the darker regions in the image (i.e. regions which have been insufficiently illuminated), so that they do not occlude the surface (Figure 1(e)).

3 Results

3.1 Visual assessment

Figure 2(a) – (c) shows endoscopic images of the brain phantom. In Figure 2(d), these images are shown texture-mapped to the surface of the phantom. The views grabbed in Figures 2(a) and (b) are oblique projections of the 3D scene, as demonstrated by the lighting in the endoscopic image itself, and by texture mapped surface. In the endoscopic images, the light intensity decreases as the distance between the surface and the endoscope tip increases (Figure 2 (a) and (b)). Because the endoscope is tilted at an oblique angle to the surface, and the surface is not necessarily planar, the edges of the viewing cone clipping the surface are often jagged (Figure 1 (d), Figure 2(d)).

All 2D views acquired by an endoscope are compressed into a circular field-of-view. However, from Figure 2 (d), it can be seen that when the endoscope is not oriented perpendicular to the surface, the circular endoscopic image is actually a compressed view of an ellipse in the case of an oblique flat surface, or a more complex geometrical shape for an arbitrary surface. The total number of virtual rays traced is equal to the number of vertices in the cutout surface (in this case less than 100 for each surface patch). In contrast, if we were to trace rays forward though every pixel in the 350x350 endoscopic texture, the number of virtual rays to be traced would be 122,500.

One of the advantages of mapping the endoscopic images onto 3-D surfaces, is that even though the original images were monoscopic, the mapped images can be ascribed depth by the underlying 3-D structure, and visualized stereoscopically. Figure 2(d) shows such a stereoscopic rendering of endoscopic images texture mapped onto the surface it originated from.

3.2 Errors

Our preliminary results demonstrate that the tracking accuracy of the POLARIS LED's is 0.3 mm, and the endoscope tip can be localized with a precision of approximately 1.0 mm. We identified anatomical landmarks in both the pre-operative CT data and on the texture mapped surfaces. If we could localize the endoscope perfectly in 3D and there were no errors in our optical model, we would expect the position of these landmarks to coincide in the two datasets, given that there is no tissue shift. We define the 3D distance between the original and texture-mapped locations of manually identified landmarks, as the texture mapping accuracy. For our phantom experiment, we found the average value of this measure to be 2.4 mm.

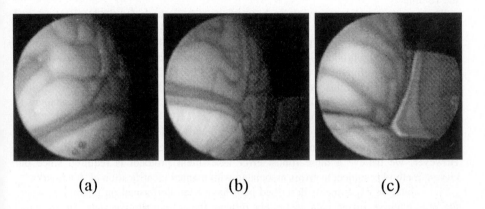

(a)   (b)   (c)

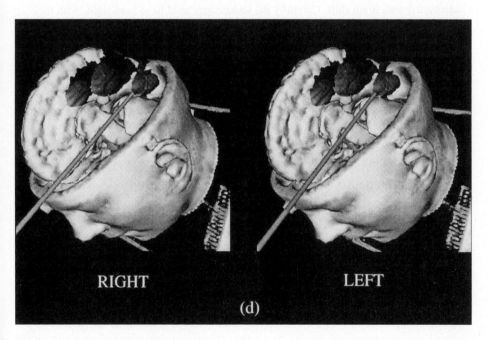

RIGHT   LEFT

(d)

Figure 2. (a) –(c) Endoscopic images of brain phantom. (d) Stereoscopic rendering of the final 3D rendering (Cross-eyed stereo).

4 Discussion

We have demonstrated preliminary results from a system that merges, in near real time, multiple tracked endoscopic views to 3D surfaces using a ray-traced texture-mapping approach. This allows both panoramic and stereoscopic visualization of the "painted" surface after the removal of the endoscope from the volume, as well as navigation within the volume.

There are many improvements to be made to our methods for tip localization, and optical calibration, and our texture mapping algorithm. From our experience, before every endoscopic procedure, the endoscope must be optically calibrated, a procedure that must be rapid and reproducible, and be able to accommodate angled endoscopes.

Further visual refinements could be made to the final 3D rendering with the mapped endoscopic images. In this work, our algorithm was applied to individual digitized video frames acquired with a tracked endoscope, but is in the process of being adapted to accommodate digitized streaming video.

Our texture mapping error is an estimate of our overall error due to inaccuracies in optical tracking, registration of physical space to image space, and our optical calibration. However, we believe a more robust error assessment protocol is required since our current method is subject to errors associated with manual identification of landmarks.

The validation experiments described here have been performed on phantoms, where the pre-operative image data accurately reflects the intra-operative state. In an actual endoscopy-assisted procedure however, we could expect additional small errors associated with brain shift. Currently the endoscopes used have been rigid, with their tip position extrapolated from the position of the tracking LED's external to the body. However, tracking technologies exist that employ optical fibre properties, ShapeTape^{TM}, (Measurand Inc., Fredericton, Canada), and electromagnetic localization principles (Biosense Inc, Seatauket, NY) that could detect the position and orientation of the tip itself. This could enable the techniques described in this paper, to be employed with flexible endoscopes.

Note that our surface painting algorithm is not limited solely to endoscopic images. It is equally applicable to mapping of video or still images of any object, acquired by any calibrated camera, to the arbitrary surfaces from which the original image emanated.

Acknowledgements

We would like to thank our colleagues Dr. Yves Starreveld and Dr. Andrew Parrent for many useful discussions; Ms. Kathleen Surry for her help with endoscope calibration; Trudell Medical for the loan of endoscopic equipment, and Nuclear Diagnostics (Stockholm, Sweden) for the use of Multimodality software for segmentation. We acknowledge the financial support of the Medical Research Council of Canada, and the Institute for Robotics and Intelligent Systems.

5 References

1. Milgram, P., Kishino, F.: IEICE Transactions on Information Systems E77-D (1994).
2. Grimson, W.E.L., Lozano-Perez, T., Wells, W.M. III, Ettinger, G.J.,White, S.J., Kikinis, R.: IEEE Transactions on Medical Imaging (1996).
3. Konen, W., Scholz, M., Tombrock, S.: Computer Aided Surgery 3 (1998) 144-148.
4. Jannin, P., Bouliou, A., Scarabin, J.M., Barillot, C., Luber, J.: SPIE Proceedings 3031 (1998) 518-526.
5. Clarkson, M.J., Rueckert, D., King, A.P, Edwards, P.J., Hill, D.L.G., Hawkes, D.J.: Proceedings MICCAI (1999) 579-588.
6. Szeliski, R., Shum ,H.Y.: Proceedings SIGGRAPH (1997) 251-258.
7. QuickTime VR: http://www.apple.com/quicktime/
8. Surround Video: http://www.bdiamond.com
9. Berci, G. (ed): Endoscopy. Appleton-Century_Crofts (1976).
10. Perneczky, A., Fries, G.: Neurosurgery 42 (1998) 219-224.
11. Perneczky, A., Fries, G.: Neurosurgery 42 1998) 226-231.
12. Auer, L.M., Auer, D.P: Neurosurgery 43 (1998) 529-548.
13. Jolesz, F.A.,. Lorensen, W.E, Shimoto, H., Atsumi, H., Nakajima, S., Kavanaugh, P., Saiviroonporn, P., Seltzer, S.E., Silverman, S.G., Philips, M., Kikinis, R.: AJR 169 (1997) 1229-1235.
14. Foley, J., van Dam, A., Feiner, S., Hughs, J.: Computer Graphics. 2nd Edition (1990) Addison Wesley.
15. Stefansic, J.D., Herline, A.J., Chapman, W.C., Galloway, R.L.: *SPIE Conference on Image Display Proceedings* 3335 (1998) 208-129.
16. Schroeder, W., Martin, K., Lorensen, B.: The Visualization Toolkit: An Object-Oriented Approach To 3D Graphics. 2nd Edition Prentice Hall (1997).
17. Tsai, R.Y.: *IEEE Journal of Robotics and Automation* RA-3 (4) (1987) 323-345.
18. Haneishi, H., Yagihashi, Y., Miyake, Y.: IEEE Transactions on Medical Imaging 14 (1995) 548-555.
19. Dey, D., Gobbi, D. G., Surry, K. J.M., Slomka, P. J., Peters, T.M.: *SPIE Conference on Image Display Proceedings* (2000) (in press).
20. Haeberli, P., Segal, M.: http://www.sgi.com/grafica/texmap/index.html (1993).

Improving Triangle Mesh Quality with SurfaceNets

P.W. de Bruin[1], F.M. Vos[2], F.H. Post[1], S.F. Frisken-Gibson[3], and
A.M. Vossepoel[2]

[1] Computer Graphics & CAD/CAM group, Faculty of Information Technology and Systems,
Delft University of Technology
[2] Pattern Recognition Group, Department of Applied Physics, Delft University of Technology
[3] MERL – a Mitsubishi Electric Research Laboratory, Cambridge, MA, USA

Abstract. Simulation of soft tissue deformation is a critical part of surgical simulation. An important method for this is finite element (FE) analysis. Models for FE analysis are typically derived by extraction of triangular surface meshes from CT or MRI image data. These meshes must fulfill requirements of accuracy, smoothness, compactness, and triangle quality. In this paper we propose new techniques for improving mesh triangle quality, based on the SurfaceNets method. Our results show that the meshes created are smooth and accurate, have good triangle quality, and fine detail is retained.

Keywords: *Surgical simulation, surface extraction, tissue deformation modelling, visualization, SurfaceNets*

1   Introduction

In recent years, endoscopic surgery has become well established practice in performing minimally-invasive surgical procedures. In training, planning, and performing procedures, pre-operative imaging such as MRI or CT can be used to provide an enhanced view of the restricted surgical field. Simulation of intra-operative tissue deformation can also be used to increase the information provided by imaging. However, accurate simulation requires patient-specific modeling of the mechanical behavior of soft tissue under the actual surgical conditions.

To derive an accurate and valid model for intra-operative simulation, we propose a five-stage process:

1. Image data acquisition (MRI, CT)
2. Image segmentation
3. Deformable tissue model generation
4. Intra-operative simulation of tissue deformation, guided by actual surgical conditions and/or intra-operative measurements conditions
5. Enhanced intra-operative visualization

In order to simulate tissue deformation, many authors have proposed finite element (FE) analysis of the relevant structures (see for example [2, 1, 8]). The FE models are commonly initialized by supervised segmentation of preoperative image data, resulting in a classification accurate to the pixel level. Using a surface extraction technique such as the Marching Cubes algorithm [9], the result is converted into a set of triangular meshes representing the surfaces of relevant organs. Such a representation can then be imported into an environment for FE analysis.

For optimal mechanical modelling and visualization, the triangular surface models should meet the following requirements :

- *Accuracy:* the representation of the organ surface geometry should be sufficiently accurate;
- *Smoothness:* the model should conform to the smooth organ boundaries. Sharp corners should be avoided as these can cause disturbing artifacts such as stress concentrations;
- *Compactness:* to achieve fast response times, the number of elements (triangles) in the model should be minimal; the resolution of the triangle mesh should be considerably lower than the medical image, with minimal loss of accuracy;
- *Triangle quality:* the shape of the triangles in the mesh should be as near as possible to equilateral to avoid FE errors and visualization artifacts.

Segmentation commonly results in a binary image (i.e., classification at pixel level). Extracting a surface from these binary data results in a triangulated surface model that does not meet all of the requirements above. The smoothness of the mesh can be poor due to quantization effects, showing ridges or terraces. Some solutions to this problem are inadequate. For example, Gaussian prefiltering of the binary image (before surface extraction) reduces accuracy, and significant anatomical detail (such as narrow ridges and clefts) may be lost, while insufficient smoothness is achieved [7].

In addition, the number of triangles generated by surface extraction may be very large. Compactness may be improved using mesh decimation techniques [5, 10], but these techniques are usually most effective with smooth meshes. Thus, smoothing of a surface mesh with minimal loss of accuracy is useful to avoid errors in FE analysis and for reducing mesh size. Exploiting the original greyscale data rather than binary segmented data can help to achieve this.

Recently, a technique called *SurfaceNets* was proposed to optimize a triangle mesh derived from binary data [6]. In this paper, the SurfaceNets method is extended to incorporate greyscale data. Several new techniques are examined and compared with Marching Cubes.

The paper is organized as follows. Section 2 briefly describes the basic SurfaceNets method, the extension to incorporate grey-scale data and new techniques for achieving smoothness, accuracy, and good triangle quality. In Section 3 these techniques are evaluated with respect to the requirements for mechanical modeling and visualization listed above. Finally, Section 4 summarizes our findings and draws conclusions.

2   Techniques

This section presents a brief explanation of the original SurfaceNet method (largely following [6]) which assumes that a binary segmentation of the original data exists. Then, two techniques will be introduced that utilize the greyscale image data during relaxation of the SurfaceNet.

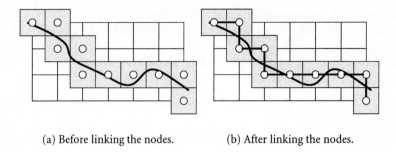

(a) Before linking the nodes.   (b) After linking the nodes.

Fig. 1. Building a SurfaceNet. The white squares represent voxels, the thick black line represents the edge of an object and the grey squares are cells with nodes represented by white circles in the center.

2.1 Generating a SurfaceNet from binary data

The goal of the SurfaceNet approach is to create a globally smooth surface that retains the fine detail present in the original data. Generation of the surface net for binary data consists of the following four steps [6]:

1. Identify nodes of the SurfaceNet;
2. Create links between the nodes;
3. Relax node positions while maintaining constraints on node movement;
4. Triangulate the SurfaceNet for visualization and FE analysis.

The first step in creating a SurfaceNet is to locate the cells that contain the surface. A cell is formed by 8 neighbouring voxel centers in the binary segmented data (Figure 1 presents the 2D case as illustration). If all eight voxels have the same binary value, then the cell is either entirely inside or entirely outside of the object. If, however, at least one of the voxels has a binary value that is different from its neighbours, then the cell is a surface cell. The net is initialized by placing a node at the center of each surface cell (step 1). Subsequently, links are created with nodes that lie in adjacent surface cells (step 2). Assuming only face connected neighbours, each node can have up to 6 links (corresponding to right, left, top, bottom, front and back neighbours). Once the SurfaceNet has been defined, each node is moved to achieve better smoothness and accuracy ("relaxation", step 3) subject to the constraint that each node must remain within its original cell. The relaxation process is described in more detail in the next section.

2.2 Improving smoothness

Once a SurfaceNet has been defined, the node positions are adjusted to improve the smoothness of the surface. This is often desirable to remove furrows and terraces due to the binary segmentation. Let us first only consider the smoothness of the net.

One way to smooth the surface is to move every node to the average position of its linked neighbours [4]. The vector $\vec{a}$ pointing from the current position of the node $\vec{p}_{\text{old}}$

to the average position is calculated as:

$$\vec{a} = \frac{1}{N} \sum_{i=1}^{N} \vec{p}_i - \vec{p}_{\text{old}} \tag{1}$$

where $\vec{p}_i$ corresponds to the position of a linked neighbour and N is the total number of neighbours of this node.

It may well be that the average position is outside the original cube and therefore diverges from the initial segmentation. To impose conformance, the relocation vector $\vec{a}$ is constrained to stay within the boundaries of the original cell by the function c (see Figure 2):

$$\vec{p}_{\text{new}} = \vec{p}_{\text{old}} + c(\vec{a}) \; . \tag{2}$$

Here, c is defined to satisfy the proper constraint of the node position such that $\vec{p}_{\text{new}}$ is always within the boundaries of the cell. Note that this approach is different from the original SurfaceNet method which simply clips the new position's x, y, and z coordinates to cell boundaries when the new position falls outside the cell.

Fig. 2. Position constraint of a node. If $\vec{p}_{\text{old}} + \vec{a}$ is outside the cell boundary, the function c is used such that $\vec{p}_{\text{old}} + c(\vec{a})$ is on the cell boundary.

The relaxation is implemented in an iterative manner by considering each node in sequence and calculating a relocation vector for that node. The SurfaceNet is updated only after each node in the net has been visited. This procedure is repeated until the number of iterations has reached a preset threshold, or when the largest relocation distance is less than a given minimum value.

2.3 Increasing accuracy using greyscale data

The technique described above ignores all greyscale information in the dataset after building the SurfaceNet. The nodes shrink-wrap around the object without trying to conform to an iso-surface of the data. This is reasonable when the binary segmentation is the best estimate of the object. However, if the object surface can be estimated to lie at an iso-surface of the image data, this iso-surface can be used to increase the accuracy of the SurfaceNet.

Let us assume that the true object surface can be obtained by drawing an iso-surface (at I_{iso}) in the original greyscale data. For instance, in many CT based applications the Marching Cubes algorithm is used to approximate the object shape in this way. By definition, at a given point the greyscale gradient vector is perpendicular to the iso-surface through that point. Thus, to enhance accuracy; a node can be displaced along the gradient vector to the iso-surface (see Figure 3(a)). This is expressed as:

$$\vec{g} = \text{SIGN}\left(I_{\text{iso}} - I(\vec{p}_{\text{old}})\right) \nabla \vec{p}_{\text{old}} \; . \tag{3}$$

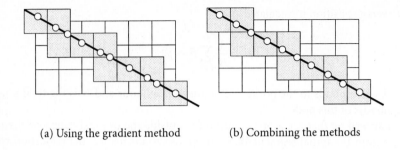

(a) Using the gradient method   (b) Combining the methods

Fig. 3. Using the gradient method for relaxation the nodes (white circles) are projected onto the iso-surface (thick line, left). The combined relaxation technique also spaces out the nodes along the iso-surface (right).

Here, SIGN is a function that returns the sign of its argument, $I(\vec{p}_{\text{old}})$ is the interpolated intensity and $\nabla \vec{p}_{\text{old}}$ is the normalized gradient vector at $\vec{p}_{\text{old}}$. The latter vector is obtained either by a central difference gradient method or by convolution with Gaussian derivatives.

The node position is updated by:

$$\vec{p}_{\text{new}} = \vec{p}_{\text{old}} + c\,(d\vec{g}) \quad . \tag{4}$$

In this equation, d is a scaling parameter representing the distance to the iso-surface. The value of d can be estimated by assuming a linear image field near the iso-surface and interpolating the greyscale values at the node and at a point sampled along the vector $\vec{g}$. As in Equation 2, c imposes a position constraint on the node to stay within the boundaries its cell.

2.4 A combined approach

Combining the methods presented in Section 2.2 and Section 2.3, we obtain a surface that fits the iso-surface of the data and is also globally smooth. To combine these features, a node should be displaced to obtain better smoothness within the iso-surface. The combination is made by first calculating the projection $\vec{a}_p$ of the averaging vector $\vec{a}$ on the plane perpendicular to the gradient $\vec{g}$ (cf. Equation 1, Equation 3):

$$\vec{a}_p = \vec{a} - \vec{g}(\vec{a} \cdot \vec{g}) \quad . \tag{5}$$

Subsequently, the combined displacement function is defined as:

$$\vec{p}_{\text{new}} = \vec{p}_{\text{old}} + c\left(\vec{a}_p + d\vec{g}\right) \quad . \tag{6}$$

This formula combines relocation towards the iso-surface with smoothing in the orthogonal plane (i.e., on the surface). This can be seen in Figure 3 where the nodes are first projected onto the line and then evenly spaced out along the line by the averaging.

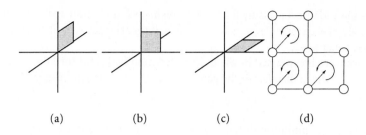

Fig. 4. Three possible configurations for a quadrilateral. For each main direction only the top right corner has to be checked for a quadrilateral (right).

Again, c ensures that the new position of each node is always within the boundaries of its original surface cell. Note that there may be some tension between the goals of a smooth SurfaceNet and one that fits the iso-surface. One of these goals can be favored over the other either by weighting the independent contributions differently or by applying them sequentially rather than simultaneously; ending with the favored goal.

2.5 Triangulation

After relaxation the SurfaceNet is triangulated to form a 3D polygonal surface. We have simplified the original triangulation process described in [6]. Instead of directly building the triangles, first quadrilaterals are identified. There are three sets of four links of a node that lie in a plane (e.g., the left, right, top, and bottom links of a node lie in a plane) (see Figure 4). In each plane the connected nodes form quadrilaterals and each node is a vertex of at most four quadrilaterals. In order to find all quadrilaterals it is sufficient to check in each plane one "corner" of a node. For example, in Figure 4(d), all quadrilaterals are found by checking the upper right region of a node.

After relaxation each quadrilateral is triangulated using either a shortest diagonal or a Delaunay criterion [4]. Either of these criteria creates triangles that result in a smoother shape than choosing a fixed configuration. The resulting triangle mesh can be rendered using standard 3D graphics techniques.

3 Results

To evaluate the relative effectiveness of the presented techniques, the SurfaceNet is compared to Marching Cubes, which is the standard iso-surface extraction tool [9]. The effectiveness of each technique will be tested against the requirements listed in Section 1. Each of these requirements is measured as follows.

- A measure expressing the local smoothness of a polygon mesh is given in [11]. As a first step, the angles α_i of all triangles around a vertex are summed. If all triangles connected to a vertex are coplanar this sum is equal to 2π. A measure of the local smoothness at a vertex is defined by $2\pi - \sum \alpha_i$, the absolute value of which is then averaged over all vertices.

- A simple and direct measure for triangle quality is found upon division of the smallest angle of each triangle by its largest angle. If the triangle is equilateral this expression is equal to 1.
- The accuracy is expressed by the unidirected modified Hausdorff distance that represents the mean distance of the generated mesh to a reference shape [3]:

$$H_{ave}(S_1, S_2) = 1/N \sum_{p \in S_1} e(p, S_2) \qquad (7)$$

where e is the minimum distance between a point and a surface, and S_1 and S_2 are two surfaces.

Using these measures, the following experiments are conducted. Two volumes, containing greyscale images of distance maps of respectively a plane and a sphere were created, where the greyscale values were stored as floats. An iso-surface is extracted using Marching Cubes (MC), a SurfaceNet with averaging (SNA) and a SurfaceNet with the combined technique (SurfaceNet with Extended Relaxation and Triangulation SNERT) as presented in Section 2. These surfaces are compared to the exact reference shape. The results of this comparison are shown in Table 1.

Table 1. Measured results on the Plane and the Sphere. Methods are Marching Cubes (MC), SurfaceNet Averaging only (SNA) and SurfaceNet with Extended Relaxation and Triangulation (SNERT). The accuracy is measured respectively at the vertices and at the centers of the faces.

	Quality		Smoothness		Accuracy (vertices)		Accuracy (face centers)	
	Plane	Sphere	Plane	Sphere	Plane	Sphere	Plane	Sphere
MC	0.64	0.54	$0.25 \ 10^{-6}$	0.0028	$5.59 \ 10^{-6}$	$2.92 \ 10^{-3}$	$4.81 \ 10^{-6}$	0.028
SNA	0.92	0.74	$16.2 \ 10^{-3}$	0.0108	0.092	0.204	0.788	0.426
SNERT	0.93	0.75	$0.15 \ 10^{-6}$	0.0028	$20.9 \ 10^{-6}$	$12.3 \ 10^{-3}$	$12.5 \ 10^{-6}$	0.043

Comparing the quality of the triangles for each method shows that both SNA and SNERT produce triangles of a higher quality than MC for the plane as well as the sphere. Also, in the case of the plane the MC and SNERT method produce a smoother (=flatter) surface than SNA. The sphere has a constant curvature that corresponds to the smoothness outcome of MC and SNERT. SNA shrinks the mesh and pulls the nodes away from the iso-surface accounting for the lower smoothness and accuracy. The accuracy of the SNERT surface is lower than MC because the nodes are placed according to the trilinearly interpolated values. However, the error at the vertices for SNERT is smaller than the error at the face centers for the Marching Cubes generated sphere.

To illustrate the effectiveness of our technique a graphical example is shown in Figure 7. Clearly, the SNERT surface is as flat as the MC surface and the triangles have higher quality. Figure 6 shows the mesh generated by MC and SNERT on a dataset containing two overlapping spheres. The average triangle quality for the Marching Cubes mesh is 0.64, for the SNERT mesh this number is equal to 0.93.

In addition to the results presented, several experiments were done on true greyscale MRI and CT data. Figure 5 shows a histogram of triangle quality for meshes generated

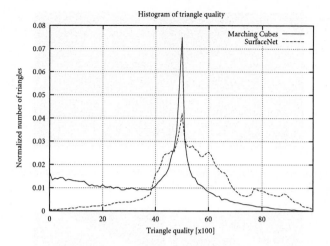

Fig. 5. Histogram of triangle quality for meshes generated using Marching Cubes and a SurfaceNet. The mesh was generated from a CT scan of an ankle. The peak at 0.5 indicates that many right triangles are generated (quality measure is smallest angle divided by largest angle of a triangle).

by MC and SNERT from a greyscale CT dataset containing part of a human ankle. It can be seen that the SNERT mesh contains less low quality triangles and contains more high quality triangles. Figure 8 shows the meshes generated from a CT-scan of a human ankle. Lastly, Figure 9 shows a close-up of the bladder extracted from a 256x256x61 MRI dataset of the abdomen of a female patient.

4   Conclusions

Finite element analysis is a standard way to simulate soft tissue deformation. For proper modelling, triangular mesh models must satisfy requirements of accuracy, smoothness and triangle quality. Several approaches proposed in the literature do not meet these requirements (e.g., Marching Cubes in combination with low pass filtering).

In this paper we extending the SurfaceNet method, and evaluated two variants. Optimization of a triangle mesh was performed by averaging vertices, stepping in the direction of the gradient to the iso-surface, and a combined approach.

From visual inspection of test objects, the meshes generated by a SurfaceNet appear to be of similar quality as those created by Marching Cubes. This is backed up by measurements. The SurfaceNet meshes are more suitable for finite element modelling as they are significantly smoother and have a low number of poor quality triangles.

We conclude that SurfaceNet creates a globally smooth surface description that retains fine detail.

Future research will focus on improving the performance of the SurfaceNets technique and developing suitable mesh reduction techniques for finite element analysis.

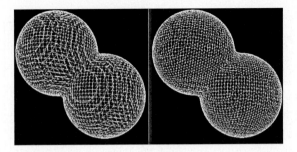

Fig. 6. Two spheres partly overlapping. Meshes generated by Marching Cubes (left) and SNERT (right). Both meshes have the same number of triangles.

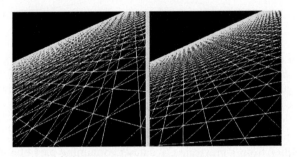

Fig. 7. Generated mesh using Marching Cubes (left) and SurfaceNets (smoothing+gradient) (right). A view of a plane is shown.

Acknowledgements

This research is part of the MISIT (Minimally Invasive Surgery and Intervention Techniques) programme of the Delft Interfaculty Research Center on Medical Engineering (DIOC-9). The work described here was largely carried out at MERL – a Mitsubishi Electric Research Laboratory in Cambridge, MA (USA).

References

1. BRO-NIELSEN, M. Finite element modelling in surgery simulation. *Proceedings of the IEEE Special Issue on Virtual & Augmented Reality in Medicine 86*, 3 (Mar. 1998), 490–503.
2. COTIN, S., DELINGETTE, H., AND AYACHE, N. Real-time elastic deformations of soft tissues for surgery simulation. *IEEE Transactions on Visualization and Computer Graphics 5*, 1 (1998), 62–73.
3. DUBUISSON, M., AND JAIN, A. A modified Hausdorff distance for object matching. In *Proceedings, 12th IAPR International Conference on Pattern Recognition, Conference A (Jerusalem, Israel, October 9–13, 1994)* (Oct. 1994), IEEE Computer Society Press, Los Alamitos, CA, 1994, 566-568, pp. 566–568.
4. FIELD, D. Laplacian smoothing and Delaunay triangulations. *Communications in Applied Numerical Methods 4*, 6 (1988), 709–712.

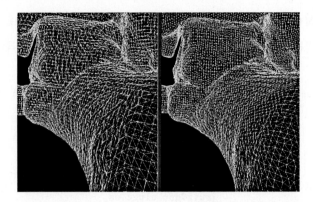

Fig. 8. Mesh generated by Marching Cubes (left) and SurfaceNet (right) on a greyscale image of an ankle. The dataset is a CT-scan with dimensions 132 × 141 × 69.

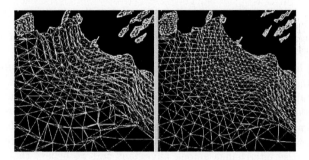

Fig. 9. View of a bladder extracted using Marching Cubes (left) and using SurfaceNets (right).

5. GARLAND, M. *Quadric-Based Polygonal Surface Simplification*. PhD thesis, School of Computer Science, Carnegie Mellon University, Pittsburgh, PA, 1999. Software and thesis available at: http://www.cs.cmu.edu/~garland/quadrics/.

6. GIBSON, S. Constrained elastic surfacenets: generating smooth surfaces from binary sampled data. In *Proceedings Medical Image Computation and Computer Assisted Interventions, MICCAI '98* (1998), pp. 888–898. http://www.merl.com/reports/TR99-24/.

7. GIBSON, S. Using distance maps for accurate surface representation in sampled volumes. In *Proceedings 1998 IEEE Symposium on Volume Visualization* (Oct. 1998), pp. 23–30. http://www.merl.com/reports/TR99-25/.

8. KOCH, R., GROSS, M., CARLS, F., VON BÜREN, D., AND FANKHAUSER, G. Simulating facial surgery using finite element models. In *Proc. ACM SIGGRAPH 96* (1996), pp. 421–428. New Orleans, USA, August 4-9 1996, Computer Graphics Proceedings.

9. LORENSEN, W., AND CLINE, H. Marching cubes: a high resolution 3D surface construction algorithm. In *Proc. ACM SIGGRAPH'87* (July 1987), pp. 163–169.

10. MONTANI, C., SCATENI, R., AND SCOPIGNO, R. Decreasing isosurface complexity via discrete fitting. Tech. Rep. xx, Istituto per l'Elaborazione dell'Informazione - Consiglio Nazionale delle Ricerche, Pisa, Italy, Dec. 1997. http://vcg.iei.pi.cnr.it.

11. VERON, P., AND LEON, J. Shape preserving polyhedral simplification with bounded error. *Computers & Graphics 22*, 5 (1998), 565–585.

diSNei: A Collaborative Environment for Medical Images Analysis and Visualization*

Carlos Alberola[1], Rubén Cárdenes[2], Marcos Martín[1], Miguel A. Martín[1],
Miguel A. Rodríguez-Florido[2], Juan Ruiz-Alzola[2,3]

[1] ETSI Telecomunicación. University of Valladolid, Spain,
caralb@tel.uva.es,
WWW home page: http://atenea.tel.uva.es
[2] Dep Señales y Comunicaciones. University of Las Palmas de Gran Canaria, Spain,
jruiz@dsc.ulpgc.es,
WWW home page: http://grtv.teleco.ulpgc.es
[3] Dep. Radiology. Harvard Medical School and Brigham & Women's Hospital, USA

Abstract. In this paper we describe our environment diSNei, a graphical tool for collaborative image analysis and visualization of models created out of slices of volume data; this application allows a number of users to simultaneous and coordinatedly analyze medical images, create graphical models, navigate through them and superimpose raw data onto the models. The application is intended to help physicians interpret data in the case that ambiguous situations may appear, by means of collaboration with other colleagues. It is therefore an integrated environment for expertise interchange among physicians and we believe that it is a powerful tool for academic purposes as well. Other outstanding application features are its being multiplatform, and, particularly, the fact that it can run on NT computers, and the support for stereo rendering so as to obtain a deep sensation of inmersion into the models.

1 Introduction

Information technologies are nowadays naturally incorporated in most of the professional and social activities; in particular, graphical computer applications ease tremendously the complex process of multidimensional data interpretation. Consequently, an important effort has been focused on this concept, in very different areas, such as geology, meteorology, chemistry and, of course, medical imaging. As far as the medical field is concerned, a number of applications have been created and reported, both in Europe[1][2] and in the US [3][4] just to mention a few. Specifically, the joint effort of MIT & and the Surgical Planning Laboratory (SPL) at Harvard Medical School has given rise to a Web-based environment consisting of Java applets running on any popular browser. This package is a framework for integration of images and textual information and

---

* This paper is a joint effort of two Spanish universities coordinated by the European research grant 1FD97-0881. Alphabetical order has been used in the author's order. Correspondence should be addressed to the last author.

it allows users to combine 3D surface models of anatomical structures, their cross-sectional slices, and textual descriptions about the structures.

Other advanced systems have been reported, as 3DSlicer [5] from SPL, ANALYZE [6], MEDx [7] and MNI [8]. All of these present more extensive applications in image-guided medicine and surgical guidance, and they incorporate powerful tools for segmentation, registration and quantitative data analysis.

All these tools, seem to work as isolated environments, in which a single user 'does the job'. However, inclussion of the possibility of collaboration seems very interesting, as it has been reported elsewhere [9] for one-dimensional signals and [10] for echocardiography.

In this paper we present our package called diSNei[1] which is an integrated environment for collaborative data analysis and visualization. This environment is totally architecture-independent, easily expandable, and though it was initially intended as a fetal-growth monitoring tool, it is transparent to the source of data to be analyzed. Apart from the classical visualization tools, it incorporates a module for symmetric collaborative work, which, to the best of our knowledge, has not been reported in fetal echography. This type of data have an inherent difficulty in interpretation (specially if data have already been acquired and no further views can be incorporated into the data base) and therefore we believe it will be of interest for diagnosis applications in unclear or ambiguous situations (in which several opinions of experts may disambiguate the situation) and academia and training as well.

In what follows, we describe the system architecture, as well as the functionality currently included in our environment. A number of snapshots taken from the application illustrate it ease of use and its graphical capabilities.

2 The diSNei Data Analysis and Visualization Environment

Our package incorporates the facilities of other more classical packages (image segmentation, generation of graphical models out of segmented data, inmersive navigation and so forth) and it also incorporates a module for computer supported collaborative work (CSCW).

Due to the fruitfull activity that characterizes this field, a major issue in diSNei is its modularity; new functions can be easily attached to the graphical user interface (GUI) just with a button, leaving the rest of the environment untouched. Portability is also a major concern, due to the great number of platforms currently in use. These two points are easily achieved by means of portable environments and programming languages such a the Visualization Toolkit (VTK) [11] for data processing and visualization and Tcl/Tk [12] for flow control and GUI design and management. VTK is an open shareware code, so new classes

---

[1] The acronym comes from the Spanish *Diseño Integrado de Segmentador y Navegador de Estructuras Internas*, i.e. integrated design of a segmenter and a visualizer of inner structures.

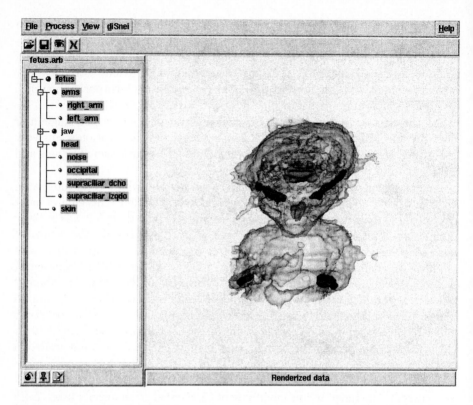

Fig. 1. An example of diSNei for some volume data (specifically, data from a fetus). The GUI shows menu buttons on top, the hierarchically-interpreted data on the left hand side and the navigation window with a semitransparently rendered view.

can be easily incorporated into the environment and made available to the community.

Figure 1 shows the diSNei's GUI. As it can be seen it basically consists of four frames, namely, the menu buttons bar, the rendering and navigation window, the hierarchical data structure of organs at disposal, and a status bar. The hierarchical data structure makes it particularly simple to select and highlight organs of interest in the navigation model. Details follow.

2.1 Image Segmentation with diSNei

An image segmentation tool is crucial for graphical medical applications since every organ must be properly identified in order to perform any further analysis. Our hypothesis is that data are originally stored as a series of parallel slices; therefore, for the sake of efficiency, our Segmentation Manager Tool (SMTool) starts working in a selected slice with classical segmentation algorithms, but then it projects the labels so far obtained onto subsequent slices to benefit from the spatial redundancy inherent in a slice-oriented imaged volume.

As we said in the Introduction, diSNei is clinically-oriented and it is initially aimed at fetal 3D ultrasound image analysis. Therefore we have divided our modules into two categories, namely, clinical and research algorithms. As far as the first category is concerned, our segmentation algorithms are simple since no sophisticated algorithms have proven their success in such an image modality. The segmentation procedure requires human intervention, but we have tried to build a flexible environment to let the operator perform simple operations comfortably. Specifically (see figure 2):

1. The segmentation procedure is governed by the concept of a session. The operator can start a session at anytime, proceed with the segmentation for as long as desired, and then save the segmentation results to disk. If the segmentation procedure is not finished, the operator can retrieve the session in exactly the same stage as it was saved; consequently, fatigue in the operator does not interfere the labelling procedure.
2. The organs and structures are segmented by simple interactions with the mouse: the operator draws a coarse template of the object to extract, and then a region growing procedure is triggered to adjust the template to the organ actual layout. Our current implementation controls the region growing method by means of a requirement of connectivity in the region, and also by keybord-introducing a range of allowable intensity pixel values. Note that the region growing may turn out to be a region shrinking procedure depending of the image data.
3. The procedure so far described applies to the current slice; however, it does not have to be repeated for every slice, but we exploit the fact that a great spatial redundancy exists in volumetric image data between two consecutive slices. This is implemented by projecting the segmentation result onto the two consecutive slices, and the triggering a new region growing procedure on the two neighbouring slices using this projection as the seed. The procedure is finished when the region eventually dissapears (i.e., no connected image pixels within the intensity bounds exists).
4. Part of the segmentation protocol is a region modification module; the user can delete regions or portions of regions at will, either in one particular section or throughout the volume. Also, new pieces of the images can be added to existing regions.

Although simple, this procedure gives enough interactivity so as to create graphical models out of 3D ecographies (the rendered images shown in the paper have been segmented with this tool by a non-expert operator). However, surfaces so created are prone to suffer degradations due to the lack of smoothing operations between slices. To alleviate this problem we have a module of research algorithms some of which have drawn interesting results [13]. Also, active contours allow the user to apply less effort in the initial sketch of the regions to be segmented. However, the proper characteristics of the ultrasound data make energy function selection an issue[14]; we are currently porting this module into the diSNei platform.

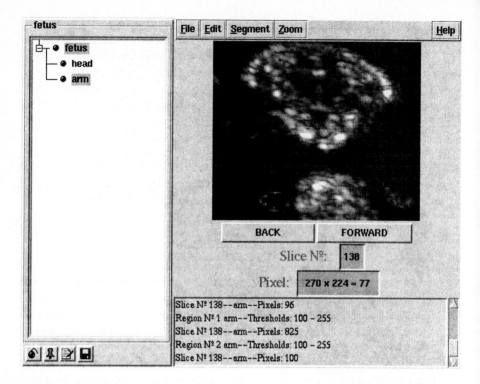

Fig. 2. Non-collaborative SMTool. The user can draw on the regions to be segmented, and the data structure is created and shown on the data tree structure on the left hand side.

Finally, the segmenter allows the user to create the hierarchical data structure we have mentioned before. It also creates the logic needed to generate graphical models out of the labelled data.

2.2 Medical Image Visualization with diSNei

Our environment diSNei incoporates advanced visualization facilities that have quite the same features as other reported browsers. However, the design has been oriented at maximizing the ergonomics and ease of use. Specifically:

- The gap between segmented (labelled) volume data and graphical model is filled in by means of an isosurface extraction module. Though the algorithm to calculate the isosurfaces is well-known [15] it is worth mentioning that this complexity is completely transparent to the end user, since the functionality is guided by the data structure that has been created in the segmentation phase. Therefore a few mouse-guided operations are the only thing to be done by the application end user.
- Once the graph model is created, the user selects which organs to represent, the colors to be used, and whether the representation should be done semi-transparent or opaque. Once again, this is done by the menus on the left-hand

Fig. 3. The stereo-rendered fetus

side of the browser (see figure 1). Functional relations can be highlighted in the graphical model by proper selection of an entire tree branch.

- A number of predefined views can be selected from a menu. This increases speed in representation since no rotations are needed to achieve positions of interest.
- Navigation throughout the graphical model is a built-in feature in our application. The user can fly through the outer surface of the graphical model so as to see inner structures from inside.
- Our current version includes stereo rendering. By using stereo glasses the feeling of inmersion is dramatically improved at no further computational cost (in a Sillicon Graphics NT workstation). The glasses are also comfortable to wear and they only need a small controlling device (placed on top of the monitor in figure 3) so no further cables are needed.
- A plane can intersect the model with axial, sagittal and coronal planes (see figure 4a) and a plane with an arbitrary position and orientation (figure 4b), The volume data are resliced according to the plane orientations and displayed. With this module, the end user can see at a glance where every slice in the raw data is actually located in 3D space.
- Data can be exported to other environments, specifically, a virtual reality modelling language (VRML) browser.

2.3 CSCW with diSNei

Human interaction is a very useful source of knowledge when it comes to solve a difficult problem. An example of this interaction are the so-called *brainstorm* meetings in which the interactors put in common ideas with the ultimate purpose to come up with an interesting overall idea created from contributions of the participants. Computer technology has made this collaboration possible by the so-called CSCW or, more generally, electronic groupware [16].

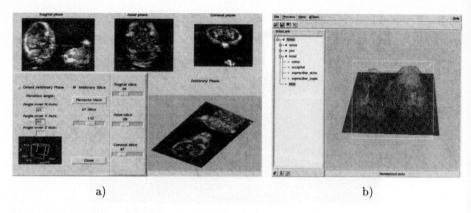

a)   b)

Fig. 4. a) Sagittal, axial and coronal planes (top) together with an arbitrarily oriented plane (bottom). Volume data are properly interpolated to get this representation. b) Graphical model with the intersected plane.

CSCW is a hot topic that is under use in many different scenarios. As far as the medical imaging field is concerned, several implementations have been reported during this decade. Most of them have to do with one or two dimensional signals [9, 17–19], which are so far the ones typically used by physicians in their daily practice. However, we adhere to Thomas Berlage [10] in believing that graphical models of volume medical data create the *common ground* that is needed for useful interactions. This is specially true in the case of fetal echographies, since there is no scanning protocol widely accepted, but every echographist has his/her proper style. Once the volume data is acquired with a traditional free-hand two dimensional probe, image interpretation is not so straightforward. Indeed, unless some provision is made, the spatial data of the scanning procedure itself is lost (i.e., the absolute position and orientation of the ultrasonic probe), so it cannot be used as an additional piece of information for cooperation.

In diSNei we are aware of how important it could be for a physician to be assisted in some cases by a colleague, so we have created a collaborative environment in which at least two such physicians may collaborate to jointly reach a proper decision. Some of the foregoing ideas have been generalized in order to simultaneous and coordinatedly be carried out by more than one user. To be more specific, several users (see figure 5) can exchange information on the echographies both by text-based messages and also by free-hand drawing on the data; the arrows in the bottom of figure 5b) give the possibility to scroll back and forth in the slices of the volume data. Collisions in synchronous actions are avoided by means of a token-grabbing procedure: the user who wants to perform any task with the shared data needs to grab the token. Once the token is grabbed, the rest of the participants in the conference become listeners. The users can also manipulate the graphical model at will, with the same access

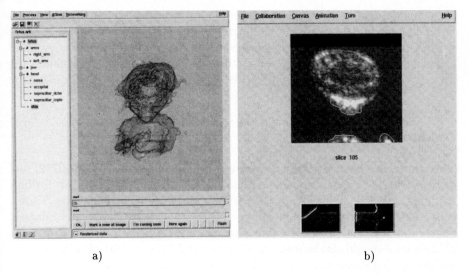

a)   b)

Fig. 5. CSCW version of diSNei. a) Browser and talk channel b) CSCW-SMTool. The arrows allow any user to scroll forward or backward in the slices of the volume data.

protocol as before. The text-based information exchange is asynchronous, so any user can communicate with the rest of the participants at any time. This communication channel is the natural way to let the users agree a respectful token grabbing procedure.

Our platform uses the GroupKit package [20] which is a Tcl/Tk application programming interface (API) that allows programmers to create group activites fast and reliably. Most of the coordination is taken care of by the API; the programmer is therefore allowed to focus on the final application.

3   Conclusions

In this paper we describe a novel collaborative environment for volume data analysis and inmersive visualization; the package currently contains an easy-to-use module for volume data segmentation and a very ergonomic stereo-rendering environment, which gives a high degree of inmersion in the graphical model. All the operations can be performed coordinatedly by several simultaneous users, regadless of their actual physical location. A steady communication channel is always available for physicians to exchange typed messages at any time.

We are currently working on different tiers to improve the system performance and capabilities. One effort is focused on developing compression algorithms for multiresolution data visualization; this module could be useful in the case that the application were to be used with a low-bandwith channel. Moreover, for clinical purposes, direct measures of organs and structures on 3D space seem mandatory. As far as the collaboration is concerned, the application is still in its infancy, but several ideas are currently under development. Some of them are the use of powerful database management tools for information retrieval and exchange, and also the implementation of a *commond ground* that includes the information of the actual scanning procedure. To this end, we are using magnetic position devices as reported in [1].

Acknowledgements

This paper has been partially supported by research grants TIC97-0772, and 1FD97-0881. The authors want to thank the students Miguel Ángel Toribios and José Carlos de la Fuente for their help in much of the programming of some collaborative aspects of diSNei and Marco Pinacho, Alberto Rivera and Eduardo Suárez for their help in the SMTool. Prof. Yannis Dimitriadis at the University of Valladolid should also be mentioned for the useful discussions with the authors and for all the on-going research on the CSCW in diSNei. Finally, the authors are indebted to Med. Dr. Arya Nabavi at University of Kiel (Germany) and Brigham & Women's Hospital (USA) for many fruitful discussions on clinical aspects of this work and to prof. Yang at Saskatchewan University (Canada) for kindly sharing with us the ultrasound data we have used in our research.

References

1. Solus Project, http://svr-www.eng.cam.ac.uk/Research/Projects/Solus.
2. Invivo Project, http://www.igd.fhg.de/teleinvivo
3. Golland et al. AnatomyBrowser: A Framework for Integration of Medical Information. *MICCAI98 (Proc of the First International Conference*, MA, Cambridge, USA, Springer Lecture Notes in Computer Science, 1998, Vol. 1679, pp.720-731.
4. Brigham & Women's Hospital and Harvard Medical School. Surgical Planning Lab, http://www.splweb.bwh.harvard.edu:8000
5. Gering et al. An Integrated Visualization System for Surgical Planning and Guidance Using Image Fusion and Interventional Imaging. In C.J. Taylor and A. Colchester (eds.): Medical Image Computing and Computer-Assisted Interventions, Lecture Notes in Computer Science, Vol. 1679. Springer-Verlag, Berlin Heidelberg, New York (1999) 809-819.
6. Mayo Clinic. ANALYZE software, http://www.mayo.edu/bin/analyze/$ANALYZE_Main.html
7. Sensor. MEDx software, http://www.sensor.com/$medx_info/medx_docs.html
8. MNI. MNI software, http://www.bic.mni.mcgill.ca/software
9. Y. Bouillon, F. Wendling, F. Bartolomei, Computer Supported Collaborative Work (CSCW) in Biomedical Signal Visualization and Processing, IEEE Trans. on Info. Tech. in Biomed. 3 (1999) 28-31.
10. T. Berlage, Augmented Reality Communication for Diagnostic Tasks in Cardiology, in Biomedical Signal Visualization and Processing, IEEE Trans. on Info. Tech. in Biomed. 2 (1998) 169-173.
11. W. Schroeder, K. Martin, B. Lorensen, The Visualization Toolkit: an object-oriented approach to 3D graphics, Prentice Hall Int., New Jersey (1998).

12. B. B. Welch, Practical Programming in Tcl/Tk, Prentice-Hall Int., New Jersey (1997).
13. R. San José, A. Rivera, M. Pinacho, C. Alberola, J. Ruiz-Alzola, A Kalman filter technique applied to surface reconstruction and visualization from noisy volume data. In E. Kramer (eds.): Ultrasonic Imaging and Signal Processing, Proceedings of the SPIE, Vol. 3982, San Diego, California (2000) 396-407.
14. M. Martín, E. Rodríguez, D. Tejada, C. Alberola, J. Ruiz-Alzola, Energy Functions for the Segmentation of Ultrasound Volume Data using Active Rays, *Proc. of the IEEE Int. Conf. On Acoustics, Speech and Signal Processing, ICASSP-2000*, Istambul, Turkey (2000) 2274-2277.
15. W. E. Lorensen and H. E. Cline, Marching Cubes: a High Resolution 3D Surface Construction Algorithm, Comp. Graph. 21 (1987) 163-169.
16. C. A. Ellis, S. J. Gibbs, G. L. Rein, Groupware: Some experiences and issues, Comm. of the ACM, 34 (1991).
17. L. Kleinholz, M Ohly, Supporting Cooperative Medicine: The Bermed Project, IEEE Multimed. Mag. (1994) 44-53.
18. L. Makris, I. Kamilatos, E. V. Kopsacheilis, M. G. Strintzis, Teleworks: A CSCW Application for Remote Medical Diagnosis Support and Teleconsultation in IEEE Trans. Inform. Technol. in Biomed, 2 (1998) 62-73.
19. E. J. Gómez, F. del Pozo, E. J. Oritz, N. Malpica, H. Rahms, A Broadband Multimedia Collaborative System for Advanced Teleradiology and Medical Imaging Diagnosis in Biomedical Signal Visualization and Processing, IEEE Trans. on Info. Tech. in Biomed. 2 (1998) 146-155.
20. GroupLab of University of Calgary, http://www.cpsc.ucalgary.ca/projects/group-lab/groupkit/

Impact of Combined Preoperative Three-Dimensional Computed Tomography and Intraoperative Real-Time Three-Dimensional Ultrasonography on Liver Surgery

Mitsuo Shimada, M.D.[1], Takayuki Hamatsu, M.D.[1], Tatsuya Rikimaru, M.D.[1],
Yo-ichi Yamashita, M.D.[1], Shinji Tanaka, M.D.[1], Ken Shirabe, M.D.[1],
Horoshi Honda, M.D.[2], Makoto Hashizume, M.D.[3], Keizo Sugimachi, M.D.[1]

[1] The Department of Surgery II, Faculty of Medicine, Kyushu University,

Fukuoka 812-8582, Japan.

[2] The Department of Radiology, Faculty of Medicine, Kyushu University,

Fukuoka 812-8582, Japan

[3] The Department of Disaster and Emergency Medicine, Graduate School of Medical

Sciences, Kyushu University, Fukuoka 812-8582, Japan

Abstract. Objective: The aim of this study was to clarify the impact of combined preoperative three dimensional computed tomography (3D-CT) imaging and intraoperative real-time 3D-ultrasonography (US) imaging on image navigation for hepatic resection. Patients and Methods: An integrated navigation system using combined preoperative 3D-CT images and intraoperative real-time 3D-US images were used with patients who underwent hepatic resection for liver tumors, including HCC, cholangiocarcinoma and metastatic liver cancer, and for donor hepatectomy for living related liver transplantation. 3D-CT imaging was made using the workstation "ZIO M900" (ZIO software, Inc., Tokyo, Japan). 3D-US was made in real-time during an operation using the ultrasonographic device SSD5500 (ALOKA, Tokyo, Japan) and the workstation "SAS 200"(ALOKA, Tokyo, Japan). Results: The 3D-CT imaging from CT during hepatic arteriography gave us details of the arteries feeding the tumor. By 3D-CT imaging from CT during arterial portography, the recognition of Glisson's branches in the liver, as well as the hepatic venous system, was much easier than conventional 2D-CT imaging. 3D-US images allowed us to understand the anatomy of the liver more easily than from 2D-US images, and the 3D-US images could reinforce in real-time preoperative information using 3D-CT images. Conclusions: Combined preoperative 3D-CT and intraoperative real-time 3D-US provides us with integrated information of liver anatomy, especially the spatial relationship between tumor and intrahepatic vessels, which can not be visualized because of their location inside of the liver. Intraoperative 3D-US imaging can reinforce preoperative excellent 3D-CT imaging. Therefore, our integrated navigation system consisting of combined 3D-CT and 3D-US is useful for liver surgery.

1. Introduction

Recent medical advances have made hepatic resections, especially for hepatocellular carcinoma (HCC), much safer than before [1-3]. However, hepatic resections remain one of the most difficult operations in general surgery. Important vessels are hard to be visualized inside the liver, as with brain. Furthermore, in surgery for liver tumors, especially HCC, most livers have chronic liver diseases such as liver cirrhosis. Therefore, the tumor must be resected judging from both anatomical and functional viewpoints. In other words, liver parenchyma including the tumor has to be resected minimally and necessarily, and is often removed in units of less than the segment classified by Couinaud [4], as systemically as possible. Several authors have reported the efficacy of preoperative surgical planning [5-8] in preoperative three-dimensional computed tomography (3D-CT), . In contrast to the great merit of 3D-CT images, the anatomy of the liver is altered in hepatic surgery because of pre-resectional mobilization of the liver from the retroperitoneum. Therefore, during hepatic resections, the efficacy of preoperative planning by 3D-CT images is reduced, and a real-time navigation method has to be developed. Intraoperative ultrasonography (US) has been routinely used during hepatic resections to better understand vessels inside the liver in real-time. Intraoperative US provides a great deal of information about tumor location and vascular anatomy in the liver, and is an indispensable procedure for hepatic resections. However, an accurate three-dimensional understanding of US images depends on the operator's skill and experience, as the original display is two-dimensional. Three-dimensional ultrasonography (3D-US) is, therefore, expected to be useful for liver surgery. However, little has been reported on the use of 3D-US during surgery, especially liver surgery [9].

The aim of this study was to clarify the impact of combined preoperative 3D-CT imaging and intraoperative real-time 3D-US imaging on image navigation for hepatic resections.

2. Patients

Up to now, we have used a combination of preoperative 3D-CT images and intraoperative real-time 3D-US images in cases that have underwent hepatic resection for liver tumors, including HCC, cholangiocarcinoma and metastatic liver cancer, and for donor hepatectomy for living related liver transplantation, at the Department of Surgery II, Kyushu University.

3. Methods

3.1 Definition of anatomy of the liver

The definition of liver segment was that of Couinaud [4], and the anatomic division of the liver of less than Couinaud's segment was defined by the classification of Takayasu et al [10].

3.2 Preoperative 3D-CT

Helical CT images of CT arteriography (CTA) and CT during arterial portography (CTAP) were obtained using an X Vigor scanner (Toshiba, Tokyo, Japan) with a section thickness of 3 mm and a pitch of 3 mm (150 mA, 130 kV, 512x512 matrix) [11]. For CTAP, 100 ml of iopamidol contrast medium (Iopamiron 150; Nihon Schering, Osaka, Japan) was administered through a catheter in the superior mesenteric artery at an estimated rate of 2.5 ml/sec, and sequential helical scanning of the liver was begun 25 seconds after biginning the injection. For CTA, 30 - 50 ml of Iopamiron 150 was administered through a catheter in the proper hepatic atery at an estimated rate of 1.5 ml/sec, and sequential helical scanning of the liver was begun 5 seconds after biginning the injection. Both 3D CTAP and 3D-CTA imaging were reconstructed from digital data of two-dimensional CT images, and 3D-CTA imaging was constructed from two-dimensional CTA images. 3D-CT imaging was quickly made using the workstation "ZIO M900" (ZIO software, Inc., Tokyo, Japan: http://www.zio.co.jp).

3.3 Intraoperative real-time 3D-US

The 3D-US was made in real-time during an operation using the ultrasonographic device SSD5500 (ALOKA, Tokyo, Japan) and the workstation "SAS 200"(ALOKA, Tokyo, Japan). The principle of the SAS 200 is to recognize the 3D-position of the US probe as a 3D field in the magnetic field using a magnetic-field generator. The 3D-US image was generated by the volume-rendering method.

3.4 Assessment

Both preoperative 3D-CT and intraoperative real-time 3D-US images were assessed by experienced hepatic surgeons and inexperienced doctors. The points of assessment were both quality of the images and accuracy of the spatial recognition of tumor and intrahepatic vessels.

4. Results

4.1 3D-CT

The 3D-CT imaging from CTA gave us a details of the arteries feeding the tumor (Fig. 1). Furthermore, the 3D-CTA image was superior to 2D-hepatic arteriography. Using 3D-CT imaging from CTAP, recognition of Glisson's branches in the liver, as well as hepatic venous system, was much easier than with conventional 2D-CT imaging (Fig.2). By adding colors to each structure, such as portal vein, hepatic artery and hepatic vein, intrahepatic tissue structures were easily visualized. Those 3D-CT images from CTA and CTAP allowed us to plan details of operations.

4.2 Real-time 3D-US

Constructing 3D-US imaging using the workstation "SAS 200" took approximately 3 min. A color doppler mode image was useful to recognize the direction of the blood flow (both inflow and outflow vessels), but the quality of 3D-US images using the color doppler mode was inferior to the images using a power doppler mode, which was excellent to better understand the whole image of the tumor itself and the surrounding structures.Using a power-doppler mode, the vascular system in the liver became more real, and the relationship between tumor and vessels were well visualized in real time. Such 3D-US images allowed us to understand the anatomy in the liver more easily than with 2D-US images (Fig. 3). With

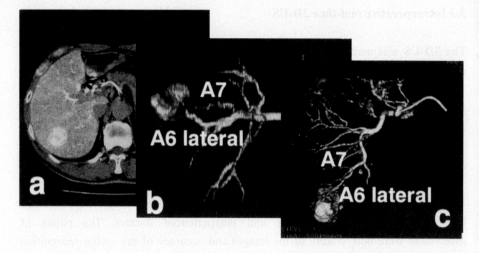

Fig. 1. Preoperative 2D- and 3D-CT images from CT arteriogaraphy (CTA). The 3D-CTA provides an excellent view of the artery feeding the tumor, and is superior to both 2D CTA and 2D-hepatic arteriography. a. 2D-CTA, b. 3D-CTA (right oblique view), c. 3D-CTA (view from foot side)

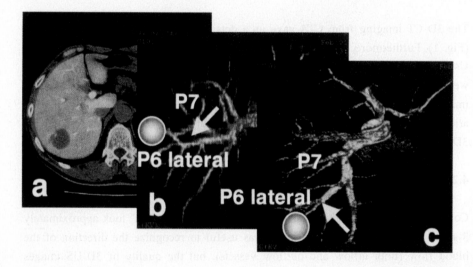

Fig. 2. Preoperative 2D- and 3D-CT images from CT during arterial portography (CTAP). The 3D-CTAP provides excellent information to understand the spatial relationship between tumor and neighboring vessels more easily (Glisson's branches and hepatic veins). An arrow indicates the dividing point, which was planned before an operation.

a. 2D-CTA, b. 3D-CTAP (right oblique view), c. 3D-CTAP (view from foot side) hyperechoic tumor, visualization of both tumor and feeding and/or draining vessels was easy. However, with iso- or hypo-echoic masses, simultaneous views of both tumor and vascular system were difficult. The 3D-US images reinforced in real-time preoperative information using 3D-CT images.

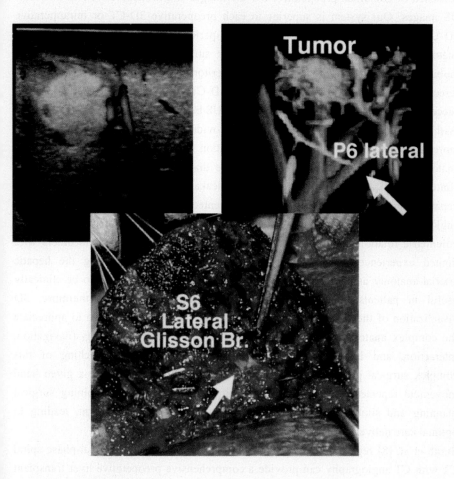

Fig. 3. Intraoperative real-time 2D-US and 3D-US images and intraoperative photography. 2D-US shows a hyperechoic mass and the feeding vessels, which are partly shown by power doppler mode. 3D-US provides excellent imaging by which the spatial relationship between tumor and neighboring vessels, including portal veins and hepatic veins, are clearly recognized.

An arrow indicates the dividing point, which was reconfirmed in real-time during an operation.

5. Discussion

This is the first report on an integrated navigation system for liver surgery, which consisted of combined preoperative 3D-CT images and intraoperative real-time 3D-US images. Our system is superior to each preoperative 3D-CT or intraoperative 3D-US. The quality of images of 3D-CT is superior to 3D-US images now, but, the anatomy of the liver is altered in hepatic surgery, because of pre-resectional mobilization of the liver from the retroperitoneum. Therefore, during hepatic resection, the information of preoperative 3D-CT images is reduced. Our system succeeded in integrating both 3D-CT and 3D-US beneficial characteristics.

With 3D-CT of liver, the 3D-CT technique provides a more accurate diagnosis and a more realistic virtual image of a tumor's location in the liver and so makes possible anatomic resection of the liver [6]. Because diagnostic errors can result in such clinical complications as postoperative bile leakage, this is a useful technique for hepatectomy, especially for sub-subsegmentectomy. Three-dimensional CT angiograms, with their global view of the anatomy and the inherent advantage of volumetric rotation of the vascular system, are useful to surgeons and others with limited experience in interpreting axial anatomy [5]. Determining the hepatic arterial anatomy using 3D-CT angiography has already been shown to be clinically useful in patients being evaluated for liver transplantation. Furthermore, 3D visualization of the organ in relation to the pathology is of great help to appreciate the complex anatomy of the liver [7]. Using virtual reality concepts (navigation, interaction, and immersion), surgical planning, training, and teaching of this complex surgical procedure may be possible. The ability to practice a given hand movement repeatedly will revolutionize surgical training, and combining surgical planning and simulation will improve the efficiency of intervention, leading to optimal care delivery.

Smith et al. [8] reported that their preliminary results suggest that dual-phase spiral CT with CT angiography can provide a comprehensive preoperative liver transplant evaluation, supplying the necessary information for patient selection and surgical planning. As a single, minimally invasive examination, this should have a marked impact on patient care by minimizing procedures and avoiding potential complications. With other uses of preoperative 3D-CT, the advantages of 3D-CT cholangiography are low level invasiveness, easily obtained images compared with those obtained with endoscopic retrograde cholangiography (ERC), good opacification, and provision of a three dimensional understanding of the biliary system, especially of the cystic duct [12]. When combined with US and routine liver function tests, 3D-CT cholangiography is considered useful to obtain information before laparoscopic cholecystectomy, as it allows the omission of ERC in many patients who have no common bile duct stone. With 3D-US in liver surgery, minimum and maximum intensity projections can show the arrangement of blood vessels (hepatic veins and portal veins) and hyperechoic regions, respectively [9]. An intensity projection from several viewpoints can reconstruct 3D imaging by cine-display. Moving a probe manually, images are taken and are processed real-time (in about 10 seconds). 3D-US was used with 24 patients undergoing hepatic resection, and allowed easy visualization of the tumors and vascular anatomy. It is an efficient and safe navigation system in liver surgery.

In conclusion, combined preoperative 3D-CT and intraoperative real-time 3D-US provides us integrated information of liver anatomy, especially the spatial relationship between tumor and intrahepatic vessels, which can not be visualized because of their location inside the liver. Intraoperative 3D-US imaging can reinforce preoperative excellent 3D-CT imaging. Therefore, our integrated navigation system consisted of combined 3D-CT and 3D-US is useful for liver surgery.

Acknowledgements

This study was supported by the Japan Society for the Promotion of Science, Research for the Future Program, Research on Robotic System in General Surgery (Project number: JSPS-RFTF 99I00902)

References

1. Shimada M, Takenaka K, Gion T, Fujiwara Y, Kajiyama K, Maeda T, Shirabe K, Nishizaki T, Yanaga K, Sugimachi K.   Prognosis of recurrent hepatocellular   carcinoma : a 10-year surgical experience in Japan. Gastroenterology   **111** (1996) 720-726.
2. Shimada M, Takenaka K, Taguchi K, Fujiwara Y, Gion T, Kajiyama K, Maeda T, Shirabe K, Yanaga K, Sugimachi K.   Prognostic factors after repeat hepatectomy for recurrent hepatocellular carcinoma. Ann Surg   **227** (1998) 80-85.

3. Shimada M, Takenaka K, Fujiwara Y, Gion T, Shirabe K, Yanaga K, Sugimachi K. Risk factors linked to postoperative morbidity in patients with hepatocellular carcinoma. Br J Surg **85** (1998) 195-198.

4. Couinaud C. Lobes et segments hepatiques; notes sur l'architecture anatomique et chirurgicale du foie. Press Med **62** (1954) 709-712.

5. Winter TC 3rd, Nghiem HV, Freeny PC, Hommeyer SC, Mack LA. Hepatic arterial anatomy: demonstration of normal supply and vascular variants with three-dimensional CT angiography. Radiographics **15** (1995) 771-80.

6. Togo S, Shimada H, Kanemura E, Shizawa R, Endo I, Takahashi T, Tanaka K. Usefulness of three-dimensional computed tomography for anatomic liver resection: sub-subsegmentectomy. Surgery **123** (1998) 73-8.

7. Marescaux J, Clement JM, Tassetti V, Koehl C, Cotin S, Russier Y, Mutter D, Delingette H, Ayache N. Virtual reality applied to hepatic surgery simulation: the next revolution. Ann Surg **228** (1998) 627-34.

8. Smith PA, Klein AS, Heath DG, Chavin K, Fishman EK. Dual-phase spiral CT angiography with volumetric 3D rendering for preoperative liver transplant evaluation: preliminary observations. J Comput Assist Tomogr **22** (1998) 868-74.

9. Shimazu M, Wakabayashi G, Ohgami M, Hiroshi H, Kitajima M. Clinical application of three dimensional ultrasound imaging as intraoperative navigation for liver surgery. Nippon Geka Gakkai Zasshi **99** (1998) 203-7.

10. Takayasu K, Moriyama N, Muramatsu Y, Shima Y, Goto H, Yamada T. Intrahepatic portal vein branches studied by percutaneous transhepatic portography. Radiology **154** (1985) 31-36.

11. Honda H, Tajima T, Kajiyama K, Kuroiwa T, Yoshimitsu K, Irie H, Abe H, Shimada M, Masuda K. Vascular changes in hepatocellular carcinoma: correlation of radiologic and pathologic findings. AJR **173** (1999) 1213-1217.

12. Kinami S, Yao T, Kurachi M, Ishizaki Y. Clinical evaluation of 3D-CT cholangiography for preoperative examination in laparoscopic cholecystectomy. J Gastroenterol **34** (1999) 111-8.

AR Navigation System for Neurosurgery

Yuichiro Akatsuka[2], Takakazu Kawamata[1], Masakazu Fujii[2], Yukihito Furuhashi[2], Akito Saito[2], Takao Shibasaki[2], Hiroshi Iseki[1], and Tomokatsu Hori[1]

[1]Dept. of Neurosurgery, Neurological Institute, Tokyo Women's Medical University
[2]Advanced Technology Research Center, Olympus Optical Co., Ltd.
Address: 2-3 Kuboyama-cho Hachioji-shi Tokyo, 192-8512 Japan
Email: y_akatsuka@ot.olympus.co.jp

Abstract. This paper presents a navigation system for an endoscope which can be used for neurosurgery. In this system, a wire frame model of a target tumor and other significant anatomical landmarks are superimposed in real-time onto live video images taken from the endoscope. The wire frame model is generated from a CT/MRI slice images. Overlaid images are simultaneously displayed in the same monitor using the picture-in-picture function so that the surgeon can concentrate on the single monitor during the surgery. The system measures the position and orientation of the patient using specially designed non-contact sensing devices mounted on the endoscpe. Based on this real-time measurement, the system displays other useful information about the navigation as well as the rendered wire frame. The accuracy of registration between the wire frame model and the actual live view is less than 2mm. We applied this AR navigation clinically in surgical resection of pituitary tumors in six cases, and verified its performance and effectiveness.

1. Introduction

Recently, the minimally invasive surgery has been playing a critical role over in the field of neurosurgery. Several attempts have been already made for this fields [1][2]. The surgeons need to carefully observe two or more different scopes and/or monitors to acquire visual information about the operational status. A prospective surgical navigator needs to provide efficient and effective navigational components as a good human interface function, as well as to provide the state-of-the-art technologies of sensing and control devices.

We developed an augmented reality (AR) navigation system especially suitable for neurosurgery[3] . This system helps the surgeons to navigate an endoscopes to the target in the patient body, while displaying efficient and effective visual information about the relative position to the target and the instrument(s) as well as surgical plans on a single monitor. First, 3D wire frame models of target tumor and other significant anatomical landmarks are generated from patient's CT/MRI slices images and are registered with the patient body coordinate frame. During the surgery, the relative position and orientation of the patient body and the surgical instruments are measured in real-time, and a wire frame model scene of the target tumor is generated and superimposed onto the live video images taken from the endoscope so that the operators can acquire navigational information.

In this paper, we will first present the system architecture and then usefulness and effectiveness of the system by showing the result of clinical test.

2. System

The system consists of an endoscope (Olympus), an optical tracking system, and a controller. The optical tracking system (I.G.T. Flashpoint 5000 –3D localizer) uses two sets of infrared LEDs to measure the position and the orientation of the endoscope respect to the patient body.

The first set of LEDs is attached on the head of the patient to measure the position and orientation of the patient head with respect to the reference frame. The second set of LEDs (shown in Fig.1 with cyan circle) is mounted on the endoscope to measure the position and orientation of the endoscope with respect to the reference frame. By taking into account the above two sets of measurements, the system calculates the relative position and orientation of the patient body and the endoscope so that the patient movement can be allowed during the surgery.

The controller (Pentium III PC) generates the wire frame model scenes of the tumor and other significant anatomical landmarks with the measured position and orientation of the endoscope as a viewpoint. The 3D wire frame models were previously reconstructed from patient's MRI data and superimposed to the scenes onto the endoscopic live images.

One good feature here is that the system takes into account the lens distortion observed in the endoscopic views. More specifically, the wire frame model scene is adjusted based on the lens distortion coefficient. Another good feature associated with the endoscope is that the system extends the wire frame scene beyond limited circular endoscopic views so that the surgeons can visually estimate the shape of the tumor beyond the viewing area.

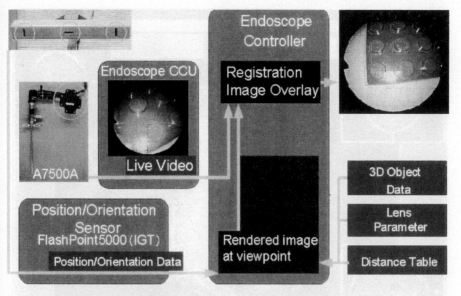

Fig. 1. AR Navigation system

In prior to the operation, the tumor and significant anatomical landmarks were reconstructed as a 3D wire frame model from patient's MRI data. On the monitor, three slice views of MRI, axial, coronal, and sagittal views, are displayed at the corners, and the position and the orientation of the endoscope are shown over slice views.

The colors of the wire frame images are controlled based on the tip position of the endoscope with respect to the target tumor. The tumor is colored "blue" in Fig.2 and "yellow" in Fig.3. This describes that the distance between the tumor and the tip position of the endoscope is less than 10mm in the case of Fig.3. The distance is also shown in the bar graph and figures at the top of the screen. we verified that the registration accuracy of the image overlay is less than 2mm[3].

3. Clinical test

We applied the AR navigation clinically in surgical resection of pituitary tumors in six cases. The patients underwent endoscope-assisted unilateral endonasal transsphenoidal surgery. Wire frame images of the anatomical structures of sphenoid sinuses, optic nerves, and internal carotid arteries, and tumors were superimposed onto the live images taken from the endoscope (Figure 2, 3). The AR navigation system indicated location of the tumors and surrounding anatomical

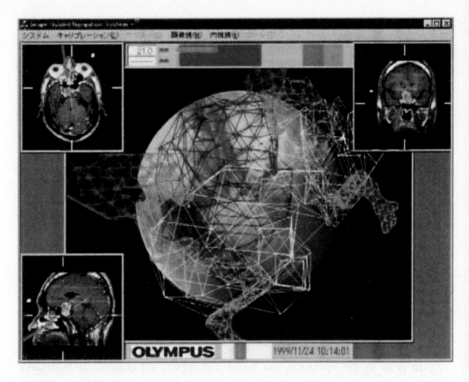

Fig. 2. Endoscopic view during surgery for pituitary tumor with superimposed images. The anterior wall of the sphenoid sinus is approached. The green, blue, light blue, and white wire frame images indicate optic nerves, tumor, internal carotid arteries, and sphenoid sinus, respectively. The upper color bar indicates reaching point of the endoscope toward the tumor. Three MRI images (axial, coronal, and sagittal) are shown on the monitor with indication of endoscopic direction (green) andendoscopic beam (yellow).

structures precisely. We could detect the location of the tumor easily. One of the most important points for safe surgery in the endonasal transsphenoidal approach is making sure the midline. The midline was always identified in the AR navigation system when we used an endoscope. Furthermore, location and direction of an endoscope itself was also indicated on MRI images on the monitor (green) with direction of endoscopic beam (yellow) (Figure 2, 3). This was very useful because neurosurgeons are sometimes confused at the direction observed in an endoscope with 30 or 70 degrees.

Intraoperative scene is demonstrated in Figure 4.

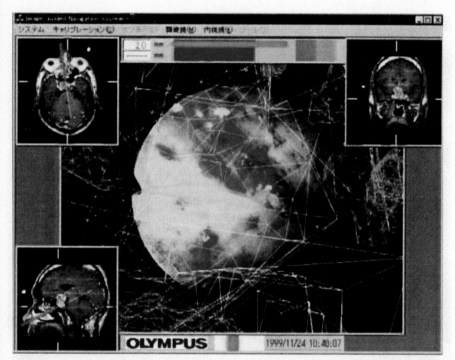

Fig. 3. Endoscopic view during surgery for pituitary tumor with superimposed images. The basal face of the sellar floor is approached following anterior sphenoidotomy. The yellow, light blue, and white wire frame images indicate tumor, internal carotid arteries, and sphenoid sinus, respectively. The upper color bar indicates reaching point of the endoscope toward the tumor. Three MRI images (axial, coronal, and sagittal) are shown on the monitor with indication of endoscopic direction (green) and endoscopic beam (yellow).

4. Conclusions and future works

The AR navigation we developed was very efficient in the endonasal transsphenoidal surgery for pituitary tumors in detecting location of tumors and surrounding anatomical structures. It was also effective to perform safe surgeries.

For further clinical application of the AR navigation system in pituitary tumor surgeries, it may be necessary to ensure the more absolute accuracy. Furthermore, it is better to set up the AR navigation system easily preoperatively.

5. Acknowledgements

This work was supported in part by Information-technology Promotion Agency, Japan.

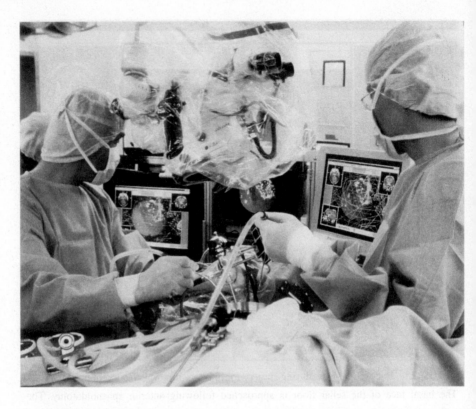

Fig. 4.   Intraoperative scene of pituitary tumor surgery. Our endonasal transsphenoidal surgery is performed using both a microsurgery and an endoscope. Monitors of endoscopic live image and AR navigation are set up.

6. References

[1] T. Dohi : Computer SurgeryandMedical Image Processing , Medical Imaging Technology Vol.12 No.5 (1994) pp 612-618

[2] H. Iseki et. al. : Medical science and virtual reality – Computer aided surgery (CAS)-, Proceedings of 73th Annual Meeting (The Japan Society of Mechanical Engineers) (1996) Narashino

[3] Y. Akatsuka et. al. :Navigation System for Neurosergery with PC Platform, Medicine Meets Virtual Reality 2000 Proceedings pp10-16(2000) Newport Beach, CA

Magneto-Optic Hybrid 3-D Sensor for Surgical Navigation

Masahiko Nakamoto[12†], Yoshinobu Sato[21‡], Yasuhiro Tamaki[3]
Hiroaki Nagano[3], Masaki Miyamoto[12], Toshihiko Sasama[12]
Morito Monden[3], and Shinichi Tamura[21]

[1] Department of Informatics and Mathematical Science
Graduate School of Engineering Science, Osaka University
[2] Division of Functional Diagnostic Imaging
[3] Department of Surgery and Clinical Oncology
Osaka University Graduate School of Medicine
†nakamoto@image.med.osaka-u.ac.jp, ‡yoshi@image.med.osaka-u.ac.jp

Abstract. The objective of the work described in this paper was to develop an accurate three dimensional (3-D) sensory system without a line-of-sight requirement for surgical navigation inside the body. Although magnetic sensors seem to be particularly suitable for this purpose, their accuracy is affected by metallic objects, which can hardly be avoided in a surgical environment. We propose a new magneto-optic hybrid 3-D sensor configuration that overcomes this limitation. Unlike previous hybrid systems, both the receiver and transmitter of the magnetic sensor are mobile, thereby permitting them to be positioned flexibly and adaptively so as to minimize inaccuracies arising from the presence of peripheral metallic objects. The 3-D position and orientation of the transmitter are measured by an optical sensor in order to accurately track the transformation between the coordinate systems of the magnetic and optical sensors. The effects of the distance between the receiver and the transmitter and their respective distances from metallic objects on the accuracy of the system were evaluated by experiments both in the laboratory and in the operating room.

1 Introduction

Although optical three-dimensional (3-D) position sensors are widely used in surgical navigation systems on account of their high degree of accuracy and acceptable speed [1]–[3], their line-of-sight requirement means that they have inherent limitations in obtaining 3-D information relating to flexible instruments and imaging devices inside the body. Magnetic 3-D position sensors [3] seem to be particularly suitable for acquiring such 3-D positions, and miniature magnetic sensors have recently been developed (for example, Biosense, Johnson & Johnson) that enable the 3-D position and orientation to be obtained even at the tip of a catheter or a flexible endoscope. However, magnetic sensors have the drawback that their accuracy is affected by metallic objects, which are inevitably in a surgical environment. Our aim is to overcome the limitations of optical and magnetic sensors by combining them in a hybrid system.

In the operating room (OR), the OR table and surgical instruments are considered to be the major causes of inaccuracies in magnetic systems. Since the OR table has a large surface area, the magnetic sensor performance is considerably affected at all time. Although surgical instruments are much smaller, they can give rise to inaccuracies when placed in the vicinity of the measurement area. Unlike the case of the OR table, distortion caused by surgical instruments varies with their arrangement during the operation. In previous work on optical and magnetic sensor hybridization [4]–[6], magnetic sensor distortion caused by metallic objects in measured 3-D positions has been corrected by using pre-calibrated data. However, since such distortion depends on the specific environment, time-consuming calibration is needed for each environment, and it is particularly difficult to calibrate for surgical instrument distortion before an operation. In addition, distortion correction is not usually carried out in 3-D orientations because of its difficulty.

In the work reported here, we designed and evaluated a new magneto-optic hybrid sensor configuration. Unlike previous configurations, which employ a fixed transmitter for the magnetic system, we arrange for the transmitter to be flexibly positioned so that the accuracy is sufficient without the need for any correction by pre-calibration. The 3-D position and orientation of the transmitter are measured by an optical sensor to accurately track the transformation between the coordinate systems of the magnetic and optical sensors. The accuracy of a magnetic sensor typically depends on the ratio of the distance between the receiver and transmitter to the distance between them and a metallic object. Hence, inaccuracies arising from the presence of metallic objects can be expected to be sufficiently reduced by adaptively arranging the transmitter so that the former distance is small while the latter is large. This strategy can also be expected to reduce the effects of surgical instruments placed in the vicinity of the measurement area.

Our goal is to provide surgeons with augmented reality visualization by utilizing 3-D ultrasound images [1]. Since the probe is usually scanned from the incision opening to the internal region, the line-of-sight requirement is a significant limitation. This is the main reason for using a magneto-optic hybrid sensor. On the other hand, an optical sensor is suitable for obtaining the position and orientation of the camera that takes video images of a patient for augmented reality. Like previous hybrid systems, our hybrid configuration allows complementary use of both types of sensors, but they are combined in a way that reduces the inaccuracy of the magnetic sensor without the need for time-consuming pre-calibration. We evaluated the accuracy of the system through experiments in the laboratory and in the operating room, itself.

2 Description of the System

2.1 Magnetic and Optical Sensors

Fastrak (Polhemus Inc., Colchester, VT), which utilizes of a transmitter and up to four receivers, was used as a magnetic sensor. The 3-D positions and orientations of the receivers can be measured with an accuracy of 0.8 mm RMS error if they are not affected by metallic objects. Fastrak's field of view is centered at the transmitter.

Optotrak (Northern Digital Inc., Waterloo, Ontario, Canada), which consists of a camera array and LED markers, was used as an optical sensor. The 3-D positions of the LED markers can be measured with an accuracy of 0.1 mm RMS if the line-of-sight requirement is satisfied. Optotrak's field of view includes a $1.3 \times 1.3 \times 1.3$ m^3 volume whose center is at a distance of 2.25 m from the center of the camera array.

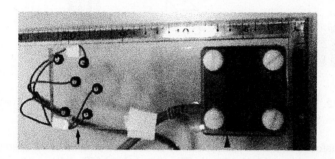

Fig. 1. Rigid body with LED markers attached to the Fastrak transmitter. Arrow: LED markers. Arrowhead: transmitter.

2.2 Magneto-Optic Integration

The Fastrak and Optotrak coordinate systems were integrated by moving Optotrak's LED markers on a rigid body attached to Fastrak transmitter (Fig. 1). Let T_{ot} be the 4×4 matrix representing the transformation from the Optotrak coordinate system to the Fastrak coordinate system, and let T_{tr} be the matrix representing the 3-D position and orientation of the receiver in the Fastrak coordinate system. By combining T_{ot} and T_{tr}, the 3-D position and orientation of the receiver in the Optotrak coordinate system, T_{or}, is estimated by

$$T_{or} = T_{ot}T_{tr}. \tag{1}$$

Figure 2 shows the transformations of the coordinate systems in our hybrid sensor. In Eq. (1), T_{tr} is directly measured and updated by Fastrak. T_{ot} is obtained by

$$T_{ot} = T_{ob}T_{bt}, \tag{2}$$

where T_{ob} is the 3-D position and orientation of the rigid body attached to the transmitter, which is directly measured and updated by Optotrak. T_{bt} is the

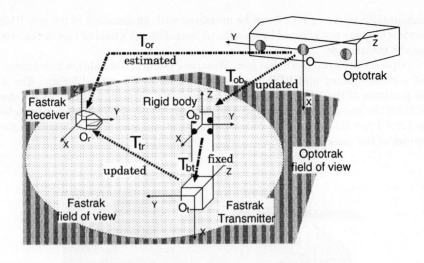

Fig. 2. Transformations between Fastrak and Optotrak coordinate systems.

fixed transformation from the rigid body to the Fastrak coordinate system, which needs to be estimated beforehand. The estimation of T_{bt} is the only calibration needed for magneto-optic integration in our system.

2.3 Calibration

Let x_m and x_o be the 3-D coordinates in the Fastrak and Optotrak coordinate systems, respectively. The relationship between them can be written as

$$x'_m = x'_o T_{ob} T_{bt}, \tag{3}$$

where x'_m and x'_o are the homogeneous coordinates of x_m and x_o, respectively. We acquire the Fastrak and Optotrak 3-D coordinates of control points uniformly distributed in the fields of view of both sensors. During the measurement of the control points, T_{ob} is simultaneously measured by Optotrak. Given a sufficient number of control points, T_{bt} is estimated using the least squares method. Data for the calibration needs to be acquired in an environment unaffected by metallic objects. Note that this calibration is performed only once, at the time when the rigid body is attached to the transmitter. Let x_h and x'_h be the 3-D coordinates measured by the hybrid sensor and its homogeneous coordinates, respectively. Using Eq. (3), x'_h is given by

$$x'_h = x'_m T_{bt}^{-1} T_{ob}^{-1}, \tag{4}$$

where x'_m is the homogeneous coordinates of the direct measurement by Fastrak. x_h provides the 3-D coordinates in the Optotrak coordinate system.

3 Experiments

3.1 Effects of Surgical Instruments

We evaluated the effects of surgical instruments on the accuracy of the magnetic sensor through laboratory experiments using steel and aluminium rods with 100 mm in length and 9 mm in diameter and surgical instrument as sources of distortion. Steel and aluminium rods were respectively selected as typical ferromagnetic and non-magnetic metals. Figures 3(a) and 3(b) show typical setups for the experiment. The receiver was fixed and its position was assumed to be the origin; the transmitter and metallic object were placed at different positions. Let D_{tr} be the position of the transmitter and D_{mr} be that of the metallic object. The effect of the metallic object was evaluated by two settings. First, the transmitter was fixed at the position $D_{tr} = -30$ cm and the position of the metallic object was varied as $D_{mr} = -20, -15, -10, -5, -3, 3, 5, 10, 15, 20$ cm. Second, the metallic object was fixed at the position $D_{mr} = 5$ cm and the position of the transmitter was varied as $D_{tr} = -25, -20, -15, -10, -5, 10, 15, 20, 25, 30$ cm. The bias was defined as $|\bar{x}_{metal} - \bar{x}_{nometal}|$, where x_{metal} and $x_{nometal}$ are the 3-D positions respectively measured with and without the metallic object in the same arrangement, and $\bar{x}$ represents the average of different measurements for the same position.

Figures 3(c) and 3(d) show the results of the experiments. Thirty measurements were made at each position; the standard deviation was less than 0.1 mm at all the positions. In the first experiment, the bias was smaller when the metallic object was further from the receiver. When $D_{tr} = -30$ cm and the metallic object was not positioned between the transmitter and the receiver, the bias was within 0.2 mm if the metal–receiver distance was more than 10 cm, hereafter, the distance from object A to object B is referred to as the A–B distance — for example, the distance from the receiver to the transmitter is termed the receiver-transmitter distance. In the second experiment, the bias was smaller when the transmitter was nearer to the receiver. When $D_{mr} = 5$ cm and the metallic object was not positioned between the transmitter and the receiver, the bias was within 0.4 mm if the transmitter–receiver distance was less than 15 cm, and the bias was around 0.6 mm if the transmitter-receiver distance was 20 cm. In both experiments, the bias was significantly smaller when the transmitter and metallic object were placed on opposite sides of the receiver, that is, when the metallic object was not positioned between the transmitter and the receiver.

3.2 Effect of the OR Table

Dual-Purpose Pen Probe for Accuracy Evaluation To evaluate the accuracy of the system, we compared the 3-D coordinates measured by the hybrid sensor with those measured by Optotrak. In order to simultaneously measure the 3-D position of the same point using the hybrid sensor and Optotrak, a dual-purpose pen probe was connected to both the Fastrak receiver and Optotrak's LEDs (Fig. 4(a)). Two aspects of the accuracy were evaluated — bias

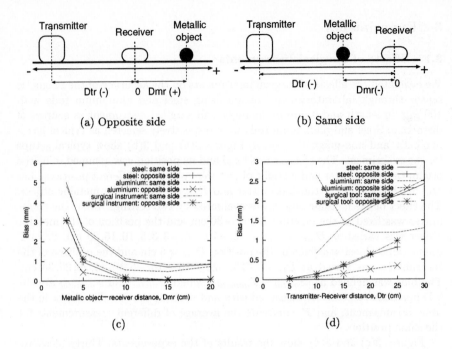

(a) Opposite side   (b) Same side

(c)   (d)

Fig. 3. Effects of surgical instruments. (a) Setup for experiments in which the transmitter is positioned on the opposite side of the receiver from the metallic object. The position of the receiver is regarded as the origin. D_{mr} is the position of a metallic object. D_{tr} is the position of the transmitter. (b) Setup for experiments in which the transmitter is positioned on the same side of the receiver as the metallic object. (c) Effects of D_{mr} on bias, when $D_{tr} = -30$ cm. (d) Effects of D_{tr} on bias, when $D_{mr} = 5$ cm.

and precision. The bias was defined as $|\bar{x}_h - \bar{x}_o|$, where x_h and x_o are the 3-D position measured by the hybrid sensor and directly by Optotrak, respectively. The precision was defined as the standard deviation in the measurements for the same point. Figures 4(b) and 4(c) show the results of the pen-probe accuracy evaluation, which was carried out in an environment unaffected by metallic objects. The precision was maintained at about 0.1 mm in all the measurement ranges. The bias increased in proportion to the distance from the transmitter to the pen probe; when the distance was within 40 cm, the bias was less than 1.0 mm.

Laboratory Experiments The effect of a metal plate on the accuracy of the hybrid sensor was evaluated in laboratory experiments. Figure 5(a) shows a typ-

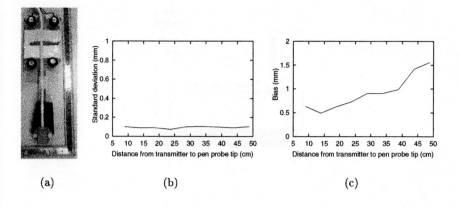

Standard deviation (mm) — Distance from transmitter to pen probe tip (cm)

Bias (mm) — Distance from transmitter to pen probe tip (cm)

(a) (b) (c)

Fig. 4. Accuracy evaluation of dual-purpose pen probe. (a) Setup of Fastrak receiver and Optotrak LEDs. (b) Precision (standard deviation) evaluation. (c) Bias evaluation.

ical experimental setup. A metal plate approximately 30×80 cm^2 was used. The transmitter and receiver were positioned above the plate at different metal plate–measurement plane distances, D_{mp}, and metal plate–transmitter distances, D_{mt}. Figure 5(b) shows the effect of D_{mt} on the bias when D_{mp} was fixed at 19 cm. The bias, $|\bar{x}_h - \bar{x}_o|$, was plotted for different values of D'_{tp}. Figure 5(c) shows the effect of the metal plate–transmitter distance D'_{tp}, when D_{mt} was fixed at 40 cm. When $D_{mt} = 40$ cm and $D_{mp} = 37$ cm, the transmitter–pen probe tip distance, D'_{tp}, needed to be not more than around 20 cm to attain a bias within 1 mm.

Operating Room Experiments The accuracy of the hybrid sensor was also evaluated in the OR environment. Figure 6(a) shows a typical experimental setup for the experiments. The transmitter was positioned at four different distances above the OR table plane. Figure 6(b) shows the effect of the transmitter–pen probe tip distance, D_{tp}, when D_{mp} was fixed at 27 cm. When the OR table–transmitter distance was 40 cm, the average bias of three points within 20 cm of D_{tp} was 2.2 mm; when the distance was 30 cm, the average bias of six points within 15 cm was 2.0 mm.

4 Discussion and Conclusions

We have proposed a new magneto-optic hybrid sensor configuration in which the 3-D position and orientation of the transmitter in a magnetic sensor system are tracked by an optical sensor, allowing the coordinate systems of the magnetic

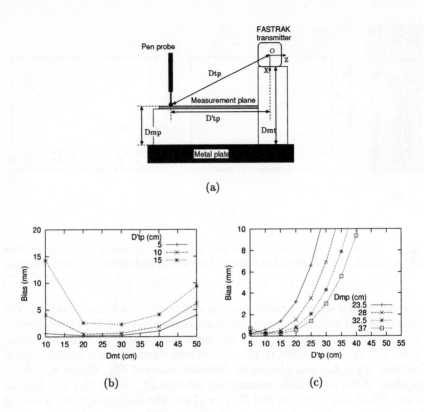

(a)

(b)   (c)

Fig. 5. Laboratory experiments for accuracy evaluation. (a) Setup for the experiments and definition of distances. D_{mt} is the distance from the metal plate to the transmitter. D_{mp} is the distance from the metal plate to the measurement plane. D_{tp} is the distance from the transmitter to the pen probe tip. D'_{tp} is the length of the D_{tp} projected onto the measurement plane. (b) Effects of D_{mt} on bias when $D_{mp} = 19$ cm. (c) Effects of D'_{tp} on bias when $D_{mt} = 40$ cm.

and optical sensors to be integrated and the transmitter to be positioned flexibly and adaptively. Experiments in the laboratory and in the operating room itself showed that inaccuracies arising from metallic objects could be reduced to acceptable levels by suitable placement of the transmitter and receivers.

Surgical instruments and the OR table are considered to be the two major sources of distortion. The experimental results on the effects of surgical instruments showed that the distortion was much smaller when the transmitter and the surgical instrument (or metallic object) were placed on opposite sides of the receiver. Thus, our hybrid system has the potential to minimize the distortion caused by surgical instruments since the placement of the transmitter can be

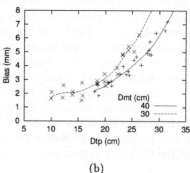

(a)   (b)

Fig. 6. Experiments in the operating room (OR). (a) Setup for experiments in the OR. Black arrow: transmitter. Black arrowhead: measurement plane. White arrow: metallic part of OR table. (b) Effects of D_{tp} on bias when $D_{mp} = 27$ cm.

changed intraoperatively. It should be noted that reducing the distortion caused by movable metallic objects such as surgical instruments is quite difficult when the pre-calibration method is used for distortion correction.

With respect to the effect of the OR table, it was found that the distance between the transmitter and the receiver needed to be relatively short to attain acceptable accuracy — for example, 20 cm to attain a bias of around 2 mm in the operating room experiment. Since acceptable accuracy is attainable within a limited area, say within a sphere with a radius of 20 cm, flexibility in transmitter placement during clinical use is essential to track a sweet spot.

A potential criticism of the system relocating the transmitter may not be easy in a crowded operating environment. Because most commercial transmitters currently available are designed to be placed at a fixed position during a series of 3-D sensing procedures, they are relatively heavy and not suitable for flexible placement in the operating environment. However, transmitters can be made lighter and more compact. For instance, the transmitter used in the InstaTrak system (Visualization Technology Inc., Woburn, MA), which is used for surgical navigation during ENT surgery, is designed to be mobile, and is thus small and much lighter. This transmitter is attached to a head adaptor similar to a pair of spectacles, and moves with the patient's head motion to track the head-centered coordinate system. Flexible placement in the operating environment would be facilitated by using such a light and compact transmitter in the hybrid system.

Another problem concerns the design of the rigid body attached to the transmitter. To attain good accuracy, this needs to be relatively large, while the distance between the transmitter and the group of LED markers mounted on the rigid body should be small. However, a large rigid body is disadvantageous in terms of flexible transmitter placement and a short distance between the group of makers and the transmitter easily causes a blind spot. In our current design, the error due to the rigid body was relatively small compared with that due to distortion arising from the proximity to metallic objects. The tradeoffs involved in rigid body design will be further examined in future work.

Our ultimate goal is to use the hybrid sensor for 3-D ultrasound imaging, and we are now evaluating the effect of the ultrasound probe on the accuracy of the system. Preliminary experimental results indicate that the effect of the probe is less serious as that of the metallic objects examined in this paper. Future work will also include a quantitative comparison of 3-D ultrasound images obtained with the Optotrak system and the hybrid sensor.

Acknowledgment This work was partly supported by the Japan Society for the Promotion of Science (JSPS Research for the Future Program, JSPS Grant-in-Aid for Scientific Research (B)(2) 12558033, and Grant-in-Aid for JSPS Fellows).

References

1. Y. Sato, M. Nakamoto, Y. Tamaki, et al. "Image guidance of breast cancer surgery using 3-D ultrasound images and augmented reality visualization". *IEEE Trans. Med. Imaging*, 17(5):681–693, 1998.
2. T. Sasama, Y. Sato, N. Sugano, K. Nakahodo, S. Yoden, T. Nishii, K. Ohzono, T. Ochi, and S. Tamura. "Accuracy evaluation in computer assisted hip surgery". In *Computer Assisted Radiology and Surgery: 13th Internaional Symposium and Exhibition (CARS'99)*, Paris, 1999.
3. A. Kato, T. Yoshimine, T. Hayakawa, Y. Tomita, T. Ikeda, M. Mitomo, K. Harada, and H. Mogami. "A frameless, armless navigational system for computer-assisted neurosurgery". *J Neurosurg*, 5(74):845–849, 1991.
4. A. State, G. Hirota, D. T. Chen, W. F. Garrett, and M. A. Livingston. "Superior augmented reality registration by integrating landmark tracking and magnetic tracking". In *SIGGRAPH '96*, pages 429–438, New Orleans, LA, 1996.
5. W. Birkfellner, F. Watzinger, F. Wanschitz, G. Enislidis, M. Truppe, R. Ewers, and H. Bergmann. "Concepts and results in the development of a hybrid tracking system for CAS". In *MICCAI '98*, number 1496, pages 343–351, Cambridge, MA, 1998.
6. W. Birkfellner, F. Watzinger, F. Wanschitz, R. Ewers, and H. Bergmann. "Calibration of tracking systems in a surgical environment". *IEEE Trans Med Imaging*, 17(5):737–742, 1998.

Computer-Assisted ENT Surgery Using Augmented Reality: Preliminary Results on the CAESAR Project

R. J. Lapeer[1], P. Chios[1], G. Alusi[2], A.D. Linney[1], M.K. Davey[2], and A.C. Tan[2]

[1] Department of Medical Physics and Bioengineering, University College London (UCL), London WC1E 6JA - UK.
[2] Institute of Laryngology and Otology (ILO), University College London (UCL), London WC1X 8EE - UK.

Abstract. The 'Computer Assisted ENT Surgery using Augmented Reality' (CAESAR) project aims to improve ENT surgical procedures through augmentation of the real operative scene during surgery: a virtual scene, which shows structures that are normally hidden to the eye of the surgeon, is superimposed onto the real scene. The main distinction of this project as opposed to previous work in the field is to create a hierarchical and stepwise implemented system which allows operations such as calibration, tracking and registration to be assessed on an individual basis. This allows us to compare different alternatives for each operation and eventually apply the best solution without interfering with the performance of other parts of the system. In this paper, we present a framework for the alignment of the objects/subject in the real and virtual operating environment before the onset of surgery, and test its performance on a phantom skull. The operations involved are thus based on a static system and include calibration of the stereo microscope and registration of the virtual patient (as reconstructed from CT data) with the real patient. The final alignment of all objects in the real and virtual operating scene is assessed by cumulating maximum errors of each individual step.

1 Introduction

Computer assisted surgery (CAS) has its roots in the field of stereotactic neurosurgery where accurate mechanical systems (stereotactic frames) were used in conjunction with CT images and powerful low cost computers [8]. The introduction of algorithms for surface and volume rendering from CT and MRI images, powerful computer graphics, the use of computer vision techniques for calibration and the commercialisation of different types of tracking devices and shape sensors have, amongst other developments, contributed to significant improvements in this field. Today, CAS has also found its way in other disciplines such as ear, nose and throat (ENT) surgery, cardiac surgery, minimal invasive surgery, maxillo-facial surgery, eye surgery and many other surgical disciplines. *Augmented reality* (AR) locates itself in between reality (the real world) and virtual reality. Imagine a line of which the two extremes comprise the real world and a totally virtual environment (see Figure 1), then augmented reality (AR) is that part of the line which lies near to the real world extremum with the predominate perception being the real world augmented by computer generated data [9].

The application of the AR concept in CAS aims to provide the surgeon with more visual information than what he can typically perceive through a 3D stereo operating microscope during surgery. Important parts of the anatomy which are not normally visible during the operation can be perceived in the augmented image. The Microscope-Assisted Guided Intervention (MAGI) system as developed at Guy's Hospital [5], which has been evaluated on six patients in a clinical experiment, augments the surgical microscope image by projecting the preoperative images back into the binocular optics of the microscope. Alternatively, the augmented image can be displayed on a separate device such as a 3D stereo display [4].

In this paper, we describe an initial setup and calibration of the participating subjects/objects in an ENT operating scene and the corresponding alignment in the virtual scene.

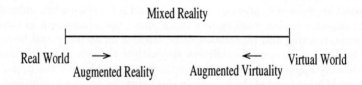

Fig. 1. The spectrum of mixed reality according to Milgram [9].

2 The alignment of the real and virtual operating scene

The following subjects/objects are included in the operating scene and their individual coordinate systems need to be referred to one another:

1. the target object, in this case the patient's head,
2. the binocular stereo operating microscope,
3. one or more surgical tools,
4. a calibration object,
5. tracking devices.

Other assets in the operating scene include the 3D augmented display, the operating table and the operating lighting equipment and of course last but not least the surgeon, but they do not require alignment!

The virtual operating scene includes the same components as mentioned above. The aim of the alignment process is to get the objects in both the real and the virtual operating scene in the same relative position as opposed to each other, before surgery begins. Since the moving objects in the scene, i.e. the subject's head, the 3D microscope and the surgical tools, are equipped with a tracking device, further changes in the individual coordinate systems as opposed to the reference or world coordinate system can be derived. For simplicity, the surgical tool(s) were excluded at this stage.

2.1 Stepwise solution to the alignment problem

Starting from an initial (random) setup to the final alignment of the components in both the real and the virtual world, requires the following operations:

Microscope calibration The 3D surgical microscope is equipped with two CCD video cameras to capture the stereoscopic images. When combined and correctly aligned, the two images form a stereoscopic image that can be displayed by an autostereoscopic 3D monitor. This calibration step was discussed in [4]. The calibration which we discuss here involves the alignment of the microscope/camera assembly, which we will call the real camera from now on, with the virtual camera[1]. The end result of this operation is that the calibration object in the real world and its exact replica in the virtual world are viewed from the same position by the real and the virtual camera, respectively.

To allow easy calibration, a microscopic pattern of a number of points[2] is required. The subject's head would obviously be ideal to use as the calibration object but not enough landmarks or points, on a typically small area as seen with the microscope, are available. Therefore, an artificial calibration object was manufactured and used. The design is briefly discussed in section 3.1. The pattern of the calibration object is used to define the world coordinate system. This implies that the calibration object has to be mounted in the operating scene and should not move during the initial alignment until all objects in both the real and virtual world are aligned.

Referring the calibration object to the subject In the first step we aligned the real and virtual camera to be in the same relative position in relation to the real and virtual calibration object respectively. However, we aim to make the camera point at the patient rather than the calibration object, thus the position of the former to the latter has to be known. We use a tracking device to achieve this goal. The receiver of a tracking device can be fixed exactly to a frame which is part of the calibration object (see Figure 2), thus its position to the origin of the world coordinate system is known. After positioning the tracking device sensor onto the calibration object, the readout of the six degrees of freedom are set to 0. A similar frame is fixed to the (real) patient's head and the sensor is placed into this frame. New readings give us the position of the patient's head relative to the world coordinate system.

---

[1] Note that this is done separately for both cameras of the stereo couple.

[2] At least five for the Tsai calibration algorithm [10] but for optimal accuracy, preferably of the order of about 100.

Registration of the real head with the virtual head The 'virtual head' is usually obtained by volume or surface rendering of a set of CT or MRI images of the (real) patient's head. In case more than one dataset is used, a preliminary registration of these datasets is required. To register the real and virtual head, we first laserscan the real head with the tracking device as positioned in the previous step. A graphical interface, allows us to roughly register the laserscanned surface with the volume-rendered CT/MRI dataset. A combined distance transform/genetic algorithm allows us to automatically optimise the registration[3]. After registration, the location of the tracking device will now also be determined on the virtual head, thus the relative position of the latter will be known as opposed to the virtual calibration object.

Tracking the cameras Although we managed to relate the coordinate systems of the patient's head, the calibration object and the cameras to one another in both the real and the virtual world, we still cannot point the camera towards the patient's head without destroying the entire alignment. Therefore we simply place a tracking device on the camera of which the readings of the six degrees of freedom are set to zero. Any change of the camera's position (e.g. to point it at a particular area of the patient's head) will allow us to correspondingly align the virtual camera and point it at the same area of the virtual head.

3 Preliminary experiments and results

At this stage, only basic lab experiments have been performed. The patient's head is simulated by a dummy which is a skull model (see Figure 3). A first series of tests included camera calibration, referring the calibration object coordinate system to the subject coordinate system and the registration of the skull model with its virtual counterpart obtained from CT images.

3.1 Microscope calibration

Calibration of the microscope or rather the calibration of each of the two CCD cameras mounted at the distal ends of the beam splitter typically involves the extraction of the extrinsic and intrinsic parameters. The *extrinsic parameters* are derived from the parameters of a Euclidean transform, i.e. the rotation and translation of a point in world coordinates into a point in camera-centered coordinates. The *intrinsic parameters* determine the projective behaviour of the camera, i.e. the principal point, which is defined as the intersection point of the optical axis and the image plane, and the camera constant or effective focal length. If radial lens distortion is considered, a number of distortion coefficients (two are usually sufficient) may be calculated as well.

---

[3] The region of the laserscanned surface, used for optimisation, should not include the tracking device for obvious reasons.

Calibration algorithms The most basic calibration algorithm is based on a *Direct Linear Transformation* [1] and has the advantage that only linear equations have to be solved. However, it typically requires at least 100 points [7], which may pose a practical problem to fit such a number into the small area of the microscope's field of view whilst preserving sufficient accuracy. A more advanced and widely used algorithm in the field of robotics and computer vision, is the Tsai calibration algorithm [10]. This algorithm includes the treatment of lens distortion (calculation of a single first-order coefficient) and gives reasonable results for a relatively small number of points (fully optimised calibration requires at least 11 data points). Furthermore, an extra intrinsic parameter, i.e. a scale factor to account for any uncertainty in the framegrabber's resampling of the horizontal scanline of the camera, is calculated.

Calibration objects The calibration object (see Figure 2) was manufactured on a CNC milling machine with an accuracy of 0.01mm. It consists of a parallelepiped base with six holes for mounting, a calibration prism and two pins for exact positioning of a Polhemus FASTRAK® tracking device. Two adjacent planes of the roof-like shaped calibration prism make an angle of 45° with the base. Each of them have a mixed pattern of calibration points of diameter 0.5mm and 1mm spacing in between points.

Fig. 2. Calibration object with attached Polhemus receiver.

Fig. 3. Skull used as the subject with attached Polhemus receiver.

Results A first series of laboratory experiments was performed using a Zeiss stereo operating microscope with two discrete magnification levels of 10 and 16, respectively. The focusing distance of the microscope's objective lens is 200mm. Figure 4a shows the image as captured by one of the two cameras. Eighty four points were used to derive the extrinsic and intrinsic camera parameters using the planar optimised Tsai calibration algorithm. Figure 4b shows the image as seen by the virtual camera. The first-order distortion coefficient, κ, was of the order 10^{-6} which means radial distortion is negligible. The average (undistorted) image plane error was 1.0±0.6 pixels and the maximum image plane error, 2.7 pixels. This corresponds with an average object space error of 0.01±0.008mm and a maximum error of 0.04mm. Considering the accuracy of the points in the calibration pattern (0.01mm), a maximum calibration error of 0.05mm can be expected.

(a) as seen by the microscope (after thresholding).

(b) as seen by the virtual camera after calibration.

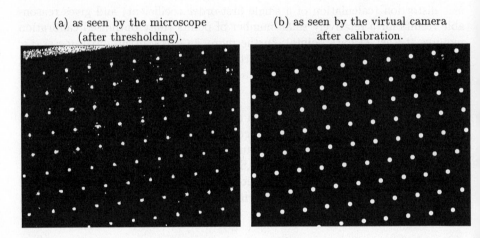

Fig. 4. Calibration pattern images: alignment of the virtual calibration object with the real calibration object - 84 points were used for calibration.

3.2   Referring the calibration object to the subject

As described in section 2.1, this step (number 2) involves the real world only. At this stage, the Polhemus FASTRAK® was used as a tracking device. The static accuracy of this magnetic tracking device is around 0.8mm. The calibration object is foreseen of two cylindrical pins upon which the receiver fits with a maximum tolerance of 0.05mm (see Figure 2). A similar device with pins is placed onto a dummy which is the skull model as shown in Figure 3 (with Polhemus receiver mounted on it). Placing the receiver, first onto the calibration object, record the readings and then replace it to the dummy, allows us to relate the coordinate systems of the dummy and the calibration object. Consideration of all the steps involved yields a maximum error of about 1mm.

3.3   Registration of the laserscanned data with CT data

The subject, in this case the skull model, was both CT scanned and laserscanned. The first operation would occur early pre-operatively, the second just before the operation starts (when the entire calibration procedure is initiated). A handheld Polhemus FastSCAN® laserscanner was used to obtain a surface of the subject.

The scan includes the tracking receiver which is fixed to the dummy. If we manage to register the laserscanned surface with the volume-rendered CT dataset, then we will know the position of the receiver on the virtual subject too, thus registering the virtual and the real subject. The algorithm to register the two datasets is based on a distance transform [3] of the voxel space which lies outside the object of interest (the virtual object). Only the points of the laserscanned surface are considered and are brought into the voxel space of the virtual object. When this is done arbitrarily, each point will fall into a region of constant distance to the virtual object surface. The ideal situation is when each point falls into the zero distance region which implies the two surfaces are registered. If we introduce the laserscanned points into the voxel space so that they are reasonably close to the object's surface, then minimising the sum of squared distances will yield an optimal fit between the surfaces. We used a genetic algorithm as the optimisation strategy. If the initial registration is sufficiently close (which can be easily achieved using a graphics interface), a global optimum will always be reached. Figure 5 shows the result before and after registration.

(a) before   (b) after

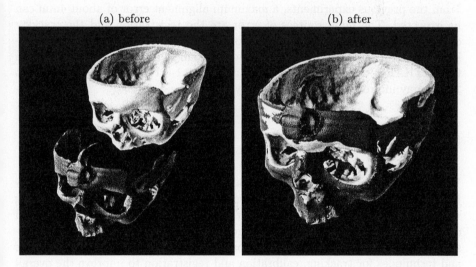

Fig. 5. Registration of the volume-rendered CT data and the laserscanned surface of the skull model.

4 Discussion

The planar and non-planar optimised Tsai calibration algorithms give stable results even for the rotational degrees of freedom. However, the solution is very sensitive to small perturbations in the input data and particularly to the accuracy by which the u,v coordinates in the image plane are determined. The accuracy improves with increasing number of calibration points. The current calibration error could be improved to sub-pixel level by better estimates of the u,v image coordinates, by using automated centre detection and using the collinearity feature to optimise centre positions.

The Polhemus FASTRAK® is a relatively inexpensive tracking device and thus suitable for initial experimentation. It proves however to be one of the main sources of error for the total accuracy of the system. Moreover, the sensitivity of the device to the introduction of any metal object in the working space may cause errors of several millimeters when used in a clinical environment [8].

The registration algorithm, based on distance transforms at voxel level, yields results with an accuracy of around 1mm. Knowing the resolution of the Polhemus FastSCAN® is also about the same order of magnitude, maximum errors of around 2mm were found.

Since the policy of the CAESAR project is to avoid any invasive solution (e.g. fiducial bone implants) for the sole purpose of tracking or registration, the placement of tracking devices on the head of the patient will be accomplished using a dental stent and/or mouthpiece [5, 6] or strapping a plastic headband around the patient's forehead or use a purpose-built head-holder [2].

From the previous experiments, a maximum alignment error of about 4mm can be expected. The main sources of error are the registration and the tracking device, which require further optimisation. For example, replacement of the Polhemus FASTRAK® with an active optical system such as the Northern Digital OPTOTRAK® with typical static accuracy of 0.1-0.15mm for distances up to 2.25m may improve the overall accuracy.

5 Conclusion

We presented a general framework to align the participating objects and subject in an ENT surgical operating environment with those in the corresponding virtual environment, the latter having the sole purpose of augmenting the real operating scene as perceived by the surgeon through the stereo operating microscope. The framework provides a solid base to experiment with different devices and techniques for tracking, calibration and registration to improve the overall accuracy of the system as each step in the framework can be optimised separately without influencing other steps. We outlined a series of initial experiments on microscope calibration and registration of the subject with its virtual counterpart. Maximum alignment errors of up to 4mm can be expected. Further experimentation should allow us to compare overlay errors with the estimated maximum error as a summation of individual steps. Although, from a statistical point of view, overlay errors would nearly always be less than the maximum possible error, the latter remains the worst case error and can only decrease by improving the methodology of the system (i.e. by minimising the number of critical steps) or the technology of the individual devices and imaging techniques! Further work will also involve the assessment of the system in a dynamic, simulated environment and, after further optimisation, in a clinical environment.

Acknowledgements

We wish to thank the staff of the CT/MRI unit at the Royal Free Hospital, London, for the acquisition of the CT data. The first and second authors would like to thank the EPSRC and the Defeating Deafness organisation for funding this project.

References

1. Y.I. Abdel-Aziz and H.M. Karara. Direct linear transformation into object space coordinates in close-range photogrammetry. In *Proceedings ASP Symposium on Close Range Photogrammetry*, pages 1–18, Urbana,Illinois,USA, 1971.
2. R.J. Bale, M. Vogele, and W. Freysinger. Minimally invasive head holder to improve the performance of frameless stereotactic surgery. *Laryngoscope*, 107:373–377, 1997.
3. G. Borgefors. Distance transformations in arbitrary dimensions. *CVGIP*, 27:321–345, 1984.
4. P. Chios, A.C. Tan, G.H. Alusi, A. Wright, G.J. Woodgate, and D. Ezra. The potential use of an autostereoscopic 3D display in microsurgery. In C. Taylor and A. Colchester, editors, *Medical Image Computing and Computer-Assisted Intervention - MICCAI'99*, volume 1679 of *Lecture Notes in Computer Science*, pages 998–1009. Springer, September 1999.
5. P.J. Edwards et al. Design and evaluation of a system for microscope-assisted guided interventions (MAGI). In C. Taylor and A. Colchester, editors, *Medical Image Computing and Computer-Assisted Intervention - MICCAI'99*, volume 1679 of *Lecture Notes in Computer Science*, pages 842–851. Springer, September 1999.
6. A.R. Gunkel, M. Vogele, A. Martin, R.J. Bale, W.F. Thumfart, and W. Freysinger. Computer-aided surgery in the petrous bone. *The Laryngoscope*, 109:1793–1799, 1999.
7. R. Klette, K. Schlüns, and A. Koschan. *Computer Vision - Three-Dimensional Data from Images*. Springer, Singapore, 1998.
8. S. Lavallée, P. Cinquin, and J. Troccaz. *Computer integrated surgery and therapy: state of the art*, volume 30 of *Technology and Informatics*, chapter 10. IOS Press, 1997.
9. P. Milgram et al. Augmented reality: A class of displays on the reality-virtuality continuum. In H. Das, editor, *SPIE Proceedings: Telemanipulator and Telepresence Technologies*, volume 2351, pages 282–292, 1994.
10. R.Y. Tsai. A versatile camera calibration technique for high-accuracy 3D machine vision metrology using off-the-shelf tv cameras and lenses. *IEEE Journal of Robotics and Automation*, 3(4):323–344, 1987.

Interventions under Video-Augmented X-ray Guidance: Application to Needle Placement

M. Mitschke[1], A. Bani-Hashemi[2], and N. Navab[2]

[1] Siemens AG Medical Engineering, Henkestr. 127, 91052 Erlangen, Germany
[2] Imaging & Visualization, Siemens Corporate Research, Princeton, NJ 08540-6632
Matthias.Mitschke@med.siemens.de, {ali,navab}@scr.siemens.com

Abstract. The camera augmented mobile C-arm (CAMC) has been introduced in [12] for the purpose of online geometrical calibration. Here, we propose its use for an augmented reality visualization. Introducing a double mirror system [11] the optical axes of both imaging systems (X-ray and optical) can be aligned. With this property both images can be merged or co-registered by only a planar transformation. This allows a real-time augmentation of X-ray and CCD camera images. We show that the needle placement procedure can be performed under this augmented reality visualization instead of fluoroscopy. Only two single X-ray images from different unknown C-arm positions are needed to align the needle to a target structure labeled by the surgeon in both X-ray images. The actual alignment is done by the surgeon while she/he sees the alignment of the needle to the target structure on *video images* co-registered with the X-ray image. Preliminary experimental results show the power of CAMC for medical augmented reality imaging. It also shows that this imaging system can provide surgeons with new possibilities for image guided surgery. In particular it reduces the X-ray exposures to both patient and physician.

1 Introduction

This paper describes an intra-operative video-augmented imaging system, introduced in [11], and presents an exemplary use of this system for precise needle placement. The system provides real-time co-registered X-ray fluoroscopy and video images. The physician could observe X-ray and video image from the same vantage point, and/or an augmented reality image merging X-ray and optical images in real-time. Assuming that the patient remains immobilized, one X-ray image may be taken, and from that point on, the procedure may be continued under video guidance This allows the surgeon to orient a surgical tool under video imaging in regard to patient's anatomy. This new imaging system has many applications, for example in orthopedic neck surgery (Fig. 1).

In this paper we first describe the design, calibration and construction of the combined video and X-ray fluoroscopy imaging system. This system is based on

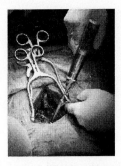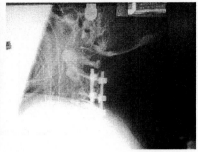

Fig. 1. Orthopedic neck surgery often requires placement of several lateral mass and pedical screws. The potential for this work is the following: Perfect registration between the X-ray and video enables the surgeon to use video to orient the surgical tools for the placement procedures. This allows the physician to use image guidance through out the procedure while taking a reduced number of X-ray images.

the camera augmented mobile C-arm, which was introduced in [12]. A CCD camera attached to the X-ray frame was used to compute the motion of the C-arm to be used for online geometrical calibration. Here, we propose to use the attached CCD camera for augmented reality visualization. Real-time blending of X-ray and video images is not possible unless both system share the same projection geometry. Introducing a double mirror system the optical axes of both imaging systems (X-ray and optical) can be aligned, see [11]. With this alignment both images can be merged or co-registered by a simple planar transformation. This allows a real-time augmented reality visualization of mixed X-ray and video images. Figure 2 illustrates the basic design of this system.

This technique of merging live camera images with X-ray fluoroscopic images bears large potential for a number of medical applications. In this paper we take the particular application of needle placement for percutaneous procedures. Minimal invasive surgical techniques are rapidly growing in popularity, among them percutaneous surgical procedures. In comparison with traditional open techniques they reduce patient recovery time and discomfort. However, during these interventions, the surgeon must rely on indirect views on the patient's anatomy provided by imaging modalities such as X-ray fluoroscopy. X-ray systems provide the surgeon with images, which are 2D projection of 3D anatomies. This is in contrast to open surgery, where the physician has a direct view (although partial) of the patient's anatomy.

Targeting deep seated anatomic structures using only X-ray projections is challenging and requires experienced physicians. Many researchers propose to register patient's pre-operative CT data with intra-operative X-ray fluoroscopic images to help the physician. Others propose to register the pre-operative CT reconstruction with the patient and an external tracking system for navigation and guidance, e.g. see [8, 1, 6]. Although a wide range of possible imaging modalities, like CT, US, MRI, and even combinations of them, is available for percutaneous procedures, X-ray fluoroscopy remains the most frequently used modality. This

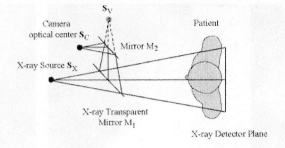

Fig. 2. Conceptual Design: A double mirror system is used to align the optical axes of X-ray and optical imaging system

may be caused by the simple fact that mobile X-ray systems and C-arm fluoroscopes are low cost, readily available in most hospitals, and they lend themselves to practical use in the operating room.

The following section introduces the design and mathematical framework of the real-time X-ray and optical image overlay (CAMC). It also describes the calibration procedure which needs to be performed only during the original construction of the CAMC system. Section 3 presents preliminary experimental results. Our new needle placement procedure is then described in section 4. The work is summarized and future work is discussed in section 5.

2 Mathematical framework for real-time X-ray and optical image overlay

2.1 General concept

Merging X-ray and optical images requires that both images are taken from the same viewpoint, which is physically impossible. Nevertheless, with the help of two mirrors the camera and the X-ray source can be positioned relative to each other such that both optical axes become aligned [11]. This is visualized in Fig.3.

Due to different focal lengths of both imaging systems it is not necessary to bring both optical centers to virtually the same point. The different images are then scaled before merging, with the scale factor determined during an offline calibration procedure.

During the C-arm motion the intrinsic geometry of the X-ray imaging system (relation between X-ray source and detector) will not be constant due to the weights of both X-ray source and detector and the thereby caused minor bending of the C-arm. This has an influence on the optical axis of the X-ray system, which is defined as the normal dropping from the X-ray source onto the detector plane. This effect makes it impossible to calibrate the system, because the camera position would have to be different for every C-arm position. To overcome this

 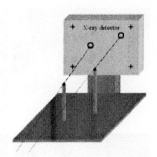

Fig. 3. X-ray source, optical camera attached to it, and the double mirror system used to align the two imaging geometry (left), and the calibration setup (right): *on-plane* markers (crosses), *off-plane* markers (spheres) and corresponding *on-plane* markers (rings) – see text for details.

limitation we use the "virtual detector" concept proposed in [12]. A small number of X-ray opaque markers are arranged on a plate that is attached to the X-ray source such that the projected markers appear on the image close to the border. A planar transformation is applied to the image to bring the projected markers to pre-defines positions. This in effect corrects for the variations in the intrinsic parameters of the X-ray system.

2.2 Setup and Calibration procedure

In order for our system to work properly as described in the previous section we need

1. Precise positioning of the camera and the two mirrors in order to align both optical axes
2. 2D-2D co-registration (in-plane rotation, translation, and scale) that corrects for the differences in intrinsic parameters of both imaging systems.

Both requirements are satisfied with a single calibration task. The two mirrors are fixed as shown in Fig.3. We use a set of marker objects that are visible to both X-ray and optical camera. Four (or more) of them are placed on the X-ray detector. They are used to estimate the planar transformation that co-registers both images in 2D[1]. Two (or more) markers are placed visible for both imaging systems with a reasonable distance (e.g. about 5cm) to the X-ray detector. Both optical axes are aligned if a single planar transformation computed from the on-plane markers also brings the projections of the off-plane markers to a match. The calibration task now is to bring the camera in a position such that this requirement is satisfied. The following algorithm is proposed:

---

[1] Here, we consider mobile C-arms equipped with a Solid State Detector with no distortion artifact. Distortion correction is necessary for traditional systems[4, 7].

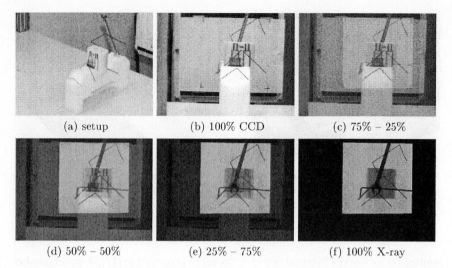

| (a) setup | (b) 100% CCD | (c) 75% – 25% |

| (d) 50% – 50% | (e) 25% – 75% | (f) 100% X-ray |

Fig. 4. Experimental results: setup (a), original video (b) and X-ray (f), and result of overlay for 3 different opacities (c-e).

1. Attach four or more markers on the X-ray detector (*on-plane markers*), and position two or more marker objects (*off-plane markers*) in the view of both imaging systems.
2. For each *off-plane* marker attach an extra marker on the detector plane such that all these pairs of markers appear superimposed in the X-ray image (see Fig.3).
3. Modify the camera's position until the pairs of *off-plane* and corresponding *on-plane* markers are also superimposed on the video image.
4. Fix the camera and estimate the planar transformation using all *on-plane* markers.

The system will then be fully calibrated, and can provide real-time overlay of X-ray and optical images. Note that this calibration procedure is done only once by the C-arm manufacturer. No extra calibration is needed before or during the operation.

3   Preliminary experimental results

Here we first want to present some results that, in general, demonstrate the power of camera augmented mobile C-arm with the double mirror system. Then we describe a simple experiments focused at a needle placement procedure in a lab setup. So far we did not conduct real experiments that would result in accuracy measures. This is due to the fact that the placement procedure depends on the person performing it. We will have to further discuss this application with surgeons in order to conduct a large number of experiments and find out how best to evaluate the accuracy of such alignment procedure.

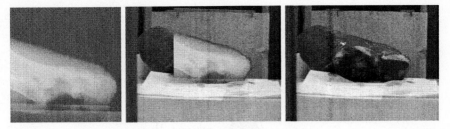

Fig. 5. Experimental results: X-ray image (left), video image (right), and overlaid optical and X-ray image of pig's knuckle (center).

3.1 Augmented reality visualization with CAMC

In this experiment we use a set of different metallic objects placed on the table as visualized in Fig.4. The objects are placed on the table such that they occlude each other. The occluded parts can then be "seen" in the combined optical and X-ray image, see Fig.4. The "visibility" can be adjusted smoothly from pure X-ray to pure optical vision.

The result of another experiment that is closer to an actual medical application is visualized in Fig.5. A pig's knuckle is positioned on the table. The X-ray image (left) shows the bone structure. In the combined optical and X-ray image (right) the bone structure can be seen on top of the optical image of the leg.

4 Needle Placement under AR visualization

4.1 Needle placement procedures – State of the art

Different needle placement techniques have been proposed in the literature. The *axial aiming technique* [5, 13] is a well established method for needle placement under X-ray fluoroscopy. The fluoroscope is positioned such that needle tip and target point become superimposed on the X-ray image during the whole insertion of the needle into the body. However, this positioning is not easy and requires much X-ray exposure. In addition, the depth of insertion cannot be observed from the axial view.

Different mechanical manipulators such as PAKY and LARS robot were introduced and used for needle injection under radiological guidance by [15, 16] and [14, 3, 2]. The design of these robots is excellent and they have proven to be precise and well adopted for their use in operating room. However, the needle placement methods they propose all require time consuming pre-operative registration procedures between the robot, imaging system and the patient's anatomy. Recently visual servoing techniques have been proposed and tested on some of these robots for automatic needle placement [9, 10]. By taking advantage of projective geometry and projective invariants the needle is aligned automatically from a small number of X-ray images taken from two different C-arm positions.

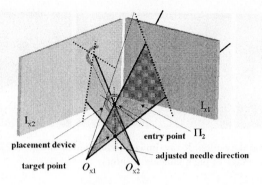

Fig. 6. 3D alignment from two perspective views: Each viewing plane includes both entry and target points as well as their projections on the respective image planes. The intersection of both planes defines the 3D orientation of the needle.

Here we show that when using our combined X-ray and video imaging system, referred to as CAMC, the necessary radiation may be decreased to a minimum of two X-ray images. Additional X-ray images might be taken to verify the accuracy of the alignments. This augmented reality imaging system can be used in conjunction with all existing methods. They can all take advantage of this system to reduce the X-ray exposure. Here we describe a method, where no motorized robotic device is used. This can be considered both as a practical method and as a proof of concept.

4.2 Placement technique

The direction of the needle will be determined from only two X-ray images from different positions of the C-arm. For each of the two views we are able to determine a plane that includes both the entry point of the needle and the target point. The orientation of the needle in 3D is then defined by the intersection of these two planes.

A plane that includes both entry and target point is the one defined by the projection of both points in the image plane and the optical center, see Fig.6. For the first view such a plane is determined. With the C-arm in the second viewing position the motion of the needle is restricted to the plane determined before. This ensures that when the needle is aligned for the second view, the final 3D alignment is accomplished.

This method requires rotations of the needle in restricted planes. A placement device such as the prototyp pictured in Fig.7 becomes necessary. Planes Π_1 and Π_3 are perpendicular to each other. Plane Π_2 can be rotated around the normal of Π_3 (axis a_3). Both planes together can be rotated around the normal of plane Π_1 (axis a_1). The needle can be rotated around the normal of plane Π_2 (axis a_2).

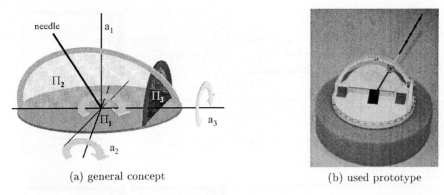

(a) general concept   (b) used prototype

Fig. 7. Placement device: General concept (a) and lab setup prototype (b).

4.3 Placement procedure

With the C-arm in the first position the goal is to orient the placement device and C-arm such that the needle, the optical(or X-ray) center, the entry point, and the target point become coplanar. This is accomplished by aligning axis a_3, which is included in plane Π_2, with the X-ray source. This can be done by moving the C-arm until a_3 is visualized as a single point on the optical image. This makes the optical center and the entry point, which lies on a_3, coplanar with Π_2. The projected plane Π_2 appears now as a line l_2, whereas a_3 is projected onto a point. The next step is then to rotate Π_2 around a_3 in order to make the target point aligned with l_2. The target point now lies on Π_2. This plane includes both entry and target point as well as the optical center. This plane is all that can be determined from the first view. The orientation of plane Π_2 relative to planes Π_1 and Π_3 is fixed and the C-arm is brought into a second position for which the optical center is not coplanar to plane Π_2. The alignment of the needle and the target in plane Π_2 is then accomplished. The needle is rotated around a_2 until entry point, target point, and the optical center become coplanar. See Fig. 6 for a visualization. Summarizing this leads to the following algorithm:

First C-arm position
1. Position needle placement device at the entry point.

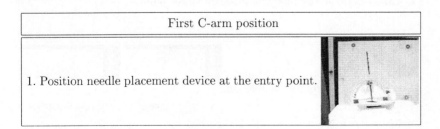

2. Align axis a_3 and camera optical center by turning placement device around axis a_1 and moving C-arm while observing the video image.

3. Take first X-ray image and select target point.

4. Merge X-ray and video image. Target point becomes visible in the video image.

5. Rotate plane Π_2 about a_3 under video control until virtual extension of needle becomes aligned with target point. Fix orientation of plane Π_2 relative to planes Π_1 and Π_3.

Second C-arm position

6. Bring C-arm into second position.

7. Take second X-ray image and select target point.

8. Merge X-ray and CCD image. Target point becomes visible in video image.

9. Rotate needle under video control around axis a_2 until virtual extension of needle becomes aligned with target point.

The needle is now completely aligned with the target in 3D. The insertion depth of the needle can be estimated using projective invariants as proposed in [9, 10].

5 Conclusion

This paper presents a simple method for semi-automatic placement of a surgical instrument such as a needle under a new augmented reality visualization system [11]. This imaging system includes an X-ray C-arm, equipped with a CCD camera that is attached to the X-ray source, and a double mirror system. The cameras and the double mirror system are installed and calibrated such the optical axes of both imaging systems are aligned. The X-ray and optical images are then co-registered using a simple planar transformation. This new imaging system results in real-time overlay of X-ray and CCD camera images. This allows the surgical operations that need intra-operative fluoroscopic imaging to be done under X-ray augmented video instead. Only a single X-ray is needed for each C-arm position. Deep-seated target structure marked in the X-ray image is by construction co-registered in the CCD camera image. The paper describes a simple interactive method to align a needle with the target under X-ray augmented video guidance. A minimum number of two X-ray images are needed for the alignment procedure. Additional X-ray images might be taken to verify the final 3D alignment from additional views. In order to evaluate the efficiency of needle placement under augmented reality visualization, one could run a series of experiments. In these experiments the physicians would first try to align the needle with the deep-seated target using X-ray fluoroscopic guidance. They would then use X-ray augmented video guidance to achieve this alignment.

Acknowledgements
We would like to thank Dr. Lee H. Riley III, M.D., at Johns Hopkins University Hospital who allowed us to observe several surgical procedures using X-ray and X-ray fluoroscopy. We also thank Bernhard Geiger for useful comments.

References

1. http://www.sofamordanek.com/fluoronav/fluoro.htm.
2. A. Bzostek, A. C. Barnes, R. Kumar, J. H. Anderson, and R. H. Taylor. A testbed system for robotically assisted percutaneous pattern therapy. In *First International Conference on Medical Image Computing and Computer-Assisted Intervention (MICCAI)*, Cambridge, UK, 1999.

3. A. Bzostek, S. Schreiner, A. C. Barnes, J. A. Cadeddu, W. W. Roberts, J. H. Anderson, R. H. Taylor, and L. Kavoussi. An automated system for precise percutaneous access of the renal collecting system. In *Proceedings of the First joint Conference on CVRMed-MRCAS*, Grenoble, France, 1997.

4. R. Fahrig, M. Moreau, and D. W. Holdsworth. Three-dimensional computer tomographic reconstruction using a C-arm mounted XRII: Correction of image intensifier distortion. *Medical Physics*, 24:1097–1106, 1997.

5. R. Greene. Transthoracic needle aspiration biopsy. In *Interventional Radiology*, Ed. C. A. Athanasoulis, R. C. Pfister and J. Saunders, pages 587–634, 1982.

6. W. E. L. Grimson, M. E. Leventon, G. Ettinger, A. Chabrerie, F. Ozlen, S. Nakajima, H. Atsumi, R. Kikinis, and P. Black. Clinical experience with a high precision image-guided neurosurgery system. In *First International Conference on Medical Image Computing and Computer-Assisted Intervention (MICCAI)*, pages 63–73, 1998.

7. E. Gronenschild. The accuracy and reproducibility of a global method to correct for geometric image distortion in the X-ray imaging chain. *Medical Physics*, 24:1875–1888, 1997.

8. S. Lavallee, P. Sautot, J. Troccaz, P. Cinquin, and P. Merloz. Computer assisted spine surgery: a technique for accurate transpedicular screw fixation using CT data and a 3D optical localizer. In *First International Symposium on Medical Robotics and Computer-Assisted Surgery (MRCAS)*, pages 315–322, Pittsburgh, PA, USA, 1994.

9. M. H. Loser, N. Navab, B. Bascle, and R. H. Taylor. Visual servoing for automatic and uncalibrated percutaneous procedures. In *Proceedings of SPIE Medical Conference*, San Diego, California, USA, February 2000.

10. N. Navab, B. Bascle, M. H. Loser, B. Geiger, and R. H. Taylor. Visual servoing for precise needle placement under x-ray fluoroscopy. In *Proceedings of the IEEE Conference on Computer Vision and Pattern Recognition*, Hilton Head Island, SC, USA, June 2000.

11. N. Navab, M. Mitschke, and A. Bani-Hashemi. Merging visible and invisible: Two camera-augmented mobile C-arm (CAMC) applications. In *IEEE International Workshop on Augmented Reality*, pages 134–141, San Francisco, CA, USA, 1999.

12. N. Navab, M. Mitschke, and O. Schütz. Camera augmented mobile C-arm (CAMC) application: 3D reconstruction using a low-cost mobile C-arm. In *Second International Conference on Medical Image Computing and Computer-Assisted Intervention (MICCAI)*, pages 688–697, Cambridge, England, 1999.

13. J. H. Newhouse and R. C. Pfister. Renal cyst puncture. In *Interventional Radiology*, Ed. C. A. Athanasoulis, R. C. Pfister and J. Saunders, pages 409–425, 1982.

14. S. Schreiner, J. H. Anderson, R. H. Taylor, J. Funda, A. Bzostek, and A. C. Barnes. A system for percutaneous delivery of treatment with a fluoroscopically guided robot. In *Proceedings of the First joint Conference on CVRMed-MRCAS*, pages 747–756, Grenoble, France, 1997.

15. D. Stoianovici, J. A. Cadeddu, R. D. Demares, H. A. Basile, R. H. Taylor, L. L. Whitcomb, W. N. Sharpe Jr., and L. Kavoussi. An efficient needle injection technique and radiological guidance method for percutaneous procedures. In *Proceedings of the First joint Conference on CVRMed-MRCAS*, Grenoble, France, 1997.

16. D. Stoianovici, L. L. Whitcomb, J. H. Anderson, R. H. Taylor, and L. R. Kavoussi. A modular surgical robotic system for image guided percutaneous procedures. In *First International Conference on Medical Image Computing and Computer-Assisted Intervention (MICCAI)*, Cambridge, MA, USA, 1998.

The Varioscope AR – A Head-Mounted Operating Microscope for Augmented Reality

Wolfgang Birkfellner[1]*, Michael Figl[1], Klaus Huber[1], Franz Watzinger[2], Felix Wanschitz[2], Rudolf Hanel[3], Arne Wagner[2], Dietmar Rafolt[1], Rolf Ewers[2], and Helmar Bergmann[1,4]

[1] Department of Biomedical Engineering and Physics
[2] Department of Oral and Maxillofacial Surgery
[3] Department of Diagnostic Radiology, Division of Osteoradiology
[4] Ludwig-Boltzmann Institute of Nuclear Medicine
All: General Hospital, University of Vienna,
Waehringer Guertel 18-20, A-1090 Vienna, Austria

Abstract. Inherent to the field of computer-aided surgery (CAS) is the necessity to handle sophisticated equipment in the operating room; an undesired side-effect of this development is the fact that the surgeon's attention is drawn from the operating field since surgical progress is partially monitored on the computer's screen. The overlay of computer-generated graphics over a real-world scene, also referred to as augmented reality (AR), provides a possibility to solve this problem. The considerable technical problems associated with this approach such as viewing of the scenery within a common focal range on the head-mounted display (HMD) or latency in display on the HMD have, however, kept AR from widespread usage in CAS. In this paper, the concept of the Varioscope AR, a lightweight head-mounted operating microscope used as a HMD, is introduced. The registration of the patient to the preoperative image data as well as preoperative planning take place on VISIT, a surgical navigation system developed at our hospital. Tracking of the HMD and stereoscopic visualisation take place on a separate POSIX.4 compliant realtime operating system running on PC hardware. While being in a very early stage of laboratory testing, we were able to overcome the technical problems described above; our work resulted in an AR visualisation system with an update rate of 6 Hz and a latency below 130 ms. First tests with 2D/3D registration have shown a match between 3D world coordinates and 2D HMD display coordinates in the range of 1.7 pixels. It integrates seamlessly into a surgical navigation system and provides a common focus for both virtual and real world objects. On the basis of our current results, we conclude that the Varioscope AR with the realtime visualisation unit is a major step towards the introduction of AR into clinical routine.

---

* e-mail: Wolfgang.Birkfellner@univie.ac.at

1 Introduction

Augmented reality (AR), the overlay of computer-generated graphics over a real world scene, has a tantalising potential for visualisation in computer-aided surgery (CAS). Several groups have tried to achieve this goal. One of the earliest approaches was the integration of the computer's monitor into the operating microscope for neuronavigation [8, 10, 11, 13, 16]. While this appears to be the most promising technology for clinical applications where an operating microscope is used (such as in neuro- or skull base surgery), the vast majority of surgical specialties does not utilise such a device, and the introduction of such an expensive system for the purpose of AR-visualisation alone appears problematic. Others have tried to use commercial head-mounted displays (HMD) [9, 18] or semi-transparent panels [7] for displaying monitor information. The major problem of this approach is the fact that a common focus of the real world scene and the computer-generated graphics cannot be achieved by the viewer's eye. A possibility to overcome this problem is the usage of miniature video cameras [12] and monitors [9]. The image from the video cameras can be merged with the computer-generated scene and displayed on the video monitors. The obvious drawback of this approach is, however, the fact that in addition to the HMD the surgeon also has to wear the miniature video cameras on the headset, and that a video generated view cannot compete with the view of a scenery as provided by an optical system alone in terms of image quality.

From our clinical experience, a number of requirements was defined for AR visualisation in CAS:

- Surgical instruments such as operating microscopes are preferrable for AR visualisation since clinical acceptance is easier to be achieved.
- Common focus for both the real-world scene and the computer-generated graphics has to be provided.
- Display latency due to lags in position measurement and rendering time requirements have to be minimised to avoid simulator sickness [2].
- The AR visualisation has to be an add-on to a normal navigation system. Sophisticated image processing such as multiplanar reformatting and volume rendering still have to be available during surgery while not overloading the scenery displayed in the HMD.
- Economic and intraoperative time expenses due to AR have to be kept to a reasonable amount.

These considerations led to the development of the Varioscope AR, a miniature head mounted operating microscope for surgical navigation; it features display of additional computer generated sceneries and communication to a surgical navigation system.

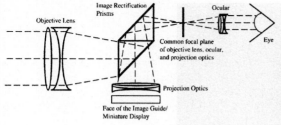

Fig. 1. The principle of image overlay in the Varioscope AR. An additional image from a miniature computer display is being projected into the focal plane of the Varioscope's objective lens. Both images can be viewed through the ocular of the Varioscope.

Working distance	300 - 600 mm
Magnification	3.6 × −7.2×
Parallax correction	Automatic
Weight	297 grams
Physical dimensions	120 mm width, 73 mm length

Table 1. Optical data of the Varioscope.

2 Materials and Methods

2.1 Visualisation Optics

The Varioscope is a head-mounted operating microscope developed and marketed by Life Optics, Vienna/Austria (*http://www.lifeoptics.com*). A list of the Varioscope's optical data can be found in Table 1. A beamsplitter together with a projection lens was inserted into the optical path by Docter Optics, Vienna/Austria. Both the image from the Varioscope's objective lens and the projection optics merge in the focal plane of the objective lens. Thus both the real world scene and the computer graphics can be viewed through the Varioscope's ocular and appear focused to the the viewer's eye (Fig. 1).

2.2 Display system

Two miniature VGA displays (AMEL HiBrite, Planar Systems, Munich/Germany, *http://www.planar.com*) with 640*480 pixel resolution and 0.75 inch display diameter were connected to a miniaturisation lens system which reduces the image from the display by a factor of 0.67. This image is being transferred to the projection optics of the Varioscope AR by means of a flexible, high resolution image guide with a resolution of 800*1000 pixels and an active area of 8*10 mm (Schott Fiber Optics, Southbridge/MA, *http://www.schottfiberoptics.com*).

2.3 HMD tracking and calibration

The HMD is being tracked by an optical tracking system (Flashpoint 5000, Image Guided Technologies, Boulder/CO, *http://www.imageguided.com*). A LED assembly was mounted to the Varioscope AR; a triaxial gyroscope (ATA ARS-09) and three accelerometers (Endevco 7290 A) are rigidly connected to the LED

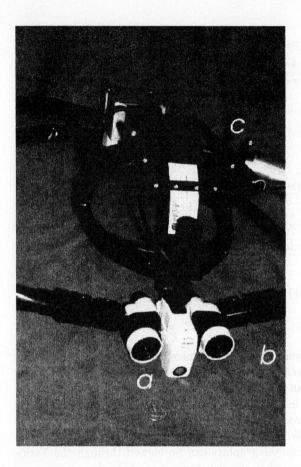

Fig. 2. The first prototype of the Varioscope AR. In addition to the normal Varioscope (a), two image guides (b) injecting a scene generated by the realtime control unit of the Varioscope are attached. Furthermore, a LED assembly (c) rigidly connected to a triaxial gyroscope and accelerometer assembly can be seen.

assembly (Fig. 2). The readings from the kinematic sensors are to be used for predictive filtering of the HMD's position through a Kalman filter [1, 14]. At the very moment, the HMD is, however, only tracked by the optical tracker.

Photogrammetric registration of the readings from the optical tracker to the actual scene to be viewed is achieved by Tsai's algorithm [17]. While this is work in progress, we have achieved first results by using a variant of Tsai's algorithm which uses a coplanar 3D world coordinate data set. These data were retrieved by aiming a crosshair displayed in th HMD at a calibration grid. From these data, six extrinsic calibration parameters (a rotation and a translation which transfers the world coordinates to the HMD's coordinate system) and two intrinsic parameters (the effective focal length of the HMD's optics and the radial distortion coeficient) are determined. Three more parameters (the center of the display relative to the optical axis and an uncertainty factor which is of no interest for this application) were either determined manually or omitted.

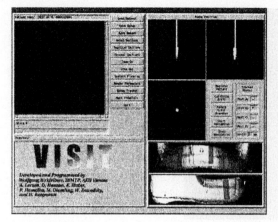

Fig. 3. A screenshot of VISIT, the non-realtime navigation system for additional visualisation on multiplanar reformatted slices and volume renderings. VISIT acquires data from the RT-control system for tool position visualisation, and delivers the data for generating the OpenGL model in the coordinate frame of the patient to the RT-control system. This OpenGL model is then visualised on the Varioscope AR (see Fig. 4).

2.4 HMD control and communication with a navigation system

The visualisation of the preoperative scene takes place on a POSIX.4 compliant real-time operating system (Lynx OS 3.0.1, *http://www.lynx.com*) running on a standard PC (Intel Pentium II Processor, 450 MHz, 128 MB RAM) with a standard Ethernet controller (3com 509B), a SCSI controller (Adaptec 2940UW), two graphics controllers (Matrox Millenium II), and an 8 channel ADC board with 200 kHz sampling rate (Pentland Systems LM1, *http://www.pentland.co.uk*). Two independent X-servers are driven by the Lynx OS. OpenGL programming was done using the Mesa 3.0 API (*http://www.mesa3d.org*) and the GLUT-toolkit [15].

The realtime system polls data from the optical tracker at an update rate of approximately 10 Hz. These data are used to render two perspective scenes on the VGA displays according to the actual position of the HMD, the patient and the surgical tool (Fig. 4). Furthermore, a navigation system is connected to the real-time system. VISIT, the system used for these experiments, was developed at our hospital (Fig. 3); it's first application is the computer aided insertion of endosteal implants in the field of cranio- and maxillofacial surgery [3–6]. The navigation system acquires data at a lower priority (approx. 1 Hz). It visualises the drill's position on obliquely reformatted slices and volume renderings. The accuracy of the system was found to be 0.9 ± 0.4 mm [4]. The preoperative planning data are sent to the realtime system after patient-to-image registration; the real-time system derives the OpenGL scene from these data (Fig. 5). Communication between the CAS-workstation and the realtime system takes place by means of POSIX.1 conformant non-canonical serial communication via the RS232 interface. The system waits for 0.1 s for a request from the CAS system; if the CAS-system does not send a request, the next position dataset is polled from the optical tracker, and the next request handler is being invoked.

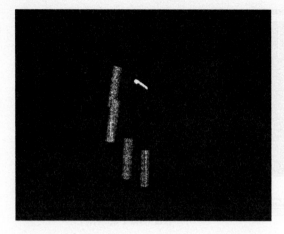

Fig. 4. The OpenGL scene as visualised on one channel of the HMD. Visible are four implants derived from the preoperative planning on the preoperative CT scan, and the drill. This scene is updated at a rate of approximately 10 Hz.

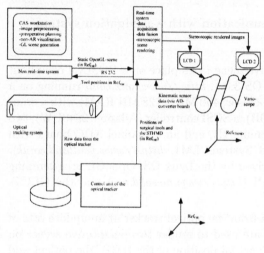

Fig. 5. The realtime control unit of the Varioscope polls data from the optical tracker at an update rate of approximately 10 Hz. The realtime system also acquires additional data from the eight channel ADC board. Two stereoscopic scenes are rendered and displayed on two miniature VGA displays connected to two independent X-servers. The navigation system (CAS workstation) acquires data at an update rate of approximately 1 Hz from the RT-system; the navigation system uses these position data to display the tool's actual position on the preoperative datasets.

2.5   First laboratory assessments

The latency in image display is a crucial problem in AR. In order to get a figure of the performance of the realtime control system of the Varioscope AR, we have analysed the single factors contributing to the overall display lag; furthermore these measurements were repeated to see whether the time lags in communication can be expected to be constant. If this is the case, predictive tracking [1] can be expected to be able to reduce the lag in display due to rapid head movements of the surgeon significantly. Measurements were taken by calling the internal timers of the Lynx OS (timer resolution: 0.01 s).

Table 2. Time requirements for data acquisition from the optical tracker and the ADC-board of the real-time workstation and the time needed for rendering the two OpenGL-scenes.

Run No.	Data Acquisition Time [s]	Rendering Time [s]
1	0.146 ± 0.02	0.04 ± 0.01
2	0.147 ± 0.02	0.04 ± 0.01
3	0.147 ± 0.02	0.04 ± 0.01
4	0.145 ± 0.02	0.04 ± 0.01
5	0.146 ± 0.02	0.04 ± 0.01
Average	0.146 ± 0.02	0.04 ± 0.01

3   Results

First of all, it turned out that common focus is easily achieved in the prototype. The HiBrite displays have turned out to be bright enough to show the OpenGL graphics in a sufficient manner without additional light sources.

The data acquisition time's repeatability was found to be in the range of the timer's resolution. Overall time requirements for accessing data from the tracker and the CAS-workstation over the serial ports was found to be 0.15 ± 0.02 s. The time needed for rendering the OpenGL scenes remained constant at 0.04 ± 0.01 s. While evaluation of the photogrammetric registration is still under progress, first experiments at a fixed focal length of 35 mm resulted in a match between measured 2D display coordinates and 3D coordinates transformed back to the HMD's display of 1.7 pixel (maximum error: 3.0 pixel).

4   Discussion

Our current results show that a head-mounted operating microscope with the capabilities of the Varioscope AR provides a very promising approach towards introduction of AR in the operating theatre. The system can handle the problems adressed in the introduction; furthermore it can easily be connected to other navigation systems since the CAS-workstation and the realtime control unit are separated. Currently, both image update rate and latency are within reasonable limits. Since the visualisation hardware can still be improved in a cost efficient manner (by usage of OpenGL accelerated graphics boards and one or more CPUs with higher computing power) we believe that the next months will bring even increased performance. Another bottleneck is the request handler for communication with the CAS-workstation; it consumes 0.1 s of computing time in the realtime control system's main event loop. This is due to the use of POSIX.1 conformant serial communication between the navigation system and the realtime-unit. A faster method using Ethernet and POSIX.4 conformant timers is currently under development.

This is a paper on work-in-progress. Therefore, the most important part of evaluation, the clinical assessment, was not yet performed. Since the basic visualisation task currently is the matching of planned implant position and actual drill position, both static and dynamic errors in photogrammetric matching of the real world view and the computer generated view are not yet crucial. The next steps include assessment of the Kalman filter's performance for improving dynamic registration, and the accuracy of the static photogrammetric registration has to be assessed. Our current work shows, however, that our approach has the potential to bring AR to a wide acceptance among surgeons in a wide variety of specialties.

5 Acknowledgment

First of all, we wish to thank M. Lehrl, W. Pesendorfer, A. Ofner, and the staff at Life Optics for their help and cooperation. The same holds true for E. Ipp and the staff at Docter Optics Vienna. We would also like to thank F. Döcke, Planar Systems, Munich, Germany. The hardware for the Varioscope AR was to a large part made and assembled by W. Piller (L.-Boltzmann Institute of Nuclear Medicine, Vienna), S. Baumgartner, A. Taubeck, and A. Gamperl (Dept. of Biomedical Engineering and Physics, Vienna). A. Larson, D. Hanson and the staff at BIR (Mayo Clinic, Rochester/MN) were of great help during the development of VISIT. AVW, the library used for VISIT's image processing capabilities, was provided courtesy of Dr. R. A. Robb. Finally, a large number of colleagues at the clinical departments at Vienna General Hospital have provided valuable input during the development phase. This work was supported by research grant P12464-MED of the Austrian Science Foundation FWF.

References

1. R. Azuma: "Predictive Tracking for Augmented Reality", PhD thesis, UNC at Chapel Hill, (1995).
2. S. Bangay, L. Preston: "An investigation into factors influencing immersion in interactive virtual reality environments", Stud Health Technol Inform 58, 43-51, (1998).
3. W. Birkfellner, P. Solar, A. Gahleitner et al.: "Computer - Aided Implant Dentistry - An Early Report", in C. Taylor, A. Colchester (eds.): "Medical Image Computing and Computer Aided Interventions - MICCAI'99", Springer LNCS 1679, 883-891, (1999).
4. W. Birkfellner, F. Wanschitz, F. Watzinger et al.: "Accuracy of a Navigation System for Computer-Aided Oral Implantology", submitted to MICCAI 2000.
5. W. Birkfellner, P. Solar, A. Gahleitner et al.: "In-vitro assessment of a registration protocol for image guided implant dentistry", Clin Oral Impl Res, in press.
6. W. Birkfellner, K. Huber, A. Larson et al.: "A modular software system for computer-aided surgery and it's first application in oral implantology", IEEE Trans Med Imaging, in press.
7. M. Blackwell, F. Morgan, A. M. DiGioia III: "Augmented reality and its future in orthopaedics", Clin Orthop 354, 111-122, (1998).
8. T. Brinker, G. Arango, J. Kaminsky et al.: "An experimental approach to image guided skull base surgery employing a microscope-based neuronavigation system" Acta Neurochir (Wien) 140(9), 883-889, (1998).

9. C. J. Calvano, M. E. Moran, L. P. Tackett, P. P. Reddy, K. E. Boyle, M. M. Pankratov: "New visualization techniques for in-utero surgery: amnioscopy with a three-dimensional head-mounted display and a computer-controlled endoscope", J Endourol 12(5), 407-410, (1998).

10. P. J. Edwards, A. P. King, D. J. Hawkes et al.: "Stereo augmented reality in the surgical microscope", Stud Health Technol Inform 62, 102-108, (1999).

11. P. J. Edwards, D. J. Hawkes, D. L. Hill et al.: "Augmentation of reality using an operating microscope for otolaryngology and neurosurgical guidance", J Image Guid Surg 1(3), 172-178, (1995).

12. H. Fuchs, M. A. Livingston, R. Raskar et al.: "Augmented Reality Visualization for Laparoscopic Surgery", in W. M. Wells, A. Colchester, S. Delp (eds.): "Medical Image Computing and Computer-Assisted Intervention - MICCAI'98", Springer LNCS 1496, 934 pp., (1998).

13. J. Haase: "Image-guided neurosurgery/ neuronavigation/the SurgiScope–reflexions on a theme", Minim Invasive Neurosurg 42(2), 53-59, (1999).

14. R. E. Kalman: "A New Approach to Linear Filtering and Prediction Problems", Transactions of the ASME - Journal of Basic Engineering, pp. 35 - 45, (1960).

15. M. J. Kilgard: "OpenGL Programming for the X Window System", Addison Wesley, (1996).

16. D. W. Roberts, J. W. Strohbehn, J. F. Hatch et al.: "A frameless stereotaxic integration of computerized tomographic imaging and the operating microscope", J Neurosurg 65(4), 545-549, (1986).

17. R. Tsai, "A verstaile camera calibration technique for high-accuracy 3D machine vision metrology using off-the-shelf TV cameras and lenses", IEEE Trans Robotics Autom, RA 3(4), 323-344, (1987).

18. A. Wagner, M. Rasse, W. Millesi et al.: "Virtual reality for orthognathic surgery: the augmented reality environment concept", J Oral Maxillofac Surg 55(5), 456-462, (1997).

An Active Hand-Held Instrument for Enhanced Microsurgical Accuracy

Wei Tech Ang, Cameron N. Riviere, and Pradeep K. Khosla

The Robotics Institute, Carnegie Mellon University, Pittsburgh, PA 15213, USA

Abstract. This paper presents the first prototype of an active hand-held instrument to sense and compensate physiological tremor and other unwanted movement during vitreoretinal microsurgery. The instrument incorporates six inertial sensors (three accelerometers and three rate gyros) to detect motion of the handle. The movement of the instrument tip in three dimensions is then obtained using appropriate kinematic calculations. The motion captured is processed to discriminate between desired and undesired components of motion. Tremor canceling will be implemented via the weighted-frequency Fourier linear combiner (WFLC) algorithm, and compensation of non-tremorous erroneous motion via an experimental neural-network technique. The instrument tip is attached to a three-degree-of-freedom parallel manipulator, actuated by three piezoelectric stacks. The actuators move the tool tip in opposition to the motion of the tremor or other erroneous motion, thereby suppressing the error. Experimental results show that the prototype is able to follow one-dimensional and three-dimensional trajectories with rms error of 2.5 μm and 11.2 μm respectively.

1 Introduction

Human limitations in positioning accuracy during micromanipulation hamper microsurgical performance. They make some procedures difficult, and some desired procedures impossible. These limitations are due primarily to small involuntary movements that are inherent in hand motion. Involuntary movement and the resulting imprecision have long been a matter of concern in microsurgery [1], and perhaps nowhere more so than in vitreoretinal microsurgery, in which manual imprecision limits both what can be done, and how well it can be done [2]. One example is the treatment of retinal vein occlusions by intraocular cannulation, a procedure that is either extremely difficult or impossible with the bare hands [3,4]. For this and other procedures, there is some degree of consensus among vitreoretinal microsurgeons that instrument-tip positioning accuracy of 10 μm is desirable [2]. This would represent an order-of-magnitude or better improvement over the capabilities of unassisted microsurgeons [5].

The most familiar type of involuntary or erroneous movement affecting microsurgery is physiological tremor [6]. Tremor is defined as any involuntary, approximately rhythmic, and roughly sinusoidal movement [7]. Physiological tremor is a type of tremor that is inherent in the movement of healthy subjects. The component of physiological tremor that is generally evident in vitreoretinal microsurgery is what Elble and Koller [7] call the "neurogenic" component: an oscillation at 8-12 Hz whose frequency is independent of the mechanical properties of the hand and arm. The resulting tool tip oscillation during vitreoretinal microsurgery is typically 50 μm peak-to-peak (p-p) or greater [8,9].

There are other significant components of erroneous motion in microsurgery. Measurements of the hand motion of surgeons have shown the significance of non-tremorous components of motion such as jerk (i.e., normal myoclonus), drift, and certain vaguely defined and poorly understood low-frequency undesired components [5]. These components are often larger than physiological tremor [5]. Therefore, in working toward positioning accuracy of 10 μm, suppressing physiological tremor is necessary but not sufficient.

Engineering approaches to the problem of increasing accuracy in microsurgery have been various. Several efforts have been based on traditional telerobotic approaches [9,10], in which filtering can be inserted between master and slave manipulators. Motion scaling can also be implemented in such systems. Taylor et al. have followed a "steady hand" approach, in which a robot and a surgeon directly manipulate the same tool [11], with the robot having high stiffness, and moving along with only those components of the manual input force that are deemed desirable. While such a system cannot scale input motion, it has advantages in terms of cost and likelihood of user acceptance. In order to further reduce cost, and to maximize ease of use, user acceptance, and compatibility with current surgical practice, the present authors are implementing accuracy enhancement within a completely hand-held tool, seeking to keep the instrument size and weight as close as possible to those of existing passive instruments. In such a device, simple lowpass filtering is inadequate; instead, the system must generate a specific estimate of undesired motion, so that the tool can deflect its own tip by an equal but opposite amount, and must do so with no time delay, so that the deflection will be in phase with the erroneous motion to be compensated. This paper presents the design, implementation, and preliminary experimental results of the first prototype of Micron, an active hand-held instrument for compensation of physiological tremor and other position errors in microsurgery. While the initial design is geared toward vitreoretinal microsurgery, the principles involved are general.

2 Design and Implementation

2.1 System Requirements

A typical vitreoretinal microsurgical instrument is 7.5 to 15 cm long and 1.0 to 1.5 cm in diameter. The intraocular shaft, roughly 3 cm long, with outer diameter of about 1 mm, is fitted with an end-effector (e.g., pick, forceps, scissor). Access to the interior of the eye is made through a sclerotomy. To be practical, an active microsurgical instrument should be of similar size to existing passive instruments. Weight must be minimized in order to avoid fatigue.

Unlike a telerobotic system, in which unwanted motion is suppressed by filtering it out, an active instrument must actually replicate the unwanted motion, generating an equal but opposite tip displacement. To achieve active error compensation, the instrument must sense its motion, distinguish desired from undesired motion in real time, and deflect its tip to nullify the undesired motion. The sensing and actuation bandwidth must be greater than 12 Hz, in order to allow canceling of physiological tremor, which is nominally 8-12 Hz [7]. In active compensation, the compensating tip deflection must be in phase with the erroneous motion, so error estimation algorithms must operate without time delay or lag. A range of tip motion of 100 μm in each of the three coordinate directions suffices for tremor canceling [9]. For canceling of certain non-tremorous types of erroneous motion, a range of motion greater than 300 μm is needed [5].

2.2 System Overview

A picture of our current system is shown in Fig. 1(a). The instrument weighs 170 g, measures 210 mm in length (including the 30 mm intraocular shaft) and has an average diameter of 22 mm. The narrowed section of the handle near the tip is contoured as an aid to securely grasping the instrument.

(a)   (b)

Fig. 1. Micron. (a) The entire active microsurgical instrument. (b) Close-up of the tip manipulator, with cover removed

An overview of the complete system is presented in Fig. 2. The current system controls the piezoelectric actuators in open loop. In the future, strain gauges will be added to sense the deflection of the actuators in order to provide closed-loop control. This feedback loop is indicated by dotted arrows in Fig. 2.

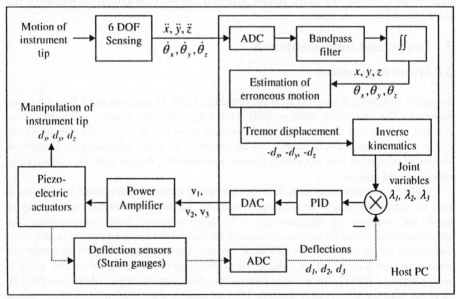

Fig. 2. Block diagram of the Micron system

The authors' previous work includes development of a motion sensing module for Micron [12], as well as algorithms for online estimation of tremor [13] and undesired non-tremorous components [14] of instrument motion. This work is summarized in Sections 2.3 and 2.4. Section 2.5 presents current work in the design, kinematics, and control of the manipulator for instrument tip deflection.

2.3 Motion Sensing

The motion-sensing module is mounted at the back end of the instrument handle, to detect translation and rotation in 6-dof [12], as seen in Figure 1(a). The sensor suite houses six inertial sensors: a CXL02LF3 tri-axial accelerometer (Crossbow Technology, Inc., San Jose, Ca.) and three CG-16D ceramic rate gyros (Tokin Corp., Tokyo). Using the data from these sensors, the three-dimensional (3-D) velocity of the instrument tip is obtained via appropriate kinematic calculations, and then integrated to obtain tip displacement. The module has been shown to estimate tremor-like oscillations (10 Hz, 30 μm p-p) with rms error of approximately 3 μm. Details of the kinematics and experimental results are presented in [12].

2.4 Error Estimation for Canceling

Tremor. Estimation of tremor will be performed by a system based on the weighted-frequency Fourier linear combiner (WFLC) algorithm [13]. The WFLC is an adaptive algorithm that estimates tremor using a dynamic sinusoidal model, estimating its time-varying frequency, amplitude, and phase online. Active canceling of physiological tremor using this algorithm has been demonstrated using a 1-dof instrument prototype. In 25 tests on hand motion recorded from eye surgeons, this technique yielded average rms amplitude reduction of 69% in the 6-16 Hz band, and average rms error reduction of 30% with respect to an off-line estimate of the tremor-free component of motion [13].

Non-tremorous Error. A neural network technique for estimation in real time of non-tremorous erroneous movement has also been developed, using the cascade learning architecture [15], which adjusts not only the values of its weights, but also the number and transfer functions of its hidden nodes. Extended Kalman filtering is used for learning [16]. This technique has been tested in simulation on recordings of vitreoretinal instrument movement, using 100 input nodes (in a tapped delay line), ten hidden nodes, and one output node. These tests resulted in an average rms error reduction of 44% [14].

2.5 Manipulator

Design. The tip of the intraocular shaft may be approximated as a point in Euclidean space. We may disregard changes in orientation of the intraocular shaft, since they will be small in any case, given the small workspace of the manipulator. This reduces the dimension of the configuration space of the manipulator to three, and simplifies the mechanical design and the online computation of inverse kinematics. A parallel manipulator design is best suited to this application because of its rigidity, compactness, and simplicity in design, as compared to a serial mechanism.

Piezoelectric actuators were chosen for their high bandwidth. The TS18-H5-202 piezoelectric stack actuator (Piezo Systems, Inc., Cambridge, Ma.) measures 5 mm x 5 mm x 18 mm, and deflects to a maximum of about 14.5 μm with an applied voltage of +100VDC. It offers good control linearity, an excellent response time of 50 μs and an actuation force of up to 840 N. A range of motion of 100 μm or greater has been achieved in each of the three coordinate directions by stacking seven piezoelectric elements to form each actuator. The response time of the piezoelectric actuator ensures the velocities in the joint space are more than adequate to map out the trajectory of the instrument tip in the workspace at the speed needed for canceling of tremor.

Fig.1(b) depicts the intraocular shaft manipulator. The 30 mm stainless steel intraocular shaft is fixed at the center of the three-legged rigid star. The three legs of the rigid star form the apexes of an equilateral triangle. The rigid star is screwed onto the "flexi-star," which has the exact same shape, by a contact pin at each of its legs. The flexi-star is a flexible thin plate made of ABS 780 thermoplastic. The flexi-star is bolted to the triangular column by three bolts close to its center, which constrains it in the three degrees of freedom that are not being driven, namely, translation in the two coordinates transverse to the long axis of the instrument, and rotation about the long axis.

The stacked piezoelectric actuators are located on the three faces of the triangular column, and sandwiched between the base star and the contact pins. When voltage is applied to the piezoelectric stacks, they expand and push against the contact pins and the base star. This deflects the three overhanging legs of the flexi-star and in turn moves the intraocular shaft on the rigid star. There is a calibration screw at each of the three legs of the base star to compensate for the manufacturing inconsistencies in the length of the piezoelectric actuators. The manipulator system fits within the main housing of the instrument handle, with an interface to the sensor suite at the back end of the handle. The specifications of the manipulator are summarized in Table 1.

Table 1. Specifications of Micron manipulator system.

	x-axis	y-axis	z-axis
Maximum tip displacement (μm)	560	560	100
Maximum tip velocity (μm/μs)	11.2	11.2	2

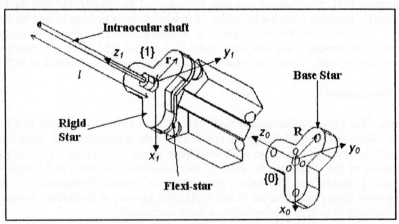

Fig. 3. Kinematic frames of the intraocular shaft manipulator of Micron.

Kinematics and Control. Since the orientation of the instrument tip in its workspace is unimportant, the dexterous workspace of the manipulator is equivalent to its reachable workspace. Moreover, this type of parallel manipulator is homeomorphic, i.e., it has a one-to-one forward and inverse mapping between its joint space and Euclidean space. It therefore has only boundary singularities and no internal singularities.

The base coordinate system $\{0\}$ is attached to the centroid of the base star, and Frame $\{1\}$ is attached to the centroid of the rigid star such that the z_1 axis aligns with the intraocular shaft, as shown in Fig. 3. Frame $\{1\}$ has the same orientation as $\{0\}$ and the origin of $\{1\}$ (x_c, y_c, z_c) has a position of $\{0, 0, z_{c0}\}$ with respect to $\{0\}$. The value R is the distance between the centroid of base star (origin of $\{0\}$) and the vertex of the equilateral triangle form by its three legs. Similarly, the value r is the distance between the centroid of the rigid star (origin of $\{1\}$) and the vertex of the equilateral triangle formed by its three legs. In this design, the ratio $\rho = r/R = 1$.

Within the small workspace for which the manipulator is intended, the constraints imposed by the design features make its kinematics essentially equivalent to those of Lee and Shah [17]. We define vectors n, o, and a to be the directional cosines of the principle axes of $\{1\}$ with respect to those of $\{0\}$.

Let the displacement of the intraocular shaft caused by tremor to be $(-d_x, -d_y, -d_z)$. Thus, the canceling displacement of the intraocular shaft (d_x, d_y, d_z) would be:

$$d_x = x_c + la_x. \tag{1}$$
$$d_y = y_c + la_y. \tag{2}$$
$$d_z = z_c - z_{c0} + l(a_z - 1). \tag{3}$$

where l is the length of the intraocular shaft. The bolts mounting the flexi-star impose the following constraints:

$$n_y = o_x. \tag{4}$$
$$X_c = \tfrac{1}{2}\rho(n_x - o_y). \tag{5}$$
$$Y_c = -n_y\rho. \tag{6}$$

where $X_c = x_c/R$; $Y_c = y_c/R$.

Let λ_1, λ_2 and λ_3 be the joint space variables or the extensions of the piezoelectric actuators. With six additional constraints imposed by the orthonormality of n, o, and a, the system of equations (1)-(6) is then solved for λ_1, λ_2 and λ_3. The actuator displacements needed for a compensation displacement of (d_x, d_y, d_z) are then:

$$\lambda_1 = \frac{R}{2}\sqrt{\left(n_x\rho + X_c - 1\right)^2 + \left(n_y\rho + Y_c\right)^2 + \left(n_z\rho + Z_c\right)^2} - z_{c0}. \tag{7}$$

$$\lambda_2 = \frac{R}{2}\sqrt{\left(-n_x\rho + \sqrt{3}o_x\rho + 2X_c + 1\right)^2 + \left(-n_y\rho + \sqrt{3}o_y\rho + 2Y_c - \sqrt{3}\right)^2 + \left(-n_z\rho + \sqrt{3}o_z\rho + 2Z_c\right)^2} - z_{c0}. \tag{8}$$

$$\lambda_3 = \frac{R}{2}\sqrt{\left(-n_x\rho - \sqrt{3}o_x\rho + 2X_c + 1\right)^2 + \left(-n_y\rho - \sqrt{3}o_y\rho + 2Y_c + \sqrt{3}\right)^2 + \left(-n_z\rho - \sqrt{3}o_z\rho + 2Z_c\right)^2} - z_{c0}. \tag{9}$$

where $Z_c = z_c/R$. Further details on the manipulator kinematics can be found in [14].

3 Manipulator Testing and Results

3.1 Testing

For testing, an infrared LED was mounted on the instrument tip, and its motion was tracked using a specialized optical tracker for micromanipulation [18]. In initial tests, Micron was driven to generate two trajectories:

1. <u>Sinusoidal 1-D motion in the long (z) axis.</u> The target trajectory was 80 μm p-p at 10.2 Hz.
2. <u>A circular motion in the transverse (x-y) plane.</u> The target trajectory had a radius of 70 μm and a frequency of 10.2 Hz. Due to the kinematics of the parallel manipulator, tracing this planar figure is actually a 3-dof task in the configuration space, and therefore offers a demonstration of the full 3-dof capabilities of the manipulator.

3.2 Results

Fig. 4 shows the results of the first test over 500 data points. The root-mean-square (rms) error for the 1-D trajectory is 2.5 μm. Fig. 5 presents the results from the 3-dof test over 0.5 s. The rms error with respect to the circular target trajectory is 11.2 μm.

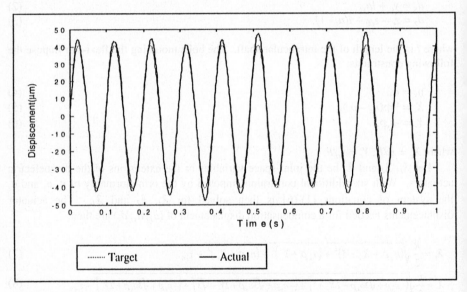

Fig. 4. Open-loop performance of Micron in axial (z-axis) trajectory generation. The dotted line depicts the predicted trajectory, the solid line the actual motion of the instrument tip.

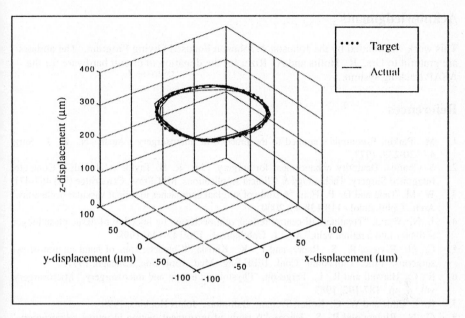

Fig. 5. Open-loop performance of Micron in 3-dof trajectory generation. Though the circle is a planar figure, tracing a circle is actually a 3-dof task in the configuration space, due to the manipulator kinematics. The dotted line depicts the predicted trajectory, the solid line the actual motion of the instrument tip.

4 Discussion

The experimental results demonstrate the ability of the manipulator to follow trajectories of the kind necessary to cancel physiological tremor. The piezoelectric actuators exhibit a certain degree of hysteresis, degrading the accuracy. Future work on Micron includes development of closed-loop control, which will minimize this source of error. Experiments will soon be performed in active error compensation, involving all components of Micron: the sensor suite, the error estimation, and the manipulator. In the long term, Micron will be redesigned for lighter weight and smaller size. This can be accomplished by using custom-designed actuators rather than off-the-shelf elements, and by continuing to exploit new developments in MEMS-based inertial sensing.

5 Conclusion

The design and preliminary testing of the first prototype of Micron, an active hand-held microsurgical instrument for accuracy enhancement, has been presented. It features six inertial sensors and three piezoelectric actuators. The manipulator is able to follow 1-D and 3-D trajectories with rms error of 2.5 µm and 11.2 µm respectively.

Acknowledgments

This work was funded by the Johnson & Johnson Focused Giving Program. The authors are grateful to Drs. R. Hollis and A. Rizzi for the donation of circuit hardware for the ASAP tracking system.

References

1. M. Patkin, Ergonomics applied to the practice of microsurgery. Austr. N. Z. J. Surg. 47:320-239, 1977.
2. S. Charles, Dexterity enhancement for surgery. In: R. H. Taylor et al. (eds.): Computer Integrated Surgery: Technology & Clinical Applications. MIT Press, Cambridge (96) 467-471.
3. W. M. Tang and D. P. Han, "A study of surgical approaches to retinal vascular occlusions." Arch. Ophthalmol. 118:138-43, 2000.
4. J. N. Weiss, "Treatment of central retinal vein occlusion by injection of tissue plasminogen activator into a retinal vein," Am. J. Ophthalmol. 126:142-144, 1998.
5. C. N. Riviere, R. S. Rader, and P. K. Khosla, "Characteristics of hand motion of eye surgeons," Proc. 19th Annu. Conf. IEEE Eng. Med. Biol. Soc., Chicago, 1997.
6. R. C. Harwell and R. L. Ferguson, "Physiologic tremor and microsurgery," Microsurgery, vol. 4, pp. 187-192, 1983.
7. R. J. Elble and W. C. Koller, Tremor. Baltimore: Johns Hopkins, 1990.
8. C. N. Riviere and P. S. Jensen, "A study of instrument motion in retinal microsurgery," submitted to 21st Annu. Conf. IEEE Eng. Med. Biol. Soc., 2000.
9. I. W. Hunter, T. D. Doukoglou, S. R. Lafontaine, P. G. Charette, L. A. Jones, M. A. Sagar, G. D. Mallinson, and P. J. Hunter, "A teleoperated microsurgical robot and associated virtual environment for eye surgery," Presence, 2:265-280, 1993.
10. P. S. Schenker, E. C. Barlow, C. D. Boswell, H. Das, S. Lee, T. R. Ohm, E. D. Paljug, G. Rodriguez, and S.T.2, "Development of a telemanipulator for dexterity enhanced microsurgery," Proc. 2nd Intl. Symp. Med. Robot. Comput. Assist. Surg., pp. 81-88, 1995.
11. R. Taylor, P. Jensen, L. Whitcomb, A. Barnes, R. Kumar, D. Stoianovici, P. Gupta, Z. Wang, E. de Juan, and L. Kavoussi, "A steady-hand robotic system for microsurgical augmentation," in: C. Taylor, A. Colchester (eds.), Medical Image Computing and Computer-Assisted Intervention--MICCAI'99. Springer, Berlin, 1999, pp. 1031-1041.
12. M. Gomez-Blanco, C. Riviere, and P. Khosla, "Sensing hand tremor in a vitreoretinal microsurgical instrument," tech. report CMU-RI-TR-99-39, Robotics Institute, Carnegie Mellon University, Pittsburgh, Pa., 1999.
13. C. N. Riviere, R. S. Rader, and N. V. Thakor, "Adaptive canceling of physiological tremor for improved precision in microsurgery," IEEE Trans. Biomed. Eng., vol. 45, pp. 839-846, 1998.
14. C. N. Riviere and P. K. Khosla, "Augmenting the human-machine interface: improving manual accuracy," Proc. IEEE Intl. Conf. Robot. Autom., Albuquerque, N.M., Apr. 20-25, 1997.
15. M. C. Nechyba and Y. Xu, "Learning and transfer of human real-time control strategies," J. Adv. Computational Intell. 1(2):137-54, 1997.
16. G. V. Puskorius and L. A. Feldkamp, "Decoupled extended Kalman filter training of feedforward layered networks," Proc. Intl. Joint Conf. Neural Networks, 1:771-7, 1991.
17. K.-M. Lee and D. K. Shah, "Kinematic analysis of a three-degrees-of-freedom in-parallel actuated manipulator," IEEE Trans. Robot. Autom., 4:354-360, 1988.
18. C. N. Riviere and P. K. Khosla, Microscale tracking of surgical instrument motion. In: C. Taylor, A. Colchester (eds.), Medical Image Computing and Computer-Assisted Intervention--MICCAI'99. Springer, Berlin, 1999, pp. 1080-1087.

A New Robotic System for Visually Controlled Percutaneous Interventions under CT Fluoroscopy

Michael H. Loser[1], Nassir Navab[2]

[1] Siemens AG, Medical Engineering, Henkestraße 127, D-91052 Erlangen, Germany
and Institute for Microsystem-Technology, University of Freiburg, Germany
michael.loser@med.siemens.de

[2] Siemens Corporate Research, 755 College Road East, Princeton, NJ 08540, USA
navab@scr.siemens.com

Abstract. Minimally invasive CT-guided interventions are an attractive option for diagnostic biopsy and localized therapy delivery. This paper describes the concept of a new prototypical robotic tool, developed in a preliminary study for radiological image-guided interventions. Its very compact and light design is optimized for usage inside a CT-gantry, where a bulky robot is inappropriate, especially together with a stout patient and long stiff instruments like biopsy needles or a trocar. Additionally, a new automatic image-guided control based on "visual servoing" is presented for automatic and uncalibrated needle placement under CT-fluoroscopy. Visual servoing is well established in the field of industrial robotics, when using CCD cameras. We adapted this approach and optimized it for CT-fluoroscopy-guided interventions. It is a simple and accurate method which requires no prior calibration or registration. Therefore, no additional sensors (infrared, laser, ultrasound, etc), no stereotactic frame and no additional calibration phantom is needed. Our technique provides accurate 3D alignment of the needle with respect to an anatomic target. A first evaluation of the robot using CT fluoroscopy showed an accuracy in needle placement of ±0.4 mm (principle accuracy) and ±1.6 mm in a small pig study. These first promising results present our method as a possible alternative to other needle placement techniques requiring cumbersome and time consuming calibration procedures.

1. Introduction

CT-guided percutaneous biopsy, drainage, and tumor ablation are widely accepted and often performed, since CT provides reproducible high-resolution images with good differentiation of soft tissue. Efficient and safe CT-guided percutaneous interventions require accurate needle placement, particularly for small or deeply situated target areas. To facilitate CT-guided interventions and to support the physician in conducting needle placement outside the CT-gantry, different guiding systems have been developed. Handheld guidance devices, consisting of a needle holder with an attached protractor and a simple level indicator, provide simple but useful support [1]. Another supporting approach for manually guided needle placement outside the CT-gantry is the use of laser guidance system [2][3][4]. Although stereotactic systems in combination with preoperative CT-imaging are commonly used in neurosurgery for localizing brain lesions, experience in other parts of the body is relatively limited [5]. Many techniques have been developed which integrate robotic guidance of end effectors with image based stereotactic procedures using a variety of registration techniques. First experiments on robot guided interventions combined with CT imaging have been performed by Kwoh *et al.* in 1985 [6]. The target applications in these early trials in using a robot for stereotactic interventions with a CT scanner were stereotactic brain surgery in combination with a head frame [7][8][9]. Glauser *et al.* also used a stereotactic head frame to register the robot and image space, but are able to perform needle placement under active CT surveillance to confirm the position of their end effector [10]. Similarly, Masamune *et al.* performed needle placement within the CT scanner and registered using a stereotactic head frame [11]. A new approach was developed by Susil *et al.* [12], who used a special localization frame fixed to a robot's end effector to guide a needle inside a CT gantry. Using the cross section of this frame in the image, they can compute the pose of the end effector holding the needle relative of the CT image. Disadvantage of this method is the need of additional space for moving this frame inside the gantry.

Our robotic prototype follows a new approach, providing a very small and compact intelligent tool together with direct CT surveillance for automatic needle alignment. We tried to imitate inside the CT-gantry what the physician is doing while manipulating the needle manually. He is visually guiding the needle - visible in the CT-image - until it is aligned with the target structure. To perform this intervention manually has basic disadvantages. Firstly, the radiation exposure for the physician especially for his hands. Because of this, he needs a special needle holder in order not to place his hands in the X-ray beam. Secondly, the manual manipulation of the needle inside the CT-gantry often forces the physician to an inconvenient posture during the intervention. With the development of our prototype, guiding the needle automatically using principles of visual servoing, we avoid these disadvantages and demonstrate a simple and intelligent tool for precise and fast needle placement under CT-fluoroscopy.

2. The robot's design

The goal was to develop an optimized needle-guiding system for automatic or remote controlled percutaneous needle alignment inside a CT gantry. The system should help to minimize radiation exposure and inconvenient posture for the physician manipulating the needle manually inside the CT gantry. Especially complicated procedures are expected to be performed with improved safety, accuracy and speed.

The resulting concept is a prototypical surgical robotic system shown in figure 1. Its very small and compact design (length 365 mm, weight 590 g) with a radiolucent distal part is optimized for precise needle manipulation inside a CT-gantry. This robot could be controlled via joystick or - as described in the next section - automatically using *visual servoing*, an image based control approach. Best performance will be achieved by using CT fluoroscopy which allows real-time imaging and is increasingly used in interventional procedures [13][14][15]. Our robotic tool consists of a very compact parallelgear that allows needle rotation around two perpendicular axes. These two rotational degrees of freedom allow us to rotate the needle in 3D around a certain rotation point

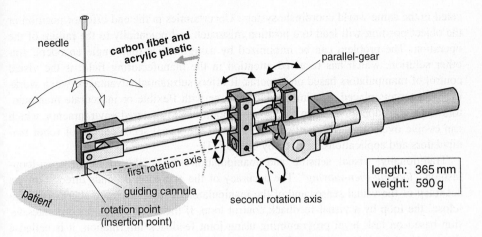

Fig. 1 The prototype has two perpendicular rotational axis that allow 3D needle-rotation around a fixed rotation point. The parallel-gear allows a very compact design for the use inside a CT-gantry. The whole system is very small and lightweight designed which provides enough space between robot and the patient, even for strongly tilted needles. All cables are integrated inside the construction.

which is defined by the geometry of the design. Two servo drives together with precise micro-encoders are responsible for accurate needle motion. The resolution of needle rotation around the two axis is ±0.04°. This parallel design allows the easy separation between a radiolucent distal part, where only carbon fiber and acrylic plastic is used (except the metallic guiding cannula), and a very rigid metallic construction containing all servo drives, encoders and limit switches. Apart from CT-guided interventions, this radiolucent distal part of the robot makes X-ray imaging (fluoroscopy) possible, where a guiding tool should not obscure patient's anatomy in the X-ray image. Additionally a special radiolucent needle transmission is integrated at the guiding cannula and allows the remote controlled needle insertion. All electric cables for motors, encoders and limit switches are integrated inside the construction. The robot can be attached to a passive arm while the electric contact to the robot is obtained by a special electric plug connection. The passive arm is a seven-degree of freedom manipulator that may be locked in the desired position. The robotic tool is a unique miniaturized radiolucent construction which provides motorized or manual needle insertion. The needle injection is powered by a small DC motor integrated in the parallel-structure of the construction. A special feature of the prototype's needle drive is that it grasps the distal end of the needle, not the needle head. This significantly reduces the unsupported length of the needle, thus minimizing the lateral flexure during injection and increasing the accuracy.

3. Visual servoing - an image based control for the robot

Most robot applications require interaction of the robot end-effector with objects in the work environment. Conventional robot control focuses on positioning the end-effector accurately in a fixed world coordinate frame. To achieve precise and reliable interaction with targets in the workspace, these objects have to be consistently and accurately located in the same world coordinate system. Uncertainties in the end-effector position or the object position will lead to a position mismatch and potentially to the failure of the operation. The problem can be minimized by using precise joint angle encoders. Another solution, which has received attention in the manufacturing field, is the visual control of manipulators based on cameras. It offers substantial advantages, when working with targets placed at unknown locations or with flexible or inaccurate manipulators. This can support accurate free-space motion in unstructured environments, which can evolve over time. An excellent overview of the visual servo control of robot manipulators and applications, is given by Corke [13].

Traditionally, visual sensing and manipulation are combined in an open-loop-fashion, *'looking-then-moving'*. The accuracy of the operation depends directly on the accuracy of the visual sensor and of the manipulator. A more accurate alternative is to 'close' the loop by a visual-feedback control loop. If the robot system incorporates vision based on task level programming using joint feedback information, it is called a *dynamic look-and-move* control. If servoing is done directly on the basis of image features without joint feedback, it is called *image based servoing* as proposed by Weiss [17]. The visual control approach that we implemented for our robotic tool using CT fluoroscopy is a *dynamic look-and-move* control and is described in section 3.2.

3.1 Visual Servoing in Medicine

A central concern of visual servoing research for industrial robotics are dynamic issues of the visual control, since the economic justification is frequently based upon cycle time. High sample rate for the vision sensor and robot control, and low latency and high-bandwidth communications are critical to a short settling time [18]. However, in medicine, the situation is radically different. In X-ray medical imaging the imaging sample rate and dose are first determined from diagnostic or clinical considerations. While in industrial robot applications a high imaging sample rate would be desirable because of dynamic issues, in X-ray imaging this could result in high radiation exposure for both patient and physician. Furthermore, guidelines for industrial robots suggest that robots should not be powered when people are in the vicinity, whereas this would be inappropriate in the OR, where robots are working very closely together with surgeons and the patient [19]. Safety aspects are the main reason why medical robots are generally moving very slowly. Thus, sophisticated dynamic visual control issues are of lower importance in medicine.

Since the use of medical robots has only begun in the last decade, there are currently not many visual servoing systems developed in the medical field. Most of these systems using visual control have been proposed in the field of minimally invasive surgery to control motion of a camera or an endoscopic instrument [20][21][22]. One recent development is a laparoscopic guiding system for an AESOP 1000 robot (Computer Motion, Goleta, CA, USA), developed by the German Aerospace Research Establishment (DLR) [23]. Using color image segmentation this system locates the tip of a laparoscope provided with a green mark. This image feature guides the laparoscopic camera so that the instrument's tip is always at a certain position in the image. Salcudean *et al.* [24] proposes a special counterbalanced robot for positioning an ultrasound probe. He uses visual servoing for robot image feature tracking in the ultrasonic image plane. A new approach for semi-automatic needle placement under X-ray fluoroscopy using visual servoing is described by Loser *et al.* [25]. To the authors knowledge, this paper presents the first experimental results on visual servoing using CT fluoroscopy.

3.2 Visual Servoing for CT fluoroscopy

In this section we describe the principle of our approach for automatic and uncalibrated needle placement with CT-fluoroscopy-guidance for two different scenarios:

(a) The target structure and the needle insertion point are located in the same CT-image plane. Therefore, the access trajectory for the needle and thus the complete insertion procedure is visible in one image plane.

(b) The target and the insertion point are located in two different CT-image planes. In case that target and insertion point could not be located in the same image plane by tilting the CT gantry, the access trajectory and thus the insertion procedure is not directly visible in the CT-image.

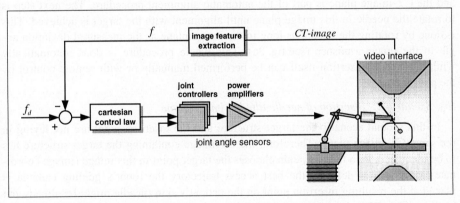

Fig. 2 Controller block diagram of visual servoing feedback loop for automatic needle alignment using CT fluoroscopy guidance.

The controller block diagram of our CT-guided visual servoing feedback loop is shown in figure 2. Using an *image based* visual control the elements of the task are specified in image feature space, not in world space. Mostly the real world task is described by one or more image space tasks [26]. For instance, the robot motion is controlled to achieve desired conditions on the image, e. g. the alignment of a needle with a target point in the scene. This means, the location of features on the image plane are directly used for feedback. In our application there are two image features to control:

1) The first image feature is the visible needle length λ in the image, which is needed to control the automatic tilting of the needle into the image plane.

2) The second image feature is the deviation angle $\Delta\phi$ between the actual and the desired needle orientation in the image. This parameter is used for the needle alignment feedback loop in the image plane.

(a) *Scenario 1: Insertion of needle in image plane*

First the robot has to be placed manually at the desired insertion point on the patients skin. The tip of the guiding cannula (rotation point) is now identical with the insertion point. In the next step the CT-table is moved automatically while imaging, until the tip of the guiding cannula is visible in the image (needle might be already put in the guide). Using image processing, this cannula is automatically detected in the image (see fig. 3a). Now the robot tilts the guiding cannula as long as it appears with maximum length in the CT-image. But the robotic system still does not know the orientation of the CT-image plane in the robot's coordinate frame. Therefore, we need a second needle position in the CT-image plane in order to define its position and orientation in robot's coordinate frame. So the robot rotates the needle around its first rotation axis (see fig. 1) with a certain amount and tilts the needle again into the image plane until its visible with maximum length. Now the CT-image plane, containing these two needle positions, is defined in the robot's coordinate frame (see fig. 3b). This registration between robot and the CT-image plane is part of the automatic alignment procedure. The next step is to rotate the needle in this image plane until alignment with the target is achieved. This is done by rotating the needle as long in the image plane as the measured deviation angle in the image vanishes (see fig. 3c). This whole procedure is done automatically while the needle insertion itself can be performed manually or with remote control by the physician.

(b) *Scenario 2: insertion of needle tilted to image plane*

In this second scenario the target structure and the insertion point are not laying in the same CT-image plane. Therefore, the CT-image containing the target structure has to be identified, then the physician chooses the target point in this image (image coordinates x_1, y_1). After defining the best access trajectory the robot's guiding cannula is placed at the resulting insertion point on the patient's skin (needle might be already put in the guide). The physician moves the CT-table until the needle tip (insertion point) becomes visible in the image (see fig. 4a). Now the image coordinates of target (x_1, y_1) and insertion-point (x_2, y_2) are defined. Since we know the CT-table translation ΔT between the target image plane and the insertion-point image plane, the complete three-dimensional settings are known in CT-image space. In the next step the robot has to be registered to the CT-image plane, as described in scenario 1 (see fig. 4b). After the image plane is known in the robot's coordinate frame the prototype can adjust the needle orientation which aligns the needle with the target. In this second scenario, needle insertion must be performed remotely by the physician. While in scenario 1 the needle insertion and deviations can be directly observed in the image, this is not possible in this second scenario. But during needle insertion the system ensures by moving the CT-table while imaging, that the needle tip is always visible as a bright dot in the CT-image. Also projected into the image is the computed desired insertion trajectory (fig. 4c), so that the physician can observe on the monitor - even in this second scenario - the deviation between desired and actual needle tip position during remote controlled insertion.

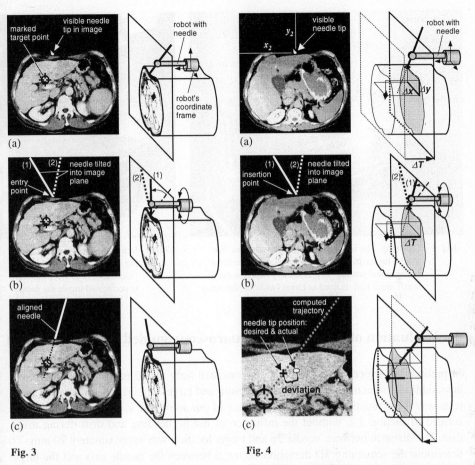

Fig. 3

The automatic needle alignment process in case that insertion point and target are located in the same CT-image plane (scenario 1).

Fig. 4

The automatic needle alignment process in case that insertion point and target are located in two different CT-image planes (scenario 2).

4. Setup of robotic prototype together with a CT scanner

We conducted our first experiments in automatic needle placement with a Somatom Plus4 CT scanner (Siemens AG, Forchheim, Germany) with CT-fluoroscopy (CARE vision) at our CT laboratory. Figure 5 shows the experimental setup. The prototype with attached needle is fixed to a passive arm. For these first experiments we used a "CT-table simulator" to perform translational motion. This simulator consists of a linear drive on which the passive arm and an acrylic plate as seat for test objects are fixed. In preliminary tests a small metal ball with a diameter of 2 mm is used as target. In a second test series we performed image guided needle placement with different pig organs (kidney, liver, lung) as phantom (see fig. 6). We used the shell of a water melon to simulate the abdominal wall (see fig. 6 and 7) and to stabilize the underlying structures.

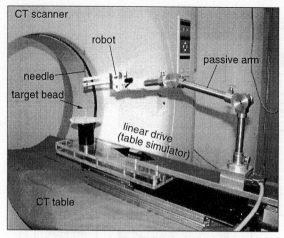

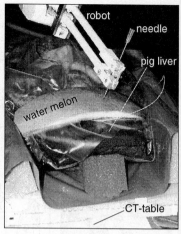

Fig. 5 CT-setup. The robot is fixed to a passive arm. We used a linear drive as "table simulator" to perform translational motion. In preliminary experiments a small metal ball is used as target (without phantom).

Fig. 6 In a second test series different pig organs are used for needle placement. Here liver puncture is performed inside the gantry.

5. Evaluation of automatic CT-fluoroscopy guided needle placement

In preliminary experiments the phantom consisted only of a 2 mm metal ball without any additional structures between rotation point and target (fig. 5). The purpose of these tests was the evaluation of principle accuracy of our automatic alignment technique described in section 3.2, without the influence of needle bending and drift during insertion. The distance between needle tip and target location was approximately 70 mm. To determine the remaining 3D deviation vector Δ between the needle axis and the target midpoint after needle alignment, we took digital images with a CCD-camera from two known viewing directions. After analyzing these two CCD-images we could precisely compute the desired deviation vector Δ for all alignment trials. The mean deviation in these preliminary experiments was about 0.4 millimeters. The authors want to emphasize that this is a kind of technical accuracy of the applied method.

In order to obtain results that are more clinically relevant, we performed several test series using different pig organs as phantom (see fig. 7). After automatic alignment, incremental manual advancement of the needle was performed, followed by intermittent CT-fluoroscopy to assess the needle position after each increment. Results are shown in table 1. We achieved a mean deviation over all series with pig organs of 1.6 mm. In these laboratory experiments we performed CT-fluoroscopy scanning continuously during the alignment process. For clinical usage together with patients we propose intermittent CT-fluoroscopy for the visual control of the needle. In conjunction with further optimization of the control loop, this would result in significantly shorter CT-fluoroscopy scan time and radiation dose. For sterilization of the robot, we propose the use of surgical drape for the whole device except the guiding cannula, which is easily detachable and sterilizable separately.

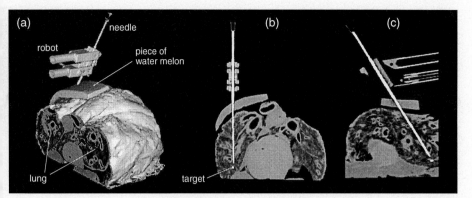

Fig. 7 Needle placement with pig lung. To determine the deviation between needle axis and target after needle insertion, we performed a high resolution spiral scan (a) and measured the deviation out of two perpendicular slices along the needle axis (b and c).

series no.	CT-fluoros- copy time	insertion depth	location	needle orienta- tion to image	mean deviation
1	39 sec	7.2 cm	kidney	tilted	1 mm
2	40 sec	7.9 cm	liver	in image plane	2 mm
3	40 sec	8.7 cm	liver	tilted	3 mm
4	38 sec	5.8 cm	lung	in image plane	1 mm
5	40 sec	7.7 cm	lung	tilted	1 mm

Tab. 1

Results of needle placement in different pig organs. We achieved a mean deviation over all series of 1.6 mm.

target: metal bead (Ø 2 mm)

6. Conclusion

In this paper we presented a new prototypical robot for image-guided interventions. With a very compact and light design, this robot is optimized for usage inside a CT gantry. Furthermore, we proposed a simple and accurate method for automatic needle placement. Using visual servoing under CT fluoroscopy guidance in two possible scenarios. The only human interaction required by the system is the choice of the needle insertion point on the patient and the manual definition of the target point on the computer display showing the CT-images. In the case of manual needle insertion this last interaction is also performed by the physician. Our method requires no additional sensors (infrared, laser, ultrasound, etc), no stereotactic frame and no prior calibration using a phantom or fiducial markers. The approach has been tested with our prototype in different experiments using pig organs as phantom. Promising first results present this method as a possible alternative to other needle placement techniques, which require cumbersome and time consuming calibration procedures.

7. References

1. A. M. Palestrant, "Comprehensive approach to CT-guided procedures with a hand-held guidance device", Radiology, vol. 174, pp. 270-272, 1990.
2. H. Ishizaka, T. Katsuya, Y. Koyama, H. Ishijima, T. Moteki, et al. "CT-guided percutaneous intervention using a simple laser director device", AJR, 170(3), pp. 745-746, 1998.

3. A. Gangi, B. Kastler, J. M. Arhan, A. Klinkert, et al. "A compact laser beam guidance system for interventional CT", J. of Comp. Assisted Tomography, 18, pp. 326-328, 1994.

4. C. Frahm, W. Kloess, H.-B. Gehl, et al., "First experiments with a new laser-guidance device for MR- and CT-guided punctures", European Radiology, vol. 5, p. 315, 1994.

5. G. Onik, P. Costello, E. Cosman, et al., "CT body stereotaxis: an aid for CT-guided biopsies", AJR, vol. 146, pp. 163-168, 1986.

6. Y. S. Kwoh, I. S. Reed, J. Y. Chen, H. M. Shao, et al., "A new computerized tomographic-aided robotic stereotaxis system", Robotics Age, vol. 7, pp.17-22, 1985.

7. R. F. Young, "Application of robotics to stereotactic neurosurgery", Neurological Research, vol. 9, pp. 123-128, 1987.

8. Y. S. Kwoh, J. Hou, E. A. Jonckeere, S. Hayati, " A Robot with improved absolute positioning accuracy for CT guided stereotactic brain surgery", IEEE Transactions on Biomedical Engineering, 35(2), pp. 153-160, 1988.

9. J. M. Darke, M. Joy, A. Goldenberg, et al. "Computer and robotic assisted resection of brain tumors", Proc. 5th Intern. Conference on Advanced Robotics, pp. 888-892., 1991.

10. D. Glauser, H. Frankenhauser, M. Epitaux, J.-L. Hefti, A. Jaccottet, "Neurosurgical robot Minerva, first results and current developments", in Proc. 2nd Symp. On MRCAS , 1995.

11. K. Masamune, et al., "A newly developed stereotactic robot with detachable driver for neurosurgery", in Proc. MICCAI 1998, pp. 215-222, 1998.

12. R. C. Susil, J. H. Anderson, R. H. Taylor, "A single image registration method for CT guided interventions", in Proc. MICCAI 1999, pp. 798-808 , 1999.

13. K. Katada, R. Kato, H. Anno, Y. Ogura, S. Koga, Y. Ida, et al., "Guidance with real-time CT fluoroscopy: early clinical experience", Radiology, 200(3), pp. 851-856, 1996.

14. J. J. Fröhlich, B. Saar, M. Hoppe, et al., "Real-time CT-fluoroscopy for guidance of percutaneous drainage procedures", Journal Vasc. Interv. Radiology, 9(5), pp. 735-740, 1998.

15. B. Daly, P. A. Templeton, " Real-time CT fluoroscopy: evolution of an interventional tool", Radiology, 211(2), pp. 309-315, 1999.

16. P. Corke, "Visual control of robot manipulators – a review", in Visual Servoing, Ed. K. Hashimoto, World Scientific, Singapore, vol. 7, pp. 1-31, 1993.

17. L. E. Weiss, A. C. Sanderson, C. P. Neumann, "Dynamic sensor-based control of robots with visual feedback", IEEE J. Robotics and Automation, vol. RA-3, pp. 404-417, 1987.

18. P. I. Corke, M. C. Good, "Dynamic effects in visual closed-loop systems", IEEE Trans. Robotics and Automation, 5(12), 1996.

19. B. Davis, "Safety of Robots in Surgery", in Proc. of 2nd Workshop on Medical Robotics, IARP , Ed. R. Dillmann et al., p. 101, 1998.

20. R. H. Taylor, J. Funda, B. Eldridge, D. Larose, et al., "A telerobotic assistant for laparoscopic surgery", IEEE Journal Engin. in Medicine and Biology, 14(3), pp. 279-288, 1995.

21. A. Casals, J. Amat, D. Prats, E. Laporte, "Vision guided robotic system for laparoscopic surgery", in Proc. Int. Conf. on Advanced Robots, pp. 33-36, 1995.

22. C. Lee, Y. F. Wang, D. R. Uecker, et al., "Image analysis for automated tracking in robot-assisted endoscopic surgery", in Proc. Int. Conf. on Pattern Recognition, pp. 88-92, 1994.

23. G.-Q. Wei, K. Arbter, G. Hirzinger, "Real-time visual servoing for laparoscopic surgery", IEEE Eng. In Medicine and Biology, 16(1), pp. 40-45, 1997.

24. S. E. Salcudean, G. Bell, S. Bachmann, et al. "Robot-assisted diagnostic ultrasound – design and feasibility experiments", in Proc. MICCAI 1999, pp. 1062-1071, 1999.

25. M. H. Loser, N. Navab, B. Bascle, R. H. Taylor, "Visual servoing for automatic and uncalibrated percutaneous procedures", in Proc. SPIE Medical Imaging, pp. 270-281, 2000.

26. D. B. Westmore, W. J. Wilson, "Direct dynamic control of a robot using an end-point mounted camera and Kalman filter position estimation", in Proc. IEEE Int. Conf. on Robotics and Automation, pp. 2376-2384, 1991.

A Miniature Instrument Tip Force Sensor for Robot/Human Cooperative Microsurgical Manipulation with Enhanced Force Feedback

Peter J. Berkelman, Louis L. Whitcomb, Russell H. Taylor, and Patrick Jensen

Engineering Research Center for Computer Integrated Surgical Systems and Technology and Microsurgery Advanced Design Lab
Johns Hopkins University
Baltimore, MD 21218*

Abstract. This paper reports the development of a new miniature force sensor to measure forces at the tip of a microsurgical instrument in three dimensions with sub-millinewton resolution. This sensor will enable enhanced force feedback during surgical intervention in which a user directly manipulates surgical instruments cooperatively with a force-reflecting robot arm. This "steady-hand" scaled force interaction enables a surgeon to sense millinewton forces between the instrument and delicate body tissues during microsurgery which would otherwise be far below the threshold of human tactile sensing. The magnified force feedback can increase the dexterity of the surgeon and improve safety by preventing large damaging forces from being exerted by the instrument. The design and analysis of the new force sensor is presented with preliminary testing and force scaling control results.

Keywords: robotics and robotic manipulators, MEMS based medical devices, clinical human computer interfaces

1 Introduction

The limits of microsurgical procedures are presently defined by the visual acuity, dexterity, and tactile sensitivity of the surgeon. We have developed a compact robot assistant manipulator arm designed to enhance human tactile sensitivity and dexterity in microsurgical operations. Our goal is to enable new procedures to be performed and to improve measurable outcomes of microsurgery.

The robotic system is based on the "steady-hand" cooperative manipulation paradigm in which the surgeon and the robot both hold a microsurgical instrument and the robot end effector responds directly to comply to the sensed manipulation forces of the surgeon's hand [1,2]. Direct hand-guided robot manipulation for surgery has been investigated by a number of authors [3–5]. Our

* We gratefully acknowledge the support of the National Science Foundation under grants IIS9801684 and EEC9731478.

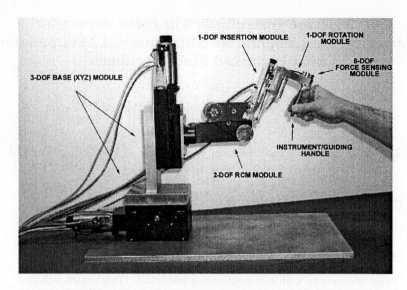

Fig. 1. Steadyhand Surgical Assistant Robot

system can significantly reduce the effects of hand tremor, hold an instrument in a fixed position as a third hand, enforce safety limits on the instrument motion, and provide amplified force reflection between the user manipulation forces and the tool tip contact forces. Our steady-hand surgical assistant robot provides a stable platform for smooth controlled motion of a microsurgical instrument with micrometer level resolution, directly controlled by the operator's manipulation forces and torques on the instrument.

The advantages of cooperative "steady-hand" manipulation over master-slave robot teleoperation systems for microsurgery such as the RAMS system [6] and Intuitive Systems' Da Vinci system [7] are direct manipulation, reduced system cost, and reduced complexity. Since the surgeon holds the actual instruments which operate on the patient to perform procedures using the steady-hand assistant, "steady-hand" assisted surgery is more similar to conventional microsurgery and would require less training by the surgeon to adapt to the system. Steady-hand manipulation requires a single robot system only, rather than the two separate robotic systems required in master-slave teleoperation with force reflection. The advantages of master-slave teleoperation systems over steady-hand manipulation for surgery are that motions can be scaled as well as forces and procedures can be performed remotely.

For precise scaled force reflection, both user manipulation and instrument tip interaction forces must be sensed independently. A commercial six-axis force sensor is used to measure the user's forces and torques on the instrument handle (ATI Industrial Automation Nano-17). At present, however, commercial force sensors are not available with the small size and high resolution necessary to measure microsurgical instrument tool tip forces. To provide the desired enhanced force feedback, we have developed a new sensor to fit inside a microsurgical instrument handle and measure the 3-D force vector at the instrument tip with sub-millinewton precision. This paper reports the design of this force sensor, its testing, and preliminary force scaling control data.

2 Steady-Hand Surgical Assistant

The steady-hand surgical assistant robot has a modular design with a 3 DOF linear translation stage module, a 2 DOF remote center of rotation module, and a final tool insertion and rotation stage [8] as shown in Figure 1. The three base stages move the end effector in the Cartesian x, y, and z directions, the remote center-of-motion arm provides rotation about x and y axes intersecting at the instrument outside of the body of the robot, and the final stage provides rotation about and translation along the axis of the surgical instrument. Separate force sensors measure user manipulation forces and torques and tool tip forces.

The actuator drives are geared for slow, precise motions with ratios of 50:1 to 200:1 in rotation and 2 mm/revolution in translation. A microsurgical instrument such as a retinal pic or needle is mounted on the robot as an end effector. The modular construction of the steadyhand surgical assistant robot allows the separate parts of the robot to easily be adapted to various different applications.

The six-axis force sensor is mounted between the robot and the end effector instrument handle to measure the forces and torques exerted by the user during manipulation. The newly developed smaller sensor between the handle and the instrument tip is used to measure the interaction forces between the instrument and its environment, such as delicate bodily tissues in microsurgery. An 8-axis DSP controller card (Motion Engineering Inc. PCX/DSP) installed in a PC provides fast low level closed-loop PID joint position control for the steadyhand robot. The position control bandwidth of the steadyhand robot is 20-25 Hz.

Since typical desired motions are less than 5 mm/sec, the robot actuators have high transmission ratios, and the environment is highly compliant, the dynamics of the manipulator can be neglected during operation. Force control is implemented as an added layer above the low-level position control by updating the desired velocity in the joint controllers as follows:

$$f_{des} = f_{handle}/C + f_{offset} \tag{1}$$

$$\dot{x}_{des} = K(f_{des} - f_{tip}) \tag{2}$$

where $\dot{x}_{des}$ is the desired end effector velocity, K is the force-to-velocity control gain, C is the force scaling factor, f_{tip} is the sensed force at the instrument tip, f_{handle} is the user manipulation force sensed on the instrument handle, and f_{offset} is the desired resting tip force, which would be set to zero during typical manipulation tasks. Given the 20 Hz closed loop joint positioning bandwidth, it is reasonable in this application to neglect servo tracking error errors and assume that $x = x_{des}$.

3 Prior Work

Microsurgical force measurement experiments were undertaken by Gupta *et al* [9], showing that typical forces on microsurgical instrument tips during retinal surgery are less than 7.5 millinewton and below the threshold of the operator's tactile sensitivity. Based on these results we conclude that microsurgeons operate using visual feedback without the influence of any tactile sensing feedback through the instruments. Measurement of the hand tremor variation in the position of handheld microsurgical instruments while being held passively and actively has been performed by Riviere *et al* [10, 11]. These studies show a limit in human tool positioning accuracy of 20-40 micrometers during microsurgery.

Preliminary results with "steady-hand" interactive force scaling are reported by Kumar *et al* [12] with motion along a single axis and using a single axis instrument tip force sensor consisting of a single strain gage pair bonded on both sides of the flattened shaft of an instrument tip in a half bridge arrangement. Force scaling experimental results during contacts with stiff wires and porcine eyes were obtained with a tool tip to user handle scaling factor of 25.

4 New Instrument Tip Force Sensor

The new 3-axis sensor measures forces at the instrument tip provided that there are no additional forces or torques exerted on the instrument shaft. The performance requirements for the new instrument tip force sensor are as follows:

Force Range:	±1.0 N
Force Resolution:	0.05 mN
Maximum Force:	5.0 N
Diameter:	12.5 mm
Instrument Tip Length:	40 mm

Due to the lever arm of the extended instrument tip, a typical four-beam cross force sensor design is much more sensitive to x and y forces at the tip than to z forces in line with the instrument axis. A compact ring-shaped sensor designed by Diddens *et al* [13] to fit inside a pen and measure writing forces, also has this drawback of anisotropic insensitivity.

To obtain a tool tip sensor with isotropic sensitivity, we have developed a double cross design with two vertically separated flexure beam crosses as suggested in [14]. The relative sensitivity of this sensor configuration in different directions varies with the vertical separation between the crosses. In this design the sensor sensitivity to axial and nonaxial forces was equalized with a 4.0 mm separation between the beam crosses. The complete force sensor is a cylinder 12.5 mm in diameter and 12.25 mm in height.

When the tip force is perpendicular to the instrument shaft, the flexure beams of the sensor are in nearly uniform tension and compression. Forces parallel to the instrument shaft cause bending strains in the beams which are maximized at the ends of each beam, however. The strain gages are bonded at the outer end

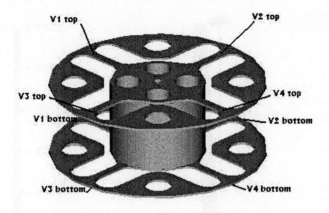

Fig. 2. Strain Gage Configuration

of each flexure beam, above the top set of beams and below the bottom set, as shown in Figure 2. The finite element static strain analysis (FEA) response of the flexure beams of the sensor (I-DEAS Master Series 7, Structural Dynamics Research Corporation) to a 1.0 N force on the instrument tip in the radial directions is shown in Figure 3, with the strain shown in the bar to the right.

Doped silicon strain gages have a strain gage factor approximately 100 times those of conventional foil gages. Silicon strain gages also have correspondingly greater thermal drift than foil gages, so that we must compensate for thermal effects during operation of the sensor. The parameters of the silicon strain gages used in the force sensor (Micron Instruments, CA) are given below:

Size:	0.56 mm
Resistance:	500 Ω
Operating Range:	±2000 $\mu\epsilon$
Maximum Strain:	±3000 $\mu\epsilon$
Gage Factor:	150

The maximum beam strains in the FEA response of Figure 3 are approximately ±500 $\mu\epsilon$, well within the operating range of the gages. The safety factors for the strain and shear yield points of the sensor beams are also in the 3-4 range. These safety factors are included to account for residual stresses arising from strain bonding, part fabrication, and sensor assembly.

The first prototype of the new sensor contains eight strain gages in four half bridges. Each gage is paired with its counterpart on the other flexure beam cross. Each pair of strain gages produces a change in voltage at the node between them proportional to the difference in strain. The instrument tip forces are calculated

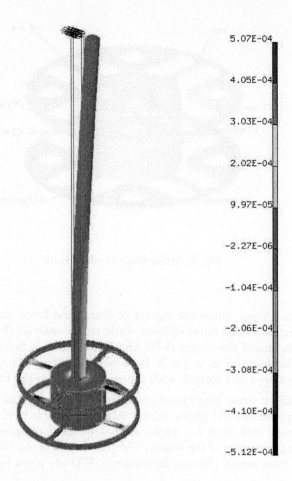

5.07E-04	
4.05E-04	
3.03E-04	
2.02E-04	
9.97E-05	
-2.27E-06	
-1.04E-04	
-2.06E-04	
-3.08E-04	
-4.10E-04	
-5.12E-04	

Fig. 3. FEA Force Sensor Strain Response to 1.0 N Radial Force

from the strain gage half bridge voltages as follows:

$$f_x \approx C(\Delta V_1 - \Delta V_3)/2 \tag{3}$$

$$f_y \approx C(\Delta V_2 - \Delta V_4)/2 \tag{4}$$

$$f_z \approx C(\Delta V_1 + \Delta V_2 + \Delta V_3 + \Delta V_4)/4 \tag{5}$$

where each ΔV refers to the change in voltage from the unloaded condition for each strain gage pair and C is a scaling factor derived from the gage factor, and sensor beam material and dimensions. To obtain accurate force measurements the sensor must be calibrated to correct for variations in strain gage mounting locations, gage resistances, and beam dimensions.

The voltage signals from the strain gage bridges are amplified by a signal amplifier (Vishay Instruments Division, 2210 A Signal Conditioning Amplifier)

Fig. 4. Fabricated Miniature Force Sensor

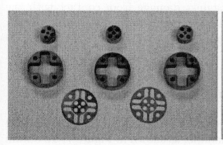 

Fig. 5. Force Sensor Components   **Fig. 6.** Flexure Plates with Strain Gages

with a gain of 100 and digitized by a 16-bit digital to analog converter (ComputerBoards Inc. PCI-DAS1602/16). The maximum resolution of the sensor at the ± 1.0 N range due to 16-bit quantization is 0.015 millinewton in the x and y directions and 0.0075 millinewton in the z direction.

The fabricated force sensor is shown in Figure 4 with a penny for scale. The internal flexure beams are visible. The component parts of the miniature force sensor, shown in Figure 5, were fabricated using wire electrical discharge machining (EDM) to achieve consistent tolerances at submillimeter dimensions and because EDM fabrication does not cause residual stresses, warping, or burrs as may be caused by conventional machining. The flexure plates and other sensor parts are assembled with standard bolts so that the instrument tips and gage plates can be easily replaced in case of damage. The flexure plates of the sensor with bonded strain gages wired to solder pads are shown in Figure 6.

Since the strain gages and flexure beams in the present sensor may be damaged by tip forces over 5 Newtons, overload limiting will be added to subsequent versions of the tool tip force sensor by placing the sensor inside the instrument handle with a preloaded spring which compresses when applied forces exceed a

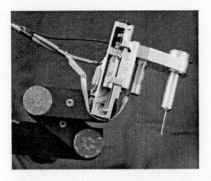

Fig. 7. Steadyhand Force Scaling Setup with New Sensor

given threshold, so that the instrument tip is compliant to excessive forces and the sensor will not be damaged. The force sensor beams and wiring will also be enclosed in a small housing to protect against damage from contact with the environment or operator. Additional strain gages can be bonded to the sensor to form full bridges instead of half bridges, improving sensitivity and greatly reducing thermal drift.

5 Calibration and Experiments

The new tip force sensor has been installed at the steadyhand robot end effector with a cylindrical tube as an instrument handle, as shown in Figure 7. To obtain sensor calibration data, 10, 20, and 50 gram weights were suspended from the tip of the instrument and the rotation stages of the robot were moved so that the sensor was loaded in the $+x$, $-x$, $+y$, $-y$, and $-z$ directions. The amplifiers that were used can internally balance the half bridge arrangements used for the strain gages, so that offsets in the strain gage signals are eliminated. The resulting collection of strain gage bridge signal data can be mapped to the forces by a linear transform which was calculated using a least squares solution:

$$
M = \begin{bmatrix}
0.0360 & 0.6568 & -0.0222 \\
-0.2711 & -0.1086 & -0.0508 \\
0.0236 & -0.2163 & -0.0206 \\
0.1878 & -0.0826 & -0.0536
\end{bmatrix}
$$

so that $SM = F$, where S is the vector of sensor gage signal voltages and F is the 3-D force vector in N. This mapping differs from the expected mapping for the sensor as described in Section 4 due to residual stresses from assembly and material and dimensional variations in the sensor components. These differences can be reduced in subsequent force sensors by assembly under better controlled conditions. The accuracy of the first fabricated sensor is within ±5% in the tested ±0.5 N range.

To test steadyhand force scaling control with the new sensor, a latex membrane was suspended horizontally with low tension in the workspace of the robot

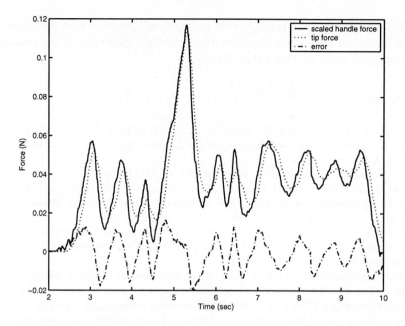

Fig. 8. Surface Contact Force Control Using Miniature Force Sensor

to simulate bodily tissue. The membrane was palpated with the tip of the force sensor instrument using steadyhand manipulation, with the instrument held by both user and robot. The handle to tip force scaling factor was 30:1 for the tip force trajectory shown in Figure 8. The forces shown in Figure 8 are the vertical components of the forces measured by the handle and instrument tip force sensors, with all gages active in the instrument tip sensor. The tip force is a smoother filtered response to the scaled handle force trajectory.

6   Conclusion

The new force sensor described will enable steadyhand enhanced feedback force scaling in all directions, not just a single axis as demonstrated in the preliminary experiment. The fidelity of millinewton force control in microsurgery can be greatly improved with the new sensor due to its high resolution. Although the sensor was designed specifically for enhanced feedback during steadyhand cooperative manipulation, use of this miniature precision force sensor will allow instrument tip forces to be measured in other microsurgical procedures.

Preliminary testing of the steadyhand surgical assistant using the tool tip force sensor is planned for application to various eye and ear microsurgical procedures, leading to clinical testing. Other application areas envisioned for the steadyhand surgical assistant system include neurosurgery and microvascular surgery. The system may also be used for applications such as MEMS assembly.

The complete system combines the accuracy of the robot with the sensitivity of the described instrument tip force sensor and the dexterity of the human for better performance of microsurgical tasks.

References

1. R. H. Taylor, J. Funda, B. Eldridge, K. Gruben, D. LaRose, S. Gomory, and M. Talamini, *A Telerobotic Assistant for Laparoscopic Surgery*, ch. 45, pp. 581–592. MIT Press, 1996.
2. R. H. Taylor, H. A. Paul, P. Kazandzides, B. D. Mittelstadt, W. Hanson, J. F. Zuhars, B. Williamson, B. L. Musits, E. Glassman, and W. L. Bargar, "An image-directed robotics system for precise orthopedic surgery," *IEEE Transactions on Robotics and Automation*, vol. 10, no. 3, pp. 261–275, 1994.
3. J. Troccaz, M. Peshkin, and B. L. Davies, "The use of localizers, robots and synergistic devices in cas," in *First Joint Conference of CVRMed and MRCAS*, (Grenoble, France), pp. 727–729, 1997.
4. S. J. Harris, W. J. Lin, K. L. Fan, R. D. Hibberd, J. Cobb, R. Middleton, and B. L. Davies, "Experiences with robotic systems for knee surgery," in *First Joint Conference of CVRMed and MRCAS*, (Grenoble, France), pp. 727–729, 1997.
5. S. C. Ho, R. D. Hibberd, and B. L. Davies, "Robot assisted knee surgery," *IEEE EMBS Special Issue on Robotics in Surgery*, pp. 292–300, April 1995.
6. S. Charles, "Dexterity enhancement for surgery," in *First International Symposium on Medical Robotics and Computer Assisted Surgery*, vol. 2, pp. 145–160, 1994.
7. G. S. Guthart and J. K. Salisbury, "The intuitive(tm) telesurgery system: Overview and application," in *IEEE International Conference on Robotics and Automation*, (San Francisco), pp. 618–621, April 2000.
8. R. H. Taylor, P. Jensen, L. Whitcomb, A. Barnes, R. Kumar, D. Stoianovici, P. Gupta, Z. Wang, E. deJuan, and L. Kavoussi, "A steady-hand robotic system for microsurgical augmentation," in *Medical Image Computing and Computer-Assisted Interventions (MICCAI)*, (Cambridge, England), pp. 1031–1041, September 1999.
9. P. K. Gupta, P. S. Jensen, and E. deJuan Jr., "Surgical forces and tactile perception during retinal microsurgery," in *Medical Image Computing and Computer-Assisted Interventions (MICCAI)*, (Cambridge, England), pp. 1218–1225, September 1999.
10. M. Gomez-Blanco, C. Riviere, and P. Khosla, "Intraoperative instrument motion sensing for microsurgery," *Proc. of the 21st Conf. of the IEEE Engineering in Medicine and Biology Society*, p. 864, October 1999.
11. C. N. Riviere, P. S. Rader, and P. K. Khosla, "Characteristics of hand motion of eye surgeons," in *Proc. of the 19th Conf. of the IEEE Engineering in Medicine and Biology Society*, (Chicago), pp. 1690–1693, October 1997.
12. R. Kumar, P. Berkelman, P. Gupta, A. Barnes, P. Jensen, L. Whitcomb, and R. Taylor, "Preliminary experiments in cooperative human/robot force control for robot assisted microsurgical manipulation," in *IEEE International Conference on Robotics and Automation*, (San Francisco), pp. 610–617, April 2000.
13. D. Diddens, D. Reynaerts, and H. van Brussel, "Design of a ring-shaped three-axis micro force/torque sensor," *Sensors and Actuators*, vol. A46, pp. 225–232, January 1995.
14. D. Reynaerts, Piers, and H. van Brussel, "A mechatronic approach to microsystem design," *IEEE/ASME Transactions on Mechatronics*, vol. 3, pp. 24–33, 1998.

Portable Virtual Endoscope System with Force and Visual Display for Insertion Training

Koji Ikuta   Koji Iritani   Junya Fukuyama

Department of Micro System Engineering, School of Engineering, Nagoya University
Furocho, Chikusa-ku, Nagoya 464-8603, Japan
TEL: +81 52-789-5024 FAX: +81 52-789-5027 E-mail: ikuta@mech.nagoya-u.ac.jp

Abstract. Drastically advanced "Portable Virtual Endoscope System" (Portable VES) with haptic ability is developed. This Portable VES can present not only computer graphical image of internal view from virtual intestine but also reaction force through the endoscope. Unlike previous VES reported in MICCAI'98, the portable VES has simpler drive mechanism with higher performance. Moreover, a large deformable dynamical model of both intestine and endoscope are formulated and implemented in the Pentium based a micro computer. Basic performance for clinical training is verified successfully by clinical surgeon.

1 Introduction

In today's medical clinics, low- invasive and non- invasive techniques of diagnosis and surgery that lighten the burden on the patient have come to be used frequently. A typical example is endoscopic diagnosis and surgery using an endoscope. However, there is no effective method of training medical personnel in these techniques, and there are reports of cases in which bodily organs have been harmed during diagnosis or surgery. To deal with this problem, in the last few years the authors have proposed and developed a virtual endoscope system (Fig. 1), with force sensation, to train young doctors and for simulation before surgery [1], [2].

In a previous report [1], we reported on functional improvements in the function of the force sensation display mechanism (VES-III) and the formulation of a real time dynamical model of an endoscope and intestine.

In the present work, we first designed and test produced a new type of force sensation display mechanism (VES-IV) that solves the problems that we had at the time of the last report, and performed a force sensation experiment. Next, in order to make possible training in endoscope insertion techniques under more realistic conditions, we developed a more sophisticated dynamical model of the endoscope and intestine. Then we displayed endoscope images and fluoroscopic images.

Finally, by linking the dynamical model and the mechanism we constructed a new system suitable for training in endoscope insertion. Here we report on an insertion experiment.

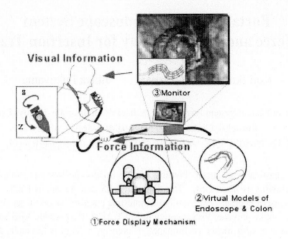

Visual Information

③Monitor

Force Information

②Virtual Models of
Endoscope & Colon

①Force Display Mechanism

Fig. 1. Concept of virtual endoscope system with force sensation

2 Design and Test Production of a Force Sensation Display Mechanism (VES-IV) Using the New Mechanism

2.1 Design and Test Production of the VES-IV

A conceptual diagram of the new VES-IV is shown in Fig. 2. Rubber balls are placed at the center of a group of 4 rollers, and motor output is transmitted only through the rubber ball to drive both linear motion ball to drive both linear motion and rotation of the endoscope (this is called "Decoupled ball drive"). The test - produced VES-IV is shown in Fig. 3.

First, we summarize the principal features of the VES-IV.
1. Small and light weight
2. Improves responsiveness
3. Increased endoscope driving force and speed
4. Improved ability to hold the endoscope
5. Easy control
Now let us explain these features in detail.
 1) As can be seen from Fig. 2 and Fig. 3, in the VES-IV the motor output is transmitted to the endoscope only through a hard rubber ball which has large surface friction; no gears are used. Therefore, compared to the VES-III, the transmission mechanism is very simple and efficient. At the same time great reductions in size and weight have been achieved.
 2) In the VES-III, the motor output transmission mechanism was complicated and had many gears, so that gear backlash adversely affected the responsiveness. In contrast, in the VES-IV, as stated above, gears are not used; the simple transmission without gears, approaching direct drive, has eliminated backlash, improving responsiveness.

3) Referring to Fig. 4(a), input T from the motor applies a downward force (1) and a moment (2) to the center of gravity of the rubber ball.Force (1) produces an increase in the vertical resistance force between the endoscope and the ball. At this time, the reaction to the frictional force between the endoscope and the ball produces the rightward force (3) on the center of gravity of the ball. This in turn increases the vertical resistance between the ball and the roller. Since the rubber ball will not slip against either the endoscope or the roller, the motor output is transmitted efficiently to the endoscope. Since the rubber ball is surrounded by 4 rollers, a similar effect occurs in all 4 directions during driving, producing an efficient driving force on the endoscope. The result is that the drive speed is 2.4 times that of the VES-III.

4) The VES-III had the problem that even within the force range expected to be encountered during the endoscope insertion operation, the endoscope slipped against the rollers, so that accurate force and position sensing could not be achieved. In the VES-IV, as shown in Fig. 4(b), the input force F produced by the endoscope insertion operation produces the rightward force (4) and the moment (5) on the center of gravity of the ball. At this time, contrary to the case discussed with regard to feature 3) above, first force(4) produces an increase in the vertical resistance between the rollers and the ball. At this time the reaction to the frictional force produces the downward force (6) on the center of gravity of the ball, increasing the vertical resistance between the endoscope and the ball, so that slippage does not occur as in 3). This effect resolves the problem that we had with VES-III producing a strong holding force without endoscope slippage even in case of a large input that exceeds the force anticipated in the endoscope insertion operation.

5) VES-IV uses 4 motors, 2 each to drive linear motion and rotation. However, since the output voltages applied to the 2 motors in each pair during driving are the same, control is as easy as it was for VES-III which had only 2 motors.

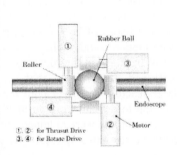

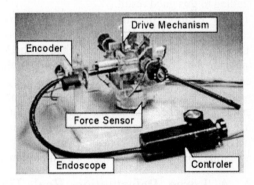

Fig. 2. Decoupled ball drive mechanism

Fig. 3. Newly developed ompact force display mechanism using ball drive mechanism (VES-IV)

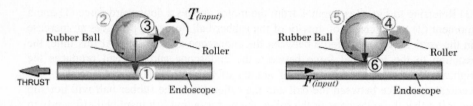

Fig. 4. Principle of a strong holding force by ball drive mechanism

2.2 The VES-IV Force Sensation Display Experiment

This system is required to provide feedback to the doctor that will provide a sensation similar to that encountered in an actual endoscope insertion operation. To determine whether it in fact does so, we did an experiment in which we used the VES-IV system to express the sensation of resistance encountered during endoscope insertion in the virtual free condition and in the virtual "spring condition" representing the resistance encountered when the endoscope contacts, for example, the intestinal wall.

In the virtual free condition, control was applied so that the reaction force on the endoscope was always 0. In the virtual free condition, the force read from a 6-axis force sensor was used to compute the theoretical position using a linear spring model; the motor output was determined from the difference between that theoretical position and the present position sensed by an encoder. The experiment results are shown in Fig. 5 and Fig. 6.

In the free condition in the forward motion direction (Fig. 5(a)), the error was held to the order of 10[gf], a very good result. In VES-III, because of the mechanical problem discussed above it was difficult to obtain adequate drive speed, so that the reaction force oscillated to some extent. In VES-IV, adequate drive speed was obtained, leading to this good result.

In the rotation direction (Fig. 5(b)), waveform oscillations were compressed compared to the VES-III but were more obvious than in the forward motion direction. The cause of this was apparently that, in the VES-IV, for reasons of layout the gear motor free of backlash that was used to drive the forward motion could not be used to drive the rotational motion, so that there was an effect of backlash in the motor gear section. In the virtual spring condition, the VES-III had the problem that the endoscope slipped against the rollers, resulting in large reaction forces in both the forward motion and rotation directions, causing the motion to deviate from an ideal straight line. In the VES-IV good results along an ideal straight line were obtained for all spring constants in both the forward motion and rotation directions (Fig. 6(a), (b)).

3. A Large Deformation Real Time Dynamical Model of an Endoscope and Intestine

3.1 Assumptions

Since this system is for the purpose of training in endoscope insertion, as the doctor inserts the endoscope, the motions of the endoscope and the intestine must be analyzed, and force sensation information fed back to the doctor in real time. For this reason, analysis using the finite element method, which provides highly accurate solutions but requires a great deal of computation time, is unsuitable. Instead, the motions of the endoscope and the intestine were modeled in real time under the following assumptions.

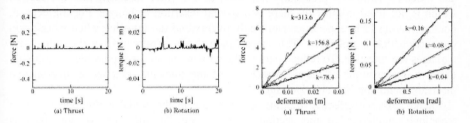

Fig. 5. Force display ability of VES-IV(free mode)

Fig. 6. Force display ability of VES-IV(elastic mode)

1. The endoscope is assumed to consist of a number of viscoelastic links, each compriding a rigid arm combined with viscoelastic components.
2. The intestine is assumed to be a viscoelastic circular cylinder of non-uniform inner diameter.
3. The reaction force is computed considering the contact force between the endoscope and the inside of the intestine.
First, we derived general equations of motion such as those that are used for manipulators. Then we neglected effects that are unimportant in considering the endoscope and intestine motions to simplify the equations.

3.2 Modeling the Endoscope

As stated, the endoscope is assumed to consist of n links, with (n-1) joints capable of viscoelastic motion. Each of the joints from the 1st joint to the (n-2)nd joint is given 2 rotational degrees of freedom; the nearest, or (n-1)st, joint is assumed to have a thrust mechanism and a rotation mechanism with the same degrees of freedom as the mechanism section. Each joint is considered to be visco-elastic. Regarding the lengths of the links, in general the segments near the distal end of the endoscope are moved frequently and for long distances, and their direction of motion determines the path of advance of the whole endoscope. For this reason, the segments were made shorter toward the distal end, to improve accuracy, and longer as the root of the endoscope was approached (Fig. 7).
The motion of the endoscope modeled as described above was assumed to be represented by the Newton - Euler equations of motion, as follows.

$$\mathbf{M}(\mathbf{q})\ddot{\mathbf{q}} + \mathbf{V}(\mathbf{q},\dot{\mathbf{q}}) = \sum_{i}^{n} \mathbf{J}_{r_i}^{T}(\mathbf{q})\mathbf{F}_i + \mathbf{T} \tag{1}$$

$$\mathbf{T} = -K[\mathbf{q}-\mathbf{q}_0] - D\dot{\mathbf{q}} \tag{2}$$

3.3 Modeling an Intestine Capable of Large Deformations

3.3.1 A Previously Reported Intestine Model
The large intestine which we are considering is assumed to consist of a fixed section and a section which is practically not fixed at all. Since insertion into the section which is not fixed is difficult, there is a special insertion technique. For this reason, doctors having little experience can easily overextend the intestine when working in the section that is not fixed. This causes pain to the patient, and in the worst case damages the intestine. To reduce pain to the patient and make it possible to smoothly perform observations of the deep recesses of the intestine, practice in this special insertion technique is very important.

To make possible effective training and practice in the actual endoscope insertion technique, modeling of the section of the large intestine that is not fixed is indispensable. Until now, models of the intestine have only considered deformation of the intestine in the radial direction, and failed to consider the large deformations of the entire intestinal tube, referred to above, that are so important in training. Therefore, we here propose a new model of the intestine that considers large deformations of the intestinal tube itself.

3.3.2 A Model That Considers Large Deformations of the Intestinal Tube Center Line
As in previous models, the center line that determines the shape of the whole intestine is represented by equation (3).

$$S(\mathbf{r}) = 0 \tag{3}$$

Designating the points into which the curve given by equation (3) is segmented by $\mathbf{r}_i$

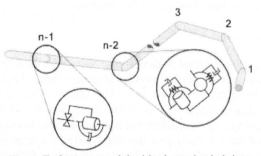

Fig. 7. Endoscope model with visco-elastic joints

(equation (4)), by using the circles of radii R_i (equation (5)) with those points as their centers, we form the 3-dimensional tube of varying diameter. These circles lie on the planes P_i to which the lines tangent to the center line of the intestinal tube at the circle centers are normal.

$$\mathbf{r}_i = \begin{bmatrix} x_{ci} & y_{ci} & z_{ci} \end{bmatrix}^T \tag{4}$$

$$R = \phi(s) \tag{5}$$

Distances from the center points are represented by the parameters. As shown in Fig. 8, deformation of the intestine is modeled by displacing the segment points from their initial positions in response to loads applied by the endoscope with a visco-elastic model having an elastic coefficient K_c and a viscous coefficient D_c.

From the load received by a point contacted by the endoscope (equivalently, the magnitude of displacement of that point) the displacements of neighboring points are determined, leading to a model for the deformation of the entire intestinal tube in response to the load. However, to simplify the deformation of the intestinal tube, it is assumed that each of the points r_i is displaced only on the corresponding plane P_i. As in previous models, deformation of the intestinal tube in the radial direction is modeled with an elastic coefficient K_d and a viscous coefficient D_d, and friction against the intestinal wall is modeled with a static friction coefficient μ_s and a dynamic friction coefficient μ_d.

3.4 Simplification of the Model

In an actual endoscope insertion operation, both the endoscope motion and deformation of the intestine are very slow. Accordingly, in equation (1) the inertial terms as well as the centrifugal and Coriolis force terms are considered to be very small compared to the weight of the endoscope and the contact force, so that they can be neglected. As a result, equation (1) is simplified as follows.

$$\mathbf{G}(\mathbf{q}) = \sum_i^n \mathbf{J}_{r_i}^T(\mathbf{q})\mathbf{F}_i - K[\mathbf{q}-\mathbf{q}_0] - D\dot{\mathbf{q}} \tag{6}$$

The motions of the endoscope and the intestine are analyzed by solving this equation of motion in short time steps. The viscous term for each movable joint has to be considered in view of the dynamical characteristics of an actual endoscope insertion tube. If these viscous terms are neglected, the time- dependent terms disappear, giving rise to unnatural behavior in which, if there is no contact force, at the next time step the joint angle returns to the balanced position.

3.5 Judgment of Contact, and Computation of the Reaction Force

To judge whether there is contact between the endoscope and the intestine, the distances between the endoscope model joints and the segment points of the intestine model are found, then the radii R_i of the intestine model at the segment points are compared geometrically.

In previous endoscope and intestine models, the intestinal tube itself did not deform, so to compute the reaction force it was sufficient to consider only contact between the endoscope and the intestine. However, this time the motion of the intestine itself is considered, so in computing the reaction force it is necessary to consider not only cases in which the endoscope contacts the intestine, but also cases in which the

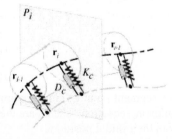

Fig. 8. Newly proposed colon model with visco-elastic suspentions

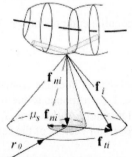

Fig. 9. Frictional cone

intestine contacts the endoscope. As in previous models, friction is judged using a friction cone. To represent the effect of friction in movement along smooth contact surfaces, the friction cone shown in Fig. 9 is used. The inside cone represents the maximum static friction force; outside the outer cone (of radius r_0) is the dynamic friction region. $\mathbf{f}_i$ is the contact force vector, with normal and tangential components $\mathbf{f}_{ni}$ and $\mathbf{f}_{ti}$, respectively.

First, the movable joints of the endoscope model and the segment points of the intestine model are given by equations (7) and (8), respectively.

$$\mathbf{E}_i = [X_i \; Y_i \; Z_i]^T \tag{7}$$

$$\mathbf{r}_i = [x_{ci} \; y_{ci} \; z_{ci}]^T \tag{8}$$

1. Case in which the endoscope contacts the intestine

In the contact judgment, from the distance between the the endoscope joints and the intestine segment points, the direction vector $\mathbf{l}_i$, for the direction in which the endoscope presses against the intestine tube, is found. In the present model, the segment point displacement, that is, whether the intestine is deformed, is determined from the magnitude of $\mathbf{l}_i$. The magnitude $|\mathbf{l}_i|$ when deformation occurs is called L_i.

If $|\mathbf{l}_i| <= L_i$, it is judged that deformation does not occur. In this case, there are reaction forces on the endoscope from deformation of the intestine in the radial direction and from the attempt of the intestine segment point to return to its initial position. These correspond to $\mathbf{f}_{ni}$ in Fig. 9. $\mathbf{f}_{ni}$ is as given by the following equation. K_c, K_d, D_c and D_d are the viscous and elastic coefficients of the intestinal tube in the radial direction and the intestinal segment points, as discussed in section 3.3.2. $\mathbf{d}_{ri}$ is the position vector of the segment point $\mathbf{r}_i$ as seen from its initial position. The 1st term represents the force due to deformation of the intestinal tube in the radial direction; the 2nd term the force exerted by an intestine segment point attempting to return to its initial position.

$$\mathbf{f}_{ni} = \left\{ - K_d \|\mathbf{l}_i\| - D_d \|\dot{\mathbf{l}}_i\| \right\} \frac{\mathbf{l}_i}{\|\mathbf{l}_i\|} + \left\{ - K_c \|\mathbf{d}_{r_i}\| - D_c \|\dot{\mathbf{d}}_{r_i}\| \right\} \frac{\mathbf{d}_{r_i}}{\|\mathbf{d}_{r_i}\|} \tag{9}$$

If $|\mathbf{l}_i| >= L_i$, the intestine deforms, so fni is given by the following equation.

$$\mathbf{f}_{ni} = \left\{ - K_c \|\mathbf{l}_i + \mathbf{d}_{r_i}\| - D_c \|\dot{\mathbf{l}}_i\| \right\} \frac{\mathbf{l}_i + \mathbf{d}_{r_i}}{\|\mathbf{l}_i + \mathbf{d}_{r_i}\|} \;) \tag{10}$$

If the endoscope contacts the intestine, the contact takes place in the fi direction. First the angle q between $\mathbf{f}_i$ and $\mathbf{f}_{ni}$ is found, then the magnitude of $\mathbf{f}_i$ becomes

$$\|\mathbf{f}_i\| = \|\mathbf{f}_{ni}\| / \cos\theta \tag{11}$$

then the tangential component $\mathbf{f}_{ti}$ is found.

$$\mathbf{f}_{ti} = \mathbf{f}_i - \mathbf{f}_{ni} \tag{12}$$

2. Case in which the intestine contacts the endoscope

In this case, $\mathbf{f}_i$ corresponds to the force of the intestine segment point trying to return to its initial position. At this time, $\mathbf{f}_i$ is given by the following equation. The $\mathbf{d}_{ri}$ are the respective position vectors of the segment points $\mathbf{r}_i$ from their initial positions.

$$\mathbf{f}_i = \left\{ -K_c \|\mathbf{d}_{r_i}\| - D_c \|\dot{\mathbf{d}}_{r_i}\| \right\} \frac{\mathbf{d}_{r_i}}{\|\mathbf{d}_{r_i}\|} \tag{13}$$

In this case, if the direction of the tangential component $\mathbf{f}_{ti}$ of $\mathbf{f}_i$ is taken to be the direction of a link that has a movable joint E_i, then the angle θ between $\mathbf{f}_i$ and $\mathbf{f}_{ti}$ is found, the magnitude of $\mathbf{f}_{ti}$ becomes:

$$\mathbf{f}_{ti} = \|\mathbf{f}_i\| \cos \theta \tag{14}$$

and the normal component $\mathbf{f}_{ni}$ is found as:

$$\mathbf{f}_{ni} = \mathbf{f}_i - \mathbf{f}_{ti} \tag{15}$$

Then, from the normal component $\mathbf{f}_{ni}$ and the tangential component $\mathbf{f}_{ti}$ of $\mathbf{f}_i$ that have been found, α, which expresses the friction condition, is found as in equation (16).

$$\alpha = \begin{cases} 0 & \|\mathbf{f}_{ti}\| < \mu_s\|\mathbf{f}_{ni}\| \\ \dfrac{\|\mathbf{f}_{ti}\| - \mu_s\|\mathbf{f}_{ni}\|}{r_0 - \mu_s\|\mathbf{f}_{ni}\|} & \mu_s\|\mathbf{f}_{ni}\| < \|\mathbf{f}_{ti}\| < r_0 \\ 1 & \|\mathbf{f}_{ti}\| > r_0 \end{cases} \tag{16}$$

Then, using α, the reaction force $\mathbf{F}_i$ on the endoscope, seen from the reference coordinate system, is computed as in the following equation. The 1st term is the static friction force component; the 2nd term is the dynamic friction force component.

$$\mathbf{F}_i = (1-\alpha)\ \mathbf{f}_i + \alpha\ (\mu_d\|\mathbf{f}_{ni}\|\ \frac{\mathbf{f}_{ti}}{\|\mathbf{f}_{ti}\|} + \mathbf{f}_{ni}) \tag{17}$$

After the $\mathbf{F}_i$ that are found in this way have been converted into the respective coordinate systems fixed in each of the movable joints using the conversion matrix, they are substituted into the endoscope model equation of motion given above.

3.6 Analysis by Simulation

Next, we performed actual analysis, using the dynamical model of an endoscope and intestine formulated as equations. Our endoscope model had 5 movable joints. As discussed in section 3.2, the 1st to 4th joints had rotating mechanisms with 2 degrees of freedom, while the 5th joint had a direct drive mechanism and a rotation mechanism with 1 degree of freedom each. The intestine model is given as a cubic center line divided into 150 segments. As discussed in section 3.3.2, each segment point was the center of a circle in the plane having a normal vector tangent to the intestine center line at the center point. In general the circles have different radii. An example of an analysis result obtained from the simulation will now be given.

We performed an analysis for the case in which an endoscope is inserted with constant force into a hypothetical intestine tube intended to model the sigmoid colon, a part of the large intestine that is practically not held in place at all. The motions of the endoscope and the intestinal tube in this case are shown in Fig. 10(a).

The reaction force toward the operator is shown in Fig. 10(b). The numbers in Fig. 10(b) have the same meanings as the corresponding numbers in Fig. 10(a).

As can be seen from the intestinal movement shown in Fig. 10(a), when the endoscope enters the sigmoid colon, the endoscope stretches the intestine considerably in the upward direction. It can be further seen from Fig. 10(b) that while this is happening the reaction force gradually increases. As can be seen particularly from (3) in Fig. 10(a), the operator seems to be inserting the endoscope at a constant rate, but as explained in section 3.3.1 at the distal end the endoscope stretches the intestine considerably, causing pain to the patient, and fails to penetrate into the deeper recesses of the intestine.

Previous intestine models, which only considered deformation of the intestine in the radial direction, gave simulation results showing that the distal end of the endoscope, while receiving a reaction force, penetrated deeply into the intestine [1]. The simulation results discussed above, on the other hand, show that when the considerable deformation of the center line, in addition to deformation in the radial direction, is considered, more realistic simulation results are obtained.

4. Improving Visual Information

4.1 Required Specifications

Doctor can obtain two kinds of visual information during endoscopic inspection into the intestine. Fig. 11 (a) is the endoscope image and (b) is the X-ray fluoroscopic image. The doctor relies primarily on the endoscope image when inserting the endo scope. An experienced doctor can judge from the color, shape and texture of the intestine. From this information and the distance that the endoscope has been inserted,

(a) Motion of endoscope model and colon model

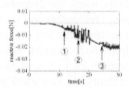

(b) Time-reactive force

Fig. 10. Simulation result

the experienced doctor forms an image of the shape of the endoscope insertion tube and is able to complete the technique in a short time without causing serious pain to the patient. It is indispensable that this endoscope image realistically shows the color, shape and texture of the intestine wall. An experienced doctor almost never uses the X-ray fluoroscopic image.

However, a young doctor cannot form an image of the shape of the endoscope insertion tube just from the endoscope image. For this purpose he uses the X-ray fluoroscopic image. It is sufficient for the doctor to know the shapes of the endoscope insertion tube and the intestine, so it is not necessary to use the X-ray fluoroscopic image to Fig. 11(b). To provide better understanding of the shape, views are obtained from different angles while the endoscope is being inserted.

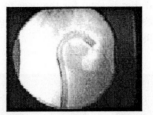

(a)Image of endoscopy   (b)Image of fluoroscopy

Fig. 11. Visual information during endoscopic operation

4.2 Three Types of Endoscope Images

As discussed in section 4.1, it is most important that an endoscope image realistically show the color, shape and texture of the intestine wall. In this system, realistic images are obtained by representing the intestinal tube model discussed in section 3.3 by a polygon model (Fig. 12(a)). The color and texture of each section of the intestine is mapped (Fig. 12(b)). In this way an experienced doct or can form an image of where the intestine the endoscope insertion tube is located. Fig. 12(c) shows an endoscope image using a wire frame. This is very effective in training beginners because it permits one to see parts of the intestine that cannot actually be seen.

Thus, it is possible to choose a method suitable for the doctor's skill level: a wire frame for training beginners, a polygon model for training intermediates, or texture mapping for training and surgery simulation for advanced physicians.

The method used can be changed while training is in progress, so that if the trainee becomes unable to form an image of the intestine it is possible to switch to a wire frame to increase the effectiveness of the training.

4.3 X-Ray Fluoroscopic Images

The two specifications required of X-ray fluoroscopic images were mentioned in section 4.1. In this system, the intestinal tube is represented by a sequence of rings (Fig. 13). This makes it easier to understand the shape of the endoscope inside the intestine and the contact situation.

Fig. 13 (a) and (b) show fluoroscopic images from different angles. To make the shape of the endoscope easier to understand, the angle of the fluoroscopic image can be varied continuously by mouse operation while training is in progress. In addition, a capability to enlarge the fluoroscopic image to show more detail is provided.

4.4 The Total Image Display

Fig. 14 shows an actual operation screen with both the endoscope image and the X-ray fluoroscopic image incorporated into the system. In the control window at the upper left, the endoscope window can be selected, training repeated, damage to the intestine displayed and the intestine and endoscope to be used in the virtual space selected. These features were all created based on the requirements of the training system.

An input method that is easy for the user to understand intuitively has been adopted. This makes it possible for even people using the system for the first time to perform the various settings and operations.

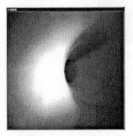
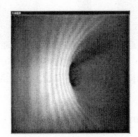
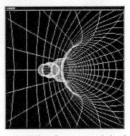

(a)Polygon model of intestine   (b)Polygon model with texture mapping of real image   (c)Wire frame model of intestine

Fig. 12. Image of endoscopy for VES-IV

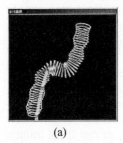
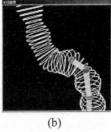

(a)   (b)

Fig. 13. CG image of fluroscopy for VES

Fig. 14. Visual information for VES

Finally, in this system it is possible to choose whether to display or hide the X-ray fluoroscopic image. It has already been noted that an experienced doctor will generally not use the X-ray fluoroscopic image. However, the present reality that young inexperienced doctors are using X-rays on the patients. In other words, the patients are being exposed to X-rays at the time of observation. By hiding the X-ray fluoroscopic image while using this system, the doctor learns how to insert the endoscope into the patient without using the X-ray fluoroscopic image, making it possible to discontinue exposing patients to X-rays. Of course, the choice of whether to display or hide the image can be changed while training is in progress.

5. Construction of an Endoscope Insertion Training System

A system incorporating and linking the components that have been discussed thus far -- the VES-V, a dynamical model of the endoscope and intestine, and a fluoroscopic image display, into a system that can be used for endoscope insertion training, was constructed (Fig. 15).When an endoscope is inserted using the VES-V, the motions of the endoscope and the intestine are analyzed in real time and the reaction force and reaction torque are transmitted to the doctor's hand. In addition, a controller similar to the controller that will be used in an actual operation is used to adjust the angulation of the distal end of the endoscope to deform the intestine in a way that will decrease the reaction force, suppressing the pain felt by the patient as much as possible and making it possible to insert the endoscope deeply into the intestine.

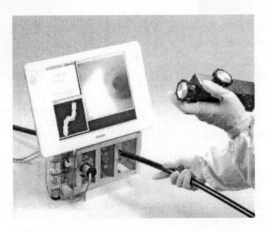

Fig. 15. Total virtual endoscope system with new force display mechanism and large deformable colon model (VES-V)

6. Portable Virtual Endoscope System

The latest version of the "portable virtual endoscope system" is shown in Fig.16. Our virtual endoscope system is constructed in three parts, a force display mechanism, virtual models of endoscope and colon, a monitor. In this portable virtual endoscpe system they are in a body of tower-type PC case. In recent years many surgical simulations are studied by various organization. But most of them isn't considered with portable ability. It is very important for training systems to be portable. Because many doctors need to practice high leveled insertion training. And the confidence of this system is progressed for using by many doctors. We made the very small force display mechanism which can be in PC case. Doctors can practice wherever there is a power supply.

As shown in Fig.17, the portable VES has a CD-ROM to read the personal data of patient's organ and the network ability also available to exchange data of the patient far from the trainee. It is easy to add new features according to demand.

7. Summary

Firstly, we clarify the problems that occur with mechanisms adopted in past research, and design and produce a new mechanism (the VES-IV) to solve those the problems that occur problems. The VES-IV was used to perform a force sensation display

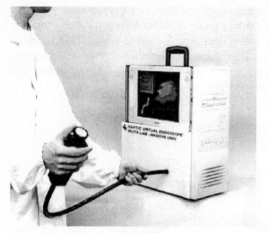

Fig. 17. Internal mechanism of Portable virtual endoscope system (VES-VI)

Fig. 16. Portable virtual endoscope system (VES-VI)

experiment, giving better results than have been obtained with past systems, confirming the effectiveness of this system. In the dynamical model, large deformation of the whole intestine that had been neglected in earlier models was considered. This model succeeded in simulating movements of the endoscope and the intestine close to those that occur in reality.

Both endoscope images and fluoroscopic images were provided, as in a real operation. To make the training system as complete as possible, several types of realistic endoscope images, and fluoroscopic images obtained from several angles, are provided.

Finally, the various elements were linked to construct a new system for realistic endoscope insertion training. And new portable training system was completed and performance was verified by the doctors.

Acknowledgment

Authors thank Mr. Shinkoh Senda and Mr. Masaki Takeichi of Nagoya University for his useful assistance for development.

References

1) K.Ikuta, M.Takeichi, T.Namiki: "Virtual Endscope System with Force Sensation",Medical Image Computting and Computer-Assisted Intervention (MICCAI 98),1998

2)K.Ikuta, M.Takeichi, T.Namiki: "Virtual Endoscope System with Force Sensation",IEEE International Conference on Robotics and Automation ICRA99, pp.1715-1721, 1999

MR Compatible Surgical Assist Robot: System Integration and Preliminary Feasibility Study

Kiyoyuki Chinzei[1], Nobuhiko Hata[2], Ferenc A. Jolesz[2], and Ron Kikinis[2]

[1] Mechanical Engineering Laboratory, AIST, MITI
1-2 Namiki, Tsukuba, 305-8564 Japan
chin@mel.go.jp (after Apr. 2001, chin@aist.go.jp)
http://www.mel.go.jp (after Apr. 2001, http://www.aist.go.jp)
[2] Department of Radiology, Brigham and Women's Hospital
Francis St. 75, Boston, MA 02115, USA
{noby, jolesz, kikinis}@bwh.harvard.edu
http://splweb.bwh.harvard.edu:8000

Abstract. A magnetic resonance (MR) compatible surgical assist robot system under preclinical evaluation is described. It is designed to coexist, and cooperate, with a surgeon, and to position and direct an axisymmetric tool, such as a laser pointer or a biopsy catheter. The main mechanical body is located above the head of the surgeon, and two rigid arms extend to the workspace. This configuration contributes to a small occupancy in the surgeon's workspace, and good MR compatibility.

The design of the robot is described. The MR compatibility is examined, and shows that there is no adverse effect on the imaging, even when the robot is in motion. Any heating effect was not evaluated, because a published study has revealed any effect is quite small. This robot system is carefully designed for safety and sterilization issues.

1 Introduction

The advantages of surgical robots and manipulators are well recognized in the clinical and technical community. Precision, accuracy, and the potential for telesurgery are the prime motivations in applying advanced robot technology in surgery [1–4]. Except for master-slave manipulators, which are controlled by human operators, surgical robots require trajectory planning, which, in practice, relies upon preoperative images. If the target organ is deformable, then the trajectory needs to be updated according to the magnitude of the deformation. Here, image-guided surgery is a natural solution.

Magnetic resonance imaging (MRI) has an excellent soft tissue discrimination, and a well-defined 3D coordinate reference system. An intraoperative MR scanner (Signa SP/i, GE Medical Systems, Milwaukee, WI, 0.5 Tesla) has been specifically designed to bring the power of MRI to the operating theater. It has a pair of parallel facing donut-shaped magnets, with an air gap of 560 mm. (Fig. 1 left) Two surgeons can stand in the gap to access the patient. In the six years to February 2000, the authors' institute has recorded more than 500 cases using theintraoperative MR scanner [5–7].

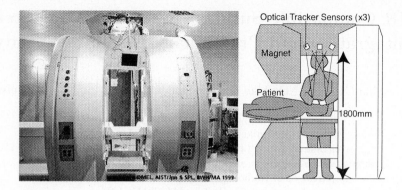

Fig. 1. The intraoperative MR scanner (left) and its profile with a patient and a medium height surgeon (right). Three visual sensors of the tracking device are installed above the imaging region. The sensor view has to be clear.

MR guidance of a surgical robot will be quite useful, as it can update the trajectory. However, an MR scanner is a highly restrictive environment for foreign objects, particularly for robots. Schenck has discussed the magnetic compatibility of MRI systems [8]. GE Medical Systems have issued a comprehensive and practical guideline regarding MR safety and compatibility [9], and Masamune has developed an MR compatible six degrees-of-freedom manipulator [10]. This was mainly built from plastics, and so suffered from a lack in rigidity. We have previously investigated the MR compatibility of mechanical parts, and have summarized MR compatible techniques, thereby showing the possibility of building a precise MR compatible mechanism [11].

In this paper, we demonstrate a unique configuration of a novel MR compatible robotic system for use in MR guided surgery. In Section 2.1, we introduce our target applications, and then define the design concept of the robot. Section 2 explains the unique configuration and the design of the robot system. In Section 3, the MR compatibility of the moving robot is validated. Safety and sterilization issues are discussed in Section 4.

2 Materials and Methods

The design of the robotic system will be introduced following the task definition. Two specific restrictions need to be overcome; (i) the layout of the robot needs to coexist, and cooperate with, the surgeon; and (ii) the design needs to be MR compatible.

2.1 Task Definition

The goal of our robot assist system is to enhance the surgeon's performance by accurate numerical control, and not to eject him or her from the surgical field.

Therefore, the system must coexist, and cooperate, with the surgeon. Minimally-invasiveness is an obvious requirement. The robot will hold a catheter or a laser pointer. In the former case, the robot will position and direct the catheter. Technically, the robot can perform the insertion; however, our current plan reserves this task for surgeon owing to ethical and legal considerations.

2.2 Specific Restrictions in the MR Environment

The intraoperative MR scanner requires two specific restrictions to the surgical robot, in addition to the standard requirements such as safety and sterilization policies.

Layout: The robot must coexist with the surgeon. However, when the patient is prepared, and the surgeons take their place, the available space for the robot is quite limited, particularly around the patient. In addition, the occlusion of the sensors of the optical tracker (Fig. 1, right) should be minimal. However, if the robot is placed some distance from the workspace, it will reduce the precision and the dynamic response.

MR compatibility: To enable the real-time tracking of the target position, the robot should be able to maneuver, even during imaging. The robot motion should not have any adverse effect on the imaging, and it should not be affected by the imaging process. This requires that the robot at least be made from paramagnetic materials. The further these are from the imaging region and the smaller volume they occupy, the better for MR compatibility. In addition, the robot should be MR safe. The MR safety of the robot requires that the machine should not unintentionally move from any magnetic attraction, and adverse electromagnetic side effects (e.g., leakage of, and heating by, eddy currents and RF pulses) should not occur.

To clear the workspace for the surgeon, the robot should be placed away from the surgical field. This also contributes to a better MR compatibility. However, there is a trade-off between space saving and mechanical performance.

2.3 Configuration and Kinematics

Figure 2 shows a schematic configuration of the robot. The actuators and the end effector are spatially separated. The main body, with all actuators, is located above the surgeon's head, and the end effector is attached at the ends of two long, rigid arms. The robot has five degrees-of-freedom, which define a polar coordinate system. Five degrees-of-freedom are sufficient to position and direct a catheter or a laser pointer, because these instruments are axisymmetric.

All axes in the main body are driven by linear motion mechanisms. The first three axes (X1, Y1, Z1) drive a rigid arm (Arm1) as well as the mechanism of remaining two axes (X2, Y2). These two axes drive the second rigid arm (Arm2). These arms are linked through two pivotal joints (P1, P2). P1 defines a point in the 3D coordinate system, and P2 travels in a 2D plane relative to P1. The

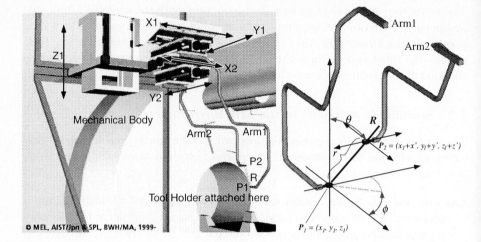

Fig. 2. The configuration of the robot. The mechanical body has 5 axes in total. All axes are driven by the linear motion mechanisms. Two rigid arms extend to the workspace. Arm 1 is actuated by X1, Y1 and Z1. Arm 2 is actuated by X2 and Y2. The ends of the arms are linked by two spherical joints P1, P2 and the sliding bar R. A tool holder is attached at P1. P1 moves in 3D space and P2 moves in a 2D plane relative to P1. The direction of the line segment between P1 and P2, r, is determined by the position of P2. Mathematically, it is a polar coordinate system whose origin is P1.

link between P1 and P2, R in Fig. 2, is fixed to the inner ring of P1, whereas there is a sliding joint between P2 and R. This is to allow the distance between P1 and P2, r, to vary according to the position of P2.

The end effector, the tool holder, is attached to P1. For simplicity, here we will assume that the position of the tool holder is at P1, and its direction of travel is parallel to that of R. This is described in a polar coordinate system whose origin is at P1 (Eq. 1)

$$x' = r \cos \phi \sin \theta$$
$$y' = r \sin \phi \sin \theta \qquad (1)$$
$$z' = r \cos \theta$$

where P1 $= (x1, y1, z1)$, P2 $= (x1 + x', y1 + y', z1 + z')$, ($z'$ is a constant), and $r^2 = x'^2 + y'^2 + z'^2$. The direction (ϕ, θ) is determined only by the relative position of P2 to P1, (x', y', z'). Therefore, the direction is independent of P1. When the tool holder is attached to P1 with an offset, its position is dependent on (ϕ, θ).

2.4 Design

The hardware of the robot was assembled by Unitek Co. (Tsukuba, Japan) and partly by Drenak Co. (Machida, Japan.) The five degrees-of-freedom main body is composed of five linear motion tables. Each table unit has a ball screw and a pair of linear guides. These are made of either stainless steel (YHD50) or beryllium-copper; both materials have low magnetic susceptibility and a hard surface that can be used as a point-touch mechanism. The ball screws and linear guides made from YHD50 were manufactured by NSK Ltd. (Tokyo, Japan), and those made from beryllium-copper were manufactured by Koyo Seiko Co., Ltd. (Osaka, Japan.) The ball screw is supported by a pair of ball bearings made from silicon nitride (Si_3N_4) ceramics, manufactured by the above companies.

A non-magnetic (piezoelectric) ultrasonic motor, USR60-S3N, (Shinsei Kogyo Corp., Tokyo, Japan) directly drives the ball screw. Its maximum rotational torque is 0.5 Nm, and its holding torque is more than 0.7 Nm. A mechanical clutch is inserted between the motor and the ball screw to allow manual motion.

Each axis has the home and limit detector, and an incremental linear encoder. The linear encoder is customely designed, using glass grating patterns manufactured by Dynamic Research Corp. (Wilmington, MA). Its resolution is currently chosen to be 0.02 mm. All signals are picked up by, and transferred via, fiber optic cables. The optoelectronic conversion circuits, as well as the other control and power circuits, are placed outside the scanner room for better noise immunity.

All parts of the robot were made from paramagnetic materials. The rigid arms, the frame structure of the vertical axis, and the attachment of the robot to the scanner were made from a titanium alloy. The frames of the horizontal axes were made from a polycarbonate resin. All the screws were made from a titanium alloy or brass, to avoid any attraction of loose screws from the magnet.

The pivotal joints, P1 and P2, were KBRM-10 Rod End Bearings (IGUS GmbH, Germany), and were all made from plastics, and the connecting link, R, was made from acrylic resin. As these are insulating materials, the leakage of the eddy current and RF pulses can be safely avoided. Each arm can be divided into three pieces. The end pieces can fit in a typical autoclave tray, whose internal dimensions are approximately $450 \times 80 \times 200$ mm.

Cooperation between the robot control, MRI, and 3D position tracking is implemented by the object distributed, server-client model [12]. Currently there are three modules: (i) a robot hardware module; (ii) a Modular Robot Control (MRC) developed at Johns Hopkins University; and (iii) Slicer3D modules (Image processing/surgical planning) [13]. The details of these are to be found in the given references.

2.5 Evaluation of MR Compatibility

We have examined the loss of homogeneity of the magnetic field, and the signal-to-noise ratio (SNR) of the image. There are several possible interactions when the robot is located in, and is maneuvering in, the MR environment. The presence of, and motion of, the robot can distort, or shift, the image by decreasing the homogeneity of the magnetic field, and these can also affect the image SNR. The leakage of the induced current can be safely neglected by the insulating devices as described above. The heaing effect has been proved to be minor as shown by the previous work [14].

A spherical phantom (diameter = 280 mm) was imaged using the imager. This contained $CuSO_4$ solution, which gave a delta-function shape resonance spectrum in an ideal, homogenous magnetic field. The inhomogeneity was defined by the diversity of the observed spectrum. The SNR was calculated using the following equation (Eq. 2).

$$SNR = Pcenter/SDcorner \ . \tag{2}$$

where $Pcenter$ is the mean value of the $40{\times}40$ pixel area at the center of the image, and $SDcorner$ is the standard deviation of the $40{\times}40$ pixel area in the lower right corner of the image. The sequence was the Spin Echo, TE/TR = 85/220 ms, and the receiver bandwidth was 62.5 kHz. These protocols are described in [9]. The robot repeated a simple Y2 axis motion, which was the most adjacent axis to the imaging region. The control data were obtained by the same phantom without the robot.

3 Results

The construction of a Mark 1 version of the robot system has already been completed. The system is currently in the software tuning, and the preclinical evaluation stages.

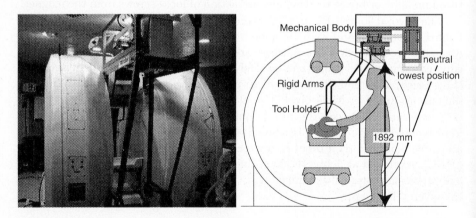

Fig. 3. Constructed robot attached with the intraoperative MR scanner (left) and the profile of the workspace (right). The moving part clears the workspace for the surgeon.

3.1 Configuration

Figure 3 shows the constructed robot installed with the intraoperative MRI and the profile of the workspace.

When the vertical axis is at the lowest position, the moving part's lowest height (except the arms and part of the vertical axis, Z1) is 1892 mm from the floor (the bottom of the Y2 axis, Fig. 3, right). The arms are bent so that they do not collide with the scanner.

3.2 MR Compatibility

The magnetic field inhomogeneity values are listed in Table 1. The inhomogeneity value was observed to be 0.53 when the robot was in motion. This is better than that of a clinically-used stereotactic frame, or of the human body itself. Therefore, the effect on the homogeneity of the magnetic field was negligible.

Table 1. Obtained inhomogeneity values. The smaller value is the greater homogeneity.

Inhomogeneity	(ppm)
Spherical phantom, without robot (baseline)	0.45
Spherical phantom, with moving robot	0.53
Spherical phantom, with an 'MR compatible' Mayfield stereotactic frame	0.9
Human volunteer	ca. 1.4

The observed signal-to-noise ratio (SNR) loss was 1.6% to 1.8%. As an SNR loss up to 10% is acceptable, the observed SNR was in the negligible range. Figure 4 shows an image of the spherical phantom with, and without, the robot.

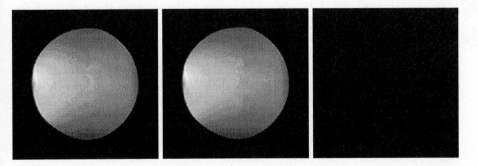

Fig. 4. Images of the spherical phantom when the robot was not installed (left) and when one axis of the robot was in motion (center). The subtraction of these images showed no shift (right).

4 Discussion

This configuration has several advantages:

- The mechanical body is positioned 1-2 m away from the center of the imaging volume. It also relaxes the MR compatibility criteria.
- The kinematics and inverse kinematics are simple. It forms a polar coordinate system, and the position and the direction can be independently determined by the position of the pivoltal joints P1 and P2. When the surgeon wants to move the end effector numerically, he or she can easily predict the motion of the robot.
- The arms and end effector are detachable, exchangeable, and sterilizable. This design also allows for the selection of a suitable arm shape, according to the desired procedure.
- The arms have no rotational joint and no actuator to sustain. Without these, the robot has the slender arms that save the precious workspace. The parallel-link mechanism has the similar advantage, however, it usually demands a large motion space. However, Koseki and the author are proposing a combination of a leverage and parallelepiped mechanisms (LPM) [15]. LPM mirrors the motion of the robot at a distance, therefore it has a potential to be a break-through to relax the specific restrictions in the MR environment.

On the other hand, the end effector needs to be light and axisymmetric, such as a catheter needle. As the needle insertion is one of common tasks in many minimally invasive procedures, this design does not lose a generality.

4.1 Safety and Sterilization Issues

In general, it is the surgeon's responsibility to avoid a collision, as there is no reliable, and universal, automated method to address this issue. Therefore, the simplicity of the kinematics, in particlar, the independency of the parameters, is an important factor in helping the surgeon to understand the motion of the robot.

Brakes are not necessary, as the motor's holding torque is strong, and the robot maintains its posture even in the event of a power cut. The noise immunity of the sensors is good, since the sensors use fiber optics to carry the signals.

The rigid arms can be divided into shorter pieces. The end piece is autoclavable. The other part of the arms, and the mechanical body, can be draped. The pivot joints and other plastics parts can be applied EOG sterilization.

4.2 An Example of Application

One of the good examples of this robot system is the needle navigation in the brachytherapy of prostate cancer. This is a minimally invasive outpatient radiotherapy that delivers an internalized radioactive source to the cancer. A number of small iodine-125 (^{125}I) radiation seeds are placed using catheters, under the guidance of MRI. The authors' institute has pioneered the use of MR guided brachytherapy of prostate cancer since 1997 [16].

This procedure implants 50 to 120 seeds by 12 to 20 catheter insertions, according to a preoperative seeding plan. It is a skilful, and hard, task for human operator, whereas it will be a good application for the robot.

4.3   Impact of the MR Guided Robotic Assist

Modern medicine has rich sources of information regarding the state of health of a patient. High quality three dimensional images (e.g., MRI) transfer the real world (patient) into the virtual world. We have developed computer models for virtual manipulation for surgical planning. In contrast, the transfer from the virtual to the real world (i.e., the operating theater) has been mostly limited to visual assist. Here, the flow of online information is unidirectional and incomplete. Surgical robots will be the means of physical assistance, and create a bi-directional flow of online information. The impact of the combination of the intraoperative MRI, and the MR compatible robot, will be even greater, because it will bring the possibility of near real-time processing of the virtual and real worlds in a bi-directional manner.

5   Conclusion

An MR compatible surgical assist robot is in the preclinical evaluation stage. It is designed to position and direct an axisymmetric tool, such as a laser pointer or a biopsy catheter. An example of its application is in MR guided brachytherapy of prostate cancer. It has a generality, because catheter insertion is a common procedure in many other operations.

To be MR compatible, and to fit into the reduced space of the intraoperative MRI, the robot has a unique configuration. The robot has five degrees-of-freedom. All the mechanical axes are in linear motions. The end effector is attached at the end of two long rigid arms. The end effector forms a polar coordinate system. The robot is made from a titanium alloy and plastic and is driven by non-magnetic ultrasonic motors. Fiber optics carry the signals from the sensor heads to circuit boards that are isolated outside the MRI room for noise immunity.

The robot shows excellent MR compatibility. Its motion does not appear to have any adverse effect on the imaging, and the robot is not affected by the imaging process. Other possible events (e.g., heating) are considered to be negligible from the isolation of the design, and from previous studies in literature.

Acknowledgment

In Japan, the project has been funded by MITI, "Study of Open MRI-Guided Diagnosis/Treatment Concurrent System", and "Research on MR compatible active holder for surgical assistance".

In the USA, the project has been funded by NSF Engineering Research Laboratory "Computer Integrated Surgical Systems and Technology" #9731748, and "MR Guided Therapy" 3P01CA67165–03S1 funded by NIH.

The authors would gracefully acknowledge the assists of Mr. Dan Kacher, Mr. Oliver Schorr and Dr. Toshikatsu Washio to conduct the experiments.

References

1. Villotte, N., Glauser, D., Flury, P., et.al.: Conception of Stereotactic Instruments for the Neurosurgical Robot Minerva. *in proc IEEE ICRA* (1992) 1089–90.
2. Taylor, R.H., Mitterlstadt, B.D., Paul, H.A., et.al.: An Image-Directed Robotic System for Precise Orthopedic Surgery. in Taylor, R.H., et.al. (eds) *Computer-Integrated Surgery : Technology and Clinical Applications,* MIT Press (1995) 379–95.
3. Sackier, J.M., Wang, Y.: Robotically Assisted Laparoscopic Surgery: from Concept to Development in Taylor, R.H., et.al. (eds) *Computer-Integrated Surgery : Technology and Clinical Applications,* MIT Press (1995) 577–80.
4. Schenker, P.S., Das, H., Ohm, T.R.: A New Robot for High Dexterity Microsurgery. in proc *CVRMed95,* Lecture Notes Computer Science, 905, Springer-Verlag (1995) 115–22.
5. Schenck, J.F., Jolesz, F.A., Roemer, P.B., et. al.: Superconducting Open-Configuration MR Imaging System for Image-Guided Therapy. *Radiology,* 195(**3**) (1995) 805–14.
6. Silverman, S.G., Collick, B.D., Figueira, M.R., et.al.: Interactive MR-guided biopsy in an open-configuration MR imaging system. *Radiology,* 197 (1995) 175–81.
7. Hata, N., Morrison, P.R., Kettenbach, J., Black, P., Kikinis, R., Jolesz, F.A.: Computer-assisted Intra-Operative Magnetic Resonance Imaging Monitoring of Interstitial Laser Therapy in the Brain: A Case Report. *J Biomedical Optics,* 3(**3**) (1998) 304–11.
8. Schenck, J.F.: The role of magnetic susceptibility in magnetic resonance imaging: MRI magnetic compatibility of the first and second kinds. *Med Phys,* 23(**6**) (1996) 815–50.
9. GE Medical Systems (ed): MR Safety and MR Compatibility: Test Guidelines for Signa SP^{TM}. www.ge.com/medical/mr/iomri/safety.htm, 1997.
10. Masamune, K., Kobayashi, E., Masutani, Y., et.al.: Development of an MRI compatible Needle Insertion Manipulator for Stereotactic Neurosurgery. *J Image Guided Surgery,* 1 (1995) 242–8.
11. Chinzei, K., Kikinis, R., Jolesz, F.A.: MR Compatibility of Mechatronic Devices: Design Criteria. in proc *MICCAI '99,* Lecture Notes in Computer Science, 1679 (1999) 1020–31.
12. Schorr, O., Hata N., Bzostek, A., Kumar, R., Burghart, C., Taylor, R.H., Kikinis, R.: Distributed Modular Computer-Integrated Surgical Robotic Systems: Architecture for Intelligent Object Distribution. in proc *MICCAI 2000* (printing), Pittsburgh PA (2000).
13. Gering D., Nabavi, A., Kikinis, R., Eric, W., Grimson, L., Hata, N., et.al.: An Integrated Visualization System for Surgical Planning and Guidance using Image Fusion and Interventional Imaging. in proc. *MICCAI '99,* Lecture Notes in Computer Science, 1679, Springer-Verlag, Berlin, Heiderberg, New York (1999) 809–19.
14. Buchili, R., Boesiger, P., Meier, D.: Heating Effects of Metallic Implants by MRI Examinations. *Magnet Reson Med,* 7 (1988) 255–61.
15. Koseki, Y., Chinzei, K., Koyachi, N., Arai, T.: Robotic Assist for MR-Guided Surgery Using Leverage and Palallelepiped Mechanism. in proc *MICCAI 2000* (printing), Pittsburgh PA (2000).
16. D'Amico, A.V., et al.: Real-time magnetic resonance image-guided interstitial brachytherapy in the treatment of select patients with clinically localized prostate cancer. *Int J Radiat Oncol Biol Phys,* 42(**3**) (1998) 507–15.

Needs Assessment for Computer-Integrated Surgery Systems

Sarah Graham[1], Russell H Taylor[1], Michael Vannier[2]

[1]Department of Computer Science
Engineering Research Center for Computer-Integrated Surgical Systems and Technology
Johns Hopkins University, Baltimore, Maryland, USA
[2]Department of Radiology
University of Iowa, Iowa City, Iowa, USA
{sarah,rht}@cs.jhu.edu, michael-vannier@uiowa.edu

Abstract. The needs of surgeons for computer-assisted systems cannot be satisfied unless their requirements and expectations are known. We determined surgeons' needs for computer-aided systems using group facilitation processes that measure customer wants and needs, including quality function deployment and Kano Analysis. A one-day workshop hosted by the CISST ERC (Computer Integrated Surgical Systems and Technology Engineering Research Center) included thirteen surgeons and eight engineers. The primary goal of the workshop was to determine medical professionals' needs and expectations at the earliest stages of research and development. Surgeons and nurses from Johns Hopkins Medical Institutions participated in the one-day event. The results include an affinity diagram, a prioritized list of wants and needs in the "voice of the customer", and a Kano Analysis of 25 question pairs that reveal exceptional features that can be included in computer-aided surgical systems.

1 Introduction

Computer-Integrated Surgery (CIS) systems will have a profound effect on 21st Century Medical Practice. A novel partnership between human surgeons and machines, made possible by advances in computing and engineering technology, will overcome many of the limitations of traditional surgery. By extending human surgeons' ability to plan and carry out interventions more accurately and less invasively, CIS will address a vital need to reduce medical costs, improve clinical outcomes, and improve the efficiency of health care delivery.

The development of CIS crucially requires close collaboration between engineering researchers, industry representatives, and clinicians who will use the systems. It is vital that research and system development be directed toward real clinical needs. As is common with emerging fields, there have been a number of "needs and research opportunities" workshops (e.g., [1-4]) directed toward different aspects of computer-integrated surgery. Typically, a mixture of engineering researchers and clinicians attends these workshops. The workshop format usually combines a series of technical presentations with breakout group discussions, followed by a report written by a subset of the participants. The workshop reported here, in contrast, used formal needs and quality assessment techniques to elicit responses from operating room personnel (surgeons, nurses, and technicians). The results provide a useful complement to the reports of the breakout-session workshops.

2 Background

Quality is a measure of the ability of a product or service's features and characteristics to satisfy given needs. The quality of computer-integrated surgery systems is judged by their users, especially surgeons and related medical professionals. We conducted a study to determine surgeons' needs using formal group facilitation processes, including the construction of an affinity diagram, one component of the "House of Quality" [5]. Kano Analysis [6] was applied to a group of invited surgeons who are interested in computer-aided surgical technologies. The purpose of this paper is to explain the process used to determine surgeons' needs for these technologies and report the results of a one-day planning session where they were applied.

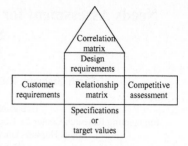

Fig. 1. Termed the "House of Quality" by Hauser and Clausing, this diagram illustrates a relationship matrix for determining users' needs and translating them into products and services.

Quality function deployment (QFD), also known as the "House of Quality" and illustrated in Figure 1, refers to the process of involving customers in the design stage of new or redesigned products. [7, 8] The "Voice of the Customer" is the first step of QFD, followed by product development, production, and sales. QFD is a requirements identification, analysis, flow-down, and tracking technique. It focuses on quality and communication to translate customer needs into product-and-process design specifics. In this study, we measured the Voice of the Customer by building an affinity diagram in a group meeting of invited surgeons and related professionals. The affinity diagram is a group decision-making technique designed to sort a large number of ideas, process variables, concepts, and opinions into naturally related groups. [9-11]

Kano Analysis is a method for extracting different types of customer requirements from survey information based on the research and publications of Noriaki Kano, a customer satisfaction researcher and member of the Japanese Union of Scientists and Engineers, sponsors of the Deming Prize. The Kano Model, shown in Figure 2, classifies product attributes and their importance based on how they are perceived by customers and their effect on customer satisfaction. The model measures the level of satisfaction with a product against consumer perceptions of attribute performance. Kano claims that attributes can be classified into three categories [12]:

1. Basic or Expected characteristics provide diminishing returns in terms of customer satisfaction. These are essential or "must" attributes of performance and do not offer any real opportunity for product differentiation. Providing *basic* attributes and meeting customer expectations for them will do little to enhance overall customer satisfaction, but removing or performing poorly on them will hurt customer satisfaction, lead to customer complaints, and possibly result in customer defections. Examples of *basic* characteristics include timely delivery of a magazine subscription, the ever-present telephone dial tone, and availability of an automatic teller machine at a bank branch.

2. Performance or Normal characteristics exhibit a linear relationship between perceptions of attribute performance and customer satisfaction. Strong performance on these "need" attributes enhances, while weak performance reduces, satisfaction with the product or service. Adding more attributes of this type to a product will also raise customer satisfaction. Examples of *performance* characteristics include the duration of a cellular telephone's rechargeable battery life and an automobile's fuel economy.

3. Excitement or Delightful characteristics are unexpected attributes that, when provided, generate disproportionately high levels of customer enthusiasm and satisfaction. When these "nice-to-have" attributes are not available in a product, it does not lead to customer dissatisfaction. Examples of *excitement* characteristics include receiving an upgraded hotel room with free breakfast for the standard rate or finding a CD player included as standard equipment on an economy car.

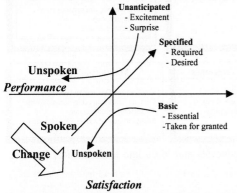

Fig. 2. The Kano Model illustrates the relationship between perceived customer need and product

Both the QFD and Kano processes are means of capturing and analyzing user input regarding the customers' needs and priorities in product development. Such formal needs assessment methods are not new to the medical arena. Both QFD and Kano Analysis have been applied in studies designed to improve patient quality-of-care and customer service in the health care community (e.g.[11, 13, 14]). In our efforts, we aim to cross professional boundaries to better define the needs of operating room personnel with respect to CIS systems and engineering efforts. It is our goal that such a forum will foster more collaboration between CIS engineers and medical professionals at all stages of the research lifecycle.

3 Method & Results

3.1 Method: QFD Analysis

The QFD process [5, 15] is comprised of several steps, each driving towards the goal of having a list of specific, prioritized features to be included in CIS products. The first step in the process is to determine a broad notion of what customers, in our case surgeons and nurses, want from the field of Computer Integrated Surgery. To do this, the customers answer a series of questions. For example, workshop participants answered questions such as, "What do you like about surgical technology as it exists today?" and "If cost were not an issue, what CIS technology or capability would you ask for and why would you want it?" The questions are designed to provoke brainstorming and to encourage descriptive, freeform responses. Multiple responses to each question are allowed and, in fact, encouraged. Each separate idea is recorded on a separate index card, and all responses are anonymous. At the end of the session, all index cards are collected.

The next step establishes categories into which the users' needs (the responses to questions in the previous step) can be classified. All index cards from the previous

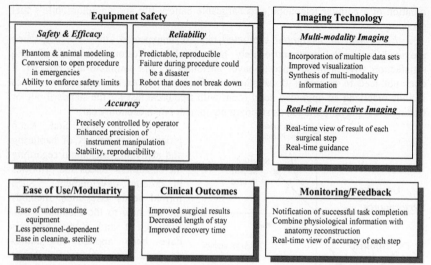

Fig. 3. An Affinity Diagram groups concepts and ideas into a hierarchical list of categories.

step are placed face-up on a table, and the participants are asked to help sort the cards into categories with common attributes. For example, cards stating "no technical support necessary" and "complicated to use" could be grouped together because they both address a system's ease-of-use. Any participant can group, ungroup, or regroup cards as he or she sees fit, and all grouping is done with minimal discussion. Grouping continues until a small, pre-determined number of categories (or card groupings) is established.

After the groupings are established, each group must be given a category name. In this step, it is important to let the users define the name for each category. This allows the users' linguistics to enter the design process from the very beginning. For each stack of index cards grouped together in the previous step, all cards in that stack are read aloud. The participants discuss the group's common characteristics and come to a consensus on what the category name should be. During this process, groupings may be combined or divided as necessary to capture all main ideas presented in the index cards. The category name of each grouping is recorded. In the workshop, a reduced set of ten groupings defined by the sorting process resulted in eighteen unique categories after the naming process was completed.

Finally, a voting process is used to prioritize the categories. Votes vary in value from 1 (feature would be nice but is unessential) to 9 (feature is absolutely essential). Each participant is given a certain number of votes of each value to distribute among the categories. The votes are tallied, and the categories are ranked from highest to lowest priority according to points received. [5, 15]

3.1.1 Results: Affinity Diagram & QFD Voting

The results of the QFD Analysis are presented in two forms. First, the affinity diagram in Figure 3 shows the grouping of customer needs into categories. The requirements listed in each category are taken from the index cards completed by workshop participants during the QFD process. A subset (8 of 18) of the categories is shown. Categories sharing common themes can be grouped under common headings, such as reliability and accuracy share the underlying theme of "Equipment

Category	Score
Tier 1	
Safety and Efficacy	99
Ease of Use / Modularity	75
Accuracy	66
Reliability	65
Tier 2	
Clinical Outcomes	49
Real-time Interactive Imaging	47
Monitoring / Feedback / Physiological Processes	38
Multi-modality Imaging	35
Tier 3	
Training and Logistical Support	33
Versatility	32
Efficiency	31
Extend Human Motor & Sensory Input or Skills	30
Tier 4	
Acceptability	25
Preoperative Planning and Procedure Practice	24
Physical Properties	21
Minimal Invasiveness	19
Tele-operation / Remote Operations	12
Cost	11

Fig. 4. The QFD Voting process yields a list of prioritized customer needs.

Safety". The subset of categories in this figure reflects the top two tiers of results obtained in the QFD voting process. These results are shown in Figure 4. The categories were prioritized based on total score received in the ranking procedure. The actual category names were determined by workshop participants and are listed in the left-hand column, and the right-hand column reflects the total score the category received.

To get a better understanding of what the surgeons' and nurses' ideas mean, the discussion about the results was open to both the workshop participants and CIS engineers who did not participate in the QFD process. The discussion proved to be lively and enlightening. A good deal of discussion focused on the top tier of results. We were surprised to find that these four features are not CIS-specific but, instead, can apply to any consumer product. From the engineers' perspective, clinical customers should assume these features, just like you assume a telephone will have a dial tone. However, our workshop participants decidedly ranked these categories at the top of their list of needs. This further illustrates the need for greater customer input during the development of CIS products. With greater communication and involvement, perhaps the medical community would come to view these 'Tier 1' features as guaranteed. The customers of our survey, however, apparently do not believe that such a guarantee currently exists.

Moving to the next tier of results, features that are more CIS-specific are found. Clinical outcomes, real-time interactive imaging, and monitoring address specific needs which surgeons deem to be of high priority in any CIS application. It is here that research initiatives can begin to find motivation from the medical arena. Knowing that medical professionals deem these features to be of high importance in CIS systems, future research initiatives can focus more directly on these areas.

We were surprised to find "hot topics" such as minimal invasiveness, remote operation, and telemedicine at the bottom of the priorities list. Surgeons explained that the definition of minimally invasive differs from procedure to procedure, and each procedure must strike a balance between being minimally invasive and giving the surgeon adequate control in the operating field. With respect to telemedicine, workshop participants were simply not focused on this as a need.

3.2 Method: Kano Analysis

Kano Analysis was used in the workshop to better understand the unspoken requirements from users. This method helps discriminate between what users say (e.g., Voice of the Customer) and what they think. [6, 9, 11, 12] This method focuses on the end-users' expectations of a product and its features. Participants are asked a series of positive and negative question pairs, 25 sets in all. Each positive question is stated in the form, "How do you feel if our product has X?" while the corresponding negative question asks, "How do you feel if our product does not have X?" Participants choose from four possible responses: (a)I like it, (b)It is normally that way, (c)I don't care, or (d)I don't like it. In our case, workshop participants answered positive/negative question pairs such as: "How do you feel if our

Responses to the Negative Question

Kano Analysis	A	B	C	D
	Like	Normal	Don't Care	Don't Like
A Like		Delightful (A,B)	Delightful (A,C)	Normal (A,D)
B Normal				Expected (B,D)
C Don't Care				Expected (C,D)
D Don't Like				

(Left axis label: Responses to the Positive Question)

Fig. 5. The combination of responses to the positive and negative Kano questions determine which category a given feature falls into: *Expected, Normal, and Delightful.*

product includes a robot?" versus "How do you feel if our product does not include a robot?" and "How do you feel if our product provides a single monitor with 2D fluoroscopy?" versus "How do you feel if our product does not provides a single monitor with 2D fluoroscopy?".

Analysis of the results considers the voters' response to both the positive and negative phrasing of the same question and categorizes the combined responses as *Delightful, Normal,* or *Expected.* [9, 10] Relating these responses to the attribute characteristic explained previously, a *Basic* characteristic can elicit an *Expected* response, a *Performance* characteristic can elicit a *Normal* response, and an *Excitement* characteristic can elicit a *Delightful* response. The correlation grid for Kano Analysis is shown in Figure 5. The grid defines response pairs associated with each category. For example, a "Like" response to the positive question combined with a "Normal" response to the corresponding negative question results in an (A,B) voting pair and is categorized as a "Delightful" response. Notice that many response combinations do not reflect any conclusive result and therefore cannot be categorized.

3.2.1 Results: Kano Analysis

Expected	Single Monitor with 2D Fluoroscopy
Normal	Automatically Move Scalpel/Needle Force Feedback Control Monitor the Therapy in Progress Operates Without Imaging Delays Submillimeter Resolution
Delightful	Instrument-tip Imaging Direct link to Imaging Scanner Voice Control Tremor-free Motion Control Intraoperative Ultrasound Imaging Uses 3D Imaging Superimposes 3D Anatomy Teleoperation Capability

Fig. 6. A sample of the results obtained from the Kano analysis process reveals a lack of established, i.e. expected, features in CIS applications.

A sample of the Kano Analysis results from the Needs and Priorities Workshop can be seen in Figure 6. A subset (14 of 25) of the features addressed is included. Column headings reflect which of the characterizations the feature received in the analysis process.

Perhaps most interesting is the fact that only one feature was decidedly categorized as "expected" in CIS technology. The surgeons and nurses who participated in the workshop assume that modern CIS products do include or use a "single monitor with 2D fluoroscopy". Several features were categorized as "normal", including force-feedback control and submillimeter resolution. Of the 25 features included in the analysis, an overwhelming majority of seventeen features were decidedly categorized as "delightful". These features, including voice control and 3D imaging, are not viewed as common parts of current CIS technology. Through these results and the discussion that followed, workshop participants confirmed their openness to CIS technology. However, participants also expressed concern that CIS has not defined and proven itself by establishing a large number of features that are viewed as "expected" in CIS products.

4 Discussion

We consider the results we obtained from the QFD and Kano techniques to be provocative while acknowledging that no single study can be conclusive. In our case, the results provide useful insight into the mind of the customer. However, a greater depth of analysis would be useful for future workshops. Several improvements could lead to more comprehensive results. One way in which improvements could be made is gathering more information about workshop participants. Specifically, it would be useful to survey participants on their current level of technology use and overall receptiveness of technology in the workplace. Results could then be interpreted with an understanding of the participants' feelings about technology. Additional improvements might be achieved by subgroup analysis. Grouping results into subgroups such as doctors, nurses, and technicians would likely reveal interesting and useful results that do not arise when the groups are analyzed as one. The difficulty in this case lies in gathering such a varied and well-balanced group of individuals from the medical profession for participation in a day-long workshop. When applying these analysis techniques, careful planning will afford the desired level of detail.

The results of both analysis techniques lead to candid discussion between the medical professionals who participated in the process and the engineers who design CIS systems. One key point that arose was a language difference between medical and engineering professionals. Key terms used during the workshop and, more importantly, in professional collaborations between the groups, carried different meanings for the two communities. For example, the engineering and medical communities represented at the workshop carried different understandings for the meaning of the term "modularity". Both groups shared the basic definition of modularity as "plug-and-play", i.e. pieces of a modular device are interchangeable. However, medical professionals extended this meaning to encompass fault-tolerance and fault-recovery capabilities. The individual units of a modular device, claimed the surgeons, should continue to operate in a safe and consistent way even if other units in the same device fail. In addition, surgeons expect modular devices to include on-the-fly recovery mechanisms. For future analysis sessions and collaborations, a common and consistent vocabulary of CIS-related terms is essential for clear communication between surgeons and engineers.

Another point of interest was the surgeons' perception of the CIS field. Medical professionals and CIS engineers, it seems, have different approaches to the design of clinical applications. The surgeons tended to focus on *existing* surgical techniques and look for ways to improve them in order to produce better clinical outcomes reduced less cost and labor. The surgeons perceived engineers as frequently focusing on *non-existing* applications, overlooking less complex, intermediate solutions that could be very effective. For example, one surgeon presented the seemingly simple task of moving a patient during surgery. A given patient must be moved multiple times during surgery and, each time, a team of medical personnel must lift the person by hand. Something as simple as a way to move patients around in the operating room, it was suggested, would be very beneficial.

5 Conclusions

A common theme in all discussions was that CIS must strike a balance between being visionary and meeting the current needs of its customers. In seeking this balance, both the QFD Voting and Kano Analysis methods proved useful for the goal of defining research initiatives and direction. By utilizing these processes, it was guaranteed that both the engineering and the medical community's needs would be fully addressed. The QFD Voting process forced the participants to work together to define and name categories of importance from a medical point-of-view, ensuring that the focus of the analysis was correctly targeted to the end-users. The Kano questions, on the other hand, were pre-determined by CIS engineers and therefore approached the issue from an engineering point-of-view.

Within the CISST ERC, the results of the Needs and Priorities Workshop have provided insight into how best to allocate its research resources to meet its customers' needs. The results also reveal the greatest challenge for CISST ERC research: current CIS technology has yet to be fully embraced by the medical community. It is the goal of those involved that the workshop proceedings permanently open doors of communication between surgeons and engineers that will prove invaluable in future collaborations and help establish CIS as a trusted partner in medical solutions.

Acknowledgements

The surgeons and nurses from the Johns Hopkins Medical Institutions who participated in the Needs and Priorities Workshop are listed in Figure 7. We appreciate their willingness to provide their opinions and expectations that serve as the basis for the needs assessment reported in this paper.

Workshop Participant	Specialty
Randy Brown, DVM, MSc	Comparative Medicine
Gaylord Clark, M.D.	Orthopedics, Hand
Theodore DeWeese, M.D.	Radiation Oncology, Urology
Frank Frassica, M.D.	Orthopedics, Bone Cancer
Tushar Goradia, M.D., Ph.D.	Neurosurgery
Brian Kuszyk, M.D.	Radiology
Byron Ladd, M.D.	Ophthalmology
Nadine Levick, M.D.	Pediatrics
Lee Riley III, M.D.	Orthopedics, Spine
Daniel Rothbaum, M.D.	Orthopedics
JoAnne Walz, D.S.N.	Neurosurgery
Keith Wiley, A.A., B.A, B.A.	Neurosurgery
David Yousem, M.D.	Neuroradiology

Fig. 7. The surgeons and nurses who participated come from a variety of specializations and represent a wide scope of customer needs.

References

[1] R. H. Taylor, G. B. Bekey, and J. Funda, "Proceedings of the NSF Workshop on Computer Assisted Surgery,", Washington, D.C., 1993.
[2] A. DiGioia and et al., presented at NSF Workshop on Medical Robotics and Computer-Assisted Medical Interventions (RCAMI), Shadyside Hospital, Pittsburgh, Pa.: Bristol, England, 1996.

[3] D. Winfield, "Final Report of the Working Group on Image Guided Therapy," USPHS Office of Women's Health and National Cancer Institute, Research Triangle Institute, Research Triangle Park April 12-14, 1999 1999.

[4] K. Cleary, "Workshop Report: Technical Requirements for Image-Guided Spine Procedures," Georgetown University Medical Center, Washington, D.C. 113, April 17-20, 1999 1999.

[5] J. B. Revelle, J. W. Moran, and C. Cox, *The QFD Handbook*: John Wiley & Sons, 1998.

[6] N. Kano, "A Perspective on Quality Activities in American Firms," *California Management Review*, pp. 12-31, 1993.

[7] Y. Akao and S. Mizuno, *QFD: The Customer-Driven Approach to Quality Planning and Deployment*. Asian Productivity Organization: Productivity Inc., 1994.

[8] Y. Akao, *Quality Function Deployment: Integrating Customer Requirements into Product Design*: Productivity Press, 1993.

[9] G. A. Churchill and C. Suprenant, "An investigation into the determinants of customer satisfaction," *Journal of Marketing Research*, vol. 19, pp. 491-504, 1982.

[10] H. K. Hunt, "Conceptualization and measurement of consumer satisfaction and dissatisfaction,". Cambridge, MA: Marketing Science Institute, 1977.

[11] B. King, "Techniques for understanding the consumer," *Quality Management in Health Care*, vol. 2, pp. 61-67, 1994.

[12] R. Jacobs, "Evaluating Satisfaction with Media Products and Services: An Attribute Based Approach," *European Media Management Review*, 1999.

[13] E. Chaplin, M. Bailey, R. Crosby, D. Gorman, X. Holland, C. Hippe, T. Hoff, D. Nawrocki, S. Pichette, and N. Thota, "Using quality function deployment to capture the voice of the customer and translate it into the voice of the provider," *Joint Commission Journal on Quality Improvement*, vol. 25, pp. 300-315, 1999.

[14] E. M. Einspruch, V. K. Omachonu, and N. G. Einspruch, "Quality function deployment: application to rehabilitation services," *International Journal of Health Care Quality Assurance*, vol. 9, pp. 42-47, 1996.

[15] L. Cohen and L. Cohen, *Quality Function Deployment: How to Make QFD Work for You (Engineering Process Improvement)*: Addison-Wesley Publishing Co., 1995.

Robotic Assist for MR-Guided Surgery Using Leverage and Parallelepiped Mechanism

Yoshihiko Koseki[1], Kiyoyuki Chinzei[1],
Noriho Koyachi[1], and Tatsuo Arai[2]

[1] Mechanical Engineering Laboratory, AIST, MITI,
1-2 Namiki, Tsukuba, Ibaraki 305-8564, Japan
{koseki, chin, koyachi}@mel.go.jp
http://www.mel.go.jp/
[2] Graduate School of Engineering Science, Osaka University,
1-3 Machikaneyama-cho, Toyonaka, Osaka 560-8531, Japan
arai@sys.es.osaka-u.ac.jp
http://www-arailab.sys.es.osaka-u.ac.jp/

Abstract. In this paper, we would propose a novel mechanism of surgical manipulator, which assists the surgeon in precise positioning and handling of surgical devices, like biopsy needle, endscope, in MR-guided surgery. This mechanism can transmit 3 translational and 3 rotational motion from the outside to the inside of MR imaging area using leverage and parallelepiped mechanism. Such a remote actuation is significantly helpful for robotic assist under MR-guided surgery because the strong magnet of MR denies the existence of magnetic and electric devices around imaging area. This mechanism also has merits of the mechanical safety and simple shape. The combination of stereotactic imaging and precise positioning would enable a less invasive surgery in brain and spine surgery.

1 Introduction

1.1 Robotic Assist for Image-Guided Surgery

The image-guided surgery is a technique to reach a surgical instrument to a target through a narrow access under observation of X-ray CT and to minimize the damage to normal tissue. The trajectory to the target is strategically decided and the treatments must be executed precisely according to the pre-operative planning. So the precise positioning is one of the keys of this operation.

A stereotactic neurosurgery is the most typical example because it has urgent requirement to avoid surgical trauma. Many researchers have proposed robotic assist [1], [2]. Although it is much difficult for a surgeon to aim manually at concealed target inside skull, a robot potentially provides precise positioning because of its numerical control referring to the coordinate frame of X-ray CT image. Not only less invasion but also robotic surgery is expected to be effective on avoiding surgeon's infection and reducing the number of surgeons.

On the other hand, the advantage of MRI in comprehending the pathobiology of soft tissue leads to an image-guided surgery under MRI, especially in brain and spine surgery [3]. However, many conventional devices and instruments cannot be used because they affect the magnetic field and causes serious noise and deformation of images. Furthermore, the magnet surrounds the patient so closely that the accessibility of surgeon is also restricted. In these days, surgeons are capable of accessing to patient due to development of interventional MRI, so-called open MRI.

Some researchers have studied robotic assist under MRI for the same motivation as that under X-ray CT [4]. This MR manipulator has much more technical difficulties because the conventional mechatronical components are not MR compatible. So MR-guided robotic surgery needs comprehensive research of image-navigation, MR compatible mechatronics, and surgical manipulator.

1.2 MR Compatible Manipulator Project

Our group started open MR compatible manipulator project at 1996 and targets to develop a surgical manipulator, which works around open MRI of GE's Signa SP^{TM}, so-called double doughnut type. We have been studying the following three topics.

The first is navigation and registration of coordinate by MRI. Hata reports the navigation method under MR [5].

The second is, the key study of this project, MR compatibility of material, shape, arrangement, and electrical design of mechatronical components. Chinzei reported the design criteria [6], defining MR compatibility and discussing mechatronical components systematically. And we temporally conclude that remote actuation results better MR compatibility.

This paper would study the third topic, mechanism design of the MR compatible manipulator. This topic has common problems to all surgical manipulators, like mechanical safety, interface safety, sterilization compatibility. Many researchers have proposed unique mechanisms for surgical manipulators [4], [7]. The MR compatible manipulator must satisfy both MR compatibility and these requirements.

To solve these problems, we would propose a novel mechanism of surgical manipulator introduced in following sections.

2 Leverage and Parallelepiped Mechanism

2.1 Mechanism

Our novel mechanism, LPM (Leverage and Parallelepiped mechanism) can transmit full set of spatial motion, 3 translational and 3 rotational motion. Its input is attached to a conventional manipulator and its output is allocated into the workspace around MR gantry.

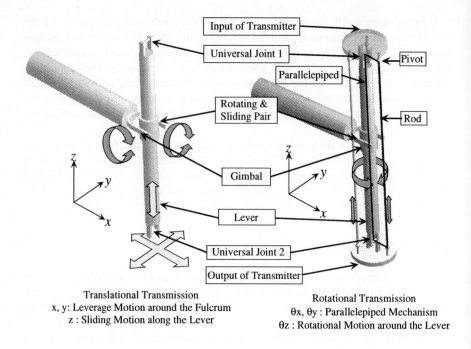

Translational Transmission
x, y: Leverage Motion around the Fulcrum
z : Sliding Motion along the Lever

Rotational Transmission
θx, θy : Parallelepiped Mechanism
θz : Rotational Motion around the Lever

Fig. 1. Leverage and Parallelepiped Mechanism

Fig. 1 schematically explains the structure and functions of LPM. The lever is supported by a gimbal via rotating and sliding pair. The gimbal is fixed on the base and works as a fulcrum of the lever.

As shown in the left of Fig. 1, the swinging motion around the fulcrum transmits the translational motion in x- and y-axes. The sliding motion along the lever allowed by the rotating and sliding pair transmits the translational motion in z-axis.

As shown in the right of Fig. 1, the rotating and sliding pair allows the rotational motion around z-axis. The input and output plates are connected with the lever via universal joint 1 and 2, respectively. The universal joints transmit the rotational motion around z-axis but never constrain the rotational motion around x- and y-axes.

Some rods joint the input and output plates via pivots. The rods are as long as the lever and are connected to the plates at relatively same positions. So the pairs of rod and lever form parallelepipeds. These parallelepiped mechanisms always keep the input and output plates in parallel and transmit rotational motion around x- and y-axes. If the rods can endure both expansion and contraction, 2 rods are necessary for 2 rotational motion and sufficient unless the lever and rods are in same plane. If the rods can endure only expansion, 3 rods are necessary and sufficient unless any 3 of the rods and the lever are in same plane. The latter case is more practical because a long rod cannot endure high pressure. If wires are used for the rods, the pivots are not necessary. Actually, our prototype has 4 wires in order to avoid confliction. (See Fig. 2(b).)

In Fig. 1, the rods are outside of the lever. In this case, the rotation range around z-axis is limited by the conflict between rods and gimbal. Therefore, our prototype has rods inside the lever as shown in Fig. 2(b).

2.2 Features

Remote Actuation Any electromagnetic actuator and a magnetic metal must not enter the neighborhood MR gantry. As for a fluid motor, actuation is free from magnet and electricity but its control system consists of magnetic valves. Even An ultrasonic motor, which doesn't include any magnetic body, makes noise to MR image, while working if it's close to MR imaging area. So the remote actuation is currently the best solution for MR manipulator, and LPM is one of the remote actuation mechanisms.

Mechanical Safety A serial link manipulator, which is widely used in factories, has potential danger of conflict between elbow joint and surgeon. Because an articular manipulator causes the motion of elbow, which is difficult for surgeon to forecast. LPM doesn't have such danger, because it's always straight. The wider workspace expands the potentiality of danger because everything in the workspace might be damaged when the manipulator runs out of control. As for LPM, the output workspace can be restrained by the input workspace. For example, if a barrier is set to the left in input workspace, the end-effector in output workspace cannot move to the right. Accordingly, LPM has potentiality of workspace control by double workspace, but this workspace control needs further study.

Simple Shape Even in robotic assist surgery, the operation is performed mainly by surgeon and robot's motion must be supervised by surgeon. Therefore, there should be left a space for surgeon(s). Because LPM has thin and straight shape and actuators and controllers are distant from the workspace, it doesn't occupy large space around the patient.

Demerits Because LPM is attached to a manipulator in serial, the backlash and deformation under load of LPM are added to that of the manipulator. So LPM potentially decreases the stiffness. As for the translational deformation and backlash except along the lever, the deformation and backlash originated in the manipulator affect to the output in proportion to the ratio of the output lever length to input lever length (See. Eq. (1), (11)). So, elasticity of LPM should be analyzed at the time of design.

The inclination of the output is limited by the limit angle of pivot, universal joints, and by conflict between the lever and rods. The translational displacement along the lever is restricted by the sliding range of sliding and rotating pair. The translational workspace changes according to the ratio of the output lever length to input lever length. So, allowable input must be analyzed at the time of design.

2.3 Current Prototype

We made a prototype of LPM manipulator for so-called double doughnut type of open MRI (See Fig. 2(a)). The purpose of this prototype is to make clear mechanical problems. So this prototype is not MR compatible but all components are designed to be exchangeable with MR compatible components. The reasons are that MR compatibility of mechatronics is studied in another paper (second topic), and that MR compatible components are very expensive. For example, titanium has good MR compatibility and biocompatibility but costs very much. For the same reason, the back drivability of manipulator is not implemented to the current prototype.

The manipulator, which drives the input of LPM, is decided to fixed-linear parallel mechanism (FLP) [8]. The FLP is one of parallel mechanisms, has 6 linear actuators fixed on the base, and has 6 middle-links, which connect between the endplate and actuators in parallel (See Fig. 2(a)). FLP has good property of remote actuation because its actuators always stay in the same positions distant from MR gantry. FLP has demerit that it has narrow workspace especially in orientation but parallelepiped mechanism also has narrow workspace in orientation. So this demerit is not bottleneck of workspace. Those are why FLP is used.

Kinematic parameters are decided by conventional Monte-Carlo method to meet the workspace of ±100[mm] in x-, y-, z-axes, and ±30[deg.] around x-, y-, z-axes.

3 Kinematics and Statics

In this section, some important kinematic and static equations of LPM are obtained. All notations are illustrated and noted in Fig. 3. The origins and power points of input and output are assumed to be located at the crossing points of universal joints, respectively.

Firstly, the forward kinematics is shown in the followings. The position and orientation of LPM's output are obtained. The inverse kinematics is omitted here.

$$\boldsymbol{p}_{output} = -l_{lever}\,\boldsymbol{q} + \boldsymbol{p}_{input} \tag{1}$$

$$\boldsymbol{R}_{output} = \boldsymbol{R}_{input} \tag{2}$$

Noting that $\boldsymbol{q} = \dfrac{(\boldsymbol{p}_{input} - \boldsymbol{p}_{fulcrum})}{|\boldsymbol{p}_{input} - \boldsymbol{p}_{fulcrum}|}$

Secondly, the forward statics and the force exerted on the fulcrum are obtained.

$$\boldsymbol{f}_{output} = -\frac{|\boldsymbol{p}_{input} - \boldsymbol{p}_{fulcrum}|}{|\boldsymbol{p}_{output} - \boldsymbol{p}_{fulcrum}|}\,\boldsymbol{f}_{input}$$
$$+\frac{l_{lever}}{|\boldsymbol{p}_{input} - \boldsymbol{p}_{fulcrum}|}\,(\boldsymbol{q} \cdot \boldsymbol{f}_{input})\,\boldsymbol{q} \tag{3}$$

Note that $\boldsymbol{a} \cdot \boldsymbol{b}$ is inner product of $\boldsymbol{a}$ and $\boldsymbol{b}$.

$$\boldsymbol{m}_{output} = \boldsymbol{m}_{input} \tag{4}$$

$$\boldsymbol{f}_{fulcrum} = \boldsymbol{f}_{input} - \boldsymbol{f}_{output} \tag{5}$$

Thirdly, infinitesimal displacements of output are obtained. They are corresponding to velocity and elastic deformation.

$$\delta\boldsymbol{p}_{output} = -\frac{\left|\boldsymbol{p}_{output} - \boldsymbol{p}_{fulcrum}\right|}{\left|\boldsymbol{p}_{input} - \boldsymbol{p}_{fulcrum}\right|}\delta\boldsymbol{p}_{input}$$
$$+\frac{l_{lever}}{\left|\boldsymbol{p}_{input} - \boldsymbol{p}_{fulcrum}\right|}\left(\boldsymbol{q} \cdot \delta\boldsymbol{p}_{input}\right)\boldsymbol{q} \tag{6}$$

$$\delta\boldsymbol{R}_{input} = \delta\boldsymbol{R}_{output} \tag{7}$$

Here, it should be noticed that the elasticity of LPM is proportional to the square of the ratio of the output lever length to input lever length. If the output lever is 2 times as long as input lever, the stiffness of LPM is 1/4 of that of input manipulator.

$$\boldsymbol{f}_{input} = -\frac{\left|\boldsymbol{p}_{output} - \boldsymbol{p}_{fulcrum}\right|}{\left|\boldsymbol{p}_{input} - \boldsymbol{p}_{fulcrum}\right|}\boldsymbol{f}_{output} \tag{8}$$
The first term of Eq. (3)

$$\delta\boldsymbol{p}_{input} = \boldsymbol{K}\,\boldsymbol{f}_{input} \tag{9}$$

$\boldsymbol{K}$ is Flexibility Matrix of the input manipulator

$$\delta\boldsymbol{p}_{output} = -\frac{\left|\boldsymbol{p}_{output} - \boldsymbol{p}_{fulcrum}\right|}{\left|\boldsymbol{p}_{input} - \boldsymbol{p}_{fulcrum}\right|}\delta\boldsymbol{p}_{input} \tag{10}$$
The first term of Eq. (6)

$$= \left(\frac{\left|\boldsymbol{p}_{output} - \boldsymbol{p}_{fulcrum}\right|}{\left|\boldsymbol{p}_{input} - \boldsymbol{p}_{fulcrum}\right|}\right)^2 \boldsymbol{K}\,\boldsymbol{f}_{output} \tag{11}$$

The following equations describe the connection between the input manipulator and input of LPM.

$$\boldsymbol{p}_{input} = \boldsymbol{p}_{endplate} + \boldsymbol{R}_{endplate}\,\boldsymbol{p}_{etoi} \tag{12}$$

$$\boldsymbol{R}_{input} = \boldsymbol{R}_{etoi}\,\boldsymbol{R}_{endplate} \tag{13}$$

$$\boldsymbol{f}_{input} = \boldsymbol{f}_{endplate} \tag{14}$$

$$\boldsymbol{m}_{input} = \boldsymbol{m}_{endplate} + \boldsymbol{p}_{etoi} \times \boldsymbol{f}_{endplate} \tag{15}$$

Note that $\boldsymbol{a} \times \boldsymbol{b}$ is outer product of $\boldsymbol{a}$ and $\boldsymbol{b}$.

The kinematics of FLP is shortly introduced here but the details are shown in [8]. The inverse kinematics can be obtained by solving the following equations individually for l_{linear_i}. The forward kinematics can be obtained by solving the following equations simultaneously for $p_{endplate}$ and $R_{endplate}$.

$$|(p_{endplate} + R_{endplate}\, p_{etol_i}) - (l_{linear_i}\, v_{linear_i} + p_{btol_i} + p_{base})| = l_{link_i} \quad (16)$$
$$\text{while } i = 1 \cdots 6$$

4 Conclusions and Future Works

In this paper, we proposed our novel and unique mechanism of surgical manipulator, leverage and parallelepiped mechanism. Its possible application is precise positioning and handling of biopsy needle, endscope and other surgical devices under MR-guided surgery. We introduced the structure and functions, discussed the merits and demerits and formulated the kinematics and statics.

The current prototype is shortly introduced but it's still under development. We are planning to test its basic experiment like precision and stiffness, and preclinical evaluations.

References

1. Burckhardt C.W., Flury P., Glauser D.: Stereotactic Brain Surgery, IEEE Engineering in Medicine and Biology Magazine, Vol. 14, No. 3, pp. 314-317, May/June 1995
2. Kwoh Y.S., Hou J., Jonckheere E., Hayati S.: A Robot with Improved Absolute Positioning Accuracy for CT Guided Stereotactic Brain Surgery, IEEE tran. on Biomedical Engineering, Vol. 35, No. 2, pp. 153-160, Feb. 1988
3. Jolesz F.A., Blumenfeld S.G.: Interventional Use of Magnetic Resonance Imaging, Magn Reson Q, 1994, Vol. 10, No. 2, pp. 85-96
4. Masamune K., Kobayashi E., et al.: Development of a MRI Compatible Needle Insertion Manipulator for Stereotactic Neurosurgery, Proc. of MRCAS'95, Baltimore MD, pp. 165-172, Nov. 1995
5. Hata et al.: Multimodality Deformable Registration of Pre- and Intraoperative Images for MRI-Guided Brain Surgery, Proc. of MICCAI'98, pp. 1067-1074, 1998
6. Chinzei K., Kikinis R., Jolesz F.: MR Compatibility of Mechatronic Devices: Design Criteria, Proc. of MICCAI'99, pp. 1020-1031, 1999
7. Taylor R., et al.: A Telerobotic Assistant for Laparoscopic Surgery, IEEE Engineering in Medicine and Biology Magazine, Vol. 14, No. 3, pp. 279-288, May/June 1995
8. Arai T., Tanikawa T., Merlet J.P., Sendai T.: Development of a New Parallel Manipulator with Fixed Linear Actuator, Proc. of Japan/USA Symposium on Flexible Automation, pp. 145-149, 1995

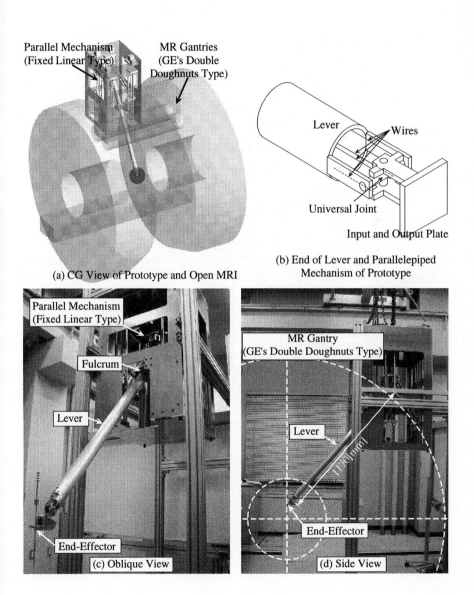

(a) CG View of Prototype and Open MRI

(b) End of Lever and Parallelepiped Mechanism of Prototype

(c) Oblique View

(d) Side View

Fig. 2. Current Prototype

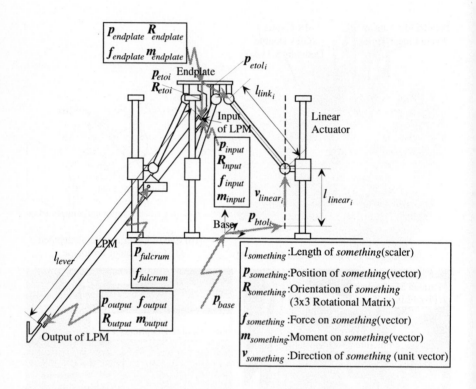

Fig. 3. Notataions in Equations

Endoscopic Robots

C. Kübler, J. Raczkowsky, H. Wörn

Institute for Process Control and Robotics, Universität Karlsruhe (TH),
D-76128 Karlsruhe, Germany
{kuebler, rkowsky, woern }@ira.uka.de

Abstract. Endoscopy is an important procedure for the diagnostic and therapy of various pathologies. We develop extensive and automated systems for this field. Due to application of these new systems, a patient is subject to considerably less strains, as opposed to prevailing commercial systems. The capability of such instruments, unlike the presently used systems, to independently follow anatomical peculiarities of the body means also a reduced risk of complications for a patient. A further advantage is that difficult to access regions deep inside the body, like the small intestine or peripheral parts of the bronchial tubes, can thus be reached.
The control is made automatic with a 6D-localisation system, a CCD camera, contact sensors, curvature sensors, and pressure sensors. This makes it possible for the doctor to control the endoscope via a human-machine interface without using his own force, and to avoid many problems of today, e.g., looping in the large intestine. The endoscope is driven in accordance with a new driving concept on the basis of fluid actuators or a magnetic probe.
By fusion of all the sensor data in a relational data model, a model of the body inside by means of supplementary image processing. Amongst other things, it make possible to do a quick postoperative checkup. The relational model has been developed specially for endoscopy. In addition, the doctor can intraoperatively see with this model what parts inside the body have already been examined. In this way, a complete examination of, e.g., bronchi can easily be proved.

1 Introduction

A new application for robotic systems is to automate the endoscopy. In our laboratory we develop a new endoscopic system which bases on a motion system for the endoscopic tip and the supply-tube of the tip. Classical the endoscope is being pushed by the physician. The insertion of endoscopes are often combined with sharp pain for the patient and danger to perforate the wall of the tube. Our systems will be used for the colon, rectum, bronchial tree, fallopian tube, etc. To build a new automated endoscopic system we use our years of experience and competence in the field of robot supported surgery and microrobotics [1] [2].

2 A new motion system

Our new endoscopic system is designed to work with fluidic elements and strong electromagnetic fields [3] or only with fluidic elements for the motion actuators. The endoscopic tip needs an extern power supply because an endoscopic capsule, which is not connected directly with the external world, cannot be used for taking up pictures. This is for the simple reason that the power requirement for complex sensors and the light source nowadays and in the near future is still too high. Not sufficient energy can be stored e.g. in-coupled (i.e. with electromagnetic induction) in the capsule. For therapy there is the need to remove several biopsies out of the body. Another approach is the improvement of classical endoscopes with steerable distal ends [4] [5]. The incoupled force for the movement of the distal end had to be redirected to the tip by the tube's wall and initiate strong pain.

The endoscope is divided into two parts: the tip and the supply-tube. The architecture of the endoscopic tip is presented in Fig. 1. The tip includes a CCD-camera, fibre optics for the light-source, a working channel and an air-/ water channel. Additionally, the tip includes a position sensor for 6D-localisation and several touch-sensors for the navigation system. The 6D-localisation system is used to get the absolute position and orientation of the tip in the tube. The function of the 6D-localisation system will be presented below.

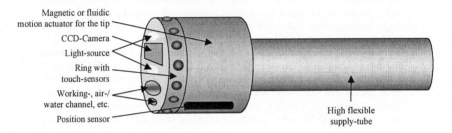

Fig. 1. Tip of the new endoscopic system

To pull a standard colonoscope out of the colon a physician applies a force between 3 - 5 N. A new motion system has to produce about the same force to pull the high flexible supply-tube in the colon. Our aim is to examine not only the colon [6] [7] but also the small intestine and other tubes. The diagnostic of the small intestine is a new challenge for the endoscopy. It is not possible to examine the complete small intestine with standard endoscopes. With our flexible approach it is possible to examine the whole small intestine. The length of the required supply-tube for the endoscopic system is between 6 - 8 m. For the intestinoscopy the required force to pull a supply-tube in the small intestine driven by the tip is higher than for colonoscopy.

For the reason that the small intestine is flexible, long and winded, the motion system is not allowed to generate the whole force to pull the 8 m long supply-tube in the small intestine. The required force has to be reduced to prevent closed loops in the intestine. Tip driven endoscopic systems require an automated intelligent supply-tube to prevent loops during pulling the endoscope in or out of the tube. Our approach bases on a distributed automated intelligent motion system with several motion actuators (see Fig. 2).

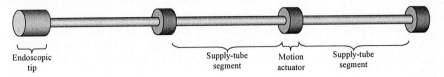

Endoscopic
tip

Supply-tube
segment

Motion
actuator

Supply-tube
segment

Fig. 2. Automated intelligent supply-tube

A design for a motion actuator bases on fluidic inch-worm locomotion principle (Fig. 3a). Every motion actuator includes min. three inner fluidic elements and a front end- and back end fluidic element. Every supply-tube's motion actuator has its own controller and the motion mechanism bases on fluidic elements for all our systems. The magnetic mechanism of a endoscopic tip cannot be used for these motion actuators because only one strong permanent magnet can be controlled individually. This magnet is placed in the tip. The endoscopic tip is controlled separately. With an ergonomic human-machine-interface (HMI) the tip and intelligent supply-tube are controlled individually.

With both force sensors connected between motion actuator and supply-tube segments the controller regulates the flow of the fluid and so the pressure in the inner fluidic elements. The reaction of a motion action is presented in Fig. 3b.

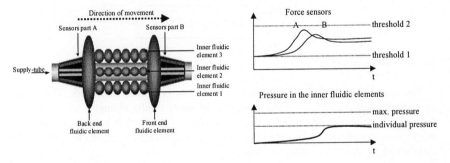

Fig. 3. a) A motion actuator and b) its force reaction

The resulting controller for every motion actuator is presented in Fig. 4. The size of the controller and the supply-tube is limited. The required fluidic pressure for all motion actuators is too high and the cycle time is too short. A triggering of every fluidic element with an own microvalve is not possible, because the fluidic elements cannot be regulated independent from outside because of the limited size of the supply tube. The solution for this problem is to trigger synchronously the corresponding fluidic elements in every motion actuator. The cycle is generated outside of the endoscope. About three fluidic pipes supply all supply-tube's motion actuators. The motion actuator's controller regulates independently the fluidic pressure resp. flow individually to control the force of every motion actuator. This mechanism is required when an actuator e.g. sticks to the wall. The following motion actuator has to decrease the force to reduce the speed to avoid a supply-tube's loop. The previous motion actuator has also to decrease the force to reduce tension between these two elements. The cycle time and the variation of the pressure for the individual control is about ten time slower than the cycle time of the fluidic elements.

Another approach is the use of shape memory alloy (SMA) for microvalves, which is used for other endoscopic robots [6]. The SMA springs open and close the shuttle of the valves. The cycle time of the valve is limited by the slow speed of the SMA spring.

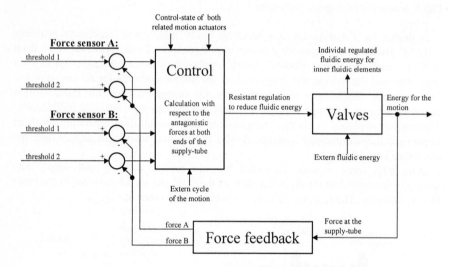

Fig. 4. Motion control

The main stages of the inch-worm locomotion is presented Fig. 5. The actual movement control of Fig. 4 inflates regular all inner fluidic elements. A bending of the motion actuator is generated with an extension of the motion actuator's controller. With miniaturised valves an individual bending is generated (Fig. 5b) to increase the stability and speed of the whole motion system.

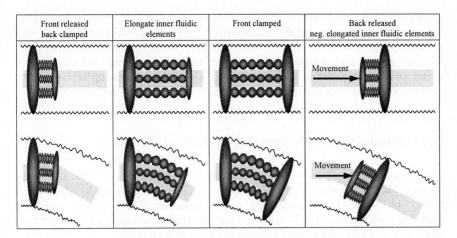

Front released back clamped	Elongate inner fluidic elements	Front clamped	Back released neg. elongated inner fluidic elements
			Movement
			Movement

Fig. 5. Locomotion stages a) without bending extension b) with bending extension

Our approach to design new microvalves, which individually throttle down the pressure in the element (Fig. 6) is to miniaturise this valves to a size, that all fluidic elements of the motion actuator have their own microvalve.

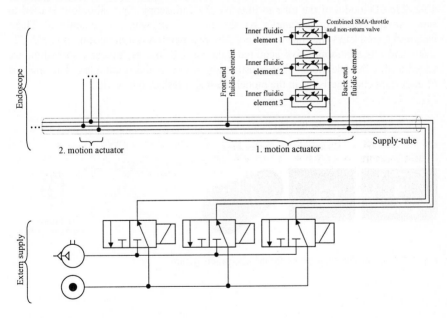

Fig. 6. Control of the motion actuators

We additionally plan to develop new mechatronic elements for the tip of the endoscope to assist the surgeon during diagnostic or therapy. One of these elements is a controllable telescope camera to support the diagnostic of small tubes. This is our first approach to examine i.e. the fallopoian tube or the peripheral bronchial tree. One or two tools have to be added to the endoscopic tip for cutting and catching biopsies. The new endoscope is a modern microrobotic system for medicine. Mainly, we want to support the diagnostic and therapy of tubes with an inner diameter between 2 and 16 mm. An extremely small inner diameter has the fallopian tube with about 1 mm.

3 An introduction in the complex navigation system

The tip of the endoscope contains a position sensor (see Fig. 1). We use a less expensive 6D-localisation system with failure detection, which is developed from our project partners. A comparable, but more expensive 6D-localisation system is produced from Biosense [8]. We used for our first tests a 6D-localisation system from Polhemus. The precision of the Polhemus system is to inaccurate.

The 6D-localisation system is necessary for the navigation system. The navigation system merges the position and orientation with the taken images (see Fig. 7) to simplify matching of successing images. The 6D-localisation system is a very important sensor for the navigationsystem. A simple rigid matching of endoscopic images and a preoperative acquired 3D-scan was presented in [9]. We try to generate a flexible 3D-model of the tube to improve the endoscopy. Special colour is used to spray small synthetic landmarks on the tube's wall which is examined. These landmarks simplifies the matching, too. The preoperative acquired data e.g. 3D-CT-scans and other intraoperative acquired data e.g. US-images, sensor information are added to this flexible model. The construction of this model is complex. The analysis of important characteristics in the model is a major task of our work.

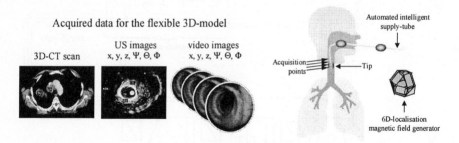

Fig. 7. Part of the data acquisition for the navigation system

4 Conclusion

A miniaturisation within the area of the endoscopy is desirable. Miniaturisation of automatic handling modules for integration into an endoscopic system is a strong demand for future development. An intelligent navigation system extends the possibilities for the physician. New endoscopic systems reduces the strains on a patient.

Acknowledgement

The research is being funded by the Ministry for Science, Research and Arts of Baden-Württemberg.

References

[1] P. Bohner, "Ein Multiagentenansatz zur Steuerung von redundanten Manipulatoren für medizinische Applikationen", VDI Verlag, Reihe 8, Nr. 675, Universität Karlsruhe (TH), Diss., 1997.

[2] H. Grabowski, "Modellbasierte Deformation medizinischer Bilder", GCA-Verlag, Universität Karlsruhe (TH), Diss., 1999.

[3] D.C. Meeker, E.H. Maslen, R.C. Ritter, F.M. Creighton, "Optimal Realization of Arbitrary Forces in a Magnetic Stereotaxis System", IEEE Transactions on Magnetics, Vol. 32, No. 2, 1996.

[4] K. Ikuta, M. Tsukamoto, S. Hirose, "Shape Memory Alloy Servo Actuator System with Electric Resistance Feedback and Application for Active Endoscope", Proc. of the 1988 IEEE Int. Conf. on Robotics and Automation, 1988.

[5] J. Szewczyk, N. Troisfontaine, Ph. Bidaud, "An Active Tubular Polyarticulated Micro-System for Flexible Endoscopes", Proc. Int. Workshop on Micro Robots Micro Machines and Systems, 1999.

[6] P. Dario, M.C. Carrozza, L. Lencioni, B. Magnani, "A Microrobot for Colonoscopy", Proc. of the Seventh Int. Symp. on Micro Machine and Human Science, 1996.

[7] A.B. Slatkin, J.W. Burdick, S.W. Grundfest, "The Development of a Robotic Endoscope", Proc. of the IEEE/RSJ/GI Int. Conf. on Intelligent Robots and Systems IROS'95, 1995.

[8] Biosense Webster, "EP Navigation", http://www.biosensewebster.com/.

[9] S.B.- Solomon, D.E. Acker, A.J. Polito, C. Wiener, A.C. Venbrux, P. White Jr., "Real-Time Bronchoscope tip Position Technology Displayed on Previously Acquired CT Images to Guide Transbronchial Needle Aspiration (TBNA)", The Cardiopulmonary and Critical Care Journal Vol. 112, No.3S, 1997.

An Augmentation System for Fine Manipulation

Rajesh Kumar[1,3], Gregory D. Hager[1,3], Aaron Barnes[2,3]
Patrick Jensen[2,3], Russell H. Taylor[1,3]

[1] Department of Computer Science
{rajesh, hager, rht}@cs.jhu.edu
[2] Wilmer Eye Institute, Johns Hopkins Medical Institutions
{abarnes,psjensen}@jhu.edu
Johns Hopkins University, Baltimore, Maryland, USA
[3] NSF Engineering Research Center
for Computer Integrated Surgical Systems and Technology
Johns Hopkins University, Baltimore, Maryland, USA

Abstract. Augmented surgical manipulation tasks can be viewed as a sequence of smaller, simpler steps driven primarily by the surgeon's input. These steps can be abstracted as controlled interaction of the tool/end-effector with the environment. The basic research problem here is performing a sequence of control primitives. In computing terms, each of the primitives is a predefined computational routine (e.g. compliant motion or some other "macro") with initiation and termination predicates. The sequencing of these primitives depends upon user control and effects of the environmental interaction. We explore a sensor driven system to perform simple manipulation tasks. The system is composed of a core set of "safe" system states and task specific states and transitions. Using the "steady hand" robot as the experimental platform we investigate using such a system.

1 Introduction

Dexterous manipulation is a key element in the speed, safety, and, ultimately, the success of most surgical interventions. The majority of surgical tasks involve hand-held tools operated using both vision and force (including both tactile, and kinesthetic) information. While most interventions use both force and vision at some level, the availability and efficacy of both varies widely. In general, during coarse, large-scale manipulation forces from the tools is an important cue, but visual information improves both the speed and facility of manipulation. In contrast fine, small-scale manipulation is often almost completely visual, as the interaction forces between the tool and the environment are imperceptible to even a trained surgeon. As demonstrated in the literature [1-5] [6], the lack of tactile information during surgical procedures probably results in them taking longer and being less accurate than if tactile information were present.

Our "steady hand" manipulation approach [7] requires tools to be held simultaneously by the user and the robot, with the robot complying to the forces applied to the tool. It provides a safe, intuitive means of addressing the above problems by *augmenting* the manipulation capabilities of a surgeon. It is safe, as the surgeon has direct control of the manipulator, and thereby his or her accustomed surgical tools. It is intuitive, as the surgeon not only directly manipulates those tools, but also receives direct force-feedback from the manipulator, thus "feeling" the manipulation much as one would during a large-scale surgical intervention. Our approach (compared to conventional tele-manipulation) is also more appealing because of its cost advantages.

While the increasing need for augmentation at micro scales provides a clear opportunity for human augmentation, it also makes it clear that *different levels of augmentation are necessary at different stages and/or scales of surgical intervention.* Open questions in the steady-hand approach (indeed, in any human-augmentation system) include 1) How might one develop a framework for human augmentation that varies its behavior in response to both the task context (e.g. scale) and the needs of the human within that context? and 2) Given such a system, does it provide "added value" to the human operator? Clearly, these two questions are closely interrelated and can only be answered through a cycle of engineering and empirical testing. In this paper, we present preliminary results based on a prototype system we have developed.

1.1 Previous Work

Some prior work exists in analogous problems such as automation of assembly tasks and vision guided control. Flexible automation of tasks for assembly (e.g.Lozano-Perez[8], Sanderson[9]) has been studied for a long period of time. Taylor [10, 11] also looked at task representations. Some existing work in methods for learning tasks focuses on determining force/position control parameters from a human worker's operation[12, 13]. Analysis of robot systems operating in tandem with humans and extraction of some information from this cooperation is an active topic of research. Kosuge [14-16] has looked at cooperative tasks. Exoskeletons, amplifying user input have been proposed by Kazerooni [17-19]. The vision community also has a body of research (e.g. Dickmanns [20]) in developing similar frameworks for vision-guided processes. Similar work also exists in space and planetary robotics, e.g. Lee [21] proposed a sensor-based architecture for planetary robotic sampling.

More recently, sophisticated systems have been developed to assist, or augment human actions in unstructured environments, especially in medicine. In medicine, they are often used to reduce human involvement in a task (i.e. to act as tool or camera holder) than for their superior manipulation abilities. However, Davies [22] and Troccaz [23] among others have devised systems that incorporate some level of integration of task information for constrained control.

2 A System For Surgical Manipulation

The following are some of the important requirements of an augmentation system:

Safety: includes identification of critical portions of the controlled task, ability to identify and/or correct faults, and redundancy. In medical procedures, the criticality of the task puts safety as the most important design consideration.

Stability: includes meeting performance specifications over time, state/condition and over the range of inputs possible.

Efficacy/Accuracy/Functionality: includes ability to perform useful function identified by the users, and the ability to perform it without significantly modifying existing processes.

Ease of Interaction: includes ability to interact with the user with conventional tools used in the process and without imposing significant restrictions on existing practice or requiring extensive training.

Interaction with the planning process, possibility of learning/teaching, and accounting/process learning are other desirable attributes. It is difficult to design an optimal solution for tasks in such a dynamic environment and the flexibility of the system to allow tuning of it performance is likely to be important. Unlike assembly environments, augmentation must seamlessly integrate with existing processes and environment. There are other significant differences between assembly and surgical environments. Safety is critical and the environment is unstructured and very dynamic.

There are a sufficiently large number of generic manipulation actions in surgical procedures that may be augmented by a single surgical assistant. Examples of such tasks are camera holding, acting as a tool guide, tool positioning, constrained and guarded motions (force constraints are especially hard to implement without augmentation and very common in practice).

2.1 Our Approach

We demonstrate our approach by choosing a simple example task that involves both fine and coarse manipulation, and is a common minimally invasive task. The task of placing a tool through a port (small incision or hole) at a surgical site is composed of the following steps: *a)* positioning the tool at the port, *b)* orienting it such that it can be inserted, *c)* insertion of the tool, *d)* adjusting the orientation of the tool towards the placement site viewing through the visual feedback device (microscope in the eye, video feedback in laparoscopy etc), *e)* approaching the site, and *f)* achieving contact.

This task has both coarse manipulation (positioning and orientation leading to the port) and fine manipulation inside the organ. In our approach, each step of the **task** is called an **action**. Actions are themselves more complex than a motion that can be directly coded. They are represented as chains of high-level subroutines **(primitives)** that are linked together by **predicates**. Primitives are composed of functions implementing sensing and manipulation routines. Predicates serve as conditions for transitions in the state graph composed of **basic system states** (for manipulation and safety), and task specific states. Predicates include both automatic sensing (e.g. contact detection) and explicit user input.(e.g. input from buttons or foot pedals).

Implementations of sensing and manipulation subroutines are the basic building blocks. These subroutines perform well-defined tasks that are robust to errors. Manipulation subroutines such as move to specified position, move relative distance, move with specified velocity, and sensing subroutines such as get current value, filter raw values, get biased value, resolve values with respect to a given Cartesian frame are examples of basic routines used here.

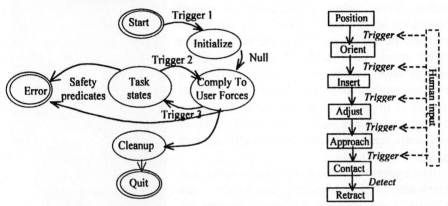

Fig. 1: Left: A Simple System Graph. System States are composed of basic states (initalize, cleanup, manipulate, error) and task specific states. Triggers Between task specific states, and initialization are user interface actions performed by the user. Data Collection is not shown. Right: Task graph for the example task, triggers between states are provided by user by pressing a foot pedal

Primitives are the next level entities. They are composed from sensing and manipulation routines. The primary primitives in "steady hand" manipulation are those that support compliant motion. A "move in compliance to forces" primitive is composed of sensing routines to obtain resolve value of the sensor, a control law (e.g. velocity proportional to force error), and manipulation routines that receive the output of the control law.

Actions are composed from primitives by specifying sensing primitives (predicates) that initiate the execution of the action. These predicates form the conditions that must be met before the primitive is executed. In the above example, the actions are composed of a single "move in compliance to forces" primitive with different parameters. Position uses the translation stages of the robot to provide XYZ positioning where as the rest of the actions (orient, insert, adjust, approach, retract) use two rotational degrees of freedom (about X,Y) and Z insertion degree of freedom about a mechanically constrained remote center of motion(RCM). In the example task, the contact action serves only as placeholder for manipulation actions that might follow in a real task.

The task representation for a given task is obtained by identifying distinct parts in the conventional approach. Given that "steady hand" manipulation imposes a sequential ordering on *planned* task execution (user only performs one action at a time with the manipulator), each of these parts can be implemented as an action. The conditions that need to be met before each action are identified. They form the predicates for that action. If an action is composed of several primitives, the process of identifying primitives is repeated for the action. Finally the user identifies safety requirements, such as limits on motion, sensory values etc for each action. This serves as a skeleton for the task graph. The task graph is then executed in a training enviornment. During execution the user may identify redundant or additional states, predicates that modify the task graph.

The system maintains a basic set of states, and predicates. These include initialization and cleanup, a manipulation set, data collection set, and safety and error checking predicates. The system basic states are sleep, and move in compliance to forces, and error states. The data collection set includes a single Dump state. Safety Predicates include workspace and force limits, and hardware and software errors.

Fig. 2: Experimental Setup: setup with the robot (*top*), a user performing free hand experiment (*middle*), and the ball with access ports and data grid (*bottom*)

3 Preliminary Experiments

We have begun to experimental studies to evaluate the effectiveness of using task-graph enhanced "steady hand" augmentation in comparison to simpler augmentation and un-augmented free hand performance. The task chosen is a constrained needle placement task that presents many of the fine manipulation difficulties encountered in eye surgery. The example task is a modified (or minimally invasive) version of the peg-in-hole task, a common task used for performance evaluation [24],[25],[26].

3.1 Experimental Environment

The experimental environment [7] consists of software and hardware components. The software consists of the machine level robot control software and the framework specific to this work. The JHU Modular Robot Control (MRC) library provides the machine level robot control functionality. The hardware consists a cooperative manipulator and augmented tools required for selected tasks. The cooperative manipulator used is the "steady hand" robot. This is a 7-degree-of-freedom manipulator with XYZ translation at the base for coarse positioning, two rotational degrees of freedom at the shoulder (the RCM linkage, [36]), and instrument insertion and rotation stages. A force sensor is built in the end-effector. This robot has a remote center of motion and an overall positional accuracy of less than 10 micrometers.

This experimental setup appears in Figure 2. To construct the target a ball was sliced and ports constructed to reflect distances similar to the eye. This ball was attached to a data surface containing 100 micrometer holes separated by 2mm. An ergonomic tool handle was mounted with a 1 mm shaft and 50 micrometer tip wire for the tool. The goal of the experiment is to touch the bottom of the hole inside the "eye" without touching the sides. Automatic electrical contact sensing is employed to detect contact between the bottom of the holes (*success*) or the sides of the hole or elsewhere on the surface of the data surface (*error*).

3.2 Task and Experimental Protocol

The selected task can be performed free hand (**Unassisted**) and with the "steady hand" robot. Two sets of augmentation parameters can be used, one using just constant gain compliant motion (**Comply**), and another using a non-linear gain scheme (**Augmented**). With non-linear gains (Figure 3), the gains for each action are modified as a function of distance to the target for that action (since the actions are positioning/orientation primitives, this predicate is easy to evaluate). The target for coarse manipulation actions is the center of port on the ball, and for fine manipulation a selected hole. The port and hole locations are taught to the robot by hand guiding it to both locations.

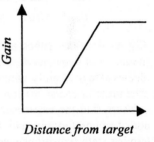

Distance from target

Fig. 3: A typical non-linear gain profile, the target is at zero.

The users are allowed unlimited training time till they are comfortable with the experimental protocol, both free hand and with the robot. The users execute each task 5 times. For both robotic and free hand experiments transitions are explicitly signaled by the user/observer (by pressing a button/pedal). Each user also evaluates the setup subjectively on ease of operation, seating comfort, and ease of viewing the target.

3.3 Results

From the current data, the success rates for this experiment (number of errors per successful try) improve significantly with the robot, and augmentation adds to the improvement. The total time for the task also decreases when the robot is used. Data from three users appears as a graph below. User 1 had the maximum training time and experience with the system, and user 2 and 3 are familiar with the system. Training time clearly affects the performance, and further experimentation is needed for analysis and evaluation. These experiments are scheduled.

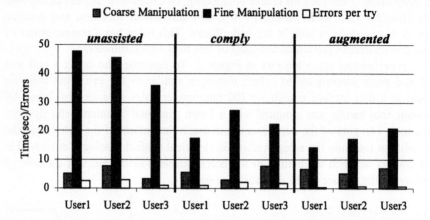

Fig 4: Average Times and errors for the experiment: (*Unassisted left, comply center, and augmented right three sets*). Errors and total time for the experiment decrease significantly with the robot, and further when non-linear gains are used. The time for coarse motion does not change significantly.

Users find the experiment challenging free hand. The seating conditions are reported as comfortable, and visibility of the target good. The use of foot pedal to determine state transitions was considered non-intuitive, but not difficult by the users. This was replaced with an observer pressing the button at the request of the user. The tip wire would bend and touch the side (when touching the bottom) of the hole with very small forces on it. This may have inflated errors. While this is a problem with all flexible instruments, we are working on improved apparatus and experimental protocol to reduce/detect this problem.

4 Discussion and Conclusions

Our system utilizes the sequential nature of the execution of augmented tasks. Note that since this manipulation uses the "steady hand" approach, this is inherent. The example task chosen above is very similar to several other common tasks in surgery, one example being placing a micropipette in a blood vessel to deliver clot-dissolving therapy. However, experimentation is needed with skilled users in conditions even closer to clinical conditions to validate any results. These experiments are also planned.

We do not describe the mechanisms for specification of tasks here. The user describes the task in a high level specification being developed. The detailed discussion is beyond the scope of this paper. Use of a foot pedal was found non-intuitive by the users. However, clinicians routinely use foot pedals and their response to the pedal may be different. Alternative mechanisms for state transitions (pause in motion, automatic detection by comparison of change in state <force,position, velocity>) are available but they are yet to be evaluated.

5 Summary

We report on a sensor driven system for augmentation based on "steady hand" manipulation. Under this approach, surgical procedures are performed by a surgeon in the same way as the convention approach. However, portions of the procedure are augmented/automated at surgeon's initiation. The system is described in detail and an implementation is discussed. A preliminary experiment and some preliminary experimental results for an example task are reported.

Acknowledgements

The authors gratefully acknowledge the support of the National Science Foundation under grant #IIS9801684, and the Engineering Research Center grant #EEC9731478. This work was also supported in part by the Johns Hopkins University internal funds.

References

[1]   D. J. Glencross, "Control of skilled movement," *Psychological Bulletin*, vol. 84, pp. 14-19, 1977.

[2]   M. Patkin, "Ergonomics applied to the practice of microsurgery.," *Aust NZ J Surg*, vol. 47, pp. 320-329, 1977.

[3]   K. R. Boff, L. Kaufman, and J. P. Thomas, *Handbook of perception and human performance*. New York, NY: John Wiley and Sons, 1986.

[4]   K. R. Boff and J. E. Lincoln, *Engineering Data Compendium: Human Perception and Performance*. Ohio: H. G. Anderson Aerospace Medical Research Laboratory., 1988.

[5]   R. D. Howe, W. J. Peine, D. A. Kontarinis, and J. S. Son, "Remote palpation technology.," *IEEE Engineering in Medicine and Biology*, vol. 14, pp. 318-323, 1995.

[6]   D. A. Kontarinis and R. D. Howe, "Tactile Display of Vibratory Information in Teleoperation and Virtual Environments," *Presence*, vol. 4, pp. 387-402, 1995.

[7]   R. H. Taylor, P. Jensen, L. L. Whitcomb, A. Barnes, R. Kumar, D. Stoianovici, P. Gupta, Z. Wang, E. d. Juan, and L. Kavoussi, " A Steady-Hand Robotic System for Microsurgical Augmentation," presented at Medical Image

Computing and Computer-Assisted Intervention - MICCAI'99, Cambridge, UK, 1999.

[8]   T. Lozano-Perez, M. T. Mason, and R. H. Taylor, "Automatic synthesis of fine-motion strategies for robots," *The International Journal of Robotics Research*, vol. 3, pp. 3--23, 1984.

[9]   A. C. Sanderson and G. Perry, "Sensor Based Robotic Assembly Systems: Research And Applications in Electronic Manufacturing," *Proceedings of the IEEE*, vol. 71, pp. 856-871, 1983.

[10]   R. Taylor, J. Korein, G. Maier, and L. Durfee, "A General Purpose Architecture for Programmable Automation Research," presented at Third Int. Symposium on Robotics Research, Chantilly, 1985.

[11]   R. H. Taylor, "The Synthesis of Manipulator Control Programs from Task Level Specifications," : Stanford University, 1976.

[12]   H. Asada and H. Izumi, "Teaching and Program Generation for the Hybrid Position/Force Control via the Measurement of Human Manipulation Tasks," *J. of the Robotics Society of Japan*, vol. 5, pp. 452-459, 1987.

[13]   H. Asada and B.-H. Yang, "Skill Acquisition from Human Experts through Pattern processing of teaching data," *Jornal of Robotics Society of Japan*, vol. 8, pp. 17-24, 1990.

[14]   K. Kosuge, Y. Fujisava, and T. Fukuda, "Control of Robot Directly Maneuvered by Operator," presented at IEEE International Conference on Robots and Systems, 1993.

[15]   K. Kosuge, H. Yoshida, and T. Fukuda, "Dynamic Control for Robot-Human Collaboration," presented at IEEE International Workshop on Robot Human Cooperation, 1993.

[16]   K. Kosuge and N. Kazamura, "Control of a robot handling an object in cooperation with a human," presented at IEEE international workshop on Robot human Communication, 1997.

[17]   H. Kazerooni, "Human/robot interaction via the transfer of power and information signals --- part i: Dynamics and control analysis," presented at Proceedings of IEEE International Conference on Robotics and Automation, 1989.

[18]   H. Kazerooni, "Human/robot interaction via the transfer of power and information signals --- part ii: Dynamics and control analysis," presented at Proceedings of IEEE International Conference on Robotics and Automation, 1989.

[19]   H. Kazerooni and G. Jenhwa, "Human extenders," *Transaction of the ASME: Journal of Dynamic Systems, Measurement and Control*, vol. 115, pp. 218-90, June, 1993.

[20]   E. A. Dickmanns, "Dynamic Vision Merging Control Engineering and AI Methods," in *The Confluence of Vision and Control*, G. D. H. e. al, Ed.: Springer-Verlag, pp. 210-229.

[21]   S. Lee and S. Ro, "Robotics with Perception and Action Nets," in *Control in Robotics and Automation: Sensor Based Integration*, N. X. B K Ghosh, T. J. Tarn, Ed. San Diego, CA: Academic Press, 1999, pp. 347-380.

[22] B. L. Davies, K. L. Fan, R. D. Hibberd, M. Jakopec, and S. J. Harris, "ACROBOT - Using Robots and Surgeons Synergistically in Knee Surgery," presented at 8th International Conference on Advanced Robotics, California, USA, 1997.

[23] J. Troccaz, M. Peshkin, and B. L. Davies, "The use of localizers, robots, and synergistic devices in CAS," presented at Proc. First Joint Conference of CVRMed and MRCAS, Grenoble, France, 1997.

[24] R. Kumar, T. Goradia, A. Barnes, P. Jensen, L. Whitcomb, D. Stoianovici, L. Auer, and R. Taylor, "Performance of Robotic Augmentation in Microsurgery-Scale Motions," presented at 2nd Int. Symposium on Medical Image Computing and Computer-Assisted Surgery, Cambridge, England, 1999.

[25] S. E. Salcudean, S. Ku, and G. Bell, "Performance measurement in scaled teleoperation for microsurgery," presented at First joint conference computer vision, virtual realtiy and robotics in medicine and medical robotics and computer-assisted surgery, Grenoble, France, 1997.

[26] H. Das, H. Zak, J. Johnson, J. Crouch, and D. Frambach, "Evaluation of a Telerobotic System to Assist Surgeons in Microsurgery," *Computer Aided Surgery*, vol. 4, pp. 15-25, 1999.

Endoscopic Robotically Assisted Coronary Artery Bypass Anastomosis on the Beating Heart: Analysis of Quality and Patency

Marco A. Zenati, M.D., Gilbert J. Burckart, Pharm.D., Bartley P. Griffith, M.D.

Division of Cardiothoracic Surgery, University of Pittsburgh Medical Center,
Pittsburgh, Pennsylvania, U.S.A.
zenatim@msx.upmc.edu

Abstract. *Background*: Construction of a coronary artery bypass anastomosis on the beating heart using a computer-enhanced telemanipulation system is feasible but its quality is unknown. *Methods*: In nine pigs the Left Internal Mammary Artery (LIMA) to the Left Anterior Descending (LAD) coronary artery anastomosis was constructed exclusively using the ZEUS System. Pressure-fixation stabilizer provided local immobilization. LIMA flow was analyzed using transit-time technology. Pigs had to survive without ischemia with proximal LAD ligated for one hour. Angiography was performed within 90 minutes of the anastomosis and graded using the Fitzgibbon Classification. After sacrifice, the anastomosis was assessed using probes. *Results*: LIMA-LAD anastomosis was successful in all cases in 22 $\pm$ 3.6 minutes. No repair stitches were necessary. LIMA flow was 21.6 $\pm$ 2.5 ml/min. Angiographic patency was 100% with grade A anastomosis in all. Eight pigs survived without ischemia for 60 min; one pig fibrillated shortly after completion of anastomosis. A 2 mm probe was passed without resistance through the anastomosis in all cases. *Conclusion*: The quality of the LIMA-LAD anastomosis performed on the beating heart using the ZEUS robotic system is excellent. Confirmation of these results with a totally endoscopic approach is warranted.

Keywords: Cardiovascular surgery; robotics and robotic manipulators; clinical system development.

1. Introduction

The ultimate goal of Minimally Invasive Coronary Surgery is to achieve a totally endoscopic ("ports only") coronary artery bypass operation on the beating heart. The high degree of difficulty associated with the performance of microvascular (2 mm) endoscopic coronary artery bypass anastomosis on the beating heart without cardiopulmonary bypass, requires dexterity enhancement through robotic technology. Several computer-enhanced telemanipulation surgical systems have been used to test the feasibility of endoscopic beating heart coronary artery bypass graft surgery (1,2). The construction of Left Internal Mammary Artery (LIMA) to Left Anterior Descending (LAD) anastomosis is the critical factor in the operation and current studies have not addressed the quality and patency of endoscopic anastomosis using robotic telemanipulation instruments. We hypothesized that the patency and quality of the LIMA to LAD anastomosis performed *exclusively* using the ZEUS Surgical

System (ComputerMotion, Goleta, CA) would be excellent, based on our previous large experience in beating heart coronary bypass surgery and our work with the ZEUS System in cadavers.

2. Methods

Nine crossbred swine of either sex weighing an average of 42.5 Kg (range 35-50 Kg) were used for the study. All animals received humane care in AAALAC, USDA registered facility in compliance with the "Principles of Laboratory Animal Care" formulated by the National Society for Medical Research and the "Guide for the Care and Use of Laboratory Animal Resources" and published by the National Institutes of Health (NIH Publication No. 85-23, revised 1985). The study protocol was approved by the Institutional Animal Care and Use Committee of the University of Pittsburgh (IACUC Approval #1002). Maintenance of general anesthesia and monitoring was performed as previously described. A median sternotomy was performed and the LIMA was harvested using skeletonization technique. Heparin (2 mg/Kg) was given intravenously and the LIMA was divided distally and prepared for anastomosis. A pressure-fixation myocardial stabilizer (OPCAB System, Genzyme or CMI-Grace II, ComputerMotion) was used. Three 5 mm ports were created on the left chest to align instruments with LAD and create a 90' angle between the instruments. The LAD was manually dissected and opened. An intracoronary shunt (Flocoil, Guidant) of appropriate size was introduced in the LAD and distal perfusion was reestablished. From this point on the procedure was performed from the ZEUS surgeon control station, located 10 feet away from the operating table. A PTFE 8-0 double armed suture with TT8 needles was used. After the completion of the anastomosis, the LAD was ligated proximally and flow was measured using transit-time technology (Transonic, Ithaca, NY). Contrast angiography was performed by injection of contrast agent (Renographin, Roche) into the LIMA. Data were expressed as mean ± standard deviation.

3. Results

All 9 LIMA-LAD anastomoses were completed successfully without need for repair stitches and using exclusively robotic telemanipulation in 22 ± 3.6 minutes (range 18-30 min). The LIMA flow after LAD ligation was 21.6 ± 2.5 ml/min. Eight animals survived for 60 minutes with the LAD ligated without ischemic EKG changes; one pig suffered from an episode of ventricular fibrillation. Angiographic patency was 100% with Fitzgibbon Grade A in all. A 2 mm Parsonnet probe was passed into the anastomosis without resistence after the animal was sacrificed. Upon gross inspection, no evidence of thrombus formation at the anastomotic site was identified.

4. Discussion

We demonstrated in a pig beating heart model that an excellent LIMA-LAD anastomosis can be constructed endoscopically using the ZEUS robotic surgical system with conventional 2D visualization and 4 degrees of freedom (DOF). Other groups (1) have suggested that a manipulator with less than 7 DOF could not attain general goal positions and orientation in space. We completed all anastomosis in 22 minutes with a 100% angiographic patency and found the ZEUS System adequate for the task.

5. References

5.1. Falk V, Diegler A, Walther T, et al. Endoscopic Coronary Artery Bypass Grafting on the Beating Heart Using a Computer Enhanced Telemanipulation System. Heart Surg Forum 1999;2:199-205
5.2. Reichenspruner H, Boehm DH, Gulbins H, et al. Robotically Assisted Endoscopic Coronary Artery Bypass Procedures without Cardiopulmonary Bypass. J Thorac Cardiovasc Surg 1999;118:960-1.

6. Acknowledgement

This work was supported by a Research Grant from the Ravitch/Hirsch Center for Minimally Invasive Surgery of the University of Pittsburgh.

Distributed Modular Computer-Integrated Surgical Robotic Systems:
Implementation Using Modular Software and Networked Systems

Andrew Bzostek[1,4], Rajesh Kumar[1,4], Nobuhiko Hata[2,4], Oliver Schorr[2,3,4],
Ron Kikinis[2,4], Russell H. Taylor[1,4]

[1]Department of Computer Science
Johns Hopkins University, Baltimore, Maryland, USA

[2]Surgical Planning Laboratory
Brigham and Women's Hospital and Harvard Medical School, Boston, MA,USA

[3]Institute of Process Control and Robotics
University of Karlsruhe, Germany

[4]Engineering Research Center for
Computer Integrated Surgical Systems and Technology

Abstract. We seek to build CIS research systems within a flexible, open architecture. In this paper, we outline our solutions to the problems of system design, construction, and integration in this environment: building distributed, modular systems for sensing, control, and processing. Based on our experience building these systems, we utilize distributed network architectures, modular software components, and intelligent object distribution to maximize flexibility while complying with the particular needs and interfaces of specific components. While a work in progress, these approaches have been integrated into several systems in development and have demonstrated significant utility, based on construction time and system flexibility. This paper will discuss the first two aspects: modular software and basic network architectures, while intelligent object distribution is presented in a companion paper. Based on the applications we have targeted and technologies we have used, this paper introduces our architecture and its components. It then presents the system's current state and finally, discusses plans for future improvements and extensions.

1 Introduction

One of the goals of our groups is to build an open architecture for CIS research, which will allow the integration of a variety of components into systems targeting various surgical applications. Software modularity and distribution of components across a standard network are two cornerstones of our work. Using modular architectures has allowed: **reuse** of research, **speed-up** of application development, and **encapsulation** of proprietary technologies. Distribution of system components on standard networks has: improved **computational capabilities** and **robustness**, as well as facilitating the support of **specialized devices**.

Based on concrete system examples, this paper presents efforts within the Center for Computer Integrated Surgical Systems and Technologies to design and implement components which use these concepts. It focuses on two major aspects of our architecture: modular software components and a basic system distribution. Presented in a companion paper [1] are our efforts to build an intelligent object distribution architecture. While at varying levels of maturity, these techniques are complementary and their integration into systems has demonstrably reduced system implementation overhead and increased flexibility.

1.1 Examples

1.1.1 Steady Hand Cooperative Manipulation

Cooperative ("Steady Hand") manipulation offers an attractive alternative to the popular master-slave teleoperation systems. Our LARS [2] test bed robots are equipped, and the "steady hand" robots [3] designed to explore this approach.

The current primary application being investigated is therapy delivery in the eye to treat vein occlusion, though future applications include other ophthalmic applications, as well as a variety of ENT and spinal surgeries. Our modular robot control (MRC) library implements control for both the LARS and Steady hand robots, as well as integrating low-level sensing.

1.1.2 Fluoroscopy-Guided Robotic Needle Placement

Under development by one of our groups, this system is designed to place needles into soft tissue organs (see fig. 1), initially targeting hepatic tumors for localized therapy delivery [4, 5]. It uses fluoroscopic imaging for guidance. Originally developed on the LARS [2] robotic platform, it currently uses the CART(Constrained Access Robotic Therapy) System [4] for manipulation. Its configuration integrates the robot and imaging, as well as force sensing and visualization. Like the steady hand system, the system's robotic platform has changed as the system has evolved. In contrast, however, further significant changes in this component are

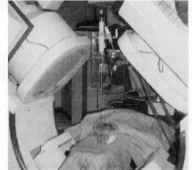

Figure 1. The Fluoroscopy-Guided Robot Needle Placement system.

anticipated. Also unlike the previous system, this approach has required the integration of imaging and a significant visualization capability. The modular construction of our robot software has allowed this system to move seamlessly between a variety of robots. Distributing the system components across a standard network has not only separated robot control from the visualization and image processing, but has also allowed the robot and imaging host computers to be specialized for their respective devices.

1.1.3 IMRI-Guided Robot-Assisted Needle Placement

An intraoperative MRI (IMRI)-guided robotic system is in development, currently targeting prostate brachytherapy needle guidance. It is discussed more thoroughly elsewhere [1, 6]. Like the previous example, this system includes a robot, an imager, and a visualization system. Integrated into the IMRI system [7] is also a tracking device which provides position information. Because of the system can acquire volumes during the procedure, a separate high-performance planning host is also used. To support these, we have extended our modular software and distributed system architectures to support trackers and to include distributed planners as well as physical devices such as robots and imagers.

1.2 Our Approach

In general, modularity and distribution are desirable in any engineered software system. Development of CIS Systems is a particularly iterative, application driven process, however. System specifications and targets are often short to medium-term and tied fairly closely to the particular task at hand. While modularity and distribution are worthy goals in abstract, their associated overhead can outweigh their benefit when only applied to isolated cases. Our approach has been to find commonality among our systems, and then try to implement this commonality reusably in second-generation systems. We've found that only after building systems which address a number of often different targets, using a variety of technologies, have we found useful patterns.

1.3 Previous Work

Complex sensing and manipulation systems are becoming common in improving surgical outcomes. These include navigation aids, including Northern Digital's Polaris optical tracking Systems and ISG's viewing wand arm-based localizer; imaging from a variety of vendors; and robotic systems such as Aesop and Zeus from Computer Motion, daVinci from Intuitive Surgical, and Robodoc from Integrated Surgical Systems. Several research groups are working on systems for a variety of clinical applications. These include Davies [8-10], and Troccaz [11], Peshkin[12], Salcudean [13], Das [14], and Sheridan [15]

As these systems evolve the need for building modular architectures is also becoming evident. Modular architectures have been proposed for industrial automation [16, 17], and integrated, propriety frameworks, such as Picker's Venue system and Surgical Navigation Network's system for component integration are becoming available.

Some open systems, such as VRPN from UNC, have also been developed, though for different domains and thus with different design specifications.

1.4 Components

Our systems use a variety of technologies to address our clinical targets. These include, but certainly are not limited to robots, imagers, trackers, and computational modeling.

Robots: Our first generation systems for cooperative manipulation and fluoroscopically-guided needle placement were built on the LARS platform [2] (see fig. 2). Next generation robots developed within the

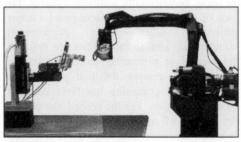

Figure 2. LARS *(right)* and the Steady Hand robot *(left)*: two generations of robots for CIS Research

center include the "steady hand robot," [3] designed for cooperative manipulation; the CART (Constrained Access Robotic Therapy) Robot [4], a platform for needle placement and experimental orthopedic work; and the RCM-PAKY System [18] designed for needle insertion for renal surgery. We are also using the MRT robot, designed for an open-configuration MR scanner, for needle guidance [19].

Within the CISST Engineering Research Center, there are also several new robotic systems suitable for particular imaging/manipulation environments are under development.

Imagers: One of our groups has focused on iMRI-guidance, the other has focused on fluoroscopy. Intraoperative MR techniques hold significant promise, [6, 20], providing unique capabilities in terms of intraoperative planning and monitoring of a procedure. Real-time fluoroscopy provides a more light-weight, real-time imaging solution. Additionally, other groups within the CISST center have now begun to use CT [21] and ultrasound as well.

Trackers: We have developed systems based on three commercially available tracking systems: Northern Digital's Optotrak, Ascension's Flock of Birds, and Image Guided Technology's Pixsys system. Integration of NDI's Polaris as well as other tracking technologies is in the initial phases.

Computational Modeling: While significant computation takes place on the computers within the OR, for very demanding tasks, it is necessary to move this work to a more powerful platform. This is particularly relevant when intraoperative dosemetry and planning is required, e.g. in IMRI-guided localized therapy. A variety of software packages, both commercial and noncommercial, are available for many of these tasks. Our systems currently use non-commercial, research packages, but this will certainly change as applications diversify, and systems move toward clinical use.

2 Overview

In this paper, we present two categories of architectures for dealing with these components: Modular software abstractions, and network distribution of devices and processing.

2.1 Modular Software

In using commercially available systems, the lack of standards for interfaces poses a major difficulty. When using experimental components, similar capabilities are required for control or processing among a class of devices. We resolve the first problem by abstracting programming interfaces common to a devices class. For the second, the control for classes of components are abstracted in layered structures (see fig. 3). These two techniques complement each other. Vendor or hardware-specific software is often available

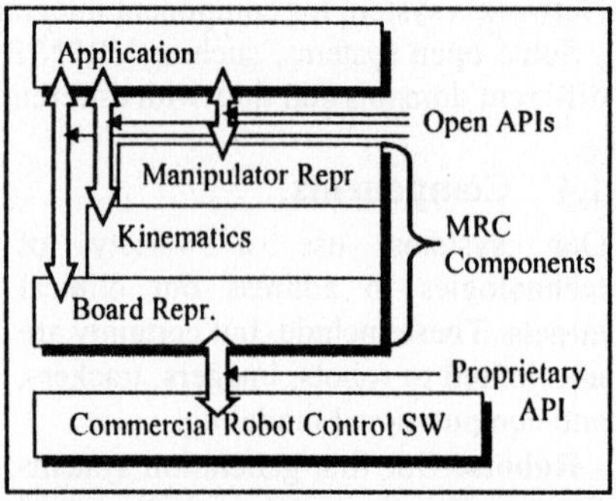

Figure 3. A portion of the modular architecture within the Modular Robot Control (MRC) Library

at levels below our general device abstractions. The appropriate layers can then be implemented around this software, and the remaining layers of our architecture are shielded from vendor implementations. We follow the same approach for our robots, trackers, planners, and to a lesser extent, general sensors.

2.2 Networked System Distribution

Monolithic systems suffer from problems of fault-sensitivity and lack of scalability in terms of computation, component integration, and system monitoring. Additionally, components of our systems have often had hardware or software-related platform restrictions. A local, OR network of computers solves many of these problems.

2.2.1 General Architecture

Our general networked system architecture uses several PC-class machines connected via a standard Ethernet LAN, though, in cases where more significant computational power is needed, it can be extended to a larger class of machines located more remotely. Within the OR, different machines are directly connected to (and responsible for) various other components (e.g. robots or trackers), though more than one component may be connected to the same host machine if

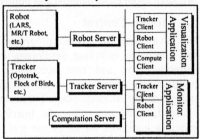

Figure 4. A prototypical implementation of our network architecture

architectures/processing allow. In general, we then call the host connected directly to device X, the X Server. Other hosts which want to use X's capabilities will then connect to the X Server as X Clients (see fig. 4). In addition to hosts for controlling application(s), clients can also include additional hosts for safety monitoring and data collection.

2.2.2 Requirements

The general architecture is not specific to a particular software implementation. Potential implementations include simple message-based communication via TCP or UDP, RCP style invocation, and distributed objects (e.g. via CORBA).

Any of these, however, will need to provide certain capabilities, and can be selected based on how well they meet these needs.

Transparent Distribution: In order to maintain maximum software flexibility the network implementation needs to make the distribution as transparent as possible. A client should not have to differentiate local vs. remote devices and should have to know as little as possible about their actual locations.

Network Performance: The network implementation should provide not only adequate bandwidth, but should minimize latency as well. Ensured quality of service will also be increasingly important as systems become more distributed.

Synchronous &. Asynchronous Communication: Depending on the distributed device and its mode of use, primary communication can be either synchronous or asynchronous. An implementation should allow for either and for their mixing within an interface.

Multi-Priority Client Support: A network implementation must thus provide the ability to not only support multiple clients, but support different levels of functionality among them. For example, a robot server should support only a single controlling client while still allowing multiple clients to connect as observers.

Safety: Although system distribution can enhance safety by adding redundancy and monitoring capabilities to the system, the components necessary to achieve this and to ensure that the distributed system maintains a high overall level of safety are not yet developed. A network implementation should at least facilitate, and ideally enhance these capabilities

3 Component Implementations

We have taken these patterns and implemented them into several different components.

3.1 Robot Control

We have chosen to abstract programming interfaces for our robots. The alternatives to this approach, including developing/using programming languages poses problems of learning curve and user acceptance. Implementations of interfaces in acceptable programming languages provide libraries that can be shared, swapped, and developed independently of each other. Furthermore, it allows the programming language to be changed, while preserving the interfaces. Our modular robot control (MRC) library provides interfaces for all of our test bed robots. The interface can be easily extended include additional manipulators and allows high level control independent of the robotic platform actually implementing the commands.

The MRC structure abstracts interfaces in layers. The lowest layers wrap hardware implementations (vendor specific sensor implementations, servo control for robots etc). This functionality is then used by logical abstractions (joints for robots, sensor interface for all sensors). The higher layers contain functionality for Cartesian control, command scheduling, and network support. This layering allows new robots and new configurations of existing robots to be quickly supported, often requiring only a change in run-time loadable configuration.

The library provides Cartesian control, forward and inverse kinematics, and joint control interfaces for our robots. It also includes implementations of force sensor, and digital and analog i/o device interfaces.

Our first implementations of robot control used remote procedure calls for network communication. However this is cumbersome, and not platform independent. The current implementation includes remote execution functionality as a part of the class interface. This also allows us to separate network implementations (socket style TCP or UDP or direct cable serial/parallel connections) from the programming interface for robot control. CORBA, an emerging standard for distributed networking, is also under active investigation [1].

Current MRC implementations support four robots (the CART robot, the Steady Hand robot, the LARS robot and the MRT robot). Implementations for several others are under development. Two motion control interfaces, the MEI motion controller, and the proprietary motion controller for LARS are supported. Support for other motion controllers is being developed. Cross platform support for Solaris and linux is being tested. We plan to support clients on most operating systems.

3.2 General Sensors

For general sensors, we have established a simple architecture which shields an application from device specific API's. In general, it uses two device representations (see fig. 5). The first, a "Physical Device Representation" (PDR), represents the device itself and, except for a very small common API, is specific to the device. The application, however, can also interact with a "Logical Device" (LD). The abstraction for LD's is that each "senses" only a single kind of data (e.g. Points, or 2D images). Specific implementations then map a common interface to specific PDR's. This abstraction layers above MRC and other device abstractions, but provides the application layer with the ability, for example, to use either a robot or a tracker, (or other position sensor) transparently as a source for position information.

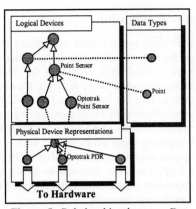

Figure 5. Relationships between Data Types, Physical Device Representations and Logical Devices within the General Sensor Architecture

PDR's for position sensing have been implemented for Optotraks, Flocks of Birds, and robots via MRC, as well as for 2D imaging via Matrox's family of image capture cards. These have been integrated into LP/PDR's and are currently in use in several systems in development. The Optotrak PDR supports network distribution using CORBA. This implementation has not yet been integrated into the architecture discussed in [1], however.

3.3 Trackers

Our architectures for robots and trackers share some significant parallels. While we view robots as maintaining information both in joint and Cartesian space, using kinematics to move between them, we view trackers as maintaining both 3D point and 6D frame (position + orientation) information, using a solver to go between. However, while the kinematics for a robot can be viewed as static, the rigid body configurations necessary for a solver are dynamic (at least run-time configurable). Additionally, solvers for pure 6DOF trackers are inherently different (frame to points, rather than points to frame), than those for point-based systems. This being said, however, many of the general principals of MRC apply here as well.

The abstract tracker library has not yet been fully implemented, though its layered structure has been used in the implementation of the Optotrak PDR.

3.4 Planners

For localized therapy, general planning systems can be posed as dosemtry planners. We have layered planners' internal information into targets and dose volumes. The Dosemtry layer takes targets to labeled dose volumes, the Optimizer layer takes prescribed dose volumes to targets. Only a prototype interface between 3D Slicer [22] and the planning software used in our fluoro-guided system [4] has been implemented.

4 Evaluation

Individual systems built using these components have been evaluated individually based on the requirements of their particular applications. Within such systems, the components have worked well, providing the required levels of performance. Direct evaluation of these methods is more difficult. However, two metrics which can be used to evaluate the architecture are integration time and the flexibility of the resulting system. A controlled comparison with other possible implementation architectures has not been done, nor is implementation of a system for the sole purpose of performance comparison likely. But, comparing these attributes to a base-line and measuring their incremental improvement as system implementations evolve are both options.

Implementation time can be easily quantified, though interpretation of the measure is dependant on environmental factors. Flexibility is harder to pin down, and is better as an incremental measure.

In two cases, the use of the techniques and components described here has demonstrably improved flexibility and/or integration time.

Prior to the integration of MRC, the fluoro-guided system was closely tied to the LARS robotic platform. With MRC's integration, however, it is able to control any of our robotic systems, clearly increasing the system's flexibility. This flexibility has sped development of the system itself, and allowed application to new imaging and manipulation environments.

The MRT robotic system provides another example. Slicer, which it uses for visualization, predates the development of the MRT robot significantly and, while it has significant 3D visualization, segmentation, registration, and measurement capabilities, it had no notion of robotics. With two days of work from two of us, however, robot control was added to Slicer using MRC. In a similar time, MRC components were implemented to control the novel MRT robot. Thus, with four day's work, three components: Slicer, MRC, and the MRT robot, which had never been connected were integrated into a working prototype. While there is no prior measure to compare this to directly, four day's work of two people is clearly significantly less than what is usually associated with the integration of robot control into new applications. Once this was accomplished, Slicer also inherited all of the flexibility associated with MRC and any application using MRC gained the ability to control the MRT robot.

5 Discussion

These architectures and implementations are a work in progress. We continue development and integration. In the long term, Slicer will serve an expanded role in systems using various kinds of image guidance, serving as a central visualization and user interaction platform for a variety of research systems.

The MRC library is being extended to include more robots, and servo controllers. The command scheduling, and networking support is also being upgraded.

The integration of imaging should also be extended. The distribution of imaging devices across the network is straightforward, although network performance will be a greater issue. An MRC-like abstraction over multiple imaging devices would also be useful, though less trivial.

Currently, device servers' functionality has been restricted—only providing computation directly related to their devices. Application specific information and processing has generally been centralized on a single host, a machine which will often also provide visualization to the user. This division is not necessarily optimal, as more distributed, agent-based techniques might better utilize the system's resources, but these are also left as future extensions.

6 Conclusion

In our efforts to build an infrastructure for CIS research which is open and flexible, modularity and distribution have proven useful, when applied appropriately. We have applied these techniques based on our experience with a variety of robotic, sensing, and computational technologies for surgical assistance. Our groups have created layered architectures for robots, trackers, planners, and general sensors as well as network implementations which allow these components to be distributed to machines connected via a standard network. Several of these components have then been implemented and integrated into systems targeting different clinical applications, including retinal surgery and cancer therapy in the liver and prostate. While the development of these systems is ongoing, our approaches have demonstrably added to the flexibility and integration time for the components we have developed. Additionally, they are enabling technologies— creating an environment which will foster integration of new technologies into the next generation of safer, smarter, more capable CIS systems.

Acknowledgements

The authors gratefully acknowledge the support of the National Science Foundation under grant #IIS9801684, the Engineering Research Center grant #EEC9731478, the NSF/Whitaker Foundation for Cost Reducing Health Care Technology Grant #5T32HL07712, and the NIH under grants P41 RR13218-01 and P01 CA67165-03. This work was also supported in part by Johns Hopkins University internal funds.

References

[1] O. Schorr, N. Hata, A. Bzostek, R. Kumar, C. Burghart, R. H. Taylor, and R. Kikinis, "Distributed Modular Computer-Integrated Surgical Robot Systems: Architecture for Intelligent Object Distribution," proc. of Medical Image Computing and Computer Assisted Interventions, Pittsburgh, PA, USA, 2000.

[2] R. H. Taylor, J. Funda, B. Eldridge, K. Gruben, D. LaRose, S. Gomory, M. Talamini MD, L. Kavoussi MD, and J. Anderson, "A Telerobotic Assistant for Laparoscopic Surgery," in IEEE EMBS Magazine Special Issue on Robotics in Surgery, 1995, pp. 279-291.

[3] R. H. Taylor, A. Barnes, R. Kumar, P. Gupta, P. Jensen, L. L. Whitcomb, E. d. Juan, D. Stoianovici, and L. Kavoussi, " A Steady-Hand Robotic System for Microsurgical Augmentation," proc. of MICCAI'99, Cambridge, UK, 1999.

[4] A. Bzostek, A. C. Barnes, R. Kumar, J. H. Anderson, and R. H. Taylor, "A Testbed System for Robotically Assisted Percutaneous Pattern Therapy," proc. MICCAI '99, pp. 1098-1107, Cambridge, UK, 1999.

[5] S. Schreiner, J. Anderson , R. Taylor , J. Funda , A. Bzostek, and A. Barnes, "A system for percutaneous delivery of treatment with a fluoroscopically-guided robot," proc. of Joint Conf. of Computer Vision, Virtual Reality, and Robotics in Medicine and Medical Robotics and Computer Surgery, Grenoble, 1997.

[6] J. F. Schenck, F. A. Jolesz, P. B. Roemer, H. E. Cline, W. E. Lorensen, R. Kikinis, S. G. Silverman, C. J. Hardy, W. D. Barber, E. T. Laskaris, and e. al, "Superconducting open-configuration MR imaging system for image-guided therapy," Radiology, v. 195, pp. 805-14, 1995.

[7] J. Schenck, F. Jolesz, P. Roemer, H. Cline, W. Lorensen, R. Kikinis, S. Silverman, C. Hardy, W. Barber, and E. Laskaris, "Superconducting open-configuration MR imaging system for image-guided therapy," Radiology, vol. 195, pp. 805-814, 1995.

[8] B. L. Davies, S. J. Harris, W. J. Lin, R. D. Hibberd, R. M. R, and J. C. Cobb, "Active Compliance in robotic surgery - the use of force control as a dynamic constraint," Proc Instn Mech Engrs, vol. 211, pp. 285-292, 1997.

[9] B. L. Davies, K. L. Fan, R. D. Hibberd, M. Jakopec, and S. J. Harris, "ACROBOT - Using Robots and Surgeons Synergistically in Knee Surgery," proc. of 8th International Conference on Advanced Robotics, California, USA, 1997.

[10] S. J. Harris, W. J. Lin, K. L. Fan, R. D. Hibberd, J. Cobb, R. Middleton, and B. L. Davies, "Experiences with robotic systems for knee surgery," proc. of Proc. First Joint Conference of CVRMed and MRCAS, pp. 757-766, Grenoble, France, 1997.

[11] J. Y. Delnondedieu and J. Troccaz, "PADyC: A Passive Arm with Dynamic Constraints - A two degree-of-freedom prototype," proc. of Proc. 2nd Int. Symp. on Medical Robotics and Computer Assisted Surgery, pp. 173-180, Baltimore, Md., 1995.

[12] J. E. Colgate, W. Wannasuphoprasit, and M. A. Peshkin, "Cobots: Robots for Collaboration with Human Operators," proc. of International Mechanical Engineering Congress and Exhibition, pp. 433-39, Atlanta, GA, USA, 1996.

[13] S. E. Salcudean, G. Bell, S. Bachmann, W. H. Zhu, P. Abolmaesumi, and P. D. Lawrence, "Robot-Assisted Diagnostic Ultrasound - Design and Feasibility Experiments," proc. of MICCAI'99, pp. 1062-1071, Cambridge, UK, 1999.

[14] H. Das, H. Zak, J. Johnson, J. Crouch, and D. Frambach, "Evaluation of a Telerobotic System to Assist Surgeons in Microsurgery," Computer Aided Surgery, vol. 4, pp. 15-25, 1999.

[15] T. B. Sheridan, "Human factors in Tele-inspection and Tele-surgery:Cooperative manipulation under Asynchronous Video and Control Feedback," proc. of MICCAI'98, pp. 368-376, Cambridge, MA, USA, 1998.

[16] A. C. Sanderson and G. Perry, "Sensor Based Robotic Assembly Systems: Research And Applications in Electronic Manufacturing," Proceedings of the IEEE, vol. 71, pp. 856-871, 1983.

[17] R. H. Taylor and D. D. Grossman, "An Integrated Robot Systems Architecture," IEEE Proceedings, 1983.

[18] D. Stoianovici, L. L. Whitcomb, J. H. Anderson, R. H. Taylor, and L. R. Kavoussi, "A Modular Surgical Robotic System for Image Guided Percutaneous Procedures," proc. of MICCAI'98, pp. 404-410, Cambridge, MA, USA, 1998.

[19] K. Chinzei, R. Kikinis, and F. A. Jolesz, "MR Compatibility of Mechatronic Devices: Design Criteria," proc. of MICCAI'99, Cambridge, UK, 1999.

[20] F. A. Jolesz, "Image-Guided Procedures and the Operating," Radiology, vol. 204, pp. 601-612, 1997.

[21] R. C. Susil, J. H. Anderson, and R. H. Taylor, "A Single Image Registration Method for CT Guided Interventions," proc. of MICCAI'99, pp. 798-808, Cambridge, UK, 1999.

[22] D. Gering, A. Nabavi, R. Kikinis, W. E. L. Grimson, N. Hata, P. Everett, F. Jolesz, and W. W. III, "An Integrated Visualization System for Surgical Planning and Guidance using Image Fusion and Interventional Imaging," proc. of MICCAI'99, Cambridge, UK, 1999.

Distributed Modular Computer-Integrated Surgical Robotic Systems: Architecture for Intelligent Object Distribution

Oliver Schorr[1,2], Nobuhiko Hata[1], Andrew Bzostek[3], Rajesh Kumar[3], Catherina Burghart[2], Russel H. Taylor[3], Ron Kikinis[1]

[1]Surgical Planning Laboratory,
Brigham and Women's Hospital and Harvard Medical School
Boston, MA, USA

[2]Institute for Process Control and Robotics
University of Karlsruhe, Germany

[3]Department of Computer Science
Johns Hopkins University, Baltimore, Maryland, USA

Abstract

This paper presents intelligent object distribution architecture to maximize the performance and intelligence of a distributed surgical robotics system and its preliminary implementation in an MR-guided surgical robot system in an open-configuration MRI scanner. The method enables networked integration of a robot control server and multiple clients with minimum engineering overhead but maximum flexibility and performance. The clients in this study include an intraoperative imager, high-performance image processing computer(s), and surgical navigation host. The first contribution of the paper is to propose the use of object distribution by common object request broker architecture (CORBA), in which a robot control object on the robot control server can be remotely but transparently invoked from the clients regardless of their hardware, operating systems, or programming language. Second, we propose a technique to achieve additional flexibility by reporting the robot configuration information, i.e. geometry and kinematics of the robot, to the clients upon connection. Third, we ensure protection against an unauthorized entity by introducing a security control host that authorized the clients' access to the robot server. In a prototype implementation of an MR-guided surgical robot system, the robot was controlled by surgical navigation software (the 3D Slicer) on a UNIX client by invoking the distributed control object on a robot control server on a PC. The method was evaluated in performance studies; and the result indicated 3.6 milliseconds for retrieving positions of the robot stages and 25.5 milliseconds to send a frame-based motion command, which are satisfactory for surgical robot control. In conclusion, the proposed method shows the potential usefulness of flexibly integrating the legacy software to a surgical robot system with minimum engineering overhead, thereby achieving highly complex and intelligent tasks in robot-assisted surgery.

1 Introduction

We have been trying to build an open architecture system for a computer-integrated robot surgery system, which can integrate component technologies into systems targeted at a variety of surgical tasks. This study aims to link surgical robot technology with medical image processing and surgical navigation technology with minimum overhead so that complex and intelligent tasks can be carried out by surgical robot assistance.

The software and hardware for surgical robot systems are developed on a per-clinical-application basis, thus limiting the flexibility of the hardware and software for reuse in multiple clinical applications. These systems usually rely on a generic computer (most likely a PC) to manage all the tasks, thereby limiting their functional expandability. For instance, real-time elastic registration can hardly be incorporated into a PC-based robot controller due to the requirement for computational heavy tasks. To overcome these problems, we present our approach of modular software architecture and networked system distribution in the associated paper [1]. The approach separates the monolithic system into an intra-operative imager host, a user interface host, and heavy computing hosts, so that each host can focus on dedicated tasks while collaboratively communicating through the network. For instance, the robot host (PC) continuously monitors actuators and sensors though real-time motion control hardware, while the user interface host (graphics workstation) interactively visualizes the status of the robot. By selecting the component hosts, one can flexibly adjust the system to application-specific requirements with minimum engineering overhead. Similar concepts can be found in intelligent manufacturing [2, 3].

While we highly appreciated the modularity and flexibility of the distributed systems in our previous studies, we also realized that the complexity of the communication using message-based commands is relatively limited. In addition, the content of commands was still designed on a per-configuration basis, and protocol definition depended on the programming language, network transport, and other factors. Therefore, the cost for updating and synchronizing a command receptor was still significant.

In order to overcome these shortcomings, we propose a method to achieve highly complex and intelligent communication within the distributed system. Specifically, the method provides the following new features: (1) the use of an object distribution mechanism by common object request broker architecture (CORBA) [4] to implement the software objects solely on the robot control server, while enabling clients to calling them in a transparent manner, (2) flexible robot configuration reporting mechanism to flexibly and automatically adjust the visualization and image-processing clients to various surgical robots; (3) a safety control mechanism to ensure security, manageability, and status reporting in the distributed system. The method is implemented and evaluated in a test-bed system, MR-guided surgical robot, in which a PC-controlled 5DOF MR-compatible robot is controlled and monitored by graphic surgical navigation and simulation software 3D Slicer on UNIX workstations.

This study is clinically significant, as it allows a surgical robot system to accomplish highly complex tasks controlled by intraoperative images, computationally heavy image processing, and graphics visualization. The method is significant technically because it achieves its clinical goal with minimum cost but maximum flexibility. The use of CORBA enables high interoperability of multiple computing hosts over different OS, hardware, and software language platforms, which encourage bringing in expertise from various research institutions by removing the barrier of software and hardware incompatibility.

2 Methods

2.1 System overview

In this study, we attempted prototype system development within the framework of an MR-guided surgical robot system[5]. The robot control objects are developed in a PC-based control server and distributed to a visualization client in a UNIX workstation. The security-monitoring host in the center of the system monitors connection to the robot control server. Other clients may be high-performance computers processing computationally heavy tasks such as elastic registration.

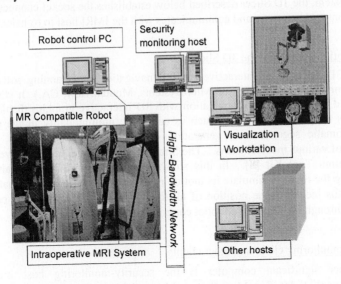

Figure 1: Schematic overview of the system

MR-compatible robot server

The center of the MR-guided surgical robot system is a MR-compatible surgical manipulator, referred as the MR/T robot hereafter, developed jointly by Mechanical Engineering Laboratory of Japan and our institution [5]. The MR/T robot has four horizontal stages and one vertical stage, each driven by ultrasonic actuators. The four stages move two arms independently, enabling the end-effector held by the two arms to have 5 degree-of-freedom DOF movements.

The ultrasonic actuators and linear optical encoders in the MR/T robot are controlled by a PC (CPU Cyrix 300 MHZ, 64MB, Windows NT4.0) using a motion control board (PCX/DSP, Motion Engineering Inc., CA). The card provides on-board PID control by a 40-MHz Analog Devices ADSP-2105 processor. Software for the robot control is built upon a modular robot control (MRC) C/C++ library providing Cartesian control, forward and inverse kinematics, and joint control interfaces for the MR/T robot. The MRC also includes implementation of a force sensor, and digital and analog input/output device interfaces; thus, expansion of the functionality of the robot can be effectively achieved with minimum cost. The library classes have a layered structure, each new layer inheriting significant functionality from its parents. Further detail on the MRC library including its modularity design can be found in [1].

Intraoperatve MRI scanner

The MR/T robot is designed to perform surgical assistance in an open-configuration intraoperative MRI (IMRI) scanner (Signa SP, GE Medical Systems, Milwaukee, WI)[6], using intraoperative MRI for monitoring and navigation. The host computer for the IMRI is a UNIX workstation (Sparc 20, Sun Microsystems, Moutainview, CA) in which a software offers a service to export real-time IMRI through an Ethernet port. In our system, the 3D Slicer described below establishes the socket connection with the IMRI front-end computer and communicates with the IMRI host to transfer the images.

Visualization client with the 3D Slicer

The 3D Slicer is an interactive surgical navigation and planning software on the UNIX workstation (Ultra30, Sun Microsystems, Moutainview, CA.) It was originally developed for neurosurgical navigation with 3D interactive graphics display. We also utilize its unique functionalities such as automatic multimodality image registration, semi-automatic segmentation, generation of 3D surface models, and quantitative analysis of various medical scans. The Slicer is coded in the Tcl/Tk [7] version of the Visualization Tool Kit [8]. In this study the role of the 3D Slicer also involves displaying the robot to simulate its motion path with respect to patient anatomy. The robot status including the position of all the stages and end-effector is continuously updated through CORBA-based robot control object invocation.

Security monitoring client and other clients

Another significant computer is the security-monitoring host on a UNIX workstation (Ultra80, Sun Microsystems, Moutainview, CA). We can also include multiple hosts for other monitoring and computing purposes, such as elastic registration for updating pre-operative images based on IMRI, which captures the deformed shaped of the target organs, or radiation planning server for planning radioactive seed planting sites for brachytherapy.

2.2 Object distribution by CORBA

Selected robot control objects for MRC are distributed to the 3D Slicer by a middleware MICO [4] which is a freely available and fully compliant implementation of the CORBA standard. MICO intercepts robot control call and finds an object that can implement the request, pass it the parameters, invoke its method, and return the results. The client does not have to be aware of where the object is located, its programming language, its OS, or any other system aspects that are not part of an object's interface. The MICO provides interoperability between applications on different machines in heterogeneous distributed environments and seamlessly interconnects multiple object systems. Figure 2: illustrates the principle of this object distribution mechanism.

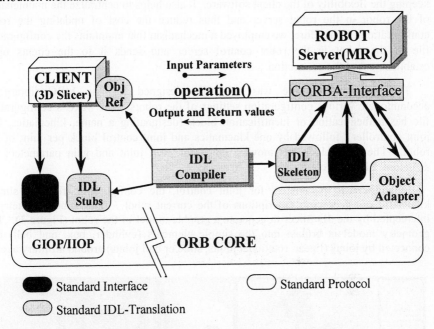

Figure 2: Schematic view of the object distribution mechanism

Objects invoked from the 3D Slicer at the robot PC server are robot hardware and software initialization, frame-level movement command, joint-level movement command, and joint-level position report. CORBA defines object interfaces via implementation language-independent hardware-independent specification, using the interface definition language (IDL). The IDL also involves definition of complex data types. In practice, MICO's IDL compiler writes "wrapper" code that translates between the standardized bus "ORB core" and the objects residing in the CORBA-interface which redirects method calls inside the MRC library.

Another unique feature of our object distribution mechanism is the Dynamic Invocation Interface (DII). Rather than hard coding the handle in the 3D Slicer using the prefixed standard IDL stub, the DII lets the 3D Slicer dynamically incorporate the interface description sent from the robot server and map it to Tcl commands.

We employed TclMico [9] that plugs into MICO's DII converts the IDL to Tcl script which can be plugged into any Tcl application at runtime. Since the 3D Slicer is coded by script language Tcl/Tk, which does not require compilation, this dynamic invocation mechanism is suitable for the 3D Slicer. In the same manner as DII, Dynamic Skeleton Interface (DSI) allows the robot server to declare accessible objects at runtime instead of fixed IDL skeleton.

2.3 Flexible robot configuration report

Minimizing the robot-dependent information in the client but managing the information in the robot control server and transferring it to the clients is suitable for keeping the flexibility of the client software. It also helps to centralize the maintenance of the robot in the robot server and thus reduce the cost of updating the robot configuration. Therefore, we employed a mechanism that maintains the configuration file of the robot in the robot control server and sends it to the clients upon establishment of the connection.

This configuration file was originally designed for the MRC library to accommodate various configuration settings of the robots. The MRC's configuration file has authentication or identification block (designating a name, kinematics, and joint controller), followed by one kinematics and joint control block per joint of the robot. The joint block has moving range for each joint and gain parameters for actuators' PID control.

In addition to parameters for joint control, the configuration file has simple kinematic and geometric descriptions of the current robot. This robot geometry is intercepted by the 3D Slicer to generate a graphic model of the robot (Figure 3). The geometry model is broken into the simple elements (cylinder, box, and ball) and connected by joints (linear, rotational joints, slide or ball joints) around arbitrary axes.

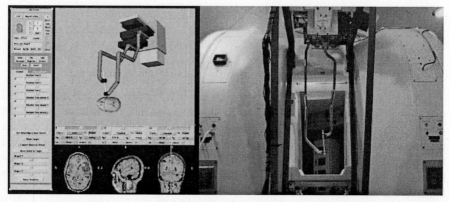

Figure 3: Integrated display of robot and anatomy data from the interventional MR scanner (left) and the robot installed in the scanner. By using object distribution (Section 2.1) and configuration distribution (Section 2.2), the 3D Slicer can control the MR-compatible robot (right) with minimum implementation cost., thus enabling the highly ~nt robot control to be coordinated with original functionalities of the 3D Slicer, ~ssing, navigation, image retrieval from the IMRI scanner, and more.

2.4 Safety-monitoring host

The safety-monitoring host implements a subset of three different tasks dedicated to increase safety and simplify communication initialization. The first task is to supervise the client's access to robot server by registering the client hosts and authorize the of passive and active control privilege to the robot control. For instance, MR/T robot limits the access to one client with active control privilege, while unlimited number of hosts can access with passive control privilege. The grounds for this access limit is that the decision for control should be finalized at the centralized in the single authorized client server, while the other client hosts should be devoted to data processor. The second task of the safety-monitoring host is to store Tcl files, interface description files, and configuration files from the robot control server, which can be later transferred to the clients upon connection. Some of the information in these files should be sent to the client when the client "greets" the client but before actual CORBA connection happens.

3 Performance test

In order to evaluate the feasibility of the method, the processing time of the prototype system was measured for the elected tasks. (Task 1) is position inquiry of five joints by one, two or three passive clients, and (Task 2) is Cartesian-level motion command of the end-effector by an active client. Task 2 involves the inverse kinematics computation in the robot control server. (Task 1) and (Task 2) were invoked 100 times from 0, 1, or 5 3D Slicer clients dedicated for (Task 1), and one client for (Task 2). Each client was connected to the 100Mbit/second fast switcher, yet was not isolated from the rest of the hospital network. Table 1 summarizes the result from the tests.

Table 1: Processing time for invoking distributed objects

# of Task 1 clienents	# of Task 2 clients	Average/STD of Task 1 time (micorosecnds)	Average/STD of Task 2 time (micorosecnds)
1	0	3373/520	N/A
2	0	3636/639	N/A
5	0	3412/399	N/A
0	1	3585/304	24764/1780
1	1	3894/637	24954/1101
5	1	3678/384	26784/1502

Though it doesn't appear in Table 1, we also found several exceptionally long processing time in a few per 100 invocations. Those processes took about 40 milliseconds for (Task 1) and 105 milliseconds for (Task 2), which are 5 to 10 times longer than average processing times.

4 Discussions

The result of the performance test indicated the feasibility of the method in computer-integrated surgical robot systems. It is conceivable that the processing of 3.6 milliseconds for passive position inquiry and 25.5 milliseconds for active motion command are acceptable for robot control using the motion control card with on-board PID control chip. The result also indicates that the number of clients accessing the robot control server does not influence the processing time.

As of the publication of this report, only the robot control objects are distributed by CORBA. Images are transferred from the IMRI scanner to the 3D Slicer through a socket connection and commanded by prefixed ASCII text, but the plan is to convert the service CORBA-based object distribution so that the image processor and other hosts can access the scanner with more complex command. An example of such task is image scanning with scanning parameter and sequence.

Added value of the object distribution using CORBA is the ability to incorporate advanced image processing software, which, due to software and hardware incompatibility, used to require significant engineering overhead to port into robot control. The currently standarized CORBA has an interface to virtually all programming language including C, C++ , Smalltalk, Java, Ada, Cobol, Visual Basic, Tcl, PL/1, LISP, Python and Perl. In our testbed prototype, we linked C/C++ with Tcl each running on a different hardware and OS platform. There is talk of making IDL compliant with XML; thus, it may be beneficial to design robot configuration description and patient-specific image and model entry to have synergy with XML to prompt the smooth transition to future XML-compliant database management systems.

5 Conclusion

We reported a method to distribute objects from robot control server to multiple clients to achieve highly intelligent and complex task by the collaborative work of automonous visualization, navigation, and image processing hosts. The method was implemented MR-guided surgical robot system in which robot control objects are transparently called from the graphical navigation software on UNIX workstation, and executed C/C++-based robot control software on PC. The evaluation test indicated the distributed objects can be remotely invocated in 3.6 milliseconds for passive monitoring and 25.5 milliseconds for active control, both of which are satisfactory for surgical robot control. The result indicated that the method is feasible to apply in distributed modular computer-integrated surgical robot systems.

6 Acknowledgement

This study is supported by National Science Foundation's Engineering Research Center (grant #EEC9731478) and NAC grant P41. The authors gratefully acknowledge the support from Dr. Chinzei of MEL for supporting MRI-compatible robot, and Mr. Gering of MIT for supporting the 3D Slicer. We also appreciate the technical contribution of Mr. Daniel Kacher of Brigham and Women's Hospital.

7 References

[1] A. Bzostek, R. Kumar, N. Hata, O. Schorr, R. Kikinis, and R. H. Taylor, "Distributed Modular Computer-Integrated Surgical Robotic Systems: Implementation using modular software and networked systems," presented at Third International Conference on Medical Robotics, Imaging And Computer Assisted Surgery, 2000.

[2] M. Lei, X. H. Yang, M. M. Tseng, and S. Z. Yang, "A CORBA-based agent-driven design for distributed intelligent manufacturing systems," *Jounal of Intelligent Manufacturing*, vol. 9, pp. 457-465, 1998.

[3] O. A. Suarez, J. L. A. Foronda, and F. M. Abreu, "Standard based framework for the development of manufacturing control systems," *International Journal of Computer Integrated Manufacturing*, vol. 11, pp. 401-415, 1998.

[4] A. Puder and K. Römer, *MICO: An Open Source CORBA Implementation*. San Francisco, CA: Morgan Kaufmann Publishers, 2000.

[5] K. Chinzei, R. Kikinis, and F. A. Jolesz, "MR Compatibility of Mechatronic Devices: Design Criteria," presented at Medical Image Computing and Computer Assisted Intervention, Cambridge, UK, 1999.

[6] J. F. Schenck, F. A. Jolesz, P. B. Roemer, H. E. Cline, W. E. Lorensen, R. Kikinis, S. G. Silverman, C. J. Hardy, W. D. Barber, E. T. Laskaris, and et al., "Superconducting open-configuration MR imaging system for image-guided therapy," *Radiology*, vol. 195, pp. 805-14, 1995.

[7] J. K. Ousterhout, *Tcl and the Tk Toolkit*. Massachusetts: Addison-wesley, 1994.

[8] W. Schroeder, K. Martin, and B. Lorensen, *The Visualization Toolkit. An object-oriented approach to 3D graphics*. Upper Saddle River, New Jersey: Prentice Hall PTR, 1996.

[9] F. Pilhofer, "TclMico", http://www.informatik.uni-frankfurt.de/~fp/Tcl/tclmico, 1999.

Motion-Based Robotic Instrument Targeting under C-Arm Fluoroscopy

Alexandru Patriciu[1,2], Dan Stoianovici, Ph.D.[1,2], Louis L. Whitcomb, Ph.D.[2],
Thomas Jarrett, M.D.[1], Dumitru Mazilu, Ph.D.[1], Alexandru Stanimir, Ph.D.[1],
Iulian Iordachita, Ph.D.[1], James Anderson, Ph.D.[4],
Russell Taylor, Ph.D.[3], Louis R. Kavoussi, M.D.[1]

Johns Hopkins University Baltimore, Maryland, USA
[1]James Buchanan Brady Urological Institute, Johns Hopkins Medical Institutions
URobotics, JHBMC-D0300, 5200 Eastern Ave., Baltimore, MD 21224, USA dss@jhu.edu
[2]Department of Mechanical Engineering, Whiting School of Engineering
[3]Department of Computer Science, Whiting School of Engineering
[4]Department of Radiology, Johns Hopkins Medical Institutions

Abstract: We present a simple and precise robot targeting method under portable x-ray fluoroscopy based on image servoing. The method is implemented for needle alignment in percutaneous procedures using the PAKY-RCM robot developed in our laboratory. Initial clinical tests apply to the access of the renal collecting system.

Previously reported methods for computer assisted instrument targeting under fluoroscopy use complex robot-image registration algorithms. These approaches use static images of fiducial markers to estimate the robot-image coordinate mapping, which is then used for targeting. In contrast, we report a new method to directly perform targeting by using a marker located on the robot/end-effector and performing fluoro-servoing under continuous imaging. Three-dimensional targeting is achieved by performing the alignment in two dissimilar views acquired at arbitrary C-Arm orientations.

The percutaneous access implementation of this method provides automated alignment of the needle towards a surgeon specified target. Needle insertion is then controlled by the surgeon using side-view fluoroscopic feedback.

The proposed method offers increased accuracy, simplicity, and repeatability. Moreover, initial clinical experiments suggest that through the use of this method access time and radiation dose have been reduced as compared to normal manual procedures.

1 Introduction

Minimally invasive and noninvasive procedures are gaining increased popularity mainly due to reduced trauma and improved recovery time. One of the main problems encountered in minimally invasive procedures is, in contrast to open procedures, a dramatic reduction in the surgeon's visual ability. Radiological, ultrasonic, and magnetic resonance imaging techniques are employed to map anatomical geometry in intraoperative procedures [5,8,12].

Portable ultrasonic and fluoroscopy units (commonly termed C-Arms) are ubiquitous in modern operating rooms. Both of these affordable imagers provide real time two-dimensional (2-D) visualization. A common impediment in using these 2-D imagers is the lack of volumetric representation necessitating extensive surgical training for correct 3-D interpretation. The problem of "retrofitting" computer image-based 3-D navigation systems on commonplace C-arms is complicated by the fact that the vast majority of portable fluoroscopy systems do not provide encoding of the C-Arm position or orientation. This creates difficulty in estimating the pose of the imager with respect to the patient, thus complicating computer assisted procedures using this image information. Many solutions have been proposed for helping surgeons in performing fluoroscopy guidance [7,8]. For example, Navab et al. proposed an efficient algorithm [16,17] allowing for the complete reconstruction of volumetric anatomy using multiple 2-D images. Simultaneously, other researchers concentrated on the development of image guidance and registration techniques for various fluoroscopy guided interventions [7,8,18,19,20].

Most image guided procedures, such as percutaneous needle access, radio, and ultrasonic ablation require targeting of a specific instrument / probe at an exact organ location. The clinical outcome of these procedures significantly relies on targeting accuracy.

To address this problem, authors and others have reported computer-assisted instrument targeting based on the development of specialized image registration algorithms. Such methods commonly use at least two images of a spatial radio-opaque marker of complex geometry or a series of one-dimensional marks distributed on a defined pattern [4]. The x-ray projection of the markers is used to estimate the instrument-image coordinate mapping, which is then used for targeting. These algorithms compute the exact position of the target with respect to the instrument and the geometrical parameters of the imager [13], such as the source position, magnification factor, etc. In these procedures distortion correction and image calibration techniques may also be required for increased accuracy [4,13].

In contrast to these "fully calibrated" approaches, our approach is "uncalibrated" in the sense that the method achieves accurate needle placement without precise camera/imager calibration. For a discussion of the advantages and disadvantages of "uncalibrated" vision methods the reader is directed to [9].

Our "uncalibrated" approach is principally motivated by the technique we have frequently observed in use by experienced surgeons in performing manual needle access under fluoroscopy. Based on this observation, we have previously developed a targeting method for percutaneous needle access based on the needle superimposition over the target, calyx of the kidney [20]. This method was implemented using our PAKY needle driver and then updated with the addition of the RCM robot [22] and GREY supporting arm [14]. The system is used clinically in our institution and has proved to offer a substantial improvement over the manual approach [6]. In this method, however, targeting is performed by the surgeon controlling the robot.

This paper reports the development of a computer-controlled image-guidance technique for automated targeting using this system. The method uses fluoro-servoing (robot control based on direct image feedback from the C-Arm) in two arbitrary image views acquired at dissimilar C-Arm orientations. Similar guidance techniques have been successfully used for industrial robot guidance based on video camera images [1,3,11, 10,15].

We present the fundamentals of this new fluoro-servoing method, its application to needle guidance using the PAKY-RCM robot, and clinical application of this system for percutaneous renal access.

2. Methods

The system comprised of a surgical robot, a PC for image processing and robot control, and a C-Arm imager, is schematically represented in Figure 1. The digital C-Arm (OEC-9600) provides x-ray images under PC command. The image is acquired using a video card (Matrox Meteor). The robot is controlled from the same PC using a real-time motion control card (PCX-DSP8, by Motion Engineering, Inc.). The fluoro-servoing algorithm controls needle orientation based on radiological feedback.

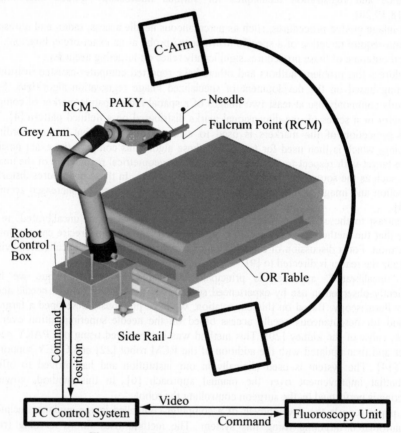

Figure 1: Schematic of system architecture

2.1 The Robotic System

The main component of the system is the PAKY-RCM robot comprising the needle driver PAKY [20] and the RCM robot [22]. PAKY (Percutaneous Access of the Kidney) is a radiolucent needle driver used to guide and actively drive a trocar needle in percutaneous access procedures. Its radiolucent design allows for unobstructed visualization of the procedure needle and anatomical target. The original PAKY driver was constructed for the "Superimposed Needle Registration Method" [20]. For the present application, PAKY was redesigned to accommodate the proposed computer-guided servoing algorithm. This was realized by providing a thinner outline in the shape of a rectangular bar as illustrated in Figure 1.

The RCM (Remote Center of Motion) is a compact robot for surgical applications that implements a fulcrum point located distal to the mechanism [20,22]. The robot can precisely orient an end-effector (i.e. surgical instrument) in space while maintaining the location of one of its points. This kinematic architecture makes it proper for applications requiring a singular entry point such as laparoscopy and percutaneous access.

The robot assembly is supported into the Grey Arm [14] mounted to the operating table. This allows for positioning and steady support of the robot in close proximity of the operated organ. The PAKY-RCM assembly is capable of orienting a needle while maintaining its tip location. This gives the ability of aiming the needle at any desired target while priory setting the skin insertion point and positioning the needle tip at that location. Only two motions are thus required for orientating the needle about the fulcrum point. The proposed targeting algorithm takes advantage of this kinematic simplicity, as presented in the following section.

2.2 Fluoro-Servoing Instrument Orientation

Fluoro servoing is a particularization of visual servoing using x-ray fluoroscopy feedback. Visual servoing is a generic name for the class of robot control algorithms using image feedback for performing positioning and tracking operations [1,3,11,10,15].

The main difficulty in performing portable fluoroscopy computer-assisted procedures is the lack of information regarding the pose of the imager with respect to the patient. As a mobile unit, the C-Arm is moved and reoriented during the procedure to satisfy surgical needs. We propose a simple and accurate algorithm for instrument (needle) targeting independent of C-Arm orientation. The algorithm uses fluoro-servoing to orient the needle about a fulcrum point located at its tip.

Aiming the needle at a desired target requires needle alignment in two dissimilar views obtained from different C-Arm orientations. That is, orienting the needle so that it extends into the target in both views. Since alignments are performed sequentially, the second alignment should not deteriorate the first.

Each alignment is performed automatically by the guidance algorithm, which corrects the needle position based on image feedback. For facilitating the automatic detection of the needle into the image the needle has been equipped with a radio-opaque spherical ball at the free end, thus providing a well-discriminated signature. A pattern-matching algorithm running on the video acquisition board is used to rapidly locate the ball marker in the x-ray image. All calculations are performed in a fixed reference frame centered at the needle tip and oriented according to the initial position of the robot.

The principle of operation is schematically represented in Figure 2, illustrating the needle supported by the PAKY driver and positioned with the tip at the skin entry point (I'). The central illustration is a 3-D representation whereas the two side views are x-ray projections of this space from View 1 and View 2 respectively, as indicated by the arrows. The figure presents the needle at different positions, which will be scanned during the two-phase alignment. In this motion the tip of the needle remains at the

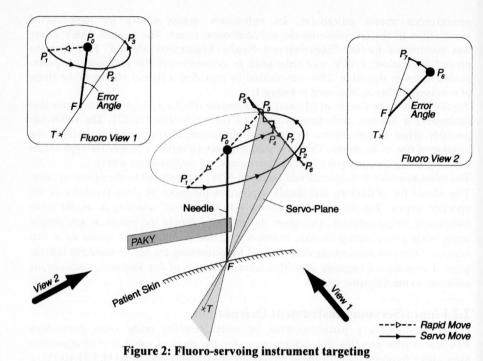

Figure 2: Fluoro-servoing instrument targeting

fulcrum point (F), while the needle end (the ball end) changes location from $P_0 \mapsto P_7$ as later described. The needle is initially located at P_0. Targeting is achieved in two steps according to the C-Arm orientations View 1 and View 2.

<u>View 1</u>: A conical needle trajectory is used to obtain an initial estimate of the relative robot-image orientation. A rapid approach move of arbitrary direction P_0P_1 places the needle on the cone. The cone angle is set so that the target is located within the space swept by the needle extension. Starting from P_1 the needle is moved on the cone using incremental steps proportional to the orientation error. The error is given by the angle $180° - \angle TFP_i$ measured in the x-ray projection (Figure 2). This proportional algorithm converges to the needle position P_2, in which the needle at P_2F points towards the target T. Continuing on the conical path a second alignment is achieved at point P_3 in a similar manner.

The plane FP_2P_3 is the initial servo-plane. This plane has the property that at any position within the plane the needle maintains View 1 target alignment. Laboratory experiments showed that the identification accuracy of this plane was not consistent, errors depending on the distance of this plane from the start point P_0. Explicitly, greater errors ware encountered when points P_2 and P_3 where closely located at a side of the cone.

To overcome this problem, while maintaining a minimal cone angle, a correction of the servo-plane was implemented, following the $P_3 \rightarrow P_0 \rightarrow P_4 \rightarrow P_5 \rightarrow P_6$ path, as follows. The needle is rapidly brought to the start position P_0 and then moved in a plane FP_0P_4 perpendicular to the servo-plane. On this path fluoro servoing is employed to achieve accurate needle alignment at point P_4. The algorithm uses the same angle-feedback proportional control. The axis FP_4 is then used as a pivot about which the servo-plane is rotated for iterative corrections. From P_4 the needle is moved in the direction of the instantaneous servo-plane towards the point P_5 and then P_6 with a prescribed angular step. Target alignment is reevaluated at each step by searching transversally, and the orientation of the servo-plane is corrected accordingly. This is achieved by adjusting the servo-plane angle through the pivot axis FP_4 with an amount proportional to the angular targeting error. This new servo plane FP_5P_6, is similar to the initial cone determined plane FP_2P_3, however, it insures that the end points P_5 and P_6 are sufficiently spaced apart to render a consistent determination of the servo-plane and also averages errors over the multiple scan points on the $P_4 \rightarrow P_5 \rightarrow P_6$ trajectory. In our algorithm the limit points P_5 and P_6 were placed at a needle angle equal to the initial cone angle measured in the servo plane.

The servo-plane insures that independent of the needle orientation within this plane the needle in properly aligned to the target in the first view. Three-dimensional targeting requires the additional determination of an axis within this plane passing the fulcrum F and the target T, as presented next.

View 2: A second view is selected by reorienting the C-Arm. The orientation of this plane is arbitrary with the restriction that in this view the servo plane does not project into a line. The best precision is achieved if the view is normal to the servo-plane.

Needle alignment is performed by servoing the needle orientation within the previously determined servo-plane based on the same angle-error feedback, as represented in the x-ray projection Figure 2, View 2. The algorithm converges to the needle position FP_7. In this orientation the target is located on the needle axis.

By using the servo-plane, the second view alignment preserves the first. Three-dimensional targeting is thus obtained by combining both 2-D alignments.

3 Results

The robotic system was adapted for the proposed method. A special design of the needle driver was implemented and integrated. The system using the servo-targeting algorithm was tested for accuracy and reliability using specially derived experiments and then clinically validated for percutaneous renal access.

3.1 Pre-Clinical Validation

For minimizing the radiation exposure during software design and evaluation the algorithm was initially developed and tested using a video camera mounted is a positioning stand. A white background and a black needle were used for achieving proper contrast. A 2mm spherical ball represented the target. Repeated tests revealed a targeting accuracy not greater than 0.5mm.

A second set of experimental tests was then performed in the operating room (Figure 3) using a digital C-Arm fluoroscope (OEC-9600). The ball-finder pattern-matching algorithm proved to be stable and robust under this imaging modality. Imager distortion was evaluated by constructing a template of equally spaced steel balls mounted on a thin radiolucent plate. For the imager used the overall distortion was found to be under 0.75 mm in a region next to the image center, this including the error of the ball-finder algorithm. Using the magnification function of the fluoroscope allowed for maintaining the field of view in the reduced distortion zone around the image center. Using a 2mm ball target located 80mm below the needle tip (fulcrum / skin entry point) the image

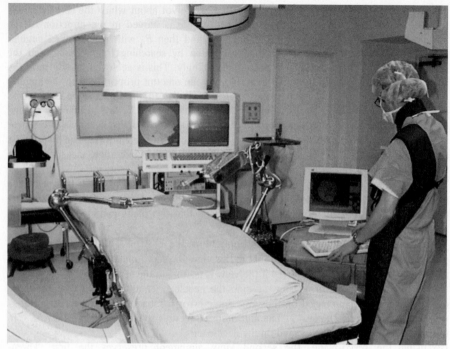

Figure 3: Experimental tests

guidance algorithm was evaluated for targeting accuracy. The maximum error in aiming the ball over 25 experiments was under 1.5mm. During the experiments the overall system comprising the robotic system, the imager, and the guidance algorithm proved to be reliable and consistently precise. The safety of the system for surgical applications is inherited from the kinematic design of the robotic component [22]. The PAKY-RCM presents decoupled orientation and needle insertion capabilities allowing for independent activation of the two stages. This insures that the needle may not be inadvertently inserted during the orientation stage and accidental reorientation may not occur during needle insertion.

The results of experimental trials, the safety of the mechanism, and the previous clinical experience using the RCM for percutaneous renal access at the Johns Hopkins Medical Institutions [6] and in numerous telesurgery transatlantic procedures [2,21] enabled us to confidently proceed with the clinical trials.

3.2 Clinical Application: Percutaneous Renal Access

In the initial application the fluoro-servoing targeting method was implemented for percutaneous renal access. A first case was recently performed in our institution (Figure 4).

For renal access the operating room and the patient are prepared as for the standard procedure. The patient is placed under total anesthesia. The fluoroscopy table is equipped with a special rigid rail. The robot is mounted onto the rail on the side of the targeted kidney and covered with a sterile bag. The needle driver is sterilized prior to the

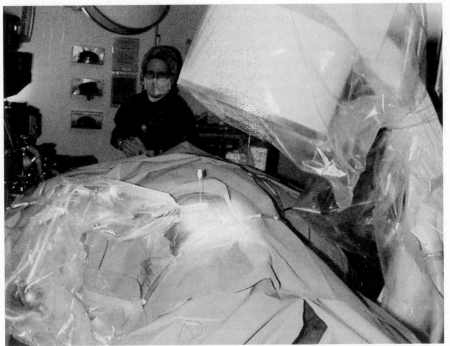

**Figure 4: Fluoro-servoing percutaneous renal access at the
Johns Hopkins Medical Institutions**

operation. As for the manual procedure the C-Arm is positioned on the opposite side of the table and all steps prior to the needle access are performed as usual. First, the urologist chooses the skin insertion point as in the manual procedure and positions the robot assembly by manipulating the passive arm such that the needle tip (located at the fulcrum point of the RCM) is located at the chosen point. The C-Arm is oriented for proper kidney and needle visibility. Then the surgeon identifies the target calyx on the PC monitor by manually selecting a point on the image. The first view needle alignment is then automatically performed. The C-Arm is then rotated to a dissimilar view in which the surgeon identifies the target again. The second needle alignment is automatically performed. Using other C-Arm orientations the surgeon verifies needle targeting and performs needle insertion under direct lateral observation.

In all steps patient respiration is shortly stopped by the anesthesiologist during the image acquisition prior to target selection and during needle insertion. The patient may breathe in all other stages including servo targeting.

In the first operation performed the kidney was accessed on the first targeting attempt in less than 10 minutes. The needle, however, needed to be slightly retracted and reinserted again, as it initially pushed the kidney aside due to tissue deflection and needle bowing. This was not caused by targeting errors, since the small retraction and reinsertion properly aimed the target. This problem was also encountered due to the fact that for this patient the target was located on a peripheral lower pole calyx.

The total radiation exposure time of the patient during this procedure was 90 seconds. With future software development this could potentially be reduced by commanding the imager to strobe activate during the servo motion. This will be approached through collaborative work with the imager manufacturer. Even in the present implementation the total radiation was significantly reduced as compared to the common manual approach. This is due to the fact that the method offers a well-defined step-by-step algorithm eliminating the need for the problematic surgeon interpretation of volumetric anatomy.

4 Conclusions

In this paper we propose an algorithm for automatic instrument orientation under C-Arm guidance, and present preliminary experimental and clinical validation results. The method represents a simple solution to a common surgical problem. Manual fluoroscopy-guided interventions are normally based on trial and error requiring considerable surgical skill and operative training. Automatic targeting has the potential to reduce the required level of surgical experience and the variability among surgeons in performing this type of procedures. Given the increasing demand for image-guided interventions and the fact that the method employs the most common type of imager available in the operating room, the proposed approach could potentially yield widespread applications.

The success experienced in the first operation gives a preliminary method validation. Additional clinical cases are scheduled in the near future and will be presented in the literature. Implementation details and surgeon interface refinements will be performed accordingly. Different surgical procedures will also be explored in urology as well as other medical fields. The method may also prove useful for applications involving similar imaging equipment such as biplanar fluoroscopy units. As for the previous application of the RCM robot we will also investigate this method for telesurgery in a remotely operated setting.

References

1. Batista,J., Araujo, H., Almeida A.T.: "Iterative multistep explicit camera calibration", IEEE Transactions on Robotics and Automation. Vol 15, No 5., October 1999, p 897.
2. Bauer J.J., Stoianovici D., Lee B.R., Bishoff J., Caddeu J.A., Whitcomb L.L., Taylor R.H., Macali S., Kavoussi L.R., (1999), "Transcontinental Telesurgical Robotic Percutaneous Renal Access: Case Study", American Telemedicine Association (ATA) conference, Salt Lake City, Utah, abstract# 18D, April 18-21, 1999, Telemedicine Journal, 5(1):27, 1999.
3. Bernadero, A., Victor, J.S., "Binocular Tracking: Integrating perception and control", IEEE Transactions on Robotics and Automation. Vol 15, No 6., December 1999
4. Bzostek, A., Schreiner, S., Barnes, A.C., Cadeddu, J.A. Roberts, W., Anderson, J.H., Taylor, R.H., Kavoussi, L.R.: "An automated system for precise percutaneous access of the renal collecting system". Lecture Notes in Computer Science, Springer-Verlag, Vol. 1205, pp.299-308, 1997.
5. Brown R.A., Roberts T.S., Osborne A.G., "Stereotaxic Frame and Computer Software for CT-Directed Neosurgical Localization", Invest Radiol, 15:308-312, 1980.

6. Cadeddu, J.A., Stoianovici, D., Chen, R.N., Moore, R.G., Kavoussi, L.R., (1998), "Stereotactic mechanical percutaneous renal access", Journal of Endourology, Vol. 12, No. 2, April 1998, p. 121-126.

7. Desbat L., Champleboux G., Fleute M., Komarek P., Mennessier C., Monteil B., Rodet T., Bessou P., Coulomb M., Ferretti G., "3D Interventional Imaging with 2D X-Ray Detectors", Medical Image Computing and Computer-Assisted Intervention, September 1999, Cambridge, England: Lecture Notes in Computer Science, Springer-Verlag, Vol. 1679, pp. 973-980, 1999.

8. Guéziec A., Kazanzides P., Williamson B., Taylor R.H., "Anatomy-Based Registration of CT-Scan and Intraoperative X-Ray Images for Guiding a Surgical Robot", IEEE Transactions on Medical Imaging, 17(5):715-728, 1998.

9. Hager, G., Hespanha, j., Dodds, Z., Morse, A.S., "What Tasks Can Be Performed with an Uncalibrated Stereo Vision System?", Submitted for review to IJCV.

10. Hager, G., Hutchinson, G., and Corke, P. A Tutorial Introduction to Visual Servo Control IEEE Transactions on Robotics and Automation, 12(5) pp. 651-670, 1996.

11. Hsu, L.; Aquino, P.L.S.: "Adaptive visual tracking with uncertain manipulator dynamics and uncalibrated camera", Proceedings of the 38th IEEE Conference on Decision and Control (1999), p. 5, vol (xvii+5325)

12. Ionescu G., Lavallée S., Demongeot J., "Automated Registration of Ultrasound with CT Images: Application to Computer Assisted Prostate Radiotherapy and Orthopedics", Medical Image Computing and Computer-Assisted Intervention, September 1999, Cambridge, England: Lecture Notes in Computer Science, Springer-Verlag, Vol. 1679, pp. 768-777, 1999

13. Jao J., Taylor, R.H., Goldberg, R.P., Kumar, R, Bzostek, A., Van Vorhis, R., Kazanzides, P., Guezniec, A., Funda, J., "A progressive Cut Refinement Scheme for Revision Total Hip Replacement Surgery Using C-arm Fluoroscopy", Lecture Notes in Computer Science, MICCAI 1999, pp. 1010-1019

14. Lerner, G., Stoianovici, D., Whitcomb, L., L., Kavoussi, L., R., (1999), "A Passive Positioning and Supporting Device for Surgical Robots and Instrumentation", Medical Image Computing and Computer-Assisted Intervention, September 1999, Cambridge, England: Lecture Notes in Computer Science, Springer-Verlag, Vol. 1679, pp. 1052-1061.

15. Molis, E., Chaumette, F., Boudet, S.: "2-1/2-D Visual Servoing", IEEE Transactions on Robotics and Automation. Vol 15, No 2., April 1999, p. 238

16. Navab, N., Bani-Hashemi, A., Nadar, M.S., Wiesent, K., Durlak, P., Brunner, T., Barth, K., Graumann, R.: "3D Reconstruction from Projection Matrices in a C-Arm Based 3D-Angiography system", 1998 MICCAI, Lecture Notes in Computer Science, Springer-Verlag, Vol. 1496, pp. 119-129, 1998.

17. Navab, N., Mitsche, M., Schutz, O.: "Camera-Augmented Mobile C-Arm (CAMC) Application: 3D Reconstruction Using a Low-Cost Mobile C-Arm", 1999 MICCAI, Lecture Notes in Computer Science, Springer-Verlag, Vol. 1679, pp. 688-705, 1999.

18. Potamiakos, P., Davies, B.L. Hilbert R.D. "Intra-operative imaging guidance for keyhole surgery methodology and calibration". Proc. First Int. Symposium on Medical Robotics and Computer Assisted Surgery, Pittsburgh, PA. P. 98-104

19. Potamiakos, P., Davies, B.L. and Hilbert R.D. "Intra-operative registration for percutaneous surgery", Proc. Second Int. Symposium on Medical Robotics and Computer Assisted Surgery, Baltimore, MD. Pp.156-164, 1995.

20. Stoianovivi, D., Cadedu, J.A., Demaree, R.D., Basile H.A., Taylor, R. Whitcomb, L.L., Sharpe, W.N.Jr., Kavoussi, L.R.:"An eficient Needle Injection Technicque and Radiological Guidance Method for Percutaneous Procedures", 1997 CVRMed-MrCas, Lecture Notes in Computer Science, Springer-Verlag, Vol. 1205, pp. 295-298, 1997.

21. Stoianovici, D., Lee, B.R., Bishoff, J.T., Micali, S., Whitcomb, L.L., Taylor, R.H., Kavoussi, L.R. (1998), "Robotic Telemanipulation for Percutaneous Renal Access", Journal of Endourology, Vol. 12, pp. S201.

22. Stoianovici, D., Witcomb, L.L., Anderson, J.H., Taylor, R.H., Kavoussi, L.R.: "A Modular Surgical Robotic System for Image Guided Percutaneous Procedures", 1998 MICCAI, Lecture Notes in Computer Science, Springer-Verlag, Vol. 1496, pp. 404-410, 1998.

Image-Based 3D Planning of Maxillofacial Distraction Procedures Including Soft Tissue Implications

Filip Schutyser[1], Johan Van Cleynenbreugel[1], Matthieu Ferrant[2],
Joseph Schoenaers[3], and Paul Suetens[1]

[1] Medical Image Computing (Radiology - ESAT/PSI), Faculties of Medicine and Engineering, University Hospital Gasthuisberg, Herestraat 49, B-3000 Leuven, Belgium

[2] Telecommunications Laboratory, Université Catholique de Louvain, Belgium

[3] Departement of Oral and Maxillofacial Surgery, University Hospitals of Leuven, Herestraat 49, B-3000 Leuven, Belgium
Filip.Schutyser@uz.kuleuven.ac.be

Abstract. Osteodistraction, a technique for new bone formation by the gradual separation of bony fragments, is a possible treatment for dysplasia of the maxillofacial skeleton. However, preoperative assessment of the optimal distraction direction, of the osteotomy trajectory, and of the influence of this procedure on the outlook of a patient, is difficult to assess preoperatively.

We report on an 3D image-based planning system for osteodistraction. Basically we adhere to a scene-based approach in which image derived visualizations and additional 3D structures are co-presented and manipulated. Osteotomy simulation with user-defined cutting trajectories, virtual distraction employing biomechanical models specific to the distraction type and evaluation tools based on a cephalometric reference frame, are available in our planning environment. According to the surgeon's findings, different choices can be redone in order to optimize the therapy. To account for soft tissue implications, skin tissue deformations are calculated using a finite element elastic model based on tetrahedral elements.

We report on results in the field of unilateral mandibular distraction. A detailed overview of the planning procedure is given by a case study.

Keywords   Image guided therapy, maxillofacial surgery simulation, osteotomy, soft tissue modelling

1 Introduction

Osteodistraction is a possible therapy for dysplasia of the maxillofacial skeleton. Although this technique has an expanding role, a number of key issues remain difficult.

Some of these issues are related to surgery actions (selection of an optimal distraction direction, definition of corresponding osteotomy paths), others to surgery outcome (prediction of facial outlook). Preoperative volumetric image data can be employed for better assessment of these questions, provided they are manipulated in an appropriate planning environment.

In this paper, an image-based planning system for maxillofacial distraction is presented. This preoperative planning environment includes three major components: osteotomy tools, distraction simulation and soft tissue deformation prediction.

Section 2 gives some background information about distraction osteogenesis and discusses current planning procedures and related research work. In section 3, our developments concerning osteotomy, distraction simulation and prediction of soft tissue deformation are explained. In section 4, examples are given of the current use of the planning system for unilateral mandibular distraction. Concluding remarks finish the paper in section 5.

2 Background

Distraction osteogenesis is a technique for new bone formation by the gradual separation of bony fragments. The method, although initially developed for limb lengthening, is now being applied in the treatment of craniofacial deformities. The technique offers considerable advantages over previous methods such as osteotomy and inlay bone grafting. Donor site morbidity is avoided, the investing soft tissue envelope is concurrently expanded, and the magnitude of the procedure is less. However, the technique is still in its infancy and requires further modification and refinement before it will be widespread accepted as a treatment in mainstream craniofacial surgery. Distraction osteogenesis is of increasing importance in craniofacial surgery [1].

Two major groups of craniofacial distractrion procedures are distuingished. Mandibular osteodistraction is applied for congenital deformities such as hemifacial microsomia (unilateral lengthening) and Nager's syndrome (bilateral lengthening) [2]. On the other hand, maxillar osteodistraction includes rapid maxillary expansion [3,4] as well as midface advancement [5].

The planning of distraction procedures currently involves evaluation of panoramic and cephalometric X rays, photographs and articulator mounted dental casts, as well as a clinical examination. Based on various measurements and using cephalometric standards, formulas can be used to compute a distraction direction [6].

Most of the cases suitable for distraction osteogenesis are problems with an important 3D effect that cannot be planned precisely with only 2D cephalometry. Furthermore, without 3D preoperative surgery planning, the direction and maximum length of distraction can only be determined during the operation itself. One should keep in mind that the amount of distraction is the sole parameter that can postoperatively be modified. Therefore, 3D preoperative planning holds a large potential.

An early attempt in craniofacial planning systems is described by Cutting et al. [7]. Based on image derived 3D landmarks, bone fragments are automatically repositioned according the Bolton normative standards. More recently, they report on optical tracking systems for the positioning of bone fragments in multisegment craniofacial surgical procedures [8].

Santler et al. [9] discuss preoperative planning using materialized models. The surgery is simulated on these models. The results are satisfactory, but disadvantages are the high extra cost and the production time for large models, and the limitation to do only one simulation per model. Moreover, all information about soft tissues is lost.

Because of the high impact of distraction therapy on the patients face, prediction of the deformation of soft tissues is highly desirable. Therefore, our work also includes a soft tissue model of the skin (i.e. the dermis and the underlying structures like fat and muscles).

Last year, we described a voxel-based approach for soft tissue modelling [14]. Although voxel-based modelling is advantageous from an image processing point of view (data remain available on a (subsampled) voxel grid), it is hard to associate biomechanical properties with the parameters of this model. Therefore, we switched our attention to models with a better biomechanical fondation.

Fung [10] reports on the biomechanical properties of living tissues. Skin tissues are called quasi-linear viscoelastic materials, meaning that these tissues show creep, relaxation and hysteresis when applying large oscillations around equilibrium, but the characteristics can be well approximated with linear viscoelasticity applying small oscillations. However, this modelling implies demanding computations which are in this application area unrealistic.

Different approaches have been investigated to reduce the computational cost. Teschner et al. describe a multilayer spring model [11], resulting in short simulation times, however no validation study is plublished. Koch et al. [12] model skin tissue as incompressible elastic tissue. They use prisms as finite elements. However, as future work, they suggest to move towards tetrahedral elements giving more topological freedom. In both cases, the meshing step, based on the approach of Waters [13], is rather tedious and error-prone.

In this paper, an isotropic homogenous elastic model for the soft tissues of the skin is presented. As a solution method, a finite element method using tetrahedral elements is chosen.

3 Methods

3.1 3D simulation environment for maxillofacial distraction

Our simulation environment is based on the ideas we have previously described in [14]. Basically we adhere to a scene-based approach in which image derived visualizations (MPR, surfaces, volume renderings) and additional 3D structures (external to the medical image volume) are co-presented and manipulated.

From CT images, bone structures and the skin surface are segmented by thresholding. To provide the patient with a realistic view of the natural complexion of his/her face, we also acquire a 3D photograph. Currently we use an "active" 3D system (projecting a pattern on the face and recovering 3D from the deformation of the pattern on the 2D picture), see [15]. The surface thus obtained is registered with the skin surface extracted from the CT using point/surface matching algorithms.

A first step in planning osteodistraction, is osteotomy simulation. The surgeon draws a trajectory on the bone surface and adds a cutting direction and depth. With this cutting blade, a virtual cutting action is performed.

Next, the distractor is planned by indicating its fixation points on the osteotomized bone parts. The orientation of the distraction line is a crucial parameter. In a virtual distraction, movements of the bone fragments are simulated using a biomechanical model specific for the distraction type.

The result is evaluated based on a cephalometric reference frame. In this simulation environment, the surgeon can decide to redo some of his previous choices, like the distraction direction or the definition of the osteotomy blade. Thanks to a menu-based approach piloting a user through the planning procedure, user-friendliness and time-efficiency are important features of this system.

3.2 Soft tissue modelling

The soft tissues are assumed to behave as an isotropic linear elastic continuum. So, the constitutive equation can be written as equation (1), with ϵ_{xx}, ϵ_{yy}, ϵ_{zz}, γ_{xy}, γ_{yz}, γ_{zx} strain components, λ en μ the Lamé constants which are a function of the Young modulus (E) and the Poisson's ratio (ν): $\lambda = \frac{E\nu}{(1+\nu)(1-2\nu)}$ en $\mu = G = \frac{E}{2(1+\nu)}$. Typical values are $E \in [2000, 4000]$ Pa and $\nu \in [0.25, 0.4]$.

$$
\begin{pmatrix} \sigma_{xx} \\ \sigma_{yy} \\ \sigma_{zz} \\ \tau_{xy} \\ \tau_{yz} \\ \tau_{zx} \end{pmatrix} = \begin{pmatrix} \lambda+2\mu & \lambda & \lambda & 0 & 0 & 0 \\ \lambda & \lambda+2\mu & \lambda & 0 & 0 & 0 \\ \lambda & \lambda & \lambda+2\mu & 0 & 0 & 0 \\ 0 & 0 & 0 & 2\mu & 0 & 0 \\ 0 & 0 & 0 & 0 & 2\mu & 0 \\ 0 & 0 & 0 & 0 & 0 & 2\mu \end{pmatrix} \begin{pmatrix} \epsilon_{xx} \\ \epsilon_{yy} \\ \epsilon_{zz} \\ \gamma_{xy} \\ \gamma_{yz} \\ \gamma_{zx} \end{pmatrix} \tag{1}
$$

As a result of the distraction simulation, displacements of bone fragments are known. These data are translated into boundary conditions for the soft tissue model. To solve the equilibrium equations (partial differential equations) with these boundary conditions, a finite element formulation is chosen. The finite elements are tetrahedra with a linear interpolation function.

Creating a volumetric tetrahedral mesh for irregular shapes, such as human organs, is a non-trivial task. Labelling soft tissues of the skin on the medical image data using semi-automatic segmentations tools, results in a volumetric description on a rectilinear voxel grid. By dividing the volume into cubes, and further subdividing the cubes into tetrahedra using the label values at the corners of the cubes, a tetrahedral mesh is built [16]. This approach enables us to automatically create tetrehadral meshes with a good level of detail. Based on this information, the mesh is deformed using our parallelized implementation of the finite element method.

4 Results: unilateral mandibular distraction

Next, we report on applying our approach in the case of unilateral mandibular distraction. This therapy is a possible treatment for jaw asymmetry in children with hemifacial microsomia. The hypoplastic side of the mandible is osteotomized and distracted (see figure 1 and 2). Difficult key issues for this procedure are the determination of a distraction direction in order to get a symmetrical face while deciding on an osteotomy trajectory that does not damage the mandibular nerve.

Constraints typical for success in unilateral mandibular distraction are incorporated in the surgical simulator. A first constraint is facial symmetry. To provide visual inspection of the asymmetry, the normal midline of the face is determined on the skin surface. In addition, we verify different distances on each side of the mandible reflecting the symmetry: distance between "incisura mandibulae – angulus mandibulae", "incisura mandibulae – protuberantia mentalis" and "angulus mandibulae – protuberantia mentalis". A second constraint concerns the degrees of freedom in the temporomandibular joint. We model this joint as a ball-and-socket joint.

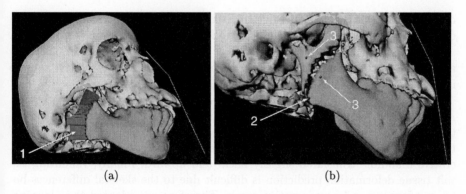

(a)   (b)

Fig. 1. This figure illustrates the planning procedure. In a first step (a) an osteotomy trajectory (1) is defined. After performing the virtual cutting action, the surgeon annotates anatomical landmarks defining important distances (2) and the distraction direction (3). Then, the surgeon can simulate and preoperatively validate the distraction. During this phase, the surgerical plan can be adapted, e.g. redefining the distraction direction and/or choosing a more suitable osteotomy trajectory.

After an optimal distraction direction and osteotomy trajectory have been defined and after a virtual distraction has been simulated, the soft tissue impli-

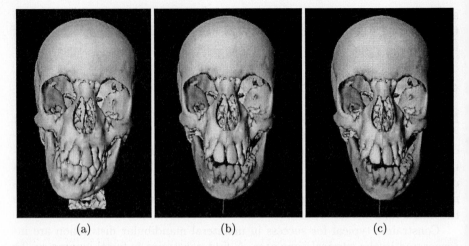

(a)   (b)   (c)

Fig. 2. After trying different possibilities for the distraction, the most suitable planning is chosen. To show the results three pictures can be compared:
- (a)   preoperative situation
- (b)   a distraction of 6 mm reaches a maximal symmetry.
- (c)   an overcorrected distraction of 9 mm is chosen to compensate for grow deficiencies.

cations are calculated. Initiated by the displacements of the bone fragments, the deformation is computed using an elastic FEM model for the skin (see figure 3).

Additionally, the preoperative plan can be transferred to the operating theatre. A customized mechanical guide can be applied on the exposed jaw bone to indicate the osteotomy and the distraction direction [17].

Postoperative CT-imaging enables us to discuss the system (see figure 4). After registration of the planned and actual shape of the mandible, we noticed a good correspondance. When comparing the postoperative position of the mandible in the temporomandibular joint with the planning, the postoperative mandible translated slightly back- and upward. Finally, validating the soft tissue deformation prediction is difficult due to the skeletal differences between planning and postoperative result. Therefore, we adapted the planning, so that the match with the postoperative skull is better. After recomputing the soft tissue deformation, and computing a distance field between both surfaces, fairly small errors were found: 1 mm $\pm$ 0.5 mm with peak values around the lips (4mm).

5   Conclusion

In this paper, we presented a planning system for planning distraction and predicting deformation of soft tissue. Key issues, such as the selection of an optimal

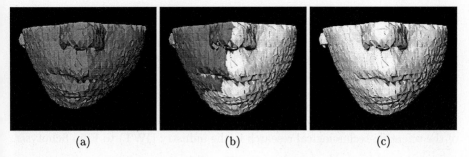

Fig. 3. This figure shows the boundary surface of the tetrahedral mesh of the soft tissues. (a) depicts the preoperative skin surface. In (b), both preoperative (dark) and expected surgical result (bright) are shown together. Figure (c) illustrates the planned result.

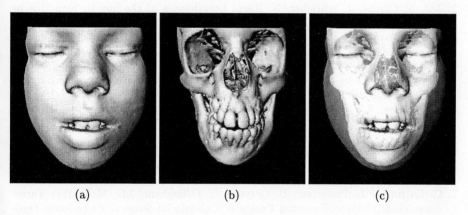

Fig. 4. From the left to the right, the postoperative skin surface, the postoperative skull and a combination of both is depicted. From postoperative CT-imaging, validation of the accuracy of our planning is possible.

distraction direction, the definition of an osteotomy, and predicting of the influence of this procedure on the outlook of a patient, are addressed in our system. Using this preoperative planning environment, trying out and validating different distractions in 3D, without large extra costs nor losing time becomes possible.

These early results of our approach opens up alleys for future work. More clinical cases and different types of distraction are planned in the near future. At the time of submission, work concerning bilateral mandibular distraction, midface advancement by sagittal (forward and downward) distraction of the maxilla are investigated. Furthermore, the soft tissue modelling will be further tested and refined. The need to incorporate incompressibility, nonlinear elastic material properties and to distinguish different types of soft tissue such as fat and muscle, are to be investigated.

Acknowledgments

The work discussed here belongs, partly to the EU-funded Brite Euram III PISA project (nr. BRPR CT97 0378), a collaboration between Materialise NV, Belgium; Philips Medical Systems BV, the Netherlands: ICS-AD; DePuy International Ltd, UK; Ceka NV, Belgium; K.U. Leuven, Belgium: ESAT/Radiology Div. Biomechanics; University of Leeds, UK: Research School of Medicine, and to a grant for research specialization from the Flemish Institute for stimulation of the scientific-technological research in the industry (IWT) to Filip Schutyser.

References

1. K. Tavakoli, K.J. Steward, M.D. Poole: Distraction Osteogenesis in Craniofacial Surgery: A Review. Annals of Plastic Surgery, Vol. 40, Nr. 1, p 88–99, January 1990
2. J.G. McCarthy, J. Schreiber, N. Karp: Lengthening of the human mandible by gradual distraction. Plast. Reconstr. Surg. 1992; 89:1–8.
3. J.A. Lehman, A.J. Haas: Surgical-orthodontic correction of transverse maxillary deficiency. Clinical Plastic Surgery, Vol. 16, p 749–775, 1989
4. C. Stromberg, J. Holm: Surgically assisted, rapid maxillary expansion in adults: a retrospective long-term follow-up study. Journal of Craniomaxillofacial Surgery, Vol. 23, p 222–227, 1995
5. J.W. Polley, A. Figueroa: Rigid External Distraction: Its Application in Cleft Maxillary Deformities. Plastic and Reconstructive Surgery, Vol. 102, Nr. 5, p 1360–1374, 1998
6. H.W. Losken, G.T. Patterson, S.A. Lazarou, T. Whitney: Planning Mandibular Distraction: a preliminary report. Cleft Palate-Craniofacial Journal, Jan, 1995, vol. 32, no. 1, p. 71–76
7. C. Cutting, F.L. Bookstein, B. Grayson, L. Fellingham, J.G. McCarthy: Three-dimensional Computer-assisted Design of Craniofacial Surgical Procedures: Optimization and Interaction with Cephalometric and CT-based Models. Plastic and Reconstructive Surgery, vol. 77:6, p. 877–885, 1986
8. C. Cutting, B. Grayson, J.G. McCarthy, C. Thorne, D. Khorramabadi, B. Haddad, R. Taylor: A Virtual Reality System for Bone Fragment Positioning in Multisegment Craniofacial Surgical Procedures. Plastic and Reconstructive Surgery, vol. 102:7, p. 2436–2443, 1998
9. G. Santler, H. Kärcher, A. Gaggl, G. Schultes, R. Mossböck: Advantage of Three-Dimensional Models in Intraoral Callus Distraction. Computer Aided Surgery, vol. 3, no. 3, p. 99–107, 1998
10. Y.C. Fung: Biomechanics: Mechanical properties of Living Tissues. 2nd edition. Ch. 7, Springer-Verlag, 1993, p. 242–320
11. M. Teschner, S. Girod, B. Girod: Optimization Approaches for Soft-Tissue Prediction in Craniofacial Surgery Simulation. Proceedings 2nd international conference on medical image computing and computer-assisted intervention - MICCAI'99, lecture notes in computer science, vol. 1679, p. 1183–1190, September 19-22, 1999
12. R.M. Koch, M.H. Gross, F.R. Carls, D.F. von Büren, G. Fankhauser, Y. Parish: Simulating Facial Surgery Using Finite Element Methods. SIGGRAPH 96 Conference Proceedings, 1996, p. 421–428
13. K. Waters: A physical model of facial tissue and muscle articulation derived from computer tomography data. SPIE Vol. 1808, Visualization in Biomedical Computing, volume 21, p. 574–583, 1992.

14. F. Schutyser, J. Van Cleynenbreugel, J. Schoenaers, G. Marchal, P. Suetens: A Simulation Environment for Maxillofacial Surgery Including Soft Tissue Implications. Proceedings 2nd international conference on medical image computing and computer-assisted intervention - MICCAI'99, lecture notes in computer science, vol. 1679, pp. 1210–1217, September 19-22, 1999

15. ShapeSnatcher, www.eyetronics.com

16. M. Ferrant, S.K. Warfield, C.R.G. Guttmann, R.V. Mulkren, F.A. Jolesz, R. Kikinis: 3D Image Matching Using a Finite Element Based Elastic Deformation Model. Proceedings 2nd international conference on medical image computing and computer-assisted intervention - MICCAI'99, lecture notes in computer science, vol. 1679, p. 202–209, September 19-22, 1999

17. F. Schutyser, J. Van Cleynenbreugel, N. Nadjmi, J. Schoenaers, P. Suetens: 3D image-based planning for unilateral mandibular distraction. Proceedings 14th international congress and exhibition on computer assisted radiology and surgery, p. 899–904, CARS 2000, June 28 - July 1 2000, San Francisco, USA

Computer-Assisted Orthognathic Surgery: Consequences of a Clinical Evaluation

G. Bettega, MD, PhD[1], F.Leitner, PhD[2], O. Raoult, PhD[3], V. Dessenne, PhD[4],
P. Cinquin, PhD[5], B. Raphael, MD, PhD[1]

1   Plastic and Maxillo-Facial Surgery Department, CHU. A. Michallon, BP 217. 38043
GRENOBLE Cedex 09. FRANCE.
E-mail: GBettega@chu-grenoble.fr
2   AESCULAP AG, Am Aesculap Platz. Postfach 40, 78000 Tuttlingen. GERMANY.
3   10 chemin du Clapero, 38360 Sassenage. FRANCE.
4   MEDIVISION, Müller Institut für Biomechanik, Durtenstrasse 35, Postfach 30, CH-3010
Bern, SWITZERLAND
5   TIMC-IMAG, institut A. Bonniot, Faculté de Médecine de Grenoble, 38700 La Tronche,
FRANCE.

Abstract: _Purpose and method:_ Computer-assisted systems, as any medical technology, have to go through several steps within an endless loop before becoming routine. The first step is the conception of a prototype, the second is its adaptation to the clinical modalities (via animals or cadavers studies), the third is the clinical validation and the last one is the ergonomics optimization preceding the industrialization.
The aim of this work is to present the evolution of a computer-assisted system built for orthognathic surgery i.e. for the surgery of the maxilla and more specifically for repositioning the mandibular condyle after sagittal split osteotomies. The system was based on three-dimensional optical localization of infrared emitting diodes. Eleven patients ("empirical group") underwent condylar repositioning using the empirical repositioning method (standard technique) and were considered controls. In ten patients ("active group") the computer-assisted system was used to replace the condyle bearing fragment in its sagittal preoperative position ; in these cases the condylar torque wasn't controlled. In the third group of ten patients ("graft group"), the computer-assisted system was used to replace the condyle in all three directions ; very often it was necessary, in this group, to fill the osteotomy gap by a bone graft. The clinical evaluation was based on four major criteria: the quality of postoperative dental occlusion, the stability of skeletal position on successive teleradiographies, the occurrence of temporo-mandibular dysfunction (TMJD), and the preservation of mandibular motion. Clinical assessment was made at 1, 3, 6 and 12 months of follow-up. _Results:_ Forty-five percent of the "empirical group" patients do not have the expected postoperative occlusion, five patients showed evidence of clinical relapse at one year, forty-five percent worsened their TMJD status, and they recovered only 63.37% of their mandibular motion amplitudes at 6 months. All the patients of the "active group" had the expected occlusion, only one patient exhibited a mild relapse and TMJD symptoms, and the average mandibular motion recovery was only 62.65% at 6 months. All the patients of the "graft

group" had a good occlusion and no relapse or TMJD. Their percentage of mandibular motion recovery was 77.58%.

Conclusion: those results confirm the utility of a condyle repositioning system. They also prove the accuracy of this computer-assisted method. This lead to an improvement of the system using a smaller localizer and a simple PC directly commanded by the surgeon with a foot-switch, without any technical support. The surgical ancillary is also simplified to reduce the time needed for the setting up of the sensors. The simplified system became a CE marked product and it is now routinely used for every orthognathic procedure in our department.

Key words: Clinical results – Surgical navigation systems – Orthognathic surgery –

1 Introduction

Computer-assisted systems, as any medical technology, have to go through several steps before becoming routine. The first step is the conception of a prototype, the second is its adaptation to the clinical modalities (via animals or cadavers studies), the third is the clinical validation and the last one is the ergonomic optimization preceding the industrialization. The aim of this work is to present the evolution of a computer-assisted system built for orthognathic surgery.

Orthognathic surgery aims to correct dentofacial deformities. The operation have been greatly improved thanks to major advances in anesthesiology and to the improvement of osteosynthesis materials. Nevertheless, many problems are still encountered, creating sources for inaccuracy and instability. Amongst these difficulties, one predominates: positioning of the condyle (the articular fragment) after osteotomy of the mandible. The quality of the procedure is closely related to the operator's experience. Joint laxity and muscular relaxation due to general anesthesia, are two factors which complicate this surgical maneuver. EPKER[1] suggests 3 main reasons to justify a precise repositioning of the condyle fragment during sagittal split osteotomies of the mandibule[1]. The first, and the main reason, is to ensure long-term stability of the surgical result. The second is to reduce noxious effects on the temporomandibular joint and notably to reduce the incidence of temporomandibular joint dysfunction (TMJD). Finally, it may improve postoperative masticatory function. VAN SICKEL[2] confirmed this in a more recent article.

The empirical method is the most widely used method to reposition the condyle fragment after a mandibular osteotomy. It consists in trying to place manually the condyle in its most superior and anterior position within the glenoid cavity[3]. As simple as it is, this method is quite dependent on the surgeon's experience. This is why many mechanical systems have been described. Today, none are really used in common practice and it is difficult to quantify the advantages brought by these repositioning systems since none have really been clinically evaluated. In 1996, we described an original computer-assisted procedure based on the three-dimensional localization optical principle of an infrared transmitter[4].

We present the clinical evaluation of this prototype. These results lead us to improve the system in a way to obtain a routinely used CE marked product. These important evolutions are presented.

2 The Clinical Validation

2.1 Material and Method

The system consists of a 3D optical localizer (Optotrak™ Northern Digital) and a workstation (DEC Alpha, OSF/1) installed on a gurney adapted to dimensions of the operating room. The optical localizer tracks 4 infrared emitting diodes assembled on a small silicon plate called "rigid body". These rigid bodies feature an attachment component making it possible to anchor them to osseous parts. Four rigid bodies are used in this application (figure 1). One is assembled on a probe, which makes it possible to define a reference mark on the patient, the 3 others are anchored on an osseous part: one on a structure not moved by the osteotomies, and used as reference (the selected site was arbitrarily the left upper orbital ridge); the two others on each condyle. Fixation to the orbit is inserted with Kirchner pins transcutaneously; an intermediary part is secured to these pins; the corresponding rigid-body can be attached or removed from this part as often as needed. Condyle anchoring is performed via a mini-plate secured with monocortical screws at the mandibular angle (figure 2); this plate is equipped with a connection which can hold a removable intermediary part, supporting the corresponding rigid-body. This plate is placed transorally using transbuccal instrumentation. Anchoring with 3 5-mm screws provides extreme stability and does not interfere with the sagittal split osteotomy. Surgery thus begins by installation of rigid bodies(figure 3). The next step consists in recording the two condyles' reference position using an intercuspidation splint established during preoperative assessment. The optical localizer determines the position of each condyle's rigid body in relation to the rigid body on the eyebrow. The rigid bodies are then removed and only the osseous anchorage remains, which completely frees the operative field. The surgeon can then perform the mandibular osteotomy of his/her choice without modifying the traditional technique. Maxillomandibular fixation is then secured and the rigid bodies are reinserted on their respective anchorage so that the surgical navigation stage may begin. The latter consists of tracking condylar displacements on computer screen while trying to make its representation overlap the reference position.

The operated patients were randomized by drawing into 3 groups. The first group named "empirical" was all the patients for whom the most experienced surgeon performed a condyle repositioning using the traditional technique, choosing the most superior and anterior position in the glenoid fossa. Computer assistance was only used at the end of surgery to measure possible displacement in relation to the reference position. In the second group named "active", the condyle was repositioned with computer assistance but only in the sagittal plane. The third group named "grafted" corresponds to all the patients whose condyles were repositioned with computer assistance but this time, the position was controlled in three dimensions. In this group, for a significant number of cases, it was necessary to maintain a gap between the various osteotomized osseous fragments by using an osseous graft or an osseous substitute in order to avoid axial or frontal displacement. Four criteria: 2 anatomical and 2 functional were used to assess outcomes. The first anatomical criterion was the evaluation of dental occlusion by comparing post osteotomy occlusion with the planned occlusion. The result is binary i.e. it is regarded as good if the patient perfectly sets in the final intercuspidation plate; it is regarded as bad if it does not. The second anatomical criterion was the stability of mandibular structures on successive cephalometric radiographs. It was studied by superposing the late postoperative cephalograms over the immediate postoperative cephalogram (before the eighth day). The degree of displacement or relapse was measured by comparing the displacement of various points along the symphysis from one cephalogram to the next. The result was regarded as good if there was no displacement, it was regarded as bad if displacement was greater than 2 mm, between 0 and 2 mm the result was labeled as dubious due to possible superposition errors. The first functional evaluation criterion was a possible temporo-mandibular joint dysfunction (TMJD). Temporomandibular joint (TMJ) related symptoms were compared before and after surgery. The results of TMJD assessment was regarded as good if the preoperative status was preserved or improved, otherwise it was regarded as bad. The last evaluation criterion was that of masticatory function assessed with a slide caliper by measuring the amplitudes of movement in opening, in protrusion, and lateral movement. The result was expressed as a percentage of preoperative amplitude. All these criteria were evaluated 1 month, 3 months, 6 months, and a year after surgery. All the patients were operated by the same surgeon and received the same postoperative care.

2.2 Results

On the whole, 31 patients were evaluated: 11 patients were included in the "empirical" group, 10 patients in the "active" group, and 10 patients in the "grafted" group. No specific complication related to the use of the system was noted.

	Right-left translation (mm)			Up-down translation (mm)			Back-front translation (mm)			frontal rotation (degrees)			axial rotation (degrees)			sagittal rotation (degrees)		
	empirical	active	grafted	empirical	active	grafted	empirical	active	grafted	empirical	active	grafted	empirical	active	grafted	empirical	active	grafted
mean	2,2	0,9	0,6	2,0	1,1	1,2	1,4	1,6	1,1	4,0	3,6	1,3	3,2	2,5	2,0	3,9	1,1	0,8
Standard deviation	1,8	0,6	0,4	1,4	1,1	0,6	1,0	1,2	0,9	2,4	2,3	1,0	2,2	2,1	1,0	3,4	1,2	0,8
minimum	0,1	0,0	0,0	0,1	0,1	0,0	0,2	0,0	0,1	0,1	0,4	0,0	0,0	0,1	0,5	0,0	0,0	0,1
maximum	6,6	1,9	1,5	4,7	4,0	2,4	3,6	4,0	3,5	7,5	7,4	4,2	8,4	9,8	3,5	11,3	4,3	1,7

Table 1: values of displacement of the condyle measured by the computer-assisted system

Table 1 summarized the mean values, standard deviation, minimum and maximum displacements of the condyle, measured by the system. The mean displacements were greater, in all directions, for the "empirical" group and the dispersion of the values was also more important. Conversely, the mean values of displacement in the "grafted" group turned around one degree or one millimeter. The standard deviation and the maximum of displacement were also lower for the "grafted" patients than in the other groups, showing a better reproducibility of repositioning. But the maximum rotation values in the frontal and axial plans (respectively 4,2 degrees and 3,5 degrees) attested the difficulties to stabilize, in these directions the condylar bearing fragment during the osteosynthesis. Nevertheless, the repositioning is still better than in the "empirical" and the "active" group. The control of the front/back translation was similar in each group.

Postoperative dental occlusion analysis showed that approximately 45 % of the "empirical" patients did not have the anticipated occlusal result (figure 4). Indeed 5 of the 11 patients required active orthodontics to correct malocclusion, although in most cases, they were not very significant. Nevertheless, only 3 of them, after 6 months of treatment, had obtained a satisfactory occlusion. All the patients of the "active" group and of the "grafted" group presented an occlusion strictly in conformity with that planned. None of them needed postoperative orthodontics. No patients in the "grafted" group presented any relapse at one year of follow-up (figure 5). One patient in the "active" group had a relatively minor relapse (about 1 mm). On the other hand, in the "empirical" group, 3 patients had a relatively minor relapse, less than 2 mm, and 2 other patients showed more than 2 mm relapse and required a second surgery. In the first case, 1 year after the first surgery because of a class III relapse, in the second case, 2 days after.

From a functional point of view, the results were comparable insofar as in the "empirical" group 5 patients had a TMJD or worsened a preexistent TMJD, even though none of them needed a particular treatment (figure 6). In the "active" group, only 1 patient worsened a preexistent TMJD whereas all the patients of the "grafted" group kept same the symptomatology of their significant difference between "grafted" patients and the other patients was noted at the sixth postoperative month follow-up, in terms of mandibular motion amplitude recovery (figure 7). Indeed, the average recovery percentage was the same for the "empirical" and "active" groups (respectively 63.37 % and 62.65 %) with a very significant variance in both groups. On the contrary, for the "grafted" group, not only the percentage of recovery was much higher (77.58 %) but the variance was also much less, demonstrating better reproductivity of the results.
Using computer assistance lead to lengthening the duration of surgery by approximately 50 minutes. This time increase is partly due to the setting up of various elements of the system (approximately 50 % of the time increase). This excludes the extra time necessary for the adaptation of the bone fragments and the graft after repositioning, in order to allow for a stable synthesis.
The scar left after insertion of the eyebrow reference was not noticeable.

2.3 Discussion

Our study featured an objective and quantified comparison between a control population (the "empirical" group) and patients treated with computer assistance. The "active" group was created to check whether the sole control of sagittal condyle repositioning was sufficient to guarantee a satisfactory clinical result. Our results confirm the worth of a condyle repositioning system, especially for anatomical results since postoperative occlusal disorders and relapse was much more significant in the "empirical" group. The comparison with the "active" group and the "grafted" group showed that the precision of sagittal repositioning was the principal factor in the quality of postoperative occlusion and skeletal stability. On the other hand, the functional result was correlated to axial and frontal rotation control, especially for the recovery of mandibular motion amplitude. Moreover, the current prototype version of the system lengthens surgery by approximately 50 minutes. Half of this lengthening is due to the setting up in of the various sensors. Another advantage of this computer-assisted system is that the interface gives the surgeon quantitative and spatial data compared to the chosen reference position. If the surgeon estimates that it is not desirable to reproduce the planned position exactly, he can vary repositioning in the direction of his choice and with the amplitude that he considers adapted.

The precision of this method reaches $\pm$ 0.5 mm and $\pm$ 0.5°. But this level of performance is one which the surgeon's hand cannot reach. In our experiment, it seemed almost impossible to reach a surgical precision below 1 millimeter, taking into account the irregularity of osseous cut and surgical constraints. In any case, clinical results show that in practice this precision is largely sufficient.

Those results confirm the utility of a condyle repositioning system. They also prove the accuracy of this computer-assisted method. But with this first version of the system the surgeon needed the help of a technician to deal with the computer. Moreover the lengthening of the procedure was incompatible with a routine application. To reach this objective it was necessary to improve and simplify the system.

3 Evolution of the Prototype

3.1 Evolution of the Navigation System

The initial system was somehow cumbersome and quite difficult to move (Figure 8A). A Unix workstation was necessary to run the application and a PC was dedicated to localizer management. Another disadvantage, for the surgeon, was his dependence to a technician during the procedure. So the system was drastically modify.

The localizer was changed for a smaller one (Polaris™ Northern Digital). Even with less accuracy, it was sufficient for our application to take into account the accuracy of the surgical procedure. All software (localizer management, and application) was concentrated into a simple PC (even a laptop) directly controlled by the surgeon through a double foot-switch (Figure 8B).

The new platform is very easy to install, it can be integrated into the operating room and managed by nurses as a classical maxillo-facial instrumentation. The great mobility of the system permits to share it with other teams.

3.2 Evolution of the Ancillary

The other ergonomic problem was the lengthening of the procedure, due essentially to fixation of condyle anchoring. The small plate was very satisfactory in a mechanical point of view. It was also very stable and it did not perturb the realization of the osteotomy. However it needed around 10 minutes installation per side. Moreover it was necessary to take off the rigid bodies during the mandibular screw fixation; so that it was impossible to control the condyle position during the osteosynthesis. This control had to be done later after re-installation of the rigid bodies. All this was time consuming (around 10 to 15 more minutes per side).

Then we replaced these plates by small endo-orally fixed vises, on the coronoid process of the mandible (Figure 9). It is simply done using a specific ancillary in less than 1 minute per side. The transbuccal approach is left free so it is possible to control condyle position during osteosynthesis with the rigid bodies in place. The total gain in time varies from 40 to 45 minutes

With this new ancillary the mean lengthening of the intervention is 10 to 15 minutes (mostly due to the insertion of the reference rigid body on the eyebrow).

The new system (Orthopilot™, BBraun Aesculap) is now adapted to routine application in orthognathic surgery. It is now used systematically in our center for this type of surgery.

References

1- Epker BN, Wylie GA.- Control of the condylar-proximal mandibular segments after sagittal split osteotomies to advance the mandibule. Oral Surg Oral Med Oral Pathol,62:613-7, 1986.
2- Van Sickel JE, Don Tiner B, Alder ME: Condylar torque as a possible cause of hypomobility after sagittal split osteotomy: report of three cases. J Oral Maxillofac Surg 55:398-402,1997.
3- Bell WH, Profitt WR, White RP: Surgical correction of dentofacial deformities, volume 2. Philadelphia, PA, Saunders, 1980,p910
4- Bettega G, Dessenne V, Cinquin P, Raphaël B. Computer assisted mandibular condyle positioning in orthognathic surgery. J. Oral Maxillofac. Surg. 54(5):553-8, 1996.

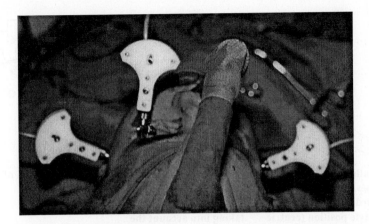

Figure 1: The fixation of the rigid-bodies on the patient. The reference rigid body is place on the left upper orbital ridge (which is not concerned by the osteotomy) and the others are place on each condylar fragment.

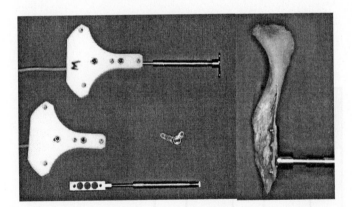

Figure 2: the condyle anchoring system. It included:
- a mini-plate fixed by three screws on the external aspect of the mandibular angle,
- a rod on which is secured the rigid body.

The connection between the rod and the plate is reversible.

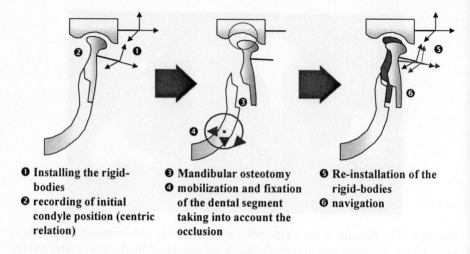

❶ Installing the rigid-bodies
❷ recording of initial condyle position (centric relation)

❸ Mandibular osteotomy
❹ mobilization and fixation of the dental segment taking into account the occlusion

❺ Re-installation of the rigid-bodies
❻ navigation

Figure 3: the algorithm of the surgical procedure

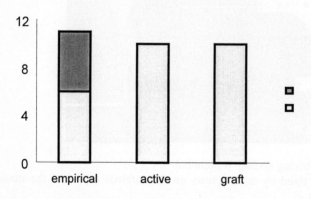

Figure 4: Quality of postoperative occlusion

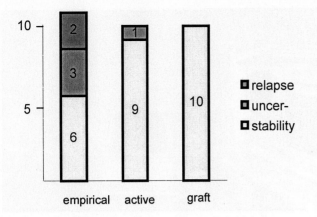

Figure 5: Stability of mandibular position

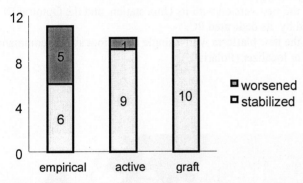

Figure 6: Occurrence of TMJ dysfunction

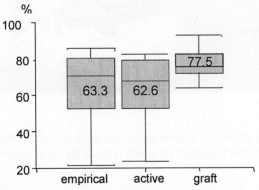

Figure 7: Recovery of mandibular movements at 6 months

Figure 8: The different prototypes
- Figure 8A: the first version with its Unix station, and the Optotrak™ commanded by its dedicated PC;
- Figure 8B: the new platform with a single PC, a foot switch command and a smaller localizer (Polaris™)

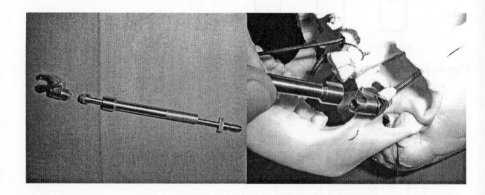

Figure 9: The new ancillary. A small vise is fixed on the coronoid process of the mandible. This vise hold the intermediate stem on which is fixed the rigid body

PC-based Virtual Reality Surgical Simulation
for Orthognathic Surgery

James Xia[1,2&3], Nabil Samman[1], Chee Kai Chua[2], Richie W.K. Yeung[1],
Dongfeng Wang[1], Steve Guofang Shen[1], Horace H.S. Ip[3], and Henk Tideman[1]

[1] Oral & Maxillofacial Surgery, Prince Philip Dental Hospital, University of Hong Kong
j.xia@ieee.org
{nsamman, htideman}@hkucc.hku.hk
{rwkyeung, dfwang, gfshen}@hkusua.hku.hk

[2] *Biomedical Engineering Research Center, School of MPE,*
Nanyang Technological University, Singapore
j.xia@ieee.org
mckchua@ntu.edu.sg

[3] Department of Computer Science, City University of Hong Kong
j.xia@ieee.org
cship@cityu.edu.hk

Abstract. Complex maxillofacial deformations continue to present challenges in analysis, planning and correction beyond modern technology. The purpose of this paper is to present a virtual reality environment for surgeons to perform virtual orthognathic surgical planning in three dimensions. A surgical planning system, "PC-based virtual reality surgical simulation for orthognathic surgery" (VRSSOS), consists of four major stages: CT data post-processing and reconstruction; 3D color facial soft tissue model generation; virtual osteotomy planning; and, soft-tissue-change simulation. The surgical planning and simulation are based on 3D CT reconstructed bone model, whereas the soft tissue prediction is based on color texture-mapped and individualized facial soft tissue model. Our approach is able to provide a quantitative osteotomy-simulated bone model and prediction of post-operative appearance with photo-realistic quality. The predicted appearance can be visualized from any arbitrary viewing point using a low-cost PC-based system.

1   Introduction

Orthognathic surgery attempts to re-establish functional and aesthetic anatomy by repositioning displaced bones or by grafting and re-contouring the deformed bone contour. Soft tissue changes usually are accompanied along with these skeletal alterations. Despite of the advance in imaging modalities, the prediction of soft-tissue-deformation upon the bone-structure-change still poses major challenges to both surgeons and computer scientists.

Three-dimensional (3D) studies in medicine started at the beginning of the 1970s. The cross-sectional imaging capability of computed tomography (CT) and three-dimensional reconstruction have led to a tremendous leap in diagnostic radiology. It can exactly record and represent the life size and shape of bone and soft tissue contour for precise bone surgery planning and simulation.

Surgical planning and simulation requires a conjunctive knowledge between medicine and computer graphics. Upon the points we presented above, the purpose of this study is to establish a virtual reality environment for orthognathic surgical planning, "PC-based virtual reality surgical planning and simulation for orthognathic surgery (VRSSOS)". It provides facilities for surgeons to make surgical plan, to simulate various orthognathic surgical procedures as they routinely performed in operating room and to predict soft tissue changes.

2 VRSSOS overview

This virtual reality workbench consists of four major functions: CT data post-processing and reconstruction; 3D color facial soft tissue model generation; virtual surgical planning and simulation; and soft-tissue-change prediction before operation (Fig.1). Surgical planning and osteotomy simulation were made on the reconstructed 3D bone structures, whereas soft tissue changes were then predicted on the individualized and color texture-mapped facial soft tissue model.

Our approach allows an immersive virtual reality surgical planning environment to be implemented on a relatively low-cost PC-based system when high-end graphics workstations were normally required. Within the VRSSOS, a surgeon can get hold of a virtual "scalpel" to operate on a "virtual" patient, to execute various surgical plans, and to visualize the predicted post-operative results (Fig.2). The system outputs orthognathic surgical plans and a 3D color photo-realistic patient's facial soft tissue model. The osteotomy-simulated bone model and predicted color model can be rotated and visualized from any point of view.

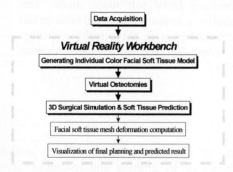

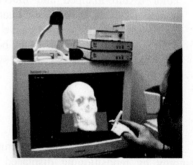

Fig. 1 System overview   Fig. 2 Virtual "patient" on VRSSOS

An IBM compatible *Pentium II®-400* personal computer (PC) was used. To create a virtual reality workbench, the virtual reality devices, such as *Logitech®* three-dimensional flying mouse system and *CrystalEyes® 2* eyewear, were also applied for this study.

The operating system was Microsoft® Windows NT™ Workstation 4.0 with Service Package 6, whereas all the programs were developed by Microsoft® Visual C++™ 6.0, OpenGL® and VTK®.

3 Clinical data acquisition

Fourteen patients with maxillofacial deformities were randomly selected from a large pool of patients awaiting surgical correction of their deformity during December 1995 to December 1998. All patients were evaluated at the Surgical-Orthodontic Joint Assessment Clinic, Oral and Maxillofacial Surgery, University of Hong Kong.

Standard cephalometric radiographs were taken with fixed 110% magnification ratio. Diagnosis and preliminary operative plan were made by the cephalometric analysis and clinical examination.

The patient's Frankfort Plane was parallel to the horizontal plane during photography. Portraits were then scanned as 24 bit images into the computer via a *Nikon LS-20* film scanner.

CT scanning was specially carried out for this study. All patients were examined by a GE® Pace CT Scanner at the Hong Kong Adventist Hospital, with a thickness of 2.0 mm and original 512x512 matrix with 16 bits. A rubber band was used to restrict the patient head's movement during CT scanning. The 2D axial slices started from the submandibular region and covered the whole head.

4 Methods

4.1 CT data post-processing and 3D unique coordinate system creation

CT data post-processing. For the true perspective of data visualization, image processing techniques were used to manipulate image contents to improve the results of subsequent processing and interpretation. The post-processing functions included: removing undesired objects, masking individual anatomical structure and the bone structure enhancement.

Segmentation was the process of classifying pixels in the image or volume. Depending on the scene complexity, it can be one of the most difficult tasks in the visualization process. Interactive segmentation was used if the *threshold filter* was not able to mask the bone in the CT dataset during the *Marching Cubes* process.

Three-dimensional reconstruction. A volumetric dataset was then prepared by using the *Marching Cubes* algorithm[1,2] from the post-processed CT raw data. For fast rendering and shading, all the succeeding operations were based on this volumetric dataset.

Unique three-dimensional coordinate system. The visualization coordinate system represented what was visible to the observer. The x, y coordinates specified location of object in a plane, whereas the z coordinate represented the depth. Horizontal Plane was defined by right and left *Porion*, and the average coordinates of the right and left *Orbitale*. Frontal Plane was perpendicular to Horizontal Plane and through the right and left *Porion*. The Midsagittal Plane was set perpendicular to the Horizontal Plane and the Frontal Plane, and through the *Nasion* (Fig.3).[3] After generating this unique coordinate system, 3D visualization and manipulation could be performed repeatedly, like cephalometric analysis.

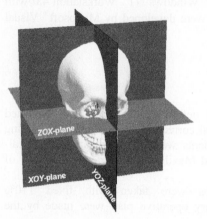

Fig. 3 3D coordinate system transformation

4.2 3D facial model generation

Two techniques were applied in this stage: individualized facial model generation from a generic mesh and color texture mapping.

Generic facial soft tissue model. Our generic facial model is a triangular mesh consisted by 2665 vertices and 5302 triangles.[3] This mesh was created from 3D bone and soft tissue which reconstructed from CT scans of a real human head. It contained the corresponding relationship of the coordinates between bone and soft tissue. Each coordinate on the bone surface had its own unique projection to the soft tissue mesh. When each point on the bone surface was changed, the corresponding coordinate of soft tissue would be changed accordingly. This allows us to simulate soft tissue deformation as a result of bone movement. Additionally, there were a series of built-in vertices representing facial outlines and features.

3D color facial model generation. The vertices controlling the facial features and outline were extracted from the 3D reconstructed soft tissues from CT scans, the generic mesh was then individualized as a texture coordinates by correspondence matching, interpolation and transformation computation based on those fiducial points.

The texture mapping was a procedure similar to pasting a picture to the surface of an object. Three digitized color portraits with the "third" dimension from reconstructed soft tissue were blended into a cartograph as a texture map.

After the cartograph was texture-mapped onto the 3D head model, a photo-realistic model of human head was generated from frontal, right and left real color portraits. It could be rotated freely and visualized from any viewpoint at a real-time speed. All the procedures described in the texture-mapping stage were fully automatically, no interactive operation was required although it involved very complex computations.

4.3 3D virtual osteotomy planning and simulation

This stage was to cut through the 3D reconstructed bone dataset with a surface and then to display interpolated data values on the surface, in order to achieve the virtual osteotomy simulation.

The bone surface was represented by millions of polygons and triangles. A fully automatic decimation computation was first applied on reconstructed bone surface to reduce the triangles up to 70%, in order to keep the balance between maintaining bone structure details, obtaining an acceptable speed of rendering and re-flash rate during cut-through the bone dataset.[4]

The data cutting operation required two pieces of information: a definition for the surface (cutting surface) and a dataset to cut (which bone to be cut). The virtual osteotomy was performed interactively and surgeon needed to decide which osteotomy would be used upon his clinical experience.[5] Fig.2 showed the cutting planes of bilateral subsigmoid osteotomies were defined by moving a 3D mouse.

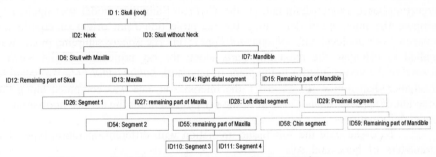

Fig. 4 *Binary Tree Structure* of simulated osteotomies

In the virtual osteotomy simulation, the *Binary Tree Structure* was used to identify the bone segments during maxillary and mandibular osteotomies[5]. Each bone segment was assigned a unique element identity (ID) automatically. A nonlinear structure, called a *tree*, consisted of *nodes*, *branches* and *leaves*. In the simulation of the operative procedures of a segmentalized Le Fort I osteotomy and bilateral vertical subsigmoid osteotomies with genioplasty, the *root* of the tree was the *skull* (ID1). Separating the *neck* from the *skull*, there were two nodes in the tree, the *neck* (ID2) and *remaining part of the skull* (ID3). Separating ID3 into ID6 and ID7, the *maxilla* (ID13) could be osteotomized from *the skull without the mandible and the neck* (ID6) – Le Fort I osteotomy. Four leaves with ID 26, 54, 110 and 111 were then assigned after the maxilla was segmentalized into four pieces. In the same principle, the *mandible* (ID7), also separated from ID3, was osteotomized as *the right distal segment* (ID14), *the left distal segment* (ID28) and *the proximal segment* (ID29) by

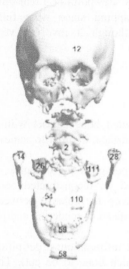

bilateral subsigmoid osteotomies. *The chin segment* (ID58) was osteotomized from *proximal segment* (ID29) by genioplasty at last (Fig.4).

After osteotomies, each bone segment with it's own ID could be manipulated separately in three dimensions, including translation, rotation, elevation, changing rotation center, displaying movement parameter, color masking, scaling, range boundaries and switching each object to visible or invisible (Fig.5).

4.4 3D soft tissue simulation

It is a soft tissue deformation scheme for simulating the soft-tissue-change according the underlay bone-movement during planning. Two soft tissue deformation algorithms were created for this scheme: the *Surface Normal-based Model*

Fig.5 osteotomized bone segments

Deformation Algorithm and the *Ray Projection-based Model Deformation Algorithm.*[6] The corresponding relationship of the vertices of triangles between the bone and the soft tissue surface was computed and saved as an intersection data file. For the ray projection-based model deformation, the *Cylinder* sampling model was applied to compute the intersection data along the ray-projection from origin of coordinate, between *Eyebrow* level and *Labiomental Fold* level. The *Sphere* sampling model was applied to calculate the remaining parts along the ray projection. If the *Surface Normal* of the vertex inferior to the plane of landmark *Labiomental Fold* was equal to or smaller than minus 10° (≤ -10°), the intersection data was computed along the direction of surface *Normal* instead of the ray projection for the surface normal-based model deformation.

To orchestrate the soft tissue prediction with different operation types, the parameters of bone and soft tissue movement ratios were transferred from intervened ratio files correspondingly.[7]

Three-dimensional color facial texture-mapping technique was applied again to regenerate a color photo-realistic facial model after soft tissue deformation.

The soft tissue changes were simply predicted by interactively and intraoperatively dragging the

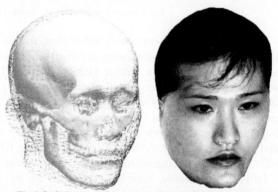

Fig.6 Soft tissue deformation after virtual osteotomies

bone segments to desired position. When the surgeon translated, rotated and elevated each bone segment, soft tissue model was correspondingly deformed and texture was re-mapped from the cartograph in a real-time speed (Fig.6). A predicted three-dimensional facial visualization with color photo-realistic quality was precisely created. Both the bone and soft tissue models could be rotated freely and visualized immediately from arbitrary observing angle with zoom-out or zoom-in. The each object of the bone and soft tissue also could be masked as visible or invisible in order to get the best visualization during the virtual operation.

5 Clinical trail

To date, 14 cases have been planned using integrated two- and three-dimensional data and analysis. Surgical planning and simulation has been performed by this VRSSOS system. The post-operative visualization is predicted before operation.

The surgeon can use different combinations of short-cut keys, two-dimensional mouse, three-dimensional flying mouse and *CrystalEye* eyewear, to perform the virtual osteotomies, to move bone segments and simulate soft tissue changes (Fig.2). The operator's clinical experience is critical for achieving a successful planning.

Based on the procedures and time requirements for analysis, six stages of treatment planning are available. A typical case with mandibular hyperplasia and mild paranasal deficiency was presented. The operative procedures were: Le Fort I osteotomy with four pieces segmentation, bilateral subsigmoid osteotomies and genioplasty.

The first stage involves data acquisition and clinical evaluation, such as CT scanning, color portraits capturing, cephalometric analysis and primary clinical assessment. This requires 2 to 7 days of effort.

The second stage involves CT raw data post-processing, including removing undesired objects, enhancement processing, masking bone structure, etc. This requires 2 to 3 hours of effort.

The third stage involves interactive visualization of the three-dimensional images, including the volumetric dataset generation, 3D coordinates geometry transformation, etc. This requires about 1 hour of effort (Fig.3).

The fourth stage of planning is to generate individual facial soft tissue model with a color texture mapping. This procedure requires about 1-2 hours.

The fifth stage of planning is to make *Decimation* computation and to perform varieties of procedures of virtual "osteotomies" in virtual reality environment. This step is only operated on the bone. It requires about 1 hour or less (Fig.5).

The sixth stage of planning is much quicker and requires about 10 to 15 minutes. This step is totally operated in the virtual reality environment with VR devices. The surgeon can move and rotate the osteotomized bone segments freely and observe the predicted result with full color visualization. The operator may also compare results between the predicted and the original facial models in three-dimensional visualization (Fig.6).

A planning and simulation session was held at the computer system with the maxillofacial team for review and final modifications. The quantitative translation and rotation data of each object could be displayed respectively. The final result of planning and simulation can be visualized from arbitrary viewpoints in the virtual reality workbench.

Follow-up was made six months after operation. A comparison, between original color portraits, predicted facial model, postoperative portraits, simulated osteotomies and bone structures after surgery, is shown in Fig.7. Fig. 8 shows another case with the same operative procedures.

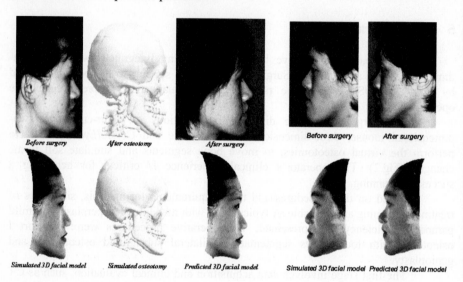

Fig. 7 Comparison between Pre-OP and Post-OP   Fig. 8 Comparison

6   Discussion

Surgical planning and simulation has wide applications, such as for orthognathic surgery, craniofacial surgery, cleft palate, trauma, distraction osteogenesis, etc. In general, surgical simulation is to perform the "surgery" in the computer and to construct a "fake" image of patient's post-surgical appearance before operation.

Three-dimensional imaging technology is now widely available. It is used to aid in the comprehension and application of volumetric data to diagnosis, planning and therapy.[8] There are several studies on surgical planning for maxillofacial and craniofacial surgery. *Altobelli* et al used the three-dimensional surface reconstruction from CT data to plan craniofacial operations.[9] Cephalometric analysis was integrated with 3D CT reconstruction to quantify the skeletal deformities and to assist in the design of the surgical procedure. Interactive techniques were developed to simulate osteotomies and skeletal movement in three dimensions on the computer-generated surface images. It took about one or two days for bone surgery simulation on an extremely powerful high-end graphic workstation. Three-dimensional soft tissue prediction based on bone movement was not included.

Stereolithographic biomodeling is a modern technology that allows 3D CT data to be used to manufacture solid plastic replicas of anatomical structures (biomodels).[10] It can produce the life-sized deformed bone structures, so that surgeon can directly observe patient's deformity and make surgical planning on the biomodel before operation. It is good for planning complex craniofacial surgery, however, it needs very expensive Stereolithography Apparatus (SLA) machine to fabricate this biomodel, and soft-tissue-change prediction is still not included.

Soft-tissue-change prediction is an important aspect in surgical planning, especially in orthognathic surgery. The orthognathic surgery is not only to correct the mal-functions and to re-locate mal-positioned bones, but also for aesthetic purpose. Currently soft tissue simulation are still mainly focused on the two-dimensional profile prediction (color video imaging techniques) combining with cephalometric analysis in recent two decades,[11,12] only a few studies involved in three-dimensional soft tissue capturing and deformation.[3,5,6,13-15] Ayoub et al designed a vision-based three-dimensional facial data capture system (photogrammetry) for the planning of maxillofacial operations.[13,14] However, the relationship between soft tissue and bone is still remaining a question.

From the methods discussed above, there is no VR-based 3D system to make a precise orthognathic surgical planning and soft tissue prediction in the personal computer. The aim of this project is to combine three-dimensional visualization, virtual reality, color video imaging technique, surgical planning and simulation together, in order to solve the bone-soft tissue relationship problem and provide surgeon an ideal virtual reality workbench for orthognathic surgical planning, simulation and soft tissue prediction in a personal computer system. The workbench is designed not only for surgical planning, but for teaching purpose as well. This study has achieved the following general objectives:

- To create a three-dimensional virtual reality workbench for surgical planning, simulation and soft tissue prediction for orthognathic surgery;
- To generate fast 3D visualization, so that the multiple virtual osteotomies can be real-time performed;
- To obtain a final product — the quantitative osteotomy-simulated 3D bone model and the precise soft-tissue-change predicted 3D facial model with color photo-realistic quality, which can be real-time visualized in three-dimensional virtual reality environment;
- All procedures can be operated in a conventional personal computer system with relatively low-cost VR devices, instead of extremely expensive high-end VR graphics workstation.

Considering the cost and radiation of whole skull CT scanning to the patient, this system is more likely to suitable for the cases with complex maxillofacial deformity, such as asymmetric deformity, cleft palate, distraction osteogenesis, trauma, where radiation and cost of CT scan need to be justified.

Future work will focus on a comparative study between predicted and postoperative models by surface anthropometry in order to verify the accuracy of the system quantitatively.

Acknowledgement

Authors thank Dr. Ron Kikinis, Associate Professor of Radiology and Director of Surgical Planning Laboratory, Brigham and Women's Hospital, Harvard Medical School, for his valuable suggestions and comments on this computer system.

References

1. Cline H.E., Lorensen W.E., Ludke S., et al: Two algorithms for the three-dimensional reconstruction of tomograms. *Med Phys.* 15: 320-327; 1988
2. Lorensen W.E. and Cline H.E.: Marching cubes: a high resolution 3D surface construction algorithm. *Comput Graphics.* 21: 163-169; 1987
3. Xia J., Wang D., Samman N., et al: Computer-assisted three-dimensional surgical planning and simulation – 3D color facial model generation. *Int J Oral Maxillofac Surg.* 29: 2-10, 2000
4. Schroeder W., Zarge J. and Lorensen W.: Decimation of triangle meshes. *Comput Graphics.* 26: 65-70, 1992
5. Xia J., Ip H.H.S., Samman N., et al: Computer-assisted three-dimensional surgical planning and simulation – 3D virtual osteotomy. *Int J Oral Maxillofac Surg.* 29: 11-7, 2000
6. Xia J., Samman N., Yeung RWK, et al: Computer-assisted three-dimensional surgical planning and simulation – 3D soft tissue planning and prediction. *Accepted by Int J Oral Maxillofac Surg*, 2000
7. Bell W.H.: Modern practice in orthognathic and reconstructive surgery. Saunders: Philadelphia, USA. *Pp*2186-96, 1992
8. Vannier M.W. and Marsh J.L.: Three-dimensional imaging, surgical planning, and image-guided therapy. *Radiol Clin North Am.* 34: 545-63, 1996
9. Altobelli D.E., Kikinis R., Mulliken J.B., et al: Computer-Assisted Three-dimensional planning in craniofacial surgery. *Plast Reconstr Surg.* 92: 576-85, 1993
10. D'Urso P.S., Barker T.M., Earwaker W.J., et al: Stereolithographic biomodelling in cranio-maxillofacial surgery: a prospective trial. *J Craniomaxillofac Surg.* 27: 30-37, 1999
11. Schultes G., Gaggl A. and Karcher H.: Accuracy of cephalometric and video imaging program Dentofacial Planner Plus in orthognathic surgical planning. *Comput Aided Surg.* 3: 108-14, 1998
12. Xia J., Wang D., Qi F., et al: Computer aided simulation system of orthognathic surgery. In *Proceedings of Eighth IEEE Symposium on Computer-based Medical Systems*, Lubbock, TX, USA: June 9-10, 1995. IEEE Computer Society Press: Los Alamitos, USA. *Pp*. 237-44, 1995
13. Ayoub A.F., Wray D., Moos K.F., et al: Three-dimensional modeling for modern diagnosis and planning in maxillofacial surgery. *Int J Adult Orthodon Orthognath Surg.* 11: 225-33, 1996
14. Ayoub A.F., Siebert P., Moos K.F., et al: A vision-based three-dimensional capture system for maxillofacial assessment and surgical planning. *Br J Oral Maxillofac Surg.* 36: 353-7, 1998
15. Keeve E., Girod S., Kikinis R., et al: Deformable modeling of facial tissue for craniofacial surgery simulation. *Comput Aided Surg.* 3: 228-38, 1998

A 3-D System for Planning and Simulating Minimally-Invasive Distraction Osteogenesis of the Facial Skeleton

Peter C. Everett[1], Edward B. Seldin[2], Maria Troulis[2], Leonard B. Kaban[2] and Ron Kikinis[1]

[1]Surgical Planning Laboratory, Brigham & Women's Hospital, Harvard Medical School, 75 Francis Street, Boston, MA, 02115, USA
{peverett,kikinis}@bwh.harvard.edu, http://spl.bwh.harvard.edu
[2]Department of Oral & Maxillofacial Surgery, Massachusetts General Hospital, Harvard Medical School, 55 Fruit Street, Boston, MA 02114, USA
{eseldin,mtroulis,lkaban}@partners.org

Abstract. Three-dimensional planning tools will enable the use of minimally-invasive distraction osteogenesis for the correction of craniomaxillofacial deformities by simulating treatment, precisely quantifying movement vectors, and aiding pre and post-treatment evaluation. Current techniques extrapolate 3D surgical movements and outcomes based on standard 2D radiographs. Surgical planning and outcome evaluation would be greatly improved by an accurate, reproducible and reliable 3D treatment planning system.   Building upon a software foundation that includes the *3D Slicer* of the Brigham & Women's Hospital, and the Visualization Toolkit (VTK) of Schroeder, Martin & Lorensen, we add algorithms that support interactive cutting of large triangulated surface models, collision detection, landmark-based registration, and cephalometric analysis. The oriented bounding-box tree (OBB tree) structure is used throughout to enhance the interactivity of selection, collision detection, and cutting. The cutting tool is notable for its generality and preservation of topological closure in the resultant models. In a retrospective case study, the collision of the proximal fragment of the distracted mandible with the skull base is detected and the resulting 3D bone movements are quantified. The distracted bone volume is computed. In prospective cases, this system will be used to compute the placement and configuration of appropriate buried distractor(s).

1   Introduction

The success of most CMF surgery depends on careful planning based on accurate diagnostic information. Most orthognathic and craniofacial surgical procedures address skeletal disproportions that are visualized in the mid-sagittal plane, hence the great utility of the lateral cephalometric x-ray in conventional treatment planning. However, for the many patients with asymmetric deformities, projection of the facial anatomy onto the mid-sagittal plane does not permit adequate evaluation and planning, necessitating 3D analysis [1]. Improved surgical techniques, such as minimally-invasive endoscopic osteotomy and placement of a buried distraction device, further heighten the need for accurate planning tools, which now must compensate for decreased visual access and decreased ability to make mid-treatment adjustments.

Traditionally, surgeons have considered translational and rotational components of a proposed correction separately. These six position parameters, three in translation and three in rotation, are sufficient to characterize the final positions of skeletal components, but do little to address the question of movement path, which is an essential characteristic of distraction osteogenesis. From two reference points, the movement path can be characterized in many cases, using Euclidian principles, as a rotation about a unique axis [3]. When applicable, this provides a way to prescribe both the treatment motion and final position.

Computer tomographic (CT) and magnetic resonance imaging of the human body has had an enormous impact in medicine during the past 20 years. Craniofacial anomalies [7] and fine anatomic details of facial traumatic injuries [14] have been accurately demonstrated with such imaging techniques. Currently, acrylic models of the skull can be created using CT based 3-D imaging and stereolithography. Surgical planning and simulation can be carried out on such models [8,9] Custom alloplastic implant prostheses can be simulated and then produced [10]. However the process of acquiring a 3-D CT scan, producing a model using stereolithography, and simulation on the model is time-consuming and very costly. Furthermore, the axis of a proposed movement can not be accurately calculated, nor can multiple "what if?" simulations be easily carried out.

Interactive surgical simulation technology [11,4] is currently being modified for use in the craniomaxillofacial skeleton. After the CT-data are acquired, bone and soft-tissue segmentation algorithms are applied, and a compression algorithm is used to decrease the size of the data set without sacrificing anatomic detail [16,5] A sytem that provides the flexibility to simulate any craniofacial skeletal operation has been [5]. This used a planar cutting tool of infinite extent and infinitesimal thickness.

Our goal was to devise a tool that overcame the limitations of prior systems, and that used a conceptual framework that was well-suited for the specific requirements of distraction osteogenesis. Specifically it should:

- Have highly interactive performance in all planning phases.
- Provide clear visualization of the relevant anatomy.
- Permit generalized cutting with realistic tool geometry.
- Create topologically closed bone surface fragments.
- Use collision detection to constrain bone fragment movements.
- Describe the planned movements in an intuitive way.
- Generate a motion plan for each distractor.
- Aid quantitative analysis via cephalometric measurement.

Our current system realizes each of these goals. It is being used for retrospective analysis of about twenty cases. The advantages of our system in cases of hemifacial microsomia are evident, and plans are in place to begin prospective cases in the coming months.

2 Methods

2.1 Surface Model Generation

Starting from a high-resolution 3D CT, we use the 3D Slicer [17] to perform semi-automatic segmentation of bone based on an appropriate radiologic threshold. Resolution limitations make false connections between upper and lower dentition, and between mandible and skull at the TMJ common. The incidence of these connections has been reduced by using tensor-controlled local structure enhancement filtering [5], but a fully automatic bone segmentation remains elusive. Therefore, an hour or two of manual refinement is typically required for segmentation.

From this segmentation, the 3D Slicer uses the marching cubes [11] method and triangular surface decimation [16] of VTK [15] to generate triangulated surface models of the facial skeleton, and the skin. The large size of these models, typically 50,000 triangles, requires a data structure that supports O(log N) triangle location, for landmark selection, collision detection, and cutting. For this we have employed an OBB-tree structure [13], and contributed enhancements to its open-source implementation in VTK.

2.2 Landmark Identification

The operator may select the cephalometric landmarks of interest either by picking points on the resliced plane images provided by the 3D Slicer, or by picking points on the surface model in the 3D window, depending on which representation makes identification simpler. We represent these landmarks visually as "radar reflector" glyphs, composed of three orthognal squares through the center of an octahedron. We introduce this representation because it has the virtue of clear visibility without obscuring the central point of interest. Picking points on the triangulated surface models is greatly sped up by intersecting the projected ray with the OBB-tree structure. By doing so, only the triangles in OBB-tree leaf nodes that intersect the ray are tested for intersection with the ray.

The selected landmarks are logically identified with the models to which they are attached, so that they are carried with any subsequent registrations or repositionings.

2.3 Data Set Alignment and Cephalometric Analysis

Cephalometric analysis requires that the scan be aligned to a standard orientation. In 2D cephalometric radiographs, this alignment is part of the x-ray procedure. In 3D

CT, the alignment of the scan with the sagittal, axial, and coronal planes is only approximate. If projected measurements are to be accurate, the CT must be aligned to the mid-sagittal and horizontal planes. We accomplish this by identifying three or more points on the mid-sagittal plane, and two or more points on the horizontal plane. Jacobi iteration on the covariance matrices of the selected landmarks provides a least-squares fit to a plane, and Gramm-Schmidt orthogonalization enforces orthogonality between the the vertical and horizontal planes.

With the CT data, surface models, and landmarks registered to standard orientation, we provide a tool for measuring cephalometric distances and angles, either in 3D, or projected onto the sagittal, coronal, or axial planes. Comparison of these measurements with a database of age-related standard values, tempered by the judgement of the surgeon in the particular case, suggests the motions necessary to treat the deformity, and the osteotomy or osteotomies that will be required.

2.4 Simulated Osteotomy

Prior tools for the simulation of craniomaxillofacial osteotomies have been limited in the types of cuts they could perform, and in the representation of the resultant bone fragments as topologically non-closed objects. We address both of these problems with the implementation of a general boolean operation engine for triangulated surface models. This engine can operate in two modes. First, it can rapidly identify collisions between two large triangulated models by colliding the two OBB-tree structures. Second, it can compute the resulting triangulated surface from the subtraction of a tool model from a bone model. This second mode achieves interactive performance by using the prior OBB-tree collision to quickly reject the large numbers of non-participating triangles in the boolean operation.

The core of the boolean engine is triangle-triangle intersection. Each triangle in a leaf node of the two OBB-trees is tested against the triangles from all intersecting nodes by the computation of "piercepoints." These are the points where the edges of one triangle cross the plane of the other. An intersection occurs if and only if there are four piercepoints, a0, a1, b0, b1, which reveal when sorted that the interiors overlap.

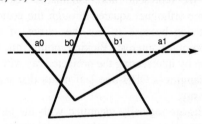

Fig. 1. Triangle intersection sorts up to 4 piercepoints, making up to 1 edge.

New edges are generated from intersecting triangles, from which are generated new triangles, which are then organized into a resultant model.

The cutting tool model is positioned with respect to the bone to be cut using a movable coordinate system actor that attaches itself to the tool model and allows dragging the tool along or around any if the three axes. Once positioned, the cut operation is applied. Additional tool positionings and cuts can be made in the case of multiple or complex osteotomies. The operator then identifies and names each resultant component.

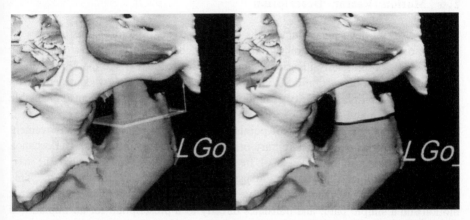

Fig. 2. The cutting tool can have any shape, has finite thickness, and caps the cut objects.

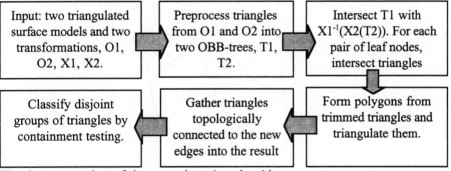

Input: two triangulated surface models and two transformations, O1, O2, X1, X2.	Preprocess triangles from O1 and O2 into two OBB-trees, T1, T2.	Intersect T1 with $X1^{-1}(X2(T2))$. For each pair of leaf nodes, intersect triangles
Classify disjoint groups of triangles by containment testing.	Gather triangles topologically connected to the new edges into the result	Form polygons from trimmed triangles and triangulate them.

Fig. 3. An overview of the general cutting algorithm

2.5 Reposition with Collision Detection

The next step in simulated treatment planning is the movement of each bone fragment to the new position. Using the same movable coordinate system interface that was used to position the cutting tool, each skeletal component is translated and rotated until visual inspection and re-computation of the cephalometric measurements from the repositioned landmarks indicate the desired correction has been achieved.

While the bone is being moved, the boolean engine is used in collision detection mode to test for collision between the moving bone and the stationary skeleton. Collision is indicated with a beep and a color change in the models that have collided. On a Sun Ultra-10 workstation with Elite3D graphics acceleration, 3-20 collision detections per second is typical, with a strong dependence on the closeness of approach and the number of triangles.

2.6 Motion Vector Description

Several different conceptual frameworks can be used to capture the six degrees of freedom of three dimensional object repositioning, each with different strengths and weaknesses. For minimally-invasive distraction osteogenesis, where the repositioning will be done by a distraction device placed by a surgeon, a primary consideration is intuitive accessibility, simplicity of motion path, and translatability to an implantable distractor prescription.

We describe the repositioning of points on bone fragments as helical movement around and along a cylinder of arbitrary orientation in three dimensions. The equation of this axis captures four degrees of freedom. The rotation angle around the axis captures the fifth, and the translation parallel to the axis captures the sixth. This representation is reasonably informative to the intuition, describes both the net motion and a simple path, and translates directly to a prescription for an implantible distractor. In the case of simple translation without rotation, the solution is degenerate, so an axis through the origin and parallel to the direction of motion is chosen.

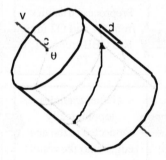

Fig. 4. Repositioning in 3-D is captured by helical movement, degenerate when $\theta = 0$.

For each osteotomy, three sets of repositioning parameters, or prescriptions, are of interest. First is the movement of the distal fragment, which in hemifacial microsomia is typically a rotation around an axis through the unaffected chondyle. Second is the movement of the proximal fragment. Third is the combined movement of the proximal fragment relative to the distal portion, which combined with placement information provides the prescriptive parameters for a buried distractor that will perform the desired repositioning when placed across the osteotomy.

Starting from the matrix representation of the bone repositioning, a 3x3 rotation matrix **R**, and a translation vector **t**, quaternion methods are used to determine the direction of the rotation axis. A quaternion [x, y, z, w] is a 4-vector whose xyz portion is along the unit axis of rotation, **v**, and whose component w is equal to the cosine of half the angle [18].

$$q = [\ v \sin(\theta/2),\ w \cos(\theta/2)\ ]$$

With the axis direction and rotation angle determined, it remains to compute a point on the axis and the movement's projection along the axis. The vector **t** is decomposed into a component along the axis, t_a, and a component perpendicular to the axis, t_p. The product $t_a \cdot v$ is the distance parameter and the latter is substituted into the following equation for finding the rotation center **c**, which is unchanged by **Rt**:

$$cR + t = c \quad \text{or} \quad c(R\text{-}I) + t = O$$

However, with any point on the axis being a solution **(R-I)** is not invertable in three dimensions. A unique solution is forced by replacing the smallest column vector of **(R-I)** with **v**, thus imposing the additional constraint $c \cdot v = 0$, making **c** the closest solution to the origin. We now have the full set of repositioning parameters: **c, v, d, θ**.

2.7 Post-Treatment Scan Registration & Evaluation

The evaluation of post-treatment scans, for comparison with pre-treatment or simulated post-treatment models, requires the registration of the corresponding non-moving anatomy. We use a rigid, landmark-based registration that performs a least-squares fit of the corresponding landmarks using singular value decomposition [19]. The most reliable non-moving landmarks vary from case to case, so the operator selects corresponding pairs of preselected cephalometric landmarks that are judged to be unchanged by the treatment interval, or by differences in patient position. By selecting a large number of widely-spaced landmarks, the errors in repeatability tend to cancel, making sub-voxel accuracy attainable.

Thus registered, the predicted and actual cephalometric quantities can be compared. Moreover, the visual superposition of the two models allows for a more general qualitative evaluation of the treatment.

3 Results

3.1 Illustrative Case

A seven year old boy with hemifacial microsomia affecting the left side, presents a shortened left mandibular ramus, which is constraining the growth of the maxilla and is interfering with bite function. We compare the actual treatment result using a conventional treatment planner and a semi-buried distractor with the simulated result of a plan created with our system.

When comparing the 3-D treatment plan with the conventional plan and its result, several advantages of the 3-D system are immediately evident. First, using the conventional planner it was not possible to accurately predict the upward movement of the proximal fragment, nor was it possible to predict that the motion-limiting collision would be the coronoid process against the skull base, rather than the hypoplastic chondyle against the fossa. Both of these predictions were readily seen in the 3-D planning process, thus avoiding an undercorrection. Second, the vectors of movement are readily available from the 3-D planner in a form that translates directly to a distraction device. Third, the ability to freely manipulate skeletal components around and along all axes, constrained by simulated bony collision, makes is easier to visualize and reach a functional and aesthetic result.

Table 1. The performance of both cutting and collision detection ranges from around 3 to 150 frames per second, thus ensuring smooth interactivity. These figures are from a Sun Ultra-10 with SunOS 5.7 and elite3D graphics acceleration.

Cutting Performance	
# of triangles	44,090 (mandible) + 12 (cutter)
OBB-tree preprocessing time (ms)	3,358 (mandible) + <1 (cutter)
cutting time (ms)	316
# of intersecting OBB leaf nodes	81
# of triangle-triangle tests	807
# of new edges	307
Collision Performance	
# of triangles	171,994 (skull) + 2,361 (prox.mandible)
collision detection time (ms)	6 to 200
# of intersecting OBB leaf nodes	0 to 35
# of triangle-triangle tests	0 to 130

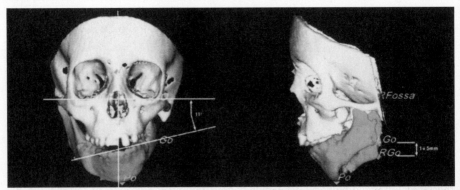

Fig. 5. Having used selected landmarks to perform least-squares alignment to the mid-sagittal and horizontal planes, A-P and L-R views allow visualization and analysis of the deformity through the measurement of distances and angles between landmarks.

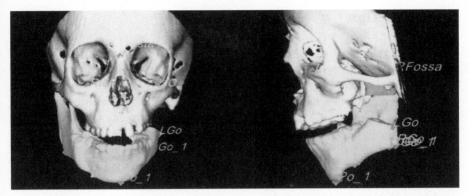

Fig. 6. After simulated cutting, the proximal fragment is moved upward, slightly leftward, and pitched forward a little to avoid the premature collision with the skull base that was detected. The distal fragment is rotated about the center of the right chondyle until functional and aesthetic constraints are satisfied.

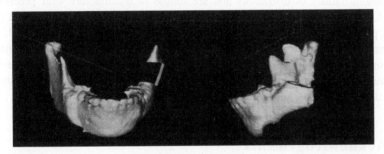

Fig. 7. The simulated post-treatment mandible in shown. The gray line is the 3-D axis of the helical motion. The object in the gap is a section of the surface along which a point at the edge of the osteotomy travels. It captures the geometry of the drive track for a buried distractor placed at that point.

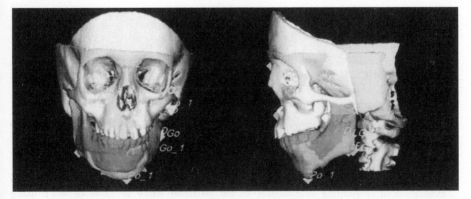

Fig. 8. Rigid registration on 6 pairs of unchanged landmarks, (3 shown), between a simulated post-treatment coronal scan and the actual post-treatment axial scan highlights the good fit in the unchanged skull (avg. distance 1.55mm), and areas where the simulated and actual treatments were different. (Bones from the actual treatment are in darker colors.)

4 Discussion

While the 3-D system presented offers a number of valuable planning tools for craniomaxillofacial surgery, it is also intended to serve as a platform for additional capabilities. It is possible to interface the 3D Slicer to a 6 degree of freedom mouse, or even a haptic interface that will provide force feedback when collision or cutting occurs, and can even be made to model the forces from soft tissue. Another important area for minimally-invasive surgery is in intraoperative navigation. One approach that has been used in another 3-D CMF treatment planning system [12] is the stereolithographic manufacture of a hole-drilling template from the CT data. However, this approach requires direct visual access to the bone surface, something that is not available with an endoscopic procedure.

By choosing a helical concept for motion planning, it was hoped that both intuitive appeal and appropriate physiologic distraction will be achieved for the majority of cases. One additional benefit of this approach is that it suggests a design for a general family of distractors that are simple to manufacture with precision and strength. Our approach is different from earlier distraction planning systems in that the desired path is computed first, and the appropriate device is selected second. Earlier work has taken the device as a given, and then computed the necessary adjustments for executing the treatment plan. It is hoped that by making the planning phase device-independent, and having a sufficiently rich family of distraction devices, the treatment plan will be simpler and more robust to execute.

5 Acknowledgements

This work has been supported by the Center for Innovative Minimally-Invasive Therapy (CIMIT).

6 References

1. Mollard B, Lavallée S and Bettega G: Computer Assisted Orthognathic Surgery. Proceedings MICCAI '98, 1-8, LNCS, Vol. 1496, Springer, Berlin, 1998
2. Altobelli D, Kikinis R, Mulliken J, Cline H, Lorensen W and Jolesz F, 1993 Computer-assisted three-dimensional planning in cranofacial surgery. Plast Reconst Surg, 92: 576-585.
3. Seldin EB, Troulis MJ, Kaban LB: Evaluation of a semiburied fixed-trajectory curvilinear distraction device in an animal model. J Oral & Maxillofac Surg., December 1999, 57(12):1442-1446
4. Richolt JA, Teschner M, Everett PC, Girod B, Millis MB, Kikinis R: Planning and evaluation of reorienting osteotomies of the proximal femur in cases of SCFE using virtual three-dimensional models. Proceedings MICCAI '98, 1-8, LNCS, Vol. 1496, Springer, Berlin, 1998
5. Keeve E, Girod S, Girod B: Interactive craniofacial surgery simulation. Proc ACM Siggraph '96, New Orleans, LA, Aug 4-9, 1996.

6. Westin CF, Sarfield SK, Bhalerao A, Mui L, Richolt JA, Kikinis R: Tensor controlled local structure enhancement of CT images for bone segmentation. Proceedings MICCAI '98, 1205-1212, LNCS, Vol. 1496, Springer, Berlin, 1998.

7. Vannier MW, marsh JL, Warren JD: Three-dimensional CT reconstruction images for craniofacial surgical planning and evaluation. Radiol 150:179, 1984.

8. Bill JS, Reuther JF, Dittmann et al: Stereolithography in oral and maxillofacial operation planning. Int J Oral Maxillofac Surg 24(1):98, 1995

9. Ayoub AF, Wray D, et al: Three-dimensional modeling for modern diagnosis and planning maxillofacial surgery. Int J Adult Orthod Orthognath Surg 11(3):225, 1996.

10. Wehmoller M., Eufinger H, Kruse D, et al: CAT by processing of computer tomography data and CAM of individually designed prosthesis. Int J Oral Maxillofac Surg 24(1):90, 1995.

11. Lorensen WE, Cline HE: Marching Cubes: A high resolution 3D surface construction algorithm. ACM Computer Graphics, 21(4):38-44, 1987.

12. Gateno J, Teichgraber JF, Aguilar E, 2000 Distraction osteogenesis: a new surgical technique for use with the multiplanar mandibular distractor. Plast Reconst Surg 105(3):883-8

13. Gottschalk S, Lin M, and Manocha D, 1996 OBB-Tree: A Hierarchical Structure for Rapid Interference Detection. Proc. ACM Siggraph '96, 171-180.

14. Levy RA, Edwards WT, Meyer JR, et al: Facial trauma and 3-D reconstructive imaging: Insufficiencies and correctives. AJNR 13:855, 1992.

15. Schroeder W, Martin K, and Lorensen W, 1997 The Visualization Toolkit. 2nd ed., Prentice-Hall, New Jersey.

16. Schroeder, W., J. Zarge, and W. Lorensen. 1992. Decimation of Triangle Meshes. Computer Graphics 26(2):65-70.

17. D. Gering, A. Nabavi, R. Kikinis, W. Eric L. Grimson, N. Hata, P. Everett, F. Jolesz, W. Wells III. An Integrated Visualization System for Surgical Planning and Guidance using Image Fusion and Interventional Imaging. Medical Image Computing and Computer-Assisted Intervention (MICCAI), Cambridge England, Sept 1999.

18. Shoemake K: Animating Rotation with Quaternion Curves, Computer Graphics, Vol. 18 No.3, July 1985, SIGGraph '85 Proc.

19. Arun KS, Huang TS, Blostein SD: Least-Squares Fitting of Two 3-D Point Sets, IEEE PAMI Vol. 9 No.3, September 1987.

Modeling for Plastic and Reconstructive Breast Surgery

David T. Chen[1], Ioannis A. Kakadiaris[1], Michael J. Miller[2],
R. Bowen Loftin[1], Charles Patrick[2]

[1] Virtual Environments Research Institute and Department of Computer Science
University of Houston, Houston, TX 77204, USA
{davechen,ioannisk,bowen}@uh.edu
[2] Laboratory for Reparative Biology & Bioengineering, Dept. of Plastic Surgery
UT M.D. Anderson Cancer Center Houston, TX 77030, USA
{mmiller,cpatrick}@mail.mdanderson.org

Abstract. In this paper, we present the modeling and estimation aspects of a virtual reality system for plastic and reconstructive breast surgery. Our system has two modes, a *model creation* mode and a *model fitting* mode. The model creation mode allows a surgeon to interactively adjust the shape of a virtual breast by varying key shape variables, analogous to the aesthetic and structural elements surgeons inherently vary manually during breast reconstruction. Our contribution is a set of global deformations with very intuitive parameters that a surgeon can apply to a generic geometric primitive in order to model the breast of his/her patient for pre-operative planning purposes and for communicating this plan to the patient. The model fitting mode allows the system to automatically fit a generic deformable model to patient specific three-dimensional breast surface measurements using a physically-based framework. We have tested the accuracy of our technique using both synthetic and real input data with very encouraging results.

1 Introduction

Virtual reality (VR) has revolutionized many scientific disciplines by providing novel methods to visualize complex data structures and by offering the means to manipulate this data in real-time in a natural way. The most promising fields for the application of VR systems include engineering, education, entertainment, military simulations and medicine. With VR-based systems surgeons are able to: navigate through the anatomy, practice established procedures, practice new procedures, learn how to use new surgical tools, and assess their progress. In particular, the application domain of existing VR applications in surgery can be broadly classified in three categories: a) education and training, b) pre-operative planning, and c) intra-operative assistance. In pre-operative planning the aim is to study patient data before surgery and to plan the best way to carry out that procedure. We are currently developing a VR system that will allow a surgeon to plan and rehearse a plastic and reconstructive breast surgery based on patient specific data.

In this paper, we present the modeling and estimation aspects of a virtual reality system for plastic and reconstructive breast surgery. Our system has two modes, a *model creation* mode and a *model fitting* mode. The model creation mode allows a surgeon to interactively adjust the shape of the breast by varying key shape variables, analogous to the aesthetic and structural elements surgeons inherently vary manually during breast reconstruction. Our contribution is a set of global deformations with very intuitive parameters that a doctor can apply to a generic geometric primitive in order to model the breast of his/her patient for pre-operative planning purposes and for communicating this plan to the patient. The model fitting mode allows the system to automatically fit a generic deformable model to patient specific three-dimensional breast surface measurements using a physically-based framework [4, 6].

The remainder of this paper is organized as follows. In Section 2 we present the motivation behind our work, and in Section 3 we present the theoretical framework of our research. In particular, in Section 3.1 we describe the basic geometric primitive employed for modeling a breast, in Section 3.2 we formulate the deformations that we have developed, and in Section 3.3 we present the methods employed for fitting a generic virtual breast model to patient-specific breast surface measurements. Finally, in Section 4 we present very encouraging results related to the adequacy of our modeling method and the accuracy of our fitting method using both synthetic and real input data.

2 Motivation

Breast size and shape are a significant part of female body image and sense of femininity. Breast deformities can occur due to cancer, congenital and traumatic causes, or the natural changes associated with aging. The psychological impact varies, but some breast deformities are a major cause of morbidity.

Breast deformities. The shape and size of the human female breast is determined by physical characteristics such as tissue volume, skin dimensions, and chest wall circumference. It is a dynamic structure that is soft and easily deformed by position, gravity, and external pressure. The configuration of the mature breast in each individual changes in response to physiologic alterations (e.g., pregnancy, menstrual cycle, etc.) and with advancing age. In addition to these natural changes, deformities can occur as a result of congenital causes, trauma, and cancer treatment.

- **Congenital breast deformities:** These include both disorders of excessively large breasts (hypermastia), small breasts (hypomastia), and breast asymmetries. Because these disorders generally appear at the time of puberty, they can cause significant emotional problems in teenagers and young adults.
- **Postpartum breast deformities:** After multiple episodes of breast-feeding, the breast atrophies and becomes ptotic in the late 30's and 40's. This is the most common reason for aesthetic breast surgery to modify the shape and nipple position.
- **Traumatic breast deformities:** These are uncommon but include sharp injuries and burns.

- **Cancer-related breast deformities:** Breast cancer is the most common cancer in women with an incidence of 11%. Deformities related to cancer treatment are the most common type. They range from complete absence of the breast to more limited contour problems resulting from breast conservation treatment (i.e., lumpectomy and radiation). Radiation can cause progressive changes in the breast over many years, making it firmer, rounded, and contracted in areas of scar.

Breast deformities can result in significant emotional and psychological morbidity. As a result, a variety of procedures have been devised to modify the shape and size of the breast.

Plastic surgery for the breast. Plastic surgery is surgery that alters the shape of tissues. Many operations have been devised to alter the female breast. Some enlarge or reduce the size the entire breast. Some alter not the size but only the shape and location of the nipple-areolar complex. Breast reconstruction recreates the entire breast (e.g., after mastectomy) using breast implants or tissue transferred from other parts of the body. Some procedures are a combination of each of these.

Specific techniques for these operations include various skin incisions and methods for removing tissue, adding tissue, inserting prosthetic devises, and fashioning the nipple/areolar complex. It is difficult to predict exactly how the breast will be changed by a specific procedure in any particular patient. How long should incisions be and where should they be placed? What size of breast implant will yield the best results? Is there enough tissue to recreate a breast that would meet the expectations of the patient? What does the patient want her breast to look like? Answering these questions is required for pre-operative planning. There is a certain amount of "trial and error" in the process. The exact result depends heavily on the experience, training, and personal artistic and surgical skills of the individual practitioner. Currently, these procedures are learned by surgeons by operating on actual patients, initially under the instruction of more senior surgeons, then later by independent experience. The patient will not know the final result until after the operation. She must trust the judgment of the surgeon to understand her needs and make many decisions without consulting her. Under these circumstances, the possibility is increased for undesirable outcomes.

Modeling and estimation. The uncertainties associated with breast surgery may be reduced by applying modeling and simulation. A virtual breast simulator may enhance the practice of breast surgery at multiple points. It enables the patient to communicate her expectations more clearly to the surgeon. It allows the surgeon to educate the patient with more accurate explanations of what can be accomplished, and after the patient encounter, it helps the surgeon plan specific aspects of the procedure to achieve the agreed upon goals. Finally, it facilitates surgical training by allowing trainees to design procedures and understand the results prior to actually performing surgery on the patient. In this paper, we limit our discussion in the modeling and estimation aspect of our VR system.

3 Theoretical Framework

In this section, the theoretical framework that will allow the analysis of a model's deformations is presented. We begin by reviewing the notation for deformable models and then we formulate the global deformations that we developed for modeling the shape of a breast.

3.1 Deformable Models: Modeling Geometry

The models used in this work are three-dimensional surface shape models. The material coordinates $\mathbf{u} = (u, v)$ of a point on these models are specified over a domain Ω. The three-dimensional position of a point w.r.t. a world coordinate system is the result of the translation and rotation of its position with respect to a non-inertial, model-centered coordinate frame ϕ. Therefore, the position of a point (with material coordinates $\mathbf{u}$) on a deformable model i at time t with respect to an inertial frame of reference Φ is given by the formula:

$$^\Phi\mathbf{x}_i(\mathbf{u}, t) = {^\Phi\mathbf{t}_i(t)} + {^\Phi_\phi\mathbf{R}_i(t)}{^\phi\mathbf{p}_i(\mathbf{u}, t)} \ , \tag{1}$$

where $^\Phi\mathbf{t}_i$ is the position of the origin O_i of the model frame ϕ_i with respect to the frame Φ (the model's translation), and $^\Phi_{\phi_i}\mathbf{R}_i$ is the matrix that encapsulates the orientation of ϕ_i with respect to Φ [4, 7]. $^{\phi_i}\mathbf{p}(\mathbf{u}, t)$ is the position of a model point with material coordinates $\mathbf{u}$ w.r.t. the model frame i and can be expressed as the sum of a reference shape $^{\phi_i}\mathbf{s}(\mathbf{u}, t)$ and a local displacement $^{\phi_i}\mathbf{d}(\mathbf{u}, t)$ as given by the formula: $^{\phi_i}\mathbf{p}(\mathbf{u}, t) = {^{\phi_i}\mathbf{s}(\mathbf{u}, t)} + {^{\phi_i}\mathbf{d}(\mathbf{u}, t)}$. The reference shape captures the salient shape features of the model and it is the result of applying global deformations $\mathbf{T}$ to a geometric primitive $\mathbf{e} = [e_x, e_y, e_z]^\mathsf{T}$. The geometric primitive $\mathbf{e}$ is defined parametrically in $\mathbf{u} \in \Omega$ and has global shape parameters $\mathbf{q_e}$. For the purposes of this research, we employ a superquadric $\mathbf{e}(u, v){:}[-\frac{\pi}{2}, \frac{\pi}{2}]x[-\pi, \pi) \to \mathbb{R}^3$, whose global shape parameters are $\mathbf{q_e} = [a_1, a_2, a_3, \epsilon_1, \epsilon_2]^\mathsf{T}$. A superquadric surface is defined by a vector sweeping a closed surface in space by varying the material coordinates u and v. The parametric equation of a superquadric is given by the formula [2, 1]:

$$\mathbf{e}(\mathbf{u}) = [a_1 C_u{}^{\epsilon_1} C_v{}^{\epsilon_2}, a_2 C_u{}^{\epsilon_1} S_v{}^{\epsilon_2}, a_3 S_u{}^{\epsilon_1}]^\mathsf{T}, \tag{2}$$

where $-\frac{\pi}{2} \leq u \leq \frac{\pi}{2}$, $-\pi \leq v \leq \pi$, $a_1, a_2, a_3 \geq 0$ are the parameters that define the superquadric size, and ϵ_1 and ϵ_2 are the "squareness" parameters in the latitude and longitude plane, respectively. To model the shape of a breast, we only need half of the (u, v) space, therefore in our case $0 \leq u \leq \frac{\pi}{2}$. In addition, in order to be able to vary the sizes of different halves of the superquadric, we employ an asymmetric superquadric with the following parameters: $\mathbf{q_e} = [a_{1b}, a_{1t}, a_{2l}, a_{2r}, a_3, \epsilon_1, \epsilon_2]^\mathsf{T}$, where a_{1b} relates to the bottom half of the superquadric ($|v| \geq \frac{\pi}{2}$), a_{1t} relates to the top half ($|v| < \frac{\pi}{2}$), a_{2l} corresponds to the left half ($v \geq 0$), and a_{2r} to the right half ($v < 0$).

The local coordinate system for our asymmetric superquadric has the x axis protruding outward through the nipple, the z axis going up, and the y axis goes to the patient's life (Fig. 1). The deformations described in this paper are all expressed in this coordinate system.

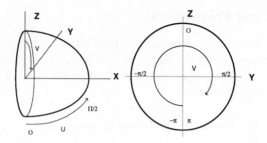

Fig. 1. Side and front views of the virtual breast model and of its associated coordinate systems.

3.2 Global Deformations

In our system, we model (using global deformations) five major features of the shape of the breast. First, the *Ptosis* deformation models the sagging that affects a breast as a subject ages. Second, the *Turn* deformation causes the breast to point towards the left or the right. Third, the *Top-Shape* deformation models the concavity/convexity of the profile of the top half of the breast. For an ideal breast shape the top half is concave, however in cases where large implants have been inserted into the breast, the top half becomes convex. Fourth, the *Flatten-Side* deformation flattens the shape of the half of the breast towards the middle of the torso. Finally, the *Turn-Top* deformation turns the top half of the breast towards the shoulder. In particular, the shape of the virtual breast model is given by $\phi_i \mathbf{s}(\mathbf{u}, t) = [s_x, s_y, s_z]^\top = \mathbf{T}(\mathbf{e}; \mathbf{q_T})$, where the global deformations $\mathbf{T}$ depend on the parameters $\mathbf{q_T}$. In the following, we provide in detail the global deformations employed for modeling the breast.

Ptosis deformation. The Ptosis deformation depends on the parameters $\mathbf{q_T} = [b_0, b_1]^\top$ and it is modeled as a quadratic function with coefficients b_0 and b_1. The deformation affects a point's vertical position (z) as a function of its depth (x). In particular, the Ptosis deformation $\mathbf{s} = \mathbf{T}_p(\mathbf{e}; b_0, b_1)$ along a centerline parallel to the z-axis of a primitive $\mathbf{e} = [e_x, e_y, e_z]^\top$ is given by:

$$s_x = e_x, \quad s_y = e_y, \quad s_z = e_z - (b_0 e_x + b_1 e_x{}^2) \ .$$

Figure 2 depicts front and side views of a deformable breast model to which Ptosis deformation has been applied.

Turn deformation. The Turn deformation causes the shape of the breast to turn to the left or to the right. It uses a quadratic function to scale the y coordinate of a model point as a function of its x coordinate. The parameters of the deformation are $\mathbf{q_T} = [c_0, c_1]^\top$, where c_0 controls the first order turning rate, and c_1 controls the second order turning. In particular, the Turn deformation $\mathbf{s} = \mathbf{T}_c(\mathbf{e}; c_0, c_1)$ along a centerline parallel to the y-axis of a primitive $\mathbf{e}$ is given by:

$$s_x = e_x, \quad s_y = e_y(c_0 e_x + c_1 e_x{}^2), \quad s_z = e_z \ .$$

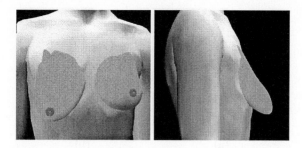

Fig. 2. Examples of Ptosis deformation with parameters $(b_0, b_1) = (0.479, -0.160)$ for the virtual patient's left breast, and $(b_0, b_1) = (0.691, 0.479)$ for the right breast.

Figure 3 depicts front and top views of a virtual patient's right breast to which turn deformation has been applied.

Top-Shape deformation. The Top-Shape deformation allows the user to modify the shape of the top half of the breast ($-\frac{\pi}{2} \leq v \leq \frac{\pi}{2}$). For such a point, we apply a polynomial that scales its z coordinate as a function of its u value (i.e., the point's longitude where the nipple is the north pole). In addition, $u' = u\frac{2}{\pi}$ and spans the range $[0, 1]$. The deformation's parameters are $\mathbf{q}_T = [s_0, t_0, s_1, t_1]^T$, where s_0 is the slope for $u' = 0$, t_0 is the curvature for $u' = 0$, s_1 is the slope for $u' = 1$, and t_1 is the curvature for $u' = 1$. The slope parameters allow the user to vary the slope of the points near the torso and near the nipple. Similarly, the curvature parameters allow the user to adjust the curvatures near the torso and near the nipple. The Top-Shape deformation $\mathbf{s} = \mathbf{T}_s(\mathbf{e}; s_0, t_0, s_1, t_1)$ is formulated as:

$$s_x = e_x, \quad s_y = e_y, \quad s_z = e_z \mathbf{f}_1(u') \ .$$

In determining the order of the polynomial and its coefficients, we seek to adhere to the following constraints. The positions of the points where the breast merges with the torso and of the nipple must stay fixed, thus $\mathbf{f}_1(0) = 1$ and $\mathbf{f}_1(1) = 1$. The following equalities must hold for the slope parameters: $\mathbf{f}_1'(0) = s_0$ and $\mathbf{f}_1'(1) = s_1$. Finally, curvature parameters should satisfy the following

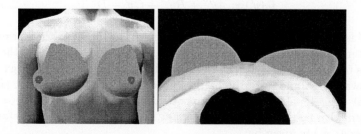

Fig. 3. Examples of Turn deformation with parameters $(c_0, c_1) = (-0.053, 0)$ for the virtual patient's left breast, and $(c_0, c_1) = (0.798, 0.213)$ for the patient's right breast.

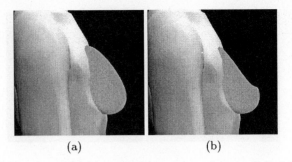

(a)   (b)

Fig. 4. Examples of Top-Shape deformation with parameters $(s_0, t_0, s_1, t_1) = (.904, -7.979, 0, 0)$ (left image) and $(s_0, t_0, s_1, t_1) = (-1.543, -6.915, 2.287, -1.064)$ (right image).

equalities: $\mathbf{f}_1''(0) = t_0$ and $\mathbf{f}_1''(1) = t_1$. With this six constraints our polynomial is a quintic with coefficients A through F as follows:

$$\mathbf{f}_1(u') = Au'^5 + Bu'^4 + Cu'^3 + Du'^2 + Eu' + F ,$$

where $A = -\frac{1}{2}t_0 - 3s_0 - 3s_1 + \frac{1}{2}t_1, B = \frac{3}{2}t_0 + 8s_0 + 7s_1 - \frac{1}{2}t_1, C = -\frac{3}{2}t_0 - 6s_0 - 4s_1 + \frac{1}{2}t_1, D = \frac{1}{2}t_0, E = s_0$, and $F = 1$. Figure 4 depicts side views of examples of the Top-Shape deformation applied to the right breast. In Fig.4(a) the slope at the torso is positive, creating a convex upper breast, while in 4(b) the slope at the nipple is negative, creating a dip near the nipple.

Flatten-Side deformation. The shape of the breast has a tendency to flatten out as it approaches the sternum. To achieve this shape, we use the Flatten-Side deformation that affects the inner half of the breast. For example, for a patient's left breast the deformation flattens the breast's right side and similarly for a patient's right breast the deformation flattens the breast's left side (as seen by the patient's point of view). The right side of the virtual breast model includes points for which $v \le 0$, while the left side includes points for which $v > 0$.

This deformation uses a cubic function that scales a point's x coordinate as a function of its horizontal position (y). The cubic is given by:

$$\mathbf{f}_2(y') = Ay'^3 + By'^2 + Cy' + D ,$$

where y' is normalized to $[0, 1]$ from the middle of the breast to the sternum. The parameters of the deformation are $\mathbf{q}_T = [g_0, g_1]^\top$, where g_0 controls the scaling of points towards the sternum, and g_1 controls the change in the scaling as points move towards the middle of the breast. Since the function should not affect the breast at the nipple, we apply the constraint $\mathbf{f}_2(0) = 1$ and $\mathbf{f}_2'(0) = 0$. For our parameters $\mathbf{f}_2(1) = g_0$ and $\mathbf{f}_2'(1) = g_1$. Thus, the Flatten-Side deformation $\mathbf{s} = \mathbf{T}_g(\mathbf{e}; g_0, g_1)$ is given by:

$$s_x = e_x, \quad s_y = e_y \mathbf{f}_2(e_{y'}), \quad s_z = e_z ,$$

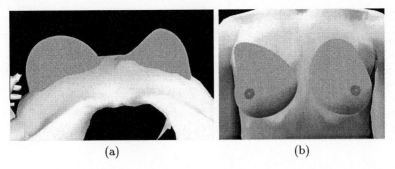

(a)   (b)

Fig. 5. (a) Example of a Flatten-Side deformation with parameters $(g_0, g_1) = (1, 0)$. (b) Example of a Turn-Top deformation with parameters $(h_0, h_1) = (0.266, 0.372)$ for the virtual patient's right breast.

where $\mathbf{f}_2(y') = Ay'^3 + By'^2 + Cy' + D$, $A = g_1 + 2 - 2g_0$, $B = -g_1 - 3 + 3g_0$, $C = 0$, and $D = 1$. Figure 5(a) depicts a top-down view of a Flatten-Side deformation applied to a virtual patient's right breast.

Turn-Top deformation. The *Turn-Top* deformation changes the shape of the top half of the breast laterally. This deformation allows the top part of the virtual breast to point towards the shoulder. The deformation uses a quadratic function to adjust a point's y coordinate as a function of its normalized z coordinate, z'. Its parameters are $\mathbf{q}_T = [h_0, h_1]^\mathsf{T}$, and it is applied to points for which $-\frac{\pi}{2} \leq v \leq \frac{\pi}{2}$. In particular, the Turn Top deformation $\mathbf{s} = \mathbf{T}_h(\mathbf{e}; h_0, h_1)$ along a centerline parallel to the y-axis of a primitive $\mathbf{e}$ is given by:

$$s_x = e_x, \quad s_y = e_y - h_0 e_z - h_1 e_{z'}{}^2, \quad s_z = e_z \ .$$

Figure 5(b) depicts a Turn-Top deformation applied to a virtual patient's right breast.

3.3 Fitting

Through the application of Lagrangian mechanics, the geometric parameters of the virtual breast deformable model, the global (parameterized) and local (free-form) deformation parameters, and the six degrees of freedom of rigid-body motion are systematically converted into generalized coordinates or dynamic degrees of freedom [7, 5]. The resulting Lagrangian equations are of the form $\dot{\mathbf{q}} + \mathbf{Kq} = \mathbf{f}_q$, for shape estimation tasks, where $\mathbf{K}$ is the stiffness matrix, $\mathbf{f}_q$ are the generalized external forces that act on the model, and $\mathbf{q}$ are the model's generalized coordinates. The damping and the stiffness matrices determine the viscoelastic properties of the deformable model. The elastic properties of the virtual breast model are being adapted in space and in time using the techniques described in [4, 6]. In physics-based shape estimation techniques, data points apply forces to the deformable model. These forces that the data apply to the model are converted to generalized 3D forces. Based on these forces the model will deform to minimize the discrepancy between the model and the data.

Table 1. The parameters for the virtual breast model depicted in Fig. 6(b)

Deformation	Parameter Values	
	Left Breast	Right Breast
Top-Shape	$(-0.532, -7.447, 1.277, 0.000)$	$(-0.532, -6.383, 1.702, 0.000)$
Ptosis	$(1.0110, 0.000)$	$(1.0110, 0.000)$
Turn	$(0.106, 0.000)$	$(0.000, 0.000)$
Flatten-Side	$(1.000, 0.000)$	$(1.000, 0.000)$
Turn-Top	$(-0.160, -1.489)$	$(0.160, 1.489)$

Table 2. The parameters for the virtual breast model depicted in Fig. 6(d)

Deformation	Parameter Values	
	Left Breast	Right Breast
Top-Shape	$(-2.234, 8.511, 1.064, -0.532)$	$(-2.234, 8.511, 1.064, -0.532)$
Ptosis	$(0.798, 0.851)$	$(0.638, 0.745)$
Turn	$(0.000, -0.053)$	$(0.213, -0.053)$
Flatten-Side	$(0.505, 0.000)$	$(0.452, 0.000)$
Turn-Top	$(-0.106, -0.638)$	$(0.319, 0.638)$

4   Experimental Results

In order to access the adequacy of the proposed global transformations, we have performed a number of experiments where a plastic surgeon constructs a virtual deformable model for a breast depicted in an image. The example images depicted in Figs. 6(a,c,e) have been randomly selected from [3]. Figures 6(b,d,f) depict the deformable models build by the surgeon. The parameters for these virtual breast models are detailed in Tables 1, 2, and 3, respectively.

Furthermore, in order to access the accuracy of the fitting, we have performed a number of shape estimation experiments with both synthetic and real data. Figure 6(g) depicts range data obtained from a subject using a Cyberware scanner, while Fig. 6(h) depicts the estimated model. The parameters for the estimated deformable breast model are detailed in Table 4.

5   Conclusion

In this paper, we have presented the modeling and shape estimation module of a VR system for plastic and reconstructive breast surgery. In particular, we presented the global deformations that enables us to model a female breast. We have presented several modeling examples along with very encouraging results from fitting a generic deformable breast model to three-dimensional data obtained using range scanning techniques.

Table 3. The parameters for the virtual breast model depicted in Fig. 6(f)

Deformation	Parameter Values	
	Left Breast	**Right Breast**
Top-Shape	$(-1.064, 7.979, 0.532, 0.000)$	$(-1.064, 7.979, 0.532, 0.000)$
Ptosis	$(0.319, 0.000)$	$(0.266, 0.106)$
Turn	$(0.266, -0.372)$	$(-0.053, 0.319)$
Flatten-Side	$(1.000, 0.000)$	$(1.000, 0.000)$
Turn-Top	$(0.319, 0.638)$	$(0, 0)$

Table 4. Parameter values for the estimated deformable model (Fig. 6(h))

Deformation	Parameter Values
Top-Shape	$(-1.543, -6.915, 1.915, -2.128)$
Ptosis	$(0.213, -0.160)$
Turn	$(0.160, 0.000)$
Flatten-Side	$(0.319, -1.489)$
Turn-Top	$(0.160, -0.372)$

References

1. A. Barr. Global and local deformations of solid primitives. *Computer Graphics*, 18(3):21–30, 1984.
2. A. H. Barr. Superquadrics and angle-preserving transformations. *IEEE Computer Graphics and Applications*, 1(1):11–23, January 1981.
3. John Bostwick III. *Plastic and Reconstructive Breast Surgery*. Quality Medical Publishing, St. Louis, Missouri, 1990.
4. I. A. Kakadiaris. *Motion-Based Part Segmentation, Shape and Motion Estimation of Multi-Part Objects: Application to Human Body Tracking*. PhD dissertation, Dept of Computer and Information Science, Univ. of Pennsylvania, Philadelphia, PA, Oct. 1996.
5. I. A. Kakadiaris and D. Metaxas. 3D Human body model acquisition from multiple views. *International Journal on Computer Vision*, 30(3):191–218, 1998.
6. D. Metaxas and I. A. Kakadiaris. Elastically adaptive deformable models. In Bernard Buxton and Roberto Cipolla, editors, *Proceedings of the Fourth European Conference on Computer Vision*, Lecture Notes in Computer Science, pages II:550–559, Cambridge, UK, April 14-18 1996.
7. D. Metaxas and D. Terzopoulos. Shape and nonrigid motion estimation through physics-based synthesis. *IEEE Transactions on Pattern Analysis and Machine Intelligence*, 15(6):580–591, June 1993.

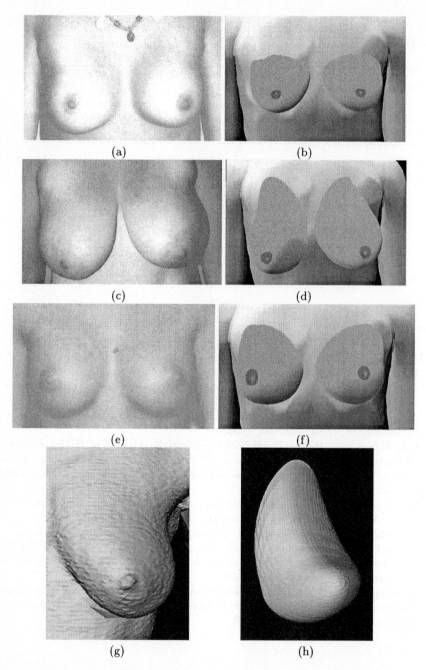

Fig. 6. (b,d,f) Examples of virtual breast models developed by a surgeon to model the female breasts depicted in (a,c,e). (g,h) Range data from a patient's breast and the estimated deformable model.

Development of 3D Measuring Techniques for the Analysis of Facial Soft Tissue Change

Z. Mao[1], P. Sebert[1] and A. F. Ayoub[2]

[1] The University of Glasgow, Glasgow, Scotland, UK
[2] Glasgow Dental Hospital and School, Glasgow, Scotland, UK

Abstract. Three-dimensional facial information is very important for assessing the influence of clef lip repair and measuring the facial growth between cleft and non-cleft children. In this paper, 3D techniques for measuring facial soft tissue change and extracting useful 3D shape information are presented. Firstly, a robust 3D registration algorithm which combines landmark-based and surface-based registration techniques is described. It uses a new surface-based registration algorithm - HICP algorithm to refine landmark-based alignment. We then describe a graphical user interface for manually extracting 3D facial landmarks. Experimental tests on both simulated surface data and real facial scans have been carried out to validate the HICP algorithm.

1   Introduction

Ora-facial clefting is the most common birth defect in the cranio-facial region. The main challenge posed by cleft lip and palate is the achievement of an early morphological and functional repair of the affected structures and to maintain the normal development of affected children. Two research projects, funded by the chief Scientists Office of the Scottish Executive and the National Lottery Board through the Cleft Lip and Palate Association, are currently being run by the Departments of Oral Surgery, Computing Science and Statistics at the University of Glasgow. The studies involve capturing multiple 3D images of both cleft and normal children at different stages throughout the first five years of life. The overall aim of the research is to undertake advanced morphometric assessment of non-cleft, cleft, and surgically-managed cleft patients to assess the influence of surgical lip repair on facial morphology. This will be achieved through the development of 3D-based surface imaging, anatomy and analysis techniques.

This paper presents a robust 3D registration technique after a brief description of the 3D imaging system for capturing facial models. The registration algorithm combines landmark-based and surface-based registration techniques and uses a modified Iterative Closest Point (ICP) [1] algorithm to refine landmark-based alignment. Facial soft tissue change in terms of surface area and soft tissue volume, before and after surgical treatment can then be measured based on the aligned models. A user-friendly graphical interface is developed for manually extracting 3D facial landmarks for statistical shape analysis [2]. The registration algorithm is validated by experimental tests on both simulated surface data and real facial scans.

2 3D facial data acquisition system

The human face is a three-dimensional object, therefore accurate 3D information of facial morpholgy is very important for auditing surgical outcome of cleft lip repair and measuring facial growth. The 3D image acquisition system employed in the research projects is the C3D [3] system developed by the collaboration between The Turing Institute and Glasgow University.

In this system, a pair of video cameras is placed at each side of the patient's face, which then takes a stereo picture. The cameras and the light source are angled at 60 degrees at a distance of 1.75 meters from the patient. A computer-controlled texture flash projector illuminates the subject with random texture pattern to facilitate stereo matching. The image capture time is about 30 millisecond and a personal computer produces the 3D facial model. Detailed system description can be found in [3]. C3D system has the following characteristics which suit this application.

- Fast facial surface capture, approximately 30 ms, especially suitable for capturing young children's faces.
- High resolution cameras: 1000 x 800 pixels.
- Accuracy of localising facial landmarks to 0.5mm.
- No exposure to harmful radiation and therefore suitable for routine use.
- Simple input operation and quick 3D display.

Fig 1 shows a cleft child's face captured by the C3D system. The 3D facial model is represented by triangle meshes or 3D surface points which can be directly used for measurement and model registration. This system has already been installed in Glasgow Dental Hospital and Yorkhill Hospital for Sick Children and the data collection is in progress.

3 Method

Techniques in image processing, computer vision and statistical shape analysis are being developed to provide the facility for measuring facial configurations in three dimensions. These techniques include registering facial models, measuring surface area and volume difference, extracting facial landmarks, performing statistical-based shape analysis and so on.

3.1 Measure facial soft tissue changes

To conduct the assessment of facial soft tissue changes, the change of surface area and soft tissue volume, before and after surgical treatment, a robust 3D registration technique has been developed to align two facial models captured at

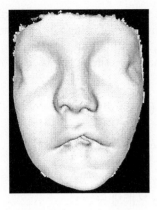
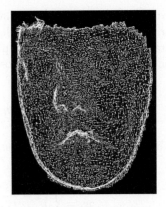

(a) Shaded facial surface.

(b) Facial surface displayed by triangle mesh.

Fig. 1. 3D cleft facial model captured by C3D system.

different times. This algorithm is a combination of landmark-based and surface-based registration techniques.

Firstly, corresponding 3D facial landmarks are manually extracted from two models using a purpose-built graphical interface (see 3.2). Landmark-based registration is then carried out to compute the relative rotation and translation between two models using a Singular Value Decomposition (SVD) approach [4]. The computed transformation is used as the initial guess in the following surface-based registration.

The surface-based registration algorithm we developed is a modified version of ICP [1] algorithm and directly uses the 3D surface points captured from the C3D system to calculate the pose parameters. It shares the similar idea to ICP algorithm which is to register a 3D point set P having N_p points $\{p_i\}, i = 1, ..., N_p$ with a second 3D point set $M = \{m_i\}, i = 1, ..., N_m$ by calculating the closest point between an individual point p_i and M based on the Euclidean distance metric. After the set of closest points has been computed, a rigid body transformation is determined which minimises the equation

$$\sum{}^2 = \sum_{i=1}^{N} \left\| m_i - (Rp_i + T)^2 \right\| \tag{1}$$

where p_i is a 3D point with $x-$, $y-$ and $z-$coordinate in the first point set and m_i is its closest point in the second point set; R represents rotation matrix and T is the translation vector. The registration $F(R, T)$ is then applied to $\{p_i\}$ and the process repeated until the change in mean square error falls below a pre-set tolerance.

The difference between our method and the original ICP algorithm is that our algorithm considers outlier problem and instead of using all the closest point pairs obtained in each iteration for pose (R, T) estimation, a weight w_i is computed for every closest point pair. Pose estimation is then computed by minimising

$$\sum{}^2 = \sum_{i=1}^{N} w_i \left\| m_i - (Rp_i + T) \right\|^2 \tag{2}$$

The method for calculating weight w_i can be expressed as

$$w_i = \begin{cases} \left[1 - \dfrac{\left\| \varepsilon_i \right\|^2}{\left(cS \right)^2} \right]^2 & if \left\| \varepsilon_i \right\| \le cS; \\ 0 & otherwise \end{cases} \tag{3}$$

where $\varepsilon_i = m_i - (Rp_i + T)$, the residual error (distance between closest point pair); S is a scale estimator that is set to median of absolute distance and c is a tuning constant which was set to 6 as suggested in [5].

Equation (3) was firstly presented by Haralick et al [5] for estimating pose parameters when the point correspondences between two data sets are known and hence we name our surface registration algorithm HICP. Our contribution in developing HICP algorithm is that we generalized the weighted least-squares techniques to iteratively locating the point correspondences and then calculating the transformation parameters and updating the data. The original design of the weighted least-squares algorithm [5] assumes that the point correspondence is known while in our algorithm a priori point correspondence is not required. Another difference from the original ICP algorithm is that a SVD algorithm [4] was used to calculate the transformation rather than using quaternions and accordingly reduces the complexity of the algorithm.

After models are properly aligned using the proposed method, soft tissue changes in surface area and volume can be measured. This registration technique can be further applied to determine facial symmetry plane [6].

3.2 Facial Landmark Extraction

A user-friendly software package has been developed to manually extract homologous facial landmarks that are required for both statistical shape analysis and approximate registration prior to applying the HICP algorithm to achieve close registration. This software is capable of displaying multiple facial models from different viewpoints at the same time. Users can easily move, zoom or rotate the facial models to a convenient position when placing landmarks. Fig 2 shows an example of the software interface. Circles indicate the positions of landmarks and vectors give the direction of surface normals at this landmark. The procedure of landmark detection can be further automated by using Active Shape Model (ASM), snakes or curvature-based algorithm.

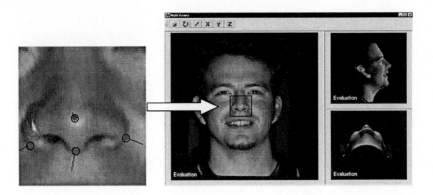

Fig. 2. Graphical interface for facial landmark extraction.

4 Experimental Results

The surface registration method – HICP algorithm has been tested using both simulated data and real facial scans. Measurement of surface volume difference was carried out on dummy facial models with known ground truth.

4.1 Test of registration algorithms with simulated surface data

Comparisons of the performance of HICP algorithm with original ICP algorithm and another modified ICP algorithm developed by Zhang [7] were made using simulated data. Zhang's registration algorithm also considers outlier problem and uses a statistical-based algorithm to computer the weight. We refer Zhang's ICP algorithm as ZICP in the following comparisons.

The simulated data used in the experiment is mathematically constructed using the function $z = xe^{-x^2-y^2}, x \in [-1, 1], y \in [-2, 2]$ with intervals of 0.1 in x-direction and 0.2 in y-direction, generating 441 3D surface points. The surface data is then scaled up by a factor of 25. This surface was chosen because it has significant variability in gradient over the interval given. Fig 3 plots the simulated surface data. A random rotation along three axes $\omega = 3.05$, $\phi = 8.2$ and $\kappa = 2.1$ and a translation of $t_x = 0.51, t_y = 0.13, t_z = -0.26$ are applied to the original surface data to produce the transformed surface (Fig 4).

The performances of the three algorithms under different conditions were compared. All the results and figures given below are repeated and averaged by 20 tries. The registration errors in rotation are measured in degree and errors in translation are given in absolute values with a comparison of data range $x \in [-25, 25]$, $y \in [-50, 50]$, $z \in [-10.7, 10, 7]$.

Comparison of performances under Gaussian noise. Zero-mean Gaussian noise with a standard deviation equal to 1 is added to the $x-$, $y-$ and

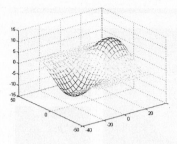
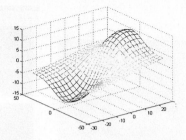

Fig. 3. The simulated original 3D surface.   **Fig. 4.** The transformed 3D surface.

$z-$coordinates of the transformed surface. The reason for chosen Gaussian noise is that the uncertainty in a 3D point reconstructed from stereo is very similar to Gaussian noise and usually modeled as Gaussian [7]. This noise is then multiplied by a scale factor c to simulate different levels of noise. C ranges from 0 to 1.2 with an incremental interval of 0.2. Fig 5 plots the registration results from the three algorithms. As we can see, they have similar performances under Gaussian noise.

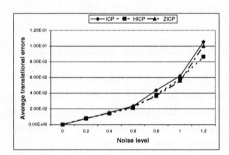
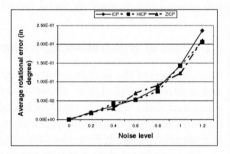

Fig. 5. Performances of three registration algorithms under Guassian noise. Translational and rotational errors as function of noise level.

Comparison of performances with outliers. In some cases only a small part of surface data is exposed to noise or has high levels of noise (we usually call it outlier) while other part of data is relatively good. To simulate such situation, uniform noise drawn from $[-2, 2]$ is added to part of the transformed data. Fig 6 shows the performances of the three registration algorithms when the percentage of outlier increases. Fig 7 plots the rates of convergence (number of iterations required) when the changes of root mean square (RMS) distance between the original surface and the transformed surface falls below a preset threshold (1e-10). In this case, our algorithm (HICP) has the best performance and the fastest convergence rate, especially when the percentage of oulier is less than 50% which is usually the case in real 3D facial scans.

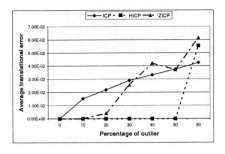
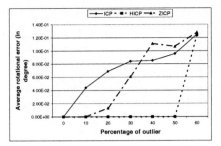

Fig. 6. Performances of the registration algorithms as a function of percentage of outlier.

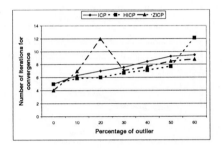

Fig. 7. The rates of the registration algorithms as a function of percentage of outlier.

Comparison of performances related to the region of overlap. The worst case in registering real surface is that the two surface patches are only partly overlapping and noise is also present. In the third test, zero-mean Gaussian noise with a standard deviation of 0.5 is added to the $x-$, $y-$ and $z-$coordinates of the transformed surface to simulate the presence of noise. At the same time, we vary the percentage of overlap between two surface patches. Fig 8 shows the performances of the three algorithms under this situation and Fig 9 plots the rate of convergence as the percentage of overlap decreases. The HICP algorithm has the best performance.

As has been shown from the above three tests using the simulated surface data, our surface registration algorithm - HICP has the overall best performance. It's especially robust in the presence of outlier and when the two surfaces to be aligned are only partially overlapped. The convergent speed of HICP is also stable and better than the other two algorithms.

4.2 Test of HICP using real facial scans

HICP algorithm is also validated using real facial scans captured at different times. Two case studies are reported.

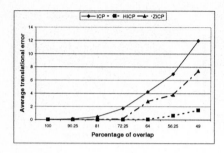
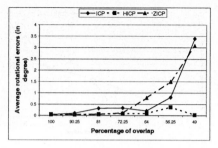

Fig. 8. Performances of the registration algorithms as a function of percentage of overlap.

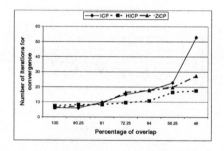

Fig. 9. The rates of the registration algorithms as a function of percentage of overlap.

Case 1: registration of the 3D facial models of the same person captured at different times.

The face of a same person was scanned using the C3D system at different positions. The 3D facial models were generated at a resolution of 5mm. Fig 10 (a) displays the two facial models before registration, one with solid surface and the other with triangle mesh. The surface patch selected for registration is displayed as meshes in Fig 10 (b). The poses of these two models were significantly different before registration, and the RMS between closest point pair was 51.056mm. After 50 iterations using HICP algorithm, RMS was reduced to 1.773mm and the aligned surfaces is shown in Fig 11.

Case 2: Measure facial growth using controlled facial models.

To simulating facial growth, a dummy facial model is scanned (Fig 12. (a)) and then material was applied to the face (Fig 12 (b)). It produces similar effect to facial growth. The ground truth for the volume added in is $11.533cm^3$. Both models were produced at a resolution of 2mm by the C3D system and the HICP algorithm was used to align them. After 28 iterations the algorithm converged at a RMS distance of 1.0mm. A Volume difference of $11.679cm^3$ was obtained based on the alignment. Comparing with the ground truth, the difference is only 1.3%.

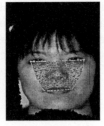
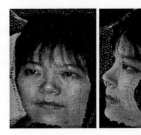

(a) Initial facial positions.

(b) Region for registration.

Fig. 11. Facial surfaces after registration.

Fig. 10. Facial surfaces before registration.

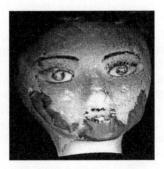

(a) before facial growth

(b) after facial growth

Fig. 12. Dummy facial model for the simulation of facial growth

5 Discussion and Further Work

In this paper, techniques for capturing 3D facial models and measuring facial soft tissue changes are presented. A new registration algorithm - HICP algorithm is developed to align 3D facial surface. This algorithm was tested and compared with other registration algorithms using simulated data. The results show that our algorithm has better performance when outlier is present or when the two surface patches are only partially overlapped. Experimental results on the real facial scans and controlled model indicate that the C3D system and the proposed measurement techniques are reliable and capable of detecting small changes in facial soft tissue morphology.

However, this work is still in its initial stage and clinical data collection is still being carried out. When data capture is completed, statistical shape analysis method, such as Procrustes analysis [8], Principal Component Analysis (PCA), Euclidean Distance Matrix Analysis (EDMA) [9] and Thin-plate Spline

(TPS) visualisation [10] will be used to measure facial growth and to establish statistical models for normal faces and deformed faces due to facial cleft. We believe that by applying 3D capture, analysis and measurement techniques, a more comprehensive understanding of the effect of facial cleft/palate repair on facial soft tissue morphology will be achieved.

References

1. Besl, P.J. and McKay, N.D. (1992) A method for registration of 3D shapes, *IEEE Transactions on Pattern Analysis and Machine Intelligence*, 14(2), 239-256.
2. Dryden, I. L. and Mardia, K. V. (1998)*Statistical shape analysis*, London, John Viley.
3. Ayoub, A. F., Siebert, P., Moos, K. F., Wray, D., Urquhart, C. and Niblett, T. B. (1998) A vision-based three-dimensional capture system for maxillofacial assessment and surgical planning, *British Journal of Oral and Maxillofacial Surgery*, 36, 353-357.
4. Arun, K.S., Huang, T.S. and Blostein, S.D. (1987) Least-squares fitting of two 3D point sets, *IEEE Transactions on Pattern Analysis and Machine Intelligence*, 9(5), 698-700.
5. Haralick, R. M., Joo, H., Lee, C. N., Zhuang, X., Vaidya, V. G. and Kim, M. B. (1989) Pose estimation from corresponding point data, *IEEE Transactions on Systems, Man and Cybernetics*, 19(60), 1426-1446.
6. Mao, Z. and Nafte, A. (1998) 3D image analysis of facial shape changes using depth from stereo, *Proceedings of Noblesse Workshop on Non-Linear Model Based Image Analysis*, 1-3 July, Glasgow, UK, Springer, pp.283-288.
7. Zhang, Z. (1994) Iterative point matching for registration of free-form curves and surfaces, *International Journal of Computer Vision*, 13(2), 119-152.
8. Rohlf, F. J. and Slice, D. (1991) Extensions of the Procrustes method for the optimal superimposition of landmarks, *Systematic Zoology*, 39(1), 40-59.
9. Lele, S. (1993) Euclidean distance matrix analysis (EDMA): estimation of mean form and mean form difference, *Mathematical Geology*, 25, 573-602.
10. Bookstein, F. L. (1989) Principal warps: thin-plate spline and the decomposition of deformations, *IEEE Transactions on Pattern Analysis and Machine Intelligence*, 11(6), 567-585.

Accuracy of a Navigation System for Computer-Aided Oral Implantology

Wolfgang Birkfellner[1]*, Felix Wanschitz[2], Franz Watzinger[2], Klaus Huber[1],
Michael Figl[1], Christian Schopper[2], Rudolf Hanel[3], Franz Kainberger[3],
Joachim Kettenbach[4], Sanda Patruta[5], Rolf Ewers[2], and Helmar Bergmann[1,6]

[1] Department of Biomedical Engineering and Physics
[2] Department of Oral and Maxillofacial Surgery
[3] Department of Diagnostic Radiology, Division of Osteoradiology
[4] Department of Diagnostic Radiology,
Division of Angiography and Interventional Radiology
[5] Dental School, University of Vienna
[6] Ludwig-Boltzmann Institute of Nuclear Medicine
General Hospital, University of Vienna,
Waehringer Guertel 18-20, A-1090 Vienna, Austria

Abstract. Placement of endosteal implants is a widespread therapy for
re-establishing full functionality in edentulous patients. As a first application of VISIT, a modular software system for research into computer-
aided surgery developed at our hospital, we have implemented a navigation system for computer-aided implant dentistry (CAID). Besides
improved accuracy, benefits of CAID include fast translation of preoperative imaging to the operating theatre and the possibility to insert
the implants without having to prepare large mucosa flaps. In this cadaver study, we have measured the overall accuracy of VISIT for inserting four intraforaminal implants in the edentulous mandible. Five cadaver mandibles were embedded into plaster. After high-resolution CT
scanning, the mandibles were registered, and the implant channels were
drilled by the surgeon. Training implants were inserted into the implant
channels, and the plaster was removed. Again, the mandibles underwent
CT scanning, and the pre- and postoperative scans were registered relative to each other. A gross registration between pre- and postoperative
scans was achieved using surface- or mutual information matching since
in some cases the fiducial markers were lost. After transformation to
a common coordinate system, the accuracy was assessed by measuring
the distance of the implant's center to the cortex of the jawbone. Average accuracy of the navigation system was found to be 0.9 ± 0.7 mm,
range $\{0.0 \ldots 3.5\}$ mm. We conclude that these results show that CAID
is an interesting novel application of computer-aided surgery superior to
conventional methods in oral surgery.

---

* e-mail: Wolfgang.Birkfellner@univie.ac.at

1 Introduction

Computer-aided insertion of endosteal implants is a rather novel application of computer-aided surgery (CAS). Basically, small titanium implants are placed in the edentulous jaw [1, 2, 7, 9]. In this cadaver study, we have assessed the precision of VISIT's module for computer-aided implant dentistry (CAID). VISIT is a modular software system for fast development of exploratory research applications in CAS developed at our hospital.

High-resolution computed tomography (CT) provides the imaging modality for preoperative planning of implant channels. This is a widespread technique in oral surgery; usually, the preoperative plan is transferred to the operating theatre by means of templates [5]. This requires an additional step for manufacturing the template after planning. Furthermore, some template techniques do also require preparation of large mucosa flaps for inserting the template in the patient's oral cavity. This is undesirable since the atrophic jaw is to a large extent vascularized by the periost over the mucosa tissue, and temporary removal of the mucosal tissue thus increases atrophy of the available bone volume.

Another interesting application of CAID is the insertion of endosteal implants in the zygoma; this allows for using endosteal implants for coverage of large skeletal defects after hemimaxillectomy in tumour patients [8]. In this case, CAS allows for accessing the zygoma for drilling an implant channel from a minimal-invasive external approach; this is not feasible in normal surgery due to the vicinity of critical anatomical structures such as the eye. In this study, we have assessed the accuracy achievable in implant placement, extending a first cadaver study on cadaver skulls for zygoma implants [4, 10].

2 Materials and Methods

2.1 Hard- and Software

VISIT, our experimental navigation system development platform, consists of an interface for communication with an optical tracking system (Flashpoint 5000, Image Guided Technologies, Boulder/CO) written in ANSI-C using the POSIX.1 standard routines for serial communication, and a platform-independent ANSI-C program. Image processing was added using AVW (Biomedical Imaging Resource, Mayo Clinic, Rochester/MN) [3]. The graphical user interface was programmed using Tcl/Tk 8.02. Currently, the program is implemented on a SGI O2 RS12000 workstation running IRIX 6.5.6. Sterilizable LED probes for tracking the patient and the surgical drill were also developed at our hospital [1].

2.2 Experimental Protocol

Three miniature osteosynthesis screws (Leibinger AG, Freiburg/Germany) were implanted into five cadaver mandibles preoperatively. The mandibles underwent

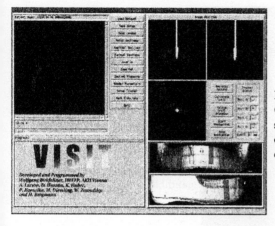

Fig. 1. Main screen of VISIT's CAID module. The position of the surgical drill is visualized using oblique reformatting and projection over volume renderings.

high resolution CT scanning (conventional CT, 0.25 *0.25*1.00 mm^3). After scanning, the mandibles were embedded into plaster so that only the remaining alveolar ridge remained visible. Registration of mandible and CT-scan was performed using a point-to-point registration algorithm [6]. Specialized tools developed for CAID [1–3] consisting of a dynamic reference frame (DRF) for tracking the patient and a probe attached to a surgical drill were used. Four intraforaminal implants were planned on the CT-scans. After drilling the implant channels, four IMZ training implants (Friatec AG, Mannheim/Germany) with 4 mm diameter and 13 mm length were inserted.

2.3 Postoperative Assessment

After implant insertion, the plastercast was removed and the mandibles underwent CT scanning again. Both the pre- and postoperative scans were interpolated to isotropic voxel size; since in some cases the fiducial markers were lost, the pre- and postoperative scans were matched either by using mutual information or by surface registration. After computing the registration, the postoperative scan was transformed to the coordinate system of the preoperative scan. The position of both the implant tip and the implant's head were then measured on both scans using axially reformatted slices. The deviation was defined as the difference between the distances between implant tip and base to the lingual and buccal cortex of the jawbone. Furthermore, the distance of the implant's center to a fiducial marker placed on the mentum was measured. All image processing besides VISIT was undertaken using Analyze$_\text{AVW}$ (BIR, Mayo Clinic, Rochester/MN) on a SGI O2 RS12000 and Analyze PC on a Intel-based PC running Windows NT 4.0.

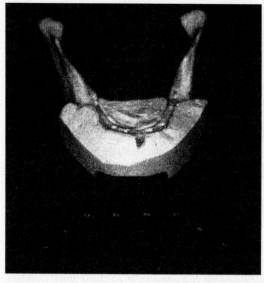

Fig. 2. A cadaver mandible. Visible is the plastercast surrounding the mandible during the surgical procedure, and the optical probe for tracking the position of the mandible during surgery.

Table 1. Average deviation between implant positions on pre- and postoperative scans as measured on five cadaver mandibles. As a measure of accuracy, the distance between the lingual and buccal cortex of the jawbone and the implant's tip (embedded in the mentum) and base (on the alveolar ridge) was measured. Furthermore, the deviation between central axis of the implant and a fiducial markers on the mentum was measured on pre- and postoperative scans and compared. All measurements in mm.

$\Delta \text{Tip}_{lingual}$	$\Delta \text{Tip}_{buccal}$	$\Delta \text{Base}_{lingual}$	$\Delta \text{Base}_{buccal}$	ΔAxis–Fiducial
1.4 ±0.7	1.4 ±0.8	0.5 ±0.4	0.5±0.3	0.9±0.7

3 Results

Average deviation from the preoperatively planned position was found to be 0.9 ± 0.7 mm. No perforation of the cortices did occur. Furthermore, the foramen mentale, the exit point of the Nervus alveolaris inferior, was not penetrated; therefore the nerve would not have been anaesthesized in a live patient despite the fact that the foramen mentale was not visible during the experiments and two of the implants were planned close to the foramina in each specimen. Additional time expense due to the navigation procedure was maintained within a few minutes as compared to the conventional procedure.

4 Discussion

There is an obvious need for transferring the preoperative plan of the implant's position to the operating theatre. This is especially the case in completely edentulous patients with highly atrophic jawbone. Typically, these patients have to undergo augmentative surgery (for instance by implanting small chips of pelvic

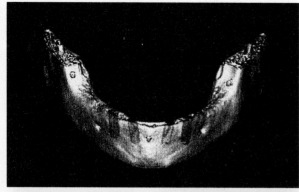

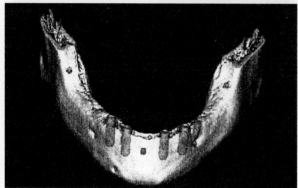

Fig. 3. Comparative view of planned (upper image) and postoperative implant position (lower image). Images were made after registration of the pre- and postoperative scan. In this case, average deviation of the planned and true implant position was found to be 0.7 mm.

bone under the mucosa of the maxillary sinus) prior to implantation, thus presenting an unphysiological anatomic situation. Preoperative CT-scanning and implant planning on the three-dimensional CT scan is an accepted therapy to make optimum implant placement sure in these cases. Typically, a template is produced to convey the preoperative plan to the operating theatre. The proposed navigation method is superior to this technique due to two reasons. First of all, the template has to be made by a dental technician, which is usually a time-consuming, cost-intensive process. Second, the template has to rest on a repeatable position in the patient's oral cavity; this can be achieved by preparing a large mucosa flap which exhibits most of the jaw's cortex. This technique is typical in implant dentistry since the flap also provides visual control for not penetrating the cortex. The mucosal flap, however, is not desirable since the jawbone's blood supply mostly stems from the periost. As said before, further atrophy of the jawbone is likely to occur due to this method.

Compared to an earlier series of cadaver studies [4, 10], we were able to improve the performance of VISIT by using newly developed LED assemblies. Taking into account the remaining mechanical problems in the surgical drill (namely the loose fit of the drill bit in the surgical drill), we conclude that accuracy achievable with the navigation system is sufficient for this type of application, especially since the experiments presented in this paper show a real-life situation;

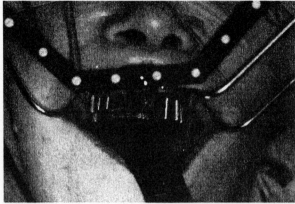

Fig. 4. Intraoperative view during the first operation with VISIT. Eight implants were placed in a completely edentulous, atrophic maxilla. Prior to the operation, the patient underwent augmentative surgery to increase the amount of bone volume available for implantation. Visible are six parallelization nails and the dynamic reference frame for tracking the maxilla.

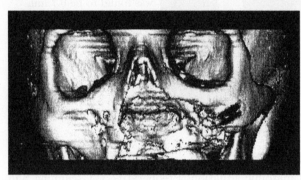

Fig. 5. Postoperative CT scan of a patient who underwent insertion of two endosteal implants in the zygoma; the maxilla and the hard palate were removed in an earlier operation due to squamous cell carcinoma of the palate. The implants were inserted for fixation of an obturator prosthesis which seals the nasopharynx.

the implant channels were drilled by a surgeon, and additional problems such as the deviation of the planned path due to hard trabecular bone or increasing deviation in the course of widening the 2 mm pilot hole to the full diameter of 4 mm ocurred. The problem of deviations due to the loose fit of the drill bit and flexion of the drill is also documented by the fact that the accuracy at the implant tip is worse than in the case of the implant base.

The reliability of the system was found to be sufficient for first intraoperative use in December 1999 (Fig. 4) in an edentulous patient. Since then, two more patients which underwent hemimaxillectomy due to squamous cell carcinoma or adenocystic caricinoma were operated. Fig 4 shows a patient after insertion of two 4 × 15 mm Friatec IMZ implants into the zygoma for placing an obturator prosthesis [4, 10]; this prosthesis seals the nasopharynx after loss of the hard palate.

5 Acknowledgment

We wish to thank W. Piller (L.-Boltzmann Institute of Nuclear Medicine, Vienna), S. Baumgartner, A. Taubeck, and A. Gamperl (Dept. of Biomedical Engineering and Physics, Vienna). The implants were provided by Schütze Dentaltechnik, Vienna. A. Larson, D. Hanson and the staff at BIR (Mayo Clinic, Rochester/MN) were of great help during the development of VISIT. AVW, the library used for VISIT's image processing capabilities, was provided courtesy of Dr. R. A. Robb. This work was supported by the Austrian Science Foundation FWF under research grant P12464-MED.

References

1. W. Birkfellner, P. Solar, A. Gahleitner et al.: "Computer - Aided Implant Dentistry - An Early Report", in A. Colchester, C. Taylor (eds.): "Medical Image Computing and Computer Aided Interventions - MICCAI'99", Springer LNCS 1679, 883-891, (1999).
2. W. Birkfellner, P. Solar, A. Gahleitner et al.: "In-vitro assessment of a registration protocol for image guided implant dentistry", Clin Oral Impl Res, in press.
3. W. Birkfellner, K. Huber, A. Larson et al.: "A modular software system for computer aided surgery and it's first application in oral implantology", IEEE Trans Med Imaging, in press.
4. W. Birkfellner, F. Watzinger, F. Wanschitz et al.: "Image Guided Insertion of Endosteal Implants in the Zygoma for Reconstructive Purposes - A Pilot Study", to appear in Proceedings of SPIE Medical Imaging 2000.
5. G. Champleboux, T. Fortin, H. Buatois et al.: "A fast, accurate and easy method to position oral implants using computed tomography", in W. M. Wells, A. Colchester, S. Delp (eds.): "Medical Image Computing and Computer-Aided Interventions - MICCAI'98", Springer LNCS 1496, 269-276, (1998).
6. B. K. P. Horn, "Closed form solution of absolute orientation using unit quaternions", J Opt Soc Am A 4(4), 629 - 642, (1987).
7. O. Ploder, A. Wagner, G. Enislidis et al.: "Computer-assisted intraoperative visualization of dental implants. Augmented reality in medicine", Radiologe 35(9), 569-572, (1995).
8. E. D. Roumanas, R. D. Nishimura, B. K. Davies et al.: "Clinical evaluation of of implants retaining edentulous maxillary obturator prostheses", J Prosthet Dent 77(2), 184 pp., (1997).
9. F. Watzinger, W. Birkfellner, F. Wanschitz et al.: "Positioning of dental implants using computer-aided navigation and an optical tracking system: case report and presentation of a new method", J Craniomaxfac Surg 27, 77-81, (1999).
10. F. Watzinger, W. Birkfellner, F. Wanschitz et al.: "Placement of endosteal implants in the zygoma after maxillectomy: a cadaver study using surgical navigation", Plast Reconstr Surg, in press.

A 3D Finite Element Model of the Face for Simulation in Plastic and Maxillo-Facial Surgery

Matthieu Chabanas and Yohan Payan

TIMC Laboratory, Faculté de Médecine de Grenoble, 38706 La Tronche, France
Matthieu.Chabanas@imag.fr

Abstract. This paper introduces a new Finite Element biomechanical model of the human face, which has been developed to be integrated into a simulator for plastic and maxillo-facial surgery. The idea is to be able to predict, from an aesthetic and functional point of view, the deformations of a patient face, resulting from repositioning of the maxillary and mandibular bone structures. This work will complete the simulator for bone-repositioning diagnosis that has been developed by the laboratory. After a description of our research project context, each step of the modeling is precisely described: the continuous and elastic structure of the skin tissues, the orthotropic muscular fibers and their insertions points, and the functional model of force generation. First results of face deformations due to muscles activations are presented. They are qualitatively compared to the functional studies provided by the literature on face muscles roles and actions.

1 Introduction

Orthognathic surgery addresses skeletal deformities of the craniofacial region in the sagittal, coronal and transverse facial planes. The correction of these deformities requires the repositioning of facial bones and their dental structures. In this framework, detailed pre-surgical planning has to be carried out to accurately determine the exact amount of movement for the jawbones and teeth. Moreover, considerations about the aesthetics of the face and the functionality of the facial muscles have to be taken into account, to try to predict the "external aspect" of the patient face, following the repositioning of the bone structures. This point is crucial from the patient point of view, but unfortunately lacks in the simulations that are driven nowadays.

This paper addresses this last point, with the introduction of a new 3D biomechanical model of the human face and its coupling with a simulator for the repositioning of cranio-facial bone structures.

2 Research Context

In the course of the European Image Guided Orthopaedic Surgery (IGOS) project, the CAMI group (Computer Assisted Medical Intervention) of TIMC Laboratory developed a prototype for a 3D simulator for bone-repositioning diagnosis during maxillo-facial surgery [1], [2]. The main idea was to provide the surgeon with a tool that offers the possibility, from patient CT scans, to plan the maxillary and mandibular osteotomies, taking into account cephalometric and orthodontic constraints. For this, a 3D virtual model of the patient skull is reconstructed and interactively used as a basis for the definition of a new 3D cephalometry and for the simulation of osteotomies. First tests were carried out on a dry skull and issues in the efficiency of this simulator were discussed [1].

The aim of this paper is to introduce a new biomechanical model of the human face that would be coupled to the software developed in the IGOS project framework, in order to offer a complete simulator for plastic and maxillo-facial surgery. To do this, our strategy will be divided into three steps:

1. A "standard" 3D model of the human face will be developed, with focuses on its anatomical, functional and physical realism. This step is the main topic of this paper.
2. Starting from patient data (CT scans and MRI), our "Mesh-Matching" algorithm [3] will be used to adapt the "standard" model to the patient morphology, thus generating a Finite Element model of the patient face.
3. This 3D model of the patient face will be attached to the virtual reconstruction of the patient skull and used to predict facial aesthetics and behavior, after bone repositioning. This step will not be described in this paper and concerns current research.

3 Biomechanical face model

3.1 Face anatomy

Facial skin has a layered structure composed of the epidermis (a superficial 0.1 mm thick layer of dead cells), dermis (0.5 - 3.5 mm thick) and hypodermis (fatty tissues connected to the skull). Many facial muscles are inserted between those skin layers and the underlying bone structure. In the case of maxillo-facial surgery, the great majority of interventions act on the upper and lower maxilla. This is the reason why our main focus for skin face modeling was dedicated to the lower part of the face, with a special emphasis on soft tissues surrounding human lips.

More than ten muscles act on lips shape, the great majority of them being pair muscles along the sagittal plane. Most of them are dilators (with a distended action like skeletal muscles), and are gathered around the lips. They all have the same kind of insertions: one into skull bone, and the other one inside the lips, onto a constrictor muscle which fibers run around the lips: the orbicularis oris (Fig. 1).

3.2 Face modeling

The first developed models of the human face focused on computer animation and were motivated by a need for external realism, from a behavioral point

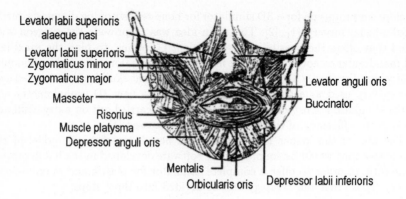

Levator labii superioris alaeque nasi
Levator labii superioris
Zygomaticus minor
Zygomaticus major
Masseter
Risorius
Muscle platysma
Depressor anguli oris
Levator anguli oris
Buccinator
Mentalis
Orbicularis oris
Depressor labii inferioris

Fig. 1. Face anatomy (from Bouchet & Cuilleret [4])

of view. Their modeling was mainly based on discrete mass-spring structures, regularly assembled inside facial tissue. In a first step, those kind of discrete models were introduced in the framework of computer assisted maxillo-facial surgery [5], [6]. Then, arguing that a precise modeling of soft tissues deformation needs a continuous description, Finite Element (FE) models were developed [7], [8], [9]. Those models were based on an automatic 3D meshing of the FE structure, on an isotropic description of tissues mechanics, and proposed, for some of them, principles for the modeling of muscular actions [9].

As we stated earlier, our strategy was to build a 3D face model that specifically addresses the simulation in the domain of plastic and maxillo-facial surgery, with a modeling framework that offers the possibility:

- to passively predict the face aesthetics after bone repositioning,
- to estimate the functional deformations of the face under muscles actions.

This last point is particularly important to ensure for example a symmetrical smiling or to guaranty no perturbation for speech production.

The first objective will induce a continuous modeling of facial tissues. For this, the partial differential equations of the linear elasticity theory were discretized through the Finite Element Method. A particular attention will be given to the definition of the 3D FE mesh, that will have to be sufficiently adapted to face anatomy: symmetry along the sagittal plane, precisions near crucial anatomical structures (nose and lips), and possibility to include muscular structures.

The second objective will need a precise modeling of muscular fibers, with a description of their course and insertion points, and with specific mechanical properties. From this point of view, and as a consequence of the complex interweaving of face muscles, we have chosen to limit the description of the deformations to the linear elasticity modeling. However, we offer the possibility to model the anatomical specificity of the muscles, i.e. their orthotropic structure (i.e. muscles reinforced by fibers).

3D Finite Element mesh We started from a 3D surface mesh of face skin [10]. This mesh was slightly modified and used to build two other surfaces, in order to model a volumetric mesh (Fig. 2). The first one is just scaled from the skin mesh, while the other one corresponds to its projection onto a standard skull. Points of this third surface are rigidly fixed, except those located around the mouth and inside the cheeks which are not fixed to the skull. A volumetric mesh, composed of two layers of material, was thus built, modeling the collagen and fatty tissues of the skin structure.

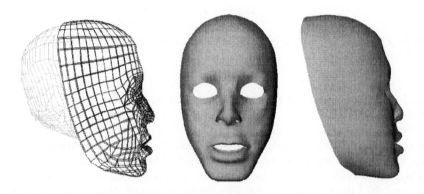

Fig. 2. Finite Element mesh (left) and rendered views (middle and right)

Muscles were defined by making some "holes" inside this global 3D mesh, and by inserting other FE meshes which elements represent the body of the muscles. The four main muscles acting on lips were inserted, with elements running in the superficial thickness of the mesh in the case of the orbicularis oris, or elements running from the extremity of the muscle, attached to the skull, to the other insertion points, fixed onto the orbicularis muscle (in the case of the zygomaticus major, risorius and depressor labii muscles). Figures 4 (left panels) plot the final 3D mesh, with the areas (dashed elements) associated to FE models of the muscles.

Mechanical characteristics As a starting point, the global face mesh, mainly describing passive collagen and fatty tissues, was modeled as an isotropic linear FE structure. The biomechanical characteristics of this structure were chosen in order to model skin tissues quasi- incompressibility and to replicate mechanical measurements made on skin. Tissues incompressibility was modeled with a Poisson's ratio value close to 0,5. Young's modulus value was chosen to match the measurements reported by Fung [11]. Those measurements were made on rabbit skin, but are, from our knowledge, the only reference about skin mechanical characteristics (they are, from this point of view, used by other authors in FE modeling [8]. The value that was retained in our simulations corresponds to the

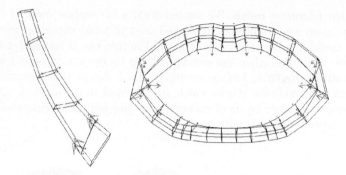

Fig. 3. Forces repartition on mesh nodes. Zygomaticus major (left) and Orbicularis oris (right)

small-deformation slope of the skin stress-strain relationship reported by Fung, i.e. 15 kPa.

Concerning muscles mechanical characteristics, we took the values reported by Duck [12], namely 6.2 kPa for a human muscle in its rest position, and 110 kPa for the same muscle when it is contracted. Those values are coherent with measurements reported by Ohayon [13] on cardiac muscles, considered as dense muscles: 30 kPa at rest and 300 kPa when activated. Moreover, due to the orthotropy of muscular fibers, we raised the Young modulus value to 110 kPa only for the modulus corresponding to the principal direction of the muscle. The Young modulus corresponding to the two orthogonal directions were maintained at the 6.2 kPa value, even when muscle is activated. Finally, as for the passive elements associated to the rest of the face, Poisson's ratio value was chosen close to 0.5, considering that muscles are mainly composed of water and thus quasi-incompressible.

4 Simulations

First simulations were carried out on the standard model to validate the deformations of the face under muscles actions, according to functional studies [14]. Each muscle was activated through a force applied in the direction of the fibers, with a level close to the measurements reported by Bunton and Weismer [15] on human tongue and lips (some Newtons in general). The Young modulus value associated to the main direction of the fiber was linearly raised with force activation, between 6.2 kPa (no activation) and 110 kPa (maximal activation). The orientations of each muscle fibers are not straight but on the contrary show a more or less important curvature among muscles. Therefore, a functional model of distributed force was adopted to simulate activation: the force is distributed on each node of the muscle FE structure, with a value proportional to the curvature. Figure 3 (left panel) shows this distribution for the zygomaticus major muscle: the global force is applied to the extremity of the muscle, but residual

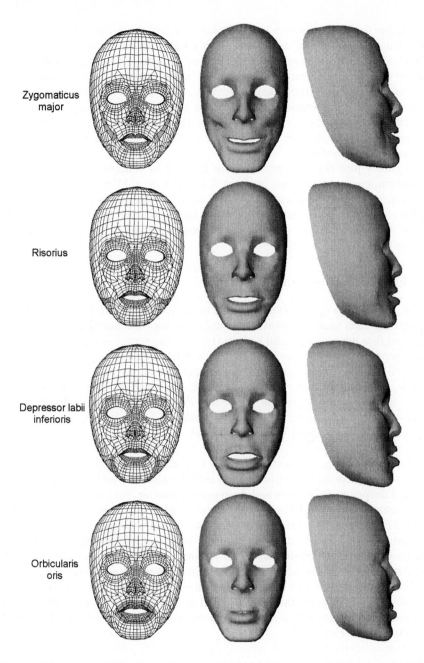

Fig. 4. Muscles modeling: geometry and actions. Left panel: muscle FE mesh (dashed elements). Middle panel: induced deformations (face). Right panel: induced deformations (profile).

forces, due to the distribution as a function of the curvature, are applied to nodes belonging to the body of the muscle, with a resulting action that will try to re-align body muscles elements. The right panel of figure 3 plots force distribution for the orbicularis oris muscle. This model is particular as it was considered as a sphincter muscle, without any extremity. Anyway, our functional distributed model allows the generation of forces on each node of the muscle, with an amplitude function of the curvature: therefore the idea is that an equilibrium can be reached if the sphincter muscle shape is close to a circle, with a curvature constant among nodes.

Figure 4 plot the face deformations induced by muscles. For each muscles activation, computation time is about 5 minutes on a 450Mhz PC to generate a 20 frames animation using quasi-static analysis. Resolution of the FE partial differential elasticity equations is provided by the Castem 2000 package. As the model was defined in the scope of linear elasticity, those figures can be reduced thanks to pre-calculus [16].

5 Discussion and perspectives

Those simulations qualitatively validate the modeling, as the face deformations induced by each muscle activation are coherent with the functional studies provided on face muscles roles and actions [14]. The zygomaticus major muscle pulls the lip corners upwards, the risorius muscle stretches the lips, while the depressor labii muscle lows lip corners. Concerning the activation of the orbicularis muscle, a lip closure gesture can be observed, while a slight backward movement of the upper lip can be seen on the profile view. This movement should not be observed once a geometric model of the teeth (with no penetration constraints for the lips) will be added to the face model.

As a perspective, the next step of our research project will consist in adding layers to complete the model, in order to include other face muscles (both superficial and deeper ones: levators and depressors, masseter, etc.) and skin epidermis. Then, quantitative validation of the modeling will have to be carried out, on a "normal" subject. The idea will be first, to measure mechanical properties of the patient (with indentation experiments for example) as well as physiological muscle properties (electromyography), and second, to confront deformations simulated by the model, with skin deformations observed on the subject. Finally, the global face model will be integrated into the simulator for bone structures repositioning, and quantitatively validated on patients (confrontations between aesthetics and functionality predicted by the model, and observations made on the patient, after surgery).

References

1. G. Bettega, Y. Payan, B. Mollard, A. Boyer, B. Raphäel, and S. Lavallée. A simulator for maxillo-facial surgery integrating cephalometry and orthodontia. *Journal of Computer Aided Surgery*, 5(4), 2000.

2. G. Bettega, Y. Payan, B. Mollard, A. Boyer, M. Chabanas, B. Raphäel, and S. Lavallée. A simulator for maxillo-facial surgery. In *Proceedings of the International Conference on Computer Assisted Radiology, CAR 2000*, 2000.

3. Payan Y. Couteau B. and Lavallée S. The Mesh-Matching algorithm : an automatic 3D mesh generator for finite element structures. *Journal of Biomechanics*, 33/8:1005–1009, 2000.

4. A. Bouchet and J. Cuilleret. *Anatomie topographique descriptive et fonctionnelle: la face, la tête et les organes des sens*. Simep ed. 1, 25, Villeurbanne, France, 1972.

5. K. Waters. Synthetic muscular contraction on facial tissue derived from computer tomography data. In R. Taylor, S. Lavallee, G. Burdea, and R. Mosges, editors, *Computer Integrated Surgery*, pages 191–200. MIT Press, Cambridge, MA, 1996.

6. E. Keeve, S. Girod, and B. Girod. Computer-Aided Craniofacial Surgery. In H.U. Lemke and al., editors, *Computer Assisted Radiology*, pages 757–763, P.O. Box 211 - 1000 AE Amsterdam - The Nederlands, 1996. Elsevier Science B.V.

7. E. Keeve, S. Girod, and B. Girod. Anatomy-Based Facial Tissue Modeling Using the Finite Element Method. In *IEEE Visualization'96*, San Francisco, USA, 1996.

8. E. Keeve, S. Girod, R. Kikinis, and B. Girod. Deformable Modeling of Facial Tissue for Craniofacial Surgery Simulation. *Journal of Computer Aided Surgery*, 3:228–238, 1998.

9. M. H. Gross R. M. Koch and A. A. Bosshard. Emotion Editing using Finite Element. *Eurographics'98*, 17(3), 1998.

10. T. Guiard-Marigny, A. Adjoudani, and C. Benoit. 3d models of the lips and jaw for visual speech synthesis. In J.P. Olive J.P.H. Van Stanten, R.W. Sproat and J. Hirschberg, editors, *Progress in Speech Synthesis*. Springer Verlag, New York, 1996.

11. Y. C. Fung. *Biomechanics: Mechanical Properties of Living Tissues*. Springer Verlag, New York, 1993.

12. F. A. Duck. *Physical properties of tissues: a comprehensive reference book*. Academic Press, London, 1990.

13. P-S. Jouk J. Ohayon, Y. Usson and H. Cai. Fibre Orientation in Human Fetal Heart and Ventricular Mechanics: A Small Perturbation Analysis. *Computer Methods in Biomedical Engineering*, 2:83–106, 1999.

14. W. J. Hardcastle. *Physiology of Speech Production*. Academic Press, London, 1976.

15. K. Bunton and G. Weismer. Evaluation of a Reiterant Force-Impulse Task in the Tongue. *Journal of Speech and Hearing Research*, 37:1020–1031, 1994.

16. S. Cotin, H. Delingette, and N. Ayache. Real time volumetric deformable models for surgery simulation. In K. Hoehne and R. Kikinis, editors, *Visualization and Biomedical Computing, VBC'96*, pages 535–540. Springer-Verlag, 1996.

Simulation of Patellofemoral Joint Reconstruction Surgery on Patient-Specific Models

Zohara A. Cohen and Gerard A. Ateshian

Department of Mechanical Engineering, Columbia University
Orthopaedic Research Laboratory, Department of Orthopaedic Surgery,
Columbia-Presbyterian Medical Center
630 West 168th St, BB1412, New York, NY 10032

Abstract. The outcome of patellofemoral joint reconstructive surgery has been found to be variable. This study presents patient-specific models of the patellofemoral joints of 13 patients diagnosed with osteoarthritis. The model is a mathematical model using quasi-static equilibrium analysis and the model inputs are patient data is acquired from a combination of two series of magnetic resonance images. Four tibial tuberosity transfer surgeries are simulated for each patient, namely: 15 mm anterior shift, 20 mm anterior shift, 8 mm anterior along with 8 mm medial shift, and 15 mm anterior along with 8 mm medial shift. The simulated surgeries produced a statistically significant decrease in mean contact stress relative to pre-surgical conditions (p=0.004), though no statistical difference was found between the different procedures. No one procedure consistently demonstrated the best outcome for all patients. In fact, certain surgical procedures increased peak stresses in some subjects while decreasing it in others. With the demonstrated variability of outcome, the utility of the model presented for surgical planning becomes apparent.

1. Introduction

1.1. Tuberosity Transfer: Variable Surgical Outcome

Clinically, patellofemoral joint (PFJ) reconstruction surgeries have met with inconsistent success. In particular, surgeries that aim to restore a lower stress level in the painful PFJ by shifting the tibial tuberosity have been reported in the clinical literature to yield variable outcomes. In 1976 Maquet introduced his surgical correction in which the tibial tuberosity, the bone protrusion where the patellar ligament inserts on the tibia, is transferred 20-25 mm anteriorly to increase the moment arm of the patellar ligament and thereby decrease the amount of quadriceps force needed to effect extension of the tibia [18]. In a modification to the Maquet procedure, Fulkerson et al. recommended a procedure in which the tibial tuberosity is cut along an antero-medial plane and then translated along that plane [12]. Such a procedure creates a medial and anterior displacement of the patellar ligament insertion without requiring a bone graft. These authors performed a cadaver study in which they combined 8 mm of medialization with both 8 and 15 mm anteriorization and found the procedure to decrease the overall contact stress, while at the same time achieving a better balance of forces between the medial and lateral aspects of the trochlear groove.

While some clinical studies of tuberosity transfer procedures show good to excellent results in as many as 80% (10-20 years follow up) [21] or 85% (1-4 years follow up) [19] of cases, others report a success rate of only 54% (6 years follow up) [13]. *In vitro* cadaver studies have also shown inconsistent results with respect to decrease in contact stress [10, 11, 17]. Generally, cadaver studies have revealed that the optimal relief of stress occurs with smaller amounts of anteriorization than Maquet had originally suggested: 10 mm according to one study by Ferrandez, et al., [11] and 12.5 mm according to Ferguson, et al., [10]. Others reported that 25 mm of anteriorization leads to an inconsistent decrease in stress [17]. Three-dimensional (3D) computer models generated from digitized cadaver data have also been used to assess the effectiveness of tibial tuberosity transfer [4, 6, 14, 20].

A limitation of these *in vitro* studies is their use of cadaver data that most likely derive from cadavers with no pathology. Nevertheless, the variable mechanical outcome observed in the cadaver studies, which are free of biological complications, indicates that the frequent failure of the clinical procedure may be due to a flaw in the mechanical basis for the procedure. Given, however, that many cases do achieve success, the mechanical benefit of the procedure may vary according to the geometry of the particular knee. With the advent of medical imaging, 3D joint models can be created from *in vivo* patient data.

1.2.   Objective of Current Study

The long-term objective of this study is to determine whether computer simulation of various tuberosity transfer procedures performed on patient-specific multibody model of the PFJ can provide guidance as to the preferred procedure for that patient. In the current study we created patient-specific multibody joint models by acquiring articular geometry, tendon insertion points and kinematic information from two sets of magnetic resonance images. The patients in the study have all been diagnosed with patellofemoral joint osteoarthritis (OA) with lateral patellar tracking and, thus, are candidates for tibial tuberosity transfer procedures. In light of the prior work regarding the optimal amount of tuberosity anteriorization, we simulated two Maquet-style tuberosity transfers, 15 mm and 20 mm anterior shift (to be referred to in this paper as M15 and M20). We also simulated two Fulkerson-style procedures, 8 mm medial - 8 mm anterior, and 15 mm anterior - 8 mm medial (F8-8 and F15-8), yielding a total of four procedures for each patient. Following each surgical simulation the state of stress in the joint was compared to the pre-surgery configuration. Based on the clinical literature, the hypothesis of the study was that the surgical outcome, assessed from certain biomechanical variables, is patient specific and that no single procedure produces superior results across the board. The long-term corollary hypothesis is that the optimum procedure for a given patient can be discerned, thereby improving the clinical outcome of patellofemoral joint reconstruction.

2.   Methods

2.1.   Acquisition of Patient Data

Thirteen patients were selected for the study because of clinical diagnosis of PFJ OA with lateral subluxation (a condition where the patella tracks on the lateral aspect of the femoral trochlea and is usually associated with dysplasia of the vastus medialis obliquus) and no prior PFJ reconstruction surgery. The surface topographies of the subchondral bone and articulating surface of the cartilage were generated by segmenting magnetic resonance images (MRI) and fitting b-spline surfaces to the segmented data. The first imaging sequence highlighted articular cartilage and was a fat suppressed spoiled gradient recalled sequence (SPGR) acquired with an extremity coil, with the knee in full extension, at a resolution of 0.55 x 0.55 x 1.5 mm^3, and requiring 9 minutes (Fig 1). The images were segmented with a semi-automated snake algorithm that has been shown to have an accuracy of 0.23 mm for topography measurement and 0.37 mm for thickness measurement [7].

A second series was acquired in the full-body coil to determine the kinematic position of the knee bones in a more functional flexed position. These images were T1 weighted with a resolution of 0.98 x 0.98 x 4.0 mm^3 and required 4.5 minutes (Fig 1). The bone contours from the extended images were registered with the flexed images to determine the correct kinematic position in which to place the high-resolution bone and cartilage surfaces [8].

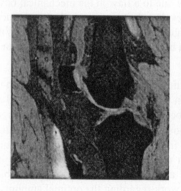 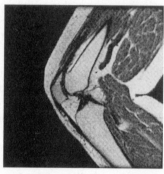

Figure 1:

Left: high-resolution image; knee in extended position

Right: large field of view image; knee in flexed position

2.2. 3D Multibody Modeling of the Patellofemoral Joint

Patellofemoral joint kinematics were simulated using a mathematical multibody model which employs quasi-static equilibrium analysis to predict the position of interacting bodies [16]. The articular layers in the model are represented by mathematical surfaces and the contact stress between them approximated by a function of the local strain. In regions of thinner cartilage, the same overlap distance produces a larger strain and, thus, an apparent stiffening of the cartilage [5]. (To read more about the algorithm for calculating overlap distance, see [3].) Furthermore, the bone is assigned a higher compressive modulus than the cartilage so that contact with the bone generates accordingly higher stresses. Ligaments are modeled as passive line elements that exert forces dependent on their length. Prescribed muscle forces are applied via tendon elements whose direction is guided by their points of origin on one bone and insertion points on other bones. For the PFJ model, the quadriceps muscles were loaded in 3 groups: vastus lateralis (VL); rectus femoris (RF) + vastus intermedius (VI) + vastus medialis longus (VML); and vastus medialis obliquus (VMO). The muscle groups were loaded in the ratio 2:3:1 following the findings in cadavers for relative physiological cross-sectional areas of the quadriceps muscles by Ahmed [1] and Farahmand [9].

2.3. Simulation of Surgery

To simulate the weakened state of the VMO for patients with lateral tracking, the portion of the total quadriceps force carried by the VMO was cut in half relative to the proportions reported in our prior cadaver calibration study [16] yielding a ratio of 4:6:1 for VL:RF+VI+VML:VMO.

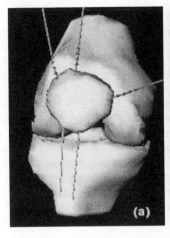 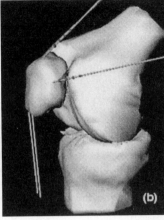

Figure 2:

a) model of female knee at 59° flexion

b) simulated tuberosity transfer (M20)

The tuberosity transfers were simulated by changing three inputs to the model. First, the force on the VL was decrease to 75% of its original value to simulate the lateral retinacular release, which typically accompanies a tuberosity transfer. Second, the insertion points on the tibia of the two line elements, which model the patellar ligament, were moved in the anterior or antero-medial directions (see Fig 2). The anatomical directions of the tibia were determined for each model using anatomic features, as described in our earlier work [15]. Finally, to maintain a constant flexion moment about the knee, the quadriceps forces were decreased to properly account for the increased moment arm of the patellar ligament resulting from the tuberosity transfer. The flexion moment prescribed to the model was derived from the peak closed-chain moment for rising from a chair, 82.2 N·m for males and 59.4 N·m for females [2]. For each simulation, including the initial pre-surgery simulation, the quadriceps forces were iteratively adjusted, keeping the ratio among them constant, until the model produced the desired flexion moment.

2.4.   Outcome Evaluation

Several biomechanical variables were available for analyzing the results of the simulated surgery. Most importantly here, the model output contour maps showing the distribution of articular contact stress across the femur and patella cartilage surfaces; implicit in the use of this variable is the assumption that elevated articular stresses cause not only cartilage wear, but also joint pain. Clinically, the explicit motivation for the various PFJ tuberosity transfer procedures is indeed to reduce the magnitude of articular contact stress. For statistical comparison analyses, the mean stress across the surface and the peak stress were employed. The model yielded other useful data including: kinematics of the patella relative to the femur, articular contact area size, contact force, moment arm of the patellar ligament about the knee's helical (flexion) axis, and medial-lateral position of the centroid of the contact area.

In the analysis of results, the variables of mean articular contact stress and peak contact stress were normalized by their pre-surgery values for the purpose of statistical comparison across different patients. The flexion angles at which patients were imaged were not necessarily the same and, thus, the applied closed-chain flexion moment varied. An ANOVA with repeated measures was performed to compare the pre-surgical and post-surgical outcomes. A pairwise comparison was performed on the least squares (LS) means to detect differences between the various surgeries.

3. Results

The articular contact stress pattern for a typical knee prior to surgery is presented in Figure 3a, where the model simulates a weak VMO and tight lateral retinaculum. This female patient was flexed at 53°, requiring an applied flexion moment of 31N·m. In this pre-surgical configuration, the mean stress is 1.36 MPa and peak stress 2.53 MPa. Following M15 and M20 anteriorization procedures, the stress patterns were altered as demonstrated in Figures 3b and 3c. The mean stress decreased by 11%, to 1.20 MPa, and by 16%, to 1.14 MPa, for the two procedures respectively. The centroid of contact shifted only 0.1 mm medially for M15 and 0.8 mm laterally for M20. The impact of an 8 mm medial shift of the tuberosity is demonstrated in Figure 3d. The mean stress decreased by a total of 18% to 1.11 MPa when the medialization was combined with the 15 mm anteriorization, while the contact area centroid shifted by 2.7 mm medially relative to the pre-surgical position, yielding a more uniformly distributed contact pattern. It can be noted that for this patient the F15-8 procedure produced the greatest decrease in stress and most centralized contact pattern, making it the best candidate procedure based on biomechanical criteria.

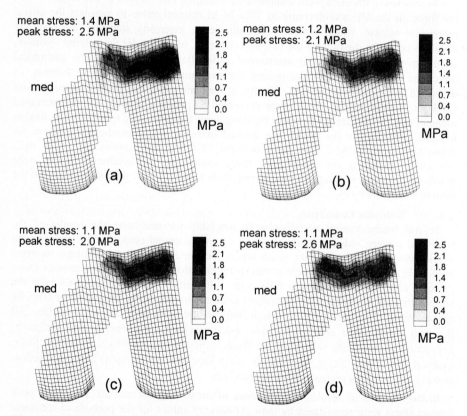

Figure 3: Patellofemoral contact stress maps for female patient with knee at 53° flexion, displayed on femoral surface: a) pre-surgery; b) M15; c) M20; and d) F15-8.

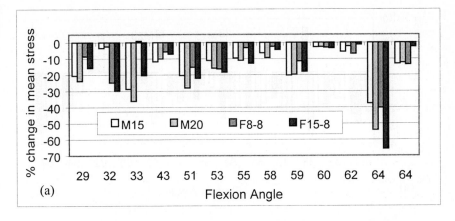

(a)

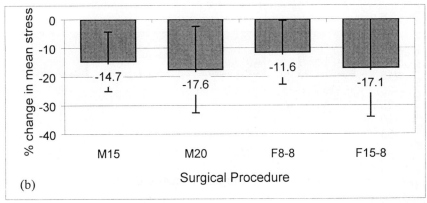

(b)

Figure 4: (a) Change in mean articular contact stress as a percentage of pre-surgery values for all 13 patients. The numbers on the horizontal axis represent the flexion angles at which the analysis was performed. (b) Average change in mean stress for n=13 from pre-surgery, p<0.0001

When looking at the results for all patients however, results are more variable (Figures 4,5,6). On average, surgery produced a statistically significant (p=0.004) decrease in mean contact stress relative to pre-surgical conditions (Figure 4b), though a Duncan's multiple range test demonstrated no difference among the various surgical procedures. When examining the patient-specific results (Figure 4a), only six of thirteen knees achieved a decrease of more than 20% of the original stress for any of the procedures, while three knees demonstrated no better than a 10% decrease. M15 achieved the best result in 2 patients, M20 in 4 patients, F8-8 in 2 patients, and F8-15 in 5 patients, thus emphasizing the variability among patients. When examining changes in peak articular contact stress relative to pre-surgical conditions (Figure 5), it is also found that, on average, surgery reduced the peak contact stress significantly (p=0.014), with Duncan's multiple range test indicating a grouping of pre-surgery and F8-8 against a grouping of M20, M15, and F15-8. However, patient-specific results indicate that certain surgical procedures may actually increase peak stresses in some subjects, as demonstrated in four of 13 patients. Importantly, increasing the amount of tuberosity transfer, e.g., from M15

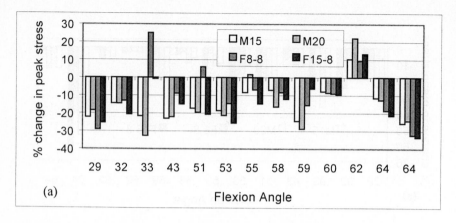

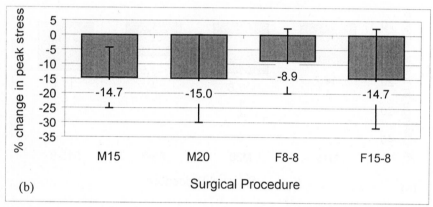

Figure 5: (a) Change in peak contact stress as a percentage of pre-surgery stress; (b) Average change in peak stress for n=13 from pre-surgery value, p<0.001

to M20 or from F8-8 to F15-8 did not necessarily lead to a decrease in mean or peak stress, emphasizing that the patient-specific articular surface topography strongly influences the potential success of these surgical procedures.

Another outcome measure of the simulated surgical procedures is the amount by which the centroid of the contact area shifted medially (Figure 6). Not surprisingly, the Fulkerson procedure (F8-8 or F15-8) produced a consistent and statistically significant medial shift of the contact area relative to pre-surgical conditions (p<.0001); however, the outcome of the Maquet procedure (M15 or M20) is less consistent, showing no difference relative to pre-surgery according to Duncan's multiple range test (p=0.227 and p=0.439).

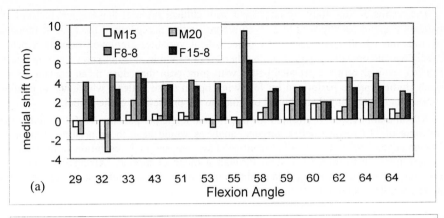

(a)

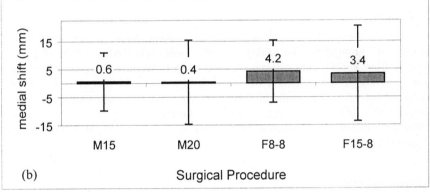

(b)

Figure 6: (a) Medial shift of stress centroid for 13 patients, 4 procedures; (b) Average shift for n=13 from pre-surgery location, p<0.0001

4. Discussion

For the 13 patients in this study the mean stress across the articular surfaces of the PFJ decreased following surgical simulation, although the magnitude of decrease varied considerably among patients. For example, comparing M20 to M15, one patient sees the mean stress decrease by an additional 17% while another sees the stress increase by 3%. Although neither the Maquet nor the Fulkerson procedure stands out as universally superior to the other, for a given patient it is possible to identify one procedure as being superior to the other. While F15-8 was the superior procedure (as measured by decrease in mean stress) for the highest number of patients, i.e. 5 of 13, M20 was a close second demonstrating superiority for 4 of 13 patients.

This model accounted for individual surface topography by using *in vivo* data to describe the articular cartilage layer and by calculating articular contact stress based on local cartilage thickness. Since the size and location of lesions as well as the general surface topography are not consistent among patients, it is not surprising that the biomechanical outcomes of the surgical simulations produced results that varied from patient to patient.

The results of this study are consistent with clinical findings that suggest that not all patients respond equally well to a given reconstructive procedure of the PFJ. It remains to ascertain whether some of the biomechanical variables employed provide a better predictive outcome of surgery than do others. For example, it is not known whether peak articular contact stress correlates better with joint pain than does mean contact stress. Such studies can be performed by following up on those patients who have undergone surgery and comparing their clinical outcome to the predictions of the model.

All simulations were performed for one flexion angle per patient. To fully predict the outcome of PFJ reconstructive surgery, analyses would have to be performed at more flexion angles. With the advent of open-MRI systems and high-resolution surface coils it is now possible to image a patient through a range of motion. Using the principles presented here, such imaging data could be used to predict the outcome of surgery for a more complete range of motion.

5. Conclusion

This study demonstrates that it is possible to perform cogent surgical simulations of corrective PFJ surgery. The analyses in this study were performed on PFJ models acquired from symptomatic patients, as opposed to cadavers with an unknown clinical history. Using the simulations it was possible to establish the best procedure for each individual patient. The results of this study are consistent with clinical findings of outcomes that differ from patient to patient. Further clinical studies are required to determine whether these surgical predictions in fact correlate with objective and subjective surgical outcomes.

Acknowledgement

We thank Mr. Peter Kung and Mr. Bryte Moa-Anderson for their help in data processing. We are grateful to The Whitaker Foundation for graduate student support.

References

1. Ahmed, A.M., Burke, D.L., Yu, A. In-vitro measurement of static pressure distribution in synovial joints--Part II: Retropatellar surface. J Biomech Eng 105 (1983) 226-36.
2. Andriacchi, T.P., Natarajan, R.N., Hurwitz, D.E. Musculoskeletal Dynamics, Locomotion, and Clinical Applications. In: Mow, V.C. and Hayes, W.C. (eds.): Basic Orthopaedic Biomechanics, Raven Press, New York, (1991) 51 -92.
3. Ateshian, G.A., Kwak, S.D., Soslowsky, L.J., Mow, V.C. A stereophotogrammetric method for determining in situ contact areas in diarthrodial joints, and a comparison with other methods. J Biomech 27 (1994) 111-24.
4. Benvenuti, J.F., Rakotomanana, L., Leyvraz, P.F., Pioletti, D.P., Heegaard, J.H., Genton, M.G. Displacements of the tibial tuberosity. Effects of the surgical parameters. Clin Orthop (1997) 224-34.
5. Cohen, Z.A., Ateshian, G.A. The Effect of Cartilage Layer Thickness on Patellofemoral Joint Kinematics and Contact Stresses. ASME Adv in Bioeng BED (1999)
6. Cohen, Z.A., Kwak, S.D., Ateshian, G.A., Blankevoort, L., Henry, J.H., Grelsamer, R.P. , et al. The effects of tibial tuberosity transfer on the patellofemoral joint: A 3-D simulation. ASME Adv in Bioeng BED 33 (1996) 387-8.
7. Cohen, Z.A., McCarthy, D.M., Kwak, S.D., Legrand, P., Fogarasi, F., Ciaccio, E.J. , et al. Knee cartilage topography, thickness, and contact areas from MRI: in-vitro calibration and in-vivo measurements. Osteoarthritis Cartilage 7 (1999) 95-109.

1085

8. Cohen, Z.A., McCarthy, D.M., Roglic, H., Henry, J.H., Rodkey, W.G., Steadman, J.R. et al. Computer-Aided Planning of Patellofemoral Joint OA Surgery: Developing Physical Models from Patient MRI.. In: Wells WM, C.A., Delp S (eds.): Medical Image Computing and Computer Assisted Intervention, Springer-Verlag, (1998) 9 -20.
9. Farahmand, F., Senavongse, W., Amis, A.A. Quantitative study of the quadriceps muscles and trochlear groove geometry related to instability of the patellofemoral joint. J Orthop Res **16** (1998) 136-43.
10. Ferguson AB, J.r., Brown, T.D., Fu, F.H., Rutkowski, R. Relief of patellofemoral contact stress by anterior displacement of the tibial tubercle. J Bone Joint Surg [Am] **61** (1979) 159-66.
11. Ferrandez, L., Usabiaga, J., Yubero, J., Sagarra, J., de No, L. An experimental study of the redistribution of patellofemoral pressures by the anterior displacement of the anterior tuberosity of the tibia. Clin Orthop (1989) 183-9.
12. Fulkerson, J.P., Becker, G.J., Meaney, J.A., Miranda, M., Folcik, M.A. Anteromedial tibial tubercle transfer without bone graft. Am J Sports Med **18** (1991) 490-6; discussion 496-7.
13. Heatley, F.W., Allen, P.R., Patrick, J.H. Tibial tubercle advancement for anterior knee pain. A temporary or permanent solution. Clin Orthop **208** (1986) 215-24.
14. Hirokawa, S. Three-dimensional mathematical model analysis of the patellofemoral joint. J Biomech **24** (1991) 659-71.
15. Kwak, S.D., Ahmad, C.S., Gardner, T.R., Grelsamer, R.P., Henry, J.H., Blankevoort, L., et al. Hamstrings and iliotibial band forces affect knee kinematics and contact pattern. J Orthop Res **18** (2000) 101-8.
16. Kwak, S.D., Blankevoort, L., Ateshian, G.A. A 3D quasi-static mathematical multibody model for diarthrodial joints. Computer Methods in Biomechanics and Biomedical Engineering (October 2000)
17. Lewallen, D.G., Riegger, C.L., Myers, E.R., Hayes, W.C. Effects of retinacular release and tibial tubercle elevation in patellofemoral degenerative joint disease. J Orthop Res **8** (1990) 856-62.
18. Maquet, P. Advancement of the tibial tuberosity. Clin Orthop (1976) 225-30.
19. Noll, B.J., Ben-Itzhak, I., Rossouw, P. Modified technique for tibial tubercle elevation with realignment for patellofemoral pain. A preliminary report. Clin Orthop **234** (1988) 178-82.
20. Roglic, H., Ateshian, G.A., Cohen, Z.A., Kwak, S.D., Gardner, T., Henry, J. , et al. A Computer Simulation of Tibial Tuberosity Elevation and Patellar Tendon Adhesions. Trans ORS **23** (1998) 617.
21. Schmid, F. The Maquet procedure in the treatment of patellofemoral osteoarthrosis. Long-term results. Clin Orthop (1993) 254-8.

A Strain-Energy Model of Passive Knee Kinematics for the Study of Surgical Implantation Strategies

E. Chen   R. E. Ellis   J. T. Bryant

Computing and Information Science   Mechanical Engineering   Surgery
Queen's University at Kingston, Ontario, Canada
Contact: ellis@cs.queensu.ca

Abstract. A mathematical model for studying the passive kinematics of condylar-type total knee prostheses can be useful in planning and performing total joint replacement. If the insertion location and neutral length of knee ligaments is known, the passive kinematics of the knee can be calculated by minimizing the strain energy stored in the ligaments in any angular configuration of the knee.

The model considered here takes into consideration the geometry of the prosthesis, patient-specific information, and operation-specific placement of the prosthesis. Based on an energy-minimization principle, this model can be used to study the kinematics of the knee joint of a patient after total joint replacement. The effect of various articular geometries, alternative surgical placements of prosthetic devices, and intraoperative ligamentous release can be simulated. The model may be useful in preoperative planning, intraoperative guidance, and the design of new prosthetic joints.

1   Introduction

One of the major goals of total knee replacement is to restore the normal function of the knee. The function of the knee after total joint replacement is heavily influenced by the design of the prosthesis and the surgical placement of the prosthetic devices [5]. In passive knee motion, where no external forces are present, the femur is kept in contact with the tibia by the tensile forces exerted by the surrounding ligaments. The geometry of the articular surfaces provides a set of feasible contact locations that determine the orientation of the knee. The interaction of the surrounding ligaments thus governs the contact conditions of the knee. It is supposed here that, at any given angulation, the contact condition that the knee would naturally assume is the contact condition that would minimize the total strain energy stored in the ligaments of the knee.

The theoretical goal of this work was to understand interactions between the geometry of the articular surfaces and the surrounding ligaments and to study how such interactions affect the overall kinematics of the knee. The passive kinematics were analyzed as the instantaneous quasi-static solution to ligament strain-energy minimization. This model required knowledge of the geometry of the articular surfaces, the insertion locations and mechanical properties of the knee ligaments, and the implanted location of the prosthetic components. The model determined the contact location between the femoral and tibial articular surfaces that minimized the strain energy stored in the ligaments of the knee. The ligament states, contact trajectories, and passive knee kinematics were analyzed to determine the effects of alternative surgical placement of the components. The knee kinematics were also animated, using standard computer graphics techniques, to provide a visualization of the passive motion of the knee.

The 3D mathematical model of the knee proposed in this paper is an extension of the 2D model proposed by Martelli *et al.*[6]. Based on the principle of energy minimization [4], the mathematical construction of our model is similar to the work by Wismans *et al.* [10] and Blankevoort *et al.* [1, 2, 7], but our model can be applied to other body joints with either prosthetic or natural articular surfaces. The motion studied is passive, which means no external forces are considered and articular surfaces are assumed to be rigid. However, extension of this model to study the dynamics of the joint and the inclusion of deformable articular surfaces is also possible.

1.1 General Description and Assumptions

Current knee prostheses are designed to have bearing surfaces that are not geometrically congruent [8] for reasons related to human biomechanics and the wear properties of the materials that are used in the prosthetic components. Non-conforming bearing surfaces introduce additional degrees of freedom in knee motion, which further complicates the problem of describing the three-dimensional knee kinematics. One example of an additional degree of freedom is sliding of the femoral component along the bearing surface of the tibial component in a given relative joint angle.

In our knee model, we include patient-specific ligament data to provide the necessary constraints for describing passive kinematics. Each ligament filament is modeled as having a particular length/strain relation, so the passive knee kinematics can be computed by finding the contact state that minimizes the total energy in the ligaments surrounding the knee. That is, each contact state will stretch the ligaments and produce internal strain energy in the filaments that compose the ligament; the contact state that minimizes strain energy is the local equilibrium to which the knee components would relax if disturbed from this equilibrium position.

2 Methods

Passive knee kinematics can be described as a series of instantaneous quasi-static solutions to energy minimization [4] of a system in which contact is ideal single-point contact. Potential energy stored in a passive knee is the sum of the energies stored in each filament of the ligaments, because no external load is present. Friction between the rigid articular surfaces is assumed to be negligible.

2.1 Relative Joint Position and Kinematic Constraints

The coordinate systems used in this model were similar to those of Blankevoort *et al.* [2] and Martelli *et al.* [6]. Two Cartesian coordinate systems were assigned to the major bones of the lower limb. The *absolute*, space-fixed coordinate system was associated with the tibia, and the *relative*, body-fixed coordinate system was associated with the femur. The Z axes were aligned with the anatomical axis of the limb, with the proximal direction being positive. The X axes were perpendicular to Z, lying in the sagittal plane with the anterior direction being the positive X direction. The Y axes were derived as $Y = Z \times X$. Without loss of generality, the origin of the absolute coordinate system was located on the mid-point of the tibial insert in the medial-lateral direction, lying on the resection plane on which the tibial component was fixed. The origin of the relative coordinate system was located on the distal cut for the femoral component, also on the mid-point of the femoral component in the medial-lateral direction. The pose of each prosthetic component was specified with respect to the associated coordinate system.

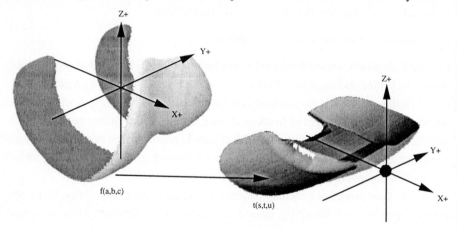

Fig. 1. Two coordinate systems were used. Motion was described as a mobile femur that was rotated and translated with respect to the fixed tibia.

Coordinate systems are related by rigid-body transformations. If $\bar{p}$ is the vector that measures the coordinates of a point in one system, then its coordinates $\bar{q}$ in another system can be expressed as $\bar{q} = T(\bar{p})$ where

$$\bar{q} = T(\bar{p}) = R(\alpha, \beta, \gamma)\bar{p} + \bar{d}$$

where R is an orthogonal rotation matrix and $\bar{d}$ is a displacement vector. Here, R is described by three successive rotations about the coordinate axes. For angles α, β, and γ, abbreviating $\sin(\alpha) = S\alpha$, $\cos(\alpha) = C\alpha$, and so on, $R(\alpha, \beta, \gamma) = R_Y(\beta)R_Z(\gamma)R_X(\alpha)$ so

$$R(\alpha, \beta, \gamma) = \begin{bmatrix} C\gamma \cdot C\beta & -S\gamma & C\gamma \cdot S\beta \\ C\alpha \cdot S\gamma \cdot C\beta + S\alpha \cdot S\beta & C\alpha \cdot C\gamma & C\alpha \cdot S\gamma \cdot S\beta - S\alpha \cdot C\beta \\ S\alpha \cdot S\gamma \cdot C\beta - C\alpha \cdot S\beta & S\alpha \cdot C\gamma & S\alpha \cdot S\gamma \cdot S\beta \end{bmatrix} \quad (1)$$

A vector is thus rotated first in flexion (about Y), then in internal-external rotation (about Z), then in varus/valgus angulation (about X), and finally linearly displaced by $\bar{d}$.

The sequence of rotations was chosen to be flexion angulation first, because flexion is the principal motion in knee kinematics. Internal-external torsion is the next largest angulation, so it was selected as the second angle in the sequence. In practice, varus angulation is very small and, placed as the final angulation, will have little effect on the product of the first two rotations.

2.2 Surface Contact

The articular geometry was obtained from 3-dimensional laser surface scanning, in which the inputs were given as a set of points with associated point normals. Assuming that the prosthetic components are ideal rigid bodies, the contact between tibial and femoral articular surfaces is at two contact points, one in the medial compartment and the other in the lateral compartment. Let the contact 4-tuple $\{C_m, C_l, \hat{C}_m, \hat{C}_l\}$ be points on the medial tibial, lateral tibial, medial femoral, and lateral femoral articular surfaces respectively, with corresponding unit normals $\{n_m, n_l, \hat{n}_m \hat{n}_l\}$. There is thus some transformation $T(\cdot)$ such that

$$
\begin{aligned}
\hat{C}_m &= T(C_m) = R(C_m) + \bar{d} \\
\hat{C}_l &= T(C_l) = R(C_l) + \bar{d} \\
\hat{n}_m &= T(n_m) = -R(n_m) \\
\hat{n}_l &= T(n_l) = -R(n_l)
\end{aligned}
\tag{2}
$$

These equations specify that ideal contact occurs when surface points coincide in space and have local normals pointing in opposite directions.

2.3 Ligament Strain Energy

The main ligaments of the knee are the anterior cruciate ligament (ACL), the posterior cruciate ligament (PCL), the medial collateral ligament (MCL), and the lateral collateral ligament (LCL). During the implantation of most total knee prostheses the ACL is resected, so it was eliminated from the model.

Ligaments were modeled as sets of independent, straight filaments. Each filament had a neutral length and stored no strain energy if the distance between the femoral and tibial insertion points was less than or equal to the neutral length. Each filament was modeled as a tension-only linear spring, so the strain energy increased quadratically with its extension and was zero in compression.

Let L_i be the instantaneous length of the ith filament, $\tilde{L}_i$ be its neutral length, and K_i be its spring constant. The strain energy of the filament was defined as

$$
E_i = \begin{cases} K_i(L_i - \tilde{L}_i)^2/\tilde{L}_i^2 & \text{if } L \geq \tilde{L}_i \\ 0 & \text{if } L < \tilde{L}_i \end{cases}
\tag{3}
$$

If F_{PCL} represents the indices of the filaments that constitute the PCL, then the PCL strain energy could be defined as

$$E_{PCL} = \sum_{i \in F_{PCL}} E_i \qquad (4)$$

Defining E_{ACL}, E_{MCL}, and E_{LCL} similarly, the governing equation of knee kinematics was to find the contact 4-tuple whose transformation, applied to the femoral insertions of the filaments, minimized the total energy of the system

$$E_{total} = E_{PCL} + E_{ACL} + E_{MCL} + E_{LCL} \qquad (5)$$

If a ligament was virtually resected during a simulated surgical procedure the stiffnesses of its filaments K_i were adjusted accordingly. For example, if the ligament was resected then all its filament stiffnesses were set to zero.

Equation 3 represents a dimensionless strain energy that is proportional to the ratio of ligament elongation with respect to its neutral length. The energy is normalized to avoid giving bias toward the collateral ligaments, which are naturally much longer than the cruciate ligaments. Because the kinematics are passive, the stiffness constant of the various ligaments are related by scale factors. As recommended in previous work [6], the PCL was given a scale factor of 3.5 times that of the collateral ligaments so that its stiffness was much greater.

2.4 Computation

The computation of knee kinematics took place in two stages, referred to as the *off-line* and *online* computations. Off-line, the set of all feasible contact 4-tuples were determined for a range of poses. Online, given the insertion locations of a given patient's ligaments and the neutral lengths, the instantaneous contact location that minimized ligament strain energy in each pose was calculated. The kinematics of the knee were then animated as a sequence of instantaneous contact locations.

The advantage of the separation of purely geometrical constraints from a patient's biomechanics was two-fold. First, the off-line computation reduced the real-time requirement in determining the patient-specific kinematics. Second, various surgical strategies, including different prosthetic placements and release of ligaments, could be rapidly simulated by the online computations of the model.

3 Validation and Results

To validate our knee model we first created a generic knee model that was symmetric in the sagittal plane. The articular geometry used was size 3 of the *Anatomic Modular Knee* (AMK) of DePuy Inc. (Warsaw, IN.). The surface geometries were found as sets of points with associated local surface normals.

Ligament data of a patient were derived from our previous work [6], with ligament insertion locations adjusted so that the MCL was the mirror image of the LCL and posterior cruciate ligament was located at the mid-distance between the collateral ligaments. The PCL, MCL, and LCL were represented by 3, 4, and 4 independent linear springs respectively.

The articular geometries were virtually implanted to a standard position with respect to our ligament data (as recommended by the manufacturer's surgical protocol and subsequently validated by a surgeon). The model was tested in 1, 575 distinct angulations. Flexion/extension angles ranged from 0° to 120° in 5° increments. Internal/external rotation angles ranged from −9° to +9° in 3° increments. Varus/valgus angle ranged from −2.0° to +2.0° in 0.5° increments. At each angulation, the contact state that minimized the total strain energy in the ligaments was noted, as were the individual ligament strains.

3.1   Ligament Strain and Joint Laxity

The instantaneous contact location for each pose was determined from the strain energy stored in the posterior cruciate ligament and collateral ligaments. The contribution from each ligament varied through knee flexion. The ligaments states of a knee with standard surgical placement of prosthetic component is plotted in Figure 2.

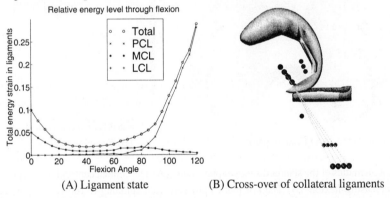

(A) Ligament state   (B) Cross-over of collateral ligaments

Fig. 2. Ligament states of the standard surgical placement. (A) The energy contribution of the ligaments through flexion, the MCL and LCL are identical here. (B) The filaments of the collateral ligament "cross over", as seen in anatomical dissection.

For the standard surgical placement of prosthetic components our model predicted that the PCL was relaxed at low flexion. The collateral ligaments in low flexion contributed the majority of the total ligament strain. As the flexion angle increased the PCL was gradually elongated; its effects on kinematics began at approximately 65° of knee flexion and in deep flexion the PCL became the predominant constraint of knee motion, which is consistent with anatomical observations. Simultaneously, the collateral ligaments gradually relaxed. At deep flexion the PCL was stretched, causing the femoral component to move posteriorly. This prediction is consistent with the results reported by Essinger *et al.* [4, Page 1234], in which the PCL fibers were shown to be elongated at 120° of flexion to as much as 1.4 times of the original length.

The range of flexion was limited by PCL strain. Based on previously validated criteria [5, 3] and equations presented in the previous section, a numerical value of 0.1 was chosen for the PCL strain that limited knee motion. For standard surgical placement of prosthetic components this corresponds to a range of postoperative motion from 0° to about 95° flexion.

The model correctly predicted that cross-over of collateral ligaments occurred at deep flexion, as shown in Figure 2(b).

3.2 Contact Path and Passive Kinematics

The contact path on the tibial component is of considerable interest to surgeons and engineers because the wearing characteristics of the polyethylene bearing surface depend on how the surface is loaded over time. Figure 3 depicts the trajectory of the contact path on the tibial bearing surface.

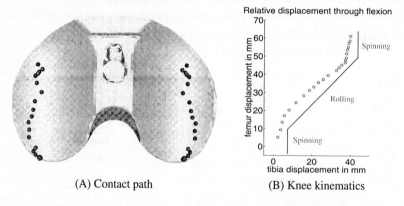

| (A) Contact path | (B) Knee kinematics |

Fig. 3. Kinematics of the standard surgical placement. (A) The contact centroid moves posteriorly with increasing flexion. (B) The kinematics proceed from spinning to rolling to spinning with increasing flexion.

For the knee with standard component placement, the tibial contact was located at the anterior center of tibial plateau in full extension. Spinning motion was observed at low flexion angles ($< 30°$). The tibial contact location traveled posteriorly as the flexion angle increased. In mid-flexion the distance between successive contact locations were almost uniform, suggesting that a rolling motion occurred. In deep flexion, the tibial contact locations were located at the posterior aspect of the tibial component. Overall the anterior to posterior movement of the contact locations through flexion/extension angulation corresponded to roll-back motion, which is observed in normal postoperative gait motion. This pattern of motion suggests that polyethylene failure, of the type associated with spinning in place, should occur at the extreme anterior and posterior margins of the prosthesis.

3.3 Alternative Surgical Implantation Strategies

Both nominal and alternative surgical implantation strategies of the tibial component were simulated. These were anterior and posterior displacement of the component, anterior and posterior sloping of the component, and a combined posterior displacement and posterior sloping. The results are tabulated in Table 1.

Table 1. Comparison of kinematics resulting from different tibial-component placements. Extended/reduced spin was spinning in place for a greater/lesser range of flexion than was observed in the nominal placement.

Tibial Placement	Kinematics during Flexion			Range of Motion
	$0° - 30°$	$30° - 65°$	$65° - 120°$	
nominal	spin	roll	spin	$0° - 95°$
anterior disp.	roll	roll	extended spin	reduced
posterior disp.	extended spin	roll	reduced spin	extended
anterior slope	spin	roll	spin	reduced
posterior slope	spin	roll	extended spin	extended
posterior disp & slope	spin	roll	spin	greatly extended

The knee with an anteriorly displaced tibial component had different knee kinematics than the standard knee. Predicted kinematics for low flexion angles was a rolling motion instead of a spinning motion. The range of motion was also reduced. This was because anteriorly displacement of tibial component effectively increased the distance between femoral and tibial PCL insertion sites, which led to a highly elongated PCL at high flexion. Similarly, an anteriorly sloped tibial component also had reduced range motion because anterior sloping increased the distance between ligament insertion sites.

Posterior displacement or sloping of the tibial component increased the range of motion, in part because such placements of tibia decreased the distance between PCL insertion sites. Of particular interest is the knee with both posteriorly displaced and posteriorly sloped tibial component; it had greatly increased range motion and the overall kinematics were still a spin-roll-spin motion.

3.4 Applications

Because of the separation of on-line computation from off-line computation, knee kinematics from full extention to deep flexion can be calculated in just a few minutes from any specific ligament configuration. This facilitates potential applications of the knee model, including:

Preoperative surgical planning: If data of a patient's knee can be obtained prior to surgery, the surgeon can preoperatively simulate the surgical procedure. This planning can include determination of the most suitable prosthesis design and appropriate surgical placement for the patient. Effects of different sizes of components and various sloping of the tibial insert can also be visualized.

Intraoperative surgical adjustment: The simulation can run fast enough that the surgeon can intraoperatively simulate modifications to a preoperative plan. For example, effects of ligament release and of alternate sloping of the tibial component can be simulated before the surgeon actually performs the tissue modifications.

Prosthesis Design: Current knee prosthesis are designed to have non-congruent bearing surfaces. The effect of different design parameters, such as the flatness of the tibial insert and radii of the femoral head, can be simulated for placement in a "standard" patient or in a suite of actual patients for whom knee data are available. The simulations can be used directly, or can act as input into a subsequent statistical model of how knee design parameters affect knee motion.

4　Conclusions

We have described the development and construction of a computational model of the kinematics of condylar-type total-knee prostheses. The model required knowledge of the geometry of articular surfaces, the insertion locations and relative mechanical properties of the knee ligaments, and the implanted location of the prosthetic components. Based on the principle of ligament strain minimization [4, 6], the computer program generated the motion of the center of contact of each condyle, the kinematics of the prosthetic joint, and the resulting ligament state from full extension to deep flexion.

The model was first validated on a prosthesis design for which the lateral profiles were is the mirror images of the medial profiles. Using the articular surfaces of an AMK prosthesis and ligament data derived from previous work [6], the model was used to simulate the spatial kinematics of a prosthetic knee. The model demonstrated that the pattern of knee motion was influenced by the design of the prosthesis and by the surgical implantation strategies of the prosthetic components. With respect to the AMK design, the simulation suggested that spinning and point-loading motions take place with a wide variety of prosthesis implantation strategies.

The model provided a unique tool for evaluating the effects of different surgical implantation strategies. The model predicted that posterior displacement or sloping of the tibial component would usually lead to improved patterns of knee kinematics and increased postoperative ranges of motion. Conversely, anterior displacement or sloping of the tibia would have adverse effects, such as excessive elongation of the PCL and point loading in deep flexion. These finding are consistent with current knowledge of the *in vivo* behavior of this prosthesis design [5, 9].

The model was validated by simulating a simplified geometry of the knee of an actual patient. The model predictions were be plausible and in accordance with published results. Further validation by *in vitro* and *in vivo* experiments are needed, particularly to consider the effects of asymmetric ligaments and asymmetric components. The model may prove useful not only to determine patient-specific implantation parameters but also to design new knee components that interact better with the biological structures of the human knee.

Acknowledgements

This research was supported in part by Communications and Information Technology Ontario, the Institute for Robotics and Intelligent Systems, and the Natural Sciences and Engineering Research Council of Canada.

References

1. L. Blankevoort, R. Huiskes, and A. de Lange. The envelope of passive knee joint motion. *Journal of Biomechanics*, 21(9):705–720, 1988.
2. L. Blankevoort, J. H. Kuiper, R. Huiskes, and H. J. Grootenboer. Articular contact in a three-dimensional model of the knee. *Journal of Biomechanics*, 24(11):1019–1031, 1991.
3. D. L. Butler, M. D. Kay, and D. C. Stouffer. Comparison of material properties in fascicle-bone units from human patellar tendon and knee ligaments. *Journal of Biomechanics*, 19(6):425–432, 1986.
4. J. R. Essinger, P. F. Leyvraz, J. H. Heegard, and D. D. Robertson. A mathematical model for the evaluation of the behaviour during flexion of condylar-type knee prostheses. *Journal of Biomechanics*, 22(11–12):1229–1241, 1989.
5. A. Garg and P. S. Walker. Prediction of total knee motion using a three-dimensional computer-graphics model. *Journal of Biomechanics*, 23(1), 1990.
6. S. Martelli, R. E. Ellis, M. Marcacci, and S. Zaffagnini. Total knee arthroplasty kinematics, computer simulation and intraoperative evaluation. *The Journal of Arthroplasty*, 13(2):145–155, February 1998.
7. T. J. A. Mommersteeg, R. Huiskes, L. Blankevoort, J. G. M. Kooloos, J. M. G. Kauer, and P. G. M. Maathuis. A global verification study of a quasi-static knee model with multi-bundle ligaments. *Journal of Biomechanics*, 29(12):1659–1664, 1996.
8. J. J. O'Connor and J. W. Goodfellow. The role of meniscal bearing vs. fixed interface in unicondylar and bicondylar arthroplasty. In V. M. Goldberg, editor, *Controversis of Total Knee Arthroplasty*, pages 27–49. Raven Press, 1991.
9. L. A. Whiteside and D. D. Amador. The effect of posterior tibial slope on knee stability after Ortholoc total knee arthroplasty. *Journal of Arthroplasty*, 3(Supplement):S51–S57, October 1988.
10. J. Wismans, F. Veldpaus, J. Janssen, A. Huson, and P. Struben. A three-dimensional mathematical model of the knee-joint. *Journal of Biomechanics*, 13(8):677–685, 1980.

Computer-Assisted Anatomical Placement of a Double-Bundle ACL through 3D-Fitting of a Statistically Generated Femoral Template Into Individual Knee Geometry

J.W.H. Luites[1a], A.B. Wymenga[1b], M. Sati[2], Y. Bourquin[2], L. Blankevoort[3], R. van der Venne[4a], J.G.M. Kooloos[4b] and H.-U. Stäubli[5]

[1a] SMK-Research, [1b] Dept. of Orthopaedic Surgery, St. Maartenskliniek,
P.O. 9011, 6500 GM Nijmegen, The Netherlands
a.wymenga@maartenskliniek.nl
[2] Maurice E. Muller Institute for Biomechanics, University of Bern, Switzerland
[3] Orthopaedic Research Center Amsterdam, Academic Medical Center,
University of Amsterdam, Amsterdam, The Netherlands
[4] Orhopaedic Research Laboratory, [4b] Dept. of Anatomy and Embryology,
University of Nijmegen, Nijmegen, The Netherlands
[5] Dept. of Orthopaedic Surgery, Surgical Clinic Tiefenauspital, Bern, Switzerland

Abstract. Femoral graft placement is an important factor in the success of ACL-reconstruction. Besides improving the accuracy of femoral tunnel placement, Computer Assisted Surgery (CAS) can be used to determine the anatomic location. This requires a 3D femoral template with the position of the anatomical ACL-center, based on endoscopical measurable landmarks. This study describes the development and application of this method. The template is generated through statistical shape analysis of the ACL-insertion, with respect to the anteromedial- (AMB) and posterolateral bundle (PLB). The data is mapped onto a cylinder and related to the intercondylar notch surface and the cartilage border on the lateral notch wall (n=33). The template was programmed in a computer-assisted system for ACL-replacement and validated. The program allows real-time tracking of the femur and interactive digitization under endoscopic control. In a wizard-like fashion the surgeon is guided through steps of acquiring the landmarks for the template alignment. The AMB- and PLB-center are accurate positioned within 1-3 mm of the anatomic insertion-centers in individual knees.

1 Introduction

The anterior cruciate ligament (ACL) consists of two functional bundles, each with its anatomic insertion area [1]. The anteromedial bundle (AMB) is most tight in flexion and the posterolateral bundle (PLB) is tighter near extension [2]. An ACL-reconstruction is aimed at obtaining normal knee laxity and kinematics. However, current ACL-reconstruction aims at the repair of only the AMB through an isometric single-bundle technique. A reconstruction with two bundles approximates normal ACL function near extension and in full flexion by repairing both AMB and PLB.

Restoration of knee kinematics also depends on accurate graft placement. Femoral attachments affect graft length changes most and incorrect positioned femoral tunnels lead to malfunctioning and graft failure [3]. Using a double-bundle technique requires anatomical placement. Different methods for anatomic femoral tunnel localization have been developed [4-6]. Besides the lack of the distinction between the AMB and PLB, most techniques are 2D approaches. Computer Assisted Surgery (CAS) can improve femoral tunnel placement accuracy [7-11], but can also be used to determine the 3D anatomical ACL-insertion center. This requires the development of an averaged template with the 3D positions of the anatomical AMB and PLB center.

1.1 Goal

Development, incorporation in a CAS system and validation of a 3D femoral template with the anatomical ACL-, AMB- and PLB centers, to improve accurate and anatomical femoral placement of a double bundle ACL-graft into the individual knee.

This work describes the computer-assisted placement of the position of the anatomical ACL-center with respect to the AMB and PLB, through 3D fitting of a statistically generated template, based on 2 endoscopical measurable landmarks into individual femoral knee geometry.

2 Methods

2.1 Development

The template is generated through statistical shape analysis of the ACL-insertion, with respect to the AMB and PLB. 33 Cadaver knees were dissected till the ligaments. During knee movement the fibers of the ACL were divided. The part that was slack in flexion, the PLB, was separated with a sharp scalpel from the fibers that were tight in that position, the AMB. The PLB tensioned near extension. After marking the outlines of both insertion sites with leaded wire, the bundles were dissected.

With an electromagnetic 3Space Fastrak system (Polhemus Navigation Sciences, USA) the 3D positions of points distributed over the insertion outlines (ACL, AMB and PLB) were measured. To relate the insertion sites to the knee geometry, two major landmarks were also 3D measured; Points on the surface of the intercondylar notch and on the cartilage bone-transition at the medial wall of the lateral femoral condyle. (Fig. 1.)

To develop the 3D template, the notch surface geometry of each femur was approximated by a cylinder. The alignment and radius of the cylinder were determined by a least-squares fit on the notch surface data. The cartilage border determined rotation about and location along the cylindrical axis. (Fig. 2.)

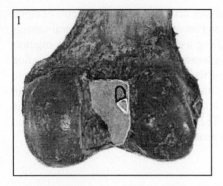

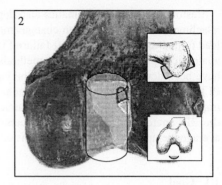

Fig. 1. Left: Dorsal view of the femoral intercondylar notch with the areas 3D digitized by Fastrak: the lateral notch surface, the cartilage border of the lateral condyle and the outlines of the AMB (*black*) and PLB (*white*) insertion sites

Fig. 2. Right: The cylindrical fit into the notch surface in dorsal view, lateral (upper inset) and notch view (lower inset)

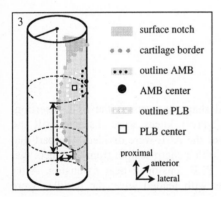

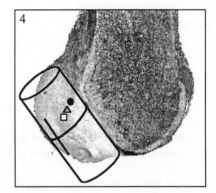

Fig. 3. Left: Ventrolateral view of the fitted cylinder with the projected data of the cartilage border and the AMB- and PLB-insertion outlines. The ACL, AMB and PLB centers were calculated. A parabolic curve was fitted through the cartilage border. The top of the parabola was defined as the origin at the cylindrical surface. From each projected point, the distance in axial (Z) and in horizontal direction at the cylindrical surface (X) (as a result of angle φ times mean cylinderradius R) was calculated in mm

Fig. 4. Right: Lateral view of a femur with dissected medial condyle. The alignment of the 3D cylinder template with projected on the surface: the origin of the coordinate system (apex of the cartilage border) and the mean position of the centers of the ACL (*gray triangle*), AMB (*black round*) and PLB (*white square*)

The cartilage-bone transition data and the calculated ACL, AMB and PLB insertion centers were superimposed on the cylinder. A 2° polynomial was fit through all cartilage-bone transition data by a least squares fit, thus resulting in a function with which the cartilage border was described as a parabola on the cylinder. The apex of the cartilage border curve, i.e. the top of the parabola was defined as the origin on the cylinder surface. The distance in axial direction (Z) between each projected point and the origin was calculated, as was the angle (φ). The distance at the cylindrical surface (X) was calculated as a result of the multiplication of the angle with the mean cylinder radius (R). (Fig. 3.)

After scaling all data to the mean cylinder radius of 10.5 mm, the positions of the 33 ACL, AMB and PLB insertion centers, relative to their origin (the apex of the cartilage border curve), were mapped onto the average cylinder. The mean positions of the centers were averaged, thus resulting in a mean femoral template. (Fig. 4)

2.3 Incorporation

The femoral template was incorporated in a developed computer assisted system developed at the M.E. Müller Institute for Biomechanics to perform intra-operative planning for ACL replacement. It allows real-time tracking of the femur and tibia and interactive digitization and labeling of anatomical structures identified under direct visual or endoscopic control [7].

2.4 Validation

The CAS system was first tested on the population of 33 knees that was used to define the template by comparing the anatomically marked bundle locations to those found by numerical template fit.

In the femora first the outlines of the insertion sites were digitized. The center of the outline was defined as the mean value of the 3D positions of all measured points. Then the developed program was applied to the femora. The obtained data of the 3D positions of the AMB and PLB centers were compared in the 3 directions (X, Y, and Z). Also the direct 3D distance between the anatomical and computer determined centers was calculated.

3 Results

3.1 Computer assisted placement

The system consists of an Optotrak system, a computer and computer-tracked devices. The Optotrak system consists of 3 camera's that track the 3D positions of LED's. These are located in the non-rigid Dynamic Reference Bases (DRB's) who will be fixed to the patients femur and tibia. This will allow real-time tracking of the bones and devices and interactive digitization.

The interface was programmed in a wizard-like fashion guiding the surgeon through the steps of acquiring the required landmarks for the template alignment. After starting the ACL-module, the first step is controlling tracking and visibility of the LED's on the DRB's on the monitor. To inform the system about the orientation of the femur and tibia, the long axis of the femur (trochantor major femur - mid condyles) and tibia (intercondylar tibia – mid ankle) have to be digitized, as is the medial-lateral axis (between the femoral epicondyles). The next step is the template alignment.

First the notch surface will be interactively digitized as a cloud of points using a computer-tracked palpation hook (Stillehook) (Fig.5.a). An algorithm uses this cloud of points to generate a 3D surface for better visualization (Fig.5.b). The same palpation hook is then used to interactively digitize the cartilage border (fig. 5.c). A cylinder fit into the notch surface through a Levenberg-Marquardt algorithm, determines template orientation. At the same time this results in the sizing of the template to the patient's knee. The apex of the cartilage border curve, as described in a cylindrical coordinate system, will find template rotation about the cylindrical axis. Since surgical orientation of the reference base on the femur is arbitrary, the system gives several different start conditions for the Levenberg-Marquardt algorithm and chooses the solution having the smallest residual. The template is then visualized as a 3D cylinder for the user to verify proper algorithm convergence and suggested tunnel locations (in ACL-, AMB- or PLB-center) are displayed as small 3D spheres (Fig. 5.d). The computer interface is then used to guide a computer-tracked awl to mark the drill hole point(s) in the patient's knee with a physical indent (Fig. 5.e+5.f) A K-wire is drilled into the marked point, after which the tunnel will be drilled. If wanted the drilling can also be done computer-assisted by using a tracked drill guide.

3.2 Validation results template

Table 1. The mean differences (±sd) in mm reported for the anatomical AMB and PLB center locations and those found by numerical template fit with CAS, in 33 tested femora.

	3D	X-direction	Y-direction	Z-direction
AMB	3.2±1.3	1.3±1.2	1.8±1.2	1.0±0.8
PLB	2.1±1.2	1.4±1.1	1.4±1.0	0.7±0.6

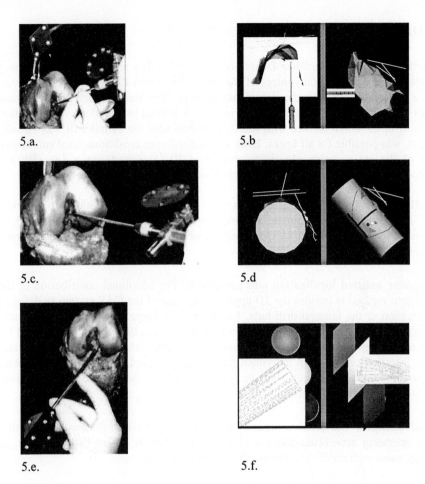

5.a.

5.b

5.c.

5.d

5.e.

5.f.

Fig. 5. Computer-assisted placement of a double bundle ACL through 3D fitting of a statistically generated femoral template into individual knee geometry. A bone block with a spliced Quadriceps tendon will be fixated in a tunnel located in the anatomic ACL-center. Both slips are positioned in the correct place with the use of CAS

5.a. Picture of a tunnel view into the femoral notch: Digitizing the notch surface with the computer-tracked Stillehook. The femur is tracked through the dynamic reference base (DRB)

5.b. Notch-view (*left*) and lateral-medial view of the notch (*right*) on the monitor-screen: The digitized notch surface is 3D visualized for better orientation. CAS calculates the cylinder-algorithm to fit the notch surface

5.c. Picture of a tunnel view into the femoral notch: Digitizing of the cartilage border at the medial notch wall of the lateral femoral condyle with the tip of the Stillehook

5.d. Notch-view (*left*) and lateral-medial view of the notch (*right*) on the monitor-screen: CAS positions the scaled template with the anatomic *gray* AMB-, *black* PLB- and *white* ACL-center

5.e. Picture of a tunnel view into the femoral notch: Marking of the drill hole location in the notch with a physical indent of the computer-tracked awl

5.f. Zoomed notch-view (*left*) and lateral-medial view of the notch (*right*) on the monitor-screen: The *white* anatomic ACL-center is the target point for the mark made by the awl, which is visualized on the CAS monitor

4 Discussion

Practical use. The wizard-like fashion of the interface is a simple and practical way of determine the insertion centers of the ACL, AMB and PLB. Algorithm convergence did not always give the right solution on the first trail on each knee. In some knees the first digitization was enough, in some it needed to be redone several times. Refining of the notch surface with additional points until the cylinder fit was visually correct, was possible for all knees. Even with several start conditions, total calculation time of the optimization algorithm was about 10-15 seconds on a Sun (Ultra 1 creator). The total extra operating time was about 10 minutes.

Application. The system was used in an in-vitro experiment with double bundle ACL-reconstruction for computer assisted anatomic femoral graft placement [12]. In the tested femora (n=11), no cartilage damage or dorsal outbreak of the drilled tunnel appeared.

Computer assisted localization and placement. The additional contribution of the developed method is besides the 3D approach, the use of the CAS system in defining the location of the femoral drill hole, based on the average knee anatomy. No extra images are needed, neither the use of fluoroscopy. CAS is therefore not only used for accurate placement, but also to reproduce the location within 1-3 mm of the anatomical insertion center of an individual patient. It is a simple method, which enables the surgeon to place the graft where he wants it (anatomic AMB, PLB or ACL center) and to use an outside-in drill technique.

Disadvantages. The disadvantages of the system are, besides the costs and (little) extra operating time (10'), the use of the DRB's, which causes extra wounds. In chronic cases with notch deformation, the method is not fit.

Accuracy. The developed method to determine the localization of the anatomical AMB and PLB center has a high accuracy (<2 mm) in the separate directions. From 3D viewpoint the centers can be placed within 1-3 mm of its anatomical center.

Future. Further biomechanical studies are needed to evaluate the difference between "isometric" and anatomic placement.

5 Conclusion

The computer-program with the template is a simple and accurate guide in computer-assisted 3D localization of the femoral drill hole and anatomical placement of a double bundle ACL-graft into individual knee geometry.

Acknowledgment. This study is supported by the AO Foundation, Switzerland.

References

1. Harner C.D., Hyun Baek, G., Vogrin, T.M., Carlin, G.J., Kashiwaguchi, S., Woo, S. L-Y.: Quantitative Analysis of Human Cruciate Ligament Insertions. J Arth Rel Surg Vol. 17 (7) (1999) 741-49

2. Girgis, F.G., Marshall, J.L., Al Monajem, A.R.S.: The cruciate ligaments of the knee joint, anatomical, functional and experimental analysis. Clin Orth Rel Res Vol. 106 (1975) 216-231

3. Amis, A.A., Jakob, R.P.: Knee Surg, Sports Traumatol, Arthrosc Vol. 6 suppl 1 (1998) S2-S12

4. Bernard, M., Hertel, P., Hornung, H., Cierpinski ,Th.: Femoral insertion of the ACL, radiographic quadrant method. Am J Knee Surg Vol. 10 (1) (1997) 14-22

5. Harner, C.D., Marks, P.H., Fu, F.H., Irrgang, J.J., Silby, M.B., Mengato, R.: Anterior cruciate ligament reconstruction: endoscopic versus two- incision technique. J Arth Rel Surg Vol. 10 (1994) 502-512

6. Lintner, D..M., Dewitt, S.E., Moseley, J.B.: Radiographic evaluation of native anterior cruciate ligament attachments and graft placement for reconstruction, A cadavaric study. Am J Sports Med Vol. 24 (1) (1996) 72-78

7. Sati, M., Stäubli, H.-U., Bourquin, Y., Kunz, M., Käsermann, S., Nolte, L.-P.: Clinical Integration of New Computer Assisted Technology for Arthroscopic ACL Replacement. In: T.DiGioia (Guest Ed.), F. Fu (Ed.): Operative Techniques in Orthopaedics, Vol. 10 (1) (January 2000) 40-49

8. Stäubli, H.-U., Käsermann, S., Kunz, M., Sati, M.: Inter-operator variance of ligament placement: Endoscopic versus. Abstr 4th Intern Symp on CAOS (1999) 17

9. Klos, T.V.S., Habets, R.J.E., Banks, A.Z., Banks, S.A., Devilee, R.J.J., Cook F.F.: Computer Assistance in Arthroscopic Anterior Cruciate Ligament Reconstruction. Clin Orth Rel Res Vol. 354 (1998) 65-69

10. Picard, F., Moody, J., Jaramaz, B., GiGioia, A.M.: Computer-assisted ACL reconstruction surgery: "High tech" does not necessarily mean "high complexity". Abstr 5th Intern Symp on CAOS (2000) 22

11. Petermann, J., Schierl, M., Heeckt, P.F., Gotzen L.: The CASPAR-system (Computer Assisted Surgery Planning And Robotics) in the reconstruction of the ACL. First follow-up results. Abstr 5th Intern Symp on CAOS (2000) 23

12. Wymenga, A.B., Luites, J.W.H., Sati, M., Bourquin, Y., Blankevoort, L., van der Venne, R., L., Kooloos, J.G.M., Stäubli, H.-U.: Computer-Assisted Anatomical Placement of a Double-Bundle ACL through 3D-Fitting of a Statistically Generated Femoral Template Into Individual Knee Geometry. Abstr 5th Intern Symp on CAOS (2000) 24

Post-operative Measurement of Acetabular Cup Position Using X-ray/CT Registration

David LaRose[1,2], Laura Cassenti[2], Branislav Jaramaz[2,3], James Moody[3], Takeo Kanade[1,2], and Anthony DiGioia[2,3]

[1] Department of Electrical and Computer Engineering, Carnegie Mellon University, Pittsburgh, PA, USA
[2] Center for Medical Robotics and Computer Assisted Surgery, Robotics Institute, Carnegie Mellon University, Pittsburgh, PA, USA
[3] Center for Medical Robotics and Computer Assisted Surgery, UPMC Shadyside Hospital, Pittsburgh, PA, USA

Abstract. This paper describes a system for measuring acetabular implant orientation following total hip replacement surgery. After a manual initialization procedure, the position of the pelvis is established relative to a pair of nearly orthogonal radiographs by automatically registering to pre-operative pelvic CT data. The pose of the cup is then recovered by projecting a 3D surface model into the two images. A phantom study is presented in which this pose is expressed relative to well defined anatomical landmarks and compared to measurements obtained using an image-guided surgery system.

1 Introduction

Following total hip replacement surgery, acetabular implant orientation is typically measured using anterior-posterior (AP) X-rays[9]. Implant orientation has been shown to be predictive of post-operative outcome, and of complications such as dislocation[6][7]. Despite recent improvements in measurement technique[3], these measurements are still unreliable due to unknown pelvic flexion at the time of X-ray acquisition.

This paper describes a system which measures acetabular implant placement with respect to the pelvis independent of pelvic flexion. The pose of the implant is initially recovered in the coordinate system of the CT volume by 2D/3D registration between planar X-ray images, the pre-operative CT volume, and a 3D surface model of the implant. This pose is then expressed relative to stable anatomic landmarks which are defined in the CT volume, providing an accurate basis for measurements of cup orientation and post-operative evaluation.

Initial results are presented in which pose estimates are generated using images of a high-density pelvis phantom. These estimates are compared with ground truth results obtained using the HipNav image-guided system for total hip replacement surgery [1].

2 Problem Description

During post-operative evaluation, the pose of the acetabular implant must be determined with respect to the pelvis. AP and lateral X-ray images are acquired with known source-to-film distances. The approximate projection centers of each X-ray image are known, but the pose of the patient is not. In particular, the images are not known to be true AP or lateral views, and each image may be acquired with the patient either lying or standing. In addition to the X-ray images, the pre-operative pelvic CT volume and a triangulated surface model of the implanted cup are available.

Although the X-ray images are not acquired simultaneously, it is assumed that the pose of the cup with respect to the pelvis does not change between acquisitions. A schematic drawing of this scenario is shown in Fig. 1(a). There are three unknown coordinate transformations in this figure: the transformation from the coordinate system of the acetabular cup to the CT coordinate system, which is labeled $^{ct}T_{cup}$; the transformation from the CT coordinate system to the coordinate system of the X-ray imager at the time of AP image acquisition, labeled $^{i0}T_{ct}$; and the transformation from the CT coordinate system to the coordinate system of the X-ray imager at the time of lateral image acquisition, labeled $^{i1}T_{ct}$. Each of these unknown coordinate transformations has six degrees of freedom, for a total of 18 unknown parameters.

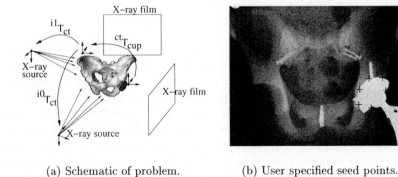

(a) Schematic of problem.   (b) User specified seed points.

Fig. 1. Schematic of problem and input image with marked points.

3 Approach

The problem is broken down into 3 steps: 1) All parameters are initialized to reasonable starting values using input from the user; 2) The pose of the pelvis is estimated in each image using an iterative intensity-based 2D/3D registration scheme; and 3) The pose of the acetabular implant is estimated with respect to the coordinate system of the CT volume by simultaneously matching the projection of of the implant surface model to contours in each X-ray image. These three steps are discussed in sections 3.1, 3.2, and 3.3 respectively.

3.1 Initialization

The 18 unknown degrees of freedom are initialized to reasonable starting values by user input. To initialize the coordinate transformations $^{i0}T_{ct}$ and $^{i1}T_{ct}$, the user indicates the approximate positions of at least three anatomical landmarks in each radiograph and then enters the corresponding 3D coordinates from the CT volume. The initial pose is estimated using a point based 2D/3D registration.

The initial pose of the acetabular cup implant is known from the pre-operative plan. In addition, the user provides the image positions of several points on the border of the cup in each image as shown in Fig. 1(b). These points are used to further constrain the position of the cup as described in Sec. 3.3.

3.2 X-ray/CT Registration

After initialization, the pose of the pelvis with respect to the X-ray imaging apparatus is recovered for each image by iterative comparison between the input images and synthetic X-ray images. These synthetic images are known as Digitally Reconstructed Radiographs (DRRs).

DRR Generation Traditional DRR generation is computationally expensive, and involves numerically integrating through the CT volume along the path which connects each pixel to the X-ray source, as shown in Fig. 2. To avoid this time consuming integration during registration, DRRs are generated using an intermediate data structure called a Transgraph [4]. The Transgraph is a 4D database of precomputed integral values, and is based on data structures used in the computer graphics area called *view-based rendering* [2][5]. DRRs can be generated over a limited range of pose parameters by interpolating among the values in the database. This interpolation procedure is much faster than conventional DRR generation and allows generation of 621x512pixel images in roughly 2 seconds on a 550 MHz Pentium III computer, without requiring clipping of the CT volume. Work is currently underway to implement the interpolation using off the shelf texture mapping hardware as described in [2]. We expect this change to yield a significant additional speedup.

Image Comparison. Since the pose of the pelvis is estimated by iterative comparison of DRRs with the input radiograph, registration accuracy depends on an appropriate choice of image comparison metric. The images are compared using a local *semi-normalized* correlation measure which works as follows:

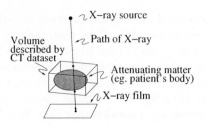

Fig. 2. Path of a single ray from radiation source to film surface. The box indicates the volume in space which is represented by the CT dataset. p1 and p2 represent the points at which the ray enters and exits this volume.

1. For each pixel in the synthetic image, define a surrounding region of interest. For the rest of this paper, we assume that the region of interest is an 11 pixel by 11 pixel window.
2. Compute the variance of the synthetic image over each 11x11 window, excepting those windows which extend past the border of the image.
3. Compute the normalized correlation between each of these 11x11 windows and the corresponding region of the input image.
4. Scale each correlation value by the associated variance, and compute the weighted average.

More precisely, we compute the local semi-normalized correlation by:

$$\widehat{C}(I_0, I_1) = \frac{\sum_{x=5}^{M-6} \sum_{y=5}^{N-6} C(I_0, I_1, x, y)}{\sum_{x=5}^{M-6} \sum_{y=5}^{N-6} (S(I_1, I_1, x, y)/K^2)} \tag{1}$$

$$C(I_0, I_1, x, y) = \frac{S(I_0, I_1, x, y)\sqrt{S(I_1, I_1 x, y)}}{K^2 \sqrt{S(I_0, I_0, x, y)}} \tag{2}$$

$$S(I_p, I_q, x, y) = K \sum_{i,j \in W(x,y)} I_0(x+i, y+j) I_1(x+i, y+j) \tag{3}$$

$$- \sum_{i,j \in W(x,y)} I_0(x+i, y+j) \sum_{i,j \in W(x,y)} I_1(x+i, y+j) \ , \tag{4}$$

where $\widehat{C}$ is the weighted average described in step 4 above, C is the scaled correlation value computed over a single region of interest, I_0 is the input X-ray image, and I_1 is the DRR. The two images are the same shape, and the constants M and N represent their width and height in pixels. $W(i,j)$ is the 11x11 region of interest surrounding pixel (i,j) and the constant K is equal to $11^2 = 121$. Note that the pixel values of I_1 depend on the estimated pose of the pelvis, although this dependence is not made explicit in the notation. In Sec. 3.2 we introduce the pose parameter vector θ which specifies this estimated pose. In order to evaluate these expressions efficiently, all of the summations within S are computed using recursive filters.

The local semi-normalized correlation function has three key advantages: first, the intensity response function of the film/digitizer need not be known, as long as it is approximately linear over small variations in irradiance; second, weighting by the variance of the synthetic image concentrates attention in those regions of the image which have high signal strength; and third, any unmodeled variations in the X-ray images (such as those caused by soft tissue deformation) are effectively filtered out, provided that they have low spatial frequency compared to the correlation window size.

Occasionally, one or more of the 11x11 regions in the input image has all pixels at the same intensity. When this happens, the normalized correlation coefficient for that window is undefined, and a value of zero is substituted.

Minimization. The local semi-normalized correlation function reaches a maximum of 1.0 when the input image and the synthetic image match each other perfectly. A non-negative function which has a theoretical minimum of zero is generated by subtracting the local semi-normalized correlation value from 1.0, and this function is minimized using the quasi-Newton algorithm of Broyden, Fletcher, Goldfarb and Shanno, as described in [8]. The minimization is run for each input image, to recover the coordinate transformations $^{i0}T_{ct}$ and $^{i1}T_{ct}$.

Gradient Computation. Quasi-Newton minimization requires that the first derivative of the objective function be computed. The free parameters in the optimization describe the pose of the pelvis with respect to the X-ray imaging apparatus, and we represent these parameters using the vector θ. Repeated application of the chain rule to equation (2) gives

$$\frac{\partial}{\partial \theta_i} \widehat{C}(I_0, I_1) = \frac{K^2 \sum_{x,y} \frac{\partial}{\partial \theta_i} C(I_0, I_1, x, y)}{\sum_{x,y} S(I_1, I_1 x, y)}$$

$$- \frac{2K^2 \left(\sum_{x,y} C(I_0, I_1, x, y) \right) \left(\sum_{x,y} S(I_1, \frac{\partial}{\partial \theta_i} I_1, x, y) \right)}{\left(\sum_{x,y} S(I_1, I_1 x, y) \right)^2}$$

$$\frac{\partial}{\partial \theta_i} C(I_0, I_1, x, y) = \frac{S(I_0, \frac{\partial}{\partial \theta_i} I_1, x, y) S(I_1, I_1, x, y) + S(I_0, I_1, x, y) S(I_1, \frac{\partial}{\partial \theta_i} I_1, x, y)}{K^2 \sqrt{S(I_0, I_1, x, y) S(I_1, I_1, x, y)}},$$

where the notation $\sum_{x=5}^{M-6} \sum_{y=5}^{N-6}$ from equation (2) has been replaced with $\sum_{x,y}$. $\frac{\partial}{\partial \theta_i} I_1$ is the the first derivative of DRR pixel intensity with respect to the parameter θ_i, and all other variables are defined as in equation (2).

In general, the derivative of DRR pixel intensity with respect to pelvis pose parameters is difficult to compute. In this case, however, the DRR pixel values are found by interpolating among the values stored in the Transgraph, and the first derivative is easily found.

3.3 Determination of Cup Position

Once the pose of the pelvis has been estimated in both images, the pose of the acetabular cup implant is estimated with respect to the coordinate system of the CT volume. This is done by simultaneously matching the projection of the implant surface model to contours in the two X-ray images.

For a given pose, the silhouette of the surface model is projected into each image, and an error measure is computed based on the image positions of the projected points. This error measure is then minimized over the parameter space of the rigid body transformation $^{\text{ct}}T_{\text{cup}}$. The first subsection below describes how the silhouette is computed, the second describes a rough point-based registration used to approximate the actual cup position, and the third describes a final minimization which increases the registration accuracy.

Silhouette Generation. To generate the silhouette of the cup, the vertices of the cup surface model are projected into each image. The projected vertices define a set of 2D triangles corresponding to the 3D triangles of the surface model. The silhouette is generated by culling those vertices which lie interior to any of the projected triangles. To speed this culling process, the projected triangles are organized into a quadtree data structure, and each vertex is compared against only those triangles which lie in or intersect its cell in the quadtree.

Approximate Solution. The pose of the cup is initially computed based on the image coordinates supplied by the user during manual initialization. We define the objective function

$$f(\gamma) = \frac{1}{|R|} \sum_{r \in R} \min_{t \in H_0(\gamma)} (\|r - t\|) + \frac{1}{|S|} \sum_{s \in S} \min_{u \in H_1(\gamma)} (\|s - u\|) , \tag{5}$$

where γ is the vector of parameters describing the rigid body transformation $^{\text{ct}}T_{\text{cup}}$, R is the set of user-supplied initialization points in the AP image, and $|R|$ is the number of points in this set. S is the set of user-supplied initialization points in the lateral image, and $|S|$ is the number of points in this set. $H_0(\gamma)$ is the set of points comprising the silhouette of the cup in the AP image at the pose specified by γ, and $H_1(\gamma)$ is the set of points comprising the silhouette in the lateral image. The notation $\|x\|$ denotes the magnitude of vector x.

The objective function in equation 5 reaches a minimum of zero when every user supplied point is exactly overlapped by one of the points on the silhouette of the projected model. In practice, this minimum is nearly met when the boundaries of the projected surface model lie close to the edges of the cup in the X-ray images. The objective function is minimized using the downhill simplex method of Nelder and Mead, as described in [8].

Refinement of Approximate Solution. The minimization above gives a good approximation to the pose of the acetabular cup. There are, however, small inaccuracies. This is because the initialization points may not lie exactly on the boundaries of the implant in the X-ray images, and because these points may not match well with the points which make up the silhouette of the projected cup. The estimate is, however, good enough to initialize a more precise search.

The gradient of pixel intensity with respect to pixel coordinates x and y is computed for each input image, and an objective function is defined

$$g(\gamma) = \frac{1}{|H_0(\gamma)|} \sum_{t \in H_0(\gamma)} \|\nabla_{x,y} I_0(t[0], t[1])\|$$

$$+ \frac{1}{|H_1(\gamma)|} \sum_{u \in H_1(\gamma)} \|\nabla_{x,y} I_0(u[0], u[1])\| \quad ,$$

where γ is the vector of parameters describing $^{ct}T_{cup}$, $|H_0|$ is the number of points in the silhouette corresponding to the AP image, and $t[0]$ and $t[1]$ are the x and y coordinates of projected vertex t. Similarly, H_1, $u[0]$ and $u[1]$ refer to the silhouette corresponding to the lateral image. $\nabla_{x,y} I_0()$ and $\nabla_{x,y} I_1()$ are the spatial gradients the two images.

4   Results

A preliminary study was conducted using phantom data to evaluate the performance of the registration algorithm. A 62mm diameter VerSys Acetabular cup (Zimmer, Inc.) was fitted to a high density pelvis phantom. The pelvis coordinate system was defined with respect to the left and right anterior iliac spines, and the left and right pubic symphysis points. These points were marked on the model with 1mm diameter fiducial markers. A CT dataset was acquired, having an intra-slice pixel spacing of approximately 0.74mm, a slice thickness of 1mm, and an inter-slice spacing of 1mm. The fiducials were identified in the CT and used to compute the coordinate transformation between the pelvis coordinate system and the CT coordinate system. In order to prevent the fiducial markers from biasing the X-ray/CT registration, an image manipulation program was used to paint out any visible markers in the subsequent X-ray images.

Four series of images were acquired. In each series, AP films were taken with an approximate source-to-film distance of 40 inches, while lateral films were taken with an approximate source-to-film distance of 72 inches. The first series involved 3 AP images and 3 lateral images. Each image was acquired from a slightly different angle, and in four of the six images, household objects were placed in the field of view in order to simulate occluding patient anatomy as shown in Fig. 3(a).

For the second series of images, a simulated torso, surrounding the pelvis, was constructed out of plastic film and filled with rice to simulate soft tissue. Small balloons were inserted into the rice to simulate the effects of bowel gas and soft tissue inhomogeneity. In these images, the difference in density between the rice and the pelvis phantom was very small, and there was not enough contrast for successful registration.

For the third and fourth series of images, the rice was replaced with oatmeal in order to improve the visibility of the pelvis. Although the contrast in these images is still quite low, there is sufficient detail for successful registration. These two series differ in the arrangement of the soft tissue and in the placement of the bowel gas. In addition, the acetabular cup was removed prior to acquisition of the fourth series, and reattached in a different orientation. As before, the AP images of each series span a range of roughly 20° in flexion, while the lateral images span a range of roughly 15° rotation around the superior-inferior axis. Typical images from the these series are shown in Fig.s 3(b) and 3(c).

Ground truth measurements were performed using the HipNav system [1]. Repeated measurements of the implant orientation for the first three series had a mean abduction of 45.2^o and a mean flexion of 10.4^o, with standard deviations of 0.11^o and 0.22^o respectively. After repositioning for the fourth series, the cup was measured to have an orientation of 52.6^o abduction and 48.9^o flexion.

All of the films were digitized and resampled to resolution of 36 dpi, giving a final image size of 621x512 pixels. The center of projection for each image was assumed to lie at pixel coordinates (310, 255.5). No further attempt was made to calibrate the X-ray imaging system. The images from each series were grouped into pairs, each consisting of one AP image and one lateral image. The registration algorithm was run using each pair as input to recover the pose of the acetabular cup implant with respect to the coordinate system of the CT. This transformation was composed with the CT-to-pelvis coordinate transformation, and flexion/abduction measurements were calculated. The outline of the cup surface model was projected into each image so that the accuracy of registration could be visually assessed.

Abduction and flexion measurements are presented in 1, and visual registration results for cup alignment are shown in Fig. 3. The poor result in case 7 is due to incorrect convergence of the X-ray/CT registration, and is discussed further in Sec. 5.

Series 1	Abduction	Flexion	Series 2	Abduction	Flexion	Series 3	Abduction	Flexion
HipNav	45.2^o	10.4^o	HipNav	45.2^o	10.4^o	HipNav	52.6^o	48.9^o
Case 1	45.3^o	13.0^o	Case 4	46.2^o	11.9^o	Case 7	59.3^o	-10.7^o
Case 2	45.6^o	13.1^o	Case 5	45.4^o	10.3^o	Case 8	52.6^o	47.0^o
Case 3	45.2^o	12.3^o	Case 6	45.8^o	10.8^o	Case 9	50.8^o	50.2^o

Table 1. The results closely reflect the nominal pose of the acetabular cup.

5 Discussion and Conclusion

These initial results are encouraging. The recovered cup orientation matches the measured ground truth to within 2° abduction and 3° flexion in all except one

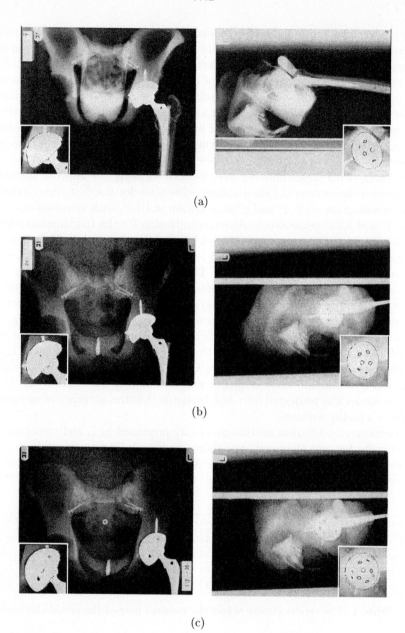

(a)

(b)

(c)

Fig. 3. (a) A pair of input images from the first series of radiographs. The inset shows recovered cup position, and a peanut butter jar is visible in each image. (b) A pair of input images from the third series of radiographs. In the lateral image, the superior boundary of the simulated torso runs almost parallel to the superior edges of the iliac crests. (c) A pair of input images from the fourth series of radiographs for which the X-ray/CT registration was inaccurate, making it impossible to simultaneously align the cup in both images. The bright line running superior-inferior in the lateral image is a lexan plate to which the pelvis is attached.

of the trial cases. RMS error among these "correctly" converged cases was 0.78^o abduction and 1.78^o flexion. The results are presented in this space since these are the measures most commonly used intra-operatively.

Although the X-ray/CT registration algorithm converges reliably for AP images, correct convergence for lateral images is somewhat sensitive to the initial conditions. This is illustrated by the failure to converge correctly in case 7 above. One possible reason for this sensitivity is that the bilateral symmetry of the pelvis leads to a pose ambiguity when the view direction is very nearly lateral. In these cases it becomes very difficult, for example, to distinguish the right iliac crest from the left iliac crest. This is especially true when the source-to-film distance is large, as is the case in this study, since a large source-to-film distance decreases foreshortening due to perspective projection.

Future work includes automating the pose initialization to eliminate the need for user input, resolving the pose ambiguity in direct lateral images, and applying the system to real patient data.

References

1. A. M. DiGioia, B. Jaramaz, M. Blackwell, D. A. Simon, F. Morgan, J. E. Moody, C. Nikou, B. D. Colgan, C. A. Aston, R. S. Labarca, E. Kischell, and T. Kanade. The otto aufranc award. image guided navigation system to intraoperatively measure acetabular implant alignment. *Clinical Orthopaedics and Related Reseearch*, 355:8–22, October 1998.
2. S. J. Gortler, R. Grzeszczuk, R. Szeliski, and M. F. Cohen. The lumigraph. In *Computer Graphics Proceedings, Annual Conference Series*, pages 43–54, 528, New Orleans, LA, USA, August 1996. ACM SIGGRAPH.
3. B. Jaramaz, C. Nikou, and T. J. Levison. Cupalign: Computer-assisted postoperative radiographic measurement of acetabular components following total hip arthroplasty. In *Proceedings of MICCAI '99*, pages 876–882, Cambridge, UK, September 1999.
4. D. LaRose, J. Bayouth, and T. Kanade. Transgraph: Interactive intensity-based 2d/3d registration of x-ray and ct data. In *Proceedings of SPIE - International Society for Optical Engineering, Medical Imaging 2000 (In press)*, San Diego, CA, USA, February 2000.
5. M. Levoy and P. Hanrahan. Light field rendering. In *Computer Graphics Proceedings, Annual Conference Series*, pages 31–42, New Orleans, LA, USA, August 1996. ACM SIGGRAPH.
6. G. E. Lewinnek, J. L. Lewis, and R. Tarr. Dislocations after total hip replacement arthroplasties. *J Bone Joint Surg*, 60A:217–220, 1978.
7. D. E. McCollum and W. J. Gray. Dislocation after total hip arthroplasty: Causes and prevention. *Clinical Orthopaedics and Related Reseearch*, 261:159–170, 1990.
8. W. H. Press, B. P. Flannery, S. A. Teukolsky, and W. T. Vetterling. *Numerical Recipes in C - The Art of Scientific Computing*. Cambridge University Press, Cambridge, England, 1988.
9. R. Sellers, D. Lyles, and L. Dorr. The effect of pelvic rotation on alpha and theta angles in total hip arthroplasty. *Contemporary Orthop.*, 17:67–69, 1988.

Intraoperative Simulation and Planning Using a Combined Acetabular and Femoral (CAF) Navigation System for Total Hip Replacement

Yoshinobu Sato[1][4], Toshihiko Sasama[4][1], Nobuhiko Sugano[2]
Kei Nakahodo[4][1], Takashi Nishii[2], Kenji Ozono[3]
Kazuo Yonenobu[2], Takahiro Ochi[3], Shinichi Tamura[1][4]

[1] Division of Functional Diagnostic Imaging
[2] Department of Orthopaedic Surgery
[3] Division of Computer Integrated Orthopaedic Surgery
Osaka University Graduate School of Medicine
[4] Department of Informatics and Mathematical Science
Graduate School of Engineering Science, Osaka University
yoshi@image.med.osaka-u.ac.jp, http://www.image.med.osaka-u.ac.jp/~yoshi

Abstract. This paper describes intraoperative simulations for the limb length and the range of motion (ROM) adjustment in total hip relacement (THR) surgery, and their utility in intraoperative planning. After implantation of the cup and stem, final adjustments can be made to the limb length and ROM by selecting the optimal combination of femoral neck and head components from the range available in a changeable modular system. The aim of this work is to provide intraoperative assistance to the surgeon in selecting the optimal component combination as well as in planning additional osteotomy to remove unwanted bone impingements and widen the safe ROM. Using the positions and orientations of the cup and stem intraoperatively measured by a combined acetabular and femur (CAF) navigation system, limb length and ROM simulations are carried out for neck and head components of various lengths and angles. These simulations provide information on limb length, the ROM, and where in a 3D model impingements will occur for each combination of components. The accuracy of the simulations was evaluated by comparison with postoperative CT data for the limb length and actually measured motions for the ROM.

1 Introduction

In total hip replacement (THR), computer assisted navigation for placement of the acetabular cup has been shown to be highly useful [1],[2],[3]. Precise alignment of the cup orientation is regarded as particularly important to reduce the possibility of complications such as dislocation, wear, and loosening. Previous navigation systems for THR have consisted of the distinctly separate stages – a preoperative planning and simulation stage and the intraoperative navigation itself [1],[2],[3]. In the preoperative stage, optimal parameters such as cup size, position, and orientation are determined through surgical planning that includes the range of motion (ROM) simulation [4] and the limb length adjustment. In the intraoperative stage, the actual position and orientation of the cup are accurately measured using an optical 3D sensor so as to execute exactly what was planned preoperatively. However, this two-stage approach is insufficient because the actual execution during the intraoperative stage often differs from the preoperative plan for the following reasons:

- Manual placement of the cup component is inherently inaccurate even if a sophisticated navigation system is utilized.
- The preoperative plan sometimes needs to be changed during the operation if conditions are encountered that were not anticipated from the preoperative images. For example, the cup position may need to be changed if the bone tissue around the preoperatively planned position is found to be unexpectedly fragile.

To address the above problems, our system incorporates additional intraoperative simulation and planning stages that are implemented after the two stages described above.

In this paper, we describe intraoperative simulations for limb length and ROM adjustment in THR. In our hospital, after implanting the acetabular cup and femoral stem, the surgeon makes final adjustments to the limb length and ROM by selecting the optimal combination of neck and head components from the range available in a changeable modular femoral head and neck system (ANCA-FIT, Cremascoli, Milan) (Fig. 1). The neck and head components vary in their length and angle (the depth of the socket in the case of the head component), and component selection has a significant effect on both the limb length and ROM. The surgeon performs additional osteotomy (bone cutting) to remove some bone regions so as to prevent the impingement of bone on bone or of the implant on bone, which would reduce the safe ROM (hereafter, the impingement of A on B is referred to as the A–B impingement). The purpose of the intraoperative simulation is to assist the surgeon in intraoperative planning with regard to selecting the optimal combination of neck and head components and determining which additional bone regions should be removed by osteotomy. In our intraoperative simulation, the limb length and ROM are simulated for neck and head components with various lengths and angles based on the positions and orientations of the acetabular cup and the femoral stem implants, which are intraoperatively measured after implantation using a combined acetabular and femur (CAF) navigation system for THR [5]. Intraoperative planning is then carried out based on the simulation results as to which neck and head components should be selected and where additional osteotomy should be performed to avoid unwanted impingements.

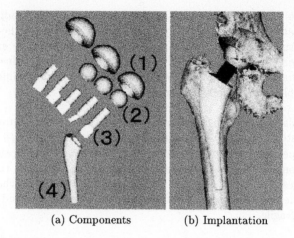

(a) Components   (b) Implantation

Fig. 1. Changeable modular femoral head and neck system. (1) Cup, (2) head, (3) neck, and (4) stem.

2 Intraoperative Simulation and Planning for Total Hip Replacement

2.1 Combined Acetabular and Femur (CAF) Navigation System

The combined acetabular and femur (CAF) navigation system employed is an Optotrak-based guidance system for THR (Optotrak, Northern Digital Inc., Waterloo, Ontario, Canada) [5]. The novel feature of this CAF system is that it provides the guidance of both pelvis and femoral sides – unlike other systems which deal with either the acetabular [1],[2],[3] or femur [6] side. This feature enables the surgeon to intraoperatively evaluate the geometric and kinematic parameters inherent in the hip joint, including limb length and ROM as addressed in this paper. Using 3D surface models of the pelvis and femur reconstructed from preoperative CT images, surface-based registration is performed by applying the iterative closest point (ICP) algorithm for both the acetabular and femoral sides. To attain a high level of accuracy on each side, we evaluated the optimal sampling areas of 3D points on clinically available bony surfaces so as to balance accuracy and invasiveness [7]. The placement of the cup and stem components is guided and measured using localizers to which Optotrak LED markers are attached (Fig. 2(a)). During the operation, the motions of both the pelvis and femur are tracked by means of rigid bodies attached to them (Fig. 2(b)). Another advantage of our CAF system in THR is that it provides guidance in osteotomy for femoral head resection on the femoral side.

2.2 Pelvis- and Femur-Centered Coordinate Systems

Using anatomical landmarks, the pelvis-centered and femur-centered coordinate systems (pelvis-CS and femur-CS) are preoperatively determined to measure the

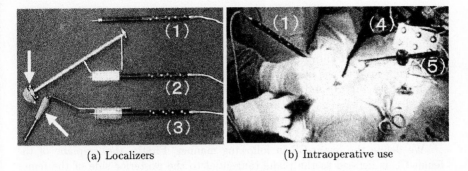

      (a) Localizers   (b) Intraoperative use

Fig. 2. Localizers and trackers with LED markers for the CAF system and their intra-operative use. (1) Optotrak pen-probe, (2) cup localizer, (3) stem localizer, (4) pelvis tracker (rigid body attached to pelvis), and (5) femur tracker (rigid body attached to femur).

geometric and kinematic parameters inherent in the hip joint of each patient. Figure 3 shows the anatomy of the hip joint. The anatomical landmarks are localized on the 3D surface model of the pelvis and femur reconstructed from the CT images. The pelvis-CS and femur-CS are based only on the 3D shape inherent in the pelvis and femur, respectively.

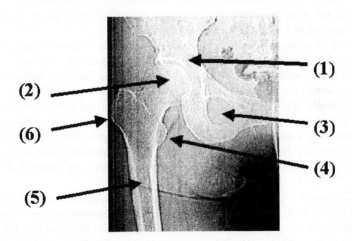

Fig. 3. Anatomy of the hip joint and anatomical landmarks. (1) Acetabulum, (2) femoral head, (3) foramen obturatum, (4) trochanter minor, (5) medullary space, and (6) trochanter major.

We define the pelvis-CS in the following manner. Firstly, the xz-plane of the pelvis-CS is defined. Consider a plane whose normal is vertical (like the top of a desk). A pelvis placed on the plane with its frontal side downward is in a stable pose supported by three points on the pelvis surface. The xz-plane of the pelvis-CS is then defined as the plane tangential to the frontal side of the pelvis as described above (Fig. 4(a)). Secondly, the xy-plane is defined as the plane tangential to the upper rims of the two holes in the pelvis (the foramina obturata, see Fig. 3) and orthogonal to the xz-plane. Finally, the yz-plane is defined as the plane passing through the center of the bridge between the left and right parts of the pelvis (the symphysis pubica) and orthogonal to both the xz-plane and the xy-plane (Fig. 4(b)).

We define the femur-CS in following manner. Firstly, the xz-plane of the femur-CS is defined as the plane tangential to the posterior side of the femur in a manner analogous to that described above for the pelvis-CS (Fig. 4(c)). Secondly, the z-axis is defined as the orthogonal projection of the medullary axis, which is the principal axis of the medullary space (see Fig. 3), on the xz-plane. Thirdly, the origin is defined as the orthogonal projection of the crown position of the trochanter minor, a prominent anatomical landmark (see Fig. 3) on the xz-plane. Finally, the xy-plane is defined as the plane passing through the trochanter minor crown and orthogonal to the z-axis (Fig. 4(d)).

2.3 Representation of Positions and Orientations of Cup and Stem

The CT image coordinate system (image-CS) and the Optotrak coordinate system (Optotrak-CS) are both unsuitable for representing the geometric and kinematic properties inherent in the hip joint. In order to measure and analyze these properties, we need to determine the positions and orientations of the cup and stem in the pelvis-CS and femur-CS, respectively.

For this purpose, we employ cup-centered and stem-centered coordinate systems (cup-CS and stem-CS) in which the 3D shapes of the cup and stem are respectively represented. We define the cup-CS in the following manner. First, the origin of the cup-CS is defined as the center of the spherical surface of the cup hemisphere. The z-axis of the cup-CS is then defined as the normal of the plane fitted to the cup rim. Since the cup implant we use is rotationally symmetric around the z-axis, the x-axis and y-axis are not determined. The stem-CS is similarly uniquely defined based on the rim shape and orientation of the socket into which the modular neck component is inserted (the details are not described here).

Let T_{io} be a 4×4 matrix representing the transformation from the image-CS to the Optotrak-CS, which is estimated during the intraoperative registration stage; let T_{oc} be the transformation from the Optotrak-CS to the cup-CS, which is estimated by combining the intraoperatively measured 3D positions of the cup localizer with the preoperative tool calibration; and let T_{pi} be the transformation from the pelvis-CS to the image-CS, which is estimated during the preoperative planning stage. What we need is T_{pc}, representing the cup position and orientation in the pelvis-CS. We obtain the transformation from the pelvis-CS to the cup-CS, T_{pc}, using the above transformations as follows:

$$T_{pc} = T_{pi}T_{io}T_{oc}. \qquad (1)$$

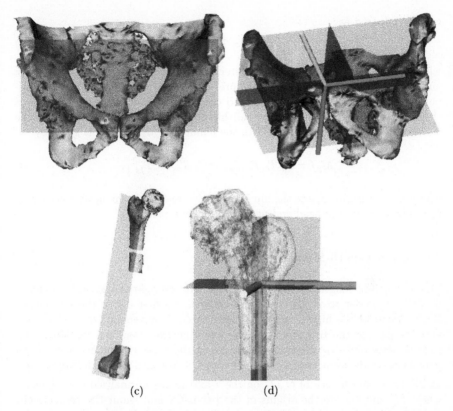

(c)   (d)

Fig. 4. Pelvis-centered and femur-centered coordinate systems (Pelvis-CS and femur CS). (a) The xz-plane with the pelvis 3D model. (b) Pelvis-centered coordinate system. (c) The xz-plane with the femur 3D model. (d) Femur-centered coordinate system.

Similarly, we obtain the transformation from the femur-CS to the stem-CS, T_{fs}, as follows:

$$T_{fs} = T_{fi}T_{io}T_{os}, \tag{2}$$

where T_{os} and T_{fi} are the transformations from the Optotrak-CS to the stem-CS and the femur-CS to the image-CS, respectively.

2.4 Representation of Hip Joint Motion

After implanting the cup and stem, the surgeon selects the most appropriate combination of neck and head components from those available in the changeable modular femoral head and neck system (Fig. 1) to adjust the limb length and ROM.

This combination determines the transformation from the cup-CS to the stem-CS, M_{cs}. M_{cs} is also a 4×4 matrix representing the transformation, which can decomposed as

$$M_{cs} = T_{cs}R_{cs}, \tag{3}$$

where T_{cs} is a fixed transformation determined by the combination of the neck and head components. R_{cs}, which represents the hip joint motion, is a variable transformation but constrained to the rotational motion whose center is the origin of the cup-CS. The overall hip joint motion is represented by the transformation from the pelvis-CS to the femur-CS, M_{pf}, which is given by

$$M_{pf} = T_{pc}T_{cs}R_{cs}T_{sf} = T_{pc}T_{cs}R_{cs}T_{fs}^{-1}, , \tag{4}$$

where R_{cs} is variable due to the hip joint motion and T_{cs} is changeable according to the combination of head and neck components selected.

2.5   Limb Length Simulation

Minimizing limb length discrepancy between left and right is of great importance in THR. In order to measure the limb length, we first define the normalized arrangement of the hip joint. In the normalized arrangement, R_{cs}, which is a variable part of the transformation representing the rotational motion of hip joint, is determined so that the directions of the three coordinate axes of the pelvis-CS are the same as those of the femur-CS. We then define the limb length as the projective length of $\overrightarrow{O_pO_f}$ on the z-axis in the normalized arrangement, where O_p and O_f are the origins of the pelvis-CS and femur-CS, respectively. This length corresponds to the distance along the z-axis between the upper rim of the foramen obturatum on the pelvis and the crown of the trochanter minor on the femur in the normalized arrangement, which are anatomical landmarks used in determining the pelvis-CS and femur-CS. The limb length simulation for testing different combinations of neck and head components is realized by changing the transformation T_{cs} (in Equation (4)), which corresponds to each combination. Since T_{pc} (from the pelvis to the cup) and T_{sf} (from the stem to the femur) needed to measure the limb length are obtained intraoperatively, this simulation is performed intraoperatively.

With respect to the other side of the leg (which is assumed not to be an implant), the limb length is preoperatively measured using the method described above. The femur-CS is determined by the same method. The pelvis-CS is common to both the left and right legs. T_{pc} (from the pelvis to the cup) and M_{cf} (from the cup to the femur) are measured from CT images, where $M_{cf} = T_{cf}R_{cf}$. The origin of the cup-CS is determined as the center of the sphere approximating the femoral head in CT images. The orientation of the cup-CS does not affect the limb length since it is measured in the normalized arrangement. R_{cf} is determined so that the the directions of the three axes of the pelvis-CS are the same as those of the femur-CS. Figure 5 shows examples of the intraoperative simulation for limb length adjustment.

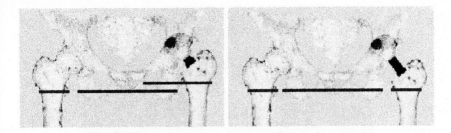

Fig. 5. Intraoperative simulation for limb length adjustment. Left: A large discrepancy between left and right is observed. Right: The discrepancy is minimized by changing the neck component.

2.6 Range of Motion (ROM) Simulation

The range of motion (ROM) is regarded as the intersection of sets of ROMs estimated on the basis of implant–implant, bone–implant, and bone–bone impingements (Fig. 6). In the hip joint motion shown in Equation (4), T_{cs} can be changed using different combinations of neck and head components. As described earlier, R_{cs} is the rotational motion whose center is the origin of the cup-CS. The range of rotational angles in R_{cs} is constrained by the impingements mentioned above, and the safe ROM can be regarded as the range of rotation angles without such impingements. R_{cs} can be decomposed into the direction $\overrightarrow{O_cO_s}$ and the rotation around $\overrightarrow{O_cO_s}$. Here, we assume that the rotation around $\overrightarrow{O_cO_s}$ is fixed and we estimate the range of the direction $\overrightarrow{O_cO_s}$ without impingements. A collision detection algorithm (V-CoLLIDE, http://www.cs.unc.edu/~geom/V_COLLIDE) is employed to find the ROM using the CT surface models of the pelvis and femur, and the CAD surface models of the implants. In the ROM simulation, the intraoperatively measured positions and orientations of the cup and stem (T_{pc} and T_{sf}) are used. The ROM simulation for testing the different combinations of neck and head components is realized by changing the transformation T_{cs} Thus, the ROM simulation results suggest the best combination of neck and head components.

Using the collision detection algorithm, the surgeon can also determine where in the 3D models collisions (impingements) occur and the extent to which they reduce the ROM. Hence, the ROM simulation results also suggest where bone should be removed by osteotomy to widen the safe ROM.

3 Results

We applied the procedures described in Section 2 to ten cases of THR at Osaka University Hospital. Postoperative CT images were obtained to compare intraoperatively measured parameters with those estimated from the postoperative images. Both the preoperative and postoperative CT images (512×512 matrix)

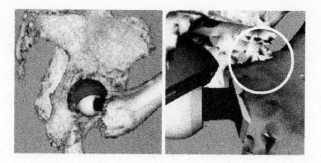

Fig. 6. Impingement of bone on bone in intraoperative ROM simulation.

were obtained with a 3-mm slice thickness, 3-mm reconstruction pitch, and 420 mm FOV.

3.1 Limb Length Simulation

We evaluated the intraoperatively estimated limb length by comparing it with that estimated using postoperative CT images. The postoperative images were processed, and the postoperative cup position, cup orientation, and limb length were measured, as follows. Firstly, the CT surface models of the pelvis and femur were reconstructed from the postoperative CT images. The postoperative pelvis model was registered to the preoperative pelvis model using the ICP algorithm and the same pelvis-CS as the preoperative one was determined. Secondly, the position and orientation of the implanted cup (the cup-CS) in the pelvis-CS were estimated by fitting the CAD model of the cup. In this way, the center of the rotation R_{cf} was found. Finally, the femur-CS in the postoperative model was defined and its orientation was aligned to the pelvis-CS so that the pelvis and femur models were positioned in the normalized arrangement. The postoperative limb length was then measured.

Table 1 shows the differences between the intraoperative and postoperative results for the cup position, cup orientation, and limb length, which taken together can be regarded as a good approximation of the overall accuracy of the method. For the cup position and orientation, the distance and difference in angle, which are absolute values, were respectively used as the error measures. For the limb length, the difference between the two results, which is a signed value, was used. The root-mean-squares (RMS) were around 3 mm and 4 degrees. Intraoperative adjustment of limb length in gradations of every 2 or 3 mm is possible by changing the neck and head component combinations. The accuracy of the limb length simulation was regarded as acceptable considering that the differences included the registration error between the preoperative and postoperative pelvis models.

3.2 ROM Simulation

Figure 7 shows the ROM results for two cases (#6 and #7). These were plotted based on the implant–implant impingement (labeled "implant" in Fig. 7) and all possible impingements (labeled "simulation"). The ROM is represented by the z-axis direction of the femur-CS in the pelvis-CS. The radial directions $0°$ and $90°$ correspond to the left and frontal directions, respectively. We also plotted the directions of the femur relative to the pelvis measured intraoperatively by actual ROM testing – $i.e.$ not by simulation – using the Optotrak system (labeled "real"). These measurements were made while the surgeon moved the leg in the frontal directions. In each case, the same combination of neck and head components was used for the actual and simulated measurements. The ranges of movement simulated intraoperatively were consistent with those actually measured.

In the case shown in Fig. 7(a), a ROM simulation was also performed with a different combination of neck and head components from that used for the actual measurement. The result, shown in the left frame, confirmed that the ROM could be markedly altered by changing the component combination.

In the case shown in Fig. 7(b), the surgeon performed additional osteotomy on the pelvic bone to remove unwanted impingements and widen the safe ROM. We intraoperatively reconstructed the bone shape resurfaced by the additional osteotomy based on a set of 3D points obtained by digitizing the bone surface using the Optotrak pen-probe [8]. ROM simulations were done using the pelvis models obtained before and after the additional osteotomy (respectively labeled "simulation 1" and "simulation 2"). The results confirmed that the ROM was widen by the additional osteotomy.

4 Discussion and Conclusions

We have described intraoperative limb length and range of motion (ROM) simulations in total hip replacement (THR) surgery and demonstrated their usefulness in intraoperative planning to select the best combination of neck and head components in a changeable modular system as well as in additional osteotomy to widen the safe ROM. In order to measure the geometric and kinematic properties inherent in the hip joint, the pelvis- and femur-centered coordinate systems were preoperatively determined. The positions and orientations of the acetabular cup and femoral stem in these coordinate systems were then obtained intraoperatively using a combined acetabular and femur (CAF) navigation system for

Table 1. Accuracy evaluation of intraoperatively measured cup position, orientation, and limb length.

Case #	1	2	3	4	5	6	7	8	9	10	RMS
Cup position (mm)	2.96	1.50	4.38	3.05	3.40	3.55	1.15	2.08	3.79	4.64	2.98
Cup orientation (degree)	4.55	2.84	3.38	1.84	6.60	4.91	1.91	2.67	3.28	1.52	4.25
Limb length (mm)	6.35	0.50	−0.84	3.53	1.11	−4.45	−0.18	3.12	−2.21	−5.11	3.40

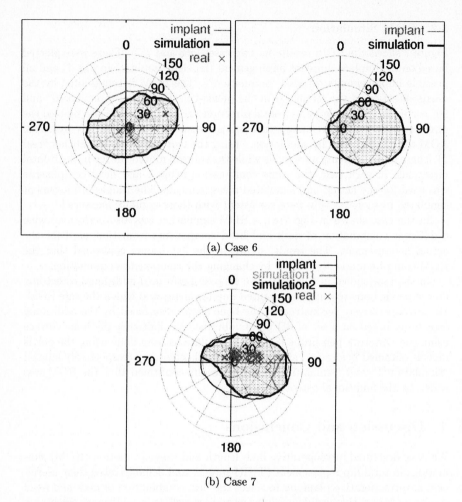

(a) Case 6

(b) Case 7

Fig. 7. Intraoperative ROM simulation results. See text for details.

utilization in the simulations based on the intraoperative conditions of the hip joint.

These intraoperative simulations offer two distinct advantages. Firstly, the most suitable combination of head and neck components can be selected based on comprehensive simulation results. This is advantageous for the surgeon because it is difficult to actually measure the motion limits in various directions or the limb length for various combinations of head and neck components during the operation. Secondly, using 3D models the surgeon can easily determine impingement locations and where additional osteotomy should be performed.

In clinical trials, the accuracy of intraoperatively estimated limb lengths was evaluated by comparison with postoperative CT images, and was found to be acceptable. The ROM obtained by intraoperative simulation was compared with actually measured motions and confirmed to be consistent with them. While the experimentally evaluated overall accuracy was acceptable, it was apparent that the overall accuracy was liable to be affected by several potential sources of error including tool calibration and registration [7] as well as the determination of the pelvis- and femur-centered coordinate systems. Furthermore, error sources on the pelvis side are combined with those on the femur side in the hip joint simulation. Hence, in future work we will consider how the theory of error propagation [9],[10] can be applied to determine the extent to which each error source might affect the overall accuracy.

Since our work is focused on the uniqueness of the geometric and kinematic parameters of the hip joint, in defining the pelvis- and femur-centered coordinate systems it was our intention that these coordinate systems should be uniquely determined based on their 3D shapes. Although the coordinate systems used in this study provided unique parameters, they should be improved so as to describe geometric and kinematic properties that can be directly related to the joint function. For example, in the present system the z-axis of the pelvis-centered coordinate system does not correspond to any functional axis of hip joint motion. In future work, such factors need to be incorporated when defining the coordinate systems [11].

Acknowledgment This work was partly supported by the Japan Society for the Promotion of Science (JSPS Research for the Future Program).

References

1. Digioia AM III, Jaramaz B, Blackwell M, et al.: Image guided navigation system to measure intraoperatively acetabular implant alignment, *Clinical Orthopaedics and Related Research*, **355**, 8-22 (1998).
2. Jaramaz B, Digioia AM III, Blackwell M, et al.: Computer assisted measurement of cup placement in total hip replacement, *Clinical Orthopaedics and Related Research*, **354**, 70-81 (1998).
3. Langlotz U, Lawrence J, Hu Q, et al.: Image guided cup placement, *Computer Assisted Radiology and Surgery (CAR'99)*, Paris, 717–721 (1999).
4. Jaramaz B, Nikou C, Digioia AM III, et al.: Effect of cup orientation and neck length in range of motion simulation, *Proc. 43rd Annual Meeting of the Orthopaedics Research Society*, 186 (1997).
5. Sugano N, Sato Y, Sasama T, et al.: Combined acetabular and femoral surgical navigation in total hip arthroplasty, *Computer Assisted Radiology and Surgery (CAR'99)*, Paris, 722–725 (1999).
6. Taylar RH, Mittelstadt BD, Paul HA, et al.: An image-directed robotic system for precise orthopaedic surgery, *IEEE Transactions on Robotics and Automation*, **10**(3), 261–275 (1994).
7. Sasama T, Sato Y, Sugano N, et al.: Accuracy evaluation in computer assisted hip surgery, *Computer Assisted Radiology and Surgery (CAR'99)*, Paris, 772–776 (1999).
8. Nakahodo K, Sasama T, Sato Y, et al.: Intraoperative update of 3-D bone model during computer navigation of pelvic osteotomies using real-time 3-D position data, *Computer Assisted Radiology and Surgery (CARS2000)*, San Francisco (2000).
9. Lea J and Peshkin M: Registration graphs: a diagramming and analysis tool for registration in computer-assisted surgery, *Computer Assisted Radiology and Surgery (CAR'99)*, Paris, 767–771 (1999).
10. Smith RC and Cheeseman P: On the representation and estimation of spatial uncertainty, *International Journal of Robotics Research*, **5**(4), 56–68 (1986).
11. Zatsiorsky VM: Kinematics of human motion, Human Kinematics Publishers Inc. (1998).

Fixatoin-Based Surgery:
A New Technique for Distal Radius Osteotomy

H. Croitoru   R. E. Ellis   C. F. Small   D. R. Pichora

Computing and Information Science   Mechanical Engineering   Surgery
Queen's University at Kingston, Canada
contact: ellis@cs.queensu.ca

Abstract. Fixation-based surgery is a new technique for achieving difficult corrections in some orthopedic procedures. The method is premised on using a fixation device, such as a fracture-fixation plate, during the alignment and distration phases of an open-wedge osteotomy. The basic idea is similar to assembly of manufactured components: pilot holes are drilled in the bone fragments, the fixation plate is attached to one fragment and, when the fragment is moved, the alignment has been achieved when the pilot holes in one fragment line up with predetermined through holes in the fixation plate.

The method has been specifically developed to address osteotomy of the distal radius to correct a malunited fracture. The method has been validated in laboratory studies. Clinical trials suggest that the method is no slower than the conventional technique, there is almost no intraoperative X-ray exposure, and that exceptionally large corrections can easily be achieved.

1 Rationale and Objectives

Fractures of the distal radius constitute about 15% of all fractures seen in the emergency room [4, 6]. The distal bone fragment may fail to realign to its proper anatomical position during healing, which can affect the alignment, kinematics, and load transfer across the wrist. Such malunions often lead to reduced strength, reduced range of motion, and pain. Correction of a malunion is performed using distal radius osteotomy (DRO). This procedure involves cutting the distal radius near its original fracture site and realigning it to restore normal function. For optimal outcome, proper planning is important to ensure that all the proper lengths and angles are restored. We use a patient-specific measure, in which the affected wrist is realigned to match the healthy wrist.

1.1 Traditional Technique

In traditional technique, AP and lateral X-ray films of the deformed wrist are taken. Osteotomy lines are drawn on the X-ray films to determine the size and shape of the bone graft [1] needed to correct the wrist to accepted radiographic indices. Intraoperatively, after the bone has been cut, fluoroscopic images are taken to determine the alignment of the distal bone fragment. This process is repeated until the desired orientation and position is obtained. The defect gap is filled with a bone graft or bone substitute. A fixation plate is contoured to the shape of the distal and proximal fragments, secured with bone screws.

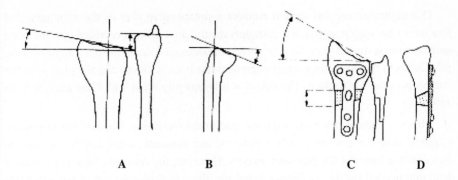

Fig. 1. Distal Radius Osteotomy. (A) A malunion typically exhibits shortening and an radial tilt of the radius. (B) Dorsal tilt may also be exhibited. (C,D) After correction, a wedge of bone fills the gap that is created by extending and realigning the distal radius.

This technique has three major limitations: poor planning of rotations in three dimensions; difficulty in achieving the desired lengthening, due to soft-tissue contracture; and a high dosage of fluoroscopic radiation to the patient and OR staff during the surgery.

1.2 Previous Computer-Assisted Techniques

To date, two studies have attempted to overcome these limitations. The BIZCAD system created by Bilić and Zdravković [2, 3, 12] used data obtained from two perpendicular X-ray films to create 3D wireframe models of the healthy and deformed radii. These models were constructed from fourteen landmarks that were marked on the X-ray films and later digitized. Despite positive outcome of seven trial cases this system limited: there were potential data-entry errors in entering the coordinates of the landmark locations, and the system was only a planning system (there was no way to inform the surgeon during the operation as to where the exact location and orientation of the osteotomy should be).

Jupiter *et al.* [9] used CAD/CAM technology and CT data to generate plastic models of the wrist to visualize and complex malunions. This approach was useful in envisioning corrections of complex malunions, but offered little help in determining the quantitative data necessary to plan the correction. Use of this method was also limited due to the high cost of creating the models.

1.3 A Fixation-Based Computer-Assisted Technique

We have developed a technique that is a fundamental departure from all previously reported techniques. By analysing the geometry of the completed osteotomy, we have determined that there are three components to a successful precedure:

Distraction of the fragment, which requires stretching soft tissues that have contracted;

Alignment of the fragment to restore the distal radius to its correct anatomical location; and

Fixation of the fragment to preserve the alignment.

Our computer-assisted system requires a preoperative plan of the alignment and fixation of an <u>uncontoured</u> plate. Intraoperatively, pilot holes are drilled into the bone and the bone is shaved to fit the plate. When the through holes in the plate align with the pilot holes in the bone, the correct alignment has been achieved and the plate is in the correct position for fixation. The defect is subsequently filled with bone graft or bone substitute.

This is a radical departure from the traditional technique. Fixation-based surgery requires computer assistance in both planning and guidance, which requires additional preoperative time for the planning process. However, the operative time is no greater than that needed for the traditional technique, there is almost no use of intraoperative X-ray fluoroscopy, and exceptionally large defects can be corrected. We have conducted an *in vitro* study that showed significant improvements over the traditional method. We have also performed four clinical cases that demonstrate the practicality of our method.

2 Materials and Methods

Five steps are involved in our procedure. Preoperatively, the patient was scanned, models were generated, and the surgical plan was formulated. Intraoperatively, the patient's anatomy was registered to the preoperative data and image-guided surgery was performed.

1. **Patient Scanning:** Both patient wrists were scanned in neutral rotation using helical CT. Scans were also acquired from the proximal radii and ulnae to help in the long bone alignment.

2. **3D Model Generation:** Voxel intensities were extracted from the scan and used to create isosurface models [5]. The models of the healthy radius and ulna were reflected to serve as a template for the correction.

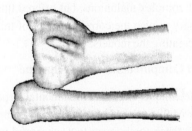

The ulna and the malunited radius

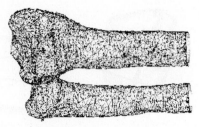

Mirror image of the healthy arm acts as a template

3. Plan Creation: The planning system used an X-Windows interface and the OpenGL 1.11 graphics library. Creating a preoperative plan involved five steps:

(a) **Initial Alignment** The entire ulna and the proximal radius of the affected wrist were aligned with the template. The affected radius demonstrated a deformity.
(b) **Virtual Osteotomy** A cutting plane was chosen by the surgeon and the isosurface model was cut into two models.

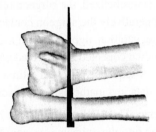

(c) **Aligning the Distal Radii** The distal radial fragment of the deformed radius was aligned with the healthy distal radius. Of particular interest were the subchondral arc that was formed between the ulna and radius and the ulnar variance.

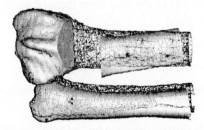

(d) **Positioning the Fixation Plate** Once the surgeon was satisfied with the new alignment of the deformed distal radius, the model of the fixation plate was placed on the models such that *in vivo* it would hold the bone fragments in place. In placing the fixation plate on the bones, it was important to verify that it lay flat on the bones and that none of the screw holes for the plate would protrude into any joints.

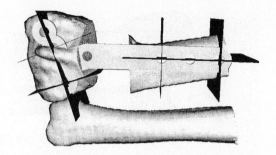

(e) **Saving the Plan** The location of the two distal and proximal drill holes for the plate were saved. The locations of the distal holes before the cutting of the bone were determined by applying the inverse transformation of the distal fragment to the drill locations.

4. **Intraoperative Registration:** We used a robust registration method to estimate the rigid-body transformation from the anatomy to the preoperative data [10]. While the patient was being anesthetized, the surgeon selected initial-estimate landmarks ("spotlights"). Intraoperatively the surgeon contacted the bone in these spotlights, which provided the registration algorithm with a good initial estimate of the registration. The surgeon then collected about 10 additional surface points to refine the registration estimate. A sequence of robust estimators were used to discard statistical outliers from the data in a mathematically disciplined manner.

5. **Image-Guided Surgery:** The 3D bone and plate models, CT data, and preoperative plan are imported into the guidance system. An OPTOTRAK 3020 (Northern Digital Inc., Waterloo, Canada) was used for tracking the surgical tools and patient movements. The sequence of the surgical procedure was:

   - Drill proximal and distal pilot holes for the fixation plate with image guidance, with the locations of the pilot holes dictated by the preoperatively planned location of the plate;
   - Cut the bone (with image guidance);
   - Affix the plate to the distal radius fragment;
   - Implant a temporary anchor screw into the radial midshaft, proximal to the plate;
   - Progressively distract the plate by "jacking" the plate against the anchor screw with a laminar spreader;
   - Align the screw holes in the plate with the pilot holes in the bone;
   - Secure the plate to the proximal fragment and remove the anchor screw; and
   - Fill the defect with autologous bone graft.

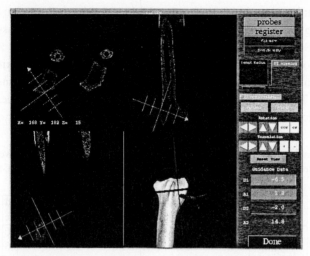

Image guidance for drilling a pilot hole for the fixation plate

2.1 A Laboratory Study

Fourteen identical polyurethane-foam models of a single deformed radius (model #1029, Pacific Research Laboratories, Bellingham, WA), one normal radius (model #1018), and two normal ulnae (model #1017) were used for the study. One normal ulna and radii were affixed in neutral rotation and scanned to represent the template for correction. A jig was constructed for the deformed radii to ensure that they would always lie in the exact location relative to the ulna.

Seven corrections were performed using the traditional procedure using fluoroscopy and seven using the computer planning and navigation system. Corrective procedures were alternated to reduce potential bias in the results.

In each of the procedures, a fresh deformed radius was fastened in the jig and infrared emitting diode (IRED) markers were attached to the distal radial bone fragment and the radial shaft. The locations of the two markers were captured before and after the procedure and used to analyze the overall correction established. The data collected from the marker locations were analyzed for the translational and rotational corrections. Mean, Standard Deviation, and range were computed for each spatial degree of freedom.

3 Results

Statistical analyses were performed using the Statistical Package for the Social Sciences (SPSS Inc., Chicago, IL). For accuracy, the mean translations and rotations were analyzed using Student's t-test; in cases where there was a significant difference in the variance between the computer-assisted and traditional group results, a Kruskal-Wallis H test for two samples (equivalent to the Mann-Whitney U test) was used to analyze the accuracy. For repeatability, standard deviations were analyzed using the two-sample F-test. Accuracy and repeatability data are presented in Table 1.

Significant increases in accuracy were observed for the total rotation angle (θ_{total}) and dorsal angulation (α, about the lateral axis, X) using the computer-assisted method ($p < 0.01$). Significant improvement in repeatability was observed in the computer-assisted method for radial/ulnar rotation (β, about the anteroposterior axis, Y, $p < 0.1$) and in radial/ulnar translations (along X, $p < 0.01$) and dorsal/volar translation (along Y, $p < 0.1$). There were no significant changes in accuracy of the other geometrical degrees of freedom, which were low in both groups.

Table 1. Accuracy and Repeatability comparisons. Errors are the difference between the planned and measured rotations and translations. The range is the difference between the largest and smallest error.

	Errors, Traditional			Errors, Computer-Assisted		
	Mean	Std. Dev.	Range	Mean	Std. Dev.	Range
x (mm)	1.9	3.8	10.7	0.4	1.2	3.0
y (mm)	-1.7	3.7	11.64	-1.0	2.1	5.3
z (mm)	-0.1	1.8	6.0	0.8	1.7	4.6
α (deg)	-10.3	3.0	9.2	-1.1	2.9	7.1
β (deg)	-0.4	5.3	13.2	0.5	3.3	9.5
γ (deg)	7.8	3.7	10.8	4.0	6.5	18.4
θ_{total} (deg)	14.0	4.0	10.9	7.7	3.5	11.12

4 A Pilot Clinical Study

Four clinical procedures have been performed at Kingston General Hospital (Kingston, Ontario, Canada) to date. In the first procedure the correction achieved was as planned, but fluoroscopic validation suggested that the planned correction was short by approximately 2 mm. This small additional correction was readily achieved intraoperatively. Subsequently, the planner was modified to provide contour rendering (which better displays concave features than does surface rendering). The next three cases all showed excellent alignment from fluoroscopic images and excellent postoperative results.

Case #4 is a typical example of the use of our system. A 50-year-old male fell while traveling in Central America in February 1999. The immediate treatment, cast application without reduction, was unsuccessful. A severe deformity resulted (30 degrees of dorsal tilt, 15 degrees less of radial inclination, 8 degrees of supination, 12mm of shortening), as shown in Figure 2. The consequence was pain, with functional impairment due to restricted motion and weakness.

The correction was planned as described above. The three views, shown in Figure 3, show the large amount of correction required to realign the deformity.

The final position and orientation of the articular radial surface was near anatomic, which is an excellent result. It is doubtful that this correction, shown in Figure 4 could have been achieved by conventional means.

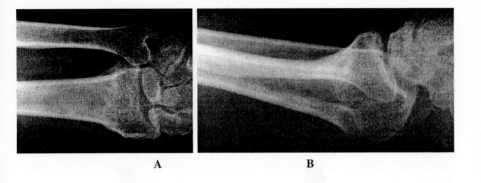

A   B

Fig. 2. Preoperative radiographs. (A) AP view of the malunited radius. Note the severe shortening and ulnar inclination (B) ML view of the malunited radius. Note the severe dorsal tilt.

5 Discussion

We report the design, implementation, and testing of a preoperative planning system that worked in conjunction with a guidance system to assist a surgeon in restoring the biomechanics and anatomy of the wrist. The ability of the system to simulate the osteotomy, and to view the expected results in solid and reduced renderings, allowed a surgeon to plan an osteotomy in a way that was far superior to the traditional film-based method. Laboratory results have shown that the system can improve the accuracy of producing a desired alignment.

Clinical trials have shown that the system is useful in improving surgical performance. An immediate benefit to the surgical team was that the amount of intraoperative radiation was reduced. Traditionally, numerous fluoroscopic images were required during the operation to determine the correct orientation of the distal radius after the cut. Using the image-based guidance system, X-ray images were needed only at the end of the procedure to verify the alignment.

Specific benefits of the planning system include the ability to perform multiple simulations of the surgical procedure preoperatively, which can be used to optimize the plan and identify potential problems during realignment (such as bone/bone contact). The planning system can also be used as a teaching tool to give medical students spatial visualization of deformities and ways of correcting them. Since the exact geometry of the created gap between the bones is known, CAD/CAM technology could be used to make artificial bone grafts, reducing the amount of trauma to the patient's body.

According to the literature reviewed, the three main anatomical measurements considered in correcting deformed distal radii are the ulnar variance, dorsal tilt, and radial tilt [8, 7, 6, 11]. It was therefore not surprising to find that there was a large standard deviation in the other geometric variables, which were translation along the lateral and A/P axes, and rotation about the pronation/supination axis. The results obtained using the computer-assisted corrections showed significant reductions in the other directions.

Despite promising experimental and clinical results, some potential pitfalls are:

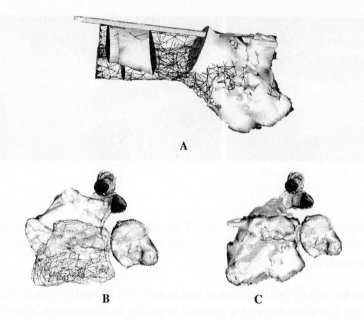

Fig. 3. Preoperative planning. (A) ML view of the planned correction, including the fixation plate. (B,C) Axial views of the original models and the planned correction.

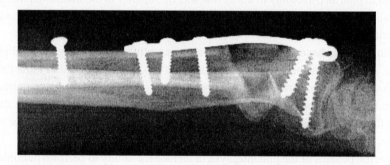

Fig. 4. Postoperative radiograph 8 weeks after surgery. The wrist has been restored to its anatomic alignment. Note the jacking screw. The bowing of the plate is due to the residual tension in the soft tissues.

- The outcome of the surgery depends on the accuracy of the plan. Poor planning may result in poor surgical outcome.
- Due to the design of the fixation plate, the fixation screws are not always perpendicular to the surface of the plate. This must be checked in the planning process.
- When drilling the pilot holes, the drill could potentially bend. Bending is not accounted for by the tracking system.
- Placement of the plate depends on how well the images are registered to the patient's bone: a poor registration could result in poor surgical outcome.

Fixation-based surgery is a method of planning and performing a surgical procedure in which the relative position of two bones is dictated by the fixation device that holds them together. For distal radius osteotomies, the holes for the screws in the fixation plate determined how the distal radius would lie relative to the proximal radius. The use of a fixation-based technique also avoids the complexity of attempting to guide the surgeon to realign a bone fragment in six degrees of freedom of correction.

Fixation-based surgery offers much to the practising hand surgeon. We are now examining ways in which the plate and fixation could be optimized for computer-assisted instumentation. The idea of fixation-based surgery might also be applicable in other surgical procedures.

Acknowledgements

This research was supported in part by Communications and Information Technology Ontario, the Institute for Robotics and Intelligent Systems, and the Natural Sciences and Engineering Research Council of Canada.

References

1. R. Bilić and V. Zdravković. Planning corrective osteotomy of the distal end of the radius. 1. Improved Method. *Unfallchirurg*, 91:571–574, 1988.
2. R. Bilić and V. Zdravković. Planning corrective osteotomy of the distal end of the radius. 2. Computer-aided planning and postoperative follow-up. *Unfallchirurg*, 91:575–580, 1988.
3. R. Bilić, V. Zdravković, and Z. Boljević. Osteotomy for deformity of the radius. Computer-assisted three-dimensional modeling. *Journal of Bone & Joint Surgery [Br]*, 76B(1):150–154, January 1994.
4. A. J. Bronstein, T. E. Trumble, and A. F. Tencer. The Effects of Distal Radius Fracture Malalignment on Forearm Rotation: A Cadaveric Study. *Journal of Hand Surgery [Am]*, 22A(2):258–262, March 1997.
5. H. E. Cline, W. E. Lorensen, S. Ludke, C. R. Crawford, and B. C. Teeter. Two algorithms for the three-dimensional reconstruction of tomograms. *Medical Physics*, 15(3):320–327, May 1988.
6. D. J. Pogue et. al. Effects of distal radius fracture malunion on wrist joint mechanics. *Journal of Hand Surgery [Am]*, 15A(5):721–727, September 1990.
7. J. Oskam et. al. Corrective osteotomy for malunion of the distal radius: The effects of concominant ulnar shortening osteotmy. *Archives of Orthopaedic & Trauma Surgery*, 115(5):219–222, September 1996.
8. D. L. Fernandez and J. B. Jupiter. *Fractures of the Distal Radius: A Practical Approach to Management*, chapter 2-5, 11. Springer Verlag, Inc., New York, 1996.
9. J. B. Jupiter, J. Ruder, and D. A. Roth. Computer-generated bone models in the planning of osteotomy of multidirectional distal radius malunions. *Journal of Hand Surgery [Am]*, 17A(3):406–415, May 1992.
10. B. Ma, R. E. Ellis, and D. J. Fleet. Spotlights: A robust method for surface-based registration in orthopedic surgery. In *Medical Image Computing and Computer-Assisted Intervention – MICCAI'99*, pages 936–944. Springer Lecture Notes in Computer Science #1496, 1999.
11. M. Porter and I. Stockley. Fractures of the Distal Radius: Intermediate and End Results in Relation to Radiologic Parameters. *Clinical Orthopedics*, 220:241–251, July 1987.
12. V. Zdravković and R. Bilić. Computer-assisted preoperative planning (CAPP) in orthopaedic surgery. *Computer Methods & Programs in Biomedicine*, 32:141–146, 1990.

A Computer-Assisted ACL Reconstruction System Assessment of Two Techniques of Graft Positioning in ACL Reconstruction

F. Picard[1], J. Moody[1], V. Martinek[3], F.Fu[3], M.Rytel[1], C. Nikou[1], R.S. LaBarca[2], B. Jaramaz[1,2], A.DiGioia[1,2].

Centers for Medical Robotics and Computer-Assisted Surgery,
[1]UPMC Shadyside Hospital, Pittsburgh, PA and
[2]Robotics Institute, Carnegie Mellon University, Pittsburgh, PA.
[3]Department of Orthopaedic Surgery, University of Pittsburgh, Pittsburgh, PA.

Abstract. KneeNav™ACL, a CT-based surgical navigation system, is used to demonstrate two fundamental requirements for a practical computer-assisted surgical device: 1) it can guide the surgeon with the accuracy required for successful graft placement, and 2) this guidance can be achieved through a simple and intuitive interface. A series of ACL reconstruction preparation procedures were performed on foambone models using traditional arthroscopic techniques. Half of the procedures were done using conventional techniques, the other half were performed with KneeNav™ACL guidance. Resulting tunnel and femoral graft site locations were measured via radiographic and computer means, and the two techniques are compared. Ease of system use is also assessed.

Keywords: Computer-assisted surgery, navigation systems, ACL surgery.

1. Introduction

The outcomes following treatment of anterior cruciate ligament injuries have been shown to be highly dependent on the surgical technique employed. Two main clinical challenges faced by surgeons are graft placement and tensioning. Outcomes following ACL reconstruction depend on the selection of the graft position and the initial graft tension. Optimizing these two factors is expected to improve outcomes and reduce long-term complications [1, 2]. Although ACL reconstruction is a routine procedure, accurate ACL graft placement remains a difficult task even using current arthroscopic techniques. Alternatively, a computer-guided surgical system could be used to intraoperatively locate ideal graft placement sites.

The aim of this study was to evaluate two techniques for guiding holes used for graft attach during ACL reconstruction. These were a traditional "free hand" method using standard guide-pin tools, and a computer-assisted surgical guidance system that augments the standard tool set. The specific goals were:

- To assess and compare the accuracy, reliability and repeatability of the two ACL graft positioning techniques.
- To test whether the use of a computer-assisted surgical guidance system can enhance a surgeon's ability to consistently achieve a predetermined (i.e. target) plan.

2. Material and Methods

2.1. Personnel, Equipment & Materiel Requirements:

The following list summarizes the main requirements for the core experiments.
• Two surgeons experienced in ACL reconstruction, with little previous exposure to computer-assisted surgical guidance systems.
• 21 sawbones knee models per surgeon. These are custom knee models complete with meniscus, collateral ligaments and PCL. All are fully enclosed in a simulated capsule of light-impervious elastic material.
• An arthroscopic station suitable for ACL reconstruction surgery included 1) a thirty degree optical device, 2) a camera plugged onto the optical device, 3) central processing unit, 4) light cable, 5) monitor, and 6) a VCR to record the experimental data.
• A traditional ACL tool set (Paramax ^{TM} ACL guide system) and a drill.
• Computer-assisted guidance system (KneeNav^{TM}), including trackers to be attached to the femur, tibia, and the Paramax^{TM} cruciate guide assembly (guide pin).

2.2. Design and Execution of the Experiment

2.2.1.General
During the experiment two surgeons independently prepared ACL tunnels on sets of sawbones knee models. Each surgeon used both traditional and computer-guided methods to attempt to place the femoral and tibial tunnels in a pre-determined orientation and location. Postoperatively, the tunnel positions and orientations were measured. In addition to direct physical measurements, computer and radiographic analysis techniques were used.

Foambones knee models were used in the experiment since they are readily available, anatomically accurate, and (to within manufacturing tolerances) physically identical. Thus, meaningful results can be obtained by comparing follow-up measurements from all sawbones test specimens. Although not truly representative of an actual surgical environment, careful test site preparation and the use of sawbones knee models provided sufficiently realistic OR conditions for the purposes of this study.

2.2.2. Preparation:
Several days before the experiment sessions each surgeon was given a sawbone knee model from which the simulated capsule was removed. On this model the surgeon indicated the exact locations of the desired entry and exit holes for ACL tunnels. This reference knee then became, in effect, that surgeon's preoperative plan for the surgical experiment. The participating surgeons used their specific reference model throughout the experiment. This gave each surgeon the advantage of working within their personal preference, but limits how the experimental data can be related across all participants.

Each reference knee's hole locations were marked with radio-opaque fiducials. The model was CT scanned using the following protocol.

All slices were 1mm thick. Inter-slice density varies depending on the location in the model:• 1 mm spacing was used in critical areas, e.g., areas used for registration and areas containing fiducials. • 3mm spacing is used in between high-density areas, and in other areas of anatomic interest. • 10mm spacing is used in all other areas (most proximal and most distal).

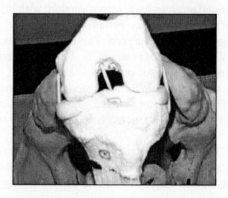

Figure 1: Sawbones model with radio-opaque fiducials.

Computer surface models were generated from the CT scan. These models were used intraoperatively with the computer-assisted surgical guidance system as well as in the post-operative analysis.

2.2.3. Randomized Procedure

The study was a randomized prospective parallel investigation comparing traditional and computer-assisted guided approaches to ACL reconstruction. Beforehand, two groups of twenty identical knee models were selected. Models within these groups were numbered respectively from 1 to 20 for the first surgeon and from 21 to 40 for the second surgeon. For each surgeon a set of 20 envelopes was prepared indicating whether the "Traditional technique" or "Computer-Guided technique" was to be performed. The 20 envelopes were combined and thoroughly mixed, thus randomizing the subsequent selection order.

During the experimental sessions the knee models were processed in numerical order. Once a model was set up, an envelope was selected and the indicated technique was used to perform the ACL tunnel placement on that model.

2.2.4. Surgical Procedure

Each surgery was performed on a new sawbones model. Each model was placed in an operative posture and secured in a fixture designed to permit the surgeon the use of a standard surgical approach. The models had fully intact capsules to minimize any external visual cues for tunnel placement.

In the traditional technique, the tibial guide-pin was arthroscopically positioned. A 10 mm bit was then used to over-drill the tibia tunnel along the guide-pin. Next, the guide-pin was used through the tibial tunnel to place the femoral tunnel entry point. The pin was drilled through the femur. The femoral tunnel was not over-drilled.

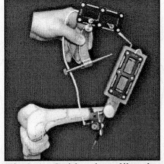

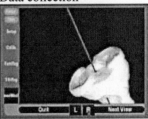

Fig 2 A: Guide-pin calibration   Fig 2 B: Data collection

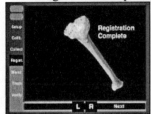

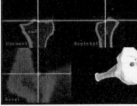

Fig 2 C: Registration.   Fig 2 D: Verification   Fig 2 E: Guide-pin
                                                              orientation

In the computer-assisted technique the following steps were performed prior to the normal surgical flow: device calibration, data registration and verification. Then, using the calibrated guide-pin, the surgeon placed the tibial and femoral tunnels as above, but under computer guidance. The arthroscope was not used during the computer technique.

A gray line (fig 2E) represented the guide-pin and gray dot marked the predetermined ACL sites (surgical goal). Three different knee views (tibial superior view, femoral inferior view and tibio-femoral sagittal view) were continuously displayed on the monitor and could be selected by the surgeon using a foot pedal control.

The surgery then resumed in a traditional manner. No notchplasty was performed.

2.2.5.   *Evaluation Protocol*

For follow-up analysis each model was carefully reregistered, and the final position of the femoral and tibial holes were recorded.

The main assessment criterion was the radiological measurement of the femoral and tibial hole positioning. The following parameters were measured from X-rays for each of the foambone models by two observers. (Figure 3): FSP = Femoral hole sagittal positioning; FCP = Femoral hole coronal positioning; TSP = tibia hole sagittal positioning; and TCP = Tibial hole coronal positioning.

In order to standardize the x-ray measurements a casting was prepared to secure the identical sawbones models in exactly the same film orientation. Each model was filmed in the AP and lateral views.

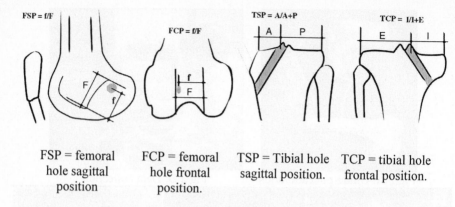

FSP = femoral hole sagittal position

FCP = femoral hole frontal position.

TSP = Tibial hole sagittal position.

TCP = tibial hole frontal position.

Figure 3: Technique of X-ray measurements.

For the follow up measurements the knee model was equipped with rigid bodies (one each for the tibia and femur) then registered in the normal "Nav" fashion. Good registration was verified. A centering rod was inserted through the tibial tunnel so that each end of the rod could be easily accessed, as illustrated in Figure 4. The ball probe tip was inserted into the distal detent, and the position was measured with respect to the tibia. The ball probe was moved to the proximal detent and a second measurement was recorded. For femoral measurements the ball probe was placed directly on each end of the femoral guide-pin tunnel. The points were collected with respect to the femoral tracker. The ball probe self-centered to the tunnel ends for most knee models. On models with oblique femoral tunnels care had to be taken to center the probe tip. Post processing software computed the intersection of the tunnel axes with the bone surface model.

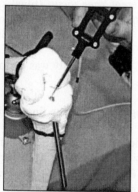
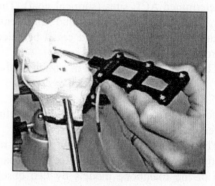

Figure 4: Computer-assisted measurement technique.

3. Results

Table I: Measurements of the bone holes positioning.

Hole position	Measurement technique	Surgeon number one		Surgeon number two	
		Scope	KneeNav	Scope	KneeNav
FCP Coronal femoral hole position.	X-ray measurement	75.7+/- 6.7	70.7 +/- 8.9	79.8 +/- 8.7	90.1 +/- 4.9
		Goal = 70.1 p= 0.08		**Goal =100** p= 0.01	
FSP Sagittal femoral hole position	X-ray measurement	67.9 +/- 3.37	68 +/- 7.13	66.9 +/- 4.41	82.5 +/- 8.4
		Goal =72 p=0.94		**Goal = 83.4** p=0.71	
TCP Coronal tibial hole position.	X-ray measurement	47.5 +/- 1.52	50.3 +/- 1.48	48 +/- 2.03	51.9 +/- 1.93
		Goal = 52.2 p=0.98		**Goal = 50.5** p= 0.86	
TSP Sagittal tibial hole position	X-ray measurement	41.1 +/- 2.04	35.1 +/- 4.5	33.5 +/- 4.21	35.5 +/- 3.04
		Goal = 39.1 p=0.02		**Goal = 36** p= 0.08	

No statistical differences between inter-observer radiological measurements have been found.

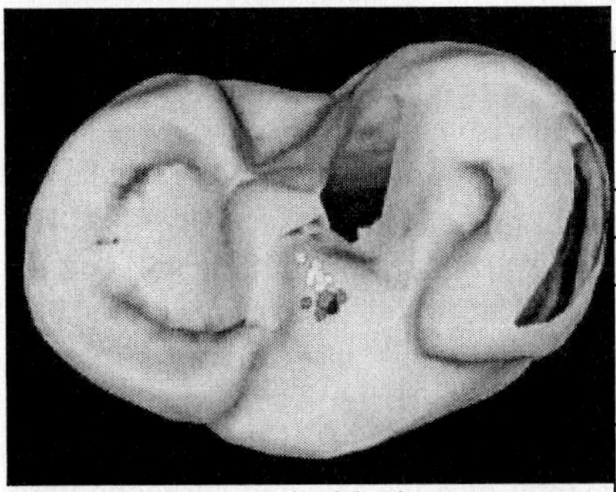

Figure 5: Target in black, computer-assisted technique in gray and traditional technique in white.

Table II: Comparison between Scope and KneeNav techniques (p=0.008)

N = 40	Scope Technique (N= 20)	KneeNav-ACL (N= 20)
Min	4.07 mm	2.05 mm
Max	10.01 mm	15.08 mm
Average distance to the goal	6.86 mm	4.6 mm
Standard Deviation	1.69 mm	2.7 mm

In order to evaluate the learning curve for surgeons using computer-assisted system, we normalized each measurement to the target values (T1, T2, T3, and T4). For each experiment appraisal we used four views: one coronal and one lateral/ sagittal view for each the femoral and the tibial holes (X_{1n}, X_{2n}, X_{3n} and X_{4n}) respectively. Figures below show additional errors to the target values for both surgeons.

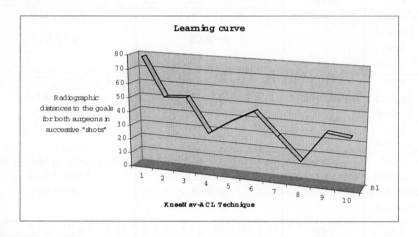

Figure 6: Learning curve. Cumulative errors to the goal during the computer-assisted surgery trial.

4. Discussion

The computer-assisted system enabled the surgeons to accurately place the tibial tunnel and femoral hole without direct visualization. Each surgeon placed the holes in accordance with his specific preoperative plan. Identical foambone knee models provided a common baseline for follow-up comparison.

However, because the surgeons did not reference a common plan the results had to be independently examined for radiographic appraisal. Neither surgeon's results showed a significant statistical difference between the traditional and computer-assisted trials using the radiographic appraisal. The initial conclusion might be that the CAS system is comparable to the traditional tools in terms of accuracy in achieving goals. However, closer examination and computer-assisted measurements suggest another explanation.

As mentioned before, to prevent experimental bias we selected surgeons with no prior CAS system experience, but who did have extensive experience in ACL reconstruction surgery. A few iterations (about 4 or 5) with the CAS system were sufficient for the surgeons to become completely comfortable with the technology, demonstrating the ease of using CAS systems. However, since each surgeon performed a total of 10 tunnel placements under CAS guidance, only half can be considered to be outside of the learning curve. This resulting sample size is not sufficient for meaningful statistical analysis.

Even so, within these small sets we see a trend beginning to emerge. Beyond the initial learning curve the hole positions tend to be more tightly clustered around the planned locations as showed on figure 5. Once the surgeons became comfortable with the CAS system they were able to achieve the planned placement more accurately than with the traditional technique.

It is interesting to note that the tunnel sites preoperatively selected by each experienced surgeon differ appreciably [1, 5, 6]. For example, the femoral hole placement (coronal/sagittal) was 70/66 for surgeon 1, and 100/91 for surgeon 2. This illustrates the difficulty in establishing a consensus regarding femoral tunnel position [1,5,6], and provides a strong argument for a guidance system that allows surgeons to select tunnel sites. This also suggests a requirement for any surgical navigation system to support customizable planning parameters [4, 5, 6]. Moreover, the surgeon must be given the option to verify and/or change his plan intraoperatively. Such capability is not currently available in robotic-based surgical systems [7].

This study demonstrated the ease with which traditional surgical tools can be adapted for use in a computer-assisted surgical system. This is an important consideration for CAS systems to gain acceptance and make the transition to widespread practical use.

It was also noted that the surgical system could provide guidance that could not be fully exploited due to limitations in the traditional instrumentation. For example, in our study the tracked tibial guide pin could be oriented such that the tibial tunnel, when over-reamed, would provide optimal access to the intercondylar notch for the subsequent femoral tunnel placement. However, this optimal orientation could not always be maintained during tibial guide pin drilling because the oblique angle between the ACL guide cannula and the tibia prevented a stable purchase of the guide. A lesser angle had to be used to ensure stable guide positioning during pin drilling. Thus, although traditional tools can easily incorporated into computer-assisted systems, instrumentation may have to be redesigned before surgeons can take full advantage of a computer-assisted system's capabilities.

The surgeons were exposed to new technical concepts related to CAS systems. Tracking issues, tool calibration, model registration, and operation of the user-interface were quickly grasped and posed no problem in the execution of the experiment [3, 4, 8]. Computer-Assisted Navigation Systems will provide viable alternatives to current surgical tools. However, CT scan is not a routine exam for ACL reconstruction and MRI models will probably more helpful in a near future. Regardless of the complexity of the underlying technology, a truly practical system must provide a simple and intuitive interface.

Acknowledgements

Chuck Cheetham from Linvatec Company, Bob Brown from Zimmer Randall Associate Inc. Company and Freddie Fu for additional support.

References

1. Frank C.B, Jackson D.W., The Science of Reconstruction of the Anterior Cruciate Ligament. J Bone J.Surg 79-A, 1556-1576, 1997.
2. Fu F.H., Harner C.D., Johnson D.L. et al., Biomechanics of knee ligaments: Basic concepts and clinical application. Chapter 14, 137-148, 1994.
3. Sati M., Deguise J.A., Drouin G., Computer Assisted Knee Surgery: Diagnostics and planning of knee surgery. Computer Aided Surgery 2: 108-123, 1997.
4. Klos T.V.S., Banks A.Z., Banks S.A., et al., Computer and Radiographic Assisted Anterior Cruciate Ligament Reconstruction of the Knee. in Computer Assisted Orthopaedic Surgery L.P. Nolte and R.Gantz Hogrefe& Huber Publishers.184-189, 1999.
5. Dessenne V., Lavallee S., Julliard R., et al., Computer Assisted Knee Anterior Cruciate Ligament Reconstruction: First Clinical Tests, Journal of Image Guided Surgery 1, 59-64, 1995.
6. Julliard R., Lavallee S., Dessenne V. Computer Assisted Reconstruction of the Anterior Cruciate Ligament. Clin.Orthop. Relat. Res. 354, 57-64, 1998.
7. Peterman J., Kober R., Heinze et al., Computer Assisted Planning and Robot Assisted Surgery in the Reconstruction of the Anterior Cruciate Ligament. Operative Techniques in Orthopaedics, 50-55, Vol 10, No 1 (January), 2000.
8. DiGioia A.M., Jaramaz B., Colgan B.D., Computer Assisted Orthopaedic Surgery, Clin. Orthop.Relat.Res., 354, 8-16, 1998.
9. Stulberg S.D, Picard F., Saragaglia D., Computer-Assisted Total Knee Arthroplasty. Operative Techniques in Orthopaedics, 25-39, Vol 10, No 1 (January), 2000.

A Classification Proposal for Computer-Assisted Knee Systems

F. Picard[1], J. Moody[1], B. Jaramaz[1,2], A. DiGioia[1,2], C. Nikou[1], R. S. LaBarca[2]

Centers for Medical Robotics and Computer-Assisted Surgery,
[1]UPMC Shadyside Hospital, Pittsburgh, PA and
[2]Robotics Institute, Carnegie Mellon University, Pittsburgh, PA.

Abstract: Computer-Assisted Surgery (CAS) combines various enabling technologies to help surgeons meet and exceed quality requirements in performing knee surgery. Emerging CAS systems for the knee already include applications for TKR, ACL reconstruction and tibial osteotomies. In this paper we propose a general scheme to classify computer-assisted knee systems currently in use and under development, with an emphasis on surgical navigation systems. A literature review and available commercialized product analyses allowed us to sort the different computer-assisted systems in several categories. We proposed a classification scheme relies upon medical criteria instead of solely technical criteria (such as localizer properties and computer specifics). New concepts in computer-assisted surgery can easily fit into this classification framework.

Keywords: Computer-assisted knee surgery systems.

1. Introduction

Knee-related operations comprise a significant percentage of the routine orthopaedic surgeries performed daily in the United States[1,2,3]. Within this group two procedures predominate: Total Knee Replacement (TKR) for degenerative conditions, and Anterior Cruciate Ligament (ACL) reconstruction for ACL tears. Approximately 300,000 knee arthroplasties and 50,000 ACL reconstructions are performed every year in the USA[4].

Making these knee procedures more accurate and less invasive would increase immediate and long-term success, thereby increasing the overall benefit to the patient. However, there remain several practical challenges to improving clinical techniques[5,6]. Computer-assisted surgery has emerged as one of the key-solutions for solving these technical challenges. Over the last few years, several computer-assisted systems have been developed.

In this paper we propose a general scheme to classify computer-assisted knee systems currently in use and under development with an emphasis on surgical navigation systems.

2. Background

Computer-assisted technologies can be loosely classified in two categories: Robotic Assistive Systems and Surgical Navigation Systems. Sub-classifications within these groups can be determined by examining the underlying principles and functionality of various systems.

1. Robotic Assistive Systems:

Robotic systems are those in which an active manipulator is used in some aspect of the surgical procedure. Robotic systems come in direct contact with the patient and must necessarily employ patient-specific models. Several robotic systems have been developed for use in knee surgery to date.

1.1. TKR surgery:

Matsen[7] and coworkers were the first authors to describe a robotic system for knee arthroplasty using active robotic templates. Approximately at the same time Kienzle and Stulberg[8] developed a robotic assisted system that utilized a preoperative CT scan and pin based registration technique. A 3-D reconstruction allowed them to define the ideal positioning and the ideal size of the prostheses. Other researchers (Davies [9], Glozman[10] and Fadda/ Marcacci/ Martelli[11]) have used similar approaches. Afterwards, these teams used either an active robot or a semi-active [9] robot in order to orient a drilling or milling tool to perform surgery according to a preoperative plan. The preliminary cadaver results showed accuracy to within 1 mm and out "degree". Following the same principles implemented by Paul and Bargar, Bauer et al[29], Kober et al.[30] described an active robot for milling bones.

1.2. ACL Reconstruction Surgery:

Two types of robotic techniques have been developed:

- An active robot for drilling the tibial and femoral tunnels has been developed by Peterman[12] and coworkers. After obtaining a CT scan with pins previously applied, the authors used a calibrated robot to drill bones in the ideal position for anchoring the ACL graft. The first clinical trial involved 25 patients. Gotte and al.[13] following the same principles have developed a similar robotic technique.

- Active robot for leg stabilization: using a robotic manipulator, Sakane[14] and al. proved the reliability of robotic technique used to provide ideal knee positioning during the ACL reconstruction on cadavers. The goal was to stabilize the knee in a good position during the surgical procedure in order to improve the location of the graft.

2. Surgical Navigation Systems

Surgical Navigation Systems essentially provide intraoperative guidance. These systems typically use optical or magnetic markers to track tools and patient anatomy. Surgical plan may be preoperatively generated, intraoperatively generated, or both (preoperative data refined intraoperatively). Several knee surgical navigation systems currently exist.

2. 1. Surgical Navigation Systems based on Preoperative Models:

2.1.1. Total Knee Arthroplasty:

KneeNav™, an image guided navigation system using optoelectronic tracking, for TKR (and also ACL reconstruction) under development in the Centers for Medical Robotics and Computer Assisted Surgery at Carnegie Mellon University and UPMC Shadyside Hospital. KneeNav relies on the same principles as the HipNav[15] system, a system successfully used for total hip replacement surgery.

A similar guided navigation system named Navitrack™[23] for TKR has been developed by Orthosoft, Inc, Montreal, Canada. This system relies upon an image-guided navigation system using electromagnetic tracking.

Other researchers have worked on applications like high tibial osteotomies, (Tso and Ellis[18]) using CT images to perform computer-assisted surgical navigation.

2.1.2. ACL reconstruction:

Sati[16] et al. proposed a system that utilizes preoperative planning based on CT Scan information. The CT scan is used for registration and 3D visualization of the patient anatomy. Using a magnetic tracking system, the computer was able to follow knee movement in real time. A graphical interface displayed the 3D image of the knee and allowed the analysis of the knee motion in real time. The system also predicts elongation properties and isometric points of the ligaments. This system is currently not in clinical use.

2.2. Surgical Guided Systems based on Intraoperative Model:

2.2.1. Total Knee Arthroplasty:

Picard and Leitner[19,24,25,31] described the first clinically used computer-assisted system for performing a complete TKR procedure. No preoperative imagery was required. The principle was to determine the mechanical axes of the limb during the surgical procedure and then to orient pre-calibrated cutting guides in reference to those axes using optoelectronic guidance. Real-time user interfaces presented the mechanical axes and the orientation of the cutting guides. The surgeon then secured the jigs using the computer guidance system and extra-medullary instrumentation. The bone was cut using traditional oscillating saws.

• Krackow[20] and coworkers developed a similar system, also without the use of preoperative imagery. An optoelectronic localizer was used to track bones and tools during the surgical procedure. Two rigid bodies (tracking markers) are fixed to the distal femur and the proximal tibia. The center of the femoral head was found by moving the femur around the hip joint. The author utilized a traditional cannulated intramedullary femoral rod oriented with computer guidance to perform the surgery.

• Kuntz, Sati, and Nolte[28] developed a similar system. For registration of the hip center they used a pivoting algorithm that does not require a reference base on the pelvis.

2.2.2. ACL reconstruction:

Dessenne and Julliard[21] used an optical tracking system to determine the ideal placement of the ACL during the usual surgical procedure, with the help of an isometry map. Using a pre-calibrated probe, the surgeon recorded anatomical data (the points on the knee surfaces) during the arthroscopic procedure, and then by moving the knee computed the relative position between femur and tibia so that define an isometry map. System was also able to calculate the position of the graft relative to the tibial and femoral holes (tension of the graft) and to the roof (impingement).

Klos and Banks[17] developed a system based on fluoroscopic images. During the surgical procedure they were able to define ideal placement and isometry of the ACL graft, using an overlap template, which coincides with the lateral fluoroscopic view. The system tracks the surgical instruments within the fluoroscopic image, and generates graphic overlays. The results of the clinical trial showed an accuracy of 1.1mm.

Fleute[22] et al: Using ten cadaver femurs authors generated a generic shape of the femur. Intraoperatively, point collections and a registration process of the distal femur allowed them to match the ideal fitting between intraoperative registration and generic shape. A statistical model was used to ideally match the femoral registration and the generic shape.

3. Results

Based on our survey of existing and developing computer-assisted technologies, we propose the classification according to the following categories[27,32]:

1- Robotic Assistive Systems are robotic devices that perform a surgical task according to some preoperative data. Three classifications have been defined in the literature:

    1.1. Active robotic system perform some surgical task, such as drilling or milling, without the direct intervention of the surgeon.

    1.2. Semi-active robotic systems augment the surgeon's control of the tool. Such a system may, for example, not directly control a saw during a resection, but may limit the depth of cut. Most systems in this category restrict a task within a pre-determined envelope. In other words they enforce constraints.

    1.3. Passive robotic system performs some part of the surgical procedure while under continuous and direct control of the surgeon but does so under the direct guidance or control of the surgeon. It may, for example be used to position a template or cutting guide block.

- Surgical Navigation Systems are computer-assisted systems that display information for orientation and guidance during the surgical procedure. They may present accurate anatomical images, simple graphics, or a combination of images (for example, icons superimposed on radiographic data). These can be loosely classified into two groups depending on their use of models:

    2.1. Preoperative-model systems rely on models generated preoperatively, usually from large data sets (CT or MRI image series, for example). They can provide a wealth of detailed information, and are typically used in conjunction with a planning system. Preoperative model system can be either patient-specific or not patient-specific.

        2.1.1. Patient-specific means that the preoperative model is set up from specific data of the patient itself. For example, the preoperative plan can be based on CT images of the patient and this data will be used as a reference during the surgical procedure.

        2.1.2. Non patient-specific means that the preoperative model is based on a generic shape or generic model. A homothetic model resulting from a generic model derived from images or digitized cadaver bones can be used during the surgical procedure.

    2.2. Intraoperative-model systems develop anatomical models through intraoperative data collection. Intraoperative model systems can be divided in two categories depending on their use of intraoperative images.

2.2.1. Image based: an intraoperative image (such as a set of coordinated fluoroscopic images[26]) are generated during the surgical procedure and used as a frame of reference.

2.2.2. Non-image based: all model information required for the task is determined from direct measurement of the bone surface (using, for instance a calibrated and tracked probe), or from direct measurement of limb dynamics (e.g., computing rotational centers from relative bone movement).

4. Discussion

Table I summarizes the classification breakdown. This classification scheme relies upon medical criteria instead of solely technical criteria (such as localizer properties and computer specifics).

Table I: Classification of Computer-Assisted Knee Systems.

Robotic Assistive Systems	Surgical Navigation Systems			
Passive Systems	Preoperative-model Systems		Intraoperative-model Systems	
Semi-active Systems	Patient Specific	Non patient specific	Intraoperative Images	No Intraoperative Images
Active Systems				

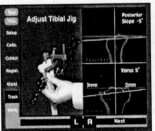

Fig. 1: Robotic Assistive System (Guided tools).   **Fig. 2:** Preoperative model Image guided model.   **Fig. 3:** Intraoperative model None Image guided based.

Conclusion

We propose a classification scheme for computer assisted systems for knee reconstructive surgery, based on a survey of systems currently in use or in development. The approach provides easy comprehension of technological requirements and the medical utility for a particular computer-assisted system. New concepts in computer-assisted surgery can easily fit into this classification framework.

References

1. Ayers D.C., Berman A.T., Duncan C.P., and al., Economic Aspects of Total Joint Replacement. AAOS Committee on Hip and Knee Arthritis, 1997.
2. Lavernia C.J., Guzman J.F., Gachupin-Garcia A., Cost Effectiveness and Quality of Life in Knee Arthroplasty. Clin. Orthop. Relat. Res., 345, 134-139, 1997.
3. Rorabeck C.H., Murray P., The Cost Benefit of Total Knee Arthroplasty. Orthopedics, 19 (9), 777-779, 1996.
4. Frank C.B, Jackson D.W., The Science of Reconstruction of the Anterior Cruciate Ligament. J. Bone J. Surg, 79-A (10), 1556-1576, 1997.
5. Ayers D.C, Dennis D.A., Johanson N.A and al., Common Complications of Total Knee Arthroplasty. J. Bone J. Surg., 79-A (2), 278-311, 1997.
6. Andersson C., Odensten M., Gillquist J., Knee function after surgical or non-surgical treatment of acute rupture of the ACL: A randomized study with a long-term follow-up period. Clin. Orthop., 264, 1991.
7. Matsen F.A., Garbini J.L., Sidles J.A, and al. Robotic Assistance in Orthopaedic Surgery. (A proof of principle using distal femoral arthroplasty). Clin. Orthop. Relat. Res., 296, 178-186, 1993.
8. Kienzle T.C., Stulberg S.D., Peshkin M. and al., A Computer-Assisted Total Knee Replacement Surgical System Using a Calibrated Robot. Orthopaedics. In computer Integrated Surgery, Cambridge, Massachusetts, The MIT Press. R.H. Taylor and al. Editor. 409-416, 1996.
9. Davies B.L., Harriss J., Lin W.J., and al., Active compliance in Robotic Surgery- The use of force control as a dynamic constraint. Journal of Engineering in Medicine, Proceedings H. of the Institution of Mechanical Engineers, UK, Volume 211, page H 4, November 1997.
10. Glozman D., Shoham M., Fischer A., Efficient registration of 3-D objects in Robotic-Assisted Surgery. Proceedings CAOS 'US, Editor UPMC Shadyside Medical Center (A.DiGioia) 248-252, 1999.
11. Fadda M., Bertelli D., Martelli S., Marcacci .M., and al., Computer Assisted Planning for Total Knee Arthroplasty. In First Joint Conference of CVRMed and MRCAS, Grenoble, France: Springer, 619-628, 1997.
12. Peterman J., Kober R., Heinze and al., Computer Assisted Planning and Robot Assisted Surgery in the Reconstruction of the Anterior Cruciate Ligament. Operative Techniques in Orthopaedics, 50-55, Vol 10, No 1(January), 2000.
13. Gotte H., Roth M., Brack C.H., and al. A new less-invasive approach to knee surgery using a vision-guided manipulator. IARP workshop on Medical Robotics, Vienna (Austria) Oct. 99-106, 1996.
14. Sakane M., Li G., Fox R., Woo L-Y., and al., The advantage of Robot – Assisted Knee Positioning for ACL Reconstruction Surgery. Advances in Bioengineering. ASME (33), 397-398, 1996.
15. DiGioia A., Jaramaz B., Blackwell M., and al., Image Guided Navigation System to Measure Intraoperatively Acetabular Implant Alignment. Clin. Orthop. and Relat. Res., 355, 8-22, 1998.
16. Sati M., DeGuise J.A., Drouin G., Computer Assisted Knee Surgery. Computer Aided Surgery 2: 108-123, 1997.
17. Klos T.V.S., Banks A.Z., Banks S.A., and al., Computer and Radiographic Assisted Anterior Cruciate Ligament Reconstruction of the Knee. In Computer Assisted Orthopaedic Surgery L.P. Nolte and R.Ganz Hogrefe& Huber Publishers.184-189, 1999.

18. Tso C.Y., Ellis R.E., Rudan J., and al., A Surgical Planning and Guidance System for High Tibial Osteotomies. In Proceedings Medical Image Computing and Computer–Assisted Intervention- MICCAI'98, 39-50, 1998.
19. Leitner F, Picard F., Minfelde R. and al. Computer-Assisted Knee Surgical Total Replacement. In First Joint Conference of CVRMed and MRCAS, Grenoble, France: Springer, 629-638, 1997.
20. Krackow K., Serpe L., Phillips M.J., and al., A New Technique for Determining Proper Mechanical Axis Alignment During Total Knee Arthroplasty. Orthopedics, 22 (7), 698-701, 1999.
21. Dessenne V., Lávallee S., Julliard R., and al., Computer Assisted Knee Anterior Cruciate Ligament Reconstruction: First Clinical Tests. Journal of Image Guided Surgery 1, 59-64, 1995.
22. Fleute M., Lavallee S., Julliard R., Incorporating a statiscally-based shape model into a system for computer-assisted anterior cruciate ligament surgery. Medical Image Analysis, Oxford University Press volume 3, 209-222, 1999.
23. Amiot L.P., Labelle H., DeGuise J.A., Sati M., and al., Computer-assisted Pedicle Screw Fixation: a feasibility Study, Spine, 10, 1208-1212, 1995.
24. Stulberg S.D, Picard F., Saragaglia D. Computer-Assisted Total Knee Arthroplasty. Operative Techniques in Orthopaedics, 25-39, Vol 10, No 1(January), 2000.
25. Delp S.L., Stulberg S.D., Davies B., Picard F., Leitner F., Computer Assisted Knee Replacement. Clin. Orthop. Relat. Res., 354, 49-56, 1998.
26. Hofstetter R., Slomczykowski M., Sati M., Nolte L.P., Fluoroscopy as an Imaging Means for Computer-Assisted Surgical Navigation. Computer Aided Surgery., 4:65-76, 1999.
27. Troccaz J. Man-Machine Interfaces in Computer-Augmented Surgery. In Computer Assisted Orthopaedic Surgery L.P. Nolte and R.Ganz Hogrefe & Huber Publishers. 53-68, 1999.
28. Kuntz M., Sati M., Nolte L-P., and al. Computer Assisted Total Knee Arthroplasty. In International Symposium on CAOS, Davos, February, 17-19, 2000.
29. Bauer A., Robot-Assisted Total Hip Replacement in Primary and Revision Cases. Techniques in Orthopaedics, 9-13, Vol 10, No 1(January), 2000.
30. Kober R., Meister D., Total Knee Replacement using the Caspar-system. Computer Assisted Total Knee Arthroplasty. In International Symposium on CAOS, Davos, February 17-19, 2000.
31. Picard F., Leitner F., Raoult O., Saragaglia D. Computer-Assisted Knee Replacement. Location of a rotational center of the knee. Total Knee Arthroplasty. In International Symposium on CAOS, Davos, February 17-19, 2000.
32. Taylor R.H., Robotics in Orthopedic Surgery, In Computer Assisted Orthopaedic Surgery (CAOS), L.P. Nolte and R.Ganz, 35-41, 1998.

A Navigation System for Computer Assisted Unicompartmental Arthroplasty

Maurilio Marcacci[2], Oliver Tonet[1], Giuseppe Megali[1], Paolo Dario[1],
Maria Chiara Carrozza[1], Laura Nofrini[2], Pier Francesco La Palombara[2]

[1] MiTech Lab, Scuola Superiore Sant'Anna, Pisa, Italy
peppe@mail-arts.sssup.it
oly@sssup.it
dario@arts.sssup.it
chiara@arts.sssup.it
http://www-mitech.sssup.it/
Biomechanics Lab, Istituti Ortopedici Rizzoli, Bologna, Italy
{M.Marcacci, L.Nofrini, F.La.Palombara}@biomec.ior.it
http://www.ior.it/biomec/

Abstract. This work presents an overview of a prototype navigation system for computer assisted unicompartmental arthroplasty. The navigation system allows the surgeon to use a minimally invasive surgical technique, solving the problems due to a restricted knee exposure with an augmented reality environment. The key feature of the system is an interactive graphical interface in which a 3D model of the patient's joint and two 2D projections (frontal and lateral) are shown together with the model of a sensorized custom surgical cutting guide. The position of the sensorized instruments are tracked in real-time by an optoelectronic localizer and reproduced in the virtual scene. The 3D model of the joint is reconstructed from a preoperative CT/MRI data set and matched to the actual anatomy by means of an ICP-based non-fiducial registration algorithm. The prototype system has been positively evaluated by a selected group of skilled orthopaedic surgeons.

1 Introduction

Unicompartmental knee arthroplasty (UKA) is a kind of intervention that offers several advantages for the treatment of single compartment joint disease. Even if in the past there has been a lack of general agreement on how successful UKA has been, recently, clinical studies have reported satisfactory results [1][2] mainly depending on a correct patient selection criteria, and they have shown that in some cases UKA can also be considered a valid alternative procedure to high tibial osteotomy [3][4].

Traditional surgical techniques for UKA require numerous and difficult to use instruments, a large exposure area and patellar dislocation to determine the correct placement of the components, and its invasiveness is similar to total knee arthroplasty one.

Recently, minimally invasive surgical techniques for UKA (RepicciII, , Miller-Galante system) have been proposed: these procedures involves smaller incision, minimal blood loss, preservation of normal tissue, reduced pain, and shortens recovery and rehabilitation times[5]. However, the reduced exposure area leads to a hard retrieval of the anatomical landmarks used to correctly position the prosthesis components, requiring considerable surgical dexterity and skill. In particular the Repicci II technique has been criticize because it relies on freehand surgery.

Augmented reality systems can be useful to overcome the limitation due to the restricted exposure enhancing the surgical skill.

We propose a new computer-assisted approach to help the surgeon to solve problems related to minimally invasive UKA. Our methodology is based on the use of an intraoperative navigation system and of a reduced set of surgical instruments. In the following paragraphs we will describe the system showing the main advantages it offers in comparison with the other existing techniques.

2 Material

2.1 Equipment

The described procedure needs the standard surgical equipment for monitoring the patient, a PC for running the navigation system (the central control unit we use is an Intergraph TDZ 2000 GX1 workstation (pentium II xeon 450 MHz, 128 MB RAM, Windows NT 4.0) with RealiZm II 3D graphics board), a commercial optical localizer (FlashPoint 5000 system, Image Guided Technologies Inc.), three sensorized sterilizable frame with four IR LEDs each and a custom cutting guide.

The optical localizer. The optical localizer is based on three cameras located on a stand placed over the surgical scene, which defines the coordinate reference frame of the operating room. The cameras detect the IR pulses emitted by LEDs placed on rigid frames fixed to the objects to track. The control unit of the localization system computes the coordinates of the LED positions by means of geometrical triangulation and determines the position and orientation of the objects present in the scene, with an accuracy better than 0.5 mm. This localizer is provided with a probe digitizer that can be used to acquire points in the working area.

In the surgical room, the optical localizer is positioned over the head of the patient so that it does not encumber the surgeon in the operating area.

The sensorized frames. Three sterilizable frames are used to track the objects in the operating room to reproduce their movements in the virtual scenary.

One frame is fastened to the femur in the distal frontal third of the bone. Another frame is fastened to the tibia in the proximal lateral third of the bone. The last frame is fixed on a custom cutting guide, designed on purpose for this surgical system (fig.1).

The cutting guide. The cutting guide is a tool consisting of a small metal block and of a bar connecting the block with the sensorised frame. The block is 4 x 2 x 1 cm and it is provided with two holes 3.2mm∅ to secure it to the bone with two pins.

The design of the new surgical tool has been outlined in collaboration with expert surgeons in order to satisfy the new surgical requirements of minimal invasiveness (fig. 1).

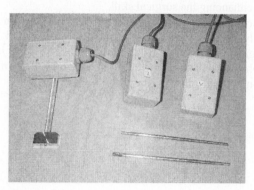

Fig. 1. The sensorized frames

2 .2 The navigation system software

The navigation system has been developed by MiTech Lab of Scuola Superiore Sant'Anna (Pisa, Italy), in collaboration with Biomechanics Lab of Istituti Ortopedici Rizzoli (Bologna, Italy)[6].

The navigation system includes a *registration module* for matching preoperative data, i.e. the virtual model of the anatomical structures, and intraoperative data, i.e. the actual patient's anatomy, a *tracking module* for real time monitoring of the position of the surgical tools and anatomical structures present in the operating room, and a *graphical human-machine interface* that displays the enhanced reality scenario based on a 3D model of patient's limb computed from CT/MRI. The navigation system has been developed in C++ language for Windows NT 4.0 platform, using Microsoft Foundation Classes for the graphical user interface (GUI) and OpenGL Optimizer for 3D visualization.

The model. The virtual models of the anatomical structures are obtained elaborating the images acquired with an accurate preoperative CT/MRI examination of the patient's limb. The examination is done with the patient in a standard position. The total number of slice is around 120 divided in two sets of different acquisition parameters: one group is 1mm thickness 1mm scan spacing from the upper side of patella to beneath the tibial spine (around 80 slices); the other group is 1mm thickness 3mm scan spacing in the adjacent areas both of the femur and of the tibia. The examination is safe (dose on critical organs <0.2μGy) and fast (time in helical

machine < 5 min). The images processing consists in a first segmentation phase performed semi-automatically with an algorithm that provides contour tracking by means of gradient-of-gaussian filtering and in a successive 3D computer reconstruction phase using the marching cube algorithm.
CAD drawings of the digitizer and of the cutting guide are also drawn in the scenario.

The registration module. The intraoperative registration module matches the coordinate system of the virtual scenario with the real position of the patient and tools in the operating room. Pre-operatively, an operator identifies, on the virtual model, areas that will be accessible during surgery and intra-operatively the surgeon acquires, with a digitizer, points in the correspondent areas on the bone. Registration is performed by means of an ICP-based non-fiducial algorithm.

The tracking module. The tracking module elaborates the information retrieved by the optical localizer to reproduce in real time the surgical environment in the virtual scenario.

The user interface. The virtual scenario of the operating room is shown on a computer display, which contains four windows: the first is the *external viewpoint* window, and it displays the surgical scene as viewed from an external viewpoint. In this window the main objects in the operating room are displayed together with the anatomical structures and the surgical tools; the view may be adjusted according to surgeon preference by changing camera position, orientation and magnification.
The remaining three windows show the anatomical components present in the scene and the trajectory of the cutting plane drawn from different point of view and in different graphical style: in the *lateral projection window* and the *frontal projection window*, the bones are shown in translucent wire-frame mode while in the *Solid 3D view* they are shown in solid mode (fig.2).

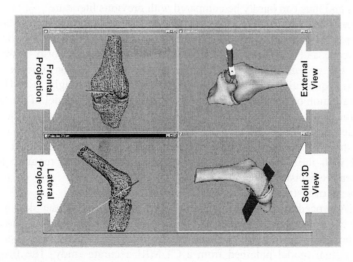

Fig. 2. The user interface

3 Description of the assisted surgical procedure

The surgeon makes a medial parapatellar skin incision of 8-10 cm, extending from the proximal border of the patella to the proximal tibial region.

After the knee exposure, the two sensorized frames are fastened to femur and tibia with two fiches.

Next step is the registration phase: the surgeon acquires, with the optical localizer's digitizer, points in the bone surface, in areas corresponding to the ones preoperatively identified on the virtual model. Using this set of points the system provides to perform the registration.

After the registration phase, the system displays the virtual scenario reproducing in real time the surgical environment: the sensorized cutting guide is shown together with the two bones.

Now the surgeon can track in real-time the placement of the cutting plane on the bone in the enhanced reality scenario, so that he can choose the desired cutting guide position and orientation with respect to the anatomical structures before the cut is effectively done.

The sensorized cutting guide is used by the surgeon to make the horizontal tibial cut and the femur first distal cut.

4 Discussion and conclusions

In this paper we have presented new minimally invasive computer-assisted procedure for UKA relying on an augmented reality navigation system and on a reduced set of easily usable surgical instruments.

The procedure that we have described is an original application of navigation systems to UKA, and thus can hardly be compared with previous literature.

Preliminary tests done to see the accuracy of the navigation system, show that the *overall system error*, defined as the average difference between coordinates in the virtual scene and the corresponding coordinates in the real world, is about 2 mm, whereas the *system update rate,* defined as the inverse of the time needed by the system to acquire all data and update the virtual scene accordingly, is 13 Hz.

Early evaluation of the surgical system done using artificial model of the human knee shows that the computer assisted UKA is as time-consuming as the traditional technique: this is probably due to the fact that the registration procedure is completed in less than 10 minutes, and the reduced and simple set of surgical instruments coupled with the enhanced field of view diminish the time required to take the intraoperative decisions.

The new set of instruments can be easily sterilised.

The UKA procedure presented allows the surgeon to reduce the skin incision size from 20 cm to 8-10 cm; the navigation system solves the problems due to a restricted knee exposure providing the surgeon with the enlarged view of the patient's limb on the 3D virtual model obtained from a CT/MRI accurate study. The first prototype system has been favorably evaluated by a selected group of skilled orthopedic surgeons. At this moment we are performing tests to evaluate the final accuracy and repeatability of the surgical decisions.

References

1. Ryd L, Boegård T, Egund N, Lindstrand A, Selvik G, Thorngren KG: Migration of the tibial component in successful unicompartmental knee arthroplasty. A clinical, radiographic and roentgen stereophotogrammetric study. Acta Orthop. Scand. 54: 408, 1983

2. Ryd L: Micromotion in knee arthroplasty. A Roentgen Stereophotogrammetric Analysis of tibial component fixation. Acta Orthop. Scand. 57 (Suppl. 220): 1, 1986

3. Lindstrand A, Stenström A, Egund N: The PCA unicompartmental knee: a 1-4 year comparison of fixation with or without cement. Acta Orthop. Scand. 59(6): 695, 1988

4. Goodfellow J.W., Kershaw C.J., D'A Benson M.K., O'Connor J.J.: The Oxford knee for unicompartmental osteoarthritis. J Bone Joint Surg [Br] 1988;70-B:692 – 701.

5. Repicci J.A., Eberle R.W.: Minimally Invasive Surgical Technique for Unicondylar Knee Arthroplasty. J South Orthop Assoc 1999;8(1):20-27

6. P. Dario, M.C. Carrozza, M. Marcacci, S. D'Attanasio, B. Magnani, O. Tonet and G. Megali, (2000) "A Novel Mechatronic Tool for Computer-Assisted Arthroscopy". IEEE Transactions on Information Technology in Biomedicine, 4(1), 15 – 29.

An Augmented Reality Navigation System for Computer Assisted Arthroscopic Surgery of the Knee

Oliver Tonet[1], Giuseppe Megali[1], Simona D'Attanasio[1], Paolo Dario[1],
Maria Chiara Carrozza[1], Maurilio Marcacci[2], Sandra Martelli[2],
Pier Francesco La Palombara[2]

[1] MiTech Lab, Scuola Superiore S. Anna, Via Carducci 40, 56127 Pisa, Italy
oly@sssup.it, peppe@mail-arts.sssup.it, dario@arts.sssup.it,
chiara@arts.sssup.it
http://www-mitech.sssup.it/

[2] Biomechanics Lab, Istituti Ortopedici Rizzoli, Bologna, Italy
{M.Marcacci, S.Martelli, F.La.Palombara}@biomec.ior.it
http://www.ior.it/biomec/

Abstract. This work presents a prototype navigation system for computer assisted arthroscopic surgery. Arthroscopic surgery is minimally invasive and ensures fast post-surgical recovery. It requires, however, considerable dexterity due to the limited 2D view of the scene and cramped operating area. Augmented reality can help the surgeon overcome these limitations. The key feature of the system presented is an interactive graphical interface. A 3D model and two 2D projections of the patient's joint and models of the arthroscope and other sensorized surgical instruments are shown. The positions of all sensorized objects, including the patient's anatomy, are tracked in real-time by an optoelectronic localizer and reproduced in the virtual scene. The field of view of the arthroscope is dynamically highlighted on the 3D model of the joint, which is reconstructed from a preoperative CT/MRI data set and matched to the actual anatomy by means of an ICP-based non-fiducial registration algorithm. The prototype system has been positively evaluated by a selected group of skilled orthopaedic surgeons. Preliminary tests on the overall accuracy of the system show that errors can be kept within 3°/3mm.

1 Introduction

Arthroscopic surgery is widely used as a routine treatment for several joint disorders. Being minimally invasive, it ensures little tissue damage and faster post-surgical recovery. Arthroscopic surgical procedures require, however, considerable dexterity and skill due to the limited 2D view of the scene and the cramped operating area. Augmented reality systems enhance the surgical skills and help the surgeon overcome these limitations [1]. A navigation system, aimed at the improvement of the current state of the art of arthroscopic surgery has been developed and tested in vitro. The system requirements have been defined in collaboration with a team of experts in the field of arthroscopy.

2 Clinical Relevance

Despite the strong social and economic importance that minimally invasive surgical techniques have achieved, still many technical problems are to be solved in order to improve the quality of the care delivered to patients. In arthroscopic surgery the narrow access imposes severe constraints on the operating workspace. Vision is limited by the small monoscopic field of view of the arthroscope and by image quality. Other sensory signals, such as tactile sense and force feedback are significantly reduced. Research efforts are currently devoted not only to restore, but also to enhance the surgeon's perceptive capabilities. The proposed augmented reality navigation system can provide guidance during the surgical procedure by integrating the images from the arthroscopic camera and other sensory data with a virtual environment.

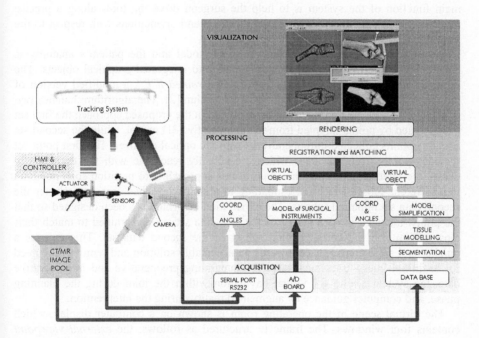

Fig. 1. A general scheme of the navigation system

3 System Overview

The system includes different intercommunicating modules. Some of them account for preoperative image processing, such as image processing modules for:

1. Segmentation of preoperative CT or MR images;
2. 3D reconstruction of the virtual model of the anatomy.

Other modules are used intraoperatively, such as:

1. The registration module for matching preoperative data, consisting in a virtual model of the anatomical structures, and intraoperative data, consisting in the patient's anatomy;
2. A tracking module for the real time monitoring of the position of the surgical tools and anatomical structures;
3. A graphical human-machine interface displaying the enhanced reality scene.

To determine the position and orientation of the objects, the system uses a commercial optical localizer which employs three cameras located on a stand placed over the surgical scene and defines the coordinate system of reference of the operating room. The cameras detect IR pulses emitted by LED's placed on rigid frames fixed to the objects to track. The control unit of the localization system computes the coordinates of the LED positions by means of geometrical triangulation and determines the precise position and orientation of the objects present in the scene. The main function of the system is to help the surgeon drive the tools along a precise trajectory and maintain the desired tool positions and orientations with respect to the anatomical structures.

An accurate registration between the joint model and the patient's anatomy is essential to blend pre- and intraoperative data into integrated graphical objects. The implemented registration technique is the *"constrained ICP"*, a modified version of the renowned Iterative Closest Point (ICP) algorithm [2]. The algorithm matches two different point sets belonging to the same object. In the proposed approach the first set is composed by points sampled from the preoperative 3D model, while the second set contains points collected intraoperatively with the optical digitizer. The first point set includes points from the areas which are actually reachable with the digitizer tip during the intervention. The innovation in this approach is to partition such point set into different subsets, each corresponding to a single detached area, and cluster the second data set in paired subsets. The registration algorithm has been designed so that the points included in each subset of the first data set are constrained to match their closest points in the corresponding subset of the second data set. This ensures a quicker and safer convergence towards an acceptable solution and removes the need for an initial guess assessment step. By integrating preoperative and intraoperative data, the system allows for virtual navigation within the joint during the planning phase, and computer guidance by augmented reality during the intervention.

The virtual scene of the operating room is shown on a computer display, which contains four windows. The frame is structured as follows: the *external viewpoint* window displays the surgical scene as viewed from an external viewpoint. This view may be adjusted according to surgeon preference by changing camera position, orientation and magnification. For external reference, the main objects in the operating room are displayed together with the anatomical structures and the surgical tools. The *lateral projection* and the *frontal projection* windows show a translucent projection of the anatomical segments and lines representing the axes of the surgical tools for trajectory control. The *arthroscope tracking light* window shows the virtual model and highlights the anatomical area included in the arthroscope field of view.

A mechatronic arthroscope has been designed, manufactured in a prototype version and successfully integrated with the navigation system. The mechatronic arthroscope has a cable-actuated servomotor-driven multi-joint mechanical structure. It is equipped with a position sensor, measuring the orientation of the tip, and a force sensor, detecting possible contact with delicate tissues in the knee. Moreover it incorporates an embedded microcontroller for sensor signal processing, motor driving and interfacing with the surgeon and/or the system control unit [3].

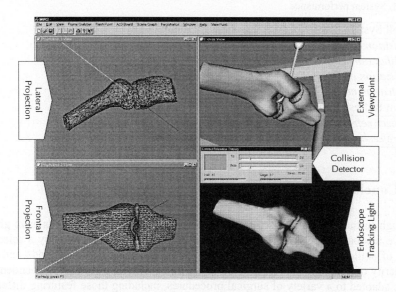

Fig. 2. The graphical interface of the navigation system

4   Tests and Results

The system has been tested using an artificial model of the human knee (Sawbones Europe AB) including femur, tibia, patella, menisci, ligaments and the surrounding soft tissues. A virtual model of the knee was obtained from a CT scan of this model, after segmentation and 3D reconstruction. The points of the anatomy database were selected on the virtual model and registered with the reference frame of the 3D localizer (FlashPoint 5000 system, Image Guided Technologies Inc.). The position of three cameras, located on a frame placed over the surgical scene, defines the coordinates of the common reference frame. The cameras detect the IR pulses emitted by LED's mounted on the tools to be tracked. The LED position is reconstructed by means of geometrical triangulation. In the experiments, a square frame was used as support for four IR LED's. The frame was mounted on the external housing of a commercial 30° arthroscope (Karl Storz Endoskope GmbH). The central control unit is an Intergraph TDZ 2000 GX1 workstation (Pentium II Xeon 450 MHz, 128 MB RAM, Windows NT 4.0) with RealiZm II 3D graphics board. Communication between the localizer and the workstation is based on a standard serial port. The navigation system has been developed in C++ language for Windows NT 4.0

platform, using Microsoft Foundation Classes for the graphical user interface (GUI) and OpenGL Optimizer for 3D visualization.

The following table summarizes the system performance parameters as resulting from the laboratory trials.

Table 1. System performance

Overall System Error[1]:	2.3 mm
Acquisition error	*1.0 mm*
Simplification error	*0.4 mm*
Localization error	*< 0.5 mm*
Registration error	*2.0 mm*
System Refresh Rate[2]:	13 Hz
Localizer refresh rate	*14 Hz*
GUI refresh rate	*40 Hz*

5   Conclusions

Although further testing is essential to validate the system, the first trials have given positive feedback. The prototype system has been favorably evaluated by a selected group of skilled orthopedic surgeons. The use of general-purpose methods and the modularity of the architecture make the navigation system application-independent. It can be adapted to a variety of surgical procedures, including those featuring different graphical user interfaces, and can be used with both ordinary surgical tools and custom instrumentation.

References

1. Satava, R.M., Jones, S.B.: Current and Future Applications of Virtual Reality for Medicine. Proceedings of the IEEE **86**(3) (1998) 484-489
2. Besl, P.J., McKay, N.D.: A Method for Registration of 3-D Shapes. IEEE Trans PAMI **14**(2) (1992) 239-256
3. Dario, P., Carrozza, M.C., Marcacci, M., D'Attanasio, S., Magnani, B., Tonet, O., Megali, G.: A Novel Mechatronic Tool for Computer-Assisted Arthroscopy. IEEE Trans ITB **4**(1) (2000) 15–29

---

[1] Mean difference between coordinates in the virtual scene and in the real environment.

[2] Inverse of the time needed by the system to acquire all data and update the virtual scene accordingly.

Repeatability and Accuracy of Bone Cutting and Ankle Digitization in Computer-Assisted Total Knee Replacements

Kevin B. Inkpen[1], Antony J. Hodgson[1], Christopher Plaskos[1], Cameron A. Shute[1] and Robert W. McGraw[2]

Departments of [1]Mechanical Engineering and [2]Orthopaedics
University of British Columbia, Vancouver, BC, Canada
Email: ahodgson@mech.ubc.ca

Abstract – In conventional total knee replacement (TKR) surgery, a significant fraction of implants have varus/valgus alignment errors large enough to reduce the lifespan of the implant, so we are developing a more accurate computer-assisted procedure aimed at reducing the standard deviation (SD) of the implant's frontal alignment to under 1°. In this study we measured the contributions to overall alignment error of two steps in our proposed procedure: ankle digitization and manual bone cutting. We introduce a new digitizing probe that quickly and robustly locates the midpoint between the ankle malleoli. Based on repeated measurements on 8 cadavers by 6 operators (318 measurements), we estimate that the new probe introduces only 0.15° (SD) of variability into the definition of the tibial mechanical axis in the frontal plane. We also measured the accuracy and repeatability with which surgeons can implement a bone cut using conventional cutting guides to see if conventional cutting techniques are sufficiently accurate. A total of 53 tibial plateau, distal and anterior/posterior (A/P) femoral cuts approximated primary and revision TKR resections. In 20 test cuts on cadaver bone made by 2 expert TKR surgeons, we found a SD of 0.37° (bias of 0.29°) in the varus/valgus difference between the guide orientation and the implant orientation before cementing, but in 23 additional cuts performed by four less-experienced surgeons, the SD was 0.83° (bias of 0.31°). Ten A/P femoral cuts showed similar trends. We conclude that, in the hands of an experienced surgeon, our current technique (based on our previously-reported non-invasive hip centre locating technique [Inkpen 1999b], robust ankle digitization and manual cutting using computer-guided cutting guides) can approach our target alignment variability goal of a SD less than 1°.

1 Introduction

We are developing computer-assisted total knee replacement (TKR) tooling that eliminates intramedullary rods and improves alignment accuracy without introducing additional imaging requirements (such as preoperative computed tomography scans) or invasive procedures (such as bone pins remote to the operating site). Varus/valgus alignment error of 3° has been strongly correlated with aseptic loosening [Jeffery 1991]. In a meta-analysis of 10 published studies involving 1373 knees, we estimate the SD of varus/valgus alignment to be 2.6° on the overall limb and 1.9° each for the distal femoral and tibial plateau cuts [Inkpen 1999a]. To ensure that upwards of 95% of all computer-assisted procedure result in overall alignment lying within this critical 3° window, it is necessary to reduce the SD of the overall procedure to approximately 1°.

In the computer assisted TKR technique developed by Leitner *et al* [Leitner 1997], the mechanical axis is defined intraoperatively using a 3D optoelectronic localizer. They find the hip centre by tracking a marker at the distal femur in a reference frame rigidly attached to the pelvis (using a bone pin) as the femur is moved through its range of motion, taking the centre of the best fitting sphere to the femoral path as the hip centre. Krakow *et al* [Krackow 1999] use a similar technique for the hip but eliminate the bone pin at the pelvis by simply holding the patient steady during the femoral motion. In early testing we have achieved precise and accurate hip centre location using a new non-invasive hip tracker and, although more testing on a variety of patients is ongoing, we estimate that the femoral mechanical axis can be located intraoperatively with a SD of 0.05° [Inkpen 1999a].

Krakow *et al* use a conventional point probe to digitize the medial and lateral malleoli, and the ankle centre point is then defined as the intersection of this malleolar axis and a digitized unit vector representing the surgeons' estimate of the best centre along the anterior portion of the ankle. Leitner, despite the complexity of the ankle joint complex, uses the same sphere-fitting algorithm applied to the hip to find the ankle centre. They track a bone pin marker inserted in the calcaneus in a coordinate frame pinned to the tibia. In an attempt to validate this determination of a kinematic ankle centre, Leitner compared the sphere fitting method with direct digitization on a plastic skeleton leg. Although not directly comparable to a human study, digitizing the mechanical centres of the skeleton showed better repeatability over the sphere fitting method (0.29° vs. 0.49° SD). We also have found poor repeatability with the motion tracking technique even when a bone pin marker is used in the calcaneus [Inkpen 1999a, Inkpen 1999b]. Even though digitization is more repeatable than motion-tracking techniques at the ankle, the contribution of variance is still large in comparison with that found at the hip. Using a conventional point probe, a user who is asked to digitize an ankle malleolus could select any point within an area of radius ~3-5 mm and still consider the point to be 'on' the malleolus. Although a careful operator may be very precise using this method, we would not expect it to be robust across different operators and patients. It is also not clear that a motion tracking and sphere fitting method without realistic weight bearing load gives a biomechanically correct ankle centre. A passively manipulated ankle acts about a single talocrural joint axis [Leardini 1999], although inversion/eversion torques generate motion at the subtalar joint, so the ankle complex can act more like a biaxial joint [van den Bogert 1994]. Given the practical problems of simulating and tracking weight bearing ankle motion intraoperatively, we propose using a robust digitizing probe to precisely locate the malleoli, allowing the traditional ankle centre estimate of the midpoint between the malleoli to be used in the computer-assisted technique.

Finally, once the patient's mechanical axis is registered, cutting planes for the desired limb alignment can be calculated and conventional cutting guides fitted with markers can be manually adjusted to the correct pose [Leitner 1997][Krackow 1999]. For cemented TKR where the surface quality from a conventional bone saw is sufficient, the question is whether errors due to conventional cutting are large enough to warrant more sophisticated bone cutting tools such as robotic guidance of the bone saw or robotic machining. One study of cuts made in balsa wood blocks shows flexion/extension SD of 0.25° with a bias of 1° towards flexing of the saw blade up and away from the guide surface [Otani 1993].

Although several studies have been made of the bone surface quality and roughness after cutting, we could not find any published studies on overall implant orientation error in the frontal plane compared to the guide setting.

2 Methods

2.1 Ankle digitizing probe design and testing

Probe Design: In making the traditional estimate of the ankle centre based on the malleoli, precision and robustness could be improved if the digitizing probe only fit onto the malleolus in a certain way, forcing the user to place the probe in a smaller range of positions. As the malleoli are roughly spherical in shape (even indistinct malleoli can often be palpated well enough to find some amount of spherical protrusion), the ankle probe (Figure 1a) contacts the anatomical feature with three 7.8 mm diameter balls spaced evenly on a common 12.8 mm radius about the probe centreline, with all 3 contact ball centres lying in a plane normal to the centreline. A plunger is free to move along the probe centreline and contacts the feature with a normal planar face. When the probe is pressed against a roughly spherical feature, the three contact balls force the centreline of the probe to pass through the sphere centre and the plunger face contact point lying on the sphere centreline is known (Figure 1b). Unless used in a completely careless way, this design repeatably guides the plunger contact point to the extreme medial and lateral surfaces of the malleoli.

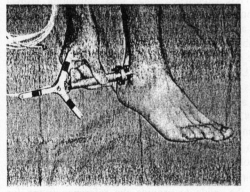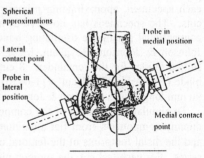

Figure 1. Ankle probe in use (a) and frontal plane view of both measurement positions (b).

Study design and data analysis: To simulate conditions in a computer-assisted TKR, sets of 21 to 30 repeated measurements of the ankle centre were made by 6 different operators on 8 different embalmed cadavers for a total of 318 ankle centre measurements. To estimate intra-observer repeatability on a variety of patients, we report the standard deviation (SD) of 30 ankle centre measurements by a single operator on 7 different embalmed cadavers. To show the probe's robustness when used by different operators, three additional operators each made 21 to 30 repeated measurements on one of these specimens and we report the total range spanning the 95% confidence limits of each individual operator. To show the relation between the characteristic ankle point found with the probe and the centre of area for the articular surface, two additional operators each made 20 measures of an additional specimen, after which we dissected the talocrural joint, digitized the tibial mortise and calculated its centre of area.

Measurement techniques: All measurements were recorded relative to a three-emitter triangular local reference frame (120 mm on a side) pinned rigidly to the proximal tibia and nominally aligned with the body planes. We used a Flashpoint 5000 localizer (Image Guided Technologies, Boulder, CO, USA) for all measurements, which has a typical accuracy of ~0.5 mm [Chassat 1998]. We removed indentations in the skin by massaging the tissue overlying the malleoli between measurements to prevent the probe contact balls from repeatedly falling into the same position.

2.2 Bone cutting errors

Study design: For this study we define bone cutting error as the difference in orientation between the cutting guide and the implant, described as two rotations. The first is in the plane of the face of the cutting guide facing the surgeon (labeled varus/valgus for distal femoral and tibial plateau cuts, and internal/external rotation for A/P femoral cuts), and the second is about a nominally mediolateral (ML) axis defined by the intersection of the front face plane and the cutting plane of the guide (labeled flexion/extension for distal femoral and tibial plateau cuts, and anterior/posterior deviation for A/P femoral cuts).

Two expert TKR surgeons, three orthopaedic residents with TKR experience and a foreign orthopaedic surgeon without TKR experience made cuts simulating typical practice on five fresh frozen human femora and eight tibia. Several cuts were made on each specimen, approximating primary and revision bone resections, for a total of 53 cuts. The specimens had no obvious abnormalities in bone quality. Cutting guides, a conventional pneumatic oscillating bone saw (CPS #1535), and saw blades supplied by Johnson & Johnson (Raynham, Mass. USA). We used open guides for 33 cuts and slotted guides for the remainder. 8 cuts were made with a 0.8 mm thick, 75 mm long blade (Synvasive 'Stablecut' #11-0470) and the remainder were made with 1 mm thick, 90 mm long blades (Johnson & Johnson #26-6050) renewed at the surgeon's request.

To simulate orientation of the components after placement on the cut bone, we made dummy components from flat aluminum plate by copying the profile of the tibial trays and the distal footprints of the femoral components in a Johnson & Johnson PFC implant series. For each test cut, the surgeon pinned the guide to the specimen and we measured the front face and cutting surfaces of the guide to provide the reference frame for reporting errors in varus/valgus (internal/external rotation for A/P cuts) and flexion/extension (anterior/posterior deviation for A/P cuts). The surgeon then made the cut and did any trimming and checking as would normally be done in TKR. We then re-measured the guide, placed a dummy implant on the cut bone surface to simulate placement of an implant with proper bone coverage, and measured the plane of the dummy implant.

Measurement techniques and data analysis: All measurements were made using the Flashpoint 5000 localizer (described above). To determine the variability associated with placing a dummy implant on a bone cut and measuring its orientation, we made 30 repeated placements and measurements on a single bone cut. The resulting SD was

<0.10° for varus/valgus errors and <0.17° for flexion/extension errors (95% confidence). For test cuts, we computed the mean and SD of all cutting errors in each direction. We are continuing to enroll surgeons in this study, so until we have a sufficient number to represent the general surgical population and until more cuts are made with slotted guides, we will not try to extract the effects of guide style, or type of cut.

3 Results

3.1 Ankle digitizing probe results

The average SD of ankle centre location for any one set of repeated measurements is 0.75 mm in the mediolateral (ML) direction and 1.12 mm in the anteroposterior (AP) direction (indicating that the new probe will give a result within ±1.9 mm (ML) and ±2.9 mm (AP) 95% of the time, Figure 2). This corresponds to an angular variability of 0.15° SD in varus/valgus and 0.2° SD in flexion/extension when registering a short (375 mm) tibial mechanical axis. The maximum range of a single operator on a single specimen is 6.5 mm (ML) and 10 mm (AP), giving a maximum tibial axis range of 1.0° varus/valgus and 1.5° flexion/extension.

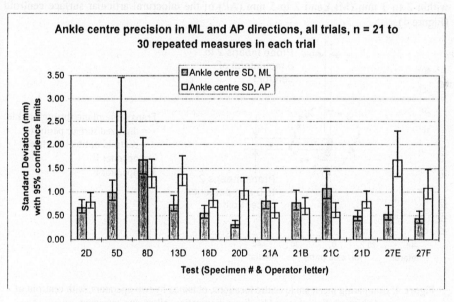

Figure 2. Range of variation in estimated ankle centre locations in the mediolateral (ML) and anteroposterior (AP) directions based on multiple specimens and operators.

A truly robust method would have no significant differences between the operator means. All means are within 0.8 mm in both AP and ML directions (Figure 3) and the range of the 95% confidence limits of the means is 1.6 mm (ML) and 1.3 mm (AP).

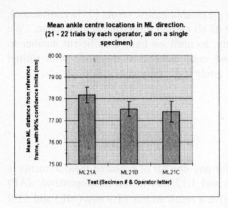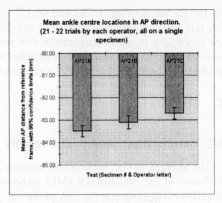

Figure 3. Interoperator differences in estimated ankle centre locations in the ML and AP directions on a single specimen (ankle centre locations are relative to a local reference frame pinned rigidly to the proximal tibia and nominally aligned with the body planes).

On the one specimen available for dissection, 2 operators both obtained mean results within 2 to 3 mm (ML) and 2 to 5 mm (AP) of the talocrural articular surface centroid (Figure 4).

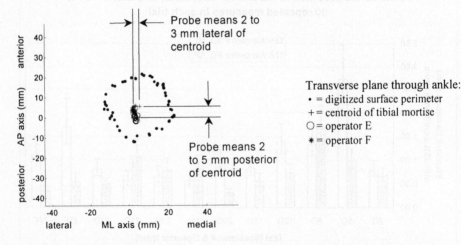

Figure 4. Comparing mean ankle centre locations, obtained by two operators, with centroid of tibial mortise, obtained by direct digitization, all on one specimen.

3.2 Bone cutting results

Both the guide movement and implant errors follow a normal distribution (checked using normal scores plots). Based on the actual variances, the test has 95% power to detect an implant alignment error of 0.4° in varus/valgus and 0.7° in flexion/extension. The mean cutting error in the frontal plane is about 0.3° degrees in the varus/valgus direction for both surgeon types (Figure 5). Precision in varus/valgus is within a SD of

0.54° for the experts and 1.25° for the less experienced surgeons (95% confidence); guide movement accounts for about 30% of the expert surgeons' variance and 34% of the less experienced surgeons' variance.

Mean 0.29° (95% CI 0.11 to 0.46)
SD 0.37° (95% CI 0.28 to 0.54)
Range 1.30°

Mean 1.03° (95% CI 0.49 to 1.58)
SD 1.16°(95% CI 0.89 to 1.70)
Range 4.65°

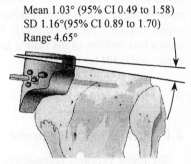

Mean 0.31° (95% CI -0.05 to 0.66)
SD 0.83° (95% CI 0.68 to 1.25)
Range 3.71°

Mean 0.74° (95% CI 0.05 to 1.42)
SD 1.58° (95% CI 1.22 to 2.24)
Range 7.02°

Figure 5. Varus/valgus (left) and flexion/extension (right) implant error results for 43 bone cuts: 20 made by 2 expert surgeons (top) and 23 by 2 less experienced surgeons (bottom).*

Mean 0.03° (95% CI -0.33 to 0.39)
SD 0.50° (95% CI 0.35 to 0.92)
Range 1.73°

Mean –0.02° (95% CI -0.47 to 0.43)
SD 0.63° (95% CI 0.43 to 1.15)
Range 2.02°

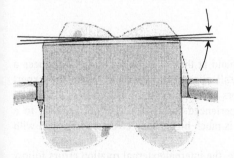

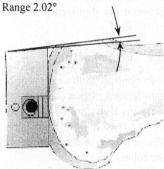

Figure 6. Internal/external rotation (left) and anterior/posterior deviation (right) implant error results for 10 bone cuts.*

In flexion/extension, there is a bias towards flexing of the saw blade up and away from the guide surface of over 1° (P = 0.0006) for the expert surgeons. The greatest flexion/extension errors in the current study occurred on large specimens where the saw blade did not reach the end of the cut before the saw body hit the guide, requiring the cut to be finished off 'freehand' at the far cortex. Precision in flexion/extension for both surgeon groups is within a SD of 2.24°, where guide movement accounts for about 37%

---

* Positive mean rotations are counter-clockwise about axes pointing out of the page.

of this variance. The measurement variance in either direction accounts for only about 1% of the observed variance.

The results of ten A/P femoral cuts executed by an expert surgeon (six cuts) and a resident (four cuts) on four femora are shown in Figure 6. Error in internal/external rotation (about the femoral mechanical axis) averages to a SD of 0.50° (0.34° for the expert and 0.67° for the resident). In the sagittal plane, angular deviation of the resected bone plane relative to the cutting guide surface resulted in a SD of 0.63° (0.70° for the expert and 0.50° for the resident).

4 Discussion

4.1 Ankle digitizing probe

We found excellent repeatability in using digitization to locate a characteristic point within the ankle joint complex. Our ankle probe only introduces a frontal plane SD of 0.15°, which compares favourably to Leitner's digitization result of 0.29°. A future repeatability study that will compare our probe to an alternative method such as using a point probe to digitize the malleoli is also required to validate the robustness of this new design. We plan to do this *in vivo* because the skin dimpling encountered in cadaveric studies artificially decreases the repeatability of locating the same points with the point probe. We are currently working on developing an ankle centre location protocol that uses the talocrural and subtalar joint axes to define a biomechanically meaningful centre. More specimens need to be measured and dissected to show that a reliable relationship between the digitized point and a biomechanically relevant ankle centre point can be established.

4.2 Bone cutting

In contrast to the excellent repeatability found at the ankle, bone cutting introduces a significantly higher contribution to the overall alignment variance. While the expert surgeons have limited their varus/valgus errors to under 1° (0.37° SD), errors as high as 2° (0.83° SD) were found with the less experienced surgeons. This corresponds to a 13mm ML error in ankle registration, which is much higher than errors experienced with the robust ankle probe.

While only ten A/P femoral cuts were made, the internal/external rotation errors follow similar trends to those found in varus/valgus. The expert surgeon's SD in internal/external rotation error is almost equal to that in varus/valgus (0.34° and 0.37° respectively) and is half that of the less experienced surgeon. However, we do not see the sagittal plane bias of the saw blade flexing up and away from the guide surface as with flexion/extension errors. This flexion/extension sagittal plane bias is similar to that found by Otani [Otani 1993]. It may be possible that the amount of bone resected in an A/P femoral cut is not enough to cause the blade to flex away from the guide surface and there is no need for any freehand trimming.

Although we have noted marked differences in repeatability in bone cutting between the two expert surgeons and the four surgeons with less TKR experience, we have not yet studied enough surgeons to make a statistically valid argument that this is the case. As mentioned above, we are continuing to enroll surgeons of various experience levels in this study and hope soon to be in a position to test the effect of experience on repeatability in bone cutting. These results are also based on the use of both slotted and open guides; here too we expect to be able to test for differences in performance when using different guide types once we have enrolled more surgeons into the study. Finally, although the surgeons who have participated have commented that the cutting process on the benchtop is essentially identical to that in the operating room, we do not yet have any OR data to support this impression. As our system moves closer to clinical implementation, we will be in a position to test this hypothesis.

4.3 Cumulative repeatability

There are numerous contributors to variability in our computer-assisted TKR procedure; these include variance and bias in estimating the hip, knee and ankle centres, setting the correct cutting guide pose, cutting the bone, and cementing the implant into place. To date, we have only measured values for the hip and ankle and for the bone cuts. Assuming that the knee measurements will contribute roughly the same variability as those at the hip and ankle, and assuming that the angular precision of locating the cutting guides will be similar to that of reading the planar probe, we estimate the overall repeatability of our proposed procedure by summing the variances of the various measurements. When we do this using the data for all six surgeons, we find that our predicted repeatability, before cementing, is roughly 1° (SD). The cutting errors, particularly for the less experienced surgeons, contribute the greatest fraction to the overall variance (~90%), so further development of this procedure should focus on the cutting issue and the precision of the cementing process, which is currently unknown.

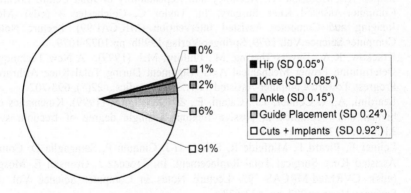

Figure 7. Contributions to overall varus/valgus variance in computer assisted TKR procedure.

5 Conclusions

We conclude that even inexperienced surgeons can likely achieve a SD within 1.5° for overall varus/valgus alignment in computer-assisted TKR using our proposed procedure consisting of a non-invasive hip centre identification technique, a robust ankle digitizing technique and a conventional cutting technique based on a computer-guided cutting guide. Expert surgeons exhibit significantly better cutting precision, so we expect their overall precision to be better (<1.2°). In flexion/extension, alignment precision will be worse due to greater cutting errors in this direction.

Acknowledgements

We thank Drs. Thomas Oxland of the Division of Orthopaedic Engineering Research, Vancouver Hospital, and Vlad Stanescu, UBC Dept. of Anatomy, for their assistance.

References

1. Chassat F, Lavallee S: Experimental Protocol of Accuracy Evaluation of 6-D Localizers for Computer-Integrated Surgery: Application to Four Optical Localizers. In: Wells, W; Colchester, A; Delp, S. (eds): Medical Imaging and Computer Assisted Intervention (MICCAI'98). Lecture Notes in Computer Science Vol. 1496, Springer-Verlag 1998, pp. 421-430.
2. Jeffery RS, Morris RW, Denham RA: Coronal Alignment After Total Knee Replacement. J. of Bone and Joint Surgery (British Ed.) 73-B, 1991, pp. 709-714.
3. Inkpen, KB: Precision and Accuracy in Computer-Assisted Total Knee Replacement. Master's thesis, Dept. of Mechanical Engineering, University of British Columbia, Vancouver BC Canada. 1999a.
4. Inkpen KB, Hodgson AJ: Accuracy and Repeatability of Joint Centre Location in Computer-Assisted Knee Surgery. In: Taylor C, Colchester A (eds): Medical Imaging and Computer Assisted Intervention (MICCAI'99). Lecture Notes in Computer Science Vol. 1679, Springer-Verlag 1999b, pp.1072-1079.
5. Krackow, K.A., Bayers-Thering, M., Phillips, M.J. (1999). A New Technique for Determining Proper Mechanical Axis Alignment During Total Knee Arthroplasty: Progress Toward Computer-Assisted TKA. Orthopedics., 22(7), 698-702.
6. Leardini, A., O'Connor, J.J., Catani, F., & Giannini, S. (1999). Kinematics of the human ankle complex in passive flexion; a single degree of freedom system. J.Biomech., 32(1), 111-118.
7. Leitner F, Picard F, Minfelde R, Schulz H-J, Cinquin P, Saragaglia D: Computer Assisted Knee Surgical Total Replacement. In: Troccaz J, Grimson E, Mosges R (eds): CVRMed-MRCAS '97. Lecture Notes in Computer Science Vol. 1205, Springer-Verlag 1997, pp. 627-638.
8. Otani T, Whiteside LA, White SE: Cutting Errors in Preparation of Femoral Components in Total Knee Arthroplasty. Journal of Arthroplasty, Vol 8, No 5, Oct. 1993 pp. 503-510.
9. Bogert, A.J. van den, Smith GD, Nigg BM. In Vivo Determination of the Anatomical Axes of the Ankle Joint Complex: an Optimization Approach, Journal of Biomechanics, Vol. 27(12), 1477 – 1488, 1994.

Double-Bundle Anatomic ACL-Reconstruction with Computer Assisted Surgery: An In-Vitro Study of the Anterior Laxity in Knees with Anatomic Double-Bundle versus Isometric Single-Bundle Reconstruction

J.W.H. Luites[1a], A.B. Wymenga[1b], L. Blankevoort[2], J.G.M. Kooloos[3] and M. Sati[4].

[1a] SMK-Research and [1b] Department Orthopaedic Surgery, St. Maartenskliniek,
P.O. 9011, 6500 GM Nijmegen, The Netherlands
a.wymenga@maartenskliniek.nl
[2] Orthopaedic Research Center Amsterdam, Academic Medical Center,
University of Amsterdam, Amsterdam, The Netherlands
[3] Department of Anatomy and Embryology,
University of Nijmegen, Nijmegen, The Netherlands
[4] Maurice E. Muller Institute for Biomechanics, University of Bern, Switzerland

Abstract. The current standard technique to reconstruct ruptured anterior cruciate ligament (ACL) is through an isometric single-bundle graft. However, the ACL has a two-bundled nature, with the anteromedial bundle (AMB) most tight in flexion and the posterolateral bundle (PLB) near extension. Therefore a double-bundle, anatomically placed, graft is expected to provide better stability over the complete range of motion. The anterior laxity was tested in various flexion angles in knees with anatomic double-bundle and isometric single-bundle reconstruction. Anatomic femoral graft placement was performed using Computer Assisted Surgery (CAS) with the AMB-graft fixated in 90° of flexion and the PLB in 20°. Near extension the double-bundle technique caused no significant laxity change compared to a 2.1 mm (34%) increase in single-bundle reconstructed knees. In 90° of flexion the laxity increase was nearly similar. Over the complete range of motion the anatomic double-bundle technique results in a better functional reconstruction, at the cost of a higher AP-error near extension.

1 Introduction

The anterior cruciate ligament (ACL) consists of two functional bundles, each with its anatomic insertion area. The anteromedial bundle (AMB) is most tight in flexion and the posterolateral bundle (PLB) is tighter near extension. An anatomical double-bundle graft restores the functionally two-bundled nature of the ACL in full flexion a near extension, where as an isometric single-bundle technique cannot [1].

1.1 Goal

This in-vitro study evaluates the anterior laxity characteristics in knees with double-bundle reconstruction, anatomically placed using Computer Assisted Surgery (CAS) by comparing to intact knees and the standard isometric single-bundle technique.

2 Methods

In 9 fresh cadaver knees, fixed in a knee loading system under 50N axial compression, the anterior-posterior (AP) position of the tibia relative to the femur was measured using an electromagnetic system [3Space Fastrak, Polhemus Navigation Sciences, USA]. This was done in various flexion angles for the unloaded joint and loaded with an anterior force (100N drawer test) on the tibia, whereby axial tibial rotation was restrained. Mean AP-error, as the result of the difference in AP-position between the unloaded intact and unloaded reconstructed knee was calculated. As was the anterior laxity, defined as the difference between the unloaded and loaded joint. Each knee was measured, in 6 conditions: 1) intact; 2) ACL deficient; 3) anatomic double-bundle reconstruction (4+5); 4) anatomic AMB-reconstruction; 5) anatomic PLB-reconstruction; 6) isometric single-bundle reconstruction. In the anatomic double-bundle, AMB- and PLB-reconstruction a Quadriceps Tendon-patellar Bone graft (QTB) with 1 bone block and two tendon slips was used. The AMB-graft was tensioned in 90° of flexion, the PLB-graft in 20°. A Bone-Patellar Tendon-Bone (BPTB) graft, tensioned in 20° of flexion, was used in the single-bundle reconstruction. Tibial tunnel positioning was performed using the PCL Oriented Placement Marking Hook™ [Arthrex Inc., USA]. Anatomic femoral graft placement was performed through CAS [SurgiGate, Medivision, Switzerland] with the use of an ACL-module [2, MEM Institute for Biomechanics, Switzerland] and a 3D femoral template [3]. One 11 mm-tunnel was drilled at the 3D position of the ACL-center in the template. A transtibial femoral guide™ [Arthrex Inc., USA] with a 7 mm offset, placed in the over-the-top position at 11 0'clock determined the isometric tunnel location.

3 Results

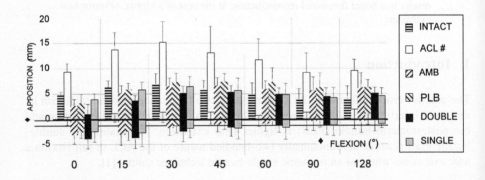

Fig. 1. Mean anterior laxity in mm, as the result of the difference between the anterior-posterior (AP) position (Y) of the unloaded joint (*lower* position) and loaded joint (upper position), in various knee flexion angles (X). In intact knees (*horizontal stripes*), ACL deficient knees (*white*) and reconstructed knees with AMB-graft (*upping stripes*), PLB-graft (*downing stripes*), double-bundle graft (*black*) and isometric single-bundle graft (*gray*)

4 Discussion

The slight posterior shift of the tibia (posterior AP-error) seems to be a characteristic that is common in all ACL-reconstruction using tendons, due to the tensioning at the time of fixation. The three single bundle reconstructions were not able to restore the normal laxity in the position in which they were tensioned and fixed. The increased posterior AP-error and lower anterior laxity in the double bundle technique, specifically up to 30° of flexion where both bundles are recruited, can be the result of using more tendon material in the double bundle graft of these reconstructions.

5 Conclusion

Double bundle anatomic ACL-reconstruction, placed with CAS and a 3D template, results in a good functional reconstruction having anterior laxity characteristics that are close to those of the original ACL over the complete range of motion.

Acknowledgment. This study is supported by the AO Foundation, Switzerland.

References

1. Radford, W.J.P., Amis, A.A.: Biomechanics of a double prosthetic ligament in the anterior cruciate deficient knee. J.B.J.S. [Br] Vol. 72-B, 6 (1990) 1038-43
2. Sati, M., Stäubli, H.-U., Bourquin, Y., Kunz, M., Käsermann, S., Nolte, L.-P.: Clinical Integration of New Computer Assisted Technology for Arthroscopic ACL Replacement. In: T.DiGioia (Guest Ed.), F. Fu (Ed.): Operative Techniques in Orthopaedics, Vol. 10, No 1(January 2000) 40-49
3. Wymenga, A.B., Luites, J.W.H., Sati, M., Bourquin, Y., Blankevoort, L., van der Venne, R., L., Kooloos, J.G.M., Stäubli, H.-U.: Computer-Assisted Anatomical Placement of a Double-Bundle ACL through 3D-Fitting of a Statistically Generated Femoral Template Into Individual Knee Geometry. Abstr 5th Intern Symp on CAOS (2000) 24

Measurement of Pelvic Tilting Angle during Total Hip Arthroplasty Using a Computer Navigation System

S.Nishihara[1], N.Sugano[2], K.Nakahodo[3], T.Sasama[3], T.Nishii[2], Y.Sato[3], S.Tamura[3], K.Yonenobu[2], H.Yoshikawa[2], and T.Ochi[1]

[1]Department of Computer Aided Orthopaedic Surgery, Osaka University Medical School, 2-2 Yamadaoka, Suita 565-0871, Osaka, Japan

[2]Department of Orthopaedic Surgery, Osaka University Medical School, 2-2 Yamadaoka, Suita 565-0871, Osaka, Japan

[3]Division of Functional Diagnostic Imaging, Osaka University Medical School, 2-2 Yamadaoka, Suita 565-0871, Osaka, Japan

Correspondence should be sent to S.Nisihihara, MD.
FAX: +81-6-6879-3559, email: nishihara@caos.med.osaka-u.ac.jp

Abstract. The purpose of this study was to measure pelvic orientation during total hip arthroplasty (THA) in a lateral decubitus position using a computer navigation system with an optical localizer (OPTOTRAK). THA was performed in 17 hips. Much attention was paid to set the patients in neutral axial rotation with the anatomical plane of the pelvis perpendicular to the operating table. After shape-based registration, pelvic orientation was tracked with light emitting diode markers fixed to the pelvis. Measurements were based on the anatomical plane. Mean movement of the pelvis from the supine position to the dislocated lateral decubitus position was 8 degrees posterior (26 posterior to 6 anterior), 3 degrees abducted (6 adduction to 20 abduction), and 4 degrees internally rotated (24 internal to 5 external). This study showed that the pelvis was not always placed in neutral axial rotation, despite the surgical plan. It was further tilted posteriorly and rotated internally when the socket was inserted. This tilt can lead to decreased socket anteversion with conventional alignment guide systems, while unknown pelvic orientation in general can cause socket malposition with such systems. Therefore, intraoperative three-dimensional measurements of pelvic orientation seems to be useful to avoid socket malposition.

1 Introduction

Socket malposition has been reported to be one of the causes of dislocation, impingement, wear, and loosening after total hip arthroplasty (THA)[3,4,7]. Socket anteversion is affected by the flexion and rotational angles of the pelvis[1]. Therefore, it is important for surgeons to know the three-dimensional orientation of the pelvis on the operating table, especially in the lateral decubitus position[2,5,8]. The position of the pelvis may change during various operative maneuvers, even if surgeons initially pay sufficient attention to neutrally align the pelvis on the operating table. The purpose of this study is to measure the orientation of the pelvis on the operating table during THA procedures using a computer navigation system with an optical sensor (OPTOTRAK), and to estimate the risk of socket malposition with conventional alignment guides.

2 Materials and Methods

Our navigation system consists of the following three steps: (1) making a computer model from preoperative CT data of the object to be used in the operation; (2) registration of the computer model to the real object; and (3) tracking and measurement of the object and operative tools during operation. Preoperatively, transverse images from the level of the superior anterior iliac spines to the level of the femoral canal isthmus are obtained using a helical CT scanner. 3D acetabular bone surface models are reconstructed from this CT data. A pelvic coordinate system is defined using the inter-teardrop line and the standard anatomic pelvic plane through the superior anterior iliac spines and the pubic tubercles. THA was performed in 17 hips with osteoarthritis using this computer navigation system. Mean age at surgery was 54 years (range, 45-68 years).

In the operating room, the optical three-dimensional position sensor (OPTOTRAK 3020, Northern Digital Inc., Waterloo, Canada) was placed at the wall caudal to the patient. The operating table was placed so that the surgical area was within 2.5 meters from the position sensor. Patients were placed in a lateral decubitus position on the operating table such that the anatomical plane of the pelvis was perpendicular to the plane of the operating table. A posterolateral approach was used. A flat plate with 6 LED markers was fixed with a rod to the iliac crest using an extraskeletal fixation system (Hoffmann system, Howmedica, Rutherford, USA).

Shape-based surface registration of the previously constructed bone models to the real objects was performed in two steps by using the ICP algorithm with the least square method. First, the surgeon digitized four surface points to provide the starting position for matching, and then 30 surface points were digitized for the final registration. After shape-based surface registration, LED markers fixed to the pelvis and femur allowed for continuous intraoperative tracking of the position and movement of these bones. Orientation of the pelvis was determined based on the anatomical plane. The pelvic orientation, as measured by flexion, abduction and axial rotation, was measured in the supine and lateral decubitus positions, as well as a dislocated position when the socket was inserted.

3 Results

The mean pelvic tilting angle in the supine position was 6 degrees anterior (standard deviation (SD): 5.8, range: 4 posterior to 16 anterior). The mean abduction angle in the supine position was 0 degrees (SD: 4.5, range: 14 adduction to 9 abduction). The mean axial rotation angle in the supine position was 1 degree (SD: 2.9, range: 6 internal to 4 external).

Mean movement of the pelvis from the supine position to the initial lateral decubitus position was 3 degrees anterior (SD: 7.1, range: 6 posterior to 22 anterior), 0 degrees abducted (SD: 4.9, range: 10 adduction to 9 abduction), and 2 degrees externally rotated (SD: 7.5, range: 17 internal to 13 external).

Mean movement of the pelvis from the initial lateral decubitus position to the dislocated position was 12 degrees posterior (SD: 7.6, range: 25 posterior to 1 anterior), 2 degrees abducted (SD: 4.4, range: 6 adduction to 11 abduction), and 7 degrees internally rotated (SD: 5.0, range: 15 internal to 1 external).

Mean movement of the pelvis from the supine position to the dislocated position was 8 degrees posterior (SD: 8.6, range: 26 posterior to 6 anterior), 3 degree abducted (SD: 6.0, range: 6 adduction to 20 abduction), and 4 degrees internally rotated (SD: 6.8, range: 24 internal to 5 external).

4 Discussion

One important factor affecting alignment of the acetabular component has been reported to be the orientation of the pelvis on the operating table[2),5),8)]. It is hypothesized that anterior tilting or internal rotation of the pelvis intraoperatively in the lateral decubitus position may lead to decreased anteversion of the acetabular component, as compared to the surgeon's intended orientation of the component. It is also hypothesized that adduction of the pelvis in the lateral decubitus position on the operating table may lead to increased abduction of the acetabular component. However, in spite of the impact of these factors on clinical outcomes, there are few reports on measuring pelvic tilting angle, abduction angle, and axial rotation angle in the lateral decubitus position on the operating table.

This study showed that the pelvis was not always placed in neutral, despite the surgeon's intention. It was further tilted posteriorly and rotated internally when the socket was inserted. Movement of the pelvis from the initial lateral decubitus position to the dislocated position was larger than that of the pelvis from the supine position to the initial lateral decubitus position.

With conventional mechanical alignment guide systems, mean 8 degrees posterior tilting of the pelvis from the supine position to the dislocated lateral decubitus position may lead to 8 degrees less anteversion of the acetabular component. Mean 3 degrees abduction of the pelvis seemed to have a small effect on the cup abduction angle, but variation of the pelvic abduction angle on the table is not insignificant. Mean 4 degrees internal rotation of the pelvis may lead to an additional decrease in the cup anteversion with conventional mechanical alignment guide systems. This study demonstrated that mean 12 degrees of anteversion may be overestimated, and the cup may be placed in less anteversion than intended, when using mechanical guide systems. The variation of the pelvic tilt on the operating table also seemed to be an issue with regard to socket malposition. Therefore, intraoperative three-dimensional measurement of pelvic orientation is necessary to place the socket without a large bias and variation.

References

1. Abel, M.F., Sutherland, D.H. : Evaluation of CT scans and 3-D reformatted images for quantitative assessment of the hip. J. Pediatr. Orthop. (1994); 14 : 48-53
2. DiGioia, A.M., Jaramaz, B., et al : Image guided navigation system to measure intraoperatively acetabular implant alignment. Clin. Orthop. (1998); 355 : 8-22

3. D'lima, D.D., Urquhart, A.G., et al : The effect of the orientation of the acetabular and femoral components on the range of motion of the hip at different head-neck ratios. J. Bone and Joint Surg. (2000); 82-A(3): 315-321

4. Kennedy, J.G., Rogers, W.B. : Effect of acetabular component orientation on recurrent dislocation, pelvic osteolysis, polyethylene wear, and component migration. J. Arthroplasty (1998) ; 13: 530-534

5. McCollum, D.E., Gray, W.J. : Dislocation after total hip arthroplasty : Causes and prevention. Clin. Orthop. (1990) ; 261: 159-170

6. Murray, D.W. : The definition and measurement of acetabular orientation. J. Bone Joint Surg. (1992) ; 75B : 228-232

7. Robinson, R.P., Simonian, P.T., et al : Joint motion and surface contact area related to component position in total hip arthroplasty. J. Bone and Joint Surg. (1997) 79-B(1): 140-14

8. Woo, R.Y.G., Morrey, B.F. : Dislocations after total hip arthroplasty. J. Bone Joint Surg. (1982) ; 64A: 1295-1306

Efficacy of Robot-Assisted Hip Stem Implantation. A Radiographic Comparison of Matched-Pair Femurs Prepared Manually and with the Robodoc® System Using an Anatomic Prosthesis

F. Gossé, K.H. Wenger, K. Knabe, C.J. Wirth

Orthopaedics Department, Hannover Medical University; Heimchenstrasse 1-7, Hannover, Germany D-30625; +49-511-535-4340; gosse@annastift.de

Abstract. The number of robots in operation in the German-speaking orthopaedic community more than doubled last year, reflecting considerable excitement among patients and doctors over the prospects of improving the accuracy of endoprosthesis surgery in both the planning and implantation stages. The purported improvement, however, adds considerable expense to the procedures although little data exist still in many applications to prove its efficacy.

The following study compares pre- and post-operative radiographs of conventionally- and robotically-prepared groups of donor femurs implanted with an anatomic hip stem to quantify their respective preservation of three anatomic parameters— leg length, anterior bow angle, and mediolateral offset. All femurs were planned blindly both with virtual fitting using 3D reconstruction from CT scans and with conventional templates.

Maintenance of leg length as well as anterior bow angle were significantly higher (p<0.05) for the robotically-prepared group than for the manually-prepared group. Maintenance of mediolateral offset, however, was significantly higher for the manually-prepared group, likely due to the common varus implantation experienced with this prosthesis type. Further studies will examine microradiographic slices in the transverse plane to quantify the gap size resulting from the two methods.

Introduction

The recent surge in the adaptation of robotic technology for surgical purposes in a sweeping range of medical fields speaks for a new era in the precision with which tumors, stenoses, and diseased joints are treated. What began over a decade ago in the neurosurgical field (1) now has found application in areas as disparate and complex as laparoscopic surgery (2), cardiovascular and ocular surgery (3), as well as orthopaedic surgery (4).

The primary goals of robotically-assisted hip replacement surgery are to improve the reliability of placement of the prosthesis stem and to optimize the amount of bony contact in the trabecular bone region proximally. The Robodoc® system from Integrated Surgical Systems (Davis, California, USA) provides an Orthodoc® computer station for planning the implantation procedure in three dimensions, based on CT scan data. The milling path and the prosthesis silhouette are displayed together in A-P, lateral, and transverse plane views, over more than a decade of available zoom range, to allow detailed planning of the placement of the prosthesis and to help determine the amount of cortical bone, if any, that will be removed in the procedure.

Currently in Germany there are over two dozen Robodoc® units in operation, with at least six different hip stems presently programmed for application. The reported functional accuracy of the robot, in deviation from plan, is +/- 0.4 mm; working in an artificially rigid environment, the deviation is much lower. Given this order of mechanical precision, clearly the critical factor for the success of the implantation lies in the planning itself and how well it is supported by the technology associated with the implantation method. Jerosch et al. (5) recently compared CCD and antertorsion angles pre- and post-operatively in two cadaveric femur groups, one prepared with manual techniques and the other with a robot. They found that the CCD angle changed by a similar amount in the two approaches, but that the antetorsion angle changed almost 10° with hand preparation, whereas with robotic preparation there was virtually no change.

In this study, our goals were to compare the preservation of three anatomical parameters complementing those of the Jerosch et al. study in manually-prepared and robotically-prepared femoral cavities implanted with an anatomical hip stem. The parameters were: 1) leg length; 2) anterior bow angle in the sagittal plane; and 3) mediolateral offset. The prosthesis used in the study is now available for manual and robotic implantation in Europe.

Methods

Ten pairs of donor femurs were obtained fresh then maintained frozen at -20°C until scanning and testing. Femurs larger or smaller than the common-sizes range currently programmed for this prosthesis were excluded before forming the final test groups with n=10 for each. A two-pin registration system was employed, with one 10-mm proximal pin set into the anteromedial aspect of the femur slightly above the level of the trochanter minor, and one 30-mm distal pin set into the medial condyle. The femurs were scanned in a Siemens Somatom +4 model using a spiral CT method. Slice thickness in the pin region was 1.0 mm, otherwise 3.0 mm.

Radiographs were obtained in the A-P and lateral views using a perpendicular block and podiatric orthosis foam to assure the stability of the femur during exposure in the upright position as well as the trueness and orthogonality of the perspectives. A long Steinmann pin, attached in length to the block, defined an anatomically-vertical axis between the block contact points on the femoral condyles and lesser trochanter. The distance between the femur and the film plane, as well as that between the film plane and source, was maintained constant. Anteterior bow was measured in the lateral radiograph, and mediolateral offset and leg length in the A-P radiograph by two authors, then averaged.

After pair-wise random assignment to the manual and robot groups, the femurs of both groups were blindly planned for operation with an Antega® anatomical hip stem prosthesis from Aesculap AG & Co. KG (Tuttlingen, Germany) by the same two authors.

Table 1 lists the distribution of sizes for the study groups, with size 13 being predominantly represented. After the one surgeon agreed with the planning of the other in all cases, the femoral cavities were appropriately prepared either manually, using the existing instrument set, or robotically. In one case a different size was planned for the left and right femurs of a pair. Additional planning was performed on all specimens using the manufacturer's template system and the radiographs.

Following cavity preparation and implantation, with all manual aspects of the various procedures performed by the senior surgeon, the femurs were again radiographed for posteroperative measurements. Anticipated cutting time of the robot was about 30 minutes, with two tool changes required to accomodate a fillet length on the anterior aspect of this particular prosthesis. Distal reaming was performed by hand, pending certification of that aspect of the software.

Statistical analysis consisted of a Kruskal-Wallis analysis of variance using a level of significance of 0.05.

Results

Leg length change in the manual group (Table 1) was 7.5 mm (SD 4.9 mm), and in the robot group, 1.3 mm (SD 3.8 mm). This difference was statistically significant (p<0.05). Anterior bow angle change in the manual group was -7.4° (SD 4.6°), and in the robot group -5.0 ° (SD 2.7 °). Mediolateral offset change in the manual group was -3.1 mm (SD 3.6 mm), and in the robot group -7.3 mm (SD 3.6 mm). Neither of these differences was significant. Actual milling time, with two tool changes, averaged about 35 minutes. The diagnostics and preparation procedures attendant on the robot increasingly became more efficient as the team gained experience with setup procedures. Still, there is an irreducible time when the surgery waits on the robot before it begins milling.

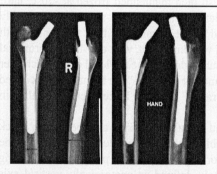

Fig. 1 shows an example of the resulting fit of the prosthesis in the two groups on radiographs. These images are intended to provide a general perspective on the outcome of the implantations and should not be taken as truly representative, since such images do not lend themselves easily to statistical treatment. Based on a clinically subjective assessment, however, it was consistently appraised among the authors through blind radiographic analysis that the robot group overall demonstrated better-fitting hip stems.

Fig. 1. Examples of robotically- (left) and manually-prepared femurs (right). Images are not intended to be viewed as results, but rather as indications of fit in some cases.

Table 1. Differences of Pre- and Post-Op in Robot and Hand Preparation Groups

	Robot					Hand			
Speci. No.	Prosth. Size	Leg Length	Medio-Lateral Offset	Anterior Bow Angle	Speci. No.	Prosth. Size	Leg Length	Medio-Lateral Offset	Anterior Bow Angle
R1	13	-1	-9	-5	H1	13	12	-9	-3
R2	13	1	-13	-3	H2	14	8	-8	-7
R3	14	0	-5	-1	H3	14	3	-6	-4
R4	13	-6	-2	-8	H4	13	0	-1	-10
R5	14	5	-5	-4	H5	13	11	-2	-6
R6	16	0	-12	-4	H6	16	0	3	-17
R7	10	1	-7	-5	H7	11	9	-3	-7
R8	13	6	-9	-5	H8	12	10	-2	-11
R9	13	0	-3	-11	H9	14	8	-2	-1
R10	13	7	-8	-4	H10	13	14	-1	-8
Avg	13.2	1.3	-7.3	-5.0	Avg	13.3	7.5	-3.1	-7.4
Std Dev	1.4	3.6	3.4	2.6	Std Dev	1.3	4.7	3.4	4.3

Table 1. Differences of Pre- and Post-Op in Robot and Hand Preparation Groups

Discussion

The use of a robot to improve accuracy of placement of a hip stem and to preserve leg anatomy is experimentally supportable, based on the results of this study, especially regarding retention of anterior bow angle and leg length. Mediolateral offset was actually larger in the robot group, but this may reflect a larger error in the manual group from the commonly-observed varus placement of the prosthesis by hand.

In the Jerosch et al. study (5), the measured CCD angle of the femur represented a combined anatomical parameter, which we attempted in the current study to separate into the more individual and indicative mediolateral offset and leg length. Using the combined parameter of CCD angle, Jerosch et al. found no differences between the two groups. In the present study, the decoupled parameter of leg length revealed differences in the performance of the two approaches, with the robot group indicating less change from the preoperative value than that of the manual group, although the ranges were similar for the two. The greater change in mediolateral offset in the robot group may reflect a relatively consistent varus malpositioning of the prosthesis using the manual technique. Further analysis is needed to determine whether this difference can be categorized as one of accuracy or misplaced precision.

A limit of this study is that the use of one surgeon discourages extrapolation to the general case. For our purposes, however, it served to control the inter-surgeon variability that otherwise could mask or distort group differences. Also, measurement of the parameters from the radiographs relied on common clinical techniques, using hand measurements with an accurate mechanical rule rather than a more exact, but less practiced digitizing technique. Further study of these groups is planned using microradiographic techniques from which a more detailed analysis of the bony contact will complement the anatomical parameters studied here.

Although it represents a costly sophistication to hip replacement procedure options, this particular surgical robotic system does provide an enhanced measure of reliability in the implantation of the anatomical hip stem used in this study.

References

1. Hefti JL, Epitaux M, Glauser D, Fankhauser H. Robotic three-dimensional positioning of a stimulation electrode in the brain. *Comput Aided Surg* 3(1):1-10, 1998.
2. Mettler L, Ibrahim M, Jonat M. One year experience working with the aid of a robotic assistant in gynaecological endoscopic surgery. *Hum Reprod* 13(10):2748-50, 1998.
3. Okada S, Tanaba Y, Kimura K, Yamaguchi H, Sato S. Thoracoscopic surgery using voice controlled robot for spontaneous pneumothorax. *Kyobu Geka* 51(7):561-3, 1998.
4. Nolte LP, Visarius H, Arm E, Langlotz F, Schwarzenbach O, Zamorano L. Computer-aided fixation of spinal implants. *J Image Guid Surg* 1(2):88-93, 1995.
5. Jerosch J, von Hasselbach C, Filler T, Peuker E, Rahgozar M, Lahmer A. Increasing the quality of preoperative planning and intraoperative application of computer-assisted systems and surgical robots –an experimental study. *Chirurg* 69 (9): 973-976, 1998.

Surgical Navigation for THR: A Report on Clinical Trial Utilizing HipNav

Timothy J. Levison[1], James E. Moody[1], Branislav Jaramaz[1,2], Constantinos Nikou[1],
Anthony M. DiGioia[1,2]

Centers for Medical Robotics and Computer Assisted Surgery,
[1]UPMC Shadyside Hospital, Pittsburgh, PA and
[2]Carnegie Mellon University, Pittsburgh, PA
{[levison,moody,branko,costa,tony]@cor.ssh.edu}

Abstract: Computer-assisted Total Hip Replacement (THR) surgery using the HipNav surgical navigational system was evaluated. This summary reports on the first 100 HipNav clinical trial patients and focuses on: 1) patient demographics, 2) post-operative clinical outcomes, 3) incision length measurements, 4) mechanical guide measures, and 5) functional pelvic tilt measurements. Results from this clinical trial have shown no system-related complications, an improvement in post-operative clinical outcomes, reductions in soft tissue dissection, an unreliability of traditional mechanical alignment guides, and a variability of pelvic orientation during functional activity.

Keywords: orthopaedics, surgical navigation systems, clinical results

1. INTRODUCTION

The use of HipNav [1] computer-assisted system in total hip replacement surgery provides the ability to plan the alignment of the acetabular cup before surgery, and accurately perform the surgery according to this preoperative plan using image guided technology. We have established a methodology to evaluate the outcomes based on preoperative, intraoperative and postoperative measurements.

2. METHODS

The HipNav system is an image guided hip navigation system used to assist surgeons in planning of the acetabular implant placement and precise alignment according to the preoperative plan. HipNav is a collaborative research effort involving investigators from Carnegie Mellon University and UPMC Shadyside Hospital in Pittsburgh, PA. The system is composed of three components: 1) a preoperative planner; 2) a kinematic simulator; and 3) an intraoperative navigation guide. The specific technical aspects of the HipNav system have been described in other published literature [1,2].

The reported results of the clinical trial involved THR patients operated with the help of the HipNav system between April 1997 and December 1999. All patients were informed of the experimental status of the system and each signed an informed consent approved by the Institutional Review Board. All patients in which the HipNav system was used were enrolled in the Total Joint Registry at UPMC Shadyside Hospital. The

Total Joint Registry collects all relevant data preoperatively, intraoperatively and at 1, 3, 6, and 12 months post-operatively, and then annually for the rest of their lives.

3. RESULTS

One hundred patients have been enrolled in the study, with eight patients receiving bilateral THR procedures, for a total of 108 total hip replacement surgeries. Patient demographics are presented in Table 1.

Table 1. HipNav Patient Demographics

Number of Patients	100 patients (8 bilateral)
Number of Procedures	108 Hips
Gender	51 Female, 49 Male
Average Age	63 years old (62.6)
Age Range	37 to 81 years
Affected Hip	48 Left, 60 Right

Outcome measurements were recorded using Harris Hip Score (HHS) measures[3]. The HHS measurement tool incorporates functional activity levels, perceived pain and range of motion measurements to compute a clinical outcome score. The preoperative HHS average for 100 patients was 51.6 (σ = 12.2). The three month HHS average was 82.1 (σ = 10.6). The six month HHS average was 88.5 (σ = 9.8). One year postoperative HHS showed an average of 93.5 (σ = 7.9).

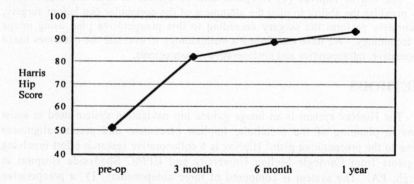

Fig 1. Change in Harris Hip Score Averages from Pre-op to 1 year Post-op following Total Hip Arthroplasty using HipNav

Results from the HipNav clinical study have also shown a reduction in the extent of soft tissue dissection on the order of 50% when compared to traditional THR techniques. Skin incisions for all 100 patients ranged between 7.3 cm to 23.8 cm, with an average length of 13.6 cm (σ = 4.2). However, beginning with the 36th patient, the surgical navigation capabilities of the HipNav system were utilized to enable the surgeon to use a less invasive surgical approach. As a consequence, the mean incision length changed from 19.6 cm (σ = 2.2) with a range of 14.8 cm to 23.8 cm for the first 35 HipNav cases, to the average incision length of 12.1 cm (σ = 3.1) with a range of 7.3 cm to 20.6 cm in patients 36 to 100.

Before the cup was placed in the preoperatively planned orientation, acetabular component orientation using traditional mechanical guides was measured by the HipNav system. The goal of the mechanical guides is to align the acetabular cup in 45° of abduction and 20° of flexion. However, results of the measured cup alignment using the mechanical guide ranged from 35° to 59° in abduction and from -26° to 33° of flexion.

Functional pelvic orientation, expressed as a change in pelvic flexion and extension, was radiographically measured in all patients. These measurements of pelvic tilt for each patient represent the differences seen as they moved from a standing to a sitting position. Results of pelvic tilt measurements showed a trend towards a neutral pelvic orientation during standing, with a range of -16° to 27° (μ= 2°, σ = 8°). However, a significant posterior tilting was seen among patients in a sitting position, with a range of pelvic tilting from 5° to -58° (μ = -35°, σ = 14°).

4. DISCUSSION

The HipNav system has been successfully utilized during the ongoing clinical trial and has been shown to provide a measurement tool that is useful in examining conventional methods of THR. HipNav also provides the surgeon with the ability to accurately align implants and to implement less invasive surgical procedures while not significantly increasing surgical time. The reduction in incision length, and thus soft tissue dissection, is expected to decrease complication rates and to increase the rate of recovery following surgery. HipNav patients have also shown an improvement in post-operative clinical outcomes measured by Harris Hip Scores. The current version of the system can now assist on the femoral side of the joint, permitting on-line leg length determination and patient specific maps of safe, stable range of motion.

5. REFERENCES

1. DiGioia AM, Jaramaz B, Blackwell M, et al: Image guided navigation system to measure intraoperatively acetabular implant alignment. Clinical Orthopaedics and Related Research, Vol. 355, pp 8-22, 1998.
2. DiGioia AM, Jaramaz B, Nikou C, et al: Surgical navigation for total hip replacement with the use of HipNav. Operative Techniques in Orthopaedics, Vol. 10, No. 1, pp 3-8, 2000.
3. Harris WH: Traumatic arthritis of the hip after dislocation and acetabular fractures: treatment by mold arthroplasty. An end-result study using a new method of result evaluation. J Bone Joint Surg 51A:737, 1969.

Description of Anatomic Coordinate Systems and Rationale for Use in an Image-Guided Total Hip Replacement System

Constantinos Nikou, Branislav Jaramaz, Anthony M. DiGioia, and Timothy J. Levison

Center for Medical Robotics and Computer-Assisted Surgery, UPMC Shadyside Hospital,
Pittsburgh, PA.
{[costa,branko,tony,levison]@cor.ssh.edu}

Abstract Lowering the risks of a surgical procedure is extremely important, especially for high-volume procedures such as total hip replacement. Significant work has been done to study total hip replacement procedures and provide the surgeon with techniques and tools to achieve better patient outcomes. Computer-assisted intervention allows surgeons to "close the loop" in medical research, allowing the surgeon to preoperatively plan, interoperatively navigate, and postoperatively analyze medical procedures, then use the results to repeat or improve the quality of future procedures. In order to expedite the cycle of planning, execution, and analysis amoung multiple research groups, standards for description, measurement, and procedure are necessary. In this work, the authors preset the coordinate systems used in their suite of computer-based tools for planning, executing, and evaluating the total hip replacement procedure. Rationales for the choices of each system are given along with experimental data which support the definitions.

1 Introduction

Total hip replacement (THR) is a very common and very successful procedure used to treat patient pain and immobility due to degenerative disease of the hip joint. Each year, about 250,000 hip replacement procedures are performed in the United States alone [1]. With such a large number of cases, even low percentages of failure are significant. Consistent failure of the prosthetic joint can require revision, which requires an extra surgical procedure, causing patient distress and incurring additional medical costs. Therefore, minimizing the risks of failure is critical.

Possible modes of failure include implant dislocation, excessive wear, loosening, and leg length discrepancy. All of these phenomena are affected by the positions of the prostheses in an artificial hip joint. Studies have shown a significantly lower risk of certain failure modes when the implants are placed within a "safe zone" [2]. However, patient-to-patient differences in anatomical structure and activity make general rules of implant placement a near impossibility.

Prior work done by the authors has attempted to lower the risk of failure by analyzing the THR procedure, recognizing its limitations, and providing a means to preoperatively plan [3], intraoperatively execute [4], and postoperatively evaluate [5] THR cases with a suite of computer-based tools known as HipNav. In this work, we describe the coordinate systems defined in HipNav and provide rationale for the choice of each.

2 Methods

2.1 Body Coordinate System

There is an inherent body coordinate frame defined by the transverse, mid-sagittal, and coronal planes of the patient. This coordinate frame is used as a global orientation reference. The body X-axis points to the left of the patient, perpendicular to the sagittal plane. The body Y- direction is normal to the transverse plane, pointing to the superior. All coordinate systems described in this work are right-handed coordinate systems, thereby defining the body Z-axis to lie normal to the coronal plane, pointing to the anterior (Figure 1).

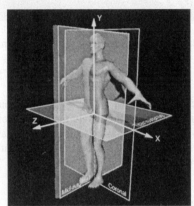

Fig. 1. Body coordinate system

2.2 Pelvic Coordinate System

The pelvic coordinate system is defined by four points on the pelvis: the maximally anterior left and right iliac spine points, and the maximally anterior left and right pubis symphysis points. The anterior spine points and the median of the pubis symphysis points define an anterior plane of the pelvis. This plane, along with the line between the iliac spine points, defines a coordinate system for the pelvis (Figure 2). The pelvic X-axis points to the patient's left, parallel with the line between the iliac spine points. The above. The pelvic Y-axis points to the superior, perpendicular to the other axes. The pelvic coordinate system is centered at the median of the iliac crest points, thereby making the definition rely wholly on the four landmark points.

It was hypothesized that the orientation of the APP in the average standing patient is roughly vertical, or parallel to the coronal (or here, the body XZ) plane. To test this hypothesis, lateral radiographs for 100 THR cases were analyzed. Rotation of the APP in the sagittal plane is easily visible in lateral radiographs. It was found that the standing position orientations showed a trend toward neutral alignment of the anterior pelvic plane, with a mean anterior rotation of 2 degrees relative to vertical (and a standard deviation of 8 degrees). We thereby can safely define "neutral standing position" (hereafter called simply "neutral position") of the pelvis to lie parallel with the body

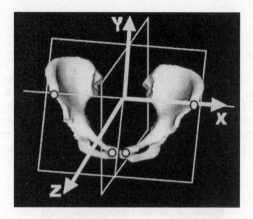

Fig. 2. Pelvic landmark points and the established coordinate frame

coordinate system. Any functional rotation (such as pelvic flexion) of the pelvis is defined relative to neutral position.

Measurements of APP orientation of sitting patients (mean posterior rotation of 35 degrees with a 14 degree standard deviation) were also done, reinforcing past studies [4, 6, 7, 8] stating the variability of functional pelvic position in patients both in and out of the operating room. This implies need of a definition of acetabular cup alignment that is independent of pelvic orientation.

2.3 Acetabular Component Coordinate System

The coordinate frame for the acetabular component is centered at the rotation center of the liner. The cup Z-axis points away from the liner opening and lies normal to the flat plane of the implant shell (Figure 3). If the implant liner is not axisymmetric (i.e. it has an extended lip), the Y-axis marks the center of the extension. The orientation of the cup is commonly quantified using one of three paramaterizations: anatomic, operative and radiographic.   The anatomic definition decribes the cup position in terms of cup abduction and anteversion.   Neutral cup position orients the cup's Z-axis pointing

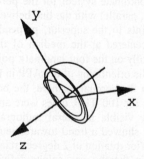

Fig. 3. Acetabular implant reference frame

Table 1. Parameterization values for identical cup orientations. As the secondary value increases (anteversion or flexion), the parameter values become less consistent.

Abduction/ Radiographic Version	Abduction/ Flexion	Abduction/ Anteversion
35 / 5	35 / 6	35 / 9
35 / 15	34 / 18	38 / 25
35 / 25	31 / 30	42 / 39
45 / 5	45 / 7	45 / 7
45 / 15	43 / 21	47 / 21
45 / 25	40 / 33	50 / 33
55 / 5	55 / 9	55 / 6
55 / 15	52 / 25	56 / 18
55 / 25	48 / 39	59 / 30

opposite the pelvic Y-axis, with the cup Y-axis pointing laterally, parallel to the pelvic X-axis. Abduction describes cup rotation from neutral around an axis through its center and parallel to the pelvic Z-axis (hereafter called pelvic Z-rotation). Anteversion in this case is a subsequent pelvic Y-rotation. The radiographic definition decribes the cup position in terms of cup abduction as above followed by cup anteversion, this time defined as cup rotation around the Y-axis of the *cup*. This definition is often used in postoperative radiographic studies because this definition of anteversion is easily measured via the view of the cup in the X-ray image, and is therefore commonly used in postoperative studies. The operative definition decribes the cup position in terms of cup abduction followed by cup "flexion" (pelvic X-rotation). Traditional mechanical cup positioning guides are set to follow the abduction/flexion alignment and are therefore preset to achieve planned alignment in terms of this operative definition. All three systems are combined with a "twist" value that represents rotation of the implant about the cup axis. This is required when studying effects of non-axisymmetric cup liners on postoperative range of motion. These parameterizations are all relative to the pelvic coordinate system, which is important for describing cup position independent of the varying functional position of the pelvis, but they are not interchangeable. As we can see in Table 1, as anteversion (or flexion) increases the parameterization values diverge significantly.

2.4 Femoral Coordinate System

Four landmark points define the femoral coordinate system: the center of joint rotation, the posterior lesser trochanter, and the posterior femoral condyle points. The posterior lesser trochanter and condyle points define a plane. (Figure 4) The projection of the vector from the condylar midpoint to the joint center in this plane defines a vector parallel to the mechanical axis of the femur. The Z-axis of the femoral frame is normal to the defined plane. The X-axis lies parallel to the plane and orthogonal to the mechanical axis. The Y-axis of the femoral frame is simply the resultant cross product. The femoral coordinate frame origin is located at the center of rotation for the femur. Therefore, because the Y-axis passes through the joint center (by definition) and, roughly, center of the knee joint, it lies along the mechanical axis of the femur.

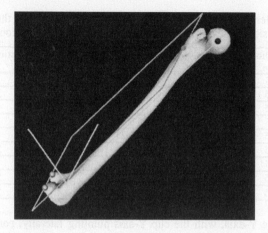

Fig. 4. Femoral landmark points which establish the femoral coordinate system

Note that because the femoral coordinate system is dependent on the center of rotation of the joint, changing the position of the joint center (which happens during THR) relative to the femur may change the orientation of the femoral frame. This change is most relevant in calculation of leg lengths and offsets, which are very important factors in determining proper component placement during THR. "Neutral position" of the femur is achieved when the coordinate frame of the femur lies parallel to the body coordinate frame. Any functional motion of the femur (discussed below in section 2.6) can be described as a rotation relative to the body coordinate system.

2.5 Femoral Component Coordinate System

The origin of the femoral implant neck coordinate frame is at the center of joint rotation, i.e. the center of the implant head. The orientation of the frame is parallel to the symmetry plane of the stem (Figure 5a). The Z-axis lies coincident with the axis of the implant neck, and the Y-axis is normal to the neck axis and in the symmetry plane

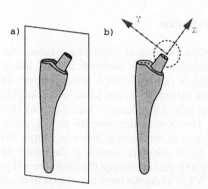

Fig. 5. a) Stem symmetry plane and b) neck coordinate frame

(Figure 5b). Orientation of the implant neck relative to the can be described as a pair of rotations in the femoral coordinate frame: abduction and anteversion (or retroversion). These orientation parameters are analogous to their namesakes in acetabular component placement, with abduction as a femoral Z-rotation from vertical, followed by anteversion (a femoral Y-rotation). By using this parameterization, studies can be conducted on the effect "combined anteversion", or the difference of anteversion values for the femoral component neck and acetabular component placements, on postoperative range of motion [9].

2.6 Function Leg Movement Definitions

For purposes of simulation and analysis, the six functional movements of the femur (flexion, extension, abduction, adduction, internal rotation and external rotation) have been given precise geometrical definitions that follow the descriptions presented by Greene and Heckman [10]. Mathematically, these motions are pure rotations in the various coordinate frames described in previous sections. Any motion of the femur can be described as an ordered set of these rotations.

Flexion and extension describe rotation of the femur about the center of joint rotation parallel to the body YZ plane. Internal and external rotation are defined as rotations about the mechanical axis of the femur (femoral Y-axis). (Figure 6).

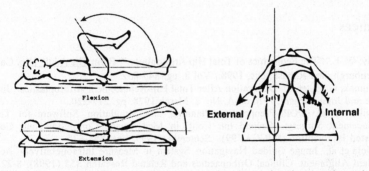

Fig. 6. Illustrations of femoral flexion, extension, internal and external rotation

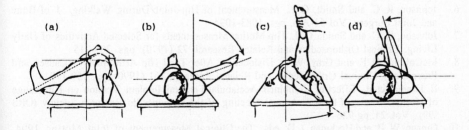

Fig. 7. Modes of abduction and adduction preceding and following flexion. (a) abduction, (b) abduction in flexion, (c) adduction, and (d) adduction in flexion

The geometric definitions of femoral abduction and adduction depend on their occurrence relative to flexion or extension in the same motion. Abduction and adduction not taking place after flexion or extension describe rotations parallel to the body XY plane. When flexion (or extension) occurs before the ab/adduction, the movements are referred to as *ab/adduction in flexion (or extension)*. In this case, they represent rotation parallel to the body XZ plane. (Figure 7).

3 Discussion

In this work, we have presented the anatomic and geometric definitions allowing quantification of results related to total hip replacement surgery. These definitions have been used in the HipNav computer-assisted THR system, and remain in use to allow repeatable results and ready application of information gained by the cases. It is important to rigorously define these coordinate systems and kinematic relationships when reporting results in order to eliminate confusion when comparing studies within and between research groups. In the future, standards for computer-assisted surgery in all disciplines will facilitate data integration allowing surgeons and engineers to improve medical practice and patient quality of life.

References

1. Healy, W. L. "The Economics of Total Hip Arthroplasty." From The Adult Hip. Callaghan, Rosenburgh, and Rubash, eds. 1998. Vol 2. pg. 845-852
2. Lewinnek, G. E., et al. Dislocation After Total Hip-Replacement Arthroplasties. Journal of Bone and Joint Surgery. Vol 60-A, No. 2, March 1978. pg. 217-220.
3. C. Nikou et al. POP: Preoperative Planning and Simulation Software for Total Hip Replacement Surgery. Second Int. Conf. in Med. Image Computing and Computer-Assisted Intervention (MICCAI'99). September 1999. pgs. 868-875
4. DiGioia et al. Image Guided Navigation System to Measure Intraoperatively Acetabular Implant Alignment. Clinical Orthopaedics and Related Research 355 (1998): 8-22
5. C. Nikou et al. CupAlign: Computer-Assisted Postoperative Radiographic Measurement of Acetabular Components Following Total Hip Arthroplasty. Second Int. Conf. in Med. Image Computing and Computer-Assisted Intervention (MICCAI'99). September 1999. pgs. 876-882
6. Johnson, R. C. and Smidt, G. L. Measurement of Hip-Joint During Walking. J. of Bone and Joint Surgery. Vol 51-A:6, pgs. 1083-1094
7. Johnson, R. C. and Smidt, G. L. Hip Motion Measurements for Selected Activities of Daily Living. Clinical Orthopaedics and Related Research 72 (1970): pgs. 205-215
8. McCollum, D. E. and Gray, W. J.: Dislocation After Total Hip Arthroplasty (Causes and Prevention). Clinical Orthopaedics and Related Research 261 (1990). pg. 159-170
9. B. Jaramaz et al. Effect of Combined Acetabular/Femoral Implant Version on Hip Range of Motion. Trans. of the 45th Annual Meeting of the Orthopaedic Research Society (ORS '99). Vol. 24. pg 926
10. Greene, W. B. and Heckman, J. D., eds. The Clinical Measurement of Joint Motion. 1994. pgs. 99-114

XIMIT – X-Ray Migration Measurement Using Implant Models and Image Templates

Kathrin Burckhardt and Gábor Székely

Computer Vision Laboratory,
Swiss Federal Institute of Technology, ETH Zürich,
[tribur|szekely]@vision.ee.ethz.ch
WWW home page: http://www.vision.ee.ethz.ch

Abstract. XIMIT is a newly developed method for precisely measuring the 2D migration of a hip socket implant using standard radiographs. 2D migration is the change of the 2D distance between implant and bone observable in the x-ray image. It is measured in order to judge the fixation of the implant in the bone. The proposed method is based on a carefully designed measurement of the 2D distance minimizing the influence of different sources of errors. The magnitude of the entering errors is minimized by estimating the implant's 3D orientation and position using the CAD model of the implant and by involving a state of the art template matching algorithm. According to the first error estimate, the new method makes a migration measurement with a standard deviation of about 0.2 mm possible.

1 Introduction

Many people suffer from coxarthrosis, the arthrosis of the hip joint. Often, the replacement of the hip by a bone implant is the only way to relieve the pain. About 400.000 artificial hip joints are implanted every year only in Europe. However, despite its success, hip replacement still involves complications. One major problem is the loosening of the cup, the acetabular part of the implant. The fixation of the cup in the bone depends on the design of the implant and on the implantation technique. If a certain kind of replacement has a high tendency to loosen, its extensive use on patients should be prevented. Therefore, the aim is to promptly and reliably find out the fixation properties of a certain cup or implantation technique.

Migration of the implant, which means the change of the cup's position in the bone, is interpreted as an objective sign for loosening. Especially early postoperative migration is supposed to give information about the (later) fixation of the implant in the bone [6, 10, 8]. In [10], it is recapitulated that the risk of a failed ingrowth is minimal if the migration is less than 1 mm and occurs only in the first year after surgery.

The various methods for measuring the migration can be classified into one 3D method and several 2D methods. The migration is assessed using time series of x-ray images, divided into reference image(s) and compare images. The methods are based on the observation of the change of the distance between the implant and certain bone-related landmarks over time. In the 3D method, RSA (Roentgen Stereophotogrammetric Analysis) [12, 6], all three components of this distance are determined. This is achieved by inserting about 7 metal markers into the bone and into the implant and simultaneously taking two radiographs from two different directions of the pelvis as well as of a calibration frame.

In the 2D methods, e.g. [11, 3, 13, 7], only the 2D distance, i.e. the components of the bone–implant distance parallel to the film plane, is observed. The 2D distance is measured in one single x-ray image, by determining a cup reference point and certain bone reference points or lines defined by bony structures. The used x-ray images are the standard radiographs of the hip replacement's follow-up study. The bony structures and the cup reference point are marked by hand using a pencil (in case of conventional radiographs) or by mouse-clicking on digitized radiographs. The magnification of the radiographic system is taken into consideration, either through the division of the 2D distance by the estimated magnification factor or through the reconstruction of the spatial situation at exposure [7]. In both cases the known radius of the head of the implant's femoral part and the measured size of this radius in the image are used. In [11], the 2D distance is corrected for pelvis rotation between the exposures by interpreting the varying distance between two bony structures in the x-ray images. In [7], the rotation is taken into consideration by evaluating only pairs of radiographs showing the pelvis in a similar orientation. To select these pairs also image distances between bony structures are compared.

In order to be useful for the predictive early postoperative migration, the migration measurement should have a precision lying in the submillimeter range. This is the case for RSA, of which a maximum standard deviation of 0.1 mm for the migration parallel to the film plane is reported [9, 6]. However, RSA is not suited for the clinical routine and thus for a large-scale examination because of the necessary implantation of markers and of the expensive technique required. The existing 2D methods, on the other side, do not require special preparation or equipment, but are probably too imprecise. Dickob *et al.* [3] evaluated radiographs showing a pelvis phantom with varying orientation to test their method and found a standard deviation of the measured migration of about 1.0 mm. Nunn *et al.* [11] and Sutherland *et al.* [13] roughly estimated the errors of their methods to be 3.0 mm and 2.5 mm, respectively. The precision of different 2D methods was theoretically estimated in [2] by considering all sources of error and by applying the principle of error propagation. The resulting standard deviations for the two components of the migration in [7], [11], [3] and [13] were found to lie in the range of 0.7 mm to 2.7 mm.

Measuring the migration using single standard radiographs is affected by the following well known sources of error: the variability of the extrinsic imaging parameters (here the position and orientation of the patient's pelvis at exposure), the unknown intrinsic parameters (here only the film–focus distance), and the lack of definite reference points in the image (here the implant reference point and bony landmarks). It is clear that, because of these sources of errors, it will hardly be possible to reach the precision of a stereo radiographic method using implanted markers. Still, 2D migration measurement can be improved. With XIMIT, its optimization is pursued in order to attain a method making the submillimeter measurement of implant migration under clinical conditions possible.

2 Minimizing the Error of 2D Migration Measurement

The method proposed here is based on two pillars: On the one side the design of a distance measurement using single radiographs which is as insensitive as possible towards the above mentioned sources of errors (Sect. 2.1). On the other side the reduction of the magnitude of the remaining relevant entering errors (Sect. 2.2). The latter is achieved by involving a 2D template matching algorithm to localize the bony structures and by using the known 3D model of the implant to determine its position and orientation at exposure (thus the acronym **X**-ray migration measurement using **I**mplant **M**odels and **I**mage **T**emplates).

2.1 Minimizing the Sensitivity towards the Sources of Error

Applying the principle of error propagation to different 2D methods for measuring migration, the partial derivatives of the measured 2D distance with respect to the sources of error were calculated [2]. These derivatives are measures for the significance of the different sources of error for the total precision. They depend on the design of the method, that means on the bony structures which are used, on the definition of the bone reference points or lines, etc. According to the conclusions in [2], the following general rules can be stated:

1. The sensitivity of the measured migration towards pelvis rotation between two exposures can be reduced by choosing bony landmarks with a small 3D distance (especially in sagittal direction) to the cup reference point.
2. The smaller the sagittal component of this 3D distance, the lower is the influence of the pelvis translation relative to the focus (i.e. the x-ray source).
3. Pelvis rotation around the sagittal axis can be neglected, if a bone reference system defined by two bony landmarks is used.
4. If the correction for magnification is considered for each x-ray image separately, the variability of the focus–film and of the focus–pelvis distance plays no role, and the influence of the rotation is reduced.

The design of the proposed method is based on these observations and on the existing migration measurement methods. It is illustrated in Fig. 1. Considering the Rules 1 and 2, the so-called teardrop figures were chosen as bony landmarks (m_1 and m_2). These figures are the projections of bony structures close to the acetabulum and with a similar sagittal position as its center. The cup reference point (**c**) is chosen be the cup center. The distance between cup and bone is represented in a coordinate system (e_x, e_y), which is defined by the line connecting the most caudal points of the teardrop figures and by the perpendicular of this line. Thus, the distance is not affected by rotation around the sagittal axis according to Observation 3. Taking Rule 4 into consideration, the measured distance is corrected for magnification through division by the magnification factor g, which is estimated for each exposure. The magnification factor corresponds to the ratio between focus–film and focus–implant distance and is determined as described in Sect. 2.2. The measured 2D distance indicated with $\mathbf{d}$ is given by

$$\mathbf{d} = \frac{1}{g} \begin{pmatrix} (\mathbf{c} - \mathbf{m}_1) \cdot \mathbf{e}_x \\ (\mathbf{c} - \mathbf{m}_1) \cdot \mathbf{e}_y \end{pmatrix} . \tag{1}$$

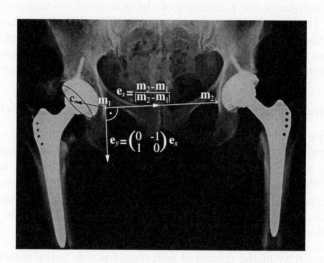

Fig. 1. The measured 2D distance in XIMIT

2.2   Minimizing the Magnitude of the Entering Errors

Using the 3D Model of the Implant. In order to precisely determine the image position of the cup and its distance to the focus, its 3D position is determined using its CAD model and the approximate focus–film constellation at exposure. The principle of "Analysis-by-Synthesis" is applied: the implant's position and orientation parameters are estimated by minimizing the difference between a synthesized and the original x-ray image of the cup.

The focus-film constellation is defined by the intrinsic parameters, which are the focus–film distance f and the intersection of the central beam with the film $\mathbf{t}$ (see Fig. 2). f is assumed to have the nominal value of the x-ray unit, and $\mathbf{t}$ is estimated to be the film center, because in the clinical practice the central beam is directed to this point.

In this focus-film system, the 3D orientation and position of the implant is first roughly estimated by the manual fitting of an ellipse to the circumference of the cup's equatorial plane. The ellipse is shown in Fig. 1. With this initial estimate the Analysis-by-Synthesis algorithm is started, searching the cup's 3D orientation and position where the difference between the synthetic and the original x-ray image of the cup is at minimum. The difference measure, chosen to be the sum of the squared intensity differences at each pixel, is minimized using a combined modified Newton and Gauss-Newton algorithm. The x-ray image is synthesized by calculating the absorption of the x-rays reaching from the assumed position of the focus to each pixel. The absorption depends on the x-rays' lengths inside the cup, which are gained by intersecting each ray with the cup's surface given by the CAD model.

To get the image position $\mathbf{c}$ of the cup center, the estimated 3D cup center is projected on the x-ray film plane using the same intrinsic parameters as used for Analysis-by-Synthesis. The magnification factor g is yielded by division of the film–focus distance by the estimated cup's z-position. Even if the assumed intrinsic parameters differ from the real ones at exposure, $\mathbf{c}$ and g are precisely determined, because the difference is automatically neutralized by an appropriate x-y- and z-translation of the cup during Analysis-by-Synthesis.

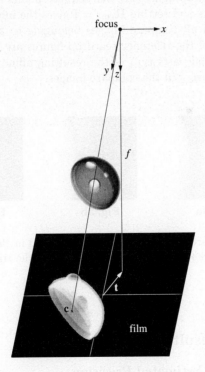

Fig. 2. Illustration of the spatial situation while synthesizing the cup's x-ray image

Locating the Bony Landmarks by Template Matching. Template matching is a general procedure to estimate the mapping between a template, i.e. a structured area in a reference image, and a patch, i.e. the corresponding area in a compare image. The mapping is expressed by a geometrical transformation (translation, rotation, scale, etc.) in the patch's space. The translational part of this transformation can be used to locate the patch in the compare image. In XIMIT, a least square matching (LSM) [1, 4] is applied, which means that the transformation parameters are estimated through a least square minimization.

Here, the template is the area containing a teardrop figure in a reference x-ray image. The patch is the area of the same teardrop figure in a compare image. The transformation working best is a translation plus rotation to compensate for rotation of pelvis around the sagittal axis. It is represented by $\mathbf{x}'_i = \mathbf{R}_{\mathrm{LSM}}\mathbf{x}_i + \mathbf{t}_{\mathrm{LSM}}$, where $\mathbf{x}_i$ and $\mathbf{x}'_i$ indicate the original and the transformed coordinates of pixel i, $\mathbf{t}_{\mathrm{LSM}}$ the translation, and $\mathbf{R}_{\mathrm{LSM}}$ the rotation matrix. The transformation's parameters are found by minimizing the sum of the squared differences between the intensities of the template at $\mathbf{x}_i$ and of the patch at $\mathbf{x}'_i$.

The template is manually defined by marking in the reference radiograph the area of the teardrop figures which is relatively independent from the pelvis orientation. In the reference as well as in the compare x-ray image, the most caudal points of the teardrop figures are marked (see points in Fig. 3a and Fig. 3b). Subtracting the points and setting $\mathbf{R}_{\mathrm{LSM}} = \mathbf{I}$ gives the initial estimate for LSM. For the evaluation of all compare images belonging to a reference image, the once marked points of the reference teardrop figures are kept. Adding to these points the translational parts $\mathbf{t}_{\mathrm{LSM}}$ of the resulting affine transformations gives the points' positions $\mathbf{m}_{1,2}$ in the compare images.

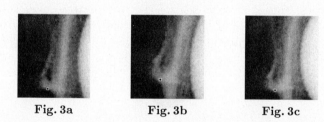

| Fig. 3a | Fig. 3b | Fig. 3c |

Fig. 3. The marked teardrop figure in the reference and in the compare x-ray image (Fig. 3a and 3b), and the compare image warped with the transformation estimated using LSM (Fig. 3c).

3 Tests and Results

3.1 Theoretically Estimated Precision

The scheme of theoretically estimating the standard deviation worked out in [2] was applied to (1). The following sums of error components resulted, in which the coefficients are the partial derivatives of **d** with respect to the variables representing the sources of error:

$$\sigma_{d_x}^2 = (0.01806\sigma_{\delta\alpha})^2 + (-0.00098\sigma_{\delta\beta})^2 + (-0.00277\sigma_{\delta\gamma})^2 + (0.00264\sigma_{\delta T_x})^2 + (-0.00020\sigma_{\delta T_z})^2 +$$
$$(0.79091\sigma_{\delta c_x})^2 + (-0.79091\sigma_{\delta m_{1x}})^2 + (-0.08648\sigma_{\delta m_{1y}})^2 + (0.08648\sigma_{\delta m_{2y}})^2 +$$
$$(-26.47220\sigma_{\delta g})^2 + (-0.00020\sigma_{\delta f})^2 \qquad (2)$$

$$\sigma_{d_y}^2 = (-0.04433\sigma_{\delta\alpha})^2 + (-0.00264\sigma_{\delta T_y})^2 + (-0.00001\sigma_{\delta T_z})^2 +$$
$$(-0.79091\sigma_{\delta c_y})^2 + (0.99603\sigma_{\delta m_{1y}})^2 + (-0.20512\sigma_{\delta m_{2y}})^2 + (-11.16060\sigma_{\delta g})^2 + (-0.00001\sigma_{\delta f})^2$$

With the same notation as in [2], $\sigma_{\delta\alpha,\delta\beta,\delta\gamma}$, $\sigma_{\delta T_{x,y,z}}$, and $\sigma_{\delta f}$ indicate the standard deviation of the pelvis orientation[1], of its position, and of the film-focus distance, respectively. The error in locating the cup center and the teardrop figures and the one in determining the correction for magnification are represented by $\sigma_{\delta c_{x,y}}$, $\sigma_{\delta m_{1,2x,y}}$, and $\sigma_{\delta g}$.

For comparison of the theoretical estimate with the experimental results, $\sigma_{\delta T_{x,y}}$ and $\sigma_{\delta\alpha,\delta\beta}$ is derived using the pelvis position and orientation parameters in the validation exposures (see next section). Calculating the standard deviations of these parameters listed in the middle column of Table 1 yields $\sigma_{\delta T_x} = 22\,\text{mm}$, $\sigma_{\delta T_y} = 27\,\text{mm}$, $\sigma_{\delta\alpha} = 4.1°$, and $\sigma_{\delta\beta} = 2.3°$. From [2], the values $\sigma_{\delta\gamma} = 3.0°$, $\sigma_{\delta T_z} = 15\,\text{mm}$, and $\sigma_{\delta f} = 100\,\text{mm}$ were adopted. LSM and Analysis-by-Synthesis were run ten times with a new definition of the template and of the initial estimates. The standard deviations $\sigma_{\delta m_{1,2x,y}} = 0.07\,\text{mm}$ and $\sigma_{\delta c_{x,y}} = 0.015\,\text{mm}$, and $\sigma_{\delta g} = 0.0008$ were derived from these tests made using several radiographs of two different pelves.

Regarding the coefficients in (2) and considering the size of these standard deviations, the sources of error having a significant influence are: the pelvis rotation around the x-axis, the error in locating the cup and the bony landmarks, the error of the correction for magnification, and less importantly also the variable pelvis x-y-position. Inserting these standard deviations in (2), the theoretical error of the measured 2D distance is: $\sigma_{d_x} = 0.11\,\text{mm}$, $\sigma_{d_y} = 0.21\,\text{mm}$.

3.2 Experimental Validation

XIMIT was validated using x-ray images of a pelvis phantom equipped with a cup and mounted on a cage. The phantom-cage system allows the simulation of implant migration in the bone and of pelvis translation in the film-focus system with a precision of about 0.05 mm. It also makes the rotation of the pelvis with a precision of about 0.05° possible.

In the first experiment, the pelvis was translated along the x- as well as along the y-axis, keeping the cup–bone distance constant. Then, the orientation of the pelvis was varied by rotation, together with the cup, around these two axes. Finally, exposures were made at different cup distances to the bone, keeping the

---

[1] $\delta\alpha$, $\delta\beta$, and $\delta\gamma$ represent the random angles of rotation around the x- (the medio-lateral), the y- (the cranio-caudal), and the z- (the sagittal) axis

position and orientation of the pelvis constant. The radiographs were scanned with a pixel resolution of 300 dpi and a gray value resolution of 12 bit.

In the Tables 1 and 2, the measured migration values are listed, i.e. the differences $\Delta\mathbf{d} = (\Delta d_x, \Delta d_y)$ between the bone–implant distance $\mathbf{d}$ measured in the reference radiograph and in the ones made under changed conditions. Their standard deviations from groundtruth are also shown.

The standard deviations in the tables mainly agree with the theoretical error estimate: From (2), similar values result, if the standard deviations $\sigma_{\delta\alpha}$, $\sigma_{\delta\beta}$, $\sigma_{\delta T_x}$, or $\sigma_{\delta T_y}$ corresponding to the unchanged position or orientation parameters are set to zero. Only the errors of d_x at a x-translation and especially the one at a rotation around the y-axis significantly differ from the theoretical result, because these motions influenced the shape of the teardrop figures. The values in Table 2 show that the measured 2D distance change in fact corresponds to the projection of the implant migration on the x-y-plane.

Table 1. Measured migration while translating and rotating the pelvis

pelvis motion	Δd_x [mm]	Δd_y [mm]
groundtruth: no migration	0.0	0.0
translation in x-direction		
-2.4 cm	0.37	0.01
-1.2 cm	0.19	-0.13
1.2 cm	-0.15	-0.06
2.4 cm	-0.16	-0.05
standard deviation	0.26	0.05
translation in y-direction		
-3.0 cm	0.01	-0.11
-1.5 cm	0.01	-0.07
1.5 cm	0.06	-0.01
3.0 cm	0.03	0.06
standard deviation	0.03	0.07
rotation around x-axis		
$-4.39\,°$	-0.00	-0.34
$-2.19\,°$	-0.09	-0.08
$2.19\,°$	0.21	-0.04
$3.29\,°$	0.04	0.33
$5.47\,°$	0.32	0.03
standard deviation	0.16	0.24
rotation around y-axis		
$-2.19\,°$	0.41	0.06
$-1.09\,°$	0.16	-0.06
$1.09\,°$	-0.18	-0.02
$2.19\,°$	-0.20	-0.01
$3.29\,°$	-0.48	0.04
standard deviation	0.35	0.05

Table 2. Measured migration at constant position and orientation of pelvis

	groundtruth		measurement	
	Δd_x [mm]	Δd_y [mm]	Δd_x [mm]	Δd_y [mm]
	0.0	0.0	−0.13	0.00
	0.0	0.0	0.02	−0.13
	0.0	0.0	−0.04	−0.08
	0.60	0.72	0.52	0.78
	1.20	1.44	1.08	1.57
	1.80	2.17	1.82	2.12
standard deviation			0.09	0.07

4 Conclusions

According to the above results, XIMIT substantially improves implant migration measurement using standard radiographs. It makes a 2D migration measurement of artificial hip sockets in the submillimeter range possible. However, achieving this precision requires the restriction of pelvis rotation between two exposures to about ±3° and ±2° around the medio-lateral and the cranio-caudal axis, respectively.

This limitation is necessary mainly because pelvis rotation alters the bony reference structures in the image, i.e. the most caudal points of the teardrop figures. At a variable orientation, the exact identification of these bony landmarks in all images is difficult. Another problem of natural bony landmarks may be the interobserver variability of their location in the image, which is not tested up to the present. Despite its automation using a template matching algorithm, the location may be influenced by the user as it depends on the definition of the template. Because of these two reasons, implanted markers as bone reference points instead of natural bony landmarks would improve XIMIT.

The pelvis rotation between two exposures may be restricted by putting a pad under the patients knees, by repeating exposures not comparable to the preceding ones, or by omitting for evaluation the follow-up series' radiographs which show the pelvis in an extraordinary orientation. These possibilities require a measure for pelvis rotation. Additional studies are planned to find such a measure by evaluating the appearance of certain bony structures. For example, the alteration at certain rotation angles of the position of the coccyx relative to the symphysis pubis or of the shape of the foramina obturatoria can be investigated. Different from the pelvis bone, the orientation of the implant is automatically estimated in XIMIT by applying Analysis-by-Synthesis. Therefore, having a quantitive measure for pelvis orientation also the rotation of the cup relative to the bone, intended to be measured already in [5], could be determined.

The proposed method is well suited for the measurement of implant migration integrated in the clinical routine. It makes reliable large-scale examinations of the fixation properties of artificial hip sockets possible. At the same time the basic algorithms can be well generalized to other anatomical regions and provide a good approach for migration measurement, for example, of knee or of shoulder implants.

Acknowledgment

This work has been supported by the grant No. 9803 of the Synos Foundation, Münsingen-Bern, Switzerland

References

[1] M. Berger. The framework of least squares template matching. Technical Report 180, Communication Technology Lab, Computer Vision Group, ETH Zürich, 1998.

[2] K. Burckhardt, Ch. Brechbühler, Ch. Gerber, J. Hodler, and Gabor Székely. Precision of distance determination using 3d to 2d projections: The error of migration measurement using x-ray images. *MedIA*, in press.

[3] M. Dickob, J. Bleher, and W. Puhl. Standardisierte Pfannenwanderungsanalyse in der Hüftendoprothetik mittels digitaler Bildverarbeitung. *Unfallchirurg*, 97:92–97, 1994.

[4] A Grün. Adaptive least squares correlation: A powerful image matching technique. *South African Journal of Photogrammetry, Remote Sensing & Cartography*, 14(3):175–187, 1985.

[5] B. Jamaraz, C. Nikou, T.J. Levison, A.M. DiGioia III, and R.S. LaBarca. CupAlign: Computer-Assisted Postoperative Radiographic Measurement of Acetabular Components Following Total Hip Arthroplasty. In *Proc. of MICCAI*, 1999.

[6] J. Kärrholm, P. Herberts, P. Hultmark, H. Malchau, B. Nivbrant, and J. Thanner. Radiostereometry of Hip Prostheses: Review of Methodology and Clinical Results. *Clin Orthop*, November(344):94–110, 1997.

[7] M. Krismer, R. Bauer, J. Tschupik, and P. Mayrhofer. EBRA: a method to measure migration of acetabular components. *J Biomechanics*, 28:1225–1236, 1995.

[8] M. Krismer, B. Stöckl, M. Fischer, R. Bauer, P. Mayrhofer, and M. Ogon. Early migration predicts late aseptic failure of hip sockets. *J Bone Joint Surg*, 78-B426, 1996.

[9] B. Mjöberg, G. Selvik, L.I. Hansson, R. Rosenqvist, and R. Önnerfält. Mechanical loosening of total hip prostheses: A radiographic and roentgen stereophotogrammetric study. *J Bone Joint Surg*, 68-B:646–652, 1986.

[10] K.G. Nilsson and J. Kärrholm. RSA in the assessment of aseptic loosening. *J Bone Joint Surg [Br]*, 78-B(1):1–2, 1996.

[11] D. Nunn, M.A.R. Freeman, P.F. Hill, and S.J.W. Evans. The measurement of migration of the acetabular component of hip prosthesis. *J Bone Joint Surg [Br]*, 71:629–631, 1989.

[12] G. Selvik. Roentgen Stereophotogrammetric Analysis. *Acta Radiologica*, 31:113–126, 1990.

[13] C.J. Sutherland, A.H. Wilde, L.S. Borden, and K.E. Marks. A Ten-Year Follow-up of One Hundred Consecutive Müller Curved-Stem Total Hip-Replacement Arthroplasties. *J Bone Joint Surg*, 64-A(7):970–982, 1982.

Construction of 3D Shape Models of Femoral Articular Cartilage Using Harmonic Maps

A. D. Brett and C. J. Taylor

Division of Imaging Science, Stopford Building, Oxford Road,
University of Manchester, Manchester M13 9PT, UK
{a.brett,ctaylor}@man.ac.uk

Abstract. A previous publication has described a method of pairwise 3D surface correspondence for the automated generation of landmarks on a *set* of examples from a class of shape [3]. However, that method did not guarantee a diffeomorphic correspondence between examples. This affected the model compactness (the ability of the model to capture shape variation in a small number of parameters) and model specificity (the fact that the model will describe shapes only within the class used for training). In this paper we describe a method of generating the pairwise correspondences using piecewise-linear harmonic maps of the surfaces which is constrained to be diffeomorphic. In particular, we are interested in producing shape models of articular cartilage. In general these models will be close to being planar discs which makes the use of harmonic mapping particularly suitable for our application. An example statistical model built using this new correspondence method is shown for the human femoral articular cartilage; a complex biological shape which demonstrates considerable variation between individuals.

1 Introduction

The Active Shape Model (ASM) has been shown to provide an accurate and robust segmentation of medical images in 3D [9]. However, construction of these models remains a difficult problem. In 2D, this construction involves the manual identification of a set of L *landmarks* $\{\mathbf{x}_i; (1 \leq i \leq L)\}$ for each of N training examples of a class of shapes. A landmark is a point which identifies a salient feature of the shape and which is present on every example of the class. Manual definition of landmarks on 2D shape has proved to be both time-consuming and subjective. The interactive identification of landmarks in 3D images or on pre-segmented surfaces is considerably more difficult and time-consuming than in the 2D case. In particular, we are interested in producing shape models of articular cartilage, the 3D surface of which has been defined by hand as a set of planar contours. Each contour represents the cartilage outline as it appears in a single 2D slice of a 3D sagittal MR image of the knee. The vertices of these sets of contours comprise the pointsets $\{D_{i,j}; (1 \leq i \leq N), (1 \leq j \leq n_{D_i})\}$. The connectivity of each pointset is generated from its contour representation using the algorithm of Geiger [8]. Our motivation for this is to be able to produce corresponded thickness maps [4] of pre-segmented cartilage from 3D MR images.

A previous publication [3] described a possible solution to the problem of automatic 3D model building. However, the resultant shape model showed folding of the surface due to some groups (triples) of landmarks being re-ordered between training examples. This was as a result of the automated landmarking algorithm not producing a *diffeomorphism* between the corresponded pairs of shape examples. A diffeomorphism is a mapping which is continuous, one-to-one, onto and differentiable. The effect of this surface folding is on the model compactness (the ability of the model to capture shape variation in a small number of parameters) and model specificity (the fact that the model will describe shapes only within the class used for training). A shape model with a surface which may fold will be both less compact and less specific. Here, we describe a method of generating point correspondences between triangulated surfaces which ensures that the correspondence is diffeomorphic. This method may be incorporated into an algorithm, which is also described here, to produce the corresponded *set* of examples required to build an ASM.

2 Related Work

Kambhamettu and Goldgof [11] and Benayoun *et al* [1] both propose methods of surface correspondence based on the minimisation of a cost function which involves the difference in the curvature of the surfaces. As pointed out by Tagare [17], curvature is a rigid invariant of shape and its applicability to general non-rigid correspondence is problematic.

Several methods of 3D shape registration have been applied to the problem of building statistical models by producing point correspondences across a training set. Joshi *et al* [10] deform a template onto hippocampal surface representations using the registration method of Christensen *et al* [5]. This non-rigid registration uses a course linear elastic matching of volumes followed by refinement by a viscous fluid transform. However, this technique is computationally expensive, requiring a massively parallel computer to solve the partial differential equations of the fluid model. Fleute and Lavalée [7] use an framework of initially matching each training example to a single template, building a mean from these matched examples, and then iteratively matching each example to the current mean and repeating until convergence. Matching is performed using the multi-resolution registration method of Szeliski and Lavalée [16]. This method deforms the volume of space embedding the surface rather than deforming the surface itself. Szekély *et al* [15] parameterise the surfaces of each of their shape examples using the method of Brechbühler *et al* [2]. Correspondence may then be established between surfaces but relies upon the choice of a parametric origin on each surface mapping and registration of the coordinate systems of these mappings by the computation of a rotation.

3 Harmonic Maps

A harmonic mapping $h : D \mapsto P$ uniquely maps a (triangulated) topological disk D in $\mathbb{R}^3$ to a polygonal region in $\mathbb{R}^2$ whilst minimizing *metric dispersion*. Metric dispersion is a measure of the extent to which small regions in D are stretched in the mapping. In addition, the mapping is independent of the triangulation of D, is differentiable on each face of D and is an *embedding* i.e. it is one-to-one, containing no 'folds' in the planar surface. Because $h : D \mapsto P$ is an embedding, the inverse h^{-1} is a parameterisation of D over P. If we imagine that D is constructed from a set of triangular rubber sheets, the harmonic map minimises the total energy E_h of this set of rubber sheets.

We construct a piecewise-linear (PL) approximation to a harmonic map using a method described by Eck *et al* [6]. This method uses an explicit integration of the function E_h over each face to reinterpret the problem as the minimisation of the energy of a set of springs, one placed along each edge of D. First we form a polygon P by mapping the boundary vertices of D onto a circle in $\mathbb{R}^2$. The vertices of P are positioned such that the polygon sides subtend angles proportional to the arc lengths of the boundary segments of D. The positions of the rest of the vertices are calculated so that the total energy of the configuration E_h is minimised. E_h can be interpreted as the energy of a configuration of springs, each spring corresponding to one edge of D:

$$E_h(\mathbf{v}) = \frac{1}{2} \sum_{\{i,j\} \in \mathrm{edges}(D)} \kappa_{i,j} \| \mathbf{v}_i - \mathbf{v}_j \|^2 \qquad (1)$$

where $\mathbf{v}_i$ is the position of the ith vertex of D. The spring constants $\kappa_{i,j}$ are computed as:

$$\kappa_{i,j} = \frac{(l_{i,k_1}^2 + l_{j,k_1}^2 - l_{i,j}^2)}{A_{i,j,k_1}} + \frac{(l_{i,k_2}^2 + l_{j,k_2}^2 - l_{i,j}^2)}{A_{i,j,k_2}} \qquad (2)$$

where $l_{i,j}$ is the length of the edge $\{i,j\}$ and $A_{i,j,k}$ is the area of the face $\{i,j,k\}$, both measured in D. Each interior edge $\{i,j\}$ is incident to two faces $\{i,j,k_1\}$ and $\{i,j,k_2\}$, the formula for spring constants associated with boundary edges has only one term.

Although $\kappa_{i,j}$ can assume negative values, function (1) is positive definite and its unique minimum can be found by solving the sparse linear least-squares problem $\nabla E_h = 0$. Following the treatment of Kanai [12], we represent the problem in the form:

$$E_h = \mathbf{V}^{\mathrm{T}} \mathbf{K} \mathbf{V} \qquad (3)$$

where $\mathbf{V} = (v_{1x}, v_{1y}, v_{2x}, v_{2y}, \dots, v_{nx}, v_{ny})$. This is a vector describing all the positions of the vertices in the map. This variable vector can then be separated into two parts, a free part $\mathbf{V}_\alpha$ and a fixed part $\mathbf{V}_\beta$ representing the boundary vertices. The matrix $\mathbf{K}$ is also divided accordingly:

$$E_h = \begin{bmatrix} \mathbf{V}_\alpha^{\mathrm{T}} & \mathbf{V}_\beta^{\mathrm{T}} \end{bmatrix} \begin{bmatrix} \mathbf{K}_{\alpha\alpha} & \mathbf{K}_{\alpha\beta} \\ \mathbf{K}_{\beta\alpha} & \mathbf{K}_{\beta\beta} \end{bmatrix} \begin{bmatrix} \mathbf{V}_\alpha \\ \mathbf{V}_\beta \end{bmatrix}. \qquad (4)$$

The minimisation of this energy term has only to be solved over the free parts $\mathbf{V}_\alpha$:

$$\nabla E_h = \frac{\partial E_h}{\partial \mathbf{V}_\alpha} = \mathbf{K}_{\alpha\alpha}\mathbf{V}_\alpha + \mathbf{K}_{\alpha\beta}\mathbf{V}_\beta = 0. \tag{5}$$

Although this is a sparse system, our triangulations and hence $\mathbf{K}$ are not large and we can solve the system using a Cholesky decomposition of the matrix - making this a computationally efficient algorithm. Examples of embedded mapping are shown in Figure 1. The mappings of these triangulated meshes took $\sim 2 - 3$ secs on a 450 MHz P-II.

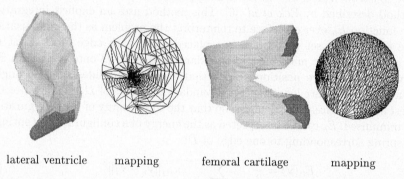

| lateral ventricle | mapping | femoral cartilage | mapping |

Fig. 1. A piecewise linear harmonic mapping of a triangulated surface of the frontal part of the left ventricle of the brain and of the upper surface of the femoral articular cartilage. In each case, on the left is a shaded representation of the original dense triangulation with approximately 1000 vertices, on the right is the embedded mapping on the unit circle.

4 Pairwise Surface Correspondence

Point correspondences between pairs of surfaces D_1 and D_2 may be generated by using their PL-harmonic maps P_1 and P_2 in a two-stage algorithm:

1. The two triangulated surfaces D_1 and D_2 are registered using a symmetric version of the ICP algorithm which operates on sparse, decimated versions of the surface triangulations. The sets of closest points used for this registration are then used to find the rotation which best registers the PL-harmonic maps P_1 and P_2.
2. The positions of the vertices on P_1 are projected onto P_2 to give a set of corresponding points on D_1 and D_2. Vertex positions on P_1 are mapped directly to D_1, the projected vertex positions on P_2 are mapped onto D_2 using a barycentric coordinate parameterisation of each triangle.

These two steps will now be described in more detail.

4.1   Registration of maps

We have used a *symmetric* version of the Iterative Closest Point (ICP) algorithm [3] to establish registration of the sparse pointset $\{D''_{1,i}\}$ with the dense pointset $\{D_{2,i}\}$ and of the sparse pointset $\{D''_{2,i}\}$ with the dense pointset $\{D_{1,i}\}$. The symmetry of this algorithm means that registration is dependant upon matching the salient features (as defined by the decimation algorithm) of *both* the triangulated surfaces.

Various metrics can be used to define the *closest* distance between point pairs. We weight the squared distance between points $\|X_i - Y_j\|^2$ by a factor of $2/(\hat{\mathbf{n}}_{X_i} \cdot \hat{\mathbf{n}}_{Y_j})$ where $\hat{\mathbf{n}}_{X_i}$ is the unit surface normal on X at point i. This encourages the correspondence of points on the surfaces which are topographically equivalent. We label the closest points to $D''_{1,i}$ from D_2 as the pointset $\{D'_{2,i}\}$, and the closest points to $D''_{2,i}$ from D_1 as the pointset $\{D'_{1,i}\}$.

We minimise the registration error such that the pose Q is determined which satisfies:

$$E_D^2 = \min_Q \left[\frac{1}{n_{D''_1}} \sum_{i=1}^{n_{D''_1}} \|Q(D''_{1,i}) - Q^{-1}(D'_{2,b_i})\|^2 + \right.$$

$$\left. \frac{1}{n_{D''_2}} \sum_{j=1}^{n_{D''_2}} \|Q(D'_{1,a_j}) - Q^{-1}(D''_{2,j})\|^2 \right] \tag{6}$$

where $Q(\mathbf{p}) = s\mathbf{R}\mathbf{p} + \mathbf{t}$, s is a scale factor, $\mathbf{R}$ is a rotation matrix and $\mathbf{t}$ is a translation. The integer values a_j index the vertices of D'_1 which are closest to each of the vertices of D''_2 indexed by j, and the integer values b_i index the vertices of D'_2 which are closest to each of the vertices of D''_1 indexed by i. The geometric information of each surface is first normalised such that the centre-of-gravity is at the origin and the mean distance of the points from the origin is 1. This provides a necessary initial approximation as input to the ICP algorithm which speeds convergence and makes the final solution more robust.

Our sparse triangulation generation algorithm we use makes use of a decimation method described by Schroeder *et al* [14]. However, we use a different distance metric which we have found to better preserve sharp edges and thin structures. The distance metric, L, is computed using Schroeder's distance to mean plane measure as:

$$L(\mathbf{v}_0) = |d(\mathbf{v}_0) - d'(\mathbf{v}_0)| \tag{7}$$

where $d(\mathbf{v}_0)$ and $d'(\mathbf{v}_0)$ are the *signed* distances of the vertex $\mathbf{v}_0$ to the mean plane of the triangle loop before and after decimation i.e. $d(\mathbf{v}_0) = \hat{\mathbf{u}} \cdot (\mathbf{v}_0 - \bar{\mathbf{x}})$, where $\hat{\mathbf{u}}$ is the unit normal to the mean plane of the triangle loop and $\bar{\mathbf{x}}$ is its centroid. See Figure. 2.

The pointset pairs $\{D''_{1,i}, D'_{2,i}\}$ and $\{D''_{2,i}, D'_{1,i}\}$ are all positions of vertices on D_1 and D_2 and hence can be mapped directly to the vertices of P_1 and P_2 as a single pointset pair $\{P'_{1,i}, P'_{2,i}\}$ where $P'_1 \subset P_1$ and $P'_2 \subset P_2$. The maps are

Fig. 2. Result of applying the decimation algorithm to a triangulated surface of the left ventricle of the brain. On the left is a shaded representation of the original dense triangulation with approximately 2000 vertices. On the right the same surface represented by 200 vertices (decimated by 90%).

now registered by finding the rotation $\mathbf{R}$ which minimises the term:

$$E_P^2 = \min_{\mathbf{R}} \sum_{i=1}^{n_{P_1'}} \|P_{1,i}' - \mathbf{R}(P_{2,i}')\|^2 \tag{8}$$

4.2 Generating point correspondences

Once the PL-harmonic maps have been registered, we can use them to give corresponding parameterisation of D_1 and D_2. For each vertex on D_1, the mapping of this vertex is found on P_1 and its position is projected onto the registered version of P_2. The projection will be in one of three positions with respect to P_2.

1. Coincident with a vertex position on P_2. In this case, the mapping to D_2 is to the vertex position on that surface.
2. Off the edge of the polygon P_2. The projected point position is simply projected onto the closest boundary edge of P_2. The mapping onto D_2 is then made using the positions of the two vertices $\mathbf{v}_1$ and $\mathbf{v}_2$ which define this closest edge and by the parameterisation of this edge, $s \in [0, 1]$, so that the mapped position is: $\mathbf{v}_m = \mathbf{v}_1 + s(\mathbf{v}_2 - \mathbf{v}_1)$ on D_2.
3. Within a triangle on P_2. In this case, the mapping onto D_2 is made using the positions of the three vertices defining the triangle and three barycentric coordinates which are used to parameterise the triangle so that the point position on D_2 is given by: $\mathbf{v}_m = u\mathbf{v}_1 + v\mathbf{v}_2 + w\mathbf{v}_3$, $u + v + w = 1$.

This algorithm generates a set of point correspondence pairs between vertex positions of D_1 and positions on the triangulated surface of D_2. By matching each example of the training set $\{D_i; (2 \leq i \leq N)\}$ to D_1, we can produce a *set* of corresponding pointsets. We can produce corresponding surfaces by combining the triangulated connectivity of D_1 with each of these pointsets.

5 Results

In order to test the pairwise correspondence algorithm, we have constructed pairs of merged 'mean' shapes, see Figure 3. The algorithms used are computationally efficient, the decimation of these ~ 600 vertex triangulations to 10 % of their original vertices takes $\sim 5-6$ secs. and the matching phase takes ~ 20 sec. on a 450 MHz P-II. A quantitative assessment of the matching and merging of two shape examples is difficult because the 'mean' of two unmatched shape examples is not defined. However, a 'good' merging should preserve the salient features of each example and produce a smooth interpolation of the representation between these. The mean shape can be generated in two ways; by mapping the vertices of D_1 onto D_2 or by mapping the vertices of D_2 onto D_1. In each case, the final shape is produced by taking the mean of the point positions of the corresponded pair and combing these with the connectivity description (triangulation) of D_1 or D_2 respectively. In Figure 3, can we can see that the means generated in these two ways are very similar.

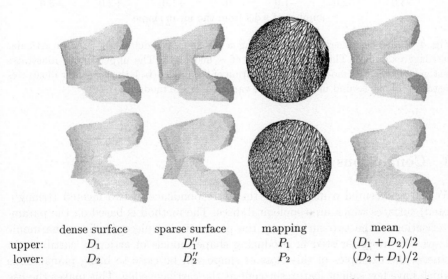

	dense surface	sparse surface	mapping	mean
upper:	D_1	D_1''	P_1	$(D_1 + D_2)/2$
lower:	D_2	D_2''	P_2	$(D_2 + D_1)/2$

Fig. 3. Building a 'mean' shape by merging two examples using surface correspondence based on a PL-harmonic mapping. The dense examples are the upper surface of the femoral articular cartilage, defined by a triangulation of ~ 600 vertices. The sparse, decimated version are defined by ~ 60 vertices. The mean shapes are constructed using mappings of all the vertices of D_1 on D_2 (upper) and mappings of all the vertices of D_2 on D_1 (lower).

We have also generated a 3D statistical model from six complex biological shapes - the human femoral articular cartilage. These have been defined by hand as contours on a series of 2D slices from 3D Magnetic Resonance images. The example shapes consisted of ~ 600 vertices. The first two modes of variation of this model are illustrated in Fig. 4, b_1 explains 50 % of the total variation, and b_2 explains 22 %. Upon inspection, we can see that the shape instances generated are legal and accurately reflect the shape variation present in the training set of six examples - indicating an accurate automated placing of landmark points. There are no tears or folds in the triangulated surfaces of the shape instances which can be caused by crossing or folding of the correspondences between examples.

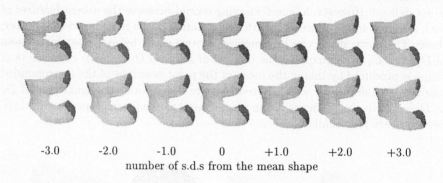

| -3.0 | -2.0 | -1.0 | 0 | +1.0 | +2.0 | +3.0 |

number of s.d.s from the mean shape

Fig. 4. Shape instances generated using a 3D shape model of six femoral articular cartilage examples. The model consists of ~ 600 points.The upper row is illustrates instances of the first mode of shape variation of the model (b_1), the lower row illustrates instances of the second mode of shape variation of the model (b_2).

6 Conclusions

We have presented a method for the correspondence of two faceted (triangulated) surfaces which are topological discs. The method is based on the parameterisation of the two surfaces by the production of piecewise-linear harmonic maps. We are interested in producing shape models of articular cartilage. In general the examples of this class of shape will be close to being planar discs which have few salient features except at the cartilage edge. This makes the use of harmonic mapping particularly suitable for our application since we assume correspondence between these edges which are the boundaries of our example triangulations.

The harmonic maps are registered by a symmetric form of the ICP algorithm to produce corresponded parameterizations of the pair of surfaces. The use of these maps in the production of corresponding pointset pairs produces a correspondence between the two triangulated surfaces which is diffeomorphic. This pairwise correspondence can be successively applied to a training set to produce a set of landmarked surface examples. We have presented sets of shape instances generated from a 3D shape model built from such a set of landmarks for a set of complex biological shapes. These models have been shown to be compact and to generate legal instances of the set of shapes. We are currently constructing and testing 3D ASMs built using these automatically generated 3D shape models.

We are aiming to use the harmonic mappings of triangulated surfaces to extend the work on automatic statistical model construction described by Kotcheff and Taylor [13]. In that work, a genetic algorithm was used to build 2D statistical models of classes of shape by minimising an objective function based upon the compactness of the resulting model. The optimisation progressed by re-parameterising the boundaries of each of the example shapes to produce a new set of corresponding boundary descriptions at each iteration of the algorithm. Our proposed extension of this work to 3D uses a PL-harmonic mapping of each example as an initial parameterisation of the surface. The model optimisation process is then one of re-parameterising each surface to refine the resulting model.

7 Acknowledgements

The authors would like to Dr. Andrew Stoddart for his contribution to the ideas described here. This work was funded by AstraZeneca Pharmaceuticals, Alderley Park, Macclesfield, Cheshire, UK.

References

1. A. Benayoun, N. Ayache, and I. Cohen. Adaptive meshes and nonrigid motion computation. In 12th International Conference on Pattern Recognition, pages 730–732, Jerusalem, Israel, 1994.
2. C. Brechbühler, G. Gerig, and O. Kübler. Parameterisation of closed surfaces for 3-D shape description. Computer Vision, Graphics and Image Processing, 61:154–170, 1995.
3. A. D. Brett and C. J. Taylor. A method of automated landmark generation for automated 3D PDM construction. Image and Vision Computing, 18(9):739–748, 2000.
4. A. D. Brett, J. C. Waterton, S. Solloway, J. E. Foster, M. C. Keen, S. Gandy, B. J. Middleton, R. A. Maciewicz, I. Watt, P. A. Dieppe, and C. J. Taylor. The measurement of focal diurnal variation in the femoral articular cartilage of the knee. In MICCAI'99, pages 328–337, Cambridge, England, Sept. 1999.
5. G. E. Christensen, S. C. Joshi, and M. Miller. Volumetric transformation of brain anatomy. IEEE Trans. Medical Image, 16:864–877, 1997.
6. M. Eck, T. DeRose, T. DuChamp, H. Hoppe, M. Lounsberry, and W. Stuetzle. Multiresolution analysis of arbitrary meshes. In Computer Graphics (SIGGRAPH '95 Proceedings), pages 173–182, 1995.
7. M. Fleute and S. Lavallée. Building a complete surface model from sparse data using statistical shape models: Application to computer assisted knee surgery. In MICCAI, pages 878–887, 1998.
8. B. Geiger. Three-dimensional modeling of human organs and its application to diagnosis and surgical planning. Technical report no. 2105, Institut National de Recherche en Informatique et Automatique, Dec. 1993.
9. A. Hill, A. Thornham, and C. J. Taylor. Model-based interpretation of 3D medical images. In J. Illingworth, editor, 4th British Machine Vison Conference, pages 339–348, Guildford, England, Sept. 1993. BMVA Press.

10. S. C. Joshi, A. Banerjee, G. E. Christensen, J. G. Csernansky, J. W. Haller, M. I. Miller, and L. Wang. Gaussian random fields on sub-manifolds for characterizing brain surfaces. In J. Duncan and G. Gindi, editors, 15th Conference on Information Processing in Medical Imaging, pages 381–386, Poulteney, VT, 1997. Springer-Verlag.

11. C. Kambhamettu and D. B. Goldgof. Point correspondence recovery in non-rigid motion. In IEEE Conference on Computer Vision and Pattern Recognition, pages 222–227, 1992.

12. T. Kanai, H. Suzuki, and F. Kimura. Three-dimensional geometric metamorphosis based on harmonic maps. The Visual Computer, 14:166–176, 1998.

13. A. C. W. Kotcheff and C. J. Taylor. Automatic construction of eigenshape models by direct optimization. Medical Image Analysis, 2(4):303–314, 1998.

14. W. J. Schroeder, J. A. Zarge, and W. E. Lorensen. Decimation of triangle meshes. Computer Graphics, 26(2):65–70, 1992.

15. G. Székely, A. Kelemen, C. Brechbühler, and G. Gerig. Segmentation of 2-d and 3-d objects from mri volume data using constrained elastic deformations of flexible fourier contour and surface models. Medical Image Analysis, 1:19–34, 1996.

16. G. Szeliski and S. Lavalée. Matching 3-D anatomical surface with non-rigid deformations using octree-splines. International Journal of Computer Vision, 18(2):171–186, 1996.

17. H. D. Tagare. Shape-based nonrigid correspondence with application to heart motion analysis. IEEE Transactions on Medical Imaging, 18(7):434–439, 1999.

Virtual Fluoroscopy: Safe Zones for Pelvic Screw Fixations

B. König[1], U. Stöckle[1], M. Sati[2], L.P. Nolte[2], N.P. Haas[1]

[1]Unfall- und Wiederherstellungschirurgie
Humboldt Universität zu Berlin, Charité, Campus Virchow Klinikum, Augustenburgerplatz 1,
13353 Berlin, Germany
E-mail: benjamin.koenig@charite.de

[2]M.E. Müller Institute for Biomechanics, Murtenstr.35, P.O. Box 30, CH- 3010 Bern,
Switzerland

Abstract. *Hypothesis*: Percutaneous pelvic screw fixations are technically demanding.
Purpose of this prospective study: To define appropriate fluoroscopy projections for the safe placement of SI screws, anterior column screw, posterior column screw and iliac screw using fluoroscopy based navigation.
Material & Methods: A conventional C-arm (Ziehm®) and two different navigation systems were used (Medivision SurgiGate® and Sofamore Danek Stealth Station®). In 6 pelvic models and 4 cadaver pelves three to four projections per screw position were defined and a total of 84 screws were inserted. Evaluation of accuracy was performed with X-ray, CT and dissection.

Results: The following fluoroscopy projections proved to be appropriate:
SI srews: Lateral, lateral 20° a.p., inlet, outlet
Ant. Column: Inlet30° ipsilateral, obturator outlet ipsilateral, obturator outlet contralateral, obturator
Post. Column: Obturator outlet ipsilateral, obturator outlet contralateral, outlet 20° contralateral
Iliac screw: Inlet 30° ipsilateral, obturator ipsilateral, obturator contralateral

The average fluoroscopy time was 6 seconds/ screw in pelvic models and 86 seconds/ screw in cadaver pelves.
In the first series with 60 screws in the pelvic models, 51 screws were inserted correctly. In five cases there was minor displacement without perforating the cortex, in four cases the cortex was perforated. In the cadaver study 19 of 24 screws were placed correctly, two were satisfactory and three were not satisfactory (one SI, ant. Column, post. Column each).
Conclusion: With a remarkable low fluoroscopy time of 6 seconds/ screw in pelvic models and 86 seconds/ procedure in cadaver pelves 83% of all pelvic screws were inserted perfectly. With further improvements in fluoroscopic image quality the method of fluoroscopy based navigation will be very helpful not only for pelvic screw fixations but for many applications in Traumatology.

1. Introduction

Percutaneous screw fixations of the pelvis are technically ambitious because of the complex anatomy of the pelvis. The screws are inserted under fluoroscopic control or

CT-guidance to stabilize pelvic ring injuries or acetabular fractures. Procedures under fluoroscopic control need rather long fluoroscopy times for the correct positioning. The drill can only be controlled in one projection at a time and at least three different projections are necessary. It is not uncommon that several drilling procedures are necessary with repetitive fluoroscopic controls in different projections. This can result in fluoroscopy times of ten minutes and more per screw.

For CT-based navigation the matching procedure can be difficult as the bony structures are covered. For each reduction maneuver the acquisition of a new data set is necessary. Therefore either the CT has to be located in the operating room or surgery has to take place in the CT. This can cause logistic problems.

The new method of fluoroscopy based navigation has the advantage that a fluoroscope is available in any orthopedic OR. The image acquisition in the needed projections is a quick procedure. A maximum of 4 different projections can be displayed on the navigation screen with direction and length of a referenced drill. According to the display on the screen the drill is inserted.

Before the clinical use of this promising new method it was necessary to perform an experimental study with pelvic models to define appropriate fluoroscopy projections for the most common percutaneous screws. In the next step the precision of screws which were inserted under fluoroscopy based navigation was to be evaluated with X-ray, CT and dissection of the pelvic models. The second part of this study was to verify the practicability of these predefined projections in human cadavers for pelvic screw fixation.

2. Material and Methods

A conventional C-arm (Ziehm®) was used for the first part of the study with the fluoroscopy based navigation module SurgiGate® of Medivision® and for screw insertion in human cadavers with the Stealth Station® navigation system of Sofamore Danek®. Optoelectronic markers are mounted on instruments, the dynamic reference base and the C-arm. Three (SurgiGate®) or two (Stealth Station®) infrared cameras assess the position of the LED's in the three dimensional space.

On pelvic models (Synthes®) appropriate fluoroscopy projections were defined for these pelvic screws:

- *Sacroiliacal screw in S1*
- *Sacroiliacal screw in S2*
- *Anterior column*
- *Posterior column*
- *Supraacetabular iliac screw*

The fluoroscopic images had to display the osseous borders of the screw pathways and important anatomic structures like the acetabulum. With respect to standard projections they had to be reproducible.

In a second trial the five screws were inserted on each side using fluoroscopy based navigation. Cannulated screws with a diameter of 7.0 mm were used. The precision of 60 screws in pelvic models was assessed by X-ray, CT and dissection. Fluoroscopy time was documented.

Afterwards in four cadavers twenty-four screws were inserted: six SI- screws in S1 and six in S2, four anterior column, one posterior column and two iliac screws. Screw position and accuracy of the system were analyzed by means of stored system settings, postoperative radiological and CT- scans as well as dissection of the models. Fluoroscopy time and radiation dose was registered.

3. Results

3.1. Definition of Appropriate Fluoroscopy Projections

Combinations of the standard pelvic projections inlet / outlet and iliac / obturator view were defined as appropriate projections.

- SI screws: Inlet, outlet, lateral, lateral 20° a.p.

- Iliac screw: Inlet 30° ipsilateral, obturator ipsilateral, obturator contralateral

- Anterior column: Inlet 30° ipsilateral, obturator outlet ipsilateral, obturator outlet contralateral, obturator

- Posterior column: Obturator outlet ipsilateral, obturator outlet contralateral, outlet 20°contralateral

3.2. Precision of the Screws

All the 60 screws in pelvic models were inserted according to the defined views. The average fluoroscopy time per screw was 6 seconds.
51 of those 60 screws (85%) were placed correctly. The postoperative X-ray correlated with the screen displays and in CT as well as in dissection there was no perforation of the cortex. In five cases (two SI and three anterior column screws) there were deviations from the screen display with acceptable screw position without joint or cortex laceration.
Two SI screws in pelvic models were misplaced with cortex perforation respectively intraforaminal placement of the screw tip. Additionally two anterior column screws were malpositioned, one with cranial perforation of the pubic ramus and one with intraarticular placement of 2 threads.

In the cadaver study 19 of 24 screws (79%) where placed correctly. Two screws where located satisfactory without perforation of cortex and three screws perforated the cortex. Those were one SI, anterior column and posterior column each.
The used projections for pelvic screws fixation proved to be good. Accuracy of the navigation system was sufficient. The number of required pictures of each projection considerable varied. On average 86 seconds of radiation time was needed per procedure.

4. Discussion

Because of the complex pelvic anatomy percutaneous screw fixations are technically difficult. Yet cannulated screws are used under fluoroscopic control. As different projections have to be controlled separately a repetitive correction of the guide wire can be necessary to achieve the correct position. This can result in fluoroscopy times of 10 minutes per screw [10].

Procedures under CT control offer high precision of the inserted screws especially when performed with CT based navigation [1,2,5]. However, in case of reduction maneuvers a new data set has to be acquired after each manipulation. Therefore either a CT is needed in the operating room or surgery has to be performed in the CT. This can cause logistic problems.

Fluoroscopy based navigation is a new method in which the fluoroscope, the patient and the instruments are referenced with optoelectronic markers. Infrared cameras assess the actual position in the three dimensional space [3,4]. There is the advantage that a fluoroscope already is available in the orthopedic operating room and is in frequent use. The necessary fluoroscopy projections are acquired only at the beginning of the procedure to be displayed on the screen simultaneously. As a prerequisite the C-arm projections have to be defined exactly for each screw position to enable a correct screw placement. In this study appropriate C-arm projections were defined for five common screw positions in the pelvis. Using these views a reduction of fluoroscopy time to 6 seconds per screw was possible in pelvic models. In the cadaver study the needed fluoroscopy time (86 seconds per procedure) was higher than in the previous series. Of course this extreme low fluoroscopy time in artificial pelvis is only possible in the experimental setup. But it is promising for the clinical use.

In this experimental study a correct placement of the screws could be achieved in 83 %. This equals the results of the conventional technique under fluoroscopic control. The misplacements in artificial pelves were in five cases minor deviations without perforation of the cortex and in four cases with perforation of the cortex for pelvic models. In the cadaver study 19 of 24 screws were placed correctly, two were satisfactory and three were not satisfactory. The reasons for the deviations from the planned position were insufficient image quality and technical failures.

A good image quality of the fluoroscope in the defined views is a mandatory prerequisite for a safe and precise placement of the drill. In six of nine misplacements in pelvic models and all three misplacements in cadavers the visualization was impaired by already inserted screws or was misinterpreted. The placement of ten percutaneous screws into one pelvis is an extreme situation with less clinical relevance. However, further improvement of the image quality of the fluoroscope is necessary as two or three percutaneous screws are realistic.

Technical failures were mainly caused by deviations of thin drills from the planned pathway. This is also known from the clinical use of the guide wire for the 7.0mm screw.

The defined C-arm projections proved to be appropriate for safe placement of percutaneous pelvic screws using fluoroscopy based navigation. With remarkable low fluoroscopy times a high precision of the screws could be achieved. With further improvements in fluoroscopic image quality the method of fluoroscopy based navigation will be very helpful not only for pelvic screw fixations but for many applications in Traumatology.

5. References

1. Ebrahim NA et al: Percutaneous computed tomography-guided stabilization of posterior pelvic fractures. Clin Orthop 1994 Oct (307): 222-8
2. Gay SB et al: Percutaneous screw fixation of acetabular fractures with CT-guidance: preliminary results of a new technique. Am J Roentgenol 1992 Apr; 158(4): 819-22
3. Hamadeh A et al: Automated 3-dimensional computed tomographic and fluoroscopic image registration. Comput Aided Surg 1999; 4(2): 65-76
4. Hofstetter R, Slomczykowski M, Sati M, Nolte P: Fluoroscopy as an imaging means for computed assisted surgical navigation. Comput Aided Surg 1999; 4(2): 65-76
5. Jacob AL et al: Posterior pelvic ring fractures: closed reduction and percutaneous CT-guided sacroiliac screw fixation. Cardiovasc Intervent Radiol 1997 Jul-Aug; 20(4): 285-94
6. Parker PJ et al: Percutaneous fluoroscopic screw fixation of acetabular fractures. Injury 1997 Nov-Dec; 28(9-10): 597-600
7. Routt ML Jr et al: Early results of percutaneous ilisacral screws placed with the patient in the supine position. J Orthop Trauma 1995 Jun; 9(3): 207-14
8. Routt ML Jr, Simonian PT: Closed reduction and percutaneous skeletal fixation of sacral fractures. Clin Orthop 1996 Aug; (329):121-8
9. Shuler TE et al: Percutaneous ilisacral screw fixation; early treatment for unstable posterior pelvic ring disruptions. J Trauma 1995 Mar; 38(3):453-8
10. Starr AJ et al: Percutaneous fixation of the columns of the acetabulum: a new technique. J Orthop Trauma 1998 Jan; 12(1):51-8

Three-Dimensional Measurement of the Femur Using Clinical Ultrasound: Developing a Basis for Image Guided Intramedullary Nail Fixation of the Femur

Dominic I. Young[1], Sean M. Staniforth[1], and Richard W. Hu[3]

[1] McCaig Centre for Joint Injuries and Arthritis Research, University of Calgary
2500 University Drive N.W., Calgary, AB, Canada T2N 1N4
diyoung@ucalgary.ca
[3] Calgary FootHills Hospital, Department of Orthopaedic Surgery
1403 29 Street N.W., Calgary, AB, Canada T2N 2T9

Abstract. Purpose: Quantify the precision and accuracy in coordinate measurements of anatomic landmarks of the femur using spatially tracked ultrasound (US) images. Establish the limits on coordinate measurement errors required for accurate determination of bone fragment alignment during intramedullary (IM) nail fixation of femoral shaft fractures. Relevance: A surgical guidance system based on a three-dimensional (3D) representation of femoral anatomy from US images would eliminate the hazard of radiation exposure and potentially increase the accuracy of IM nailing procedures. Summary: Fiducial spheres (dia. 6.3mm) were embedded in a plastic femur to mark anatomic landmarks. The femur was suspended in a water tank and could be rotated about its long axis. An US probe was mounted to a track above the femur. Images were collected at 5mm increments along the anterior, posterior, lateral and medial aspects. After the US experiment, fiducial centroid locations (x, y, z–coordinates) were measured in a coordinate measuring machine (CCM). Reconstructed fiducial positions from US images were compared to the CMM data to assess precision and accuracy. A numerical model relating errors in landmark coordinate measurements to rigid body alignment was implemented. The mean precision (std-dev.) in fiducial coordinate measurements was 1.69mm. Mean and maximum errors in fiducial positions were 17.65mm and 58.01mm, respectively. At the observed level of accuracy in coordinate measurements, the model predicted rigid body rotation errors of 3.4 (SD = 2.4)° and translation errors of 4.7 (SD = 3.2)mm. A proof-of-concept has been demonstrated in the use of clinical US to obtain a quantitative description of femoral anatomy in a 3D framework. The model of error limits provided a basis for assessing the capability of a tracked US system in the context of a clinical criterion for rotational alignment (anteversion angle). Accuracy requirements for landmark coordinate measurements were at the limits of the capability of the current US tracking system.

1 Introduction

Closed IM nail fixation is the treatment of choice for adult femoral shaft fractures. Compared to an open technique, which is complicated by a large surgical incision, closed fracture fixation reduces the risk of infection at the fracture site by using a small incision at the proximal end of the femur to introduce the IM nail [1]. A C-arm image intensifier (fluoroscope) is used to provide radiographic images that guide the placement of the IM nail and positioning of bone fragments during the procedure.

Alignment of the two bone fragments may be described by translations along and rotations about three axes corresponding to the anterior-posterior (A-P), lateral-medial (L-M) and proximal-distal (P-D) directions, respectively. The fractured ends of the bone must be aligned in the A-P and L-M directions for IM nail insertion. Proper rotational alignment about the P-D axis, called femoral anteversion, is particularly critical for a successful outcome. Femoral anteversion is characterised by the anteversion (AV) angle, commonly defined as the angle between a tangent plane to the posterior aspect of the femoral condyles and the longitudinal axis of the femoral neck [2]. Many patients tolerate some torsion, but efforts should be made to reduce and stabilize the femoral shaft fracture with an anteversion angle difference of less than 15° compared to the contralateral leg [3].

Without the image intensifier, current methods and results of femur fixation would not be possible. However, there are two important concerns regarding this technique. First, there is a potential for prolonged exposure of the operating room personnel and the patient to dangerous levels of radiation [4], [5], [6], [7]. Due to the high frequency of femur fractures there is a risk of repetitive exposure of the surgeon to radiation. Second, mal-reduction problems and complications such as errors in length/rotation and improper fixation have been associated with the procedure [3]. Even with the aid of the fluoroscope, considerable skill is required to mentally reconstruct the three dimensional (3D) positions of the bone fragments and surgical tools, relying on non-continuous two dimensional (2D) X-ray images to confirm their positions.

Pre-operative computed tomography (CT) scans are the basis for many image guided surgical systems in use or under development [8], [9], [10]. Studies have described the use of virtual 3D bone surfaces constructed from CT images to aid surgeons in orthopaedic procedures [11]. Recently, intra-operative US has been studied for its ability to locate and register fiducial markers during interactive image-guided neurosurgery [12], [13]. The use of intra-operative US has been demonstrated in providing 2D images for monitoring the alignment of bone fragments in simple or minimally comminuted fractures that required unlocked nailing of the femur [14]. A surgical guidance system based on US technology would avoid the hazard of ionising radiation and offer the potential for real-time, intra-operative, imaging of the patient's anatomy.

The present study was undertaken as a first step in the development of a guided surgical system for IM nail fixation based on US imaging. The main purpose was to demonstrate the application of a clinical US system to the problem of obtaining a 3D description of the bony anatomy of the femur. The first objective was to quantify the precision and accuracy of landmark coordinate measurements made from tracked US images. The second objective was to establish limits on the coordinate measurement errors required to obtain the degree of accuracy necessary for determining the alignment between proximal and distal fragments of the femur during IM nailing procedures.

2 Methods

2.1 Preparation and image acquisition

A plastic model femur (length 455mm) was fitted with 26 fiducial markers placed at anatomical landmarks at the proximal and distal ends and along the mid-shaft. Fiducial markers consisted of nylon spheres (diameter 6.3mm) that were embedded in small divots drilled into the surface of the specimen. Plastic rods were inserted at both ends of the specimen, allowing for rotation about its longitudinal axis when mounted in a Perspex frame specifically designed for this purpose (Fig. 1).

The specimen and support frame were placed inside a water tank so that the rotational axis of the specimen was parallel to the long sides of the tank and roughly 100mm below the surface of the water. A clamp was placed across two aluminum rails fixed to the top of the tank. An US probe (SSD-900 Diagnostic Ultrasound System with 7.5MHz Probe, Aloka Corporation, Japan) was placed in the clamp so that it could slide along the length of the specimen parallel to the axis of rotation with the transducer head just below the surface of the water (Fig. 1).

The axial position of the US probe was measured (within ±2.5mm) using a ruler fixed to one of the aluminum rails. The rotational position of the specimen could be set at 90° intervals, allowing the presentation of each of its four surfaces (anterior, posterior, lateral and medial) to the probe. The probe was oriented perpendicularly to the rotational axis of the specimen with the US image-plane oriented to obtain axial cross-sections of the femoral surface in US images (Figs. 1, 2). Video output from the US system was connected to a capture board (LG-3 Scientific Frame Grabber, Scion Corporation, USA) controlled by Scion Image software (Scion Corporation, USA) running on a PC.

Sequential axial images of a single randomly assigned surface were captured as the US probe was translated along the rails at 5mm increments. The specimen was then rotated to capture images of the remaining three surfaces in random order. Five trials each comprising four sets of surface images (anterior, posterior, lateral and medial) were completed. US images captured from the video display were stored in TIF format as 8-bit grayscale images.

2.2 Analysis of US images

Each image from a particular series (e.g., anterior series of trial 1) was examined in sequence and compared with the physical specimen. The pixel coordinates of the centre of any visible fiducial marker were identified and then labelled according to the fiducial's location on the specimen. US images were also used to identify the location of the centre of rotation (C_R), of the specimen based on the centres of the plastic rods inserted in the specimen.

2.3 Reconstruction of fiducial 3D coordinates

The second part of the analysis consisted of reconstructing the 3D locations of fiducials in a coordinate system based on the physical arrangement of the experimental set up (Fig. 1). The experimental coordinate system (ECS) was defined so that its origin was fixed at the centre of the proximal rotation support of the specimen. The x,z–plane was parallel to the short sides of the tank with the y–axis pointed distally and parallel to the specimen's axis of rotation. The x–, y– and z–axes of the ECS corresponded roughly to the L-M, P-D and A-P axes of the model femur. The coordinate collection analysis produced the x– and z–positions (horizontal and vertical, respectively) of fiducials in terms of image coordinates. The x,z–positions were scaled to millimetres (US image resolution 0.20mm-pixel^{-1}). Because the specimen was rotated in 90° intervals to image all four of its surfaces, the x– and z–coordinates of fiducials identified in the posterior, lateral or medial views were reoriented in the ECS using a rotational transformation of 180°, -90° or 90° about C_R, respectively. Fiducial x,z–coordinates were combined with the y–coordinates (axial) to produce five reconstructed data sets (x_{US}, y_{US}, z_{US}) of 26 markers in terms of x,y,z–coordinates in the ECS.

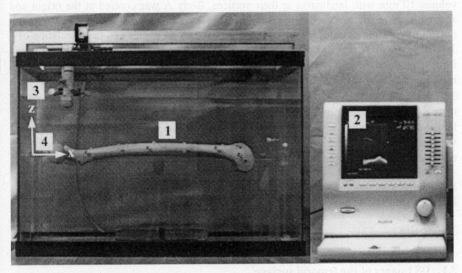

Figure 1. Experimental set-up. 1) Specimen (visible fiducials: lesser trochanter, medial epicondyle and femoral shaft markers). 2) Clinical US system (note the axial cross section displayed on the monitor). 3) US probe. 4) Experimental coordinate system (ECS)

2.4 Precision and accuracy

Mean and standard deviation (SD), denoted by σ, for the five measurements of x,y,z–coordinates were calculated for each fiducial. Precision, ρ, was quantified by the norm of the SDs:

$$\rho = \sqrt{\sigma_x^2 + \sigma_y^2 + \sigma_z^2}.$$

After the US experiment, the specimen and Perspex frame were fixed rigidly to the working surface of a coordinate measuring machine, or CMM (Mitutoyo MTI Corp., Japan, precision ±0.050mm) and the 3D coordinates of the centroids of the 26 fiducial markers were measured. This calibrated data set (x_{CMM}, y_{CMM}, z_{CMM}) was used to quantify the accuracy of the reconstructed data sets. The five sets of 26 reconstructed fiducial coordinates (x_{US}, y_{US}, z_{US}) were averaged and compared to the calibrated data set. Errors in x-, y- and z-coordinates were calculated as $(x_{CMM} - x_{US})$, $(y_{CMM} - y_{US})$ and $(z_{CMM} - z_{US})$, respectively. Error in marker position, Δ, was defined as the norm of the x-,y- and z-errors:

$$\Delta = \sqrt{(x_{CMM} - x_{US})^2 + (y_{CMM} - y_{US})^2 + (z_{CMM} - z_{US})^2} \ .$$

2.5 Modelling accuracy requirements

To answer the second objective of the study, a simple model relating errors in landmark coordinate measurements to rigid body alignment (relative rotation and translation) was implemented. The model consisted of two 3D cubes (A and B) of volume 10^6mm with landmarks at their vertices. Body A was centred at the origin and aligned with the global coordinate system (GCS). Body B was rotated and translated with respect to A. The rotation consisted of three equal Cardan angles and the translation of a 3D vector $d = [100\ 100\ 100]$mm. Landmark x,y,z-coordinates on A and B were perturbed from their ideal positions by 100 sets of normally distributed random errors of mean magnitudes 0.5, 5, 15, 25 and 50mm. For five different Cardanic rotations (5°, 10°, 15°, 20° and 25°) of B, the rigid body transformations between A and B (a 3×3 rotation matrix, M, and a 3D translation vector t) were determined for the perturbed landmark locations [15]. Relative rotations (Cardan angles, α, β and γ) about and translations (t_x, t_y, t_z) along the x- ,y- and z-axes between A and B were resolved from the perturbed transformations and compared to the ideal rotations and translations.

The fiducial coordinate collection analysis, 3D reconstruction of fiducial coordinates and accuracy modelling were carried out in Matlab5.3 (The MathWorks Inc., USA) on a PC. Precision and accuracy analysis was carried out in Excel97 (Microsoft Corp., USA).

3 Results

3.1 US images of the femoral surface

The "bone surface" of the specimen was clearly visible in all of the images. Large anatomic features (e.g., lesser trochanter, patellar notch and inter-condylar fossa) were easily identified. Finer details such as the 6mm fiducial markers could also be resolved in the images. Occasionally, reflected sound waves reduced the sharpness of the surface boundary but this did not interfere with the reliable identification of fiducials (Fig. 2).

3.2 Precision and accuracy

The femur's centre of rotation, C_R, was identified very reliably from the US images. There were usually from four to eight measurements of C_R from each series of images. The largest SDs obtained in the horizontal and vertical components of C_R were 1.1 and 1.5 pixels, respectively. Based on the image resolution, this variation was less than 0.5mm. Precision, ρ, in coordinate measurements from the five reconstructed data sets was found to be very high with mean and maximum values of 1.69mm and 3.73mm, respectively. The mean error, Δ, was 17.65mm and the maximum error was 58.01mm. Errors in coordinate values were also examined as separate components: Δ_x ($x_{CMM} - x_{US}$),

Δ_y ($y_{CMM} - y_{US}$) and Δ_z ($z_{CMM} - z_{US}$). The mean values of Δ_x, Δ_y and Δ_z were 14.76, 2.05 and 9.49mm, respectively. The comparatively larger errors in the x– and z–coordinates, corresponding to the horizontal and vertical directions in the US images, were largely responsible for the low accuracy of the fiducial location measurements.

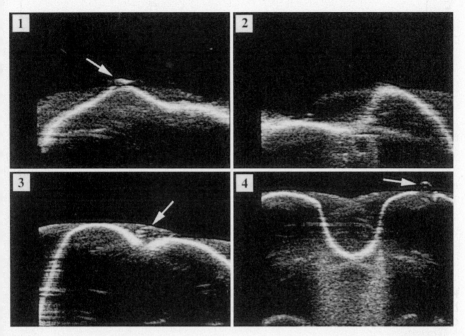

Figure 2. US images illustrating surface features of the proximal and distal ends of the femur. *Fiducials are marked with arrows.* 1) Anterior view of greater trochanter. 2) Posterior view of the greater trochanter and portion of the femoral neck. 3) Anterior view of the distal femur. 4) Posterior view of the femoral condyles

3.3 Accuracy requirements

As expected, errors in rotation (α, β and γ) and translation (t_x, t_y, t_z) increased with the magnitude of landmark coordinate errors. Larger errors were observed for larger Cardanic rotations at each level of perturbation, however, alignment errors were not strongly dependent on the amount of rotation. The results are presented for values of β and t_y, at a rotation of 15° (Fig. 3). The largest mean error in fiducial coordinates obtained in the coordinate collection analysis was $\Delta_x = 14.76$mm. At this level, the predicted rigid body alignment errors were 3.4 (SD = 2.4)° and 4.7 (SD = 3.2)mm for β and t_y, respectively. The largest amount of perturbation applied in the model (50mm) corresponded to the maximum observed errors in fiducial coordinates. At this level, the predicted alignment errors were 11.2 (SD = 8.8)° and 15.5 (SD = 10.8)mm for β and t_y, respectively.

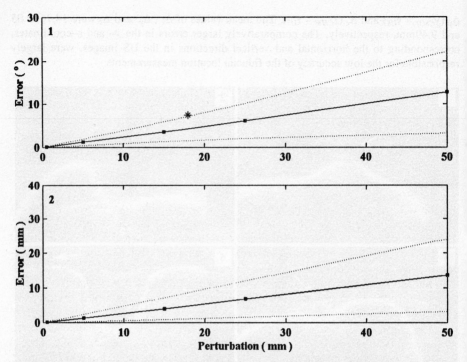

Figure 3. Accuracy model results for a nominal rotation $\beta = 15°$ and translation $t_y = 100$mm. *Dashed lines indicate SD in the error results (n = 100 perturbations).* 1) Errors in rotation, β. *Clinical accuracy criterion marked with an asterisk.* 2) Errors in translation, t_y

4 Discussion

The main purpose of this investigation was to demonstrate the application of clinical US in measuring the relative positions of anatomic landmarks on the femur. Considering the simplicity of the "US tracking system", the precision (ρ) in fiducial coordinate measurements was very high (within ±2mm, on average).

A successful IM nail fixation procedure depends on the accurate assessement of the angular and linear position of the bone fragments. For example, a 15° criterion for AV angle [3] indicates an error limit of ±7.5° for assessments of relative rotation between the proximal and distal fragments about the P-D axis. In order to achieve this accuracy, the errors in anatomic landmark coordinates (Δ_x, etc.) must be within ±18mm, according to the model of accuracy requirements (Fig. 3-1). The mean error in fiducial coordinate measurements ($\Delta = 17.65$mm) compares well to the accuracy requirements based on the 15° criterion. However, it should be noted that alignment measurements depend on the accuracy in individual landmark coordinates. Therefore the range of fiducial coordinate errors of up to 58.01mm were outside the limits of the 15° criterion.

Fixing the orientation of the US transducer provided a means of "tracking" its position and reconstructing the 3D coordinates (x_{US}, y_{US}, z_{US}) of fiducials identified in US images. Coordinate errors were thought to be related to three assumptions that were made in order to execute the reconstruction: i) the US image plane was parallel to the x,z-plane of the ECS; ii) the centre of rotation of the specimen was parallel to the y-axis of the ECS; iii) the location of the centre of rotation, C_R, of points on the specimen's surface was identical for all images. The limitation of the third assumption was indicated by a "wobbling" observed at the proximal end of the specimen caused by misalignment of the rotation supports. The x- and z-coordinates depended on a fiducial's location with respect to C_R. Larger errors in x- and z-coordinates were likely caused by this motion. By comparison, errors in the y-coordinates were within the range of 2.0 to 5.0mm, which was consistent with the estimated accuracy in the axial position measurements (±2.5mm).

5 Conclusion

Results of the US imaging protocol demonstrated that fine details of femoral anatomy can be resolved and precisely located in a 3D framework. We have demonstrated an important proof-of-concept in the use of clinical US in providing a quantitative description of femoral anatomy. Our model of error limits provided a basis for assessing the capability of a tracked US system in the context of a clinical criterion for rotational assessment. The present system produced coordinate measurement errors that were within the limits indicated by the accuracy model. However, the range in coordinate errors needs to be reduced.

A plastic specimen in a water tank was an ideal model for producing sharp US images of the femoral surface. In a human limb, the acoustical properties of skin, fat and muscle as well as the presence of other bones at the hip and knee joints would likely affect the quality of US images and obscure details that were clearly visible in the experiment. A more physiologically representative model needs to be incorporated into future development of the system.

Currently, work is being done to incorporate optical tracking technology into the project. This will provide accurate spatial tracking of US images obtained using "free hand" scanning of the femur. In the long term, the goal is to develop a system that will give the surgeon an intuitive visual display of the alignment parameters relevant to IM nail fixation, a system that benefits from the safe and real time imaging capabilities of clinical US.

References

1. Mooney V., Claudi B.F.: Fractures of the shaft of the femur. In: Rockwood, C.A. and Green, D.P. (eds): Fractures in Adults. J.B. Lippincott Company, Philidelphia, (1984) 1396-1398
2. Lausten G.S., Jorgeson F., Boesen J: Measurement of anteversion of the femoral neck: ultrasound and computerised tomography compared. J. Bone Joint Surg. (Br). 75-B (1989) 799-803
3. Braten M., Terjesen T., Rossvoll I.: Torsional deformity after intramedullary nailing of femoral shaft fractures: Measurement of anteversion angles in 110 patients. J.Bone Joint Surg. (Br) 75(5) (1993) 7990-8003

4. Levin P.E., Schoen R.W., Browner B.D.: Radiation exposure to the surgeon during closed interlocking intramedullary nailing. J. Bone and Joint Surg. (Am) **69**(5) (1987) 761-766

5. Coetzee J.C., van der Merwe E.J.: Exposure of surgeons-in-training to radiation during intramedullary fixation of femoral shaft fractures. S. Afr. Med. J. **81**(6) (1992) 312-314

6. Muller L.P., Suffner J., Wenda K., Mohr W., Rommens P.M.: Radiation exposure to the hands and the thyroid of the surgeon during intramedullary nailing. Injury. **29**(6) (1998) 461-468

7. Smith G.L., Wakeman R., Briggs T.W.: Radiation exposure of orthopaedic trainees: quantifying the risk. J Royal College Surg. Edinburgh. **41**(2) (1996) 132-134

8. Vannier M.W., Marsh J.L.: Three-dimensional imaging, surgical planning, and image-guided therapy. Radiol. Clin. North. Am. 34(3) (1996) 545-563

9. Glossop N., Hu R.: Effect of registration method on clinical accuracy of image guided pedicle screw surgery. Computer Assisted Radiology and Surgery, Proceedings of the 11th International Symposium and Exhibition, Berlin, Germany (1997) 884-888

10. Hu R., Glossop N., Steven D., Randle J.: Accuracy of image guided placement of iliosacral lag screws. 1st Joint CVRMed/MrCAS, Proceedings, Grenoble, France (1997) 593-596

11. Joskowicz L., Milgrom C., Simkin A., Tockus L., Yaniv Z.: FRACAS: a system for computer-aided image-guided long bone fracture surgery. Comput. Aided Surg. 3(6) (1998) 271-288

12. Hata N., Dohi T., Iseki H., Takakura K.: Development of a frameless and armless stereotactic neuronavigation system with ultrasonographic registration. Neurosurgery. 41(3) (1997) 608-613

13. Schreiner S., Galloway R.L., Lewis J.T., Bass W.A., Muratore D.M.: An ultrasonic approach to localization of fiducial markers for interactive, image-guided neurosurgery – Part II: Implementation and automation. IEEE Transactions on Biomedical Engineering. 45(5) (1998) 631-641

14. Mahaisavariya B., Laupattarakasem W.: Ultrasound or image intensifier for closed femoral nailing. J. Bone Joint Surg. (Br) 75(1) (1993) 66-68

15. Soederqvist I., Wedin P.A. Determining the movement of the skeleton using well-configured markers. J. Biomech. 26 (1993) 1473-1477

Percutaneous Computer Assisted Iliosacral Screwing: Clinical Validation

Lionel Carrat [a], Jérome Tonetti [b], M.D,
Philippe Merloz [c], M.D, Jocelyne Troccaz[a], Ph.D.
e-mail : lionel.carrat@imag.fr

[a] TIMC Laboratory, Faculté de médecine, I.A.B., 38706 La Tronche (France)
[b] Anatomy Laboratory, Pr Chirossel, University Joseph Fourier 38043 Grenoble
c Orthopaedic Department, CHU A. Michallon - BP 217 - 38043 Grenoble

Abstract. This paper describes the clinical validation of an image-guided system for the percutaneous placement of iliosacral screws. The goals of the approach are to decrease surgical complications, with a percutaneous technique, and to increase the accuracy and security of screw positioning thanks to a computer assisted system. Pre-operative planning is performed on CT-scan images and a 3D model is built. During surgery, tools are tracked with an optical localizer. An ultrasound acquisition is performed and images are segmented to obtain 3D intra-operative data that are registered with the CT-scan 3D model. The surgeon is assisted during drilling and screwing processes with re-sliced CT-scan images displayed on the computer screen and comparison between pre-operative planning and tools position. The system was validated in a cadaver study [1]. The clinical validation has then started and four patients have been successfully instrumented.
Keywords : orthopaedics, clinical results, ultrasound-based registration.

1. Introduction

Unstable pelvic ring fractures with complete posterior ring disruption (see Fig. 1A) require a surgical fixation as described in [2]. The non-operative treatment is not adequate to stabilize the closed reduction in order to prevent hemipelvic ascension or pseudarthrosis of the fracture. The surgical procedure consists in drilling two holes and introducing two screws through the iliosacral joint into the sacral ala and the body of the first sacral vertebra (see Fig. 1B). The trajectory of the drill must avoid the close anatomic elements[3] that are the first sacral nerve root in the S1-S2 foramen, the lumbosacral nerve trunc, the roots of the cauda equina, the iliac vessels and the peritonial cavity.

Two surgical procedures are currently used.

The first one is open reduction described by several authors [4,5,6]. In order to secure the procedure, empirical reference marks must be taken. As a consequence, a large surgical exposure of the sacrum and the posterior part of the fossa iliaca lateral must be done. Complications like blood loss, hematoma and wound infection are not rare after such exposure because of the perineum's proximity.

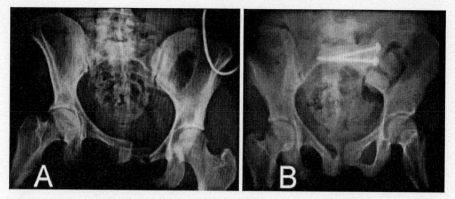

Fig. 1 : (A) Unstable pelvic ring fracture with complete disruption of the posterior arch. (B) Example of iliosacral screw placement to obtain durable fixation and to prevent non-union.

Percutaneous technique using fluoroscopy or computer tomography [5,7,8] minimizes the rate of wound complications. However this technique requires a great experience of the surgeon[3] and many C-arm manipulation. Furthermore, digestive gases can make the radioscopic vision difficult.

This paper presents a computer assisted method. The main goals are to increase accuracy, thanks to the computer assistance, and to decrease the rate of wound complications, by using a percutaneous procedure. Our approach is based on computer assisted surgery using pre-operative CT-scan model registered with ultrasound data of the patient during surgery.

Clinical validation has started at the end of February 2000. Up to now four patients have been instrumented, three with two screws on one side of the pelvis, one with two screws on each side of the pelvic bone (because of arthrodesis of the two iliosacral joints). This clinical validation follows a cadavers study described in [1]. The major modification concerns the ultrasound acquisition. On cadavers, it was performed on the iliac crest, fossa iliaca lateral, ischiatic process and posterior side of the sacrum. To avoid problem of relative motion at fracture site, we are now working only on posterior part of the sacrum thanks to a smaller probe.

2. Material and Methods

Our protocol is divided into three steps.

2.1. Pre-operative CT-Scan Segmentation and Surgical Planning

A CT-scan acquisition of nearly 50 images with a 3mm slice thickness is performed with a spiral CT GE Hispeed Advantage. The examination field includes the whole sacrum. A semi-automatic segmentation of the bone is performed in order to have a CT-based 3D model of the pelvis.

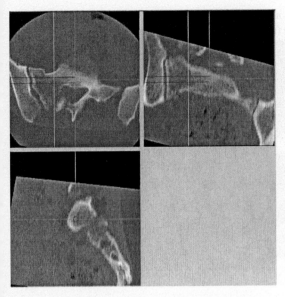

Fig. 2 : Surgical planning on CT-scan images.

Then, the surgeon defines the optimal placement of screws using an interactive interface that allows the computation of arbitrary planes orientation in the volume of CT-scan images (see Fig. 2). Each screw is described by an entry point, a direction, and a diameter. Two screw positions are defined : one from supraforaminal S1-S2 superior into the S1 vertebra's body, one from supraforaminal S1-S2 inferior into the S1 body. The screw tips are defined on the mid-sagittal line of the sacrum. The planning allows accurate determi-nation of each screw length

2.2 Intra-operative Ultrasound Acquisition and Registration

During surgery, a 6D optical localizer (Polaris, NDI – Toronto) is used. It locates the position and orientation of wireless customized rigid-bodies equipped with reflective markers (TIMC, laboratory). Rigid-bodies are fixed on tools used by the surgeon (see Fig. 3) :
- a standard ultrasound probe (A) of 75 MHz frequency with an examination field of 2 cm width and 5 cm depth
- a linear tool (B) used for registration validity checking and drilling trajectory approach.
- a standard surgical power drill (C).
- a linear tool guide used during drilling and screwing (D).

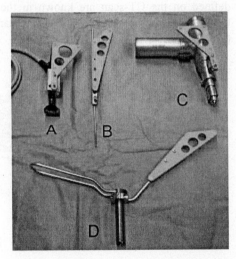

Fig. 3 : Tools used during surgery

A reference rigid-body is firmly fixed on the sacrum or in the posterior iliac crest on the opposite side of the fracture. It defines the intra-operative reference coordinate system that is used during the whole surgery.

As described in [9], we are using a standard ultrasound probe to image the bone and soft tissue interface of the sacrum (see Fig. 4). The patient is in prone position and a sterile transonic gel is used. Nearly 40 ultrasound images are acquired. Then, a segmen-tation procedure enables us to build 2D curves that belong to the bone surface on each image (see Fig. 5).

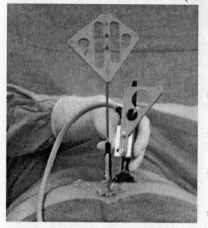
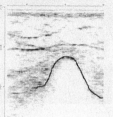

Fig. 4 : US acquisition. **Fig. 5 :** US segmentation.

Using the 6D location and the calibration of ultrasound probe, those curves are converted into a cloud of 3D points referenced in the intra-operative coordinate system. These points are registered with the 3D CT-scan model of the pelvis (see Fig. 6) using surface based registration[14]. As a result, the optimal screw positions defined on the CT-scan are known in the intra-operative coordinate system. This technique avoid the insertion of feducial markers into the bones when constructing the pre-operative model.

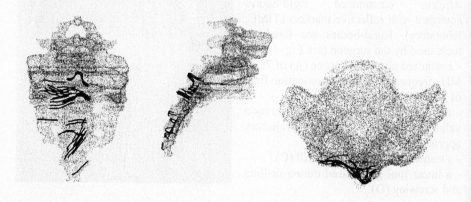

Fig. 6 : Registration between the CT-scan 3D model (in gray) and the 3D data obtained with the ultrasound segmentation (in black) .

2.3 Passive Drilling and Screwing Guidance

The passive guidance process can then be started using a real-time navigation assistance. The first step consists in the registration validation. The surgeon uses the linear tool and put it on specific anatomical points. The tool axis and extremity are tracked and corresponding images are displayed. The surgeon is then able to check if displayed images really correspond to the tool position. Once the registration result is validated, the surgeon localizes the planned entry point and direction of the drilling trajectory (see Fig. 7).

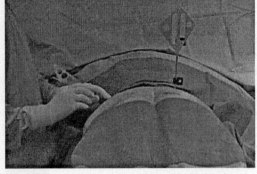

Fig. 7 : Navigation with the linear tool.

The linear tool is positioned at the entry point of the bone and used to make a small hole. This step is done in order to find the trajectory position in a less invasive way than with the drill guide, that has a bigger diameter. Furthermore, this hole will be used to prevent unexpected motion of the drill tip on the bone at the beginning of the drilling process.

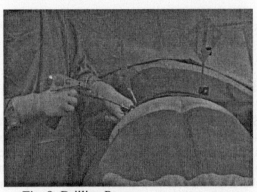

Fig. 8: Drilling Process.

The second step consists in finding the entry point with the drill guide. Like the linear tool, its axis and extremity have been calibrated. Once the entry point and direction of the trajectory have been found, the drilling process is started (see Fig.8). The power-drill calibration allows the monitoring of the drill depth penetration and the comparison with the maximum depth penetration allowed by the CT-scan planning. Before crossing the S1-S2 foramen, drill trajectory is checked thanks to one x-ray image (see Fig. 9) as proposed in [10].

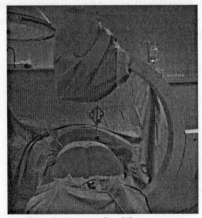

Fig. 9: X-ray checking.

At the end of the drilling process, the surgeon takes the power drill off but leaves the drill guide in the correct position. The third step is the screwing. Thanks to a tool device, the surgeon can insert the screw driver within the guide. As a consequence, the screw is guided into the hole whereas without assistance, it can be difficult and takes several minutes.

During each step, real-time tool tracking and re-sliced images are displayed with tool position shown on CT-scan images. In addition, a cross-hair alignment system allows an easy comparison between the planned trajectory and the real position of the tool (see Fig. 10).

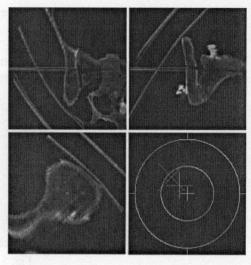

Fig. 10: Computer screen display

Thanks to this assisted guidance, 6.5mm diameter screws for cancellous bones (Synthes inc, Etupe, France) are easily inserted in the pelvic bone.

3. Clinical Validation

Since the end of February, four pelvis have been instrumented with a total of 10 screws. X-ray exposition time during surgery is recorded. This value is important, and should decrease thanks to the computer assistance in comparison with percutaneous surgery under fluoroscopy. X-ray images are acquired to check the correct position of the drill and screw. Ultrasound acquisition, segmentation and registration time are also recorded. They are additional steps to a classical surgery and directly influence the surgical time. After surgery, a post-operative CT-scan is performed in planes positioned along and perpendicular to each screw axis by report to the topography of the lumbosacral trunk and the S1 root.

These images are used to take measurement on screw positions (see Fig. 11). A is the distance from the screw to the anterior cortex of the sacrum. B is the distance from the screw to the spinal canal or to the foramen. C is the distance from the tip of the screw to the mid-sagittal line of the sacrum. A and B values are measured if the screw is in the bone.

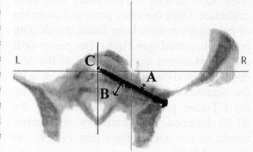

Fig. 11 : Method of screw position measurement.

We have established a security score that depends on A and B values such as :
If A or B equal 0, security score is 0% because one anatomical limit is reached.
If A = B, security score is 100% because the screw is accurately centered.

		A (mm)	B (mm)	C (mm)	Score (%)
Patient 1	Sup	10.3	3	0	45
	Inf	3.6	7.2	0	67
Patient 2	Sup	8	3.5	3.5	61
	Inf	5.5	2	5.5	53
Patient 2	Sup	3.5	4.5	4.6	88
	Inf	3	5.5	5	71
Patient 3	Sup	10.4	5.2	1.5	67
	Inf	8.7	4.3	0	66
Patient 4	Sup	0	4.5	0	0
	Inf	3	3	0	100

Table 1 : Measurements.

Values reported in Table 1 show that all screws were in the correct position and preserve the close anatomical elements. Anatomical limit has been reached only one time

Average x-ray irradiation time measured during surgery has a value of 0.35 minutes and ultrasound acquisition, segmentation and registration has an average value of 24 minutes. All those values are recorded in an observation book in which blood loss, complications, prospective follow-up with neurological evaluation and self-evaluation of pain at 7 days, 45 days and 3 months are also reported. Current status of the validation shows that the four patients have an excellent neurological status.

4. Discussion

A study on percutaneous surgery under fluoroscopy has been made on 30 patients. Comparing those results with the ones obtained with the computer assisted technique allows us to draw table2. However, result tendencies have to be confirmed on a larger number of cases.

The first result of this table concerns the number of outside bone screws: 28 % with classical technique and 0 % with the computer assisted technique. On these outside bone screws, one over three has directly induce a neurological lesion. That represents 9.3 % of screws inserted with the classical method.

With the computer assisted technique, the x-ray irradiation time has been divided by three. Moreover, this duration should decrease with the experience of the surgeon and its confidence in the system.

	Under fluoroscopy	Computer assisted
Number of patients	30	4
Number of screws	43	10
Number of outside bone screws	12 (28%)	0 (0%)
Neurological lesion due to screw	4 (9.3 %)	0 (0%)
X-ray exposition time (min)	1.07 min	0.35 min
Average security score (%)	47.1 %	61.8 %

Table 2 : Comparison between classical and computer assisted techniques.

The major drawback of our system is the time needed for ultrasound acquisition, segmentation and registration. The increase time of surgery was nearly 25 minutes. The ultrasound acquisition time will probably decrease with the surgeon's learning. Research is currently performed to develop an automatic segmentation and registration of ultrasound data with 3D model. This will decrease the time and the possible inaccuracy of ultrasound segmentation due to images quality.

Cadaver study has shown registration average root mean squares error of 1.5 degrees and 2.6 mm. Cadaver study average security score was 52.8 %. We have obtained 61.8 % for the clinical validation. This accuracy increasing may be due to ultrasound images quality that was bad on cadaver because of deshydratation.

5. Conclusion

In comparison with the percutaneous technique under fluoroscopy, the computer assisted technique allows the real-time navigation in CT-scan images and divide the x-ray exposition by three. The ultrasound based registration with CT-scan model takes advantage on feducial markers registration [11, 12] and method used for spine surgery [13, 14, 15] because of wound complications decreasing.

The percutaneous technique described in this paper is secure enough to implant 2 screws of 6.5 mm diameter through the iliosacral joint. The clinical validation has started at the end of February 2000 and 4 pelvic have already been successfully instrumented with 10 screws. Clinical validation with computer assisted technique is going on.

References

1. Tonetti J, Carrat L, Lavallee S, Pittet L, Cinquin Ph, Merloz Ph and Chirossel JP. Percutaneous iliosacral screw placement using image guided techniques. Clin. Orthop. (354), p. 103-110, 1998.
2. Waddel JP, Pennal GF, Tile M and Garside H. Pelvic disruption : Assessment and classification. Clin. Orthop. (151), p. 12-21, 1978.
3. Freese J, Templeman D, Schmidt A and Weisman I. Proximity of iliosacral screws to neurovascular structures after internal fixation. Clin. Orthop. (329), p. 194-198, 1996.
4. Letournel E. Pelvic fractures. Rev. Chir. Orthop. (10), p 145-148, 1978.
5. Matta JM and Saucedo T. Internal fixation of pelvic fracture. Clin. Orthop. (242), p 83-97, 1998.
6. Tile M and Pennal GF. Pelvic disruption : Principe of management. Clin. Orthop. (151), p 56-64, 1980.
7. Xoombs RJ, Jackson WT, Ebraheim NA, Russin JJ and Holiday B. Percutaneous computer-tomography stabilization of pelvic fractures. Preliminary report. Journal Orthop. Trauma. (1), p 197-204, 1987.
8. Routt ML and Simonian PT, Closed reduction and percutaneous skeletal fixation of sacral fractures. Clin. Orthop. (329), p121-128, 1996.
9. Barbe C, Troccaz J, Mazier B and lavallee S. Using 2.5D echography in computer assisted spine surgery. Engineering in Medicine and Biology Society Proceedings. San Diego, Institute of Electrical and Electronics Engineers Inc p 160-161, 1993.
10. Routt ML, Simonian PT, Agnew SG and Mann FA. Radiographic recognition of the sacral ala slope for optimal placement of iliosacral screws: a cadaveric and clinical study. J Orhtop. Trauma.(10:3),p 171-173, 1996.
11. Jacob AL. Computer assistance in pelvic and acetabular fractures. In L. Nolte editor, CAOS 96, Bern (CH), 1996. Muller Institute.
12. Hu R, Glossop N, Steven D and Randle J. Accuracy of image guided placement of iliosacral lag screws. In Troccaz J and al. Editors, CVRMed-MRCAS'97 Proc., LNCS Series 1205, p 593-596, Springer, 1997.
13. Lavallee S, Sautot P, Troccaz J, Cinquin P and Merloz P. Computer Assisted Spine Surgery : a technique for accurate transpedicular screw fixation using CT data and a 3D optical localizer. Journ. Of Image guided Surgery (1), p 65-73, 1995.
14. Lavallee S, Szeliski R and Brunie L. Anatomy-based registration of 3D medical images, range images, X-ray projections, 3D models using Octree-Splines. In Taylor R, Lavallee S, Burdea G and Mosges R, editors, Computer Integrated Surgery, p 115-143, MIT Press, Cambridge, MA, 1996.
15. Merloz P, Tonetti J, Eid A, Faure C, Lavallee S, Troccaz J, Sautot P, Hamadeh A and Cinquin P. Computer assisted spine surgery. Clin. Orthop. And Related Research (337) p 86-96, 1997.

Milling versus Sawing: Comparison of Temperature Elevation and Clinical Performance During Bone Cutting

A. Malvisi, P. Vendruscolo, F. Morici, S. Martelli, M. Marcacci.

Andrea Malvisi Lab. Biomeccanica via di Barbiano 1/10, 40136 Bologna, Italy.
Tel.:+39 051 6366520 - Fax:+39 051 583789
E-mail: a.malvisi@biomec.ior.it - URL: http://www.ior.it/biomec/

Abstract. A fundamental requirement for the success of an implant process is the quality of bone resection, mainly related to temperature elevation during cutting and to the accuracy of the resection. Existing surgical saws have largely been investigated, but appear still unable to maintain the temperature elevation within acceptable parameters. In this paper we analyse an alternative approach to realising large bone resections, studying the performance of the milling technique. The temperature elevation during milling is carefully analysed with different surgical tools, and an overall estimation of the accuracy and usability of this technique is discussed. The results of this study prove that milling can be used without the need for cooling procedures (T < 46°C), within an acceptable surgical time (~ 4 min for knee resections), and with negligible fatigue to the user.

Introduction

In orthopaedic surgery large resections are usually performed using surgical saws. This technique involves the use of oscillating air-powered saws, with special materials for sterilisation and a special design for fast and ergonomic use. Modern high-speed cutting machines have a shortened operating time, but a major drawback of such equipment is the tissue damage due to the release of energy. Resulting resections have a satisfactory quality in most cases, but some problems remain in controlling the final accuracy of the cut and the temperature elevation during bone machining.

Several analyses have shown that commercially available surgical saws are statistically unable to constrain the temperature elevation below 44°C to 47°C, which is the critical limit range for heat-induced bone necrosis [1] [2]. Despite the influence of mechanical and geometrical blade features on the saws' performance [1] [3] [4] [5], the temperature registered during bone sawing is generally very high and can reach even 450°C [5]. Only internally cooled saw prototypes [6] can lower the cutting temperature to a reasonable level in order to keep a normal osteointegration and prevent anomalous bone modelling.

The attempt to overcome the temperature rise during cutting led us to investigate a different approach to realising bone resections, based on the use of a high-speed power milling tool installed at the end of a robot arm with five degrees of freedom.

Our study, performed on animal limbs, investigates the performance and usability of this technique to realise knee resections.

In this study we have considered the other critical factors influencing the possible use of the milling technique for surgical interventions: firstly the temperature elevation and propagation during cutting, and secondly the time requirements and the usability in a surgical environment. Therefore this work completes an original investigation on a new technique for bone resections in orthopaedic interventions.

Materials & Methods

We studied 15 knees defrosted, a few weeks the subject animal was killed in the normal way, to insure a sufficient level of hydration. The limbs were locked in a fixed position with a clamping device. Bones appeared, on manual inspection, to be within the normal range of individual variability of shape, size, bone density, and cartilage thickness, and had an average resection surface of 1925.5 mm^2. To simulate the in vivo conditions bones were heated to a temperature of between 34°C and 37°C at the bone core , by a heating element located far enough from the milled surface (more than 450 mm) not to be affected by the temperature changes produced by the machining.

The temperature measured at the surface of the bone before the beginning of the milling was from 1°C to 3°C lower than the temperature of the bone core, as in real conditions.

The room temperature was maintained at between 23°C and 25°C.

The measuring instrumentation consisted of two thermocouples: an E-type (nickel-chromium vs. copper-nickel) thermocouple specially shaped for penetration probes with a measurement range of 0-250°C , and a J-type (iron vs. copper-nickel) exposed junction thermocouple with a measurement range of 20-700°C.

Both devices were used to monitor the bone temperature: the E-type thermocouple was set 1mm under the cut and recorded the temperature of the milled surface, while the J-type was set 45mm under the cut and recorded the temperature of the core bone (Fig. 1).

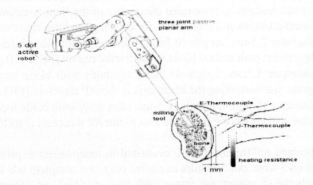

Fig. 1. (a) Experimental setup and equipment. The robot was used to localize both the resection plane and the thermocouples holes. E-type measures surface temperature, J-type measures core bone temperature. The cutting was executed manually using the planar jig.

The accurate preparation of the holes for the two thermocouples was guaranteed by the use of a robot as location pointer, to drill the holes and also to hold a planar guide for the milling machine in the planned position. The position accuracy of the experiment set-up was ±0.4mm. The two thermocouples were manually lodged in the relevant holes and cemented with a special thermo-conductive compound (Tech Spray™ Silicon Free Heat Sink Compound) to fix their position, assure good thermal contact between the bone and the device, and to prevent air infiltration into the hole.

Voltage values were read on the junction of the thermocouple every 0.25sec using an HP 34970A Data Acquisition/Switch unit equipped with a built-in cold reference junction, to monitor precisely the variations of temperature as a function of time.

The milling was carried out using MIDAS REX (Forth Worth, TX, USA) standard dissecting tools with a rotation velocity of between 70,000 and 100,000 rpm. During the tests no measures were taken to cool the milled area.

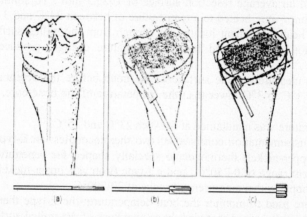

Fig. 2. Milling tools and relative cutting trajectory to realize a planar resection. The cutting trajectory is exemplified for the tibia cut. (a) M10 (b)M16 (c)M15

We studied 3 types of pneumatic milling cutters with different shapes. In order to exploit their typical features and optimize the quality of the final resections they were used with different attitudes and trajectories for carrying out knee resections:

• **M10 tool:** diameter 2.5mm, length 40.5mm, thin body with blade located along the body following a spiral path and no blade in the frontal region (FIG2 (a)).

• **M16 tool:** diameter 6.5mm, length 45.0mm, tall body with blade located in front and lateral regions, used orienting the blade axis in the AP direction (FIG2 (b)).

• **M15 tool:** diameter 12.5 mm, length 18.5mm, slim body with blade located in front and lateral region, used orienting the blade axis in the AP direction (FIG2(c)).

The three mentioned milling tools were evaluated in independent experiments. Each tool was tested on 5 tibia and 5 femurs to realise only one complete sets of resections for knee prosthesis (5 planes on femur and one on tibia), as recommended by HOWMEDICA, to get correct performances. Resection were realised manually by the surgeon, using the milling machine on the plane constrained by the robotic jig.

The temperature was monitored during and after each cut until cooling process was over.

The time needed to finish resections was measured by an external observer. After each test the surgeon filled in a form prepared for the ergonomic estimation of the tool including a subjective evaluation of the feasibility of the access, the amount of debris, and the fatigue during the task.

Results

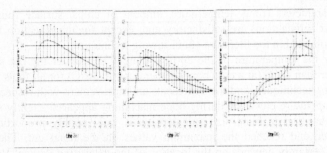

Fig. 3. Bone temperature (1mm under the surface) during milling. The profile is the mean value for 5 knees and bars show 95% confidence interval (a) M10 (b)M16 (c)M15 curve.

The temperature behaviour during milling is reported in Fig. 3 for the different tools, as an average on 5 knees. For a clearer comparison of the tools' performance the temperature profile is reported for the tibial cut where the time needed for the complete resection was very similar (±1sec.) for all knees and did not require normalisation. The temperature profiles for the 5 femoral cuts are similar, with different time ranges according to the resection size.Fig. 4 reports the time needed to perform a complete milling on an average resection surface of 3260 mm^2 with the three tools.

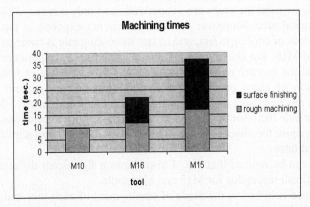

Fig. 4. Milling time for the different tools. The reported time for each tool is computed as the mean value of the tibial resection on 5 knees with an average resection size of 3260 mm^2.

The above mentioned parameters(mean cooling rate, maximum temperature during the milling, milling time) are the basics of the comparison of the three milling tools.

Discussion

Our results have shown the potential benefits deriving from a milling technique instead of sawing.

All main parameters (time, temperature, amount of chips, surface roughness and flatness, user satisfaction) are satisfactory and, except for the resection time, better than those collected in the literature about sawing techniques.

In particular the temperature of the cortical bone 1 mm under the cut surface was always lower than 46°C (at a core bone temperature of 36°C) for all tools without the use of cooling techniques, and this represents an improvement with respect to sawing [5]. The temperature elevation during bone machining appeared slightly sensitive to individual variation in bone size and density, it was roughly proportional to the resection area and reached the maximum values on the tibial cut. The measured temperature certainly depends on the location of the measuring device. Variations in the temperature are here measured 1mm from the milled surface, which showed a significant temperature elevation curve

The time necessary to mill knee resection is usually higher than the time necessary for sawing, because milling needs time consuming trajectories of the tools and often repeated machining of the surface. However it is largely under the critical exposure time for bone tissue injury [7] for all the tools. Therefore the global temperature-exposition rate of milling is better than sawing.

A more accurate analysis of the reported result should take into account the different performance of the tools tested for total knee replacement, which can be important to optimize specific applications.

We can notice that the highest temperature produced by the M10 tool with respect to the M15 and M16 is also related to the presence of chips that are retained near the cut bone, and are known to contain 60-70% of the total heat produced in metal-cutting operations [8] and commonly extended to bone cutting. On the contrary when the M15 and 16 are used the chips are «vaporized» and only a reduced part of them falls into the surgical area. Moreover the M10 mill is not exposed to the airflow with a consequent loss of cooling effect, and in fact its cooling rate is twice as slow as that of the M15 and M16. We also expect that the M15 and M16 temperature results gain an advantage in the in-vitro experiments, because of the absence of soft tissue making the dispersion of chips produced easier. All these factors explain why the bone temperature reached using the M10 is higher than that using the M15/M16.

On the other side the resecting time with the M10 is much shorter than with the M15 and M16, because the absence of a rough phase for the M10 tool significantly reduces the requested time.

However it can be noticed that Fig. 4 also shows a significant difference in the time required to finish resections for M15 and M16 tools.

This difference is linked to the mode of employment of each tool and its geometrical and mechanical features.

For the M16 tool the reduced frontal section and a more noticeable frontal rake angle, reduced bending effects, allowing a resection thickness equal to tool diameter and resulted in faster and easier use with compared to the M15, although the resection trajectories are similar.

The M15 has a larger frontal section than the M16 and did not work properly if the resection thickness was greater than 75% of its diameter, because the forces generated by the friction reduce dramatically the penetration factor of the frontal blade and required more effort by the user to counter bending effect and vibrations. As a consequence rough machining with the M15 tool needed more time than with the M16. Moreover the reduced lateral section of the M15 with respect to the M16 implies a larger number of passages and therefore increase the time for the refining phase, and produces the presence of an anomalous feature in the temperature graph of M15, with respect to the graph of M10 and M16: two temperature peaks, one for the rough machining and one for the refining.

In summary these results showed that the M10 is the best mill for TKR, because its temperature-exposition rate is optimal.

It should be noticed that the general recommendation for the drill process is to use a low feed rate [9] [10], but with our prototype system, we have not reported any drawbacks using milling techniques with tool speed at 100000 rpm. By contrast, we have found advantages using a higher feeding rate in accordance with some authors [11] [12]. The M10 performance results were the best and only chip production was worse than for the M15 and M16. As a general consideration, it can be noticed that during milling the surface remains polished, and chips do not accumulate between bone and blade like in sawing, but are dispersed by rotational movement around the working area.

Although further experiments with human bones should be performed in future, the used technique needed very little training (2/3 sessions before optimal confidence).

Conclusions

In this paper we have analysed the temperature elevation and the usability of the milling technique for performing bone resections (in particular, in total knee replacement).

Our results show that all tested tools are able to finish any resection on tibia and femur (from 1320 mm^2 to 4960 mm^2) with the temperature of the cortical bone, 1mm under the cut, lower than 46°C without using cooling techniques. Moreover the exposition time is largely under the critical value for bone tissue injury and the surgeon's degree of satisfaction is good. These results appear much better than those obtained with commercial saws or even new prototypes proposed in literature. Even if further investigation is probably necessary to evaluate the histological consequences of milling, the reported results provide encouraging indications for a possible introduction of new power tools and the milling technique into orthopaedic surgical practice as an alternative to standard equipment. The use of this technique can improve the overall quality of bone resection in TKR [13], but also reduce the risk of necrosis and could be used without the need for a cooling device, which is, at present, mandatory.

For TKR In particular, the tested high speed tools have also shown the benefit of reduced friction and control force. The results indicated that the M10 was the most suitable tool because it was simpler and faster to use than the M15 and M16 and still provided an acceptable temperature elevation during milling.

The use of this tool and this cutting technique can be particularly suitable for robot-assisted TKR because it has low vibrating feedback on the robot interface, negligible force control and does not require additional devices (such as milling methods or change of tool/approach).

Bibliography

[1] Krause, W.R., Bradbury D.W., Kelly J.E., *Temperature elevations in orthopaedic cutting operations*, J. Biomechanics, Vol. 15 No. 4, pp. 267-275 1982.

[2] Larsen S.T. , Lyd L. *Temperature elevation during knee arthroplasty*, Acta Orthop Scandinava , 1989; 60(4): 439-442

[3] Klip EJ., *Sawing in bone.*, Report no. 1, Departement of orthopaedic Surgery and the Institute of Technology, Linkoping, Sweden, 1976.

[4] Krause, W.R., *Bone cutting: Mechanical and thermal effects.* Proc. Bull. Hosp. Joint Dis. 38, 5-7, 1977.

[5] Larsen S.T. , Lyd L., Lindstrand, A. *Temperature influence in different orthopaedic saw blades*, J of arhoplasty, Vol. 7 No. 1, march 1992.

[6] Larsen S.T. , Lyd L., Lindstrand, A. *On the problem of heating generation in bone cutting*, J Bone and Joint Surgery, vol. 73-B, No. 1, Jan. 1991

[7] Eriksson AR, Albrektsson T, Albrektsson B, *Heat caused by drilling cortical bone: temperature measured in vivo in patients and animals*, Acta Orthop Scand, 1984; 55:629-31.

[8] Schmidt A.O., *Heat in metal cutting p.128 In Machining Theory and practice.* American Society of Metals, Cleveland, 1950

[9] Tetsch P. *Development of raised temperature after osteotomies*, J Maxillofac Surg, 1974; 2 (2-3): 141-145.

[10] Lavelle C., Wedgwood D. *Effect of internal irrigation on frictional het generated from bone milling*, J Oral Surg 1980, 38 (7): 499-503

[11] Matthews LS, Hirsh C. *Temperature measured in human cortical bone when drilling*, J Bone Joint Surg, 1972; 54(2):297-308.

[12] Krause W, Bradbury DW, Kelly JE, Lunceford EM. *Temperature elevations in orthopaedic cutting operations*, J Biomech, 15:267, 1982

[13] Fadda M., Marcacci M., Toksvig-Larsen S., Wang T., Meneghello R.: *Improving accuracy of bone resections using robotics tool holder and a high speed milling cutting tool.* Journal of Medical Engineering & Technology,22(6), 280–284, November/December 1998.

Author Index

Lecture Notes in Computer Science

For information about Vols. 1–1865
please contact your bookseller or Springer-Verlag